PARKINSON'S DISEASE

PARKINSON'S DISEASE: MOLECULAR AND THERAPEUTIC INSIGHTS FROM MODEL SYSTEMS

EDITED BY

RICHARD NASS AND SERGE PRZEDBORSKI

AMSTERDAM • BOSTON • HEIDELBERG • LONDON • NEW YORK • OXFORD
PARIS • SAN DIEGO • SAN FRANCISCO • SINGAPORE • SYDNEY • TOKYO
Academic Press is an imprint of Elsevier

Academic Press is an imprint of Elsevier
360, Park Avenue South, New York, NY-10010-1700.
525 B Street, Suite 1900, San Diego, California 92101-4495, USA
30 Corporate Drive, Suite 400, Burlington, MA 01803, USA
32 Jamestown Road, London NW1 7BY, UK
Radar weg 29, PO Box 211, 1000 AE Amsterdam, The Netherlands

First Edition 2008

Library of Congress Cataloguing in Publication Data
A catalogue record for this book is available from the Library of Congress

British Library Cataloguing in Publication Data
A catalogue record for this book is available from the British Library

ISBN: 978-0-12-374028-1

For information on all Elsevier Academic Press Publication
visit our Web site at www.elsevierdirect.com

Typeset by Charon Tec Ltd., A Macmillan Company. (www.macmillansolutions.com)

Printed and bound by CPI Group (UK) Ltd, Croydon, CR0 4YY

Transferred to Digital Print 2011

Working together to grow
libraries in developing countries

www.elsevier.com | www.bookaid.org | www.sabre.org

ELSEVIER   BOOK AID
International   Sabre Foundation

CONTENTS

III   RODENT TOXIN MODELS   133

10   Rodent Toxin Models of PD: An Overview   135

VINCENT RIES AND ROBERT E. BURKE

11   The MPTP Mouse Model of Parkinson's Disease: The True, the False, and the Unknown   147

VERNICE JACKSON-LEWIS AND SERGE PRZEDBORSKI

20 Viral Vectors: A Potent Approach to Generate Genetic Models of Parkinson's Disease 269

BERNARD L. SCHNEIDER, MERET N. GAUGLER AND PATRICK AEBISCHER

21 Environmental Explorations of Parkinson's Disease using Rodent Genetic Models 285

MARIE-FRANCOISE CHESSELET AND PIERRE-OLIVIER FERNAGUT

22 α-Synuclein, CSPα, SNAREs and Neuroprotection *in vivo* 295

SREEGANGA CHANDRA AND THOMAS C. SÜDHOF

23 Insights from Zebrafish PD Models and Their Potentials for Identifying Novel Drug Targets and Therapeutic Compounds 309

SU GUO

V MULTICELLULAR INVERTEBRATE MODELS   321

24 Parkinson's Disease: Insights from Invertebrates   323

ELLEN B. PENNEY AND BRIAN D. MCCABE

25 *Drosophila* Models for Parkinson's Disease Research   335

NANCY M. BONINI

29 PC12 Cells as a Model for Parkinson's Disease Research 375

CRISTINA MALAGELADA AND LLOYD A. GREENE

30 Dissociated Mesencephalic Cultures: A Research Tool to Model Dopaminergic Cell Death in Parkinson's Disease 389

MYRIAM ESCOBAR-KHONDIKER, DAMIEN TOULORGE, SERGE GUERREIRO, ETIENNE C. HIRSCH AND PATRICK P. MICHEL

34 The Role of the Foxa2 Gene in the Birth and Death of Dopamine Neurons 449

RAJA KITTAPPA, WENDY CHANG AND RONALD MCKAY

35 Embryonic Stem Cell-Based Models of Parkinson's Disease 461

MARK J. TOMISHIMA AND LORENZ STUDER

43 The Use of Cell-Free Systems to Characterize Parkinson's Disease-Related Gene Products 597

JEAN-CHRISTOPHE ROCHET AND JEREMY L. SCHIELER

▮ LIST OF CONTRIBUTORS

Numbers in parentheses indicate the pages on which the authors' contributions begin

Patrick Aebischer (269) Brain and Mind Institute, Ecole Polytechnique Fédérale de Lausanne (EPFL), Switzerland

Garnik Akopian (105) Department of Neurology, and the Andrus Gerontology Center, University of Southern California, Los Angeles, CA, USA

Karen L. Allendoerfer (505) Whitehead Institute for Biomedical Research, and Howard Hughes Medical Institute, 9 Cambridge Center, Cambridge, MA

Vellareddy Anantharam (475) Parkinson's Disorder Research Laboratory, Iowa Center for Advanced Neurotoxicology, Department of Biomedical Sciences, Iowa State University, Ames, IA, USA

D. W. Anderson (87) Department of Pathology, Anatomy and Cell Biology, Thomas Jefferson University, Philadelphia, PA, USA

Ranjita Betarbet (195) Center for Neurodegenerative Diseases, Emory University, Atlanta, GA, USA

Erwan Bezard (65) CNRS UMR 5227, Universite Victor Segalen – Bordeaux 2, Bordeaux, France

Nancy M. Bonini (335) Department of Biology, Howard Hughes Medical Institute, University of Pennsylvania, Philadelphia, PA, USA

Adam T. Braithwaite (9) Department of Neuroscience, Mayo Clinic College of Medicine, Jacksonville, FL, USA

Robert E. Burke (135) Departments of Neurology and Pathology, The College of Physicians and Surgeons Columbia University Black Building, Room 306 650 West 168th Street New York, NY 10032 USA

xxv

Sreeganga Chandra (295) Program in Cellular Neuroscience, Neurodegeneration and Repair, Yale University, New Haven, CT, USA, Department of Neuroscience, Howard Hughes Medical Institute, UT Southwestern Medical Center, Dallas, TX, USA

Wendy Chang (449) Laboratory of Molecular Biology, National Institutes of Neurological Disorders, Stroke, Bethesda, MD, USA

Marie-Francoise Chesselet (285) Departments of Neurology and Neurobiology, David Geffen School of Medicine, UCLA, CA, USA

Yaping Chu (77) Department of Neurological Sciences, Rush University Medical Center, Chicago, IL, USA

Mark R. Cookson (423) Cell Biology and Gene Expression Unit, Laboratory of Neurogenetics, National Institute on Aging, Bethesda, MD, USA

Ana Maria Cuervo (409) Department of Anatomy and Structural Biology, Marion Bessin Liver Research Center, Institute for Aging Research, Albert Einstein College of Medicine, Bronx, NY, USA

William Dauer (221) Columbia University, Departments of Neurology and Pharmacology

E. Decamp (87) Departments of Pathology, Anatomy and Cell Biology, Thomas Jefferson University, Philadelphia, PA, USA

Ted M. Dawson (225) Institute for Cell Engineering, Johns Hopkins University School of Medicine; Department of Neuroscience, Johns Hopkins University School of Medicine, Baltimore, MD, USA

Valina L. Dawson (225) Institute for Cell Engineering, Department of Neurology, Department of Neuroscience, Department of Physiology, Johns Hopkins University School of Medicine, Baltimore, MD, USA

M. R. DeLong (55) Department of Neurology, Emory University School of Medicine, Atlanta, GA, USA

Donato A. Di Monte (207) The Parkinson's Institute, Sunnyvale, CA, USA

Dennis W. Dickson (35) Mayo Clinic, Jacksonville, FL, USA

Robert H. Edwards (237) Departments of Neurology and Physiology, University of California, San Francisco, CA, USA

David Eliezer (575) Department of Biochemistry and Program in Structural Biology, Weill Medical College of Cornell University, New York, NY, USA

Myriam Escobar-Khondiker (389) INSERM Unité Mixte de Recherche S679, Experimental Neurology and Therapeutics, Paris, France; Université Pierre et Marie Curie-Paris 6, Paris, France; Experimental Neurology, Philipps University, Marburg, Germany

Stanley Fahn (3) Department of Neurology, Columbia University, College of Physicians and Surgeons, New York, NY

Matthew J. Farrer (9) Department of Neuroscience, Mayo Clinic College of Medicine, Jacksonville, Florida, USA

Howard J. Federoff (545) Department of Neuroscience; The Department of Neurology, Georgetown University Medical Center, Washington, DC, USA

Pierre-Olivier Fernagut (285) Departments of Neurology and Neurobiology, David Geffen School of Medicine, UCLA, CA, USA

Doris L. Fortin (237) Department of Molecular and Cell Biology, University of California, Berkeley, CA, USA

Don Marshall Gash (51) Department of Anatomy and Neurobiology and the Morris K. Udall Parkinson's Disease Research Center of Excellence, College of Medicine, University of Kentucky, Lexington, KY, USA

Meret N. Gaugler (269) Brain and Mind Institute, Ecole Polytechnique Fédérale de Lausanne (EPFL), Switzerland

Dwight C. German (159) Department of Psychiatry, University of Texas, Southwestern Medical School, Dallas, TX, USA

J. Timothy Greenamyre (195) Pittsburgh Institute for Neurodegenerative Diseases, University of Pittsburgh, Pittsburgh, PA, USA

Lloyd A. Greene (375) Department of Pathology and Neurobiology, Columbia University, New York, NY, USA

Elisa Greggio (423) Cell Biology and Gene Expression Unit, Laboratory of Neurogenetics, National Institute on Aging, Bethesda, MD, USA

Serge Guerreiro (389) INSERM Unité Mixte de Recherche S679, Experimental Neurology and Therapeutics; Université Pierre et Marie Curie-Paris 6, Paris, France

Su Guo (309) Department of Biopharmaceutical Sciences, Programs in Biological Sciences and Human Genetics, University of California San Francisco, CA, USA

Etienne C. Hirsch (389) INSERM Unité Mixte de Recherche S679, Experimental Neurology and Therapeutics; Université Pierre et Marie Curie-Paris 6, Paris, France

Vernice Jackson-Lewis (147) Departments of Neurology, Columbia University, New York, NY, USA

Michael W. Jakowec (105) Department of Neurology, and the Andrus Gerontology Center, University of Southern California, Los Angeles, CA, USA

Huajun Jin (475) Parkinson's Disorder Research Laboratory, Iowa Center for Advanced Neurotoxicology, Department of Biomedical Sciences, Iowa State University, Ames, IA, USA

Anumantha G. Kanthasamy (475) Parkinson's Disorder Research Laboratory, Iowa Center for Advanced Neurotoxicology, Department of Biomedical Sciences, Iowa State University, Ames, IA, USA

Arthi Kanthasamy (475) Parkinson's Disorder Research Laboratory, Iowa Center for Advanced Neurotoxicology, Department of Biomedical Sciences, Iowa State University, Ames, IA, USA

Susmita Kaushik (409) Department of Anatomy and Structural Biology, Marion Bessin Liver Research Center, Institute for Aging Research, Albert Einstein College of Medicine, Bronx, NY, USA

Raja Kittappa (449) Laboratory of Molecular Biology, National Institutes of Neurological Disorders, Stroke, Bethesda, MD, USA

Jeffrey H. Kordower (77) Department of Neurological Sciences, Rush University Medical Center, Chicago, IL, USA

Rehana K. Leak (173) Pittsburgh Institute for Neurodegenerative Diseases, Department of Neurology, University of Pittsburgh, Pittsburgh, PA, USA

Sue-Ann Lee (433) Gladstone Institute of Neurological Disease, University of California, San Francisco, CA, USA; Biomedical Sciences Program, University of California, San Francisco, CA, USA

Patrick Lewis (423) Cell Biology and Gene Expression Unit, Laboratory of Neurogenetics, National Institute on Aging, Bethesda, MD, USA

Susan Lindquist (505) Whitehead Institute for Biomedical Research, and Howard Hughes Medical Institute, 9 Cambridge Center, Cambridge, MA

Xiao-Hong Lu (247) Center for Neurobehavioral Genetics, Semel Institute for Neuroscience and Human Behavior; Department of Psychiatry and Biobehavioral Sciences, Brain Research Institute, David Geffen School of Medicine; at UCLA, Los Angeles, CA, USA

Kathleen A. Maguire-Zeiss (545) Department of Neuroscience, Georgetown University Medical Center, Washington, DC, USA

Cristina Malagelada (375) Department of Pathology and Neurobiology, Columbia University, New York, NY, USA

Brian D. Mccabe (323) Department of Physiology and Cellular Biophysics and Department of Neuroscience, Center for Motor Neuron Biology and Disease, Columbia University, New York, NY, USA

Ronald McKay (449) Laboratory of Molecular Biology, National Institutes of Neurological Disorders, Stroke, Bethesda, MD, USA

Patrick P. Michel (389) INSERM Unité Mixte de Recherche S679, Experimental Neurology and Therapeutics; Université Pierre et Marie Curie-Paris 6, Paris, France; Centre de Recherche Pierre Fabre, Castres, France

Brit Mollenhauer (559) Center for Neurologic Diseases, Brigham and Women's Hospital, Harvard Medical School, Boston, MA, USA; Current Address: Paracelsus Elena Klinik Kassel, Germany, Europe

Paul J. Muchowski (433) Gladstone Institute of Neurological Disease; Biomedical Sciences Program; Department of Biochemistry and Biophysics; Department of Neurology, University of California, San Francisco, CA, USA

Richard Nass (347, 361) Department of Pharmacology and Toxicology, Center for Environmental Health, and Stark Neuroscience Research Institute, Indiana University School of Medicine, Indianapolis, IN, USA

Venu M. Nemani (237) Departments of Neurology and Physiology, University of California, San Francisco, CA, USA

Ellen B. Penney (323) Department of Physiology and Cellular Biophysics and Department of Neuroscience, Center for Motor Neuron Biology and Disease, Columbia University, New York, NY, USA

Leonard Petrucelli (35) Mayo Clinic, Jacksonville, FL, USA

Giselle M. Petzinger (105) Department of Neurology, and the Andrus Gerontology Center, University of Southern California, Los Angeles, CA, USA

Serge Przedborski (147) Departments of Neurology, Columbia University, Pathology and Cell Biology, Columbia University, New York, NY, USA

Stephen Rayport (491) Department of Psychiatry, Columbia University, Department of Molecular Therapeutics, NYS Psychiatric Institute, New York, NY, USA

Vincent Ries (135) Department of Neurology Philipps University Marburg, Rudolf-Bultmann-Strasse 8, 35033 Marburg, Germany e-mail: ries@med.uni-marburg.de

Jean-Christophe Rochet (597) Department of Medicinal Chemistry and Molecular Pharmacology, Purdue University, West Lafayette, IN, USA

Owen A. Ross (9) Department of Neuroscience, Mayo Clinic College of Medicine, Jacksonville, Florida, USA

Vicente Sancenon (433) Gladstone Institute of Neurological Disease, University of California, San Francisco, CA, USA

Jeremy L. Schieler (597) Department of Medicinal Chemistry and Molecular Pharmacology, Purdue University, West Lafayette, IN, USA

Michael G. Schlossmacher (559) Center for Neurologic Diseases, Brigham and Women's Hospital, Harvard Medical School, Boston, MA, USA, Neuroscience Institute East – OHRI; University of Ottawa, Ottawa, ON, Canada

J. S. Schneider (87) Department of Pathology, Anatomy and Cell Biology, Thomas Jefferson University, Philadelphia, PA, USA

Bernard L. Schneider (269) Brain and Mind Institute, Ecole Polytechnique Fédérale de Lausanne (EPFL), Switzerland

Raja S. Settivari (347) Department of Pharmacology and Toxicology, Center for Environmental Health, and Stark Neuroscience Research Institute, Indiana University School of Medicine, Indianapolis, IN, USA

Patricia K. Sonsalla (159) Department of Neurology, UMDNJ-Robert Wood Johnson Medical School, Piscataway, NJ, USA

Lorenz Studer (461) Developmental Biology Program, Sloan-Kettering Institute, New York, NY, USA

Linhui Julie Su (505) Whitehead Institute for Biomedical Research, and Howard Hughes Medical Institute, 9 Cambridge Center, Cambridge, MA

Xiaomin Su (545) Department of Microbiology and Immunology, University of Rochester School of Medicine and Dentistry, Rochester, NY, USA, Current Address: Department of Neurosurgery, University of California San Francisco, San Francisco, CA, USA

Thomas C. Südhof (295) Department of Neuroscience, Howard Hughes Medical Institute, UT Southwestern Medical Center, Dallas, TX, USA

David Sulzer (491) Departments of Neurology, Psychiatry, & Pharmacology, Columbia University; Department of Molecular Therapeutics, NYS Psychiatric Institute, New York, NY, USA

Faneng Sun (475) Parkinson's Disorder Research Laboratory, Iowa Center for Advanced Neurotoxicology, Department of Biomedical Sciences, Iowa State University, Ames, IA, USA

D. James Surmeier (371) Department of Physiology, Feinberg School of Medicine, Northwestern University, Chicago, IL, USA

Daniel M. Togasaki (105) Department of Neurology, and the Andrus Gerontology Center, University of Southern California, Los Angeles, CA, USA

Mark J. Tomishima (461) SKI Stem Cell Research Facility, Developmental Biology Program, Sloan-Kettering Institute, New York, NY, USA

Damien Toulorge (389) INSERM Unité Mixte de Recherche S679, Experimental Neurology and Therapeutics; Université Pierre et Marie Curie-Paris 6, Paris, France

Louis-Eric Trudeau (491) Departments of Pharmacology, Psychiatry and Physiology, CNS Research Group, Faculty of Medicine, Université de Montréal, Montréal, Québec, Canada

John P. Walsh (105) Department of Neurology, and the Andrus Gerontology Center, University of Southern California, Los Angeles, CA, USA

T. Wichmann (55) Yerkes National Primate Research Center, Emory University School of Medicine, Atlanta, GA, USA; Department of Neurology, Emory University School of Medicine, Atlanta, GA, USA

Esther Wong (409) Department of Anatomy and Structural Biology, Marion Bessin Liver Research Center, Institute for Aging Research, Albert Einstein College of Medicine, Bronx, NY, USA

X. William Yang (247) Center for Neurobehavioral Genetics, Semel Institute for Neuroscience and Human Behavior; Department of Psychiatry and Biobehavioral Sciences, Brain Research Institute, David Geffen School of Medicine; at UCLA, Los Angeles, CA, USA

Mingyao Ying (225) Institute for Cell Engineering, Johns Hopkins University School of Medicine; Department of Neurology, Johns Hopkins University School of Medicine, Baltimore, MD, USA

Gail D. Zeevalk (159) Department of Neurology, UMDNJ-Robert Wood Johnson Medical School Piscataway, NJ, USA

Danhui Zhang (475) Parkinson's Disorder Research Laboratory, Iowa Center for Advanced Neurotoxicology, Department of Biomedical Sciences, Iowa State University, Ames, IA, USA

Zhiming Zhang (51) Department of Anatomy and Neurobiology and the Morris K. Udall Parkinson's Disease Research Center of Excellence, College of Medicine, University of Kentucky, Lexington, KY, USA

Chun Zhou (539) Department of Neurology, Columbia University, New York, NY, USA

Michael J. Zigmond (173) Pittsburgh Institute for Neurodegenerative Diseases, Department of Neurology, University of Pittsburgh, Pittsburgh, PA, USA

PREFACE

Parkinson's disease (PD) is a devastating neurodegenerative disorder that results in a wide range of motor and non-motor deficits that includes tremors, bradykinesia, rigidity, cardiovascular andgastrointestinal abnormalities, cognitive dysfunction, and depression. It is the second most prevalent neurodegenerative disease and affects an estimated 6 million people worldwide, with projections suggesting a twofold increase within 25 years. PD is largely characterized by the irreversible loss of dopamine (DA) neurons, although it is becoming increasingly clear that other neurotransmitter systems are likely involved in the pathogenesis. Recent studies suggest that the origin of the cellular dysfunction and neurodegeneration likely has both genetic and environmental component, yet the specific molecular determinants involved in the idiopathic disease have remained elusive. Currently there is no cure for PD, nor is there a therapeutic intervention that will slow the progression of the disease.

The identification of novel molecular targets and therapeutic leads to combat this debilitating illness has been severely limited due to the lack of experimental model systems that completely recapitulate the disease state. Several PD models accurately reproduce one or several phenotypic or biochemical characteristics of the pathogenesis, but fail to reflect other essential hallmarks of the disorder. Other experimental systems, while activating putative PD-associated molecular pathways and forming cellular responses reminiscent of the disease, may not by virtue of their *in vitro* nature recapitulate signaling cascades important in the disease progression. Since there is not a single PD model that completely reproduces the *in vivo* neurodysfunction and degeneration, individual researchers are left with the arduous task of eventually interpreting their scientific findings not only within their experimental system but also within the context of other model systems in order to develop a unifying theme of the pathogenesis. These challenges can be amplified when the exquisite strengths or limitations of a particular model may be known in detail only to those researchers intimately involved with the specific experimental model system.

Largely for these reasons we decided to undertake this highly relevant and important project. This book is unique in that a virtual tour-de-force of scientific experts review and analyze current PD-associated studies that have been generated within a particular model system. Moreover, these experts critically evaluate the advantages and limitations of the model in identifying and characterizing the molecular components and pathways involved in disease. The book progresses from clinical studies to non-human primates, to rodent toxin and genetic models, through invertebrates, yeast, and *in vitro* studies. Each chapter also gives a critical assessment on how these models have the potential to identify novel therapeutic targets and lead compounds. All figures may also be viewed in color at URL: www.elsevierdirect.com/companions/ISBN_ 9780123740281/. We hope that these thoughtful reviews combined with the critical analysis of the specific model systems will provide both experts and newcomers to the field of Parkinson's disease research greater insight into the utility these models have in exploring the molecular basis and mechanisms of PD-associated pathologies and drug discovery.

We also envision this volume as a useful resource to those involved in or considering studies involving other neurodegenerative diseases.Since some issues with experimental models are universal, the book should prove useful to an investigator determining whether a particular model system would be more useful or appropriate in addressing a key issue involved in cell death pathways or mitochondrial dysfunction regardless of the neurodegenerative disease being studied.

Finally we would like to express our gratitude to all the authors and contributors to this unique and exciting project. We are honored to have worked with these individuals, all of whom are among the most creative, talented, and thoughtful PD investigators in the world. Another significant influence on this project was an extraordinarily exciting and intimate Banbury Conference entitled "*Parkinson's Disease: Insights from Genetic and Toxin Models*" held in Cold Spring Harbor, New York, in May 2006. The interactions of those who attended provided an invigorating atmosphere that provoked exciting and thoughtful discussions on the current state of PD research and the potential that these model systems have in elucidating the molecular basis of the disorder and in identifying novel therapeutic targets. We hope that this book will encourage similar passionate discussions and provide a useful resource on the utility of model systems in identifying molecular mechanisms involved in neurodegeneration and facilitation of therapeutic target and lead discovery.

Richard Nass
Serge Przedborski

◼ FOREWORD

In May 2007 I had the pleasure of participating in a meeting in Goteborg, Sweden celebrating the 50th anniversary of the publication by Arvid Carlsson of his seminal paper announcing the identification of dopamine as a unique putative neurotransmitter. Dopamine had long been known to exist as a precursor to norepinephrine. Carlsson had recently returned to Sweden following postdoctoral training at the NIH in Bethesda where he learned to employ the newly fabricated Aminco-Bowman spectrophotofluorometer in which one elicits fluorescence of small molecules by exciting them at specific wavelengths of ultraviolet light. The machine had been successfully employed to monitor serotonin concentrations in the brain. Carlsson adapted it to measure dopamine as well as norephinephrine and reported vastly higher levels of dopamine than norepinephrine in the corpus striatum. He suggested that in this part of the brain dopamine might serve a unique function, perhaps as a neurotransmitter. Soon thereafter Oleh Hornykiewicz reported depletion of dopamine in the caudate of patients with Parkinson's Disease (PD) and detected fleeting improvement in parkinsonian patients receiving intravenous injections of DOPA. George Cotzias tested various regimens of orally administered D,L-DOPA in parkinsonian patients and obtained therapeutic responses despite substantial nausea. When he was able to obtain adequate amounts of the pure isomer L-DOPA, he reported in 1967 dramatic therapeutic benefit which initiated the modern era of PD research and treatment.

The 1960s were the golden era of catecholamine research commencing with the landmark studies of Julie Axelrod showing that re-uptake of norepinephrine by sympathetic nerve endings accounted for its synaptic inactivation. 6-Hydroxydopamine was developed as an agent selectively accumulated by the catecholamine transport system where it would be oxidized to a quinone and destroy cells permitting selective destruction of norepinephrine and dopamine neurons. Urban Ungerstedt at the Karolinska Institute utilized unilateral injections of 6-hydroxydopamine in the brain, depleting dopamine unilaterally in the caudate and eliciting a motor imbalance so that the rats rotated when they walked and one could quantify the extent of neuronal damage by the number of rotations. Thus was born one of the first and finest models of dopamine neuronal destruction.

Another giant step forward in animal models of PD emerged in the 1980s with the discovery that opiate addicts in San Francisco, attempting to self-administer a derivative of meperidine, developed acutely profound parkinsonian symptoms reflecting highly selective destruction of dopamine neurons by a contaminant of the opiate preparation, MPTP (1-methyl-4-phenyl-1,2,3,6-tetrahydropyridine). In contrast to 6-hydroxydopamine, MPTP bore no resemblance to dopamine. Thus, it was a surprising revelation when my MD-PhD student Jonathan Javitch elucidated the mechanism for selective destruction of dopamine neurons. MPP$^+$, generated by the action of monoamine oxidase upon MPTP, is accumulated robustly by the dopamine transport system to levels of about 10,000 times higher than the external medium, and, as a free radical, it destroys the neurons. The use of MPTP in rodents, monkeys and in catecholamine cells in tissue culture, especially PC12 cells, has provided the most long-lasting, heuristic model for studying dopamine deficiency in PD.

Emergence of the MPTP model coincided and interfaced with evidence for environmental sources of PD. MPTP bears a close resemblance to paraquat, a component of herbicides used by the American military to deforest jungles during the Vietnam war. Substances resembling MPTP abound in the industrial workplace. Several published reports described a lower incidence of PD in agricultural communities that didn't employ herbicides. To this day the role of environmental toxins in the disease is not altogether clear. Some environmental influences had always been implied by the existence of post-encephalitic PD following the great influenza epidemic of 1918.

Genetic influences in PD had never been thought to be prominent. However, in recent decades mutations of specific genes have been pinpointed in familial forms of PD and these may teach us much about the pathophysiology of the idiopathic condition. Most extensively studied have been mutations in α-synuclein. The pathogenic forms of α-synuclein cause misfolding of the protein reminiscent of the misfolding that is associated with amyloid plaques in Alzheimer's Disease. Numerous other genetic forms of the disease have been elucidated, such as those involving leucine-rich repeat kinase 2 (LRRK2), DJ-1, which may be an atypical peroxiredoxin-like peroxidase, and parkin. Like α-synuclein, parkin, and ubiquitin-3-ligase, can aggregate in cells. Parkin ubiquitinates itself promoting its own degradation, and familial mutations of parkin with reduced ubiquitin ligase activity elicit the aggregations of Lewy bodies. These Lewy bodies include α-synuclein, parkin, and synphilin. Mice, drosophila, and C. elegans containing mutant forms of these genes have provided valuable models for PD research. Yeast have been a useful tool to evaluate the misfolding of proteins in Parkinson's and other diseases.

Drs. Nass and Przedborski have assembled chapters dealing with diverse models of PD ranging from test tube to primates. All too often, edited volumes are out of date by the time of publication and provide little beyond what one can assimilate from individual published papers. The present volume, by contrast, represents a valuable contribution of utility to a specialist as well as to anyone wanting to understand how to approach the study of this ubiquitous neurodegenerative condition. The authors are all leading investigators in their areas. Research reported is very much up to date. The chapters are clearly presented and well integrated. All-in-all, a most valuable contribution.

Solomon H. Snyder, MD
Johns Hopkins University
Department of Neuroscience
725 N.Wolfe Street, WBSB 813
Baltimore, MD 21205-2185
Tel.: (410) 955-3204
Fax: (410) 955-3623
E-mail: ssnyder@jhmi.edu

I

CLINICAL OVERVIEW

1

CLINICAL ASPECTS OF PARKINSON DISEASE

STANLEY FAHN

Department of Neurology, Columbia University,
College of Physicians and Surgeons,
New York, NY

CLINICAL FEATURES OF PARKINSON DISEASE

Parkinsonism is defined as any combination of six specific, independent motoric features: tremor-at-rest, bradykinesia, rigidity, loss of postural reflexes, flexed posture, and the freezing phenomenon (where the feet are transiently "glued" to the ground) (Table 1.1). Not all six of these cardinal features need be present, but at least two should be before the diagnosis of parkinsonism is made, with at least one of them being tremor-at-rest or bradykinesia. Parkinson disease (PD) is the type of parkinsonism that is most commonly encountered by the general clinician; it is also the one on which much research has been expended and the one we know the most about. The great majority of cases of PD are sporadic, without any other family members being affected. One of the great advances in the last decade is that several gene mutations have been discovered to cause PD. But these monogenetic causes do not explain the great majority of sporadic cases. Environmental factors of an unknown nature and the combination of genetic and environmental risk factors are considered to play a role in the etiology of PD.

The symptoms of PD begin insidiously and worsen gradually over time. They typically begin on one side of the body, rather than bilaterally, before eventually spreading to involve the other side. The most common initial symptom recognized by the

TABLE 1.1 Cardinal motor features of parkinsonism

Tremor-at-rest
Bradykinesia/hypokinesia/akinesia
Rigidity
Flexed posture of neck, trunk, and limbs
Loss of postural reflexes
Freezing of gait

patient is tremor of a hand or foot when that limb is at rest, called tremor-at-rest or resting tremor with a frequency of about 4 Hz. The hand is the more common site. The tremor can be intermittent at the beginning, being present only in stressful situations. Later, tremor tends to be more constant. Tremor is not present in everyone with PD. When absent, the initial symptom is a reduced arm swing or a decreased stride length and speed when walking.

Although PD can develop at any age, it is most common in older adults, with a peak age at onset around 60 years. The prevalence and incidence increases with age, with a lifetime risk of about 2%. A positive family history doubles the risk of developing PD to 4%. Twin studies indicate that PD with an onset under the age of 50 years is more likely to have a genetic relationship than for patients with an older age at onset.

Of the six cardinal motor symptoms, three occur early in the course of the illness, and three occur later. The early symptoms and signs are rest tremor, bradykinesia, and rigidity. Bradykinesia is slowness and reduced amplitude of movement. Features of limb bradykinesia are a smaller and slower of handwriting, difficulty shaving, brushing teeth, and putting on make-up. Walking becomes slow, with decreased arm swing and with a tendency to shuffle feet. Difficulty arising from a deep chair, getting out of automobiles, and turning in bed are symptoms of truncal bradykinesia. Rigidity of muscles is detected by the examiner when he/she moves the patient's limbs, neck or shoulders and experiences increased resistance. There is often a ratchet-like feel to the muscles, so-called cogwheel rigidity.

PD is a neurodegenerative disease, and these early motoric symptoms appear to be related to striatal dopamine deficiency due to loss of dopaminergic neurons in the substantia nigra pars compacts, which sends axons to the striatum. Whereas these early features of PD usually respond to medication that activate striatal dopamine receptors (such as levodopa and dopamine agonists), the three later motoric symptoms of flexed posture, loss of postural reflexes and freezing of gait do not. This lack of response suggests that these late features of PD are the result of nondopaminergic effects. Moreover, increasing bradykinesia that is not responsive to levodopa also appears as the disease worsens. All these intractable symptoms lead to disability. These symptoms lead to impaired walking and balance, resulting in falls and eventually the need to use a walker or a wheelchair.

While the motor symptoms of PD dominate the clinical picture – and even define the parkinsonian syndrome – most patients with PD have other features that have been classified as *nonmotor*. These include bradyphrenia (slowness in mental function), decreased motivation and apathy, dementia, fatigue, depression, anxiety, sleep disturbances (fragmented sleep and REM sleep behavior disorder), constipation, bladder and other autonomic disturbances (sexual, gastrointestinal), and sensory complaints. Sensory symptoms include pain, numbness, tingling, and burning in the affected limbs occur in about 40% of patients. Dementia is associated with age, and has been reported to occur in over 70% of patients with PD eventually.

Patients with PD can live 20 or more years, depending on the age at onset; the mortality rate is about 1.5 times that of normal individuals the same age. Death in PD is usually due to some concurrent unrelated illness or due to the effects of decreased mobility, aspiration, or increased falling with subsequent physical injury.

DIAGNOSIS AND DIFFERENTIAL DIAGNOSIS

There are other causes of the motor features of parkinsonism besides PD, and these need to be considered in the differential diagnosis when an individual presents with these signs and symptoms. These parkinsonian disorders can be divided into four categories (Table 1.2). The major category that presents difficulty in diagnosis is the Parkinson-plus disorders. Like PD, these are also neurodegenerative with loss of dopamine neurons in the substantia nigra, but they have other nerve cell loss that results in additional clinical features. These Parkinson-plus disorders have a worse prognosis. Not only do they respond poorly or not at all to levodopa, they have a higher mortality rate. While it may be difficult to distinguish between PD and Parkinson-plus syndromes in the early stages of the illness, with disease progression over time, the clinical distinctions of the Parkinson-plus disorders become more apparent with the development of other neurological findings, such as cerebellar ataxia, loss of downward ocular movements, and autonomic dysfunction (e.g., postural hypotension, loss of bladder control, and impotence).

There are no practical diagnostic laboratory tests for PD, and the diagnosis rests on the clinical features or by excluding some of the other causes of parkinsonism. The research tool of fluorodopa positron emission tomography (PET) measures levodopa uptake into dopamine nerve terminals, and this shows a decline of about 5% per year of the striatal uptake. A similar result is seen using ligands for the dopamine transporter, either by PET or by single photon emission computed tomography (SPECT); these ligands also label the dopamine nerve terminals. All these neuroimaging techniques reveal decreased dopaminergic nerve terminals in the striatum in both PD and the Parkinson-plus syndromes, and do not distinguish between them. A substantial response to levodopa is most helpful in the differential diagnosis, indicating presynaptic dopamine deficiency with intact postsynaptic dopamine receptors, features typical for PD. Also helpful to distinguish PD from Parkinson-plus syndromes is an asymmetric onset of symptoms (because the other disorders usually

▰ **TABLE 1.2 Classification of the parkinsonian states**

I. Primary parkinsonism (Parkinson disease)
    Sporadic
    Known genetic etiology

II. Secondary parkinsonism (environmental etiology)
    A. Drugs
        1. Dopamine receptor blockers (most commonly antipsychotic medications)
        2. Dopamine storage depletors (reserpine)
    B. Postencephalitic
    C. Toxins – Mn, CO, MPTP, cyanide
    D. Vascular
    E. Brain tumors
    F. Head trauma
    G. Normal pressure hydrocephalus

III. Parkinsonism-Plus Syndromes
    A. Progressive supranuclear palsy (PSP)
    B. Multiple system atrophy (MSA)
    C. Cortical-basal ganglionic degeneration (CBGD)
    D. Diffuse Lewy body disease (DLBD)
    E. Parkinson–dementia–ALS complex of Guam
    F. Progressive pallidal atrophy

IV. Heredodegenerative disorders
    A. Alzheimer's disease
    B. Wilson disease
    C. Huntington disease
    D. Frontotemporal dementia on chromosome 17
    E. X-linked dystonia–parkinsonism (in Filipino men; known as lubag)

present with bilateral signs) and the presence of rest tremor (which is usually absent in Parkinson-plus disorders).

The development of dementia in a patient with parkinsonism remains a difficult differential diagnosis. If the patient's parkinsonian features did not respond to levodopa, the diagnosis is likely to be Alzheimer's disease, which can occasionally present with parkinsonism. If the presenting parkinsonism responded to levodopa, and the patient developed dementia over time, the diagnosis could be either PD Dementia (PDD) or diffuse Lewy body disease (DLBD), also known as Dementia with Lewy Bodies (DLB). The nosologic distinction is less of substance and more of useful categorization. The term PDD is used if the symptoms of PD have been present for at least 1 year before dementia develops. The term DLBD is used if the symptoms of PD have been present less than 1 year before onset

of dementia, or if dementia presents the onset of parkinsonism. A major feature of PDD and DLBD is the presence of hallucinations. Without hallucinations, other types of dementias should be considered, including vascular disease, Alzheimer's disease, and frontotemporal dementia. DLBD is a condition where Lewy bodies are present in the cerebral cortex as well as in the brainstem nuclei. The heredodegenerative disease, known as frontotemporal dementia, is an autosomal dominant disorder due to mutations of the tau gene or the progranulin gene on chromosome 17; the full syndrome presents with dementia, loss of inhibition, parkinsonism, and sometimes muscle wasting. PDD is associated with aging and increased duration of PD. The prevalence of PDD is about 20%, but the likelihood of developing dementia eventually in a patient with PD is much greater, with the highest estimate around 78%.

Some adults may develop a more benign form of PD, in which the symptoms respond to very low dosage levodopa, and the disease does not worsen severely with time. This form is usually due to the autosomal dominant disorder known as dopa-responsive dystonia, which typically begins in childhood as a dystonia. But when it starts in adult life, it can present with parkinsonism. There is no neuronal degeneration. The pathogenesis is due to a biochemical deficiency involving dopamine synthesis. The gene defect is for an enzyme (GTP cyclohydrolase I) required to synthesize the cofactor for tyrosine hydroxylase activity, the crucial rate-limiting first step in the synthesis of dopamine and norepinephrine. Infantile parkinsonism is due to the autosomal recessive deficiency of tyrosine hydroxylase, another cause of a biochemical dopamine deficiency disorder. Young-onset PD (less than 40 years of age, but some use a cut-off of 50 years) usually worsens more slowly than those with older onset. But these younger-onset patients are more likely to develop motor complications from levodopa therapy (see below).

TREATMENT

Treatment of patients with PD can be divided into three major categories: medications, surgery, and physical therapy (to keep the joints and muscles loose and to learn techniques of better gait and balance). Medications are the mainstay of therapy, but brain surgery known as deep brain stimulation (DBS) can be appropriate for selected patients with advanced disease and complications from medications.

Dopamine replacement therapy is the major medical approach to treating PD, and a variety of dopaminergic agents are available. The most powerful drug is levodopa, the immediate precursor of dopamine. Levodopa, an amino acid, can enter the brain, whereas dopamine is blocked from entering by the blood–brain barrier. Levodopa is usually administered combined with a peripheral decarboxylase inhibitor (carbidopa and benserazide) to prevent formation of dopamine in the peripheral tissues, thereby increasing levodopa's bioavailability and also markedly reducing gastrointestinal side effects.

Although levodopa is the most effective drug to treat the symptoms of PD, about 60% of patients develop troublesome complications of disabling response fluctuations ("wearing-off" effect) and dyskinesias (abnormal involuntary movements) after 5 years of levodopa therapy, and younger patients (less than 60 years of age) are particularly prone to develop these problems even sooner. Thus, younger patients are often started with a dopamine agonist rather than levodopa since these drugs are much less likely to cause these motor complications, but they are less effective in reversing the symptoms of PD.

Other drugs that influence brain dopamine are also used. Besides being metabolized by aromatic amino acid decarboxylase (commonly known as dopa decarboxylase), levodopa is also metabolized by catechol-O-methyltransferase (COMT) to form 3-O-methyldopa. COMT inhibitors (entacapone and tolcapone) can be added to levodopa to extend the plasma half-life of levodopa, thereby prolonging the duration of action of each dose of levodopa. They can often reduce motor fluctuations, that is, increase "on" time and reduce "off" time. Because they enhance levodopa's efficacy, these agents can increase dyskinesias and the dosage of levodopa may need to be lowered.

Drugs that inhibit monoamine oxidase (MAO) can also prolong the therapeutic half-life of levodopa. There are two isoforms, MAO-A and MAO-B. Inhibiting MAO-A is more dangerous because such inhibitors can cause hypertensive crises if a high amount of tyramine in food is ingested. Tyramine, normally metabolized in the gut by MAO-A activity, can act as a false neurotransmitter and displace norepinephrine in the peripheral blood vessels, leading to a transient high blood pressure that can result in headache or cerebral hemorrhage. But MAO-B inhibition is safe, and such inhibitors can prolong levodopa's action when transformed into brain dopamine. Two MAO-B inhibitors are selegiline and rasagiline. These drugs have also been shown to slow the clinical progression of PD.

After levodopa, the next most powerful drugs in treating PD symptoms are the dopamine agonists. Several of these are available. The ergot compounds of pergolide, bromocriptine, and cabergoline have the potential to induce fibrosis (cardiac valvulopathy and retroperitoneal, pleuropulmonary and pericardial fibrosis, so these agents are not recommended, and indeed pergolide has been withdrawn from the US market. Pramipexole and ropinirole are non-ergolines and they appear to be equally effective at therapeutic levels. Dopamine agonists are more likely to cause hallucinations, confusion, and psychosis, especially in the elderly, than levodopa. Thus, it is safer to utilize levodopa in patients over the age of 70 years. On the other hand, clinical trials have shown dopamine agonists are less likely to produce dyskinesias and the wearing-off phenomena than levodopa. These trials also showed that levodopa provides greater symptomatic benefit than do dopamine agonists. Other problems more likely to occur with dopamine agonists that levodopa are sudden sleep attacks, including falling asleep at the wheel, daytime drowsiness, ankle edema, and impulse control problems such as hypersexuality and compulsive gambling, shopping and eating. Apomorphine may be the most powerful dopamine agonist, but it needs to be injected subcutaneously (or taken sublingually in Europe). It is used to provide faster relief to overcome a deep "off" state.

Amantadine has several actions; it has antimuscarinic effects, but more importantly it can activate release of dopamine from nerve terminals, block dopamine uptake into the nerve terminals, and block glutamate NMDA receptors. Its dopaminergic actions make it a useful drug to relieve symptoms in about two-thirds of patients, but it can induce livedo reticularis, ankle edema, visual hallucinations, and confusion. Its antiglutamatergic action is useful in reducing the severity of levodopa-induced dyskinesias, and in fact, is the only known effective antidyskinetic agent. The dose of amantadine for its anti-PD effect is usually 100 mg twice daily, but its antidyskinetic effect requires higher dosages, usually 300–400 mg/day. Unfortunately, the antidyskinetic effect tends to lessen over time. The elderly do not tolerate amantadine well because of mental adverse effects of confusion and hallucinations.

Nondopaminergic agents are useful to treat both motor and nonmotor symptoms in PD. Antimuscarinic drugs have been widely used since the 1950s, but these are much less effective than the dopaminergic agents, including amantadine. Because of sensitivity to memory impairment and hallucinations in the elderly population, antimuscarinics should be avoided in patients over the age of 70 years. They can reduce the severity of tremor. Coenzyme Q10 is currently being tested in a controlled clinical trial to determine if it has protective effects. Creatine is being tested in a clinical trial to see if it slows clinical worsening in patients who are already on a stable dose of levodopa.

Deep Brain Stimulation (DBS): Stereotaxic DBS has replaced the older technique of lesioning in the brain because the latter is more risky for inducing permanent neurological deficits. With DBS, the parameters of stimulation, such as voltage and frequency, can be adjusted, and the electrodes could be removed if required. However, DBS is more costly, and frequent adjustments of the stimulators are usually needed. The location of the stereotaxic target is a major factor that needs to be individualized for each patient. The subthalamic nucleus (STN) is the favored target for this reduces bradykinesia and tremor, allowing for a reduction of levodopa dosage, thus reducing the severity of dyskinesias as well. The internal segment of the globus pallidus is a more satisfactory target for controlling choreic and dystonic dyskinesias, which in turn would allow a higher dose of levodopa to be used to control the major symptoms of PD. The thalamus, particularly the ventral intermediate nucleus, is the target most successful for controlling tremor, but this target does not eliminate bradykinesia as well as the STN does, so the thalamus is not a preferred choice today. Surgical procedures for patients with PD are best performed at specialty centers with an experienced team of a neurosurgeon, neurophysiologist to monitor the target during the operative procedure, and neurologist to program the stimulators. The patient needs close follow-up to adjust the stimulator settings to their optimum. Patients with cognitive decline should not have DBS because cognition can be further impaired. Also, intractable symptoms of freezing of gait, loss of postural reflexes, and falling are not benefited. The major benefits of STN stimulation are those symptoms that respond to levodopa. Adverse effects include surgical complications, mechanical problems with the stimulator and leads to the electrodes, infections attacking any of the inserted hardware, and neurologic and behavioral changes. The latter include troubles with speech, dystonic postures, depression, suicide attempts, and cognitive decline. The best candidates are younger patients who can tolerate the penetration of the brain and who have uncontrollable motor fluctuations and dyskinesias.

REFERENCES

Fahn, S., and Jankovic, J. (2007). *Principles and Practice of Movement Disorders.* Churchill Livingstone Elsevier, Philadelphia. This latest textbook devoted to the field of movement disorders contains chapters detailing all the clinical and basic aspects of Parkinson disease. This is an excellent source to obtain additional information about this disease.

Chaudhuri, K. R., Healy, D. G., and Schapira, A. H. (2006). Non-motor symptoms of Parkinson's disease: Diagnosis and management. *Lancet Neurol* 5, 235–245. This review article provides a thorough discussion of the nonmotor features occurring in patients with Parkinson disease and also provides a listing of available quantitative rating scales to measure many of these.

Palhagen, S., Heinonen, E., Hagglund, J., Kaugesaar, T., Maki-Ikola, O., and Palm, R. (2006). Swedish Parkinson Study Group. Selegiline slows the progression of the symptoms of Parkinson disease. *Neurology* 66(8), 1200–1206, Apr 25. This article reports the results of a controlled clinical trial over 7 years showing that sustained use of selegiline before and after the addition of levodopa is superior to placebo in the ultimate clinical severity of Parkinson disease, and even requires lower amounts of levodopa to achieve this benefit.

Suchowersky, O., Reich, S., Perlmutter, J., Zesiewicz, T., Gronseth, G., and Weiner, W. J. (2006). Quality Standards Subcommittee of the American Academy of Neurology. *Neurology* 66(7), 968–975, Apr 11 Practice Parameter: Diagnosis and prognosis of new onset Parkinson disease (an evidence-based review): Report of the Quality Standards Subcommittee of the American Academy of Neurology. This Practice Parameter from the American Academy of Neurology provides guidance on distinguishing Parkinson disease (PD) from other forms of parkinsonism. Early falls, poor response to levodopa, symmetry of motor manifestations, lack of tremor, and early autonomic dysfunction are suggestive of these other parkinsonian syndromes.

Pahwa, R., Factor, S. A., Lyons, K. E., Ondo, W. G., Gronseth, G., Bronte-Stewart, H., Hallett, M., Miyasaki, J., Stevens, J., and Weiner, W. J. (2006). Quality Standards Subcommittee of the American Academy of Neurology. *Neurology* 66(7), 983–995, Apr 11 Practice Parameter: Treatment of Parkinson disease with motor fluctuations and dyskinesia (an evidence-based review): Report of the Quality Standards Subcommittee of the American Academy of Neurology. This Practice Parameter from the American Academy of Neurology provides evidence-based medical and surgical treatment recommendations for patients with Parkinson disease with levodopa-induced motor fluctuations and dyskinesia.

Miyasaki, J. M., Shannon, K., Voon, V., Ravina, B., Kleiner-Fisman, G., Anderson, K., Shulman, L. M., Gronseth, G., and Weiner, W. J. (2006). Quality Standards Subcommittee of the American Academy of Neurology. *Neurology* 66(7), 996–1002, Apr 11 Practice Parameter: Evaluation and treatment of depression, psychosis, and dementia in Parkinson disease (an evidence-based review): Report of the Quality Standards Subcommittee of the American Academy of Neurology. This Practice Parameter from the American Academy of Neurology provides evidence-based treatment recommendations for patients with Parkinson disease with behavioral and cognitive problems, such as depression, psychosis, and dementia.

Krack, P., Batir, A., Van Blercom, N., Chabardes, S., Fraix, V., Ardouin, C., Koudsie, A., Limousin, P. D., Benazzouz, A., LeBas, J. F., Benabid, A. L., and Pollak, P. (2003). Five-year follow-up of bilateral stimulation of the subthalamic nucleus in advanced Parkinson's disease. *N Engl J Med* 349(20), 1925–1934, Nov 13. This is a report of a 5-year prospective study of the first 49 consecutive patients these investigators treated with bilateral stimulation of the subthalamic nucleus. Compared with base line, the patients' motor function (excluding speech which did not improve) while off medication improved by 54%. However, the disease progressed as reflected by a worsening of akinesia, speech, postural stability, and freezing of gait scores while on-medication over the 5 years. The dose of dopaminergic treatment and the duration and severity of levodopa-induced dyskinesia were reduced.

Deuschl, G., Schade-Brittinger, C., Krack, P., Volkmann, J., Schäfer, H., Bötzel, K., Daniels, C., Deutschländer, A., Dillmann, U., Eisner, W., Gruber, D., Hamel, W., Herzog, J., Hilker, R., Klebe, S., Kloss, M., Koy, J., Krause, M., Kupsch, A., Lorenz, D., Lorenzl, S., Mehdorn, H. M., Moringlane, J. R., Oertel, W., Pinsker, M. O., Reichmann, H., Reuss, A., Schneider, G. H., Schnitzler, A., Steude, U., Sturm, V., Timmermann, L., Tronnier, V., Trottenberg, T., Wojtecki, L., Wolf, E., Poewe, W., and Voges, J. (2006). German Parkinson Study Group, Neurostimulation Section. A randomized trial of deep-brain stimulation for Parkinson's disease. *N Engl J Med* 355(9), 896–908, Aug 31 This is a report of an unblinded randomized-pairs trial, comparing changes from baseline to 6 months in the quality of life and the severity of motor symptoms without medication in patients with advanced Parkinson disease undergoing stimulation of the subthalamic nucleus or medical treatment. Neurostimulation resulted in improved quality of life and motor scores (without medication). There was also a lower dose of dopaminergic medication with less dyskinesia. Serious adverse events were more common with neurostimulation than with medication alone.

2

GENETICS OF PARKINSON'S DISEASE

OWEN A. ROSS, ADAM T. BRAITHWAITE AND MATTHEW J. FARRER

Department of Neuroscience, Mayo Clinic College of Medicine, Jacksonville, Florida, USA

GENETICS OF FAMILIAL PARKINSON'S DISEASE

As described by the English physician James Parkinson (1755–1824) in "*An Essay on the Shaking Palsy,*" parkinsonism is clinically characterized by the triad of tremor, rigidity and bradykinesia (Parkinson, 2002). Parkinson's disease (PD; the term coined by French neurologist Jean-Martin Charcot) is the most common cause of parkinsonism and the second most prevalent neurodegenerative disorder after Alzheimer's disease (AD). Historically PD was thought to have no genetic basis and epidemiological data appeared to support this view. Cross-sectional studies suggested that either there is no genetic basis, or that it is only evident in early-onset PD (age of onset <50 years) (Tanner *et al.*, 1999; Wirdefeldt *et al.*, 2004), although to date twin studies have been underpowered to refute incompletely penetrant genetic causes of PD (Simon *et al.*, 2002). Differing disease concordance rates between monozygotic and dizygotic twins in longitudinal studies including those using 18F-dopa positron emission tomography (PET) do support heritability in PD (Piccini *et al.*, 1999). In fact, many clinical reports noted familial aggregation of parkinsonism and a family history of disease is the second most significant risk factor after age (Semchuk *et al.*, 1993).

Contention regarding the importance of genetics in PD was challenged by the identification of several large pedigrees in which parkinsonism appears to have a monogenic, Mendelian pattern of inheritance; either autosomal dominant, autosomal recessive or X-linked. Analyses of PD families with classical linkage methods have nominated 13 regions of the human genome and pathogenic mutations have been identified in several genes at these loci (Table 2.1). Most recently, heterozygous mutations in *leucine-rich repeat kinase 2 (LRRK2)* were found to cause both late-onset, autosomal dominant familial forms of disease and sporadic PD. Herein, we provide an overview of the genetics of PD and how these recent discoveries are changing long-held beliefs. This knowledge provides an understanding of the etiology of PD and may be functionally translated into patient therapy.

α-Synuclein (PARK1/4)

The first gene implicated in familial parkinsonism was α-*synuclein* (*SNCA*; OMIM*163890). In 1996, genome-wide linkage analysis in an Italian kindred mapped autosomal dominant parkinsonism to a locus on chromosome 4q21 (*PARK1*) containing the *SNCA* gene (Polymeropoulos *et al.*, 1996). The missense mutation 209G > A in *SNCA*; which leads to the amino acid substitution Ala53Thr, was identified in the original Italian and three other families (Polymeropoulos *et al.*, 1997). Interestingly this mutation is a reversion back to the mouse sequence, and was therefore potentially

TABLE 2.1   Genes/PARK loci

Gene	Locus	Inheritance	Mutations	Age of onset	Clinical features
SNCA	PARK1/4 (4q21)	AD	Ala30Pro, Glu46Lys and Ala53Thr substitutions; genomic duplications and triplications	Early onset; 38–65 years (duplications); 24–48 years (triplications)	Progressive, levodopa responsive parkinsonism with autonomic dysfunction and dementia; hallucinations and cognitive decline in Glu46Lys cases, rapid progression in Ala53Thr and triplication cases
Parkin	PARK2 (6q25.2–27)	AR	Homozygous/compound heterozygous missense (>57) and exonic deletion/duplication/triplication mutations	<45 years (range 16–72)	Parkinsonism with slow progression, responsive to low doses of levodopa, with early dyskinesias, diurnal fluctuation and sleep benefit
Unknown	PARK3 (2p13)	AD	N/A	N/A	Typical parkinsonism
UCH-L1	PARK5 (4p14)	AD	Possible linkage to Ile93Met substitution	55–58 years	Typical parkinsonism
PINK1	PARK6 (1p35–36)	AR	Missense and exon-deletion mutations	20–40 years	Slow progression, responsive to low doses of levodopa, some with dyskinesias
DJ-1	PARK7 (1p36)	AR	Homozygous missense (Leu166Pro) and deletion (delEx1–5) mutations, and compound heterozygotes	20–40 years	Slow progression, levodopa responsive, occasionally with dystonia, psychiatric disturbance; rare compound heterozygotes have dementia, amyotrophy
LRRK2	PARK8 (12q12)	AD	Many substitutions, notably Arg1441Cys/Gly, Tyr1699Cys, Gly2019Ser and Ile2020Thr	50–70 years (range 32–79)	Predominantly levodopa responsive parkinsonism, occasionally with dystonia, gaze palsy, dementia and amyotrophy
ATP13A2	PARK9 (1p36)	AR	Homozygous/compound heterozygous deletions	12–18 years	Levodopa responsive parkinsonism, with pyramidal degeneration, upward gaze palsy, spasticity and dementia
Unknown	PARK10 (1p32)	N/A	N/A	N/A	Typical parkinsonism
Unknown	PARK11 (2q36–37)	N/A	N/A	N/A	Typical parkinsonism
Unknown	PARK12 (Xq21–25)	X-linked	N/A	N/A	
HTRA2	PARK13 (2p12)	AD	Heterozygous Gly399Ser	49–77 years	Typical parkinsonism

AD = autosomal dominant; AR = autosomal recessive; SN = substantia nigra; N/A = not available; MND = motor neuron disease (amyotrophic lateral sclerosis).

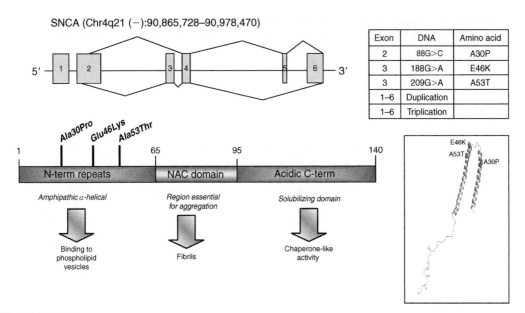

SNCA (Chr4q21 (−):90,865,728–90,978,470)

Exon	DNA	Amino acid
2	88G>C	A30P
3	188G>A	E46K
3	209G>A	A53T
1–6	Duplication	
1–6	Triplication	

FIGURE 2.1 α-Synuclein gene and protein in PD.

a polymorphism linked with the pathogenic variant. However, two further pathogenic mutations, 88G > C (Ala30Pro) and 188G > A (Glu46Lys), were later identified in German and Spanish families respectively, confirming the role of *SNCA* variants in disease (Kruger *et al.*, 1998; Zarranz *et al.*, 2004).

SNCA spans 117kb and has 6 exons encoding a 140 amino acid protein (Figure 2.1), whose aggregates form the pathognomonic brain lesions of the synucleinopathies: "Lewy bodies" and "Lewy neurites" (Spillantini *et al.*, 1997; Giasson *et al.*, 2000). The *SNCA* gene is alternately spliced, with 112 (exon 5) and 126 (exon 3) amino acid protein isoforms also predicted. Sequencing of the 6 exons of *SNCA* in patients from 30 European and American Caucasian kindreds failed to reveal further pathogenic variants (Vaughan *et al.*, 1998), and missense mutations of *SNCA* appear to be a very infrequent cause of parkinsonism (Farrer *et al.*, 1998).

Genomic multiplications of regions encompassing *SNCA* were identified in families with parkinsonism and subsequently dementia, consistent with a *post-mortem* diagnosis of diffuse Lewy body disease (DLBD). *SNCA* triplications have been found in an American (Singleton *et al.*, 2003) and a Swedish-American kindred (Farrer *et al.*, 2004),

and duplications in two French (Chartier-Harlin *et al.*, 2004; Ibanez *et al.*, 2004), one Italian, one Swedish (Fuchs *et al.*, 2007) and two Japanese families (Nishioka *et al.*, 2006). These multiplications cause an increase in *SNCA* expression, with higher levels of α-synuclein mRNA and protein found in brain tissue (Farrer *et al.*, 2004). Clinical and pathological data from these patients are consistent with diagnoses of PD, parkinsonism with dementia (PDD) and dementia with Lewy bodies (DLB) and suggests a continuum of Lewy body disorders (Braak *et al.*, 2003).

Screening of a large number of patients has shown that multiplication of *SCNA* is a rare cause of parkinsonism (Johnson *et al.*, 2004; Lockhart *et al.*, 2004); however, this mechanism highlights the link between overexpression of wild-type *SNCA* at varying levels and distinctly different ages of onset, severity and progression of disease (Farrer *et al.*, 2004). *SNCA* duplication causes disease at a later age which progresses slowly and is reminiscent of typical idiopathic PD (Chartier-Harlin *et al.*, 2004; Ibanez *et al.*, 2004), while patients with *SNCA* triplication present at an earlier age with a more aggressive form of PD, with dementia and autonomic dysfunction (Muenter *et al.*, 1998; Farrer *et al.*, 2004; Fuchs *et al.*, 2007).

Although only a limited number of *SNCA* multiplication families have been identified, these findings have provided a novel pathomechanistic insight into the possible role of *SNCA* in disease manifestation. The hypothesis generated is that overexpression of wild-type α-synuclein protein is sufficient to cause parkinsonism. The true function of α-synuclein protein remains elusive although a number of putative roles have been postulated in vesicle dynamics, through phospholipase D2 and tyrosine hydroxylase inhibition, as well as chaperone activities (Farrer, 2006). Twofold expression causes disease onset at approximately 50 years of age, therefore a subtle difference in expression profile over the lifetime of the individual may affect their susceptibility to PD. This also provides a potential therapeutic target, by means of *SNCA* expression knock-down.

Parkin (PARK2)

Autosomal recessive forms of familial juvenile parkinsonism (AR-JP), typically present with an onset before age 40 years (Takahashi *et al.*, 1994; Ishikawa and Tsuji, 1996). In 1997, a locus was mapped to chromosome 6q25.2–27 (*PARK2*; OMIM*602544) and the causative gene was identified as *parkin* (*PRKN*) (Matsumine *et al.*, 1997; Kitada *et al.*, 1998). The gene contains 12 exons (Figure 2.2) separated by large intronic regions and spanning more than 1.53 Mb (Kitada *et al.*, 1998; West *et al.*, 2001). The open reading frame encodes a 465 amino acid protein with a molecular weight of 52 kDa. The protein has a modular structure, containing an N-terminal ubiquitin-like domain (UBL), a central linker region, and a C-terminal

cassette of two really interesting new gene (RING) domains, separated by an in-between-RING (IBR) domain.

The *PRKN* mutations initially found were large homozygous deletions of one and five exons, respectively (Kitada *et al.*, 1998). Since then a multitude of mutations have been identified (Table 2.2), including deletions, duplications and triplications of exons, frameshift mutations and point mutations (missense, nonsense and splicing site mutations) (Mata *et al.*, 2004). A novel mutation was recently identified in a French family and a patient of German origin, in which there were compound heterozygous deletions of the *PRKN* promoter and exon 3, and homozygous deletion of the promoter, respectively (Lesage *et al.*, 2007). The majority of mutations discovered to date however are localized within the RING-IBR-RING domain of the protein, in particular in the first RING domain, implying this region is essential to protein function. There is no true mutational hotspot, as these have been found in each domain of parkin. Parkin is reported to act as E3-ubiquitin ligase and play a functional role in the proteasomal degradation and receptor trafficking (Shimura *et al.*, 2000; Fallon *et al.*, 2006).

Parkin mutations usually occur either as homozygous or compound heterozygous mutations (with different mutations in both alleles). Nevertheless, several cases have been published in which, despite extensive screening, only one of the alleles appears to be mutated (Farrer *et al.*, 2001; West *et al.*, 2002a). Given the size and complexity of the genomic structure of *PRKN*, a disease-associated mutation in the seemingly unaffected allele may have been missed in these cases; for example, in a non-translated region with an effect on expression.

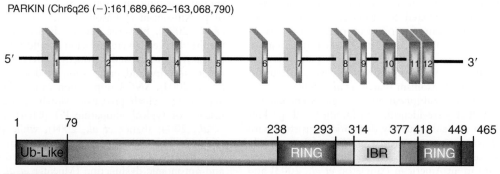

FIGURE 2.2 Structure of *PARKIN* gene and protein.

TABLE 2.2 Non-synonymous coding variants in *PARKIN*

Exon	rs#	cDNA	Amino acid	Domain
1		1A > T	new start at codon 80	Ub-like
2		43G > A	V15M	Ub-like
2		97C > T	R33X	Ub-like
2		98G > A	R33Q	Ub-like
2		101_102delAG	Q34fsX38	Ub-like
2		101delA	Q34fsX43	Ub-like
2		110C > T	P37L	Ub-like
2		125G > C	R42P	Ub-like
2		136G > C	A46P	Ub-like
2		154delA	N52fsX80	Ub-like
2		167T > A	V56E	Ub-like
3		220_221insGT	W74fsX81	Ub-like
3		235G > T	E79X	Ub-like
3		245C > A	A82E	Ub-like
3		300G > C	Q100H	
3		337_376del	L112fsX163	
4		483A > T	K161N	
4	rs1801474	500G > A	S167N	
4		518C > T	T173M	
5	rs9456735	574A > C	M192L	
5		574A > G	M192V	
6		632A > G	K211R	
6		633A > T	K211N	
6		635G > A	C212Y	
6		719C > G	T240R	RING1
6		719C > T	T240M	RING1
6		730G > A	V244I	RING1
7		758G > A	C253Y	RING1
7		766C > T	R256C	RING1
7		804T > A	C268X	RING1
7		813A > T	R271S	RING1
7	rs34424986	823C > T	R275W	RING1

(Continued)

TABLE 2.2 Continued

Exon	rs#	cDNA	Amino acid	Domain
7		838G > A	D280N	RING1
7		850G > C	G284R	RING1
7		865T > G	C289G	RING1
7		871delG	A291fsX297	RING1
8		931C > T	Q311X	
9		970delT	V324fsX434	IBR
9		971delC	V324fsX434	IBR
9		982G > A	G328E	IBR
9		1000C > T	R334C	IBR
9		1015G > T	A339S	IBR
9		1041–1042delGA	Q347fsX368	IBR
9		1046–1047delAA	K349fsX368	IBR
9		1051A > C	T351P	IBR
10		1096C > T	R366W	IBR
10	rs1801582	1138G > C	V380L	
11		1175_1176delGA	R392fsX394	
11	rs1801334	1180G > A	D394N	
11		1193G > A	A398T	
11		1204C > T	R402C	
11		1225G > T	E409X	
11		1244C > A	T415N	
11		1284insA	N428fsX568	RING2
12		1289G > A	G430D	RING2
12		1292G > T	C431F	RING2
12		1310C > T	P437L	RING2
12		1321T > C	C441R	RING2
12		1334G > A	W445X	RING2
12		1358G > A	W453X	

Alternatively it has been suggested that heterozygous mutations causing haploinsufficiency of the *PRKN* gene may constitute a risk factor for PD (West *et al.*, 2002a). However, two studies have refuted this hypothesis by showing that single mutations and polymorphisms in *PRKN* are at similar frequency in patients and controls (Lincoln *et al.*, 2003; Kay *et al.*, 2007).

PINK1 (Chr1p36 (+): 20,832,535–20,850,591)

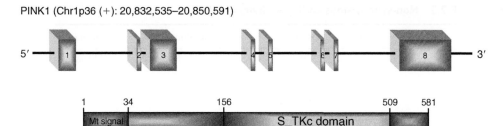

FIGURE 2.3 Structure of *PINK1* gene and protein.

For almost a decade, and until the identification of *LRRK2*, mutations in *PRKN* were apparently the most common cause of familial forms of PD. In a large European study of patients with an age of onset <45 years (or with an affected sibling showing such an early onset), *PRKN* mutations were identified in up to 50% of familial cases, and in 18% of sporadic disease (Abbas *et al.*, 1999; Lucking *et al.*, 2000; Kann *et al.*, 2002). However, the prevalence of parkinsonism <45 years at onset accounts for only 1% of idiopathic PD.

Ubiquitin Carboxyl-Terminal Hydrolase L1 (PARK5)

Ubiquitin carboxyl-terminal hydrolase L1 (UCH-L1; OMIM*191342) is a gene with nine coding exons, spanning 10 kb at chromosome 4p14. The role of the 212 amino acid protein coded for by *UCH-L1* in PD remains contentious. A single missense mutation, 277C > G (Ile93Met) was reported in two affected siblings from a German kindred presenting clinically with typical PD (Leroy *et al.*, 1998). However, no other carriers of the Ile93Met substitution have been identified, although another substitution Ser18Tyr has been associated with protective effect as discussed later in the chapter (Lincoln *et al.*, 1999). A second rare variant was identified in French families (Met124Leu) but it did not segregate with PD (Farrer *et al.*, 2000).

PTEN-Induced Kinase 1 (PARK6)

PTEN-induced putative kinase 1 (PINK1; OMIM*608309) maps to chromosome 1p36. The gene contains 8 exons, spanning 18 kb and encoding a 581 amino acid protein with a predicted mass of 63 kDa. The PINK1 protein includes a 34 amino acid mitochondrial targeting motif and a highly conserved protein kinase domain (residues 156–509) that shows a high degree of homology to the serine/threonine kinases (S_TKc) of the calcium/calmodulin family (Figure 2.3). Valente and colleagues (2001, 2002, 2004a) reported that mutations in this gene are the cause of recessive early-onset parkinsonism in families previously linked to the *PARK6* locus.

To date, several mutations in *PINK1* have been found in different populations (Table 2.3). The first pathogenic substitutions identified were two G > A transitions in nucleotide 926 (Gly309Asp) and 1311 (Trp437X), in a Spanish and two Italian families, respectively (Valente *et al.*, 2004a). Valente and colleagues (2004b) also found mutations in two early-onset sporadic patients suggesting that this gene may play a role in the more frequent sporadic forms of PD. A heterozygous deletion of the entire *PINK1* gene and a splice site mutation on the remaining copy has been recently described (Marongiu *et al.*, 2007). More variants, including single heterozygous mutations have been found in several studies (Healy *et al.*, 2004; Valente *et al.*, 2004b), supporting the notion that heterozygous mutations may be a risk factor for the development of PD (Abou-Sleiman *et al.*, 2006; Djarmati *et al.*, 2006). Alternatively, other mutations such as exon rearrangements could be present with more complex, compound inheritance. Although PINK1 function is not known yet, its localization in the mitochondria and similar structure to serine/threonine kinases suggest that it may phosphorylate mitochondrial proteins in response to cellular stress, protecting against mitochondrial dysfunction. Most mutations in *PINK1* fall within the kinase domain, and probably cause loss of function in patients with a recessively inherited forms of PD (Sim *et al.*, 2006).

TABLE 2.3   Non-synonymous coding variants in *PINK1*

Exon	rs#	cDNA	Amino acid	Domain
1		70–101del	K24fsX54	MTSignal
1		203_204GC > CT	R68P	
1		275G > T	C92F	
1		373T > G	C125G	
2		440G > A	R147H	
2		502G > C	A168P	S_TKc
2		587C > T	P196L	S_TKc
2	rs35802484	588C > T	P196S	S_TKc
2	rs17852513	627C > G	P209A	S_TKc
2	rs34677717	628C > T	P209L	S_TKc
2		650C > A	A217D	S_TKc
3		692A > G	E231G	S_TKc
3		715C > T	Q239X	S_TKc
3		718G > A	E240K	S_TKc
3		736C > T	R246X	S_TKc
3		774C > A	Y258X	S_TKc
4		802C > G	L268V	S_TKc
4	rs28940284	813C > A	H271Q	S_TKc
4		836G > A	R279H	S_TKc
4		838G > A	A280T	S_TKc
4		887C > T	P296L	S_TKc
4	rs7349186	914C > T	P305L	S_TKc
4		926G > A	G309D	S_TKc
4		938C > T	T313M	S_TKc
4		949G > A	V317I	S_TKc
4		952A > T	M318L	S_TKc
5		1015G > A	A339T	S_TKc
5	rs3738136	1018G > A	A340T	S_TKc
5	rs35813094	1023G > A	M341I	S_TKc
5	rs28940285	1040T > C	L347P	S_TKc
5		1106T > C	L369P	S_TKc
6	rs34203620	1130G > T	C377F	S_TKc
6	ss69356471	1147G > A	A383T	S_TKc

(*Continued*)

TABLE 2.3   Continued

Exon	rs#	cDNA	Amino acid	Domain
6		1157G > C	G386A	S_TKc
6		1163T > C	C388R	S_TKc
6		1196C > T	P399L	S_TKc
6		1220G > A	R407Q	S_TKc
6		1226G > T	G409V	S_TKc
6	ss69356473	1231G > A	G411S	S_TKc
6		1250A > G	E417G	S_TKc
6		1273C > T	P425S	S_TKc
7		1291T > C	Y431H	S_TKc
7		1311G > A	W437X	S_TKc
7		1325T > C	I442T	S_TKc
7		1352A > G	N451S	S_TKc
7	ss69356475	1366C > T	Q456X	S_TKc
7		1382T > G	L461S	S_TKc
7		1391G > A	R464H	S_TKc
7	ss69356476	1426G > A	E476K	S_TKc
7	rs34416410	1429T > A	S477T	S_TKc
7		1466T > C	L489P	S_TKc
7	rs34208370	1474C > T	R492X	S_TKc
8		1493C > T	P498L	S_TKc
8		1502G > A	R501P	S_TKc
8		1558delG	L519fsX522	
8	rs1043424	1562A > C	N521T	
8		1573G > A	D525N	
8		1573_1574 insTTAG	D525fsX562	
8		1602_1603 ins CAA	534_535insQ	
8		1647–1650del	C549fsX553	
8		1723T > C	C575R	
8		1745G > T	X582Leu	

Oncogene DJ-1 (PARK7)

DJ-1 (OMIM*602533) was identified as a causal gene for early-onset autosomal recessive PD in 2003 (Bonifati *et al.*, 2003b). The *PARK7* locus was first localized to chromosome 1p36 in two families from a genetically isolated community in the Netherlands (van Duijn *et al.*, 2001) and later confirmed in two

DJ-1 (Chr1p36 (+): 7,944,321–7,967,925)

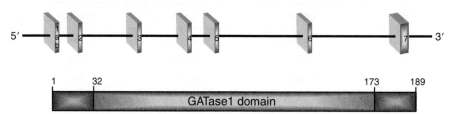

FIGURE 2.4 Structure of *DJ-1* gene and protein.

TABLE 2.4 **Non-synonymous coding variants in *DJ-1***

Exon	rs#	cDNA	Amino acid	Domain
1		c.56delC c.57G > A	T19fsX4	
1		78G > A	M26I	
2		115G > T	A39S	GATase1
2		192G > C	E64D	GATase1
4		310G > A	A104T	GATase1
6		446A > C	D149A	GATase1
6		487G > A	E163K	GATase1
6		497T > C	L166P	GATase1

independent families (Bonifati *et al.*, 2002). The *DJ-1* gene contains 8 exons spanning 24 kb, encoding a 189 amino acid (20 kDa) protein (Figure 2.4). The first 2 exons (1A and 1B) are noncoding and alternatively spliced (Bonifati *et al.*, 2003b).

A large homozygous chromosomal deletion removing 14 kb; including exons 1–5 and 4 kb of sequence upstream from the start of translation site, was identified in one Dutch kindred, while in an Italian family, affected individuals were homozygous for a 497T > C (Leu166Pro) point mutation (Bonifati *et al.*, 2003b). Additional missense mutations have been discovered in patients (Table 2.4). These pathogenic mutations account for only very few patients with early-onset PD. Although some heterozygous mutations have been identified in early-onset disease, segregation has not been demonstrated (Abou-Sleiman *et al.*, 2003; Hague *et al.*, 2003). During the course of the mutation screenings, some noncoding variations were identified throughout the gene, including an 18 bp insertion/deletion

in the promoter (Hague *et al.*, 2003). Further analysis showed that this specific polymorphism does not appear to be a significant risk factor for late-onset PD in an English population (Morris *et al.*, 2003).

Expression of the DJ-1 protein has been shown to be ubiquitous, particularly in liver, skeletal muscle and kidney. In the brain it is also ubiquitous, with higher levels in subcortical regions, which are more affected in PD (Bonifati *et al.*, 2003b). Unlike α-synuclein, DJ-1 is not an essential component of the Lewy bodies and Lewy neurites, and antibodies raised to the protein best label tangles in Pick's disease (Bonifati *et al.*, 2003b; Bandopadhyay *et al.*, 2004; Rizzu *et al.*, 2004). DJ-1 also has been found to shift to a more acidic isoform after treatment of cells with the herbicide paraquat, causing α-synuclein upregulation and aggregation by oxidizing proteins (Mitsumoto and Nakagawa, 2001; Manning-Bog *et al.*, 2002). Oxidative conditions induce a modification of DJ-1, supporting the hypothesis that DJ-1 is an oxidative stress

LRRK2 (Chr12q12 (+):38,905,081–39,049,353)

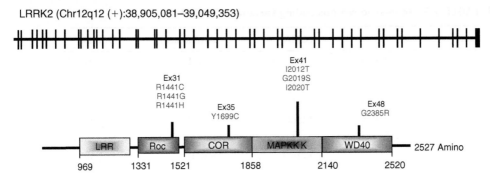

FIGURE 2.5   Structure of *LRRK2* gene and protein.

sensor within cells (Canet-Aviles *et al.*, 2004). Studies of the dopaminergic system in DJ-1-deficient mice have suggested an essential role for DJ-1 in dopaminergic physiology and D2-receptor-mediated functions (Goldberg *et al.*, 2005). The DJ-1 protein is localized to mitochondria, at least in a proportion of transfected cells, suggesting that DJ-1 can be targeted to the mitochondrion under certain conditions and protect against neuronal death (Bonifati *et al.*, 2003a; Miller *et al.*, 2003). Thus, DJ-1 indicates a further link between mitochondrial impairment and the pathogenesis of PD. Interestingly, patients in one early-onset recessive PD family have been reported to harbor a single missense mutation in each of the *DJ-1* and *PINK1* genes (Tang *et al.*, 2006).

Leucine-Rich Repeat Kinase 2 (PARK8)

Mutations in *leucine-rich repeat kinase 2* (*LRRK2*; OMIM*609007) have most recently been identified in PD. Located at chromosome 12q12 (*PARK8*), the gene spans 144 kb and contains 51 exons which encode a 2,527 amino acid protein (Figure 2.5). In 2002, the locus was mapped in a Japanese kindred with autosomal dominant inheritance (Funayama *et al.*, 2002). Linkage to the *PARK8* locus was also found in several European families, confirming the locus and highlighting the global significance of *LRRK2* in PD (Zimprich *et al.*, 2004b). Identification and sequencing of the gene initially revealed six pathogenic mutations (Paisan-Ruiz *et al.*, 2004; Zimprich *et al.*, 2004a).

A seventh mutation, 6055G > A (Gly2019Ser) was also identified (Di Fonzo *et al.*, 2005; Gilks *et al.*, 2005; Kachergus *et al.*, 2005; Nichols *et al.*, 2005), and has been found to be the most common

pathogenic cause of PD, accounting for upto 5% of familial cases in Caucasians (Di Fonzo *et al.*, 2005). This is even higher in Ashkenazi Jews and North African Arab populations, where the Gly2019Ser substitution accounts for between 14% and 30% of idiopathic PD (Lesage *et al.*, 2006; Ozelius *et al.*, 2006). While disease penetrance in Gly2019Ser patients increases from 17% at years of age to 85% at 70 years (Kachergus *et al.*, 2005), the age of onset is variable. There does not appear to be any variation in clinical phenotype between heterozygous and homozygous Gly2019Ser carriers, presumably because the protein is activated by autophosphorylation (see below) (Ishihara *et al.*, 2006; Lesage *et al.*, 2006). The Lrrk2 G2019S variant appears to be indistinguishable both clinically and pathologically from typical sporadic PD (Aasly *et al.*, 2005; Ross *et al.*, 2006).

An increasing number of *LRRK2* variants are being identified (Table 2.5) and a number of these are being nominated as pathogenic (Farrer *et al.*, 2006a). However, for the most part these variants are being observed in either very small kindreds/sib-pairs or in sporadic PD patients with no/limited recorded family histories. Given this high number of putatively pathogenic variants large collaborative studies are required to identify those which are PD related and those that are rare benign polymorphisms. These studies are crucial and may help elucidate the functional relevance of the protein and specific domains, and may demonstrate an even more important role for *LRRK2* in PD.

The 286 kDa Lrrk2 protein may be associated with lipid-rafts or localized to the cell cytoplasm, and is expressed in most brain regions (Gloeckner *et al.*, 2006). It is a member of the ROCO family and

TABLE 2.5   Non-synonymous coding variants in *LRRK2*

Exon	rs#	cDNA	Amino acid	Domain
4	rs33995463	356T > C	L119P	N-terminal
6		632C > T	A211V	N-terminal
7	rs28365216	713A > T	N238I	N-terminal
11	rs34594498	1256C > T	A419V	N-terminal
14	rs35328937	1561A > G	R521G	N-terminal
14		1630 A > G	K544E	N-terminal
14	rs7308720	1653C > C	N551K	N-terminal
18	rs10878307	2167A > G	I723V	N-terminal
19	rs34410987	2264C > T	P755L	N-terminal
19	rs35173587	2378G > T	R793M	N-terminal
21		2789A > G	Q930R	N-terminal
22	rs17519916	2830G > T	D944Y	N-terminal
24		3200G > A	R1067Q	LRR
24		3287C > G	S1096C	LRR
25	rs34805604	3364A > G	I1122V	LRR
25		3451G > A	A1151T	LRR
27		3683G > C	S1228T	LRR
28	rs4640000	3784C > G	P1262A	LRR
29	rs17466213	4111A > G	I1371V	Roc
29	rs28365226	4125C > A	D1375E	Roc
30	rs7133914	4193G > A	R1398H	Roc
31	rs33939927	4321C > T	R1441C	Roc
31	rs33939927	4321C > G	R1441G	Roc
31	rs34995376	4322G > A	R1441H	Roc
31		4324G > C	A1442P	Roc
32	rs35507033	4541G > A	R1514Q	COR
32	rs33958906	4624C > T	P1542S	COR
32	rs17491187	4666C > A	L1556I	COR
33	rs721710	4793T > A	V1598E	COR
34	rs33949390	4883G > C	R1628P	COR
34	rs35303786	4937T > C	M1646T	COR
34	rs11564148	4939T > A	S1647T	COR
35	rs35801418	5096A > G	Y1699C	COR
36	rs11564176	5173C > T	R1725STOP	COR
38		5605A > G	M1869V	COR

(Continued)

■ **TABLE 2.5** **Continued**

Exon	rs#	cDNA	Amino acid	Domain
38	rs35602796	5606T > C	M1869T	COR
38		5620G > T	E1874STOP	COR
39		5822G > A	R1941H	MAPKKK
41		6016T > C	Y2006H	MAPKKK
41	rs34015634	6035T > C	I2012T	MAPKKK
41	rs34637584	6055G > A	G2019S	MAPKKK
41	rs35870237	6059T > C	I2020T	MAPKKK
41		6091A > T	T2031S	MAPKKK
42	rs33995883	6241A > G	N2081D	MAPKKK
43	rs12423862	6356C > T	P2119L	MAPKKK
44	rs35658131	6566A > G	Y2189C	WD40
46	rs12581902	6782A > T	N2261I	WD40
48		7067C > T	T2356I	WD40
48	rs34778348	7153G > A	G2385R	WD40
49	rs3761863	7190C > T	T2397M	WD40

contains five domains which are highly conserved among humans, mice and rats (Zimprich *et al.*, 2004a). These are the leucine-rich repeats, a WD40 domain, a Roc (Ras of complex) domain containing a GTPase, a COR (C-terminal of Roc) domain and an MAPKKK (mitogen-activated protein kinase kinase kinase) domain (Mata *et al.*, 2006).

The kinase activity of Lrrk2 may be integral in the pathomechanistic effects of several autosomal dominantly inherited *LRRK2* mutations. The Arg1441Cys, Gly2019Ser and Ile2020Thr substitutions may enhance the Lrrk2 kinase activity, suggesting a gain-of-function mechanism of pathogenesis (West *et al.*, 2005; Gloeckner *et al.*, 2006) in *LRRK2* PD patients. It has been shown that GTP binding at the Roc domain, and subsequent phosphorylation, is necessary for Lrrk2 kinase activity (Ito *et al.*, 2007) and therefore substitutions such as Arg1441Cys in the GTPase domain may have a pathogenic effect on the GTP hydrolysis and downstream kinase activity (Lewis *et al.*, 2007). The actin-binding ERM proteins (ezrin, radixin and moesin) have been shown to be phosphorylated by Lrrk2 (Jaleel *et al.*, 2007). These proteins anchor the actin-cytoskeleton to the plasma membrane, and have been implicated in the regulation of neurite outgrowth (Paglini *et al.*, 1998). Overexpression of wild type and mutant Lrrk2 causes reduced neurite length and branching in primary neuronal cell cultures (MacLeod *et al.*, 2006), whereas knockout of the homologous protein in *C. elegans* causes defects in vesicle trafficking and in dendritic versus axonal cargo sorting. Progress is fast but the pathomechanistic role for Lrrk2 in processes leading to PD remains to be elucidated.

ATPase Type 13A2 (PARK9)

Kufor–Rakeb syndrome (KRS) is a rare form of recessively inherited, juvenile-onset parkinsonism which is responsive to levodopa. Patients present with additional symptoms atypical of parkinsonism including pyramidal degeneration, upward gaze palsy, spasticity and dementia. KRS was first described in a consanguineous Jordanian family in 1994 (Najim al-Din *et al.*, 1994) and the disease was later mapped to chromosome 1p36

(Hampshire *et al.*, 2001). The causative gene in this region was found to be *ATPase type 13A2* (*ATP13A2*) when a compound heterozygous deletion (exon 26: delC3057; G1019*fs*X1021) and splice site (exon 13: 1306 + 5G > A) mutation was found in affected members of a large Chilean family. A homozygous 22 bp duplication (exon 16: 1632_1653dup22 Leu552*fs*X788) was also discovered in affected members of the original Jordanian family (Ramirez *et al.*, 2006).

The *ATP13A2* gene encodes 29 exons and spans 29 kb. The mutations found in the two families lead to loss of function in the 1,180 amino acid ATP13A2 protein, and it has been suggested there may be a link between aggregation of mutant protein in the endoplasmic reticulum and proteasomal or lysosomal dysfunction (Ramirez *et al.*, 2006). A homozygous missense mutation (exon 15: 1510G > C; Gly504Arg) has been reported in one juvenile parkinsonism patient with disease onset at age 12 years and it is possible that this variant is the cause of disease (Di Fonzo *et al.*, 2007). Di Fonzo *et al.* also identified two further heterozygous variants (exon 2: 35C > T; Thr12Met and exon 16: 1597G > A; Gly533Arg) in young-onset PD patients although the effect of homozygous/heterozygous variants in this gene on PD are yet to be determined.

OMI/HTRA Serine Peptidase 2 (PARK13)

Strauss and colleagues (2005) have reported the presence of mutations within the *OMI/HTRA2* (OMIM*606441) gene in PD patients. *OMI/HTRA2* is located within the *PARK3* linkage region on chromosome 2p13. It spans 3.8 kb and has 8 exons encoding a 458 amino acid protein that is reported to function as a serine protease (Suzuki *et al.*, 2001).

A missense mutation 1195G > A producing the amino acid substitution Gly399Ser in the protein was observed in four sporadic PD patients and was not seen in 370 healthy controls. The four Gly399Ser patients share a haplotype, suggesting a common ancestor (Strauss *et al.*, 2005). The Gly399Ser substitution is located in the PDZ binding domain of the protein and may result in decreased protease activity. A second 421G > T variant that results in an Ala141Ser substitution was reported to associate with increased risk of disease and was observed in 6.2% of PD patients (*n* = 414) and only 3% of

controls (*n* = 331). As with *UCH-L1*; the role of the *OMI/HTRA2* gene in PD remains contentious and necessitates validation by independent groups.

PARKINSONISM-PLUS GENES

Mutations in several genes have been implicated in parkinsonism-plus syndromes, including frontotemporal dementia and parkinsonism (FTDP), progressive supranuclear palsy (PSP) and dystonia parkinsonism. Herein we discuss four examples of dominantly inherited mutations in *microtubule-associated protein tau* (*MAPT*), *progranulin* (*PGRN*), *GTP cyclohydrolase I* (*GCH1*) and *Na⁺/K⁺-ATPase α3 subunit* (*ATP1A3*).

Microtubule-Associated Protein Tau

In 1998 mutations in *MAPT* were identified in rare families with autosomal dominant frontotemporal dementia and parkinsonism linked to chromosome 17 (FTDP17) (Hutton *et al.*, 1998; Poorkaj *et al.*, 1998; Spillantini *et al.*, 1998). Tau is encoded within the *MAPT* locus on chromosome 17q21. The *MAPT* gene (OMIM#157140) consists of 16 exons spanning ~140 kb of genomic DNA (Poorkaj *et al.*, 2001) and three major spliced transcripts of 2, 6 and 8 kb (Goedert *et al.*, 1998). Numerous dominant coding substitutions including exon 10 splice site mutations have been described in FTDP17 (Hutton, 2001). In adult human brain, additional heterogeneity is due to alternative splicing of exons 2, 3 and 10, which gives rise to 6 tau isoforms (352–441 amino acids) (Goedert *et al.*, 1989; Kosik *et al.*, 1989). Tau's normal function in the brain is to modulate the assembly, dynamic behavior and spatial organization of microtubules by binding to tubulin (Dickson *et al.*, 1996; Delacourte and Buee, 1997).

The 1.3 Mb *MAPT* locus is comprised of two extended linkage disequilibrium (LD) haplotypes, previously designated H1 and H2 (Baker *et al.*, 1999). Genetic studies report significant association of the more common H1 haplotype with increased risk for the sporadic parkinsonian disorders including progressive supranuclear palsy (OMIM#601104) and PD (Baker *et al.*, 1999; Skipper *et al.*, 2004). The finding of an association between *MAPT* H1 and PD has now been supported by many studies although the functional mechanism, whether increased expression or exon 10 splicing, has yet to be elucidated (Zabetian *et al.*, 2007).

Progranulin

There were a number of FTDP17-linked families that do not harbor *MAPT* mutations. They also did not demonstrate tau pathology, but had TDP43/ubiquitin positive inclusions (Neumann *et al.*, 2006). Eight years after the identification of *MAPT* mutations by Hutton and colleagues, the same laboratory identified mutations in the *PGRN* gene (OMIM*138945) as responsible for disease in FTDP17-tau negative families (Baker *et al.*, 2006). *PGRN* is located ~2Mb centromeric of *MAPT* on chromosome 17q21; it spans 3.7kb and has 12 coding exons and 1 noncoding exon. The original finding was confirmed by Cruts and colleagues(2006), and since the initial reports over 30 nonsense, splice site and frameshift mutations have been described (Gass *et al.*, 2006).

It appears that *PGRN* mutations result in null alleles that are degraded by nonsense-mediated decay of the mutant mRNA (Baker *et al.*, 2006). The pathomechanism behind *PGRN*-associated FTDP17 is unclear. The *PGRN* protein is reported to play a role in tissue development, inflammation and wound repair (He and Bateman, 2003; He *et al.*, 2003). The majority of *PGRN* FTDP17 patients develop parkinsonian symptoms; mainly rigidity and bradykinesia, and they exhibit age-dependent penetrance similar to *LRRK2*-associated parkinsonism.

An online database (http://www.molgen.ua.ac.be/ADMutations) collects the currently known mutations and polymorphisms in *MAPT* and *PGRN* and other genes involved in AD and FTD.

GTP Cyclohydrolase I

Dopa-responsive dystonia (DRD) is an autosomal dominant primary dystonia with a dramatic therapeutic response to levodopa in most patients, making it hard to differentiate it from juvenile PD (Nemeth, 2002). In fact some apparent DRD families have been found to carry mutations in the *PRKN* gene (Tassin *et al.*, 2000). *GCH1* (OMIM*600225) at chromosome 14q22.1–22.2, is responsible for the biosynthesis of tetrahydrobiopterin (BH4), a rate-limiting enzyme in dopamine synthesis. Reduced activity of *GCH1* in DRD patients is thought to cause symptoms by dopamine depletion, explaining why there is a pronounced therapeutic effect by

levodopa (Nemeth, 2002). Approximately 100 mutations have been found in all 6 exons of the *GCH1* gene (http://www.bh4.org/BH4_databases_biomdb2.asp) which are the underlying cause of disease in more than half of DRD patients (Bandmann *et al.*, 1996; Bandmann *et al.*, 1998; Tassin *et al.*, 2000).

Na$^+$/K$^+$-ATPase α3 Subunit

Rapid-onset dystonia parkinsonism (RDP) is a very rare condition in which dystonic symptoms develop in late adolescence or early adulthood, in a short period of time (hours to weeks), usually accompanied by signs of parkinsonism. This sudden onset is often associated with physical or emotional stress (Dobyns *et al.*, 1993). The disease was linked to a region of chromosome 19q13 in several families (Kramer *et al.*, 1999; Pittock *et al.*, 2000; Kamm *et al.*, 2004). More recently six different mutations were identified in seven unrelated families in the gene *ATP1A3* (OMIM*182350) (Cannon, 2004; de Carvalho Aguiar *et al.*, 2004). The *ATP1A3* gene spans 25kb, with 23 exons coding for a 1,013 amino acid protein. This protein belongs to a group of pumps that catalyze active transport of cations across cell membranes and maintain ionic gradients through hydrolysis of ATP. The α subunit is the catalytic one and three isoforms (α1–3) are expressed in the nervous system (McGrail *et al.*, 1991). All six mutations have been predicted to result in either loss of activity or loss of folding stability, or both. How this results in the acquisition of dystonic and/or parkinsonism symptoms after stressful events may be related to an inability to keep up with a high demand for ion transport activity (Cannon, 2004; de Carvalho Aguiar *et al.*, 2004).

SPORADIC PD, GENETIC ASSOCIATIONS AND FAMILIAL GENES

Given that only ~10–15% of parkinsonism patients report a family history of disease, some have questioned the relevance of the familial genes described to the much more prevalent sporadic form of PD. Over the last decade many association studies have been performed in the hope of identifying genetic variants that are common within the population (>1%) yet lead to an increased risk of developing PD. However, association studies performed using "candidate" genes have often been problematic, with conflicting results and a lack of replication for the vast majority. To help catalog and aid meta-analytical studies a website has recently been created to gather the data from these types of study in a similar format to that used in AD forum, the address of which is *www.PDGene.org*. In this section we will discuss how PD research has moved into the era of genome-wide association (GWA) studies and the level of success achieved.

GWA studies are proclaimed as the latest, and perhaps the most powerful, tool in the mapping of causative/modifying genetic loci in complex disorders (Risch, 2000). The first such GWA study in PD was reported using a two-tiered approach with an initial screen of ~200,000 SNPs in a cohort of 443 discordant sib-pairs (Tier 1) (Maraganore et al., 2005). A matched unrelated patient–control series ($n = 332$) was then used in the replication study (Tier 2; 3,035 SNPs) to assess the influence of SNPs identified in Tier 1 ($p < 0.01$) and a number of hypothesis-based SNPs selected in previously identified PARK loci or genes previously associated with PD. Maraganore and colleagues (2005) reported the 11 most significant SNPs that generated the lowest overall P-values from the combined analysis of the two tiers. However, independent studies by our group and others have not been able to replicate the association of the nominated SNPs (Elbaz et al., 2006; Farrer et al., 2006b).

Fung and colleagues (2006) at the NIH generated a full dataset of the second GWA study which has been publicly released. Access to this empirical dataset may allow the development of novel methods of analysis benefiting both the PD genetics and the field of GWA studies. While both GWA studies have proposed several PD susceptibility loci, their reproducibility remains questionable. A limitation may be current marker densities, sample sizes, disease heterogeneity and the differential effects of variants across populations, as well as population substructures. However, with reducing costs, high-throughput microarrays for both extensive gene expression and association studies still hold promise. Evidence of common disease-associated alleles has previously been obtained in dominantly inherited linkage-derived genes.

Only limited association has been observed between common variants in genes linked with recessive forms of familial PD. Genetic associations between idiopathic PD and several polymorphisms in the promoter of the *PRKN* gene have been assayed, with contrary results in the case of the –258 T/G SNP (West et al., 2002b; Ross et al., 2007b) and no association in the case of other SNPs (Mata et al., 2002). As mentioned previously, it has also been proposed that heterozygous *PRKN* mutations may cause haploinsufficiency and comprise a risk variant for sporadic PD. However, this effect is disputed by conflicting evidence, and currently the exact nature of the role of *PRKN* heterozygous mutations and other variants in idiopathic disease is not clear (Kay et al., 2007). Common variants in both *DJ-1* and *PINK-1* have been associated within subgroups of PD patients and these results are still to be validated. Valente and colleagues (2004b) found that heterozygous mutations in *PINK1* appear to be a risk factor for early-onset PD, with 5% of patients ($n = 100$) and only 1% of controls ($n = 200$) carrying them. Maraganore and colleagues (2004b) published an association study with four common polymorphisms of *DJ-1* showing no association with PD overall, but with two SNPs showing association with PD in female patients.

The problem of reproducibility is highlighted by a missense polymorphism, 53C > A (Ser18Tyr) which was discovered in the *UCH-L1* gene (Lincoln et al., 1999) and case–control association studies showed a reduced susceptibility to PD for the carriers of this variant (Maraganore et al., 1999). The frequency of this variant is 14–20% in Caucasians (Maraganore et al., 1999) and be as high as 50% in Japanese and Chinese populations (Satoh and Kuroda, 2001; Momose et al., 2002) suggesting that it may have a major effect on population attributal risk. Maraganore and colleagues performed a collaborative pooled analysis of data from 11 published studies of the *UCH-L1* Ser18Tyr variant and PD and demonstrated a positive association overall (Maraganore et al., 2004a), however these findings were refuted in another large association study (Healy et al., 2006).

Common variants in the promoter and 3′-region of *SNCA*, including the repeat element Rep1, have also been associated with overexpression and increased risk of sporadic disease (Pals et al., 2004; Mueller et al., 2005). A Japanese case–control association study of candidate genes also identified multiple SNPs in the *SNCA* gene as a susceptibility factors for sporadic PD (Mizuta et al., 2006). These multiple associations may be due to independent factors in the gene or as a result of the relatively high level of LD across the *SNCA* gene. This scenario means the recent meta-analysis that suggested the Rep1 allele size effects "risk," may actually represent an alternate variant located elsewhere in the gene in LD (Maraganore *et al.*, 2006). However, the consistent association of variants at the *SNCA* locus would indicate that a functional variant/s that effect a differential risk, perhaps through regulation of gene expression, are still to be identified.

The greatest evidence to date that common variants within the general population can lead to increased risk of PD has been identified with the *LRRK2* gene. As we previously described, the Lrrk2 Gly2019Ser substitution has demonstrated reduced penetrance and is therefore observed in a number of sporadic patients (~1% of North American Caucasians) (Kachergus *et al.*, 2005). As Lrrk2 Gly2019Ser has been identified in a few elderly controls, it may be considered a genetic risk factor rather than a causal pathogenic mutation for PD. Age is still the greatest risk factor, even for Lrrk2-parkinsonism.

Recent *LRRK2* association studies in Caucasian (German/American) populations did not identify any association with PD (Biskup *et al.*, 2005; Paisan-Ruiz *et al.*, 2005; Paisan-Ruiz *et al.*, 2006). However, Skipper and colleagues (2005) reported an intronic SNP association within a Singapore population. Even more convincingly, though the role of genetics in sporadic PD is the recent discovery of the Lrrk2 variant Gly2385Arg in Asian populations. Lrrk2 Gly2385Arg was originally identified as a putatively pathogenic substitution by Mata and colleagues (2005), and was subsequently found to be a risk factor associated with both familial and sporadic PD in four studies (Di Fonzo *et al.*, 2006; Fung *et al.*, 2006; Farrer *et al.*, 2007; Funayama *et al.*, 2007). The higher prevalence of Gly2385Arg in Asian populations, in contrast to that of Gly2019Ser in Western populations, highlights the importance of careful choice of sample groups for association studies. The effects of studying a highly heterogeneous sample from North America may not be as notable, as in a population isolate, although results obtained may be less generalizable (Ross *et al.*, 2007a).

PERSPECTIVES

Genetic discoveries provide the basis for functional experiments, and the development of etiologic rather than purely symptomatic models of disease. The insights nominate novel targets and pathways for therapeutic intervention. Genetic findings in parkinsonism argue for common and convergent molecular mechanisms in dopaminergic neurodegeneration. Many mechanisms have been implicated, including vesicular transport, cytoskeletal dynamics, neurite growth and arborization, mitochondrial function and the ubiquitin–proteasome system. However, environmental and stochastic factors also have a role to play as age remains the greatest risk factor in PD, even in apparent monogenic families. As advances are made in high throughput and large-scale molecular genetic technologies and genotyping platforms further genes will be identified. With the future potential to target disease prevention in pre-symptomatic genetic risk carriers, the resolution of the "PD genome" will direct individualized diagnosis, medicines and ultimately may facilitate a cure.

ACKNOWLEDGEMENTS

Mayo Clinic Jacksonville is a Morris K. Udall Parkinson's Disease Research Center of Excellence (NINDS P50 #NS40256). We would like to thank all those who have contributed to our research, particularly the patients and their families.

REFERENCES

Aasly, J. O., Toft, M., Fernandez-Mata, I., Kachergus, J., Hulihan, M., White, L. R., and Farrer, M. (2005). Clinical features of LRRK2-associated Parkinson's disease in central Norway. *Ann Neurol* 57, 762–765.

Abbas, N., Lücking, C. B., Ricard, S., Dürr, A., Bonifati, V., DeMichele, G., Bouley, S., Vaughan, J. R., Gasser, T., Marconi, R., Broussolle, E., Brefel-Courbon, C., Harhangi, B. S., Oostra, B. A., Fabrizio, E., Bohme, G. A., Pradier, L., Wood, N. W., Filla, A., Meco, G., Denèfle, P., Agid, Y., and Brice, A. (1999). A wide variety of mutations in the *parkin* gene are responsible for autosomal recessive parkinsonism in Europe. *Hum Mol Genet* 8, 567–574.

Abou-Sleiman, P. M., Healy, D. G., Quinn, N., Lees, A. J., and Wood, N. W. (2003). The role of pathogenic DJ-1 mutations in Parkinson's disease. *Ann Neurol* 54, 283–286.

Abou-Sleiman, P. M., Muqit, M. M., McDonald, N. Q., Yang, Y. X., Gandhi, S., Healy, D. G., Harvey, K., Harvey, R. J., Deas, E., Bhatia, K., Quinn, N., Lees, A., Latchman, D. S., and Wood, N. W. (2006). A heterozygous effect for PINK1 mutations in Parkinson's disease? *Ann Neurol* 60, 414–419.

Bagade, S., Allen, N.C., Tanzi, R., Bertram, L. The PDGene Database, Alzheimer Research Forum, Available at http://www.pdgene.org/

Baker, M., Litvan, I., Houlden, H., Adamson, J., Dickson, D., Perez-Tur, J., Hardy, J., Lynch, T., Bigio, E., and Hutton, M. (1999). Association of an extended haplotype in the tau gene with progressive supranuclear palsy. *Hum Mol Genet* 8, 711–715.

Baker, M., Mackenzie, I. R., Pickering-Brown, S. M., Gass, J., Rademakers, R., Lindholm, C., Snowden, J., Adamson, J., Sadovnick, A. D., Rollinson, S., Cannon, A., Dwosh, E., Neary, D., Melquist, S., Richardson, A., Dickson, D., Berger, Z., Eriksen, J., Robinson, T., Zehr, C., Dickey, C. A., Crook, R., McGowan, E., Mann, D., Boeve, B., Feldman, H., and Hutton, M. (2006). Mutations in progranulin cause tau-negative frontotemporal dementia linked to chromosome 17. *Nature* 442, 916–919.

Bandmann, O., Daniel, S., Marsden, C. D., Wood, N. W., and Harding, A. E. (1996). The GTP-cyclohydrolase I gene in atypical parkinsonian patients: A clinico-genetic study. *J Neurol Sci* 141, 27–32.

Bandmann, O., Valente, E. M., Holmans, P., Surtees, R. A., Walters, J. H., Wevers, R. A., Marsden, C. D., and Wood, N. W. (1998). Dopa-responsive dystonia: A clinical and molecular genetic study. *Ann Neurol* 44, 649–656.

Bandopadhyay, R., Kingsbury, A. E., Cookson, M. R., Reid, A. R., Evans, I. M., Hope, A. D., Pittman, A. M., Lashley, T., Canet-Aviles, R., Miller, D. W., McLendon, C., Strand, C., Leonard, A. J., Abou-Sleiman, P. M., Healy, D. G., Ariga, H., Wood, N. W., de Silva, R., Revesz, T., Hardy, J. A., and Lees, A. J. (2004). The expression of DJ-1 (PARK7) in normal human CNS and idiopathic Parkinson's disease. *Brain* 127, 420–430.

Biskup, S., Mueller, J. C., Sharma, M., Lichtner, P., Zimprich, A., Berg, D., Wullner, U., Illig, T., Meitinger, T., and Gasser, T. (2005). Common variants of LRRK2 are not associated with sporadic Parkinson's disease. *Ann Neurol* 58, 905–908.

Bonifati, V., Breedveld, G. J., Squitieri, F., Vanacore, N., Brustenghi, P., Harhangi, B. S., Montagna, P., Cannella, M., Fabbrini, G., Rizzu, P., van Duijn, C. M., Oostra, B. A., Meco, G., and Heutink, P. (2002). Localization of autosomal recessive early-onset parkinsonism to chromosome 1p36 (PARK7) in an independent dataset. *Ann Neurol* 51, 253–256.

Bonifati, V., Rizzu, P., van Baren, M. J., Schaap, O., Breedveld, G. J., Krieger, E., Dekker, M. C., Squitieri, F., Ibanez, P., Joosse, M., van Dongen, J. W., Vanacore, N., van Swieten, J. C., Brice, A., Meco, G., van Duijn, C. M., Oostra, B. A., and Heutink, P. (2003a). Mutations in the DJ-1 gene associated with autosomal recessive early-onset parkinsonism. *Science* 299, 256–259.

Bonifati, V., Rizzu, P., van Baren, M. J., Schaap, O., Breedveld, G. J., Krieger, E., Dekker, M. C., Squitieri, F., Ibanez, P., Joosse, M., van Dongen, J. W., Vanacore, N., van Swieten, J. C., Brice, A., Meco, G., van Duijn, C. M., Oostra, B. A., and Heutink, P. (2003b). Mutations in the DJ-1 gene associated with autosomal recessive early-onset parkinsonism. *Science* 299, 256–259.

Braak, H., Del Tredici, K., Rub, U., de Vos, R. A., Jansen Steur, E. N., and Braak, E. (2003). Staging of brain pathology related to sporadic Parkinson's disease. *Neurobiol Aging* 24, 197–211.

Canet-Aviles, R. M., Wilson, M. A., Miller, D. W., Ahmad, R., McLendon, C., Bandyopadhyay, S., Baptista, M. J., Ringe, D., Petsko, G. A., and Cookson, M. R. (2004). The Parkinson's disease protein DJ-1 is neuroprotective due to cysteine-sulfinic acid-driven mitochondrial localization. *Proc Natl Acad Sci USA* 101, 9103–9108.

Cannon, S. C. (2004). Paying the price at the pump: Dystonia from mutations in a Na+/K+-ATPase. *Neuron* 43, 153–154.

Chartier-Harlin, M. C., Kachergus, J., Roumier, C., Mouroux, V., Douay, X., Lincoln, S., Levecque, C., Larvor, L., Andrieux, J., Hulihan, M., Waucquier, N., Defebvre, L., Amouyel, P., Farrer, M., and Destee, A. (2004). Alpha-synuclein locus duplication as a cause of familial Parkinson's disease. *Lancet* 364, 1167–1169.

Cruts, M., Gijselinck, I., van der Zee, J., Engelborghs, S., Wils, H., Pirici, D., Rademakers, R., Vandenberghe, R., Dermaut, B., Martin, J. J., van Duijn, C., Peeters, K., Sciot, R., Santens, P., De Pooter, T., Mattheijssens, M., Van den Broeck, M., Cuijt, I., Vennekens, K., De Deyn, P. P., Kumar-Singh, S., and Van Broeckhoven, C. (2006). Null mutations in progranulin cause ubiquitin-positive frontotemporal dementia linked to chromosome 17q21. *Nature* 442, 920–924.

de Carvalho Aguiar, P., Sweadner, K. J., Penniston, J. T., Zaremba, J., Liu, L., Caton, M., Linazasoro, G., Borg, M., Tijssen, M. A., Bressman, S. B., Dobyns, W. B., Brashear, A., and Ozelius, L. J. (2004). Mutations in the Na+/K+-ATPase alpha3 gene ATP1A3 are associated with rapid-onset dystonia parkinsonism. *Neuron* 43, 169–175.

Delacourte, A., and Buee, L. (1997). Normal and pathological Tau proteins as factors for microtubule assembly. *Int Rev Cytol* 171, 167–224.

Di Fonzo, A., Rohe, C. F., Ferreira, J., Chien, H. F., Vacca, L., Stocchi, F., Guedes, L., Fabrizio, E., Manfredi, M., Vanacore, N., Goldwurm, S., Breedveld, G., Sampaio, C., Meco, G., Barbosa, E., Oostra, B. A., and Bonifati, V. (2005). A frequent LRRK2 gene mutation associated with autosomal dominant Parkinson's disease. *Lancet* 365, 412–415.

Di Fonzo, A., Wu-Chou, Y. H., Lu, C. S., van Doeselaar, M., Simons, E. J., Rohe, C. F., Chang, H. C., Chen, R. S.,

Weng, Y. H., Vanacore, N., Breedveld, G. J., Oostra, B. A., and Bonifati, V. (2006). A common missense variant in the LRRK2 gene, Gly2385Arg, associated with Parkinson's disease risk in Taiwan. *Neurogenetics* 7, 133–138.

Di Fonzo, A., Chien, H. F., Socal, M., Giraudo, S., Tassorelli, C., Iliceto, G., Fabbrini, G., Marconi, R., Fincati, E., Abbruzzese, G., Marini, P., Squitieri, F., Horstink, M. W., Montagna, P., Libera, A. D., Stocchi, F., Goldwurm, S., Ferreira, J. J., Meco, G., Martignoni, E., Lopiano, L., Jardim, L. B., Oostra, B. A., Barbosa, E. R., and Bonifati, V. (2007). ATP13A2 missense mutations in juvenile parkinsonism and young onset Parkinson disease. *Neurology* 68, 1557–1562.

Dickson, D. W., Feany, M. B., Yen, S. H., Mattiace, L. A., and Davies, P. (1996). Cytoskeletal pathology in non-Alzheimer degenerative dementia: New lesions in diffuse Lewy body disease, Pick's disease, and corticobasal degeneration. *J Neural Transm Suppl* 47, 31–46.

Djarmati, A., Hedrich, K., Svetel, M., Lohnau, T., Schwinger, E., Romac, S., Pramstaller, P. P., Kostic, V., and Klein, C. (2006). Heterozygous PINK1 mutations: A susceptibility factor for Parkinson disease? *Mov Disord* 21, 1526–1530.

Dobyns, W. B., Ozelius, L. J., Kramer, P. L., Brashear, A., Farlow, M. R., Perry, T. R., Walsh, L. E., Kasarskis, E. J., Butler, I. J., and Breakefield, X. O. (1993). Rapid-onset dystonia-parkinsonism. *Neurology* 43, 2596–2602.

Elbaz, A., Nelson, L. M., Payami, H., Ioannidis, J. P., Fiske, B. K., Annesi, G., Carmine Belin, A., Factor, S. A., Ferrarese, C., Hadjigeorgiou, G. M., Higgins, D. S., Kawakami, H., Kruger, R., Marder, K. S., Mayeux, R. P., Mellick, G. D., Nutt, J. G., Ritz, B., Samii, A., Tanner, C. M., Van Broeckhoven, C., Van Den Eeden, S. K., Wirdefeldt, K., Zabetian, C. P., Dehem, M., Montimurro, J. S., Southwick, A., Myers, R. M., and Trikalinos, T. A. (2006). Lack of replication of thirteen single-nucleotide polymorphisms implicated in Parkinson's disease: A large-scale international study. *Lancet Neurol* 5, 917–923.

Fallon, L., Belanger, C. M., Corera, A. T., Kontogiannea, M., Regan-Klapisz, E., Moreau, F., Voortman, J., Haber, M., Rouleau, G., Thorarinsdottir, T., Brice, A., van Bergen En Henegouwen, P. M., and Fon, E. A. (2006). A regulated interaction with the UIM protein Eps15 implicates parkin in EGF receptor trafficking and PI(3)K-Akt signalling. *Nat Cell Biol* 8, 834–842.

Farrer, M., Wavrant-De Vrieze, F., Crook, R., Boles, L., Perez-Tur, J., Hardy, J., Johnson, W. G., Steele, J., Maraganore, D., Gwinn, K., and Lynch, T. (1998). Low frequency of alpha-synuclein mutations in familial Parkinson's disease. *Ann Neurol* 43, 394–397.

Farrer, M., Destee, T., Becquet, E., Wavrant-De Vrieze, F., Mouroux, V., Richard, F., Defebvre, L., Lincoln, S., Hardy, J., Amouyel, P., and Chartier-Harlin, M. C. (2000). Linkage exclusion in French families with probable Parkinson's disease. *Mov Disord* 15, 1075–1083.

Farrer, M., Chan, P., Chen, R., Tan, L., Lincoln, S., Hernandez, D., Forno, L., Gwinn-Hardy, K., Petrucelli, L., Hussey, J., Singleton, A., Tanner, C., Hardy, J., and Langston, J. W. (2001). Lewy bodies and parkinsonism in families with parkin mutations. *Ann Neurol* 50, 293–300.

Farrer, M., Kachergus, J., Forno, L., Lincoln, S., Wang, D. S., Hulihan, M., Maraganore, D., Gwinn-Hardy, K., Wszolek, Z., Dickson, D., and Langston, J. W. (2004). Comparison of kindreds with parkinsonism and alpha-synuclein genomic multiplications. *Ann Neurol* 55, 174–179.

Farrer, M., Ross, O. A., and Stone, J. T. (2006a). LRRK2-related Parkinson's disease in GeneReviews at GeneTests: Medical Genetics Information Resource [database online]. Copyright, University of Washington, Seattle, 1997–2006. Available at http://www.genetests.org.

Farrer, M. J. (2006). Genetics of Parkinson disease: Paradigm shifts and future prospects. *Nat Rev Genet* 7, 306–318.

Farrer, M. J., Haugarvoll, K., Ross, O. A., Stone, J. T., Milkovic, N. M., Cobb, S. A., Whittle, A. J., Lincoln, S. J., Hulihan, M. M., Heckman, M. G., White, L. R., Aasly, J. O., Gibson, J. M., Gosal, D., Lynch, T., Wszolek, Z. K., Uitti, R. J., and Toft, M. (2006b). Genome-wide association, Parkinson disease, and PARK10. *Am J Hum Genet* 78, 1084–1088, author reply 1092-4.

Farrer, M. J., Stone, J. T., Lin, C. H., Dachsel, J. C., Hulihan, M. M., Haugarvoll, K., Ross, O. A., and Wu, R. M. (2007). Lrrk2 G2385R is an ancestral risk factor for Parkinson's disease in Asia. *Parkinsonism Relat Disord* 13, 89–92.

Fuchs, J., Nilsson, C., Kachergus, J., Munz, M., Larsson, E. M., Schule, B., Langston, J. W., Middleton, F. A., Ross, O. A., Hulihan, M., Gasser, T., and Farrer, M. J. (2007). Phenotypic variation in a large Swedish pedigree due to SNCA duplication and triplication. *Neurology* 68, 916–922.

Funayama, M., Hasegawa, K., Kowa, H., Saito, M., Tsuji, S., and Obata, F. (2002). A new locus for Parkinson's disease (PARK8) maps to chromosome 12p11.2-q13.1. *Ann Neurol* 51, 296–301.

Funayama, M., Li, Y., Tomiyama, H., Yoshino, H., Imamichi, Y., Yamamoto, M., Murata, M., Toda, T., Mizuno, Y., and Hattori, N. (2007). Leucine-rich repeat kinase 2 G2385R variant is a risk factor for Parkinson disease in Asian population. *Neuroreport* 18, 273–275.

Fung, H. C., Scholz, S., Matarin, M., Simon-Sanchez, J., Hernandez, D., Britton, A., Gibbs, J. R., Langefeld, C.,

Stiegert, M. L., Schymick, J., Okun, M. S., Mandel, R. J., Fernandez, H. H., Foote, K. D., Rodriguez, R. L., Peckham, E., De Vrieze, F. W., Gwinn-Hardy, K., Hardy, J. A., and Singleton, A. (2006). Genome-wide genotyping in Parkinson's disease and neurologically normal controls: First stage analysis and public release of data. *Lancet Neurol* 5, 911–916.

Gass, J., Cannon, A., Mackenzie, I. R., Boeve, B., Baker, M., Adamson, J., Crook, R., Melquist, S., Kuntz, K., Petersen, R., Josephs, K., Pickering-Brown, S. M., Graff-Radford, N., Uitti, R., Dickson, D., Wszolek, Z., Gonzalez, J., Beach, T. G., Bigio, E., Johnson, N., Weintraub, S., Mesulam, M., White, C. L., 3rd, Woodruff, B., Caselli, R., Hsiung, G. Y., Feldman, H., Knopman, D., Hutton, M., and Rademakers, R. (2006). Mutations in progranulin are a major cause of ubiquitin-positive frontotemporal lobar degeneration. *Hum Mol Genet* 15, 2988–3001.

Giasson, B. I., Duda, J. E., Murray, I. V., Chen, Q., Souza, J. M., Hurtig, H. I., Ischiropoulos, H., Trojanowski, J. Q., and Lee, V. M. (2000). Oxidative damage linked to neurodegeneration by selective alpha-synuclein nitration in synucleinopathy lesions. *Science* 290, 985–989.

Gilks, W. P., Abou-Sleiman, P. M., Gandhi, S., Jain, S., Singleton, A., Lees, A. J., Shaw, K., Bhatia, K. P., Bonifati, V., Quinn, N. P., Lynch, J., Healy, D. G., Holton, J. L., Revesz, T., and Wood, N. W. (2005). A common LRRK2 mutation in idiopathic Parkinson's disease. *Lancet* 365, 415–416.

Gloeckner, C. J., Kinkl, N., Schumacher, A., Braun, R. J., O'Neill, E., Meitinger, T., Kolch, W., Prokisch, H., and Ueffing, M. (2006). The Parkinson disease causing LRRK2 mutation I2020T is associated with increased kinase activity. *Hum Mol Genet* 15, 223–232.

Goedert, M., Spillantini, M. G., Jakes, R., Rutherford, D., and Crowther, R. A. (1989). Multiple isoforms of human microtubule-associated protein tau: Sequences and localization in neurofibrillary tangles of Alzheimer's disease. *Neuron* 3, 519–526.

Goedert, M., Crowther, R. A., and Spillantini, M. G. (1998). Tau mutations cause frontotemporal dementias. *Neuron* 21, 955–958.

Goldberg, M. S., Pisani, A., Haburcak, M., Vortherms, T. A., Kitada, T., Costa, C., Tong, Y., Martella, G., Tscherter, A., Martins, A., Bernardi, G., Roth, B. L., Pothos, E. N., Calabresi, P., and Shen, J. (2005). Nigrostriatal dopaminergic deficits and hypokinesia caused by inactivation of the familial Parkinsonism-linked gene DJ-1. *Neuron* 45, 489–496.

Hague, S., Rogaeva, E., Hernandez, D., Gulick, C., Singleton, A., Hanson, M., Johnson, J., Weiser, R., Gallardo, M., Ravina, B., Gwinn-Hardy, K., Crawley, A., St George-Hyslop, P. H., Lang, A. E., Heutink, P., Bonifati, V., and Hardy, J. (2003). Early-onset Parkinson's disease caused by a compound hetero-zygous DJ-1 mutation. *Ann Neurol* 54, 271–274.

Hampshire, D. J., Roberts, E., Crow, Y., Bond, J., Mubaidin, A., Wriekat, A. L., Al-Din, A., and Woods, C. G. (2001). Kufor-Rakeb syndrome, pallido-pyramidal degeneration with supranuclear upgaze paresis and dementia, maps to 1p36. *J Med Genet* 38, 680–682.

He, Z., and Bateman, A. (2003). Progranulin (granulin-epithelin precursor, PC-cell-derived growth factor, acrogranin) mediates tissue repair and tumorigenesis. *J Mol Med* 81, 600–612.

He, Z., Ong, C. H., Halper, J., and Bateman, A. (2003). Progranulin is a mediator of the wound response. *Nat Med* 9, 225–229.

Healy, D. G., Abou-Sleiman, P. M., Gibson, J. M., Ross, O. A., Jain, S., Gandhi, S., Gosal, D., Muqit, M. M., Wood, N. W., and Lynch, T. (2004). PINK1 (PARK6) associated Parkinson disease in Ireland. *Neurology* 63, 1486–1488.

Healy, D. G., Abou-Sleiman, P. M., Casas, J. P., Ahmadi, K. R., Lynch, T., Gandhi, S., Muqit, M. M., Foltynie, T., Barker, R., Bhatia, K. P., Quinn, N. P., Lees, A. J., Gibson, J. M., Holton, J. L., Revesz, T., Goldstein, D. B., and Wood, N. W. (2006). UCHL-1 is not a Parkinson's disease susceptibility gene. *Ann Neurol* 59, 627–633.

Hutton, M. (2001). Missense and splice site mutations in tau associated with FTDP-17: Multiple pathogenic mechanisms. *Neurology* 56, S21–S25.

Hutton, M., Lendon, C. L., Rizzu, P., Baker, M., Froelich, S., Houlden, H., Pickering-Brown, S., Chakraverty, S., Isaacs, A., Grover, A., Hackett, J., Adamson, J., Lincoln, S., Dickson, D., Davies, P., Petersen, R. C., Stevens, M., de Graaff, E., Wauters, E., van Baren, J., Hillebrand, M., Joosse, M., Kwon, J. M., Nowotny, P., Heutink, P. et al. (1998). Association of missense and 5'-splice-site mutations in tau with the inherited dementia FTDP-17. *Nature* 393, 702–705.

Ibanez, P., Bonnet, A. M., Debarges, B., Lohmann, E., Tison, F., Pollak, P., Agid, Y., Durr, A., and Brice, A. (2004). Causal relation between alpha-synuclein gene duplication and familial Parkinson's disease. *Lancet* 364, 1169–1171.

Ishihara, L., Warren, L., Gibson, R., Amouri, R., Lesage, S., Durr, A., Tazir, M., Wszolek, Z. K., Uitti, R. J., Nichols, W. C., Griffith, A., Hattori, N., Leppert, D., Watts, R., Zabetian, C. P., Foroud, T. M., Farrer, M. J., Brice, A., Middleton, L., and Hentati, F. (2006). Clinical features of Parkinson disease patients with homozygous leucine-rich repeat kinase 2 G2019S mutations. *Arch Neurol* 63, 1250–1254.

Ishikawa, A., and Tsuji, S. (1996). Clinical analysis of 17 patients in 12 Japanese families with autosomal-recessive type juvenile parkinsonism. *Neurology* 47, 160–166.

Ito, G., Okai, T., Fujino, G., Takeda, K., Ichijo, H., Katada, T., and Iwatsubo, T. (2007). GTP binding is essential to the protein kinase activity of LRRK2, a causative gene product for familial Parkinson's disease. *Biochemistry* 46, 1380–1388.

Jaleel, M., Nichols, R. J., Deak, M., Campbell, D. G., Gillardon, F., Knebel, A., and Alessi, D. R. (2007). LRRK2 phosphorylates moesin at Thr558; characterisation of how Parkinson's disease mutants affect kinase activity. *Biochem J.*

Johnson, J., Hague, S. M., Hanson, M., Gibson, A., Wilson, K. E., Evans, E. W., Singleton, A. A., McInerney-Leo, A., Nussbaum, R. L., Hernandez, D. G., Gallardo, M., McKeith, I. G., Burn, D. J., Ryu, M., Hellstrom, O., Ravina, B., Eerola, J., Perry, R. H., Jaros, E., Tienari, P., Weiser, R., Gwinn-Hardy, K., Morris, C. M., Hardy, J., and Singleton, A. B. (2004). SNCA multiplication is not a common cause of Parkinson disease or dementia with Lewy bodies. *Neurology* 63, 554–556.

Kachergus, J., Mata, I. F., Hulihan, M., Taylor, J. P., Lincoln, S., Aasly, J., Gibson, J. M., Ross, O. A., Lynch, T., Wiley, J., Payami, H., Nutt, J., Maraganore, D. M., Czyzewski, K., Styczynska, M., Wszolek, Z. K., Farrer, M. J., and Toft, M. (2005). Identification of a novel LRRK2 mutation linked to autosomal dominant parkinsonism: Evidence of a common founder across European populations. *Am J Hum Genet* 76, 672–680.

Kamm, C., Leung, J., Joseph, S., Dobyns, W. B., Brashear, A., Breakefield, X. O., and Ozelius, L. J. (2004). Refined linkage to the RDP/DYT12 locus on 19q13.2 and evaluation of GRIK5 as a candidate gene. *Mov Disord* 19, 845–847.

Kann, M., Jacobs, H., Mohrmann, K., Schumacher, K., Hedrich, K., Garrels, J., Wiegers, K., Schwinger, E., Pramstaller, P. P., Breakefield, X. O., Ozelius, L. J., Vieregge, P., and Klein, C. (2002). Role of parkin mutations in 111 community-based patients with early-onset parkinsonism. *Ann Neurol* 51, 621–625.

Kay, D. M., Moran, D., Moses, L., Poorkaj, P., Zabetian, C. P., Nutt, J., Factor, S. A., Yu, C. E., Montimurro, J. S., Keefe, R. G., Schellenberg, G. D., and Payami, H. (2007). Heterozygous parkin point mutations are as common in control subjects as in Parkinson's patients. *Ann Neurol* 61, 47–54.

Kitada, T., Asakawa, S., Hattori, N., Matsumine, H., Yamamura, Y., Minoshima, S., Yokochi, M., Mizuno, Y., and Shimizu, N. (1998). Mutations in the parkin gene cause autosomal recessive juvenile parkinsonism. *Nature* 392, 605–608.

Kosik, K. S., Orecchio, L. D., Bakalis, S., and Neve, R. L. (1989). Developmentally regulated expression of specific tau sequences. *Neuron* 2, 1389–1397.

Kramer, P. L., Mineta, M., Klein, C., Schilling, K., de Leon, D., Farlow, M. R., Breakefield, X. O., Bressman, S. B., Dobyns, W. B., Ozelius, L. J., and Brashear, A. (1999). Rapid-onset dystonia-parkinsonism: Linkage to chromosome 19q13. *Ann Neurol* 46, 176–182.

Kruger, R., Kuhn, W., Muller, T., Woitalla, D., Graeber, M., Kosel, S., Przuntek, H., Epplen, J. T., Schols, L., and Riess, O. (1998). Ala30Pro mutation in the gene encoding alpha-synuclein in Parkinson's disease. *Nat Genet* 18, 106–108.

Leroy, E., Boyer, R., Auburger, G., Leube, B., Ulm, G., Mezey, E., Harta, G., Brownstein, M. J., Jonnalagada, S., Chernova, T., Dehejia, A., Lavedan, C., Gasser, T., Steinbach, P. J., Wilkinson, K. D., and Polymeropoulos, M. H. (1998). The ubiquitin pathway in Parkinson's disease. *Nature* 395, 451–452.

Lesage, S., Durr, A., Tazir, M., Lohmann, E., Leutenegger, A. L., Janin, S., Pollak, P., and Brice, A. (2006). LRRK2 G2019S as a cause of Parkinson's disease in North African Arabs. *N Engl J Med* 354, 422–423.

Lesage, S., Magali, P., Lohmann, E., Lacomblez, L., Teive, H., Janin, S., Cousin, P. Y., Durr, A., and Brice, A. (2007). Deletion of the parkin and PACRG gene promoter in early-onset parkinsonism. *Hum Mutat* 28, 27–32.

Lewis, P. A., Greggio, E., Beilina, A., Jain, S., Baker, A., and Cookson, M. R. (2007). The R1441C mutation of LRRK2 disrupts GTP hydrolysis. *Biochem Biophys Res Commun* 357, 668–671.

Lincoln, S., Vaughan, J., Wood, N., Baker, M., Adamson, J., Gwinn-Hardy, K., Lynch, T., Hardy, J., and Farrer, M. (1999). Low frequency of pathogenic mutations in the ubiquitin carboxy-terminal hydrolase gene in familial Parkinson's disease. *Neuroreport* 10, 427–429.

Lincoln, S. J., Maraganore, D. M., Lesnick, T. G., Bounds, R., de Andrade, M., Bower, J. H., Hardy, J. A., and Farrer, M. J. (2003). Parkin variants in North American Parkinson's disease: Cases and controls. *Mov Disord* 18, 1306–1311.

Lockhart, P. J., Kachergus, J., Lincoln, S., Hulihan, M., Bisceglio, G., Thomas, N., Dickson, D., and Farrer, M. J. (2004). Multiplication of the alpha-synuclein gene is not a common disease mechanism in Lewy body disease. *J Mol Neurosci* 24, 337–342.

Lucking, C. B., Durr, A., Bonifati, V., Vaughan, J., De Michele, G., Gasser, T., Harhangi, B. S., Meco, G., Denefle, P., Wood, N. W., Agid, Y., and Brice, A. (2000). Association between early-onset Parkinson's disease and mutations in the parkin gene. French Parkinson's Disease Genetics Study Group. *N Engl J Med* 342, 1560–1567.

MacLeod, D., Dowman, J., Hammond, R., Leete, T., Inoue, K., and Abeliovich, A. (2006). The familial Parkinsonism gene LRRK2 regulates neurite process morphology. *Neuron* 52, 587–593.

Manning-Bog, A. B., McCormack, A. L., Li, J., Uversky, V. N., Fink, A. L., and Di Monte, D. A. (2002). The herbicide paraquat causes up-regulation and aggregation of alpha-synuclein in mice: Paraquat and alpha-synuclein. *J Biol Chem* 277, 1641–1644.

Maraganore, D. M., Farrer, M. J., Hardy, J. A., Lincoln, S. J., McDonnell, S. K., and Rocca, W. A. (1999). Case-control study of the ubiquitin carboxy-terminal hydrolase L1 gene in Parkinson's disease. *Neurology* 53, 1858–1860.

Maraganore, D. M., Lesnick, T. G., Elbaz, A., Chartier-Harlin, M. C., Gasser, T., Kruger, R., Hattori, N., Mellick, G. D., Quattrone, A., Satoh, J., Toda, T., Wang, J., Ioannidis, J. P., de Andrade, M., and Rocca, W. A. (2004a). UCHL1 is a Parkinson's disease susceptibility gene. *Ann Neurol* 55, 512–521.

Maraganore, D. M., Wilkes, K., Lesnick, T. G., Strain, K. J., de Andrade, M., Rocca, W. A., Bower, J. H., Ahlskog, J. E., Lincoln, S., and Farrer, M. J. (2004b). A limited role for DJ1 in Parkinson disease susceptibility. *Neurology* 63, 550–553.

Maraganore, D. M., de Andrade, M., Elbaz, A., Farrer, M. J., Ioannidis, J. P., Kruger, R., Rocca, W. A., Schneider, N. K., Lesnick, T. G., Lincoln, S. J., Hulihan, M. M., Aasly, J. O., Ashizawa, T., Chartier-Harlin, M. C., Checkoway, H., Ferrarese, C., Hadjigeorgiou, G., Hattori, N., Kawakami, H., Lambert, J. C., Lynch, T., Mellick, G. D., Papapetropoulos, S., Parsian, A., Quattrone, A., Riess, O., Tan, E. K., and Van Broeckhoven, C. (2006). Collaborative analysis of alpha-synuclein gene promoter variability and Parkinson disease. *JAMA* 296, 661–670.

Maraganore, D. M., de Andrade, M., Lesnick, T. G., Strain, K. J., Farrer, M. J., Rocca, W. A., Pant, P. V., Frazer, K. A., Cox, D. R., and Ballinger, D. G. (2005). High-resolution whole-genome association study of Parkinson disease. *Am J Hum Genet* 77, 685–693.

Marongiu, R., Brancati, F., Antonini, A., Ialongo, T., Ceccarini, C., Scarciolla, O., Capalbo, A., Benti, R., Pezzoli, G., Dallapiccola, B., Goldwurm, S., and Valente, E. M. (2007). Whole gene deletion and splicing mutations expand the PINK1 genotypic spectrum. *Hum Mutat* 28, 98.

Mata, I. F., Alvarez, V., Garcia-Moreira, V., Guisasola, L. M., Ribacoba, R., Salvador, C., Blazquez, M., Sarmiento, R. G., Lahoz, C. H., Menes, B. B., and Garcia, E. C. (2002). Single-nucleotide polymorphisms in the promoter region of the PARKIN gene and Parkinson's disease. *Neurosci Lett* 329, 149–152.

Mata, I. F., Lockhart, P. J., and Farrer, M. J. (2004). Parkin genetics: One model for Parkinson's disease. *Hum Mol Genet* 13(Spec No 1), R127–R133.

Mata, I. F., Kachergus, J. M., Taylor, J. P., Lincoln, S., Aasly, J., Lynch, T., Hulihan, M. M., Cobb, S. A., Wu, R. M., Lu, C. S., Lahoz, C., Wszolek, Z. K., and Farrer, M. J. (2005). Lrrk2 pathogenic substitutions in Parkinson's disease. *Neurogenetics* 6, 171–177.

Mata, I. F., Wedemeyer, W. J., Farrer, M. J., Taylor, J. P., and Gallo, K. A. (2006). LRRK2 in Parkinson's disease: Protein domains and functional insights. *Trends Neurosci* 29, 286–293.

Matsumine, H., Saito, M., Shimoda-Matsubayashi, S., Tanaka, H., Ishikawa, A., Nakagawa-Hattori, Y., Yokochi, M., Kobayashi, T., Igarashi, S., Takano, H., Sanpei, K., Koike, R., Mori, H., Kondo, T., Mizutani, Y., Schaffer, A. A., Yamamura, Y., Nakamura, S., Kuzuhara, S., Tsuji, S., and Mizuno, Y. (1997). Localization of a gene for an autosomal recessive form of juvenile Parkinsonism to chromosome 6q25.2-27. *Am J Hum Genet* 60, 588–596.

McGrail, K. M., Phillips, J. M., and Sweadner, K. J. (1991). Immunofluorescent localization of three Na,K-ATPase isozymes in the rat central nervous system: Both neurons and glia can express more than one Na,K-ATPase. *J Neurosci* 11, 381–391.

Miller, D. W., Ahmad, R., Hague, S., Baptista, M. J., Canet-Aviles, R., McLendon, C., Carter, D. M., Zhu, P. P., Stadler, J., Chandran, J., Klinefelter, G. R., Blackstone, C., and Cookson, M. R. (2003). L166P mutant DJ-1, causative for recessive Parkinson's disease, is degraded through the ubiquitin-proteasome system. *J Biol Chem* 278, 36588–36595.

Mitsumoto, A., and Nakagawa, Y. (2001). DJ-1 is an indicator for endogenous reactive oxygen species elicited by endotoxin. *Free Radic Res* 35, 885–893.

Mizuta, I., Satake, W., Nakabayashi, Y., Ito, C., Suzuki, S., Momose, Y., Nagai, Y., Oka, A., Inoko, H., Fukae, J., Saito, Y., Sawabe, M., Murayama, S., Yamamoto, M., Hattori, N., Murata, M., and Toda, T. (2006). Multiple candidate gene analysis identifies alpha-synuclein as a susceptibility gene for sporadic Parkinson's disease. *Hum Mol Genet* 15, 1151–1158.

Momose, Y., Murata, M., Kobayashi, K., Tachikawa, M., Nakabayashi, Y., Kanazawa, I., and Toda, T. (2002). Association studies of multiple candidate genes for Parkinson's disease using single nucleotide polymorphisms. *Ann Neurol* 51, 133–136.

Morris, C. M., O'Brien, K. K., Gibson, A. M., Hardy, J. A., and Singleton, A. B. (2003). Polymorphism in the human DJ-1 gene is not associated with sporadic dementia with Lewy bodies or Parkinson's disease. *Neurosci Lett* 352, 151–153.

Mueller, J. C., Fuchs, J., Hofer, A., Zimprich, A., Lichtner, P., Illig, T., Berg, D., Wullner, U., Meitinger, T., and Gasser, T. (2005). Multiple regions of alpha-synuclein are associated with Parkinson's disease. *Ann Neurol* 57, 535–541.

Muenter, M. D., Forno, L. S., Hornykiewicz, O., Kish, S. J., Maraganore, D. M., Caselli, R. J., Okazaki, H., Howard, F. M., Jr., Snow, B. J., and Calne, D. B. (1998). Hereditary form of parkinsonism – dementia. *Ann Neurol* 43, 768–781.

Najim al-Din, A. S., Wriekat, A., Mubaidin, A., Dasouki, M., and Hiari, M. (1994). Pallido-pyramidal degeneration, supranuclear upgaze paresis and dementia: Kufor-Rakeb syndrome. *Acta Neurol Scand* 89, 347–352.

Nemeth, A. H. (2002). The genetics of primary dystonias and related disorders. *Brain* 125, 695–721.

Neumann, M., Sampathu, D. M., Kwong, L. K., Truax, A. C., Micsenyi, M. C., Chou, T. T., Bruce, J., Schuck, T., Grossman, M., Clark, C. M., McCluskey, L. F., Miller, B. L., Masliah, E., Mackenzie, I. R., Feldman, H., Feiden, W., Kretzschmar, H. A., Trojanowski, J. Q., and Lee, V. M. (2006). Ubiquitinated TDP-43 in frontotemporal lobar degeneration and amyotrophic lateral sclerosis. *Science* 314, 130–133.

Nichols, W. C., Pankratz, N., Hernandez, D., Paisan-Ruiz, C., Jain, S., Halter, C. A., Michaels, V. E., Reed, T., Rudolph, A., Shults, C. W., Singleton, A., and Foroud, T. (2005). Genetic screening for a single common LRRK2 mutation in familial Parkinson's disease. *Lancet* 365, 410–412.

Nishioka, K., Hayashi, S., Farrer, M. J., Singleton, A. B., Yoshino, H., Imai, H., Kitami, T., Sato, K., Kuroda, R., Tomiyama, H., Mizoguchi, K., Murata, M., Toda, T., Imoto, I., Inazawa, J., Mizuno, Y., and Hattori, N. (2006). Clinical heterogeneity of alpha-synuclein gene duplication in Parkinson's disease. *Ann Neurol* 59, 298–309.

Ozelius, L. J., Senthil, G., Saunders-Pullman, R., Ohmann, E., Deligtisch, A., Tagliati, M., Hunt, A. L., Klein, C., Henick, B., Hailpern, S. M., Lipton, R. B., Soto-Valencia, J., Risch, N., and Bressman, S. B. (2006). LRRK2 G2019S as a cause of Parkinson's disease in Ashkenazi Jews. *N Engl J Med* 354, 424–425.

Paglini, G., Kunda, P., Quiroga, S., Kosik, K., and Caceres, A. (1998). Suppression of radixin and moesin alters growth cone morphology, motility, and process formation in primary cultured neurons. *J Cell Biol* 143, 443–455.

Paisan-Ruiz, C., Jain, S., Evans, E. W., Gilks, W. P., Simon, J., van der Brug, M., Lopez de Munain, A., Aparicio, S., Gil, A. M., Khan, N., Johnson, J., Martinez, J. R., Nicholl, D., Carrera, I. M., Pena, A. S., de Silva, R., Lees, A., Marti-Masso, J. F., Perez-Tur, J., Wood, N. W., and Singleton, A. B. (2004). Cloning of the gene containing mutations that cause PARK8-linked Parkinson's disease. *Neuron* 44, 595–600.

Paisan-Ruiz, C., Lang, A. E., Kawarai, T., Sato, C., Salehi-Rad, S., Fisman, G. K., Al-Khairallah, T., St George-Hyslop, P., Singleton, A., and Rogaeva, E. (2005). LRRK2 gene in Parkinson disease: Mutation analysis and case control association study. *Neurology* 65, 696–700.

Paisan-Ruiz, C., Evans, E. W., Jain, S., Xiromerisiou, G., Gibbs, J. R., Eerola, J., Gourbali, V., Hellstrom, O., Duckworth, J., Papadimitriou, A., Tienari, P. J., Hadjigeorgiou, G. M., and Singleton, A. B. (2006). Testing association between LRRK2 and Parkinson's disease and investigating linkage disequilibrium. *J Med Genet* 43, e9.

Pals, P., Lincoln, S., Manning, J., Heckman, M., Skipper, L., Hulihan, M., Van den Broeck, M., De Pooter, T., Cras, P., Crook, J., Van Broeckhoven, C., and Farrer, M. J. (2004). Alpha-synuclein promoter confers susceptibility to Parkinson's disease. *Ann Neurol* 56, 591–595.

Parkinson, J. (2002). An essay on the shaking palsy. 1817. *J Neuropsychiatry Clin Neurosci* 14, 223–236, discussion 222.

Piccini, P., Burn, D. J., Ceravolo, R., Maraganore, D., and Brooks, D. J. (1999). The role of inheritance in sporadic Parkinson's disease: Evidence from a longitudinal study of dopaminergic function in twins. *Ann Neurol* 45, 577–582.

Pittock, S. J., Joyce, C., O'Keane, V., Hugle, B., Hardiman, M. O., Brett, F., Green, A. J., Barton, D. E., King, M. D., and Webb, D. W. (2000). Rapid-onset dystonia-parkinsonism: A clinical and genetic analysis of a new kindred. *Neurology* 55, 991–995.

Polymeropoulos, M. H., Higgins, J. J., Golbe, L. I., Johnson, W. G., Ide, S. E., Di Iorio, G., Sanges, G., Stenroos, E. S., Pho, L. T., Schaffer, A. A., Lazzarini, A. M., Nussbaum, R. L., and Duvoisin, R. C. (1996). Mapping of a gene for Parkinson's disease to chromosome 4q21-q23. *Science* 274, 1197–1199.

Polymeropoulos, M. H., Lavedan, C., Leroy, E., Ide, S. E., Dehejia, A., Dutra, A., Pike, B., Root, H., Rubenstein, J., Boyer, R., Stenroos, E. S., Chandrasekharappa, S., Athanassiadou, A., Papapetropoulos, T., Johnson, W. G., Lazzarini, A. M., Duvoisin, R. C., Di Iorio, G., Golbe, L. I., and Nussbaum, R. L. (1997). Mutation in the alpha-synuclein gene identified in families with Parkinson's disease. *Science* 276, 2045–2047.

Poorkaj, P., Bird, T. D., Wijsman, E., Nemens, E., Garruto, R. M., Anderson, L., Andreadis, A., Wiederholt, W. C., Raskind, M., and Schellenberg, G. D. (1998). Tau is a candidate gene for chromosome 17 frontotemporal dementia. *Ann Neurol* **43**, 815–825.

Poorkaj, P., Kas, A., D'Souza, I., Zhou, Y., Pham, Q., Stone, M., Olson, M. V., and Schellenberg, G. D. (2001). A genomic sequence analysis of the mouse and human microtubule-associated protein tau. *Mamm Genome* **12**, 700–712.

Ramirez, A., Heimbach, A., Grundemann, J., Stiller, B., Hampshire, D., Cid, L. P., Goebel, I., Mubaidin, A. F., Wriekat, A. L., Roeper, J., Al-Din, A., Hillmer, A. M., Karsak, M., Liss, B., Woods, C. G., Behrens, M. I., and Kubisch, C. (2006). Hereditary parkinsonism with dementia is caused by mutations in ATP13A2, encoding a lysosomal type 5 P-type ATPase. *Nat Genet* **38**, 1184–1191.

Risch, N. J. (2000). Searching for genetic determinants in the new millennium. *Nature* **405**, 847–856.

Rizzu, P., Hinkle, D. A., Zhukareva, V., Bonifati, V., Severijnen, L. A., Martinez, D., Ravid, R., Kamphorst, W., Eberwine, J. H., Lee, V. M., Trojanowski, J. Q., and Heutink, P. (2004). DJ-1 colocalizes with tau inclusions: A link between parkinsonism and dementia. *Ann Neurol* **55**, 113–118.

Ross, O. A., Toft, M., Whittle, A. J., Johnson, J. L., Papapetropoulos, S., Mash, D. C., Litvan, I., Gordon, M. F., Wszolek, Z. K., Farrer, M. J., and Dickson, D. W. (2006). Lrrk2 and Lewy body disease. *Ann Neurol* **59**, 388–393.

Ross, O. A., Farrer, M. J., and Wu, R. M. (2007a). Common variants in Parkinson's disease. *Mov Disord* **22**, 899–900.

Ross, O. A., Haugarvoll, K., Stone, J. T., Heckman, M. G., White, L. R., Aasly, J. O., Mark Gibson, J., Lynch, T., Wszolek, Z. K., Uitti, R. J., and Farrer, M. J. (2007b). Lack of evidence for association of Parkin promoter polymorphism (PRKN-258) with increased risk of Parkinson's disease. *Parkinsonism Relat Disord,* **13**, 386–388.

Satoh, J., and Kuroda, Y. (2001). A polymorphic variation of serine to tyrosine at codon 18 in the ubiquitin C-terminal hydrolase-L1 gene is associated with a reduced risk of sporadic Parkinson's disease in a Japanese population. *J Neurol Sci* **189**, 113–117.

Semchuk, K. M., Love, E. J., and Lee, R. G. (1993). Parkinson's disease: A test of the multifactorial etiologic hypothesis. *Neurology* **43**, 1173–1180.

Shimura, H., Hattori, N., Kubo, S., Mizuno, Y., Asakawa, S., Minoshima, S., Shimizu, N., Iwai, K., Chiba, T., Tanaka, K., and Suzuki, T. (2000). Familial Parkinson disease gene product, parkin, is a ubiquitin-protein ligase. *Nat Genet* **25**, 302–305.

Sim, C. H., Lio, D. S., Mok, S. S., Masters, C. L., Hill, A. F., Culvenor, J. G., and Cheng, H. C. (2006). C-terminal truncation and Parkinson's disease-associated mutations down-regulate the protein serine/threonine

kinase activity of PTEN-induced kinase-1. *Hum Mol Genet* **15**, 3251–3262.

Simon, D. K., Lin, M. T., and Pascual-Leone, A. (2002). "Nature versus nurture" and incompletely penetrant mutations. *J Neurol Neurosurg Psychiatry* **72**, 686–689.

Singleton, A. B., Farrer, M., Johnson, J., Singleton, A., Hague, S., Kachergus, J., Hulihan, M., Peuralinna, T., Dutra, A., Nussbaum, R., Lincoln, S., Crawley, A., Hanson, M., Maraganore, D., Adler, C., Cookson, M. R., Muenter, M., Baptista, M., Miller, D., Blancato, J., Hardy, J., and Gwinn-Hardy, K. (2003). Alpha-synuclein locus triplication causes Parkinson's disease. *Science* **302**, 841.

Skipper, L., Wilkes, K., Toft, M., Baker, M., Lincoln, S., Hulihan, M., Ross, O. A., Hutton, M., Aasly, J., and Farrer, M. (2004). Linkage disequilibrium and association of MAPT H1 in Parkinson disease. *Am J Hum Genet* **75**, 669–677.

Skipper, L., Li, Y., Bonnard, C., Pavanni, R., Yih, Y., Chua, E., Sung, W. K., Tan, L., Wong, M. C., Tan, E. K., and Liu, J. (2005). Comprehensive evaluation of common genetic variation within LRRK2 reveals evidence for association with sporadic Parkinson's disease. *Hum Mol Genet* **14**, 3549–3556.

Spillantini, M. G., Schmidt, M. L., Lee, V. M., Trojanowski, J. Q., Jakes, R., and Goedert, M. (1997). Alpha-synuclein in Lewy bodies. *Nature* **388**, 839–840.

Spillantini, M. G., Murrell, J. R., Goedert, M., Farlow, M. R., Klug, A., and Ghetti, B. (1998). Mutation in the tau gene in familial multiple system tauopathy with presenile dementia. *Proc Natl Acad Sci USA* **95**, 7737–7741.

Strauss, K. M., Martins, L. M., Plun-Favreau, H., Marx, F. P., Kautzmann, S., Berg, D., Gasser, T., Wszolek, Z., Muller, T., Bornemann, A., Wolburg, H., Downward, J., Riess, O., Schulz, J. B., and Kruger, R. (2005). Loss of function mutations in the gene encoding Omi/HtrA2 in Parkinson's disease. *Hum Mol Genet* **14**, 2099–2111.

Suzuki, Y., Imai, Y., Nakayama, H., Takahashi, K., Takio, K., and Takahashi, R. (2001). A serine protease, HtrA2, is released from the mitochondria and interacts with XIAP, inducing cell death. *Mol Cell* **8**, 613–621.

Takahashi, H., Ohama, E., Suzuki, S., Horikawa, Y., Ishikawa, A., Morita, T., Tsuji, S., and Ikuta, F. (1994). Familial juvenile parkinsonism: Clinical and pathologic study in a family. *Neurology* **44**, 437–441.

Tang, B., Xiong, H., Sun, P., Zhang, Y., Wang, D., Hu, Z., Zhu, Z., Ma, H., Pan, Q., Xia, J. H., Xia, K., and Zhang, Z. (2006). Association of PINK1 and DJ-1 confers digenic inheritance of early-onset Parkinson's disease. *Hum Mol Genet* **15**, 1816–1825.

Tanner, C. M., Ottman, R., Goldman, S. M., Ellenberg, J., Chan, P., Mayeux, R., and Langston, J. W. (1999).

Parkinson disease in twins: An etiologic study. *JAMA* **281**, 341–346.

Tassin, J., Durr, A., Bonnet, A. M., Gil, R., Vidailhet, M., Lucking, C. B., Goas, J. Y., Durif, F., Abada, M., Echenne, B., Motte, J., Lagueny, A., Lacomblez, L., Jedynak, P., Bartholome, B., Agid, Y., and Brice, A. (2000). Levodopa-responsive dystonia. GTP cyclohydrolase I or parkin mutations? *Brain* **123**(Pt 6), 1112–1121.

Valente, E. M., Bentivoglio, A. R., Dixon, P. H., Ferraris, A., Ialongo, T., Frontali, M., Albanese, A., and Wood, N. W. (2001). Localization of a novel locus for autosomal recessive early-onset parkinsonism, PARK6, on human chromosome 1p35-p36. *Am J Hum Genet* **68**, 895–900.

Valente, E. M., Brancati, F., Ferraris, A., Graham, E. A., Davis, M. B., Breteler, M. M., Gasser, T., Bonifati, V., Bentivoglio, A. R., De Michele, G., Durr, A., Cortelli, P., Wassilowsky, D., Harhangi, B. S., Rawal, N., Caputo, V., Filla, A., Meco, G., Oostra, B. A., Brice, A., Albanese, A., Dallapiccola, B., and Wood, N. W. (2002). Park6-linked parkinsonism occurs in several European families. *Ann Neurol* **51**, 14–18.

Valente, E. M., Abou-Sleiman, P. M., Caputo, V., Muqit, M. M., Harvey, K., Gispert, S., Ali, Z., Del Turco, D., Bentivoglio, A. R., Healy, D. G., Albanese, A., Nussbaum, R., Gonzalez-Maldonado, R., Deller, T., Salvi, S., Cortelli, P., Gilks, W. P., Latchman, D. S., Harvey, R. J., Dallapiccola, B., Auburger, G., and Wood, N. W. (2004a). Hereditary early-onset Parkinson's disease caused by mutations in PINK1. *Science* **15**, 15.

Valente, E. M., Salvi, S., Ialongo, T., Marongiu, R., Elia, A. E., Caputo, V., Romito, L., Albanese, A., Dallapiccola, B., and Bentivoglio, A. R. (2004b). PINK1 mutations are associated with sporadic early-onset parkinsonism. *Ann Neurol* **56**, 336–341.

van Duijn, C. M., Dekker, M. C., Bonifati, V., Galjaard, R. J., Houwing-Duistermaat, J. J., Snijders, P. J., Testers, L., Breedveld, G. J., Horstink, M., Sandkuijl, L. A., van Swieten, J. C., Oostra, B. A., and Heutink, P. (2001). Park7, a novel locus for autosomal recessive early-onset parkinsonism, on chromosome 1p36. *Am J Hum Genet* **69**, 629–634.

Vaughan, J. R., Farrer, M. J., Wszolek, Z. K., Gasser, T., Durr, A., Agid, Y., Bonifati, V., DeMichele, G., Volpe, G., Lincoln, S., Breteler, M., Meco, G., Brice, A., Marsden, C. D., Hardy, J., and Wood, N. W. (1998). Sequencing of the alpha-synuclein gene in a large series of cases of familial Parkinson's disease fails to reveal any further mutations. The European Consortium on Genetic Susceptibility in Parkinson's Disease (GSPD). *Hum Mol Genet* **7**, 751–753.

West, A., Farrer, M., Petrucelli, L., Cookson, M., Lockhart, P., and Hardy, J. (2001). Identification and characterization of the human parkin gene promoter. *J Neurochem* **78**, 1146–1152.

West, A., Periquet, M., Lincoln, S., Lucking, C. B., Nicholl, D., Bonifati, V., Rawal, N., Gasser, T., Lohmann, E., Deleuze, J. F., Maraganore, D., Levey, A., Wood, N., Durr, A., Hardy, J., Brice, A., and Farrer, M. (2002a). Complex relationship between Parkin mutations and Parkinson disease. *Am J Med Genet* **114**, 584–591.

West, A. B., Maraganore, D., Crook, J., Lesnick, T., Lockhart, P. J., Wilkes, K. M., Kapatos, G., Hardy, J. A., and Farrer, M. J. (2002b). Functional association of the parkin gene promoter with idiopathic Parkinson's disease. *Hum Mol Genet* **11**, 2787–2792.

West, A. B., Moore, D. J., Biskup, S., Bugayenko, A., Smith, W. W., Ross, C. A., Dawson, V. L., and Dawson, T. M. (2005). Parkinson's disease-associated mutations in leucine-rich repeat kinase 2 augment kinase activity. *Proc Natl Acad Sci USA* **102**, 16842–16847.

Wirdefeldt, K., Gatz, M., Schalling, M., and Pedersen, N. L. (2004). No evidence for heritability of Parkinson disease in Swedish twins. *Neurology* **63**, 305–311.

Zabetian, C. P., Hutter, C. M., Factor, S. A., Nutt, J. G., Higgins, D. S., Griffith, A., Roberts, J. W., Leis, B. C., Kay, D. M., Yearout, D., Montimurro, J. S., Edwards, K. L., Samii, A., and Payami, H. (2007). Association analysis of MAPT H1 haplotype and subhaplotypes in Parkinson's disease. *Ann Neurol* **62**, 137–144.

Zarranz, J. J., Alegre, J., Gomez-Esteban, J. C., Lezcano, E., Ros, R., Ampuero, I., Vidal, L., Hoenicka, J., Rodriguez, O., Atares, B., Llorens, V., Gomez Tortosa, E., del Ser, T., Munoz, D. G., and de Yebenes, J. G. (2004). The new mutation, E46K, of alpha-synuclein causes Parkinson and Lewy body dementia. *Ann Neurol* **55**, 164–173.

Zimprich, A., Biskup, S., Leitner, P., Lichtner, P., Farrer, M., Lincoln, S., Kachergus, J., Hulihan, M., Uitti, R. J., Calne, D. B., Stoessl, A. J., Pfeiffer, R. F., Patenge, N., Carbajal, I. C., Vieregge, P., Asmus, F., Muller-Myhsok, B., Dickson, D. W., Meitinger, T., Strom, T. M., Wszolek, Z. K., and Gasser, T. (2004a). Mutations in LRRK2 cause autosomal-dominant parkinsonism with pleomorphic pathology. *Neuron* **44**, 601–607.

Zimprich, A., Muller-Myhsok, B., Farrer, M., Leitner, P., Sharma, M., Hulihan, M., Lockhart, P., Strongosky, A., Kachergus, J., Calne, D. B., Stoessl, J., Uitti, R. J., Pfeiffer, R. F., Trenkwalder, C., Homann, N., Ott, E., Wenzel, K., Asmus, F., Hardy, J., Wszolek, Z., and Gasser, T. (2004b). The PARK8 locus in autosomal dominant parkinsonism: Confirmation of linkage and further delineation of the disease-containing interval. *Am J Hum Genet* **74**, 11–19.

3

NEUROPATHOLOGY OF PARKINSON'S DISEASE[1]

LEONARD PETRUCELLI AND DENNIS W. DICKSON

Mayo Clinic, Jacksonville, FL, USA

INTRODUCTION

Parkinson's disease (PD) is clinically characterized by a motor syndrome featuring bradykinesia,
tremor, rigidity, and postural instability; however,
non-motor manifestations, including autonomic and
cognitive dysfunction, are increasingly recognized
as being part of the wider clinical syndrome. The
histopathology of PD is characterized by a loss of
dopaminergic neurons in the substantia nigra (SN)
that project to the striatum. Disorders that produce
clinical parkinsonism share this common feature.
In PD, neurons that degenerate, accumulate cytoplasmic inclusion bodies composed of α-synuclein
(Spillantini *et al.*, 1997); these inclusions are referred
to as Lewy bodies (LBs) (Figure 3.1). Most other
parkinsonian disorders, such as progressive supranuclear palsy and multiple system atrophy, are not
associated with LBs. In addition, LBs are detected
in brains of individuals with a range of other non-
parkinsonian clinical syndromes, such as dementia,
psychosis, and dysautonomia. This chapter focuses
on the neuropathology of Lewy body parkinsonism.

GROSS AND MICROSCOPIC FINDINGS IN PD

In patients with PD the external appearance of the
brain is most often unremarkable or it shows atrophy

that is not different from that expected for the age
of the individual. In contrast, when the brain is
sectioned, there is loss of the expected dark black
pigmentation in the SN and locus coeruleus (LC)
(Figure 3.2). The dark pigment in these two nuclei
corresponds to neuronal cytoplasmic neuromelanin, which accumulates in an age-related manner
and is first detected consistently only after the second decade of life (Graham, 1979). Loss of pigment
correlates with neuronal loss and increases in proportion to the severity and duration of parkinsonism (Halliday *et al.*, 2005).

Neuronal loss in the SN is not uniform. The
ventrolateral tier of neurons in the pars compacta
(A9) is selectively vulnerable in parkinsonian disorders. In cases of long duration neuronal loss in the
ventrolateral SN may be almost total (Figure 3.3). In
contrast, there is evidence that dorsal tier neurons may be vulnerable to age-related neuronal
loss (Gibb and Lees, 1991) and the medial neuronal groups (e.g., ventral tegmental region or A10)
may be lost in patients with dementia (Rinne *et al.*,
1989). Age-associated neuronal loss is never as
severe as that detected in PD.

In addition to the SN, neuronal loss is readily
apparent with routine histologic methods in other
vulnerable nuclei (Figure 3.3). Like the SN, neuronal loss in the LC in long standing PD cases may
be almost total. Other nuclei that show neuronal
loss that is not usually as severe as the SN and LC
are the basal nucleus of Meynert and dorsal motor
nucleus of the vagus (Figure 3.3). In all of these
regions neuronal loss is accompanied by LBs as well
as α-synuclein-immunoreactive neuritic processes

[1] Supported by NIH grants P50-NS40256, P50-AG25711,
P50-AG16574, P01-AG17216, and P01-AG03949.

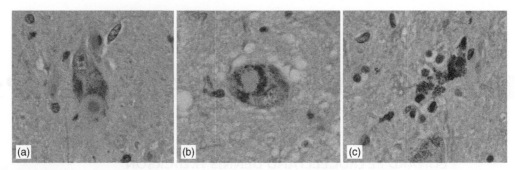

FIGURE 3.1 Lewy bodies (a) are hyaline intracytoplasmic inclusions that are found in neuromelanin containing neurons in the SN. A related inclusion is the "pale body" (b), which is less compact and considered by some to represent a precursor to classic hyaline-type LBs. In the end, neurons in the SN die and undergo phagocytosis, a process referred to as neuronophagia (c) (a, b, c ×400).

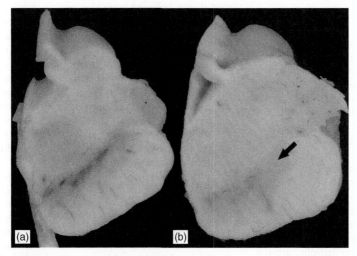

FIGURE 3.2 Transverse sections of the midbrain at the level of the third nerve in an elderly normal individual (a) and a patient with long standing PD (b). The major difference is loss of neuromelanin pigment that is most marked in the lateral parts of the SN (arrow).

(Figure 3.4). In regions with the most severe neuronal loss, LBs may be sparse. Hence, for these regions there is a poor correlation between LB density and disease duration. In contrast to LBs, at least a few α-synuclein-immunoreactive neuritic processes are usually apparent even in advanced disease.

On routine histologic stains, only a subset of α-synuclein-immunoreactive neurites can be detected, while others are essentially invisible (Figure 3.4). The histologically visible processes are called "intraneuritic Lewy bodies," and the neuritic processes that are only visible with immunohistochemical methods are called "Lewy neurites" (LNs) (Figure 3.4) (Braak *et al.*, 1999). In terms of absolute amount of α-synuclein pathology, the burden in neuritic pathology is far greater than in LBs and this correlates better with clinical features (Churchyard and Lees, 1997). In addition to the major nuclei that show neuronal loss and LBs, LNs are also detected in areas that are not typically vulnerable to classical LBs, such as the hippocampus (Dickson *et al.*, 1991), amygdala (Braak *et al.*, 1994), neocortex, (Irizarry *et al.*, 1998) and olfactory bulb (Del Tredici *et al.*, 2002) (Figure 3.5).

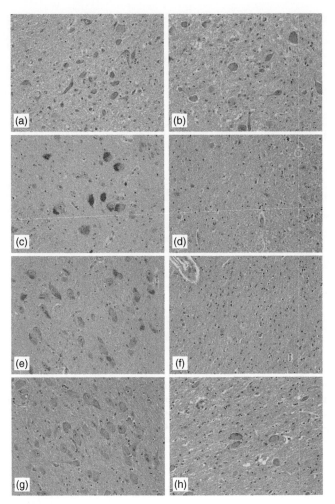

FIGURE 3.3 Sections of vulnerable nuclei in an elderly normal individual (a, c, e, and g) and a patient with long standing PD (b, d, f, and h). Neuronal loss in PD is selective and affects the dorsal motor nucleus of the vagus (b), locus coeruleus (d), substantia nigra, (f) and basal nucleus of Meynert (h). Neuronal loss is almost total in both the LC and the SN (d and f) compared to the normal (c and e). Neuronal loss is substantial in the dorsal motor nucleus of the vagus and the basal nucleus of Meynert, but not total (all figures are ×200).

Neuronal degeneration and loss is also accompanied by reactive changes in astrocytes and microglia. Microglia express markers of activation, such as the class II major histocompatibility antigen HLA-DR (McGeer *et al.*, 1988). Neurons in the SN and LC occasionally show evidence of phagocytosis by microglia, which is called neuronophagia (Figure 3.1). A hallmark of neuronal loss in the SN and LC is the presence of neuromelanin pigment in the cytoplasm of microglia. In PD patients with long disease duration, even this pigment is lost as microglia migrate to blood vessels and exit the brain along with the neuromelanin pigment. Additional pathologic findings in the SN, but not in other areas vulnerable to neurodegeneration, include axonal spheroids (Figure 3.6) and hemosiderin granules. Eosinophilic intranuclear inclusions, so-called Marinesco bodies, are common in the SN.

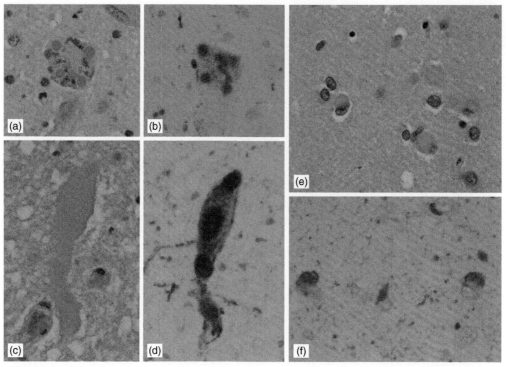

FIGURE 3.4   LBs contain α-synuclein. Typical hyaline inclusions, sometimes more than one within the same neuron (a), show α-synuclein immunoreactivity (b). Swollen neuronal processes, most notably in the dorsal motor nucleus of the vagus (c), also contain compact dense α-synuclein-immunoreactive inclusions (so-called "intraneuritic Lewy bodies") (d). In the cortex less well defined cytoplasmic inclusions, so-called "cortical Lewy bodies," (e) also have α-synuclein immunoreactivity (f). Note also the presence of fine neuritic processes (all figures are ×400).

Although they can be found in the brains of normal elderly subjects, one study found that they might be associated with dopaminergic pathology (Beach *et al.*, 2004).

As noted above, neuronal loss in PD is not limited to dopaminergic neurons of the SN but also occurs in cholinergic neurons in the basal nucleus of Meynert (Whitehouse *et al.*, 1983; Nakano and Hirano, 1984) and the pedunculopontine nucleus (Zweig *et al.*, 1989), noradrenergic neurons in the LC (Mann *et al.*, 1983; Jellinger, 1991), and serotonergic neurons of the raphe nuclei (Halliday *et al.*, 1990). Clearly neurochemical deficits in PD include not only dopamine, but also noradrenalin, serotonin, and acetylcholine. In addition to these neurotransmitter deficits, decreases in glutamatergic neurons in the cortex and hypothalamic neuropeptides have also been reported in PD (Jellinger and Mizuno, 2003).

NEUROPATHOLOGIC STAGING OF PD

Recently, Braak and coworkers (2004) proposed a pathologic staging system for PD. In this system, early pathologic manifestations of PD are detected in the dorsal motor nucleus of the vagus and anterior olfactory nucleus, with subsequent spread of the pathology to LC, SN, and basal forebrain. In the final stages, pathology extends to the neocortex, particularly limbic cortices and multimodal association cortices of frontal and temporal lobes (Table 3.1). This progression of pathology in the cortex is similar to that originally described by Kosaka and coworkers for Lewy body dementia (Katsuse *et al.*, 2003). The Braak PD staging system has been verified in autopsy studies of normal controls and patients with disorders associated with LBs (Dickson *et al.*, 2006). Although the staging

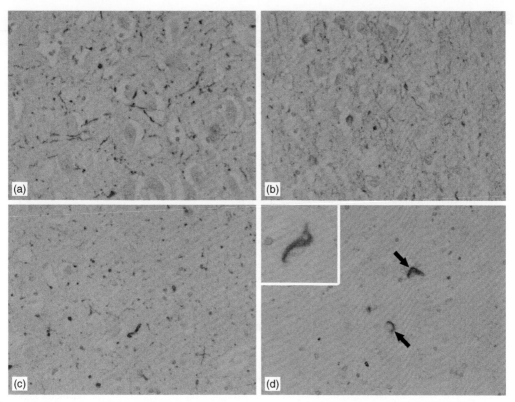

FIGURE 3.5 LNs are abundant in the hippocampus CA2 sector (a), but also present in the anterior olfactory nucleus in the olfactory bulb (b) and the amygdala (c). In the latter location many of the neurites have a dot-like morphology (c). In the midbrain and basal ganglia (d) in addition to dot-like and short neurites, α-synuclein-immunoreactive glial cytoplasmic inclusions (arrows) are not uncommon (inset) (all figures are ×400).

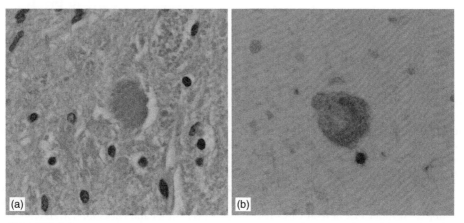

FIGURE 3.6 Granular swollen axons, "axonal spheroids," are common in the SN and globus pallidus in PD (a). Spheroids are immuno-positive for α-synuclein (b) (a and b ×400).

TABLE 3.1   Distribution of α-synuclein pathology in PD

Anatomical region	α-Synuclein pathology	Neuronal loss	Braak PD stage
Autonomic nervous system			
Sympathetic ganglia	LN, LB	+	0
Gastroesophageal	LN	−	0
Cardiac	LN	−	0
Adrenal	LN	−	0
Olfactory bulb			
Anterior olfactory nucleus	LN	++	1
Medulla			
Dorsal motor nucleus of the vagus	LB, iLB	++	1
Pons			
Locus coeruleus	LB, LN	+++	2
Midbrain			
Substantia nigra (A9)	LB, LN, GCI, Sph	+++	3
Basal forebrain			
Basal nucleus of Meynert	cLB, iLB	++	3
Amygdala	cLB, LN,	−	4
Medial temporal lobe			
Hippocampus (CA2)	LN	−	4
Hippocampus (CA4)	cLB	−	4
Convexity neocortex			
Frontal cortex	cLB, LN	−	5
Parietal cortex	cLB, LN	−	6
Basal ganglia			
Putamen	LN, GCI	−	5–6
Globus pallidus	Sph	−	5–6

LB = Lewy bodies; LN = Lewy neurites; GCI = glial cytoplasmic inclusions; Sph = spheroids; neuronal loss: +++ = severe; ++ = moderate; + = documented; − = negative or not documented.

system fits in most cases, it does not fit all disorders associated with LBs. The staging scheme does not have relevance to LBs that occur in the setting of Alzheimer's disease (AD) (Dickson *et al.*, 2006), where many cases have LBs confined to the amygdala (Uchikado *et al.*, 2006). In a large cross-sectional study of 904 brains, Parkkinen and coworkers (2003, 2005a) used α-synuclein immunohistochemistry to study the frequency and distribution of LBs in brains of subjects without PD. Several cases had LBs in the SN without the involvement of DMN, a finding also noted by Jellinger (2003).

Involvement of the spinal cord intermediolateral column is relatively early in the disease process, possibly after the dorsal motor nucleus of the vagus, but before the SN (Klos *et al.*, 2006). In addition some individuals with possibly preclinical PD have LBs in autonomic ganglia (Saito *et al.*, 2003). The basal ganglia have α-synuclein pathology, mostly LNs, in many cases of PD and evidence suggests that it is involved relatively late in the disease process (Tsuboi and Dickson, 2005).

If the staging system proposed by Braak is correct, non-motor, non-dopaminergic symptoms should precede motor symptoms (Langston, 2006). For example, one might predict that early PD may be characterized by autonomic dysfunction, olfactory dysfunction, sleep disorder, and depression, given the roles of lower brainstem monoaminergic nuclei and spinal and enteric ganglia in those processes (Langston, 2006). Epidemiologic studies indicate that autonomic symptoms precede clinical PD by as much as 12 years (Abbott *et al.*, 2001). One might speculate that the pathology responsible for this syndrome is LBs in autonomic nuclei of the lower brainstem and spinal cord as well as the enteric plexus (Braak *et al.*, 2006). Another clinical syndrome that may be a harbinger of PD is rapid eye movement (REM) behavior disorder (RBD), a condition that appears several years before PD (Schenck *et al.*, 1996). The RBD syndrome appears to have its anatomic origins in lower brainstem nuclei (Boeve *et al.*, 2007) that are consistently affected in LB dementia. Olfactory dysfunction is common in PD (Hawkes *et al.*, 1997), and it may precede overt motor symptoms (Berendse *et al.*, 2001). The later stages of PD are associated with involvement of limbic cortices and multimodal association cortices and finally unimodal association cortices and primary cortices. These later stages of PD involve mainly cognitive and psychiatric symptoms similar to those found in LB dementia (McKeith *et al.*, 2004).

MOLECULAR BIOLOGY OF LBS AND NEURONAL LOSS IN PD

Classical LBs have a hyaline appearance on routine histologic stains (e.g., hematoxylin and eosin), whereas α-synuclein-immunoreactive inclusions in less vulnerable neuronal populations, such as neurons in the amygdala and the neocortex, are pale staining and poorly circumscribed. These lesions are referred to as cortical LBs (Ikeda *et al.*, 1978). Another related pale-staining neuronal cytoplasmic inclusion found in pigmented brainstem neurons of the SN and LC is the pale body (Figure 3.1) (Pappolla *et al.*, 1988; Dale *et al.*, 1992). Evidence suggests that cortical LBs and pale bodies might be early cytologic alterations that precede the classical LB, so-called "pre-LBs." In some cases of PD with severe pathology, hyaline-type inclusions consistent with classical LBs can be detected in the amygdala and cortex, particularly the limbic cortex. Although most of the α-synuclein-immunoreactive cytopathology in PD is within neurons, careful inspection often reveals small α-synuclein-immunoreactive glia, particularly in the midbrain and basal ganglia (Figure 3.5) (Wakabayashi and Takahashi, 1996; Wakabayashi *et al.*, 2000). The significance of glial pathology is unknown, but fibrillary glial inclusions composed of α-synuclein are the hallmark of multiple system atrophy, a parkinsonian disorder with a distinct pattern on neuronal and glial α-synuclein pathology (Dickson *et al.*, 1999).

At the ultrastructural level, LBs are composed of dense granular material and straight filaments that are approximately 10–15 nm in diameter (Forno, 1969; Tiller-Borcich and Forno, 1988; Galloway *et al.*, 1992). Similar filaments have been created in the test tube by using recombinant α-synuclein, which is normally an unfolded and structureless protein (Conway *et al.*, 2000; Crowther *et al.*, 2000). This fact, combined with the immunolocalization of α-synuclein to the filaments in tissue sections by electron microscopy, indicates that the filaments in LBs are probably derived from aggregates of α-synuclein that have an abnormal conformation. Furthermore, the presence of α-synuclein in cytoplasmic inclusions indicates that it undergoes aberrant cytologic localization, since it is normally a protein enriched in presynaptic terminals (Irizarry *et al.*, 1996). The factors that cause the abnormal conformation remain to be determined, but several post-translational modifications, including phosphorylation, truncation, and oxidative damage have been implicated (Dickson, 2001).

In addition to α-synuclein other proteins have been localized to LBs mostly through immunohistochemical methods. Of the many proteins only a few have been reported in more than one study and these include neurofilament (Galvin *et al.*, 1997), ubiquitin, (Kuzuhara *et al.*, 1988) and the ubiquitin-binding protein p62 (Kuusisto *et al.*, 2003). A subset (<50%) of LBs and LNs shows immunoreactivity with antibodies to the microtubule-associated protein tau (Ishizawa *et al.*, 2003), which is the molecular determinant of neurofibrillary tangles found in AD and other disorders. LNs and intraneuritic LBs have the same immunoreactivity profile as LBs. Biochemical studies of purified LBs indicate that in addition to α-synuclein (Baba *et al.*, 1998), they may contain other proteins, such as neurofilament (Pollanen *et al.*, 1993; Galvin *et al.*, 1997).

α-Synuclein is a 140-amino acid protein with an estimated molecular weight of 19 kDa. It is highly enriched in presynaptic terminals throughout the central nervous system (Irizarry et al., 1996; Takeda et al., 1998; Beyer, 2006). It shares homology with two other proteins, β-synuclein and γ-synuclein which are both expressed in the nervous system (Galvin et al., 2001). Sequence analysis revealed the following three protein domains in α-synuclein: a highly conserved amino-terminal region (where all known coding region mutations that cause PD reside [Polymeropoulos et al., 1997; Kruger et al., 1998; Zarranz et al., 2004]) that may have lipid-binding, apolipoprotein-like properties, and an α-helical secondary structure (Bussell and Eliezer, 2003); an internal hydrophobic domain that is critical for self aggregation and fibril formation (Giasson et al., 2001); and a carboxyl-terminal acidic region that is rich in glutamate and aspartate residues and is the site of phosphorylation (Saito et al., 2003). In solution, α-synuclein is a natively unfolded protein, but it becomes more structured when it binds lipids (Perrin et al., 2000). In LBs and LNs as well as in other glial inclusions, it assumes a β-sheet structure similar to amyloid (Conway et al., 2000), a structural change associated with the increased propensity of α-synuclein to aggregate. In contrast to α-synuclein, β-synuclein has sequence differences in the hydrophobic domain that make it incapable of aggregating (Giasson et al., 2001). In fact, when expressed simultaneously, β-synuclein may have anti-aggregating effects on α-synuclein (Hashimoto et al., 2001).

The known mutations in the coding region of the gene for α-synuclein generate α-synuclein species that have greater toxicity than wild-type α-synuclein (Moussa et al., 2004; Jiang et al., 2007). Truncated species of α-synuclein are higher in patients with PD compared to normal controls (Li et al., 2005). Other mutations in the gene for α-synuclein, such as gene multiplications (Singleton et al., 2003; Chartier-Harlin et al., 2004; Fuchs et al., 2007), do not alter the sequence but lead to overexpression of α-synuclein. It remains to be determined how excessive α-synuclein leads to toxicity, but protein aggregation is likely critical (Periquet et al., 2007). In sporadic PD, defective protein degradation may lead to critical levels of α-synuclein concentration that favor aggregation within the cell. Both proteasomal and non-proteasomal (e.g., calpain [Mishizen-Eberz et al., 2005]) pathways may be involved in proteolysis of α-synuclein, and evidence suggests that proteasomal functions may be impaired in PD (McNaught and Jenner, 2001).

The proteasomal pathway is implicated because it interfaces with protein ubiquitination, and pathologic α-synuclein is ubiquitinated. Thus, deficiencies in the ubiquitin–proteasomal processing of α-synuclein may lead to its accumulation to critical levels that lead to aggregation.

The normal function of α-synuclein is unknown, but structural analysis suggests that it may function as a chaperone. It shares sequence homology with the family of 14-3-3 chaperone molecules (Ostrerova et al., 1999) and may modulate of the activity of tyrosine hydroxylase, the rate-limiting enzyme in dopamine synthesis (Perez et al., 2002). It has also been suggested that α-synuclein plays a role in synaptic vesicle trafficking in dopaminergic neurons (Murphy et al., 2000). It undergoes phosphorylation, and as noted above phosphorylated forms are enriched in LBs (Saito et al., 2003), but it is not known whether phosphorylation is necessary for lesion formation (Chen and Feany, 2005). The physiologic function of phosphorylation and the kinases responsible (Ellis et al., 2001; Lee et al., 2004) are not known with certainty.

Whereas aggregated forms of α-synuclein may be toxic to neurons, there is evidence that normal α-synuclein may have neuroprotective properties (Kim et al., 2004). In cell culture studies α-synuclein expression is associated with a more differentiated state and relative resistance to apoptotic stimuli (Sidhu et al., 2004). In PD the mechanisms implicated in neuronal death include both programmed cell death (i.e., apoptosis) and lysosome-mediated (i.e., autophagic) cell death, but most of that evidence is based on animal or cell culture models. There is little evidence for a particular type of neuronal death in human PD brains (Anglade et al., 1997; Jellinger and Stadelmann, 2000), and both apoptosis and autophagy are pleiotropic in nature. For example, both protective and toxic effects have been attributed to autophagy (Rubinsztein et al., 2005).

LBS IN OTHER DISORDERS

Incidental LBs

LBs are found in about 10% of brains from normal elderly individuals over age 65 years (Forno, 1969). These cases may represent the earliest stages of PD, and the distribution of LBs and the non-motor clinical manifestations (e.g., RBD) in some cases seem to favor this argument (Uchiyama et al.,

1995; Iwanaga *et al.*, 1999; Del Tredici *et al.*, 2002; Jellinger, 2003; Parkkinen *et al.*, 2003; Bloch *et al.*, 2006; Klos *et al.*, 2006). In particular, such cases have LBs, albeit in small numbers and not accompanied by neuronal loss or gliosis, in brain regions that are vulnerable to pathology in full-blown PD. Given the lack of overt parkinsonism, such cases have been referred to as being "incidental." It is not known whether these individuals, who may or may not have non-motor prodromal features of PD, would have eventually progressed to PD, but preliminary evidence favors this hypothesis (Dickson *et al.*, 2008).

Pure Autonomic Failure

When the involvement of the autonomic nervous system in PD and prodromal PD was investigated, it was found that some individuals with pure autonomic failure have LB pathology at autopsy (Arai *et al.*, 2000). In those cases, LBs were detected in brain and autonomic ganglia and LNs in sympathetic nerve fibers in epicardium and peri-adrenal tissues (Hague *et al.*, 1997). It is of interest that PD and individuals with incidental LBS may also have adrenal α-synuclein pathology (Fumimura *et al.*, 2007).

Dementia with LBs

Dementia with Lewy bodies (DLB) is a clinicopathologic entity with a specific constellation of clinical features, including cognitive impairment, visual hallucinations, fluctuating cognition, and spontaneous extrapyramidal signs (McKeith *et al.*, 2004). Other common clinical features are RBD, severe neuroleptic sensitivity, and reduced striatal dopamine transporter activity on functional neuroimaging. The initial neuropathologic criterion for DLB was the presence of LBs in the brain of a patient with a clinical history of dementia (McKeith *et al.*, 1996), but this criterion was not rigorous and led to poor clinicopathologic correlations, since many cases also have varying degrees of AD pathology. The criteria for DLB were refined to emphasize that the likelihood of DLB is directly related to the severity of LBs and is inversely related to the severity of AD pathology (McKeith *et al.*, 2005). Severity of LB pathology is based on the distribution of LBs – brainstem predominant LBs, LBs in mostly limbic cortices, and LBs in multimodal association cortices. Severity

of AD-type pathology is assessed using methods to score the severity of cortical plaques and the topography of neurofibrillary degeneration. This revision to the neuropathologic criteria was based on prevailing evidence that the more severe the AD-type pathology, the less likely the patient would present with the DLB clinical syndrome, even if there were diffuse cortical LBs (Lopez *et al.*, 2002).

Dementia in PD

Formal clinical guidelines for the diagnosis of PD-related dementia have only recently been formulated (Emre *et al.*, 2007). Pathological findings that likely account for dementia in PD are mixed, but most recent evidence suggests that dementia is related to cortical and subcortical LB pathology, particularly affecting monoaminergic and cholinergic nuclei that project to the cortex (Whitehouse *et al.*, 1983; Nakano and Hirano, 1984). While some cases have coexistent AD, in many cases the Alzheimer-type pathology is insufficient to warrant confident diagnosis of AD. Although virtually all brains of patients with PD have a few cortical LBs, they are usually neither widespread nor numerous. Several recent studies have shown, however, that cortical LBs are numerous and widespread in cases of PD with dementia (Hurtig *et al.*, 2000; Apaydin *et al.*, 2002; Aarsland *et al.*, 2005) and that the density of cortical LBs and LNs, especially in medial temporal lobe structures (Churchyard and Lees, 1997), correlates with the severity of dementia. There are exceptions to this, with occasional reports of clinically normal patients with many LBs (Parkkinen *et al.*, 2005a, b).

RELEVANCE OF NEUROPATHOLOGY TO PD MODELS

Modeling PD is important for understanding the pathogenesis of the disease and for developing therapeutics. Neuropathologic findings may be brought to bear in considering what should be modeled. It is very unlikely that any model will be able to capture all aspects of neuropathology and choices need to be made on what feature is ultimately the most significant. Many of the current animal models focus on one of the two aspects of the pathology of PD – either α-synuclein pathology or pathology in the nigrostriatal dopaminergic pathway. Cellular and transgenic animal models tend to focus on the former, while toxin models tend to be more appropriate to the latter. Both approaches have their strengths and weaknesses. The challenge of animal models is to accurately replicate not only the morphologic aspects of the disease, but also the selective vulnerability and neurodegeneration. None of the available models shows neurodegeneration in all of the brain regions that are commonly affected in PD. As pathologic studies of PD increasingly focus on changes occurring outside of the nigrostriatal pathway, especially in autonomic and peripheral nervous systems in preclinical PD and in cortico-limbic areas in late stages of PD where cognitive problems become increasingly frequent, the challenge of animal models to recapitulate these features will need to be explored. An essential neuropathologic feature of PD is not only α-synuclein pathology, but also neuronal loss and gliosis. Additional studies in humans are needed to determine the relative contribution of cell stress (e.g., oxidative and endoplasmic reticulum stresses), protein degradation (e.g., ubiquitin–proteasome and lysosomal systems) and neuroinflammation to selective death of specific populations of neurons. Ultimately, better understanding of the disease process in humans will lead to better models of PD.

REFERENCES

Aarsland, D., Perry, R., Brown, A., Larsen, J. P., and Ballard, C. (2005). Neuropathology of dementia in Parkinson's disease: A prospective, community-based study. *Ann Neurol* 5, 773–776.

Abbott, R. D., Petrovitch, H., White, L. R., Masaki, K. H., Tanner, C. M., Curb, J. D., Grandinetti, A., Blanchette, P. L., Popper, J. S., and Ross, G. W. (2001). Frequency of bowel movements and the future risk of Parkinson's disease. *Neurology* 3, 456–462.

Anglade, P., Vyas, S., Javoy-Agid, F., Herrero, M. T., Michel, P. P., Marquez, J., Mouatt-Prigent, A., Ruberg, M., Hirsch, E. C., and Agid, Y. (1997). Apoptosis and autophagy in nigral neurons of patients with Parkinson's disease. *Histol Histopathol* 1, 25–31.

Apaydin, H., Ahlskog, J. E., Parisi, J. E., Boeve, B. F., and Dickson, D. W. (2002). Parkinson disease neuropathology: Later-developing dementia and loss of the levodopa response. *Arch Neurol* 1, 102–112.

Arai, K., Kato, N., Kashiwado, K., and Hattori, T. (2000). Pure autonomic failure in association with human alpha-synucleinopathy. *Neurosci Lett* 2–3, 171–173.

Baba, M., Nakajo, S., Tu, P. H., Tomita, T., Nakaya, K., Lee, V. M., Trojanowski, J. Q., and Iwatsubo, T. (1998). Aggregation of alpha-synuclein in Lewy bodies of sporadic Parkinson's disease and dementia with Lewy bodies. *Am J Pathol* 4, 879–884.

Beach, T. G., Walker, D. G., Sue, L. I., Newell, A., Adler, C. C., and Joyce, J. N. (2004). Substantia nigra Marinesco bodies are associated with decreased striatal expression of dopaminergic markers. *J Neuropathol Exp Neurol* 4, 329–337.

Berendse, H. W., Booij, J., Francot, C. M., Bergmans, P. L., Hijman, R., Stoof, J. C., and Wolters, E. C. (2001). Subclinical dopaminergic dysfunction in asymptomatic Parkinson's disease patients' relatives with a decreased sense of smell. *Ann Neurol* 1, 34–41.

Beyer, K. (2006). Alpha-synuclein structure, posttranslational modification and alternative splicing as aggregation enhancers. *Acta Neuropathol (Berl)* 3, 237–251.

Bloch, A., Probst, A., Bissig, H., Adams, H., and Tolnay, M. (2006). Alpha-synuclein pathology of the spinal and peripheral autonomic nervous system in neurologically unimpaired elderly subjects. *Neuropathol Appl Neurobiol* 3, 284–295.

Boeve, B. F., Silber, M. H., Saper, C. B., Ferman, T. J., Dickson, D. W., Parisi, J. E., Benarroch, E. E., Ahlskog, J. E., Smith, G. E., Caselli, R. C., Tippman-Peikert, M., Olson, E. J., Lin, S. C., Young, T., Wszolek, Z., Schenck, C. H., Mahowald, M. W., Castillo, P. R., Del Tredici, K., and Braak, H. (2007). Pathophysiology of REM sleep behaviour disorder and relevance to neurodegenerative disease. *Brain* Pt 11, 2770–2788.

Braak, H., Braak, E., Yilmazer, D., de Vos, R. A., Jansen, E. N., Bohl, J., and Jellinger, K. (1994). Amygdala pathology in Parkinson's disease. *Acta Neuropathol (Berl)* 6, 493–500.

Braak, H., Sandmann-Keil, D., Gai, W., and Braak, E. (1999). Extensive axonal Lewy neurites in Parkinson's disease: A novel pathological feature revealed by alpha-synuclein immunocytochemistry. *Neurosci Lett* 1, 67–69.

Braak, H., Ghebremedhin, E., Rub, U., Bratzke, H., and Del Tredici, K. (2004). Stages in the development of Parkinson's disease-related pathology. *Cell Tissue Res* 1, 121–134.

Braak, H., de Vos, R. A., Bohl, J., and Del Tredici, K. (2006). Gastric alpha-synuclein immunoreactive inclusions in Meissner's and Auerbach's plexuses in cases staged for Parkinson's disease-related brain pathology. *Neurosci Lett* 1, 67–72.

Bussell, R., Jr., and Eliezer, D. (2003). A structural and functional role for 11-mer repeats in alpha-synuclein and other exchangeable lipid binding proteins. *J Mol Biol* 4, 763–778.

Chartier-Harlin, M. C., Kachergus, J., Roumier, C., Mouroux, V., Douay, X., Lincoln, S., Levecque, C., Larvor, L., Andrieux, J., Hulihan, M., Waucquier, N.,

Defebvre, L., Amouyel, P., Farrer, M., and Destee, A. (2004). Alpha-synuclein locus duplication as a cause of familial Parkinson's disease. *Lancet* **9440**, 1167–1169.

Chen, L., and Feany, M. B. (2005). Alpha-synuclein phosphorylation controls neurotoxicity and inclusion formation in a Drosophila model of Parkinson disease. *Nat Neurosci* **5**, 657–663.

Churchyard, A., and Lees, A. J. (1997). The relationship between dementia and direct involvement of the hippocampus and amygdala in Parkinson's disease. *Neurology* **6**, 1570–1576.

Conway, K. A., Harper, J. D., and Lansbury, P. T., Jr. (2000). Fibrils formed *in vitro* from alpha-synuclein and two mutant forms linked to Parkinson's disease are typical amyloid. *Biochemistry* **10**, 2552–2563.

Crowther, R. A., Daniel, S. E., and Goedert, M. (2000). Characterisation of isolated alpha-synuclein filaments from substantia nigra of Parkinson's disease brain. *Neurosci Lett* **2**, 128–130.

Dale, G. E., Probst, A., Luthert, P., Martin, J., Anderton, B. H., and Leigh, P. N. (1992). Relationships between Lewy bodies and pale bodies in Parkinson's disease. *Acta Neuropathol (Berl)* **5**, 525–529.

Del Tredici, K., Rub, U., De Vos, R. A., Bohl, J. R., and Braak, H. (2002). Where does Parkinson disease pathology begin in the brain? *J Neuropathol Exp Neurol* **5**, 413–426.

Dickson, D. W. (2001). Alpha-synuclein and the Lewy body disorders. *Curr Opin Neurol* **4**, 423–432.

Dickson, D. W., Ruan, D., Crystal, H., Mark, M. H., Davies, P., Kress, Y., and Yen, S. H. (1991). Hippocampal degeneration differentiates diffuse Lewy body disease (DLBD) from Alzheimer's disease: Light and electron microscopic immunocytochemistry of CA2-3 neurites specific to DLBD. *Neurology* **9**, 1402–1409.

Dickson, D. W., Lin, W., Liu, W. K., and Yen, S. H. (1999). Multiple system atrophy: A sporadic synucleinopathy. *Brain Pathol* **4**, 721–732.

Dickson, D. W., Uchikado, H., klos, K. J., Josephs, K. A., Boeve, B., and Ahlskog, J. E. (2006). A critical review of the Braak staging scheme for Parkinson's disease. *Mov Disord* **Suppl 15**, S559.

Dickson, D. W., Fujishiro, H., Delledonne, A., Menke, J., Ahmed, Z., Klos, K. J., Josephs, K. A., Frigerio, R., Burnett, M., Parisi, J. E., and Ahlskog, J. E. (2008). Evidence that incidental Lewy body disease is presymptomatic Parkinson's disease. *Acta Neuropathol* **115**, 437–444.

Ellis, C. E., Schwartzberg, P. L., Grider, T. L., Fink, D. W., and Nussbaum, R. L. (2001). Alpha-synuclein is phosphorylated by members of the Src family of protein-tyrosine kinases. *J Biol Chem* **6**, 3879–3884.

Emre, M., Aarsland, D., Brown, R., Burn, D. J., Duyckaerts, C., Mizuno, Y., Broe, G. A., Cummings, J.,

Dickson, D. W., Gauthier, S., Goldman, J., Goetz, C., Korczyn, A., Lees, A., Levy, R., Litvan, I., McKeith, I., Olanow, W., Poewe, W., Quinn, N., Sampaio, C., Tolosa, E., and Dubois, B. (2007). Clinical diagnostic criteria for dementia associated with Parkinson's disease. *Mov Disord* **12**, 1689–1707, Quiz 1837.

Forno, L. S. (1969). Concentric hyalin intraneuronal inclusions of Lewy type in the brains of elderly persons (50 incidental cases): Relationship to parkinsonism. *J Am Geriatr Soc* **6**, 557–575.

Fuchs, J., Nilsson, C., Kachergus, J., Munz, M., Larsson, E. M., Schule, B., Langston, J. W., Middleton, F. A., Ross, O. A., Hulihan, M., Gasser, T., and Farrer, M. J. (2007). Phenotypic variation in a large Swedish pedigree due to SNCA duplication and triplication. *Neurology* **12**, 916–922.

Fumimura, Y., Ikemura, M., Saito, Y., Sengoku, R., Kanemaru, K., Sawabe, M., Arai, T., Ito, G., Iwatsubo, T., Fukayama, M., Mizusawa, H., and Murayama, S. (2007). Analysis of the adrenal gland is useful for evaluating pathology of the peripheral autonomic nervous system in Lewy body disease. *J Neuropathol Exp Neurol* **5**, 354–362.

Galloway, P. G., Mulvihill, P., and Perry, G. (1992). Filaments of Lewy bodies contain insoluble cytoskeletal elements. *Am J Pathol* **4**, 809–822.

Galvin, J. E., Lee, V. M., Baba, M., Mann, D. M., Dickson, D. W., Yamaguchi, H., Schmidt, M. L., Iwatsubo, T., and Trojanowski, J. Q. (1997). Monoclonal antibodies to purified cortical Lewy bodies recognize the mid-size neurofilament subunit. *Ann Neurol* **4**, 595–603.

Galvin, J. E., Schuck, T. M., Lee, V. M., and Trojanowski, J. Q. (2001). Differential expression and distribution of alpha-, beta-, and gamma-synuclein in the developing human substantia nigra. *Exp Neurol* **2**, 347–355.

Giasson, B. I., Murray, I. V., Trojanowski, J. Q., and Lee, V. M. (2001). A hydrophobic stretch of 12 amino acid residues in the middle of alpha-synuclein is essential for filament assembly. *J Biol Chem* **4**, 2380–2386.

Gibb, W. R., and Lees, A. J. (1991). Anatomy, pigmentation, ventral and dorsal subpopulations of the substantia nigra, and differential cell death in Parkinson's disease. *J Neurol Neurosurg Psychiatry* **5**, 388–396.

Graham, D. G. (1979). On the origin and significance of neuromelanin. *Arch Pathol Lab Med* **7**, 359–362.

Hague, K., Lento, P., Morgello, S., Caro, S., and Kaufmann, H. (1997). The distribution of Lewy bodies in pure autonomic failure: Autopsy findings and review of the literature. *Acta Neuropathol (Berl)* **2**, 192–196.

Halliday, G. M., Blumbergs, P. C., Cotton, R. G., Blessing, W. W., and Geffen, L. B. (1990). Loss of brainstem serotonin- and substance P-containing neurons in Parkinson's disease. *Brain Res* **1**, 104–107.

Halliday, G. M., Ophof, A., Broe, M., Jensen, P. H., Kettle, E., Fedorow, H., Cartwright, M. I., Griffiths, F. M., Shepherd, C. E., and Double, K. L. (2005). Alpha-synuclein redistributes to neuromelanin lipid in the substantia nigra early in Parkinson's disease. *Brain* **Pt 11**, 2654–2664.

Hashimoto, M., Rockenstein, E., Mante, M., Mallory, M., and Masliah, E. (2001). Beta-synuclein inhibits alpha-synuclein aggregation: A possible role as an anti-parkinsonian factor. *Neuron* **2**, 213–223.

Hawkes, C. H., Shephard, B. C., and Daniel, S. E. (1997). Olfactory dysfunction in Parkinson's disease. *J Neurol Neurosurg Psychiatry* **5**, 436–446.

Hurtig, H. I., Trojanowski, J. Q., Galvin, J., Ewbank, D., Schmidt, M. L., Lee, V. M., Clark, C. M., Glosser, G., Stern, M. B., Gollomp, S. M., and Arnold, S. E. (2000). Alpha-synuclein cortical Lewy bodies correlate with dementia in Parkinson's disease. *Neurology* **10**, 1916–1921.

Ikeda, K., Ikeda, S., Yoshimura, T., Kato, H., and Namba, M. (1978). Idiopathic Parkinsonism with Lewy-type inclusions in cerebral cortex. A case report. *Acta Neuropathol (Berl)* **2**, 165–168.

Irizarry, M. C., Kim, T. W., McNamara, M., Tanzi, R. E., George, J. M., Clayton, D. F., and Hyman, B. T. (1996). Characterization of the precursor protein of the non-A beta component of senile plaques (NACP) in the human central nervous system. *J Neuropathol Exp Neurol* **8**, 889–895.

Irizarry, M. C., Growdon, W., Gomez-Isla, T., Newell, K., George, J. M., Clayton, D. F., and Hyman, B. T. (1998). Nigral and cortical Lewy bodies and dystrophic nigral neurites in Parkinson's disease and cortical Lewy body disease contain alpha-synuclein immunoreactivity. *J Neuropathol Exp Neurol* **4**, 334–337.

Ishizawa, T., Mattila, P., Davies, P., Wang, D., and Dickson, D. W. (2003). Colocalization of tau and alpha-synuclein epitopes in Lewy bodies. *J Neuropathol Exp Neurol* **4**, 389–397.

Iwanaga, K., Wakabayashi, K., Yoshimoto, M., Tomita, I., Satoh, H., Takashima, H., Satoh, A., Seto, M., Tsujihata, M., and Takahashi, H. (1999). Lewy body-type degeneration in cardiac plexus in Parkinson's and incidental Lewy body diseases. *Neurology* **6**, 1269–1271.

Jellinger, K. A. (1991). Pathology of Parkinson's disease. Changes other than the nigrostriatal pathway. *Mol Chem Neuropathol* **3**, 153–197.

Jellinger, K. A. (2003). Alpha-synuclein pathology in Parkinson's and Alzheimer's disease brain: Incidence and topographic distribution – A pilot study. *Acta Neuropathol (Berl)* **3**, 191–201.

Jellinger, K. A., and Stadelmann, C. (2000). Mechanisms of cell death in neurodegenerative disorders. *J Neural Transm Suppl*, 95–114.

Jellinger, K., and Mizuno, Y. (2003). Parkinson's disease. In *Neurodegeneration: The Molecular Pathology of Dementia and Movement Disorders* (D. W. Dickson, Ed.), pp. 159–187. International Society of Neuropathology, Basel, Switzerland.

Jiang, H., Wu, Y. C., Nakamura, M., Liang, Y., Tanaka, Y., Holmes, S., Dawson, V. L., Dawson, T. M., Ross, C. A., and Smith, W. W. (2007). Parkinson's disease genetic mutations increase cell susceptibility to stress: Mutant alpha-synuclein enhances H(2)O(2)- and Sin-1-induced cell death. *Neurobiol Aging* **29**, 1709–1717.

Katsuse, O., Iseki, E., Marui, W., and Kosaka, K. (2003). Developmental stages of cortical Lewy bodies and their relation to axonal transport blockage in brains of patients with dementia with Lewy bodies. *J Neurol Sci* **1–2**, 29–35.

Kim, T. D., Choi, E., Rhim, H., Paik, S. R., and Yang, C. H. (2004). Alpha-synuclein has structural and functional similarities to small heat shock proteins. *Biochem Biophys Res Commun* **4**, 1352–1359.

Klos, K. J., Ahlskog, J. E., Josephs, K. A., Apaydin, H., Parisi, J. E., Boeve, B. F., DeLucia, M. W., and Dickson, D. W. (2006). Alpha-synuclein pathology in the spinal cords of neurologically asymptomatic aged individuals. *Neurology* **7**, 1100–1102.

Kruger, R., Kuhn, W., Muller, T., Woitalla, D., Graeber, M., Kosel, S., Przuntek, H., Epplen, J. T., Schols, L., and Riess, O. (1998). Ala30Pro mutation in the gene encoding alpha-synuclein in Parkinson's disease. *Nat Genet* **2**, 106–108.

Kuusisto, E., Parkkinen, L., and Alafuzoff, I. (2003). Morphogenesis of Lewy bodies: Dissimilar incorporation of alpha-synuclein, ubiquitin, and p62. *J Neuropathol Exp Neurol* **12**, 1241–1253.

Kuzuhara, S., Mori, H., Izumiyama, N., Yoshimura, M., and Ihara, Y. (1988). Lewy bodies are ubiquitinated. A light and electron microscopic immunocytochemical study. *Acta Neuropathol (Berl)* **4**, 345–353.

Langston, J. W. (2006). The Parkinson's complex: Parkinsonism is just the tip of the iceberg. *Ann Neurol* **4**, 591–596.

Lee, G., Tanaka, M., Park, K., Lee, S. S., Kim, Y. M., Junn, E., Lee, S. H., and Mouradian, M. M. (2004). Casein kinase II-mediated phosphorylation regulates alpha-synuclein/synphilin-1 interaction and inclusion body formation. *J Biol Chem* **8**, 6834–6839.

Li, W., West, N., Colla, E., Pletnikova, O., Troncoso, J. C., Marsh, L., Dawson, T. M., Jakala, P., Hartmann, T., Price, D. L., and Lee, M. K. (2005). Aggregation promoting C-terminal truncation of alpha-synuclein is a normal cellular process and is enhanced by the familial Parkinson's disease-linked mutations. *Proc Natl Acad Sci USA* **6**, 2162–2167.

Lopez, O. L., Becker, J. T., Kaufer, D. I., Hamilton, R. L., Sweet, R. A., Klunk, W., and DeKosky, S. T. (2002).

Research evaluation and prospective diagnosis of dementia with Lewy bodies. *Arch Neurol* 1, 43–46.

Mann, D. M., Yates, P. O., and Hawkes, J. (1983). The pathology of the human locus ceruleus. *Clin Neuropathol* 1, 1–7.

McGeer, P. L., Itagaki, S., Boyes, B. E., and McGeer, E. G. (1988). Reactive microglia are positive for HLA-DR in the substantia nigra of Parkinson's and Alzheimer's disease brains. *Neurology* 8, 1285–1291.

McKeith, I. G., Galasko, D., Kosaka, K., Perry, E. K., Dickson, D. W., Hansen, L. A., Salmon, D. P., Lowe, J., Mirra, S. S., Byrne, E. J., Lennox, G., Quinn, N. P., Edwardson, J. A., Ince, P. G., Bergeron, C., Burns, A., Miller, B. L., Lovestone, S., Collerton, D., Jansen, E. N., Ballard, C., de Vos, R. A., Wilcock, G. K., Jellinger, K. A., and Perry, R. H. (1996). Consensus guidelines for the clinical and pathologic diagnosis of dementia with Lewy bodies (DLB): Report of the consortium on DLB international workshop. *Neurology* 5, 1113–1124.

McKeith, I., Mintzer, J., Aarsland, D., Burn, D., Chiu, H., Cohen-Mansfield, J., Dickson, D., Dubois, B., Duda, J. E., Feldman, H., Gauthier, S., Halliday, G., Lawlor, B., Lippa, C., Lopez, O. L., Carlos Machado, J., O'Brien, J., Playfer, J., and Reid, W. (2004). Dementia with Lewy bodies. *Lancet Neurol* 1, 19–28.

McKeith, I. G., Dickson, D. W., Lowe, J., Emre, M., O'Brien, J. T., Feldman, H., Cummings, J., Duda, J. E., Lippa, C., Perry, E. K., Aarsland, D., Arai, H., Ballard, C. G., Boeve, B., Burn, D. J., Costa, D., Del Ser, T., Dubois, B., Galasko, D., Gauthier, S., Goetz, C. G., Gomez-Tortosa, E., Halliday, G., Hansen, L. A., Hardy, J., Iwatsubo, T., Kalaria, R. N., Kaufer, D., Kenny, R. A., Korczyn, A., Kosaka, K., Lee, V. M., Lees, A., Litvan, I., Londos, E., Lopez, O. L., Minoshima, S., Mizuno, Y., Molina, J. A., Mukaetova-Ladinska, E. B., Pasquier, F., Perry, R. H., Schulz, J. B., Trojanowski, J. Q., and Yamada, M. (2005). Diagnosis and management of dementia with Lewy bodies: Third report of the DLB Consortium. *Neurology* 12, 1863–1872.

McNaught, K. S., and Jenner, P. (2001). Proteasomal function is impaired in substantia nigra in Parkinson's disease. *Neurosci Lett* 3, 191–194.

Mishizen-Eberz, A. J., Norris, E. H., Giasson, B. I., Hodara, R., Ischiropoulos, H., Lee, V. M., Trojanowski, J. Q., and Lynch, D. R. (2005). Cleavage of alpha-synuclein by calpain: Potential role in degradation of fibrillized and nitrated species of alpha-synuclein. *Biochemistry* 21, 7818–7829.

Moussa, C. E., Wersinger, C., Tomita, Y., and Sidhu, A. (2004). Differential cytotoxicity of human wild type and mutant alpha-synuclein in human neuroblastoma SH-SY5Y cells in the presence of dopamine. *Biochemistry* 18, 5539–5550.

Murphy, D. D., Rueter, S. M., Trojanowski, J. Q., and Lee, V. M. (2000). Synucleins are developmentally expressed, and alpha-synuclein regulates the size of the presynaptic vesicular pool in primary hippocampal neurons. *J Neurosci* 9, 3214–3220.

Nakano, I., and Hirano, A. (1984). Parkinson's disease: Neuron loss in the nucleus basalis without concomitant Alzheimer's disease. *Ann Neurol* 5, 415–418.

Ostrerova, N., Petrucelli, L., Farrer, M., Mehta, N., Choi, P., Hardy, J., and Wolozin, B. (1999). Alpha-synuclein shares physical and functional homology with 14-3-3 proteins. *J Neurosci* 14, 5782–5791.

Pappolla, M. A., Shank, D. L., Alzofon, J., and Dudley, A. W. (1988). Colloid (hyaline) inclusion bodies in the central nervous system: Their presence in the substantia nigra is diagnostic of Parkinson's disease. *Hum Pathol* 1, 27–31.

Parkkinen, L., Soininen, H., and Alafuzoff, I. (2003). Regional distribution of alpha-synuclein pathology in unimpaired aging and Alzheimer disease. *J Neuropathol Exp Neurol* 4, 363–367.

Parkkinen, L., Kauppinen, T., Pirttila, T., Autere, J. M., and Alafuzoff, I. (2005a). Alpha-synuclein pathology does not predict extrapyramidal symptoms or dementia. *Ann Neurol* 1, 82–91.

Parkkinen, L., Pirttila, T., Tervahauta, M., and Alafuzoff, I. (2005b). Widespread and abundant alpha-synuclein pathology in a neurologically unimpaired subject. *Neuropathology* 4, 304–314.

Perez, R. G., Waymire, J. C., Lin, E., Liu, J. J., Guo, F., and Zigmond, M. J. (2002). A role for alpha-synuclein in the regulation of dopamine biosynthesis. *J Neurosci* 8, 3090–3099.

Periquet, M., Fulga, T., Myllykangas, L., Schlossmacher, M. G., and Feany, M. B. (2007). Aggregated alpha-synuclein mediates dopaminergic neurotoxicity *in vivo*. *J Neurosci* 12, 3338–3346.

Perrin, R. J., Woods, W. S., Clayton, D. F., and George, J. M. (2000). Interaction of human alpha-synuclein and Parkinson's disease variants with phospholipids. Structural analysis using site-directed mutagenesis. *J Biol Chem* 44, 34393–34398.

Pollanen, M. S., Bergeron, C., and Weyer, L. (1993). Deposition of detergent-resistant neurofilaments into Lewy body fibrils. *Brain Res* 1, 121–124.

Polymeropoulos, M. H., Lavedan, C., Leroy, E., Ide, S. E., Dehejia, A., Dutra, A., Pike, B., Root, H., Rubenstein, J., Boyer, R., Stenroos, E. S., Chandrasekharappa, S., Athanassiadou, A., Papapetropoulos, T., Johnson, W. G., Lazzarini, A. M., Duvoisin, R. C., Di Iorio, G., Golbe, L. I., and Nussbaum, R. L. (1997). Mutation in the alpha-synuclein gene identified in families with Parkinson's disease. *Science* 5321, 2045–2047.

Rinne, J. O., Rummukainen, J., Paljarvi, L., and Rinne, U. K. (1989). Dementia in Parkinson's disease is related to neuronal loss in the medial substantia nigra. *Ann Neurol* 1, 47–50.

Rubinsztein, D. C., DiFiglia, M., Heintz, N., Nixon, R. A., Qin, Z. H., Ravikumar, B., Stefanis, L., and Tolkovsky, A. (2005). Autophagy and its possible roles in nervous system diseases, damage and repair. *Autophagy* 1, 11–22.

Saito, Y., Kawashima, A., Ruberu, N. N., Fujiwara, H., Koyama, S., Sawabe, M., Arai, T., Nagura, H., Yamanouchi, H., Hasegawa, M., Iwatsubo, T., and Murayama, S. (2003). Accumulation of phosphorylated alpha-synuclein in aging human brain. *J Neuropathol Exp Neurol* 6, 644–654.

Schenck, C. H., Bundlie, S. R., and Mahowald, M. W. (1996). Delayed emergence of a parkinsonian disorder in 38% of 29 older men initially diagnosed with idiopathic rapid eye movement sleep behaviour disorder. *Neurology* 2, 388–393.

Sidhu, A., Wersinger, C., Moussa, C. E., and Vernier, P. (2004). The role of alpha-synuclein in both neuroprotection and neurodegeneration. *Ann NY Acad Sci*, 250–270.

Singleton, A. B., Farrer, M., Johnson, J., Singleton, A., Hague, S., Kachergus, J., Hulihan, M., Peuralinna, T., Dutra, A., Nussbaum, R., Lincoln, S., Crawley, A., Hanson, M., Maraganore, D., Adler, C., Cookson, M. R., Muenter, M., Baptista, M., Miller, D., Blancato, J., Hardy, J., and Gwinn-Hardy, K. (2003). Alpha-synuclein locus triplication causes Parkinson's disease. *Science* 5646, 841.

Spillantini, M. G., Schmidt, M. L., Lee, V. M., Trojanowski, J. Q., Jakes, R., and Goedert, M. (1997). Alpha-synuclein in Lewy bodies. *Nature* 6645, 839–840.

Takeda, A., Hashimoto, M., Mallory, M., Sundsumo, M., Hansen, L., Sisk, A., and Masliah, E. (1998). Abnormal distribution of the non-A beta component of Alzheimer's disease amyloid precursor/alpha-synuclein in Lewy body disease as revealed by proteinase K and formic acid pretreatment. *Lab Invest* 9, 1169–1177.

Tiller-Borcich, J. K., and Forno, L. S. (1988). Parkinson's disease and dementia with neuronal inclusions in the cerebral cortex: Lewy bodies or Pick bodies? *J Neuropathol Exp Neurol* 5, 526–535.

Tsuboi, Y., and Dickson, D. W. (2005). Dementia with Lewy bodies and Parkinson's disease with dementia: Are they different? *Parkinsonism Relat Disord* 45, S47–S51.

Uchikado, H., Lin, W. L., DeLucia, M. W., and Dickson, D. W. (2006). Alzheimer disease with amygdala Lewy bodies: A distinct form of alpha-synucleinopathy. *J Neuropathol Exp Neurol* 7, 685–697.

Uchiyama, M., Isse, K., Tanaka, K., Yokota, N., Hamamoto, M., Aida, S., Ito, Y., Yoshimura, M., and Okawa, M. (1995). Incidental Lewy body disease in a patient with REM sleep behavior disorder. *Neurology* 4, 709–712.

Wakabayashi, K., and Takahashi, H. (1996). Gallyas-positive, tau-negative glial inclusions in Parkinson's disease midbrain. *Neurosci Lett* 2–3, 133–136.

Wakabayashi, K., Hayashi, S., Yoshimoto, M., Kudo, H., and Takahashi, H. (2000). NACP/alpha-synuclein-positive filamentous inclusions in astrocytes and oligodendrocytes of Parkinson's disease brains. *Acta Neuropathol (Berl)* 1, 14–20.

Whitehouse, P. J., Hedreen, J. C., White, C. L., 3rd, and Price, D. L. (1983). Basal forebrain neurons in the dementia of Parkinson disease. *Ann Neurol* 3, 243–248.

Zarranz, J. J., Alegre, J., Gomez-Esteban, J. C., Lezcano, E., Ros, R., Ampuero, I., Vidal, L., Hoenicka, J., Rodriguez, O., Atares, B., Llorens, V., Gomez Tortosa, E., del Ser, T., Munoz, D. G., and de Yebenes, J. G. (2004). The new mutation, E46K, of alpha-synuclein causes Parkinson and Lewy body dementia. *Ann Neurol* 2, 164–173.

Zweig, R. M., Jankel, W. R., Hedreen, J. C., Mayeux, R., and Price, D. L. (1989). The pedunculopontine nucleus in Parkinson's disease. *Ann Neurol* 1, 41–46.

NON-HUMAN PRIMATE MODELS

II

NON-HUMAN PRIMATE
MODELS

4

INTRODUCTION TO: NEUROTOXIN-BASED NONHUMAN PRIMATE MODELS OF PARKINSON'S DISEASE

ZHIMING ZHANG AND DON MARSHALL GASH

Department of Anatomy and Neurobiology and the Morris K. Udall Parkinson's Disease Research Center of Excellence, College of Medicine, University of Kentucky, Lexington, KY, USA

The chapters in this section review extensive experimentation by very talented research groups to model Parkinson's disease (PD) in nonhuman primates via neurotoxin-induced lesions. Most recent studies have focused on the neurotoxin 1-methyl-4-phenyl-1,2,3,6-tetrahydropyridine (MPTP). The specific neurotoxic actions of MPTP are produced when it is metabolized by monoamine oxidase B into 1-methyl-4-phenylpyridinium (MPP+), a complex I mitochondrial neurotoxin with relative specificity for dopamine neurons in the substantia nigra (see Langston *et al.*, 1983; Nicklas *et al.*, 1987; and Richardson *et al.*, 2007). In the 1980s, MPTP was a major contaminate of a street drug being sold in Northern California. Some drug abusers exposed to MPTP developed movement dysfunctions closely mirroring PD, while others displayed mild parkinsonian features (Tetrud *et al.*, 1989). In species including mice and nonhuman primates where monoamine oxidase B is found in the brain, systemically administered MPTP crosses the blood brain barrier and is converted into MPP+. As the following chapters demonstrate, MPP+ toxicity selectively and robustly lesions the nigrostriatal dopamine system, replicating many features of PD. Even as MPTP/MPP+ toxicity remarkably closely mimics components of the disease, depending upon

the protocol used in nonhuman primates, the results do not fully represent the entity described by James Parkinson in 1817 and seen in patients in movement disorder clinics today. As *Erwan Bezard* emphasizes in his chapter, the ultimate goal is to create clinically relevant models that lead to a better understanding of the disease processes and improved clinical treatments. In this regard, the five chapters demonstrate that details are important in modeling PD, even with a potent dopaminergic neurotoxin like MPTP.

There is an important dichotomy in the PD mimicry produced by different MPTP models. Understanding what is faithfully modeled in different MPTP protocols and what elements of idiopathic PD are missing provide important insights into the human disease process. On the positive side, administration of MPTP to nonhuman primates invariably targets the nigrostriatal dopamine pathway (see Table 4.1), a primary neural system that degenerates in PD. Some models using MPTP however fail to replicate other key features of PD. As pointed out by *Chu and Kordower*, "a number of novel therapeutic strategies that have been successful when tested in young parkinsonian (i.e., MPTP-treated) monkeys have failed to pass muster when tested clinically in patients with PD." Age is a critical factor in idiopathic PD and failure to take

■■■ **TABLE 4.1 Primary Parkinson's disease models**

Model	Features
Clinical Diagnosis Criteria	1. Manifestation of at least two of three cardinal signs: tremor at rest, rigidity and bradykinesia, in the absence of a secondary cause such as known neuroleptic of toxin exposure. 2. Demonstrate a clear cut response to levodopa or a dopamine agonist.

Nigrostriatal Dopamine System Pathology
(e.g., see Fearnley and Lees, 1991; Lee *et al.*, 2004; Bohnen *et al.*, 2006)
	1. Degeneration and loss of dopamine neurons in the substantia nigra.
	  • 50% loss in early stages of PD
	  • 80% loss in late stages of PD
	2. Dopamine innervation and dopamine levels markedly lower in the caudate nucleus and putamen.
	  • 60–80% dopamine loss in earlier PD
	  • 99% loss in end-stage PD

Progressive Multisystem PNS and CNS Degeneration (Braak Hypothesis) (e.g., see Braak *et al.*, 2004; Hawkes *et al.*, 2007)	1. Long prodromal period with olfactory and vagal dysfunctions, sleep disturbances. 2. Early events include olfactory bulb and enteric NS pathology. 3. Mid-stage begins with nigrostriatal dopamine system pathology and PD symptoms. 4. Later stages include neocortical and thalamic pathology/dysfunctions. Progression can be mapped by Lewy bodies and α-synuclein accumulation.
Principal Risk Factors (e.g., see Lang and Lazano, 1998; Dawson and Dawson, 2003; Gash *et al.*, 2008)	1. *Age*: Disease onset is after age 40 in 90–95% of cases. 2. *Genetic factors*: The list keeps growing and includes Parkin, Pink-1, DJ-1 LRRK-2 and α-synuclein. 3. *Environmental factors*: The list keeps growing here also, includes MPTP, pesticides and trichloroethylene.

it into account can be a fatal flaw in experimental design.

The PD features modeled by MPTP toxicity are detailed in the five chapters on nonhuman primates. *Petzinger et al.* provide a brief history of MPTP and its use in animal models. As they discuss, a major strength of MPTP models of PD is the close replication of nigrostriatal dopamine system pathology. The chapters by *Bezard* and *Schneider, Anderson and Decamp* discuss the dopaminergic and non-dopaminergic effects from MPTP toxicity. The chapter by *Wichmann and DeLong* describes how the neural circuitry affected by MPTP-induced degeneration of the nigrostriatal dopamine system can be mapped and analyzed. In these cases, MPTP-induced parkinsonism has been used productively to better understand the neurobiology and neurophysiology of degenerative processes.

The other aspect of MPTP modeling, the PD features that are not replicated in the human disease,

as previously mentioned is equally important. As *Chu and Kordower* emphasize, age is an essential component to consider in modeling PD. In retrospect, this is not surprising as age is the single most predictive risk factor in idiopathic PD. The onset of PD is after age 40 in 90–95% of the cases with disease incidence and prevalence increasing nearly exponentially with age thereafter (Lang and Lozano, 1998; Silver, 2006). A major problem encountered in using young animals in PD-related research is occasional spontaneous behavioral recovery regardless of the route for MPTP administration (Crossman *et al.*, 1987; Schneider *et al.*, 1987; Moratalla *et al.*, 1992; Albanese *et al.*, 1993; Ovadia *et al.*, 1995; Ross and Pickart, 2004; Mounayar *et al.*, 2007). For example, Mounayar *et al.* (2007) reported spontaneous behavioral recovery in 4–6 year old young vervet monkeys receiving multiple intramuscular injections of MPTP. In retrospect, it seems obvious that data on neuroprotective studies in young MPTP-treated animals might not be predictive of outcomes in patients in their 60s and 70s. Based on factors such as life expectancy, age at puberty and brain volume, 1 year in rhesus monkeys equals approximately 3 years of human life (Andersen *et al.*, 1999), suggesting that studies using 4–6 year old macaques are modeling 12–18 year old humans, and not the target age group for idiopathic PD.

The rationale for using aged primates for PD-related research is compelling. Firstly, age-associated decline of motor activity is seen in both humans and nonhuman primates. Zhang *et al.* (2000) reported that movement dysfunction scores rose significantly with increasing age, and were also accompanied with two or more parkinsonian signs in 20% of tested mid-aged (12–17 years old) monkeys and 36% of aged (21–27 years old) monkeys. Slowing performance times of fine-motor hand tasks correlated also significantly with increasing age (Zhang *et al.*, 2000). Secondly, age-associated downregulation of dopaminergic phenotype occurs in the nigrostriatal system in both humans and nonhuman primates. As extensively discussed by *Chu and Kordower*, age-related declines in motor function are associated with a decrease of nigrostriatal dopaminergic function including decrease of dopamine and its metabolites, downregulation of TH expression and loss of dopaminergic phenotype in substantia nigra dopamine neurons. Thirdly, an age-associated increase of α-synuclein, a risk factor in PD, occurs in both humans and nonhuman primates. For example, α-synuclein was almost undetectable in nigral dopamine neurons in young monkeys, while the number of detectable α-synuclein immunoreactive neurons increased by 169% and 215% in mid-aged and aged monkeys (Yavich *et al.*, 2004; Totterdell and Meredith, 2005). These data illustrate the inherent advantages of using older animals to replicate age-associated declines in the nigrostriatal dopamine system in idiopathic PD patients.

While advanced age is important in animal models, middle-aged monkeys may be better suited for some studies than older animals. Recently, we completed a study designed to produce a stable, MPTP-induced nonhuman primate model of early PD (Ding *et al.*, in review). The goal was to develop a better model for testing neuroprotective and restorative treatments for PD. The new model was created by (i) lowering the dose of MPTP, (ii) directly infusing MPTP into the brain via the internal carotid artery, and (iii) using only late middle-aged rhesus monkeys

(i.e. 16–19 years old). A total of 27 female rhesus monkeys were tested for up to a year following treatment. All animals showed stable parkinsonian features lasting for up to 12 months per behavioral evaluation. Compared with late-stage PD animals, postmortem analysis demonstrated that more dopaminergic neurons remained in the substantia nigra pars compacta, and more dopaminergic fibers were found in the striatum. In addition, tissue levels of striatal dopamine and its metabolites were also higher. Our results support that a milder, but stable PD model can be produced in the middle-aged rhesus monkeys. These results are important in that they indicate more readily available middle-aged nonhuman human primates can be effectively used in PD models, rather than scarce older monkeys.

Another concern about MPTP models is that they do not appear to replicate the long prodromal period preceding the expression of parkinsonian signs in many patients. Braak and his associates (e.g., see Braak *et al.*, 2004; Hawkes, *et al.*, 2007) posit that early events in the development of PD begin with pathological changes in the olfactory bulb and enteric nervous system. They believe these changes are instigated by environmental agents and initiate a progressive degenerative process that gradually involves more of the brainstem and eventually the nigrostriatal dopamine system. The ability for MPTP to adequately model the prodromal stage of PD has not been fully investigated, but it may be the Achilles heel of this dopamine neurotoxin and explains important differences between MPTP models and the real disease.

CONCLUSIONS

The five chapters on nonhuman primates provide valuable insights into various MPTP-lesioned models, highlighting strengths and weaknesses of each model system. The chapters also report the latest results from the authors' laboratories. One is left with the sense that the widely used MPTP model has served the neuroscience community well over the past two decades since its discovery. But, there is also a sense that studies on MPTP have been fully exploited and that it is time to explore new approaches to better model idiopathic PD. One prominent challenge is to adequately model the prodromal stage of the disease and identify the molecular basis for early pathological changes in neural circuitry that progresses to eventually compromise the nigrostriatal dopamine system.

ACKNOWLEDGEMENTS

We would like to thank our colleagues including Alexander Blandford, Hamed Haghnazar, Eric Forman and Avalon Sandoval for their support in preparing the manuscript. Special thanks go to Dr. Greg A. Gerhardt for his meaningful discussion. This work was supported by NIH RO1 NS050242 (ZM) and NIH center grants NS39783 (GAG) and AG013494 (DMG).

REFERENCES

Albanese, A., Granata, R., Gregori, B. et al. (1993). Chronic administration of 1-methyl-4-phenyl-1,2,3,6-tetrahydropyridine to monkeys: behavioural, morphological and biochemical correlates. *Neuroscience* **55**, 823–832.

Andersen, A. H., Zhang, Z., Zhang, M. et al. (1999). Age-associated changes in rhesus CNS composition identified by MRI. *Brain Res.* **829**, 90–98.

Bohnen, N. I., Albin, R. L., Koeppe, R. A. et al. (2006 Sep.). Positron emission tomography of monoaminergic vesicular binding in aging and Parkinson disease. *J Cereb Blood Flow Metab.* **26**(9), 1198–1212.

Braak, H., Ghebremedhin, E., Rüb, U. et al. (2004). Stages in the development of Parkinson's disease-related pathology. *Cell Tissue Res.* **318**, 121–134.

Crossman, A. R., Clarke, C. E., Boyce, S. et al. (1987). MPTP-induced parkinsonism in the monkey: neurochemical pathology, complications of treatment and pathophysiological mechanisms. *Can J Neurol Sci.* **14**, 428–435.

Dawson, T. M., and Dawson, V. L. (2003). Molecular pathways of neurodegeneration in Parkinson's disease. *Science* **302**, 819–822.

Ding, F., Luan, L., Ai, Y. et al. (2008). Development of a stable, early stage unilateral model of Parkinson's disease in middle-aged rhesus monkeys. *Exp Neurol.*, (in press).

Fearnley, J. M., and Lees, A. J. (1991 Oct.). Ageing and Parkinson's disease: substantia nigra regional selectivity. *Brain* **114**(Pt 5), 2283–2301.

Gash, D. M., Rutland, K., Hudson, N. L. et al. (2008 Feb.). Trichloroethylene: Parkinsonism and complex 1 mitochondrial neurotoxicity. *Ann Neurol.* **63**(2), 184–192.

Hawkes, C. H., Del Tredici, K., and Braak, H. (2007). Parkinson's disease: a dual-hit hypothesis. *Neuropathol Appl Neurobiol.* **33**, 599–614.

Lang, A. E., and Lozano, A. M. Parkinson's disease. First of two parts. N Engl J Med.

Langston, J. W., Ballard, P., Tetrud, J. W., and Irwin, I. (1983). Chronic Parkinsonism in humans due to a product of meperidine-analog synthesis. *Science* **219**, 979–980.

Lee, C. S., Schulzer, M., de la Fuente-Fernández, R. et al. (2004 Dec.). Lack of regional selectivity during the progression of Parkinson disease: implications for pathogenesis. *Arch Neurol.* **61**(12), 1920–1925.

Moratalla, R., Quinn, B., DeLanney, L. E. et al. (1992). Differential vulnerability of primate caudate-putamen and striosome-matrix dopamine systems to the neurotoxic effects of 1-methyl-4-phenyl-1,2,3,6-tetrahydropyridine. *Proc Natl Acad Sci U S A* **89**, 3859–3863.

Mounayar, S., Boulet, S., Tande, D. et al. (2007). A new model to study compensatory mechanisms in MPTP-treated monkeys exhibiting recovery. *Brain* **130**, 2898–2914.

Nicklas, W. J., Youngster, S. K., Kindt, M. V., and Heikkila, R. E. (1987). MPTP, MPP+ and mitochondrial function. *Life Sci.* **40**, 721–729.

Ovadia, A., Zhang, Z., and Gash, D. M. (1995). Increased susceptibility to MPTP toxicity in middle-aged rhesus monkeys. *Neurobiol Aging* **16**, 931–937.

Richardson, J. R., Caudle, W. M., Guillot, T. S. et al. (2007). Obligatory role for complex I inhibition in the dopaminergic neurotoxicity of 1-methyl-4-phenyl-1,2,3,6-tetrahydropyridine (MPTP). *Toxicol Sci.* **95**, 196–204.

Ross, C. A., and Pickart, C. M. (2004). The ubiquitin-proteasome pathway in Parkinson's disease and other neurodegenerative diseases. *Trends Cell Biol.* **14**, 703–711.

Schneider, J. S., Yuwiler, A., and Markham, C. H. (1987). Selective loss of subpopulations of ventral mesencephalic dopaminergic neurons in the monkey following exposure to MPTP. *Brain Res.* **411**, 144–150.

Silver, D. (2006). Impact of functional age on the use of dopamine agonists in patients with Parkinson disease. *Neurologist* **12**, 214–223.

Tetrud, J. W., Langston, J. W., Garbe, P. L., and Ruttenber, A. J. (1989). Mild parkinsonism in persons exposed to 1-methyl-4-phenyl-1,2,3,6-tetrahydropyridine (MPTP). *Neurology* **39**, 1483–1487.

Totterdell, S., and Meredith, G. E. (2005). Localization of alpha-synuclein to identified fibers and synapses in the normal mouse brain. *Neuroscience* **135**, 907–913.

Yavich, L., Tanila, H., Vepsäläinen, S., and Jäkälä, P. (2004). Role of alpha-synuclein in presynaptic dopamine recruitment. *J Neurosci* **24**, 11165–11170.

Zhang, Z., Andersen, A. H., Smith, C. D. et al. (2000). Motor slowing and parkinsonian signs in aging rhesus. *J Gerontology: Biol Sci.* **55A**, B473–B480.

5

SYSTEMS LEVEL PHYSIOLOGY OF THE BASAL GANGLIA, AND PATHOPHYSIOLOGY OF PARKINSON'S DISEASE

T. WICHMANN[1,2] AND M. R. DeLONG[2]

[1]*Yerkes National Primate Research Center, Emory University School of Medicine, Atlanta, GA, USA*
[2]*Department of Neurology, Emory University School of Medicine, Atlanta, GA, USA*

INTRODUCTION

Parkinsonism is a syndrome characterized by movement abnormalities, such as akinesia (poverty of movement and impaired movement initiation), bradykinesia (slowness of movement), muscular rigidity and tremor at rest. These signs have long been recognized to result from the loss of dopaminergic cells in the substantia nigra pars compacta (SNc), and of their projections to the striatum. Consequently, most symptomatic therapies for parkinsonism are aimed at replacing dopamine. Recent studies have shown that non-dopaminergic neurons also degenerate in Parkinson's disease, accounting for a host of non-dopaminergic features of the disease, such as depression, dementia, sleep, olfactory and balance problems (Braak *et al.*, 2003). While the dopamine-dependent motor signs dominate the early phases of Parkinson's disease, the non-dopaminergic signs and symptoms are frequently more disabling than the motor dysfunction in advanced stages of the disease.

There is no single best animal model for all facets of the disease. The available animal models can be grossly grouped into two categories, that is, models that mimic the neuronal degeneration process that occurs in the parkinsonian brain, and models that replicate the loss of the dopaminergic nigrostriatal projection in Parkinson's disease. Since the discovery

that 1-methyl-4-phenyl-1,2,3,6-tetrahydropyridine (MPTP) is toxic to dopaminergic neurons in the SNc (Langston *et al.*, 1983), and that this agent can be used to induce parkinsonism in primates (Burns *et al.*, 1983), animal models using this compound have become an important tool to examine the functional consequences of dopamine loss in the basal ganglia. Other widely used toxins that induce lesions specific for dopaminergic cells are 6-hydroxy dopamine (6-OHDA; Ungerstedt, 1968) and several pesticides, including rotenone (Betarbet *et al.*, 2000). Studies in these models have greatly helped us to understand the pathophysiology of Parkinson's disease, and to develop new therapies directed at the motor dysfunction in parkinsonism. In this chapter, we will summarize the circuit anatomy of the basal ganglia, followed by a description of some of the research into the pathophysiology of parkinsonism.

CIRCUIT ANATOMY OF THE BASAL GANGLIA

General Composition of Basal Ganglia Circuits

The general anatomy of the basal ganglia and their connections with cortex and brainstem are shown

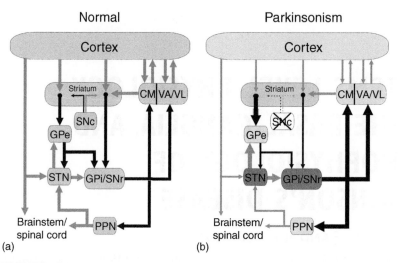

FIGURE 5.1 (a) Anatomical structure of the basal ganglia circuitry, and (b) changes in the activity of basal ganglia nuclei associated with the development of parkinsonism. Gray arrows indicate excitatory (glutamatergic) connections, black arrows indicate inhibitory (GABAergic) connections. Changes in thickness of the arrows indicate changes in firing rates. Abbreviations: GPe, external pallidal segment; STN, subthalamic nucleus; GPi, internal pallidal segment; SNr, substantia nigra pars reticulata; SNc, substantia nigra pars compacta; PPN, pedunculopontine nucleus; CM, centromedian nucleus of the thalamus; VA/VL, ventral anterior and ventrolateral nucleus of the thalamus.

in Figure 5.1a ("Normal"). The term basal ganglia encompasses the striatum, the external and internal segments of the globus pallidus (GPe, GPi, respectively), the SNc and the pars reticulata of the substantia nigra (SNr), and the subthalamic nucleus (STN). These structures are components of a system of larger cortico-subcortical pathways which remain largely segregated throughout their subcortical course. Depending on their cortical area of origin, and the basal ganglia region occupied by them, these pathways subserve different functions. In each of these loops, the striatum and STN serve as "input" stages of the basal ganglia, whereas GPi and SNr serve as "output" stages, sending inhibitory projections to thalamus and brainstem. The projections from the basal ganglia to the thalamus are central in current functional schemes regarding motor and non-motor functions of the basal ganglia. They will be described in more detail below. Interactions between the basal ganglia and brainstem structures such as the pedunculopontine nucleus (PPN) are also integral to basal ganglia function, both in terms of modulating basal ganglia activity, and in transmitting basal ganglia output to brainstem areas which are important for gait, balance and other functions (e.g., Mena-Segovia *et al.*, 2004).

The motor signs of parkinsonism appear to depend primarily on abnormalities in the "motor circuit" of the basal ganglia. This circuit arises from the precentral motor fields, and engages the "motor" area of the striatum (the putamen), "motor" portions of GPe, STN and GPi/SNr, as well as parts of the ventrolateral thalamus and brainstem. The physiologic functions of this circuit appear to include roles in the scaling of kinematic movement parameters (amplitude or velocity of movement), the initiation or execution of internally triggered movements, sequencing of movements, switching between movements or motor programs and roles in procedural learning.

Cell Types and Basic Neurochemistry

The striatum is composed of several cell types. The majority of cells are GABAergic medium spiny neurons, which receive glutamatergic inputs from cortex and thalamus, and project to both segments of the globus pallidus (see Figure 5.1 and below). In addition, the striatum contains different classes of interneurons, including cholinergic interneurons and several classes of GABAergic interneurons (Tepper *et al.*, 2004). While small in number, these interneurons may contribute significantly to the functions of the striatum. Thus, together with dopaminergic inputs, cholinergic interneurons may be involved in reward processing, while "fast-spiking," parvalbumin-positive interneurons receive cortical inputs and project to neighboring medium spiny neurons, providing the substrate for "center-surround"-type inhibition affecting striatal outputs (Mallet et al., 2005).

The other basal ganglia structures are more homogeneous in their neuronal composition. GPe, GPi and SNr largely consist of tonically active GABAergic neurons, while the STN consists of glutamatergic cells. The SNc is a dopaminergic nucleus, and provides inputs to all of the other basal ganglia structures (see below).

Direct and Indirect Pathways

In terms of the intrinsic circuitry of the basal ganglia (Figure 5.1), two major projection systems between input and output structures have been described: a monosynaptic inhibitory (GABAergic) "direct" pathway between striatum and GPi/SNr, and a polysynaptic "indirect" pathway which involves GPe and STN. In the indirect pathway, the projections between the striatum and the GPe and between the GPe and the STN are inhibitory (GABAergic), while the STN–GPi/SNr pathway is excitatory (glutamatergic). In addition, a reciprocal inhibitory connection between both segments of the globus pallidus which circumvents the STN is also considered as part of the indirect pathway.

In addition to their projection targets, direct and indirect pathways also differ in several other important aspects. Thus, the two populations of striatal projection neurons that give rise to direct or indirect pathways appear to be innervated by different populations of cortical neurons (e.g., Wilson, 1987; Lei et al., 2004). Furthermore, direct pathway neurons in the striatum receive stronger monosynaptic inputs from the intralaminar nuclei of the thalamus (the centromedian and the parafascicular nucleus, CM/Pf) than indirect pathway neurons, while inputs from other thalamic nuclei, such as VA/VL projections to the striatum are not specific for the direct or indirect pathways (Smith et al., 2004). While both pathways are GABAergic, they differ in terms of associated co-transmitters. The indirect pathway carries enkephalin as a co-transmitter, while the direct pathway expresses substance P. The function of these co-transmitters remains unclear.

Another difference between the direct and indirect pathways is that the adenosine A2A-receptor is preferentially found on the striatal neurons that give rise to the indirect pathway.

Finally, the direct and indirect pathways are likely to differ substantially in terms of their functional effects of their activation. Activation of the direct pathway by cortical inputs may inhibit the activity of the output nuclei GPi and SNr, and thereby disinhibit thalamocortical projection neurons. In contrast, activation of striatal neurons that give rise to the indirect pathway may have a net excitatory effect on GPi/SNr activity and would thereby act to inhibit thalamocortical neurons (Alexander and Crutcher, 1990). The functional difference between direct and indirect pathways is central to almost all of the current models of basal ganglia function. A balance between direct and indirect pathways actions may be essential to regulate basal ganglia output. In terms of motor functions, a preponderance of indirect pathway activity would result in over-inhibition of thalamostriatal projections, while the opposite would be true for activation of the direct pathway.

Functions of Dopamine

An additional, functionally highly relevant difference between the two striatal output pathways is the fact that the activity of the direct and indirect pathways appears to be differentially modulated by dopamine, released from terminals of the nigrostriatal projection, which arises from neurons in the SNc. The majority of dopaminergic terminals terminate on the neck of dendritic spines of medium spiny neurons in the striatum. The same dendritic spines also receive the corticostriatal projection, as well as the majority of terminals of the thalamostriatal projection (Raju et al., 2006). This anatomical arrangement has given rise to the concept that one of the major functions of striatal dopamine may be to regulate the effectiveness of synaptic transmission reaching dendritic spines.

While some striatal neurons express both, D1- and D2-receptors (Aizman et al., 2000), most of these receptors are differentially expressed (Gerfen et al., 1990). Direct pathway neurons have been shown to preferentially express D1-receptors, while indirect pathway neurons express preferentially D2-receptors (Gerfen et al., 1990). Activation of D1-receptors is thought to gate (activate) corticostriatal transmission at spines of direct pathway neurons, while activation of D2-receptors would act to inhibit the spread of activation at spines of medium spiny neurons belonging to the indirect pathway. In addition, D2-receptors also function as presynaptic autoreceptors, regulating the dopamine release from dopaminergic terminals.

Through its differential actions on the direct and indirect pathways, striatal dopamine release may substantially alter the balance of activity between direct and indirect pathways. Increased release of dopamine may activate the striatal neurons that give rise to the direct pathway and inactivate those neurons that give rise to the indirect pathway. Mediated to the output nuclei of the basal ganglia, both changes would eventually result in a reduction of basal ganglia output from GPi/SNr to the thalamus. Reduced release of dopamine, as is found in parkinsonism, would result in disinhibition of the indirect pathway, and reduced facilitation of the direct pathway, leading to increased basal ganglia output. Dopaminergic transmission in the striatum may also have other direct or indirect functions, for instance, regulating plasticity at corticostriatal synapses (e.g., Pisani et al., 2005), or maintaining their anatomical integrity (Day et al., 2006).

There is also evidence that dopamine is released at extrastriatal sites, including the pallidum, STN and SNr. In the pallidum and STN, dopamine release occurs at dopaminergic axon terminals, while dopamine release in the substantia nigra is an example of dendritic transmitter release. Both D1- and D2-receptors exist at all of these locations. The functions of dopamine at these extrastriatal sites are far less understood than those of striatal dopamine release, but they are suspected to have substantial effects on basal ganglia output. For instance, dopamine release in GPi and SNr may preferentially activate D1-receptors on terminals of the striatonigral projection, which, in turn, may result in increased GABA release from such terminals in the SNr, and subsequent inhibition of the activity of SNr neurons (e.g., Trevitt et al., 2002).

PATHOPHYSIOLOGY OF PARKINSONISM

One of the principal areas of research that has greatly benefited from the use of the animal models of parkinsonism is the elucidation of the pathophysiologic consequences of dopamine depletion in primates. Many of these studies were carried out in MPTP-treated primates and our discussion will mostly focus on these experiments. Activity changes in the basal ganglia were first documented in biochemical studies, which indicated that in MPTP-induced parkinsonism in primates the metabolic activity is increased in both pallidal segments (e.g., Crossman et al., 1985), likely due to increased activity of the striatum–GPe connection and the STN–GPi pathway, or increased activity via the projections from the STN to both pallidal segments. Later, it was shown with microelectrode recordings of neuronal activity, that parkinsonism in primates is associated with reduced tonic neuronal discharge in GPe, and increased mean discharge rates in the STN and GPi, as compared to normal controls (Miller and DeLong, 1987; Filion et al., 1988; Bergman et al., 1994; Boraud et al., 1998). These findings have given rise to a "rate-based" model of parkinsonian pathophysiology (Albin et al., 1989) in which striatal dopamine depletion leads to increased activity of striatal neurons that project to GPe, suppressing neuronal discharge in GPe and resulting in disinhibition of STN and GPi via the indirect pathway. Loss of dopamine in the striatum is also postulated to lead to reduced activity via the inhibitory direct pathway. The classic model of MPTP-induced changes in activity is shown in Figure 5.1b ("Parkinsonism").

Although not yet directly shown in the MPTP-treated primate, the centromedian nucleus of the thalamus (CM) and the PPN may also be involved in the development of parkinsonism. Increased inhibition of the thalamus may reduce CM activity, which may then contribute to the reduction of activity in the direct pathway via its projections to the striatum. An involvement of reduced activity in the PPN in the development of parkinsonian signs is suggested by the fact that lesions of this nucleus in normal monkeys can lead to a hemiparkinsonian syndrome, possibly by reducing excitation of SNc neurons by input from the PPN, or through mechanisms independent of the dopaminergic system (e.g., Kojima et al., 1997; Nandi et al., 2002; Mena-Segovia et al., 2004). Recently, this finding has been exploited in attempts to treat parkinsonism by deep brain stimulation (DBS) directed at the PPN. Preliminary studies PPN–DBS have been promising, even in patients who have responded poorly to established antiparkinsonian treatments such as levodopa or DBS of the STN (Stefani et al., 2007).

Strong evidence for the importance of increased basal ganglia output in the development of parkinsonian motor signs comes from lesion stud-ies. Lesions of the STN in MPTP-treated primates reverse all of the cardinal motor signs of parkin-sonism (Bergman *et al.*, 1990). More recently, this finding has been replicated in parkinsonian patients (e.g., Gill and Heywood, 1997; Alvarez *et al.*, 2005). Stereotactic lesions of the motor portion of GPi are also known to be effective against all major parkinsonian motor signs and against drug-induced dyskinesias in humans and MPTP-treated monkeys (Laitinen, 1995; Baron *et al.*, 1996). PET studies in patients undergoing such pallidotomy proce-dures have shown that frontal motor areas whose metabolic activity was reduced in the parkinsonian state became more active after the lesion (Ceballos-Bauman *et al.*, 1994; Dogali *et al.*, 1995; Grafton *et al.*, 1995; Samuel *et al.*, 1997).

Despite its predictive value in terms of new therapies for parkinsonism, the aforementioned rate-based model of the pathophysiology of par-kinsonism fails to explain many findings of recent studies in parkinsonian subjects. For instance, lesions of the motor thalamus that abolish tha-lamocortical output from the motor circuit do not result in akinesia/bradykinesia and, conversely, GPi lesions do not result in excessive movement, as would be expected if the rate model was entirely true. There is also evidence that electrical stimu-lation of the STN in the MPTP monkey model of parkinsonism increases the mean discharge rates of neurons in GPi without worsening parkinsonism (Hashimoto *et al.*, 2003). Because of such discrep-ancies, the attention of researchers has increasingly turned toward the prominent changes in discharge patterns that accompany the rate changes in par-kinsonism. These changes are now seen to be cru-cial for the development of specific parkinsonian motor signs (e.g., see, the examples of such firing pattern changes in Figure 5.2, with a summary of pattern changes presented in Figure 5.3).

One of the earliest pattern changes described in MPTP-treated animals was that the responses of individual neurons to passive limb manipulations in STN, GPi and thalamus (Miller and DeLong, 1987; Filion *et al.*, 1988; Bergman *et al.*, 1994) occur more often, are more pronounced, have wid-ened receptive fields, and are more often inhibitory (Boraud *et al.*, 2000) than under normal conditions. Cross-correlation studies have also revealed that a substantial proportion of neighboring neurons in globus pallidus, STN and cortex discharge in unison in MPTP-treated primates (Bergman *et al.*,

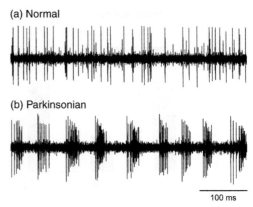

(a) Normal

(b) Parkinsonian

100 ms

FIGURE 5.2 Examples of traces of neuronal activity, recorded with standard extracellular electro-physiologic recording methods in a monkey before (a) and after (b) induction of parkinsonism with MPTP. The animal was awake and sitting quietly in a primate chair during the recordings. The traces of these dif-ferent neurons are each 1-s long. Treatment with MPTP resulted in an increased firing rate, and obvious changes in firing patters in this animal.

1994; Nini *et al.*, 1995; Raz *et al.*, 2000; Raz *et al.*, 2001; Goldberg *et al.*, 2002; Heimer *et al.*, 2002). This is in contrast to the virtual absence of synchro-nized discharge of such neurons in normal monkeys (e.g., Wichmann *et al.*, 1994; Jaeger *et al.*, 1995). Finally, the proportion of cells in STN, GPi and SNr which discharge in oscillatory or non-oscilla-tory bursts is greatly increased in the parkinsonian state, and the temporal structure of burst- and peri-burst discharges is altered (discussed in Wichmann and Soares, 2006).

The potentially increased neuronal synchrony in parkinsonism has recently been further explored through recordings of local field potentials in the basal ganglia of human patients with the disor-der, using implanted DBS electrodes for recording (Brown and Williams, 2005). The origin of local field potentials recorded with these electrodes in the basal ganglia is not certain, but, as in other brain regions, it is likely that they represent compound electrical activities, which are generated at synapses, in dendrites or are influenced by neuronal spik-ing activity. Importantly, local field potentials are strongly influenced by the synchronicity of neuronal and other activities close to the recording electrode. The local field potential recordings in parkinsonian humans have demonstrated prominent oscillatory

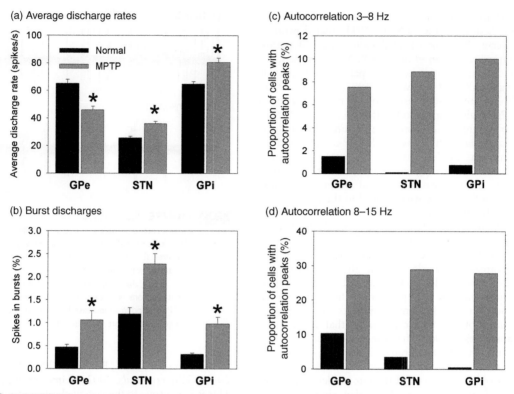

FIGURE 5.3 Changes in (a) discharge rates and (b) burst discharges (proportion of spikes occurring in bursts; and changes in the proportion of cells with oscillatory discharge in the (c) 3–8 Hz range and (d) 8–15 Hz ranges. Data from Soares *et al.* (2004).

activity in the beta band (10–25 Hz) in both the STN and the GPi. The increased beta-band activity was reduced by treatment with dopaminergic medications, in favor of activity at higher frequencies (Brown *et al.*, 2001), and similarly modulated in GPi by high frequency stimulation in the STN (Brown *et al.*, 2004). Brown and colleagues have argued that Parkinson's may result from a replacement of the normal gamma-band oscillations with abnormal beta-band oscillations.

Interesting new studies have suggested that dopamine depletion may also affect corticostriatal transmission in non-traditional ways. Thus, it has been suggested that dopamine loss in the striatum may induce substantial morphologic changes in the striatal neurons that receive dopaminergic innervation. For instance, the spine density of striatal neurons is reduced in dopamine-depleted animals (Ingham *et al.*, 1989) and human PD patients (Stephens *et al.*, 2005), affecting preferentially those

neurons that give rise to the indirect pathway (Day *et al.*, 2006). In addition, recent (rodent) studies have indicated that the sources of cortical inputs to the direct and indirect pathways may differ (Lei *et al.*, 2004), and that dopamine depletion may differentially alter the activity along these specific corticostriatal projections (Mallet *et al.*, 2006). These studies suggest that the activity changes at corticostriatal synapses may be far more complex (and enduring) than previously thought, and that changes in cortical activity may play a role in the proposed differential activity changes along the direct and indirect pathways. Thus far, such studies have only been done in rodents, and, given the known differences in cortical and basal ganglia anatomy between rodents and primates, it will be important to test whether they can be verified in the MPTP model or primate parkinsonism.

The effects of dopamine replacement have also been studied in detail in MPTP-treated monkeys.

Such studies have demonstrated that dopamine agonists lower the increased GPi firing rate (Papa *et al.*, 1999), and de-correlate the hypersynchronous basal ganglia activities (Heimer *et al.*, 2002), although a complete normalization of firing patterns does not occur (Heimer *et al.*, 2006). Furthermore, electrophysiologic studies have also demonstrated that drug-induced dyskinesias, a frequent side effect of dopaminergic therapy in Parkinson's disease, are associated with strong reductions, or suppression of GPi firing rates (Papa *et al.*, 1999).

CONCLUSION

Physiologic studies in the primate MPTP model of parkinsonism have been invaluable for the understanding of the pathophysiologic basis of the disorder and instrumental in the development of current models of the pathophysiology of parkinsonism. In addition, the MPTP-treated primate has been used for trials of medications or neurosurgical interventions in parkinsonism, and to model side effects of such treatments, such as drug-induced dyskinesias. These studies have been very important in the development of new therapies. For instance, the resurgence of neurosurgical therapies for Parkinson's disease, and the choice of the STN as a primary target for such treatments in the early 1990 was largely motivated by the finding that parkinsonism could be ameliorated by STN lesions. It is likely that the MPTP model of primate parkinsonism will continue to be useful in the development of new symptomatic treatments in this disorder. The success of the use of this model in the development of neuroprotective treatments is less certain, because the mechanism of cell death in many forms of human parkinsonism may differ from that induced by MPTP in primates, and because human Parkinson's disease clearly affects multiple cell types that are not affected by MPTP.

REFERENCES

Aizman, O., Brismar, H., Uhlen, P., Zettergren, E., Levey, A. I., Forssberg, H., Greengard, P., and Aperia, A. (2000). Anatomical and physiological evidence for D1 and D2 dopamine receptor colocalization in neostriatal neurons. *Nat Neurosci* **3**, 226–230.

Albin, R. L., Young, A. B., and Penney, J. B. (1989). The functional anatomy of basal ganglia disorders. *Trends Neurosci* **12**, 366–375.

Alexander, G. E., and Crutcher, M. D. (1990). Functional architecture of basal ganglia circuits: Neural substrates of parallel processing. *Trends Neurosci* **13**, 266–271.

Alvarez, L., Macias, R., Lopez, G., Alvarez, E., Pavon, N., Rodriguez-Oroz, M. C., Juncos, J. L., Maragoto, C., Guridi, J., Litvan, I., Tolosa, E. S., Koller, W., Vitek, J., DeLong, M. R., and Obeso, J. A. (2005). Bilateral subthalamotomy in Parkinson's disease: Initial and long-term response. *Brain* **128**, 570–583.

Baron, M. S., Vitek, J. L., Bakay, R. A. E., Green, J., Kaneoke, Y., Hashimoto, T., Turner, R. S., Woodard, J. L., Cole, S. A., McDonald, W. M., and DeLong, M. R. (1996). Treatment of advanced Parkinson's disease by GPi pallidotomy: 1 year pilot-study results. *Ann Neurol* **40**, 355–366.

Bergman, H., Wichmann, T., and DeLong, M. R. (1990). Reversal of experimental parkinsonism by lesions of the subthalamic nucleus. *Science* **249**, 1436–1438.

Bergman, H., Wichmann, T., Karmon, B., and DeLong, M. R. (1994). The primate subthalamic nucleus. II. Neuronal activity in the MPTP model of parkinsonism. *J Neurophysiol* **72**, 507–520.

Betarbet, R., Sherer, T. B., MacKenzie, G., Garcia-Osuna, M., Panov, A. V., and Greenamyre, J. T. (2000). Chronic systemic pesticide exposure reproduces features of Parkinson's disease. *Nat Neurosci* **3**, 1301–1306.

Boraud, T., Bezard, E., Guehl, D., Bioulac, B., and Gross, C. (1998). Effects of L-DOPA on neuronal activity of the globus pallidus externalis (GPe) and globus pallidus internalis (GPi) in the MPTP-treated monkey. *Brain Res* **787**, 157–160.

Boraud, T., Bezard, E., Bioulac, B., and Gross, C. E. (2000). Ratio of inhibited-to-activated pallidal neurons decreases dramatically during passive limb movement in the MPTP-treated monkey. *J Neurophysiol* **83**, 1760–1763.

Braak, H., Del Tredici, K., Rub, U., de Vos, R. A., Jansen Steur, E. N., and Braak, E. (2003). Staging of brain pathology related to sporadic Parkinson's disease. *Neurobiol Aging* **24**, 197–211.

Brown, P., and Williams, D. (2005). Basal ganglia local field potential activity: Character and functional significance in the human. *Clin Neurophysiol* **116**, 2510–2519.

Brown, P., Oliviero, A., Mazzone, P., Insola, A., Tonali, P., and Di Lazzaro, V. (2001). Dopamine dependency of oscillations between subthalamic nucleus and pallidum in Parkinson's disease. *J Neurosci* **21**, 1033–1038.

Brown, P., Mazzone, P., Oliviero, A., Altibrandi, M. G., Pilato, F., Tonali, P. A., and Di Lazzaro, V. (2004).

Effects of stimulation of the subthalamic area on oscillatory pallidal activity in Parkinson's disease. *Exp Neurol* **188**, 480–490.

Burns, R. S., Chiueh, C. C., Markey, S. P., Ebert, M. H., Jacobowitz, D. M., and Kopin, I. J. (1983). A primate model of parkinsonism: Selective destruction of dopaminergic neurons in the pars compacta of the substantia nigra by N-methyl-4-phenyl-1,2,3,6-tetrahydropyridine. *Proc Natl Acad Sci USA* **80**, 4546–4550.

Ceballos-Bauman, A. O., Obeso, J. A., Vitek, J. L., DeLong, M. R., Bakay, R., Linaasoro, G., and Brooks, D. J. (1994). Restoration of thalamocortical activity after posteroventrolateral pallidotomy in Parkinson's disease. *Lancet* **344**, 814.

Crossman, A. R., Mitchell, I. J., and Sambrook, M. A. (1985). Regional brain uptake of 2-deoxyglucose in N-methyl-4-phenyl-1,2,3,6-tetrahydropyridine (MPTP)-induced parkinsonism in the macaque monkey. *Neuropharmacology* **24**, 587–591.

Day, M., Wang, Z., Ding, J., An, X., Ingham, C. A., Shering, A. F., Wokosin, D., Ilijic, E., Sun, Z., Sampson, A. R., Mugnaini, E., Deutch, A. Y., Sesack, S. R., Arbuthnott, G. W., and Surmeier, D. J. (2006). Selective elimination of glutamatergic synapses on striatopallidal neurons in Parkinson disease models. *Nat Neurosci* **9**, 251–259.

Dogali, M., Fazzini, E., Kolodny, E., Eidelberg, D., Sterio, D., Devinsky, O., and Beric, A. (1995). Stereotactic ventral pallidotomy for Parkinson's disease. *Neurology* **45**, 753–761.

Filion, M., Tremblay, L., and Bedard, P. J. (1988). Abnormal influences of passive limb movement on the activity of globus pallidus neurons in parkinsonian monkeys. *Brain Res* **444**, 165–176.

Gerfen, C. R., Engber, T. M., Mahan, L. C., Susel, Z., Chase, T. N., Monsma, F. J., Jr., and Sibley, D. R. (1990). D1 and D2 dopamine receptor-regulated gene expression of striatonigral and striatopallidal neurons. *Science* **250**, 1429–1432.

Gill, S. S., and Heywood, P. (1997). Bilateral dorsolateral subthalamotomy for advanced Parkinson's disease [letter]. *Lancet* **350**, 1224.

Goldberg, J. A., Boraud, T., Maraton, S., Haber, S. N., Vaadia, E., and Bergman, H. (2002). Enhanced synchrony among primary motor cortex neurons in the 1-methyl-4-phenyl-1,2,3,6-tetrahydropyridine primate model of Parkinson's disease. *J Neurosci* **22**, 4639–4653.

Grafton, S., Waters, C., Sutton, J., Lew, M., and Couldwell, W. (1995). Pallidotomy increases activity of motor association cortex in Parkinson's disease: A positron emission tomographic study. *Ann Neurol* **37**, 776–783.

Hashimoto, T., Elder, C. M., Okun, M. S., Patrick, S. K., and Vitek, J. L. (2003). Stimulation of the subthalamic nucleus changes the firing pattern of pallidal neurons. *J Neurosci* **23**, 1916–1923.

Heimer, G., Bar-Gad, I., Goldberg, J. A., and Bergman, H. (2002). Dopamine replacement therapy reverses abnormal synchronization of pallidal neurons in the 1-methyl-4-phenyl-1,2,3,6-tetrahydropyridine primate model of parkinsonism. *J Neurosci* **22**, 7850–7855.

Heimer, G., Rivlin-Etzion, M., Bar-Gad, I., Goldberg, J. A., Haber, S. N., and Bergman, H. (2006). Dopamine replacement therapy does not restore the full spectrum of normal pallidal activity in the 1-methyl-4-phenyl-1,2,3,6-tetra-hydropyridine primate model of parkinsonism. *J Neurosci* **26**, 8101–8114.

Ingham, C. A., Hood, S. H., and Arbuthnott, G. W. (1989). Spine density on neostriatal neurones changes with 6-hydroxydopamine lesions and with age. *Brain Res* **503**, 334–338.

Jaeger, D., Gilman, S., and Aldridge, J. W. (1995). Neuronal activity in the striatum and pallidum of primates related to the execution of externally cued reaching movements. *Brain Res* **694**, 111–127.

Kojima, J., Yamaji, Y., Matsumura, M., Nambu, A., Inase, M., Tokuno, H., Takada, M., and Imai, H. (1997). Excitotoxic lesions of the pedunculopontine tegmental nucleus produce contralateral hemiparkinsonism in the monkey. *Neurosci Lett* **226**, 111–114.

Laitinen, L. V. (1995). Pallidotomy for Parkinson's disease. *Neurosurg Clin N Am* **6**, 105–112.

Langston, J. W., Ballard, P., Tetrud, J. W., and Irwin, I. (1983). Chronic parkinsonism in humans due to a product of meperidine-analog synthesis. *Science* **219**, 979–980.

Lei, W., Jiao, Y., Del Mar, N., and Reiner, A. (2004). Evidence for differential cortical input to direct pathway versus indirect pathway striatal projection neurons in rats. *J Neurosci* **24**, 8289–8299.

Mallet, N., Le Moine, C., Charpier, S., and Gonon, F. (2005). Feedforward inhibition of projection neurons by fast-spiking GABA interneurons in the rat striatum *in vivo*. *J Neurosci* **25**, 3857–3869.

Mallet, N., Ballion, B., Le Moine, C., and Gonon, F. (2006). Cortical inputs and GABA interneurons imbalance projection neurons in the striatum of parkinsonian rats. *J Neurosci* **26**, 3875–3884.

Mena-Segovia, J., Bolam, J. P., and Magill, P. J. (2004). Pedunculopontine nucleus and basal ganglia: Distant relatives or part of the same family? *Trends Neurosci* **27**, 585–588.

Miller, W. C., and DeLong, M. R. (1987). Altered tonic activity of neurons in the globus pallidus and subthalamic nucleus in the primate MPTP model of parkinsonism. In *The Basal Ganglia II* (M. B. Carpenter, and A. Jayaraman, Eds.), pp. 415–427. Plenum Press, New York.

Nandi, D., Aziz, T. Z., Giladi, N., Winter, J., and Stein, J. F. (2002). Reversal of akinesia in experimental parkinsonism by GABA antagonist microinjections in the pedunculopontine nucleus. *Brain* 125, 2418–2430.

Nini, A., Feingold, A., Slovin, H., and Bergman, H. (1995). Neurons in the globus pallidus do not show correlated activity in the normal monkey, but phase-locked oscillations appear in the MPTP model of parkinsonism. *J Neurophysiol* 74, 1800–1805.

Papa, S. M., Desimone, R., Fiorani, M., and Oldfield, E. H. (1999). Internal globus pallidus discharge is nearly suppressed during levodopa-induced dyskinesias. *Ann Neurol* 46, 732–738.

Pisani, A., Centonze, D., Bernardi, G., and Calabresi, P. (2005). Striatal synaptic plasticity: Implications for motor learning and Parkinson's disease. *Mov Disord* 20, 395–402.

Raju, D. V., Shah, D. J., Wright, T. M., Hall, R. A., and Smith, Y. (2006). Differential synaptology of vGluT2-containing thalamostriatal afferents between the patch and matrix compartments in rats. *J Comp Neurol* 499, 231–243.

Raz, A., Vaadia, E., and Bergman, H. (2000). Firing patterns and correlations of spontaneous discharge of pallidal neurons in the normal and the tremulous 1-methyl-4-phenyl-1,2,3,6-tetrahydropyridine vervet model of parkinsonism. *J Neurosci* 20, 8559–8571.

Raz, A., Frechter-Mazar, V., Feingold, A., Abeles, M., Vaadia, E., and Bergman, H. (2001). Activity of pallidal and striatal tonically active neurons is correlated in MPTP-treated monkeys but not in normal monkeys. *J Neurosci* 21, RC128.

Samuel, M., Ceballos-Baumann, A. O., Turjanski, N., Boecker, H., Gorospe, A., Linazasoro, G., Holmes, A. P., DeLong, M. R., Vitek, J. L., Thomas, D. G., Quinn, N. P., Obeso, J. A., and Brooks, D. J. (1997). Pallidotomy in Parkinson's disease increases supplementary motor area and prefrontal activation during performance of volitional movements an H2(15)O PET study. *Brain* 120, 1301–1313.

Smith, Y., Raju, D. V., Pare, J.-F., and Sidibe, M. (2004). The thalamostriatal system: A highly specific network of the basal ganglia circuitry. *Trends Neurosci* 27, 520–527.

Soares, J., Kliem, M. A., Betarbet, R., Greenamyre, J. T., Yamamoto, B., and Wichmann, T. (2004). Role of external pallidal segment in primate parkinsonism: Comparison of the effects of MPTP-induced parkinsonism and lesions of the external pallidal segment. *J Neurosci* 24, 6417–6426.

Stefani, A., Lozano, A. M., Peppe, A., Stanzione, P., Galati, S., Tropepi, D., Pierantozzi, M., Brusa, L., Scarnati, E., and Mazzone, P. (2007). Bilateral deep brain stimulation of the pedunculopontine and subthalamic nuclei in severe Parkinson's disease. *Brain*.

Stephens, B., Mueller, A. J., Shering, A. F., Hood, S. H., Taggart, P., Arbuthnott, G. W., Bell, J. E., Kilford, L., Kingsbury, A. E., Daniel, S. E., and Ingham, C. A. (2005). Evidence of a breakdown of corticostriatal connections in Parkinson's disease. *Neuroscience* 132, 741–754.

Tepper, J. M., Koos, T., and Wilson, C. J. (2004). GABAergic microcircuits in the neostriatum. *Trends Neurosci* 27, 662–669.

Trevitt, T., Carlson, B., Correa, M., Keene, A., Morales, M., and Salamone, D. (2002). Interactions between dopamine D1 receptors and gamma-aminobutyric acid mechanisms in substantia nigra pars reticulata of the rat: Neurochemical and behavioral studies. *Psychopharmacology* 159, 229–237.

Ungerstedt, U. (1968). 6-Hydroxy-dopamine induced degeneration of central monoamine neurons. *Eur J Pharmacol* 5, 107–110.

Wichmann, T., and Soares, J. (2006). Neuronal firing before and after burst discharges in the monkey basal ganglia is predictably patterned in the normal state and altered in parkinsonism. *J Neurophysiol* 95, 2120–2133.

Wichmann, T., Bergman, H., and DeLong, M. R. (1994). The primate subthalamic nucleus. I. Functional properties in intact animals. *J Neurophysiol* 72, 494–506.

Wilson, C. J. (1987). Morphology and synaptic connections of crossed corticostriatal neurons in the rat. *J Comp Neurol* 263, 567–580.

6

NEUROPROTECTION FOR PARKINSON'S DISEASE: CLINICALLY DRIVEN EXPERIMENTAL DESIGN IN NON-HUMAN PRIMATES

ERWAN BEZARD

CNRS UMR 5227, Universite Victor Segalen – Bordeaux 2, Bordeaux, France

In contrast to other neurodegenerative conditions, there is relatively good symptomatic therapy for Parkinson's disease (PD). However, there is no proven therapy yet that prevents cell death or restores sick neurons to a normal state. Interventions that can slow or halt the progression of PD remain a crucial unfulfilled need. The current proliferation of candidate drugs may bring more angst than excitement when the time comes to choose the few ones to be tested in PD patients (Ravina *et al.*, 2003). Indeed, the selection most likely still relies on data from 1-methyl-4-phenyl-1,2,3,6-tetrahydropyridine (MPTP)-induced or other neurotoxin-induced mouse models of PD, since genetic animal models of PD and other neurodegenerative disorders, which may have related pathogenic mechanisms, are just beginning to yield important discoveries.

Current animal models of PD allow assessment of the efficacy of drugs that exert symptomatic effects on PD motor abnormalities (Bezard *et al.*, 1998). With all these models, a reliable degeneration of substantia nigra pars compacta (SNc) neurons can be achieved. However, they are limited by their acute character (i.e., a single systemic or local injection of a toxin) that triggers a degeneration of SNc neurons that does not necessarily reflect the ongoing degenerative process in PD patients. Thus, in order to mimic the slow degenerative process

occurring in PD, more chronic regimen of intoxication, featuring either constant or repeated exposure to a single toxin/agent or combination of toxins/agents, seems better suited to assess putative neuroprotective effects of any given substance. Moreover, in humans, the onset of PD motor abnormalities occurs when SNc neuronal death exceeds 50–60% and DA nerve terminal death in the striatum exceeds 70–80% (Bernheimer *et al.*, 1973). PD patients are therefore likely to receive a neuroprotective agent only after diagnosis as no available predictive biomarker exists yet. Most of the current animal experiments do not reconcile this observation because the tested agent is usually applied prophylactically (i.e., weeks before) or together with the neurotoxin in acute or semi-chronic neurotoxin-induced models. Thus, we propose that, in order to assess drugs in a similar setting to clinical trials, the tested substance has to be applied after the loss of nigral DA-containing neurons exceeds the critical threshold responsible for the onset of PD motor abnormalities. Examples of such rather puzzling designs are numerous and basically, almost all the literature on neuroprotection falls in this category.

We do consider that, even though the neurotoxin models are not perfect, it is possible to make the most of them with respect to this key question of the screening of potential disease-modifying agents.

What we propose to call *a clinically driven design* should ideally satisfy the following three criteria. First, the chosen animal model should recapitulate several features of sporadic PD, including its progressiveness as we posit that specific cell death mechanisms or pathways would play a pivotal role only in this context and are somewhat masked in acute designs. Second, administration of the drug candidate should begin once neurodegeneration has started or from a pre-defined level of DA neuronal loss in order to mimic the clinical setting. Third, final proof of efficacy should be obtained from non-human primate models and not limited to rodents because it is likely that complex cell death mechanisms may differ in rodents and primates. Although being an opinion, this is supported by the differences in anatomo-functional organization of the SNc in primates (Hirsch *et al.*, 1997; Damier *et al.*, 1999b) and by fundamental physiological differences such as the presence of neuromelanin in primates/humans (McCormack *et al.*, 2004; Purisai *et al.*, 2005) but not in rodents, an index for critical role of microautophagy in maintaining SNc neurons in primates (Martinez-Vicente and Cuervo, 2007).

A CHRONIC (STEPWISE) MPTP PRIMATE MODEL

With this in mind, we have developed a chronic (stepwise) MPTP non-human primate model that mimic the different stages of the disease (Bezard *et al.*, 1997), although in a much shorter time frame (Figure 6.1a). In this model, macaques are treated daily with a low dose of MPTP (0.2 mg/kg i.v.) until parkinsonian motor abnormalities appear and reach an empirically defined threshold of behavioral impairment. As the model develops, animals exhibit a reduction in spontaneous movements and an increased muscle tone of the limbs, adopt a flexed posture, and have a decrease vocalization. Their movements become less accurate, for example when reaching for fruit, and there are occasional episodes of freezing. Both postural tremor and some resting tremor are observed. After the motor abnormalities have developed, animals pass through a critical period of very severe parkinsonism (Bezard *et al.*, 2001d) that stabilizes progressively over 2 months. Macaques then remain parkinsonian for years without recovery.

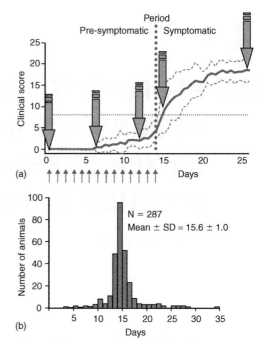

FIGURE 6.1 (a) Evolution of mean daily clinical scores (solid line) ± standard deviation (dashed lines) while daily injecting MPTP 0.2 mg/kg i.v. (small arrows) until reaching a score of 8 on our clinical scale (horizontal dashed line) (Imbert *et al.*, 2000). The dashed vertical line symbolizes the transition between the pre-symptomatic and the symptomatic periods. Large arrows correspond to time points at which animals were killed in order to characterize kinetics of neurodegeneration (see Figure 6.2; at D0, D6, D12, D15 and D25). (b) Historical record of the number of MPTP injection days required for reaching the clinical score of 8 (Imbert *et al.*, 2000). It shows the reproducibility of the model in developing symptoms.

Besides those motor abnormalities classically described in all MPTP monkey model, an unexpected but critical add-on of this model is its reproducibility (Figure 6.1b). Indeed, while the primate models have been plagued by huge variability in response to the toxin (Bédard *et al.*, 1992; Bezard *et al.*, 1998), we observe on the contrary an amazing reproducibility (Figure 6.1b) in probably the largest sample size ever produced with the very same regimen. We cannot, however, attribute this reproducibility to a given factor; that is, is this the consequence of the intoxication regimen or of the unique homogeneity of the monkey population (F2-bred female *Macaca fascicularis* or *Macaca mulatta* with a very low variability in weight and age).

TIME COURSE OF NIGROSTRIATAL DEGENERATION AFTER CHRONIC MPTP TREATMENT

The kinetics of nigrostriatal degeneration in this model and the critical thresholds associated with the motor abnormalities appearance were further validated by assessing striatal DA content and metabolism, DA transporter (DAT) binding in striatal sections, striatal DA receptor (DAR; D_1-like and D_2-like subtypes) binding and the number of both tyrosine hydroxylase immunoreactive (TH-IR) and Nissl-stained neurons in the SNc (Bezard *et al.*, 2001d). At day 13.2, the calculated day for motor abnormalities manifestation (day 1 being the first day of MPTP administration), $56.8 \pm 6.3\%$ of TH-IR and $75.2 \pm 6.2\%$ of Nissl-stained neurons remained in the SNc (Figure 6.2a; Table 6.1). In parallel, DAT binding in striatal sections (Figure 6.2b; Table 6.1) and DA content decreased to $19.7 \pm 4.9\%$ and $18.2 \pm 5.6\%$ of day 0 values while striatal D1 and D2 binding were similar to day 0 values (Bezard *et al.*, 2001d) (Figure 6.2d; Table 6.1). Besides the DAT binding in striatal sections, sequential ^{123}I-PE2I single photon emission computed tomography (SPECT) acquisitions was performed allowing monitoring of disease progression within the same animal by assessing striatal DAT binding in living animals (Figure 6.2c). In agreement with the DAT binding in striatal sections, the mean distribution volume calculated according to Logan's graphical method was significantly decreased from day 6 onward, that is when animals presented clinically normal (Prunier *et al.*, 2003).

As shown in Figure 6.2a, within the presymptomatic period, the decrease of DAT binding in striatal sections was more pronounced than the degeneration of TH-IR neurons, which in turn was dramatically accelerated after the onset of PD motor abnormalities between day 13.2 and day 25. This dissociation suggests that within the applied experimental protocol SNc neurons may loose the functional integrity of their terminals in early stages while their cell bodies remain unaffected as reflected by the dramatic decrease in striatal DAT binding and unaltered TH expression.

A mechanism to explain this dissociation may be the so-called "dying back", that is, an axon of a compromised neuron progressively degenerates over weeks or even months, beginning distally and then spreading toward the cell body (Raff *et al.*, 2002). These authors hypothesized that "dying back" may be due to an activation of a self-destruct program in the distal parts of an axon in response to an external stressor. It must be noted, however, that disease process might well start first in cell body but that the axon dies first. Subsequently, the nature, extent and time course may determine whether the neuron undergoes apoptotic cell death or activates a caspase independent, axonal self-destruct program to disconnect the axon from its target cell. However, with the onset of PD motor abnormalities the degeneration of TH-IR neurons is accelerated. Changes in DA content and DAT binding in striatal sections are linearly correlated and are both markers of ongoing degeneration of striatal DA terminals ($y = 1.71 + 1.08x$, $r = 0.92$, $p < 0.05$) (Bezard *et al.*, 2001d). The relationship between striatal DA turn-over and DAT binding in striatal sections is characterized by a logarithmic correlation ($y = 6.11 - 2.59 \log(x)$, $r = 0.84$, $p < 0.05$) (Bezard *et al.*, 2001d). Although there is a dramatic decrease of DAT binding in striatal sections, DA turn-over increases significantly only at D25, that is in full-stage parkinsonism. This underlines the need for pronounced DA depletion before any increase in DA turn-over can be observed (Bernheimer *et al.*, 1973; Elsworth *et al.*, 2000). The time course of D_2 binding cannot be described by a simple equation as well. The relationship between D_2- and DAT binding in striatal sections is quadratic ($y = 524 - 5.54x + 0.04x^2$, $r = 0.75$, $p < 0.05$) (Bezard *et al.*, 2001d). This implies a synergistic action of two first order processes. Since D_2 DAR are located pre- and postsynaptically the initial decrease in D_2 DAR binding mainly reflects the disappearance of striatal DA terminals as indicated by decreasing DAT binding in striatal sections, whereas the subsequent increase represents an upregulation of postsynaptic D_2 DAR.

HOW DOES THAT FIT WITH HUMAN PD?

According to clinical studies in human PD, motor abnormalities appearance would require a 70–80% loss of striatal terminals, a 70–90% depletion of

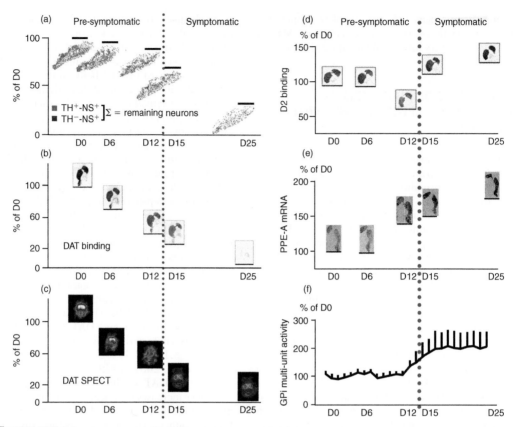

FIGURE 6.2   Time course of nigrostriatal degeneration. The dashed red line indicates the transition between the pre-symptomatic and symptomatic period in these animals. (a) Examples of cell counting maps showing the typical patterns of degeneration in the SNc ($n = 5$ at D0, D6, D12, D15 and D25) (Bezard et al., 2001d). The number of TH-IR and Nissl-stained neurons in the SNc were counted in one representative section corresponding to a median plane of the SNc by one examiner blinded to the experimental condition. TH-IR neurons are shown in red whereas the blue symbols represent the Nissl-stained cells that were not TH positive. The horizontal line above each map indicates the mean percentage of surviving cells (i.e., Nissl stained). Note the selective disappearance of the dorsal tier of the SNc with time. (b) Representative examples of DAT binding autoradiographs showing the progression of striatal denervation at the caudal level of the same animals (Bezard et al., 2001d). Note the homogenous degeneration and the severe lesion in the D25 group. The horizontal line under each example indicates the mean percentage of DAT binding in striatal sections. Non-specific binding is shown on the bottom left corner of the figure (c) Representative examples of ^{123}I-PE2I SPECT as a marker of DAT binding in living animals during disease progression between D0 and D25 ($n = 2$) (Prunier et al., 2003). In agreement with the DAT binding in striatal sections, there is a homogenous degeneration and severe lesion in the D25 group. The inferior border of each image corresponds to the mean percentage of striatal ^{123}I-PE2I binding potential as determined by Logan's graphical method. Changes affecting D_2 DAR binding (d) (Bezard et al., 2001d), PPE-A mRNA expression (e) of the same animals (Bezard et al., 2001 c) and GPi multi-unit electrophysiological activity (f) in two additional animals (Bezard et al., 1999). D_2 DAR binding (a) and PPE-A mRNA (c) autoradiographs have been taken at the rostral and caudal level of the striatum, respectively.

striatal DA as well as a 50–60% loss of nigral DA neurons (Bernheimer et al., 1973; Riederer and Wuketich, 1976). In the present chronic MPTP model, the depletion of striatal markers fits with

these predictions while the nigral threshold is lower than expected. Fearnley and Lees, however, determined a threshold of 31% DA cell loss in human PD (1991) while German and colleagues (1989)

TABLE 6.1 Time course of nigrostriatal degeneration after chronic MPTP treatment between day 0 and day 25 in the SNc (TH-IR, Nissl staining) and the putamen (DAT binding in striatal sections, DA content, DA metabolites/DA ratio, D_1 and D_2 binding) according to Bezard et al. (2001)

	D0	D6	D12	D13.2	D15	D25
TH-IR	100	82.4	62.5	56.8	52.5	14.2
Nissl staining	100	97.9	93.3	75.2	71.6	34.7
DAT binding	100	69.8	40.6	19.7	18.3	2.3
DA content	100	58.3	41.8	18.2	14.2	2.0
DA metabolites/DA ratio	100	130.4	148.3	188.2	232.5	559.3
D_1-like binding	100	103.7	95.5	98.3	101.2	91.4
D_2-like binding	100	95.7	68.8	93.3	114.9	140.6

Values are displayed as percent of day 0. Mean onset of PD motor abnormalities occurs at day 13.2. The given numbers for day 13.2 were calculated from best-fit regression equations

reported a decrease of 46% in mildly symptomatic MPTP-treated *M. fascicularis*. In accordance with previous reports (German *et al.*, 1989), the general gradient loss we observed started off in the whole dorsal tier of the SNc and thereafter spread to its ventral tier, where only few TH-IR neurons remained detectable in the fully parkinsonian state (D25, Figure 6.2a) (German *et al.*, 1996). This suggests that the present dynamic MPTP model would correspond to human PD in terms of similar thresholds for motor abnormalities appearance. In accordance with neuropathological findings in human MPTP-induced parkinsonism (Langston *et al.*, 1999), but in contrast to patients with PD, no Lewy bodies were found in surviving SNc neurons. However, the present model is not intended to determine mechanisms of cellular degeneration of PD, but to correlate biochemical and behavioral parameters with a defined extend of nigrostriatal degeneration.

ADDITIONAL *IN VIVO* AND *EX VIVO* MARKERS

Besides the in-depth characterization of the kinetics of nigrostriatal degeneration using *in vivo* and *ex vivo* markers, we have also studied the pathophysiological consequences of the denervation progression upon basal ganglia physiology. Those markers can be used for defining the level of imbalance in the transcriptional, biochemical and/or electrophysiological basal ganglia activity. Two examples can be given for illustrating this assertion. γ-Aminobutyric acid (GABA)-utilizing efferents from the striatum to the globus pallidus pars externalis (GPe) also use enkephalins, derived from the precursor pre-proenkephalin-A (PPE-A) as co-transmitters. A wealth of evidence suggests that the activity of both the GABAergic and the enkephalinergic components of this pathway is increased in the parkinsonian MPTP-treated monkey (Herrero *et al.*, 1995; Levy *et al.*, 1995; Morissette *et al.*, 1999). Therefore increased PPE-A mRNA expression is a classic marker of the DA-depleted brain. In our model, we have shown that PPE-A mRNA levels are elevated before the appearance of parkinsonian motor abnormalities (Bezard *et al.*, 2001c) (Figure 6.2e). Importantly, this upregulation is restricted to motor regions of the basal ganglia circuitry. The increased PPE-A mRNA expression observed in asymptomatic but DA-depleted animals provides a key tool for assessing the relative degree of impairment of the basal ganglia activity in the context of a neuroprotective drug.

In the same way, the metabolic (Bezard et al., 2001a, 2003) and electrophysiological brain activity (Bezard *et al.*, 1999, 2001b) has been characterized in this model. Figure 6.2f, for instance, shows the changes in the globus pallidus pars internalis firing frequency over time and unravels that the increased pathological activity actually begins before the appearance of motor abnormalities. Last, but not least, we monitored the transcriptional fluctuations in the substantia nigra using Affymetrix microarrays (Bassilana *et al.*, 2005). Although the overall number of regulated transcripts was relatively modest at the different time points, five clusters exhibiting different profiles were defined using a hierarchical clustering algorithm. Besides the identity of these regulated clusters of genes (Bassilana *et al.*, 2005), the approach offers the possibility to then assess the normalizing action of a drug upon midbrain neuron expression profile.

In conclusion, all these approaches can be then used for monitoring the effect of a neuroprotective drug. Indeed, these various markers should be used in order to not simply rely upon the clinical assessment of macaque motor behavior, but instead to ground the protection of DA neurons through functional independent outcome measures that directly or indirectly reflect the proper functioning of basal ganglia.

A CLINICALLY RELEVANT DESIGN THAT TAKES ADVANTAGE OF THE MODEL

We propose to use the above-described model for applying the so-called "clinically relevant" design for assessing potentially neuroprotective drugs in a similar setting to clinical trials, that is, the tested substance would be applied after the loss of nigral DA-containing neurons exceeds the critical threshold responsible for the onset of PD motor abnormalities.

In our model, drugs can be administered either prophylactically (before MPTP starts), together with MPTP with enough administrations to obtain 24 h/day optimal plasmatic concentration (Figure 6.3; dashed line) (e.g., Diguet *et al.*, 2004; Bezard *et al.*, 2006) or few days after the MPTP has begun (Figure 6.3; punctiform line) (e.g., Bezard *et al.*, 2006; Scheller *et al.*, 2007). Such "delayed application" is usually initiated at a time (day 7–8) when there is 30% neuronal death in the SNc and nerve terminal loss in the striatum is 40%, that is, compatible with the clinical situation where early symptomatic patients would receive such a treatment (e.g., Diguet *et al.*, 2004; Bezard *et al.*, 2006; Scheller *et al.*, 2007). Delayed appearance of motor abnormalities (Figure 6.3) is suggestive of neuroprotection

but should be supported by post-mortem measures made from animals killed at different key time points (Figure 6.3; arrows) and/or by *in vivo* imaging of DAT (Figure 6.3 – right) or of any validated metabolical/biochemical/transcriptomical marker (see above for the examples). Combination of approaches is however preferred to, once again, mimic as closely as possible the clinical situation. Using such experimental designs, only few drug candidates have shown a clinically relevant efficacy (i.e., the ability to delay the appearance of clinical motor abnormalities associated with a rescue or protection of remaining DA-containing neurons). Some drug candidates, previously shown to be efficacious in acute models when applied before or together with the neurotoxin, either failed or had deleterious effects in chronic models (E. Bezard, unpublished observation). The most impressive deleterious effect has been obtained with minocycline, one of the twelve selected compounds by the National Institute of Neurological Disorders and Stroke (USA) for further study in human patients (Group, 2006). Indeed we have tested minocycline in two phenotypic models of neurodegenerative conditions, a non-human primate model of PD and a mouse model of Huntington's disease. In these two models that fulfill most of the above-described criteria, minocycline induced a worsening of the motor abnormalities paralleled by an increased extent of lesion (Diguet *et al.*, 2004).

Although we have developed a chronic (stepwise) MPTP non-human primate model that we use to study PD progression (Bezard *et al.*, 2001d; Prunier *et al.*, 2003) and validate neuroprotective drugs (Bezard *et al.*, 2006), we do not think that it should be the gold standard model. Although this model is useful, as it allows the tested drug to be administered from a given threshold of lesion (i.e., the cornerstone of our rationale and the minimal prerequisite of any experiment aimed at validating the protective feature of a drug), it still uses a neurotoxin, MPTP, the mechanisms of action of which might be different from sporadic PD neurodegeneration. In addition, it is still a stepwise model as it does not seem to trigger an ongoing degenerative process. Several features of human PD are still lacking in this model such as the Lewy body inclusions, the widespread degeneration of non-monoaminergic neurons, etc. Recent refinement of the MPTP regimen of administration has shown both progression and presence of Lewy body-like inclusions in the SNc of mice (Fornai *et al.*, 2005). Such an

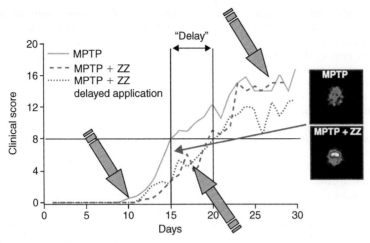

FIGURE 6.3 Example of a clinically driven experimental design in the MPTP macaque model. Data are the mean of daily clinical scores in the course of the intoxication protocol. All animals are daily treated with MPTP, from Day 1 onward, until they individually reach a score of 8 on the clinical rating scale. In addition to MPTP, animals receive either vehicle (solid line), tested drug (zz) from day 1 onward (dashed line) or tested drug (zz) from day 8 onward (punctiform line – delayed application). The only pertinent parameter is the day of reaching a score of 8 on the clinical scores. "Delay" denotes a significant difference in reaching this predefined threshold between zz-treated groups and vehicle-treated group. In this theoretical case, both concomitant and delayed application of the zz drug proved to be efficacious. Since animals are treated with MPTP until they reach that score of 8, they ALL become parkinsonian. Thus the putative difference in clinical scores once syndrome is stabilized does not reflect necessarily a protection effect. This emphasizes the fact that only the day for symptom reaching a score of 8 must be taken as the relevant parameter in the behavioral part of the experiment. DAT SPECT (bottom left examples) could be performed at specific time points of the intoxication protocol to maximize the data collection and to differentiate between symptomatic-like from neuroprotective-like activity of a given drug. Finally, these experiments should then be associated to post-mortem measures at different time points (arrows). A difference in the extent of nigrostriatal denervation at stages 2 and 3 would be indicative of neuroprotection.

exciting result rejuvenates the neurotoxin approach and further stress the need for improved experimental designs. It remains, however, to demonstrate the efficacy of given compounds to stop or interfere with the MPTP-induced cell death with this specific regimen of administration (Fornai *et al.*, 2005). As strategies aiming at interfering with alpha-synuclein aggregation are being developed, such a model might offer the opportunity to discriminate between neuroprotection and anti-aggregation as the pathological role for aggregation is still debated. Other toxins, or even genetically engineered animals, should also be used provided that the basic concept of coming closer to the clinical situation is respected. In this respect, our hope certainly lies in the successful development of conditional mutations inducing progressive neuronal death. At the moment, the alpha-synuclein overexpressing lines display at best a small (albeit significant)

dopaminergic neuronal death in the SNc (for review, see Fernagut and Chesselet, 2004). The models, such as the doubly mutated alpha-synuclein under TH promoter (Thiruchelvam *et al.*, 2004) or the overexpressing wild-type alpha-synuclein under the Thy1 promoter (Rockenstein *et al.*, 2002), recapitulate some features of synucleinopathies and would thus be useful to study the potential pathogenic role of alpha-synuclein and its molecular partners, applying the same experimental design (Chesselet, 2007). The same remark applies to the more recently developed parkin knockout (e.g., see Goldberg *et al.*, 2003; Perez and Palmiter, 2005) and DJ-1 knockout mice (e.g., see Goldberg *et al.*, 2005; Kim *et al.*, 2005), which showed no dopaminergic neuronal death. Characterization of the PINK1 or LRRK2 mutations in mice is still awaiting. However, these mouse lines consistently showed impaired locomotion (e.g., see Fernagut and

Chesselet, 2004) and possibly aberrant dopamine transmission (Goldberg *et al.*, 2005). This raises the possibility to test drugs that would "protect" from these impairments rather than protect dopaminergic neurons *stricto sensu* with experimental designs in keeping with the current proposal.

We do consider, however, that the validation process must include a non-human primate model, whatever the specifics of the model are, provided it fulfills the different above-mentioned criteria. The anatomical organization of the nigrostriatal pathway is fundamentally different in rodents and primates as only primates resembles the specific pattern of striatal innervation by specific subsets of dopaminergic neurons. Moreover, the complex intrinsic organization of the human SNc (Damier *et al.*, 1999a, b) is better approached by non-human primates than by rodents (German *et al.*, 1983) rendering possible to identify specific subpopulations of more susceptible or resistant neurons. Such differences in cellular organization and possibly in microenvironment are likely to translate in differences in mechanisms of cell death or in weighing differently the different lethal factors (Hirsch *et al.*, 1997; Hirsch, 1999). Finally, it is the only model into which you can easily correlate the behavioral and anatomical outcome measures.

To date no compound is marketed or has even been proved efficacious in human PD in spite of a myriad of preclinical studies and of a significant number of clinical trials as all proved to be inconclusive. The number of drugs emerging from academic or industry research aiming at slowing down neuronal death far exceeds those for symptomatic relief calling for extreme caution in selecting the drug deserving to be tested in human patients and thus for further preclinical validation. Preclinical research should focus on the development of more reliable chronic models and apply putative neuroprotective drugs after nigral cell death exceeds the threshold that accounts for the onset of PD motor abnormalities as drugs would thus have to rescue compromised neurons instead of blocking acute toxicity. We here acknowledged very openly the shortcomings of our models, not to blame ourselves but to emphasize how aware we are of their limitation. Probably as problematic than the models, the clinical trial do suffer as well from drawbacks that make us uncertain of our current ability to soundly demonstrate the protective capacity of a drug, a fact that explain why the regulatory authorities suspiciously consider the field. Among those drawbacks,

some are inherent to the disease itself: the heterogeneity of parkinsonian syndromes and the heterogeneity of clinical expression. Others come from lack of careful investigations such as the lack of thorough pharmacologic and pharmacokinetics studies, therefore not allowing knowing precisely if a given drug effectively penetrates the brain at relevant concentrations. Other drawbacks come from the paucity of the tools used for monitoring such a "protection." Several human studies have investigated putative neuroprotective effects of DAR agonists on nigrostriatal degeneration with imaging techniques. These imaging techniques assess either DA terminal function or DAT activity instead of directly measuring the number of remaining SNc neurons. Thus, the major limitation in interpreting these results is that the difference in loss of $[^{123}I]b$-CIT (a DAT ligand whose binding is proportional to the number of DA terminals present) and ^{18}F-dopa uptake (a marker of DA terminal function) in DAR agonist-treated versus L-dopa-treated groups might be caused by a pharmacological interaction rather than slowed or accelerated neuronal degeneration (Marek *et al.*, 2003). Another critical issue of current imaging studies is the wash-out period at the end of the trial. If a tested agent is neuroprotective, one would expect a persistent difference in surrogate markers, even when the drug is removed. This raises the question of whether it is ethical to remove all medication in patients with PD? And if so, for how long? At present, the Food and Drug Administration does not accept a primary endpoint for clinical trials based on imaging techniques.

The principal endpoint of current clinical trials to assess putative neuroprotective effects is the reversal of PD motor abnormalities as measured by the UPDRS. Although this rating scale has a low interobserver variability (Martinez-Martin *et al.*, 1994), the use of this endpoint is tricky because any agent that improves PD motor abnormalities would already be considered neuroprotective. To add to the confusion, the ELLDOPA (Earlier versus Later **L-DOPA**) trial, conceived to determine whether L-dopa alters the natural course of PD, showed contradictory results between clinical and imaging data. After 2 weeks of wash-out, patients that received 600 mg/day L-dopa were still clinically improved over baseline and in comparison with the placebo group. However, although the behavioral analysis suggested an improvement of the L-dopa-treated patients, the DAT imaging, using $[^{123}I]b$-CIT, showed a further loss in the same patients (Fahn,

2005, 2006). The behavioral and imaging endpoints are thus contradictory in the ELLDOPA study. A second measure to determine a putative neuroprotective effect of a tested agent is the additional need for L-dopa treatment in case of insufficient motor abnormalities control. According to this measure, it is expected that the faster a patient would need additional L-dopa, the faster would be the degeneration of nigral DA-containing neurons and the progression of the disease. However, any symptomatic effect would confound the results of this very subjective endpoint. This explains why the putative neuroprotective capacity of the monoamine oxidase B inhibitor selegiline remains controversial because of its symptomatic effect (Shoulson, 1998).

In view of these methodological limitations of clinical trials, it is now even more important that putative neuroprotective compounds should be tested in clinically driven preclinical studies. Resolution of these issues is of critical importance to convince (i) the regulatory authorities of the reliability of our preclinical experimental designs/models and therefore of their predictability and (ii) the governmental agencies and/or pharmaceutical companies to expend the hundreds of millions of dollars necessary to bring a new drug to market.

DISCLOSURE

EB is Non-Executive Director of Motac Neuroscience, Manchester, UK.

REFERENCES

Bassilana, F., Mace, N., Li, Q., Stutzmann, J. M., Gross, C. E., Pradier, L., Benavides, J., Ménager, J., and Bezard, E. (2005). Unraveling substantia nigra sequential gene expression in a progressive MPTP-lesioned macaque model of Parkinson's disease. *Neurobiol Dis* **20**, 93–103.

Bédard, P. J., Boucher, R., Gomez-Mancilla, B., and Blanchette, P. (1992). Primate models of Parkinson's disease. In *Animal models of neurological disease, I: Neurodegenerative diseases* (A. A. Boulton, G. B. Baker, and R. F. Butterworth, Eds.), pp. 159–173. Humana press, Totowa.

Bernheimer, H., Birkmayer, W., Hornykiewicz, O., Jellinger, K., and Seitelberger, F. (1973). Brain dopamine and the syndromes of Parkinson and Huntington. Clinical, morphological and neurochemical correlations. *J Neurol Sci* **20**, 415–455.

Bezard, E., Imbert, C., Deloire, X., Bioulac, B., and Gross, C. (1997). A chronic MPTP model reproducing the slow evolution of Parkinson's disease: Evolution of motor symptoms in the monkey. *Brain Res* **766**, 107–112.

Bezard, E., Imbert, C., and Gross, C. E. (1998). Experimental models of Parkinson's disease: From the static to the dynamic. *Rev Neurosci* **9**, 71–90.

Bezard, E., Boraud, T., Bioulac, B., and Gross, C. (1999). Involvement of the subthalamic nucleus in glutamatergic compensatory mechanisms. *Eur J Neurosci* **11**, 2167–2170.

Bezard, E., Crossman, A. R., Gross, C. E., and Brotchie, J. M. (2001a). Structures outside the basal ganglia may compensate for dopamine loss in the presymptomatic stages of Parkinson's disease. *Faseb J*, 10.1096:fj.00-0637fje.

Bezard, E., Boraud, T., Bioulac, B., and Gross, C. (2001b). Evolution of the multiunit activity of basal ganglia in the course of a dynamic experimental parkinsonism. In *The Basal ganglia VI* (A. Graybiel, M. Delong, and S. T. Kitaï, Eds.), pp. 107–116. Kluwer Academic Publishers, Norwell.

Bezard, E., Ravenscroft, P., Gross, C. E., Crossman, A. R., and Brotchie, J. M. (2001c). Upregulation of striatal preproenkephalin gene expression occurs before the appearance of parkinsonian signs in 1-methyl-4-phenyl-1,2,3,6-tetrahydropyridine monkeys. *Neurobiol Dis* **8**, 343–350.

Bezard, E., Dovero, S., Prunier, C., Ravenscroft, P., Chalon, S., Guilloteau, D., Bioulac, B., Brotchie, J. M., and Gross, C. E. (2001d). Relationship between the appearance of symptoms and the level of nigrostriatal degeneration in a progressive MPTP-lesioned macaque model of Parkinson's disease. *J Neurosci* **21**, 6853–6861.

Bezard, E., Gross, C. E., and Brotchie, J. M. (2003). Presymptomatic compensation in Parkinson's disease is not dopamine-mediated. *Trends Neurosci* **26**, 215–221.

Bezard, E., Gerlach, I., Moratalla, R., Gross, C. E., and Jork, R. (2006). 5-HT1A receptor agonist-mediated protection from MPTP toxicity in mouse and macaque models of Parkinson's disease. *Neurobiol Dis* **23**, 77–86.

Chesselet, M. F. (2007). *In vivo* alpha-synuclein overexpression in rodents: A useful model of Parkinson's disease? *Exp Neurol*.

Damier, P., Hirsch, E. C., Agid, Y., and Graybiel, A. M. (1999a). The substantia nigra of the human brain-I. Nigrosomes and the nigral matrix, a compartmental organization based on calbindin D-28K immunohistochemistry. *Brain* **122**, 1421–1436.

Damier, P., Hirsch, E. C., Agid, Y., and Graybiel, A. M. (1999b). The substantia nigra of the human brain-II. Patterns of loss of dopamine-containing neurons in Parkinson's disease. *Brain* 122, 1437–1448.

Diguet, E., Fernagut, P. O., Wei, X., Du, Y., Rouland, R., Gross, C., Bezard, E., and Tison, F. (2004). Deleterious effects of minocycline in animal models of Parkinson's disease and Huntington's disease. *Eur J Neurosci* 19, 3266–3276.

Elsworth, J. D., Taylor, J. R., Sladek, J. R., Collier, T. J., Redmond, D. E., and Roth, R. H. (2000). Striatal dopaminergic correlates of stable parkinsonism and degree of recovery in Old-World primates one year after MPTP treatment. *Neuroscience* 95, 399–408.

Fahn, S. (2005). Does levodopa slow or hasten the rate of progression of Parkinson's disease? *J Neurol* 252(Suppl 4), IV37–IV42.

Fahn, S. (2006). A new look at levodopa based on the ELLDOPA study. *J Neural Transm Suppl*, 419–426.

Fearnley, J. M., and Lees, A. J. (1991). Ageing and Parkinson's disease: substantia nigra regional selectivity. *Brain* 114, 2283–2301.

Fernagut, P. O., and Chesselet, M. F. (2004). Alpha-synuclein and transgenic mouse models. *Neurobiol Dis* 17, 123–130.

Fornai, F., Schluter, O. M., Lenzi, P., Gesi, M., Ruffoli, R., Ferrucci, M., Lazzeri, G., Busceti, C. L., Pontarelli, F., Battaglia, G., Pellegrini, A., Nicoletti, F., Ruggieri, S., Paparelli, A., and Sudhof, T. C. (2005). Parkinson-like syndrome induced by continuous MPTP infusion: Convergent roles of the ubiquitin-proteasome system and alpha-synuclein. *Proc Natl Acad Sci USA* 102, 3413–3418.

German, D. C., Schlusselberg, D. S., and Woodward, D. J. (1983). Three-dimensional computer reconstruction of midbrain dopaminergic neuronal populations: From mouse to man. *J Neural Transm* 57, 243–254.

German, D. C., Manaye, K., Smith, W. K., Woodward, D. J., and Saper, C. B. (1989). Midbrain dopaminergic cell loss in Parkinson's disease: Computer visualization. *Ann Neurol* 26, 507–514.

German, D. C., Nelson, E. L., Liang, C.-L., Speciale, S. G., Sinton, C. M., and Sonsalla, P. K. (1996). The neurotoxin MPTP causes degeneration of specific nucleus A8, A9, and A10 dopaminergic neurons in the mouse. *Neurodegeneration* 5, 299–312.

Goldberg, M. S., Fleming, S. M., Palacino, J. J., Cepeda, C., Lam, H. A., Bhatnagar, A., Meloni, E. G., Wu, N., Ackerson, L. C., Klapstein, G. J., Gajendiran, M., Roth, B. L., Chesselet, M. F., Maidment, N. T., Levine, M. S., and Shen, J. (2003). Parkin-deficient mice exhibit nigrostriatal deficits but not loss of dopaminergic neurons. *J Biol Chem* 278, 43628–43635.

Goldberg, M. S., Pisani, A., Haburcak, M., Vortherms, T. A., Kitada, T., Costa, C., Tong, Y., Martella, G., Tscherter, A., Martins, A., Bernardi, G., Roth, B. L., Pothos, E. N., Calabresi, P., and Shen, J. (2005). Nigrostriatal dopaminergic deficits and hypokinesia caused by inactivation of the familial Parkinsonism-linked gene DJ-1. *Neuron* 45, 489–496.

Group, P. S. (2006). A randomized, double-blind, futility clinical trial of creatine and minocycline in early Parkinson disease. *Neurology* 66, 664–671.

Herrero, M. T., Augood, S. J., Hirsch, E. C., Javoy-Agid, F., Luquin, M. R., Agid, Y., Obeso, J. A., and Emson, P. C. (1995). Effects of L-Dopa on preproenkephalin and preprotachykinin gene expression in the MPTP-treated monkey striatum. *Neuroscience* 68, 1189–1198.

Hirsch, E. C. (1999). Mechanism and consequences of nerve cell death in Parkinson's disease. *J Neural Transm Suppl*, 127–137.

Hirsch, E. C., Faucheux, B., Damier, P., Mouatt-Prigent, A., and Agid, Y. (1997). Neuronal vulnerability in Parkinson's disease. *J Neural Transm* 50, 79–88.

Imbert, C., Bezard, E., Guitraud, S., Boraud, T., and Gross, C. E. (2000). Comparison between eight clinical rating scales used for the assessment of MPTP-induced parkinsonism in the macaque monkey. *J Neurosci Meth* 96, 71–76.

Kim, R. H., Smith, P. D., Aleyasin, H., Hayley, S., Mount, M. P., Pownall, S., Wakeham, A., You-Ten, A. J., Kalia, S. K., Horne, P., Westaway, D., Lozano, A. M., Anisman, H., Park, D. S., and Mak, T. W. (2005). Hypersensitivity of DJ-1-deficient mice to 1-methyl-4-phenyl-1,2,3,6-tetrahydropyridine (MPTP) and oxidative stress. *Proc Natl Acad Sci USA* 102, 5215–5220.

Langston, J. W., Forno, L. S., Tetrud, J., Reeves, A. G., Kaplan, J. A., and Karluk, D. (1999). Evidence of active nerve cell degeneration in the substantia nigra of humans years after 1-methyl-4-phenyl-1,2,3,6-tetrahydropyridine exposure. *Ann Neurol* 46, 598–605.

Levy, R., Herrero, M.-T., Ruberg, M., Villares, J., Faucheux, B., Guridi, J., Guillen, J., Luquin, M. R., Javoy-Agid, F., Obeso, J. A., Agid, Y., and Hirsch, E. C. (1995). Effects of nigrostriatal denervation and L-dopa therapy on the GABAergic neurons of the striatum in MPTP-treated monkeys and Parkinson's disease: An *in situ* hybridization study of GAD$_{67}$- mRNA. *Eur J Neurosci* 7, 1199–1209.

Marek, K., Jennings, D., and Seibyl, J. (2003). Dopamine agonists and Parkinson's disease progression: What can we learn from neuroimaging studies. *Ann Neurol* 53(Suppl 3), S160–S166, discussion S166-169.

Martinez-Martin, P., Gil-Nagel, A., Gracia, L. M., Gomez, J. B., Martinez-Sarries, J., and Bermejo, F. group tcm (1994). Unified Parkinson's disease rating scale characteristics and structure. *Mov Disord* 9, 76–83.

Martinez-Vicente, M., and Cuervo, A. M. (2007). Autophagy and neurodegeneration: when the cleaning crew goes on strike. *Lancet Neurol* 6, 352–361.

McCormack, A. L., Di Monte, D. A., Delfani, K., Irwin, I., DeLanney, L. E., Langston, W. J., and Janson, A. M. (2004). Aging of the nigrostriatal system in the squirrel monkey. *J Comp Neurol* 471, 387–395.

Morissette, M., Grondin, R., Goulet, M., Bedard, P. J., and Di Paolo, T. (1999). Differential regulation of striatal preproenkephalin and preprotachykinin mRNA levels in MPTP-lesioned monkeys chronically treated with dopamine D-1 or D-2 receptor agonists. *J Neurochem* 72, 682–692.

Perez, F. A., and Palmiter, R. D. (2005). Parkin-deficient mice are not a robust model of parkinsonism. *Proc Natl Acad Sci USA* 102, 2174–2179.

Prunier, C., Bezard, E., Mantzarides, M., Besnard, J. C., Baulieu, J. L., Gross, C. E., Guilloteau, D., and Chalon, S. (2003). Presymptomatic diagnosis of experimental parkinsonism with 123I-PE2I SPECT. *Neuroimage* 19, 810–816.

Purisai, M. G., McCormack, A. L., Langston, W. J., Johnston, L. C., and Di Monte, D. A. (2005). Alpha-synuclein expression in the substantia nigra of MPTP-lesioned non-human primates. *Neurobiol Dis* 20, 898–906.

Raff, M. C., Whitmore, A. V., and Finn, J. T. (2002). Axonal self-destruction and neurodegeneration. *Science* 296, 868–871.

Ravina, B. M., Fagan, S. C., Hart, R. G., Hovinga, C. A., Murphy, D. D., Dawson, T. M., and Marler, J. R. (2003). Neuroprotective agents for clinical trials in Parkinson's disease: A systematic assessment. *Neurology* 60, 1234–1240.

Riederer, P., and Wuketich, S. (1976). Time course of nigrostriatal degeneration in Parkinson's disease. *J Neural Transm* 38, 277–301.

Rockenstein, E., Mallory, M., Hashimoto, M., Song, D., Shults, C. W., Lang, I., and Masliah, E. (2002). Differential neuropathological alterations in transgenic mice expressing alpha-synuclein from the platelet-derived growth factor and Thy-1 promoters. *J Neurosci Res* 68, 568–578.

Scheller, D., Chan, P., Li, Q., Wu, T., Zhang, R., Guan, L., Ravenscroft, P., Guigoni, C., Crossman, A. R., Hill, M., and Bezard, E. (2007). Rotigotine treatment partially protects from MPTP toxicity in a progressive macaque model of Parkinson's disease. *Exp Neurol* 203, 415–422.

Shoulson, I. (1998). DATATOP: A decade of neuroprotective inquiry. Parkinson Study Group. Deprenyl and tocopherol antioxidative therapy of parkinsonism. *Ann Neurol* 44, S160–S166.

Thiruchelvam, M. J., Powers, J. M., Cory-Slechta, D. A., and Richfield, E. K. (2004). Risk factors for dopaminergic neuron loss in human alpha-synuclein transgenic mice. *Eur J Neurosci* 19, 845–854.

7

THE USE OF AGED MONKEYS TO STUDY PD: IMPORTANT ROLES IN PATHOGENESIS AND EXPERIMENTAL THERAPEUTICS

YAPING CHU AND JEFFREY H. KORDOWER

Department of Neurological Sciences, Rush University Medical Center, Chicago, IL, USA

The etiology of Parkinson's disease (PD) is multifactorial, with genetics, aging, environmental agents all playing etiological roles. Of these, age is clearly the greatest risk factor for the development of PD. The onset of PD is less common before 50 years of age and its incidence and prevalence increase nearexponentially with age thereafter (Ross *et al.*, 2004; Silver, 2006). However, the molecular and neural mechanisms responsible for the age-enhanced disease vulnerability remain unknown. Many pathological phenomena integral in PD pathogenesis are present in normal aging. In particular, a loss of a dopaminergic phenotype is consistently seen in normal aging (Beal, 1992; Levy *et al.*, 2005; Biskup and Moore, 2006; Chu and Kordower, 2007). This is especially notable since striatal dopamine insufficiency is the key pathological feature responsible for most of the cardinal signs seen in PD. However, it is clear that aging is just a component of PD pathogenesis and other factors, for example genetics or exposure to toxic events, may be superimposed upon age-related losses in striatal dopamine and these, in combination, drive striatal dopamine insufficiency to a level that causes symptoms (Di Monte *et al.*, 2002; Jellinger, 2004; Chu and Kordower, 2007).

To understand disease pathogenesis and to best test novel therapeutic strategies, it is important to employ models that mimic the critical aspects of PD. An increasing use of invertebrates has provided many new insights into aging processes, especially regarding possible longevity genes (Roth *et al.*, 2004). Given the complexity of human physiology, however, models more phylogenetically similar to humans are needed. Although rodents remain the most widely used animal model for PD, they display little or modest age-related changes in nigrostriatal dopamine (Irwin *et al.*, 1992; Pasinetti *et al.*, 1992; Dawson *et al.*, 2002) and do not display age-related loss of TH-immunoreactivity in nigral cells (McNeill and Koek, 1990). Monkeys are closer to humans phylogenetically and display a motor repertoire similar to humans. As a function of age, monkey display crucial pathological and behavioral features that mimics PD (Roth *et al.*, 2004; Emborg *et al.*, 2006). Thus, monkey models provide an important behavioral and morphological association with PD and provide an important resource to evaluate disease pathogenesis and a forum to test new symptomatic therapeutic strategies. It is notable that a number of novel therapeutic strategies that have been successful when tested in young parkinsonian (i.e., MPTP treated) monkeys have failed to past muster when tested clinically in patients with PD. We hypothesize that this failure may result in part from failing to test these therapies in the aged parkinsonian nervous system.

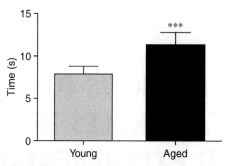

FIGURE 7.1 A pickup test to measure the speed of movement showing the increased latency (i.e., decreased motor function) in aged monkeys relative to young (***$p < 0.0001$).

AGE-RELATED DECLINE OF MOTOR ACTIVITY IN PRIMATES: RELATIONSHIP TO PD

Motor activity in rhesus monkeys gradually decline with aging advance (Emborg et al., 1998; Roth et al., 2004; Levy et al., 2005). This is especially true with fine motor skills (Figure 7.1). The profile of motor decline in aged rhesus monkeys (Grondin et al., 2003; Collier et al., 2007) is remarkably similar to normal human aging (McCormack et al., 2004; Roth et al., 2004). Qualitative observations of home cage activity by young monkeys are characterized by highly active movements and intense interactions with their environment. In contrast, aged monkeys display a stooped posture and are less interactive with their environment. Some aged monkeys displayed significant home cage bradykinesia and deficits in gait. However, other pathologic motor deficits that are seen in PD such as freezing are not observed. Young monkeys move about the cage freely by using all parts of the cage while the general activity dramatically declining with advancing age (Moscrip et al., 2000). In young monkeys, accelerometers quantifying home cage activity reveal high levels of activity were seen during each day with diminished activity during the night portion of the day/night cycle. In aged monkeys, there is a dramatic reduction in activity is seen during the day with levels declining by three times in aged monkeys compared with young monkeys (Emborg et al., 1998).

The speed of movement also undergoes age-related decline in the rhesus monkey. In previous studies, our lab has used an objective hand reach measure quantify the speed of movement in nonhuman primates. Aged monkeys displayed much slower movements (a decline in speed of 39%) when compared with young monkeys (Figure 7.1). However, both young and aged monkeys learned to perform this task relatively quickly, taking 15–20 sessions to achieve asymptotic performance indicating that young and aged monkeys acquire the task in a similar fashion suggesting that issues in motivation and cognition do not influence the performance of the task and the bradykinesia displayed by aged monkeys are due to deficits purely in the motor domain.

Using a clinical rating scale (CRS) analogous to the Unified Parkinson's Disease Rating Scale (UPDRS), we also demonstrated that aged monkeys displayed clinical signs of motor deficits relative to young cohorts. Using CRS (0–32 point scale; 0 corresponding to normal scoring and 32 corresponding to severe disability), young monkeys consistently rate a score of zero. They are fast and with steady movements and did not show any neurologic impairment. In contrast, aged monkeys are bradykinetic in a manner similar to that seen with parkinsonian monkeys. They can, on occasion display tremor but this is almost always of the action type and not like a parkinsonian resting tremor. The CRS for the aged monkeys ranged from 3.0 to 9.5 (Emborg et al., 1998), the latter of which is in the parkinsonian range.

These data indicate that motor deficiencies of aging in monkeys are remarkably similar to human aging. Aged monkeys may provide a useful animal model to study age-related motor dysfunction in human and test new drugs for treatment of PD.

AGE-RELATED DOWNREGULATION OF DOPAMINERGIC PHENOTYPE IN NIGROSTRIATAL SYSTEM: RELATIONSHIP TO PD

Age-related declines in motor function are associated with a decrease of nigrostriatal dopaminergic function. Several investigators have compared the alterations on nigrostriatal dopamine between young and aged monkeys. Age-related declines striatal dopamine (DA), homovanillic acid (HVA) and 3,4-dihydroxyphenylacetic acid (DOPAC) are well documented in monkeys (Irwin et al., 1994; Gerhardt et al., 2002; McCormack et al., 2004; Collier et al., 2007). Critically, dopamine concentrations in putamen are significantly decreased in middle-aged and aged monkeys compared with the young animals while significant differences in dopamine levels are not seen in the caudate nucleus. This is important because in PD, there is preferential loss of dopamine in the putamen relative to the caudate nucleus. Using amphetamine-evoked dopamine release in substantia nigra, a significant age-related decrease of stimulated dopamine levels is seen both middle-aged and aged monkeys as compared to the young animals (Gerhardt et al., 2002) as well.

Dopaminergic presynaptic markers such as dopamine transporter (DAT) and type-2 vesicular monoamine transporter (VMAT2) also display age-related declines in nonhuman primates (DeJesus et al., 2002; Harada et al., 2002). Analysis of PET images of [^{11}C] β-CFT reveal age-related declines in the striatal DAT availability supporting the concept that presynaptic dopaminergic function is declining in nonhuman primates as a function of age (Harada et al., 2002). Immunohistochemical experiments reveal that there is a reduction in the number of DAT-ir neurons and the intensity of the DAT-ir neuropil in aged monkeys as compared with young (Emborg et al., 1998). The effects of aging on striatal DAT was also examined in a cohort of healthy human subjects. Similar to what is seen in monkeys, striatal DAT was inversely correlated with age, declining in 6.6% per decade (van Dyck et al., 2002).

In aged monkeys, the decline of striatal dopamine is associated with a downregulation of nigral TH expression (Gerhardt et al., 2002; McCormack et al., 2004; Collier et al., 2007). However, stereological estimates of the total number of nigral neurons are unchanged with age (Pakkenberg et al., 1995; Collier et al., 2007). Thus the biochemical and immunohistochemical changes that are seen in aged monkeys are the result of age-related changes in dopaminergic phenotypic and not frank neuronal degeneration (Pakkenberg et al., 1995; McCormack et al., 2004; Collier et al., 2007). The phenotypic change seen in aged monkey is similar to what has been seen in human aging (Chu et al., 2002, 2006). We have found that the number of human nigral neurons, as counted stereologically using neuromelanin as a marker, is unchanged as a function of age. However, the numbers of nigral neurons that express markers of dopamine synthesis such as TH or Nurr1 significantly diminish with age (Chu et al., 2002, 2006).

In PD, clearly there is a loss of melanin containing neurons and this differentiates it from normal aging. However, there is a clear mismatch between the number of melanin neurons that are lost in PD and the number of TH-immunopositive neurons are lost. Even at end stage, there is about a 50% loss of melanin containing neurons in PD but about an 80% loss of TH-immunopositive neurons (Chu et al., 2006). Thus there is a significant loss of phenotype in nigral neurons that occurs as part of the disease process, similar to what is seen in normal aging. Analysis of activation of the transcription factor NF-κB, which is associated with oxidative stress-induced apoptosis, revealed a 70-fold increase in the number of NF-κB nuclei in remaining nigral neurons of PD cases relative to age-matched control subjects (Hunot et al., 1997). Taken together, these data suggest that a loss of dopaminergic phenotype in PD may be one of the earliest pathological events in the cascade of processes which ultimately results in nigral neurodegeneration and this loss of phenotype in PD may be an extension of a normally occurring age-related process. These data support the concept that aged monkeys provide for a good model for specific processes involved in PD pathogenesis, namely loss of dopaminergic phenotype.

AGE-RELATED INCREASE OF α-SYNUCLEIN: RELATIONSHIP TO PD

It remains unclear as to the molecular mechanisms that are responsible for the loss of a dopaminergic phenotype that occurs in normal aging and PD. We have recently put forth the hypothesis that the loss of dopaminergic phenotype is the result of an accumulation of the synaptic protein α-synuclein (Chu and Kordower, 2007). Toward this end, we have examined the effects of aging on α-synuclein expression within nigral perikarya in rhesus monkeys and in humans (Lo et al., 2002; Chu and Kordower, 2007). In young monkeys, α-synuclein was almost undetectable within nigral perikarya and α-synuclein-immunoreactivity is restricted to fibers and terminals in keeping with this protein's well-established role in synaptic processes (Yavich et al., 2004; Totterdell and Meredith, 2005). With increasing age, intense α-synuclein accumulates within the soma of neurons distributed throughout the substantia nigra pars compacta. The number of detectable α-synuclein immunoreactive neurons increases by 169% and 215% in middle-aged (14–23 years) and aged monkeys (>24 years) relative to young monkeys (<13 years), respectively (Figure 7.2a). In contrast, the intensity and density of α-synuclein immunoreactive neuropil in striatum and substantia nigra is unaltered across the three age groups attesting

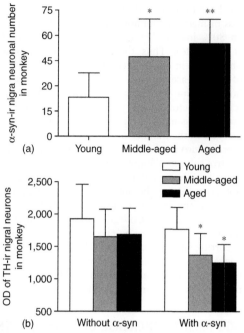

(a)

(b)

FIGURE 7.2 (a) The number of α-synuclein-ir nigral neurons significantly increases in monkeys as a function of age (*$p < 0.05$, **$p < 0.01$ compared with young group). (b) The optical density (OD) of TH-ir nigral neurons was significantly decreased in nigral neurons with detectable α-synuclein-ir when compared with the OD of TH-ir nigral neurons without detectable α-synuclein-ir profiles (*$p < 0.05$ compared with young group). Modified from Chu and Kordower (2007) with permission.

to the specificity of changes observed within nigral perikarya. The age-related increases in α-synuclein within the substantia nigra also stand in contrast to what is observed within the adjacent ventral tegmental area. In this region, α-synuclein-immunoreactivity was rarely observed with only an occasional immunoreactive cell seen in the aged group, indicating that the age-related accumulation of α-synuclein is specific for the vulnerable region of PD and is rarely expressed within an adjacent dopaminergic region that is resistant to degeneration. On the contrary to age-related increase of α-synuclein, the β-synuclein expression, a non-pathogenic analog of α-synuclein (Ohtake *et al.*, 2004), is unchanged as a function of aging, again supporting the concept that the age-associated expression in α-synuclein is a specific pathological event.

While it is important to demonstrate the associations between (1) aging and motor dysfunction; (2) aging and nigrostriatal dopamine depletion; (3) motor dysfunction nigrostriatal depletion; and (4) aging and α-synuclein accumulation, it is critical to demonstrate and association between α-synuclein accumulation and a loss of dopamine phenotype for our hypothesis to have merit. In this regard, we have found that *only* the neurons within the nigra that display increases in α-synuclein expression display age-related decreases of TH within nigral perikarya (Figure 7.2b). Nigral neurons, regardless of the age of the individual, display normal levels of TH expression if they display normal levels of α-synuclein. In contrast, if a cell has increased levels of α-synuclein, as what occurs in normal aging, then the cell virtually always has diminished levels of TH. These findings are similar to what we have observed in aged humans again supporting the relationship between aged monkeys and humans with nigrostriatal insufficiency. Relative to young individuals, there is a significant increase in the intensity of α-synuclein-immunoreactivity in middle-aged and aged humans within individual nigral neurons. The number of nigral perikarya with detectable α-synuclein also increases as a function of age (Figure 7.3a). Critically, this age-related increase of cellular α-synuclein level was strongly associated with a loss of TH (Figure 7.3b), one of the earliest cellular manifestations seen in the substantia nigra in PD. Indeed, as seen in aged monkeys, the age-related decreases in TH was only observed within nigral neurons displaying detectable α-synuclein and adjacent neurons without detectable synuclein expression displayed stable levels of TH (Figures 7.2b and 7.3b). These age-related decreases in nigral TH likely mediate the loss of striatal dopamine that has well been documented by others (McCormack *et al.*, 2004; Collier *et al.*, 2007) and supports the concept that the age-related loss of dopamine phenotype is secondary to accumulations of cytosolic α-synuclein within aging nigral neurons.

The data from aged humans and monkeys are complementary to what we observed in patients with PD (Chu *et al.*, 2006). Even though aggregation of α-synuclein has become a hallmark pathological feature in PD, even at end stage, the number of nigral cells expressing α-synuclein positive aggregates is less than nigral cells expressing cytosolic increases in α-synuclein. Cytosolic α-synuclein is the type that is seen in aged monkeys and humans and is associated with decreased dopaminergic

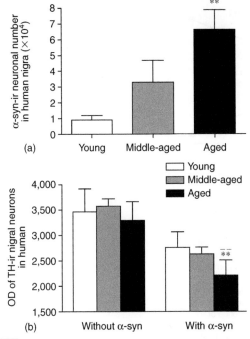

(a)

(b)

FIGURE 7.3 (a) In human nigra, the number of α-synuclein-ir nigral neurons significantly increases as a function of age (**$p < 0.01$). (b) The optical density (OD) of TH-ir nigral neurons was significantly decreased in the neurons with α-synuclein-ir as compared with the neurons without detectable α-synuclein-ir (**$p < 0.01$ compared with young group, -- $p < 0.05$ compared with middle-aged group). Modified from Chu *et al.* (2007) with permission.

phenotype. It is interesting that we have seen that TH expression is virtually undetectable in nigral neurons with α-synuclein immunoreactive inclusions (Chu *et al.*, 2006) suggesting that the age-related increases in α-synuclein may be a precursor to α-synuclein aggregation.

How do the age-related increases in non-aggregated α-synuclein potentially model the symptoms observed in patients with PD? Based upon publish findings (Chu *et al.*, 2006; Chu and Kordower, 2007), we hypothesize that the age-related decreases in nigrostriatal dopamine are mediated by increases in non-aggregated α-synuclein. This accumulation may be due to a suboptimal, but still functioning. The increase in α-synuclein is never of a sufficient magnitude to drive dopamine levels past a threshold that would engender the cardinal

signs and symptoms of PD (Figure 7.4). In PD, for reasons that still remain to be determined, we hypothesize that the lysosomal burden of the nigral neuron is overwhelmed; the age-related accumulation of α-synuclein becomes further intensified and misfolded to form an inclusion. These events cause cells to completely lose their dopaminergic phenotype; dopamine levels pass a critical threshold, and cause symptoms to emerge (Figure 7.4).

AGED MONKEYS AS A MODEL OF PD: RELATIONSHIP TO FUNCTION

Do aged monkeys model all aspects of PD? Likely not. Is PD a unitary disease? Likely not as well. Most patients experience the motor symptoms of PD later in life. There are differences in the clinical course of PD patients with older onset relative to young-onset disease. Patients with early-onset disease display a much less malignant progression of disease but display greater motor complications. Patients with old-onset disease display a more malignant course with fewer treatment-related side effects. What is the reason for this dichotomy? We hypothesize that it is due to age-related differences in compensation. We treated young, middle-aged and aged monkeys with MPTP, so that each age group displayed similar levels of behavioral dysfunction. Upon sacrifice, the residual nigrostriatal system in young monkeys displayed robust compensation as indicated by a HVA/DA ratio (Figure 7.5). The level of compensation seen in middle-aged monkeys was not significantly different from their young counterparts, but the standard errors in this group were very large suggesting that some monkeys were beginning to decompensate while others were not. Finally, there was no compensation in the aged monkeys given MPTP. We hypothesize that the failure for aged monkeys to compensate may explain why old-onset PD patients display a more malignant course of progression while their younger counterparts, that can compensate, display a less malignant course. These data provide a strong rationale for the use of aged moneys in PD experiments to better mimic the residual nigrostriatal system in old-onset PD patients, which are the majority of the cases. These data may also explain why novel therapeutic strategies that have only been tested in young parkinsonian monkeys fail to be effective when tested clinically in PD patients.

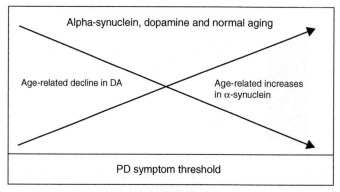

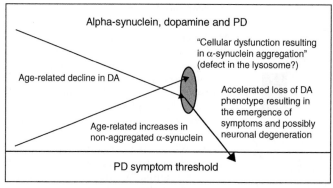

FIGURE 7.4 Hypothetical process by which α-synuclein influences nigrostriatal function during normal aging and PD. (top) During normal aging, soluble α-synuclein accumulates within nigral perikarya and this causes a phenotypic downregulation of dopamine. The level of dopamine insufficiency is not severe enough to induce the cardinal symptoms of PD. (bottom) In PD, the same process occurs, but at some point, lysosomal function is overwhelmed, the progressive loss of dopaminergic function becomes more severe, and the magnitude of dopamine dysfunction is sufficient for the cardinal symptoms of PD to occur. Used with permission from Chu *et al.* (2007).

SO WHY HAS NOT AGING PROCESSES BECOME A MORE CENTRAL FACTOR IN PD STUDIES?

There are two main reasons why aging processes have not received their due attention as a manifestation of PD. One is theoretical and the other practical. The theoretical problem is that for many years it has been assumed that the pattern of nigrostriatal degeneration is different in aging and PD. Fearnley and Lees (1991) initially reported that loss of TH phenotype in normal aging occurs in the dorsal tier of the substantia nigra while PD degeneration occurs in the ventral tier. These data are inaccurate

and are based upon a single section analysis through the caudal substantia nigra. When modern stereological probes are employed (Kanaan *et al.*, 2007), aged primates display loss of TH in a pattern virtually identical to that seen in PD. Secondly, Kish *et al.* (1992) reported that in normal aging, there is a similar loss of dopamine in the caudate and putamen while there is a preferential loss of dopamine in the putamen in PD. These data are often over interpreted. There are consistent losses of dopamine in the caudate in PD (Wang et al., 2007), the degree of which is just less than what is seen in PD. It remains possible that there are similar decreases in the caudate nucleus and putamen as

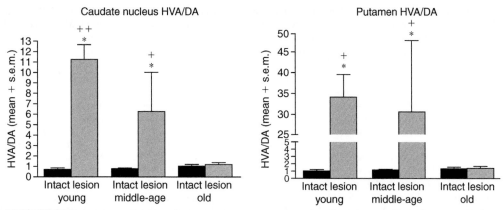

FIGURE 7.5 With aging process, the capacity for nigrostriatal neurons to compensate in response to MPTP insult is significantly lost in the oldest monkeys. After MPTP exposure, a large increase in the HVA/DA ratio is seen in young monkeys indicating their ability to compensate for lost nigrostriatal function. Although middle-aged monkeys do not show a statistically significant difference from young monkeys, the standard errors are much greater indicating that some middle-aged animals can compensate while others cannot. In contrast, no aged monkeys can compensate for the loss of nigrostriatal function (Caudate: $*p < 0.005$ for comparison of intact to lesioned hemispheres, $+p < 0.009$ for comparison of middle-age lesion to old lesion, $++ p < 0.04$ for comparison of young lesion to both middle-age and old lesion. Putamen: $*p < 0.005$ for comparison of intact to lesioned hemispheres, $+p < 0.003$ for comparison of young and middle-age lesion to old lesion). Modified from Collier *et al.* (2007) with permission.

a function of normal aging that is exacerbated by some unknown factor in the putamen in PD. Under these conditions, aging remains a precursor of PD with regard to loss of nigrostriatal function. Lastly, the failure for aging to be a central theme in PD research is also a practical problem. Aged rodents are expensive and difficult to get and aged nonhuman primates are even more rare, expensive and difficult to maintain. Still, these practical problems should not prevent science from proceeding along what may be a critical path in the understanding of disease pathogenesis and the evaluation of novel therapeutic strategies.

CONCLUSIONS

Taken together, these data illustrate that aged monkeys provide a unique model to study specific aspects of PD pathogenesis. Aged monkeys display motor deficits that are associated with a loss of dopamine phenotype independent of the loss of nigral neurons. This occurs early in the pathogenic cascade in PD. The loss of phenotype seen in aged monkeys is associated specifically with the accumulation of α-synuclein, a phenomenon also associated

with the loss of dopaminergic markers in PD. Finally, the nigrostriatal system in aged monkeys treated with MPTP fail to compensate for the loss of dopaminergic function in a manner similar to that seen in young monkeys treated with MPTP. This may explain the different clinical course seen in individuals with old-onset PD compared to young MPTP-treated monkeys.

REFERENCES

Beal, M. F. (1992). Does impairment of energy metabolism result in excitotoxic neuronal death in neurodegenerative illnesses? *Ann Neurol* **31**, 119–130.

Biskup, S., and Moore, D. J. (2006). Detrimental deletions: Mitochondria, aging and Parkinson's disease. *Bioessays* **28**, 963–967.

Chu, Y., Kompoliti, K., Cochran, E. J. et al. (2002). Age-related decreases in Nurr1 immunoreactivity in the human substantia nigra. *J Comp Neurol* **450**, 203–214.

Chu, Y., Le, W., Kompoliti, K. et al. (2006). Nurr1 in Parkinson's disease and related disorders. *J Comp Neurol* **494**, 495–514.

Chu, Y., and Kordower, J. H. (2007). Age-associated increases of alpha-synuclein in monkeys and humans are associated with nigrostriatal dopamine depletion: Is this the target for Parkinson's disease? *Neurobiol Dis* 25, 134–149.

Collier, T. J., Lipton, J., Daley, B. F. et al. (2007). Aging-related changes in the nigrostriatal dopamine system and the response to MPTP in nonhuman primates: Diminished compensatory mechanisms as a prelude to parkinsonism. *Neurobiol Dis* 26, 56–65.

Dawson, T., Mandir, A., Lee, M. et al. (2002). Animal models of PD: Pieces of the same puzzle? *Neuron* 35, 219–222.

DeJesus, O. T., Shelton, S. E., Roberts, A. D. et al. (2002). Effect of tetrabenazine on the striatal uptake of exogenous L-DOPA *in vivo*: A PET study in young and aged rhesus monkeys. *Synapse* 44, 246–251.

Di Monte, D. A., Lavasani, M., Manning-Bog, A. B. et al. (2002). Environmental factors in Parkinson's disease. *Neurotoxicology* 23, 487–502.

Emborg, M. E., Moirano, J., Schafernak, K. T. et al. (2006). Basal ganglia lesions after MPTP administration in rhesus monkeys. *Neurobiol Dis* 23, 281–289.

Emborg, M. E., Ma, S. Y., Mufson, E. J. et al. (1998). Age-related declines in nigral neuronal function correlate with motor impairments in rhesus monkeys. *J Comp Neurol.* 401, 253–265.

Fearnley, J. M., and Lee, A. J. (1991). Ageing and Parkinson's disease: Substantia nigra regional selectivity. *Brain* 114, 2283–2301.

Gerhardt, G. A., Cass, W. A., Yi, A. et al. (2002). Changes in somatodendritic but not terminal dopamine regulation in aged rhesus monkeys. *J Neurochem* 80, 168–177.

Grondin, R., Zhang, Z., Yi, A. et al. (2003). Intracranial delivery of proteins and peptides as a therapy for neurodegenerative diseases. *Prog Drug Res* 61, 101–123.

Jellinger, K. A. (2004). Lewy body-related alpha-synucleinopathy in the aged human brain. *J Neural Transm* 111, 1219–1235.

Harada, N., Nishiyama, S., Satoh, K. et al. (2002). Age-related changes in the striatal dopaminergic system in the living brain: A multiparametric PET study in conscious monkeys. *Synapse* 45, 38–45.

Hunot, S., Brugg, B., Ricard, D. et al. (1997). Nuclear translocation of NF-κB is increased in dopaminergic neurons of patients with Parkinson disease. *Proc Natl Acad Sci USA* 94, 7531–7536.

Irwin, I., DeLanney, L. E., McNeill, T. et al. (1994). Aging and the nigrostriatal dopamine system: A non-human primate study. *Neurodegeneration* 3, 251–265.

Irwin, R. P., Nutt, J. G., Woodward, W. R. et al. (1992). Pharmacodynamics of the hypotensive effect of levodopa in parkinsonian patients. *Clin Neuropharmacol* 15, 365–374.

Kanaan, N. M., Kordower, J. H., and Collier, T. J. (2007). Age-related accumulation of Marinesco bodies and lipofuscin in rhesus monkey midbrain dopamine neurons: Relevance to selective neuronal vulnerability. *J Comp Neurol* 502, 683–700.

Kish, S. J., Shannak, K., Rajput, A. et al. (1992). Aging produces a specific pattern of striatal dopamine loss: Implications for the etiology of idiopathic Parkinson's disease. *J Neurochem* 58, 642–648.

Levy, G., Louis, E. D., Cote, L. et al. (2005). Contribution of aging to the severity of different motor signs in Parkinson disease. *Arch Neurol* 62, 467–472.

Lo, Bianco, C., Ridet, J. L., Schneider, B. L. et al. (2002). Alpha-synucleinopathy and selective dopaminergic neuron loss in a rat lentiviral-based model of Parkinson's disease. *Proc Natl Acad Sci USA* 99, 10813–10818.

McCormack, A. L., Di Monte, D. A., Delfani, K. et al. (2004). Aging of the nigrostriatal system in the squirrel monkey. *J Comp Neurol* 471, 387–395.

McNeill, T. H., and Koek, L. L. (1990). Differential effects of advancing age on neurotransmitter cell loss in the substantia nigra and striatum of C57BL/6N mice. *Brain Res* 521, 107–117.

Moscrip, T. D., Ingram, D. K., Lane, M. A. et al. (2000). Locomotor activity in female rhesus monkeys: Assessment of age and calorie restriction effects. *J Gerontol A Biol Sci Med Sci* 55, 373–380.

Ohtake, H., Limprasert, P., Fan, Y. et al. (2004). Beta-synuclein gene alterations in dementia with Lewy bodies. *Neurology* 63, 805–811.

Pakkenberg, H., Andersen, B. B., Burns, R. S. et al. (1995). A stereological study of substantia nigra in young and old rhesus monkeys. *Brain Res* 693, 201–206.

Pasinetti, G. M., Osterburg, H. H., Kelly, A. B. et al. (1992). Slow changes of tyrosine hydroxylase gene expression in dopaminergic brain neurons after neurotoxin lesioning: A model for neuron aging. *Brain Res Mol Brain Res* 13, 63–73.

Ross, G. W., Petrovitch, H., Abbott, R. D. et al. (2004). Parkinsonian signs and substantia nigra neuron density in decedents elders without PD. *Ann Neurol* 56, 532–539.

Roth, G. S., Mattison, J. A., Ottinger, M. A. et al. (2004). Aging in rhesus monkeys: Relevance to human health interventions. *Science* 305, 1423–1426.

Silver, D. (2006). Impact of functional age on the use of dopamine agonists in patients with Parkinson disease. *Neurologist* 12, 214–223.

Totterdell, S., and Meredith, G. E. (2005). Localization of alpha-synuclein to identified fibers and synapses in the normal mouse brain. *Neuroscience* 135, 907–913.

van Dyck, C. H., Seibyl, J. P., Malison, R. T. et al. (2002). Age-related decline in dopamine transporters: Analysis of striatal subregions, nonlinear effects. *Am J Geriatr Psychiatry* 10, 36–43.

Wang, J., Zuo, C. T., Jiang, Y. P. et al. (2007). 18F-FP-CIT PET imaging and SPM analysis of dopamine transporters in Parkinson's disease in various Hoehn & Yahr stages. *J Neurol* 254, 185–190.

Yavich, L., Tanila, H., Vepsalainen, S. et al. (2004). Role of alpha-synuclein in presynaptic dopamine recruitment. *J Neurosci* 24, 11165–11170.

8

1-METHYL-4-PHENYL-1,2,3,6-TETRAHYDROPYRIDINE-INDUCED MAMMALIAN MODELS OF PARKINSON'S DISEASE: POTENTIAL USES AND MISUSES OF ACUTE, SUB-ACUTE, AND CHRONIC MODELS

J. S. SCHNEIDER, D. W. ANDERSON AND E. DECAMP

Department of Pathology, Anatomy and Cell Biology, Thomas Jefferson University, Philadelphia, PA, USA

INTRODUCTION

The use of animal models to gain insight into human diseases and the potential treatment thereof has been extremely valuable and has led to many substantial advances in human medicine. As opposed to isolated *in vitro* models, the use of a living animal as a model system is more productive for studying multi-factorial diseases or illnesses that affect complex systems. This has been particularly true of the use of animal models to understand the pathology and physiology of central nervous system disorders, including epilepsy, stroke, and neurodegenerative disorders. Key among the neurodegenerative disorders which has perhaps benefited the most from the development and availability of animal models is Parkinson's disease (PD).

PD is a uniquely human disorder (i.e., does not appear to occur spontaneously in animal populations) and the etiology of PD remains unknown. Thus, no animal model of PD can ever duplicate the exact disease that occurs in man. However, the neuropathology that defines PD, that is, degeneration of nigrostriatal dopaminergic neurons, can be reproducibly induced in a variety of animal species by a number of different means, allowing for study of the role of dopamine (DA) in a variety of behaviors and the impact of loss of DA on brain function and the behavioral repertoire of the species under study. Damage to the nigrostriatal DA system, with a resulting DA-deficiency syndrome, has been produced by mechanical lesion of ascending nigrostriatal nerve fibers (Brecknell *et al.*, 1995), electrolytic lesion to the substantia nigra (SN) (Donaldson *et al.*, 1976), and administration of various toxins including reserpine (to deplete brain catecholamines) (Carlsson, 1966), methamphetamine (Seiden and Ricaurte, 1987), 6-hydroxydopamine (either into the SN, striatum, or medial forebrain bundle) (Deumens *et al.*, 2002), rotenone (Betarbet *et al.*, 2000), paraquat (McCormack *et al.*, 2002), and 1-methyl-4-phenyl-1,2,3,6-tetrahydropyridine

(MPTP) (Langston *et al.*, 1983; Chiba *et al.*, 1984; Heikkila *et al.*, 1984; Langston *et al.*, 1984). While all of these models, and particularly the newer toxin-induced models, have merits, none is ideal nor are any of the commonly employed models universally suited for all types of studies. In order for a particular animal model to be valuable it should possess certain qualities: it should accurately model the disease or pathology of interest; allow for inferences to be drawn concerning the human counterpart of the model; and the system under study should have the necessary biochemical and anatomical organization so as to make the model a reliable substitute for the target species (i.e., human).

Since the discovery of MPTP in the early 1980s its ability to destroy nigrostriatal dopaminergic (DAergic) neurons after systemic administration and cause a Parkinson-like disorder in a number of different species (Langston *et al.*, 1984; Schneider *et al.*, 1986; Hekkila *et al.*, 1989), it has become one of the most widely used (and sometimes misused) animal models of PD. In non-human primates, MPTP-induced parkinsonism is characterized by clinical features that are remarkably similar to human PD (Schneider *et al.*, 1987; Schneider *et al.*, 1988; Schneider, 1989). Thus, this model has provided an important system in which to test new therapeutic strategies for the control of PD symptoms. The most commonly employed model is produced following acute administration of relatively high doses of MPTP and symptoms develop rapidly, failing to mimic the progressive nature of the human disease. However, slowly progressive models have been developed (Schneider and Kovelowski, 1990; Pope-Coleman and Schneider, 1998; Pope-Coleman *et al.*, 2000) but are very costly and labor-intensive to produce. Although significant loss of nigrostriatal DA occurs regardless of whether the toxin is administered acutely or chronically (Schneider *et al.*, 1987; Schneider, 1990), there are significant differences in some non-DAergic systems between the two models that may influence the development of therapies aimed at targets downstream from the DA system. In a slowly progressive disease such as PD, there are undoubtedly changes that occur in non-DAergic systems over the course of the disease as the brain attempts to compensate for the loss of DA. These changes, which cannot be reproduced in an acute model, may ultimately be very important for developing symptomatic therapies, as well as for developing therapies that counteract side effects from long-term DAergic

therapy (such as end of dose wearing off and levodopa-induced dyskinesias). For example, monkeys made acutely parkinsonian develop levodopa-induced dyskinesias very rapidly, whereas animals with a more chronic, slowly evolving parkinsonism develop dyskinesias only after many months of levodopa exposure (Schneider *et al.*, 2003). Although both types of animals develop phenotypically similar dyskinesias, there might be different mechanisms driving the dyskinesias in these animals, which might cause them to respond differently to experimental manipulations. In PD patients, levodopa-induced dyskinesias develop on a backdrop of a slowly progressive, chronic disease. Attempting to reproduce this aspect of the disease in a rapid onset, acute disorder may not faithfully model the complex physiological alterations that may lead to development of these problems and depending on the drug target, this may affect the ability of preclinical studies to predict clinical efficacy.

MPTP-induced models in rodents are also not without their problems. The mouse MPTP model has been used primarily to study pathogenic mechanisms contributing to PD and for development of neuroprotective strategies. Many iterations of the mouse MPTP model have been produced in last two decades but by and large, the most commonly used model is one that uses "acute" administration of toxin (i.e., multiple MPTP doses in a single day). Other models employ "sub-acute" administration of toxin (typically 2 doses per day over a 5-day period) or "chronic" administration protocols (over approximately 1 month). What is rarely taken into account in studies employing these models is that the cell death processes are likely quite different depending on a number of variables, including the frequency and duration of toxin exposure, and may or may not have relevance to what occurs in the brain of a PD patient. A number of different mechanisms contributing to cell death have been suggested to occur in the acute mouse MPTP model. However, despite exquisite neuroprotection produced in this model using a variety of compounds, there has been no success in translating these findings to the clinic. One possible reason for this is that the acute MPTP model employed, in which rapid cell death ensues, may not reflect the time-dependent, complex sequence or cascade of events that occurs in a slowly evolving neurodegenerative disease such as PD. Thus, using this model alone for development of putative neuroprotective strategies may not be productive.

In the remainder of this chapter, we will discuss in more detail the similarities and differences between acute and more chronic MPTP models in mice and monkeys, present our views on the advantages and disadvantages of each of these models and their relevance for understanding PD and for developing effective therapeutics for this disease.

MPTP MOUSE MODELS OF PARKINSONISM

The mouse model of parkinsonism induced by MPTP administration is probably the most commonly used Parkinson model today. Several factors contribute to the widespread use of this model including: (i) the availability and relatively low cost of mice; and (ii) the speed with which studies can be completed. However, these same factors that promote the widespread use of this model also led to several misconceptions, which are reviewed by Jackson-Lewis and Przedborski elsewhere in this book, and contributed to its misuse. This latter issue will be discussed below.

One area in which there is significant promise but also perhaps where there has been the greatest potential misuse of the mouse MPTP model is in the area of neuroprotection. As in other disorders such as stroke (1999), no single compound found to be an effective neuroprotective agent in the mouse MPTP model has yet transferred successfully to the clinic. Is the mouse MPTP model just not a good predictive model for neuroprotection or has the model not been used in the most efficient way in order to predict clinical efficacy? Clearly, no MPTP-induced model of parkinsonism can reproduce the cause(s) of idiopathic PD. However, in order to improve the prognostic value of this model for development of neuroprotective strategies, the model and its various iterations need to be more clearly understood and more carefully utilized.

Following the discovery of MPTP, mechanistic studies showed that its active metabolite, the 1-methyl-4-phenylpyridinium ion (MPP^+), was a potent mitochondrial poison, inhibiting complex I of the mitochondrial electron transport chain (Ramsay and Singer, 1986; Chan et al., 1991; Beal, 1992), resulting in alterations of intra- and extra-cellular Ca^{2+}, K^+, and Na^+ concentrations with the potential for causing either apoptosis (Hartley et al., 1994;

Tatton and Kish, 1997) or necrosis (Jackson-Lewis et al., 1995; Przedborski and Vila, 2003). After initial studies documented the toxicity of MPTP for striatal DA levels and SN DA neurons (Heikkila et al., 1984; Melamed et al., 1985; Switzer III, 1985), Sonsalla and Heikkila (1986) showed that different toxic effects of MPTP on the mouse DA system could be obtained depending on the dose, dosing interval, animal strain, and even vendor of origin of the mice used (Sonsalla and Heikkila, 1988). "Acute" administration of MPTP (4 injections of 20 mg/kg, given at 2-h intervals, a time frame based upon the half-life of MPP^+ within the mouse brain [Markey et al., 1984]) in 1 day produced an 80% decrease in striatal DA levels, whereas 2 injections per day of the same 20 mg/kg dose of MPTP given at 6-h intervals for 2 days only produced a 42% depletion in striatal DA (Sonsalla and Heikkila, 1986). Even though the total cumulative dose of MPTP administered in both paradigms was the same, these results showed that not all MPTP-induced models were the same and that care needed to be taken to use the same strain, dose, and dosing parameters when attempting to reproduce the results of others. These early studies paved the way for the use of the "acute" MPTP mouse model, primarily using the C57Bl6 strain (shown to be the most sensitive to the toxin [Sonsalla and Heikkila, 1988]) as a rapid and if used carefully, a reasonably reproducible experimental animal model.

One important problem associated with the acute MPTP mouse model that is infrequently reported in the literature is a relatively high mortality rate. In response to this problem, several groups reported "sub-acute" MPTP administration paradigms (2 injections of MPTP [20–30 mg/kg] for 5–10 days, with cumulative doses ranging from 150 to 300 mg/kg [Ricaurte et al., 1986; Schneider and Denaro, 1988; Seniuk et al., 1990]) that resulted in reduced mortality while still producing a substantial striatal DA depletion (Ricaurte et al., 1986; Schneider and Denaro, 1988), albeit with more limited SN cell loss than achieved with acute toxin administration (Ricaurte et al., 1986).

Important differences between acute and sub-acute MPTP mouse models extend beyond just the amount of striatal DA depletion or SN cell loss achieved. While both acute and sub-acute toxin administration paradigms seemingly rely upon the same initial mechanism of neurotoxicity, the mode of cell death produced in each case appears to be fundamentally different. Administration of MPTP can produce either primarily apoptotic or necrotic cell death (Hartley *et al.*, 1994; Jackson-Lewis *et al.*, 1995; Tatton and Kish, 1997; Przedborski and Vila, 2003). The way in which DA neurons die appears to be related to the method of MPTP administration. Following acute administration of MPTP, DA neuron death seems to proceed primarily via non-apoptotic mechanisms (necrosis) (Jackson-Lewis *et al.*, 1995; Przedborski and Vila, 2003); following sub-acute MPTP administration cells appear to die primarily via apoptotic mechanisms (Tatton and Kish, 1997). Although the precise mechanisms driving the two forms of cell death resulting from different MPTP exposures are not completely understood, they appear to relate at least in part to the kinetics of MPTP conversion to MPP^+ and the interval between toxin exposures.

MPTP administration which can result in two distinct forms of cell death can be an advantage or disadvantage of the models, depending on how the models are used. While this would seem to be beneficial for developing putative neuroprotective strategies for PD, relatively few studies have taken advantage of this aspect of the models in the search for effective therapeutics to slow the progression of PD. Since the mechanisms contributing to DA cell death in human PD are complex, multifactorial and not completely known, selection of a single MPTP administration paradigm for use in neuroprotection studies may provide potentially misleading evidence for or against neuroprotection (depending upon which model has been chosen for study) leading to unrealized efficacy in the clinic.

SIMILARITIES AND DIFFERENCES BETWEEN ACUTE AND SUB-ACUTE MPTP MOUSE MODELS

Using young, mature male C57Bl6 mice (8 weeks old), we found that a significant striatal DA depletion can be produced following either acute MPTP administration (4 injections of 20 mg/kg each, at 2-h intervals in 1 day) or sub-acute MPTP administration (2 injections of 20 mg/kg each day, at 4-h intervals for 5 consecutive days) (Figure 8.1). While the loss of striatal DA in the acute model was significantly greater than in the sub-acute model, it was associated with cumulative doses of MPTP that were less than half of that used in the sub-acute model. Also, the loss of striatal DA appears to be almost immediate in both acute and sub-acute MPTP models

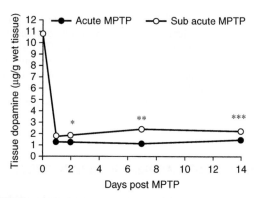

FIGURE 8.1 A comparison of the changes in striatal DA levels over time, in acute and sub-acute MPTP-lesioned mice. Acute MPTP administration produced a greater reduction in DA concentration when compared with sub-acute MPTP administration at all post-lesion time points, with significant differences between the two models apparent at 2 (*$p < 0.01$), 7 (**$p < 0.001$), and 14 days (***$p < 0.001$).

while the loss of DA neurons is a slower process that continues for up to 14 days regardless of the method of MPTP administration (Figure 8.2).

Other neuropathological findings distinguish the acute and the sub-acute models. Considerable attention has been paid to the potential role of microglia in initiating and/or perpetuating the DAergic pathology of PD. Immediately following acute administration of MPTP there is a significant increase in the number of activated microglia detected in both the striatum and the SN (Kurkowska-Jastrzebska *et al.*, 1999; Dehmer *et al.*, 2000; Furuya *et al.*, 2004) that coincides with expression of inducible nitric oxide synthase (iNOS) (Liberatore *et al.*, 1999). This has been taken to suggest that microglial activation may contribute to the degenerative processes that ensue following acute MPTP administration. The role of microglial activation in the damage that occurs following sub-acute MPTP administration is less clear and has been studied less extensively. Furuya *et al.* (2004) assessed microglial activation following acute (4 injections of 20 mg/kg each, at 2-h intervals in 1 day) and sub-acute (30 mg/kg once per day for 5 days) MPTP administration, but only examined brains at a single time point, 48-h, following MPTP administration. Acute MPTP-treated animals showed similar microglial activation as described by others (Kurkowska-Jastrzebska *et al.*, 1999; Liberatore *et al.*, 1999), but animals that received

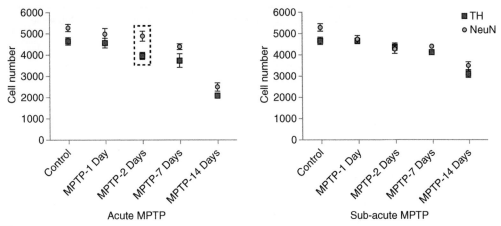

FIGURE 8.2 Effects of acute or sub-acute MPTP administration on substantia nigra pars compacta (SNc) neurons. Cell numbers were estimated by unbiased stereological analysis of the region spanning the entire rostro-caudal extent of SNc. Loss of both TH- and NeuN-immunoreactive cells in the two MPTP administration paradigms was significantly different between the models only at 2 days after acute MPTP administration (highlighted by box, $p < 0.01$), possibly indicating a loss of TH phenotype or expression preceding actual cell loss.

sub-acute MPTP administration showed minimal microglial activation in either the striatum or the SN (Furuya *et al.*, 2004) despite a significant DAergic lesion. These data might suggest that acute and sub-acute MPTP administration result in different types of cell death with only acute toxin administration leading to microglial activation. If this is true and models what happens in PD, it might suggest that there may be cell death processes in PD that do not involve microglial activation and thus, inhibiting microglia could be futile as a neuroprotective strategy in such instances.

Studies performed in our laboratory have begun to address this issue and have shown that while both acute and sub-acute MPTP administration in mature male C57Bl6 mice leads to microglial activation in the striatum and the SN (Anderson and Schneider, 2006) (Figure 8.3), acute MPTP exposure leads to a much earlier occurring and more robust microglial response than does sub-acute toxin administration. Thus, microglial activation may play a role in the degenerative process in both models but the nature of the role played by microglia might be different in the different models.

Reports that microglial activation following MPTP administration plays a role in driving cell death (primarily based on data obtained using the acute MPTP model), has led to the study of microglial inhibitors and anti-inflammatory compounds

as potential neuroprotective agents for PD. To date more than 15 different anti-inflammatory/anti-microglial pharmacotherapies have been investigated for their potential to reduce the neurodegeneration associated with either acute or sub-acute MPTP administration or 6-OHDA-induced parkinsonism (for a partial listing see [Hirsch *et al.*, 2003]). Minocycline, a semi-synthetic second-generation tetracycline with anti-inflammatory and microglial inhibitory activity (Ryan and Ashley, 1998; Du *et al.*, 2001; Wu *et al.*, 2002; Yang *et al.*, 2003; Thomas and Le, 2004), has been the most studied of these agents. Using an acute MPTP model in young C57Bl/6 mice, Du *et al.* (2001) showed that oral administration of minocycline produced a significant sparing of striatal DA levels and DAergic cell loss in the SN. Wu *et al.* (2002) carried out similar experiments also using an acute MPTP model and showed that minocycline inhibited SN and striatal microglial activation and produced a dose-dependent sparing of DAergic neurons in the ventral midbrain, as well as TH expression in the striatum. In contrast to these findings, Yang *et al.* (2003) reported that minocycline enhanced DAergic cell loss while still inhibiting microglial activation. In this latter study, minocycline was administered twice daily instead of once, it was administered pre- and post-MPTP treatment instead of only prior to *or* after MPTP treatment, and different routes of administration (i.p. or p.o.)

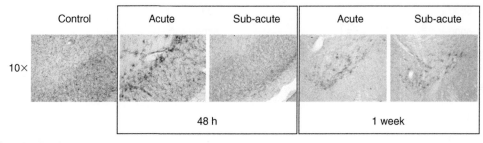

FIGURE 8.3   Microglial activation in the SN of acute and sub-acute MPTP-lesioned mice. Note the reactive microglia (MAC1[+]) at 48 h after toxin exposure in the acute MPTP condition, with no visible activation in animals with sub-acute exposure. At 1-week post-MPTP, activated microglia are seen in the SNc in both acute and sub-acute MPTP-treated animals. Acute MPTP-treated animals have reduced numbers of activated microglia at 1 week post-MPTP compared with the number of activated microglia observed at 48 h.

were assessed. Although an acute MPTP administration paradigm was used, Yang *et al.* (2003) found that in all experimental groups, minocycline pre-treatment (i.p. or p.o.) significantly exacerbated MPTP-induced striatal DA depletion and enhanced DAergic cell loss in the SN even though minocycline inhibited microglial activation in this region.

These data illustrate that there is still much to be learned about the role of activated microglia in MPTP-induced neurodegeneration and that the choice of model may be a critical factor in study outcome. Studies with minocycline in our laboratory suggest that when this drug is administered to either acute or sub-acute MPTP-treated mice, minocycline does not completely inhibit the microglial response in either group, and may only delay the onset of microglial activation (Anderson and Schneider, 2006) (Figure 8.4).

Neither the acute nor the sub-acute MPTP model is better than the other for modeling the role of microglial activation in DA neuron degeneration or for evaluating potential neuroprotective strategies based on microglial inhibition. They are simply different. Neither model can recapitulate what occurs in the PD brain, but can only model different aspects of the pathophysiology. However, when data from both models are considered, the clinical potential of minocycline as a neuroprotective agent for PD is not promising. A recent clinical trial determined that it was non-futile to continue to study minocycline as a neuroprotective agent in PD, despite no significant differences in the Unified Parkinson's Disease Rating Scale (UPDRS) scores of PD patients that received 12 months of minocycline

compared to placebo-treated controls (although this study admittedly was not powered to detect clinically significant group differences) (Galpern, *et al.*, 2006).

Would screening agents in multiple mouse MPTP models have better predictive validity for clinically effective neuroprotective agents for PD? Using sub-acute MPTP exposure paradigms, the non-ergoline DA agonist pramipexole (PPX) has been shown to be an effective neuroprotective agent (Zou *et al.*, 2000; Anderson *et al.*, 2001). However, there was no protective effect of PPX on either striatal DA levels or DA neurons in acute MPTP-treated animals (Anderson *et al.*, 2006). Clinically, although PPX use led to less decline over time in a surrogate imaging measure of the integrity of striatal DAergic terminals, there was no clinical evidence of neuroprotection in PD patients (Parkinson Study Group, 2002). These findings suggest that data from neither the acute nor the sub-acute MPTP models *alone* may have predictive value for neuroprotective efficacy in the clinical setting and that multiple models may need to be used to better assess preclinical efficacy of putative neuroprotective compounds. Perhaps those agents with broad neuroprotective efficacy in both models may be more likely to have clinical efficacy. Only time will tell if the screening of putative neuroprotective agents in both the acute and the sub-acute (or perhaps chronic) MPTP mouse models will have greater predictive value for clinical efficacy for neuroprotection in PD. However, even with the limitations of the MPTP mouse model of PD, it remains (if properly used) a valuable model in which to screen putative disease modifying therapies.

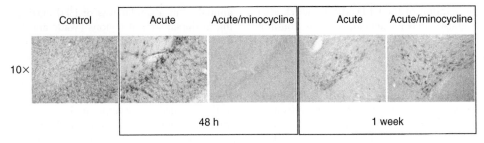

FIGURE 8.4 Microglial activation in the SN of acute MPTP-treated mice co-administered minocycline. Reactive microglia cells (MAC1⁺ immunoreactive cells) are visible throughout the SN of acute MPTP-treated animals at both 48h and 1 week. Co-administration of minocycline with MPTP inhibited microglial activation for up to 48h, but not at 1 week following the last MPTP injection.

MPTP MONKEY MODELS

Administration of MPTP to non-human primates, regardless of the specific manner in which the toxin is administered, results in the development of parkinsonian symptoms that closely mimic many of the clinical signs of PD (Crossman *et al.*, 1987; Schneider, 1989; Bezard *et al.*, 1998; Jenner, 2003a, b; Blanchet *et al.*, 2004). In contrast to mouse MPTP models, the ability to reproduce the main pathological defect of PD (i.e., DA neuron death and loss of striatal DA) as well as reproduce many of the motor and non-motor symptoms of PD in the monkey has resulted in a model system that allows for assessment of potential therapies aimed at remediating these symptoms, as well as perhaps protecting against their development. Parkinsonian monkeys also develop treatment-related side effects to symptomatic therapies (i.e., levodopa-induced dyskinesias [LIDs]), albeit in a very compressed time frame compared to PD patients, further providing a model system in which to develop strategies to enhance the efficacy or decrease the side effect profile of currently used therapies.

Although the most common monkey model involves the induction of parkinsonism by using relatively high doses of MPTP administered over a short period of time (resulting in rapid onset of symptoms), other more chronic models with slowly progressing symptoms have been developed (Schneider and Pope-Coleman, 1995; Meissner *et al.*, 2003). Over a number of years, we developed chronic low dose (CLD) MPTP-induced models of early and more advanced parkinsonism. Although the latter model results in animals with motor symptoms similar to those in animals with acute-onset parkinsonism, there appear to be fundamental differences in brain chemistry and pathology in the two models, as well as potentially different responses to drug therapies. In the following paragraphs, we will discuss the similarities and differences between acute and chronic MPTP-induced Parkinson models in monkeys and discuss some of the advantages and disadvantages of these model systems.

The first non-human primate model of MPTP-induced parkinsonism was produced by acute exposure to relatively high doses of MPTP which resulted in rapid onset of symptoms (Burns *et al.*, 1983). In these (and subsequent) studies of acutely parkinsonian monkeys, animals received doses of MPTP ranging from 0.33 to 0.75 mg/kg administered intravenously (i.v.) or intramuscularly (i.m.) over periods of time ranging from 9 to 32 days. In contrast, the initial CLD MPTP administration protocol developed by us involved administration of MPTP doses ranging from 0.05 to 0.15 mg/kg, i.v., over several months (range 4–7 months) until cognitive deficits appeared and animals showed no gross Parkinson-like motor deficits. The cognitive deficits developed by these animals are analogous to those experienced by PD patients (Schneider and Kovelowski, 1990) and are mostly frontal-like in nature, consisting of impaired performance on attention, executive functioning, and working memory tasks (Schneider *et al.*, 1988; Schneider and Kovelowski, 1990; Schneider and Roeltgen, 1993; Decamp and Schneider, 2004). These animals, with cognitive deficits and minimal or no motor deficits, are considered to be a model of early-stage PD.

In one iteration of the CLD MPTP model, after initial cognitive deficits appear, animals continue to receive increasingly higher doses of MPTP for approximately 6 months or until overt parkinsonian motor symptoms such as bradykinesia, impaired balance, decreased dexterity, rigidity, and stooped posture appear. Whether animals were made acutely parkinsonian and studied 1–3 months after MPTP exposure or had slowly progressing symptoms and were studied 10–12 months after MPTP exposure, all animals had similar parkinsonian motor symptoms at the end of their study period (i.e., Parkinson rating [rating scale of 0–41, with 41 being most severe] for acute animals = 28 ± 4; rating for chronic animals = 24 ± 5) (Schneider *et al.*, 1999). Thus, monkeys may show similar Parkinson-like motor symptoms regardless of whether they received acute or chronic administration of MPTP or developed symptoms rapidly over a few weeks or slowly over several months. However, considering the potential for adaptive plasticity in the mammalian brain, the physiological and chemical changes associated with a severe acute injury (as occurs with the rapid and massive destruction of SN DA neurons associated with acute exposure to high doses of MPTP) might be quite different from the processes associated with a lower level, more chronic insult that eventually leads to a similar level of SN injury and symptom expression. These differences could have important implications for the use of these models in the drug discovery process for neuroprotective or symptomatic therapies.

BIOCHEMISTRY OF ACUTE AND CHRONIC PARKINSONISM IN MONKEYS: SIMILARITIES AND DIFFERENCES

Striatal Dopaminergic Innervation

Using [^{3}H] mazindol binding and quantitative autoradiography to assess the integrity of the presynaptic DAergic system, we found no differences in striatal mazindol binding between animals made acutely or chronically parkinsonian, as long as the animals had similar degrees of motor dysfunction (Wade and Schneider, 2004). However, animals that received chronic administration of MPTP but had only cognitive deficits (essentially motor asymptomatic) had more [^{3}H] mazindol binding in ventral striatal regions, compared to symptomatic

animals, but a similar loss of [^{3}H] mazindol binding in dorsal striatal regions (Wade and Schneider, 2004). These findings were interpreted as suggesting the possible presence of compensatory responses involving a larger than previously appreciated contribution of the ventral striatal DAergic innervation to the maintenance of motor function in the lesioned but asymptomatic animals, perhaps involving volume transmission of DA released from relatively preserved ventral striatal terminals to the dorsal striatum, as we described in a different MPTP model (Schneider *et al.*, 1994).

Striatal DA D1 and D2 Receptors

We also used quantitative autoradiography to assess [^{3}H]SCH-23390 binding to DA D1 receptors and [^{3}H] spiperone binding to DA D2 receptors in the striatum of acutely parkinsonian monkeys with relatively brief post-lesion survival times (1–3 months) and chronically parkinsonian monkeys with extended post-lesion survival times (>8 months) (Decamp *et al.*, 1999). Despite similar symptoms and similar degrees of striatal denervation (as assessed by [^{3}H] mazindol binding), only acutely parkinsonian animals had significantly increased striatal D1 receptor binding (in most striatal subregions) and increased D2 receptor binding (in lateral subregions only), compared to normal and chronically parkinsonian animals. These results suggest that either the mode of administration of MPTP and/or the duration of parkinsonism changes in DA underlie observed receptor expression.

Preproenkephalin Gene Expression

Prepronkephalin (PPE) mRNA is co-expressed in striatal GABAergic output neurons forming the indirect striatopallidal projection system (Parent and Hazrati, 1995). These neurons primarily innervate the external segment of the globus pallidus (GP). Numerous reports have described an increase in striatal PPE mRNA expression, particularly in dorsolateral sensorimotor striatal territory, in various models of parkinsonism (Young III *et al.*, 1986; Normand *et al.*, 1988; Gerfen *et al.*, 1990) including MPTP-treated animals (Augood *et al.*, 1989; Asselin *et al.*, 1994; Herrero *et al.*, 1995). These studies all utilized animals made acutely parkinsonian (i.e., animals with rapid onset and short symptom duration).

In an attempt to evaluate the influence of time to develop symptoms and duration of symptoms on striatal PPE gene expression, we studied animals made either acutely or chronically parkinsonian, as described earlier (Schneider *et al.*, 1999). Both groups of animals had similar degrees of motor impairment and similar degrees of striatal DA denervation at the time of euthanasia. A significant increase in striatal PPE mRNA expression was seen in the sensorimotor striatal region in acutely parkinsonian animals (animals with short duration MPTP exposure [1 month] and brief [1–3 months] post-lesion survival), similar to that described by others (Asselin *et al.*, 1994; Herrero *et al.*, 1995; Nisbet *et al.*, 1995). In contrast, animals with a slowly progressing parkinsonism (MPTP exposure over 4–6 months) and long survival after symptom onset (>8 months) or animals that received acute MPTP exposure (over approximately 1 month) but had an extended post-lesion survival time (>8 months), showed either no increase in striatal PPE mRNA expression or a decrease of gene expression below normal levels. These results show that despite similarity in symptom expression and extent of DA lesion, animals with different forms of parkinsonism have drastically different striatal PPE gene expression, apparently linked to differences in duration of symptoms. The large increase in PPE signal seen in acutely parkinsonian animals seems to be a transient phenomenon unrelated to symptom expression. The possibility exists that an increase in PPE expression may also occur early in the degenerative process in chronically parkinsonian animals but this has not yet been established.

Preprotachykinin Gene expression

Substance P (SP) is co-expressed by GABAergic medium spiny neurons of the direct striatopallidal output system that primarily innervates the internal segment of the GP and the SN pars reticulata. Expression of mRNA for preprotachykinin (PPT), the precursor of SP, is decreased in response to DAergic lesions in a variety of model systems (Gerfen *et al.*, 1990; Herrero *et al.*, 1995; Morissette *et al.*, 1999). Here again, most of these data were obtained from animals made acutely parkinsonian and with brief post-lesion survival times. However, using the same rationale as for the studies of PPE gene expression discussed above, we evaluated the influence of time to develop symptoms and duration of symptoms on striatal PPT gene expression in animals made either acutely or chronically parkinsonian.

In contrast to our findings in the PPE gene expression studies, both acutely and chronically parkinsonian animals with similar symptoms and similar degrees of DA denervation had similarly decreased striatal PPT mRNA expression regardless of duration of MPTP exposure, duration of the parkinsonism or extent of DA denervation (Wade and Schneider, 2004). Interestingly, chronic MPTP-treated animals with cognitive but no motor deficits did not show any significant changes in striatal PPT expression (Wade and Schneider, 2001).

Nicotinic Acetylcholine Receptors

We have recently examined cortical and sub-cortical nicotinic acetylcholine receptor (nAChR) binding in different groups of MPTP-exposed monkeys: those with chronic low dose MPTP exposure and cognitive but no motor deficits; those with chronic MPTP exposure leading to typical parkinsonian motor deficits; and those with acute MPTP exposure and parkinsonian motor deficits (Kulak and Schneider, 2004; Kulak and Schneider, 2007). In motor asymptomatic animals, binding of $[^{125}I]$ α-bungarotoxin to α7 nAChRs was increased in the supplementary motor area, primary motor cortex and throughout the putamen, with no changes in receptor binding in cortical or sub-cortical areas related primarily to cognitive functioning. In contrast, acutely or chronically parkinsonian monkeys had increased $[^{125}I]$ α-bungarotoxin binding only in the dorsolateral putamen (Kulak and Schneider, 2004). It was suggested that the increase in α7 nAChR binding seen in these motor-asymptomatic animals may represent a compensatory mechanism that may help to maintain sensorimotor functioning in these animals. In another study (Kulak and Schneider, 2007), differences in $[^{125}I]$epibatidine binding to β2*–β4* nAChRs and $[^{125}I]$A85380 binding to β2* nAChRs in motor symptomatic and asymptomatic animals was investigated. Both groups had similar decreases (70–80%) in β2* and β4* nAChR binding in the caudate and putamen. Cognitively impaired, motor-asymptomatic animals had no significant changes in $[^{125}I]$epibatidine binding to β2*–β4* nAChRs and $[^{125}I]$A85380 binding to β2* nAChRs in cognition-related cortical regions such as Broadman's area 46, orbitofrontal cortex, anterior cingulate sulcus, and hippocampus. In contrast, motor-impaired animals had decreases in β2* and β4* nAChR in the principal sulcus (40–60%), anterior cingulate sulcus (30–55%), and orbitofrontal cortex (30–41%) (Kulak and Schneider, 2007). Thus, we found significant differences in β2* and β4* nAChR expression, particularly in the cortex, in monkeys with different forms of parkinsonism. These data suggest that therapeutic strategies based on nAChR agonist administration that might improve cognition in early PD may, due to a changing nAChR profile with advancing disease, have diminished efficacy on the same symptoms in more advanced PD.

Microglial Activation

The possible role of inflammation and activated microglia in the initiation and/or perpetuation of the neuropathology of PD has received considerable attention recently. While a number of studies, mostly using rodent models of parkinsonism (Kurkowska-Jastrzebska *et al.*, 1999; Liberatore *et al.*, 1999; Wu *et al.*, 2002; Furuya *et al.*, 2004), have described microglial activation the SN and/or striatum in response to MPTP exposure, there have been few studies that have examined this in non-human primate models of parkinsonism. In one study (Hurley *et al.*, 2003), monkeys administered MPTP (0.15 mg/kg; 2 to 3 times a week) and euthanized soon (4–5 weeks) after development of motor symptoms had an intense microglial activation in the SN, nigrostriatal tract and in both segments of the GP but not in the striatum. Silver staining showed degeneration in the SN, GP, and striatum. However, due to the short post-MPTP survival time, it is not possible to differentiate an acute degenerative process from any potential long-term residual process stimulated by the MPTP-induced injury. In another study, animals were euthanized

after a long period of time (5–14 years) following MPTP administration (either unilateral carotid infusion of MPTP that induced motor symptoms or intravenous infusion of MPTP that did not result in motor symptoms) (McGeer *et al.*, 2003). In these animals, whether symptomatic or asymptomatic, reactive microglia were observed along with a severe loss of DAergic SN neurons. The small number of animals examined, presence of complicating factors (such as diabetic coma), the lack of details on the MPTP administration, and no assessment of the presence or absence of an ongoing neurodegenerative process make the findings from this study difficult to interpret. Unfortunately, these and other studies (Goto *et al.*, 1996; Barcia *et al.*, 2004) in monkeys have not been able to clarify the relationship between microglial activation and the perpetuation of a degenerative process in MPTP-exposed monkeys.

Studies in our laboratory have begun to examine this issue by examining tissue derived from monkeys that received acute or chronic MPTP administration and had short or long post-lesion survival times (Schneider *et al.*, 2006). In preliminary studies, we have assessed TH expression, microglial activation (with CD11b immunohistochemistry), and evidence for active neurodegeneration (silver staining) in the SN in animals with different forms of parkinsonism (Table 8.1). Animals acutely treated with MPTP and with short survival times (up to 3 months post-MPTP administration) had moderate TH cell loss in the SN, intense microglial activation, and evidence of ongoing degeneration at the time of euthanasia. Animals with acute-onset parkinsonism but long-term survival (over a year post-MPTP administration) had no evidence of activated microglia or ongoing neurodegeneration

TABLE 8.1   Effects of different types of MPTP administration and survival times on the number of TH-positive neurons, activated microglia, and silver-stained neurons in the substantia nigra in non-human primates

	Normal	Acute – STS	Acute – LTS	CLD Asymp	CLD Symp
TH-positive neurons	+++	++	+	++	+
Activated microglial	–	+++	–	++	+++
Silver-stained neurons	–	+++	–	++	+++

Number of cells observed per section: − = no cells observed; + = low; ++ = intermediate; +++ = high; STS = short-term survival; LTS = long-term survival; CLD Asymp = chronic low dose, motor asymptomatic; CLD Symp = chronic low dose, motor symptomatic.

in the SN. In contrast, activated microglia and evidence of ongoing degeneration were observed in animals with slowly evolving parkinsonism and long post-MPTP survival times. In chronic low dose motor-asymptomatic MPTP-treated animals, activated microglia, and silver stained neurons were also observed in the SN.

These results suggest that acute MPTP administration in monkeys may result in self-limiting degenerative and inflammatory processes. In these animals, we found no evidence for a long-lasting microglial activation that might participate in perpetuation of neural degeneration. In contrast, in animals with a long-term, low-level insult to SN DA neurons, there was evidence of chronic microglial activation and ongoing neurodegeneration even years after the last toxin exposure, suggesting again that the mechanisms underlying the degenerative processes may be different in different models of MPTP-induced parkinsonism. Thus, investigating inflammatory processes related to DA neuron degeneration or testing anti-inflammatory therapies only on acutely parkinsonian monkeys, in which there is possibly a different type/duration of inflammatory process than in chronically parkinsonian monkeys, may not enhance our understanding of the role of microglia in perpetuation of DA neuron degeneration nor provide the best model in which to develop therapies aimed at altering the ongoing degeneration that occurs in PD.

DO DIFFERENT MODELS RESPOND DIFFERENTLY TO ANTI-PARKINSONIAN THERAPIES?

As we have discussed thus far, there are a number of pathological and neurochemical differences in various brain regions depending on the period of time over which the MPTP was administered to monkeys and the amount of post-lesion survival time (Table 8.2). Consequently, the availability of different models may allow for a better understanding of the pathophysiology of symptom generation as well as greater insight into the possible mechanisms involved in disease progression. Can these different models also be used enhance the development of symptomatic therapies for PD?

Response to Levodopa and Development of Levodopa-Induced Dyskinesias

Levodopa (L-dopa), administered with a peripheral dopa decarboxylase inhibitor, is still the gold standard

TABLE 8.2 Neurochemical similarities and differences between acute, chronic asymptomatic and symptomatic MPTP-treated monkeys

	CLD MPTP	Acute MPTP	Reference
Mazindol binding	Decreased	Decreased	Wade and Schneider (2004)
DA receptors	Normal	Increased D1	Decamp et al. (1999)
Preproenkephalin mRNA	Normal or below normal	Increased	Schneider et al. (1999)
Preprotachykinin mRNA	Decreased	Decreased	Wade and Schneider (2001); Wade and Schneider (2004)
nAChR subunit: α7	Asymptomatic: Increased in supplementary motor area, primary motor cortex and putamen; Symptomatic: Increased in dorsolateral putamen only	Increased in dorsolateral putamen only	Kulak and Schneider (2004)
nAChR subunit: β2/β4	Asymptomatic: Decreased in caudate and putamen; Symptomatic: Decreased in and caudate putamen and in cognition-related cortical regions	NA	Kulak and Schneider (2007)

CLD = chronic low dose

for symptomatic relief of the symptoms of PD. However, many patients develop response fluctuations and side effects such as dyskinesias after several years of L-dopa use (Sweet and McDowell, 1975). Different patient sub-populations have different susceptibilities for developing L-dopa-induced dyskinesias (LIDs). Patients with early-onset PD (Kostic *et al.*, 1991), patients treated with higher doses of L-dopa (Mouradian *et al.*, 1990), patients with a longer treatment duration (Miyawaki *et al.*, 1997) and patients who started L-dopa treatment early in the course of their PD (Caraceni *et al.*, 1991) have a higher likelihood of developing LIDs. MPTP-treated monkeys also develop LIDs, however, as with other studies using MPTP-treated monkeys, most dyskinesia studies have utilized animals made acutely parkinsonian and with early initiation of DAergic therapy following the appearance of symptoms. This is different from what occurs in patients, where symptoms develop slowly and there is often a lag of several years between diagnosis and initiation of L-dopa therapy. Thus, we recently examined the extent to which the duration of MPTP exposure and the duration of symptoms prior to the onset of DAergic therapy might effect the development of LIDs (Schneider *et al.*, 2003).

Four groups of animals were studied: acutely parkinsonian with short symptom duration (mean 105 days) prior to L-dopa administration; acutely parkinsonian with long symptom duration prior to start of L-dopa therapy (mean 817 days); chronically parkinsonian with long symptom duration (1,000 days) prior to start of L-dopa administration; and animals that developed cognitive deficits and no motor deficits after chronic MPTP exposure (106 ± 32 days), were then exposed to high doses of MPTP to induce motor symptoms (26 ± 8 day exposure period) and then began receiving L-dopa shortly after developing motor symptoms (mean 52.5 days).

Although all animals had similar parkinsonian symptoms and similar therapeutic responses to L-dopa, the propensity to develop dyskinesias differed significantly between the groups. The acutely parkinsonian animals with rapid symptom onset and short symptom duration prior to receiving L-dopa developed LIDs very quickly (mean 19.7 ± 3 days). Acutely parkinsonian animals that remained symptomatic for a longer period of time prior to starting L-dopa therapy failed to develop LIDs after 279 ± 13 consecutive days of L-dopa dosing. An animal that received chronic low dose MPTP administration over 369 days, developed stable

parkinsonism and was maintained for 1,000 days prior to initiation of L-dopa therapy developed dyskinesia after 146 days of L-dopa therapy. Animals initially motor asymptomatic after prolonged low dose MPTP exposure but with acute onset of motor symptoms after switching to high dose MPTP exposure and short duration of motor symptoms prior to L-dopa administration developed dystonia after 80.5 ± 18.5 days of L-dopa administration but failed to develop choreiform dyskinesias after 130 ± 10 days (Schneider *et al.*, 2003).

These results suggest that the rapidity with which the lesion occurs and symptoms develop and the duration of symptoms prior to exposure to L-dopa may play important roles in the pathophysiology of LIDs. While these different propensities for developing LIDs are obviously related to physiological differences in the brains of these animals, the precise mechanisms underlying these differences in LID development are not presently known. However, these results reinforce the point that the model used for a particular study needs to be chosen carefully depending upon the question being addressed by that study. There have been numerous studies performed evaluating potential therapies for inhibiting LIDs. These studies have all utilized monkeys that were made acutely parkinsonian, had short duration of symptoms prior to initiation of L-dopa therapy and developed LIDs rapidly after starting L-dopa therapy (Tye *et al.*, 1989; Grondin *et al.*, 2000; Hadj Tahar *et al.*, 2000; Bibbiani *et al.*, 2001; Iravani *et al.*, 2001; Bezard *et al.*, 2003, 2004; Chassain *et al.*, 2003). Yet, despite promising results from these studies, there has been little success in translating preclinical findings to the clinic and it is not yet known if the use of a model that more closely approximates the clinical condition (i.e., slow onset of symptoms, long duration of symptoms prior to L-dopa initiation, and slow onset of dyskinesia) would better predict clinical outcome. For example, Sarizotan, a full agonist at serotonin $5\text{-}HT_{1a}$ receptors with high affinity for DA D3 and D4 receptors, was studied preclinically in monkeys with acute MPTP-induced parkinsonism and rapid development of LIDs (Bibbiani *et al.*, 2001), where it was found to have significant beneficial effects on LIDs. Yet, Phase II and Phase III clinical trials with this drug failed to show significant effects on dyskinesia, compared to placebo, in advanced PD patients (Goetz *et al.*, 2007, Merck KGaA, 2007). Whether a different preclinical outcome would have been achieved using a different model is not known.

CONCLUSIONS

The information presented in this chapter is not meant to suggest that one model of MPTP-induced parkinsonism, whether in mice or in monkeys, is better than any other. The MPTP-induced models of parkinsonism have proven their worth over decades of use and currently, no other models exist that are superior in terms of reproducing the pathology and symptomatology of PD. We suggest that there is still much to be learned about these models and with a better understanding of their various iterations, investigators should be able to choose which model or models best suit the particular research problem under study.

REFERENCES

Anderson, D. W., Schneider, J. S. (2006). Microglial activation and perpetuation of neurodegeneration in MPTP-treated mice are dependent upon toxin administration protocol, Society for Neuroscience, Vol. Program No. 75.4. Neuroscience Meeting Planner, Atlanta, GA.

Anderson, D. W., Neavin, T., Smith, J. A., and Schneider, J. S. (2001). Neuroprotective effects of pramipexole in young and aged MPTP-treated mice. *Brain Res* **905**, 44–53.

Anderson, D. W., Bradbury, K. A., and Schneider, J. S. (2006). Neuroprotection in Parkinson models varies with toxin administration protocol. *Eur J Neurosci* **24**, 3174–3182.

Asselin, M.-C., Soghomonian, J. J., Cote, P. Y., and Parent, A. (1994). Striatal changes in preproenkephalin mRNA levels in parkinsonian monkeys. *Neuroreport* **5**, 2137–2140.

Augood, S. J., Emson, P. C., Mitchell, I. J., Boyce, S., Clarke, C. E., and Crossman, A. (1989). Cellular localization of enkephalin gene expression in MPTP-treated cynomologus monkeys. *Mol Brain Res* **6**, 85–92.

Barcia, C., Sanchez Bahillo, A., Fernandez-Villalba, E., Bautista, V., Poza, Y. P. M., Fernandez-Barreiro, A., Hirsch, E. C., and Herrero, M. T. (2004). Evidence of active microglia in substantia nigra pars compacta of parkinsonian monkeys 1 year after MPTP exposure. *Glia* **46**, 402–409.

Beal, M. F. (1992). Does impairment of energy metabolism result in excitotoxic neuronal death in neurodegenerative illnesses? *Ann Neurol* **31**, 119–130.

Betarbet, R., Sherer, T. B., MacKenzie, G., Garcia-Osuna, M., Panov, A. V., and Greenamyre, J. T. (2000). Chronic systemic pesticide exposure reproduces features of Parkinson's disease. *Nat Neurosci* **3**, 1301–1306.

Bezard, E., Imbert, C., and Gross, C. E. (1998). Experimental models of Parkinson's disease: From the static to the dynamic. *Rev Neurosci* **9**, 71–90.

Bezard, E., Ferry, S., Mach, U., Stark, H., Leriche, L., Boraud, T., Gross, C., and Sokoloff, P. (2003). Attenuation of levodopa-induced dyskinesia by normalizing dopamine D3 receptor function. *Nat Med* **9**, 762–767.

Bezard, E., Hill, M. P., Crossman, A. R., Brotchie, J. M., Michel, A., Grimee, R., and Klitgaard, H. (2004). Levetiracetam improves choreic levodopa-induced dyskinesia in the MPTP-treated macaque. *Eur J Pharmacol* **485**, 159–164.

Bibbiani, F., Oh, J. D., and Chase, T. N. (2001). Serotonin 5HT1A agonist improves motoc complications in rodent and primate parkinsonian models. *Neurology* **57**, 1829–1834.

Blanchet, P. J., Calon, F., Morissette, M., Hadj Tahar, A., Belanger, N., Samadi, P., Grondin, R., Gregoire, L., Meltzer, L., Di Paolo, T., and Bedard, P. J. (2004). Relevance of the MPTP primate model in the study of dyskinesia priming mechanisms. *Parkinsonism Relat Disord* **10**, 297–304.

Brecknell, J. E., Dunnett, S. B., and Fawcett, J. W. (1995). A quantitative study of cell death in the substantia nigra following a mechanical lesion of the medial forebrain bundle. *Neuroscience* **64**, 219–227.

Burns, R. S., Chiueh, C. C., Markey, S. P., Ebert, M. H., Jacobowitz, D. M., and Kopin, I. J. (1983). A primate model for parkinsonism: Selective destruction of dopaminergic neurons in the pars compacta of the substantia nigra by N-methyl-4-phenyl-1,2,3,6-tetrahydropyridine. *Proc Natl Acad Sci USA* **80**, 4546–4550.

Caraceni, T., Scigliano, G., and Musicco, M. (1991). The occurrence of motor fluctuations in parkinsonian patients treated long term with levodopa: Role of early treatment and disease progression. *Neurology* **41**, 380–384.

Carlsson, A. (1966). A pharmacological depletion of catecholamine stores. In *Modification of Sympathetic Function* (F. F. Shideman, Ed.), Vol. 18, pp. 541–549. The Williams and Wilkins Company, New York.

Chan, P., DeLanney, L. E., Irwin, I., Langston, J. W., and Di Monte, D. (1991). Rapid ATP loss caused by 1-methyl-4-phenyl-1,2,3,6-tetrahydropyridine in mouse brain. *J Neurochem* **57**, 348–351.

Chassain, C., Eschalier, A., and Durif, F. (2003). Antidyskinetic effect of magnesium sulfate in MPTP-lesioned monkeys. *Exp Neurol* **182**, 490–496.

Chiba, K., Trevor, A., and Castagnoli, N. (1984). Metabolism of the neurotoxic tertiary amine MPTP, by brain monoamine oxidase. *Biochem Biophys Res Commun* **120**, 574.

Crossman, A. R., Clarke, C. E., Boyce, S., Robertson, R. G., and Sambrook, M. A. (1987). MPTP-induced parkinsonism in the monkey: Neurochemical pathology, complications of treatment and pathophysiological mechanisms. *Can J Neurol Sci* **14**, 428–435.

Decamp, E., Wade, T., and Schneider, J. S. (1999). Differential regulation of striatal dopamine D1 and D2 receptors in acute and chronic parkinsonian monkeys. *Brain Res* **847**, 134–138.

Decamp, E., and Schneider, J. S. (2004). Attention and executive function deficits in chronic low-dose MPTP-treated non-human primates. *Eur J Neurosci.* **20**, 1371–1378.

Dehmer, T., Lindenau, J., Haid, S., Dichgans, J., and Schulz, J. B. (2000). Deficiency of inducible nitric oxide synthase protects against MPTP toxicity *in vivo*. *J Neurochem* **74**, 2213–2216.

Deumens, R., Blokland, A., and Prickaerts, J. (2002). Modeling Parkinson's disease in rats: An evaluation of 6-OHDA lesions of the nigrostriatal pathway. *Exp Neurol* **175**, 303–317.

Donaldson, I., Dolphin, A., Jenner, P., Marsden, C. D., and Pycock, C. (1976). The roles of noradrenaline and dopamine in contraversive circling behaviour seen after unilateral electrolytic lesions of the locus coeruleus. *Eur J Pharmacol* **39**, 179–191.

Du, Y., Ma, Z., Lin, S., Dodel, R. C., Gao, F., Bales, K. R., Triarhou, L. C., Chernet, E., Perry, K. W., Nelson, D. L., Luecke, S., Phebus, L. A., Bymaster, F. P., and Paul, S. M. (2001). Minocycline prevents nigrostriatal dopaminergic neurodegeneration in the MPTP model of Parkinson's disease. *Proc Natl Acad Sci USA* **98**, 14669–14674.

Furuya, T., Hayakawa, H., Yamada, M., Yoshimi, K., Hisahara, S., Miura, M., Mizuno, Y., and Mochizuki, H. (2004). Caspase-11 mediates inflammatory dopaminergic cell death in the 1-methyl-4-phenyl-1,2,3,6-tetrahydropyridine mouse model of Parkinson's disease. *J Neurosci* **24**, 1865–1872.

Galpern, W. R. for The NINDS NET-PD Investigators (2006). A randomized, double-blind, futility clinical trial of creatine and minocycline in early Parkinson disease. *Neurology* **66**, 664–671.

Gerfen, C. R., Engber, T. M., Mahan, L. C., Susel, Z., Chase, T. N., Monsma, F. J., and Sibley, D. R. (1990). D-1 and D-2 dopamine receptor-regulated gene expression of striatonigral and striatopallidal neurons. *Science* **250**, 1429–1432.

Goetz, C. G., Damier, P., Hicking, C., Laska, E., Muller, T., Olanow, C. W., Rascol, O., Russ, H. (2007). Sarizotan as a treatment for dyskinesias in Parkinson's disease: A double-blind placebo controlled trial *Mov. Disord* **22**, 179–186.

Goto, K., Mochizuki, H., Imai, H., Akiyama, H., and Mizuno, Y. (1996). An immuno-histochemical study of ferritin in 1-methyl-4-phenyl-1,2,3,6-tetrahydropyridine

(MPTP)-induced hemiparkinsonian monkeys. *Brain Res* **724**, 125–128.

Grondin, R., Hadj Tahar, A., Doan, V. D., Ladure, P., and Bedard, P. J. (2000). Noradrenoceptor antagonism with idazoxan improves l-dopa-induced dyskinesias in MPTP monkeys. *Naunyn Schmiedebergs Arch Pharmacol* **361**, 181–186.

Hadj Tahar, A., Belanger, N., Bangassoro, E., Gregoire, L., and Bedard, P. J. (2000). Antidyskinetic effect of JL-18, a clozapine analog, in parkinsonian monkeys. *Eur J Pharmacol* **399**, 183–186.

Hartley, A., Stone, J. M., Heron, C., Cooper, J. M., and Schapira, A. H. (1994). Complex I Inhibitors induce dose-dependant apoptosis in PC12 cells: relevance to Parkinson's disease. *J Neurochem* **63**, 1987–1990.

Heikkila, R. E., Hess, A., and Duvoisin, R. C. (1984). Dopaminergic neurotoxicity of 1-methyl-4-phenyl-1,2,5,6-tetrahydropyridine in mice. *Science* **224**, 1451–1453.

Hekkila, R., Seiber, B.-A., Manzino, L., and Sonsalla, P. K. (1989). Some features of the nigrostriatal dopaminergic neurotoxin 1-methyl-4-phenyl-1,2,3,6-tetrahydropyridine (MPTP) in the mouse. *Mol Chem Neuropathol* **10**, 171.

Herrero, M. T., Augood, S. J., Hirsch, E. C., Javoy-Agid, F., Luquin, R., Agid, Y., Obeso, J. A., and Emson, P. C. (1995). Effects of l-dopa on preproenkephalin and preprotachykinin gene expression in the MPTP-treated monkey striatum. *Neuroscience* **68**, 1189–1198.

Hirsch, E. C., Breidert, T., Rousselet, E., Hunot, S., Hartmann, A., and Michel, P. P. (2003). The role of glial reaction and inflammation in Parkinson's disease. *Ann N Y Acad Sci* **991**, 214–228.

Hurley, S. D., O"Banion, M. K., Song, D. D., Arana, F. S., Olschowka, J. A., and Haber, S. N. (2003). Microglial response is poorly correlated with neurodegeneration following chronic, low-dose MPTP administration in monkeys. *Exptl Neurol.* **184**, 659–668.

Iravani, M. M., Costa, S., Jackson, M. J., Tel, B. C., Cannizzaro, C., Pearce, R. K., and Jenner, P. (2001). GDNF reverses priming for dyskinesia in MPTP-treated, l-DOPA-primed common marmosets. *Eur J Neurosci* **13**, 597–608.

Jackson-Lewis, V., Jakowec, M., Burke, R. E., and Przedborski, S. (1995). Time course and morphology of dopaminergic neuronal death caused by the neurotoxin 1-methyl-4-phenyl-1,2,3,6-tetrahydropyridine. *Neurodegeneration* **4**, 257–269.

Jenner, P. (2003a). The contribution of the MPTP-treated primate model to the development of new treatment strategies for Parkinson's disease. *Parkinsonism Relat Disord* **9**, 131–137.

Jenner, P. (2003b). The MPTP-treated primate as a model of motor complications in PD: Primate model of motor complications. *Neurology* **61**, S4–S11.

Kostic, V., Przedborski, S., Flaster, E., and Sternic, N. (1991). Early development of levodopa-induced dys-kinesias and response fluctuations in young-onset Parkinson's disease. *Neurology* **41**, 202–205.

Kulak, J. M., and Schneider, J. S. (2004). Differences in alpha7 nicotinic acetylcholine receptor binding in motor symptomatic and asymptomatic MPTP-treated monkeys. *Brain Res* **999**, 193–202.

Kulak, J. M., Fan, Hand Schneider, J. S. (2007). β2*, β4* and B4 nicotinic acetylcholine receptor expression changes with progressive parkinsonism in non-human primates. *Neurobiol Dis* **27**, 312–319.

Kurkowska-Jastrzebska, I., Wronska, A., Kohutnicka, M., Czlonkowski, A., and Czlonkowska, A. (1999). The inflammatory reaction following 1-methyl-4-phenyl-1,2,3,6-tetrahydropyridine intoxication in mouse. *Exp Neurol* **156**, 50–61.

Langston, J. W., Ballard, P. A., Tetrud, J. W., and Irwin, I. (1983). Chronic parkinsonism in humans due to a product of merperidine-analog synthesis. *Science* **219**, 979–980.

Langston, J. W., Forno, L. S., Rebert, C. S., and Irwin, I. (1984). Selective nigral toxicity after systemic admin-istration of 1-methyl-4-phenyl-1,2,3,6-tetrahydropy-ridine (MPTP) in the squirrel monkey. *Brain Res* **292**, 390–394.

Liberatore, G. T., Jackson-Lewis, V., Vukosavic, S., Mandir, A. S., Vila, M., McAiliffe, W. G., Dawson, V. L., Dawson, T. M., and Przedborski, S. (1999). Inducible nitric oxide synthase stimulates dopaminergic neurodegeneration in the MPTP model of Parkinson disease. *Nat Med* **5**, 1403–1409.

Markey, S. P., Johannessen, J. N., Chiueh, C. C., Burns, R. S., and Herkenham, M. A. (1984). Intraneuronal generation of a pyridinium metabolite may cause drug-induced parkinsonism. *Nature* **311**, 464–467.

McCormack, A. L., Thiruchelvam, M., Manning-Bog, A. B., Thiffault, C., Langston, J. W., Cory-Slechta, D. A., and Di Monte, D. A. (2002). Environmental risk factors and Parkinson's disease: Selective degen-eration of nigral dopaminergic neurons caused by the herbicide paraquat. *Neurobiol Dis* **10**, 119–127.

McGeer, P. L., Schwab, C., Parent, A., and Doudet, D. (2003). Presence of reactive microglia in monkey sub-stantia nigra years after 1-methyl-4-phenyl-1,2,3,6-tetrahydropyridine administration. *Ann Neurol* **54**, 599–604.

Meissner, W., Prunier, C., Guilloteau, D., Chalon, S., Gross, C., and Bezard, E. (2003). Time-course of nigrostriatal degeneration in a progressive MPTP-lesioned macaque model of Parkinson's disease. *Mol Neurobiol* **28**, 209–218.

Melamed, E., Rosenthal, J., Globus, M., Cohen, O., Frucht, Y., and Uzzan, A. (1985). Mesolimbic dopaminergic neurons are not spred by MPTP neuro-toxicity in mice. *Eur J Pharmacol* **114**, 97–100.

Merck KGaA: Sarizotan phase III studies did not meet primary efficacy endpoint (2007). http://me.merck.de/ n/A720D2336B86597DC1257196001CCA99/$FILE/ Sarizo-e.pdf.

Miyawaki, E., Lyons, K., Pahwa, R., Troster, A. I., Hubble, J., Smith, D., Busenbark, K., McGuire, D., Michalek, D., and Koller, W. C. (1997). Motor com-plications of chronic levodopa therapy in Parkinson's disease. *Clin Neuropharmacol* **20**, 523–530.

Morissette, M., Grondin, R., Goulet, M., Bedard, P. J., and Di Paolo, T. (1999). Differential regulation of striatal preproenkephalin and preprotachykinin mRNA levels in MPTP-lesioned monkeys chronically treated with dopamine D1 or D2 receptor agonists. *J Neurochem* **72**, 682–692.

Mouradian, M. M., Heuser, I. J., Baronti, F., and Chase, T. N. (1990). Modification of central dopaminergic mecha-nisms by continuous levodopa therapy for advanced Parkinson's disease. *Ann Neurol* **27**, 18–23.

Nisbet, A. P., Foster, O. J. F., Kingsbury, A., Eve, D. J., Daniel, S. E., Marsden, C. D., and Lees, A. J. (1995). Preproenkephalin and preprotachykinin messenger RNA expression in normal human basal ganglia and in Parkinson's disease. *Neuroscience* **66**, 361–376.

Normand, E., Popovici, T., Onteniente, B., Fellman, D., Pieter-Tonneau, D., Auffrey, C., and Bloch, B. (1988). Dopaminergic neurons of the substantia nigra modu-late preproenkephalin A gene expression in rat striatal neurons. *Brain Res* **439**, 39–46.

Parent, A., and Hazrati, L.-N. (1995). Functional anatomy of the basal ganglia. I. The cortico-basal ganglia-thalamo-cortical loop. *Brain Res Rev* **20**, 91–127.

Parkinson Study Group (2002). Dopamine transporter brain imaging to assess the effects of pramipexole vs levodopa on Parkinson disease progression. *JAMA* **287**, 1653–1661.

Pope-Coleman, A., and Schneider, J. S. (1998). Effects of chronic GM1 ganglioside treatment on cognitive and motor deficits in a slowly progressing model of parkinsonism in non-human primates. *Restor Neurol Neurosci* **12**, 255–266.

Pope-Coleman, A., Tinker, J. P., and Schneider, J. S. (2000). Effects of GM1 ganglioside treatment on pre- and postsynaptic dopaminergic markers in the striatum of parkinsonian monkeys. *Synapse* **36**, 120–128.

Przedborski, S., and Vila, M. (2003). The 1-methyl-4-phe-nyl-1,2,3,6-tetrahydropyridine mouse model: A tool to explore the pathogenesis of Parkinson's disease. *Ann NY Acad Sci* **991**, 189–198.

Ramsay, R. R., and Singer, T. P. (1986). Energy-depend-ent uptake of N-methyl-4-phenylpyridinium, the neurotoxic metabolite of 1-methyl-4-phenyl-1,2,3,

6-tetrahydropyridine, by mitochondria. *J Biol Chem* **261**, 7585–7587.

Ricaurte, G. A., Langston, J. W., Delanney, L. E., Irwin, I., Peroutka, S. J., and Forno, L. S. (1986). Fate of nigrostriatal neurons in young mature mice given 1-methyl-4-phenyl-1,2,3,6-tetrahydropyridine: A neurochemical and morphological reassessment. *Brain Res* **376**, 117–124.

Roundtable, S. T. A. I. (1999). Recommendations for standards regarding preclinical neuroprotective and restorative drug development. *Stroke* **Vol 30**, 2752–2758.

Ryan, M. E., and Ashley, R. A. (1998). How do tetracyclines work? *Adv Dent Res* **12**, 149–151.

Schneider, J. S. (1989). Levodopa-induced dyskinesias in parkinsonian monkeys: Relationship to extent of nigrostriatal damage. *Pharmacol Biochem Behav* **34**, 193–196.

Schneider, J. S. (1990). Chronic exposure to low doses of MPTP. II. Neurochemical and pathological consequences in cognitively-impaired, motor asymptomatic monkeys. *Brain Res* **534**, 25–36.

Schneider, J. S., and Denaro, F. J. (1988). Astrocytic responses to the dopaminergic neurotoxin 1-methyl-4-phenyl-1,2,3,6-tetrahydropyridine (MPTP) in cat and mouse brain. *J Neuropathol Exp Neurol* **47**, 452–458.

Schneider, J. S., and Kovelowski, C. J. (1990). Chronic exposure to low doses of MPTP. I. Cognitive deficits in motor asymptomatic monkeys. *Brain Res* **519**, 122–128.

Schneider, J. S., and Pope-Coleman, A. (1995). Cognitive deficits precede motor deficits in a slowly progressing model of parkinsonism in the monkey. *Neurodegeneration* **4**, 245–255.

Schneider, J. S., and Roeltgen, D. P. (1993). Delayed matching-to-sample, object retrieval, and discrimination reversal deficits in chronic low dose MPTP-treated monkeys. *Brain Res* **615**, 351–354.

Schneider, J. S., Yuwiler, A., and Markham, C. H. (1986). Production of a Parkinson-like syndrome in the cat with N-methyl-4-phenyl-1,2,3,6-tetrahydropyridine (MPTP): Behavior, histology, and biochemistry. *Exp Neurol* **91**, 293–307.

Schneider, J. S., Yuwiler, A., and Markham, C. H. (1987). Selective loss of subpopulations of ventral mesencephalic dopaminergic neurons in the monkey following exposure to MPTP. *Brain Res* **411**, 144–150.

Schneider, J. S., Unguez, G., Yuwiler, A., Berg, S. C., and Markham, C. H. (1988). Deficits in operant behaviour in monkeys with N-methyl-4-phenyl-1,2,3,6-tetrahydropyridine (MPTP). *Brain* **111**, 1265–1285.

Schneider, J. S., Rothblat, D. S., and DiStefano, L. (1994). Volume transmission of dopamine over large distances may contribute to recovery from experimental parkinsonism. *Brain Res* **643**, 86–91.

Schneider, J. S., Decamp, E., and Wade, T. (1999). Striatal preproenkephalin gene expression is upregulated in acute but not chronic parkinsonian monkeys: Implications for the contribution of the indirect striatopallidal circuit to parkinsonian symptomatology. *J Neurosci* **19**, 6643–6649.

Schneider, J. S., Gonczi, H., and Decamp, E. (2003). Development of levodopa-induced dyskinesias in parkinsonian monkeys may depend upon rate of symptom onset and/or duration of symptoms. *Brain Res* **990**, 38–44.

Schneider, J. S., Kulak, J. M., and Anderson, D. W. (2006). *Microglial Activation and Perpetuation of Neurodegeneration in MPTP-Treated Primates Is Dependent Upon Toxin Administration Protocol.* Society for Neuroscience, Atlanta, GA, pp. 75.3/GG17.

Seiden, L. S., and Ricaurte, G. A. (1987). Neurotoxicity of methamphetamine and related drugs. In *Psychopharmacology: The Third Generation of Progress* (H. Y. Meltzer, Ed.), pp. 359–366. Raven, New York.

Seniuk, N. A., Tatton, W. G., and Greenwood, C. E. (1990). Dose-dependent destruction of the coeruleus-cortical and nigral-striatal projections by MPTP. *Brain Res* **527**, 7–20.

Sonsalla, P. K., and Heikkila, R. E. (1986). The influence of dose and dosing interval in MPTP-induced dopaminergic neurotoxicity in mice. *Eur J Pharmacol* **129**, 339–345.

Sonsalla, P. K., and Heikkila, R. E. (1988). Neurotoxic effects of 1-methyl-4-phenyl-1,2,3,6-tetrahydropyridine (MPTP) and methamphetamine in several strains of mice. *Prog Neuropsychopharmacol Biol Psychiatry.* **12**, 345–354.

Sweet, R. D., and McDowell, F. H. (1975). Five years' treatment of Parkinson's disease with levodopa. Therapeutic results and survival of 100 patients. *Ann Intern Med* **83**, 456–463.

Switzer III, R. C. (1985). Argyrophilic degeneration of substantia nigra neurons in C57 mice due to 1-methyl-4-phenyl-1,2,3,6-tetrahydropyridine (MPTP). *Society for Neuroscience, Abstr* **11**, 992.

Tatton, N. A., and Kish, S. J. (1997). *In situ* detection of apoptotic nuclei in the substantia nigra compacta of 1-methyl-4-phenyl-1,2,3,6-tetrahydropyridine-treated mice using terminal deoxynucleotidyl transferase labelling and acridine orange staining. *Neuroscience* **77**, 1037–1048.

Thomas, M., and Le, W. D. (2004). Minocycline: Neuroprotective mechanisms in Parkinson's disease. *Curr Pharm Des* **10**, 679–686.

Tye, S. J., Rupniak, N. M., Naruse, T., Miyaji, M., and Iversen, S. D. (1989). NB-355: A novel prodrug for l-DOPA with reduced risk for peak-dose dyskinesias in MPTP-treated squirrel monkeys. *Clin Neuropharmacol* **12**, 393–403.

Wade, T. V., and Schneider, J. S. (2001). Expression of striatal preprotachykinin mRNA in symptomatic and asymptomatic 1-methyl-4-phenyl-1,2,3,6-tetrahydro-pyridine-exposed monkeys is related to parkinsonian motor signs. *J Neurosci* **21**, 4901–4907.

Wade, T. V., and Schneider, J. S. (2004). Striatal prepro-tachykinin gene expression reflects parkinsonian signs. *Neuroreport* **15**, 2481–2484.

Wu, D. C., Jackson-Lewis, V., Vila, M., Tieu, K., Teismann, P., Vadseth, C., Choi, D.-K., Ischiropoulos, H., and Przedborski, S. (2002). Blockase of microglial activation is neuroprotective in the 1-methyl-4-phenyl-1,2,3,6-tetrahydropyridine mouse model of Parkinson disease. *J Neurosci* **22**, 1763–1771.

Yang, L., Sugama, S., Chirichigno, J. W., Gregorio, J., Lorenzl, S., Shin, D. H., Browne, S. E., Shimizu, Y.,

Joh, T. H., Beal, M. F., and Albers, D. S. (2003). Minocycline enhances MPTP toxicity to dopaminergic neurons. *J Neurosci Res* **74**, 278–285.

Young III, W. S., Bonner, T. I., and Brann, M. R. (1986). Mesencephalic dopaminergic neurons regulate the expression of neuropeptide mRNAs in the rat fore-brain. *Proc Natl Acad Sci USA* **83**, 9827–9831.

Zou, L.-L., Xu, J., Jankovic, J., He, Y., Appel, S. H., and Le, W.-D. (2000). Pramipexole inhibits lipid peroxida-tion and reduces injury in the substantia nigra indued by the dopaminergic neurotoxin 1-methyl-4-phenyl-1,2,3,6-tetrahydropyridine in C57Bl/6 mice. *Neurosci Lett* **281**, 167–170.

9
NON-HUMAN PRIMATE MODELS OF PARKINSON'S DISEASE AND EXPERIMENTAL THERAPEUTICS

GISELLE M. PETZINGER, DANIEL M. TOGASAKI, GARNIK AKOPIAN,
JOHN P. WALSH AND MICHAEL W. JAKOWEC

*Department of Neurology, and the Andrus Gerontology Center, University of Southern California,
Los Angeles, CA, USA*

INTRODUCTION

The non-human primate serves as an important model for understanding the pathophysiology of the basal ganglia, evaluating new treatment modalities for neurodegenerative disorders affecting this region, especially Parkinson's disease (PD), and provides a valuable tool for discovery of new therapeutic targets that may lead to a cure for PD. The non-human primate model generated through either neurotoxicant or surgical lesioning has been most commonly used for experimental therapeutic studies in PD, in particular for identifying new symptomatic strategies primarily targeting the dopaminergic system, as well as those neurotransmitter systems known to modulate dopamine (including serotonin, glutamate, adenosine, acetylcholine, endocannabinoid, and noradrenalin). The model has also been extremely valuable in providing important insights into the pathophysiology and treatment of levodopa-induced dyskinesia, a disabling complication of long-term levodopa use in PD. In addition, the model has provided fertile ground for investigating innovative therapeutic approaches such as gene therapy using vector delivery systems, tissue transplantation, neurotrophic factor delivery, and deep brain stimulation (DBS). Finally, the 1-methyl-4-phenyl-1,2,3,6-tetrahydropyridine (MPTP)-lesioned non-human primate model has been valuable for understanding basal ganglia function in both the normal and lesioned state by providing insights into the role of neurotransmitters and circuitry in motor learning, motor control, and synaptic function that will guide the development of new therapeutic approaches for PD and related disorders (Israel and Bergman, 2008). The strength of the non-human primate as a model of PD lies in its similarities to the human condition for (i) clinical phenomenology, (ii) response to dopamine replacement therapy, and (iii) neuroanatomy. These strengths have been confirmed by 25 years of investigations that have yielded valuable insights toward the treatment of PD and normal basal ganglia function.

Although the non-human primate model has been important in identifying many new treatments for PD, including pharmacological targeting of non-dopaminergic neurotransmitter systems, vector-based gene therapy, and tissue transplant approaches, a number of these preclinical studies, did not accurately predict efficacy in clinical trials. There are many factors that may contribute to this lack of a translational success including (i) limitations in our knowledge regarding basal ganglia function in both the normal and diseased state; (ii) limited understanding of the pharmacokinetics and bioavailability of novel compounds; (iii) poorly elucidated adverse effects, including cardiovascular and cognitive changes; (iv) limited understanding of molecular mechanisms of the treatment intervention; and (v) differences between preclinical and clinical study design. Still, studies have shown that the non-human primate model addresses many important questions in addition to efficacy, including safety and tolerability, and technical issues addressing feasibility. More importantly, the model can be used to understand the underlying molecular mechanisms responsible for the success or failure of any new therapy, and consequently becomes a valuable conduit in bi-directional translational medicine from bench to bedside and bedside to bench. The value of information gained from studies using the non-human primate model of PD resides in understanding the details and differences of any one of the several distinct non-human primate models that may be used in the experimental paradigm. Important parameters that may vary amongst the many different non-human primate models include the lesioning regimen, behavioral assessments, and the pathological, neurochemical and molecular features of the model, some of which may be dynamic and display altered plasticity weeks to months post-lesioning.

The primary goal of this chapter is to review commonly used non-human primate models for PD, including (i) the behavioral assessments used to measure parkinsonian motor features or levodopa-related abnormal involuntary movements; (ii) the different model types and subtypes; (iii) the use of these models in pharmacological and non-pharmacological experimental therapeutics; and (iv) the impact of these models in understanding basal ganglia function and physiology. While the vast majority of non-human primate models of PD utilize the neurotoxicant MPTP, other neurotoxicants with similar selectivity for midbrain dopaminergic neurons, such as 6-hydroxydopamine, methamphetamine (METH), and proteasome inhibitors have been utilized. The development of many of these primate models has often been preceded by studies in rodents, which provided the foundation and basic understanding of these neurotoxicants. By understanding the details of each model, the investigator can appreciate more fully the individual strengths and limitations of the specific models, and can therefore select the most appropriate model for the question under investigation. This will then lead to improved study design, clearly defined outcome measures, and more comprehensive interpretation of preclinical studies.

ASSESSMENT OF MOTOR BEHAVIOR IN NON-HUMAN PRIMATE MODELS OF PD

Studies evaluating pharmacological and non-pharmacological agents in non-human primate models require the observation and quantification of motor behavior(s) associated with the parkinsonian state. While the primary motor feature is bradykinesia, these animal models may also demonstrate balance problems, tremor, freezing, and changes in posture. Most models also display the motor complications related to levodopa therapy, specifically dyskinesia (chorea and dystonia), and in some cases wearing-off phenomena. Several approaches have been used to quantify motor activity related to either the parkinsonian state or levodopa-related dyskinesias, including (i) clinical rating scales (CRSs); (ii) cages designed to measure animal movement (typically by having the animal break infrared beams traversing the cage); (iii) personal activity monitors affixed to the animals (often based on a small accelerometer); and (iv) digitized video monitoring systems. Each of these approaches has advantages and disadvantages.

Clinical Rating Scales

Since the initial reports of the utility of MPTP in non-human primate, a number of different CRSs have been published. These scales vary with respect to the species of the model and behaviors that are readily observed after MPTP administration. The design of a CRS is often developed by comprehensive observations of the full behavioral repertoire of the animal model. Most rating scales are then often refined to highlight those behavioral features that have a human clinical correlate as assessed in the Unified Parkinson's Disease Rating Scale (UPDRS). In general, many of these published scales for the non-human primate models have been tested for inter-rater reliability, validated against other behavioral measures, and have been demonstrated to be sensitive enough to capture response to intervention. For example, in the squirrel monkey, CRSs have been developed that have been shown to be sensitive to the degree of MPTP-lesioning and to capture behavioral features that are responsive to levodopa replacement therapy (Petzinger *et al.*, 2001). Rating scales have also been devised to evaluate the presence and intensity of motor complications in a number of different non-human primate models including the squirrel monkey and the marmoset (Petzinger *et al.*,

2001; Tan *et al.*, 2002; Jenner, 2003). These scales also vary within the different models. For example some scales distinguish between the types (chorea versus dystonia), and distribution (generalized versus focal) of dyskinesia, while others do not.

The main disadvantages of CRSs are that their application is often labor-intensive, and rating assignments are subjective and require extensive training. In addition, while many scales attempt to highlight features of animal behavior that resemble human parkinsonism, some behavioral features included in the rating scales are unique to a specific model, and do not have a direct human correlate. For example, action tremor, which is not a classical clinical feature of PD, is frequently observed in the squirrel monkey after MPTP administration and is often included in the clinical rating of these animals. In addition, hypokinesia or a general poverty of movement of an animal in its environment is another behavioral feature that is often observed and rated, but has no direct correlation with the UPDRS rating scale in human PD. The degree to which MPTP-induced behavioral features should be included in a CRS, despite the lack of a direct human clinical correlate, is an issue of debate.

Automated Behavioral Observation Methods

Another approach to measuring parkinsonian motor behavior is through the application of activity monitors either applied to the home cage or attached to the primate itself. Cage-based activity monitors are most commonly grounded in technologies similar to those used in rodent studies where movements are counted when a subject breaks infrared beams traversing the cage. These techniques distinguish between repetitive breaking of a single beam and consecutive breaks of different beams, thus allowing the apparatus to distinguish translational movement of the animal through space, which provides a measure generally accepted to represent locomotor activity. A recent technique utilizes telemetric methods to detect a probe implanted in the subcutaneous tissue of the animal to monitor locomotion. Personal activity monitors, such as the small accelerometers manufactured by Minimitter Company (Bend, OR), can be attached to the subject either using a collar, jacket, or straps. Most non-human primates require training or habituation to adapt to these encumbrances. These personal monitors measure general motor activity, so that the output encompasses both translational movement through space and movements performed with the animal stationary in the cage. Distinguishing between the different movement types, however, is impossible from the raw data from this device alone.

Newer types of activity measures have been derived from the advances in computer-based video technologies. These methods span a wide spectrum. At the low end are ones requiring minimal expense but provide little or no technical support and yield relatively crude measures. In contrast, there are other, commercially produced systems that can track each of the animals' limbs separately, recording position and velocity, and then provide sophisticated analyses of specific movement subtypes, with the trade-off being that acquiring these systems requires substantial financial investment.

Each of these systems has advantages and disadvantages. Choosing which method to use for measuring activity depends upon the specific requirements of the study design. In many cases, complementary methods are combined. For example, an automated system is often used in conjunction with a CRS. We suspect that, in the future, a popular approach will be the simultaneous application of a video-based computerized method with ratings performed on the video images.

One possible consideration in selecting a behavioral monitoring method is its ability to be applied repeatedly in a serial fashion to generate a time course. Because behavior is highly variable, the additional data points gathered in constructing a time course allows for much greater power in the statistical analyses. Furthermore, it is often difficult to know *a priori* how long after drug administration the peak effect will occur: making a series of repeated measurements throughout the entire duration of the behavioral change provides a more accurate assessment than making one measurement at a single time point when the peak effect is presumed to occur. This assumes even greater importance in studies using multiple drugs, leading to complex pharmacokinetic interactions. For example, using one drug to suppress levodopa-induced dyskinesias could lead to uncertainty about what aspect of the behavior might be affected: the duration of the dyskinesias might be shortened, the severity of the peak effect might be lessened, or both might occur. Automated methods in home cages have an obvious advantage for making repeated measurements over an extended period of time. In contrast, clinical ratings may be challenging when performed at cage side for long periods of time. However, clinical rating methods can be applied in conjunction with video recording over an extended period of time with animals in their home cages.

In contrast, the more objective methods of motor assessment in home cages using mechanical (accelerometers) and electronic (infrared beams) technology provide quantitative data on movement in space that can undergo parametric statistical analyses. However, these methods do not often target different subtypes of movement, or specific body regions. The more recent video-based systems with computerized analysis can have advantages over both clinical and automatic cage monitors by providing sophisticated quantitative analyses while simultaneously distinguishing different types of movements.

Data Analysis

Behavioral experiments tend to produce large amounts of data requiring thorough and sometimes complex statistical analysis. These analyses can be performed using currently available statistical software packages, such as SPSS (SPSS Inc., Chicago, IL) or SAS/STAT (SAS America Inc., Cary, NC). For example, the statistical analysis using time as a variable requires the application of repeated measures analysis of variance for parametric variables, but may require more sophisticated analysis for non-parametric variables (Togasaki *et al.*, 2005). Some studies may require correlation analysis to compare one behavioral analysis to another such as the improvement in levodopa-induced dyskinesias compared with the change in the severity of parkinsonism (Hsu *et al.*, 2004). Researchers can benefit from the inclusion of statisticians as a component of a study not only for the analysis of data but also for the early stages of the experimental design.

NON-MOTOR FEATURES IN MPTP-LESIONED NON-HUMAN PRIMATES

A potentially valuable application of the MPTP-lesioned non-human primate is in elucidating the anatomical sites and neurotransmitter systems involved in non-motor behavior. Although the substantia nigra pars compacta (SNpc) is the most vulnerable region affected by MPTP, there are varying degrees of injury and/or cell death in other brain

areas including the ventral tegmental area (VTA), retrorubal field, nucleus Basalis of Meynert, raphe nucleus, and locus ceruleus (Mitchell *et al.*, 1985a; Forno, 1986a, b). The fact that these anatomical regions mediate non-motor behavior and that MPTP can alter their neurotransmitter systems (including serotoninergic, cholinergic, and noradrenergic) suggests that non-motor features may be observed in these models (Namura *et al.*, 1987; Friedman and Mytilineou, 1990; Perez-Otano *et al.*, 1991). Unfortunately, there have been a limited number of studies in the MPTP-lesioned non-human primate examining non-motor features, which include cognitive impairment (executive function), sleep disorders, and affective behaviors, such as depression and anxiety (Schneider and Pope-Coleman, 1995; Decamp and Schneider, 2004). For example, cognitive changes, including reduced executive function and attention, are evident with early parkinsonian features after chronic low-dose MPTP administration, which might be associated with injury of extrastriatal dopaminergic pathways to the prefrontal cortex (Schneider, 1990; Schneider and Kovelowski, 1990; Schneider *et al.*, 1995; Slovin *et al.*, 1999a, b; Decamp and Schneider, 2004). In addition, other cognitive deficits in working memory, which might reflect dopaminergic injury to the hippocampus, can be observed (Decamp *et al.*, 2004). Alterations in sleep organization patterns have also been observed in MPTP-lesioned monkeys, with more severe disturbances in animals receiving a greater amount of MPTP (Almirall *et al.*, 1999).

THE MPTP-LESIONED NON-HUMAN PRIMATE MODEL

The most common model of PD in the non-human primate is generated by the administration of the neurotoxicant MPTP. The major strengths of this model include (i) parkinsonian features that closely resemble the human condition; (ii) the robust response to dopaminergic therapy; (iii) manifestation of levodopa-induced motor complications including dyskinesia and wearing-off; and (iv) neuroanatomical and physiological features that are similar to that in humans. This model has been well validated over the last 25 years as reflected in its ability to successfully predict the efficacy of drug treatments in patients with PD. In the following sections we present the history of MPTP, features of the MPTP non-human primate model, the different subtypes of models generated using various lesioning regimens, and finally, examples of the utility of the model for evaluating new therapeutic agents and strategies, and for providing insights into basal ganglia function in the normal and parkinsonian states.

A BRIEF RETROSPECTIVE OF MPTP IN MONKEYS

Immediately following its identification in humans, MPTP was administered to both rodents and non-human primates and some of the most valuable animal models of PD were established. MPTP-lesioned non-human primate models encompass a wide spectrum of species including the squirrel monkey (*Saimiri sciureus*) (Langston *et al.*, 1984), long-tailed macaque or cynomologus (*Macaca fascicularis*) (Mitchell *et al.*, 1985), rhesus macaque (*Macaca mulatta*) (Burns *et al.*, 1983; Chiueh, 1984b; Markey *et al.*, 1984), Japanese macaque (*Macaca fuscata*) (Crossman *et al.*, 1985; Jenner *et al.*, 1986), bonnet monkey (*Macaca radiata*) (Freed *et al.*, 1988), owl monkey (*Aotus trivirgatus*) (Collins and Neafsey, 1985), baboon (*Papio papio*) (Hantraye *et al.*, 1993; Moerlein *et al.*, 1986), African green monkey or vervet (*Chlorocebus aethiops* formerly *Ceropithecus aethiops*) (Taylor *et al.*, 1994), pigtail macaque (*Macaca nemestrina*) (Chefer *et al.*, 2007), and common marmoset (*Callithrix jacchus*) (Jenner *et al.*, 1986). The administration of MPTP to the non-human primate results in parkinsonian symptoms including bradykinesia, postural instability, freezing, stooped posture, and rigidity. Although postural and action tremors have been observed in many species after MPTP treatment, a resting tremor, characteristic of PD, is less commonly documented (Tetrud *et al.*, 1986; Hantraye *et al.*, 1993; Raz *et al.*, 2000).

The mechanism of MPTP toxicity has been thoroughly investigated. The meperidine analog MPTP is converted to 1-methyl-4-pyridinium (MPP^+) by monoamine oxidase B. MPP^+ acts as a substrate of the dopamine transporter (DAT) and is selectively taken up by the dopaminergic cells of the SNpc, leading to the inhibition of mitochondrial complex I, the depletion of ATP, and cell death. In mice and non-human primates MPTP selectively destroys dopaminergic neurons of the SNpc, the same neurons affected in PD (Langston *et al.*, 1984; Forno *et al.*, 1993; Jackson-Lewis *et al.*, 1995). Similar to

PD other catecholaminergic neurons, such as those in the VTA and locus coeruleus, may be affected albeit to a lesser degree (Mitchell *et al.*, 1985a; Forno, 1986b; Forno, 1996). In addition, dopamine depletion occurs in both the putamen and caudate nucleus. Whether the putamen or caudate nucleus is preferentially lesioned may depend on animal species and regimen of MPTP administration (Ricaurte *et al.*, 1986; Kalivas *et al.*, 1989; Bezard *et al.*, 2000).

Unlike PD, Lewy Bodies have not been consistently documented. Animal age and route of MPTP administration are two factors that may influence the development of Lewy Bodies. Specifically, eosinophilic inclusions (resembling Lewy Bodies) have been described in aged MPTP-lesioned squirrel and cynomologus monkeys (Forno, 1986a; Kiatipattanasakul *et al.*, 2000). In humans, autosomal recessive juvenile parkinsonism (AR-JP) due to mutations in the *parkin* gene do not develop Lewy Bodies (Shimura *et al.*, 2000) whereas Lewy Bodies are common (and serve as the pathological hallmark) in idiopathic PD in the aged brain (Jellinger, 2001). The time course of MPTP-induced neurodegeneration is rapid, and therefore, represents a major difference from idiopathic PD, which is a chronic progressive disease manifesting over several years. Modifications in the MPTP-lesioning regimen, such as chronic treatment in the presence of probenecid or the delivery of alpha-synuclein via stereotactic targeting of expression vectors have been reported to promote the formation of intracellular protein aggregates reminiscent of Lewy Bodies (Meredith *et al.*, 2002; Eslamboli *et al.*, 2007). The requirement of Lewy Body formation in animal models is an issue of debate since they might be neurotoxic or they might be neuroprotective (Calne *et al.*, 2004; Bodner *et al.*, 2006).

Following the administration of MPTP, the non-human primate progresses through acute (hours), sub-acute (days), and chronic (weeks) behavioral phases of toxicity that are due to the peripheral and central effects of MPTP. The acute phase occurs within minutes after MPTP administration and is characterized by sedation, and a hyper-adrenergic state. This state may also include hyper-salivation, emesis, exaggerated startle, seizure-like activity, and dystonic posturing of trunk and limbs (Jenner *et al.*, 1986; German *et al.*, 1988; Jenner and Marsden, 1988; Petzinger and Langston, 1998; Irwin *et al.*, 1990). The sub-acute phase generally occurs within hours and persists for several days and may be due to the peripheral actions of MPTP on the autonomic nervous system and peripheral organs such as the liver, kidney, and heart (see below) (Petzinger and Langston, 1998). Weight loss, altered blood pressure, and hypothermia may occur, requiring oral gastoral tube feeding and placement in an incubator to stabilize body temperature. In addition, elevated liver transaminases and creatinine phosphokinase may develop; reflecting impaired liver function and muscle breakdown. Behaviorally, these animals may appear prostrate and cognitively impaired. Occasionally, animals may demonstrate self-injurious behavior such as finger biting and hyper-flexion of the neck and trunk with head banging. Assessment of parkinsonian features may be confounded by alterations in the general health of the animal. The chronic phase starts within days to weeks after MPTP administration. It is characterized by the stabilization of body weight and temperature as well as the normalization of blood chemistries such as hepatic enzymes. Parkinsonian features clearly emerge and remain stable for weeks to months or longer. Similar to PD, the MPTP-lesioned non-human primate responds to traditional anti-parkinsonian therapies such as levodopa and dopamine receptor agonists; in some cases severely lesioned animals may require such an intervention to sustain survival. The degree of behavioral stability may be predicted in part by the initial degree of behavioral impairment as observed between the subacute and the chronic phases. Animals with greater behavioral impairments require a longer period of time for recovery. Behavioral improvement after MPTP administration has been reported in most species of non-human primates (see below). In general the behavioral response to MPTP-lesioning may vary both within and between species. One must be cautious in simply extrapolating findings from one species or strain to another. For example, many, but not all, strains of mice are sensitive to the effects of MPTP, whereas rats are almost completely resistant (Riachi, 1988; Zuddas et al., 1994). This variability due to age and species phylogeny also applies to the non-human primate. For example, Old World monkeys (such as rhesus, Macaca mulatta or African Green, Chlorocebus aethiops) tend to be more sensitive to MPTP administration than New World monkeys (such as the squirrel monkey, Saimiri sciureus or marmoset, Callithrix jacchus) (Rose et al., 1993; Ovadia et al., 1995; Gerlach and Reiderer, 1996). Also, within a species, younger animals tend to be more resistant to the effects of MPTP and are more likely to recover from mild lesions compared to older

animals (Ovadia et al., 1995; Collier et al., 2007). There are a number of different reasons that account for the variability in MPTP sensitivity including its metabolism (especially in the liver), ability to cross the blood–brain barrier, conversion of MPTP to the toxic form MPP$^+$ in the brain primarily by astrocytic monoamine oxidase B, uptake into dopaminergic neurons through the DAT, and distribution within cells to target mitochondrial energy depletion and to mediate cell death (Dauer and Przedborski, 2003; Jackson-Lewis and Smeyne, 2005).

THE SUBTYPES OF THE MPTP-LESION NON-HUMAN PRIMATE MODEL

Administrating MPTP through a number of different regimens has led to the development of several distinct models of parkinsonism in the non-human primate that vary in both behavioral and pathological features. Each model is characterized by unique behavioral and neurochemical features, which should be taken into consideration when addressing specific scientific objectives or designing studies. As a result, numerous studies addressing a variety of hypothesis have been carried out in these different models. These studies include the investigation of basal ganglia function including motor behavioral recovery, mechanisms of motor complications, and cognitive impairment. These models also help in the evaluation of new treatment modalities including pharmacological agents, cell transplantation, DBS, and novel neuroprotective and restorative strategies. In some models (such as the intracarotid delivery of MPTP) there is profound striatal dopamine depletion and denervation with few or almost no dopaminergic axons or terminals remaining. This model provides an optimal setting to test fetal tissue grafting since the presence of any tyrosine hydroxylase positive axons or sprouting cells would be due to surviving transplanted tissue or its influence on intrinsic striatal neurons. Other models, such as mild systemic delivery of MPTP, have less extensive dopamine depletion and only partial denervation with a moderate number of dopaminergic axons and terminals remaining. This partially denervated model best resembles mildly to moderately affected PD patients. Therefore, sufficient dopaminergic neurons and axons as well as compensatory mechanisms are likely to be present. Growth factors (inducing sprouting) or neuroprotective factors (promoting cell survival) are best evaluated in this situation. The most commonly used MPTP-lesioning paradigms in non-human primate models include (1) the systemic-lesioned model, (2) hemi-lesioned, (3) bilateral intracarotid, (4) over-lesioned, and (5) low-dose chronic. The following sections briefly highlight the features of these different models. It should be kept in mind that these models are delineated based on the mode of MPTP delivery rather than the species of non-human primate utilized. Other factors will influence these models due to the degree of lesioning including the age, subspecies, and sex of the animals used.

(1) In the *systemic-lesioned model*, MPTP is administered via intramuscular, intravenous, intraperitoneal, or subcutaneous injection (Waters *et al.*, 1987; Eidelberg *et al.*, 1986; Tetrud and Langston, 1989; Elsworth *et al.*, 1990). This model is most common with smaller non-human primates such as the squirrel monkey and marmoset, which can tolerate modest levels of MPTP. Old World monkeys tend to display a high degree of sensitivity toward MPTP and if utilizing systemic administration small doses are typically used and are spread out over several weeks. Large doses result in severe akinesia and may require intensive veterinarian intervention including prolonged hand feeding to rescue animals. In addition, large doses can result in increased risk of animal death from the effects of MPTP and MPP^+ on peripheral organs. Systemic administration of MPTP leads to bilateral depletion of striatal dopamine and nigrostriatal cell death. One feature of this model is that the degree of lesioning can be titrated by adjusting the MPTP concentration administered resulting in a range (mild to severe) of parkinsonian symptoms. The presence of clinical asymmetry in motor features is common with one side more severely affected, but this feature may be subtle. Levodopa or dopamine agonist administration leads to the reversal of all behavioral signs of parkinsonism in a dose-dependent fashion. After several days to weeks of levodopa administration, animals develop reproducible motor complications. The principal advantage of this model is that the behavioral syndrome closely resembles the clinical features of idiopathic PD. The systemic model has partial dopaminergic denervation bilaterally and probably best represents the degree of loss seen in all stages of PD including end-stage disease, where some dopaminergic neurons are still present. This model is well suited for therapeutics that interact with remaining dopaminergic neurons including growth factors, neuroprotective agents, and dopamine agonists. The easily reproducible dyskinesia in this model allows for extensive investigation of its underlying mechanism and treatment (see below). Disadvantages of this model include spontaneous recovery in mildly affected animals, while severely affected animals may require extensive veterinary care and dopamine supplementation, specifically in the early period following MPTP-lesioning. In most cases early interventive care is necessary to overcome the systemic effects of MPTP on peripheral organ systems including the liver, kidneys, and heart which are typically transient (see section on systemic effects of MPTP below).

(2) The *hemi-parkinsonian or hemi-lesioned model* involves administration of MPTP via unilateral intracarotid infusion and has been used to induce a hemi-parkinsonian state in the primate (Bankiewicz *et al.*, 1986). The surgical delivery of MPTP goes directly to the brain, avoiding the systemic effects of MPTP as well as potential xenobiotic metabolism in the liver influencing its bioavailability. The rapid metabolism of MPTP to MPP^+ in the brain may account for the localized toxicity to the hemisphere ipsilateral to the infusion and in some cases can induce stroke-like lesions and necrosis within the basal ganglia (Emborg *et al.*, 2006). Motor impairments appear primarily on the contralateral side. Hemi-neglect, manifested by a delayed motor reaction time, also develops on the contralateral side. In addition, spontaneous ipsilateral rotation may develop. Levodopa administration reverses the parkinsonian symptoms and induces contralateral rotation. SNpc neurodegeneration and striatal dopamine depletion (greater than 99%) on the ipsilateral side to the injection is more extensive than in the systemic model. The degree of unilateral lesioning in this model is dose-dependent. Major advantages of this model include (i) the ability of animals to feed and maintain themselves without supportive care; (ii) the availability of the relatively unaffected limb on the ipsilateral side to serve as a control; and (iii) the utility of the dopamine-induced rotation for pharmacological testing. In addition, due to the absence of dopaminergic innervation in the striatum, the hemi-lesioned model is well suited for examining neuronal sprouting of transplanted cells or tissue. A disadvantage of this model is that only a subset of parkinsonian features are evident and are restricted to one side of the body, a situation never seen in idiopathic PD.

(3) The *bilateral intracarotid model* employs an intracarotid injection of MPTP followed several months later by another intracarotid injection on the opposite side (Smith *et al.*, 1993). This model combines the less debilitating features of the carotid model with bilateral clinical features, a situation more closely resembling idiopathic PD. The advantage of this model is its prolonged stability and limited inter-animal variability. Similar to the hemi-lesioned model, where there is extensive striatal dopamine depletion and denervation, the bilateral intracarotid model is well suited for evaluation of transplanted tissue or vector infusion. However, levodopa administration may result in only partial improvement of parkinsonian motor features and food retrieval tasks. This can be a disadvantage since high doses of test drug may be needed to demonstrate efficacy, increasing the risk for medication-related adverse side effects.

(4) The *over-lesioned model* is a novel approach to MPTP-lesioning and involves the administration of MPTP via intracarotid infusion followed by a systemic MPTP injection (Eberling *et al.*, 1998). This model is characterized by severe dopamine depletion ipsilateral to the MPTP-carotid infusion and a partial depletion on the contralateral side due to the systemic MPTP injection. Consequently, animals are still able to maintain themselves as they retain one relatively intact side. The behavioral deficits consist of asymmetric parkinsonian features. The more severely parkinsonian side is contralateral to the intracarotid injection. Levodopa produces a dose-dependent improvement in behavioral features. The complications of levodopa therapy, however, such as dyskinesia have not been as consistently observed. This model combines some of the advantages of both the systemic and intracarotid MPTP models, including stability. This model is suitable for both transplant studies, utilizing the more depleted side, and neuro-regeneration with growth factors, utilizing the partially depleted side where dopaminergic neurons still remain.

(5) The *chronic low dose model* consists of intravenous injections of very low dose of MPTP administration over a 5-to 13-month period (Bezard *et al.*, 1997; Slovin *et al.*, 1999a, b; Decamp and Schneider, 2004). Rather than primarily targeting motor behavior this model is characterized by cognitive deficits consistent with frontal lobe dysfunction reminiscent of PD or normal-aged monkeys. These animals have impaired attention and short-term memory processes and perform poorly in tasks of delayed response or delayed alternation. Since gross parkinsonian motor symptoms are essentially absent at least in the early stages, this model is well adapted for studying cognitive deficits analogous to those that accompany idiopathic PD.

THE ADVERSE EFFECTS OF MPTP-LESIONING

In addition to its motor effects, MPTP clearly has additional central nervous system effects that may include hypothermia and seizures. The systemic administration of MPTP may also lead to deleterious peripheral effects especially on systemic organs that must be taken into consideration when designing therapeutic studies since they could lead to adverse effects influencing animal behavior, alterations in drug bioavailability, or alterations in drug–target interactions. For example, the peripheral conversion of MPTP to MPP^+, especially in the liver, can result in short-term changes in liver metabolism. Alterations within the liver can influence subsequent MPTP administration sessions (especially those occurring within the first week) and may alter xenobiotic metabolism of MPTP, MPP^+, or the therapeutic drug of interest. This point is illustrated in Figure 9.1. The level of serum MPP^+ was determined by high performance liquid chromatography HPLC analysis after each of six successive subcutaneous MPTP injections at 2-week intervals in the squirrel monkey. Data collected at post-lesioning day 1, 4, and 10 after each injection demonstrates altered serum MPP^+ levels indicating that the peripheral conversion of MPTP to MPP^+ is occurring in a successively higher rate with each injection. Since MPP^+ itself does not cross the blood–brain barrier, there is subsequent reduced bioavailability of the toxin to the brain and therefore less MPTP-mediated cell death. Our analysis of brain tissue early in the injection regimen suggests that the majority of MPTP-induced cell loss of midbrain dopaminergic neurons occurs within the period of the first three injections.

Systemic insult in the non-human primate may also influence motor and non-motor behavior, independent of the central brain lesion itself, since animals that are sick due to MPTP may become very quiescent and disengaged. Furthermore, the peripheral effects on systemic organs, especially the heart, kidneys, and liver, are often responsible for the

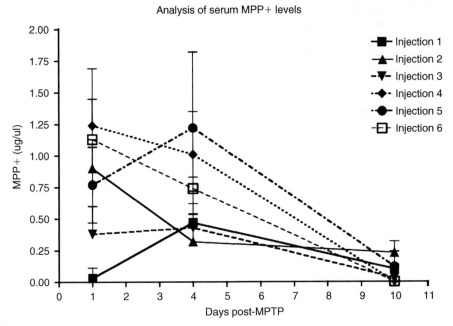

FIGURE 9.1 Altered levels of serum MPP$^+$ levels with successive injections of MPTP. Squirrel monkeys were administered a series of six injections of MPTP (i.p., 2.0 mg/kg free-base, 2 weeks between each injection). Blood was collected at days 1, 4, and 10 after each injection of MPTP and the level of MPP$^+$ determined by HPLC analysis. These data demonstrate that with successive injections of MPTP there is an increase in the serum level of MPP$^+$ indicating systemic conversion of MPTP to MPP$^+$ most likely due to induction of metabolic enzymes in peripheral organs especially the liver. Since MPP$^+$ cannot cross the blood–brain barrier the degree of lesioning in the brain is reduced with later injections of MPTP.

death of animals following MPTP administration. Death can occur within hours of MPTP administration indicating an immediate organ failure, or within the first week due to the inability of animals to eat or drink for themselves. In addition, animals can become cachectic, and can show general wasting and decline without proper intervention. Supportive intervention to avoid this adverse effect includes feeding lesioned non-human primates by gavage with an enriched diet, injection of subcutaneous fluids, or the administration of levodopa to promote movement. Investigators must be cognizant of using dopamine replacement therapy for rescue treatment since such intervention may influence experimental outcome measures.

To address the potential adverse effects of MPTP, we studied issues of the systemic effects of MPTP to gain insight into interventions that would result in greater animal survival. Squirrel monkeys were administered MPTP (in a series of six subcutaneous

injections of 2 mg/kg, free-base, 2 weeks apart) and were given a comprehensive physical examination 1, 4, and 10 days after each injection. The results are summarized in Table 9.1. After the first injection of MPTP, there were significant alterations in a number of physiologic parameters including elevated liver transaminases and elevated creatinine phosphokinase, indicative of liver damage and muscle breakdown, respectively. By the second injection, there was a significant decrease in body weight, which was cumulative with each subsequent injection and tended not to recover without intervention (gavage feeding). Greater than 20% loss of body weight is a significant predictor of animal death. Evidence of hepatocellular toxicity persisted for several weeks after the final MPTP injection. In addition, animals had hypothermia beginning 48 h after lesioning and persisting for up to 10 days after the last MPTP injection. The pathophysiology of these effects may be directly related to MPTP

TABLE 9.1   Systemic effects of MPTP in the non-human primate. Squirrel monkeys received a series of six injections of MPTP (s.c., 2.0 mg/kg, free-base) with 2 weeks between injections. At days 1, 4, and 10 after each of the six MPTP injections animals were subjected to a comprehensive physical exam that included body weight, heart rate, blood pressure, core body temperature, blood cell counts, and comprehensive blood chemistry. Data are grouped according to either the MPTP exposure early (injections 1–3) or late (injections 4–6) as well as post-lesion time early (1 day) and late (10 days). These data demonstrate the variation in systemic parameters at different stage of lesioning

Measures	Change			
	Time		Injection #	
	Early	Late	Early	Late
Cardiovascular				
Heart rate	↑	↓	...	...
Blood pressure	↓	↓	↓	↓
Body weight	↓	↓	↓	↓
Body temperature	↓	↑	↓	↓
Blood cell counts				
White blood Cells	↑	↓	↓	↑
Hemoglobin	↓	↓	↓	↓
Reticulocytes	↑	↓	↓	↑
Hematocrit	↓	↓	↓	↓
Blood chemistry				
Creatine phosphokinase	↑	↑	↑	↑
Alanine transaminase	↑	↑	↑	↑
Aspartate transaminase	↑	↑	↑	↑
Alkaline phosphatase	↑	↑	↑	↑
Creatine	...	...	...	...
Blood urea nitrogen	↓	↓	↓	↓
Sodium	...	...	...	...
Potassium	...	↑	...	↓
Chloride	...	...	...	...

Early time = Day 1; Late time = Day 10; Early injection = 1, 2, 3; Late injection = 4, 5, 6
↑ = Generally increased compared to baseline
↓ = Generally decreased compared to baseline
... = Similar to baseline

itself and/or its metabolites and their adverse effects may persist for several weeks. Overall, body weight and white blood cell count were the key predictors of mortality and should be monitored during MPTP administration. Supplemental caloric intake may be helpful in improving survival. These studies highlight the systemic effects of MPTP on animal models that should be taken into consideration during the design of any pharmacological study. Some researchers may be interested in details of MPTP toxicity and safety as it pertains to risks of exposure to researchers. Using good laboratory practice and general safety procedures MPTP can be utilized with little risk. Researchers may refer to specific technical reviews for additional information (Przedborski et al., 2001; Jackson-Lewis and Przedborski, 2007)

TESTING PHARMACOLOGICAL THERAPIES FOR NEUROPROTECTIVE AND SYMPTOMATIC BENEFIT

The MPTP non-human primate model can be used to test compounds that may provide neuroprotective or symptomatic benefit. In neuroprotective studies, it is extremely valuable to understand the mechanism(s) and time course of toxin-induced nigrostriatal dopaminergic cell death. This information is critical in determining the timing of drug administration relative to toxin exposure, and in providing possible explanations to neuroprotection that may include toxin bioavailability. Neuroprotective interventions are typically started before or during the lesioning phase, and may therefore vary depending on the animal species and timing of toxin-induced cell death. For example, in the non-human primate, such as the squirrel monkey, the half-life of MPTP/MPP^+ clearance is approximately 11 h (Irwin, 1985; Irwin et al., 1990). In contrast, the half-life of MPTP/MPP^+ clearance in the C57BL/6 mouse is approximately 3 h and the time course of nigrostriatal dopaminergic neuron cell death is complete by day 3 post-injection (Irwin, 1989; Jackson-Lewis et al., 1995). Drug-related neuroprotection study may be due to the attenuation of cell death by directly supporting midbrain dopaminergic neurons to promote survival (as seen with neurotrophic factors), or it may be due to reduced insult to the brain by altering the bioavailability of toxin. For example, mechanisms by which a drug may effect MPTP/MPP^+ bioavailability include (i) inhibition of monoamine oxidase B, which would suppress the conversion of MPTP to MPP^+ and (ii) alterations in the pattern of expression of proteins involved in MPP^+ uptake and storage, such as the DAT and vesicular monoamine transporter-2 (VMAT-2), respectively (Gainetdinov, 1997; Staal, 2000). Finally, given that many species of non-human primates demonstrate spontaneous behavioral recovery after MPTP administration, it is important to include lesion-only control animals to assess and compare behavioral recovery that is intrinsic to the model versus that due to the neuroprotective agent (Elsworth, 1989; Petzinger et al., 2006).

The MPTP-lesioned non-human primate model is especially useful for evaluating new symptomatic treatments of PD. The primary utility of the model has been to test compounds for symptomatic relief of motor deficits, including bradykinesia, balance impairment, and freezing. Because cognitive and affective behaviors in the MPTP-lesioned non-human primate have not yet been well characterized, and have limited scales to evaluate treatment, few studies have targeted non-motor features in this model. The testing of symptomatic drug therapies often involves the use of an experimental therapeutic colony of non-human primates that have been rendered parkinsonian after MPTP administration, and have been followed for behavioral stability for weeks to months. The experimental therapeutic colony may undergo testing of numerous and a diverse list of symptomatic therapy that may include drugs that directly or indirectly affect dopamine neurotransmission. These drugs include compounds that act directly on dopamine receptor subtypes within the striatum, including dopamine agonists, or compounds that target other neurotransmitter systems such as adenosine or glutamate that may enhance dopamine neurotransmission either directly (by acting on dopamine receptors themselves) or indirectly (by affecting other dopaminergic parameters such as release or downstream effector pathways). Given the repetitive use of the experimental colony in drug testing, investigators must be cognizant of potential long-lasting effects due to drug exposure that could influence subsequent studies.

The MPTP-lesioned non-human primate has also been valuable in testing potential neurorestorative therapies in PD. The goal of neurorestoration is to re-establish basal ganglia function and improve behavior. Neurorestorative studies include gene/protein delivery via stereotactic targeting or transplantation of genetically engineered or stem/progenitor cells. The different MPTP-lesioned non-human primate models provide contrasting templates to evaluate such interventions. For example, the systemic lesion model typically results in a residual degree of midbrain dopaminergic neuron survival and a partial degree of striatal innervation (Petzinger et al., 2006), that may serve as a template to test neurotrophic factors including glia-derived neurotrophic factor (GDNF) (Kordower et al., 2000; Oiwa et al., 2006). In contrast, the intracarotid lesion paradigm, analogous to the 6-OHDA-lesioned rat, has a near complete depletion of nigrostriatal dopaminergic neurons and their axonal projections. In this model, the lesioned basal ganglia may serve as a "blank slate" where the recovery of any dopamine function may be attributed directly to the transplantation of stem/progenitor cells (Taylor *et al.*, 1991; Sortwell *et al.*, 1998). Specific details of different neurorestorative therapeutic approaches can be found in other related chapters in this book.

The MPTP-lesioned non-human primate model can also help us better understand the potential properties of pharmacological treatments already in clinical use. For example, an interest in our laboratory is to elucidate the underlying mechanisms of intrinsic motor recovery in the squirrel monkey following systemic lesioning with MPTP (Petzinger *et al.*, 2006). It is hypothesized that dopamine could, in fact, act as a neurotrophic factor helping to maintain the integrity of the basal ganglia (Borta and Hoglinger, 2007). We are interested in knowing if dopamine replacement therapy with either levodopa or a dopamine agonist could provide benefit beyond purely symptomatic improvement. In a set of experiments performed in our labs, MPTP-lesioned squirrel monkeys treated with the dopamine agonist pramipexole, and to a lesser extent those treated with levodopa, had higher levels of striatal dopaminergic markers including dopamine and tyrosine hydroxylase, and amphetamine-induced dopamine release than parkinsonian animals treated with saline alone. This occurred despite similar degrees of cell loss based on SNpc counts. These data suggest that dopamine replacement therapy may have a beneficial effect not only on symptomatic treatment of parkinsonian features but also may influence neuroplasticity in the injured basal ganglia. Pharmacological treatment of motor symptoms targeting dopamine replacement may have an analogous effect in patients with PD.

DYSKINESIA AND MOTOR COMPLICATIONS IN NON-HUMAN PRIMATES

In PD, levodopa therapy in many patients leads to the development of motor complications, typically after a few years. The underlying pathogenesis of these complications remains obscure (Vitek and Giroux, 2000; Blanchet *et al.*, 2004; Brotchie, 2005). For levodopa-induced dyskinesias, abnormal involuntary movements induced by levodopa, studies have implicated several neurotransmitter systems, especially the dopaminergic and the glutamatergic

systems. As in patients with PD, the parkinsonian non-human primate also develops complications of levodopa therapy, with both motor fluctuations (wearing-off) and levodopa-induced dyskinesias. In this model, animals develop movements that are abnormal (i.e., they differ phenomenologically from movements that are typically present), but it is impossible to determine whether they are involuntary since we cannot ask the animal its intent in making the movements. We can only judge whether the movements appear purposeful and use this to infer the degree of voluntary control. In any case, the movements bear a striking resemblance to levodopa-induced dyskinesias observed in patients with PD. For example, dyskinetic movements in the MPTP-lesioned squirrel monkey or marmoset involve all four limbs and the trunk, with choreoathetoid movements that develop a few minutes after a dose of levodopa and last 3–4h. The time course for the movements corresponds to the time course for reversal of MPTP-induced bradykinesia. Observation of the phenomenology of the movements has led to rating scales that have been developed by a number of different groups for squirrel monkeys, marmosets, and macaques (Brotchie and Fox, 1999; Petzinger *et al.*, 2001; Chassain *et al.*, 2001; Tan *et al.*, 2002).

The effect of pharmacologic agents upon levodopa-induced dyskinesias have been studies for a variety of agents, including D2 dopamine receptor agonists (Calon *et al.*, 1995; Smith *et al.*, 2006), D3 dopamine receptor partial agonists (Hsu *et al.*, 2004; Smith *et al.*, 2006), dopamine receptor antagonists (Andringa *et al.*, 1999), A2A-adenosine receptor antagonists (Kanda *et al.*, 1998; Blandini, 2003), opioid receptor agents (Fox *et al.*, 2002; Samadi *et al.*, 2003; Samadi *et al.*, 2004; Cox *et al.*, 2007), and glutamate receptor antagonists (Papa and Chase, 1996; Verhagen-Metman *et al.*, 1998; Samadi *et al.*, 2007). Unfortunately, the search for an effective suppressor of levodopa-induced dyskinesias has had limited success, although some studies have provided clues for a useful therapeutic strategy; for example, minimizing wide fluctuations in the delivery of levodopa to the striatum. Although the underlying mechanism of levodopa-induced dyskinesias is unknown, electrophysiological, neurochemical, molecular, and neuro-imaging studies in non-human primate models suggest that the pulsatile delivery of levodopa may lead to a variety of changes in the post-synaptic cell and in other regions of the basal ganglia that are further downstream, giving rise to levodopa-induced dyskinesias. These changes could include (i) changes in the neuronal firing rate and pattern of the globus pallidus and subthalamic nucleus (STN); (ii) enhancement of D1 and/or D2 dopamine receptor mediated signal transduction pathways; (iii) super-sensitivity of the D2 receptor; (iv) alterations in the phosphorylation state or subcellular localization of glutamate receptors; (v) modifications in dopamine receptor subtypes and their functional links; and (vi) enhancement of opioid-peptide-mediated neurotransmission (Bedard et al., 1992; Papa and Chase, 1996; Bezard et al., 2001; Hurley et al., 2001; Calon et al., 2002).

In designing studies to examine potential treatments for suppressing levodopa-induced dyskinesias, it is necessary to examine the effect on both the dyskinesias and the parkinsonism, as mentioned above. A drug that improves dyskinesias at the expense of worsening parkinsonism will be of limited utility as a treatment in patients. It is also important to keep in mind that dyskinesias differ phenomenologically and possibly mechanistically in different primate species. For example, such a difference might exist for the occurrence of facial dyskinesias, which have been observed in Old World monkeys but not in New World monkeys (Petzinger et al., 2001; Tan et al., 2002).

The presence of a nigral lesion has long been considered a necessary prerequisite for the development of levodopa-induced dyskinesias with the intensity dependent on the degree of lesioning (Di Monte et al., 2000; Schneider et al., 2003; Kuoppamaki et al., 2007). Recent studies have challenged this dogma and it has been reported that non-human primates without any dopaminergic lesions can manifest levodopa-induced dyskinesias. For example, when administered sufficiently large doses of levodopa, dyskinesias can develop in squirrel monkeys within a few days (Togasaki et al., 2001), and in marmosets within 8 weeks (Pearce et al., 2001). The relatively high doses of levodopa administered to these animals may serve to exhaust the buffering capacity of the dopaminergic terminals within the striatum and accentuate the pulsatile deliver of dopamine to the post-synaptic receptors of the normal animal, thus giving rise to dyskinesias.

INTRINSIC NEUROPLASTICITY AND BEHAVIORAL RECOVERY

Understanding the molecular mechanisms underlying behavioral recovery in the non-human primate may provide insights into neuroplasticity of the brain after injury, help identify new therapeutic targets for treatment of PD, and provide an opportunity to elucidate basal ganglia function (Zigmond et al., 1990; Zigmond, 1997; Bezard and Gross, 1998; Jakowec et al., 2003; Jakowec et al., 2004). Behavioral recovery after MPTP-induced parkinsonism has been reported in both New World and Old World non-human primates (Eidelberg et al., 1986; Kurlan et al., 1991; Scotcher et al., 1991; Schneider et al., 1994, 1995; Cruikshank and Weinberger, 1996; Petzinger and Langston, 1998; Oiwa et al., 2003; Petzinger et al., 2006). The degree and time course of behavioral recovery is dependent on age, species, and mode of MPTP administration (Albanese et al., 1993; Ovadia et al., 1995; Taylor et al., 1997; Petzinger and Langston, 1998). In general, severely lesioned animals are less likely to recover than mildly lesioned animals, and intracarotid injection models are less likely to recover than systemic models (Taylor et al., 1997). For example, squirrel monkeys rendered severely parkinsonian by a series of six subcutaneous injections of MPTP over a 12-week period recover motor behavior within several months of their last injection (Petzinger et al., 2006). Figure 9.2 demonstrates differences in the time course of recovery of motor behavior depending on the degree of MPTP-lesioning. In contrast to systemic MPTP-lesioning, non-human primates administered MPTP via intracarotid injection tend not to recover motor behavior and may display stable parkinsonian features for months to years after lesioning. These differences in recovery between the various models likely reflect differences in the degree of midbrain dopaminergic cell loss. With systemic administration of MPTP, a proportion of SNpc dopaminergic neurons are spared (between 40% and 60%), whereas in the intracarotid paradigm almost no midbrain dopaminergic neurons remain ipsilateral to the site of injection. In fact, intracarotid lesioning can be so intense as to cause necrosis and stroke in the ipsilateral striatum (Emborg et al., 2006).

Studies investigating the mechanisms of recovery in these models have shown (i) alterations in dopamine biosynthesis (tyrosine hydroxylase) and metabolism (increased turnover); (ii) altered regulation of DAT expression and function; (iii) sprouting and branching of tyrosine hydroxylase fibers; (iv) alterations of other neurotransmitter systems including glutamate and serotonin; and (v) alterations of signal transduction pathways in both the

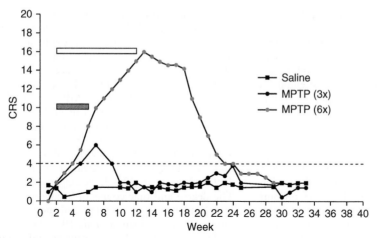

FIGURE 9.2   Time course in motor recovery in the MPTP-lesioned non-human primate model. Squirrel monkeys (N = 6 per group) were administered saline, three or six injections of MPTP (2.0 mg/kg free-base, 2 weeks between injections). A CRS was administered to monitor parkinsonian features (Petzinger *et al.*, 2006). A score greater than 4 is considered the threshold for parkinsonian motor features. Animals receiving three injections of MPTP displayed mild transient parkinsonian features for only a few weeks while those receiving six injections showed moderate to severe parkinsonian features and showed full motor behavioral recovery 12 weeks after the last injection of MPTP. The open and gray bars represent the time frame of 6 and 3 injections of MPTP, respectively.

direct (D1 dopamine receptor) and indirect (D2 dopamine receptor) pathways (Chiueh, 1984a; Eidelberg *et al.*, 1986; Mori *et al.*, 1988; Nishi *et al.*, 1989; Rose *et al.*, 1989; Morgan *et al.*, 1991; Russ *et al.*, 1991; Cruz-Sanchez *et al.*, 1993; Frohna *et al.*, 1995; Bezard and Gross, 1998; Ho and Blum, 1998; Mitsumoto *et al.*, 1998; Rozas *et al.*, 1998; Rothblat *et al.*, 2001; Wade *et al.*, 2001; Jakowec *et al.*, 2004). Interestingly, the return of striatal dopamine is incomplete despite full motor recovery, although more pronounced return in the ventral striatum compared to the dorsal regions has been reported in several different MPTP-lesioned species (Elsworth *et al.*, 2000; Petzinger *et al.*, 2006). In the squirrel monkey we found that dopamine levels in tissue homogenate increased from 0.7% to 1.6% of baseline in the dorsal putamen at 6 weeks (when animals are moderately parkinsonian) and 9 months (when animals are fully recovered from motor impairment) (Petzinger *et al.*, 2006). However, in the ventral putamen dopamine levels at the same time points were 8.7% and 28.0% of baseline, respectively. One explanation may be that the ventral dopaminergic system is less sensitive to MPTP toxicity (Moratalla *et al.*, 1992). This raises the possibility that the ventral striatum, with a higher dopamine return than the dorsal, may

play a role in behavioral recovery and one means may be through the diffusion of dopamine into the dorsal denervated regions (Schneider *et al.*, 1994). Even with insufficient total dopamine return, studies in the rodent models have shown that when dopamine loss is less than 80%, homeostatic mechanisms can lead to complete normalization of extracellular levels of dopamine; however, when dopamine depletion exceeds 80% there is only partial normalization (Robinson and Wishaw, 1988; Altar and Marien, 1989; Abercrombie *et al.*, 1990; Castaneda *et al.*, 1990). In our studies we also observed a dynamic change in protein expression in the caudate nucleus and putamen, where animals at 9 months after lesioning compared to animals at 6 weeks, showed increased levels of tyrosine hydroxylase and DAT protein that was more dramatic in the ventral than dorsal regions of the basal ganglia. Studies suggest that there are many pre- and post-synaptic molecular changes that occur as a consequence of dopamine denervation/dysfunction which may contribute to behavioral recovery and/or play a role in other phenomenology such as susceptibility to levodopa-induced dyskinesias. The fact that there are molecular and neurochemical changes that occur in a time course fashion in these models underscores the importance of considering the time from lesioning as an important parameter in study design. Another cautionary point is that many of these molecular changes may take place in a time course fashion even in animals that do not display overt behavioral recovery. This raises the issue that studies should consider time since lesioning as a potential factor in influencing outcome measures when examining molecular and physiological changes.

ELECTROPHYSIOLOGICAL STUDIES OF BASAL GANGLIA FUNCTION IN THE NON-HUMAN PRIMATE MODEL OF PD

Electrophysiological studies in the normal and MPTP-lesioned non-human primate have provided valuable information regarding basal ganglia physiology and pathophysiology of PD and have led to the identification and testing of new therapeutic interventions. The utility of the MPTP-lesioned non-human primate is due in part to the following strengths: (i) the non-human primate shares similar basal ganglia structure and circuitry with humans and (ii) the MPTP-lesioned model demonstrates analogous pathological and clinical characteristics to idiopathic PD (Langston *et al.*, 1983). For example, electrophysiological studies in the MPTP-lesioned non-human primate have provided evidence for increased activity in both the STN and globus pallidus and have suggested that abnormalities in the circuitry between nuclei of the basal ganglia underlie parkinsonian features. These findings supported the hypothesis that inactivation of the STN alleviates the parkinsonian features in the non-human primate and eventually led to the use of DBS surgery of the STN as a treatment for PD (Israel and Bergman 2008).

Electrophysiological studies in MPTP-lesioned non-human primate have also been valuable in examining alterations in the physiological properties of neurons and their molecular modulators within the basal ganglia and have identified new pharmacological targets. These studies have included the investigation of dopamine, glutamate, γ-aminobutyric acid (GABA), nicotine, and adenosine and their respective receptors, their influence on the physiological properties of medium spiny neurons of the striatum, and their effects on motor (including dyskinesia) and cognitive behavior. For example, neurophysiological studies have implicated overactivity

at corticostriatal synapses, due to altered glutamatergic neurotransmission, as one underlying mechanism for the development of motor impairment and levodopa related motor complications in PD. (Konitsiotis *et al.*, 2000; Wichmann and DeLong, 2003; Soares *et al.*, 2004). These findings have supported the clinical use of glutamatergic antagonists, such as amantadine, for the treatment of levodopa-induced dyskinesia in PD, and the concept that glutamatergic antagonists with receptor and anatomical specificity may provide a future therapeutic intervention for symptomatic treatment.

Electrophysiological studies in our labs, using the MPTP-lesioned squirrel monkey, have shown changes in α-amino-3-hydroxy-5-methyl-4-isoxazolepropionate (AMPA) and GABA mediated synaptic neurotransmission that may account for excessive excitatory corticostriatal drive. For these studies, we administered MPTP in a series of six subcutaneous injections of 2.0 mg/kg (free-base) every 2 weeks for a total of 12 mg/kg. Whole brains were harvested at either 6 weeks (when animals are parkinsonian) or 9 months (when animals are motorically recovered) after the last injection of MPTP and striatal synaptic function was examined in coronal *in vitro* brain slices. We found that the input/output relationship was greater for AMPA-receptor-mediated synaptic currents at 6 weeks after MPTP-lesioning compared to saline control using whole cell voltage clamp. The relative strength of $GABA_A$-receptor-mediated synaptic inhibition versus AMPA-receptor-mediated synaptic excitation response was calculated as the I_{GABA-A} changes in the I_{GABA-A}/I_{AMPA} ratio. Interestingly, we also found a reduced I_{GABA-A}/I_{AMPA} ratio 6 weeks after MPTP. These GABAergic inhibition that we and others have observed may play an important role in facilitating the synchrony and oscillatory patterns of discharge found throughout the basal ganglia motor circuit in MPTP-treated akinetic primates (Raz *et al.*, 1996; Raz *et al.*, 2001; Goldberg *et al.*, 2002). Analysis of animals 9 months after MPTP administration suggests that there is normalization of corticostriatal hyperactivity when animals demonstrate full behavioral recovery. Specifically we found the input/output ratio for AMPA-receptor-mediated synaptic responses and the I_{GABA-A}/I_{AMPA} ratio returned back to control levels (Figure 9.3). These observations are in agreement with the view that excessive glutamatergic corticostriatal synaptic function may be a contributing factor to the behavioral pathology of PD (Konitsiotis *et al.*,

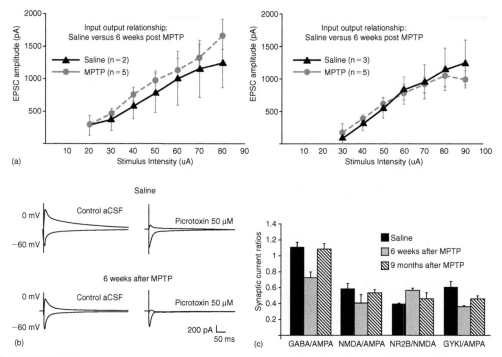

FIGURE 9.3 Electrophysiological evidence for MPTP-induced changes in synaptic transmission. (a) Input (stimulus intensity applied to corpus callosum) – output (excitatory post-synaptic current or EPSC amplitude) relationships were determined for cortico-putamen synapses using whole cell voltage clamp with GABA$_A$ receptor blocked by picrotoxin. Note a greater tendency for larger amplitude EPSCs at 6 weeks post-MPTP (left panel), with a return to normal amplitude EPSCs by 9 months post-MPTP (right panel). (b) Example of synaptic currents recorded from a putamen neuron in response to corpus callosum stimulation before and after addition of picrotoxin. Synaptic currents were evoked in cells clamped at membrane potentials of −60 mV to maximize AMPA current activation and 0 mV to maximize GABA$_A$ current activation. (c) Ratios of synaptic currents illustrate a shift in synaptic function. At 6-week post-MPTP there was a reduction in the GABA$_A$/AMPA ratio, a reduction in the NMDA/AMPA ratio, an increase in the NR2B/total NMDA ratio, and a decrease in the GYKI 52466 sensitive/CNQX sensitive AMPA ratio. Interestingly, these trends returned to saline control levels by 9 months post-MPTP.

2000; Muriel *et al.*, 2001). Future studies will exam whether changes in glutamatergic drive in fully recovered animals differentially impacts corticostriatal synapses in direct versus indirect basal ganglia pathways, as has been reported in the parkinsonian state (Wichmann and DeLong, 2003; Day *et al.*, 2006). This could represent an additional consideration for therapeutic targeting.

The AMPA and N-methyl-D-aspartate (NMDA) receptors play a key role in determining the physiological properties of medium spiny neurons of the striatum. In the MPTP-lesioned non-human primate we observed changes in the pharmacological profile of AMPA and NMDA receptors, which are

consistent with previously reported molecular studies in the dopamine denervated striatum (Betarbet *et al.*, 2000; Betarbet *et al.*, 2004; Nash *et al.*, 2004; Hallett *et al.*, 2005; Hurley *et al.*, 2005). For example as shown in Figure 9.3, animals examined 6 weeks post-MPTP-lesioning displayed (i) a decrease in the I_{NMDA}/I_{AMPA} ratio; (ii) an alteration in the NMDA receptor subunit composition as indicated by increased sensitivity to the selective NMDA-NR2B antagonist CP-101606; and (iii) an alteration in AMPA-receptor-mediated synaptic responses, as indicated by changes in the sensitivity to the selective AMPA receptor antagonist, GYKI-52466 compared to saline control animals (Ruel *et al.*, 2002; Nash *et al.*, 2004). Following behavioral recovery at 9 months post-MPTP-lesioning, there was a normalization of NMDA and AMPA receptor function toward saline control (Figure 9.3).

The glutamatergic corticostriatal and the dopaminergic nigrostriatal system are important mediators of synaptic plasticity, termed long-term depression (LTD) and long-term potentiation (LTP), within the basal ganglia (Centonze *et al.*, 2001; Reynolds and Wickens, 2002; Mahon *et al.*, 2004; Picconi *et al.*, 2005). Electrophysiological studies in our lab, using saline control squirrel monkeys, have shown that the induction of long-term synaptic plasticity at corticostriatal synapses is region specific, with LTP being induced in more medial regions and LTD in more lateral regions. These findings agree with previous reports from the rodent model of PD (Partridge *et al.*, 2000; Smith *et al.*, 2001). Studies in the rat have shown a loss of synaptic plasticity after 6-OHDA administration, which we have observed in the MPTP-lesioned mouse model, 1–2 weeks after neurotoxicant exposure (Calabresi *et al.*, 1992; Centonze *et al.*, 1999; Kreitzer and Malenka, 2007). Presently, there is little known regarding alterations in synaptic plasticity immediately following MPTP-lesioning in the non-human primate.

Analysis of the expression of synaptic plasticity in the squirrel monkey 9 months after MPTP-lesioning has shown that LTD and LTP expression is evident. In the same animals used for analysis of glutamate neurotransmission above, we observed a dramatic and permanent decrease in dopamine release as measured by fast-scan cyclic voltammetry (Cragg, 2003) (Figure 9.4). This finding is in agreement with previous reports examining dopamine function in the squirrel monkey using HPLC (Petzinger *et al.*, 2006). The expression of dopamine-dependent forms of LTP we observed in the dopamine-depleted squirrel monkey suggest that an adaptation may occur in the expression and/or sensitivity of both D1 and D2 dopamine receptors (Centonze *et al*, 2001; Reynolds and Wickens, 2002; Mahon *et al.*, 2004; Picconi *et al.*, 2005). Preliminary studies in our lab have shown that LTD expression at lateral cortico-putamen synapses from the 9-month MPTP-lesioned squirrel monkey is D2 dopamine receptor dependent, since this effect is blocked by the D2 dopamine receptor antagonist *l*-sulpiride. In addition, use of *l*-sulpiride results in the unexpected expression of LTP in lateral synapses (Figure 9.4). Our findings are consistent

with the literature, where D1 and D2 dopamine receptors have been shown to play an important role in LTP and LTD, respectively (Calabresi *et al.*, 1992; Centonze *et al.*, 1999; Wang *et al.*, 2006). Taken together, these data suggest behavioral recovery from MPTP exposure in the squirrel monkey may be due at least in part to compensatory increases in the sensitivity of dopamine receptors, which enables the normal and expected expression of long-term plasticity at corticostriatal synapses. The studies outlined above demonstrate how the MPTP-lesioned non-human primate model provides valuable insights regarding the role that neurotransmitters and their respective signaling pathways play in modulating the electrophysiological properties of basal ganglia neurons. These findings are serving to identify new therapeutic targets for treatment of PD.

OTHER NEUROTOXICANTS IN NON-HUMAN PRIMATES

This chapter has focused on MPTP as the neurotoxicant to generate parkinsonism in the non-human primate since this is the most common regimen. Other neurotoxicants such as 6-hydroxydopamine, METH, proteasome inhibitors, and pesticides, while most often used in rodent models have had limited utility in non-human primates. The following sections highlight some features of these models.

6 Hydroxydopamine

6-hydroxydopamine (6-OHDA or 2,4,5-trihydroxyphenylethylamine) is a specific catecholaminergic neurotoxin structurally analogous to both dopamine and noradrenalin. In addition to the rat, other species including the non-human primate (specifically the marmoset) have served as models for 6-OHDA lesioning (Annett *et al.*, 1992; Eslamboli, 2005). Lesioning in non-human primates provides for the analysis of behaviors not observed in the rat, such as targeting and retrieval tasks of the arm and hand. This model, however, has not gained popularity for non-human primates because the toxin must be delivered directly in the vicinity of the dopamine cells by intracerebral injections. This is much more difficult method than administering MPTP systemically.

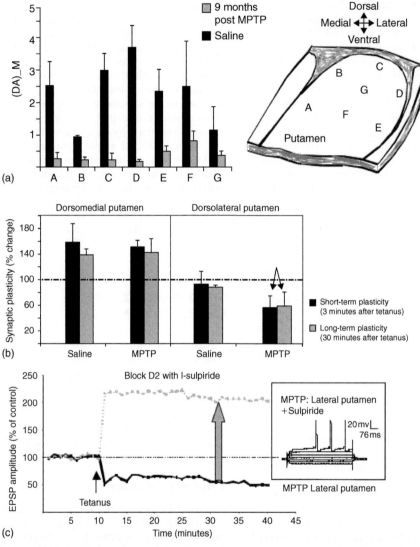

FIGURE 9.4 Changes in dopamine release and synaptic plasticity in the MPTP-lesioned non-human primate. (a) Fast-scan cyclic voltammetry revealed regional differences in evoked dopamine release in the putamen that is markedly reduced even after 9 months post-MPTP when animals are motorically recovered. Letters in the graph correspond to putamen brain slice sites. (b) Comparison of short-term (3 min post-tetanus) and long-term (30 min post-tetanus) synaptic plasticity at corticostriatal synapses from saline and 9 post-MPTP-lesioning. Intracellular recording of EPSPs Excitatory Post-Synaptic potential evoked at cortico-putamen synapses was used to monitor changes in strength induced by tetanic activation of the corpus callosum. Medial cortico-putamen synapses produced short- and long-term potentiation (LTP) in both groups. Lateral cortico-putamen synapses expressed short- and long-term depression (LTD) in saline and MPTP exposed monkeys, but the MPTP group tended toward greater LTD. (c) Example of LTD induced at lateral cortico-putamen synapses and tetanic activation of lateral cortico-putamen synapses in the presence of the D2 dopamine receptor antagonist l-sulpiride that enabled the expression of profound LTP. Insert shows the response of the putamen neuron recorded during the l-sulpiride experiment to current injection, which is typical of a medium spiny projection neuron.

Methamphetamine

Amphetamine and its derivatives (including METH, N-methyl-beta-phenylisopropylamine) lead to long-lasting depletion of both dopamine and serotonin when administered to rodents and non-human primates including vervet, macaques, squirrel monkeys, and baboons (Ricaurte et al., 1980; Ricaurte et al., 1982; Villemagne et al., 1998; Davidson et al., 2001; Czoty et al, 2004). METH, one of the most potent of these derivatives, is typically administered in a series of small intramuscular or oral doses from 0.1 to 2 mg/kg and leads to dose-dependent terminal degeneration of dopaminergic neurons in the caudate nucleus and putamen, nucleus accumbens, and neocortex. Despite the severe depletion of striatal dopamine, the motor behavioral alterations seen in rodents and non-human primates tend to be transient and subtle.

In contrast to MPTP, which destroys nigrostriatal dopaminergic neurons and their terminals, METH administration spares axonal trunks and soma of SNpc and VTA neurons targeting terminals found within the caudate nucleus and putamen (Kim et al., 2000). Depending on the species and dosing regimen of METH, the effects of lesioning involves a spectrum from axonal degeneration to suppression of markers of nigrostriatal neuron integrity including tyrosine hydroxylase, DAT, and VMAT-2 proteins. The fact that these markers can be differentially affected by METH indicates that phenotypic suppression in dopaminergic neurons is a significant feature of METH exposure. In general, the effects of severe METH lesioning are long lasting. Interestingly, there is evidence of recovery of dopaminergic system depending on the METH regimen and species used (Harvey et al., 2000a). Studies employing PET-imaging in conjunction with histological analysis of markers of the dopaminergic system have demonstrated that re-establishment of the nigrostriatal system occurs which probably involves a combination of re-innervation (neuronal sprouting) and return of previously suppressed TH and DAT protein expression (Melega et al., 1997; Harvey et al., 2000b).

Similar to studies with MPTP, METH administration demonstrates the dynamic neuroplasticity of the nigrostriatal system and its ability to respond to neurotoxic injury. The administration of METH to adult animals has played an important role in testing the molecular and biochemical mechanisms underlying dopaminergic and serotonergic neuronal axonal degeneration especially the role of free radicals and glutamate neurotransmission. Understanding these mechanisms has led to the testing of different neuroprotective therapeutic modalities. An advantage of the METH model over MPTP is that the serotonergic and dopaminergic systems can be lesioned in utero during the early stages of the development of these neurotransmitter systems. Such studies have indicated that there is a tremendous degree of architectural rearrangement that occurs within the dopaminergic and serotonergic systems of injured animals as they develop. These changes may lead to altered behavior in the adult animal (Frost and Cadet, 2000). And finally, in light of the fact that METH and other substituted amphetamines (including methylenedioxymethamphetamine (MDMA) "ecstasy") are major drugs of abuse in our society, animal models have provided a means to understand the mechanisms of brain injury with these toxic compounds and to determine the long-lasting effects of these drugs including if humans who abuse METH are prone to develop parkinsonism (McCann et al., 1998; Guilarte, 2001; Paulus et al., 2002).

Proteasome Inhibition

It has been proposed that inhibition of the ubiquitin proteasome system (UPS) can lead to the inability to remove toxic protein moieties, accumulation of protein aggregates, neuronal dysfunction, and cell death (Petrucelli and Dawson, 2004; Tanaka et al., 2004). However, thus far modeling PD via a systemic administration of proteasome inhibitor has produced unreliable and irreproducible results (Kordower et al., 2006; Manning-Bog et al., 2006; Beal and Lang, 2006; Bove et al., 2006).

CONCLUSION

To successfully translate findings in the laboratory to the clinical setting it is critical that novel experimental therapeutic approaches be evaluated in the non-human primate. Studies in normal animals can evaluate safety and tolerability issues, while studies in models of PD can test potential efficacy of pharmacological, surgical, and molecular approaches. Despite the fact that neurotoxicant models do not replicate all the pathological features seen in patients with PD, studies in non-human primate models can help us better understand the human condition by providing a template to test hypotheses. For example, understanding why a therapeutic approach shown to be efficacious in rodent models but fails in patients with PD can be addressed in non-human primates and may reveal previously unknown or unrecognized features of the disease. The non-human primate serves as an important link bridging the phylogenetic continuum between rodents and *Homo sapiens*. In finding new therapeutic modalities for the treatment of neurological disorders such as PD many researchers feel that the non-human primate provides an essential model to validate findings in the preclinical phase.

ACKNOWLEDGEMENTS

We would like to acknowledge the generous support of the Parkinson's Disease Foundation, Team Parkinson LA, and the Whittier PD Education Group, NIH RO1 NS44327-1 and RO1 AG21937, U.S. Army NETRP (Grant # W81XWH-04-1-0444). A special thanks to Friends of the USC Parkinson's Disease Research Group including George and Mary Lou Boone, and Walter and Susan Doniger. We thank our colleagues including Elizabeth Hogg, Pablo Arevalo, Jon VanLeeuen, Marta Vuckovic, Erlinda Kirkman, and Charles Meshul for their insightful discussions. Special thanks to Nicolaus, Pascal, and Dominique for their support.

REFERENCES

Abercrombie, E. D., Bonatz, A. E., and Zigmond, M. J. (1990). Effects of L-DOPA on extracellular dopamine in striatum of normal and 6-hydroxydopamione-treated rats. *Brain Res* 525, 36–44.

Albanese, A., Granata, R., Gregori, B., Piccardi, M. P., Colosimo, C., and Tonali, P. (1993). Chronic administration of 1-methyl-4-phenyl-1,2,3,6-tetrahydropyridine to monkeys: Behavioural, morphological and biochemical correlates. *Neuroscience* 55, 823–832.

Almirall, H., Pigarev, I., de la Calzada, M. D., Pigareva, M., Herrero, M. T., and Sagales, T. (1999). Nocturnal sleep structure and temperature slope in MPTP treated monkeys. *J Neural Transm* 106, 1125–1134.

Altar, C. A., and Marien, M. R. (1989). Preservation of dopamine release in the denervated striatum. *Neurosci Lett* 96, 329–334.

Andringa, G., Stoof, J. C., and Cools, A. R. (1999). Subchronic administration of the dopamine D(1) antagonist SKF 83959 in bilaterally MPTP-treated rhesus monkeys: Stable therapeutic effects and wearing-off dyskinesia. *Psychopharmacology (Berl)* 146, 328–334.

Annett, L. E., Rogers, D. C., Hernandez, T. D., and Dunnett, S. B. (1992). Behavioral analysis of unilateral monoamine depletion in the marmoset. *Brain* 115, 825–856.

Bankiewicz, K. S., Oldfield, E. H., Chiueh, C. C., Doppman, J. L., Jacobowitz, D. M., and Kopin, I. J. (1986). Hemiparkinsonism in monkeys after unilateral internal carotid infusion of 1-methyl-4-phenyl-1,2,3,6-tetrahydropyridine. *Life Sci* 39, 7–16.

Beal, F., and Lang, A. (2006). The proteasomal inhibition model of Parkinson's disease: "Boon or bust"? *Ann Neurol* 60, 158–161.

Bedard, P. J., Mancilla, B. G., Blanchette, P., Gagnon, C., and Di Paolo, T. (1992). Levodopa-induced dyskinesia: Facts and fancy. What does the MPTP monkey model tell us? *Can J Neurol Sci* 19, 134–137.

Betarbet, R., Porter, R. H., and Greenamyre, J. T. (2000). GluR1 glutamate receptor subunit is regulated differentially in the primate basal ganglia following nigrostriatal dopamine denervation. *J Neurochem* 74, 1166–1174.

Betarbet, R., Poisik, O., Sherer, T. B., and Greenamyre, J. T. (2004). Differential expression and ser897 phosphorylation of striatal N-methyl-D-aspartate receptor subunit NR1 in animal models of Parkinson's disease. *Exp Neurol* 187, 76–85.

Bezard, E., and Gross, C. (1998). Compensatory mechanisms in experimental and human parkinsonism: Towards a dynamic approach. *Prog Neurobiol* 55, 96–116.

Bezard, E., Imbert, C., Deloire, X., Bioulac, B., and Gross, C. E. (1997). A chronic MPTP model reproducing the slow evolution of Parkinson's disease: Evolution of motor symptoms in the monkey. *Brain Res* 766, 107–112.

Bezard, E., Dovero, S., Imbert, C., Boraud, T., and Gross, C. E. (2000). Spontaneous long-term compensatory dopaminergic sprouting in MPTP-treated mice. *Synapse* 38, 363–368.

Bezard, E., Brotchie, J. M., and Gross, C. E. (2001). Pathophysiology of levodopa-induced dyskinesia: potential for new therapies. *Nat Rev Neurosci* 2, 577–588.

Blanchet, P. J., Calon, F., Morissette, M., Tahar, A. H., Belanger, N., Samadi, P., Grondin, R., Gregoire, L., Meltzer, L., Paolo, T. D., and Bedard, P. J. (2004). Relevance of the MPTP primate model in the study of dyskinesia priming mechanisms. *Parkinsonism Relat Disord* 10, 297–304.

Blandini, F. (2003). Adenosine receptors and L-DOPA-induced dyskinesia in Parkinson's disease: Potential targets for a new therapeutic approach. *Exp Neurol* **184**, 556–560.

Bodner, R. A., Outeiro, T. F., Altmann, S., Maxwell, M. M., Cho, S. H., Hyman, B. T., McLean, P. J., Young, A. B., Housman, D. E., and Kazantsev, A. G. (2006). Pharmacological promotion of inclusion formation: A therapeutic approach for Huntington's and Parkinson's diseases. *Proc Natl Acad Sci USA* **103**, 4246–4251.

Borta, A., and Hoglinger, G. U. (2007). Dopamine and adult neurogenesis. *J Neurochem* **100**, 587–595.

Bove, J., Zhou, C., Jackson-Lewis, V., Taylor, J., Chu, Y., Rideout, H. J., Wu, D. C., Kordower, J. H., Petrucelli, L., and Przedborski, S. (2006). Proteasome inhibition and Parkinson's disease modeling. *Ann Neurol* **60**, 260–264.

Brotchie, J. M. (2005). Nondopaminergic mechanisms in levodopa-induced dyskinesia. *Mov Disord* **20**, 919–931.

Brotchie, J. M., and Fox, S. H. (1999). Quantitative assessment of dyskinesia in subhuman primates. *Mov Disord* **14**, 40–47.

Burns, R. S., Chiueh, C. C., Markey, S. P., Ebert, M. H., Jacobowitz, D. M., and Kopin, I. J. (1983). A primate model of parkinsonism: Selective destruction of dopaminergic neurons in the pars compacta of the substantia nigra by N-methyl-4-phenyl-1,2,3,6-tetrahydropyridine. *Proc Natl Acad Sci USA* **80**, 4546–4550.

Calabresi, P., Maj, R., Pisani, A., Mercuri, N. B., and Bernardi, G. (1992). Long-term synaptic depression in the striatum: Physiological and pharmacological characterization. *J Neurosci* **12**, 424–4233.

Calne, D. B., and Mizuno, Y. (2004). The neuromythology of Parkinson's Disease. *Parkinsonism Relat Disord* **10**, 319.

Calon, F., Goulet, M., Blanchet, P. J., Martel, J. C., Piercey, F., Bedard, P. J., and Di Paolo, T. (1995). Levodopa or D2 agonists induce dyskinesia in MPTP monkeys: Correlation with changes in dopamine and GABA A receptors in the striatopallidal complex. *Brain Res* **680**, 43–52.

Calon, F., Morissette, M., Ghribi, O., Goulet, M., Grondin, R., Blanchet, P. J., Bedard, P. J., and Di Paolo, T. (2002). Alteration of glutamate receptors in the striatum of dyskinetic 1-methyl-4-phenyl-1,2,3,6-tetrahydropyridine-treated monkeys following dopamine agonist treatment. *Prog Neuropsychopharmacol Biol Psychiatr* **26**, 127–138.

Castaneda, E., Whishaw, I. Q., and Robinson, T. E. (1990). Changes in striatal dopamine neurotransmission assessed with microdialysis following recovery from a bilateral 6-OHDA lesion: Variation as a function of lesion size. *J Neurosci* **10**, 1847–1854.

Centonze, D., Gubellini, P., Picconi, B., Calabresi, P., Giacomini, P., and Bernardi, G. (1999). Unilateral dopamine denervation blocks corticostriatal LTP. *J Neurophysiol* **82**, 3575–3579.

Centonze, D., Picconi, B., Gubellini, P., Bernardi, G., and Calabresi, P. (2001). Dopaminergic control of synaptic plasticity in the dorsal striatum. *Eur J Neurosci* **13**, 1071–1077.

Chassain, C., Eschalier, A., and Durif, F. (2001). Assessment of motor behavior using a video system and a clinical rating scale in parkinsonian monkeys lesioned by MPTP. *J Neurosci Method* **111**, 9–16.

Chefer, S. I., Kimes, A. S., Matochik, J. A., Horti, A. G., Kurian, V., Shumway, D., Domino, E. F., London, E. D., and Mukhin, A. G. (2007). Estimation of D2-like Receptor Occupancy by Dopamine in the Putamen of Hemiparkinsonian Monkeys. *Neuropsychopharmacology.*

Chiueh, C. C., Markey, S. P., Burns, R. S., Johannessen, J. N., Pert, A., and Kopin, I. J. (1984a). Neurochemical and behavioral effects of systemic and intranigral administration of N-methyl-4-phenyl-1,2,3,6-tetrahydropyridine in the rat. *Eur J Pharmacol* **100**, 189–194.

Chiueh, C. C., Markey, S. P., Burns, R. S., Johannessen, J. N., Jacobowitz, D. M., and Kopin, I. J. (1984b). Selective neurotoxic effects of N-methyl-4-phenyl-1,2,3,6-tetrahydropyridine (MPTP) in subhuman primates and man: A new animal model of Parkinson's disease. *Psychopharmac Bull* **20**, 548–553.

Collier, T. J., Lipton, J., Daley, B. F., Palfi, S., Chu, Y., Sortwell, C., Bakay, R. A., Sladek, J. R., Jr., and Kordower, J. H. (2007). Aging-related changes in the nigrostriatal dopamine system and the response to MPTP in nonhuman primates: Diminished compensatory mechanisms as a prelude to parkinsonism. *Neurobiol Dis* **26**, 56–65.

Collins, M. A., and Neafsey, E. J. (1985). Beta-carboline analogues of N-methyl-4-phenyl-1,2,5,6-tetrahydropyridine (MPTP): Endogenous factors underlying idiopathic parkinsonism? *Neurosci Lett* **55**, 179–184.

Cox, H., Togasaki, D. M., Chen, L., Langston, J. W., Di Monte, D. A., and Quik, M. (2007). The selective kappa-opioid receptor agonist U50,488 reduces l-dopa-induced dyskinesias but worsens parkinsonism in MPTP-treated primates. *Exp Neurol.*

Cragg, S. J. (2003). Variable dopamine release probability and short-term plasticity between functional domains of the primate striatum. *Trends Neurosci* **23**, 4378–4385.

Crossman, A. R., Mitchell, I. J., and Sambrook, M. A. (1985). Regional brain uptake of 2-deoxyglucose in N-methyl-4-phenyl-1,2,3,6-tetrahydropyridine (MPTP)-induced parkinsonism in the macaque monkey. *Neuropharmacology* **24**, 587–591.

Cruikshank, S. J., and Weinberger, N. M. (1996). Evidence for the Hebbian hypothesis in experience-dependent physiological plasticity of neocortex: A critical review. *Brain Res Rev* 22, 191–228.

Cruz-Sanchez, F. F., Cardozo, A., Ambrosio, S., Tolosa, E., and Mahy, N. (1993). Plasticity of the nigrostriatal system in MPTP-treated mice. A biochemical and morphological correlation. *Mol Chem Neuropathol* 19, 163–176.

Czoty, P. W., Makriyannis, A., and Bergman, J. (2004). Methamphetamine discrimination and *in vivo* microdialysis in squirrel monkeys. *Psychopharmacology (Berl)* 175, 170–178.

Dauer, W., and Przedborski, S. (2003). Parkinson's disease: Mechanisms and models. *Neuron* 39, 889–909.

Davidson, C., Gow, A. J., Lee, M. K., and Ellinwood, E. H. (2001). Methamphetamine neurotoxicity: Necrotic and apoptotic mechanisms and relevance to human abuse and treatment. *Brain Res Brain Res Rev* 36, 1–22.

Day, M., Wang, Z., Ding, J., An, X., Ingham, C. A., Shering, A. F., Wokosin, D., Ilijic, E., Sun, Z., Sampson, A. R., Mugnaini, E., Deutch, A. Y., Sesack, S. R., Arbuthnott, G. W., and Surmeier, D. J. (2006). Selective elimination of glutamatergic synapses on striatopallidal neurons in Parkinson disease models. *Nat Neurosci* 9, 251–259.

Decamp, E., and Schneider, J. S. (2004). Attention and executive function deficits in chronic low-dose MPTP-treated non-human primates. *Eur J Neurosci* 20, 1371–1378.

Decamp, E., Tinker, J. P., and Schneider, J. S. (2004). Attentional cueing reverses deficits in spatial working memory task performance in chronic low dose MPTP-treated monkeys. *Behav Brain Res* 152, 259–262.

Di Monte, D. A., McCormack, A., Petzinger, G., Janson, A. M., Quik, M., and Langston, W. J. (2000). Relationship among nigrostriatal denervation, parkinsonism, and dyskinesias in the MPTP primate model. *Mov Disord* 15, 459–466.

Eberling, J. L., Jagust, W., Taylor, S., Bringas, J., Pivirotto, P., VanBrocklin, H. F., and Bankiewicz, K. S. (1998). A novel MPTP primate model of Parkinson's disease: Neurochemical and clinical changes. *Brain Res* 805, 259–262.

Eidelberg, E., Brooks, B. A., Morgan, W. W., Walden, J. G., and Kokemoor, R. H. (1986). Variability and functional recovery in the N-methyl-4-phenyl-1,2,3,6-tetrahydropyridine model of parkinsonism in monkeys. *Neurosci* 18, 817–822.

Elsworth, J. D., Deutch, A. Y. et al. (1989). Symptomatic and asymptomatic 1-methyl-4-phenyl-1,2,3,6-tetrahydropyridine-treated primates: biochemical changes in striatal regions. *Neuroscience* 33(2), 323–331.

Elsworth, J. D., Deutch, A. Y., Redmond, D. E., Sladek, J. R., and Roth, R. H. (1990). MPTP-induced parkinsonism: Relative changes in dopamine concentration in subregions of substantia nigra, ventral tegmental

area and retrorubal field of symptomatic and asymptomatic vervet monkeys. *Brain Res* 513, 320–324.

Elsworth, J. D., Taylor, J. R., Sladek, J. R., Collier, T. J., Redmond, D. E., and Roth, R. H. (2000). Striatal dopaminergic correlates of stable parkinsonism and degree of recovery in old-world primates one year after MPTP treatment. *Neurosci* 95, 399–408.

Emborg, M. E., Moirano, J., Schafernak, K. T., Moirano, M., Evans, M., Konecny, T., Roitberg, B., Ambarish, P., Mangubat, E., Ma, Y., Eidelberg, D., Holden, J., Kordower, J. H., and Leestma, J. E. (2006). Basal ganglia lesions after MPTP administration in rhesus monkeys. *Neurobiol Dis* 23, 281–289.

Eslamboli, A. (2005). Marmoset monkey models of Parkinson's disease: Which model, when and why? *Brain Res Bull* 68, 140–149.

Eslamboli, A., Romero-Ramos, M., Burger, C., Bjorklund, T., Muzyczka, N., Mandel, R. J., Baker, H., Ridley, R. M., and Kirik, D. (2007). Long-term consequences of human alpha-synuclein overexpression in the primate ventral midbrain. *Brain* 130, 799–815.

Forno, L. S. (1996). Neuropathology of Parkinson's disease. *J Neuropath Exp Neurol* 55, 259–272.

Forno, L. S., DeLanney, L. E., Irwin, I., and Langston, J. W. (1986a). Neuropathology of MPTP-treated monkeys. Comparison with the neuropathology of human idiopathic Parkinson's Disease. In *MPTP: A Neurotoxin Producing a Parkinsonian Syndrome* (S. P. Markey, N. J. Castagnoli, A. J. Trevor, and I. J. Kopin, Eds.), pp. 119–139. Academic Press.

Forno, L. S., Langston, J. W., DeLanney, L. E., Irwin, I., and Ricaurte, G. A. (1986b). Locus ceruleus lesions and eosinophilic inclusions in MPTP-treated monkeys. *Ann Neurol* 20, 449–455.

Forno, L. S., DeLanney, L. E., Irwin, I., and Langston, J. W. (1993). Similarities and differences between MPTP-induced parkinsonsim and Parkinson's disease. Neuropathologic considerations. *Adv Neurol* 60, 600–608.

Fox, S. H., Henry, B., Hill, M., Crossman, A., and Brotchie, J. (2002). Stimulation of cannabinoid receptors reduces levodopa-induced dyskinesia in the MPTP-lesioned nonhuman primate model of Parkinson's disease. *Mov Disord* 17, 1180–1187.

Freed, C. R., Richards, J. B., Sabol, K. E., and Reite, M. L. (1988). Fetal substantia nigra transplants lead to dopamine cell replacement and behavioral improvement in Bonnet monkeys with MPTP induced parkinsonism. In *Jackson DMPharmacology and Functional Regulation of Dopaminergic Neurons* (P. M. Beart, and G. N. Woodruff, Eds.), pp. 353–360. MacMillan Press, London.

Friedman, L. K., and Mytilineou, C. (1990). Neurochemical and toxic effects of 1-methyl-4-phenyl-1,2,3,6-tetrahydropyridine and 1-methyl-4-

phenylpyridine to rat serotonin neurons in dissociated cell cultures. *J Pharmacol Exp Ther* 253, 892–898.

Frohna, P. A., Rothblat, D. S., Joyce, J. N., and Schneider, J. S. (1995). Alterations in dopamine uptake sites and D1 and D2 receptors in cats symptomatic for and recovered from experimental parkinsonism. *Synapse* 19, 46–55.

Frost, D. O., and Cadet, J. L. (2000). Effects of methamphetamine-induced neurotoxicity on the development of neural circuitry: A hypothesis. *Brain Res Brain Res Rev* 34, 103–118.

Gainetdinov, R. R., Fumagalli, F. et al. (1997). Dopamine transporter is required for in vivo MPTP neurotoxicity: evidence from mice lacking the transporter. *J Neurochem.* 69(3), 1322–1325.

Gerlach, M., and Reiderer, P. (1996). Animal models of Parkinson's disease: An empirical comparison with the phenomenology of the disease in man. *J Neural Transm* 103, 987–1041.

German, D., Dubach, M., Askari, S., Speciale, S., and Bowden, D. (1988). 1-Methyl-4-phenyl-1,2,3,6-tetrahydropyridine-induced parkinsonian syndrome in Macca fasicularis: Which midbrain dopaminergic neurons are lost? *Neurosci* 24, 161–174.

Goldberg, J. A., Boraud, T., Maraton, S., Haber, S. N., Vaadia, E., and Bergman, H. (2002). Enhanced synchrony among primary motor cortex neurons in the 1-methyl-4-phenyl-1,2,3,6-tetrahydropyridine primate model of Parkinson's disease. *J Neurosci* 22, 4639–4653.

Guilarte, T. R. (2001). Is methamphetamine abuse a risk factor in parkinsonism? *Neurotoxicol* 22, 725–731.

Hallett, P. J., Dunah, A. W., Ravenscroft, P., Zhou, S., Bezard, E., Crossman, A. R., Brotchie, J. M., and Standaert, D. G. (2005). Alterations of striatal NMDA receptor subunits associated with the development of dyskinesia in the MPTP-lesioned primate model of Parkinson's disease. *Neuropharmacology* 48, 503–516.

Hantraye, P., Varastet, M., Peschanski, M., Riche, D., Cesaro, P., Willer, J. C., and Maziere, M. (1993). Stable parkinsonian syndrome and uneven loss of striatal dopamine fibres following chronic MPTP administration in baboons. *Neurosci* 53, 169–178.

Harvey, D. C., Lacan, G., and Melega, W. P. (2000a). Regional heterogeneity of dopaminergic deficits in vervet monkey striatum and substantia nigra after methamphetamine exposure. *Exp Brain Res* 133, 349–358.

Harvey, D. C., Lacan, G., Tanious, S. P., and Melega, W. P. (2000b). Recovery from methamphetamine induced long-term nigrostriatal dopaminergic deficits without substantia nigra cell loss. *Brain Res* 871, 259–270.

Ho, A., and Blum, M. (1998). Induction of interleukin-1 associated with compensatory dopaminergic sprouting in the denervated striatum of young mice: Model of aging and neurodegenerative disease. *J Neurosci* 18, 5614–5629.

Hsu, A., Togasaki, D. M., Bezard, E., Sokoloff, P., Langston, J. W., Di Monte, D. A., and Quik, M. (2004). Effect of the D3 Dopamine Receptor Partial Agonist BP897 on L-dopa-Induced Dyskinesias and Parkinsonism in Squirrel Monkeys. *J Pharmacol Exp Ther* 311, 770–777.

Hurley, M. J., Mash, D. C., and Jenner, P. (2001). Dopamine D1 receptor expression in human basal ganglia and changes in Parkinson's disease. *Brain Res Mol Brain Res* 87, 271–279.

Hurley, M. J., Jackson, M. J., Smith, L. A., Rose, S., and Jenner, P. (2005). Immunoautoradiographic analysis of NMDA receptor subunits and associated postsynaptic density proteins in the brain of dyskinetic MPTP-treated common marmosets. *Eur J Neurosci* 21, 3240–3250.

Imbert, C., Bezard, E., Guitraud, S., Boraud, T., and Gross, C. E. (2000). Comparison of eight clinical rating scales used for the assessment of MPTP-induced parkinsonism in the Macaque monkey. *J Neurosci Method* 96, 71–76.

Irwin, I., and Langston, J. W. (1985). Selective accumulation of MPP+ in the substantianigra: a key to neurotoxicity? *Life Sci.* 36(3), 207–212.

Irwin, I., DeLanney, L. E., Forno, L. S., Finnegan, K. T., Di Monte, D., and Langston, J. W. (1990). The evolution of nigrostriatal neurochemical changes in the MPTP-treated squirrel monkey. *Brain Res* 531, 242–252.

Israel, Z., and Bergman, H. (2008). Pathophysiology of the basal ganglia and movement disorders: From animal models to human clinical applications. *NeurosciBiobehav Rev.* 32(3), 367–377.

Jackson-Lewis, V., and Smeyne, R. J. (2005). MPTP and SNpc DA neuronal vulnerability: Role of dopamine, superoxide and nitric oxide in neurotoxicity. Minireview. *Neurotox Res* 7, 193–202.

Jackson-Lewis, V., and Przedborski, S. (2007). Protocol for the MPTP mouse model of Parkinson's disease. *Nat Protoc* 2, 141–151.

Jackson-Lewis, V., Jakowec, M., Burke, R. E., and Przedborski, S. (1995). Time course and morphology of dopaminergic neuronal death caused by the neurotoxin 1-methyl-4-phenyl-1,2,3,6-tetrahydropyridine. *Neurodegen* 4, 257–269.

Jakowec, M. W., Fisher, B., Nixon, K., Hogg, L., Meshul, C., Bremmer, S., McNeill, T., and Petzinger, G. M. (2003). Neuroplasticity in the MPTP-lesioned mouse and non-human primate. *Ann NY Acad Sci* 991, 298–301.

Jakowec, M. W., Nixon, K., Hogg, L., McNeill, T., and Petzinger, G. M. (2004). Tyrosine hydroxylase and dopamine transporter expression following 1-methyl-4-phenyl-1,2,3,6-tetrahydropyridine-induced neurodegeneration in the mouse nigrostriatal pathway. *J Neurosci Res* 76, 539–550.

Jellinger, K. A. (2001). The pathology of Parkinson's disease. *Adv Neurol* **86**, 55–72.

Jenner, P. (2003). The contribution of the MPTP-treated primate model to the development of new treatment strategies for Parkinson's disease. *Parkinsonism Relat. Disord.* **9**(3), 131–137.

Jenner, P., and Marsden, C. D. (1988). MPTP-induced parkinsonism as an experimental model of Parkinson's disease. In *Parkinson's Disease and Movement Disorders* (J. Jankovic, and E. Tolosa, Eds.), pp. 37–48. Urban and Schwarzenberg, Inc, Baltimore.

Jenner, P., Rose, S., Nomoto, M., and Marsden, C. D. (1986). MPTP-induced parkinsonism in the common marmoset: Behavioral and biochemical effects. *Adv Neurol* **45**, 183–186.

Kalivas, P. W., Duffy, P., and Barrow, J. (1989). Regulation of the mesocortocolimbic dopamine system by glutamic acid receptor subtypes. *J Pharmacol Exp Therap* **251**, 378–387.

Kanda, T., Jackson, M. J., Smith, L. A., Pearce, R. K., Nakamura, J., Kase, H., Kuwana, Y., and Jenner, P. (1998). Adenosine A2A antagonist: A novel antiparkinsonian agent that does not provoke dyskinesia in parkinsonian monkeys. *Ann Neurol* **43**, 507–513.

Kiatipattanasakul, W., Nakayama, H., Yongsiri, S., Chotiapisitkul, S., Nakamura, S., Kojima, H., and Doi, K. (2000). Abnormal neuronal and glial argyrophilic fibrillary structures in the brain of an aged albino cynomolgus monkey (Macaca fascicularis). *Acta Neuropathol (Berl)* **100**, 580–586.

Kim, B. G., Shin, D. H., Jeon, G. S., Seo, J. H., Kim, Y. W., Jeon, B. S., and Cho, S. S. (2000). Relative sparing of calretinin containing neurons in the substantia nigra of 6-OHDA treated rat Parkinsonian model. *Brain Res* **855**, 162–165.

Konitsiotis, S., Blanchet, P. J., Verhagen, L., Lamers, E., and Chase, T. N. (2000). AMPA receptor blockade improves levodopa-induced dyskinesia in MPTP monkeys. *Neurology* **54**, 1589–1595.

Kordower, J. H., Emborg, M. E., Bloch, J., Ma, S. Y., Chu, Y., Leventhal, L., McBride, J., Chen, E. Y., Palfi, S., Roitberg, B. Z., Brown, W. D., Holden, J. E., Pyzalski, R., Taylor, M. D., Carvey, P., Ling, Z., Trono, D., Hantraye, P., Deglon, N., and Aebischer, P. (2000). Neurodegeneration prevented by lentiviral vector delivery of GDNF in primate models of Parkinson's disease. *Science* **290**, 767–773.

Kordower, J. H., Kanaan, N. M., Chu, Y., Suresh Babu, R., Stansell, J., 3rd, Terpstra, B. T., Sortwell, C. E., Steece-Collier, K., and Collier, T. J. (2006). Failure of proteasome inhibitor administration to provide a model of Parkinson's disease in rats and monkeys. *Ann Neurol* **60**, 264–268.

Kreitzer, A. C., and Malenka, R. C. (2007). Endocannabinoid-mediated rescue of striatal LTD and motor deficits in Parkinson's disease models. *Nature* **445**, 643–647.

Kuoppamaki, M., Al-Barghouthy, G., Jackson, M. J., Smith, L. A., Quinn, N., and Jenner, P. (2007). L-dopa dose and the duration and severity of dyskinesia in primed MPTP-treated primates. *J Neural Transm*.

Kurlan, R., Kim, M. H., and Gash, D. M. (1991). The time course and magnitude of spontaneous recovery of parkinsonism produced by intracarotid administration of 1-methyl-4-phenyl-1,2,3,6-tetrahydropyridine to monkeys. *Ann Neurol* **29**, 677–679.

Langston, J. W., Ballard, P., Tetrud, J. W., and Irwin, I. (1983). Chronic parkinsonism in humans due to a product of meperidine-analog synthesis. *Science* **219**, 979–980.

Langston, J. W., Forno, S., Rebert, C. S., and Irwin, I. (1984). Selective nigral toxicity after systemic administration of 1-methyl-4-phenyl-1,2,5,6,-tetrahydropyridine (MPTP) in the squirrel monkey. *Brain Res* **292**, 390–394.

Mahon, S., Deniau, J. M., and Charpier, S. (2004). Corticostriatal plasticity: Life after the depression. *Trends Neurosci* **27**, 460–467.

Manning-Bog, A. B., Reaney, S. H., Chou, V. P., Johnston, L. C., McCormack, A. L., Johnston, J., Langston, J. W., and Di Monte, D. A. (2006). Lack of nigrostriatal pathology in a rat model of proteasome inhibition. *Ann Neurol* **60**, 256–260.

Markey, S. P., Johannessen, J. N., Chiueh, C. C., Burns, R. S., and Herkenham, M. A. (1984). Intraneuronal generation of a pyridinium metabolite may cause drug-induced parkinsonism. *Nature* **311**, 464–467.

McCann, U. D., Wong, D. F., Yokoi, F., Villemagne, V., Dannals, R. F., and Ricaurte, G. A. (1998). Reduced striatal dopamine transporter density in abstinent methamphetamine and methcathinone users: Evidence from positron emission tomography studies with [11C]WIN-35,428. *J Neurosci* **18**, 8417–8422.

Melega, W. P., Raleigh, M. J., Stout, D. B., Lacan, G., Huang, S.-C., and Phelps, M. E. (1997). Recovery of striatal dopamine function after acute amphetamine- and methamphetamine-induced neurotoxicity in the vervet monkey. *Brain Res* **766**, 113–120.

Meredith, G. E., Totterdell, S., Petroske, E., Santa Cruz, K., Callison, R. C., and Lau, Y. S. (2002). Lysosomal malfunction accompanies alpha-synuclein aggregation in a progressive mouse model of Parkinson's disease. *Brain Res* **956**, 156–165.

Mitchell, I. J., Cross, A. J., Sambrook, M. A., and Crossman, A. R. (1985a). Sites of the neurotoxic action of 1-methyl-4-phenyl-1,2,3,6-tetrahydropyridine in the monkey include the ventral tegmental area and the locus coeruleus. *Neurosci Lett* **61**, 195–200.

Mitsumoto, Y., Watanabe, A., Mori, A., and Koga, N. (1998). Spontaneous regeneration of nigrostriatal dopaminergic neurons in MPTP-treated C57BL/6 mice. *Biochem Biophys Res Comm* **248**, 660–663.

Moerlein, S. M., Stocklin, G., Pawlik, G., Wienhard, K., and Heiss, W. D. (1986). Regional cerebral pharmacokinetics of the dopaminergic neurotoxin 1-methyl-4-phenyl-1,2,3,6-tetrahydropyridine as examined by positron emission tomography in a baboon is altered by tranylcypromine. *Neurosci Lett* **66**, 205–209.

Moratalla, R., Quinn, B., DeLanney, L. E., Irwin, I., Langston, J. W., and Graybiel, A. M. (1992). Differential vulnerability of primate caudate-putament and striosime-matrix dopamie systems to the neurotoxic effects of 1-methyl-4-1,2,3,6-tetrahydropyridine. *Proc Natl Acad Sci USA* **89**, 3859–3863.

Morgan, S., Nomikos, G., and Huston, J. P. (1991). Changes in the nigrostriatal projections associated with recovery from lesion-induced behavioral asymmetry. *Behav Brain Res* **46**, 157–165.

Mori, S., Fujitake, J., Kuno, S., and Sano, Y. (1988). Immunohistochemical evaluation of the neurotoxic effects of 1-methyl-4-phenyl-1,2,3,6-tetrahydropyridine (MPTP) on dopaminergic nigrostriatal neurons of young adult mice using dopamine and tyrosine hydroxylase antibodies. *Neurosci Lett* **90**, 57–62.

Muriel, M. P., Agid, Y., and Hirsch, E. (2001). Plasticity of afferent fibers to striatal neurons bearing D1 dopamine receptors in Parkinson's disease. *Mov Disord* **16**, 435–441.

Namura, I., Douillet, P., Sun, C. J., Pert, A., Cohen, R. M., and Chiueh, C. C. (1987). MPP+ (1-methyl-4-phenylpyridine) is a neurotoxin to dopamine-, norepinephrine- and serotonin-containing neurons. *Eur J Pharmacol* **136**, 31–37.

Nash, J. E., Ravenscroft, P., McGuire, S., Crossman, A. R., Menniti, F. S., and Brotchie, J. M. (2004). The NR2B-selective NMDA receptor antagonist CP-101,606 exacerbates L-DOPA-induced dyskinesia and provides mild potentiation of anti-parkinsonian effects of L-DOPA in the MPTP-lesioned marmoset model of Parkinson's disease. *Exp Neurol* **188**, 471–479.

Nishi, K., Kondo, T., and Narabayashi, H. (1989). Difference in recovery patterns of striatal dopamine content, tyrosine hydroxylase activity and total biopterin content after 1-methyl-4-phenyl-1,2,3,6-tetrahydropyridine (MPTP) administration: A comparison of young and older mice. *Brain Res* **489**, 157–162.

Oiwa, Y., Nakai, K., and Itakura, T. (2006). Histological effects of intraputaminal infusion of glial cell line-derived neurotrophic factor in Parkinson disease model macaque monkeys. *Neurol Med Chir (Tokyo)* **46**, 267–275, discussion 275–266.

Oiwa, Y., Eberling, J. L., Nagy, D., Pivirotto, P., Emborg, M. E., and Bankiewicz, K. S. (2003). Overlesioned hemiparkinsonian non human primate model: Correlation between clinical, neurochemical and histochemical changes. *Front Biosci* **8**, 155–166.

Ovadia, A., Zhang, Z., and Gash, D. M. (1995). Increased susceptibility to MPTP toxicity in middle-aged rhesus monkeys. *Neurobiol Aging* **16**, 931–937.

Papa, S. M., and Chase, T. N. (1996). Levodopa-induced dyskinesias improved by a glutamate antagonist in parkinsonian monkeys. *Ann Neurol* **39**, 574–578.

Partridge, J. G., Tang, K. C., and Lovinger, D. M. (2000). Regional and postnatal heterogeneity of activity-dependent long-term changes in synaptic efficacy in the dorsal striatum. *J Neurophysiol* **84**, 1422–1429.

Paulus, M. P., Hozack, N. E., Zauscher, B. E., Frank, L., Brown, G. G., Braff, D. L., and Schuckit, M. A. (2002). Behavioral and functional neuroimaging evidence for prefrontal dysfunction in methamphetamine-dependent subjects. *Neuropsychopharmacology* **26**, 53–65.

Pearce, R. K., Heikkila, M., Linden, I. B., and Jenner, P. (2001). L-Dopa induces dyskinesia in normal monkeys: Behavioural and pharmacokinetic observations. *Psychopharmacology (Berl)* **156**, 402–409.

Perez-Otano, I., Herrero, M. T., Oset, C., De Ceballos, M. L., Luquin, M. R., Obeso, J. A., and Del Rio, J. (1991). Extensive loss of brain dopamine and serotonin induced by chronic administration of MPTP in the marmoset. *Brain Res* **567**, 127–132.

Petrucelli, L., and Dawson, T. M. (2004). Mechanism of neurodegenerative disease: Role of the ubiquitin proteasome system. *Ann Med* **36**, 315–320.

Petzinger, G. M., and Langston, J. W. (1998). The MPTP-lesioned non-human primate: A model for Parkinson's disease. In *Advances in Neurodegenerative Disease. Volume I: Parkinson's Disease* (J. Marwah, and H. Teitelbbaum, Eds.), pp. 113–148. Prominent Press, Scottsdale, AZ.

Petzinger, G. M., Quik, M., Ivashina, E., Jakowec, M. W., Jakubiak, M., Di Monte, D., and Langston, J. W. (2001). Reliability and validity of a new global dyskinesia rating scale in MPTP-lesioned non-human primate. *Mov Disord* **16**, 202–207.

Petzinger, G. M., Fisher, B. E., Hogg, E., Abernathy, A., Arevalo, P., Nixon, K., and Jakowec, M. W. (2006). Behavioral Recovery in the MPTP (1-methyl-4-phenyl-1,2,3,6-tetrahydropyridine)-lesioned Squirrel Monkey (Saimiri sciureus): Analysis of Striatal Dopamine and the Expression of Tyrosine Hydroxylase and Dopamine Transporter Proteins. *J Neuorsci Res* **83**, 332–347.

Picconi, B., Pisani, A., Barone, I., Bonsi, P., Centonze, D., Bernardi, G., and Calabresi, P. (2005). Pathological synaptic plasticity in the striatum: Implications for Parkinson's disease. *Neurotoxicology* **26**, 779–783.

Przedborski, S., Jackson-Lewis, V., Naini, A. B., Jakowec, M., Petzinger, G., Miller, R., and Akram, M. (2001). The parkinsonian toxin 1-methyl-4-phenyl-1,2,3,6-tetrahydropyridine (MPTP): A technical review of its utility and safety. *J Neurochem* **76**, 1265–1274.

Raz, A., Vaadia, E., and Bergman, H. (2000). Firing patterns and correlations of spontaneous discharge of pallidal neurons in the normal and the tremulous 1-methyl-4-phenyl-1,2,3,6-tetrahydropyridine vervet model of parkinsonism. *J Neurosci* **20**, 8559–8571.

Raz, A., Feingold, A., Zelanskaya, V., Vaadia, E., and Bergman, H. (1996). Neuronal synchronization of tonically active neurons in the striatum of normal and parkinsonian primates. *J Neurophysiol* 76, 2083–2088.

Raz, A., Frechter-Mazar, V., Feingold, A., Abeles, M., Vaadia, E., and Bergman, H. (2001). Activity of pallidal and striatal tonically active neurons is correlated in MPTP-treated monkeys but not in normal monkeys. *J Neurosci* 21, RC128.

Reynolds, J. N., and Wickens, J. R. (2002). Dopamine-dependent plasticity of corticostriatal synapses. *Neural Netw* 15, 507–521.

Riachi, N. J., Harik, S. I., Kalaria, R. N., and Sayre, L. M. (1988). On the mechanisms underlying 1-methyl-4-phenyl-1,2,3,6-tetrahydropyridine neurotoxicity. II. Susceptibility among mammalian species correlates with the toxin's metabolic patterns in brain microvessels and liver. *J Pharmacol Exp Ther* 244, 443–448.

Ricaurte, G. A., Schuster, C. R., and Seiden, L. S. (1980). Long-term effects of repeated methylamphetamine administration on dopamine and serotonin neurons in the rat brain: A regional study. *Brain Res* 193, 153–163.

Ricaurte, G. A., Guillery, R. W., Seiden, L. S., Schuster, C. R., and Moore, R. Y. (1982). Dopamine nerve terminal degeneration produced by high doses of methylamphetamine in the rat brain. *Brain Res* 235, 93–103.

Ricaurte, G. A., Langston, J. W., DeLanney, L. E., Irwin, I., Peroutka, S. J., and Forno, L. S. (1986). Fate of nigrostriatal neurons in young mature mice given 1-methyl-4-phenyl-1,2,3,6-tetrahydropyridine: A neurochemical and morphological reassessment. *Brain Res* 376, 117–124.

Robinson, T. E., and Wishaw, I. Q. (1988). Normalization of extracellular dopamine in striatum following recovery from a partial unilateral 6-OHDA lesion of the substantia nigra: a microdialysis study in freely moving rats. *Brain Res* 450, 209–224.

Rose, S., Nomoto, M., Kelly, E., Kilpatrick, G., Jenner, P., and Marsden, C. D. (1989). Increased caudate dopamine turnover may contribute to the recovery of motor function in marmosets treated with the dopaminergic neurotoxin MPTP. *Neurosci Lett* 101, 305–310.

Rose, S., Nomoto, M., Jackson, E. A., Gibb, W. R. G., Jaehnig, P., Jenner, P., and Marsden, C. D. (1993). Age-related effects of 1-methyl-4-phenyl-,2,3,6-tetrahydropyridine treatment of common marmosets. *Eur J Pharm* 230, 177–185.

Rothblat, D. S., Schroeder, J. A., and Schneider, J. S. (2001). Tyrosine hydroxylase and dopamine transporter expression in residual dopaminergic neurons: Potential contributors to spontaneous recovery from experimental parkinsonism. *J Neurosci Res* 65, 254–266.

Rozas, G., Liste, I., Guerra, M. J., and Labandeira-Garcia, J. L. (1998). Sprouting of the serotonergic afferents into striatum after selective lesion

of the dopaminergic system by MPTP in adult mice. *Neurosci Lett* 245, 151–154.

Ruel, J., Guitton, M. J., and Puell, J. L. (2002). Negative allosteric modulation of AMPA-preferring receptors by the selective isomer GYKI 53784 (LY303070), a specific non-competitive AMPA antagonist. *CNS Drug Rev* 8, 235–254.

Russ, H., Mihatsch, W., Gerlach, M., Riederer, P., and Przuntek, H. (1991). Neurochemical and behavioural features induced by chronic low dose treatment with 1-methyl-4-phenyl-1,2,3,6-tetrahydropyridine (MPTP) in the common marmoset: Implications for Parkinson's disease? *Neurosci Lett* 123, 115–118.

Samadi, P., Gregoire, L., and Bedard, P. J. (2003). Opioid antagonists increase the dyskinetic response to dopaminergic agents in parkinsonian monkeys: Interaction between dopamine and opioid systems. *Neuropharmacology* 45, 954–963.

Samadi, P., Gregoire, L., and Bedard, P. J. (2004). The opioid agonist morphine decreases the dyskinetic response to dopaminergic agents in parkinsonian monkeys. *Neurobiol Dis* 16, 246–253.

Samadi, P., Gregoire, L., Morissette, M., Calon, F., Hadj Tahar, A., Dridi, M., Belanger, N., Meltzer, L. T., Bedard, P. J., and Di Paolo, T. (2007). mGluR5 metabotropic glutamate receptors and dyskinesias in MPTP monkeys. *Neurobiol Aging*.

Schneider, J. S. (1990a). Chronic exposure to low doses of MPTP. II. Neurochemical and pathological consequences in cognitively-impaired, motor asymptomatic monkeys. *Brain Res* 534, 25–36.

Schneider, J. S., and Kovelowski, C. J. (1990b). Chronic exposure to low doses of MPTP. I. Cognitive deficits in motor asymptomatic monkeys. *Brain Res* 519, 122–128.

Schneider, J. S., and Pope-Coleman, A. (1995). Cognitive deficits precede motor deficits in a slowly progressing model of parkinsonism in the monkey. *Neurodegeneration* 4, 245–255.

Schneider, J. S., Rothblat, D. S., and DiStefano, L. (1994). Volume transmission of dopamine over large distances may contribute to recovery from experimental parkinsonism. *Brain Res* 643, 86–91.

Schneider, J. S., Lidsky, T. I., Hawks, T., Mazziotta, J. C., and Hoffman, J. M. (1995). Differential recovery of volitional motor function, lateralized cognitive function, dopamine-agonist-induced rotation and dopaminergic parameters in monkeys made hemi-parkinsonian by intracarotid MPTP infusion. *Brain Res* 672, 112–127.

Schneider, J. S., Gonczi, H., and Decamp, E. (2003). Development of levodopa-induced dyskinesias in parkinsonian monkeys may depend upon rate of symptom onset and/or duration of symptoms. *Brain Res* 990, 38–44.

Scotcher, K. P., Irwin, I., DeLanney, L. E., Langston, J. W., and Di Monte, D. (1991). Mechanism of accumulation of the 1-methyl-4-phenyl-phenylpyridinium species into mouse brain synaptosomes. *J Neurochem* **56**, 1602–1607.

Shimura, H., Hattori, N., Kubo, S., Mizuno, Y., Asakawa, S., Minoshima, S., Shimizu, N., Iwai, K., Chiba, T., Tanaka, K., and Suzuki, T. (2000). Familial Parkinson disease gene product, parkin, is a ubiquitin-protein ligase. *Nat Genet* **25**, 302–305.

Slovin, H., Abeles, M., Vaadia, E., Haalman, I., Prut, Y., and Bergman, H. (1999a). Frontal cognitive impairments and saccadic deficits in low-dose MPTP-treated monkeys. *J Neurophysiol* **81**, 858–874.

Slovin, H., Abeles, M., Vaadia, E., Haalman, I., Prut, Y., and Bergman, H. (1999b). Frontal cognitive impairments and saccadic deficits in low-dose MPTP-treated monkeys. *J Neurophysiol* **81**, 858–874.

Smith, R., Zhang, Z., Kurlan, R., McDermott, M., and Gash, D. (1993). Developing a stable bilateral model of parkinsonism in rhesus monkeys. *Neuroscience* **52**, 7–16.

Smith, R., Musleh, W., Akopian, W., Buckwalter, G., and Walsh, J. P. (2001). Regional differences in the expression of corticostriatal synaptic plasticity. *Neuroscience* **106**, 95–101.

Smith, L. A., Jackson, M. J., Johnston, L., Kuoppamaki, M., Rose, S., Al-Barghouthy, G., Del Signore, S., and Jenner, P. (2006). Switching from levodopa to the long-acting dopamine D2/D3 agonist piribedil reduces the expression of dyskinesia while maintaining effective motor activity in MPTP-treated primates. *Clin Neuropharmacol* **29**, 112–125.

Sortwell, C. E., Blanchard, B. C., Collier, T. J., Elsworth, J. D., Taylor, J. R., Roth, R. H., Redmond, D. E., and Sladek, J. R. (1998). Pattern of synaptophysin immunoreactivity within mesencephalic grafts following transplantation in a parkinsonian primate model. *Brain Res* **791**, 117–124.

Staal, R., and Sonsalla, P. (2000). Inhibition of brain vesicular monoamine transporter (VMAT2) enhances 1-methyl-4-phenylpyridinium neurotoxicity in vivo in rat striata. *J Pharmacol Exp Ther.* **293**(2), 336–342.

Tan, L. C., Protell, P. H., Langston, J. W., and Togasaki, D. M. (2002). The hyperkinetic abnormal movements scale: A tool for measuring levodopa-induced abnormal movements in squirrel monkeys. *Mov Disord* **17**, 902–909.

Tanaka, K., Suzuki, T., Hattori, N., and Mizuno, Y. (2004). Ubiquitin, proteasome and parkin. *Biochim Biophys Acta* **1695**, 235–247.

Taylor, J. R., Elsworth, J. D., Roth, R. H., Sladek, J. R., Collier, T. J., and Redmond, D. E. (1991). Grafting of fetal substantia nigra to striatum reverses behavioral deficits induced by MPTP in primates: A comparison with other types of grafts as controls. *Exp Brain Res* **85**, 335–348.

Taylor, J. R., Elsworth, J. D., Roth, R. H., Sladek, J. R., and Redmond, D. E. (1994). Behavioral effects of

MPTP administration in the vervet monkey. A primate model of Parkinson's disease. In *Toxin-Induced Models of Neurological Disorders* (M. L. Woodruff, and A. J. Nonneman, Eds.), pp. 139–174. Plenum Press, New York.

Taylor, J. R., Elsworth, J. D., Roth, R. H., Sladek, J. R., and Redmond, D. E. (1997). Severe long-term 1-methyl-4-phenyl-1,2,3,6-tetrahydropyridine-induced parkinsonism in the vervet monkey (Cercopithecus aethiops sabaeus). *Neuroscience* **81**, 745–755.

Tetrud, J. W., and Langston, J. W. (1989). MPTP-induced parkinsonism as a model for Parkinson's disease. *Acta Neurol Scand* **126**, 35–40.

Tetrud, J. W., Langston, J. W., Redmond, D. E., Roth, R. H., Sladek, J. R., and Angel, R. W. (1986). MPTP-induced tremor in human and non-human primates. *Neurology* **36**(Suppl. 1), 308.

Togasaki, D. M., Tan, L., Protell, P., Di Monte, D. A., Quik, M., and Langston, J. W. (2001). Levodopa induces dyskinesias in normal squirrel monkeys. *Ann Neurol* **50**, 254–257.

Togasaki, D. M., Protell, P., Tan, L. C., Langston, J. W., Di Monte, D. A., and Quik, M. (2005). Dyskinesias in normal squirrel monkeys induced by nomifensine and levodopa. *Neuropharmacology* **48**, 398–405.

Verhagen Metman, L., Del Dotto, P., Blanchet, P. J., van den Munckhof, P., and Chase, T. N. (1998). Blockade of glutamatergic transmission as treatment for dyskinesias and motor fluctuations in Parkinson's disease. *Amino Acids* **14**, 75–82.

Villemagne, V., Yuan, J., Wong, D. F., Dannals, R. F., Hatzidimitriou, G., Mathews, W. B., Ravert, H. T., Musachio, J., McCann, U. D., and Ricaurte, G. A. (1998). Brain dopamine neurotoxicity in baboons treated with doses of methamphetamine comparable to those recreationally abused by humans: Evidence from [11C]WIN-35,428 positron emission tomography studies and direct *in vitro* determinations. *J Neurosci* **18**, 419–427.

Vitek, J. L., and Giroux, M. (2000). Physiology of hypokinetic and hyperkinetic movement disorders: Model for dyskinesia. *Ann Neurol* **47**, S131–S140.

Wade, T. V., Rothblat, D. S., and Schneider, J. S. (2001). Changes in striatal dopamine D3 receptor regulation during expression of and recovery from MPTP-induced parkinsonism. *Brain Res* **905**, 111–119.

Wang, Z., Kai, L., Day, M., Ronesi, J., Yin, H. H., Ding, J., Tkatch, T., Lovinger, D. M., and Surmeier, D. J. (2006). Dopaminergic control of corticostriatal long-term synaptic depression in medium spiny neurons is mediated by cholinergic interneurons. *Neuron* **50**, 443–452.

Waters, C. M., Hunt, S. P., Jenner, P., and Marsden, C. D. (1987). An immunohistochemical study of the acute and long-term effects of 1-methyl-4-phenyl-1,2,3,6-tetrahydropyridine in the marmoset. *Neuroscience* **23**, 1025–1039.

Wichmann, T., and DeLong, M. R. (2003). Pathophysiology of Parkinson's Disease: The MPTP primate model of the human disorder. *Ann NY Acad Sci* **991**, 199–216.

Yamamoto, B. K., and Zhu, W. (1998). The effects of methamphetamine on the production of free radicals and oxidative stress. *J Pharmacol Exp Ther* **287**, 107–114.

Zigmond, M. J. (1997). Do compensatory processes underlie the preclinical phase of neurodegenerative disease? Insights from an animal model of parkinsonism. *Neurobiol Dis* **4**, 247–253.

Zigmond, M. J., Abercrombie, E. D., Berger, T. W., Grace, A. A., and Sticker, E. M. (1990). Compensations after lesions of central dopaminergic neurons: Some clinical and basic implications. *Trends Neurosci* **13**, 290–295.

Zuddas, A., Fascetti, F., Corsini, G. U., and Piccardi, M. P. (1994). In brown Norway rats, MPP+ is accumulated in the nigrostriatal dopaminergic terminals but is not neurotoxic: A model of natural resistance to MPTP toxicity. *Exp Neurol* **127**, 54–61.

RODENT TOXIN MODELS

10
RODENT TOXIN MODELS OF PD: AN OVERVIEW

VINCENT RIES AND ROBERT E. BURKE

Department of Neurology Philipps University Marburg, Rudolf-Bultmann-Strasse 8, 35033 Marburg, Germany e-mail: ries@med.uni-marburg.de

Departments of Neurology and Pathology, The College of Physicians and Surgeons Columbia University Black Building, Room 306 650 West 168th Street New York, NY 10032 USA

INTRODUCTION

The rodent toxin models of Parkinson's disease (PD), are among the earliest developed, they remain in wide use today. While a number of toxin models exist, our introductory overview will focus primarily on the models presented in detail in this book: the 6-hydroxydopamine (6-OHDA) 1-methyl-4-phenyl-1,2,3, 6-tetrahydropyridine (MPTP), rotenone, and paraquat models. We will not touch upon the recently described proteasome inhibitor-induced model (McNaught *et al.*, 2004), because at this writing, considerable controversy exists over its reproducibility. McNaught and colleagues reported a new model of PD in adult rats using systemic exposure to proteasome inhibitors. Although two recent reports confirm some aspects of the original finding (Schapira *et al.*, 2006; Zeng *et al.*, 2006), work in four other laboratories failed to replicate the ability of proteasome inhibitors to induce nigrostriatal degeneration (Bove *et al.*, 2006; Kordower *et al.*, 2006; Manning-Bog *et al.*, 2006). The interested reader is referred to an excellent recent review of the proteasome inhibitor model (Beal and Lang, 2006).

With so much being learned today about the genetic basis of PD, and with widespread availability of modern techniques of transgenesis, one might ask, why continue to study neurotoxin models? The principal reason is that to date, only these neurotoxin models reliably produce neurodegeneration in dopaminergic neurons of the substantia nigra (SN). For reasons that are unknown, there has been no model developed to date, based on transgenesis, which recapitulates this essential pathology of the disease.

These observations of course immediately raise what has become a thorny contemporary issue: Is it essential to observe neurodegeneration in dopamine neurons of the SN in order to have a useful model of PD? After all, Braak *et al.* (2003) have suggested that abnormal synuclein deposition in lower brainstem regions may be the first sign of the disease, and Langston (2006) has suggested that the earliest signs of the disease may be non-motor. What comprises a useful "model of PD" depends on the question that you are trying to answer. If you wish to ask, "How does excess expression of synuclein kill neurons?", then to address that question you do not necessarily need a model in which dopamine neurons die. However, if one accepts that the clinical signs of parkinsonism, in conjunction with loss of SN dopamine neurons, remain the essential criteria to diagnose PD, that is, they are both necessary and sufficient, then the question may be: "What molecular and cellular processes can bring about the degeneration of adult dopamine neurons of the SN?" To attempt to address such questions today, one must use neurotoxin models, because only these reliably and robustly bring about degeneration of these neurons. It is only these models that induce an essential defining feature of PD: loss of dopamine neurons of the SN. Simply put, in the absence of loss of these neurons, given our present state of knowledge of the disease, it cannot be diagnosed.

Parkinson's Disease: molecular and therapeutic insights from model systems

The effect of neurotoxins can be studied not only in rodents, but also in invertebrates, such as *Drosophila melanogaster* and *Caenorhabditis elegans* (*C. elegans*), in which powerful genetic tools can be applied. In recent years, *Drosophila* has been used as a model of several neurodegenerative diseases (Bilen and Bonini, 2005), including a genetic model of PD based on directed expression of human alpha-synuclein in the *Drosophila* brain (Feany and Bender, 2000; Auluck *et al.*, 2002). As an advantage, this invertebrate offers a large panel of genetic approaches and allows the rapid screening of potential therapeutic drugs. Only a few laboratories used toxin-induced approaches in flies. Coulom and Birman (2004) showed that rotenone-treated flies present a selective loss of dopaminergic neurons in the brain and severe locomotor impairments. Furthermore, the addition of L-dopa to the feeding medium rescued the motor deficits but did not prevent dopaminergic cell death, as in human PD patients. The antioxidant N-acetyl-5-methoxy-tryptamine (melatonin) prevented both rotenone-induced behavioral defects and neuronal loss. There has also been a recent report on paraquat-treated flies, showing behavioral symptoms and loss of specific dopaminergic neurons (Chaudhuri *et al.*, 2007). In addition, the effect of dopamine regulating genes on the susceptibility to paraquat-induced oxidative damage has been studied, including those that regulate tetrahydrobiopterin. Protection against the neurotoxicity of paraquat is conferred by mutations that elevate dopamine pathway function. In *C. elegans* a model of neurotoxin-induced (6-OHDA) dopamine neuron death has been established (Nass *et al.*, 2002). 6-OHDA toxicity was found to depend on the function of the plasma membrane transporter for dopamine (DAT). The use of these models and insights derived from them are presented in detail in Part V in this book.

While invertebrate models are powerful tools, it remains important to use rodent models of PD. The potential importance of species differences and different system properties of the central nervous system in the pathogenesis of disease cannot be ignored. Despite the conservation of fundamentally important basic cellular processes between, for example, *Drosophila* and mammals, there remain substantial differences between the organisms such that observations in *Drosophila* will not generalize to human disease. Furthermore, the strategies used to make transgenic flies might not model pathology comparable to that seen in humans, owing to

inappropriate levels of expression that will result in the activation of non-relevant pathways. Some cellular pathways in higher organisms utilize molecules that either do not have homologs or have minimal homolog representation in the fly (Muqit and Feany, 2002) and *C. elegans*. The caspase family of proteases, for example, are key molecules in the execution of neuronal apoptosis. Fourteen caspases have been described in human. In *Drosophila*, seven caspases have been identified and shown to be required for cell death induced by various stimuli. Although there are four caspase-like proteins in *C. elegans*, CED-3 is the only one that has been shown to be required for apoptosis (reviewed by Kumar, 2007). Invertebrate models also lack utility to address the mechanisms of behavioral and motor abnormalities that are prominent in neurodegenerative diseases. For example, the anatomical and physiological differences between the fly and vertebrate motor systems will severely limit the contribution of these models to questions related to how the clinical manifestations of PD are caused by the loss of specific populations of dopamine neurons. In conclusion, for studies related to neuron death, and with a reasonable likelihood of direct relevance to human PD, rodent toxin models remain indispensable.

AN OVERVIEW OF THE NEUROTOXIN MODELS

6-OHDA

The noradrenaline analog 6-OHDA (2,4,5-trihydroxy phenylethylamine) was introduced as a catecholaminergic neurotoxin 40 years ago (Tranzer and Thoenen, 1967). It was the first catecholaminergic neurotoxin to be discovered, and it remains one of the best characterized. It was used to induce the first animal model of PD to be associated with dopamine neuron death in the SN pars compacta (Ungerstedt, 1968). Ever since, 6-OHDA has been extensively used for both *in vitro* and *in vivo* studies.

Due to its structural similarities with dopamine and norepinephrine, 6-OHDA is efficiently taken up and accumulated by neurons that have catecholaminergic plasma membrane transporters for dopamine and norepinephrine, thus accounting for its relatively selective toxicity for monoaminergic neurons (Luthman *et al.*, 1989). Evidence suggests that within neurons its cytotoxic actions are associated with its ease of autoxidation. This is a complex process with simultaneous formation of reactive oxygen species (ROS), like hydrogen peroxide (Heikkila and Cohen, 1971) and quinones (Saner and Thoenen, 1971). These chemical reactions have been reviewed recently by Przedborski and Ischiropoulos (2005).

As 6-OHDA is unable to cross the blood–brain barrier (BBB), for the induction of PD models it must be injected directly into the brain stereotactically. Administered by systemic injection, 6-OHDA causes a chemical sympathectomy by damaging the peripheral nervous system (Jonsson, 1983). To target the nigrostriatal dopaminergic pathway it must be injected directly into the SN, the median forebrain bundle (MFB) or the striatum. After 6-OHDA injections into the SN or the MFB, dopamine neurons start degenerating within 24 h and die without apoptotic morphology (Jeon et al., 1995). When injected into the striatum, however, 6-OHDA causes a more progressive, retrograde degeneration of the nigrostriatal system over a period of weeks (Sauer and Oertel, 1994; Przedborski et al., 1995). In none of these models does 6-OHDA lead to the formation of cytoplasmic inclusions (Lewy bodies). Typically, 6-OHDA is injected unilaterally. The unilateral nigrostriatal lesion results in an asymmetric and quantifiable circling behavior, which can be induced by systemic administration of either L-dopa, dopamine receptor agonists or dopamine-releasing drugs (Ungerstedt and Arbuthnott, 1970; Hefti et al., 1980; Iancu et al., 2005). For morphological and biochemical studies the contralateral side can serve as an internal control. Animals are usually not lesioned bilaterally with 6-OHDA, because they require intensive nursing care (Cenci et al., 2002), and often die primarily due to the occurrence of marked aphagia and adipsia (Ungerstedt, 1971).

MPTP

In 1982, young drug users in California inadvertently injected MPTP, that was produced during the illicit synthesis of 1-methyl-4-phenyl-4-propion-oxypiperidine (MPPP), an analog of the narcotic meperidine (Demerol) (Langston et al., 1983). This injection resulted in clinical symptoms remarkably similar to PD in humans (Langston et al., 1999; Przedborski and Vila, 2003). After systemic administration, MPTP crosses the BBB and is metabolized in astrocytes to its active metabolite, 1-methyl-4-phenylpyridinium (MPP$^+$). MPP$^+$ is selectively taken up by the DAT, as well as by the norepinephrine and serotonin transporters (Javitch et al., 1985; Bezard et al., 1999). Inside neurons, MPP$^+$ binds to the vesicular monoamine transporter-2 (VMAT2), and is translocated into synaptic vesicles (Liu et al., 1992). This sequestration of toxin appears to protect cells from neurodegeneration. It can remain in the cytosol to interact with cytosolic enzymes, and it can also be concentrated within the mitochondria, its likely site of action. Within the mitochondria, MPP$^+$ impairs oxidative phosphorylation by inhibiting complex I of the mitochondrial electron transport chain (Nicklas et al., 1985; Mizuno et al., 1987). This inhibition leads to a decrease in tissue ATP levels especially in the striatum and ventral midbrain of mice. However, the fact that in vivo MPTP causes only a transient 20% reduction in ATP levels (Chan et al. 1991) argues against the ATP deficit being the only factor of MPTP-induced cell death. Another consequence of complex I inhibition by MPP$^+$ is an increased production of ROS. Incubation of MPTP with brain mitochondria resulted in an oxygen-dependent formation of ROS (Rossetti et al., 1988). In another study, it was also shown that the degree of complex I inhibition is proportional to the amount of superoxide radicals produced (Hasegawa et al., 1990). The MPP$^+$-related ROS production also contributes to MPTP-induced cell death is further underlined by the fact that modulation of mitochondrial ROS scavengers, such as manganese superoxide dismutase, affects MPTP-induced neurotoxicity in mice (Klivenyi et al., 1998; Andreassen et al., 2001). ATP depletion and ROS overproduction appear to occur soon after MPTP injection, correlating poorly with the time course of neuronal death in vivo (Jackson-Lewis et al., 1995). There is evidence that rather than killing the cells, alterations in ATP synthesis and ROS production are triggering cell-death-related molecular pathways which lead to the demise of the intoxicated neurons (Przedborski and Vila, 2003). Thus far Lewy bodies have not been convincingly observed in MPTP-induced parkinsonism (Forno et al., 1993).

MPTP is usually systemically administered (subcutaneous, intraperitoneal, intravenous or intramuscular). Susceptibility to MPTP varies across species and strains of animals. Rats are relatively resistant to toxicity and mouse strains vary markedly in their sensitivity. Therefore, MPTP is now mainly used in specific strains of mice, such as C57BL/6 mice. The MPTP mouse model has been used extensively to explore the molecular mechanisms of dopamine neuron degeneration. Mice treated with MPTP do not, however, develop persistent and progressive motor symptoms (Przedborski et al., 2001).

Administration of MPTP to non-human primates has been used extensively to assess the behavioral and physiological consequences of SN dopamine neuron loss. MPTP-treated monkeys have also been used to test a variety of experimental therapeutic modalities, including new drugs, gene therapies and cell replacement approaches. In fact, assessment of a new treatment modality in MPTP-treated primates is considered to be the "gold standard" assessment before proceeding to clinical trials in humans. Monkeys, when lesioned bilaterally, often exhibit a generalized parkinsonian syndrome, so that an accompanying application of L-dopa is required to allow the animals to eat and drink adequately (Petzinger and Langston, 1998). Unilateral intracarotid infusion of MPTP in non-human primates causes mostly symptoms on one side, which enables the animals to maintain normal nutrition without medication (Przedborski et al., 1991).

Dose and schedule of MPTP administration substantially influence the time course and character of neurodegeneration. The most common model of MPTP toxicity utilizes an acute dosing regimen of multiple doses on a single day (Jackson-Lewis and Przedborski, 2007). In this model apoptosis is not observed (Jackson-Lewis et al., 1995). A chronic dosing regimen of MPTP in mice (30 mg/day for 5 days) induces cell death by apoptosis (Tatton and Kish, 1997). With this regimen, Bax is upregulated in dopamine neurons of the SN pars compacta, whereas Bcl-2, an anti-apoptotic protein, is down-regulated (Vila et al., 2001). Consistent with these findings, Bax knockout animals and mice overexpressing Bcl-2 are significantly protected against MPTP toxicity. Nevertheless, in Bcl-2 transgenic mice neuroprotection is greater in the acute MPTP model rather than a chronic regimen (Yang et al., 1998). Apoptotic cell death has also been described recently in a chronic MPTP/probenecid mouse model of PD (Novikova et al., 2006).

In the acute mouse model, there is evidence for inflammation (Kurkowska-Jastrzebska et al., 1999). This is of interest because inflammation is also observed in neuropathological studies in patients after exposure to MPTP (Langston et al., 1999). It has been shown that cyclooxygenase-2 (COX-2) expression is induced within dopamine neurons of the SN pars compacta in postmortem PD specimens and in the acute MPTP mouse model of PD during the destruction of the nigrostriatal pathway. COX-2 ablation and inhibition attenuate MPTP-induced nigrostriatal dopaminergic neurodegeneration by mitigating oxidative damage (Teismann et al., 2003).

Rotenone

Rotenone represents one of the most recently developed toxin models of PD in rodents (Betarbet et al., 2000). It is widely used as an insecticide and fish poison. Since it is extremely lipophilic, rotenone crosses biological membranes easily and gains access to all organs (Talpade et al., 2000); it does not depend on specific plasma membrane transporters for cell entry. Like MPTP, rotenone is a high-affinity, specific inhibitor of complex I, and in mitochondria it impairs oxidative phosphorylation (Schuler and Casida, 2001). It does, however, have other biological effects; it also inhibits the formation of microtubules from tubulin (Brinkley et al., 1974; Marshall and Himes, 1978).

Betarbet et al. (2000) reported that a systemic (i.v.) administration of rotenone in rats causes a selective degeneration of the nigrostriatal dopaminergic system. However, subsequent studies by other investigators have reported a more widespread neurotoxicity than originally proposed (Höglinger et al., 2003; Lapointe et al., 2004). In addition, the extent of the dopaminergic axonal lesion within the striatum of rats appears highly variable (Betarbet et al., 2000; Sherer et al., 2003; Lapointe et al., 2004; Zhu et al., 2004). Furthermore, degeneration of intrinsic striatal neurons, such as cholinergic neurons, has been reported (Höglinger et al., 2003; Lapointe et al., 2004; Zhu et al., 2004). Behaviorally, rotenone-exposed rats are hypokinetic with a flexed posture and in some cases show rigidity. However, in some rotenone-infused rats without a dopaminergic lesion, a similar set of motor abnormalities has been observed (Sherer et al., 2003). Others have also observed that the extent of dopaminergic damage does not correlate with motor behavior in individual rats (Fleming et al., 2004). In contrast to the 6-OHDA and MPTP models, the rotenone model has the interesting feature that some of the dopamine neurons of the SN show the formation of ubiquitin- and alpha-synuclein-positive inclusions (Betarbet et al., 2000).

Paraquat

The herbicide paraquat (1,1'-dimethyl-4,4'-bipyridinium) provides another toxin model of PD. Exposure to paraquat may increase the risk for PD according to epidemiological studies (Liou et al., 1997). It does not cross the BBB easily (Shimizu et al., 2001), but there is evidence of damage to the brain seen in individuals who died from paraquat intoxication, including generalized edema and hemorrhages, both subependymal and subarachnoid (Grant et al., 1980). Paraquat has structural similarity to MPP$^+$. Its mode of action is believed to involve the formation of superoxide radicals (Day et al., 1999; Przedborski and Ischiropoulos, 2005). Systemic administration of paraquat causes a dose-dependent loss of nigral dopamine neurons and its projections in mice, accompanied by reduced motor activity (Brooks et al., 1999; McCormack et al., 2002), although Thiruchelvam et al. (2000) initially failed to observe a significant reduction of dopamine neurons in paraquat-treated mice. Like the rotenone model, the paraquat model has the attractive feature that it induces intracellular inclusions in nigral neurons that are positive for alpha-synuclein. In addition it induces elevation of levels of alpha-synuclein both in the frontal cortex and ventral midbrain (Manning-Bog et al., 2002).

The fungicide manganese ethylenebisdithiocarbamate, maneb, and paraquat are used in overlapping geographical areas. Maneb decreases locomotor activity (Morato et al., 1989) and potentiates MPTP toxicity (Takahashi et al., 1989). Combined exposure to paraquat and maneb has a greater effect on the nigrostriatal dopaminergic system than either compound alone (Thiruchelvam et al., 2000).

WHAT ARE THE STRENGTHS OF THE NEUROTOXIN MODELS?

6-OHDA

Unlike the other neurotoxins used to induce PD models, 6-OHDA is an endogenous metabolite of dopamine (Kostrzewa and Jacobowitz, 1974), and it has been detected in the human caudate nucleus (Curtius et al., 1974). Thus, to the extent that oxidative metabolism of endogenous dopamine may play a role in human PD (reviewed by Fahn and Sulzer, 2004), 6-OHDA may model this process.

An attractive feature of the intrastriatal 6-OHDA model is that it causes a progressive retrograde neuron loss over a period of weeks (Sauer and Oertel, 1994; Przedborski et al., 1995), and in this respect mimics the human disease. Another advantage of the intrastriatal 6-OHDA model is that, among models of PD associated with dopamine neuron death in the SN, it has been demonstrated unequivocally to induce apoptosis, both in a developmental setting as well as in the adult rat brain (Marti et al., 1997; Marti et al., 2002). Thus, the processes of dopaminergic cell death in this model can be interpreted unequivocally in terms of the large body of knowledge about the canonical pathways of programmed cell death. Such a clear and straightforward basis of interpretation does not exist, for example, for the acute MPTP model, in which apoptosis is not observed (Jackson-Lewis et al., 1995), and the activation of caspases has not been demonstrated at the cellular level in dopamine neurons.

An advantage of all 6-OHDA models is that they allow a comprehensive analysis in terms of morphology, biochemical parameters and, importantly, behavior. As mentioned above, the unilateral 6-OHDA lesion model of PD is most widely used, and it causes a hemiparkinsonian syndrome with asymmetries of body posture and contralateral sensorimotor deficits. After administration of dopamine-releasing drugs, animals will show a typical turning behavior, ipsiversive to the lesion, due to the greater release of dopamine on the intact side. On the other hand, postsynaptic dopamine receptor supersensitivity on the lesioned side (Pycock, 1980) results in contraversive rotations after treatment with L-dopa or dopamine receptor agonists (Ungerstedt, 1976; Hefti et al., 1980; Iancu et al., 2005). These simple and objective tests, applicable to both mice and rats, allow quantification of the unilateral lesion. Such quantification has been a valuable tool to monitor the properties of new antiparkinsonian drugs (Jiang et al., 1993), the functional dopaminergic efficacy of transplantation (Brundin et al., 1986; Björklund et al., 2002) and gene therapies (Bensadoun et al., 2000; Kirik et al., 2002; Ries et al., 2006).

Furthermore, in contrast to the other neurotoxin models, intracerebral injection of 6-OHDA into the nigrostriatal pathway has been shown to induce a profound loss of dopaminergic neurons in the SN pars compacta, usually of 80% or more. Thus, among current neurotoxin models, it provides the most stringent test of a novel therapeutic modality to protect or restore the nigrostriatal system. The 6-OHDA model has been used to make numerous seminal observations in the neurobiology of dopamine neurons, including the first demonstration of apoptosis in a PD model (Marti *et al.*, 1997) and the observation that dopamine depletion impairs precursor cell proliferation in the adult rat brain (Höglinger *et al.*, 2004).

MPTP

Among the neurotoxins used to model PD, only exposure to MPTP produces a syndrome in humans that resembles PD. In a study of patients with MPTP-induced parkinsonism, the clinical signs included bradykinesia, rigidity, postural and resting tremor, flexed posture, gait disturbance, loss of postural reflexes, loss of facial expression, drooling, speech disturbance, micrographia and seborrhea (Burns *et al.*, 1985). Also "freezing" episodes have been observed. Furthermore, in MPTP-intoxicated humans, the response to L-dopa and the complications of long-term therapy are almost identical to those seen in PD. As in PD patients, most of these patients developed L-dopa-induced motor complications, like dyskinesia (Langston and Ballard, 1984).

MPP$^+$ is a specific inhibitor of the nicotinamide adenine dinucleotide (NADH)–ubiquinone reductase (complex I). A postmortem study in PD patients on the activity of the mitochondrial respiratory chain in the SN revealed a specific defect of complex I in these patients. The similarity of this finding to the effects produced in animal models of PD by MPP$^+$ suggests that complex I deficiency in PD may be related to the primary disease process (Schapira *et al.*, 1989).

Over the years, the MPTP model has been used in a number of different animal species to recapitulate the hallmarks of PD, the loss of dopamine neurons in the SN and dopamine terminals from the striatum, especially in mice (Heikkila *et al.*, 1989) and non-human primates (Burns *et al.*, 1983). As mentioned above, the model of MPTP-treated primates is considered to be the "gold standard" in the assessment of novel therapeutic strategies. For example, it was demonstrated in non-human primates treated with MPTP that delivery of glial cell line-derived neurotrophic factor (GDNF) reverses functional deficits and prevents nigrostriatal degeneration (Kordower *et al.*, 2000). Also, electrophysiological studies in parkinsonian primates

have shown that increased activity of the subthalamic nucleus (STN) plays an important role in the development of characteristic motor symptoms in PD. Electrical stimulation of the STN markedly improves these symptoms (Bergman *et al.*, 1990). This finding lead to the treatment of PD patients by deep brain stimulation of the STN (Limousin *et al.*, 1998). Thus, the MPTP model of PD is considered to be one of the best experimental models of this common neurodegenerative disease.

Rotenone

Rotenone is widely used around the world as an insecticide and fish poison (Hisata, 2002). Epidemiological data suggest that PD is associated with occupational exposure to herbicides and insecticides (Gorell *et al.*, 1998).

Even if the pattern of neuron loss remains controversial, as discussed above, the preponderance of pathology is in the nigrostriatal system (Betarbet *et al.*, 2000; Höglinger *et al.*, 2003; Lapointe *et al.*, 2004). Evidence that rotenone-induced neurodegeneration spreads beyond the dopaminergic system is not necessarily contradictory, since degeneration in PD is not confined to dopamine neurons, according to new staging studies (Braak *et al.*, 2003). Furthermore, L-dopa appears to reverse motor deficits caused by chronic administration of rotenone (Alam and Schmidt, 2004). Motor behavior, however, seems to be variable (Sherer *et al.*, 2003; Fleming *et al.*, 2004).

The rotenone model was the first model to link an environmental toxin to the pathogenesis of PD. The formation of cytoplasmic inclusions further supports its potential relevance to the human disease. These inclusions share some of the features of Lewy bodies characteristic of PD (Spillantini *et al.*, 1998), containing ubiquitin and alpha-synuclein, and by electron microscopy, they appear to be composed of a dense core surrounded by fibrillar elements (Betarbet *et al.*, 2000). Therefore, the rotenone model can offer the opportunity to study how an environmental toxin interacts with genes, like alpha-synuclein, known to be involved in the pathogenesis of PD.

Paraquat

A number of case–control studies have suggested an association between exposure to the potent herbicide paraquat and an increased risk for PD (Hertzman *et al.*, 1990; Semchuk *et al.*, 1993; Liou *et al.*, 1997; Gorell *et al.*, 1998). Interestingly, paraquat shows structural similarity to MPP^+, however, it does not cross the BBB easily.

McCormack *et al.* (2002) demonstrated selective loss of dopaminergic neurons in the SN of paraquat-treated mice, counting nigral dopamine neurons by stereology for the first time. Despite the loss of nigral neurons, the neurochemical changes caused by paraquat in mouse striatum are relatively modest. In terms of the molecular mechanism of cell death, activation of the c-jun N-terminal kinase pathway seems to be involved (Peng *et al.*, 2004). Based on the hypothesis that paraquat-induced toxicity is associated with excessive oxidative stress, several studies used antioxidant enzymes to modulate the level of ROS *in vivo* (Peng *et al.*, 2005; Thiruchelvam *et al.*, 2005, Choi *et al.*, 2006), showing protection against paraquat-induced dopamine neuron death.

Similar to rotenone, paraquat exposure may also represent a model for the formation of alpha-synuclein-containing inclusions. In addition, Manning-Bog *et al.* (2002) reported an upregulation of alpha-synuclein by paraquat in both the frontal cortex and ventral midbrain of mice. The fact that the paraquat model ties in dopamine neuron death with upregulation and aggregation of alpha-synuclein could make it a valuable tool in resembling pathology characteristic of PD.

WHAT ARE THE WEAKNESSES OF THESE MODELS?

The chief weakness of all neurotoxin-based models is that their actual relationship to the molecular mechanisms of neurodegeneration in human disease remains unknown. Among the neurotoxins used, only MPTP is clearly linked to a form of human parkinsonism. The discovery of genetic defects, that have been associated with familial forms of PD, will allow the generation of new animal models on the basis of "proven" causes of human disease, as presented in Part IV in this book. As more is learned about the interactions between environmental and genetic factors in the cause of PD, useful models may emerge that are based on both known genetic causes and neurotoxin approaches.

A weakness of some of these neurotoxin models is that they are entirely based on acute injury, for example, the acute MPTP model (Table 10.1). The relevance of these acute models to the pathogenesis of human disease seems less certain than models which recapitulate the chronic, progressive nature of the disease.

TABLE 10.1   Characteristics of rodent toxin models of Parkinson's disease

Model	Route	Morphology of cell death	Lewy body-like inclusions	Motor deficits/symptoms
6-Hydroxydopamine	Intranigral	Non-apoptotic	No	Unilateral: quantifiable rotational behavior
	MFB	Non-apoptotic	No	Bilateral: parkinsonian motor syndrome
	Intrastriatal	Apoptotic and non-apoptotic	No	
MPTP	Systemic (i.p.) – acute	Non-apoptotic	No	No stable motor deficit
	Systemic (i.p.) – chronic	Apoptotic and non-apoptotic	(No)	No stable motor deficit
Rotenone	Systemic (i.v.)	Non-apoptotic	Yes	Highly variable parkinsonian motor syndrome
Paraquat	Systemic (i.p.)	(Apoptotic and) non-apoptotic	Yes	Reduced locomotor activity

MFP – Median Forebrain bundle, i.p – intraperitoneal, i.v – intra venous

With increasing awareness that the pathology of PD is not confined to dopamine neurons, there is growing concern that some of these models are focused exclusively on dopamine neurons. In this regard, the rotenone and paraquat models do induce the formation of Lewy body-like inclusions that are immunoreactive for both ubiquitin and alpha-synuclein, and thus model this important feature of the disease.

Just as these models have had their successes, they have also had their possible failures. The primate MPTP model, for example, predicted neuroprotective capability for CEP-1347/KT-7515, an inhibitor of c-jun N-terminal kinase activation. CEP-1347 attenuated the MPTP-mediated loss of nigrostriatal dopaminergic neurons in mice (Saporito *et al.*, 1999) as well as in primates. In addition, in the primate study, CEP-1347 was highly effective in preventing the development of behavioral parkinsonism due to chronic MPTP administration (Saporito *et al.*, 2002). However, in a double-blind Phase 2/3 trial in 800 PD patients, CEP-1347 failed to modify disease progression. While there are a number of possible explanations for this negative result, including an inability of the oral doses in the trial to achieve kinase inhibition in patients, nevertheless there is concern that the MPTP primate model failed to predict this negative result.

CONCLUSION

This overview summarized the most salient features of the four rodent toxin models of PD to be presented in detail in this book. The 6-OHDA and the MPTP models have been available for decades and are the most widely used and best characterized toxin models. But, all of the models have advantages and shortcomings as discussed above. The goal of the particular investigation and the questions to be answered should determine which model will be used.

REFERENCES

Alam, M., and Schmidt, W. J. (2004). L-DOPA reverses the hypokinetic behaviour and rigidity in rotenone-treated rats. *Behav Brain Res* 153, 439–446.

Andreassen, O. A., Ferrante, R. J., Dedeoglu, A., Albers, D. W., Klivenyi, P., Carlson, E. J., Epstein, C. J., and Beal, M. F. (2001). Mice with a partial deficiency of manganese superoxide dismutase show increased vulnerability to the mitochondrial toxins malonate, 3-nitropropionic acid, and MPTP. *Exp Neurol* 167, 189–195.

Auluck, P. K., Chan, H. Y., Trojanowski, J. Q., Lee, V. M.-Y., and Bonini, N. M. (2002). Chaperone suppression of α-synuclein toxicity in a *Drosophila* model for Parkinson's disease. *Science* 295, 865–868.

Beal, F., and Lang, A. (2006). The proteasomal inhibition model of Parkinson's disease: "Boon or bust"? *Ann Neurol* 60, 158–161.

Bensadoun, J. C., Deglon, N., Tseng, J. L., Ridet, J. L., Zurn, A. D., and Aebischer, P. (2000). Lentiviral vectors as a gene delivery system in the mouse midbrain: Cellular and behavioral improvements in a 6-OHDA model of Parkinson's disease using GDNF. *Exp Neurol* 164, 15–24.

Bergman, H., Wichmann, T., and DeLong, M. R. (1990). Reversal of experimental parkinsonism by lesions of the subthalamic nucleus. *Science* 249, 1436–1438.

Betarbet, R., Sherer, T. B., MacKenzie, G., Garcia-Osuna, M., Panov, A. V., and Greenamyre, J. T. (2000). Chronic systemic pesticide exposure reproduces features of Parkinson's disease. *Nat Neurosci* 3, 1301–1306.

Bezard, E., Gross, C. E., Fournier, M. C., Dovero, S., Bloch, B., and Jaber, M. (1999). Absence of MPTP-induced neuronal death in mice lacking the dopamine transporter. *Exp Neurol* 155, 268–273.

Bilen, J., and Bonini, N. M. (2005). *Drosophila* as a model for human neurodegenerative disease. *Annu Rev Genet* 39, 153–171.

Björklund, L. M., Sanchez-Pernaute, R., Chung, S., Andersson, T., Chen, I. Y., McNaught, K. S., Brownell, A. L., Jenkins, B. G., Wahlestedt, C., Kim, K. S., and Isacson, O. (2002). Embryonic stem cells develop into functional dopaminergic neurons after transplantation in a Parkinson rat model. *Proc Natl Acad Sci USA* 99, 2344–2349.

Bove, J., Zhou, C., Jackson-Lewis, V., Taylor, J., Chu, Y., Rideout, H. J., Wu, D. C., Kordower, J. H., Petrucelli, L., and Przedborski, S. (2006). Proteasome inhibition and Parkinson's disease modeling. *Ann Neurol* 60, 260–264.

Braak, H., Del Tredici, K., Rub, U., de Vos, R. A., Jansen Steur, E. N., and Braak, E. (2003). Staging of brain pathology related to sporadic Parkinson's disease. *Neurobiol Aging* 24, 197–211.

Brinkley, B. R., Barham, S. S., Barranco, S. C., and Fuller, G. M. (1974). Rotenone inhibition of spindle microtubule assembly in mammalian cells. *Exp Cell Res* 85, 41–46.

Brooks, A. I., Chadwick, C. A., Gelbard, H. A., Cory-Slechta, D. A., and Federoff, H. J. (1999). Paraquat elicited neurobehavioral syndrome caused by dopaminergic neuron loss. *Brain Res* 823, 1–10.

Brundin, P., Isacson, O., Gage, F. H., Prochiantz, A., and Bjorklund, A. (1986). The rotating 6-hydroxydopamine-lesioned mouse as a model for assessing functional effects of neuronal grafting. *Brain Res* **366**, 346–349.

Burns, R. S., Chiueh, C. C., Markey, S. P., Ebert, M. H., Jacobowitz, D. M., and Kopin, I. J. (1983). A primate model of parkinsonism: Selective destruction of dopaminergic neurons in the pars compacta of the substantia nigra by N-methyl-4-phenyl-1,2,3,6-tetrahydropyridine. *Proc Natl Acad Sci USA* **80**, 4546–4550.

Burns, R. S., LeWitt, P. A., Ebert, M. H., Pakkenberg, H., and Kopin, I. J. (1985). The clinical syndrome of striatal dopamine deficiency. Parkinsonism induced by 1-methyl-4-phenyl-1,2,3,6-tetrahydropyridine (MPTP). *N Engl J Med* **312**, 1418–1421.

Cenci, M. A., Whishaw, I. Q., and Schallert, T. (2002). Animal models of neurological deficits: How relevant is the rat? *Nat Rev Neurosci* **3**, 574–579.

Chan, P., DeLanney, L. E., Irwin, I., Langston, J. W., and Di Monte, D. (1991). Rapid ATP loss caused by 1-methyl-4-phenyl-1,2,3,6-tetrahydropyridine in mouse brain. *J Neurochem* **57**, 348–351.

Chaudhuri, A., Bowling, K., Funderburk, C., Lawal, H., Inamdar, A., Wang, Z., and O'Donnell, J. M. (2007). Interaction of genetic and environmental factors in a *Drosophila* parkinsonism model. *J Neurosci* **27**, 2457–2467.

Choi, H. S., An, J. J., Kim, S. Y., Lee, S. H., Kim, D. W., Yoo, K. Y., Won, M. H., Kang, T. C., Kwon, H. J., Kang, J. H., Cho, S. W., Kwon, O. S., Park, J., Eum, W. S., and Choi, S. Y. (2006). PEP-1-SOD fusion protein efficiently protects against paraquat-induced dopaminergic neuron damage in a Parkinson disease mouse model. *Free Radic Biol Med* **41**, 1058–1068.

Coulom, H., and Birman, S. (2004). Chronic exposure to rotenone models sporadic Parkinson's disease in *Drosophila melanogaster*. *J Neurosci* **24**, 10993–10998.

Curtius, H. C., Wolfensberger, M., Steinmann, B., Redweik, U., and Siegfried, J. (1974). Mass fragmentography of dopamine and 6-hydroxydopamine. Application to the determination of dopamine in human brain biopsies from the caudate nucleus. *J Chromatogr* **99**, 529–540.

Day, B. J., Patel, M., Calavetta, L., Chang, L. Y., and Stamler, J. S. (1999). A mechanism of paraquat toxicity involving nitric oxide synthase. *Proc Natl Acad Sci USA* **96**, 12760–12765.

Fahn, S., and Sulzer, D. (2004). Neurodegeneration and neuroprotection in Parkinson disease. *NeuroRx* **1**, 139–154.

Feany, M. B., and Bender, W. W. (2000). A *Drosophila* model of Parkinson's disease. *Nature* **404**, 394–398.

Fleming, S. M., Zhu, C., Fernagut, P. O., Mehta, A., Dicarlo, C. D., Seaman, R. L., and Chesselet, M. F.

(2004). Behavioral and immunohistochemical effects of chronic intravenous and subcutaneous infusions of varying doses of rotenone. *Exp Neurol* **187**, 418–429.

Forno, L. S., DeLanney, L. E., Irwin, I., and Langston, J. W. (1993). Similarities and differences between MPTP-induced parkinsonism and Parkinson's disease. Neuropathologic considerations. *Adv Neurol* **60**, 600–608.

Gorell, J. M., Johnson, C. C., Rybicki, B. A., Peterson, E. L., and Richardson, R. J. (1998). The risk of Parkinson's disease with exposure to pesticides, farming, well water, and rural living. *Neurology* **50**, 1346–1350.

Grant, H., Lantos, P. L., and Parkinson, C. (1980). Cerebral damage in paraquat poisoning. *Histopathology* **4**, 185–195.

Hasegawa, E., Takeshige, K., Oishi, T., Murai, Y., and Minakami, S. (1990). 1-Methyl-4-phenylpyridinium (MPP$^+$) induces NADH-dependent superoxide formation and enhances NADH-dependent lipid peroxidation in bovine heart submitochondrial particles. *Biochem Biophys Res Commun* **170**, 1049–1055.

Hefti, F., Melamed, E., Sahakian, B. J., and Wurtman, R. J. (1980). Circling behavior in rats with partial, unilateral nigro-striatal lesions: Effect of amphetamine, apomorphine, and DOPA. *Pharmacol Biochem Behav* **12**, 185–188.

Heikkila, R., and Cohen, G. (1971). Inhibition of biogenic amine uptake by hydrogen peroxide: A mechanism for toxic effects of 6-hydroxydopamine. *Science* **172**, 1257–1258.

Heikkila, R. E., Sieber, B. A., Manzino, L., and Sonsalla, P. K. (1989). Some features of the nigrostriatal dopaminergic neurotoxin 1-methyl-4-phenyl-1,2,3,6-tetrahydropyridine (MPTP) in the mouse. *Mol Chem Neuropathol* **10**, 171–183.

Hertzman, C., Wiens, M., Bowering, D., Snow, B., and Calne, D. (1990). Parkinson's disease: A case–control study of occupational and environmental risk factors. *Am J Ind Med* **17**, 349–355.

Hisata, J. (2002). Final supplemental environmental impact statement. Lake and stream rehabilitation: Rotenone use and health risks (Washington Department of Fish and Wildlife).

Höglinger, G. U., Feger, J., Annick, P., Michel, P. P., Karine, P., Champy, P., Ruberg, M., Oertel, W. H., and Hirsch, E. C. (2003). Chronic systemic complex I inhibition induces a hypokinetic multisystem degeneration in rats. *J Neurochem* **84**, 1–12.

Höglinger, G. U., Rizk, P., Muriel, M. P., Duyckaerts, C., Oertel, W. H., Caille, I., and Hirsch, E. C. (2004). Dopamine depletion impairs precursor cell proliferation in Parkinson disease. *Nat Neurosci* **7**, 726–735.

Iancu, R., Mohapel, P., Brundin, P., and Paul, G. (2005). Behavioral characterization of a unilateral

6-OHDA-lesion model of Parkinson's disease in mice. *Behav Brain Res* **162**, 1–10.

Jackson-Lewis, V., and Przedborski, S. (2007). Protocol for the MPTP mouse model of Parkinson's disease. *Nat Protoc* **2**, 141–151.

Jackson-Lewis, V., Jakowec, M., Burke, R. E., and Przedborski, S. (1995). Time course and morphology of dopaminergic neuronal death caused by the neurotoxin 1-methyl-4-phenyl-1,2,3,6-tetrahydropyridine. *Neurodegeneration* **4**, 257–269.

Javitch, J. A., D'Amato, R. J., Strittmatter, S. M., and Snyder, S. H. (1985). Parkinsonism-inducing neurotoxin, N-methyl-4-phenyl-1,2,3,6-tetrahydropyridine: Uptake of the metabolite N-methyl-4-phenylpyridinium by dopamine neurons explains selective toxicity. *Proc Natl Acad Sci USA* **82**, 2173–2177.

Jeon, B. S., Jackson-Lewis, V., and Burke, R. E. (1995). 6-Hydroxydopamine lesion of the rat substantia nigra: Time course and morphology of cell death. *Neurodegeneration* **4**, 131–137.

Jiang, H., Jackson-Lewis, V., Muthane, U., Dollison, A., Ferreira, M., Espinosa, A., Parsons, B., and Przedborski, S. (1993). Adenosine receptor antagonists potentiate dopamine receptor agonist-induced rotational behavior in 6-hydroxydopamine-lesioned rats. *Brain Res* **613**, 347–351.

Jonsson, G. (1983). Chemical lesioning techniques: Monoamine neurotoxins. In *Handbook of Chemical Neuroanatomy. Methods in Chemical Neuroanatomy, 1st edn* (A. Bjorklund, and T. Hokfelt, Eds.), Vol. 1, pp. 463–507. Elsevier Science Publishers B.V, Amsterdam.

Kirik, D., Georgievska, B., Burger, C., Winkler, C., Muzyczka, N., Mandel, R. J., and Bjorklund, A. (2002). Reversal of motor impairments in parkinsonian rats by continuous intrastriatal delivery of L-dopa using rAAV-mediated gene transfer. *Proc Natl Acad Sci USA* **99**, 4708–4713.

Klivenyi, P., St Clair, D., Wermer, M., Yen, H. C., Oberley, T., Yang, L., and Beal, M. F. (1998). Manganese superoxide dismutase overexpression attenuates MPTP toxicity. *Neurobiol Dis* **5**, 253–258.

Kordower, J. H., Emborg, M. E., Bloch, J., Ma, S. Y., Chu, Y., Leventhal, L., McBride, J., Chen, E. Y., Palfi, S., Roitberg, B. Z., Brown, W. D., Holden, J. E., Pyzalski, R., Taylor, M. D., Carvey, P., Ling, Z., Trono, D., Hantraye, P., Deglon, N., and Aebischer, P. (2000). Neurodegeneration prevented by lentiviral vector delivery of GDNF in primate models of Parkinson's disease. *Science* **290**, 767–773.

Kordower, J. H., Kanaan, N. M., Chu, Y., Suresh Babu, R., Stansell, J., III, Terpstra, B. T., Sortwell, C. E., Steece-Collier, K., and Collier, T. J. (2006). Failure of proteasome inhibitors to provide a model of Parkinson's disease in rats and monkeys. *Ann Neurol* **60**, 264–268.

Kostrzewa, R. M., and Jacobowitz, D. M. (1974). Pharmacological actions of 6-hydroxydopamine. *Pharmacol Rev* **26**, 199–288.

Kumar, S. (2007). Caspase function in programmed cell death. *Cell Death Differ* **14**, 32–43.

Kurkowska-Jastrzebska, I., Wronska, A., Kohutnicka, M., Członkowski, A., and Członkowska, A. (1999). The inflammatory reaction following 1-methyl-4-phenyl-1,2,3,6-tetrahydropyridine intoxication in mouse. *Exp Neurol* **156**, 50–61.

Langston, J. W. (2006). The Parkinson's complex: Parkinsonism is just the tip of the iceberg. *Ann Neurol* **59**, 591–596.

Langston, J. W., and Ballard, P. (1984). Parkinsonism induced by 1-methyl-4-phenyl-1,2,3,6-tetrahydropyridine (MPTP): Implications for treatment and the pathogenesis of Parkinson's disease. *Can J Neurol Sci* **11**, 160–165.

Langston, J. W., Ballard, P., and Irwin, I. (1983). Chronic parkinsonism in humans due to a product of meperidine-analog synthesis. *Science* **219**, 979–980.

Langston, J. W., Forno, L. S., Tetrud, J., Reeves, A. G., Kaplan, J. A., and Karluk, D. (1999). Evidence of active nerve cell degeneration in the substantia nigra of humans years after 1-methyl-4-phenyl-1,2,3,6-tetrahydropyridine exposure. *Ann Neurol* **46**, 598–605.

Lapointe, N., St-Hilaire, M., Martinoli, M. G., Blanchet, J., Gould, P., Rouillard, C., and Cicchetti, F. (2004). Rotenone induces non-specific central nervous system and systemic toxicity. *FASEB J* **18**, 717–719.

Limousin, P., Krack, P., Pollak, P., Benazzouz, A., Ardouin, C., Hoffmann, D., and Benabid, A. L. (1998). Electrical stimulation of the subthalamic nucleus in advanced Parkinson's disease. *N Engl J Med* **339**, 1105–1111.

Liou, H. H., Tsai, M. C., Chen, C. J., Jeng, J. S., Chang, Y. C., Chen, S. Y., and Chen, R. C. (1997). Environmental risk factors and Parkinson's disease: A case–control study in Taiwan. *Neurology* **48**, 1583–1588.

Liu, Y., Roghani, A., and Edwards, R. H. (1992). Gene transfer of a reserpine-sensitive mechanism of resistance to N-methyl-4-phenyl-pyridinium. *Proc Natl Acad Sci USA* **89**, 9074–9078.

Luthman, J., Fredriksson, A., Sundstrom, E., Jonsson, G., and Archer, T. (1989). Selective lesion of central dopamine or noradrenaline neuron systems in the neonatal rat: Motor behavior and monoamine alterations at adult stage. *Behav Brain Res* **33**, 267–277.

Manning-Bog, A. B., McCormack, A. L., Li, J., Uversky, V. N., Fink, A. L., and Di Monte, D. A. (2002). The herbicide paraquat causes up-regulation and aggregation of alpha-synuclein in mice: Paraquat and alpha-synuclein. *J Biol Chem* **277**, 1641–1644.

Manning-Bog, A. B., Reaney, S. H., Chou, V. P., Johnston, L. C., McCormack, A. L., Johnston, J.,

Langston, J. W., and Di Monte, D. A. (2006). Lack of nigrostriatal pathology in a rat model of proteasome inhibition. *Ann Neurol* 60, 256–260.

Marshall, L. E., and Himes, R. H. (1978). Rotenone inhibition of tubulin selfassembly. *Biochim Biophys Acta* 543, 590–594.

Marti, M. J., James, C. J., Oo, T. F., Kelly, W. J., and Burke, R. E. (1997). Early developmental destruction of terminals in the striatal target induces apoptosis in dopamine neurons of the substantia nigra. *J Neurosci* 17, 2030–2039.

Marti, M. J., Saura, J., Burke, R. E., Jackson-Lewis, V., Jimenez, A., Bonastre, M., and Tolosa, E. (2002). Striatal 6-hydroxydopamine induces apoptosis of nigral neurons in the adult rat. *Brain Res* 958, 185–191.

McCormack, A. L., Thiruchelvam, M., Manning-Bog, A. B., Thiffault, C., Langston, J. W., Cory-Slechta, D. A., and Di Monte, D. A. (2002). Environmental risk factors and Parkinson's disease: Selective degeneration of nigral dopaminergic neurons caused by the herbicide paraquat. *Neurobiol Dis* 10, 119–127.

McNaught, K. S., Perl, D. P., Brownell, A. L., and Olanow, C. W. (2004). Systemic exposure to proteasome inhibitors causes a progressive model of Parkinson's disease. *Ann Neurol* 56, 149–162.

Mizuno, Y., Sone, N., and Saitoh, T. (1987). Effects of 1-methyl-4-phenyl-1,2,3,6-tetrahydropyridine and 1-methyl-4-phenylpyridinium ion on activities of the enzymes in the electron transport system in mouse brain. *J Neurochem* 48, 1787–1793.

Morato, G. S., Lemos, T., and Takahashi, R. N. (1989). Acute exposure to maneb alters some behavioral functions in the mouse. *Neurotoxicol Teratol* 11, 421–425.

Muqit, M. M., and Feany, M. B. (2002). Modelling neurodegenerative diseases in *Drosophila*: A fruitful approach? *Nat Rev Neurosci* 3, 237–243.

Nass, R., Hall, D. H., Miller, D. M., III, and Blakely, R. D. (2002). Neurotoxin-induced degeneration of dopamine neurons in *Caenorhabditis elegans*. *Proc Natl Acad Sci USA* 99, 3264–3269.

Nicklas, W. J., Vyas, I., and Heikkila, R. E. (1985). Inhibition of NADH-linked oxidation in brain mitochondria by MPP^+, a metabolite of the neurotoxin MPTP. *Life Sci* 36, 2503–2508.

Novikova, L., Garris, B. L., Garris, D. R., and Lau, Y. S. (2006). Early signs of neuronal apoptosis in the substantia nigra pars compacta of the progressive neurodegenerative mouse 1-methyl-4-phenyl-1,2,3,6-tetrahydropyridine/probenecid model of Parkinson's disease. *Neuroscience* 140, 67–76.

Peng, J., Mao, X. O., Stevenson, F. F., Hsu, M., and Andersen, J. K. (2004). The herbicide paraquat induces dopaminergic nigral apoptosis through sustained activation of the JNK pathway. *J Biol Chem* 279, 32626–32632.

Peng, J., Stevenson, F. F., Doctrow, S. R., and Andersen, J. K. (2005). Superoxide dismutase/catalase mimetics are neuroprotective against selective paraquat-mediated dopaminergic neuron death in the substantial nigra: Implications for Parkinson disease. *J Biol Chem* 280, 29194–29198.

Petzinger, G. M., and Langston, J. W. (1998). The MPTP-lesioned non human primate: A model in Parkinson's disease. In *Advances in Neurodegenerative Disorders. Parkinson's Disease* (J. Marwah , and H. Teitelbaum, Eds.), pp. 113–148. Prominent, Scottsdale.

Przedborski, S., and Ischiropoulos, H. (2005). Reactive oxygen and nitrogen species: Weapons of neuronal destruction in models of Parkinson's disease. *Antioxid Redox Signaling* 7, 685–693.

Przedborski, S., and Vila, M. (2003). The 1-methy-4-phenyl-1,2,3,6-tetrahydropyridine mouse model. A tool to explore the pathogenesis of Parkinson's disease. *Ann NY Acad Sci* 991, 189–198.

Przedborski, S., Jackson-Lewis, V., Popilskis, S., Kostic, V., Levivier, M., Fahn, S., and Cadet, J. L. (1991). Unilateral MPTP-induced parkinsonism in monkeys. A quantitative autoradiographic study of dopamine D1 and D2 receptors and re-uptake sites. *Neurochirurgie* 37, 377–382.

Przedborski, S., Levivier, M., Jiang, H., Ferreira, M., Jackson-Lewis, V., Donaldson, D., and Togasaki, D. M. (1995). Dose-dependent lesions of the dopaminergic nigrostriatal pathway induced by intrastriatal injection of 6-hydroxydopamine. *Neuroscience* 67, 631–647.

Przedborski, S., Jackson-Lewis, V., Naini, A. B., Jakowec, M., Petzinger, G., Miller, R., and Akram, M. (2001). The parkinsonian toxin 1-methyl-4-phenyl-1,2,3,6-tetrahydropyridine (MPTP): A technical review of its utility and safety. *J Neurochem* 76, 1265–1274.

Pycock, C. J. (1980). Turning behaviour in animals. *Neuroscience* 5, 461–514.

Ries, V., Henchcliffe, C., Kareva, T., Rzhetskaya, M., Bland, R., During, M. J., Kholodilov, N., and Burke, R. E. (2006). Oncoprotein Akt/PKB induces trophic effects in murine models of Parkinson's disease. *Proc Natl Acad Sci USA* 103, 18757–18762.

Rossetti, Z. L., Sotgiu, A., Sharp, D. E., Hadjiconstantinou, M., and Neff, N. H. (1988). 1-Methyl-4-phenyl-1,2,3,6-tetrahydropyridine (MPTP) and free radicals *in vitro*. *Biochem Pharmacol* 37, 4573–4574.

Saner, A., and Thoenen, H. (1971). Model experiments on the molecular mechanism of action of 6-hydroxydopamine. *Mol Pharmacol* 7, 147–154.

Saporito, M. S., Brown, E. M., Miller, M. S., and Carswell, S. (1999). CEP-1347/KT-7515, an inhibitor of c-jun N-terminal kinase activation, attenuates the 1-methyl-4-phenyl tetrahydropyridine-mediated loss of nigrostriatal dopaminergic neurons *in vivo*. *J Pharmacol Exp Ther* 288, 421–427.

Saporito, M. S., Hudkins, R. L., and Maroney, A. C. (2002). Discovery of CEP-1347/KT-7515, an inhibitor of the JNK/SAPK pathway for the treatment of neurodegenerative diseases. *Prog Med Chem* 40, 23–62.

Sauer, H., and Oertel, W. H. (1994). Progressive degeneration of nigrostriatal dopamine neurons following intrastriatal terminal lesions with 6-hydroxydopamine: A combined retrograde tracing and immunocytochemical study in the rat. *Neuroscience* 59, 401–415.

Schapira, A. H., Cooper, J. M., Dexter, D., Jenner, P., Clark, J. B., and Marsden, C. D. (1989). Mitochondrial complex I deficiency in Parkinson's disease. *Lancet* 1, 1269.

Schapira, A. H., Cleeter, M. W., Muddle, J. R., Workman, J. M., Cooper, J. M., and King, R. H. (2006). Proteasomal inhibition causes loss of nigral tyrosine hydroxylase neurons. *Ann Neurol* 60, 253–255.

Schuler, F., and Casida, J. E. (2001). Functional coupling of PSST and ND1 subunits in NADH:ubiquinone oxidoreductase established by photoaffinity labeling. *Biochim Biophys Acta* 1506, 79–87.

Semchuk, K. M., Love, E. J., and Lee, R. G. (1993). Parkinson's disease: A test of the multifactorial etiologic hypothesis. *Neurology* 43, 1173–1180.

Sherer, T. B., Kim, J. H., Betarbet, R., and Greenamyre, J. T. (2003). Subcutaneous rotenone exposure causes highly selective dopaminergic degeneration and alpha-synuclein aggregation. *Exp Neurol* 179, 9–16.

Shimizu, K., Ohtaki, K., Matsubara, K., Aoyama, K., Uezono, T., Saito, O., Suno, M., Ogawa, K., Hayase, N., Kimura, K., and Shiono, H. (2001). Carrier-mediated processes in blood–brain barrier penetration and neural uptake of paraquat. *Brain Res* 906, 135–142.

Spillantini, M. G., Crowther, R. A., Jakes, R., Hasegawa, M., and Goedert, M. (1998). Alpha-synuclein in filamentous inclusions of Lewy bodies from Parkinson's disease and dementia with Lewy bodies. *Proc Natl Acad Sci USA* 95, 6469–6473.

Takahashi, R. N., Rogerio, R., and Zanin, M. (1989). Maneb enhances MPTP neurotoxicity in mice. *Res Commun Chem Pathol Pharmacol* 66, 167–170.

Talpade, D. J., Greene, J. G., Higgins, D. S., Jr., and Greenamyre, J. T. (2000). *In vivo* labeling of mitochondrial complex I (NADH:ubiquinone oxidoreductase) in rat brain using [³H]dihydrorotenone. *J Neurochem* 75, 2611–2621.

Tatton, N. A., and Kish, S. J. (1997). *In situ* detection of apoptotic nuclei in the substantia nigra compacta of 1-methyl-4-phenyl-1,2,3,6-tetrahydropyridine-treated mice using terminal deoxynucleotidyl transferase labelling and acridine orange staining. *Neuroscience* 77, 1037–1048.

Teismann, P., Tieu, K., Choi, D. K., Wu, D. C., Naini, A., Hunot, S., Vila, M., Jackson-Lewis, V., and Przedborski, S. (2003). Cyclooxygenase-2 is instrumental in Parkinson's disease neurodegeneration. *Proc Natl Acad Sci USA* 100, 5473–5478.

Thiruchelvam, M., Richfield, E. K., Baggs, R. B., Tank, A. W., and Cory-Slechta, D. A. (2000). The nigrostriatal dopaminergic system as a preferential target of repeated exposures to combined paraquat and maneb: Implications for Parkinson's disease. *J Neurosci* 20, 9207–9214.

Thiruchelvam, M., Prokopenko, O., Cory-Slechta, D. A., Richfield, E. K., Buckley, B., and Mirochnitchenko, O. (2005). Overexpression of superoxide dismutase or glutathione peroxidase protects against the paraquat + maneb-induced Parkinson disease phenotype. *J Biol Chem* 280, 22530–22539.

Tranzer, J. P., and Thoenen, H. (1967). Ultramorphologische Veränderungen der sympathischen Nervenendigungen der Katze nach Vorbehandlung mit 5- und 6-Hydroxy-Dopamin. *Naunyn Schmiedebergs Arch Exp Pathol Pharmakol* 257, 73–75.

Ungerstedt, U. (1968). 6-Hydroxydopamine induced degeneration of central monoamine neurons. *Eur J Pharmacol* 5, 107–110.

Ungerstedt, U. (1971). Adipsia and aphagia after 6-hydroxydopamine induced degeneration of the nigro-striatal dopamine system. *Acta Physiol Scand Suppl* 367, 95–122.

Ungerstedt, U. (1976). 6-Hydroxydopamine-induced degeneration of the nigrostriatal dopamine pathway: The turning syndrome. *Pharmacol Ther* 2, 37–40.

Ungerstedt, U., and Arbuthnott, G. (1970). Quantitative recording of rotational behaviour in rats after 6-hydroxydopamine lesions of the nigrostriatal dopamine system. *Brain Res* 24, 485–493.

Vila, M., Jackson-Lewis, V., Vukosavic, S., Djaldetti, R., Liberatore, R., Offen, D., Korsmeyer, S. J., and Przedborski, S. (2001). Bax ablation prevents dopaminergic neurodegeneration in the 1-methyl-4-phenyl-1,2,3,6-tetrahydropyridine mouse model of Parkinson's disease. *Proc Natl Acad Sci USA* 98, 2837–2842.

Yang, L., Matthews, R. T., Schultz, J. B., Klockgether, T., Liao, A. W., Martinou, J. C., Penney, J. B., Hyman, B. T., and Beal, M. F. (1998). 1-methyl-4-phenyl-1,2,3,6-tetrahydropyridine neurotoxicity is attenuated in mice overexpressing Bcl-2. *J Neurosci* 18, 8145–8152.

Zeng, B. Y., Bukhatwa, S., Hikima, A., Rose, S., and Jenner, P. (2006). Reproducible nigral cell loss after systemic proteasomal inhibitor administration to rats. *Ann Neurol* 60, 248–252.

Zhu, C., Vourc'h, P., Fernagut, P. O., Fleming, S. M., Lacan, S., Dicarlo, C. D., Seaman, R. L., and Chesselet, M. F. (2004). Variable effects of chronic subcutaneous administration of rotenone on striatal histology. *J Comp Neurol* 478, 418–426.

11

THE MPTP MOUSE MODEL OF PARKINSON'S DISEASE: THE TRUE, THE FALSE, AND THE UNKNOWN

VERNICE JACKSON-LEWIS[1] AND SERGE PRZEDBORSKI[1,2]

[1]*Departments of Neurology, Columbia University, New York, NY, USA*
[2]*Pathology and Cell Biology, Columbia University, New York, NY, USA*

INTRODUCTION

Parkinson's disease (PD) appears to be the second most frequent degenerative disorder of the nervous system in elderly next to the dementia of Alzheimer's. In part, because a of number of celebrities have publicly acknowledged being affected with PD, its cardinal features which include tremor at rest, muscle rigidity, slowness of voluntary movement, and postural instability (Fahn and Przedborski, 2005) are by now universally known. Although PD neuropathology comprises a number of different neurotransmitter pathways, the key motor abnormalities exhibited by PD patients are due mainly, though not exclusively, to the degeneration of the ascending nigrostriatal pathway and the ensuing deficit in brain dopamine (Dauer and Przedborski, 2003). As with several other prominent neurodegenerative disorders, PD is essentially a sporadic condition whose etiology is unknown and whose pathogenesis is only partially understood (Dauer and Przedborski, 2003). In attempts to shed light on both of these issues, PD investigators rely heavily on a host of experimental model systems ranging from unicellular to multicellular organisms and from invertebrates to vertebrates in which some but never all of the PD hallmarks are recapitulated. Many of these different and useful models are reviewed in this book and will thus not be discussed here. Among the various means of inducing PD-like lesions, a variety of toxins of uncertain relevance to the cause of PD have been used to destroy dopaminergic neurons (Przedborski and Tieu, 2005). As pointed out elsewhere, the popularity of such an approach stems form the premise that dopaminergic neurons may have a stereotyped death cascade that can be activated by a range of insults (Jackson-Lewis and Przedborski, 2007).

Among the various and commonly used dopaminergic neurotoxins, MPTP (1-methyl-4-phenyl-1,2,3, 6-tetrahydropyridine) has received the lion's share of attention for the past few decades for at least three reasons. First of all, MPTP is the only known dopaminergic neurotoxin capable of causing a clinical picture in both humans and monkeys (Langston *et al.*, 1983) that is indistinguishable from PD. Second, while its handling requires a series of precautions (Przedborski *et al.*, 2001), the use of MPTP is rather straightforward since it typically does not require any particular piece of equipment such as a stereotaxic frame nor does it require surgery on live animals as needed for 6-hydroxydopamine (Sotelo *et al.*, 1973) or rotenone (Betarbet *et al.*, 2000). And finally, MPTP produces a reliable and reproducible lesion of the nigrostriatal dopaminergic pathway after its systemic administration, which is often not the case for other publicized poisons (Bove *et al.*, 2006). While MPTP monkeys remain the gold standard for the pre-clinical testing of new therapies for PD, most of the studies geared toward unraveling the mechanisms underlying the demise

█ TABLE 11.1   Popular assumptions about the MPTP mouse model of PD and their validity

	Assumption	Evaluation
1	MPTP is neurotoxic *per se*	False
2	DAT explains MPTP neurotoxic specificity	False
3	MPTP causes a pure dopaminergic neurodegeneration	False
4	MPTP causes an acute death of dopaminergic neurons	Probably true
5	MPTP does not produce Lewy bodies	Not always true
6	MPTP does not cause parkinsonism in mice	True
7	Mice lack neuromelanin	True
8	Age does not alter mouse susceptibility to MPTP	False
9	Different sensitivity to MPTP among mouse strains	True for the loss of striatal dopamine levels Unproven for the loss of nigral neuron numbers

Abbreviations: DAT = dopamine transporter; MPTP = 1-methyl-4-phenyl-1,2,3,6-tetrahydropyridine; PD = Parkinson's disease.

of dopaminergic neurons have been performed primarily in mice. However, over the years, there have been tendencies toward applying uncritically a considerable series of assumptions which have often undermined either the sound interpretations of the data generated in MPTP mice or have even discouraged investigators from using this valuable model. The purpose of this chapter is to evaluate some of the most pervasive of these assumptions and to discuss their truthfulness. To facilitate this analysis, we have subdivided the selected notions about the MPTP mouse model into six central assumptions, to be evaluated individually in the following sections (Table 11.1).

ASSUMPTION 1: MPTP IS NEUROTOXIC

While this assumption may be a question of semantics, it is false, for in, reality, MPTP is not neurotoxic *per se*. MPTP is a pro-toxin which must undergo a complex, multistep activation process prior to acquiring its neurotoxic properties (Dauer and Przedborski, 2003). As indicated by its octanol/water partition coefficient of 15.6 (Riachi *et al.*, 1989), MPTP is a highly lipophilic molecule. This enables MPTP to readily permeate lipid bilayer membranes and thus, to cross the blood–brain barrier freely. Consequently, MPTP can accumulate in the brain parenchyma in a matter of minutes after its systemic administration (Markey *et al.*, 1984), but as is, it will not destroy any neurons. Instead, once in the brain, MPTP must be rapidly converted into 1-methyl-4-phenylpyridinium (MPP^+) via a monoamine oxidase

(MAO) dependent reaction (Chiba *et al.*, 1984) in order to become neurotoxic. This transformation requires two biochemical steps. During the first reaction, MPTP undergoes a two electron oxidation catalyzed by MAO, which gives rise to the unstable intermediate 1-methyl-4-phenyl-2,3-dihydropyridinium ($MPDP^+$) (Castagnoli, Jr. *et al.*, 1985). As for the second reaction, there is some consensus about the fact that $MPDP^+$ is further oxidized non-enzymatically to form MPP^+, but there is still disagreement as to how this event occurs. It was shown that, once formed, $MPDP^+$ can quickly undergo spontaneous disproportionation to MPP^+ and MPTP, at least *in vitro* (Chiba *et al.*, 1985; Fritz *et al.*, 1985). However, the *in vivo* relevance of this observation has been questioned by several experts since the study was done at millimolar concentrations. Alternatively, it has been suggested that the oxidation of $MPDP^+$ to MPP^+ could be mediated by superoxide radicals (Castagnoli, Jr. *et al.*, 1985); this suggestion would explain why transgenic mice expressing high levels of superoxide dismutase were found more resistant to MPTP (Przedborski *et al.*, 1992) and why dopaminergic neurons, in which the auto-oxidation of dopamine takes place, generate superoxide radicals (Cohen, 1984), may be particularly susceptible to MPTP.

In most animals, there are two MAO isoforms, namely MAO-A and MAO-B and the use of selective MAO inhibitors has shed light onto important aspects of MPTP bioactivation. For instance, since the MAO-B antagonist deprenyl, but not the MAO-A antagonist clorgyline blocks the conversion of

MPTP to MPP$^+$ *in vitro*, Chiba *et al.* have correctly inferred that MAO-B is pivotal in this reaction (Chiba *et al.*, 1984). Also important to note is the fact that since the cellular distribution of MAO-B in the brain (Westlund *et al.*, 1985, 1988; Kitahama *et al.*, 1991) is primarily found in astrocytes and serotonergic neurons, and not in dopaminergic neurons, it is believed that the conversion of MPTP to MPDP$^+$ takes place in these non-dopaminergic cells. Even in living animals, it is now well-established that the MAO-B-dependent bioconversion of MPTP into MPP$^+$ is critical to its neurotoxicity as elegantly shown by targeting MAO-B either genetically using knockout MAO-B mice (Grimsby *et al.*, 1997), or pharmacologically using a variety of antagonists such as deprenyl (Heikkila *et al.*, 1984; Da Prada *et al.*, 1987). Although all known neurotoxic analogs of MPTP must be activated to become neurotoxic, not all rely on MAO-B as some like 2'-methyl-MPTP rely on MAO-A (Kindt *et al.*, 1988) for its bioactivation and deleterious effects on dopaminergic neurons.

ASSUMPTION 2: DOPAMINE TRANSPORTERS EXPLAIN MPTP NEUROTOXIC SPECIFICITY

The pervasive notion that MPTP-induced toxicity is specific to dopaminergic neurons because of their specific expression of the dopamine transporter (DAT) is false and probably emanates from one of the most serious misinterpretations of the data available on MPTP neurotoxicity. Once formed, MPP$^+$ is presumably released from glial and serotonergic cells into the extracellular space prior to entering the neighboring neurons. In contrast to MPTP, MPP$^+$ has an octanol/water partition coefficient of only 0.09 (Riachi *et al.*, 1989), which indicates that, while being a lipophilic cation, MPP$^+$ is unable to effectively cross cellular lipid bilayer membranes. Instead, its entry into adjacent neurons depends on specialized plasma membrane carriers. It has been known since the mid-1980s that MPP$^+$ binds with high affinity to DAT (Javitch and Snyder, 1984; Javitch *et al.*, 1985). Unfortunately the titles of these two otherwise landmark papers, namely "Uptake of MPP$^+$ by dopamine neurons explains selectivity of parkinsonism-inducing neurotoxin, MPTP" (Javitch and Snyder, 1984) and "Parkinsonism-inducing neurotoxin, N-methyl-4-phenyl-1,2,3, 6-tetrahydropyridine: uptake of the metabolite N-methyl-4-phenylpyridine

by dopamine neurons explains selective toxicity" (Javitch *et al.*, 1985) are likely responsible for the misunderstanding discussed herein. While in both papers, the authors show that MPP$^+$ exhibits a Kd of 170 nM for its uptake into dopaminergic neurons, they also clearly show that MPP$^+$ harbors a Kd of 65 nM for its uptake into norepinephrine neurons. Less known is the fact that MPP$^+$ also binds avidly to other carrier systems such as that for serotonin (Heikkila *et al.*, 1985a) and the so-called organic cation/extraneuronal catecholamine transporters (Kristufek *et al.*, 2002; Shang *et al.*, 2003) which, among other monoamines, transport histamine (Gasser *et al.*, 2006). Collectively, these data indicate that MPP$^+$ can gain access to a wide variety of neurons and not solely or even preferentially to dopaminergic neurons. That being said, if one studies the damage to the dopaminergic neurons only – which is often the case – then, it is correct to say that MPP$^+$ entry via DAT is mandatory for the killing of these specific neurons (Javitch *et al.*, 1985; Bezard *et al.*, 1999). However, the real issue here is not whether DAT is needed to kill dopaminergic neurons, but rather to know whether MPTP's preference for killing dopaminergic neurons is due to DAT, as many researchers are tempted to believe. The answer to this critical question is unambiguously, no! Indeed, an analysis of the literature clearly indicates that despite expressing DAT, not all dopaminergic cell groups are equally affected by MPTP (Javitch *et al.*, 1985; Seniuk *et al.*, 1990; Jackson-Lewis *et al.*, 1995). For instance, although dopaminergic neurons in the ventral tegmental area seem to express greater numbers of DAT than those dopaminergic neurons in the substantia nigra, the former are consistently less damaged by the systemic administration of MPTP than the latter (Seniuk *et al.*, 1990; Muthane *et al.*, 1994; Jackson-Lewis *et al.*, 1995). Furthermore, noradrenergic and serotonergic neurons, that can be found, respectively, in locus coeruleus and in raphe nuclei, do take up MPP$^+$ (Heikkila *et al.*, 1985a; Javitch *et al.*, 1985) and are susceptible to it (Namura *et al.*, 1987), although they are never damaged as much as the dopaminergic neurons in the substantia nigra (Seniuk *et al.*, 1990). Finally, similar to noradrenergic neurons in the brain, adrenomedullary chromaffin cells are only minimally affected by MPTP even though large quantities of MPP$^+$ are found in the gland after peripheral administration to animals (Johannessen *et al.*, 1986). From all of the above, it can be emphatically stated that the preferential toxicity of MPTP to substantia nigral dopaminergic neurons cannot be explained merely by either the presence of DAT or its level of expression.

ASSUMPTION 3: MPTP CAUSES A PURE DOPAMINERGIC NEURODEGENERATION IN MICE

For many, the idea that MPTP causes a pure dopaminergic neurodegeneration in mice challenges the validity of using the MPTP mouse model to study PD. Although the degeneration of the nigrostriatal pathway in PD is a critical feature of the disease, there are neuropathological changes that extend well beyond dopaminergic neurons (reviewed by Hornykiewicz and Kish, 1987). As emphasized by Dauer and Przedborski (2003) in postmortem brains from PD patients, histological alterations are found in the noradrenergic (locus coeruleus), serotonergic (raphe), and cholinergic (nucleus basalis of Meynert, dorsal motor nucleus of vagus) systems, as well as in the cerebral cortex (especially cingulate and entorhinal cortices), olfactory bulb, and autonomic nervous system. Is there compelling evidence that a comparable multi-systemic neuropathology exists in the MPTP mouse model? As indicated above, aside from the lesion in the nigrostriatal pathway, the mesolimbic and the mesocortical systems are also affected as is the locus coeruleus (Seniuk *et al.*, 1990). Damage to the serotoninergic terminals have also been reported, although without evidence of overt cell body loss in raphe nuclei (Gupta *et al.*, 1984; Namura *et al.*, 1987). The use of FluoroJade, a general fluorescent marker for dying cells, has disclosed loci of dying cells in unexpected areas of the central nervous system (CNS) such as the midline and intralaminar thalamic nuclei (Schmued *et al.*, 1997) and we have found evidence of DNA damage, used as a marker for cellular lesions in hippocampal neurons (Chesselet and Przedborski, unpublished observation). Furthermore, as in PD, several studies have also documented alterations in the sympathetic system in MPTP mice (Fuller *et al.*, 1984; Takatsu *et al.*, 2002). In contrast, unlike in PD, we cannot offer an unambiguous answer as to whether or not the nucleus basalis of Meynert, the nucleus dorsal motor nucleus of vagus, the cingulate cortex or the entorhinal cortex are affected in mice after MPTP administration. Perhaps, the neuropathology of the MPTP mouse should be revisited now that many effective and reliable regimens of MPTP have been validated (Jackson-Lewis and Przedborski, 2007) and that refined quantitative morphological methods such as Stereology can be used.

In light of the above, it is thus clear that the notion of MPTP damaging only the dopaminergic system is false. Retrospectively, it is very possible that this wrong idea originated from the fact that the majority of the MPTP studies in mice, including our own, have unfortunately reported on histological and biochemical changes at the level of the substantia nigra and the striatum, ignoring, most of the time, all other brain regions. Although the absence of references to non-nigrostriatal damage in all of these papers should have been regarded as no more than evidence of omission, it has become commonplace to use this lack of information as evidence of the absence of non-nigrostriatal damage. From a practical standpoint, this misunderstanding may have contributed to the small amount of progress made toward the understanding of the physiopathology and consequently the treatment of the non-dopaminergic features of PD including depression, dementia, and constipation by diverting the interest of MPTP researchers away from the non-nigrostriatal pathways. On the other hand, if one accepts the view that the degenerative process of PD is driven by a similar cascade of deleterious events in all neurons, then it becomes somewhat immaterial which type of neuron one studies as long as one can accurately and reproducibly identify the neuron of interest as well as its molecular changes.

ASSUMPTION 4: MPTP MICE DEPART FROM PD ON SEVERAL IMPORTANT ASPECTS

As pointed out in the introduction, MPTP, like the other available experimental models of PD, is no more than a model. None are exactly recapitulating all of the key features of PD. Therefore, it is correct to say that the MPTP mouse model departs from PD on several important aspects. All investigators must thus be aware of all of these shortcomings and must carefully evaluate, prior to any study, whether or not these issues may have any bearing on the experimental design or the interpretation of the anticipated results. Among the imperfections of the MPTP mouse model of PD, some aspects pertain to the use of MPTP and thus are common to all living organisms intoxicated with this compound and some pertain to the use of mice.

Toxin-Related Shortcomings of the MPTP Mouse Model of PD

The two common issues raised in using MPTP to provoke PD-like neurodegeneration include the rate of neuronal death following MPTP administration and the lack of Lewy bodies in the lesioned mice.

Rate of Neuronal Death

Following the classical 1-day regimen of four injections of 20 mg/kg free base MPTP, the death of dopaminergic neurons begins within 12 h after the last injection and is completed by 4–5 days post-injections (Jackson-Lewis *et al.*, 1995). Although the time course of dopaminergic neurodegeneration after the many other published regimens of MPTP injections is often not known, it is reasonable to believe that the decay of the nigrostriatal pathway may be consistently as swift. In contrast, in PD, the protracted worsening of the patient's clinical condition suggests a slow progressive loss of the nigrostriatal pathway. In the MPTP model, several investigators including ourselves have shown that during the active phase of neurodegeneration, a large number of neurons are succumbing at any given time (Jackson-Lewis *et al.*, 1995). Conversely, in PD, it is believed that during the active phase of neurodegeneration only a small number of neurons are succumbing at any given time. If correct, this difference may account for the striking difference in the apparent "course of the disease" seen in the MPTP mouse model and in PD. Correlatively, there may be no difference in the speed at which a given dopaminergic neuron dies in the MPTP mouse model and in PD. Until proven otherwise, one cannot exclude that in the former, many neurons are dying quickly over a short period of time whereas in the latter, very few neurons are dying quickly over a prolonged period of time. Therefore, if one experiment relies on what happens to the nigrostriatal dopaminergic neurons at the population level, then using the MPTP mouse model might be problematic. However, if it relies on what happens to the nigrostriatal dopaminergic neurons at the cellular level, then, based on current literature, we fail to see why should the use of the MPTP mouse model to, for example, study the pathogenesis of PD or test neuroprotective strategies, not be valid.

Lack of Lewy Bodies

Aside from a loss of neurons, the neuropathology of PD is characterized by the presence of proteinaceous inclusions called Lewy bodies (Shults, 2006). These intraneuronal inclusions are not pathognomonic for PD, but, with only a few exceptions (see below), all cases of PD will show them (Shults, 2006). In contrast, in almost all postmortem analyses of MPTP intoxicated humans, monkeys, or mice, Lewy bodies could never be unequivocally documented. It has been thus commonly thought that MPTP could not recruit the proper molecular machinery indispensable for the formation of these proteinaceous inclusions; this view has been often used by the detractors of the MPTP model to challenge its relevance to PD. However, it may be that the lack of Lewy bodies in MPTP tissue samples is due less to any molecular reason rather than to the rate of injury. Indeed, until recently, all cases of either accidental or purposely induced MPTP intoxications occurred after an acute administration of the toxin (Langston *et al.*, 1999; Jackson-Lewis and Przedborski, 2007). In contrast, intraneuronal proteinaceous inclusions have been documented in mice after infusion of MPTP with osmotic pumps (Fornai *et al.*, 2005) and after the co-administration of MPTP with probenecid which, supposedly, should have slowed the clearance of the toxin (Meredith *et al.*, 2002). Although one may argue that these observations must be independently confirmed and whether or not these inclusions are true Lewy bodies, both studies raise the hypothesis that when MPTP injury is more sustained, Lewy body-like inclusions can form. In support of this speculation are the observations that the mitochondrial poison rotenone, which operates presumably very much like MPP^+, when acutely administered to rodents is not apparently associated with Lewy body-like inclusions (Heikkila *et al.*, 1985b; Ferrante *et al.*, 1997) whereas when chronically infused, striking synuclein-positive intraneuronal inclusions can be observed (Betarbet *et al.*, 2000).

In our opinion, a more fundamental issue is to know whether or not MPTP can cause Lewy body formation rather than whether or not the formation of these inclusions is mandatory for a model of PD to be accepted as such. For the past decade, we have become aware of cases of familial PD linked to mutations in the genes encoding for the E3 ubiquitin ligase *parkin* or in the kinase *Leucine repeat rich kinase-2*, in which postmortem examinations have revealed a typical loss of nigrostriatal dopaminergic neurons and yet in the absence of Lewy bodies (Takahashi *et al.*, 1994; Zimprich *et al.*, 2004). Although, at the very least, it is true that having an *in vivo* model of PD with Lewy bodies may be crucial to elucidating the pathobiology underlying their formation, it is unclear how important these inclusions are in the neurodegenerative process and whether such a feature is an essential attribute for a model of PD to have.

Mouse-Related Shortcomings of the MPTP Mouse Model of PD

The two common issues raised in using mice as a model organism to study PD include the behavioral manifestations they develop after MPTP administration and their lack of neuromelanin.

Behavioral Abnormalities in MPTP Mice

There is no doubt that a behavioral correlate of the nigrostriatal dopaminergic pathway degeneration in MPTP mice would be desirable, but can one expect it to parallel the motor deficits of PD. In other words, can we be sure that rodents do develop parkinsonism. In reality, no one knows with certainty whether, for instance, a reduction of spontaneous motor activity or some type of limb trembling in mice is a murine equivalent to the PD bradykinesia or resting tremor. Unfortunately, it is not infrequent to read in the PD modeling literature, authors who use casually and misleadingly descriptors borrowed from the clinical field to depict behavioral alterations in rodents. As discussed elsewhere (Dauer and Przedborski, 2003), in the MPTP mouse model, it would be more prudent to hunt for rodent-specific behavioral changes that involve striatal functions, such as habituation to a novel environment or the ability to learn a stimulus-response paradigm. Furthermore, since the motor system organization differs in rodents and humans, we believe that the value of a particular behavioral phenotype depends more upon its relationship to striatal dopamine function rather than to its apparent similarity to PD. Furthermore, we take the position that any behavior in mice claimed to result from striatal dopamine deficiency should improve with dopamine replacement. Previously, in a collaborative work, striking motor alterations were found in MPTP mice, but these could not be alleviated by a proper exogenous dopaminergic stimulation (Smith et al., 2003). While the observed motor alterations are interesting, clearly it would be erroneous to attribute them to the neurodegeneration of the nigrostriatal pathway and the ensuing deficit in brain dopamine simply because they emerged after MPTP administration.

As indicated above, MPP^+ can gain access to a host of different cell types and, even if mainly dopaminergic neurons are actually dying, all of these cells are intoxicated to some extent giving rise to a transient state of generalized toxicosis. Depending on the regimen and the dose administered, this state of generalized toxicosis can be more or less severe and associated with a variety of manifestations that can emerge as early as a few hours after the first injection and that include death, prostration, loss of spontaneous motor activity and coordination as well as a significant decrease in drinking, eating, and grooming behaviors. In the case of the 1-day 4 × 20mg/kg MPTP regimen, these behavioral alterations can last for 1–2 days after the fourth injection of MPTP and, by 3–4 days, the MPTP mice have recovered normal spontaneous behaviors. Accordingly, it is essential that any behavioral testing be done at a timeframe away from the early phase of toxicosis and that any behavioral testing done soon after the last injection (Sedelis et al., 2001) be taken with circumspection *a fortiory* if not confirmed at least a week later. Using a computerized treadmill gait analysis system (Mouse Specifics Inc., Boston, MA), we found distinct gait alterations in MPTP mice displaying >85% deficit in striatal dopamine which could be reversed by L-DOPA, 10 days after the last injection (Jackson-Lewis and Przedborski, unpublished observation). However, MPTP mice with even a 70% reduction in striatal dopamine did not differ from the normal control mice on any of the gait parameters assessed 10 days after the last injection, making this type of sophisticated test of limited usefulness. Chesselet et al. have validated a battery of novel, simple behavior tests in genetic models of PD (Fleming and Chesselet, 2006). It will thus be interesting to evaluate these simple tests in MPTP mice to see if a better quantitative relationship can be found between the degree of striatal dopamine loss and the magnitude of behavioral impairment.

Lack of Neuromelanin

As reviewed by Fedorow et al. (2005) and Zecca et al. (2001), neuromelanin is a naturally occurring dark pigment found in specific populations of catecholaminergic neurons in the brain. The neuromelanin granules are found in the cytoplasm of primarily brainstem dopaminergic and noradrenergic neurons, mostly at the level of the ventral midbrain (Bogerts, 1981; Saper and Petito, 1982). Although little is still known about its biology, neuromelanin granules are composed of a core of pheomelanin decorated at their surface with a coat of eumelanin (Bush et al., 2006). In the context of PD, the loss of neuromelanin-containing neurons represents a basic pathological diagnostic criterion for the disease, and an inverse relationship between the amount of neuromelanin contained within the midbrain dopaminergic neurons and the relative vulnerability of these neurons to PD has been reported (Hirsch *et al.*, 1988). These observations suggest a role for neuromelanin in neurodegeneration. Among several possible mechanisms put forward over the years, the ability of neuromelanin to interact with transition metals, especially iron, and to mediate intracellular oxidative mechanisms has received particular attention (Fedorow *et al.*, 2005). A series of studies on neuromelanin and MPTP (D'Amato *et al.*, 1986, 1987a, b, c) have also led to the attractive hypothesis that this pigment, by binding to toxins such as MPP^+, could play the role of an internal toxic reservoir. In the case of MPTP, during the phase of intoxication, neuromelanin would be saturated with MPP^+ and then it would progressively release MPP^+, subjecting the cell to a long-lasting toxic insult. This scenario found particular appeal among researchers attached to the idea that the etiology of PD is, at least in part, environmental (Dauer and Przedborski, 2003). The reason for discussing the question of neuromelanin here resides in the fact that while this pigment is abundant in the adult human brainstem, few other species have neuromelanin in the brain (Marsden, 1961; Barden and Levine, 1983). Even more relevant to this chapter is the absence of neuromelanin in all of the commonly used laboratory animals, including mice (Marsden, 1961; Barden and Levine, 1983). Because all of these animals have dopamine and noradrenaline neurons, albeit they do not have neuromelanized neurons, the formation of this pigment cannot be an inevitable consequence of catecholamine metabolism. This fact raises the troublesome possibility that the molecular make-up of primates and rodents may diverge with respect to some critical aspects that determine the susceptibility of catecholaminergic neurons to the PD disease process.

ASSUMPTION 5: THERE IS NO EFFECT OF AGE ON MPTP SUSCEPTIBILITY

PD is considered as a prototypical example of human disease of the aging brain, mainly because its incidence increases dramatically with age (Fahn and Przedborski, 2005). Even in familial cases of PD, which typically have an earlier onset compared to their sporadic counterpart, the clinical expression of these rare forms of PD still emerges during adulthood (Vila and Przedborski, 2004). Notably, brain imaging studies with positron emission tomography suggest that the preclinical stage of PD corresponds to only 4.5 years (Moeller and Eidelberg, 1997). Thus, the

occurrence of PD motor abnormalities in most cases after the age of 40 cannot merely be explained by a slow demise of dopaminergic neurons from birth. Instead, the most parsimonious explanation for the adult onset of PD is to assume that the neurodegenerative process starts at some point during life either because of a slow accumulation of toxic molecules (e.g., misfolded proteins) or a progressive decrease of protective mechanisms (e.g., antioxidant) enabling the ongoing disease process to reach a pathological threshold only at sometime late in life. In the context of MPTP, several studies have consistently shown that older mice are more sensitive to the toxin than younger mice (Jarvis and Wagner, 1985; Gupta *et al.*, 1986; Irwin *et al.*, 1992; Tatton *et al.*, 1992; Ali *et al.*, 1993). Although the basis for this age-related effect in mice remains unknown, it clearly supports the idea that, with age, the nigrostriatal dopaminergic pathway becomes more susceptible to toxic insults such as MPTP and perhaps to the PD disease process. In light of this, it is legitimate to wonder whether researchers in the field take this factor into account in their experimental designs. Most of the time and most of us do not, since we typically use 6–10-week-old mice for our MPTP studies, which would correspond to young adults in humans probably under the age of 40. The rationale for using under-aged mice is unfortunately not scientific, but rather practical and economical as older mice are not readily available and cost significantly more than their younger counterparts. Perhaps, the compromise may be to continue to do the ground work in young mice and to encourage investigators to confirm their most salient findings in older mice.

ASSUMPTION 6: DIFFERENT MOUSE STRAINS EXHIBIT DIFFERENT SENSITIVITIES TO MPTP

A host of factors such as age, gender, weight, and genetic background have been reported to influence both the reproducibility and the extent of the MPTP lesion in mice. Information about the extensive and systematic assessment of these different factors in mice can be found in the following references (Heikkila *et al.*, 1989; Giovanni *et al.*, 1991, 1994a, b; Miller *et al.*, 1998; Hamre *et al.*, 1999; Staal and Sonsalla, 2000). Among these important factors it is nonetheless worth discussing here the issue-genetic background. Indeed, it has been known for decades that different strains of mice (and even within a given strain obtained from different vendors) can exhibit strikingly distinct sensitivity to MPTP (Heikkila *et al.*, 1989). This differential sensitivity has been reported to be transmissible as an autosomal-dominant trait (Hamre *et al.*, 1999). However, a careful examination of the above-cited studies revealed that most of these conclusions have been obtained by assessing biochemical indices (e.g., striatal dopamine levels), and not structural indices (e.g., substantia nigra dopaminergic neuron numbers). In a ongoing study, we were astounded to find that among eight different lines of inbred mice injected with the same regimen of MPTP, while the loss of striatal dopamine levels did indeed vary among the different lines ranging from 10% to 90%, there were no significant differences in the proportion of substantia nigra dopaminergic neuronal loss (Jackson-Lewis and Przedborski, unpublished observation). If confirmed, this would argue that all common inbred lines are equally sensitive to MPTP, a notion that may have far-reaching implications for many researchers who often use engineered mice from different genetic backgrounds for their MPTP work.

In this section, it may also be worth discussing two additional points related to neuronal death. First, often the quantification of the substantia nigra dopaminergic neurons is determined by counting neurons within a defined anatomical area and labeled by a phenotypic marker such as tyrosine hydroxylase or DAT. However, these useful markers can readily be downregulated in intoxicated neurons which may skew the counting. This issue can be avoided by counting substantia nigra dopaminergic neurons at a time-frame away from intoxication, to allow enough time for surviving neurons to recover a normal expression of tyrosine hydroxylase or of DAT. Non-phenotypic-based methods can also be used. For instance, the intrastriatal injection of retrograde tracers such as FluoroGold has been successfully used in toxic models of PD for this purpose (Sauer and Oertel, 1994; Mandel *et al.*, 1997; Anderson *et al.*, 2001). While technically challenging, this is quite an elegant procedure for quantifying the number of surviving substantia nigra neurons irrespective of their expression of tyrosine hydroxylase or DAT.

A second issue pertains to the notion that MPTP does not kill neurons nor does it provoke a stable lesion. In the earlier days of MPTP, various regimens were used and some might have been subliminal in that they may have been sufficient to intoxicate neurons and transiently downregulate the expression of tyrosine hydroxylase, but actually not to kill any neurons. While we have no experimental data to support this view, we can state emphatically that with the current common regimen, such as the 1 day 4×20 mg/kg dosing regimen, neurons are dying as demonstrated by silver staining (Jackson-Lewis *et al.*, 1995) or FluoroJade labeling (Schmued *et al.*, 1997; Schmued and Hopkins, 2000). In keeping with this, we do not have any evidence that once such lesion is completed, there is any recovery as far as the number of neurons in the substantia nigra is concerned (Jackson-Lewis *et al.*, 1995; Bezard *et al.*, 1997, 2000). Could it be that striatal innervation recovers over time in these MPTP-lesioned mice? This is entirely possible as some degree of striatal reinnervation could arise from the spared mesolimbic and, to a lesser extent, nigrostriatal projections (Bezard *et al.*, 2000). This may be the reason why some investigators have reported a recovery of striatal dopaminergic levels over time in MPTP mice (Bohn *et al.*, 1987; Bezard *et al.*, 2000). Therefore, based on the current regimens of MPTP, there is indisputable evidence that the toxin kills dopaminergic neurons and causes a stable loss of dopaminergic cell bodies in the substantia nigra, but one cannot exclude that some reinnervation (i.e., collateral sprouting) occurs in the striatum.

CONCLUSION

This chapter has provided a detailed discussion on the MPTP mouse model of PD. By focusing on several selected and controversial questions, we have shown that many of the popular objections made about this model are false (Table 1). This chapter also gave us the opportunity to stress some of the important limitations of this model and to discuss how these may impact the interpretations of one's data. Like all of the other models of PD, the MPTP mouse model is not perfect, but we believe that the secret to using this model successfully is to make sure that it is suitable for the question being investigated. Safety and technical issues related to its utilization and actual experimental protocols have not been reviewed here but can be found elsewhere (Przedborski *et al.*, 2001; Jackson-Lewis and Przedborski, 2007).

ACKNOWLEDGEMENTS

The authors wish to thank Dr. Delphine Prou for her suggestions on the manuscript and acknowledge the support of the US Department of Defense Grant DAMD 17–03–1, the

Parkinson's Disease Foundation, the Muscular Dystrophy Association/Wings–over–Wall Street, and the US National Institutes of Health Grants NS42269, NS38370, and NS11766, AG 21617, for this work.

REFERENCES

Ali, S. F., David, S. N., and Newport, G. D. (1993). Age-related susceptibility to MPTP-induced neurotoxicity in mice. *Neurotoxicology* **14**, 29–34.

Anderson, D. W., Neavin, T., Smith, J. A., and Schneider, J. S. (2001). Neuroprotective effects of pramipexole in young and aged MPTP-treated mice. *Brain Res* **905**, 44–53.

Barden, H., and Levine, S. (1983). Histochemical observations on rodent brain melanin. *Brain Res Bull* **10**, 847–851.

Betarbet, R., Sherer, T. B., MacKenzie, G., Garcia-Osuna, M., Panov, A. V., and Greenamyre, J. T. (2000). Chronic systemic pesticide exposure reproduces features of Parkinson's disease. *Nat Neurosci* **3**, 1301–1306.

Bezard, E., Dovero, S., Bioulac, B., and Gross, C. (1997). Effects of different schedules of MPTP administration on dopaminergic neurodegeneration in mice. *Exp Neurol* **148**, 288–292.

Bezard, E., Gross, C. E., Fournier, M. C., Dovero, S., Bloch, B., and Jaber, M. (1999). Absence of MPTP-induced neuronal death in mice lacking the dopamine transporter. *Exp Neurol* **155**, 268–273.

Bezard, E., Dovero, S., Imbert, C., Boraud, T., and Gross, C. E. (2000). Spontaneous long-term compensatory dopaminergic sprouting in MPTP-treated mice. *Synapse* **38**, 363–368.

Bogerts, B. (1981). A brainstem atlas of catecholaminergic neurons in man, using melanin as a natural marker. *J Comp Neurol* **197**, 63–80.

Bohn, M. C., Cupit, L., Marciano, F., and Gash, D. M. (1987). Adrenal medulla grafts enhance recovery of striatal dopaminergic fibers. *Science* **237**, 913–916.

Bove, J., Zhou, C., Jackson-Lewis, V., Taylor, J., Chu, Y., Rideout, H. J., Wu, D. C., Kordower, J. H., Petrucelli, L., and Przedborski, S. (2006). Proteasome inhibition and Parkinson's disease modeling. *Ann Neurol* **60**, 260–264.

Bush, W. D., Garguilo, J., Zucca, F. A., Albertini, A., Zecca, L., Edwards, G. S., Nemanich, R. J., and Simon, J. D. (2006). The surface oxidation potential of human neuromelanin reveals a spherical architecture with a pheomelanin core and a eumelanin surface. *Proc Natl Acad Sci USA* **103**, 14785–14789.

Castagnoli, N., Jr., Chiba, K., and Trevor, A. J. (1985). Potential bioactivation pathways for the neurotoxin 1-methyl-4-phenyl-1,2,3,6-tetrahydropyridine (MPTP). *Life Sci* **36**, 225–230.

Chiba, K., Trevor, A., and Castagnoli, N., Jr. (1984). Metabolism of the neurotoxic tertiary amine, MPTP, by brain monoamine oxidase. *Biochem Biophys Res Commun* **120**, 574–578.

Chiba, K., Peterson, L. A., Castagnoli, K. P., Trevor, A. J., and Castagnoli, N., Jr. (1985). Studies on the molecular mechanism of bioactivation of the selective nigrostriatal toxin 1-methyl-4-phenyl-1,2,3,6-tetrahydropyridine. *Drug Metab Dispos* **13**, 342–347.

Cohen, G. (1984). Oxy-radical toxicity in catecholamine neurons. *Neurotoxicology* **5**, 77–82.

D'Amato, R. J., Lipman, Z. P., and Snyder, S. H. (1986). Selectivity of the parkinsonian neurotoxin MPTP: Toxic metabolite MPP$^+$ binds to neuromelanin. *Science* **231**, 987–989.

D'Amato, R. J., Alexander, G. M., Schwartzman, R. J., Kitt, C. A., Price, D. L., and Snyder, S. H. (1987a). Evidence for neuromelanin involvement in MPTP-induced neurotoxicity. *Nature* **327**, 324–326.

D'Amato, R. J., Alexander, G. M., Schwartzman, R. J., Kitt, C. A., Price, D. L., and Snyder, S. H. (1987b). Neuromelanin: A role in MPTP-induced neurotoxicity. *Life Sci* **40**, 705–712.

D'Amato, R. J., Benham, D. F., and Snyder, S. H. (1987c). Characterization of the binding of N-methyl-4-phenylpyridine, the toxic metabolite of the parkinsonian neurotoxin N-methyl-4-phenyl-1,2,3,6-tetrahydropyridine, to neuromelanin. *J Neurochem* **48**, 653–658.

Da Prada, M., Kettler, R., Keller, H. H., Bonetti, E. P., and Imhof, R. (1987). Ro 16-6491: A new reversible and highly selective MAO-B inhibitor protects mice from the dopaminergic neurotoxicity of MPTP. *Adv Neurol* **45**, 175–178.

Dauer, W., and Przedborski, S. (2003). Parkinson's disease: Mechanisms and models. *Neuron* **39**, 889–909.

Fahn, S., and Przedborski, S. (2005). Parkinsonism. In *Merritt's Neurology* (L. P. Rowland, Ed.), 11th edn., pp. 828–846. Lippincott Williams & Wilkins, New York.

Fedorow, H., Tribl, F., Halliday, G., Gerlach, M., Riederer, P., and Double, K. L. (2005). Neuromelanin in human dopamine neurons: Comparison with peripheral melanins and relevance to Parkinson's disease. *Prog Neurobiol* **75**, 109–124.

Ferrante, R. J., Schulz, J. B., Kowall, N. W., and Beal, M. F. (1997). Systemic administration of rotenone produces selective damage in the striatum and globus pallidus, but not in the substantia nigra. *Brain Res* **753**, 157–162.

Fleming, S. M., and Chesselet, M. F. (2006). Behavioral phenotypes and pharmacology in genetic mouse models of Parkinsonism. *Behav Pharmacol* **17**, 383–391.

Fornai, F., Schluter, O. M., Lenzi, P., Gesi, M., Ruffoli, R., Ferrucci, M., Lazzeri, G., Busceti, C. L., Pontarelli, F., Battaglia, G., Pellegrini, A., Nicoletti, F., Ruggieri, S., Paparelli, A., and Sudhof, T. C. (2005). Parkinson-like syndrome induced by continuous MPTP infusion: Convergent roles of the ubiquitin- proteasome system and {alpha}-synuclein. *Proc Natl Acad Sci USA* **102**, 3413–3418.

Fritz, R. R., Abell, C. W., Patel, N. T., Gessner, W., and Brossi, A. (1985). Metabolism of the neurotoxin in MPTP by human liver monoamine oxidase B. *FEBS Lett* **186**, 224–228.

Fuller, R. W., Hahn, R. A., Snoddy, H. D., and Wikel, J. H. (1984). Depletion of cardiac norepinephrine in rats and mice by 1-methyl- 4-phenyl-1,2,3,6-tetrahydropyridine (MPTP). *Biochem Pharmacol* **19**, 2957–2960.

Gasser, P. J., Lowry, C. A., and Orchinik, M. (2006). Corticosterone-sensitive monoamine transport in the rat dorsomedial hypothalamus: Potential role for organic cation transporter 3 in stress-induced modulation of monoaminergic neurotransmission. *J Neurosci* **26**, 8758–8766.

Giovanni, A., Sieber, B. A., Heikkila, R. E., and Sonsalla, P. K. (1991). Correlation between the neostriatal content of the 1-methyl-4- phenylpyridinium species and dopaminergic neurotoxicity following 1-methyl-4-phenyl-1,2,3,6-tetrahydropyridine administration to several strains of mice. *J Pharmacol Exp Ther* **257**, 691–697.

Giovanni, A., Sieber, B.-A., Heikkila, R. E., and Sonsalla, P. K. (1994a). Studies on species sensitivity to the dopaminergic neurotoxin 1-methyl-4-phenyl-1,2,3,6-tetrahydropyridine. Part 1: Systemic administration. *J Pharmacol Exp Ther* **270**, 1000–1007.

Giovanni, A., Sonsalla, P. K., and Heikkila, R. E. (1994b). Studies on species sensitivity to the dopaminergic neurotoxin 1-methyl-4-phenyl-1,2,3,6-tetrahydropyridine. Part 2: Central administration of 1-methyl-4-phenylpyridinium. *J Pharmacol Exp Ther* **270**, 1008–1014.

Grimsby, J., Toth, M., Chen, K., Kumazawa, T., Klaidman, L., Adams, J. D., Karoum, F., Gal, J., and Shih, J. C. (1997). Increased stress response and beta-phenylethylamine in MAOB-deficient mice. *Nat Genet* **17**, 206–210.

Gupta, M., Felten, D. L., and Gash, D. M. (1984). MPTP alters central catecholamine neurons in addition to the nigrostriatal system. *Brain Res Bull* **13**, 737–742.

Gupta, M., Gupta, B. K., Thomas, R., Bruemmer, V., Sladek, J. R., Jr, and Felten, D. L. (1986). Aged mice are more sensitive to 1-methyl-4-phenyl-1,2,3,6-tetrahydropyridine treatment than young adults. *Neurosci Lett* **70**, 326–331.

Hamre, K., Tharp, R., Poon, K., Xiong, X., and Smeyne, R. J. (1999). Differential strain susceptibility following 1-methyl-4-phenyl-1,2,3,6-tetrahydropyridine (MPTP) administration acts in an autosomal dominant fashion: Quantitative analysis in seven strains of Mus musculus. *Brain Res* **828**, 91–103.

Heikkila, R. E., Manzino, L., Cabbat, F. S., and Duvoisin, R. C. (1984). Protection against the dopaminergic neurotoxicity of 1-methyl-4-phenyl-1,2,3,6-tetrahydropyridine by monoamine oxidase inhibitors. *Nature* **311**, 467–469.

Heikkila, R. E., Manzino, L., Cabbat, F. S., and Duvoisin, R. C. (1985a). Effects of 1-methyl-4-phenyl-1,2,3,6-tetrahydropyridine (MPTP) and several of its analogues on the dopaminergic nigrostriatal pathway in mice. *Neurosci Lett* **58**, 133–137.

Heikkila, R. E., Nicklas, W. J., Vyas, I., and Duvoisin, R. C. (1985b). Dopaminergic toxicity of rotenone and the 1-methyl-4- phenylpyridinium ion after their stereotaxic administration to rats: Implication for the mechanism of 1-methyl-4-phenyl-1,2,3,6- tetrahydropyridine toxicity. *Neurosci Lett* **62**, 389–394.

Heikkila, R. E., Sieber, B. A., Manzino, L., and Sonsalla, P. K. (1989). Some features of the nigrostriatal dopaminergic neurotoxin 1- methyl-4-phenyl-1,2,3,6-tetrahydropyridine (MPTP) in the mouse. *Mol Chemic Neuropathol* **10**, 171–183.

Hirsch, E., Graybiel, A. M., and Agid, Y. A. (1988). Melanized dopaminergic neurons are differentially susceptible to degeneration in Parkinson's disease. *Nature* **334**, 345–348.

Hornykiewicz and Kish (1987). Biochemical pathophysiology of Parkinson's disease. In Parkinson's Disease (Yahr, M; Bergmann, K. J., eds) pp. 19–34, Raven Press, 1987.

Irwin, I., Finnegan, K. T., DeLanney, L. E., Di Monte, D., and Langston, J. W. (1992). The relationships between aging, monoamine oxidase, striatal dopamine and the effects of MPTP in C57BL/6 mice: A critical reassessment. *Brain Res* **572**, 224–231.

Jackson-Lewis, V., and Przedborski, S. (2007). Protocol for the MPTP mouse model of Parkinson's disease. *Nat Protocols* **2**, 141–151.

Jackson-Lewis, V., Jakowec, M., Burke, R. E., and Przedborski, S. (1995). Time course and morphology of dopaminergic neuronal death caused by the neurotoxin 1-methyl-4-phenyl-1,2,3,6-tetrahydropyridine. *Neurodegeneration* **4**, 257–269.

Jarvis, M. F., and Wagner, G. C. (1985). Age-dependent effects of 1-methyl-4-phenyl-1,2,3,4-tetrahydropyridine (MPTP). *Neuropharmacology* **24**, 581–583.

Javitch, J. A., and Snyder, S. H. (1984). Uptake of MPP$^+$ by dopamine neurons explains selectivity of parkinsonism-inducing neurotoxin, MPTP. *Eur J Pharmacol* **106**, 455–456.

Javitch, J. A., D'Amato, R. J., Strittmatter, S. M., and Snyder, S. H. (1985). Parkinsonism-inducing neurotoxin, N-methyl-4-phenyl-1,2,3,6-tetrahydropyridine: Uptake of the metabolite N-methyl-4-phenylpyridinium by dopamine neurons explain selective toxicity. *Proc Natl Acad Sci USA* **82**, 2173–2177.

Johannessen, J. N., Chiueh, C. C., Herkenham, M., and Markey, C. J. (1986). Relationship of the *in vivo* metabolism of MPTP to toxicity. In *MPTP: A Neurotoxin Producing a Parkinsonian Syndrome* (S. P. Markey, N. Castagnoli, Jr., A. J. Trevor, and I. J. Kopin, Eds.), pp. 173–190. Academic Press, Orlando.

Kindt, M. V., Youngster, S. K., Sonsalla, P. K., Duvoisin, R. C., and Heikkila, R. E. (1988). Role for monoamine oxidase-A (MAO-A) in the bioactivation and nigrostriatal dopaminergic neurotoxicity of the MPTP analog, 2'Me- MPTP. *Eur J Pharmacol* **146**, 313–318.

Kitahama, K., Denney, R. M., Maeda, T., and Jouvet, M. (1991). Distribution of type B monoamine oxidase immunoreactivity in the cat brain with reference to enzyme histochemistry. *Neuroscience* **44**, 185–204.

Kristufek, D., Rudorfer, W., Pifl, C., and Huck, S. (2002). Organic cation transporter mRNA and function in the rat superior cervical ganglion. *J Physiol* **543**, 117–134.

Langston, J. W., Ballard, P., and Irwin, I. (1983). Chronic parkinsonism in humans due to a product of meperidine-analog synthesis. *Science* **219**, 979–980.

Langston, J. W., Forno, L. S., Tetrud, J., Reeves, A. G., Kaplan, J. A., and Karluk, D. (1999). Evidence of active nerve cell degeneration in the substantia nigra of humans years after 1-methyl-4-phenyl-1,2,3,6-tetrahydropyridine exposure. *Ann Neurol* **46**, 598–605.

Mandel, R. J., Spratt, S. K., Snyder, R. O., and Leff, S. E. (1997). Midbrain injection of recombinant adeno-associated virus encoding rat glial cell line-derived neurotrophic factor protects nigral neurons in a progressive 6-hydroxydopamine-induced degeneration model of Parkinson's disease in rats. *Proc Natl Acad Sci USA* **94**, 14083–14088.

Markey, S. P., Johannessen, J. N., Chiueh, C. C., Burns, R. S., and Herkenham, M. A. (1984). Intraneuronal generation of a pyridinium metabolite may cause druginduced parkinsonism. *Nature* **311**, 464–467.

Marsden, C. D. (1961). Pigmentation in the nucleus substantiae nigrae of mammals. *J Anat* **95**, 256–261.

Meredith, G. E., Totterdell, S., Petroske, E., Santa, C. K., Callison, R. C., Jr., and Lau, Y. S. (2002). Lysosomal malfunction accompanies alpha-synuclein aggregation in a progressive mouse model of Parkinson's disease. *Brain Res* **956**, 156–165.

Miller, D. B., Ali, S. F., O'Callaghan, J. P., and Laws, S. C. (1998). The impact of gender and estrogen on striatal dopaminergic neurotoxicity. *Ann NY Acad Sci* **844**, 153–165.

Moeller, J. R., and Eidelberg, D. (1997). Divergent expression of regional metabolic topographies in Parkinson's disease and normal ageing. *Brain* **120**, 2197–2206.

Muthane, U., Ramsay, K. A., Jiang, H., Jackson-Lewis, V., Donaldson, D., Fernando, S., Ferreira, M., and Przedborski, S. (1994). Differences in nigral neuron number and sensitivity to 1-methyl- 4-phenyl-1,2,3, 6-tetrahydropyridine in C57/bl and CD-1 mice. *Exp Neurol* **126**, 195–204.

Namura, I., Douillet, P., Sun, C. J., Pert, A., Cohen, R. M., and Chiueh, C. C. (1987). MPP$^+$ (1-methyl-4-phenylpyridine) is a neurotoxin to dopamine-, norepinephrine- and serotonin-containing neurons. *Eur J Pharmacol* **136**, 31–37.

Przedborski, S., and Tieu, K. (2005). Toxic animal models. In *Neurodegenerative Diseases: Neurobiology, Pathogenesis and Therapeutics* (M. F. Beal, A. E. Lang, and A. Ludolph, Eds.), pp. 196–221. Cambridge, New York.

Przedborski, S., Kostic, V., Jackson-Lewis, V., Naini, A. B., Simonetti, S., Fahn, S., Carlson, E., Epstein, C. J., and Cadet, J. L. (1992). Transgenic mice with increased Cu/Zn-superoxide dismutase activity are resistant to N-methyl-4-phenyl-1,2,3,6-tetrahydropyridine-induced neurotoxicity. *J Neurosci* **12**, 1658–1667.

Przedborski, S., Jackson-Lewis, V., Naini, A., Jakowec, M., Petzinger, G., Miller, R., and Akram, M. (2001). The parkinsonian toxin 1-methyl-4-phenyl-1,2,3,6-tetrahydropyridine (MPTP): A technical review of its utility and safety. *J Neurochem* **76**, 1265–1274.

Riachi, N. J., LaManna, J. C., and Harik, S. I. (1989). Entry of 1-methyl-4-phenyl-1,2,3,6-tetrahydropyridine into the rat brain. *J Pharmacol Exp Ther* **249**, 744–748.

Saper, C. B., and Petito, C. K. (1982). Correspondence of melanin-pigmented neurons in human brain with A1–A14 catecholamine cell groups. *Brain* **105**, 87–101.

Sauer, H., and Oertel, W. H. (1994). Progressive degeneration of nigrostriatal dopamine neurons following intrastriatal terminal lesions with 6-hydroxydopamine: A combined retrograde tracing and immunocytochemical study in the rat. *Neuroscience* **59**, 401–415.

Schmued, L. C., and Hopkins, K. J. (2000). Fluoro-Jade: Novel fluorochromes for detecting toxicant-induced neuronal degeneration. *Toxicol Pathol* **28**, 91–99.

Schmued, L. C., Albertson, C., and Slikker, W., Jr. (1997). Fluoro-Jade: A novel fluorochrome for the sensitive and reliable histochemical localization of neuronal degeneration. *Brain Res* **751**, 37–46.

Sedelis, M., Schwarting, R. K., and Huston, J. P. (2001). Behavioral phenotyping of the MPTP mouse model of Parkinson's disease. *Behav Brain Res* **125**, 109–125.

Seniuk, N. A., Tatton, W. G., and Greenwood, C. E. (1990). Dose-dependent destruction of the coeruleus-cortical and nigral- striatal projections by MPTP. *Brain Res* **527**, 7–20.

Shang, T., Uihlein, A. V., Van, A. J., Kalyanaraman, B., and Hillard, C. J. (2003). 1-Methyl-4-phenylpyridinium accumulates in cerebellar granule neurons via organic cation transporter 3. *J Neurochem* **85**, 358–367.

Shults, C. W. (2006). Lewy bodies. *Proc Natl Acad Sci USA* **103**, 1661–1668.

Smith, P. D., Crocker, S. J., Jackson-Lewis, V., Jordan-Sciutto, K. L., Hayley, S., Mount, M. P., O'Hare, M. J., Callaghan, S., Slack, R. S., Przedborski, S., Anisman, H., and Park, D. S. (2003). Cyclin-dependent kinase 5 is a mediator of dopaminergic neuron loss in a mouse model of Parkinson's disease. *Proc Natl Acad Sci USA* 100, 13650–13655.

Sotelo, C., Javoy, F., Agid, Y., and Glowinski, J. (1973). Injection of 6-hydroxydopamine in the substantia nigra of the rat. I. Morphological study. *Brain Res* 58, 269–290.

Staal, R. G., and Sonsalla, P. K. (2000). Inhibition of brain vesicular monoamine transporter (VMAT2) enhances 1-methyl-4-phenylpyridinium neurotoxicity *in vivo* in rat striata. *J Pharmacol Exp Ther* 293, 336–342.

Takahashi, H., Ohama, E., Suzuki, S., Horikawa, Y., Ishikawa, A., Morita, T., Tsuji, S., and Ikuta, F. (1994). Familial juvenile parkinsonism: Clinical and pathologic study in a family. *Neurology* 44, 437–441.

Takatsu, H., Wada, H., Maekawa, N., Takemura, M., Saito, K., and Fujiwara, H. (2002). Significant reduction of 125 I-meta-iodobenzylguanidine accumulation directly caused by 1-methyl-4-phenyl-1,2,3,6-tetrahydroxypyridine, a toxic agent for inducing experimental Parkinson's disease. *Nucl Med Commun* 23, 161–166.

Tatton, W. G., Greenwood, C. E., Seniuk, N. A., and Salo, P. T. (1992). Interactions between MPTP-induced and age-related neuronal death in a murine model of Parkinson's disease. *Can J Neurol Sci* 19, 124–133.

Vila, M., and Przedborski, S. (2004). Genetic clues to the pathogenesis of Parkinson's disease. *Nat Med* 10(Suppl), S58–S62.

Westlund, K. N., Denney, R. M., Kochersperger, L. M., Rose, R. M., and Abell, C. W. (1985). Distinct monoamine oxidase A and B populations in primate brain. *Science* 230, 181–183.

Westlund, K. N., Denney, R. M., Rose, R. M., and Abell, C. W. (1988). Localization of distinct monoamine oxidase B cell populations in human brainstem. *Neuroscience* 25, 439–456.

Zecca, L., Tampellini, D., Gerlach, M., Riederer, P., Fariello, R. G., and Sulzer, D. (2001). Substantia nigra neuromelanin: Structure, synthesis, and molecular behaviour. *Mol Pathol* 54, 414–418.

Zimprich, A., Biskup, S., Leitner, P., Lichtner, P., Farrer, M., Lincoln, S., Kachergus, J., Hulihan, M., Uitti, R. J., Calne, D. B., Stoessl, A. J., Pfeiffer, R. F., Patenge, N., Carbajal, I. C., Vieregge, P., Asmus, F., Muller-Myhsok, B., Dickson, D. W., Meitinger, T., Strom, T. M., Wszolek, Z. K., and Gasser, T. (2004). Mutations in LRRK2 cause autosomal-dominant parkinsonism with pleomorphic pathology. *Neuron* 44, 601–607.

12

ACUTE AND CHRONIC ADMINISTRATION OF 1-METHYL-4-PHENYLPYRIDINIUM

PATRICIA K. SONSALLA[1], GAIL D. ZEEVALK[1] AND DWIGHT C. GERMAN[2]

[1]Department of Neurology, UMDNJ-Robert Wood Johnson Medical School Piscataway, NJ, USA
[2]Department of Psychiatry, University of Texas Southwestern Medical School, Dallas, TX, USA

INTRODUCTION

Parkinson's disease (PD), a progressive neurodegenerative disorder which afflicts primarily the aged population, presents with predominantly motor-related symptoms due to the loss of nigrostriatal dopamine (DA) neurons. While other neuronal populations are also affected in PD, it is the extensive loss of DA neurons that leads to profound immobility and eventually the cause of death in many patients. It is the DA neurons that are the primary target in experimental models of PD. Other notable features of PD pathology are the presence of intracellular Lewy bodies which contain many proteins including α-synuclein and ubiquitin, and an active microgliosis.

Various animal models of PD are used to research the disease, seeking clues to mechanisms of neurodegeneration and for testing efficacy of pharmacological agents for symptomatic relief and for neuroprotection. An ideal animal model would be the one which mimics all aspects of the disease. Unfortunately, such an animal model does not exist. Although there is no perfect model, several experimental PD animal models have been developed and have provided much information on the processes associated with neurodegeneration of DA neurons. Many of the presently used models are based on the acute or subacute administration of neurotoxicants such as 1-methyl-4-phenyl-1,2,3,6-tetrahydropyridine (MPTP), 1-methyl-4-phenylpyridinium (MPP^+), 6-hydroxydopamine (6-OHDA), or malonate using dosing and administration conditions that cause a fairly rapid death of DA neurons. One particular advantage of the acute models is in the evaluation of dynamic brain changes that occur during a toxic insult to the DA neurons (e.g., imposed metabolic stress). In the later part of the chapter, we will describe two of these models, the acute MPP^+ model and the acute malonate model, and their utility in the examination of impaired energy metabolism on DA homeostasis and for pharmacological testing of potential neuroprotective compounds. While the acute models have provided much insight into potential mechanisms underlying cell death, they do not mimic the progressive nature of PD nor present with all of the features of PD pathology. Progressive experimental models for PD are needed in order to provide a more accurate representation of the progressive nature of the disease, the contribution of factors and processes involved in neurodegeneration, especially those potentially involved in the later stages of the disease, and the ability of therapeutic intervention to slow the degeneration of neurons. We have recently developed the chronic MPP^+ model and will focus our attention on this model.

CHRONIC MPP$^+$ MODEL

Many current models utilize the mouse as the species of choice for PD models because it is readily amenable to genetic manipulation. An argument can be made, however, for the need for animal models that use different insults as well as utilize a variety of different species. This is particularly important for the evaluation of potential therapeutics as testing in different models and across species will likely best predict clinical success or conversely, unmask potential problems. We chose an outbred rather than inbred strain of rat (Sprague Dawley) with the view in mind that success for a potential therapeutic at the laboratory level must translate into a positive outcome across a widely variant genetic population. MPTP/MPP$^+$ has been used for acute models of PD since the early–mid-1980s and is well known for its preferential targeting of DA neurons due to its selective uptake of the active metabolite MPP$^+$ via the DA transporter (DAT). While the acute MPTP/ MPP$^+$ model has provided numerous insights into PD pathology and etiology, chronic, low-level perturbation of mitochondrial and cellular function may better reproduce the events important to the degeneration of neurons in the disease. For these reasons, as well as others discussed below, a chronic model of central delivery of MPP$^+$ in rats was used for model development (Yazdani et al., 2006). In establishing the model, we sought to achieve the following criteria: (1) progressive damage to DA neurons with histological and neurochemical correlates to PD, (2) low inter-animal variability in neuropathology, (3) continuous chronic low dose delivery of MPP$^+$, (4) minimal animal handling to reduce stress-related confounds, (5) low mortality, and (6) animals capable of maintaining normal weight and with minimal motor impairment.

In order to satisfy requirements of continuous consistent drug delivery and minimal animal handling, MPP$^+$ was delivered via an Alzet osmotic minipump (DURECT Corp, Cupertino, CA, model 2ML4), which delivers 2 μL of agent per hour for 28 days. Continuous delivery avoids the peaks and troughs of drug concentration observed with injection protocols. This latter approach may better represent a multi-hit paradigm rather then replicate a chronic level of dysfunction as may exist in PD patients with mitochondrial defects. In the anesthetized rat, the pump is inserted under the skin in the back of the rat below the neck. Attached to the pump via tubing is a cannula that is stereotaxically placed for delivery of MPP$^+$ into the left lateral ventricle (icv). Central delivery into the anterior portion of the lateral ventricle was done to minimize perfusion to the opposite hemisphere and produce unilateral lesioning. For chronic studies, unilateral lesioning allows the animal to eat and drink normally avoiding the complications of aphagia and adipsia that can occur in a bilaterally lesioned animal. In the chronic MPP$^+$ model, rates of weight gain between the control and the MPP$^+$ animals were similar. Final weights ($\pm$ SD) at the end of the treatment were 452 ± 16 g and 415 ± 12 g, in saline and MPP$^+$-treated animals, respectively. The final difference in weights at the end of 28 days was due to a lag period in weight gain for the MPP$^+$ group. Chronic delivery icv rather than direct delivery into the striatum was chosen as the introduction of 2 μL of fluid per hour into tissue over the course of 28 days could potentially produce non-specific damage.

Also at issue, and an advantage to the current model as well as models that provide continuous drug delivery (Betarbet et al., 2000; Fornai et al., 2005), is that once the cannula and pump placement is made, animal handling is minimized. Models that require multiple injections over the course of weeks (Petroske et al., 2001; Thiruchelvam et al., 2003) may add an unwanted stress component to the paradigm. Stress has been suggested to contribute to the loss of DA neurons in PD (Smith et al., 2002). In animal models, stress can lead to neuronal loss (Sapolsky, 1992) or enhance toxin-induced neuronal damage (Urakami et al., 1988; Smith et al., 2002). While it is unclear at present how stress impacts on the various delivery approaches in chronic models, it should be a consideration for all model development.

Dosing and Selectivity of Damage

While the selective targeting of catecholaminergic neurons by MPTP/MPP$^+$ is well established, non-selective damage can occur with high doses, particularly intracerebral infusions of MPP$^+$. To examine the dose of MPP$^+$ and selectivity of damage in the chronic model, MPP$^+$ (free base) was delivered icv in rats, at 0.086–0.86 mg/kg/day for 28 days. Striatal levels of serotonin (5-HT), γ-aminobutyric acid (GABA) and glutamate were examined for selectivity of damage. A dose-dependent loss of striatal DA was observed with an IC$_{50}$ of 0.125 mg/kg/day (Figure 12.1). No loss of striatal DA was observed

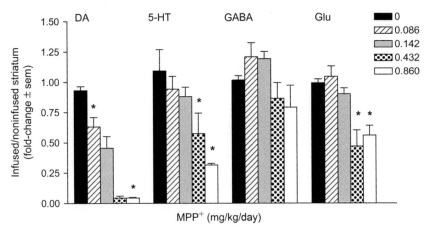

FIGURE 12.1   Dose-response effect of chronic icv MPP$^+$ infusion on striatal content of DA, serotonin (5HT), GABA and glutamate. After 28 days of MPP$^+$ infusion into the left cerebral ventricle, striatal contents of DA, 5HT, GABA, and glutamate were measured. Results (mean ± SEM from 4–5 rats/group) are presented in ratios of neurotransmitter content in left to right striata. No significant changes in neurotransmitter levels were observed in the right striata at any of the MPP$^+$ doses tested (data not shown). *Statistically different ($p < 0.05$) from vehicle controls. Reprinted from *Exp Neurol*, **200** (1); Yazdani *et al.* (2006). Rat model of Parkinson's disease: Chronic central delivery of MPP$^+$, with permission from Elsevier.

on the contralateral side at any of the doses tested (data not shown). At the lower doses tested (0.086 and 0.142 mg/kg/day), DA on the infused side was 63% and 47% of the respective contralateral side with no significant difference between left and right sides for 5-HT, GABA or glutamate. At the higher doses tested (0.432 and 0.860 mg/kg/day), there was >90% loss of DA but 5-HT and glutamate were also significantly reduced on the infused side. GABA showed a trend towards lower levels on the infused side, but results were not statistically significant. Immunoreactivity for tyrosine hydroxylase (TH) demonstrated a predominant loss of striatal TH in the dorsomedial striatum on the side of infusion (Figure 12.2), likely due to the proximity of the region to the infused ventricle. Immunostaining for TH on the contralateral side was similar to controls. Some loss of the cholinergic marker choline acetyltransferase was observed, but this was restricted to the area immediately adjacent to the infused ventricle, whereas choline acetyltransferase immunostaining throughout the rest of the striatum was normal (Figure 12.2c). Thus, selectivity of damage to the nigrostriatal DA population can be achieved with chronic administration of between 0.086 and 0.142 mg/kg/day MPP$^+$. Inclusion bodies immunopositive for ubiquitin (Figure 12.2d and

Figure 12.3) and α-synuclein (not shown) were also observed in the striatum near the infused ventricle. Increases in the DOPAC/DA ratio and in lactate levels in the striatum were observed in animals receiving 0.142 mg/kg/day or more MPP$^+$, whereas these parameters on the contralateral side for all MPP$^+$ concentrations were similar to controls. Respectively, these findings indicate an increase in DA turnover in the presence of MPP$^+$ and an increase in the anaerobic metabolism of glucose to produce lactate likely to compensate for the impairment in aerobic metabolism caused by MPP$^+$.

In the substantia nigra (SN) *pars compacta*, there was a marked loss of TH-immunoreactive neurons noted throughout all levels of the nigra (Figure 12.4). Stereological counts of TH-positive cells were decreased by 35% on the side of infusion (8892 ± 611 and 5811 ± 1014, mean ± SD, for vehicle controls and animals treated with 0.142 mg/kg/day MPP$^+$ for 28 days, respectively). Cell death rather then reversible loss of neuronal phenotype was supported by the finding that there was a similar number of Nissl stained SN neurons that were TH-negative in the control and MPP$^+$-treated groups (193 ± 31 and 201 ± 46, mean ± SD, respectively) and by the large number of neurons staining in the SN with Neurosilver (Yazdani *et al.*,

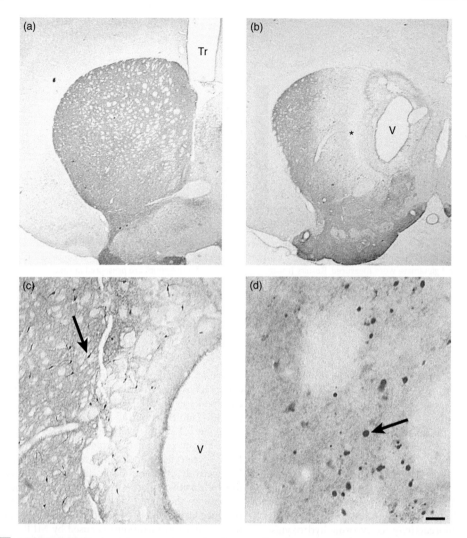

FIGURE 12.2   Striatal neuropathology after 28 days of icv infusion of MPP$^+$ (0.142 mg/kg/day). TH immunostaining in striatum of rat infused with (a) vehicle or (b) MPP$^+$. Tr = cannula track; V = ventricle. Note loss of TH immunostaining in dorsomedial region of striatum in MPP$^+$-treated rat. (c) Staining for striatal choline acetyltransferase. Note that staining for choline acetyltransferase is normal in nerve endings and cell bodies (arrow) except for a small region adjacent to the cerebral ventricle that shows depletion. (d) Presence of ubiquitin stained inclusion bodies. Inclusion bodies (arrow) were found immediately adjacent to the lateral ventricle close to where the MPP$^+$ was infused. Scale bar in panels a and b = 450 μm, panel c = 150 μm and in panel d = 14 μm. Reprinted from *Exp Neurol,* **200** (1); Yazdani *et al.* (2006). Rat model of Parkinson's disease: Chronic central delivery of 1-methyl-4-phenylpyridinium (MPP$^+$), with permission from Elsevier.

2006) indicating that they were undergoing neurodegeneration. Microglial activation as evidenced by OX-42 immunoreactivity was found in the SN (Figure 12.4e) and interestingly was observed both ipsilateral and contralateral to the infused side.

Progressive Degeneration

An important feature of a chronic model of neurodegeneration is that it recapitulates the feature of progressive and ongoing damage once the insult

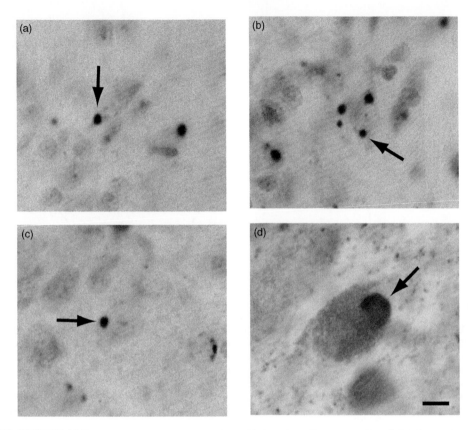

FIGURE 12.3 Striatal inclusion bodies are found 28 days after ICV infusion of MPP$^+$ (0.142 mg/kg/day). Striatal sections were stained for ubiquitin (black) and counterstained with cresyl violet. Inclusion bodies in (a–c) are depicted by arrows. A ubiquitin stained Lewy body in a nigral DA neuron from a patient with PD is shown in (d). Note the presence of neuromelanin pigment in the DA neuron. Scale bar = 10 μm. Reprinted from *Exp Neurol*, **200** (1), Yazdani *et al.* (2006). Rat model of Parkinson's disease: Chronic central delivery of 1-methyl-4-phenylpyridinium (MPP$^+$), with permission from Elsevier.

has been removed or eliminated. To address this, some animals were allowed to recover an additional 14 days after completion of MPP$^+$ infusion. These animals showed further loss of TH-positive cell numbers (3114 ± 620 neurons) when compared with animals examined immediately following 28 days of MPP$^+$ administration (5811 ± 1014 neurons). These initial findings are promising and support that this model results in the continued loss of nigral DA neurons. However, further verification of this and examination out to longer time intervals after completion of the insult are needed in order to confirm that a set of events has been triggered

that result in the continued loss of neurons in the region.

A progressive model offers the opportunity to test whether a potential neuroprotective strategy halts disease progression rather than or in addition to protecting against the initial insult. At the time of diagnosis of PD, it is estimated that there has been a loss of approximately 50% of SN neurons and 70–80% of striatal DA. The exact mechanisms underlying cell damage in PD are not clearly understood and it is possible that the mechanisms associated with the initial insult and ensuing damage as compared with the progressive degeneration

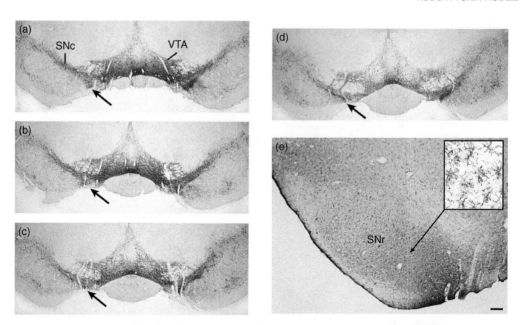

FIGURE 12.4 Loss of DA neurons and presence of microgliosis in SN of a rat chronically infused with MPP$^+$ (0.142 mg/kg/day). SN sections from rostral (a) to caudal (d) illustrate the loss of TH staining in fibers and cell bodies on the side ipsilateral to the ICV infusion of MPP$^+$ as compared to the contralateral side. Note prominent loss of medial SN pars compacta TH-positive neurons (arrows). VTA = ventral tegmental area. Scale bar = 300 μm. (e) Activation of microglia in SN of MPP$^+$-treated rat is illustrated by immunostaining with the microglial marker OX-24. Insert is a higher power view. Scale = 130 μm; insert scale = 30 μm. Reprinted from *Exp Neurol*, **200** (1), Yazdani *et al.* (2006). Rat model of Parkinson's disease: Chronic central delivery of 1-methyl-4-phenylpyridinium (MPP$^+$), with permission from Elsevier.

are not identical and may require different strategies for abrogation.

Inclusion Bodies and Abnormal Mitochondria

While α-synuclein and ubiquitin inclusion bodies were observed in the striatum, they were not observed in the SN in animals that received MPP$^+$ for 28 days or 28 days plus 14 days of recovery. Lewy bodies are a hallmark feature of PD and are observed in catecholaminergic neurons in the SN and in other brain regions. Whether their accumulation in cells is a protective mechanism to remove unwanted and potentially toxic protein aggregates or contribute to the degenerative process is at present enigmatic. Inclusion bodies have been observed in the nigral DA neurons in the chronic rat rotenone model (Betarbet *et al.*, 2000), in rats treated chronically with proteasomal inhibitors

(Fornai *et al.*, 2003) and in mice and non-human primates chronically treated with MPTP (Forno *et al.*, 1988; Meredith *et al.*, 2002). Genetic models of PD have not produced Lewy bodies in the population of neurons that are affected in the disease. The observation of inclusion bodies in the chronic MPP$^+$-treated rat striatum indicates the potential for inclusion formation in this model but may require longer periods following MPP$^+$ administration in order for their formation in more physiologically relevant regions. Although inclusion bodies were not seen in the nigra with the chronic MPP$^+$ model, examination with electron microscopy showed the presence of numerous abnormal mitochondria in nigral DA neurons (Figure 12.5). Dendritic DA terminals identified by DAT immunoreactivity showed many swollen mitochondria with broken cristae. Electron translucent or electron dense inclusions were frequently observed within mitochondria. Lewy bodies have been shown to contain degenerating mitochondria at their central

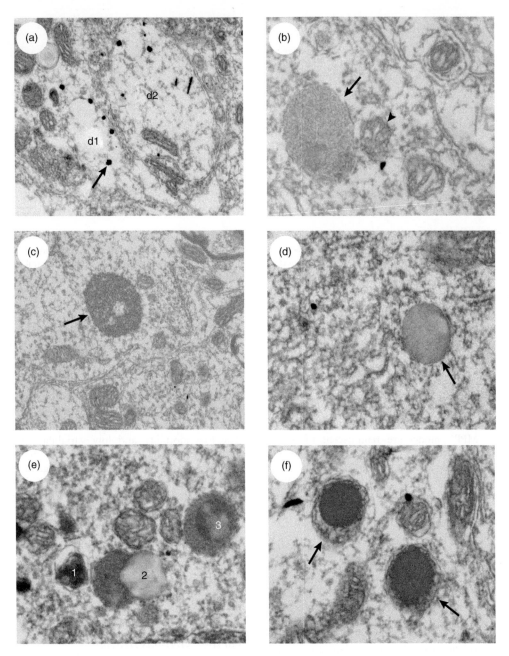

FIGURE 12.5 Abnormal mitochondria in nigral DA neurons of a rat chronically infused with MPP$^+$. Electron micrographs are from a representative rat treated with MPP$^+$ (0.142 mg/kg/day) for 28 days followed by a 2-week recovery period. (a) Two DA dendrites (d1 and d2) are identified using an antibody to the DAT. Arrow points to one of the DAT-labeled gold particles. (b) Arrow points to a very large mitochondrion located next to a normal sized mitochondrion (arrowhead). (c) An enlarged mitochondrion (arrow) which contains a clear space within its center. Note normal mitochondria in lower half of the field. (d) Mitochondrion with condensation particles. (e) Electron dense (1) and translucent (2, 3) mitochondria. (f) Mitochondria with electron dense accumulation (arrows).

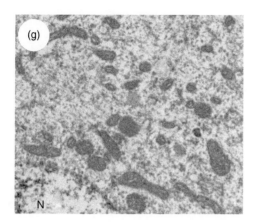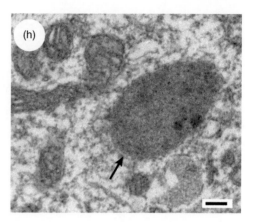

FIGURE 12.5 (continued) (g) Normal appearing mitochondria in the red nucleus. (h) Rare mitochondrion (arrow) in the red nucleus exhibit swelling. Scale bar: a, b, d–f = 0.125 μm; C = 0.25 μm; G = 0.5 μm; and h = 0.175 μm. Reprinted from *Exp Neurol*, Vol **200** (1); Yazdani *et al.* (2006). Rat model of Parkinson's disease: Chronic central delivery of 1-methyl-4-phenylpyridinium (MPP⁺), with permission from Elsevier.

core (Forno, 1986; Roy and Wolman, 1969; Gai *et al.*, 2000). Macroautophagy and the lysosomal pathway clear damaged mitochondria from the cell (Rodriguez-Enriquez *et al.*, 2004). Recent studies implicate autophagy in PD pathology (Cuervo *et al.*, 2004). It has been proposed that as mature lysosomes break down, their contents may form the seed for Lewy bodies (Meredith *et al.*, 2004). The mitochondrial inclusions observed in the chronic MPP⁺ model are reminiscent of abnormal mitochondria observed in other models of PD, that is, following acute MPTP in mouse and monkey (Tanaka *et al.*, 1988; Mizukawa *et al.*, 1990), in mice chronically treated with MPTP (Fornai *et al.*, 2005), in animals that over-express human α-synuclein (Song *et al.*, 2004) and in cybrids from PD patients (Trimmer *et al.*, 2000). It is intriguing to speculate that these abnormal mitochondria eventually contribute to Lewy body formation. A model that produces consistent inclusions would be a useful tool to investigate Lewy body formation and to determine whether cells containing Lewy bodies are lost early or late in the degenerative process. This could help provide insight into the question of the protective versus damaging nature of the Lewy body.

Variability and Mortality

Of concern with several of the chronic models of mitochondrial impairment is the high mortality associated with the model (Betarbet *et al.*, 2000; F. Fornai, personal communications). A high mortality rate makes a model unreliable for therapeutic investigation as a "survival of the fittest" phenomenon may confound findings. No mortality was observed with chronic central delivery of MPP⁺ at any of the doses tested. Variability as monitored by the loss of striatal DA at the end of 28 days ranged between 10% and 25% with the different doses of MPP⁺. The low mortality and low variability associated with the chronic, central MPP⁺ model makes it an attractive model for therapeutic testing. Indeed, we have recently used this model to test whether elevation of brain levels of glutathione with its ethyl ester derivative could provide protection when administered concurrently with MPP⁺ (Zeevalk *et al.*, 2007). Central delivery of 0.142 mg/kg/day of MPP⁺ for 28 days reduced striatal DA content by 70%. Administration of 10 mg/kg/day glutathione ethyl ester delivered via Alzet pump concurrent with MPP⁺ provided partial protection with DA levels being reduced by only 49%. (Figure 12.6).

Further Characterization

The chronic central MPP⁺ model provides a potentially promising model of PD. It produces selective and reproducible damage to the nigrostriatal DA system with consistent drug delivery while minimizing animal handling. Animals feed and drink normally eliminating the confound of poor nutritional

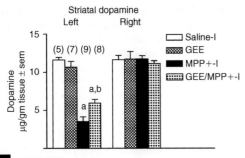

FIGURE 12.6 Glutathione monoethyl ester (GEE) partially protects against striatal DA loss produced by chronic ICV infusion of MPP$^+$. Rats received ICV infusions of vehicle, GEE (10 mg/kg/day), MPP$^+$ (0.142 mg/kg/day), or GEE plus MPP$^+$ for 28 days delivered from osmotic pumps. Results are the mean striatal DA ± SEM (5–9 rats/group). [a]Significantly different from vehicle control; [b]significantly different from MPP$^+$ group. Reprinted from *Exp Neurol*, **203** (2); Zeevalk et al. (2007). Characterization of intracellular elevation of glutathione (GSH) with GEE and GSH in brain and neuronal cultures: Relevance to Parkinson's disease, with permission from Elsevier.

status. Importantly, loss of nigrostriatal TH-positive neurons continues to occur after cessation of delivery of the toxicant. Further characterization of several aspects of the model, however, is needed to strengthen and validate its usefulness for mechanistic and pharmacological investigations. The evaluation of neuropathology was initially restricted to the nigrostriatal system and it would be of interest to determine if other brain regions such as the locus coeruleus or dorsal raphe also show cell loss. The continuous delivery of MPTP to mice via an osmotic pump produced loss of locus coeruleus noradrenergic neurons as well as damage to TH-positive neurons (Fornai et al., 2005).

In the chronic rat MPP$^+$ model, α-synuclein and ubiquitin positive inclusions were found in striatal cells but not in nigral neurons. Several months were needed before inclusions were observed in the MPTP-treated monkey (Forno et al., 1988) or mouse (Meredith et al., 2002; Meredith et al., 2004). Since our observations only extend out to 42 days from the start of the MPP$^+$ infusion, it is possible that examination of the animals at longer times after the insult would reveal the accumulation of inclusion bodies. Validation of the progressive loss of nigral neurons once toxicant administration has ceased is also needed as well as a more detailed temporal

evaluation of when damage is first observed in relation to the progressive or reversible/irreversible nature of the damage.

Microglial activation was observed on both the side of infusion as well as the contralateral side. Within the time frame of the study, the contralateral side did not show loss of striatal DA or TH-positive nigral neurons. Microglial activation has been implicated in the ongoing damage in PD (Boka et al., 1994; McGeer et al., 1988; Hunot and Hirsch, 2003; McGeer and McGeer, 2004; Gerhard et al., 2006). Thus, this model could be of use to examine whether chronic microglial activation over time results in damage to the nigrostriatal system *per se*. Longitudinal studies which characterize the onset of neuronal changes and microgliosis will help to define the interrelationship between these two processes in propagating neurodegeneration. Additionally, studies in which microgliosis is modulated will help to determine the extent of the contribution of this process in neuronal damage.

The utility of the chronic MPP$^+$ model for therapeutic testing was recently demonstrated (Zeevalk et al., 2007), but it would be important to determine if potential therapeutics such as glutathione ethyl ester or other compounds found protective when administered before or concurrently with neurotoxicants would still be beneficial if administered after the initial insult. To date, there have been no studies in animal models that have examined therapeutics in this way, although this is precisely the stage in the disease at which therapy is instituted. This model is particularly well suited for these kinds of studies.

Limitations of the Model

Every model possesses characteristics that are advantageous while at the same time limit its utility. Depending on the question that is to be addressed, one model may be better suited than another. The above discussion addresses the many features of the chronic rat MPP$^+$ model that make it attractive for a chronic model of PD, but as with all models, it has limitations. Chronic mitochondrial dysfunction with MPP$^+$ can provide insight into mechanistic issues of cell damage associated with this type of perturbation, but cannot address the equally challenging issue of selectivity for the neurons that are damaged. The complex I inhibitor rotenone does not limit its action to catecholaminergic neurons but is relatively selective for the nigrostriatal DA system

(Betarbet *et al.*, 2000), although this is not without controversy (Hoglinger *et al.*, 2003; Lapointe *et al.*, 2004). Regardless, rotenone appears to produce a hierarchy of damage within the central nervous system (CNS) and understanding the underlying reasons for this would provide important information on cell vulnerability in relation to complex I function. On the other hand, the high mortality and variability in the rotenone model limits its usefulness for pharmacological studies.

The issue of inclusion body formation in this model has been addressed above. The appearance of α-synuclein and ubiquitin positive inclusion bodies in nigral neurons has not been forthcoming in most pharmacological and genetic models of PD. While inclusion bodies have been observed in the rat rotenone model (Betarbet *et al.*, 2000), the chronic MPTP mouse model (Fornai *et al.*, 2005), and the MPTP plus probenecid model (Meredith *et al.*, 2002), these are fairly rare in appearance and/or require long periods of time before they are evident. To date, neither acute nor chronic rodent models of PD satisfactorily recapitulate this aspect of the neuropathology of PD.

Lastly, while chronic models are by nature a low-throughput tool for investigative purposes, the technical requirements for their use should be readily amenable to most laboratories. This is in general true for the chronic rat MPP$^+$ model, but does require surgical skill and training in stereotaxic placement of the brain cannula. In the rat, icv placement is relatively straightforward and with experience it can be done with 100% accuracy.

ACUTE INTRACEREBRAL INFUSIONS OF MPP$^+$ OR MALONATE AS PD MODELS

Acute models of PD, while lacking some of the pathological features seen in brains of PD patients or in progressive PD models, have nonetheless provided much valuable insight into many of the mechanisms and processes associated with the degeneration of DA neurons. Several neurotoxicants have been used (e.g., MPTP, 6-hydroxydopamine), and we will focus our discussion on two mitochondrial inhibitors with which we have considerable experience. These are MPP$^+$, an inhibitor of Complex I, and malonate, a reversible inhibitor of succinate dehydrogenase that impairs Complex II function. MPP$^+$ is more selective toward DA neurons than is malonate because it is a substrate for the DAT and thus is concentrated

within DA nerve terminals and cell bodies relative to other neurons or cells. In contrast, malonate is not selectively accumulated in DA neurons and therefore its effect represents a more generalized metabolic impairment as may be observed in PD patients with mitochondrial defects that are not limited to DA neurons. While malonate does not damage exclusively DA neurons, it does preferentially damage them at low concentrations, especially in mice (Zeevalk *et al.*, 1997; Moy *et al.*, 2000). We and/or others have documented striatal DA nerve terminal degeneration, retrograde loss of nigral DA neurons, and microgliosis following the direct single infusion of MPP$^+$ or malonate into the striatum of rats or mice (Giovanni *et al.*, 1994; Zeevalk *et al.*, 1997; Ferger *et al.*, 1999; Moy *et al.*, 2000; Xia *et al.*, 2001; Miwa *et al.*, 2004).

Acute Intrastriatal or Intranigral Infusions of Metabolic Inhibitors

Under anesthesia and using stereotaxic surgical techniques, a cannula is placed into the brain region of interest (striatum or SN of the rat; striatum in the mouse). Several days later, a needle is inserted into the cannula to the desired depth within the brain region and malonate (1–4 μmols) or MPP$^+$ (10–100 μg) is infused in a volume of 1–2 μL over several minutes after which the needle is left in place for 5–10 min to prevent the fluid from flowing up the needle track. Alternatively, the neurotoxicant may be infused into the brain while the animal is under anesthesia and in the stereotaxic apparatus. While infusion into the awake animal is somewhat more challenging, it eliminates any potential complication that might arise from the presence of the anesthetic agent during neurotoxicant administration.

The acute malonate or MPP$^+$ model has been very beneficial in identifying various pharmacological agents capable of preventing or attenuating damage due to metabolic stress. Some of the drugs found to be protective include DA transport inhibitors (Moy *et al.*, 2007), adenosine A2$_A$ receptor antagonists (Alfinito *et al.*, 2003), and tetrahydroisoquinoline (Lorenc-Koci *et al.*, 2005). It should be noted that in all of these studies, the agent under investigation was administered before and/or concurrent with exposure to the metabolic inhibitor, indicating their efficacy in protecting against the toxic insult. As discussed above under the chronic MPP$^+$ model, a crucial question is whether a potential neuroprotective strategy can in fact stop or retard neurodegeneration once the process has begun. It would be difficult to test pharmacological agents for their ability to reverse neurodegeneration in the acute models as the window for such therapeutic intervention is likely to be very narrow given the rapidity with which neurodegeneration occurs. In this aspect, the model is less attractive than a chronic progressive model for assessment of therapeutic efficacy in halting disease progression. However, one advantage of the acute model for therapeutic testing, especially for agents administered systemically, is that with the intracerebral route of neurotoxicant administration, non-specific drug interactions or alterations in the pharmacokinetics of the test compound or the neurotoxicant is dramatically reduced.

We have also used dual cannula placements for direct delivery of MPP$^+$ or malonate into one brain region (striatum) and infusion of drugs into another region (SN). This technique was instrumental in identifying the importance of nigral, but not striatal, N-methyl-D-aspartate (NMDA), and adenosinergic A2$_A$ receptors in mediating DA neurodegeneration due to striatal stresses (Zeevalk et al., 2000; Alfinito et al., 2003). While systemic administration of antagonists at these receptors had indicated their involvement in neuronal damage, it was only with the use of the dual cannula procedure that the exact brain locations of those receptors mediating protection were identified. These observations were surprising but they reveal the elaborate network of communication among the basal ganglia nuclei and additionally demonstrate the feasibility of the dual cannula approach to addressing questions of how metabolic stress in one brain region can impact on function in other regions.

Intrastriatal MPP$^+$ or Malonate Delivery via Striatal Microdialysis

One advantage of the acute intracerebral infusion method is that it provides a unique model in which to assess the dynamic changes occurring during the localized metabolic stress created directly within the brain region into which the mitochondrial inhibitors have been infused. For this type of study, MPP$^+$ or malonate is generally perfused through the dialysis probe while simultaneously collecting the dialysate for neurochemical measurements. Using this microdialysis technique, the acute infusion of the mitochondrial inhibitor has been shown to cause a rapid release of DA as well as other neurotransmitters such as GABA and glutamate (Rollema et al., 1986; Giovanni et al., 1994; Ferger et al., 1999; Moy et al.,

2000; Xia et al., 2001; Moy et al., 2007), an immediate increase in free radical formation (Obata, 2006), and elevations in lactic acid (Rollema et al., 1988), effects that are directly or indirectly due to inhibition of mitochondria and aerobic energy production. The microdialysis model was instrumental for illustrating the importance of vesicles as storage sites for MPP$^+$ and the involvement of DA in contributing to neurodegeneration during metabolic stress created by malonate (Moy et al., 2000; Moy et al., 2007).

One question raised by the disturbance in DA homeostasis caused by MPP$^+$ or malonate is whether the released DA contributes to neurodegeneration. An ongoing concern in PD research is whether DA itself propagates neurodegeneration. The possibility that DA may undergo oxidation to reactive compounds and could fuel neurodegeneration is of significant concern especially since replacement therapy with L-DOPA is the mainstay of PD treatment. Several groups including ours have pursued this question in vivo using microdialysis techniques and the acute administration of mitochondrial inhibitors to examine their effects on DA homeostasis. Results of these studies show that the striatal infusion of MPP$^+$ or malonate in rats or mice causes DA release, part of which is due to the reversal of the plasma membrane DAT which normally functions to accumulate DA from the extracellular space (Giovanni et al., 1994; Ferger et al., 1999; Moy et al., 2000; Staal and Sonsalla, 2000; Xia et al., 2001; Moy et al., 2007). The importance of DA and DA release in contributing to damage of DA nerve terminals as well as other populations of striatal neurons under conditions of a metabolic stress is demonstrated by the protection afforded by depleting DA or blocking DA release during the stress (Maragos et al., 1998; Jakel and Maragos, 2000; Moy et al., 2000; Moy et al., 2007). While more research is needed to define the relationship between metabolic stress, DA homeostasis and neurodegeneration, the acute PD models, when coupled with microdialysis, will continue to provide insight into these dynamic interactions occurring during metabolic stresses that ultimately lead to DA cell degeneration.

Intrastriatal Infusions of Metabolic Inhibitors Coupled with Microdialysis in the SN

We have also begun to explore the function of basal ganglia nuclei under conditions of a metabolic stress localized to the striatum. In these studies, rats are implanted with a cannula in the striatum and in the SN. Microdialysis probes are placed into the nigral cannula the night before experimentation. The next morning, microdialysis samples are collected prior to, during and after the delivery of malonate or MPP$^+$ into the striatum via the implanted striatal cannula. In preliminary experiments, we have found that the striatal metabolic stress causes an increase in extracellular glutamate within the SN, an effect which appears to exert an excitotoxic action on the nigral DA neurons. While these studies are still very preliminary, they nonetheless demonstrate the feasibility of this approach to examining the dynamics of brain region interactions to striatal metabolic stresses.

Advantages and Limitations of Acute Intracerebral Infusions as Models of PD

There are many advantages of the acute intracerebral MPP$^+$ or malonate model, although certainly there are several limitations as well. As with the chronic MPP$^+$ model, single unilateral infusions produce lesions essentially confined to one side of the brain. This is advantageous as it eliminates or reduces the complications associated with bilateral lesions (e.g., impaired eating, drinking, and motility). Striatal infusions with malonate or MPP$^+$ produce not only nerve terminal damage, but also the loss of nigral DA cell bodies and nigral microgliosis, allowing for assessment of temporal changes occurring in the SN with retrograde cell degeneration in the absence of an invasive procedure in the SN. Because neurodegeneration is produced by a single injection of the metabolic inhibitor, the model provides an observation window in which relatively defined and temporal events can be evaluated (e.g., onset of microgliosis relevant to degeneration of neurons). Intracerebral infusions also bypass the potential interference of systemically administered test agents which could alter the pharmacokinetics or brain distribution of the neurotoxicants that require brain penetrance and metabolism before producing their neurotoxic actions (e.g., MPTP). Moreover, localized infusions of the mitochondrial inhibitors into the striatum or the SN provide the opportunity to produce a very confined area of metabolic impairment, and when coupled with microdialysis, allow for the assessment of dynamic changes in that brain region as well as in other basal ganglia nuclei that occur during conditions of metabolic stress. The model is also useful for initial screening of drugs for therapeutic efficacy in protection against metabolic stresses, although testing in progressive models is likely to be more predictive of their ultimate efficacy in human PD patients.

A major limitation of the models is that they are acute models and therefore do not reflect neurodegeneration under chronic progressive conditions as occur in the human disease state. Additionally, the technique of infusion is invasive, disrupting the blood–brain barrier, at least temporarily, and requires stereotaxic surgery for implementation. While striatal MPP$^+$ infusion, at appropriate doses, is selective for DA nerve terminals and neurons, malonate infusions, particularly in the rat, produces non-selective damage to striatal neurons as well. As such, it is not a good model for testing therapeutic agents aimed at replacing DA function; the MPP$^+$-treated animal would be a much better model for such testing purposes.

CONCLUSIONS

Two models of PD have been described. One is created by a chronic, continuous low dose administration of MPP$^+$ into the cerebral ventricle. The other model results from a single acute administration of MPP$^+$ or malonate into the striatum. The choice of which model to use is very dependent on the research question. The chronic MPP$^+$ model is progressive in nature, presents with neuropathology seen in PD brain, exhibits low inter-animal variability, no mortality, and because of the unilateral lesion, the animals are essentially healthy. While the model requires further characterization, it nonetheless has demonstrated its utility for pharmacological testing of neuroprotective agents. Moreover, because of the progressive nature of degeneration in the model, it can be used for testing therapeutic approaches targeted at the many cellular processes involved in degeneration. The single acute MPP$^+$ or malonate PD model has proven to be especially beneficial when coupled with other injection formats or microdialysis to examine dynamic changes in brain regions subjected to a metabolic stress.

REFERENCES

Alfinito, P. D., Wang, S. P., Manzino, L., Rijhsinghani, S., Zeevalk, G. D., and Sonsalla, P. K. (2003). Adenosinergic protection of dopaminergic and GABAergic neurons against mitochondrial inhibition through receptors located in the substantia nigra and striatum, respectively. *J Neurosci* 23, 10982–10987.

Betarbet, R., Sherer, T. B., MacKenzie, G., Garcia-Osuna, M., Panov, A. V., and Greenamyre, J. T. (2000). Chronic systemic pesticide exposure reproduces features of Parkinson's disease. *Nat Neurosci* 3, 1301–1306.

Boka, G., Anglade, P., Wallach, D., Javoy-Agid, F., Agid, Y., and Hirsch, E. C. (1994). Immunocytochemical analysis of tumor necrosis factor and its receptors in Parkinson's disease. *Neurosci Lett* 172, 151–154.

Cuervo, A. M., Stefanis, L., Fredenburg, R., Lansbury, P. T., and Sulzer, D. (2004). Impaired degradation of mutant alpha-synuclein by chaperone-mediated autophagy. *Science* 305, 1292–1295.

Ferger, B., Eberhardt, O., Teismann, P., de Groote, C., and Schulz, J. B. (1999). Malonate-induced generation of reactive oxygen species in rat striatum depends on dopamine release but not on NMDA receptor activation. *J Neurochem* 73, 1329–1332.

Fornai, F., Lenzi, P., Gesi, M., Ferrucci, M., Lazzeri, G., Busceti, C. L., Ruffoli, R., Soldani, P., Ruggieri, S., Alessandri, M. G., and Paparelli, A. (2003). Fine structure and biochemical mechanisms underlying nigrostriatal inclusions and cell death after proteasome inhibition. *J Neurosci* 23, 8955–8966.

Fornai, F., Schluter, O. M., Lenzi, P., Gesi, M., Ruffoli, R., Ferrucci, M., Lazzeri, G., Busceti, C. L., Pontarelli, F., Battaglia, G., Pellegrini, A., Nicoletti, F., Ruggieri, S., Paparelli, A., and Sudhof, T. C. (2005). Parkinson-like syndrome induced by continuous MPTP infusion: Convergent roles of the ubiquitin–proteasome system and alpha-synuclein. *Proc Natl Acad Sci USA* 102, 3413–3418.

Forno, L. S. (1986). Lewy bodies. *New Engl J Med* 314, 122.

Forno, L. S., Langston, J. W., DeLanney, L. E., and Irwin, I. (1988). An electron microscopic study of MPTP-induced inclusion bodies in an old monkey. *Brain Res* 448, 150–157.

Gai, W. P., Yuan, H. X., Li, X. Q., Power, J. T., Blumbergs, P. C., and Jensen, P. H. (2000). *In situ* and *in vitro* study of colocalization and segregation of alpha-synuclein, ubiquitin, and lipids in Lewy bodies. *Exp Neurol* 166, 324–333.

Gerhard, A., Pavese, N., Hotton, G., Turkheimer, F., Es, M., Hammers, A., Eggert, K., Oertel, W., Banati, R. B., and Brooks, D. J. (2006). *In vivo* imaging of microglial activation with [11C](R)-PK11195 PET in idiopathic Parkinson's disease. *Neurobiol Dis* 21, 404–412.

Giovanni, A., Sonsalla, P. K., and Heikkila, R. E. (1994). Studies on species sensitivity to the dopaminergic neurotoxin 1-methyl-4-phenyl-1,2,3,6-tetrahydropyridine. Part

2: Central administration of 1-methyl-4-phenylpyridinium. *J Pharmacol Exp Therapeut* 270, 1008–1014.

Hoglinger, G. U., Feger, J., Prigent, A., Michel, P. P., Parain, K., Champy, P., Ruberg, M., Oertel, W. H., and Hirsch, E. C. (2003). Chronic systemic complex I inhibition induces a hypokinetic multisystem degeneration in rats. *J Neurochem* 84, 491–502.

Hunot, S., and Hirsch, E. C. (2003). Neuroinflammatory processes in Parkinson's disease. *Ann Neurol* 53(Suppl 3), S49–S58, discussion S58–S60.

Jakel, R. J., and Maragos, W. F. (2000). Neuronal cell death in Huntington's disease: A potential role for dopamine. *Trends Neurosci* 23, 239–245.

Lapointe, N., St-Hilaire, M., Martinoli, M. G., Blanchet, J., Gould, P., Rouillard, C., and Cicchetti, F. (2004). Rotenone induces non-specific central nervous system and systemic toxicity. *Faseb J* 18, 717–719.

Lorenc-Koci, E., Golembiowska, K., and Wardas, J. (2005). 1,2,3,4-Tetrahydroisoquinoline protects terminals of dopaminergic neurons in the striatum against the malonate-induced neurotoxicity. *Brain Res* 1051, 145–154.

Maragos, W. F., Jakel, R. J., Pang, Z., and Geddes, J. W. (1998). 6-Hydroxydopamine injections into the nigrostriatal pathway attenuate striatal malonate and 3-nitropropionic acid lesions. *Expe Neurol* 154, 637–644.

McGeer, P. L., Itagaki, S., Boyes, B. E., and McGeer, E. G. (1988). Reactive microglia are positive for HLA-DR in the substantia nigra of Parkinson's and Alzheimer's disease brains. *Neurology* 38, 1285–1291.

McGeer, P. L., and McGeer, E. G. (2004). Inflammation and neurodegeneration in Parkinson's disease. *Parkinsonism Relat Disord* 10(Suppl 1), S3–S7.

Meredith, G. E., Totterdell, S., Petroske, E., Santa Cruz, K., Callison, R. C., Jr., and Lau, Y. S. (2002). Lysosomal malfunction accompanies alpha-synuclein aggregation in a progressive mouse model of Parkinson's disease. *Brain Res* 956, 156–165.

Meredith, G. E., Halliday, G. M., and Totterdell, S. (2004). A critical review of the development and importance of proteinaceous aggregates in animal models of Parkinson's disease: New insights into Lewy body formation. *Parkinsonism Relat Disord* 10, 191–202.

Miwa, H., Kubo, T., Morita, S., Nakanishi, I., and Kondo, T. (2004). Oxidative stress and microglial activation in substantia nigra following striatal MPP+. *Neuroreport* 15, 1039–1044.

Mizukawa, K., Sora, Y. H., and Ogawa, N. (1990). Ultrastructural changes of the substantia nigra, ventral tegmental area and striatum in 1-methyl-4-phenyl-1,2,3,6-tetrahydropyridine (MPTP)-treated mice. *Res Comm Chem Pathol Pharmacol* 67, 307–320.

Moy, L. Y., Zeevalk, G. D., and Sonsalla, P. K. (2000). Role for dopamine in malonate-induced damage *in vivo* in striatum and *in vitro* in mesencephalic cultures. *J Neurochem* 74, 1656–1665.

Moy, L. Y., Wang, S. P., and Sonsalla, P. K. (2007). Mitochondrial stress-induced dopamine efflux and neuronal damage by malonate involves the dopamine transporter. *J Pharmacol Exp Therapuet* 320, 747–756.

Obata, T. (2006). Imidaprilat, an angiotensin-converting enzyme inhibitor exerts neuroprotective effect via decreasing dopamine efflux and hydroxyl radical generation induced by bisphenol A and MPP$^+$ in rat striatum. *Brain Res* 1071, 250–253.

Petroske, E., Meredith, G. E., Callen, S., Totterdell, S., and Lau, Y. S. (2001). Mouse model of Parkinsonism: A comparison between subacute MPTP and chronic MPTP/probenecid treatment. *Neuroscience* 106, 589–601.

Rodriguez-Enriquez, S., He, L., and Lemasters, J. J. (2004). Role of mitochondrial permeability transition pores in mitochondrial autophagy. *Int J Biochem Cell Biol* 36, 2463–2472.

Rollema, H., Damsma, G., Horn, A. S., De Vries, J. B., and Westerink, B. H. (1986). Brain dialysis in conscious rats reveals an instantaneous massive release of striatal dopamine in response to MPP$^+$. *Eur J Pharmacol* 126, 345–346.

Rollema, H., Kuhr, W. G., Kranenborg, G., De Vries, J., and Van den Berg, C. (1988). MPP$^+$-induced efflux of dopamine and lactate from rat striatum have similar time courses as shown by *in vivo* brain dialysis. *J Pharmacol Exp Therapeut* 245, 858–866.

Roy, S., and Wolman, L. (1969). Ultrastructural observations in Parkinsonism. *J Pathol* 99, 39–44.

Sapolsky, R. M. (1992). *Stress, the Aging Brain, and the Mechanisms of Neuron Death.* MIT Press, Cambridge, MA.

Smith, A. D., Castro, S. L., and Zigmond, M. J. (2002). Stress-induced Parkinson's disease: A working hypothesis. *Physiol Behav* 77, 527–531.

Song, D. D., Shults, C. W., Sisk, A., Rockenstein, E., and Masliah, E. (2004). Enhanced substantia nigra mitochondrial pathology in human alpha-synuclein transgenic mice after treatment with MPTP. [see comment]. *Exp Neurol* 186, 158–172.

Staal, R. G., and Sonsalla, P. K. (2000). Inhibition of brain vesicular monoamine transporter (VMAT2) enhances 1-methyl-4-phenylpyridinium neurotoxicity *in vivo* in rat striata. *J Pharmacol Exp Therapeut* 293, 336–342.

Tanaka, J., Nakamura, H., Honda, S., Takada, K., and Kato, S. (1988). Neuropathological study on 1-methyl-4-phenyl-1,2,3,6-tetrahydropyridine of the crab-eating monkey. *Acta Neuropathol* 75, 370–376.

Thiruchelvam, M., McCormack, A., Richfield, E. K., Baggs, R. B., Tank, A. W., Di Monte, D. A., and Cory-Slechta, D. A. (2003). Age-related irreversible progressive nigrostriatal dopaminergic neurotoxicity in the paraquat and maneb model of the Parkinson's disease phenotype. *Eur J Neurosci* 18, 589–600.

Trimmer, P. A., Swerdlow, R. H., Parks, J. K., Keeney, P., Bennett, J. P., Jr., Miller, S. W., Davis, R. E., and Parker, W. D., Jr. (2000). Abnormal mitochondrial morphology in sporadic Parkinson's and Alzheimer's disease cybrid cell lines. *Exp Neurol* 162, 37–50.

Urakami, K., Masaki, N., Shimoda, K., Nishikawa, S., and Takahashi, K. (1988). Increase of striatal dopamine turnover by stress in MPTP-treated mice. *Clin Neuropharmacol* 11, 360–368.

Xia, X. G., Schmidt, N., Teismann, P., Ferger, B., and Schulz, J. B. (2001). Dopamine mediates striatal malonate toxicity via dopamine transporter-dependent generation of reactive oxygen species and D2 but not D1 receptor activation. *J Neurochem* 79, 63–70.

Yazdani, U., German, D. C., Liang, C. L., Manzino, L., Sonsalla, P. K., and Zeevalk, G. D. (2006). Rat model of Parkinson's disease: Chronic central delivery of 1-methyl-4-phenylpyridinium (MPP$^+$). *Exp Neurol* 200, 172–183.

Zeevalk, G. D., Manzino, L., Hoppe, J., and Sonsalla, P. (1997). *In vivo* vulnerability of dopamine neurons to inhibition of energy metabolism. *Eur J Pharmacol* 320, 111–119.

Zeevalk, G. D., Manzino, L., and Sonsalla, P. K. (2000). NMDA receptors modulate dopamine loss due to energy impairment in the substantia nigra but not striatum. *Exp Neurol* 161, 638–646.

Zeevalk, G. D., Manzino, L., Sonsalla, P. K., and Bernard, L. P. (2007). Characterization of intracellular elevation of glutathione (GSH) with glutathione monoethyl ester and GSH in brain and neuronal cultures: Relevance to Parkinson's disease. *Exp Neurol* 203, 512–520.

13

ENDOGENOUS DEFENSES THAT PROTECT DOPAMINE NEURONS: STUDIES WITH 6-OHDA MODELS OF PARKINSON'S DISEASE

REHANA K. LEAK AND MICHAEL J. ZIGMOND

Pittsburgh Institute for Neurodegenerative Diseases, Department of Neurology, University of Pittsburgh, Pittsburgh, PA, USA

Parkinson's disease (PD) is a progressive disease that affects only a minority of individuals and generally does not appear until advanced age. Within the brain, the neuropathology begins in the medulla, pons, and olfactory bulb, spreading gradually to regions of the mesencephalon, and eventually affecting portions of the telencephalon (Braak *et al.*, 2003a, b). However, it is the loss of dopamine (DA) neurons in the substantia nigra (SN) that is believed to underlie many of the motor symptoms of PD (Schapira *et al.*, 2006) and has received the most attention. Here, too, the degeneration is progressive, possibly occurring over one to two decades (Marttila and Rinne, 1991).

There are several possible explanations for the progressive nature of PD. For example, it may be a result of the gradual spread of a toxin within the brain, perhaps in transneuronal fashion (Braak *et al.*, 2003a, b). It is also possible that the death of one neuron triggers the death of another, perhaps as a result of an intermediary inflammatory event (Liu and Hong, 2003; McGeer and McGeer, 2004; Jenner and Olanow, 2006; Kim and Joh, 2006). A third possibility is that remaining neurons compensate for the loss of their neighbors by increasing synthesis and release of the transmitter DA,

eventually causing lethal levels of oxidative stress from DA metabolites such as hydrogen peroxide and DA quinone in a self-perpetuating cycle (Zigmond *et al.*, 2002). Alternatively, and as discussed below, brain regions may simply vary in their inherent susceptibility to injury and progress at different rates toward eventual demise.

This review focuses on another mystery, why PD affects a relatively small number of individuals (100–300 per 100,000) and then usually only after the fifth or sixth decade of life. What determines who succumbs to PD and why is age the major risk factor? There is research pointing to oxidative stress, mitochondrial damage, aggregated proteins, and proteasome dysfunction in the illness, which we briefly review below. However, these secondary events presumably follow an unknown primary cause. In a small percentage of cases, that cause is purely genetic in origin. However, in a majority of cases, there is no strong pattern of inheritance and thus environmental factors must play a significant role. It is possible that exposure to an environmental factor is limited to a small percentage of individuals, and that this provides an explanation for the low prevalence. If so, then the high correlation between age and incidence of PD might be explained by a lengthy period of exposure. In this review we offer another explanation for the prevalence and age-dependence of PD. We suggest that many individuals do not develop PD because they have a strong *defensive stress response*, which we define as a protective reaction to what would otherwise be toxic cellular stress. We further hypothesize that in people who do develop the illness, this defensive stress response – while not adequate to prevent the disease – serves to keep the neurodegenerative process at bay for many years. Finally, we suggest that although a response to minor stress may be defensive and thereby serve to "precondition" an individual against further insults, a more intense stress might actually sensitize an individual to a potential toxin. We will review endogenous defenses and then focus upon stress responses initiated in common models of PD, those involving cells and animals exposed to the catecholaminergic neurotoxin 6-hydroxydopamine (6-OHDA), while also highlighting some of the advantages and limitations of 6-OHDA.

ENDOGENOUS DEFENSES

The organism has several ways to defend itself against stress, and it seems likely that such defenses normally act to prevent or at least delay neurodegeneration. In the following section, we will briefly review some major defense systems.

Antioxidant Systems

Among the most prominent of endogenous defenses are the *antioxidant systems*, including but not limited to the superoxide dismutases (SODs), catalase, glutathione (GSH), and related enzymes such as GSH peroxidase and GSH reductase, serving to reduce reactive oxygen species (ROS). These molecules are relevant to aging and neurodegenerative diseases because oxidative stress is thought to underlie much of the pathology in both these conditions, and it is widely believed that amelioration of ROS toxicity will be beneficial (Varadarajan et al., 2000; Floyd and Hensley, 2002; Barnham et al., 2004; Mariani et al., 2005). Although oxidative stress is unlikely to be the sole precipitating cause of PD, it appears to be an early event in its pathogenesis (Dauer and Przedborski, 2003). This is supported by the observation of a reduction in GSH in postmortem SN from patients with incidental Lewy body disease, thought to represent early PD (Dexter et al., 1994). Other indices of oxidative damage in the SN of patients with PD include excess lipid peroxidation and increased deposition of iron (Olanow and Tatton, 1999; Przedborski and Jackson-Lewis, 2000). Mitochondria are damaged in PD, leading to an impairment of oxidative phosphorylation and attendant oxidative stress (Wallace et al., 1992; Ebadi et al., 2001). Mitochondrial MnSOD activity is increased in postmortem SN of PD patients, perhaps reflecting a defensive response to increased oxidative stress (Saggu et al., 1989).

The relationship of these antioxidant defenses to the neurodegeneration that occurs with age is not clear. Some studies suggest that antioxidant enzymes such as the SODs actually *increase* with age, perhaps as a defensive response to slow accrual of damage (Hussain et al., 1995; Erraji-Benchekroun et al., 2005; Barreiro et al., 2006; Lambertucci et al., 2006; Yoon et al., 2006). However, other studies show that antioxidant capacity is reduced with age, which is reflected in a decline in GSH levels (Liu and Dickinson, 2003; Siqueira et al., 2005; Williams and Chung, 2006). It is possible that age-related increases in basal levels of antioxidant enzymes have a dark side, reducing the capacity for further increases when a more severe stress is encountered (Collier et al., 2005). In any event, apparent contradictions in the literature probably indicate that basal and stress-induced defensive responses vary depending on stressor, defense system in question, age of the organism, cellular context (stressed or unstressed), and brain region.

Protein-Folding Chaperones

An equally important system of defenses are *chaperones*, serving to ensure the proper folding of many proteins, and originally identified as highly responsive to heat shock (Ritossa, 1996). *Heat shock proteins* are chaperones that help refold misfolded proteins, prevent aggregation, and negatively regulate apoptotic pathways (Beere, 2001; Beere and Green, 2001; Nollen and Morimoto, 2002). They are induced upon cellular damage, especially the accumulation of misfolded proteins, and are an important feature of the *stress response*, a term often used to describe changes in gene expression caused by exposure to deleterious agents (Welch, 1992). Heat shock proteins are critical for proper function of the ubiquitin–proteasome system (see below) (Ciechanover et al., 2000). One of the best studied examples of such proteins is heat shock protein 70, shown to be protective in several PD models (Duan and Mattson, 1999; Auluck and Bonini, 2002; Auluck et al., 2002; Quigney et al., 2003; Shen et al., 2005).

Heat shock proteins and their alteration with age were the subject of several recent studies, but there are conflicting data as to whether levels increase or decrease with age (Lee et al., 2000; Jiang et al., 2001; Morrow and Tanguay, 2003; Taylor and Starnes, 2003; Tandara et al., 2006). Several heat shock proteins (Hsp 40, 27, 59, 70, and constitutive Hsp 70) were reported to increase in the aged central nervous system (Lee et al., 2000; Lu et al., 2004), perhaps as a defensive measure against cumulative damage. As with the changes in antioxidant systems, it is likely that the effect of age upon heat shock proteins varies according to the brain region and chaperone in question. It is noteworthy that aged cells show a reduced stress induction of heat shock proteins (Locke and Tanguay, 1996; Volloch et al., 1998; Verbeke et al., 2001; Tandara et al., 2006). It has therefore been hypothesized that loss of the heat shock response might underlie several types of age-related degeneration, including that of PD (Meriin and Sherman, 2005).

Proteasomes and Autophagy as Endogenous Defense Systems

A third defense lies in the *proteasome*, a large barrel-shaped catalytic complex comprising 1% of total cellular protein that, with the help of heat shock proteins, engages in proteolysis of oxidized, misfolded, and aggregated proteins (Davies, 2001; Goldberg, 2003). This abundant particle generates 3–20 residue peptides that are subsequently hydrolyzed to amino acids by cytosolic peptidases. The importance of the proteasome in basic cellular homeostasis has generated a large body of work with proteasome inhibitors that are specific for the proteasome to varying degrees (Lee and Goldberg, 1998; Kisselev and Goldberg, 2001). Several observations suggest a role of the proteasome in PD: proteasome function is decreased in sporadic PD (McNaught et al., 2002; McNaught et al., 2003; Blandini et al., 2006), aggregated proteins accumulate within the Lewy body inclusions, and mutations in various components of the ubiquitin–proteasome system lead to familial PD

(McNaught, 2004; Olanow and McNaught, 2006). Importantly, proteasome function is also reported to be inhibited with age (Conconi and Friguet, 1997; Stolzing and Grune, 2001; Carrard et al., 2002; Zeng et al., 2005). Therefore, uncovering ways to keep the proteasome system functional and healthy has the potential to influence the pathologies that occur both in normal aging and in neurodegenerative diseases.

Autophagy and the lysosomal pathways constitute yet another form of defense mediating protein degradation (Larsen and Sulzer, 2002; Cuervo, 2004; Kaushik and Cuervo, 2006). Autophagy occurs under basal conditions and is induced in response to stressors such as starvation. Macroautophagy involves the formation of double membrane-bound autophagosomes to engulf large portions of cytosol or organelles and subsequent fusion with lysosomes for degradation by acidic hydrolases (Rubinsztein, 2006). In chaperone-mediated autophagy, proteins are more selectively targeted for degradation and delivered for translocation into lysosomes (Massey et al., 2004). Inhibition of autophagy increases protein aggregation and cell death, whereas induction of autophagy with rapamycin is protective (Ravikumar et al., 2002, 2006; Berger et al. 2006). Loss of autophagy causes neurodegeneration in mice, in some cases even without the expression of aggregate-prone proteins and in spite of normal proteasome function (Hara et al., 2006; Komatsu et al., 2006), confirming the importance of basal autophagy as a quality control system (Mizushima and Klionsky, 2007). Larger scale autophagy might also mediate cell death in PD models, however, as inhibition of autophagy protects from MPP^+ (Chu et al., 2007). Autophagic profiles have been observed in ultrastructural studies of human SN from postmortem PD brains (Anglade et al., 1997; Zhu et al., 2003) and alterations in autophagy in the brain regions affected in various Braak stages of PD are likely to be an important subject for future studies. Age would be an additional important variable to consider as autophagic defenses fall with age (Cuervo and Dice, 1998).

Trophic Factors as Endogenous Defense

Finally, *trophic factors* can be viewed as another endogenous defense triggered by cellular stress, as numerous studies show an upregulation of trophic factors with injury (Bar *et al.*, 1998; Satake *et al.*, 2000; Saavedra *et al.*, 2006; Bella *et al.*, 2007). One of the most potent neurotrophic factors for DA neurons is glial cell line-derived neurotrophic factor (GDNF). Early studies of DA neurons in culture indicated the presence of a survival factor (Engele and Bohn, 1991), which was soon identified as GDNF (Lin *et al.*, 1993). GDNF expression rapidly falls off as development proceeds (Stromberg *et al.*, 1993; Blum and Weickert, 1995; Choi-Lundberg and Bohn, 1995). However, there are instances in which aspects of recovery of function recapitulate ontogeny (Cramer and Chopp, 2000), and there are several reports that GDNF expression increases following injury (Liberatore *et al.*, 1997; Naveilhan *et al.*, 1997; Sakurai *et al.*, 1999; Wei *et al.*, 2000; Smith *et al.*, 2003; Saavedra *et al.*, 2006). Increased GDNF, whether through the addition of exogenous GDNF or a viral vector containing the GDNF gene, protects DA neurons against the neurotoxic effects of 6-OHDA. This has been shown both *in vivo* and *in vitro* (Hoffer *et al.*, 1994; Kearns and Gash, 1995; Akerud *et al.*, 1999; Gong *et al.*, 1999; Kramer *et al.*, 1999; Schatz *et al.*, 1999; Kirik *et al.*, 2000; Kozlowski *et al.*, 2000) and we recently confirmed these effects in animal models (A. D. Cohen, M. J. Zigmond, and A. D. Smith, unpublished observations; N. Lindgren, R. K. Leak, A. D. Smith, and M. J. Zigmond, unpublished observations), MN9D cells (Ugarte *et al.*, 2003), and primary DA neurons (Ding *et al.*, 2004). In addition to 6-OHDA, the neuroprotective effects of GDNF extend to other insults and other cell types (McAlhany *et al.*, 1997; Kordower *et al.*, 2000; Grondin *et al.*, 2002), and have also been shown to ameliorate age-related DA dysfunction (Maswood *et al.*, 2002; Grondin *et al.*, 2003).

In the next section, we will further discuss the evidence that DA-depleting lesions increase GDNF levels, suggesting that *endogenous* GDNF is neuroprotective. On the other hand, there are conflicting data on whether GDNF is altered in PD; some reports indicate that levels are reduced (Chauhan *et al.*, 2001), or increased (Backman *et al.*, 2006), whereas yet others fail to observe a change (Mogi *et al.*, 2001). As in the case of contradictions regarding changes in antioxidant defenses and heat shock proteins, age, precise brain region, and the stage of the illness may be a critical factor explaining differences across studies. For example, a close analysis of the temporal progression of trophic factor alterations in PD with Braak staging as a variable has not yet been performed.

Section Summary

In summary, there are several endogenous defense systems available for cells to mount self-protective responses to cellular injury. In most individuals each of these defenses may reduce the damage that accrues with age or in response to environmental insults. However, the size of the defensive responses might be blunted as these individuals age or in individuals with PD and other neurodegenerative conditions so that cell death can progress, albeit slowly.

6-OHDA AS A MODEL OF PD

The accumulation of oxidative damage to biomolecules with age has long been considered a hallmark of neurodegeneration. Mimicking ROS and its consequences with 6-OHDA therefore remains a useful tool in the study of PD, even decades after its inception (Zigmond and Keefe, 1997). In this section we describe the use of 6-OHDA to mimic the DA deficiency of PD, describing both the strengths and weaknesses of this model.

Characteristics of 6-OHDA

6-OHDA is a structural analog of DA that is highly electroactive and oxidizes to form a variety of cytotoxic compounds, including 6-OHDA quinone and hydrogen peroxide (Figure 13.1). The toxin is transported into central and peripheral catecholaminergic terminals by their high affinity catecholamine transport systems. In this way, its effects are relatively specific to these monoaminergic neurons. The effects of 6-OHDA can be further limited to DA neurons by (a) pretreating the animals with desipramine to block uptake into noradrenergic neurons, or (b) the selective application of 6-OHDA into a brain area where neurons or their terminals are primarily dopaminergic, such as SN or striatum, respectively.

6-OHDA does not readily cross the blood–brain barrier and must be stereotaxically infused into the brain in order to model DA cell loss. Once in the brain, 6-OHDA appears to kill neurons from within through rapid formation of oxidized products (Kostrzewa and Jacobowitz, 1974; Przedborski and Ischiropoulos, 2005). Evidence for this mechanism of action includes the capacity of transporter inhibitors and antioxidants to block the toxic effects of 6-OHDA (Hou *et al.*, 1997; Pong *et al.*, 2000; Gonzalez-Hernandez *et al.*, 2004).

Injection Protocols

There are several protocols for intracerebral application of 6-OHDA. Some of the earliest studies employed *intraventricular infusions*. The amount of catecholamine loss in these studies was determined by 6-OHDA dose. When applied in sufficiently high dose, intraventricular 6-OHDA produces a bilateral loss of striatal DA (>95%) requiring intensive animal nursing care due to imminent aphagia and adipsia (Ungerstedt, 1971b; Zigmond and Stricker, 1973; Rowland and Stricker, 1982; Sakai and Gash, 1994). In contrast, smaller DA depletions produce few gross motor changes, perhaps reflecting compensatory responses to such lesions, while motor deficits can be precipitated by exposing animals to stressors such as pain and glucoprivation (Stricker and Zigmond, 1974; Snyder *et al.*, 1985). The observation that motor deficits are associated with very large depletions of DA or severe stress in the rat 6-OHDA model is somewhat analogous to observations made in PD.

A second approach to the delivery of 6-OHDA is the *intraparenchymal route* (Figure 13.1). This has

several advantages. Whereas intraventricular injections produce a widespread, bilateral loss of DA, intraparenchymal injections are used to target a particular subfield of DA cells, axons, and/or terminals and lesions can be made unilaterally. This prevents the development of aphagia and adipsia and produces motor asymmetries that are readily measured (Schallert *et al.*, 2000). The unilateral lesion model is especially useful in that it results in asymmetrical turning behavior in response to both direct and indirect DA agonists such as apomorphine and amphetamine, respectively, thereby allowing a rapid assessment of lesion size (Ungerstedt and Arbuthnott, 1970; Ungerstedt, 1971a; Przedborski *et al.*, 1995). Furthermore, the contralateral side of the brain is unlesioned and available as an internal control. For example, assays of DA levels have often been expressed as a fraction of levels on the contralateral side. The contralateral hemisphere is particularly useful in immunohistochemical studies, where both hemispheres are exposed to the same solutions during the perfusion and assay procedures. On the other hand, caution must be exercised with this approach as some changes in the contralateral side have been reported (Schallert *et al.*, 1983; Warenycia and McKenzie, 1987; Robinson and Whishaw, 1988; Kozlowski *et al.*, 2004).

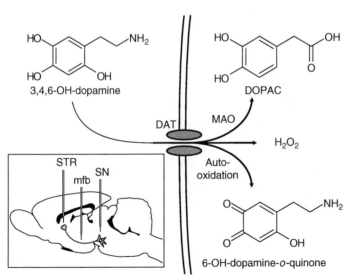

FIGURE 13.1 Mechanism of action and sites of application of 6-OHDA. 6-OHDA is concentrated within the intracellular milieu by the high affinity dopamine transporter (DAT) and oxidizes to form hydrogen peroxide, 6-OHDA quinone, or dihydroxyphenylacetic acid (DOPAC) by action of monoamine oxidase (MAO). Inset: 6-OHDA is typically applied to one of three sites – the medial forebrain bundle (mfb), striatum (STR), or substantia nigra (SN). The Paxinos rat brain is shown from a sagittal view (Paxinos and Watson, 2006).

For unilateral lesions of DA projections, 6-OHDA is typically applied to one of three sites, the DA terminal field of the striatum, the axons and terminals in the medial forebrain bundle in the lateral hypothalamus, or directly to the DA cell bodies in the SN. Of these paradigms, the *striatal* application of 6-OHDA produces the slowest loss of cell bodies over the course of 1 month (Sauer and Oertel, 1994), allowing a time window for therapeutic applications to take effect. The striatal model results in apoptotic morphology peaking between days 7 and 10 (Marti *et al.*, 2002), a mode of cell death consistent with the observation that caspase inhibitors reduce cell death by intrastriatal 6-OHDA (Cutillas *et al.*, 1999) and with *in vitro* 6-OHDA studies of caspases 9 and 3 (Liang *et al.*, 2004). However, these studies are not consistent with the recent report of a lack of activation of these caspase after intrastriatal injections (Ebert *et al.*, 2007). Striatal placements may also cause the least amount of non-specific damage, perhaps because the density of high affinity DA uptake sites results in rapid clearance of the toxin from extracellular fluid.

Quantification of the loss of DA terminals within the striatum after an intrastriatal lesion can be accomplished by immunohistochemistry with an antibody to a marker of DA terminals such as tyrosine hydroxylase (TH), or by HPLC measurement of DA levels in the region of 6-OHDA administration. The area of loss of TH immunoreactivity in the striatum can be quantified relatively rapidly planimetrically

(Figure 13.2). On the other hand, a single striatal injections of 6-OHDA typically affects only a relatively small subset of DA cells in the SN. Thus, to quantify this cell body loss, it is best to pre-label the affected region with a retrograde tracer (Sauer and Oertel, 1994; Kozlowski *et al.*, 2000). The loss of DA neurons can then be quantified by counting the number of retrogradely labeled cells, often done by stereological techniques. An alternative to increase the effect of striatal lesions is to infuse multiple striatal sites (Kirik *et al.*, 1998), a procedure which takes considerably more time.

With the application of 6-OHDA into the *medial forebrain bundle* or *SN*, cell death is much more acute than striatal application, and typically affects a larger swath of nigral neurons. Although this may lead to more non-specific damage, the great majority of DA neurons can be destroyed in this way (Hokfelt and Ungerstedt, 1973; Jackson *et al.*, 2003) and there is no subsequent need for careful dissection of the affected region of SN or striatum, as required for the one injection intrastriatal lesion paradigm. Furthermore, direct application of 6-OHDA into the medial forebrain bundle or SN leads to more readily detectable behavioral deficits. As in the case of striatal lesions, the mode of cell death following medial forebrain bundle lesions is thought to be apoptotic (He *et al.*, 2000; Zuch *et al.*, 2000). In contrast, SN injections are reported to produce a non-apoptotic cell death (Jeon *et al.*, 1995).

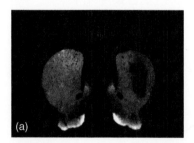 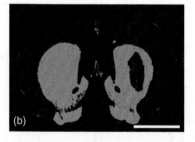

FIGURE 13.2 Quantification of striatal loss of TH immunoreactivity. (a) A coronal section of a rat infused into the striatum unilaterally with 3 ?g 6-OHDA, perfused after 7 days, immunostained with an infrared secondary against mouse anti-TH, and scanned with an Odyssey Infrared Imager (Li-Cor Biosciences, Lincoln, NA). The grayscale images have a bit depth of 16, allowing 65,536 levels of gray, increasing the linear range and sensitivity of the system. The relative lack of background in the infrared range also increases sensitivity. The image in (b) was thresholded in MetaMorph (Molecular Devices, Sunnyvale, CA) relative to the fluorescent intensity in the contralateral control hemisphere. Thus, the area of loss of TH (striatal area lacking fluorescent label in thresholded image) shows greater than 75% loss of staining relative to the contralateral, intact striatum. This technique nullifies variations in immunohistochemical staining between animals. A blind observer then traces the lesion area in sections centered around the needle track. Scale bar = 3 mm.

A strength of the 6-OHDA model is that it roughly recapitulates the topographical loss of DA neurons typical of PD, affecting SN more than ventral tegmental area (VTA) at any one time point. With striatal lesions, one would of course expect only SN neurons to be affected, particularly with lesions placed in the dorsolateral striatum (Fallon and Moore, 1978; Gerfen et al., 1987). This outcome would not necessarily be expected of medial forebrain bundle lesions since VTA axons within the lateral hypothalamus are in close enough proximity to SN axons. Likewise, in cultures of ventral mesencephalon containing both SN and VTA, 6-OHDA might be expected to show little specificity. Instead, we have shown that the loss of TH positive cells is greater in SN than VTA in vivo (Smith et al., 2003), and as will be noted in the next section, this is even true with in vitro studies of DA neurons (Ding et al., 2004).

The relative resilience of VTA holds true in MPTP and rotenone models of PD (Liang et al., 1996; Betarbet et al., 2000) as well as in a recent PD model involving genetically induced respiratory chain dysfunctions (Ekstrand et al., 2007). Although there is some debate as to whether VTA sequesters MPP^+ and 6-OHDA better than SN and is thereby less vulnerable, this cannot explain the results with rotenone and respiratory defect models. Instead, SN neurons appear to be truly more vulnerable than VTA neurons, providing a parallel to their relative susceptibility in PD (Braak et al., 2003b).

In Vitro Application of 6-OHDA

6-OHDA can be effectively used to produce in vitro models of DA depletion. The choice of in vitro model will, of course, influence outcome, and investigators are encouraged to verify their in vitro observations with different models (Falkenburger and Schulz, 2006). Cells that have been used in conjunction with 6-OHDA include PC12 (Ryu et al., 2005), SH-SY5Y (Nakamura et al., 2006), MN9D (Choi et al., 1999), B65 (Chalovich et al., 2006), and primary neurons from prenatal and neonatal rat pups (Ding et al., 2004; Signore et al., 2006).

The issue of non-specific damage is more critical here than in the case of in vivo studies as the volume of extracellular fluid greatly exceeds that of the DA cells and thus 6-OHDA is only slowly and incompletely cleared. Therefore, caution must be exercised in leaving this easily oxidized compound in cell culture media for long periods. Typically in vitro investigations using 6-OHDA keep the toxin

in the media for many hours and are actually measuring the effects of 6-OHDA plus its oxidized breakdown products (such as hydrogen peroxide) within media. This would result in non-specific entry into all cells present in the culture and damage to the outer surface of cells, as well, in contrast to the specific intracellular damage to catecholamine neurons in vivo. This confound has been discussed at length by Clement et al. (2002), and is exemplified by our observation that 6-OHDA is completely oxidized within a few minutes in standard culturing conditions (Figure 13.3). Thus, our procedure is to make 6-OHDA up in a vehicle containing ascorbic acid (0.15%) and DETAPAC (10 mM) that has been flushed with nitrogen and is stored only briefly, on ice, and shielded from light. (We follow similar precautions when using 6-OHDA in vivo, though we do not add DETAPAC.) This solution is then added as a 10× stock to the culture medium for limited intervals only (15–30 min), after which media is fully replaced. In this way, we can achieve a concentration-dependent loss of DA cells, with little or no loss of GABAergic neighbors in primary midbrain cultures (Figure 13.4) (Ding et al., 2004). Even with these precautions, however, in some experiments we see no toxicity, while in others we see excessive toxicity, and these experiments must be discarded. Given the potential confound of non-specific damage, when using 6-OHDA to study the impact of intracellular ROS in vitro it is critical to have control wells that contain no 6-OHDA as well as 6-OHDA alone and to show that the effects under investigation can be abolished with inhibitors of the DA transporter such as nomifensine. A complimentary approach is to compare changes in DA neurons with those in other adjacent GABA neurons.

The tendency for 6-OHDA to auto-oxidize is not limited to culturing. Indeed, 6-OHDA can oxidize even as a dry powder and the extent of its purity appears to vary greatly depending on the source, the batch, and the age of the material. For this reason, we store the 6-OHDA as a dry powder at −80°C, carefully examine the color of the powder (it should be white or slightly off-white) as well as monitor the color of the solution once prepared (it should remain colorless).

Two types of quantification of DA cell loss are typically performed when studying primary DA neurons in vitro. One method is to count the small number of cells that are TH immunoreactive in such cultures (typically <5%), and the other is to use tritiated DA uptake through the high affinity DA

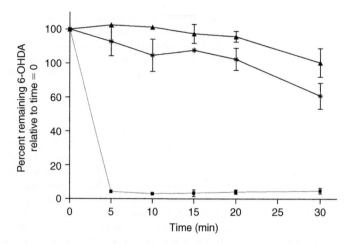

FIGURE 13.3 6-OHDA degradation as a function of time and different vehicles. Samples were prepared with 6-OHDA (100 μM) in media alone (■), media plus 0.015% ascorbic acid (*), or vehicle containing ascorbic acid, DETAPAC, and flushed with nitrogen (▲). Samples were incubated for indicated times at 37°C and analyzed by HPLC. Data represent the mean and SEM of two or three replicates for each condition. Reprinted with permission from Ding *et al.* (2004).

transporter as a rough measure of DA cell number. Albeit not perfect measures of cell death, they can be put to good use in primary culture models (Ding *et al.*, 2004; Signore *et al.*, 2006).

Loss of Phenotype

We and others have observed that changes in the phenotype of DA neurons do not always reflect the loss of the DA neurons themselves. For example, Dr. Annie Cohen in our laboratory has observed that 2 weeks after intrastriatal injection of 6-OHDA, the loss of TH immunoreactivity in the SN greatly exceeds the actual loss of DA neurons as measured by retrograde tract tracing, suggesting that DA neurons initially respond to oxidative stress by a downregulation of TH expression. Likewise, she observed that GDNF could protect DA neurons from 6-OHDA, but that full restoration of the phenotype of those cells took up to 8 weeks (A. D. Cohen, M. J. Zigmond, and A. D. Smith, unpublished observations). Clearly, using TH immunoreactivity to quantify DA cell loss has its limitations, and this approach is best combined with other indices, such as a retrograde tracer, the use of a stain for Nissl substance, or neuromelanin itself (Jackson-Lewis *et al.*, 1995; McCormack *et al.*, 2004).

Strengths of Model

6-OHDA, now in use for some 30 years to model PD, has both strengths and weaknesses (Table 13.1). These must all be borne in mind when using 6-OHDA to test novel therapeutic approaches such as new drug targets and potential drug leads. One of the strengths of 6-OHDA is that when used with care it can produce a relatively specific loss of the DA neurons of the SN, and in so doing recapitulates one of the key features of PD, the presence of a toxic level of oxidative stress within these neurons. Moreover, employing 6-OHDA to produce models of DA depletion allows one to make use of a vast body of literature accumulated over the past three decades (Zigmond and Keefe, 1997). 6-OHDA has been successfully used in tissue culture and in studies of a wide range of animal species, from worms to monkeys. The popularity of 6-OHDA animal models can be attributed to the uniformity of the lesions, the low inter-animal variability, the ease with which unilaterally lesioned animals can be quantified for motor asymmetries, the presence of a contralateral "control" hemisphere when injected unilaterally, and the avoidance of animal nursing care even with large unilateral lesions.

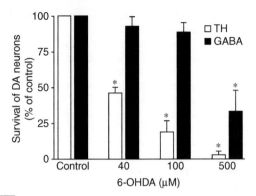

FIGURE 13.4 The effects of 15-min exposure to 40, 100, and 500 μM 6-OHDA on the survival of DA neurons. After 3–4 days in vitro, cultures were exposed to vehicle or 6-OHDA. After 48 h, cells were fixed and immunostained for TH and GABA. Note the specific loss of DA neurons, but not of GABA neurons at lower 6-OHDA concentrations. Data represent the mean and SEM for 8–10 experiments, performed in duplicate or triplicate. *$p < 0.01$ versus control (0 μM 6-OHDA), Bonferroni post hoc following ANOVA. Representative photomicrographs in color of TH versus GABA cells are shown in Ding et al. (2004), and this figure is reprinted with permission.

TABLE 13.1 The strengths and weaknesses of 6-OHDA models of Parkinson's disease

Strengths
Specific loss of dopamine neurons
Can be used *in vitro* and in a broad range of animal species
Relatively uniform lesions with low inter-animal variability
Easy quantification of motor deficits in animal model
Unilateral lesions *in vivo* permits use of contralateral hemisphere as control

Weaknesses
Models only one type of cell loss
Models only one type of cellular stress
Lack of Lewy body-like inclusions
Causes relatively rapid cell death
Possible non-specific damage due to rapid oxidation and/or high concentrations

Weaknesses of Model

There also are several weaknesses in modeling PD with 6-OHDA application to the nigrostriatal system. We will mention four. First, the human illnesses spans various regions in the central and peripheral nervous systems, ranging from an early involvement of the dorsal motor nucleus of the vagus in the brainstem to the cerebral cortex in end stages of the disease (Braak *et al.* 2003a, b). The symptomatic relief provided by L-dopa against many of the motor deficits confirms the importance of developing therapies that protect the dopaminergic system from degeneration. However, patients suffer from a constellation of affective and autonomic symptoms which also reduce the quality of their lives, particularly as the illness progresses. For testing therapeutic interventions against these troublesome symptoms, devising an effective PD model that spans these non-motor regions as well as mesencephalic motor neurons will be required.

Second, a key pathological hallmark of PD is the presence of inclusions. However, although it has been reported that 6-OHDA can increase the accumulation of ubiquitin-conjugated proteins (Elkon *et al.*, 2001), inclusions have not been observed in the acute 6-OHDA rodent model. It remains to be seen if low-dose 6-OHDA infused chronically will induce inclusion formation as has been shown for MPTP (Fornai *et al.*, 2005).

A third weakness is the relative lack of progression in the 6-OHDA model. This is less true for the striatal model than those involving injections within the lateral hypothalamus or SN, since the loss of DA cells progresses over weeks after striatal 6-OHDA (Sauer and Oertel, 1994), whereas the loss of cells after the other treatments appears to occur within a few days. An 8- to 16-week period for the rat may translate into as many as 8 years in human terms assuming a life expectancy of 2.5 years for a rat and 90 years for a human. However, even this rate of progression is probably fast when compared to the human disease. Moreover, it applies only to the DA cell bodies; the loss of DA terminals might be very rapid. Attempts have been made to give multiple small doses of 6-OHDA over time (Fleming *et al.*, 2005), but this may not be directly comparable to the progressive loss of DA neurons in PD. One can imagine that with multiple doses of 6-OHDA, there is a series of acute apoptotic events, whereas the death of DA neurons in

PD is gradual at the level of the individual cell as well as of the whole population.

Fourth, the 6-OHDA model only encompasses one of the cellular dysfunctions thought to be present in PD, oxidative stress. PD is likely to involve multiple system failures, and may even represent more than one illness (Jenner and Olanow, 2006).

Thus, 6-OHDA remains a one-dimensional model of a complex human illness, and if one is to use 6-OHDA to model PD – as we ourselves have done for many years – it is essential to use it with care and be cognizant of its limitations. This is particularly critical when using 6-OHDA to test the efficacy of various neuroprotective strategies. If we limit ourselves to this model alone, we will only know if potential candidates protect DA neurons, as well as only know if they protect from the consequences of one type and time course of cellular stress – rapid oxidative toxicity.

SELF-PROTECTION IN RESPONSE TO 6-OHDA

Response of Neurotrophic Factors to 6-OHDA-Induced Injury

As noted above, trophic factors may serve to defend against neural injury. We have observed, for example, that GDNF levels increased immediately following the administration of 6-OHDA into the medial forebrain bundle (Smith *et al.*, 2003). In our studies the increase of GDNF with 6-OHDA was modest (50%) and could only be detected within the first 3 days after the administration of 6-OHDA (Smith *et al.*, 2003), On the other hand, increases in GDNF protein and in GDNF mRNA have been both larger and longer lasting in other studies (Zhou *et al.*, 2000; Nakajima *et al.*, 2001; Yurek and Fletcher-Turner, 2001). An even greater range of results have been reported with MPTP, with some investigators finding increases in GDNF message and others finding no change (Tang *et al.*, 1998; Inoue *et al.*, 1999; Collier *et al.*, 2005). To some extent these differences in results may reflect the difference between monitoring mRNA and protein, which need not always correlate. But clearly, there are variables determining the GDNF response to the loss of DA that are not yet understood.

The increases in GDNF protein and message that have been observed with injury to DA neurons are likely to function as a self-defensive measure. This is supported by findings in primary SN cultures that inhibiting the upregulation of GDNF by oxidative stress with blocking antibodies exacerbates ROS toxicity (Saavedra *et al.*, 2006).

We and others have noted a reduction of GDNF receptors with 6-OHDA lesions, (Smith *et al.*, 2003; Kozlowski *et al.*, 2004), which may reflect a loss of the DA neurons on which the receptors are located. However, an early decrease in Ret without a concomitant decrease in GFRα has been observed after treatment with MPTP (Hirata and Kiuchi, 2007), which suggests the possibility of an active down-regulation of the protein. Moreover, others have noted transient increases in c-Ret and GFRα1 after 6-OHDA followed by decreases in these proteins (Marco *et al.*, 2002). Thus, changes in receptor levels for GDNF need to be systematically examined as a function of dose and time of injury in studies of endogenous defenses.

Another factor that influences outcome in trophic factor studies is age, as it has been reported that BDNF and GDNF are increased with the injury of 6-OHDA in the denervated striatum in young but not old animals (Yurek and Fletcher-Turner, 2000, 2001). Further, in monkeys there was a basal increase in striatal trophic activity with age, whereas the increase with additional MPTP injury failed to occur in tissue from old animals (Collier *et al.*, 2005). In short, the trophic factor response to various doses of insults, as a function of age of organism and topographic subregion, with close attention to temporal kinetics, all remain important subjects for future inquiry.

Preconditioning-Induced Protection

One explanation for the progressive nature of PD is that although the primary insult is present for years, resilient neurons respond to this with adaptive defenses to stave off degeneration for some time. If some anatomical regions are better equipped to handle stressors than others, this could result in the topographical specificity of many neurodegenerative disorders. Minor, sublethal stressors have long been known to confer protection against subsequent, otherwise lethal stressors, a phenomenon referred to as *preconditioning* or *tolerance* (Dawson and Dawson, 2000; Dirnagl *et al.*, 2003; Arumugam *et al.*, 2006). The majority of preconditioning studies employ ischemia, in which exposure to short ischemic episodes produces robust neuroprotection against subsequent longer ischemic exposures in a variety of organs, including heart and brain (Murry *et al.*, 1986; Kitagawa *et al.*, 1990). Preconditioning also has been hypothesized to be the mechanism whereby dietary restriction increases lifespan (Prolla and Mattson, 2001; Arumugam *et al.*, 2006). Preconditioning-induced defenses are thought to decrease with age (Fenton *et al.*, 2000; Schulman *et al.*, 2001; Taylor and Starnes, 2003; He *et al.*, 2005; Juhaszova *et al.*, 2005). As resistance to stress is thought to correlate with longevity (Verbeke *et al.*, 2001; Arumugam *et al.*, 2006), studies of changes in defensive responses to minor stress with age and in age-related diseases are warranted.

In ischemia, two types of preconditioning have been described. Protection occurs within minutes or hours of the sublethal episode (*rapid preconditioning*), or within hours to days of the first insult (*delayed preconditioning*). The former depends primarily on post-translational modifications of proteins and is short lived in nature, whereas the latter time frame also allows for changes in protein expression and can be sustained for weeks (Eisen *et al.*, 2004; Tsai *et al.*, 2004; Pasupathy and Homer-Vanniasinkam, 2005).

Although the literature is replete with studies using small ischemic episodes as a preconditioning stimulus, a few studies have examined preconditioning or at least stress-induced protection in models of PD. For example, using cell lines, protection was reported by heat shock against MPP^+ (Quigney *et al.*, 2003), $FeSO_4$ or xanthine/xanthine oxidase against rotenone toxicity (Tai and Truong, 2002), and hydrogen peroxide against DA (Xiao-Qing *et al.*, 2005). In addition, non-lethal serum deprivation can protect SH-SY5Y cells against MPP^+ (Andoh *et al.*, 2002). In animal models of PD, protection has been observed with dietary restriction or 2-deoxyglucose (Duan and Mattson, 1999), mechanical injury (Przedborski *et al.*, 1991), and low doses of thrombin (Cannon *et al.*, 2005). More evidence of defensive responses to injury by DA neurons comes from observations that multiple escalating intrastriatal injections of 6-OHDA are needed to produce stable behavioral deficits, and that rats can tolerate a much larger dose of toxin when preceded by smaller escalating doses (Fleming *et al.*, 2005). The latter effect has been attributed to preconditioning. Collectively, these observations argue for the presence of defensive stress responses of direct relevance to PD.

We have recently observed further evidence of preconditioning in a DA cell line, MN9D (Leak *et al.*, 2006). We found that low, sublethal concentrations

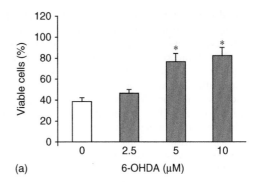

(a)

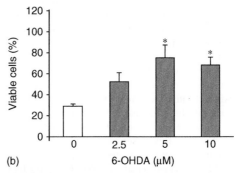

(b)

FIGURE 13.5 Preconditioning MN9D cells with sublethal doses of 6-OHDA protects them against subsequent, otherwise lethal doses of 6-OHDA. (a) MN9D cells were pretreated with indicated concentrations of 6-OHDA or vehicle 6 h prior to post-treatment with 50 μM 6-OHDA and Hoechst stained 24 h thereafter. All pretreatments were 20 min and post-treatments were 30 min in duration. Cell counts of viable nuclei are shown as a percentage of controls treated twice with vehicle in the same manner. (b) MN9D cells were preconditioned with indicated sublethal 6-OHDA for 20 min 6 h before a 24 h challenge with 1 μM MG-132. Cells were stained with Hoechst reagent 24 h after initiation of MG-132 treatment. Data are presented as means and SEM of three independent experiments. *$p < 0.05$ versus vehicle, Bonferroni *post hoc* following ANOVA. Reprinted with permission from Leak *et al.* (2006).

of 6-OHDA (5–10 μM) protected against subsequent application of high concentrations of the same toxin (50–100 μM) delivered 6 h later (Figure 13.5). The protection was dose responsive, as is typical of preconditioning. We also observed that sublethal 6-OHDA protected against otherwise lethal doses of the proteasome inhibitor MG-132, a display of *cross-tolerance*, that is, protection by a low level of one cellular stressor against a different stressor. Protection was evident with the Hoechst nuclear stain for condensed nuclei and clumped chromatin, but was not evident with other assays such as those for ATP levels and lactate dehydrogenase release. The preservation of viable nuclei with preconditioning, but not of ATP levels was consistent with previous findings in this cell line. For example, we also observed that the neurotrophic factor GDNF protected MN9D cells against 6-OHDA as measured by the Hoechst reagent, but did not protect against loss of membrane integrity as measured by Trypan blue uptake (Ugarte *et al.*, 2003). Such results led us to conclude that upon preconditioning, MN9D cells enter a quiescent state, focusing on survival alone. We do not know if the cells would eventually recover fully or if cell death was merely delayed. A recent study of ours using chronic treatments with sublethal concentrations of the proteasome inhibitor MG132 serves to illustrate that a long treatment (greater than 2 weeks) with low doses of this toxin can elicit functional, long-lasting protection against 6-OHDA in PC12 cells (Leak et al., 2008). The paradigm used was not a classic, acute preconditioning paradigm, but also elicited defensive responses to chronic stress, such as higher levels of SOD enzymes, Hsp 70, and catalase. Our RNA interference experiments demonstrated that the protection against oxidative stress was mediated by higher levels of the antioxidant enzyme CuZn SOD."

Oxidative challenges present themselves continually to neurons as a consequence of cellular metabolism. Rather than always have detrimental consequences on neuronal function, might low levels of oxidative stress resulting from such metabolic events be beneficial, at least to otherwise healthy neurons? DA neurons produce a continual level of oxidative stress due to the formation of ROS via both auto-oxidation and deamination of their neurotransmitter. Does this make them more or less vulnerable to insult? In support of the latter, we have found that α-methyl-*p*-tyrosine, when applied to MN9D cells to effectively decrease their basal DA levels, exacerbated subsequent 6-OHDA toxicity, as measured by the Hoechst nuclear stain (Figure 13.6) (E. Lin and M. Zigmond, unpublished observations). The self-protective activation of phospho-ERK1/2 by 6-OHDA (see below) was also reduced by α-methyl-*p*-tyrosine. In these

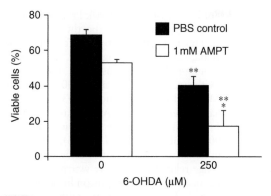

FIGURE 13.6 Decreasing DA levels in MN9D cells exacerbates 6-OHDA toxicity. MN9D cells were treated with PBS or 1 mM α-methyl-*p*-tyrosine (AMPT) for 24 h prior to a 30 min treatment with 250 μM 6-OHDA and stained for viable nuclei with the Hoechst reagent 24 h thereafter. Viable cells are expressed as live/total. The effect upon basal viability was not statistically significant. This treatment dropped DA levels down by 93.5%, as assessed by HPLC. *$p < 0.05$ versus PBS. **$p < 0.05$ versus 0 μM 6-OHDA, Bonferroni *post hoc* following ANOVA.

cells, therefore, normal levels of DA appear to be protective. Whether endogenous DA generates preconditioning levels of oxidative stress *in vivo* is not known but is a future direction of research in our laboratory.

It is possible that diseased neurons cannot mount endogenous neuroprotection in the form of preconditioning. For example, one might speculate that stressors which can precondition DA neurons in healthy individuals cannot do so in a patient with PD and that this leaves these cells vulnerable even to sublethal injury. Such a predisposition could result from genetic or environmental factors.

Studies utilizing mild stress as a means to uncover endogenous defenses have not been used extensively in PD studies and open the possibility for future studies. For example, fibroblasts could be harvested from patients and tested for defensive responses to minor stressors such as heat shock or low levels of ROS. This might determine whether PD patients are deficient in their response to cellular stress. Such a response could also be examined as a function of duration of the illness and clinical staging criteria. We emphasize here that the long-term goal of preconditioning studies is to boost the

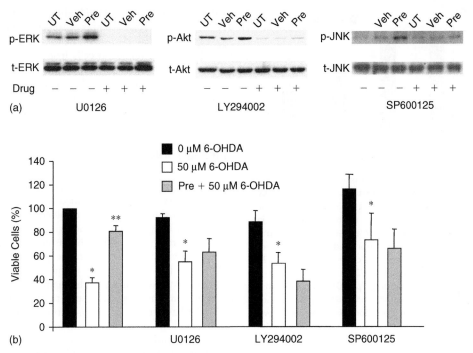

FIGURE 13.7 Sublethal 6-OHDA activates the kinases ERK1/2, Akt, and JNK and inhibitors of all three kinases blocks preconditioning-induced protection in MN9D cells. (a) Western blotting data shows ERK1/2, Akt, and JNK phosphokinase and total kinase levels in untreated cells (UT) or 15 min after the onset of treatment with vehicle (Veh) or 5 μM 6-OHDA (Pre), with or without phospho-ERK (p-ERK) inhibitor U0126 (5 μM), p-Akt inhibitor LY294002 (10 μM) and p-JNK inhibitor SP600125 (5 μM). Blots are representative of three independent experiments. Other time points examined showed no changes. (b) Effect of p-ERK, p-Akt, and p-JNK inhibition on preconditioning. MN9D cells were preconditioned with sublethal 6-OHDA (5 μM; Pre) for 20 min in the absence or presence of U0126, LY294002, or SP600125, as done for Westerns. Six hours later, cells were post-treated with 50 μM 6-OHDA for 30 min. Cells were stained with Hoechst reagent 24 h after post-treatment. Data are presented as means ± SEM of five independent experiments. *$p < 0.05$ compared to vehicle, represented by previous black bar. **$p < 0.05$ compared to non-preconditioned cells, represented by previous white bar, Bonferroni *post hoc* following ANOVA. Reprinted with permission from Leak *et al.* (2006).

endogenous defensive responses in patients pharmacologically without the stress itself.

Possible Mechanisms Underlying Defensive Responses to 6-OHDA Injury

In MN9D cells, preconditioning produced post-translational modifications, activating the kinases ERK1/2, Akt, and JNK (Figure 13.7). Inhibitors of all three kinases blocked protection, suggesting that the changes played a causal role in the 6-OHDA-induced preconditioning of MN9D cells (Leak *et al.*, 2006). Although JNK is typically a pro-death

molecule, there is evidence in the ischemia literature that JNK mediates a protective preconditioning effect (Nakano *et al.*, 2000; Sato *et al.*, 2000; Fryer *et al.*, 2001). Akt and ERK, in contrast, are viewed as pro-survival molecules in most paradigms.

Even when high-dose challenges are applied, there is evidence that cells can sometimes attempt to limit injury to themselves and their neighbors by mounting endogenous defenses. For example, we observed that high concentrations of 6-OHDA rapidly activated ERK1/2 in MN9D cells and transient exposure to U0126, an inhibitor of ERK, exacerbated 6-OHDA-induced toxicity (Figure 13.8) (Lin *et al.*, 2008). We have seen analogous

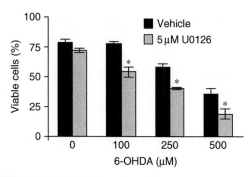

FIGURE 13.8 Decreasing phospho-ERK1/2 levels with U0126 exacerbates 6-OHDA toxicity. MN9D cells were treated with U0126 1h prior to and during 6-OHDA treatment, resulting in decreased cell viability at all 6-OHDA concentrations. Viable cells are expressed as a fraction of total (live plus dead). This concentration of U0126 decreased phospho-ERK1/2 to undetectable levels by Western blotting. *$p < 0.05$, significant difference between $0\,\mu M$ and $5\,\mu M$ U0126 at each 6-OHDA concentration; Bonferroni *post hoc* following ANOVA. Reprinted from Lin *et al.* (2008) with permission.

results when studying the effects of oxidative stress in HEK293 cells (L. Li, M. J. Zigmond, and A. K. F. Liou, unpublished observations). In our studies with MN9D cells, 6-OHDA also activated phospho-CREB and U0126 abolished its activation. This suggested that CREB activation was downstream of the rapid ERK peak. U0126 also increased ROS production and activated caspase-3, suggesting that phospho-ERK acted both to reduce the source of damage and apoptotic sequelae. Preliminary data suggested that high doses of 6-OHDA also activated Akt in MN9D cells. We have evidence that a high dose of 6-OHDA applied to the medial forebrain bundle elicits early ERK1/2 activation within nigral DA neurons (S. Castro, A. Smith, and M. Zigmond, unpublished observations). Collectively these data suggest that cells can sometimes undergo self-defensive responses even in the face of high level, possibly lethal insults. This may serve both to delay cell death and to limit injury to neighbors while cells are undergoing programmed cell death processes.

In addition to serving a protective function, cellular stress may also sensitize neurons to further stress. For example, proteasome inhibition has been shown to sensitize primary DA cultures to oxidative stress and MPP⁺ (Hoglinger *et al.*, 2003; Mytilineou *et al.*, 2004). This destructive type of a stress response may also serve the overall protective role of removing damaged cells from the organism as a whole. The nature of the stress response to injury, whether defensive or destructive, is likely to be highly paradigm and dose-dependent (Lin *et al.*, 1998) and deserves further detailed study.

SUMMARY AND CONCLUSIONS

In summary, many insights into endogenous defenses against injury have been gained by the use of 6-OHDA as a means to deliver sublethal as well as lethal ROS. These 6-OHDA studies are still limited in their direct relevance to PD. The insult only involves one of several likely factors in the precipitation of DA neuron loss and does not mimic either the very slow loss of those neurons or the many other accompanying neuropathologies. Despite these weaknesses, 6-OHDA has been useful in evaluating the therapeutic potential of compounds such as GDNF in high-throughput and inexpensive manner and has been useful in answering some fundamental questions about cell biology.

As with sublethal ischemic episodes, oxidative stress produced by 6-OHDA has the capacity to elicit self-protective responses. Whether the human brain acts to limit injury in a similar manner and whether this defense contributes to the progressive nature of neurodegeneration are not yet known. Insight into endogenous defenses against stress may have clinical implications for a broad range of illnesses. The added complication of a loss of defensive responses to injury with advancing age has not always been accounted for in studies with PD models, but further characterization of age-related changes in stress-induced defenses may help explain why age is the major risk factor for PD.

REFERENCES

Akerud, P., Alberch, J., Eketjall, S., Wagner, J., and Arenas, E. (1999). Differential effects of glial cell line-derived neurotrophic factor and neurturin on developing and adult substantia nigra dopaminergic neurons. *J Neurochem* **73**, 70–78.

Andoh, T., Chock, P. B., and Chiueh, C. C. (2002). Preconditioning-mediated neuroprotection: Role of nitric oxide, cGMP, and new protein expression. *Ann NY Acad Sci* **962**, 1–7.

Anglade, P., Vyas, S., Javoy-Agid, F., Herrero, M. T., Michel, P. P., Marquez, J., Mouatt-Prigent, A., Ruberg, M., Hirsch, E. C., and Agid, Y. (1997). Apoptosis and autophagy in nigral neurons of patients with Parkinson's disease. *Histol Histopathol* **12**, 25–31.

Arumugam, T. V., Gleichmann, M., Tang, S. C., and Mattson, M. P. (2006). Hormesis/preconditioning mechanisms, the nervous system and aging. *Ageing Res Rev* **5**, 165–178.

Auluck, P. K., and Bonini, N. M. (2002). Pharmacological prevention of Parkinson disease in *Drosophila*. *Nat Med* **8**, 1185–1186.

Auluck, P. K., Chan, H. Y., Trojanowski, J. Q., Lee, V. M., and Bonini, N. M. (2002). Chaperone suppression of alpha-synuclein toxicity in a *Drosophila* model for Parkinson's disease. *Science* **295**, 865–868.

Backman, C. M., Shan, L., Zhang, Y. J., Hoffer, B. J., Leonard, S., Troncoso, J. C., Vonsatel, P., and Tomac, A. C. (2006). Gene expression patterns for GDNF and its receptors in the human putamen affected by Parkinson's disease: A real-time PCR study. *Mol Cell Endocrinol* **252**, 160–166.

Bar, K. J., Saldanha, G. J., Kennedy, A. J., Facer, P., Birch, R., Carlstedt, T., and Anand, P. (1998). GDNF and its receptor component Ret in injured human nerves and dorsal root ganglia. *Neuroreport* **9**, 43–47.

Barnham, K. J., Masters, C. L., and Bush, A. I. (2004). Neurodegenerative diseases and oxidative stress. *Nat Rev Drug Discov* **3**, 205–214.

Barreiro, E., Coronell, C., Lavina, B., Ramirez-Sarmiento, A., Orozco-Levi, M., and Gea, J. (2006). Aging, sex differences, and oxidative stress in human respiratory and limb muscles. *Free Radic Biol Med* **41**, 797–809.

Beere, H. M. (2001). Stressed to death: Regulation of apoptotic signaling pathways by the heat shock proteins. *Sci STKE* **2001**, RE1.

Beere, H. M., and Green, D. R. (2001). Stress management – heat shock protein-70 and the regulation of apoptosis. *Trends Cell Biol* **11**, 6–10.

Bella, A. J., Lin, G., Garcia, M. M., Tantiwongse, K., Brant, W. O., Lin, C. S., and Lue, T. F. (2007). Upregulation of penile brain-derived neurotrophic factor (BDNF) and activation of the JAK/STAT signalling pathway in the major pelvic ganglion of the rat after cavernous nerve transection. *Eur Urol* **52**, 574–580.

Berger, Z., Ravikumar, B., Menzies, F. M., Oroz, L. G., Underwood, B. R., Pangalos, M. N., Schmitt, I., Wullner, U., Evert, B. O., O'Kane, C. J., and Rubinsztein, D. C. (2006). Rapamycin alleviates toxicity of different aggregate-prone proteins. *Hum Mol Genet* **15**, 433–442.

Betarbet, R., Sherer, T. B., MacKenzie, G., Garcia-Osuna, M., Panov, A. V., and Greenamyre, J. T. (2000). Chronic systemic pesticide exposure reproduces features of Parkinson's disease. *Nat Neurosci* **3**, 1301–1306.

Blandini, F., Sinforiani, E., Pacchetti, C., Samuele, A., Bazzini, E., Zangaglia, R., Nappi, G., and Martignoni, E. (2006). Peripheral proteasome and caspase activity in Parkinson disease and Alzheimer disease. *Neurology* **66**, 529–534.

Blum, M., and Weickert, C. S. (1995). GDNF mRNA expression in normal postnatal development, aging, and in Weaver mutant mice. *Neurobiol Aging* **16**, 925–929.

Braak, H., Rub, U., Gai, W. P., and Del Tredici, K. (2003a). Idiopathic Parkinson's disease: Possible routes by which vulnerable neuronal types may be subject to neuroinvasion by an unknown pathogen. *J Neural Transm* **110**, 517–536.

Braak, H., Del Tredici, K., Rub, U., de Vos, R. A., Jansen Steur, E. N., and Braak, E. (2003b). Staging of brain pathology related to sporadic Parkinson's disease. *Neurobiol Aging* **24**, 197–211.

Cannon, J. R., Keep, R. F., Hua, Y., Richardson, R. J., Schallert, T., and Xi, G. (2005). Thrombin preconditioning provides protection in a 6-hydroxydopamine Parkinson's disease model. *Neurosci Lett* **373**, 189–194.

Carrard, G., Bulteau, A. L., Petropoulos, I., and Friguet, B. (2002). Impairment of proteasome structure and function in aging. *Int J Biochem Cell Biol* **34**, 1461–1474.

Chalovich, E. M., Zhu, J. H., Caltagarone, J., Bowser, R., and Chu, C. T. (2006). Functional repression of cAMP response element in 6-hydroxydopamine-treated neuronal cells. *J Biol Chem* **281**, 17870–17881.

Chauhan, N. B., Siegel, G. J., and Lee, J. M. (2001). Depletion of glial cell line-derived neurotrophic factor in substantia nigra neurons of Parkinson's disease brain. *J Chem Neuroanat* **21**, 277–288.

Choi-Lundberg, D. L., and Bohn, M. C. (1995). Ontogeny and distribution of glial cell line-derived neurotrophic factor (GDNF) mRNA in rat. *Brain Res Dev Brain Res* **85**, 80–88.

Choi, W. S., Yoon, S. Y., Oh, T. H., Choi, E. J., O'Malley, K. L., and Oh, Y. J. (1999). Two distinct mechanisms are involved in 6-hydroxydopamine- and MPP$^+$-induced dopaminergic neuronal cell death: Role of caspases, ROS, and JNK. *J Neurosci Res* **57**, 86–94.

Chu, C. T., Zhu, J., and Dagda, R. (2007). Beclin 1-independent pathway of damage-induced mitophagy and autophagic stress: Implications for neurodegeneration and cell death. *Autophagy* **3**, 663–666.

Ciechanover, A., Orian, A., and Schwartz, A. L. (2000). The ubiquitin-mediated proteolytic pathway: Mode of action and clinical implications. *J Cell Biochem* **34**, 40–51.

Clement, M. V., Long, L. H., Ramalingam, J., and Halliwell, B. (2002). The cytotoxicity of dopamine may be an artefact of cell culture. *J Neurochem* **81**, 414–421.

Collier, T. J., Dung Ling, Z., Carvey, P. M., Fletcher-Turner, A., Yurek, D. M., Sladek, J. R., Jr., and Kordower, J. H. (2005). Striatal trophic factor activity in aging monkeys with unilateral MPTP-induced parkinsonism. *Exp Neurol* **191(Suppl 1)**, S60–S67.

Conconi, M., and Friguet, B. (1997). Proteasome inactivation upon aging and on oxidation-effect of HSP 90. *Mol Biol Rep* **24**, 45–50.

Cramer, S. C., and Chopp, M. (2000). Recovery recapitulates ontogeny. *Trends Neurosci* **23**, 265–271.

Cuervo, A. M. (2004). Autophagy: In sickness and in health. *Trends Cell Biol* **14**, 70–77.

Cuervo, A. M., and Dice, J. F. (1998). How do intracellular proteolytic systems change with age? *Front Biosci* **3**, d25–d43.

Cutillas, B., Espejo, M., Gil, J., Ferrer, I., and Ambrosio, S. (1999). Caspase inhibition protects nigral neurons against 6-OHDA-induced retrograde degeneration. *Neuroreport* **10**, 2605–2608.

Dauer, W., and Przedborski, S. (2003). Parkinson's disease: Mechanisms and models. *Neuron* **39**, 889–909.

Davies, K. J. (2001). Degradation of oxidized proteins by the 20S proteasome. *Biochimie* **83**, 301–310.

Dawson, V. L., and Dawson, T. M. (2000). Neuronal ischaemic preconditioning. *Trends Pharmacol Sci* **21**, 423–424.

Dexter, D. T., Sian, J., Rose, S., Hindmarsh, J. G., Mann, V. M., Cooper, J. M., Wells, F. R., Daniel, S. E., Lees, A. J., Schapira, A. H. *et al.* (1994). Indices of oxidative stress and mitochondrial function in individuals with incidental Lewy body disease. *Ann Neurol* **35**, 38–44.

Ding, Y. M., Jaumotte, J. D., Signore, A. P., and Zigmond, M. J. (2004). Effects of 6-hydroxydopamine on primary cultures of substantia nigra: specific damage to dopamine neurons and the impact of glial cell line-derived neurotrophic factor. *J Neurochem* **89**, 776–787.

Dirnagl, U., Simon, R. P., and Hallenbeck, J. M. (2003). Ischemic tolerance and endogenous neuroprotection. *Trends Neurosci* **26**, 248–254.

Duan, W., and Mattson, M. P. (1999). Dietary restriction and 2-deoxyglucose administration improve behavioral outcome and reduce degeneration of dopaminergic neurons in models of Parkinson's disease. *J Neurosci Res* **57**, 195–206.

Ebadi, M., Govitrapong, P., Sharma, S., Muralikrishnan, D., Shavali, S., Pellett, L., Schafer, R., Albano, C., and Eken, J. (2001). Ubiquinone (coenzyme q10) and mitochondria in oxidative stress of Parkinson's disease. *Biol Signals Recept* **10**, 224–253.

Ebert, A. D., Hann, H. J., and Bohn, M. C. (2007). Progressive degeneration of dopamine neurons in 6-hydroxydopamine rat model of Parkinson's disease does not involve activation of caspase-9 and caspase-3. *J Neurosci Res.*

Eisen, A., Fisman, E. Z., Rubenfire, M., Freimark, D., McKechnie, R., Tenenbaum, A., Motro, M., and Adler, Y. (2004). Ischemic preconditioning: Nearly two decades of research. A comprehensive review. *Atherosclerosis* **172**, 201–210.

Ekstrand, M. I., Terzioglu, M., Galter, D., Zhu, S., Hofstetter, C., Lindqvist, E., Thams, S., Bergstrand, A., Hansson, F. S., Trifunovic, A., Hoffer, B., Cullheim, S., Mohammed, A. H., Olson, L., and Larsson, N. G. (2007). Progressive parkinsonism in mice with respiratory-chain-deficient dopamine neurons. *Proc Natl Acad Sci USA.*

Elkon, H., Melamed, E., and Offen, D. (2001). 6-Hydroxydopamine increases ubiquitin-conjugates and protein degradation: Implications for the pathogenesis of Parkinson's disease. *Cell Mol Neurobiol* **21**, 771–781.

Engele, J., and Bohn, M. C. (1991). The neurotrophic effects of fibroblast growth factors on dopaminergic neurons *in vitro* are mediated by mesencephalic glia. *J Neurosci* **11**, 3070–3078.

Erraji-Benchekroun, L., Underwood, M. D., Arango, V., Galfalvy, H., Pavlidis, P., Smyrniotopoulos, P., Mann, J. J., and Sibille, E. (2005). Molecular aging in human prefrontal cortex is selective and continuous throughout adult life. *Biol Psychiatry* **57**, 549–558.

Fallon, J. H., and Moore, R. Y. (1978). Catecholamine innervation of the basal forebrain. IV. Topography of the dopamine projection to the basal forebrain and neostriatum. *J Comp Neurol* **180**, 545–580.

Fenton, R. A., Dickson, E. W., Meyer, T. E., and Dobson, J. G., Jr. (2000). Aging reduces the cardioprotective effect of ischemic preconditioning in the rat heart. *J Mol Cell Cardiol* **32**, 1371–1375.

Fleming, S. M., Delville, Y., and Schallert, T. (2005). An intermittent, controlled-rate, slow progressive degeneration model of Parkinson's disease: Antiparkinson effects of Sinemet and protective effects of methylphenidate. *Behav Brain Res* **156**, 201–213.

Fleming, S. M., Delville, Y., and Schallert, T. (2005). An intermittent, controlled-rate, slow progressive degeneration model of Parkinson's disease: Antiparkinson effects of Sinemet and protective effects of methylphenidate. *Behav Brain Res* **156**, 201–213.

Falkenburger, B. H. and Schulz, J. B. (2006) Limitations of cellular models in Parkinson's disease research. *J. Neural Transm. Suppl.* **70**, 261–268.

Floyd, R. A., and Hensley, K. (2002). Oxidative stress in brain aging. Implications for therapeutics of neurodegenerative diseases. *Neurobiol Aging* **23**, 795–807.

Fornai, F., Schluter, O. M., Lenzi, P., Gesi, M., Ruffoli, R., Ferrucci, M., Lazzeri, G., Busceti, C. L., Pontarelli, F., Battaglia, G., Pellegrini, A., Nicoletti, F., Ruggieri, S., Paparelli, A., and Sudhof, T. C. (2005). Parkinson-like syndrome induced by continuous MPTP infusion: Convergent roles of the ubiquitin–proteasome system and alpha-synuclein. *Proc Natl Acad Sci USA* **102**, 3413–3418.

Fryer, R. M., Patel, H. H., Hsu, A. K., and Gross, G. J. (2001). Stress-activated protein kinase phosphorylation during cardioprotection in the ischemic myocardium. *Am J Physiol Heart Circ Physiol* **281**, H1184–H1192.

Gerfen, C. R., Herkenham, M., and Thibault, J. (1987). The neostriatal mosaic: II. Patch- and matrix-directed mesostriatal dopaminergic and non-dopaminergic systems. *J Neurosci* **7**, 3915–3934.

Goldberg, A. L. (2003). Protein degradation and protection against misfolded or damaged proteins. *Nature* **426**, 895–899.

Gong, L., Wyatt, R. J., Baker, I., and Masserano, J. M. (1999). Brain-derived and glial cell line-derived neurotrophic factors protect a catecholaminergic cell line from dopamine-induced cell death. *Neurosci Lett* **263**, 153–156.

Gonzalez-Hernandez, T., Barroso-Chinea, P., De La Cruz Muros, I., Del Mar Perez-Delgado, M., and Rodriguez, M. (2004). Expression of dopamine and vesicular monoamine transporters and differential vulnerability of mesostriatal dopaminergic neurons. *J Comp Neurol* **479**, 198–215.

Grondin, R., Zhang, Z., Yi, A., Cass, W. A., Maswood, N., Andersen, A. H., Elsberry, D. D., Klein, M. C., Gerhardt, G. A., and Gash, D. M. (2002). Chronic, controlled GDNF infusion promotes structural and functional recovery in advanced parkinsonian monkeys. *Brain* **125**, 2191–2201.

Grondin, R., Cass, W. A., Zhang, Z., Stanford, J. A., Gash, D. M., and Gerhardt, G. A. (2003). Glial cell line-derived neurotrophic factor increases stimulus-evoked dopamine release and motor speed in aged rhesus monkeys. *J Neurosci* **23**, 1974–1980.

Hara, T., Nakamura, K., Matsui, M., Yamamoto, A., Nakahara, Y., Suzuki-Migishima, R., Yokoyama, M., Mishima, K., Saito, I., Okano, H., and Mizushima, N. (2006). Suppression of basal autophagy in neural cells causes neurodegenerative disease in mice. *Nature* **441**, 885–889.

He, Y., Lee, T., and Leong, S. K. (2000). 6-Hydroxydopamine induced apoptosis of dopaminergic cells in the rat substantia nigra. *Brain Res* **858**, 163–166.

He, Z., Crook, J. E., Meschia, J. F., Brott, T. G., Dickson, D. W., and McKinney, M. (2005). Aging blunts ischemic-preconditioning-induced neuroprotection following transient global ischemia in rats. *Curr Neurovasc Res* **2**, 365–374.

Hirata, Y., and Kiuchi, K. (2007). Rapid down-regulation of Ret following exposure of dopaminergic neurons to neurotoxins. *J Neurochem* **102**, 1606–1613.

Hoffer, B. J., Hoffman, A., Bowenkamp, K., Huettl, P., Hudson, J., Martin, D., Lin, L. F., and Gerhardt, G. A. (1994). Glial cell line-derived neurotrophic factor reverses toxin-induced injury to midbrain dopaminergic neurons *in vivo*. *Neurosci Lett* **182**, 107–111.

Hoglinger, G. U., Carrard, G., Michel, P. P., Medja, F., Lombes, A., Ruberg, M., Friguet, B., and Hirsch, E. C. (2003). Dysfunction of mitochondrial complex I and the proteasome: Interactions between two biochemical deficits in a cellular model of Parkinson's disease. *J Neurochem* **86**, 1297–1307.

Hokfelt, T., and Ungerstedt, U. (1973). Specificity of 6-hydroxydopamine induced degeneration of central monoamine neurones: an electron and fluorescence microscopic study with special reference to intracerebral injection on the nigro-striatal dopamine system. *Brain Res* **60**, 269–297.

Hou, J. G., Cohen, G., and Mytilineou, C. (1997). Basic fibroblast growth factor stimulation of glial cells protects dopamine neurons from 6-hydroxydopamine toxicity: Involvement of the glutathione system. *J Neurochem* **69**, 76–83.

Hussain, S., Slikker, W., Jr., and Ali, S. F. (1995). Age-related changes in antioxidant enzymes, superoxide dismutase, catalase, glutathione peroxidase and glutathione in different regions of mouse brain. *Int J Dev Neurosci* **13**, 811–817.

Inoue, T., Tsui, J., Wong, N., Wong, S. Y., Suzuki, F., and Kwok, Y. N. (1999). Expression of glial cell line-derived neurotrophic factor and its mRNA in the nigrostriatal pathway following MPTP treatment. *Brain Res* **826**, 306–308.

Jackson-Lewis, V., Jakowec, M., Burke, R. E., and Przedborski, S. (1995). Time course and morphology of dopaminergic neuronal death caused by the neurotoxin 1-methyl-4-phenyl-1,2,3,6-tetrahydropyridine. *Neurodegeneration* **4**, 257–269.

Jackson, A. L., Bartz, S. R., Schelter, J., Kobayashi, S. V., Burchard, J., Mao, M., Li, B., Cavet, G., and Linsley, P. S. (2003). Expression profiling reveals off-target gene regulation by RNAi. *Nat Biotechnol* **21**, 635–637.

Jenner, P., and Olanow, C. W. (2006). The pathogenesis of cell death in Parkinson's disease. *Neurology* **66**, S24–S36.

Jeon, B. S., Jackson-Lewis, V., and Burke, R. E. (1995). 6-Hydroxydopamine lesion of the rat substantia nigra: Time course and morphology of cell death. *Neurodegeneration* **4**, 131–137.

Jiang, C. H., Tsien, J. Z., Schultz, P. G., and Hu, Y. (2001). The effects of aging on gene expression in the hypothalamus and cortex of mice. *Proc Natl Acad Sci USA* **98**, 1930–1934.

Juhaszova, M., Rabuel, C., Zorov, D. B., Lakatta, E. G., and Sollott, S. J. (2005). Protection in the aged heart: Preventing the heart-break of old age? *Cardiovasc Res* 66, 233–244.

Kaushik, S., and Cuervo, A. M. (2006). Autophagy as a cell-repair mechanism: Activation of chaperone-mediated autophagy during oxidative stress. *Mol Aspect Med* 27, 444–454.

Kearns, C. M., and Gash, D. M. (1995). GDNF protects nigral dopamine neurons against 6-hydroxydopamine *in vivo*. *Brain Res* 672, 104–111.

Kim, Y. S., and Joh, T. H. (2006). Microglia, major player in the brain inflammation: Their roles in the pathogenesis of Parkinson's disease. *Exp Mol Med* 38, 333–347.

Kirik, D., Rosenblad, C., and Bjorklund, A. (1998). Characterization of behavioral and neurodegenerative changes following partial lesions of the nigrostriatal dopamine system induced by intrastriatal 6-hydroxydopamine in the rat. *Exp Neurol* 152, 259–277.

Kirik, D., Rosenblad, C., and Bjorklund, A. (2000). Preservation of a functional nigrostriatal dopamine pathway by GDNF in the intrastriatal 6-OHDA lesion model depends on the site of administration of the trophic factor. *Eur J Neurosci* 12, 3871–3882.

Kisselev, A. F., and Goldberg, A. L. (2001). Proteasome inhibitors: From research tools to drug candidates. *Chem Biol* 8, 739–758.

Kitagawa, K., Matsumoto, M., Tagaya, M., Hata, R., Ueda, H., Niinobe, M., Handa, N., Fukunaga, R., Kimura, K., Mikoshiba, K. *et al.* (1990). "Ischemic tolerance" phenomenon found in the brain. *Brain Res* 528, 21–24.

Komatsu, M., Waguri, S., Chiba, T., Murata, S., Iwata, J., Tanida, I., Ueno, T., Koike, M., Uchiyama, Y., Kominami, E., and Tanaka, K. (2006). Loss of autophagy in the central nervous system causes neurodegeneration in mice. *Nature* 441, 880–884.

Kordower, J. H., Emborg, M. E., Bloch, J., Ma, S. Y., Chu, Y., Leventhal, L., McBride, J., Chen, E. Y., Palfi, S., Roitberg, B. Z., Brown, W. D., Holden, J. E., Pyzalski, R., Taylor, M. D., Carvey, P., Ling, Z., Trono, D., Hantraye, P., Deglon, N., and Aebischer, P. (2000). Neurodegeneration prevented by lentiviral vector delivery of GDNF in primate models of Parkinson's disease. *Science* 290, 767–773.

Kostrzewa, R. M., and Jacobowitz, D. M. (1974). Pharmacological actions of 6-hydroxydopamine. *Pharmacol Rev* 26, 199–288.

Kozlowski, D. A., Connor, B., Tillerson, J. L., Schallert, T., and Bohn, M. C. (2000). Delivery of a GDNF gene into the substantia nigra after a progressive 6-OHDA lesion maintains functional nigrostriatal connections. *Exp Neurol* 166, 1–15.

Kozlowski, D. A., Miljan, E. A., Bremer, E. G., Harrod, C. G., Gerin, C., Connor, B., George, D., Larson, B., and

Bohn, M. C. (2004). Quantitative analyses of GFRalpha-1 and GFRalpha-2 mRNAs and tyrosine hydroxylase protein in the nigrostriatal system reveal bilateral compensatory changes following unilateral 6-OHDA lesions in the rat. *Brain Res* 1016, 170–181.

Kramer, B. C., Goldman, A. D., and Mytilineou, C. (1999). Glial cell line derived neurotrophic factor promotes the recovery of dopamine neurons damaged by 6-hydroxydopamine *in vitro*. *Brain Res* 851, 221–227.

Lambertucci, R. H., Levada-Pires, A. C., Rossoni, L. V., Curi, R., and Pithon-Curi, T. C. (2006). Effects of aerobic exercise training on antioxidant enzyme activities and mRNA levels in soleus muscle from young and aged rats. *Mech Ageing Dev*.

Larsen, K. E., and Sulzer, D. (2002). Autophagy in neurons: A review. *Histol Histopathol* 17, 897–908.

Leak, R. K., Liou, A. K., and Zigmond, M. J. (2006). Effect of sublethal 6-hydroxydopamine on the response to subsequent oxidative stress in dopaminergic cells: Evidence for preconditioning. *J Neurochem* 99, 1151–1163.

Leak, R. K., Zigmond, M. J., and Liou, A. K. F. (2008). Adaptation to chronic MG132 reduces oxidative toxicity by a CuZnSOD-dependent mechanism. *Journal of Neurochemistry*, E-pub ahead of print.

Lee, C. K., Weindruch, R., and Prolla, T. A. (2000). Gene-expression profile of the ageing brain in mice. *Nat Genet* 25, 294–297.

Lee, D. H., and Goldberg, A. L. (1998). Proteasome inhibitors: Valuable new tools for cell biologists. *Trends Cell Biol* 8, 397–403.

Liang, C. L., Sinton, C. M., Sonsalla, P. K., and German, D. C. (1996). Midbrain dopaminergic neurons in the mouse that contain calbindin-D28k exhibit reduced vulnerability to MPTP-induced neurodegeneration. *Neurodegeneration* 5, 313–318.

Liang, Q., Liou, A. K., Ding, Y., Cao, G., Xiao, X., Perez, R. G., and Chen, J. (2004). 6-Hydroxydopamine induces dopaminergic cell degeneration via a caspase-9-mediated apoptotic pathway that is attenuated by caspase-9dn expression. *J Neurosci Res* 77, 747–761.

Liberatore, G. T., Wong, J. Y., Porritt, M. J., Donnan, G. A., and Howells, D. W. (1997). Expression of glial cell line-derived neurotrophic factor (GDNF) mRNA following mechanical injury to mouse striatum. *Neuroreport* 8, 3097–3101.

Lin, E., Cavanaugh, J. E., Leak, R. K., Perez, R. G., and Zigmond, M. J. (2008). Rapid activation of ERK by 6-hydroxydopamine promotes survival of dopaminergic cells. *J Neurosci Res* 86, 108–117.

Lin, K. I., Baraban, J. M., and Ratan, R. R. (1998). Inhibition versus induction of apoptosis by proteasome inhibitors depends on concentration. *Cell Death Differ* 5, 577–583.

Lin, L. F., Doherty, D. H., Lile, J. D., Bektesh, S., and Collins, F. (1993). GDNF: A glial cell line-derived neurotrophic factor for midbrain dopaminergic neurons. *Science* 260, 1130–1132.

Liu, B., and Hong, J. S. (2003). Role of microglia in inflammation-mediated neurodegenerative diseases: Mechanisms and strategies for therapeutic intervention. *J Pharmacol Exp Ther* 304, 1–7.

Liu, R. M., and Dickinson, D. A. (2003). Decreased synthetic capacity underlies the age-associated decline in glutathione content in Fisher 344 rats. *Antioxid Redox Signal* 5, 529–536.

Locke, M., and Tanguay, R. M. (1996). Diminished heat shock response in the aged myocardium. *Cell Stress Chaperones* 1, 251–260.

Lu, T., Pan, Y., Kao, S. Y., Li, C., Kohane, I., Chan, J., and Yankner, B. A. (2004). Gene regulation and DNA damage in the ageing human brain. *Nature* 429, 883–891.

Marco, S., Saura, J., Perez-Navarro, E., Jose Marti, M., Tolosa, E., and Alberch, J. (2002). Regulation of c-Ret, GFRalpha1, and GFRalpha2 in the substantia nigra pars compacta in a rat model of Parkinson's disease. *J Neurobiol* 52, 343–351.

Mariani, E., Polidori, M. C., Cherubini, A., and Mecocci, P. (2005). Oxidative stress in brain aging, neurodegenerative and vascular diseases: An overview. *J Chromatogr* 827, 65–75.

Marti, M. J., Saura, J., Burke, R. E., Jackson-Lewis, V., Jimenez, A., Bonastre, M., and Tolosa, E. (2002). Striatal 6-hydroxydopamine induces apoptosis of nigral neurons in the adult rat. *Brain Res* 958, 185–191.

Marttila, R. J., and Rinne, U. K. (1991). Progression and survival in Parkinson's disease. *Acta Neurol Scand* 136, 24–28.

Massey, A., Kiffin, R., and Cuervo, A. M. (2004). Pathophysiology of chaperone-mediated autophagy. *Int J Biochem Cell Biol* 36, 2420–2434.

Maswood, N., Grondin, R., Zhang, Z., Stanford, J. A., Surgener, S. P., Gash, D. M., and Gerhardt, G. A. (2002). Effects of chronic intraputamenal infusion of glial cell line-derived neurotrophic factor (GDNF) in aged Rhesus monkeys. *Neurobiol Aging* 23, 881–889.

McAlhany, R. E., Jr., West, J. R., and Miranda, R. C. (1997). Glial-derived neurotrophic factor rescues calbindin-D28k-immunoreactive neurons in alcohol-treated cerebellar explant cultures. *J Neurobiol* 33, 835–847.

McCormack, A. L., Di Monte, D. A., Delfani, K., Irwin, I., DeLanney, L. E., Langston, W. J., and Janson, A. M. (2004). Aging of the nigrostriatal system in the squirrel monkey. *J Comp Neurol* 471, 387–395.

McGeer, P. L., and McGeer, E. G. (2004). Inflammation and neurodegeneration in Parkinson's disease. *Parkinsonism Relat Disord* 10(Suppl 1), S3–S7.

McNaught, K. S. (2004). Proteolytic dysfunction in neurodegenerative disorders. *Int Rev Neurobiol* 62, 95–119.

McNaught, K. S., Belizaire, R., Jenner, P., Olanow, C. W., and Isacson, O. (2002). Selective loss of 20S proteasome

alpha-subunits in the substantia nigra pars compacta in Parkinson's disease. *Neurosci Lett* 326, 155–158.

McNaught, K. S., Belizaire, R., Isacson, O., Jenner, P., and Olanow, C. W. (2003). Altered proteasomal function in sporadic Parkinson's disease. *Exp Neurol* 179, 38–46.

Meriin, A. B., and Sherman, M. Y. (2005). Role of molecular chaperones in neurodegenerative disorders. *Int J Hyperthermia* 21, 403–419.

Mizushima, N., and Klionsky, D. J. (2007). Protein turnover via autophagy: Implications for metabolism. *Annu Rev Nutr*.

Mogi, M., Togari, A., Kondo, T., Mizuno, Y., Kogure, O., Kuno, S., Ichinose, H., and Nagatsu, T. (2001). Glial cell line-derived neurotrophic factor in the substantia nigra from control and parkinsonian brains. *Neurosci Lett* 300, 179–181.

Morrow, G., and Tanguay, R. M. (2003). Heat shock proteins and aging in *Drosophila melanogaster*. *Sem Cell Dev Biol* 14, 291–299.

Murry, C. E., Jennings, R. B., and Reimer, K. A. (1986). Preconditioning with ischemia: A delay of lethal cell injury in ischemic myocardium. *Circulation* 74, 1124–1136.

Mytilineou, C., McNaught, K. S., Shashidharan, P., Yabut, J., Baptiste, R. J., Parnandi, A., and Olanow, C. W. (2004). Inhibition of proteasome activity sensitizes dopamine neurons to protein alterations and oxidative stress. *J Neural Transm* 111, 1237–1251.

Nakajima, K., Hida, H., Shimano, Y., Fujimoto, I., Hashitani, T., Kumazaki, M., Sakurai, T., and Nishino, H. (2001). GDNF is a major component of trophic activity in DA-depleted striatum for survival and neurite extension of DAergic neurons. *Brain Res* 916, 76–84.

Nakamura, M., Yamada, M., Ohsawa, T., Morisawa, H., Nishine, T., Nishimura, O., and Toda, T. (2006). Phosphoproteomic profiling of human SH-SY5Y neuroblastoma cells during response to 6-hydroxydopamine-induced oxidative stress. *Biochim Biophys Acta* 1763, 977–989.

Nakano, A., Baines, C. P., Kim, S. O., Pelech, S. L., Downey, J. M., Cohen, M. V., and Critz, S. D. (2000). Ischemic preconditioning activates MAPKAPK2 in the isolated rabbit heart: Evidence for involvement of p38 MAPK. *Circ Res* 86, 144–151.

Naveilhan, P., ElShamy, W. M., and Ernfors, P. (1997). Differential regulation of mRNAs for GDNF and its receptors Ret and GDNFR alpha after sciatic nerve lesion in the mouse. *Eur J Neurosci* 9, 1450–1460.

Nollen, E. A., and Morimoto, R. I. (2002). Chaperoning signaling pathways: Molecular chaperones as stress-sensing "heat shock" proteins. *J Cell Sci* 115, 2809–2816.

Olanow, C. W., and McNaught, K. S. (2006). Ubiquitin–proteasome system and Parkinson's disease. *Mov Disord* 21, 1806–1823.

Olanow, C. W., and Tatton, W. G. (1999). Etiology and pathogenesis of Parkinson's disease. *Annu Rev Neurosci* **22**, 123–144.

Pasupathy, S., and Homer-Vanniasinkam, S. (2005). Ischaemic preconditioning protects against ischaemia/reperfusion injury: Emerging concepts. *Eur J Vasc Endovasc Surg* **29**, 106–115.

Paxinos, G., and Watson, C. (2006). *The Rat Brain in Stereotaxic Coordinates*, 6th edn. Academic Press. San Diego.

Pong, K., Doctrow, S. R., and Baudry, M. (2000). Prevention of 1-methyl-4-phenylpyridinium- and 6-hydroxydopamine-induced nitration of tyrosine hydroxylase and neurotoxicity by EUK-134, a superoxide dismutase and catalase mimetic, in cultured dopaminergic neurons. *Brain Res* **881**, 182–189.

Prolla, T. A., and Mattson, M. P. (2001). Molecular mechanisms of brain aging and neurodegenerative disorders: Lessons from dietary restriction. *Trends Neurosci* **24**, S21–S31.

Przedborski, S., and Ischiropoulos, H. (2005). Reactive oxygen and nitrogen species: Weapons of neuronal destruction in models of Parkinson's disease. *Antioxid Redox Signal* **7**, 685–693.

Przedborski, S., and Jackson-Lewis, V. (2000). *ROS and Parkinson's disease: A view to a kill*. Marcel Dekker, Inc, New York, pp 273–290.

Przedborski, S., Levivier, M., Kostic, V., Jackson-Lewis, V., Dollison, A., Gash, D. M., Fahn, S., and Cadet, J. L. (1991). Sham transplantation protects against 6-hydroxydopamine-induced dopaminergic toxicity in rats: Behavioral and morphological evidence. *Brain Res* **550**, 231–238.

Przedborski, S., Levivier, M., Jiang, H., Ferreira, M., Jackson-Lewis, V., Donaldson, D., and Togasaki, D. M. (1995). Dose-dependent lesions of the dopaminergic nigrostriatal pathway induced by intrastriatal injection of 6-hydroxydopamine. *Neuroscience* **67**, 631–647.

Quigney, D. J., Gorman, A. M., and Samali, A. (2003). Heat shock protects PC12 cells against MPP+ toxicity. *Brain Res* **993**, 133–139.

Ravikumar, B., Duden, R., and Rubinsztein, D. C. (2002). Aggregate-prone proteins with polyglutamine and polyalanine expansions are degraded by autophagy. *Hum Mol Genet* **11**, 1107–1117.

Ravikumar, B., Berger, Z., Vacher, C., O'Kane, C. J., and Rubinsztein, D. C. (2006). Rapamycin pre-treatment protects against apoptosis. *Hum Mol Genet* **15**, 1209–1216.

Ritossa, F. (1996). Discovery of the heat shock response. *Cell Stress Chaperones* **1**, 97–98.

Robinson, T. E., and Whishaw, I. Q. (1988). Normalization of extracellular dopamine in striatum following recovery from a partial unilateral 6-OHDA lesion of the substantia nigra: A microdialysis study in freely moving rats. *Brain Res* **450**, 209–224.

Rowland, N., and Stricker, E. M. (1982). Effects of dopamine-depleting brain lesions on experimental hyperphagia in rats. *Physiol Behav* **28**, 271–277.

Rubinsztein, D. C. (2006). The roles of intracellular protein-degradation pathways in neurodegeneration. *Nature* **443**, 780–786.

Ryu, E. J., Angelastro, J. M., and Greene, L. A. (2005). Analysis of gene expression changes in a cellular model of Parkinson disease. *Neurobiol Dis* **18**, 54–74.

Saavedra, A., Baltazar, G., Santos, P., Carvalho, C. M., and Duarte, E. P. (2006). Selective injury to dopaminergic neurons up-regulates GDNF in substantia nigra postnatal cell cultures: Role of neuron-glia crosstalk. *Neurobiol Dis* **23**, 533–542.

Saggu, H., Cooksey, J., Dexter, D., Wells, F. R., Lees, A., Jenner, P., and Marsden, C. D. (1989). A selective increase in particulate superoxide dismutase activity in parkinsonian substantia nigra. *J Neurochem* **53**, 692–697.

Sakai, K., and Gash, D. M. (1994). Effect of bilateral 6-OHDA lesions of the substantia nigra on locomotor activity in the rat. *Brain Res* **633**, 144–150.

Sakurai, M., Hayashi, T., Abe, K., Yaginuma, G., Meguro, T., Itoyama, Y., and Tabayashi, K. (1999). Induction of glial cell line-derived neurotrophic factor and c-ret proto-oncogene-like immunoreactivity in rabbit spinal cord after transient ischemia. *Neurosci Lett* **276**, 123–126.

Satake, K., Matsuyama, Y., Kamiya, M., Kawakami, H., Iwata, H., Adachi, K., and Kiuchi, K. (2000). Up-regulation of glial cell line-derived neurotrophic factor (GDNF) following traumatic spinal cord injury. *Neuroreport* **11**, 3877–3881.

Sato, M., Cordis, G. A., Maulik, N., and Das, D. K. (2000). SAPKs regulation of ischemic preconditioning. *Am J Physiol Heart Circ Physiol* **279**, H901–H907.

Sauer, H., and Oertel, W. H. (1994). Progressive degeneration of nigrostriatal dopamine neurons following intrastriatal terminal lesions with 6-hydroxydopamine: A combined retrograde tracing and immunocytochemical study in the rat. *Neuroscience* **59**, 401–415.

Schallert, T., Upchurch, M., Wilcox, R. E., and Vaughn, D. M. (1983). Posture-independent sensorimotor analysis of inter-hemispheric receptor asymmetries in neostriatum. *Pharmacol Biochem Behav* **18**, 753–759.

Schallert, T., Fleming, S. M., Leasure, J. L., Tillerson, J. L., and Bland, S. T. (2000). CNS plasticity and assessment of forelimb sensorimotor outcome in unilateral rat models of stroke, cortical ablation, parkinsonism and spinal cord injury. *Neuropharmacology* **39**, 777–787.

Schapira, A. H., Bezard, E., Brotchie, J., Calon, F., Collingridge, G. L., Ferger, B., Hengerer, B.,

Hirsch, E., Jenner, P., Le Novere, N., Obeso, J. A., Schwarzschild, M. A., Spampinato, U., and Davidai, G. (2006). Novel pharmacological targets for the treatment of Parkinson's disease. *Nat Rev Drug Discov* **5**, 845–854.

Schatz, D. S., Kaufmann, W. A., Saria, A., and Humpel, C. (1999). Dopamine neurons in a simple GDNF-treated meso-striatal organotypic co-culture model. *Exp Brain Res* **127**, 270–278.

Schulman, D., Latchman, D. S., and Yellon, D. M. (2001). Effect of aging on the ability of preconditioning to protect rat hearts from ischemia-reperfusion injury. *Am J Physiol Heart Circ Physiol* **281**, H1630–H1636.

Shen, H. Y., He, J. C., Wang, Y., Huang, Q. Y., and Chen, J. F. (2005). Geldanamycin induces heat shock protein 70 and protects against MPTP-induced dopaminergic neurotoxicity in mice. *J Biol Chem* **280**, 39962–39969.

Signore, A. P., Weng, Z., Hastings, T., Van Laar, A. D., Liang, Q., Lee, Y. J., and Chen, J. (2006). Erythropoietin protects against 6-hydroxydopamine-induced dopaminergic cell death. *J Neurochem* **96**, 428–443.

Siqueira, I. R., Fochesatto, C., de Andrade, A., Santos, M., Hagen, M., Bello-Klein, A., and Netto, C. A. (2005). Total antioxidant capacity is impaired in different structures from aged rat brain. *Int J Dev Neurosci* **23**, 663–671.

Smith, A. D., Antion, M., Zigmond, M. J., and Austin, M. C. (2003). Effect of 6-hydroxydopamine on striatal GDNF and nigral GFRalpha1 and RET mRNAs in the adult rat. *Brain Res Mol Brain Res* **117**, 129–138.

Snyder, A. M., Stricker, E. M., and Zigmond, M. J. (1985). Stress-induced neurological impairments in an animal model of parkinsonism. *Ann Neurol* **18**, 544–551.

Stolzing, A., and Grune, T. (2001). The proteasome and its function in the ageing process. *Clin Exp Dermatol* **26**, 566–572.

Stricker, E. M., and Zigmond, M. J. (1974). Effects on homeostasis of intraventricular injections of 6-hydroxydopamine in rats. *J Comp Physiol Psychol* **86**, 973–994.

Stromberg, I., Bjorklund, L., Johansson, M., Tomac, A., Collins, F., Olson, L., Hoffer, B., and Humpel, C. (1993). Glial cell line-derived neurotrophic factor is expressed in the developing but not adult striatum and stimulates developing dopamine neurons *in vivo*. *Exp Neurol* **124**, 401–412.

Tai, K. K., and Truong, D. D. (2002). Activation of adenosine triphosphate-sensitive potassium channels confers protection against rotenone-induced cell death: Therapeutic implications for Parkinson's disease. *J Neurosci Res* **69**, 559–566.

Tandara, A. A., Kloeters, O., Kim, I., Mogford, J. E., and Mustoe, T. A. (2006). Age effect on HSP70: Decreased

resistance to ischemic and oxidative stress in HDF. *J Surg Res* **132**, 32–39.

Tang, Y. P., Ma, Y. L., Chao, C. C., Chen, K. Y., and Lee, E. H. (1998). Enhanced glial cell line-derived neurotrophic factor mRNA expression upon (-)-deprenyl and melatonin treatments. *J Neurosci Res* **53**, 593–604.

Taylor, R. P., and Starnes, J. W. (2003). Age, cell signalling and cardioprotection. *Acta Physiol Scand* **178**, 107–116.

Tsai, B. M., Wang, M., March, K. L., Turrentine, M. W., Brown, J. W., and Meldrum, D. R. (2004). Preconditioning: Evolution of basic mechanisms to potential therapeutic strategies. *Shock* **21**, 195–209.

Ugarte, S. D., Lin, E., Klann, E., Zigmond, M. J., and Perez, R. G. (2003). Effects of GDNF on 6-OHDA-induced death in a dopaminergic cell line: Modulation by inhibitors of PI3 kinase and MEK. *J Neurosci Res* **73**, 105–112.

Ungerstedt, U. (1971a). Postsynaptic supersensitivity after 6-hydroxy-dopamine induced degeneration of the nigro-striatal dopamine system. *Acta Physiol Scand Suppl* **367**, 69–93.

Ungerstedt, U. (1971b). Adipsia and aphagia after 6-hydroxydopamine induced degeneration of the nigro-striatal dopamine system. *Acta Physiol Scand Suppl* **367**, 95–122.

Ungerstedt, U., and Arbuthnott, G. W. (1970). Quantitative recording of rotational behavior in rats after 6-hydroxy-dopamine lesions of the nigrostriatal dopamine system. *Brain Res* **24**, 485–493.

Varadarajan, S., Yatin, S., Aksenova, M., and Butterfield, D. A. (2000). Review: Alzheimer's amyloid beta-peptide-associated free radical oxidative stress and neurotoxicity. *J Struct Biol* **130**, 184–208.

Verbeke, P., Fonager, J., Clark, B. F., and Rattan, S. I. (2001). Heat shock response and ageing: Mechanisms and applications. *Cell Biol Int* **25**, 845–857.

Volloch, V., Mosser, D. D., Massie, B., and Sherman, M. Y. (1998). Reduced thermotolerance in aged cells results from a loss of an hsp72-mediated control of JNK signaling pathway. *Cell Stress Chaperones* **3**, 265–271.

Wallace, D. C., Shoffner, J. M., Watts, R. L., Juncos, J. L., and Torroni, A. (1992). Mitochondrial oxidative phosphorylation defects in Parkinson's disease. *Ann Neurol* **32**, 113–114.

Warenycia, M. W., and McKenzie, G. M. (1987). Activation of striatal neurons by dexamphetamine is antagonized by degeneration of striatal dopaminergic terminals. *J Neural Transm* **70**, 217–232.

Wei, G., Wu, G., and Cao, X. (2000). Dynamic expression of glial cell line-derived neurotrophic factor after cerebral ischemia. *Neuroreport* **11**, 1177–1183.

Welch, W. J. (1992). Mammalian stress response: Cell physiology, structure/function of stress proteins, and implications for medicine and disease. *Physiol Rev* **72**, 1063–1081.

Williams, W. M., and Chung, Y. W. (2006). Evidence for an age-related attenuation of cerebral microvascular antioxidant response to oxidative stress. *Life Sci* **79**, 1638–1644.

Xiao-Qing, T., Jun-Li, Z., Yu, C., Jian-Qiang, F., and Pei-Xi, C. (2005). Hydrogen peroxide preconditioning protects PC12 cells against apoptosis induced by dopamine. *Life Sci* **78**, 61–66.

Yoon, D. K., Yoo, K. Y., Hwang, I. K., Lee, J. J., Kim, J. H., Kang, T. C., and Won, M. H. (2006). Comparative study on Cu,Zn-SOD immunoreactivity and protein levels in the adult and aged hippocampal CA1 region after ischemia-reperfusion. *Brain Res* **1092**, 214–219.

Yurek, D. M., and Fletcher-Turner, A. (2000). Lesion-induced increase of BDNF is greater in the striatum of young versus old rat brain. *Exp Neurol* **161**, 392–396.

Yurek, D. M., and Fletcher-Turner, A. (2001). Differential expression of GDNF, BDNF, and NT-3 in the aging nigrostriatal system following a neurotoxic lesion. *Brain Res* **891**, 228–235.

Zeng, B. Y., Medhurst, A. D., Jackson, M., Rose, S., and Jenner, P. (2005). Proteasomal activity in brain differs between species and brain regions and changes with age. *Mech Ageing Dev* **126**, 760–766.

Zhou, J., Yu, Y., Tang, Z., Shen, Y., and Xu, L. (2000). Differential expression of mRNAs of GDNF family

in the striatum following 6-OHDA-induced lesion. *Neuroreport* **11**, 3289–3293.

Zhu, J. H., Guo, F., Shelburne, J., Watkins, S., and Chu, C. T. (2003). Localization of phosphorylated ERK/MAP kinases to mitochondria and autophagosomes in Lewy body diseases. *Brain Pathol* **13**, 473–481.

Zigmond, M. J., and Keefe, K. (1997). 6-Hydroxydopamine as a tool for studying catecholamines in adult animals: Lessons from the neostriatum. In *Highly Selective Neurotoxins: Basic and Clinical Applications* (R. M. Kostrzewa, Ed.), pp. 75–108. Humana Press, Totowa.

Zigmond, M. J., and Stricker, E. M. (1973). Recovery of feeding and drinking by rats after intraventricular 6-hydroxydopamine or lateral hypothalamic lesions. *Science* **182**, 717–720.

Zigmond, M. J., Hastings, T. G., and Perez, R. G. (2002). Increased dopamine turnover after partial loss of dopaminergic neurons: Compensation or toxicity? *Parkinsonism Relat Disord* **8**, 389–393.

Zuch, C. L., Nordstroem, V. K., Briedrick, L. A., Hoernig, G. R., Granholm, A. C., and Bickford, P. C. (2000). Time course of degenerative alterations in nigral dopaminergic neurons following a 6-hydroxydopamine lesion. *J Comp Neurol* **427**, 440–454.

14

COMPLEX I INHIBITION, ROTENONE AND PARKINSON'S DISEASE

RANJITA BETARBET[1] AND J. TIMOTHY GREENAMYRE[2]

[1]Center for Neurodegenerative Diseases, Emory University, Atlanta, GA, USA
[2]Pittsburgh Institute for Neurodegenerative Diseases, University of Pittsburgh, Pittsburgh, PA, USA

INTRODUCTION

Parkinson's disease (PD) is the most common neurodegenerative movement disorder whose cardinal clinical features can be found in Chapter 1. As explained in Chapter 3, PD neuropathology is characterized by the degeneration of all of the ascending dopaminergic pathways in the CNS, albeit to variable degrees. That being said, it is important to stress, however, that PD neurodegeneration is not restricted to the dopaminergic systems, as widespread neuronal loss can be detected in other catecholaminergic and non-catecholaminergic nuclei (Braak *et al.*, 2004). In addition, surviving neurons in the substantia nigra pars compacta (SNc) contain cytoplasmic, proteinaceous, α-synuclein-positive inclusions, called Lewy bodies (Shults, 2006). Together, degeneration of the nigrostriatal, dopaminergic pathway and the presence of Lewy bodies define PD pathologically (Gibb and Lees, 1989; Klockgether and Turski, 1989), even though it appears today that PD may be a multifaceted disorder (Hely *et al.*, 2005).

The etiology of the common, sporadic form of PD is not fully understood; however, it is believed that PD is caused by an interaction of environmental factors and genetic susceptibilities. Furthermore, genetic analyses, epidemiologic studies, neuropathological investigations and new experimental models of PD have provided new insights into the pathogenesis of PD with several cellular pathways converging on neurodegeneration. Among the most recent ways to model PD in rodents, there is a use of a well-known mitochondrial poison, i.e. rotenone. In this chapter we will provide the reader with general information about rotenone and review the body of work in which rotenone has been used to model PD. As we will see, this model has allowed reaching significant insights into the PD pathogenesis. However, this model bears some specific technical challenges and, although it offers clear advantages, it also has some interesting shortcomings. All of these different aspects of the rotenone model will be discussed herein.

MITOCHONDRIAL DEFECTS, OXIDATIVE STRESS AND PD

Before discussing the use of rotenone, a mitochondrial toxin, to model PD, it is essential to review briefly the association of mitochondrial dysfunction to PD. For decades, mitochondrial respiratory defect has been implicated in PD pathogenesis. One of the first observations linking a mitochondrial problem to dopaminergic neurodegeneration is the development of a profound, irreversible parkinsonian syndrome in humans intoxicated with 1-methyl-4-phenyl-1,2,3,6-tetrahydropyridine (MPTP) (Langston *et al.*, 1983). The toxicity of 1-methyl-4-phenylpyridinium (MPP+), the active metabolite of MPTP, was found to kill dopaminergic neurons by first being taken up by the dopamine transporter (Javitch *et al.*, 1985) and then by concentrating inside the mitochondria where MPP$^+$ inhibits complex I of the electron transport chain (ETC) (Nicklas *et al.*, 1985; Tipton and Singer, 1993).

The fact that a complex I toxin was able to produce in humans a parkinsonian syndrome that was astonishingly similar to typical "idiopathic" PD (Langston, 1996) suggested that defects in complex I may be associated with pathogenesis of typical PD. Consistent with this view, several investigators have subsequently reported modest but reproducible reductions in complex I activity in brain tissues from PD patients including the substantia nigra (Schapira *et al.*, 1989). Interestingly, it appears that complex I defect in PD is *systemic*, affecting tissues outside the brain as well. Indeed, reduced complex I activity has been found, not only in brain, but also in the platelets, muscle and fibroblasts of patients with PD (Parker *et al.*, 1989; Krige *et al.*, 1992; Yoshino *et al.*, 1992; Haas *et al.*, 1995). On average, the loss of activity is about 25% and statistically significant, leading to a general agreement that typical PD is associated with a partial, systemic deficit in complex I activity.

The main function of the mitochondrial respiration is to produce ATP. Thus, it is tantalizing to content that a reduction in complex I activity may cause a deficit in ATP leading to a cellular energy crisis. However, partial inhibition of complex I is also known to enhance reactive oxygen species (ROS) production and contribute to oxidative stress (Sherer *et al.*, 2001). *Oxidative stress* is a term used to describe the steady state level of oxidative damage in a cell, tissue or organ, caused by reactive free radical species. Oxidative stress is caused by an imbalance between production of reactive free radical species and ability to detoxify the reactive intermediates or repair the resulting damage (Lotharius and Brundin, 2002). Free radicals, including ROS, peroxynitrite, superoxide and peroxy radicals, are molecular species with unpaired electrons and therefore are highly reactive. These reactive species can cause oxidative damage by reacting with DNA, lipids and proteins and changing their structural conformation that is unsuitable or toxic for a living system. Exogenous sources of free radicals are numerous and include environmental pollutants and toxicants, ionizing radiation and various kinds of infection. Endogenous sources are normal and essential cellular reactions including energy production from mitochondria as mentioned previously, detoxification reactions involving the liver cytochrome P-450 enzyme system and

dopamine metabolism in dopaminergic neurons and nerve terminals that are primary targets in PD. Dopaminergic neurons are believed to exist in constant state of oxidative stress due to production of peroxy radicals and ROS during dopamine metabolism and dopamine auto-oxidation (Lotharius and Brundin, 2002). Given that partial complex I inhibition is associated with PD and dopaminergic neurons are primary targets in PD, it can be assumed that oxidative stress may also have an important role in PD pathogenesis. Indeed, oxidative-stress-related changes have been reported in brains of PD patients (Jenner, 1998). Analysis of postmortem brain specimens has revealed oxidative damage to proteins (carbonyls and nitrotyrosine residues), lipids (products of peroxidation) and DNA (8′-hydroxy-deoxyguanosine) in the substantia nigra (Dexter *et al.*, 1989; Yoritaka *et al.*, 1996; Jenner, 1998). Genetic studies have further strengthened the association between mitochondria and PD pathogenesis. The identification of mutations in phosphatase and tensin homolog (PTEN)-induced kinase 1 (PINK1), a mitochondrial kinase (Valente *et al.*, 2004), parkin mutations and related mitochondrial pathology (Greene *et al.*, 2003), and DJ-1, a possible redox sensor, (Bonifati *et al.*, 2003; Hague *et al.*, 2003) all provided strong evidence that mitochondrial dysfunction and oxidative stress might have a primarily role in pathogenesis of PD.

Mitochondrial impairments also have a central role in most known neuronal cell death pathways, including excitotoxicity, caspase-dependent and caspase-independent apoptosis, necrosis and inflammation-induced injuries (Dunnett and Bjorklund, 1999; Dawson and Dawson, 2003; Hunot and Hirsch, 2003). *Since mitochondrial dysfunction appears to be a good candidate as a causative factor in PD pathogenesis, rotenone, a potent mitochondrial toxin, is being extensively used to investigate the role of mitochondrial complex I inhibition in PD pathogenesis.*

WHAT IS ROTENONE?

Rotenone as an Agricultural Chemical

Rotenone occurs naturally in roots and stems of several tropical and subtropical plant species especially those belonging to the genus *Lonchocarpus* or *Derris*. For example, the root extracts from Cubé plant (*Lonchocarpu utilis*) or Barbasco

(*Lonchocarpu urucu*), referred to as Cubé resin, are enriched in rotenone. Duboisia plants also contain rotenone and have been used in South America and by Australian aboriginals for poisoning fish for food for centuries (Maslin, 2001; Hisata, 2002). Commercially today rotenone is used as a selective and non-specific broad-spectrum insecticide, used in home garden for insect control, for lice and tick control on pets and as a piscicide for fish eradications as a part of water management (Greenamyre *et al.*, 2001). Rotenone and rotenoids have been used as crop pesticide since 1848 when they were applied to plants to control leaf-eating caterpillars (Maslin, 2001), whereas today it is used to spray on home-grown tomato plants, etc. The use of roots of certain species of *Lonchocarpus* and *Derris* was patented in 1912, when it was established that the active compounds were rotenoids with the main insecticide being rotenone. Rotenone is sold today in dispersible powder, emulsified concentrate and wettable powder formulations (Maslin, 2001). Rotenone is classified by the World Health Organization as being moderately hazardous. The LD50 for rats (the amount of the chemical lethal to one-half of experimental animals) is between 132 and 1500 mg/kg. The acute oral toxicity of rotenone is moderate for mammals, but there is a wide variation between species. It is less toxic for the mouse and hamster than for the rat; the pig seems to be especially sensitive. Rotenone is believed to be moderately toxic to humans with an oral lethal dose estimated from 300 to 500 mg/kg. Epidemiological studies have implicated residence in a rural environment and the related exposure to herbicides and pesticides with an elevated risk of PD (Tanner *et al.*, 1999). Chronic exposure to rotenone is however unlikely to cause serious CNS damage because it is very unstable, lasting only a few days in lakes (Hisata, 2002). Indeed, rotenone is rapidly broken down in soil and water: its half-life in both is between 1 and 3 days. Nearly all its toxicity is lost in 5–6 days of spring sunlight, or 2–3 days of summer sunlight. It does not readily leach from soil and it is not expected to be a groundwater pollutant. Nevertheless, it remains a prototypical model for potential PD-related environmental poisons.

Rotenone as an Inhibitor of Mitochondrial Complex I

Rotenone is highly lipophilic and thus readily gains access to all organs including the brain. After a single intravenous (i.v.) injection, rotenone reaches maximal concentration in the CNS within 15 min and decays to about half of this level in approximately 100 min (Talpade *et al.*, 2000). Its brain distribution is heterogeneous paralleling regional differences in oxidative metabolism (Talpade *et al.*, 2000). Rotenone also freely crosses all cellular membranes and can accumulate in sub-cellular organelles such as mitochondria. As indicated earlier, mitochondria are intracellular organelles in eukaryotic cells, which produce energy in the form of ATP. Production of ATP involves a series of oxidative-reduction reactions to produce a proton gradient that is used to synthesize ATP from electron donors such as NADH (reduced nicotinamide adenine dinucleotide). This process involves the coordinated activity of five enzymes (complexes I–V) of the inner mitochondrial membrane that constitute the ETC. Rotenone works by interfering with the ETC in mitochondria. More specifically, it inhibits the transfer of electrons from Fe–S centers in complex I to ubiquinone, thus inhibiting the oxidation of NADH-linked substrates (Horgan *et al.*, 1968; Schuler and Casida, 2001). This prevents NADH from being converted into usable cellular energy – ATP. The catalytic activity of complex I is, indeed, defined by its inhibition with rotenone. Also mentioned earlier, in addition to ATP production, the ETC is also involved in generation of ROS. During mitochondrial metabolism, molecular oxygen is reduced to water at complex IV of the ETC. The 1–2% of oxygen not reduced at complex IV gets reduced non-enzymatically to ROS such as superoxide () and hydrogen peroxide, by electrons that leak from sites in the ETC including at complex I. Partial inhibition of complex I, as by rotenone, can therefore also enhance ROS production and oxidative stress (Sherer *et al.*, 2001). Thus, rotenone-induced complex I inhibition can decrease ATP production and increase free radical generation, the combination of which can be fatal for the cell.

THE ROTENONE MODEL

Previous History

As previously mentioned, MPTP causes acute and permanent parkinsonian syndrome (Langston *et al.*, 1983). The toxicity of MPP^+, the toxic metabolite of MPTP, was found to be specific for dopaminergic neurons due to its ability to be taken up by the dopamine transporter (Javitch and Snyder, 1984). In addition, MPP$^+$ was found to be mitochondrial toxin that inhibits complex I of the ETC (Nicklas et al., 1985; Tipton and Singer, 1993). To confirm the mechanism of MPTP toxicity and compare the effects of a known complex I inhibitor rotenone, on dopaminergic cells, Heikkila et al. (1985) stereotaxically administered MPP$^+$ and rotenone to the median forebrain bundle in *rats*. Since rats are resistant to MPTP toxicity (Przedborski et al., 2001), MPP$^+$ and rotenone were stereotaxically administered. Both MPP$^+$ and rotenone resulted in degeneration of the dopaminergic nigrostriatal pathway with extensive dopamine depletion in the striatum. In another *in vitro* experiment, the effects of rotenone and MPP$^+$ on mitochondrial oxidation were determined in mitochondrial rat brain preparations. These experiments demonstrated that MPP$^+$, like rotenone, markedly inhibited NADH-linked respiration although having no effect on succinate oxidation, thereby demonstrating its specific effects on complex I. Taken together, the data from both *in vivo* and *in vitro* studies showed that MPP$^+$ and rotenone had similar effects on the dopaminergic nigrostriatal pathway and on mitochondrial preparations though the effects of rotenone were more potent. The authors concluded that the mechanism for neurotoxic effects of MPTP may be caused by the deleterious effects of MPP$^+$ on aerobic respiration, following selective inhibition of complex I (Heikkila et al., 1985). Two decades later these results were confirmed by Saravanan et al. (2005).

During the same time it was also demonstrated that MPTP is a substrate for the dopamine transporter, which is selectively expressed in the dopaminergic neurons (Javitch and Snyder, 1984). Therefore when administered systemically in monkeys and mice, MPTP could selectively target the dopaminergic system (Przedborski et al., 2001). Thus MPTP administration did not result in systemic and mild, complex I inhibition as reported in PD patients. The question then was *could systemic inhibition of complex I, by systemic administration of rotenone, cause selective degeneration of dopaminergic nigrostriatal pathway?*

To address this question, Ferrante et al. (1997) studied the effects of continuous i.v. administration of rotenone in rats by Alzet pumps. Rotenone, being lipophilic, can easily cross the blood–brain barrier and systemically inhibit complex I, unlike MPP$^+$, that depends on the dopamine transporter for access to dopaminergic neurons (Javitch and Snyder, 1984). Sprague-Dawley rats were used for these studies. Alzet pumps, pre-filled with rotenone, were connected to the jugular vein via a 22-gauge angiocatheter. Thus rotenone, dissolved in a mixture of 1:1 DMSO and polyethylenimine, was continuously administered i.v. for 7–9 days at a rate of 10–18 mg/kg/day. The rotenone-treated rats demonstrated behavioral symptoms, including akinesia and rigidity after 10 days of treatment (Ferrante et al., 1997). Histologic evaluation of brains from rotenone-treated rats demonstrated gliosis and degeneration in the striatum and globus pallidus with no apparent cell loss in the substantia nigra. The authors concluded that the selective vulnerability of the substantia nigra neurons to MPTP neurotoxicity was not merely due to inhibition of complex I but due to selective uptake of MPP$^+$, the active metabolite of MPTP, by the synaptic dopamine transporter, especially since blockers of dopamine transporter prevent MPTP neurotoxicity (Javitch et al., 1985).

Keeping in mind the potential role of mitochondrial dysfunction in nigrostriatal degeneration and PD pathogenesis, Thiffault et al. (2000) investigated the effects of acute, subcutaneous injection of rotenone, as a single dose of 15 mg/kg or multiple doses of 1.5 mg/kg, on the levels of striatal dopamine and its metabolites in mice. Thus, administered in mice, rotenone, while causing fatality, failed to affect striatal dopamine levels; though the dopamine turnover was increased. The authors concluded that rotenone is not capable of causing overt dopaminergic toxicity under the paradigm used in their studies.

At the time Ferrante et al. (1997) published their report on the effects of rotenone, the Greenamyre laboratory was also extensively investigating the dose effects of *systemic administration* of complex I inhibitor – rotenone. The aim of the project was to develop a model of PD with systemic yet mild complex I inhibition, selective degeneration of the dopaminergic nigrostriatal pathway with formation of α-synuclein-rich cytoplasmic inclusions in the surviving dopaminergic neurons. The premise, thus, was to determine a dose of rotenone that causes partial inhibition of complex I, and yet does not completely prevent ATP production. At the end of the year 2000, we reported that chronic, systemic and low doses of rotenone causes systemic complex I inhibition and yet selective degeneration of the nigrostriatal pathway with formation of synuclein-rich, fibrillar, cytoplasmic inclusions, thus demonstrating that by fine-tuning rotenone delivery regimen, systemic complex I inhibition could result in PD-like pathology in rats (Betarbet et al., 2000).

Rotenone Administration

To imitate low levels of exposure to an environmental toxin that can cause systemic inhibition of mitochondrial complex I during a normal human life span, Greenamyre and colleagues exposed Sprague-Dawley and Lewis rats to systemic and chronic, low levels of rotenone via an intrajugular cannula attached to a subcutaneous osmotic minipump (Betarbet et al., 2000). Lewis rats developed more consistent lesions and were therefore used exclusively for further studies. At first the doses of rotenone used ranged from 1 to 12 mg/kg/day. At high doses rotenone produced systemic cardiovascular toxicity and non-specific brain lesions similar to that observed by others (Ferrante et al., 1997; Lapointe et al., 2004). Downward titration of rotenone dosing resulted in less systemic toxicity and more specific nigrostriatal dopaminergic degeneration. The optimal dose for inducing PD-like pathology was determined to be 2–3 mg/kg/day. It is important to note at this point that even at the "optimal dose," only 30–50% of Lewis rats demonstrated PD-like pathology. Nonetheless, systemic, low levels of rotenone administration reproduced pathological features characteristic of PD (Betarbet et al., 2000). Intrajugular cannulation surgeries were however labor-intensive and increased the risk of post-surgical complications. Therefore an alternative route was developed to administer rotenone. Instead of cannulation and vascular administration, rotenone was administered by subcutaneously placed osmotic minipumps that released rotenone into the body cavity (Betarbet et al., 2000; Sherer et al., 2003b). Subcutaneous administration of rotenone, at 2–3 mg/kg/day in Lewis rats, also produced selective nigrostriatal dopaminergic lesions as previously reported (Betarbet et al., 2000). Since the initial studies daily administration of rotenone as an emulsion in sunflower oil was also able to produce degeneration of nigral dopaminergic neurons in Sprague-Dawley rats (Alam and Schmidt, 2002).

Characteristics of Rotenone-Induced Toxicity

Systemic Complex I Inhibition

Consistent with its ability to cross biological membranes easily, chronic, low doses of systemic rotenone infusion resulted in uniform inhibition of complex I throughout the rat brain. [3H] dihydrorotenone binding to complex I in brain was reduced by approximately 75%. We have estimated that this 75% inhibition of specific binding should translate to be 20–30 nM of free rotenone in the brain. Rotenone infusion did not have any effect on the enzymatic activities of complexes II and IV, as analyzed histochemically (Betarbet et al., 2000). This uniform complex I inhibition induced by rotenone was unlike the effects of MPTP, which selectively inhibits complex I in dopaminergic neurons due to the dependence of MPP^+, the active metabolite of MPTP, on the dopamine transporter (Storch et al., 2004).

Oxidative Stress

Rotenone-induced complex I inhibition resulted in increased oxidative stress, both in vitro in neuroblastoma cells (Sherer et al., 2002) and in vivo in rats (Sherer et al., 2003c), as suggested by increased levels of protein carbonyls, a marker for oxidative stress. It was observed that the toxicity in neuroblastoma cells, chronically exposed to low levels of rotenone, was mainly due to increased levels of oxidative stress and minimally due to ATP depletion. Furthermore, rotenone-induced toxicity in cells was attenuated by prior treatment with α-tocopherol, a known antioxidant, confirming that the toxic action of rotenone was via oxidative stress (Sherer et al., 2003c). Rotenone-infused rats also demonstrated increased levels of oxidative stress most notably in dopaminergic regions including the striatum, ventral midbrain and the olfactory bulb.

Oxidative-stress-related recessive DJ-1 mutations are associated with an early-onset form of parkinsonism in human patients (Bonifati et al., 2003). Interestingly, chronic rotenone exposure, both in vitro and in vivo, resulted in oxidative modifications of DJ-1 protein by a shift in pI toward a more acidic form and redistribution of cytosolic DJ-1 (Betarbet et al., 2006). It is suggested that DJ-1 is normally neuroprotective while mutations or oxidative modifications can reduce its neuroprotective effects (Mitsumoto and Nakagawa, 2001; Mitsumoto et al., 2001). Thus rotenone-induced toxicity appears to be strongly associated with oxidative stress, which in turn has been strongly implicated in PD pathogenesis (Jenner, 1998).

Microglial Activation

PD neuropathology also shows striatal and nigral microgliosis and, to a lesser extent, astrocytosis.

Microglial activation in the striatum and substantia nigra was detected in rotenone-infused rats (Sherer et al., 2003a; Zhu et al., 2004). Enlarged microglia with short, stubby processes were detected prior to dopaminergic lesions. Rotenone-induced microglial activation was less pronounced in the cortex and in rats that did not develop a striatal lesion. Microglia are the brain's resident immune cells and are activated in response to immunological stimuli and/or neuronal injuries. They are known to produce potentially neurotoxic ROS that probably add to the oxidative stress reported in PD (Liberatore et al., 1999; Gao et al., 2002).

α-Synuclein-Positive Cytoplasmic Inclusions

As mentioned above, α-synuclein-positive intraneuronal inclusions called Lewy bodies are a hallmark of PD pathology. Intraneuronal α-synuclein-positive inclusions in the substantia nigra were also detected in rotenone-infused rats (Betarbet et al., 2000; Hoglinger et al., 2003b; Sherer et al., 2003b). These cytoplasmic inclusions were positive for ubiquitin and appeared as "pale eosinophilic" inclusions with hematoxylin and eosin staining. On electron microscopy examination they typically displayed an ultrastructural appearance reminiscent of Lewy bodies, in that they had a homogenous dense core surrounded by fibrillar elements (Betarbet et al., 2000). Biochemical analysis confirmed the accumulation and aggregation of α-synuclein in rotenone-infused rats. Western immunoblotting demonstrated significant and selective increases in α-synuclein levels in the ventral midbrain regions as well as higher molecular weight bands ($\sim$ 30 and 52 kDa) in addition to the 19 kDa α-synuclein band (Betarbet et al., 2006). Notably, the α-synuclein-related changes were observed only in the ventral midbrain in all rotenone-infused, both lesioned and non-lesioned rats. In the striatum punctate α-synuclein accumulation was detected in regions that were devoid of dopaminergic structures (Betarbet et al., 2006), very similar to that observed in Alzheimer's disease/Lewy body dementia patients (Duda et al., 2002).

Proteasomal Dysfunction

A dysfunctional ubiquitin–proteasome system (UPS) was also a consequence of rotenone-induced complex I inhibition. Ubiquitin-independent proteasomal enzymatic activities were significantly and *selectively reduced in the ventral midbrain regions* in rotenone-infused rats with striatal lesions. Furthermore, ubiquitin conjugated proteins, an indicator of proteins marked for degradation, were markedly increased only in ventral midbrain suggesting impairment of ubiquitin-dependent proteasome degradation pathway (Betarbet et al., 2006).

Impairment of proteasomal function could be due to complex I inhibition-induced changes in bioenergetics such as ATP production and/or complex I inhibition-induced increase in free radicals production resulting in oxidatively damaged proteins. Acute, *in vitro* studies using ventral mesencephalic primary cultures implied that rotenone-induced impairment of proteasomal function is primarily due to ATP depletion and not from free radical production (Hoglinger et al., 2003a). However chronic exposure to low levels of rotenone, while minimally affecting bioenergetics, significantly increases the levels of oxidative stress, which may have a greater role in neuronal degeneration (Sherer et al., 2002; Sherer et al., 2003c; Betarbet et al., 2006). It is possible that increased levels of oxidatively damaged proteins, observed following rotenone infusion/treatment, could impair proteasomal pathway, by either "clogging-up" the UPS or by oxidatively modifying the proteasomal subunits themselves. In fact Shamoto-Nagai et al. (2003) have reported that complex I inhibition with rotenone in neuroblastoma SH-SY5Y cells reduced proteasomal activity through increased production of oxidatively modified proteins including oxidative modification of the proteasome itself. Thus, it appears that increased oxidative stress could inhibit proteasomal function and eventually lead to neuronal degeneration.

Selective Nigrostriatal Dopaminergic Degeneration

Interestingly, despite this uniform complex I inhibition, rotenone caused selective degeneration of the nigrostriatal dopaminergic pathway (Betarbet et al., 2000; Alam and Schmidt, 2002; Sherer et al., 2003b; Zhu et al., 2004). Immunocytochemistry for tyrosine hydroxylase (TH), a rate limiting enzyme involved in the synthesis of dopamine, is often used as a phenotypic marker for dopaminergic neurons. This assay demonstrated a sticking regional reduction in TH immunostaining in the striatum of at least of some (see later) of the rotenone-treated animals (Betarbet et al., 2000). Other dopaminergic markers including dopamine transporter and vesicular monoamine transporter type 2 confirmed the striatal lesions (Zhu et al., 2004). Staining for cell death markers such as silver and fluoro-jade B verified that the reduction of dopaminergic phenotypic markers in the striatum was due to degeneration of dopaminergic terminals. That being said, the striatal dopaminergic lesions exhibited some striking variability: partial or focal; located in the central or dorsolateral region of the anterior striatum; or diffused and spread out to involve most of the motor striatum. Interestingly, even when the lesion was severe, there was relative sparing of dopaminergic fibers in the medial aspects of the striatum, nucleus accumbens and olfactory tubercle (Betarbet et al., 2000), areas that are relatively spared in idiopathic PD as well.

Neurodegeneration was also evident in the dopaminergic neurons of the substantia nigra. Animals that had partial striatal lesions had dopaminergic neurons in the substantia nigra that looked relatively normal while animals with extensive striatal lesions had obvious reductions in nigral TH-positive neurons. Silver staining demonstrated signs of degenerating nigral neurons with silver deposits in their cell bodies and processes, even in animals that had normal looking TH-positive cells and partial striatal lesions. Rats that had severe striatal lesions exhibited more extensive signs of degeneration in the nigral neurons.

Both TH immunocytochemistry and silver staining demonstrated retrograde degeneration of nigral dopaminergic neurons following rotenone exposure; degeneration began at the terminals in the striatum where the effects were more severe compared to the nigral neurons (Hoglinger et al., 2003b; Zhu et al., 2004; Betarbet et al., 2006). Quantitative analysis of dopamine levels has also shown extensive deficiency in the striatum (Alam and Schmidt, 2002), similar to postmortem analysis of brains from PD patients that have shown more extensive loss of striatal dopamine compared to substantia nigra (Hornykiewicz, 1966). Furthermore, neurons in the lateral and ventral tiers of the substantia nigra appeared to be more vulnerable to systemic rotenone infusion, very similar to the pattern of neuronal vulnerability observed in idiopathic PD. Despite the loss of TH immunoreactivity in the substantia nigra, dopaminergic neurons of the ventral tegmental area were spared as in PD. Also, similar to PD, noradrenergic neurons of the locus coeruleus were also susceptible to rotenone toxicity (Betarbet et al., 2003; Hoglinger et al., 2003b; Zhu et al., 2004).

Despite profound loss of presynaptic dopaminergic terminals in the striatum, the postsynaptic striatal neurons remained intact. In majority of the rotenone-infused rats, striatal neurons were minimally affected as observed with Nissl stain and the pan-neuronal marker, NeuN, and with immunocytochemistry for various striatal phenotypic markers such as DARPP32 (dopamine and cAMP-regulated phosphoprotein), GAD (glutamic acid decarboxylase), nNOS (neuronal nitric oxide synthase) and histochemistry for AchE (acetylcholinesterase). There was an exception to this rule however. One or two rats with acute focal lesions had a necrotic core and showed evidence of striatal cell loss (Na et al., 2003; Zhu et al., 2004). Rotenone toxicity also had minimal effects on neurons of other brain regions, including globus pallidus and subthalamic nucleus confirmed with silver staining (Sherer et al., 2003b). These data further supported the nigrostriatal dopaminergic selectivity of rotenone-induced neurodegeneration.

Behavior

Reduced striatal dopaminergic activity is known to cause parkinsonian symptoms in humans such as rigidity and bradykinesia (Klockgether, 2004). Reminiscent of these PD manifestations, rats treated with rotenone exhibit motor behavior abnormalities such as reduced exploratory behavior, rigidity and even catalepsy which is typical of neuroleptic-induced parkinsonism; catalepsy is defined as a state of marked loss of voluntary motion in which the limbs remain in whatever position they are placed. The use of catalepsy tests revealed a significant increase in this abnormal motor behavior in rotenone-treated rats as compared to control, vehicle-treated rats (Alam and Schmidt, 2002). In addition, rotenone-treated rats displayed significant decline in locomotor activities including active sitting, rearing and line crossing behavior (Alam and Schmidt, 2002). Rigidity and reduced spontaneous motor activity as well as flexed posture were previously reported in rotenone-exposed rats as well (Betarbet et al., 2000; Sherer et al., 2003b). Some of the rotenone-infused rats also developed not only severe rigidity, but also spontaneously shaking paws; while no electrophysiological characterization of this tremulous movement has been done, could this be a rat equivalent of the PD resting tremor? It is important to note that we have not demonstrated that these behavioral changes respond to levadopa as expected in PD.

Variability in the Rotenone Model

The initial studies with rotenone numerous reports have either confirmed (Alam and Schmidt, 2002;

Zhu et al., 2004) or questioned (Hoglinger et al., 2003b; Lapointe et al., 2004) the selectivity of rotenone-induced degeneration of the nigrostriatal dopaminergic pathway. These differences could be due to a "small window" for rotenone's action that results in selective neurodegeneration. There exists a threshold for every drug beyond which they have non-specific or "side effects." For rotenone this threshold appears to be very small – some animals have an acute response while some are not affected by rotenone at all and yet there are some rats that develop very characteristic features of PD. At high doses, as shown by Ferrante et al. (1997), rotenone can have non-specific effects.

The variability observed in rotenone-induced toxicity range from none to nearly complete striatal dopaminergic lesions (Ferrante et al., 1997; Betarbet et al., 2000; Hoglinger et al., 2003b; Sherer et al., 2003b; Zhu et al., 2004) is interesting. This variability clearly demonstrates the individual susceptibility to complex I inhibition in rats which could be due to genetic differences and/or differences in the ability to metabolize environmental toxins (Uversky, 2004), similar to individual differences in humans that may determine ones susceptibility to develop PD.

Use of Rotenone to Model PD in Other Species

Rotenone has been successfully used to model PD in various species including fly and snail.

Interestingly many PD features have been recapitulated in flies, *Drosophila melanogaster* (Coulom and Birman, 2004) exposed to rotenone. Following several days of sublethal doses (125–500 μm) of rotenone exposure, flies developed locomotor impairments that increased with the dose of rotenone. Immunocytochemistry studies demonstrated a significant and selective loss of dopaminergic neurons in the brain clusters. Levodopa into the feeding medium rescued the behavioral deficits but not neuronal loss implying that locomotor deficits are due to loss of dopaminergic neurons. In contrast, the antioxidant melatonin alleviated both behavioral deficits and neuronal loss suggesting a major role for oxidative stress in neuronal degeneration. The study provides a new model to study PD pathogenesis and to screen therapeutic drugs.

Rotenone was found to be toxic to another invertebrate, the pond snail *Lymnaea stagnalis*. Chronis exposure to 0.5 μm (7 days) rotenone resulted in progressive and irreversible decrease in spontaneous locomotion and feeding as well as in a loss of TH immunoreactivity in dopaminergic RPeD1 neurons in this invertebrate suggesting that rotenone affected the dopaminergic system at neuronal and behavioral levels (Vehovszky et al., 2007).

The use of rotenone to model PD in mice was an obvious follow-up to the rat studies due to the availability of numerous genetic strains of mice that could enable the investigation of interaction of genetic and environmental factors in PD pathogenesis. However, reports so far suggest that mice may respond differently to rotenone. Initial studies demonstrated that mice were more resistant to rotenone (Betarbet et al., 2005). Chronic and systemic exposure to rotenone, at 2–4 times of the dose used in rats, the mice demonstrated loss of nigral neurons without loss of dopaminergic terminals in the striatum. However, repeated oral administration of very high doses (0.25–30 mg/kg) of rotenone induced specific nigrostriatal dopaminergic neurodegeneration with motor deficits and upregulation of α-synuclein expression in surviving nigral neurons (Inden et al., 2007) though chronic inhalation of rotenone (Rojo et al., 2007) or systemic administration in oil (dose 5 mg/kg for 45 days; Richter et al. (2007)) did not induce a PD-like phenotype.

Preliminary studies in non-human primates demonstrated that route and dose of rotenone administration could determine the specificity of rotenone-induced toxicity (Greenamyre et al., 2004). A cumulative dose of 582 mg/kg of rotenone administered via subcutaneously implanted minipumps produced parkinsonian pathology in a macaque (Macaca mulatta) including loss of TH immunoreactivity in the striatum and degenerating cells in the nigra whereas a cumulative dose of 1034 mg/kg of rotenone from slow releasing rotenone tablets and subcutaneous injections resulted in non-specific striatal degeneration in addition to loss of dopaminergic terminals.

Advantages and Disadvantages of Using Rotenone to Model PD

Though chronic and systemic inhibition of complex I using the mitochondrial and environmental toxin rotenone, reproduced PD-like phenotype in rats, this was observed in only a proportion and not all the rats exposed to this toxin. This is a disadvantage for a model to be used for drug evaluation studies. However, this model is crucial for studying the underlying mechanisms involved in the progression and evolution of PD pathogenesis that appears to be a multisystems disorder (Hely *et al.*, 2005) affecting the gastrointestinal tract (Braak *et al.*, 2006) and various brain regions including higher-order cognitive areas (Braak *et al.*, 2004) and not just restricted to the nigrostriatal dopaminergic pathway. The fact that some rats are spared from rotenone toxicity will also enable the investigation of individual factors that may provide resistance to rotenone toxicity and therefore have a therapeutic potential for curing PD.

Converging Mechanisms

The rotenone model of parkinsonism, while substantiating the involvement of mitochondrial dysfunction and environmental exposures in PD pathogenesis, demonstrates that chronic, low-grade complex I inhibition results in (a) increased oxidative stress and oxidative-stress-induced DJ-1 modification and cellular redistribution, (b) accumulation and aggregation of α-synuclein, (c) proteasomal dysfunction and (d) selective degeneration of the dopaminergic nigrostriatal pathway. Mitochondrial and proteasomal dysfunction, α-synuclein aggregation and oxidative stress are all mechanisms strongly implicated in PD pathogenesis (Dawson and Dawson, 2003). Moreover mechanistic analysis of the rotenone models suggests a sequence of events starting with complex I inhibition, followed in turn by oxidative stress, α-synuclein accumulation and proteasomal inhibition, leading ultimately to neurodegeneration. At each step, the effects become more regionally restricted such that systemic complex I inhibition eventually results in highly selective degeneration of the nigrostriatal pathway. Although it is difficult to demonstrate this sequence of events in PD patients, available data suggest that this could be the case: (i) a systemic but moderate levels of complex I inhibition (Schapira *et al.*, 1989); (ii) generalized but anatomically more restricted increase in carbonyl levels (Jenner, 1998); (iii) more restricted, yet early (prior to degeneration) synuclein aggregations in targeted regions (Braak *et al.*, 2004) and finally (iv) proteasomal dysfunction in the substantia nigra (McNaught and Jenner, 2001). This study raises hope for therapeutic interventions that target convergent key mechanisms, such as complex I defects or oxidative stress, which can in turn influence other divergent aspects

of PD pathogenesis, such as DJ-1, α-synuclein and UPS function (Greenamyre and Hastings, 2004).

REFERENCES

Alam, M., and Schmidt, W. J. (2002). Rotenone destroys dopaminergic neurons and induces parkinsonian symptoms in rats. *Behav Brain Res* **136**, 317–324.

Betarbet, R., Sherer, T. B., MacKenzie, G., Garcia-Osuna, M., Panov, A. V., and Greenamyre, J. T. (2000). Chronic systemic pesticide exposure reproduces features of Parkinson's disease [In Process Citation]. *Nat Neurosci* **3**, 1301–1306.

Betarbet, R., Sherer, T. B., Lund, S., and Greenamyre, J. T. (2003). Rotenone models of Parkinson's disease: Altered proteasomal activity following sustained inhibition of complex I. *Society for Neuroscience Abstracts*.

Betarbet, R., Na, H. M., Taylor, G., Bryant, M., and Greenamyre, J. T. (2005). Rotenone-induced substantia nigra pathology in mice. *Society for Neuroscience Abstracts*.

Betarbet, R., Canet-Aviles, R. M., Sherer, T. B., Mastroberardino, P. G., McLendon, C., Kim, J. H., Lund, S., Na, H. M., Taylor, G., Bence, N. F., Kopito, R., Seo, B. B., Yagi, T., Yagi, A., Klinefelter, G., Cookson, M. R., and Greenamyre, J. T. (2006). Intersecting pathways to neurodegeneration in Parkinson's disease: Effects of the pesticide rotenone on DJ-1, alpha-synuclein, and the ubiquitin–proteasome system. *Neurobiol Dis* **22**, 404–420.

Bonifati, V., Rizzu, P., van Baren, M. J., Schaap, O., Breedveld, G. J., Krieger, E., Dekker, M. C., Squitieri, F., Ibanez, P., Joosse, M., van Dongen, J. W., Vanacore, N., van Swieten, J. C., Brice, A., Meco, G., van Duijn, C. M., Oostra, B. A., and Heutink, P. (2003). Mutations in the DJ-1 gene associated with autosomal recessive early-onset parkinsonism. *Science* **299**, 256–259.

Braak, H., Ghebremedhin, E., Rub, U., Bratzke, H., and Del Tredici, K. (2004). Stages in the development of Parkinson's disease-related pathology. *Cell Tissue Res* **318**, 121–134.

Braak, H., de Vos, R. A., Bohl, J., and Del Tredici, K. (2006). Gastric alpha-synuclein immunoreactive inclusions in Meissner's and Auerbach's plexuses in cases staged for Parkinson's disease-related brain pathology. *Neurosci Lett* **396**, 67–72.

Coulom, H., and Birman, S. (2004). Chronic exposure to rotenone models sporadic Parkinson's disease in *Drosophila melanogaster*. *J Neurosci* **24**, 10993–10998.

Dawson, T. M., and Dawson, V. L. (2003). Molecular pathways of neurodegeneration in Parkinson's disease. *Science* **302**, 819–822.

Dexter, D. T., Carter, C. J., Wells, F. R., Javoy-Agid, F., Agid, Y., Lees, A., Jenner, P., and Marsden, C. D. (1989). Basal lipid peroxidation in substantia nigra is increased in Parkinson's disease. *J Neurochem* **52**, 381–389.

Duda, J. E., Giasson, B. I., Mabon, M. E., Lee, V. M., and Trojanowski, J. Q. (2002). Novel antibodies to synuclein show abundant striatal pathology in Lewy body diseases. *Ann Neurol* **52**, 205–210.

Dunnett, S. B., and Bjorklund, A. (1999). Prospects for new restorative and neuroprotective treatments in Parkinson's disease. *Nature* **399**, A32–aA39.

Ferrante, R. J., Schulz, J. B., Kowall, N. W., and Beal, M. F. (1997). Systemic administration of rotenone produces selective damage in the striatum and globus pallidus, but not in the substantia nigra. *Brain Res* **753**, 157–162.

Gao, H. M., Jiang, J., Wilson, B., Zhang, W., Hong, J. S., and Liu, B. (2002). Microglial activation-mediated delayed and progressive degeneration of rat nigral dopaminergic neurons: Relevance to Parkinson's disease. *J Neurochem* **81**, 1285–1297.

Gibb, W. R., and Lees, A. J. (1989). The significance of the Lewy body in the diagnosis of idiopathic Parkinson's disease. *Neuropathol Appl Neurobiol* **15**, 27–44.

Greenamyre, J. T., and Hastings, T. G. (2004). Biomedicine. Parkinson's – divergent causes, convergent mechanisms. *Science* **304**, 1120–1122.

Greenamyre, J. T., Sherer, T. B., Betarbet, R., and Panov, A. V. (2001). Complex I and Parkinson's disease. *IUBMB Life* **52**, 135–141.

Greenamyre, J. T., Nichols, C. J., Na, H. M., Martin, B., Postupna, N., Reinhardt, J., Anderson, M., and Betarbet, R. (2004). Rotenone-induced parkinsonian pathology in non-human primates. *Society for Neuroscience Abstracts*.

Greene, J. C., Whitworth, A. J., Kuo, I., Andrews, L. A., Feany, M. B., and Pallanck, L. J. (2003). Mitochondrial pathology and apoptotic muscle degeneration in *Drosophila* parkin mutants. *Proc Natl Acad Sci USA* **100**, 4078–4083.

Haas, R. H., Nasirian, F., Nakano, K., Ward, D., Pay, M., Hill, R., and Shults, C. W. (1995). Low platelet mitochondrial complex I and complex II/III activity in early untreated Parkinson's disease. *Ann Neurol* **37**, 714–722.

Hague, S., Rogaeva, E., Hernandez, D., Gulick, C., Singleton, A., Hanson, M., Johnson, J., Weiser, R., Gallardo, M., Ravina, B., Gwinn-Hardy, K., Crawley, A., St George-Hyslop, P. H., Lang, A. E., Heutink, P., Bonifati, V., Hardy, J., and Singleton, A. (2003). Early-onset Parkinson's disease caused by a compound heterozygous DJ-1 mutation. *Ann Neurol* **54**, 271–274.

Heikkila, R. E., Nicklas, W. J., and Duvoisin, R. C. (1985). Dopaminergic toxicity after the stereotaxic administration of the 1-methyl-4-phenylpyridinium ion (MPP$^+$) to rats. *Neurosci Lett* **59**, 135–140.

Hely, M. A., Morris, J. G., Reid, W. G., and Trafficante, R. (2005). Sydney Multicenter Study of Parkinson's disease: Non-l-dopa-responsive problems dominate at 15 years. *Mov Disord* **20**, 190–199.

Hisata, J. (2002). Lake and stream rehabilitation: Rotenone use and health risks. *Final Supplement of the Environmental Impact Statement*.

Hoglinger, G. U., Carrard, G., Michel, P. P., Medja, F., Lombes, A., Ruberg, M., Friguet, B., and Hirsch, E. C. (2003a). Dysfunction of mitochondrial complex I and the proteasome: interactions between two biochemical deficits in a cellular model of Parkinson's disease. *J Neurochem* **86**, 1297–1307.

Hoglinger, G. U., Feger, J., Prigent, A., Michel, P. P., Parain, K., Champy, P., Ruberg, M., Oertel, W. H., and Hirsch, E. C. (2003b). Chronic systemic complex I inhibition induces a hypokinetic multisystem degeneration in rats. *J Neurochem* **84**, 491–502.

Horgan, D. J., Ohno, H., and Singer, T. P. (1968). Studies on the respiratory chain-linked reduced nicotinamide adenine dinucleotide dehydrogenase. XV. Interactions of piericidin with the mitochondrial respiratory chain. *J Biol Chem* **243**, 5967–5976.

Hornykiewicz, O. (1966). Dopamine (3-hydroxytyramine) and brain function. *Pharmacol Rev* **18**, 925–964.

Hunot, S., and Hirsch, E. C. (2003). Neuroinflammatory processes in Parkinson's disease. *Ann Neurol* **53(Suppl 3)**, S49–sS58, discussion S58–S60.

Inden, M., Kitamura, Y., Takeuchi, H., Yanagida, T., Takata, K., Kobayashi, Y., Taniguchi, T., Yoshimoto, K., Kaneko, M., Okuma, Y., Taira, T., Ariga, H., and Shimohama, S. (2007). Neurodegeneration of mouse nigrostriatal dopaminergic system induced by repeated oral administration of rotenone is prevented by 4-phenylbutyrate, a chemical chaperone. *J Neurochem* **101**, 1491–1504.

Javitch, J. A., and Snyder, S. H. (1984). Uptake of MPP(+) by dopamine neurons explains selectivity of parkinsonism-inducing neurotoxin, MPTP. *Eur J Pharmacol* **106**, 455–456.

Javitch, J. A., D'Amato, R. J., Strittmatter, S. M., and Snyder, S. H. (1985). Parkinsonism-inducing neurotoxin, N-methyl-4-phenyl-1,2,3,6-tetrahydropyridine: Uptake of the metabolite N-methyl-4-phenylpyridine by dopamine neurons explains selective toxicity. *Proc Natl Acad Sci USA* **82**, 2173–2177.

Jenner, P. (1998). Oxidative mechanisms in nigral cell death in Parkinson's disease. *Mov Disord* **13**, 24–34.

Klockgether, T. (2004). Parkinson's disease: Clinical aspects. *Cell Tissue Res* **318**, 115–120.

Klockgether, T., and Turski, L. (1989). Excitatory amino acids and the basal ganglia: Implications for the therapy of Parkinson's disease. *Trends Neurosci* 12, 285–286.

Krige, D., Carroll, M. T., Cooper, J. M., Marsden, C. D., and Schapira, A. H. (1992). Platelet mitochondrial function in Parkinson's disease. The Royal Kings and Queens Parkinson Disease Research Group. *Ann Neurol* 32, 782–788.

Langston, J. W. (1996). The etiology of Parkinson's disease with emphasis on the MPTP story. *Neurology* 47, S153–sS160.

Langston, J. W., Ballard, P., Tetrud, J. W., and Irwin, I. (1983). Chronic Parkinsonism in humans due to a product of meperidine-analog synthesis. *Science* 219, 979–980.

Lapointe, N., St-Hilaire, M., Martinoli, M. G., Blanchet, J., Gould, P., Rouillard, C., and Cicchetti, F. (2004). Rotenone induces non-specific central nervous system and systemic toxicity. *FASEB J* 18, 717–719.

Liberatore, G. T., Jackson-Lewis, V., Vukosavic, S., Mandir, A. S., Vila, M., McAuliffe, W. G., Dawson, V. L., Dawson, T. M., and Przedborski, S. (1999). Inducible nitric oxide synthase stimulates dopaminergic neurodegeneration in the MPTP model of Parkinson disease. *Nat Med* 5, 1403–1409.

Lotharius, J., and Brundin, P. (2002). Pathogenesis of Parkinson's disease: Dopamine, vesicles and alpha-synuclein. *Nat Rev Neurosci* 3, 932–942.

Maslin, P. (2001). Rotenone. *Pesticide News* 54, 20–21.

McNaught, K. S., and Jenner, P. (2001). Proteasomal function is impaired in substantia nigra in Parkinson's disease. *Neurosci Lett* 297, 191–194.

Mitsumoto, A., and Nakagawa, Y. (2001). DJ-1 is an indicator for endogenous reactive oxygen species elicited by endotoxin. *Free Radic Res* 35, 885–893.

Mitsumoto, A., Nakagawa, Y., Takeuchi, A., Okawa, K., Iwamatsu, A., and Takanezawa, Y. (2001). Oxidized forms of peroxiredoxins and DJ-1 on two-dimensional gels increased in response to sublethal levels of paraquat. *Free Radic Res* 35, 301–310.

Na, H. M., Betarbet, R., Kim, J. H., Sherer, T. B., and Greenamyre, J. T. (2003). Rotenone models of Parkinson's disease selectively destroys striatal dopaminergic terminals and spares postsynaptic striatal neurons. *Society for Neuroscience Abstracts*.

Nicklas, W. J., Vyas, I., and Heikkila, R. E. (1985). Inhibition of NADH-linked oxidation in brain mitochondria by 1-methyl-4-phenyl-pyridine, a metabolite of the neurotoxin, 1-methyl-4-phenyl- 1,2,5,6-tetrahydropyridine. *Life Sci* 36, 2503–2508.

Parker, W. D., Jr, Boyson, S. J., and Parks, J. K. (1989). Abnormalities of the electron transport chain in idiopathic Parkinson's disease. *Ann Neurol* 26, 719–723.

Przedborski, S., Jackson-Lewis, V., Naini, A. B., Jakowec, M., Petzinger, G., Miller, R., and Akram, M. (2001). The parkinsonian toxin 1-methyl-4-phenyl-1,2,3,6-tetrahydropyridine (MPTP): A technical review of its utility and safety. *J Neurochem* 76, 1265–1274.

Richter, F., Hamann, M., and Richter, A. (2007). Chronic rotenone treatment induces behavioral effects but no pathological signs of parkinsonism in mice. *J Neurosci Res* 85, 681–691.

Rojo, A. I., Cavada, C., de Sagarra, M. R., and Cuadrado, A. (2007). Chronic inhalation of rotenone or paraquat does not induce Parkinson's disease symptoms in mice or rats. *Exp Neurol* 208, 120–126.

Saravanan, K. S., Sindhu, K. M., and Mohanakumar, K. P. (2005). Acute intranigral infusion of rotenone in rats causes progressive biochemical lesions in the striatum similar to Parkinson's disease. *Brain Res* 1049, 147–155.

Schapira, A. H., Cooper, J. M., Dexter, D., Jenner, P., Clark, J. B., and Marsden, C. D. (1989). Mitochondrial complex I deficiency in Parkinson's disease [letter] [see comments]. *Lancet* 1, 1269.

Schuler, F., and Casida, J. E. (2001). Functional coupling of PSST and ND1 subunits in NADH:ubiquinone oxidoreductase established by photoaffinity labeling. *Biochim Biophys Acta* 1506, 79–87.

Shamoto-Nagai, M., Maruyama, W., Kato, Y., Isobe, K., Tanaka, M., Naoi, M., and Osawa, T. (2003). An inhibitor of mitochondrial complex I, rotenone, inactivates proteasome by oxidative modification and induces aggregation of oxidized proteins in SH-SY5Y cells. *J Neurosci Res* 74, 589–597.

Sherer, T. B., Betarbet, R., and Greenamyre, J. T. (2001). Pesticides and Parkinson's disease. *ScientificWorldJournal* 1, 207–208.

Sherer, T. B., Betarbet, R., Stout, A. K., Lund, S., Baptista, M., Panov, A. V., Cookson, M. R., and Greenamyre, J. T. (2002). An *in vitro* model of Parkinson's disease: Linking mitochondrial impairment to altered alpha-synuclein metabolism and oxidative damage. *J Neurosci* 22, 7006–7015.

Sherer, T. B., Betarbet, R., Kim, J. H., and Greenamyre, J. T. (2003a). Selective microglial activation in the rat rotenone model of Parkinson's disease. *Neurosci Lett* 341, 87–90.

Sherer, T. B., Kim, J. H., Betarbet, R., and Greenamyre, J. T. (2003b). Subcutaneous rotenone exposure causes highly selective dopaminergic degeneration and alpha-synuclein aggregation. *Exp Neurol* 179, 9–16.

Sherer, T. B., Betarbet, R., Testa, C. M., Seo, B. B., Richardson, J. R., Kim, J. H., Miller, G. W., Yagi, T., Matsuno-Yagi, A., and Greenamyre, J. T. (2003c). Mechanism of toxicity in rotenone models of Parkinson's disease. *J Neurosci* 23, 10756–10764.

Shults, C. W. (2006). Lewy bodies. *Proc Natl Acad Sci USA* **103**, 1661–1668.

Storch, A., Ludolph, A. C., and Schwarz, J. (2004). Dopamine transporter: Involvement in selective dopaminergic neurotoxicity and degeneration. *J Neural Transm* **111**, 1267–1286.

Talpade, D. J., Greene, J. G., Higgins, D. S., Jr, and Greenamyre, J. T. (2000). *In vivo* labeling of mitochondrial complex I (NADH:ubiquinone oxidoreductase) in rat brain using [(3)H]dihydrorotenone. *J Neurochem* **75**, 2611–2621.

Tanner, C. M., Ottman, R., Goldman, S. M., Ellenberg, J., Chan, P., Mayeux, R., and Langston, J. W. (1999). Parkinson disease in twins: An etiologic study. *JAMA* **281**, 341–346.

Thiffault, C., Langston, J. W., and Di Monte, D. A. (2000). Increased striatal dopamine turnover following acute administration of rotenone to mice. *Brain Res* **885**, 283–288.

Tipton, K. F., and Singer, T. P. (1993). Advances in our understanding of the mechanisms of the neurotoxicity of MPTP and related compounds. *J Neurochem* **61**, 1191–1206.

Uversky, V. N. (2004). Neurotoxicant-induced animal models of Parkinson's disease: Understanding the role of rotenone, maneb and paraquat in neurodegeneration. *Cell Tissue Res* **318**, 225–241.

Valente, E. M., Abou-Sleiman, P. M., Caputo, V., Muqit, M. M., Harvey, K., Gispert, S., Ali, Z., Del Turco, D., Bentivoglio, A. R., Healy, D. G., Albanese, A., Nussbaum, R., Gonzalez-Maldonado, R., Deller, T., Salvi, S., Cortelli, P., Gilks, W. P., Latchman, D. S., Harvey, R. J., Dallapiccola, B., Auburger, G., and Wood, N. W. (2004). Hereditary early-onset Parkinson's disease caused by mutations in PINK1. *Science* **304**, 1158–1160.

Vehovszky, A., Szabo, H., Hiripi, L., Elliott, C. J., and Hernadi, L. (2007). Behavioural and neural deficits induced by rotenone in the pond snail *Lymnaea stagnalis*. A possible model for Parkinson's disease in an invertebrate. *Eur J Neurosci* **25**, 2123–2130.

Yoritaka, A., Hattori, N., Uchida, K., Tanaka, M., Stadtman, E. R., and Mizuno, Y. (1996). Immunohistochemical detection of 4-hydroxynonenal protein adducts in Parkinson disease. *Proc Natl Acad Sci USA* **93**, 2696–2701.

Yoshino, H., Nakagawa-Hattori, Y., Kondo, T., and Mizuno, Y. (1992). Mitochondrial complex I and II activities of lymphocytes and platelets in Parkinson's disease. *J Neural Transm Park Dis Dement Sect* **4**, 27–34.

Zhu, C., Vourc'h, P., Fernagut, P. O., Fleming, S. M., Lacan, S., Dicarlo, C. D., Seaman, R. L., and Chesselet, M. F. (2004). Variable effects of chronic subcutaneous administration of rotenone on striatal histology. *J Comp Neurol* **478**, 418–426.

15

PARAQUAT-INDUCED NEURODEGENERATION: A MODEL OF PARKINSON'S DISEASE RISK FACTORS

DONATO A. DI MONTE

The Parkinson's Institute, Sunnyvale, CA, USA

Experimental models of Parkinson's disease (PD), using both toxic and genetic manipulations, have often aimed at reproducing disease signs and pathology as closely as possible and to a degree of severity that mimics the profound lesion underlying overt PD. It is noteworthy, however, that models yielding a subtler phenotype that reproduce some, but not all, features of PD and are characterized by milder pathology also play a significant role in experimental research. They could more faithfully reflect early events in the disease process and conditions that do not necessarily lead to full-blown PD but rather predispose to it. In other words, they may represent valuable paradigms of disease risk factors.

Experimental work using models of PD risk factors has broad implications. For instance, toxin-induced and transgenic models can help elucidate the complex interactions between environmental and genetic determinants that most likely underlie disease pathogenesis. Results from these studies can also provide clues on the nature of specific risk factors (e.g., classes of chemicals that target the nigrostriatal system) and the mechanisms that characterize their action. Finally, experimental data, together with clinical and epidemiological evidence, may ultimately lead to the development of preventive strategies, including genetic screening and environmental regulations, that could identify individuals at risk, limit the consequences of risk exposures and allow for early neuroprotective intervention.

Several paradigms of neuronal injury and dysfunction that model the putative effects of PD risk factors have been described over the past few years. Examples of toxin-induced lesions include the administration of iron to newborn mice which, once they reach an older age, develop mild nigrostriatal degeneration (Kaur *et al.*, 2006). From the genetic standpoint, transgenic mouse models characterized by altered expression of PD-linked genes have been shown to develop abnormalities in dopaminergic neurotransmission in the absence of overt nigrostriatal degeneration (Goldberg *et al.*, 2003, 2005). These transgenic animals could therefore represent valuable tools for studies on preclinical deficits and predisposing factors in PD. The purpose of this chapter is to describe another model of PD risk factors generated by exposure to the herbicide paraquat. Features of this model will be reviewed with the intent of highlighting mechanisms of likely relevance to the role of environmental exposures and gene–environment interactions in PD.

THE PARAQUAT MODEL

Interest in the neurotoxic effects of paraquat and their possible relevance to PD was originally spurred by two considerations. First, after the discovery of a parkinsonian syndrome caused by 1-methyl-4-phenyl-1,2,3,6-tetrahydropyridine (MPTP) in humans and non-human primates, it was noted that the chemical structure of paraquat resembles that of MPP^+, the fully oxidized metabolite that mediates MPTP neurotoxicity (Snyder and D'Amato, 1985; Bocchetta and Corsini, 1986). Second, a potential role of paraquat in PD was supported by epidemiological evidence. Not only has a list of more than 20 studies reported an increased disease risk associated with pesticide exposure, but a few of these studies have specifically suggested paraquat as a potential culprit (Liou et al., 1997; Di Monte 2003; Kamel et al., 2007). While stimulating interest in the relationship between paraquat and PD, these considerations also became matter of a spirited debate. Reasons for skepticism included the fact that, as a charged compound, paraquat would not be expected to cross the blood–brain barrier efficiently and thus cause significant neurotoxicity (Koller, 1986; Hart, 1987). Furthermore, despite their chemical similarity, paraquat and MPP^+ may not necessarily be alike when it comes to their toxic properties and mechanisms of cell death (Di Monte et al., 1986; Ramachandiran et al., 2007). Finally, not all epidemiological data are consistent with the conclusion that pesticide exposure represents a risk factor for PD, and the implication of paraquat by retrospective population-based studies may be a consequence of methodological artifacts such as a recall bias (Nuti et al., 2004; Li et al., 2005).

Initial work in animal models contributed to making the issue of paraquat neurotoxicity a controversial one. For example, Perry et al. (1986) reported lack of changes in striatal dopamine content even after repeated subcutaneous injections of mice with paraquat. In contrast, Brooks et al. (1999) provided evidence that systemic paraquat administration damaged the mouse nigrostriatal system and caused a reduction of ambulatory activity. A study published in 2002 represents a turning point in this controversy (McCormack et al., 2002). For the first time, this study assessed and quantified the neurodegenerative effects of paraquat in the mouse substantia nigra pars compacta using state-of-the-art stereological techniques. It also addressed the critical question of selective toxicity by comparing paraquat-induced changes in neuronal number in the substantia nigra versus other brain regions. Results showed that a regimen of three intraperitoneal injections of the herbicide separated by 1-week intervals caused a dose-dependent degeneration of nigral dopaminergic neurons that was accompanied by an astrocytic and microglial response. Despite this loss of cell bodies, striatal dopamine levels remained relatively unchanged, likely as a result of compensatory mechanisms. Other neuronal populations and, in particular, GABAergic cells in the substantia nigra pars reticulata and cholinergic cells in the hippocampus were not affected by paraquat, consistent with a selective action.

Subsequent work has confirmed this ability of paraquat to kill nigral dopaminergic neurons in mice as well as rats (Table 15.1) (Peng et al., 2004; Cicchetti et al., 2005; Ossowska et al., 2005). They have also furthered the characterization of this model and elucidated important steps in the sequence of events leading to neurotoxicity. The initial concern that paraquat may not gain access into the brain has been addressed by several studies showing detectable levels of the herbicide in the central nervous system

TABLE 15.1 Paraquat administration consistently decreases the number of dopaminergic neurons in the mouse substantia nigra pars compacta

Treatment	Number of TH-positive neurons	Number of Nissl-stained neurons
Saline ($n = 30$)	$12,263 \pm 128$	$15,614 \pm 200$
Paraquat ($n = 30$)	$8,960 \pm 125^*$ (−27%)	$12,165 \pm 215^*$ (−22%)

Data were pooled from different sets of experiments in which mice received three injections of saline or paraquat 1 week apart and were sacrificed 7 days after the third injection. The number of tyrosine hydroxylase(TH)-immunoreactive cells and Nissl-stained neurons was estimated in the substantia nigra pars compacta using stereological techniques. Values are the mean of bilateral counts $\pm$ SEM.

$^*p < 0.001$ as compared to the corresponding saline control group.

(CNS) after its systemic administration to experimental animals (Corasaniti *et al.*, 1992; Shimizu *et al.*, 2001; Barlow *et al.*, 2003). Particularly relevant is the finding that a specific uptake system, that is, the neutral amino acid transporter (System L carrier), mediates or contributes to the transport of paraquat across the blood–brain barrier (Shimizu *et al.*, 2001). Based on this finding, it has been possible to protect against the neurodegenerative effects of paraquat in mice using specific amino acids that are substrates for the System L carrier, such as L-valine, L-phenylalanine or L-dopa (McCormack and Di Monte, 2003). By competing for the same transporter, these agents inhibit the access of paraquat into the CNS and thus prevent its neurotoxicity. Once paraquat enters the brain, damage to nigral dopaminergic neurons is accompanied by evidence of apoptosis that, as reported by Peng *et al.* (2004), is likely triggered by the c-Jun N-terminal kinase (JNK) signal transduction pathway.

Recent investigations into the features of paraquat-induced neurodegeneration have also provided interesting clues on the relevance of this model to PD. Aging represents an unequivocal risk factor for PD. It is therefore noteworthy that older mice as well as older rats display enhanced vulnerability to paraquat (McCormack *et al.*, 2002; Thiruchelvam *et al.*, 2003; Saint-Pierre *et al.*, 2006). In PD, not all dopaminergic neurons are equally susceptible to degeneration. In particular, dopaminergic cells in the ventral tegmental area are relatively less affected than neurons in the substantia nigra pars compacta (Hirsch *et al.*, 1988). Also, dopaminergic neurons containing the calcium-binding protein calbindin-D_{28K} are relatively spared in PD (German *et al.*, 1992). The pattern of paraquat-induced neurodegeneration reveals intriguing similarities with these PD features since (i) dopaminergic neurons are killed in the mouse substantia nigra pars compacta but not in the ventral tegmental area and (ii) midbrain dopaminergic cells immunoreactive for calbindin-D_{28K} are resistant to the herbicide (McCormack *et al.*, 2006). Other features of paraquat neurotoxicity are more directly pertinent to mechanisms of PD risk factors and will be discussed in the following sections.

EXPOSURE PARADIGMS

The time-course of neurodegeneration triggered by repeated injections of paraquat to mice has revealed a surprising pattern. One might have expected to see a progressive exacerbation of the lesion after each toxic administration. Instead, when the number of nigral dopaminergic neurons was compared in animals treated with 1–3 weekly paraquat injections, no cell loss occurred after the first exposure, a significant decrease followed the second administration, and the third injection did not produce further degeneration (McCormack *et al.*, 2005). A likely explanation for these findings is that an initial paraquat exposure, though unable to cause significant neuronal loss, acts as a "priming" event that enhances neuronal vulnerability to a subsequent challenge. Maximal toxicity (25–30% decrease in dopaminergic cell number) is reached within 4–5 days after the second paraquat injection. The lack of further degeneration, even when mice are subjected to a third exposure, is compatible with at least two interesting interpretations. Neurons that survive two toxic challenges may develop defense responses that allow them to resist a third paraquat treatment. Alternatively or in addition, paraquat-induced degeneration may target a particularly vulnerable sub-population of dopaminergic neurons that is mostly depleted after two consecutive exposures. The significance of these findings from the standpoint of the relationship between toxic insults, nigrostriatal degeneration and PD is that they emphasize the importance of exposure paradigms and toxin–tissue interactions. Based on this paraquat model, the neurodegenerative consequences of environmental exposures depend upon the sequence of toxic challenges and the interplay between neuronal susceptibility, damaging effects and protective responses.

MICROGLIAL ACTIVATION

The pattern of cell loss caused by sequential paraquat administrations, as described above, provides a valuable model for studying mechanisms that predispose dopaminergic neurons to damage and thus enhance their risk for degeneration. A recent study has identified microglial activation as one of these mechanisms (Purisai *et al.*, 2007). A single exposure to paraquat induces a time-dependent increase in the number of cells with morphological, immunocytochemical and biochemical characteristics of activated microglia in the mouse substantia nigra. If a second paraquat injection is administered

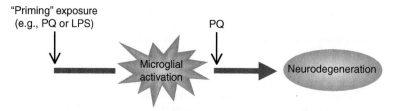

FIGURE 15.1   Role of microglia in paraquat-induced neurodegeneration. Microglial activation, which can be induced by an initial paraquat exposure or via administration of pro-inflammatory agents (lipopolysaccharide, LPS), enhances the vulnerability of dopaminergic cells to injury ("priming" effect). Once reactive microglia are present, paraquat exposure triggers neurodegeneration.

at the time of maximal microglial response (7 days after the initial insult), it immediately triggers neuro-degenerative effects. That microglia play an important role in this toxic model is not only suggested by this temporal association, but also indicated by experiments showing a causal relationship between microglial activation and paraquat-induced neuronal loss. When treatment with the anti-inflammatory drug minocycline blocked the microglial response elicited by an initial paraquat exposure, no dopaminergic cell death was observed even after a second toxic challenge (Purisai *et al.*, 2007). On the other hand, if a single paraquat treatment was preceded by pharmacological activation of microglia using lipopolysaccharide, this single exposure became capable of triggering dopaminergic cell loss.

These findings are not the first to implicate inflammatory processes and, in particular, microglial activation in the pathogenesis of neurodegeneration in animal models of PD and PD itself (Wu *et al.*, 2002; McGeer and McGeer, 2004; Saint-Pierre *et al.*, 2006). They highlight, however, an interesting relationship between neuroinflammation and toxic exposures, and reveal a new mechanism by which microglial activation could act as a risk factor for dopaminergic cell death in PD. A pre-existing inflammatory process can amplify the damaging effects of toxic insults and, in the presence of reactive microglia, dopaminergic neurons become dramatically more susceptible to toxicant-induced injury (Figure 15.1).

REDOX CYCLING

Upon activation, microglial cells are capable of producing a number of potentially harmful molecules, and this property has been suggested to explain at

least in part their role in neurodegenerative processes and models of PD (Wu *et al.*, 2002, 2003; Dringen, 2005). For example, superoxide anion and nitric oxide are formed via the catalytic activity of microglial NADPH oxidase and nitric oxide synthase, respectively. Similar mechanisms could underlie the ability of microglia to enhance tissue vulnerability in the paraquat model. In addition, however, the priming effect described in the previous sections may arise from specific microglia–paraquat interactions. Paraquat is known to undergo a one-electron reduction that generates a cation radical and is catalyzed by a variety of cellular diaphorases (Bus and Gibson, 1984; Clejan and Cederbaum, 1989; Day *et al.*, 1999). Electron transfer from this radical onto molecular oxygen regenerates the parent herbicide while producing, at the same time, superoxide anion. Reiteration of this cycle of redox reactions leads to the formation of significant amounts of reactive oxygen species (ROS) and could ultimately result in cytotoxicity. *In vitro* work using primary mesencephalic cultures has shown that redox cycling reactions are capable of killing dopaminergic neurons (Bonneh-Barkay *et al.*, 2005a). As importantly, experiments with microglial cultures have revealed that these cells promote the redox cycling of paraquat through its reduction by NADPH oxidase and nitric oxide synthase (Bonneh-Barkay *et al.*, 2005b). Thus, the role of microglia in paraquat-induced neurotoxicity in mice is compatible with the following scenario. An initial toxicant exposure causes mild injury and a microglial response in the absence of overt neurodegeneration. If, however, a second paraquat challenge occurs at the time of microglial activation, the contribution of these cells to redox cycling reactions could enhance ROS production, overwhelm neuronal defense mechanisms and trigger cell death (Figure 15.2).

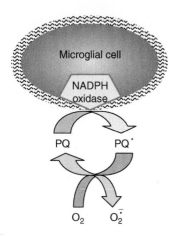

FIGURE 15.2 Mechanism of formation of ROS after paraquat exposure. Upon microglial activation, membrane-bound NADPH oxidase can catalyze the redox cycling of paraquat, generating significant amounts of deleterious superoxide anion. This sequence of toxic events is supported by *in vitro* evidence and may also characterize paraquat-induced neurotoxicity in the mouse model.

The involvement of microglia-mediated redox cycling in the *in vivo* mouse model of paraquat-induced dopaminergic cell loss remains to be demonstrated. Nonetheless, present evidence indicates that reactions catalyzed by NADPH oxidase, which may include paraquat reduction, play an important role in this model since transgenic mice lacking functional NADPH oxidase are resistant to paraquat neurotoxicity (Purisai *et al.*, 2007). Further investigation into the relationship between toxicant exposures, redox cycling and nigrostriatal damage bears significant implications for neurodegenerative processes targeting dopaminergic neurons. The ability to redox cycle is shared by a variety of naturally occurring and synthetic compounds, including quinones and bipyridyl pesticides such as paraquat and diquat (Frank *et al.*, 1987; O'Brien, 1991). This property could, therefore, characterize an entire class of toxic agents that warrant special consideration as potential PD risk factors.

OXIDATIVE STRESS

If oxidative reactions involving microglial cells are directly linked to neuronal demise in the paraquat mouse model, neurodegeneration in this model should be accompanied by evidence of oxidative stress. Results from several studies support this premise. The sequence of cell loss caused by repeated paraquat exposures was found to be similar to the pattern of oxidative neuronal injury (McCormack *et al.*, 2005). Oxidative stress, as assessed by immunocytochemical evidence of lipid peroxidation, was minimal after a single paraquat exposure in the absence of overt neurodegeneration. It became very pronounced, however, after a second toxic challenge in parallel to the onset and progression of neuronal loss. When this loss was completed, lipid peroxidation also receded, and a third paraquat administration given at this time caused neither further degeneration nor oxidative stress. A relationship between microglial activation, oxidative stress and paraquat-induced toxicity is further indicated by the finding that co-administration of the herbicide with minocycline decreased the number of reactive microglial cells, prevented lipid peroxidation and protected against dopaminergic cell degeneration (Purisai *et al.*, 2007).

Results with transgenic mice are also consistent with a role of oxidative stress in killing neurons after paraquat exposure. Mice overexpressing ferritin, Cu, Zn superoxide dismutase (SOD) or glutathione peroxidase were completely spared from paraquat-induced neurodegeneration, most likely because of their decreased vulnerability to oxidative challenges (McCormack *et al.*, 2005; Thiruchelvam *et al.*, 2005). Similarly, dopaminergic cell death triggered by paraquat in normal (i.e., non-transgenic) mice was counteracted by systemic administration of a synthetic SOD/catalase mimetic or a PEP-1–SOD fusion protein (Peng *et al.*, 2005; Choi *et al.*, 2006). In summary, data in the paraquat model underscore the importance of oxidative stress as a mechanism of nigrostriatal degeneration. A relationship appears to exist between susceptibility to oxidative damage and vulnerability to neurodegenerative processes. Even among sub-populations of midbrain dopaminergic cells, neurons that are relatively resistant to degeneration (i.e., calbindin-containing cells) are also more resistant to toxicant-induced oxidative stress (McCormack *et al.*, 2006). As a corollary to this association, PD risk may be particularly affected by genetic or environmental factors that perturb the balance between pro- and anti-oxidant mechanisms in the CNS.

CO-EXPOSURE PARADIGMS

Humans are likely to be exposed to complex mixtures of chemicals in their residential as well as occupational environments and, for this reason, experimental models of toxic interactions may more closely replicate pathogenetic processes pertinent to human diseases. Paraquat neurotoxicity represents an excellent example of neuronal injury that can be synergistically enhanced by chemical interactions. In particular, co-administration of paraquat with the fungicide manganese ethylenebisdithiocarbamate (*maneb*) has been shown to result in behavioral changes (e.g., reduced motor activity) and neurochemical and pathological alterations of the nigrostriatal system that were more pronounced than those caused by paraquat alone (Thiruchelvam *et al.*, 2000, 2003). A likely mechanism underlying this effect is the ability of maneb to modify the biodisposition and to increase the concentration of paraquat in various tissues including the brain (Barlow *et al.*, 2003). For instance, paraquat concentration in the striatum and midbrain was doubled when mice were co-injected with maneb.

Chemical interactions are most often seen as events that exacerbate toxic outcomes. It is also possible, however, that the damaging properties of a compound may actually be alleviated by co-exposure paradigms, as indicated by a recent report in mice treated with paraquat and nicotine (Khwaja *et al.*, 2007). Administration of nicotine via drinking water prior to and during paraquat exposure partially blocked nigral dopaminergic cell death. This protective effect of nicotine against paraquat neurotoxicity is especially intriguing in view of the widely documented inverse correlation between PD risk and cigarette smoking (Quik, 2004). Together with findings in other PD models (Quik *et al.*, 2006), paraquat–nicotine interactions support a direct role of nicotine as a neuroprotective agent and raise the possibility that nicotine exposure may reduce the vulnerability of dopaminergic neurons to toxic challenges.

DEVELOPMENTAL EXPOSURE

A few studies to date have assessed the long-term consequences of early-in-life paraquat exposure. In 1993, Fredriksson *et al.* reported behavioral and neurochemical changes in adult mice that had

been treated orally with the herbicide on post-natal days 10 and 11. Significant hypoactivity characterized these animals at 60 and 120 days of age and, at the latter time point, measurements of striatal neurotransmitters showed a significant decline of dopamine but not serotonin. A more recent investigation was aimed at determining whether neonatal administration of paraquat, alone or in combination with maneb, affected nigrostriatal integrity in mature mice, and whether early-in-life exposure enhanced the susceptibility of the nigrostriatal system to subsequent toxic challenges (Thiruchelvam *et al.*, 2002). Animals were exposed via intraperitoneal injections from post-natal days 5–19 and were allowed to survive to 8 months of age. A decrease in striatal dopamine and nigral cell number was observed in 8-month old animals that had received neonatal treatment with paraquat alone. The severity of this lesion was enhanced, however, in mice exposed to both paraquat and maneb. Furthermore, if animals were treated with paraquat early-in-life and then re-challenged at 8 months, nigrostriatal damage was greater than in mice exposed to the herbicide only once as adult.

Administration of paraquat or maneb to pregnant mice (gestational days 10–17) did not result in significant behavioral changes in offspring (Barlow *et al.*, 2004). Interestingly, however, male offspring of maneb-treated pregnant animals displayed enhanced susceptibility to paraquat, as reflected by reduced locomotion, striatal dopamine depletion and loss of nigral dopaminergic cells, when exposed at adult age. Taken together, these findings support the notion that developmental exposures can cause long-lasting effects on the nigrostriatal dopaminergic system. Early toxic events could not only induce a permanent lesion, but also predispose to damage from subsequent insults, thus acting as risk factors for later disease progression.

GENE–TOXICANT INTERACTIONS

Genetic studies over the past few years have identified several genes that are associated with familial forms of the disease but may also be implicated in the pathogenesis of sporadic PD. At least two of these genes encoding the proteins α-synuclein and DJ-1 appear to play a role in the paraquat model of nigrostriatal injury. The precise mechanisms that underlie the involvement of α-synuclein in the pathogenesis of human parkinsonism are still relatively unknown. Both clinical and experimental evidence, however, strongly suggests that two factors, not mutually exclusive, contribute to α-synuclein-dependent pathology. The first is an increased expression of the protein, the second its propensity to aggregate and to form intracellular inclusions. Indeed, higher levels of α-synuclein due to gene multiplication are associated with familial autosomal parkinsonism, and α-synuclein-containing inclusions (Lewy bodies and Lewy neurites) represent a pathological hallmark of PD (Spillantini et al., 1998; Singleton et al., 2003; Farrer et al., 2004). An intriguing mechanism by which levels of α-synuclein could be enhanced even in the absence of genetic modifications is suggested by experimental models of toxic exposures. Similar to other chemicals that damage the nigrostriatal system, paraquat administration to mice is accompanied by a dramatic upregulation of α-synuclein (Vila et al., 2000; Manning-Bog et al., 2002). This effect, common to different paradigms of toxic challenges, could be part of the normal neuronal response to injury but, under particular circumstances (e.g., persistent insults leading to prolonged protein upregulation), may ultimately contribute to cellular demise (Di Monte, 2003).

The relationship between α-synuclein, Lewy bodies and PD has prompted investigations into mechanisms that promote changes in protein conformation and aggregation. One of these mechanisms is α-synuclein–chemical interactions. Experiments in vitro have shown that paraquat shares with other specific pesticides the ability to accelerate the aggregation of α-synuclein into fibrils (Uversky et al., 2001). Furthermore, following the administration of paraquat to mice, protein misfolding and aggregation are indicated by the intraneuronal accumulation of thioflavin S-positive structures that are immunoreactive for α-synuclein (Manning-Bog et al., 2002). Thus, data in the paraquat model suggest that, by enhancing the level and promoting the aggregation of α-synuclein, toxicant exposures could contribute to the pathogenetic role of this protein in PD.

Mutations in the DJ-1 gene resulting in a loss of protein function are associated with autosomal recessive parkinsonism (Bonifati et al., 2003). Because of the likely involvement of DJ-1 in the cellular response to oxidative stress, it is not surprising that interactions between DJ-1 and paraquat have been explored in a variety of in vitro and animal models (Mitsumoto et al., 2001; Canet-Aviles et al., 2004; Taira et al., 2004). Particularly interesting are the results of studies in Drosophila showing that a loss of DJ-1 function enhances the sensitivity of flies to paraquat toxicity and, vice versa, DJ-1 upregulation within dopaminergic neurons confers protection against paraquat injury (Menzies et al., 2005; Meulener et al., 2005). An initial report of the effects of paraquat in transgenic mice lacking DJ-1 did not find increased damage to nigrostriatal neurons (Goldberg et al., 2005). However, given the abundant evidence in support of a protective role of DJ-1 against oxidative stress, further work on DJ-1–paraquat interactions in the mouse model is warranted. Overall, experimental models based on paraquat exposure represent valuable tools for studies on gene–environment interactions that will help elucidate how specific genetic modifications modulate the effects of toxic challenges in the context of neurodegenerative processes.

CONCLUSIONS

Taken together, data accumulated over the past few years amply justify the addition of paraquat to the list of toxins that damage the nigrostriatal system and therefore reproduce neurodegenerative changes relevant to PD. Whether or not paraquat itself acts as a risk factor and contributes to disease development in humans remains to be established. Nevertheless, as a model of environmental exposure targeting the nigrostriatal system, paraquat administration has already enhanced our understanding of toxic processes/mechanisms that could underlie the action of PD risk factors and the pathogenesis of neurodegeneration in PD. Furthermore, mechanistic studies using the paraquat model bear important implications for the development of PD therapeutics. Current findings strongly support, for example, that modulation of microglial response, enhancement of neuronal defense mechanisms against oxidative damage and prevention of the accumulation and aggregation of α-synuclein all represent promising avenues for neuroprotective intervention in PD.

A limitation of the current paradigms of paraquat-induced neurotoxicity is the difficulty in reproducing severe nigrostriatal damage. For this reason, paraquat administration will be unlikely to replace other toxin-induced models of PD-like pathology (e.g., MPTP or 6-hydroxydopamine).

These models, however, complement each other very well, and together provide a better opportunity to test pathogenetic hypotheses and therapeutic avenues pertinent to PD. The choice to use a certain model should be made based on the relevance of its features to specific experimental question(s). Paraquat toxicity, as emphasized in this review chapter, could be a model of choice for (i) investigating pre-clinical lesions, (ii) elucidating mechanisms of neuronal vulnerability and (iii) testing agents for neuroprotection against moderate nigrostriatal injury. It is also reasonable to anticipate that, given the availability of a variety of paradigms of nigrostriatal degeneration, an increasing number of future studies will utilize parallel testing using two or more experimental models. This approach might be particularly effective for screening putative therapeutics since, for example, pharmacological agents that counteract neurotoxicity in the paraquat as well as other models of dopaminergic cell death could be considered better candidates for neuroprotection in PD.

ACKNOWLEDGEMENTS

The author's work is supported by grants from the National Institute of Environmental Health Sciences (ES10806 and ES12077) and the Backus Foundation. The author thanks Dr. Sarah A. Jewell and Dr. Alison L. McCormack for their comments on this chapter. The assistance of Ms. Kirsten Thompson in the preparation of this manuscript is greatly appreciated.

REFERENCES

Barlow, B. K., Thiruchelvam, M. J., Bennice, L., Cory-Slechta, D. A., Ballatori, N., and Richfield, E. K. (2003). Increased synaptosomal dopamine content and brain concentration of paraquat produced by selective dithiocarbamates. *J Neurochem* **85**, 1075–1086.

Barlow, B. K., Richfield, E. K., Cory-Slechta, D. A., and Thiruchelvam, M. (2004). A fetal risk factor for Parkinson's disease. *Dev Neurosci* **26**, 11–23.

Bocchetta, A., and Corsini, G. U. (1986). Parkinson's disease and pesticides. *Lancet* **2**, 1163.

Bonifati, V., Rizzu, P., van Baren, M. J., Schaap, O., Breedveld, G. J., Krieger, E., Dekker, M. C., Squitieri, F., Ibanez, P., Joosse, M., van Dongen, J. W., Vanacore, N.,

van Swieten, J. C., Brice, A., Meco, G., van Duijn, C. M., Oostra, B. A., and Heutink, P. (2003). Mutations in the DJ-1 gene associated with autosomal recessive early-onset parkinsonism. *Science* **299**, 256–259.

Bonneh-Barkay, D., Langston, J. W., and Di Monte, D. A. (2005a). Toxicity of redox cycling pesticides in primary mesencephalic cultures. *Antioxid Redox Signal* **7**, 649–653.

Bonneh-Barkay, D., Reaney, S. H., Langston, J. W., and Di Monte, D. A. (2005b). Redox cycling of the herbicide paraquat in microglial cultures. *Brain Res Mol Brain Res* **134**, 52–56.

Brooks, A. I., Chadwick, C. A., Gelbard, H. A., Cory-Slechta, D. A., and Federoff, H. J. (1999). Paraquat elicited neurobehavioral syndrome caused by dopaminergic neuron loss. *Brain Res* **823**, 1–10.

Bus, J. S., and Gibson, J. E. (1984). Paraquat: Model for oxidant-initiated toxicity. *Environ Health Perspect* **55**, 37–46.

Canet-Aviles, R. M., Wilson, M. A., Miller, D. W., Ahmad, R., McLendon, C., Bandyopadhyay, S., Baptista, M. J., Ringe, D., Petsko, G. A., and Cookson, M. R. (2004). The Parkinson's disease protein DJ-1 is neuroprotective due to cysteine-sulfinic acid-driven mitochondrial localization. *Proc Natl Acad Sci USA* **101**, 9103–9108.

Choi, H. S., An, J. J., Kim, S. Y., Lee, S. H., Kim, D. W., Yoo, K. Y., Won, M. H., Kang, T. C., Kwon, H. J., Kang, J. H., Cho, S. W., Kwon, O. S., Park, J., Eum, W. S., and Choi, S. Y. (2006). PEP-1–SOD fusion protein efficiently protects against paraquat-induced dopaminergic neuron damage in a Parkinson disease mouse model. *Free Radic Biol Med* **41**, 1058–1068.

Cicchetti, F., Lapointe, N., Roberge-Tremblay, A., Saint-Pierre, M., Jimenez, L., Ficke, B. W., and Gross, R. E. (2005). Systemic exposure to paraquat and maneb models early Parkinson's disease in young adult rats. *Neurobiol Dis* **20**, 360–371.

Clejan, L., and Cederbaum, A. I. (1989). Synergistic interactions between NADPH-cytochrome P-450 reductase, paraquat, and iron in the generation of active oxygen radicals. *Biochem Pharmacol* **38**, 1779–1786.

Corasaniti, M. T., Strongoli, M. C., Pisanelli, A., Bruno, P., Rotiroti, D., Nappi, G., and Nistico, G. (1992). Distribution of paraquat into the brain after its systemic injection in rats. *Funct Neurol* **7**, 51–56.

Day, B. J., Patel, M., Calavetta, L., Chang, L. Y., and Stamler, J. S. (1999). A mechanism of paraquat toxicity involving nitric oxide synthase. *Proc Natl Acad Sci USA* **96**, 12760–12765.

Di Monte, D. A. (2003). The environment and Parkinson's disease: Is the nigrostriatal system preferentially targeted by neurotoxins? *Lancet Neurol* **2**, 531–538.

Di Monte, D. A., Sandy, M. S., Ekstrom, G., and Smith, M. T. (1986). Comparative studies on the mechanisms of paraquat and 1-methyl-4-phenylpyridine (MPP$^+$) cytotoxicity. *Biochem Biophys Res Commun* **137**, 303–309.

Dringen, R. (2005). Oxidative and antioxidative potential of brain microglial cells. *Antioxid Redox Signal* **7**, 1223–1233.

Farrer, M., Kachergus, J., Forno, L., Lincoln, S., Wang, D. S., Hulihan, M., Maraganore, D., Gwinn-Hardy, K., Wszolek, Z., Dickson, D., and Langston, J. W. (2004). Comparison of kindreds with parkinsonism an α-synuclein genomic multiplications. *Ann Neurol* **55**, 174–179.

Frank, D. M., Arora, P. K., Blumer, J. L., and Sayre, L. M. (1987). Model study on the bioreduction of paraquat, MPP$^+$, and analogs. Evidence against a "redox cycling" mechanism in MPTP neurotoxicity. *Biochem Biophys Res Commun* **147**, 1095–1104.

Fredriksson, A., Fredriksson, M., and Eriksson, P. (1993). Neonatal exposure to paraquat or MPTP induces permanent changes in striatum dopamine and behavior in adult mice. *Toxicol Appl Pharmacol* **122**, 258–264.

German, D. C., Manaye, K. F., Sonsalla, P. K., and Brooks, B. A. (1992). Midbrain dopaminergic cell loss in Parkinson's disease and MPTP-induced parkinsonism: Sparing of calbindin-D28k-containing cells. *Ann NY Acad Sci* **648**, 42–62.

Goldberg, M. S., Fleming, S. M., Palacino, J. J., Cepeda, C., Lam, H. A., Bhatnagar, A., Meloni, E. G., Wu, N., Ackerson, L. C., Klapstein, G. J., Gajendiran, M., Roth, B. L., Chesselet, M. F., Maidment, N. T., Levine, M. S., and Shen, J. (2003). Parkin-deficient mice exhibit nigrostriatal deficits but not loss of dopaminergic neurons. *J Biol Chem* **278**, 43628–43635.

Goldberg, M. S., Pisani, A., Haburcak, M., Vortherms, T. A., Kitada, T., Costa, C., Tong, Y., Martella, G., Tscherter, A., Martins, A., Bernardi, G., Roth, B. L., Pothos, E. N., Calabresi, P., and Shen, J. (2005). Nigrostriatal dopaminergic deficits and hypokinesia caused by inactivation of the familial Parkinsonism-linked gene DJ-1. *Neuron* **45**, 489–496.

Hart, T. B. (1987). Parkinson's disease and pesticides. *Lancet* **1**, 38.

Hirsch, E., Graybiel, A. M., and Agid, Y. A. (1988). Melanized dopaminergic neurons are differentially susceptible to degeneration in Parkinson's disease. *Nature* **334**, 345–348.

Kamel, F., Tanner, C., Umbach, D., Hoppin, J., Alavanja, M., Blair, A., Comyns, K., Goldman, S., Korell, M., Langston, J., Ross, G., and Sadler, D. (2007). Pesticide exposure and self-reported Parkinson's disease in the Agricultural Health Study. *Am J Epidemiol* **165**, 364–374.

Kaur, D., Peng, J., Chinta, S. J., Rajagopalan, S., Di Monte, D. A., Cherny, R. A., and Andersen, J. K. (2006). Increased murine neonatal iron intake results

in Parkinson-like neurodegeneration with age. *Neurobiol Aging*, 2007 June **28**(6), 907–913.

Khwaja, M., McCormack, A., McIntosh, J. M., Di Monte, D. A. and Quik, M. (2007). Nicotine partially protects against paraquat-induced nigrostriatal damage in mice; link to α6β2* nAChRs. *J Neurochem* **100**, 180–190.

Koller, W. C. (1986). Paraquat and Parkinson's disease. *Neurology* **36**, 1147.

Li, A. A., Mink, P. J., McIntosh, L. J., Teta, M. J., and Finley, B. (2005). Evaluation of epidemiologic and animal data associating pesticides with Parkinson's disease. *J Occup Environ Med* **47**, 1059–1087.

Liou, H. H., Tsai, M. C., Chen, C. J., Jeng, J. S., Chang, Y. C., Chen, S. Y., and Chen, R. C. (1997). Environmental risk factors and Parkinson's disease: A case–control study in Taiwan. *Neurology* **48**, 1583–1588.

Manning-Bog, A. B., McCormack, A. L., Li, J., Uversky, V. N., Fink, A. L., and Di Monte, D. A. (2002). The herbicide paraquat causes up-regulation and aggregation of α-synuclein in mice. *J Biol Chem* **277**, 1641–1644.

McCormack, A. L., and Di Monte, D. A. (2003). Effects of L-dopa and other amino acids against paraquat-induced nigrostriatal degeneration. *J Neurochem* **85**, 82–86.

McCormack, A. L., Thiruchelvam, M., Manning-Bog, A. B., Thiffault, C., Langston, J. W., Cory-Slechta, D. A., and Di Monte, D. A. (2002). Environmental risk factors and Parkinson's disease: Selective degeneration of nigral dopaminergic neurons caused by the herbicide paraquat. *Neurobiol Dis* **10**, 119–127.

McCormack, A. L., Atienza, J. G., Johnston, L. C., Andersen, J. K., Vu, S., and Di Monte, D. A. (2005). Role of oxidative stress in paraquat-induced dopaminergic cell degeneration. *J Neurochem* **93**, 1030–1037.

McCormack, A. L., Atienza, J. G., Langston, J. W., and Di Monte, D. A. (2006). Decreased susceptibility to oxidative stress underlies the resistance of specific dopaminergic cell populations to paraquat-induced degeneration. *Neuroscience* **141**, 929–937.

McGeer, P. L., and McGeer, E. G. (2004). Inflammation and neurodegeneration in Parkinson's disease. *Parkinsonism Relat Disord* **10**(Suppl 1), S3–S7.

Menzies, F. M., Yenisetti, S. C., and Min, K. T. (2005). Roles of *Drosophila* DJ-1 in survival of dopaminergic neurons and oxidative stress. *Curr Biol* **15**, 1578–1582.

Meulener, M., Whitworth, A. J., Armstrong-Gold, C. E., Rizzu, P., Heutink, P., Wes, P. D., Pallanck, L. J., and Bonini, N. M. (2005). Drosophila DJ-1 mutants are selectively sensitive to environmental toxins associated with Parkinson's disease. *Curr Biol* **15**, 1572–1577.

Mitsumoto, A., Nakagawa, Y., Takeuchi, A., Okawa, K., Iwamatsu, A., and Takanezawa, Y. (2001). Oxidized forms of peroxiredoxins and DJ-1 on two-dimensional gels increased in response to sublethal levels of paraquat. *Free Radic Res* **35**, 301–310.

Nuti, A., Ceravolo, R., Dell'Agnello, G., Gambaccini, G., Bellini, G., Kiferle, L., Rossi, C., Logi, C., and Bonuccelli, U. (2004). Environmental factors and Parkinson's disease: A case–control study in the Tuscany region of Italy. *Parkinsonism Relat Disord* **10**, 481–485.

O'Brien, P. J. (1991). Molecular mechanisms of quinone cytotoxicity. *Chem Biol Interact* **80**, 1–41.

Ossowska, K., Wardas, J., Smialowska, M., Kuter, K., Lenda, T., Wieronska, J. M., Zieba, B., Nowak, P., Dabrowska, J., Bortel, A., Kwiecinski, A., and Wolfarth, S. (2005). A slowly developing dysfunction of dopaminergic nigrostriatal neurons induced by long-term paraquat administration in rats: An animal model of preclinical stages of Parkinson's disease? *Eur J Neurosci* **22**, 1294–1304.

Peng, J., Mao, X. O., Stevenson, F. F., Hsu, M., and Andersen, J. K. (2004). The herbicide paraquat induces dopaminergic nigral apoptosis through sustained activation of the JNK pathway. *J Biol Chem* **279**, 32626–32632.

Peng, J., Stevenson, F. F., Doctrow, S. R., and Andersen, J. K. (2005). Superoxide dismutase/catalase mimetics are neuroprotective against selective paraquat-mediated dopaminergic neuron death in the substantia nigra: Implications for Parkinson disease. *J Biol Chem* **280**, 29194–29198.

Perry, T. L., Yong, V. W., Wall, R. A., and Jones, K. (1986). Paraquat and two endogenous analogues of the neurotoxic substance N-methyl-4-phenyl-1,2,3,6-tetrahydropyridine do not damage dopaminergic nigrostriatal neurons in the mouse. *Neurosci Lett* **69**, 285–289.

Purisai, M. G., McCormack, A. L., Cumine, S., Li, J., Isla, M. Z., and Di Monte, D. A. (2007). Microglial activation as a priming event leading to paraquat-induced dopaminergic cell degeneration. *Neurobiol Dis* **25**, 392–400.

Quik, M. (2004). Smoking, nicotine, and Parkinson's disease. *Trends Neurosci* **27**, 561–568.

Quik, M., Parameswaran, N., McCallum, S. E., Bordia, T., Bao, S., McCormack, A., Kim, A., Tyndale, R. F., Langston, J. W., and Di Monte, D. A. (2006). Chronic oral nicotine treatment protects against striatal degeneration in MPTP-treated primates. *J Neurochem* **98**, 1866–1875.

Ramachandiran, S., Hansen, J. M., Jones, D. P., Richardson, J. R., and Miller, G. W. (2007). Divergent mechanisms of paraquat, MPP$^+$, and rotenone toxicity: Oxidation of thioredoxin and caspase-3 activation. *Toxicol Sci* **95**, 163–171.

Saint-Pierre, M., Tremblay, M. E., Sik, A., Gross, R. E., and Cicchetti, F. (2006). Temporal effects of paraquat/maneb on microglial activation and dopamine neuronal loss in older rats. *J Neurochem* **98**, 760–772.

Shimizu, J., Ohtaki, K., Matsubara, K., Aoyama, K., Uezono, T., Saito, O., Suno, M., Ogawa, K., Hayase, N., Kimura, K., and Shiono, H. (2001). Carrier-mediated processes in blood–brain barrier penetration and neural uptake of paraquat. *Brain Res* **906**, 135–142.

Singleton, A. B., Farrer, M., Johnson, J., Singleton, A., Hague, S., Kachergus, J., Hulihan, M., Peuralinna, T., Dutra, A., Nussbaum, R., Lincoln, S., Crawley, A., Hanson, M., Maraganore, D., Adler, C., Cookson, M. R., Muenter, M., Baptista, M., Miller, D., Blancato, J., Hardy, J., and Gwinn-Hardy, K. (2003). α-Synuclein locus triplication causes Parkinson's disease. *Science* **302**, 841.

Snyder, S. H., and D'Amato, R. J. (1985). Predicting Parkinson's disease. *Nature* **317**, 198–199.

Spillantini, M. G., Crowther, R. A., Jakes, R., Hasegawa, M., and Goedert, M. (1998). α-Synuclein in filamentous inclusions in Lewy bodies from Parkinson's disease and dementia with Lewy bodies. *Proc Natl Acad Sci USA* **95**, 6469–6473.

Taira, T., Saito, Y., Niki, T., Iguchi-Ariga, S. M., Takahashi, K., and Ariga, H. (2004). DJ-1 has a role in antioxidative stress to prevent cell death. *EMBO Rep* **5**, 213–218.

Thiruchelvam, M., Richfield, E. K., Baggs, R. B., Tank, A. W., and Cory-Slechta, D. A. (2000). The nigrostriatal dopaminergic system as a preferential target of repeated exposures to combined paraquat and maneb: Implications for Parkinson's disease. *J Neurosci* **20**, 9207–9214.

Thiruchelvam, M., Richfield, E. K., Goodman, B. M., Baggs, R. B., and Cory-Slechta, D. A. (2002). Developmental exposure to the pesticides paraquat and maneb and the Parkinson's disease phenotype. *Neurotoxicology* **23**, 621–633.

Thiruchelvam, M., McCormack, A. L., Richfield, E. K., Baggs, R. B., Tank, A. W., Di Monte, D. A., and Cory-Slechta, D. A. (2003). Age-related irreversible progressive nigrostriatal dopaminergic neurotoxicity in the paraquat and maneb model of Parkinson's disease phenotype. *Eur J Neurosci* **18**, 589–600.

Thiruchelvam, M., Prokopenko, O., Cory-Slechta, D. A., Richfield, E. K., Buckley, B., and Mirochnitchenko, O. (2005). Overexpression of superoxide dismutase or glutathione peroxidase protects against the paraquat + maneb-induced Parkinson disease phenotype. *J Biol Chem* **280**, 22530–22539.

Uversky, V. N., Li, J., and Fink, A. L. (2001). Pesticides directly accelerate the rate of α-synuclein fibril formation: A possible factor in Parkinson's disease. *FEBS Lett* **500**, 105–108.

Vila, M., Vukosavic, S., Jackson-Lewis, V., Neystat, M., Jakowec, M., and Przedborski, S. (2000). α-Synuclein up-regulation in substantia nigra dopaminergic neurons following administration of the parkinsonian toxin MPTP. *J Neurochem* **74**, 721–729.

Wu, D. C., Jackson-Lewis, V., Vila, M., Tieu, K., Teismann, P., Vadseth, C., Choi, D. K., Ischiropoulos, H., and Przedborski, S. (2002).

Blockade of microglial activation is neuroprotective in the 1-methyl-4-phenyl-1,2,3,6-tetrahydropyridine mouse model of Parkinson disease. *J Neurosci* **22**, 1763–1771.

Wu, D. C., Teismann, P., Tieu, K., Vila, M., Jackson-Lewis, V., Ischiropoulos, H., and Przedborski, S. (2003). NADPH oxidase mediates oxidative stress in the 1-methyl-4-phenyl-1,2,3,6-tetrahydropyridine model of Parkinson's disease. *Proc Natl Acad Sci USA* **100**, 6145–6150.

IV

RODENT AND OTHER VERTEBRATE GENETIC MODELS

VI

RODENT AND OTHER VERTEBRATE GENETIC MODELS

16

OVERVIEW: RODENT AND FISH MODELS OF PARKINSON'S DISEASE

WILLIAM DAUER

Columbia University, Departments of Neurology and Pharmacology

INTRODUCTION

Using Parkinson's disease (PD) genes to generate models of the disease is an active and fast-moving area that is providing new insights into PD gene function and dysfunction. The intense interest in this area is evident from a rapidly expanding literature that has largely developed along two general themes. The largest body of work uses traditional modeling approaches – transgenic and knockout mouse technology, and cell-based models – and attempts to recapitulate neurodegeneration and understand the molecular events underlying this central event. A smaller body of work has similar goals, but is focused primarily on bringing new technologies to bear on PD modeling. The following chapters highlight both of these areas and emphasize the many interesting directions this research is taking.

An important question to consider when reading the following chapters is: How much will these genes tell us about typical (idiopathic) PD? This is an appropriate concern, given the fact that only approximately 5% of all PD is believed to result from Mendelian-type mutations. Indeed, in amyotrophic lateral sclerosis research, the mechanistic link between inherited (e.g., SOD1 mutations) and idiopathic disease remains uncertain, despite many years of effort. In PD, however, solid evidence exists to implicate synuclein in idiopathic PD, since it appears to be an abundant component of Lewy bodies (Spillantini *et al.*, 1997), regardless of the genetic status of the affected individual. It has been more difficult to link the recessively inherited genes to idiopathic disease. Nevertheless, the phenotypic similarity between the inherited and idiopathic forms of the disease is striking, both at the clinical and pathological levels. Moreover, polymorphisms in these PD genes have been linked to genetic susceptibility to idiopathic PD, further strengthening the rationale for this work.

A common concern voiced about PD models is that none have thus far recapitulated all of the key features of the disease. Ideally, a mouse model of PD would exhibit the adult onset of relatively specific and progressive dopamine neuron degeneration, and an accompanying behavioral phenotype that is responsive to dopamine replacement therapy. However, many factors – including neuroanatomic differences between mice and men and the compressed time frame of pathologic progression in mice – may preclude the "perfect" model. A more reasonable requirement, perhaps, is that these PD genes be employed in model systems that recapitulate some important feature of the disease, such as protein aggregation or the signaling effects of disease mutations. Information from these more limited models can give valuable information about gene function and dysfunction and, considered together, can lead to insights into disease pathogenesis. Comparing the cell-based model described by Fortin and Edwards to the transgenic mouse models reviewed by Dawson and colleagues nicely illustrates the different contributions provided by disparate models of the same gene (synuclein).

SYNUCLEIN BIOLOGY

Synuclein-based cell and mouse models have been most commonly studied because of the clear relationship of synuclein to idiopathic PD, and because multiple groups have had some success reproducing neuropathological changes similar to those found in PD. The chapter by Ted Dawson and colleagues extensively reviews the multiple alpha-synuclein-expressing transgenic mouse models. These models clearly show that overexpression of synuclein produces neuronal atrophy, dystrophic neurites and astrocytosis that are accompanied by insoluble inclusions of synuclein reminiscent of (though distinct from) Lewy bodies. Notably, none of the models reviewed by Dawson and colleagues show degeneration of dopamine neurons, a feature that has yet to be modeled in transgenic synuclein mice, or in knockout mice for recessive familial PD genes (Parkin, DJ-1 or PINK1 (Goldberg *et al.*, 2003; Kim *et al.*, 2005; Kitada *et al.*, 2007)). However, introduction of synuclein-overexpressing viral vectors into the substantia nigra, reviewed by Aebischer and colleagues, does lead to dopamine neurodegeneration. This work, largely performed in rats, raises the possibility that species differences or factors such as when during development the exogenous protein is introduced, may be critical to dopamine neuron viability.

Other models have been exploited in an attempt to develop a mechanistic understanding of synuclein cellular function and the effect of pathogenic missense mutations. The chapter by Chandra and Sudhof reviews an intriguing link between synuclein and cysteine string protein (CSP). CSP is a synaptic chaperone (Chamberlain and Burgoyne, 2000), and mice deficient for CSP develop widespread neurodegeneration and death (Fernandez-Chacon *et al.*, 2004). Based on the notion that decreasing chaperone function might enhance the aggregation and toxicity of synuclein, these investigators crossed synuclein transgenic mice to CSP mice. Surprisingly, overexpression of synuclein profoundly suppressed CSP-dependent neurodegeneration, while intercrossing CSP and synuclein null mice worsened the CSP phenotype. These findings, reviewed in detail in the chapter, lead Chandra and Sudhof to suggest a potential neuroprotective role for synuclein, and that the normal and the pathological role of synuclein are distinct. Synuclein has also been shown to be an inhibitor synaptic vesicle recycling (Abeliovich *et al.*, 2000). Thus, an alternative possibility raised by these data is that synuclein overexpression slows down synaptic vesicle cycling, thereby reducing the need for the chaperone activity provided by CSP. Indeed

a protective role for synuclein is difficult to reconcile with the fact that duplication or triplication of the synuclein locus causes PD (Singleton *et al.*, 2003).

While additional work is required to unravel the underlying mechanisms for the CSP-synuclein genetic interaction, this study emphasizes the fact that synuclein, which is enriched at presynaptic terminals, plays an important yet not clearly defined role in modulating neurotransmission. The chapter by Edwards and colleagues focuses on the role of synuclein within the presynaptic nerve terminal. In addition to reviewing the relevant literature on this topic, they detail their studies showing that the binding of synuclein to synaptic vesicles is dependent on neuronal activity. Interestingly, their work shows that enhanced synaptic activity causes synuclein to dissociate from membranes and diffuse away from the synaptic bouton, raising the possibility that the amount of synuclein in a given terminal is a marker of recent activity. Intriguingly, knockout models for parkin, PINK1 and DJ-1 all show abnormalities synaptic transmission (Nakamura and Edwards, 2007), raising the possibility that synaptic dysfunction may contribute to cellular demise in PD. Future work will be needed to define whether synaptic function promotes cell death, or is a consequence of other toxic events provoked by manipulating these PD genes.

NOVEL MODELING STRATEGIES

The discovery of multiple familial PD genes has also led investigators to develop novel methods for studying gene function. A particularly exciting advance is the development transgenic mice using very large stretches of genomic DNA known as bacterial artificial chromosomes (BACs). Yang and colleagues detail the development of this method, which enables investigators to develop transgenic models that recapitulate the normal spatial and temporal expression pattern, in contrast to traditional transgenesis, which provides far less predictable and physiologically relevant results. While mice are the most commonly used animal model, they are time-consuming to generate and costly to maintain. A cheaper and rapid method of PD model generation may ultimately be possible using zebrafish, small (~2 inch) fish that have a complex neuroanatomy including dopaminergic neurons that project to forebrain and control motor activity. This system, reviewed in the chapter by Su Guo, has many unique advantages, including the possibility of forward genetic screens. In addition, because their skin and gills are permeable to water soluble chemicals, zebrafish are also amenable to high throughput drug screens. These advantages may enable investigators to uncover novel regulators of PD gene functions in a vertebrate model, a feat not currently practical using genetically modified mice.

SUMMARY

The analysis of familial PD genes is still relatively young, having been launched with the discovery of synuclein in 1996. The models reviewed in the following chapters should thus be viewed as the "tip of the iceberg" of a rapidly expanding and active research area. Indeed, although modeling recessive PD genes in mice has been challenging, modeling these genes in drosophila has recently led to significant progress (Clark *et al.*, 2006; Park *et al.*, 2006), which may now be able to be translated back into the mouse system. Thus, advances made in different animal model and cell-based systems are beginning to synergize, and will likely lead to second- and third-generation models that will further our understanding of the biology of these genes, and their potential relationship to one another.

REFERENCES

Abeliovich, A., Schmitz, Y., Farinas, I., Choi-Lundberg, D., Ho, W. H., Castillo, P. E., Shinsky, N., Verdugo, J. M., Armanini, M., Ryan, A. *et al.* (2000). Mice lacking alpha-synuclein display functional deficits in the nigrostriatal dopamine system. *Neuron* **25**, 239–252.

Chamberlain, L. H., and Burgoyne, R. D. (2000). Cysteine-string protein: The chaperone at the synapse. *J Neurochem* **74**, 1781–1789.

Clark, I. E., Dodson, M. W., Jiang, C., Cao, J. H., Huh, J. R., Seol, J. H., Yoo, S. J., Hay, B. A., and Guo, M. (2006). Drosophila pink1 is required for mitochondrial function and interacts genetically with parkin. *Nature* **441**, 1162–1166.

Fernandez-Chacon, R., Wolfel, M., Nishimune, H., Tabares, L., Schmitz, F., Castellano-Munoz, M., Rosenmund, C., Montesinos, M. L., Sanes, J. R., Schneggenburger, R. *et al.* (2004). The synaptic vesicle protein CSP alpha prevents presynaptic degeneration. *Neuron* **42**, 237–251.

Goldberg, M. S., Fleming, S. M., Palacino, J. J., Cepeda, C., Lam, H. A., Bhatnagar, A., Meloni, E. G., Wu, N., Ackerson, L. C., Klapstein, G. J. *et al.* (2003). Parkin-deficient mice exhibit nigrostriatal deficits but not loss of dopaminergic neurons. *J Biol Chem* **278**, 43628–43635.

Kim, R. H., Smith, P. D., Aleyasin, H., Hayley, S., Mount, M. P., Pownall, S., Wakeham, A., You-Ten, A. J., Kalia, S. K., Horne, P. *et al.* (2005). Hypersensitivity of DJ-1-deficient mice to 1-methyl-4-phenyl-1,2,3,6-tetrahydropyrindine (MPTP) and oxidative stress. *Proc Natl Acad Sci USA* **102**, 5215–5220.

Kitada, T., Pisani, A., Porter, D. R., Yamaguchi, H., Tscherter, A., Martella, G., Bonsi, P., Zhang, C., Pothos, E. N., and Shen, J. (2007). Impaired dopamine release and synaptic plasticity in the striatum of PINK1-deficient mice. *Proc Natl Acad Sci USA* **104**, 11441–11446.

Nakamura, K., and Edwards, R. H. (2007). Physiology versus pathology in Parkinson's disease. *Proc Natl Acad Sci USA* **104**, 11867–11868.

Park, J., Lee, S. B., Lee, S., Kim, Y., Song, S., Kim, S., Bae, E., Kim, J., Shong, M., Kim, J. M. *et al.* (2006). Mitochondrial dysfunction in drosophila PINK1 mutants is complemented by parkin. *Nature* **441**, 1157–1161.

Singleton, A. B., Farrer, M., Johnson, J., Singleton, A., Hague, S., Kachergus, J., Hulihan, M., Peuralinna, T., Dutra, A., Nussbaum, R. *et al.* (2003). Alpha-Synuclein locus triplication causes Parkinson's disease. *Science* **302**, 841.

Spillantini, M. G., Schmidt, M. L., Lee, V. M., Trojanowski, J. Q., Jakes, R., and Goedert, M. (1997). Alpha-synuclein in Lewy bodies. *Nature* **388**, 839–840.

17

GENETIC MODELS OF FAMILIAL PARKINSON'S DISEASE

MINGYAO YING[1,2], VALINA L. DAWSON[1–4] AND TED M. DAWSON[1–3]

[1]Institute for Cell Engineering, Johns Hopkins University School of Medicine, Baltimore, MD, USA
[2]Department of Neurology, Johns Hopkins University School of Medicine, Baltimore, MD, USA
[3]Department of Neuroscience, Johns Hopkins University School of Medicine, Baltimore, MD, USA
[4]Department of Physiology, Johns Hopkins University School of Medicine, Baltimore, MD, USA

INTRODUCTION

Since the discovery of mutations in α-synuclein as the first genetic cause of Parkinson's disease (PD), there has been an explosion of new mechanistic insights into the pathogenesis of PD. These new insights have come from the investigations of how mutations in α-synuclein cause PD, as well as the identification of other genes linked to familial PD and the molecular investigations that followed the identification of these genes. In Chapter 4 Farrer and colleagues review the various genetic contributions to PD. Five genes have been clearly linked to PD, and a number of other genes or genetic linkages have been identified that may cause PD (Table 17.1) (also see Chapter 4).

AUTOSOMAL DOMINANT INHERITED PD

α-Synuclein

Mutations in the gene encoding α-synuclein were the first genetic mutations linked to PD (Polymeropoulos et al., 1996). The first mutation consisted of an alanine to threonine substitution (A53T) (Polymeropoulos et al., 1997). Two other point mutations including an alanine to proline substitution (A30P) and a glutamine to lysine substitution (E46K) also cause autosomal-dominant forms of the disease (Kruger et al., 1998; Zarranz et al., 2004). Simple overexpression of wild-type (non-mutated) α-synuclein also causes autosomal-dominant forms of the disease, since triplication and gene duplication of α-synuclein leads to PD (Singleton et al., 2003; Chartier-Harlin et al., 2004; Hope et al., 2004; Nishioka et al., 2006; Winkler et al., 2007). An increased risk of PD is also associated with polymorphisms within the α-synuclein promoter (Maraganore et al., 2006; Winkler et al., 2007). Thus, it is thought that there is an increased risk of developing PD based on the relative levels of α-synuclein. Patients with higher levels of α-synuclein have a greater risk of developing PD and those patients with lower levels have a smaller risk. The α-synuclein appears to play an important role in the pathogenesis of both sporadic and inherited forms of PD, since α-synuclein appears to be the major structural component of the Lewy body

■■■ **TABLE 17.1   Loci and genes linked to PD**

Locus	Gene	Inheritance	Phenotype	References
PARK 1 and 4	α-Synuclein	AD	PD/DLB, early onset, rapid progression	(Polymeropoulos et al., 1997; Singleton et al., 2003)
PARK 2	Parkin	AR	PD, early onset, slow progression	(Kitada et al., 1998)
PARK 3	Unknown	AD	Typical PD, sometimes dementia	(Gasser et al., 1998)
PARK 5	UCH-L1	Unclear	Typical PD	(Leroy et al., 1998)
PARK 6	PINK1	AR	PD, early onset, slow progression	(Valente et al., 2004)
PARK 7	DJ-1	AR	PD, early onset, slow progression	(Bonifati et al., 2003)
PARK 8	LRRK2	AD	Typical PD, late onset	(Funayama et al., 2002; Zimprich et al., 2004)
PARK 9	ATP13A2	AR	Kufor-Rakeb Syndrome	(Ramirez et al., 2006)
PARK 10	Unknown	Unclear	Typical PD	(Hicks et al., 2002)
PARK 11	Unknown	Unclear	Typical PD	(Pankratz et al., 2003b)
PARK12	Unknown	Unclear	Typical PD	(Pankratz et al., 2003a)
PARK13	Omi/HHA2	AD	Typical PD	(Strauss et al., 2005)

AD = autosomal dominant; AR = autosomal recessive; DLB = dementia with Lewy bodies.

(Spillantini et al., 1997; Spillantini et al., 1998) (also see Chapter 02). In this chapter we will review vertebrate α-synuclein transgenic mouse models. These models have led to substantial advances in understanding the role of α-synuclein in pathogenic processes that are relevant to PD.

Leucine Rich Repeat Kinase-2

Mutations in the gene coding for leucine rich repeat kinase-2 (LRRK2) also lead to autosomal-dominant PD (Paisan-Ruiz et al., 2004; Zimprich et al., 2004). LRRK2 is a large protein containing multiple functional domains including a GTPase and kinase domain (West et al., 2005). LRRK2 is widely distributed throughout the brain including localization to the nigrostriatal system in both rodents and human brain (Biskup et al., 2006; Higashi et al., 2007a, b). Dopamine-containing neurons of the substantia nigra pars compacta also contain LRRK2 and it is also localized to membranous and vesicular structures suggesting that it may be involved in vesicular transport (Biskup et al., 2006). Mutations in LRRK2 that cause familial PD lead to enhanced kinase activity and cell death in cellular models (West et al., 2005; Smith et al., 2006; Greggio et al., 2007). Inhibition of the kinase or GTPase activity attenuates LRRK2 toxicity, suggesting that LRRK2 kinase and GTPase inhibitors might be attractive new agents for the treatment of PD (Smith et al., 2006; West et al., 2007). Most cases of LRRK2-related disease show typical late and asymmetric onset of PD symptoms similar to α-synuclein-related PD (Paisan-Ruiz et al., 2004; Zimprich et al., 2004). Examination of brains from LRRK2 patients shows that the majority of affected patients demonstrate typical PD pathology including cell loss of the substantia nigra and locus coeruleus with Lewy bodies, although other pathologic findings have been reported (Zimprich et al., 2004; Rajput et al., 2006; Ross et al., 2006; Whaley et al., 2006). In Chapter 32 Cookson and colleagues provide a detailed review of LRRK2's function and potential role in PD. No transgenic or knockout vertebrate models of LRRK2 have been reported as of 2007.

AUTOSOMAL RECESSIVE INHERITED PD

Parkin

Three genes have been linked to autosomal-recessive PD. The first autosomal-recessive-linked PD gene identified was parkin (Kitada *et al.*, 1998). Parkin is an ubiquitin E3 ligase, where mutations in parkin lead to loss of parkin's function and accumulation of toxic substrates (Shimura *et al.*, 2000; Zhang *et al.*, 2000). Knockout of parkin and PINK1 in mice does not lead to dramatic abnormalities in the dopaminergic system (Goldberg *et al.*, 2003; Von Coelln *et al.*, 2004; Perez and Palmiter, 2005), however one line of mice in which the catalytic domain of parkin was removed have loss of norepinephrine-containing locus coeruleus neurons (Von Coelln *et al.*, 2004). A number of putative substrates for parkin have been identified, but it seems that the *a*minoacyl-tRNA *s*ynthetase (ARS) *i*nteracting *m*ultifunctional *p*rotein type 2 (AIMP2) (p38/JTV-1), and the *f*ar upstream element-*b*inding *p*rotein-1 (FBP-1) are authentic parkin substrates since they accumulate in patients with mutations in parkin and in parkin knockout mice (Corti *et al.*, 2003; Ko *et al.*, 2005; Ko *et al.*, 2006). Moreover they accumulate in sporadic PD, which is characterized by impaired parkin activity (Chung *et al.*, 2004). Accumulation of these substrates may contribute to the pathogenesis of PD due to mutations in parkin.

DJ-I

The second autosomal-recessive PD gene identified was DJ-1 (Bonifati *et al.*, 2003). DJ-1 is thought to function as an antioxidant protein as well as a redox-sensitive chaperone (Bonifati *et al.*, 2003; Shendelman *et al.*, 2004; Meulener *et al.*, 2006). Familial-associated mutations in DJ-1 disrupt its function through decreasing its ability to properly function through impairment in its dimerization (Olzmann *et al.*, 2004; Moore *et al.*, 2005b). Knockout of DJ-1 revealed that it functions as an atypical peroxiredoxin-like peroxidase leading to a deficit in scavenging mitochondrial H_2O_2 (Andres-Mateos *et al.*, 2007).

PINKI

The third autosomal-recessive-linked gene that was identified is PTEN-induced kinase (PINK-1)

(Valente *et al.*, 2004). PINK1 is a mitochondrial-associated kinase where familial-associated mutations are thought to disrupt its kinase activity (Beilina *et al.*, 2005; Silvestri *et al.*, 2005). PINK1-deficient mice also do not have dramatic abnormalities in the dopaminergic system (Kitada *et al.*, 2007). Knockout of parkin and PINK1 leads to similar mitochondrial abnormalities in *drosophila* with muscle degeneration and genetic studies indicate that they function in the same pathway (Greene *et al.*, 2003; Clark *et al.*, 2006; Park *et al.*, 2006).

α-SYNUCLEIN TRANSGENIC MOUSE MODELS

A variety of transgenic mice models have been developed that overexpress human α-synuclein (Table 17.2). Many of these models recapitulate α-synucleinopathy-induced neurodegeneration. The α-synucleinopathies represent a large spectrum of diverse and related neurodegenerative diseases, including PD (Lee and Trojanowski, 2006). Although, none of these transgenic α-synuclein mice accurately model all the pathologic features of sporadic PD, their importance as models to study the pathogenesis of PD cannot be underestimated.

Ideal Animal Models of PD?

An ideal animal model of PD would recapitulate most, if not all, the features of sporadic PD (Dawson *et al.*, 2002). PD usually begins in late adulthood with degeneration of non-dopaminergic systems beginning first, followed by loss of dopamine neurons within the substantia nigra pars compacta (SNc), and followed by degeneration in cortical and limbic structures (Del Tredici *et al.*, 2002; Braak *et al.*, 2003; Savitt *et al.*, 2006). The degeneration of dopamine neurons leads to the characteristic motoric dysfunction in PD, including rigidity, rest tremor, and slowness of movement. The degeneration in non-dopamine neurons is also a prominent feature leading to cognitive disturbances (Dubois *et al.*, 1990; Aarsland *et al.*, 1996; Dubois and Pillon, 1997; Del Tredici *et al.*, 2002; Braak *et al.*, 2003). The disease is relentlessly progressive and it is characterized by the presence of Lewy bodies and Lewy neurites (Lewy, 1912;

■ **TABLE 17.2   Mouse models of α-synucleinopathy**

Promoter	Human α-synuclein	Motor deficits	Life span	Neurodegeneration	LB-like inclusions	References
mThy-1	WT, A53T	++	ND	Neurite	++	(van der Putten et al., 2000)
mThy-1	A30P	+++ (H)	Reduced	Neurite	+++	(Neumann et al., 2002)
mThy-1	WT (m and h), A30P*, A53T*	+++	Reduced	Neuron loss	ND	(Chandra et al., 2005)
mPrp	WT, A30P, A53T*	+++ (H)	Reduced	Neurite	+++	(Giasson et al., 2002)
mPrp	WT, A30P, A53T*	+++	Reduced	Neurite, neuron loss	+++	(Lee et al., 2002; Martin et al., 2006)
mPrp	WT, A53T*	+	ND	Neurite	–	(Gispert et al., 2003)
Hamster Prp	WT, A30P*, A53T	+++	Reduced	ND	–	(Gomez-Isla et al., 2003)
PDGF-β	WT	+	–	Loss of DA terminals	+	(Hashimoto et al., 2003; Masliah et al., 2000)
Rat TH (4,5 kb)	WT, A30P, A53T	ND	–	–	–	(Matsuoka et al., 2001)
Rat TH (9 kb)	WT, A30P+A53T*	+	–	Loss of TH+ neurons in SNpc	–	(Richfield et al., 2002; Thiruchelvam et al., 2004)
Rat TH (9 kb)	Truncated (1-120)	+	ND	Reduced striatal DA level	+	(Tofaris et al., 2006)
Proteolipid	WT	ND	–	ND	+(Oligodendrocytes)	(Kahle et al., 2002)
pCNP	WT	+	–	Neurite, Neuron and oligodendrocyte loss	++(Oligodendrocytes)	(Yazawa et al., 2005)
pMBP	WT	++		Neurite, reduced TH level	++ (Oligodendrocytes)	(Shults et al., 2005)

Asterick (*) indicates that multiple lines were examined. Phenotypes +++ = severe, ++ = moderate, += Mild, – = none. ND = not determined. H– = homozygous lines, Prp = prion promoter.

Forno, 1996). Lewy bodies and Lewy neurites are eosinophilic cytoplasmic proteinaceous inclusions that contain a variety of proteins with α-synuclein being the major structural component (Spillantini et al., 1997; Spillantini et al., 1998). Lewy bodies at the ultrastructural level are composed of 10–16 nm filaments that radiate from a central core (Galloway et al., 1992). Other key features of PD are the selective loss or deficits in the activity of mitochondrial complex 1 and ubiquitin proteasomal dysfunction (Dawson and Dawson, 2003).

Thy-1 Promoter and Murine Prion Promoter α-Synuclein Transgenic Mice

As noted above, an ideal model for PD would recapitulate all the features of sporadic PD. Despite the fact that none of these α-synuclein transgenic mice fully recapitulate sporadic PD, they none-the-less represent powerful tools to investigate the toxicity of α-synuclein in vivo and the factors that modulate α-synuclein pathogenesis. Of the various α-synuclein transgenic mice that have been generated over the years, the models using the murine Thy-1 promoter and the murine prion promoter recapitulate most if not all the pathogenic features of PD, except for degeneration of dopamine neurons (van der Putten *et al.*, 2000; Giasson *et al.*, 2002; Lee *et al.*, 2002; Neumann *et al.*, 2002; Gispert *et al.*, 2003; Martin *et al.*, 2006). Both the Thy-1 and PrP promoters are pan-neuronal promoters that drive very high levels of expression in most neuronal populations. Transgenic mice overexpressing α-synuclein under either promoter develop α-synucleinopathy in neurons and overlapping brain regions with predominate pathology in spinal motor neurons, deep cerebellar nuclei, pontine reticular nuclei and the red nucleus. Thus, these neuronal populations are particularly vulnerable to α-synuclein-induced neurodegeneration in mice (van der Putten *et al.*, 2000; Giasson *et al.*, 2002; Lee *et al.*, 2002; Neumann *et al.*, 2002; Gispert *et al.*, 2003; Martin *et al.*, 2006). Interestingly, Braak and colleagues have suggested that similar brain stem structures are also preferentially vulnerable to the degenerative effects of α-synuclein in humans (Braak *et al.*, 2000). Thus, these models may recapitulate the early pathogenic features of PD, particularly with regards to the most vulnerable neuronal populations. These models are progressive and age-dependent. The mouse models exhibit mitochondrial dysfunction (Martin *et al.*, 2006), and both the mouse prion promoter and Thy-1 models are the only models with fibrillar α-synuclein inclusions in neurons. Thus, of all the α-synuclein models made to date, these two models utilizing different pan-neuronal promoters most closely recapitulate the pathogenic features of PD, particularly with regards to α-synuclein-induced degeneration.

Murine Prion Promoter Models of α-Synucleinopathies

Two independent groups have generated lines of transgenic mice expressing human α-synuclein under the direction of the murine prion promoter (Giasson *et al.*, 2002; Lee *et al.*, 2002). Mice expressing human A53T mutant α-synuclein develop an adult onset progressive neurodegenerative disorder characterized by ataxia, dystonia and reduced mobility (Giasson *et al.*, 2002; Lee *et al.*, 2002). Mid-life adult mice (8–16 months of age) exhibit neurologic symptoms and disease severity correlates with the expression level of mutant α-synuclein. Mice expressing A30P human α-synuclein or mice

expressing wild-type human α-synuclein at comparable or higher levels than the A53T α-synuclein generally do not develop a neurologic phenotype. Thus, human α-synuclein carrying the A53T mutation appears to have greater pathogenic potential than the other human α-synuclein variants. The human A53T α-synuclein transgenic mice exhibit abnormal neuronal accumulation of α-synuclein and ubiquitin. These inclusions are distributed throughout the nervous system, but they are abundant in subcortical regions such as the pons/medulla, cerebellar nuclei, spinal cord and midbrain. Thioflavin S-staining and immunoelectron microscopy indicates that these inclusion are fibrillar (Giasson *et al.*, 2002; Lee *et al.*, 2002). While no obvious pathology was observed in the striatonigral pathway, including normal striatal dopamine levels and normal dopamine neuron number (Giasson *et al.*, 2002), the human A53T α-synuclein mice develop significant hyperactivity that is dependent on dopaminergic neurotransmission due to increased sensitivity to dopamine D-1 receptors (Unger *et al.*, 2006). Thus, there are functional alterations in dopaminergic neurotransmission despite the lack of overt dopaminergic pathology. These mice exhibit overt neurodegeneration in non-dopaminergic structures since the most severely affected areas had positive TUNEL-staining, markers of caspase activation and loss of neurons (Martin *et al.*, 2006). The neuropathology in the human A53T α-synuclein transgenic mice is associated with increased levels of insoluble α-synuclein that consists of full length α-synuclein, truncated α-synuclein and higher molecular weight species of α-synuclein (Giasson *et al.*, 2002; Lee *et al.*, 2002). The overall pattern of α-synuclein biochemical abnormalities observed in the transgenic mice is very similar to that found in human α-synucleinopathies (Baba *et al.*, 1998; Li *et al.*, 2005). Overall, the murine prion promoter human α-synuclein transgenic mouse model recapitulates most of the features of human α-synucleinopathies (Table 17.2).

Modulation of α-Synuclein Pathology in α-Synuclein Transgenic Models

α-Synuclein is an abundant presynaptic phosphoprotein that is highly conserved. It exhibits an unstructured shape in solution, however in PD it aggregates into filaments and becomes insoluble. These insoluble filaments are hyperphosphorylated and ubiquitinated (Fujiwara *et al.*, 2002; Hasegawa *et al.*, 2002; Kahle *et al.*, 2002). Abnormal aggregation of α-synuclein into these toxic misfolded forms likely contributes to the neuronal death due to the missense mutated proteins and overexpression of wild-type α-synuclein in PD (Moore *et al.*, 2005a; Lee and Trojanowski, 2006). A variety of factors can influence the aggregation and folding of α-synuclein, including phosphorylation, mitochondrial and proteasomal dysfunction, oxidative and nitrosative stress and dopamine (Cookson, 2005). Unidentified synucleinases also process and cleave α-synuclein at its C terminus (Li *et al.*, 2005). The propensity of α-synuclein to oligomerize correlates with disease severity and truncation of α-synuclein (Li *et al.*, 2005). The toxic forms of α-synuclein appear to be the protofibril and fibril forms and the creation and stabilization of these forms, either due to the cellular context or by familial-associated mutations, may be the central pathogenic mechanism of α-synuclein-induced degeneration (Lee and Trojanowski, 2006). Thus, animal models which possess fibrillar inclusions of α-synuclein, such as the mouse prion promoter or Thy-1 transgenic models allows one to fully test the full pathogenic range of α-synuclein-induced degeneration.

With the availability of α-synuclein transgenic mouse models, investigators are able to address the question how α-synucleinopathy could be modulated *in vivo*. These experiments could lead to novel therapeutic strategies targeting these modulators. Masliah and colleagues showed that in PDGF promoter driven α-synuclein transgenic mice that have non-fibrillar inclusions (Masliah *et al.*, 2000) that, in α-synuclein and human APP (amyloid precursor protein) double transgenic mice, non-fibrillar α-synuclein inclusions were converted to fibrillar inclusions (Masliah *et al.*, 2001). The increase in fibrillar α-synuclein was associated with more synaptic degeneration and increased loss of dopaminergic markers in these double transgenic mice. Another potential modulator of α-synucleinopathy is β-synuclein. In vitro studies have shown that β-synuclein can inhibit aggregation of α-synuclein (Uversky *et al.*, 2002). In α-synuclein and β-synuclein double transgenic mice, β-synuclein ameliorated motor deficits, neurodegenerative alterations, and neuronal α-synuclein accumulation seen in α-synuclein transgenic mice (Hashimoto *et al.*, 2001). These results indicate that β-synuclein might function as a α-synucleinopathy inhibitor.

Recent studies suggest that α-synuclein cooperates with the cysteine-string protein (CSP) as a synaptic vesicle chaperone as mice overexpressing α-synuclein rescue the degenerative phenotype of deletion of the CSPα protein. Interesting these protective effects of α-synuclein seem to require the phospholipid-binding activity of α-synuclein. Thus, α-synuclein works with CSPs to protect nerve terminals against injury (Chandra *et al.*, 2005). The absence of parkin does not influence the pathology of the A53T α-synuclein transgenic mice since crossing parkin knockout mice to the A53T α-synuclein transgenic mice has no effect on multiple indices (Von Coelln *et al.*, 2006).

Some studies have investigated the effects of environmental factors that can modulate α-synucleinopathies (Norris *et al.*, 2007). Human α-synuclein A53T transgenic mice driven by mouse prion promoter that were treated with paraquat and maneb showed drastically increased α-synucleinopathy throughout the central nervous system. These abnormalities may be linked to oxidative stress induced modification of α-synuclein.

Utility of Models of α-Synucleinopathies for Drug Discovery

Models such as the murine Thy-1 or murine prion promoter α-synuclein transgenic mouse models are particularly relevant models that can be used to explore a variety of research questions in PD. These models have tremendous advantages over other models in PD research since they recapitulate most if not all the pathogenic features of PD. The only limitation of these models is the absence of dopaminergic neurodegeneration, but it is likely that α-synuclein injures neurons in similar ways. These are the only models that exhibit the full range of α-synuclein pathology and thus are suitable to answer most if not all questions related to α-synuclein pathogenesis.

CONCLUSIONS

Refinement of these models so that they exhibit progressive degeneration of dopamine neurons is an active area of investigation by numerous laboratories. New models of α-synucleinopathy-induced progressive degeneration of dopamine neurons will be important, but they should not hinder or distract from the importance of the current existing models. These α-synuclein transgenic models can be used to explore the molecular basis of cell death induced by α-synuclein. Further exploration of potential environmental contributions to PD pathogenesis can be explored and most importantly these models can be used to test and validate potential therapeutic targets. The most salient example of this is the use of immunotherapy, in which immunization of transgenic mice with α-synuclein has been shown to dramatically attenuate the disease phenotype in transgenic mice (Masliah *et al.*, 2005). It is likely as this approach gains greater acceptance and utility by the PD community that a number of interesting new therapeutic targets will emerge from the study of these mice.

ACKNOWLEDGEMENTS

This work was supported by the Morris K. Udall Parkinson's Disease Research Center of Excellence and National Institutes of Health-National Institute of Neurological Disorders and Stroke grants (NS 38377) and the Michael J. Fox Foundation. T.M.D. is the Leonard and Madlyn Abramson Professor in Neurodegenerative diseases.

REFERENCES

Aarsland, D., Tandberg, E., Larsen, J. P., and Cummings, J. L. (1996). Frequency of dementia in Parkinson disease. *Arch Neurol* 53, 538–542.

Andres-Mateos, E., Perier, C., Zhang, L., Blanchard-Fillion, B., Greco, T. M., Thomas, B., Ko, H. S., Sasaki, M., Ischiropoulos, H., Przedborski, S., Dawson, T. M., and Dawson, V. L. (2007). DJ-1 gene deletion reveals that DJ-1 is an atypical peroxiredoxin-like peroxidase. *Proc Natl Acad Sci USA* 104, 14807–14812.

Baba, M., Nakajo, S., Tu, P. H., Tomita, T., Nakaya, K., Lee, V. M., Trojanowski, J. Q., and Iwatsubo, T. (1998). Aggregation of alpha-synuclein in Lewy bodies of sporadic Parkinson's disease and dementia with Lewy bodies. *Am J Pathol* 152, 879–884.

Beilina, A., Van Der Brug, M., Ahmad, R., Kesavapany, S., Miller, D. W., Petsko, G. A., and Cookson, M. R. (2005). Mutations in PTEN-induced putative kinase 1 associated with recessive parkinsonism have differential effects on protein stability. *Proc Natl Acad Sci USA* 102, 5703–5708.

Biskup, S., Moore, D. J., Celsi, F., Higashi, S., West, A. B., Andrabi, S. A., Kurkinen, K., Yu, S. W., Savitt, J. M., Waldvogel, H. J., Faull, R. L., Emson, P. C., Torp, R., Ottersen, O. P., Dawson, T. M., and Dawson, V. L. (2006). Localization of LRRK2 to membranous and vesicular structures in mammalian brain. *Ann Neurol* 60, 557–569.

Bonifati, V., Rizzu, P., van Baren, M. J., Schaap, O., Breedveld, G. J., Krieger, E., Dekker, M. C., Squitieri, F., Ibanez, P., Joosse, M., van Dongen, J. W., Vanacore, N., van Swieten, J. C., Brice, A., Meco, G., van Duijn, C. M., Oostra, B. A., and Heutink, P. (2003). Mutations in the DJ-1 gene associated with autosomal recessive early-onset parkinsonism. *Science* 299, 256–259.

Braak, H., Del Tredici, K., Rub, U., de Vos, R. A., Jansen Steur, E. N., and Braak, E. (2003). Staging of brain pathology related to sporadic Parkinson's disease. *Neurobiol Aging* 24, 197–211.

Braak, H., Rub, U., Sandmann-Keil, D., Gai, W. P., de Vos, R. A., Jansen Steur, E. N., Arai, K., and Braak, E. (2000). Parkinson's disease: Affection of brain stem nuclei controlling premotor and motor neurons of the somatomotor system. *Acta Neuropathol (Berl)* 99, 489–495.

Chandra, S., Gallardo, G., Fernandez-Chacon, R., Schluter, O. M., and Sudhof, T. C. (2005). Alpha-synuclein cooperates with CSPalpha in preventing neurodegeneration. *Cell* 123, 383–396.

Chartier-Harlin, M. C., Kachergus, J., Roumier, C., Mouroux, V., Douay, X., Lincoln, S., Levecque, C., Larvor, L., Andrieux, J., Hulihan, M., Waucquier, N., Defebvre, L., Amouyel, P., Farrer, M., and Destee, A. (2004). Alpha-synuclein locus duplication as a cause of familial Parkinson's disease. *Lancet* 364, 1167–1169.

Chung, K. K., Thomas, B., Li, X., Pletnikova, O., Troncoso, J. C., Marsh, L., Dawson, V. L., and Dawson, T. M. (2004). S-nitrosylation of parkin regulates ubiquitination and compromises parkin's protective function. *Science* 304, 1328–1331.

Clark, I. E., Dodson, M. W., Jiang, C., Cao, J. H., Huh, J. R., Seol, J. H., Yoo, S. J., Hay, B. A., and Guo, M. (2006). Drosophila pink1 is required for mitochondrial function and interacts genetically with parkin. *Nature* 441, 1162–1166.

Cookson, M. R. (2005). The biochemistry of Parkinson's disease. *Annu Rev Biochem* 74, 29–52.

Corti, O., Hampe, C., Koutnikova, H., Darios, F., Jacquier, S., Prigent, A., Robinson, J. C., Pradier, L., Ruberg, M., Mirande, M., Hirsch, E., Rooney, T., Fournier, A., and Brice, A. (2003). The p38 subunit of the aminoacyl-tRNA synthetase complex is a Parkin substrate: Linking protein biosynthesis and neurode-generation. *Hum Mol Genet* 12, 1427–1437.

Dawson, T., Mandir, A., and Lee, M. (2002). Animal models of PD: Pieces of the same puzzle? *Neuron* 35, 219–222.

Dawson, T. M., and Dawson, V. L. (2003). Molecular pathways of neurodegeneration in Parkinson's disease. *Science* 302, 819–822.

Del Tredici, K., Rub, U., De Vos, R. A., Bohl, J. R., and Braak, H. (2002). Where does parkinson disease pathology begin in the brain? *J Neuropathol Exp Neurol* 61, 413–426.

Dubois, B., and Pillon, B. (1997). Cognitive deficits in Parkinson's disease. *J Neurol* 244, 2–8.

Dubois, B., Pillon, B., Sternic, N., Lhermitte, F., and Agid, Y. (1990). Age-induced cognitive disturbances in Parkinson's disease. *Neurology* 40, 38–41.

Forno, L. S. (1996). Neuropathology of Parkinson's disease. *J Neuropathol Exp Neurol* 55, 259–272.

Fujiwara, H., Hasegawa, M., Dohmae, N., Kawashima, A., Masliah, E., Goldberg, M. S., Shen, J., Takio, K., and Iwatsubo, T. (2002). Alpha-Synuclein is phosphorylated in synucleinopathy lesions. *Nat Cell Biol* 4, 160–164.

Funayama, M., Hasegawa, K., Kowa, H., Saito, M., Tsuji, S., and Obata, F. (2002). A new locus for Parkinson's disease (PARK8) maps to chromosome 12p11.2-q13.1. *Ann Neurol* 51, 296–301.

Galloway, P. G., Mulvihill, P., and Perry, G. (1992). Filaments of Lewy bodies contain insoluble cytoskeletal elements. *Am J Pathol* 140, 809–822.

Gasser, T., Muller-Myhsok, B., Wszolek, Z. K., Oehlmann, R., Calne, D. B., Bonifati, V., Bereznai, B., Fabrizio, E., Vieregge, P., and Horstmann, R. D. (1998). A susceptibility locus for Parkinson's disease maps to chromosome 2p13. *Nat Genet* 18, 262–265.

Giasson, B. I., Duda, J. E., Quinn, S. M., Zhang, B., Trojanowski, J. Q., and Lee, V. M. (2002). Neuronal alpha-synucleinopathy with severe movement disorder in mice expressing A53T human alpha-synuclein. *Neuron* 34, 521–533.

Gispert, S., Del Turco, D., Garrett, L., Chen, A., Bernard, D. J., Hamm-Clement, J., Korf, H. W., Deller, T., Braak, H., Auburger, G., and Nussbaum, R. L. (2003). Transgenic mice expressing mutant A53T human alpha-synuclein show neuronal dysfunction in the absence of aggregate formation. *Mol Cell Neurosci* 24, 419–429.

Goldberg, M. S., Fleming, S. M., Palacino, J. J., Cepeda, C., Lam, H. A., Bhatnagar, A., Meloni, E. G., Wu, N., Ackerson, L. C., Klapstein, G. J., Gajendiran, M., Roth, B. L., Chesselet, M. F., Maidment, N. T., Levine, M. S., and Shen, J. (2003). Parkin-deficient mice exhibit nigrostriatal deficits but not loss of dopaminergic neurons. *J Biol Chem* 278, 43628–43635.

Gomez-Isla, T., Irizarry, M. C., Mariash, A., Cheung, B., Soto, O., Schrump, S., Sondel, J., Kotilinek, L., Day, J., Schwarzschild, M. A., Cha, J. H., Newell, K., Miller, D. W., Ueda, K., Young, A. B., Hyman, B. T., and Ashe, K. H. (2003). Motor dysfunction and gliosis with preserved dopaminergic markers in human alpha-synuclein A30P transgenic mice. *Neurobiol Aging* 24, 245–258.

Greene, J. C., Whitworth, A. J., Kuo, I., Andrews, L. A., Feany, M. B., and Pallanck, L. J. (2003). Mitochondrial pathology and apoptotic muscle degeneration in Drosophila parkin mutants. *Proc Natl Acad Sci USA* 100, 4078–4083.

Greggio, E., Lewis, P. A., van der Brug, M. P., Ahmad, R., Kaganovich, A., Ding, J., Beilina, A., Baker, A. K., and Cookson, M. R. (2007). Mutations in LRRK2/dardarin associated with Parkinson disease are more toxic than equivalent mutations in the homologous kinase LRRK1. *J Neurochem* 102, 93–102.

Hasegawa, M., Fujiwara, H., Nonaka, T., Wakabayashi, K., Takahashi, H., Lee, V. M., Trojanowski, J. Q., Mann, D., and Iwatsubo, T. (2002). Phosphorylated alpha-synuclein is ubiquitinated in alpha-synucleinopathy lesions. *J Biol Chem* 277, 49071–49076.

Hashimoto, M., Rockenstein, E., Mante, M., Mallory, M., and Masliah, E. (2001). Beta-Synuclein inhibits alpha-synuclein aggregation: A possible role as an antiparkinsonian factor. *Neuron* 32, 213–223.

Hashimoto, M., Rockenstein, E., and Masliah, E. (2003). Transgenic models of alpha-synuclein pathology: Past, present, and future. *Ann NY Acad Sci* 991, 171–188.

Hicks, A. A., Petursson, H., Jonsson, T., Stefansson, H., Johannsdottir, H. S., Sainz, J., Frigge, M. L., Kong, A., Gulcher, J. R., Stefansson, K., and Sveinbjornsdottir, S. (2002). A susceptibility gene for late-onset idiopathic Parkinson's disease. *Ann Neurol* 52, 549–555.

Higashi, S., Biskup, S., West, A. B., Trinkaus, D., Dawson, V. L., Faull, R. L., Waldvogel, H. J., Arai, H., Dawson, T. M., Moore, D. J., and Emson, P. C. (2007a). Localization of Parkinson's disease-associated LRRK2 in normal and pathological human brain. *Brain Res* 1155, 208–219.

Higashi, S., Moore, D. J., Colebrooke, R. E., Biskup, S., Dawson, V. L., Arai, H., Dawson, T. M., and Emson, P. C. (2007b). Expression and localization of Parkinson's disease-associated leucine-rich repeat kinase 2 in the mouse brain. *J Neurochem* 100, 368–381.

Hope, A. D., Myhre, R., Kachergus, J., Lincoln, S., Bisceglio, G., Hulihan, M., and Farrer, M. J. (2004). Alpha-synuclein missense and multiplication mutations in autosomal dominant Parkinson's disease. *Neurosci Lett* 367, 97–100.

Kahle, P. J., Neumann, M., Ozmen, L., Muller, V., Jacobsen, H., Spooren, W., Fuss, B., Mallon, B.,

Macklin, W. B., Fujiwara, H., Hasegawa, M., Iwatsubo, T., Kretzschmar, H. A., and Haass, C. (2002). Hyperphosphorylation and insolubility of alpha-synuclein in transgenic mouse oligodendrocytes. *EMBO Rep* 3, 583–588.

Kitada, T., Asakawa, S., Hattori, N., Matsumine, H., Yamamura, Y., Minoshima, S., Yokochi, M., Mizuno, Y., and Shimizu, N. (1998). Mutations in the parkin gene cause autosomal recessive juvenile parkinsonism. *Nature* 392, 605–608.

Kitada, T., Pisani, A., Porter, D. R., Yamaguchi, H., Tscherter, A., Martella, G., Bonsi, P., Zhang, C., Pothos, E. N., and Shen, J. (2007). Impaired dopamine release and synaptic plasticity in the striatum of PINK1-deficient mice. *Proc Natl Acad Sci USA* 104, 11441–11446.

Ko, H. S., Kim, S. W., Sriram, S. R., Dawson, V. L., and Dawson, T. M. (2006). Identification of far upstream element-binding protein-1 as an authentic Parkin substrate. *J Biol Chem* 281, 16193–16196.

Ko, H. S., von Coelln, R., Sriram, S. R., Kim, S. W., Chung, K. K., Pletnikova, O., Troncoso, J., Johnson, B., Saffary, R., Goh, E. L., Song, H., Park, B. J., Kim, M. J., Kim, S., Dawson, V. L., and Dawson, T. M. (2005). Accumulation of the authentic parkin substrate aminoacyl-tRNA synthetase cofactor, p38/JTV-1, leads to catecholaminergic cell death. *J Neurosci* 25, 7968–7978.

Kruger, R., Kuhn, W., Muller, T., Woitalla, D., Graeber, M., Kosel, S., Przuntek, H., Epplen, J. T., Schols, L., and Riess, O. (1998). Ala30Pro mutation in the gene encoding alpha-synuclein in Parkinson's disease. *Nat Genet* 18, 106–108.

Lee, M. K., Stirling, W., Xu, Y., Xu, X., Qui, D., Mandir, A. S., Dawson, T. M., Copeland, N. G., Jenkins, N. A., and Price, D. L. (2002). Human alpha-synuclein-harboring familial Parkinson's disease-linked Ala-53 --> Thr mutation causes neurodegenerative disease with alpha-synuclein aggregation in transgenic mice. *Proc Natl Acad Sci USA* 99, 8968–8973.

Lee, V. M., and Trojanowski, J. Q. (2006). Mechanisms of Parkinson's disease linked to pathological alpha-synuclein: New targets for drug discovery. *Neuron* 52, 33–38.

Leroy, E., Boyer, R., Auburger, G., Leube, B., Ulm, G., Mezey, E., Harta, G., Brownstein, M. J., Jonnalagada, S., Chernova, T., Dehejia, A., Lavedan, C., Gasser, T., Steinbach, P. J., Wilkinson, K. D., and Polymeropoulos, M. H. (1998). The ubiquitin pathway in Parkinson's disease. *Nature* 395, 451–452.

Lewy, F. H. (1912). *Handbuch der Neurologie Band III* (M. Lewandowski, Ed.), Springer-Verlag, Berlin, pp. 920–933.

Li, W., West, N., Colla, E., Pletnikova, O., Troncoso, J. C., Marsh, L., Dawson, T. M., Jakala, P., Hartmann, T.,

Price, D. L., and Lee, M. K. (2005). Aggregation promoting C-terminal truncation of alpha-synuclein is a normal cellular process and is enhanced by the familial Parkinson's disease-linked mutations. *Proc Natl Acad Sci USA* 102, 2162–2167.

Maraganore, D. M., de Andrade, M., Elbaz, A., Farrer, M. J., Ioannidis, J. P., Kruger, R., Rocca, W. A., Schneider, N. K., Lesnick, T. G., Lincoln, S. J., Hulihan, M. M., Aasly, J. O., Ashizawa, T., Chartier-Harlin, M. C., Checkoway, H., Ferrarese, C., Hadjigeorgiou, G., Hattori, N., Kawakami, H., Lambert, J. C., Lynch, T., Mellick, G. D., Papapetropoulos, S., Parsian, A., Quattrone, A., Riess, O., Tan, E. K., and Van Broeckhoven, C. (2006). Collaborative analysis of alpha-synuclein gene promoter variability and Parkinson disease. *JAMA* 296, 661–670.

Martin, L. J., Pan, Y., Price, A. C., Sterling, W., Copeland, N. G., Jenkins, N. A., Price, D. L., and Lee, M. K. (2006). Parkinson's disease alpha-synuclein transgenic mice develop neuronal mitochondrial degeneration and cell death. *J Neurosci* 26, 41–50.

Masliah, E., Rockenstein, E., Adame, A., Alford, M., Crews, L., Hashimoto, M., Seubert, P., Lee, M., Goldstein, J., Chilcote, T., Games, D., and Schenk, D. (2005). Effects of alpha-synuclein immunization in a mouse model of Parkinson's disease. *Neuron* 46, 857–868.

Masliah, E., Rockenstein, E., Veinbergs, I., Mallory, M., Hashimoto, M., Takeda, A., Sagara, Y., Sisk, A., and Mucke, L. (2000). Dopaminergic loss and inclusion body formation in alpha-synuclein mice: Implications for neurodegenerative disorders. *Science* 287, 1265–1269.

Masliah, E., Rockenstein, E., Veinbergs, I., Sagara, Y., Mallory, M., Hashimoto, M., and Mucke, L. (2001). Beta-amyloid peptides enhance alpha-synuclein accumulation and neuronal deficits in a transgenic mouse model linking Alzheimer's disease and Parkinson's disease. *Proc Natl Acad Sci USA* 98, 12245–12250.

Matsuoka, Y., Vila, M., Lincoln, S., McCormack, A., Picciano, M., LaFrancois, J., Yu, X., Dickson, D., Langston, W. J., McGowan, E., Farrer, M., Hardy, J., Duff, K., Przedborski, S., and Di Monte, D. A. (2001). Lack of nigral pathology in transgenic mice expressing human alpha-synuclein driven by the tyrosine hydroxylase promoter. *Neurobiol Dis* 8, 535–539.

Meulener, M. C., Xu, K., Thomson, L., Ischiropoulos, H., and Bonini, N. M. (2006). Mutational analysis of DJ-1 in Drosophila implicates functional inactivation by oxidative damage and aging. *Proc Natl Acad Sci USA* 103, 12517–12522.

Moore, D. J., West, A. B., Dawson, V. L., and Dawson, T. M. (2005a). Molecular pathophysiology of Parkinson's disease. *Annu Rev Neurosci* 28, 57–87.

Moore, D. J., Zhang, L., Troncoso, J., Lee, M. K., Hattori, N., Mizuno, Y., Dawson, T. M., and Dawson, V. L. (2005b). Association of DJ-1 and parkin mediated by pathogenic DJ-1 mutations and oxidative stress. *Hum Mol Genet* **14**, 71–84.

Neumann, M., Kahle, P. J., Giasson, B. I., Ozmen, L., Borroni, E., Spooren, W., Muller, V., Odoy, S., Fujiwara, H., Hasegawa, M., Iwatsubo, T., Trojanowski, J. Q., Kretzschmar, H. A., and Haass, C. (2002). Misfolded proteinase K-resistant hyperphosphorylated alpha-synuclein in aged transgenic mice with locomotor deterioration and in human alpha-synucleinopathies. *J Clin Invest* **110**, 1429–1439.

Nishioka, K., Hayashi, S., Farrer, M. J., Singleton, A. B., Yoshino, H., Imai, H., Kitami, T., Sato, K., Kuroda, R., Tomiyama, H., Mizoguchi, K., Murata, M., Toda, T., Imoto, I., Inazawa, J., Mizuno, Y., and Hattori, N. (2006). Clinical heterogeneity of alpha-synuclein gene duplication in Parkinson's disease. *Ann Neurol* **59**, 298–309.

Norris, E. H., Uryu, K., Leight, S., Giasson, B. I., Trojanowski, J. Q., and Lee, V. M. (2007). Pesticide exposure exacerbates alpha-synucleinopathy in an A53T transgenic mouse model. *Am J Pathol* **170**, 658–666.

Olzmann, J. A., Brown, K., Wilkinson, K. D., Rees, H. D., Huai, Q., Ke, H., Levey, A. I., Li, L., and Chin, L. S. (2004). Familial Parkinson's disease-associated L166P mutation disrupts DJ-1 protein folding and function. *J Biol Chem* **279**, 8506–8515.

Paisan-Ruiz, C., Jain, S., Evans, E. W., Gilks, W. P., Simon, J., van der Brug, M., Lopez de Munain, A., Aparicio, S., Gil, A. M., Khan, N., Johnson, J., Martinez, J. R., Nicholl, D., Carrera, I. M., Pena, A. S., de Silva, R., Lees, A., Marti-Masso, J. F., Perez-Tur, J., Wood, N. W., and Singleton, A. B. (2004). Cloning of the gene containing mutations that cause PARK8-linked Parkinson's disease. *Neuron* **44**, 595–600.

Pankratz, N., Nichols, W. C., Uniacke, S. K., Halter, C., Murrell, J., Rudolph, A., Shults, C. W., Conneally, P. M., and Foroud, T. (2003a). Genome-wide linkage analysis and evidence of gene-by-gene interactions in a sample of 362 multiplex Parkinson disease families. *Hum Mol Genet* **12**, 2599–2608.

Pankratz, N., Nichols, W. C., Uniacke, S. K., Halter, C., Rudolph, A., Shults, C., Conneally, P. M., and Foroud, T. (2003b). Significant linkage of Parkinson disease to chromosome 2q36-37. *Am J Hum Genet* **72**, 1053–1057.

Park, J., Lee, S. B., Lee, S., Kim, Y., Song, S., Kim, S., Bae, E., Kim, J., Shong, M., Kim, J. M., and Chung, J. (2006). Mitochondrial dysfunction in Drosophila PINK1 mutants is complemented by parkin. *Nature* **441**, 1157–1161.

Perez, F. A., and Palmiter, R. D. (2005). Parkin-deficient mice are not a robust model of parkinsonism. *Proc Natl Acad Sci USA* **102**, 2174–2179.

Polymeropoulos, M. H., Higgins, J. J., Golbe, L. I., Johnson, W. G., Ide, S. E., Di Iorio, G., Sanges, G., Stenroos, E. S., Pho, L. T., Schaffer, A. A., Lazzarini, A. M., Nussbaum, R. L., and Duvoisin, R. C. (1996). Mapping of a gene for Parkinson's disease to chromosome 4q21-q23. *Science* **274**, 1197–1199.

Polymeropoulos, M. H., Lavedan, C., Leroy, E., Ide, S. E., Dehejia, A., Dutra, A., Pike, B., Root, H., Rubenstein, J., Boyer, R., Stenroos, E. S., Chandrasekharappa, S., Athanassiadou, A., Papapetropoulos, T., Johnson, W. G., Lazzarini, A. M., Duvoisin, R. C., Di Iorio, G., Golbe, L. I., and Nussbaum, R. L. (1997). Mutation in the alpha-synuclein gene identified in families with Parkinson's disease. *Science* **276**, 2045–2047.

Rajput, A., Dickson, D. W., Robinson, C. A., Ross, O. A., Dachsel, J. C., Lincoln, S. J., Cobb, S. A., Rajput, M. L., and Farrer, M. J. (2006). Parkinsonism, Lrrk2 G2019S, and tau neuropathology. *Neurology* **67**, 1506–1508.

Ramirez, A., Heimbach, A., Grundemann, J., Stiller, B., Hampshire, D., Cid, L. P., Goebel, I., Mubaidin, A. F., Wriekat, A. L., Roeper, J., Al-Din, A., Hillmer, A. M., Karsak, M., Liss, B., Woods, C. G., Behrens, M. I., and Kubisch, C. (2006). Hereditary parkinsonism with dementia is caused by mutations in ATP13A2, encoding a lysosomal type 5 P-type ATPase. *Nat Genet* **38**, 1184–1191.

Richfield, E. K., Thiruchelvam, M. J., Cory-Slechta, D. A., Wuertzer, C., Gainetdinov, R. R., Caron, M. G., Di Monte, D. A., and Federoff, H. J. (2002). Behavioral and neurochemical effects of wild-type and mutated human alpha-synuclein in transgenic mice. *Exp Neurol* **175**, 35–48.

Ross, O. A., Toft, M., Whittle, A. J., Johnson, J. L., Papapetropoulos, S., Mash, D. C., Litvan, I., Gordon, M. F., Wszolek, Z. K., Farrer, M. J., and Dickson, D. W. (2006). Lrrk2 and Lewy body disease. *Ann Neurol* **59**, 388–393.

Savitt, J. M., Dawson, V. L., and Dawson, T. M. (2006). Diagnosis and treatment of Parkinson disease: Molecules to medicine. *J Clin Invest* **116**, 1744–1754.

Shendelman, S., Jonason, A., Martinat, C., Leete, T., and Abeliovich, A. (2004). DJ-1 is a redox-dependent molecular chaperone that inhibits alpha-synuclein aggregate formation. *PLoS Biol* **2**, e362.

Shimura, H., Hattori, N., Kubo, S., Mizuno, Y., Asakawa, S., Minoshima, S., Shimizu, N., Iwai, K., Chiba, T., Tanaka, K., and Suzuki, T. (2000). Familial Parkinson disease gene product, parkin, is a ubiquitin-protein ligase. *Nat Genet* **25**, 302–305.

Shults, C. W., Rockenstein, E., Crews, L., Adame, A., Mante, M., Larrea, G., Hashimoto, M., Song, D., Iwatsubo, T., Tsuboi, K., and Masliah, E. (2005). Neurological and neurodegenerative alterations in a transgenic mouse model expressing human alpha-synuclein under oligodendrocyte promoter: Implications for multiple system atrophy. *J Neurosci* **25**, 10689–10699.

Silvestri, L., Caputo, V., Bellacchio, E., Atorino, L., Dallapiccola, B., Valente, E. M., and Casari, G. (2005). Mitochondrial import and enzymatic activity of PINK1 mutants associated to recessive parkinsonism. *Hum Mol Genet* **14**, 3477–3492.

Singleton, A. B., Farrer, M., Johnson, J., Singleton, A., Hague, S., Kachergus, J., Hulihan, M., Peuralinna, T., Dutra, A., Nussbaum, R., Lincoln, S., Crawley, A., Hanson, M., Maraganore, D., Adler, C., Cookson, M. R., Muenter, M., Baptista, M., Miller, D., Blancato, J., Hardy, J., and Gwinn-Hardy, K. (2003). Alpha-Synuclein locus triplication causes Parkinson's disease. *Science* **302**, 841.

Smith, W. W., Pei, Z., Jiang, H., Dawson, V. L., Dawson, T. M., and Ross, C. A. (2006). Kinase activity of mutant LRRK2 mediates neuronal toxicity. *Nat Neurosci* **9**, 1231–1233.

Spillantini, M. G., Crowther, R. A., Jakes, R., Hasegawa, M., and Goedert, M. (1998). Alpha-Synuclein in filamentous inclusions of Lewy bodies from Parkinson's disease and dementia with lewy bodies. *Proc Natl Acad Sci USA* **95**, 6469–6473.

Spillantini, M. G., Schmidt, M. L., Lee, V. M., Trojanowski, J. Q., Jakes, R., and Goedert, M. (1997). Alpha-synuclein in Lewy bodies. *Nature* **388**, 839–840.

Strauss, K. M., Martins, L. M., Plun-Favreau, H., Marx, F. P., Kautzmann, S., Berg, D., Gasser, T., Wszolek, Z., Muller, T., Bornemann, A., Wolburg, H., Downward, J., Riess, O., Schulz, J. B., and Kruger, R. (2005). Loss of function mutations in the gene encoding Omi/HtrA2 in Parkinson's disease. *Hum Mol Genet* **14**, 2099–2111.

Thiruchelvam, M. J., Powers, J. M., Cory-Slechta, D. A., and Richfield, E. K. (2004). Risk factors for dopaminergic neuron loss in human alpha-synuclein transgenic mice. *Eur J Neurosci* **19**, 845–854.

Tofaris, G. K., Garcia Reitbock, P., Humby, T., Lambourne, S. L., O'Connell, M., Ghetti, B., Gossage, H., Emson, P. C., Wilkinson, L. S., Goedert, M., and Spillantini, M. G. (2006). Pathological changes in dopaminergic nerve cells of the substantia nigra and olfactory bulb in mice transgenic for truncated human alpha-synuclein(1–120): Implications for Lewy body disorders. *J Neurosci* **26**, 3942–3950.

Unger, E. L., Eve, D. J., Perez, X. A., Reichenbach, D. K., Xu, Y., Lee, M. K., and Andrews, A. M. (2006). Locomotor hyperactivity and alterations in dopamine neurotransmission are associated with overexpression of A53T mutant human alpha-synuclein in mice. *Neurobiol Dis* **21**, 431–443.

Uversky, V. N., Li, J., Souillac, P., Millett, I. S., Doniach, S., Jakes, R., Goedert, M., and Fink, A. L. (2002). Biophysical properties of the synucleins and their propensities to fibrillate: Inhibition of alpha-synuclein assembly by beta- and gamma-synucleins. *J Biol Chem* **277**, 11970–11978.

Valente, E. M., Abou-Sleiman, P. M., Caputo, V., Muqit, M. M., Harvey, K., Gispert, S., Ali, Z., Del Turco, D., Bentivoglio, A. R., Healy, D. G., Albanese, A., Nussbaum, R., Gonzalez-Maldonado, R., Deller, T., Salvi, S., Cortelli, P., Gilks, W. P., Latchman, D. S., Harvey, R. J., Dallapiccola, B., Auburger, G., and Wood, N. W. (2004). Hereditary early-onset Parkinson's disease caused by mutations in PINK1. *Science* **304**, 1158–1160.

van der Putten, H., Wiederhold, K. H., Probst, A., Barbieri, S., Mistl, C., Danner, S., Kauffmann, S., Hofele, K., Spooren, W. P., Ruegg, M. A., Lin, S., Caroni, P., Sommer, B., Tolnay, M., and Bilbe, G. (2000). Neuropathology in mice expressing human alpha-synuclein. *J Neurosci* **20**, 6021–6029.

Von Coelln, R., Thomas, B., Andrabi, S. A., Lim, K. L., Savitt, J. M., Saffary, R., Stirling, W., Bruno, K., Hess, E. J., Lee, M. K., Dawson, V. L., and Dawson, T. M. (2006). Inclusion body formation and neurodegeneration are parkin independent in a mouse model of alpha-synucleinopathy. *J Neurosci* **26**, 3685–3696.

Von Coelln, R., Thomas, B., Savitt, J. M., Lim, K. L., Sasaki, M., Hess, E. J., Dawson, V. L., and Dawson, T. M. (2004). Loss of locus coeruleus neurons and reduced startle in parkin null mice. *Proc Natl Acad Sci USA* **101**, 10744–10749.

West, A. B., Moore, D. J., Biskup, S., Bugayenko, A., Smith, W. W., Ross, C. A., Dawson, V. L., and Dawson, T. M. (2005). Parkinson's disease-associated mutations in leucine-rich repeat kinase 2 augment kinase activity. *Proc Natl Acad Sci USA* **102**, 16842–16847.

West, A. B., Moore, D. J., Choi, C., Andrabi, S. A., Li, X., Dikeman, D., Biskup, S., Zhang, Z., Lim, K. L., Dawson, V. L., and Dawson, T. M. (2007). Parkinson's disease-associated mutations in LRRK2 link enhanced GTP-binding and kinase activities to neuronal toxicity. *Hum Mol Genet* **16**, 223–232.

Whaley, N. R., Uitti, R. J., Dickson, D. W., Farrer, M. J., and Wszolek, Z. K. (2006). Clinical and pathologic features of families with LRRK2-associated Parkinson's disease. *J Neural Transm Suppl*, 221–229.

Winkler, S., Hagenah, J., Lincoln, S., Heckman, M., Haugarvoll, K., Lohmann-Hedrich, K., Kostic, V., Farrer, M., and Klein, C. (2007). Alpha-synuclein and Parkinson disease susceptibility. *Neurology*.

Yazawa, I., Giasson, B. I., Sasaki, R., Zhang, B., Joyce, S., Uryu, K., Trojanowski, J. Q., and Lee, V. M. (2005). Mouse model of multiple system atrophy alpha-synuclein expression in oligodendrocytes causes glial and neuronal degeneration. *Neuron* **45**, 847–859.

Zarranz, J. J., Alegre, J., Gomez-Esteban, J. C., Lezcano, E., Ros, R., Ampuero, I., Vidal, L., Hoenicka, J., Rodriguez, O., Atares, B., Llorens, V., Gomez Tortosa, E., del Ser, T., Munoz, D. G., and de Yebenes, J. G. (2004). The new mutation, E46K, of alpha-synuclein causes Parkinson and Lewy body dementia. *Ann Neurol* 55, 164–173.

Zhang, Y., Gao, J., Chung, K. K., Huang, H., Dawson, V. L., and Dawson, T. M. (2000). Parkin functions as an E2-dependent ubiquitin- protein ligase and promotes the degradation of the synaptic vesicle-associated protein, CDCrel-1. *Proc Natl Acad Sci USA* 97, 13354–13359.

Zimprich, A., Biskup, S., Leitner, P., Lichtner, P., Farrer, M., Lincoln, S., Kachergus, J., Hulihan, M., Uitti, R. J., Calne, D. B., Stoessl, A. J., Pfeiffer, R. F., Patenge, N., Carbajal, I. C., Vieregge, P., Asmus, F., Muller-Myhsok, B., Dickson, D. W., Meitinger, T., Strom, T. M., Wszolek, Z. K., and Gasser, T. (2004). Mutations in LRRK2 cause autosomal-dominant parkinsonism with pleomorphic pathology. *Neuron* 44, 601–607.

18

THE DYNAMICS OF α-SYNUCLEIN AT THE NERVE TERMINAL

DORIS L. FORTIN[1], VENU M. NEMANI[2] AND ROBERT H. EDWARDS[2]

[1]Department of Molecular and Cell Biology, University of California, Berkeley, CA, USA
[2]Departments of Neurology and Physiology, University of California, San Francisco, CA, USA

INTRODUCTION

Considerable evidence has implicated α-synuclein in the pathogenesis of Parkinson's disease (PD). Mutations in the gene encoding α-synuclein, largely single amino acid substitutions (A30P, A53T and E46K), have been linked to rare cases of familial PD (Polymeropoulos *et al.*, 1997; Kruger *et al.*, 1998; Zarranz *et al.*, 2004). An increase in the dosage of the wild-type α-synuclein gene, due to duplication or triplication of the chromosomal locus, also appears to cause the disease (Singleton *et al.*, 2003; Chartier-Harlin *et al.*, 2004; Ibanez *et al.*, 2004). In addition, genetic variation in the promoter of the α-synuclein gene, which may lead to changes in α-synuclein expression, has been associated with sporadic cases of PD (Pals *et al.*, 2004). Even in the absence of genetic changes, α-synuclein is an abundant protein component of Lewy bodies and dystrophic neurites (Spillantini *et al.*, 1997, 1998; Galvin *et al.*, 1999), further supporting an important role for α-synuclein in most, if not all, cases of PD. Despite considerable research, the precise molecular mechanisms underlying the role of α-synuclein in PD and indeed the normal function of the protein remain unclear.

As it can be seen in this book, experimental models of PD can adopt various forms. Typically PD models try to recapitulate either the neuropathological or clinical hallmarks of the disease.

Nonetheless, we believe that PD models should not be restricted to this conception as useful models may also be developed to acquire insight into the normal neurobiology of known etiologic factors. The apparent central role of alpha-syuclein in the pathogenesis of PD makes this protein not only an attractive therapeutic target, but underscores the need to understand its role in normal and pathologic physiology.

WHAT DO WE KNOW ABOUT SYNUCLEIN?

α-Synuclein is a small neuronal protein implicated in synaptic plasticity, neurotransmitter release and synaptic vesicle recycling (George *et al.*, 1995; Abeliovich *et al.*, 2000; Murphy *et al.*, 2000; Cabin *et al.*, 2002; Liu *et al.*, 2004; Yavich *et al.*, 2004; Larsen *et al.*, 2006). Consistent with these proposed functions, α-synuclein localizes to the presynaptic terminal of mature neurons (Maroteaux *et al.*, 1988; Withers *et al.*, 1997; Murphy *et al.*, 2000). However, the expression of α-synuclein and its accumulation at synapses is delayed compared to other synaptic proteins (Withers *et al.*, 1997; Murphy *et al.*, 2000), suggesting that α-synuclein is not involved in synapse formation but rather in the maintenance or regulation of existing synapses. A direct association with synaptic vesicles was the basis for the original discovery of α-synuclein (Maroteaux *et al.*, 1988).

However, standard biochemical fractionation methods typically fail to isolate α-synuclein in particulate or membrane fractions and rather indicate that α-synuclein behaves as a soluble protein of the nerve terminal (Jakes et al., 1994; George et al., 1995; Iwai, 2000; Kahle et al., 2000). This suggests that the interaction between α-synuclein and synaptic vesicles is weak and easily reversible. Since α-synuclein does not contain a transmembrane domain or a consensus sequence for lipid anchor, it presumably relies on direct protein/protein or protein/lipid interactions to achieve its synaptic enrichment *in vivo*.

MEMBRANE BINDING *IN VITRO*

In vitro, α-synuclein binds directly to artificial membranes, especially small liposomes containing acidic phospholipid (Davidson et al., 1998; Jo et al., 2000; Eliezer et al., 2001; Chandra et al., 2003; Kubo et al., 2005). Upon binding to liposomes, α-synuclein undergoes a major conformational change, from relatively unstructured (Weinreb et al., 1996) to α-helical, forming an amphipathic α-helix that mediates binding to membranes (Davidson et al., 1998; Eliezer et al., 2001). The N-terminus of α-synuclein contains seven imperfect KTKEGV repeats that form the core of the phospholipid-binding domain, with individual lysines predicted to interact directly with negatively charged phospholipid head groups. The direct association between lysines and negatively charged head groups presumably accounts for the marked preference for membranes containing acidic phospholipid (Davidson et al., 1998; Eliezer et al., 2001; Bussell and Eliezer, 2003; Bisaglia et al., 2006). A reported preference for small liposomes, which display highly curved membranes, may reflect the particular topology of the amphipathic α-helix adopted by α-synuclein (Bussell and Eliezer, 2003; Chandra et al., 2003; Rhoades et al., 2006).

The amphipathic nature of the α-helix formed by α-synuclein in the presence of membranes further suggests that hydrophobic contacts between the protein and fatty acid side chains play an important role in the interaction of α-synuclein and membranes (Davidson et al., 1998; Bisaglia et al., 2005; Kim et al., 2006). This is consistent with the stability of α-synuclein/membrane complexes in high salt concentration (Davidson et al., 1998; Li et al., 2001; Lee et al., 2002; Ramakrishnan et al., 2003).

In addition, it has been shown that α-synuclein interacts directly with polyunsaturated fatty acids (Perrin et al., 2001; Sharon et al., 2001, 2003), and suggested that the protein shares features with fatty acid-binding proteins (Sharon et al., 2001 but also see Lucke et al., 2006). A prominent role for hydrophobic interactions is also supported by the influence of phase separation, that is, the formation of microdomains known as lipid rafts, in the binding of α-synuclein to different membranes (Fortin et al., 2004; Kubo et al., 2005). In contrast, recent work has led to the proposal that the role of lipid rafts may be limited to the concentration of acidic phospholipids with α-synuclein binding to membranes via electrostatic interactions exclusively (Rhoades et al., 2006). However, the use of large liposomes in these studies precludes a direct comparison with previously published data obtained using highly curved membranes. Nonetheless, these findings suggest that α-synuclein utilizes two different modalities for association with membranes that can be independently recruited depending on the particular type of membranes encountered.

PD-Associated Mutations and Membrane Binding

The PD-associated mutations A30P, A53T and E46K are located within the region of α-synuclein that forms the amphipathic α-helix and may therefore influence the interaction of the protein with artificial membranes. Indeed, it was shown that A30P and to a lesser extent A53T, disrupt the helical propensity in the N-terminus of free monomeric α-synuclein (Bussell and Eliezer, 2001). Residual helical propensity, which can be found in natively unfolded proteins, may be important in early intramolecular protein folding events that are necessary for membrane binding. Surprisingly, both A30P and A53T do not significantly influence the structure of lipid-associated α-synuclein (Bussell and Eliezer, 2004), suggesting that membrane binding overcomes A30P- and A53T-induced changes in the structure of soluble α-synuclein. Binding to artificial membranes is increased by the E46K mutation (Choi et al., 2004), unchanged for A53T and may or may not be affected by A30P (Perrin et al., 2000; Jo et al., 2002; Bussell and Eliezer, 2004). In contrast, several studies have demonstrated that A30P greatly diminishes binding of α-synuclein to lipid rafts isolated from cells (Fortin et al., 2004; Kubo et al., 2005) and disrupts association with native membranes such as synaptic vesicles, retrograde transport vesicles and yeast membranes (Jensen *et al.*, 1998; Cole *et al.*, 2002; Outeiro and Lindquist, 2003; Kim *et al.*, 2006; Wislet-Gendebien *et al.*, 2006). Binding of α-synuclein to native membranes may thus involve more stringent requirements than binding to artificial membranes, accounting for the unequivocal demonstration that A30P impairs association with membranes.

MEMBRANE INTERACTIONS *IN VIVO*

Synaptic Localization of α-Synuclein: Role of Membranes

The expression of α-synuclein occurs relatively early during neuronal development and its enrichment at the synapse has been described extensively (Withers *et al.*, 1997; Murphy *et al.*, 2000). Overexpression of α-synuclein as a fusion to the green fluorescent protein (GFP) does not disrupt the synaptic enrichment of endogenous α-synuclein in cultured hippocampal neurons (Fortin *et al.*, 2004). Furthermore, GFP α-synuclein fusion proteins localize to the nerve terminal (Fortin *et al.*, 2004, 2005; Specht *et al.*, 2005), making them a useful tool to study the protein in a cellular context. Mutational analysis has revealed that the domain responsible for synaptic localization is contained within amino acids 1–102 of α-synuclein (Specht *et al.*, 2005), similar to what has been reported for the interaction with liposomes containing acidic phospholipids *in vitro* (Perrin *et al.*, 2000). Acute or chronic disruption of lipid rafts reduces the synaptic enrichment of GFP-α-synuclein by redistributing the protein to the axon (Fortin *et al.*, 2004), supporting a role for membrane microdomains in the synaptic localization of α-synuclein. Similarly, the A30P mutation, which has been shown to disrupt interaction with raft-like liposomes (Kubo *et al.*, 2005), also disrupts the steady-state localization of GFP-α-synuclein in transfected neurons (Fortin *et al.*, 2004). Thus, the synaptic localization of α-synuclein appears to rely on the direct interaction with highly curved membranes containing microdomains enriched in acidic phospholipids. At the synapse, these membranes are likely to be synaptic vesicles. However, it has recently been suggested that transmembrane proteins dominate the external structure of synaptic vesicles, possibly restricting access to phospholipid

from the cytoplasm (Takamori *et al.*, 2006). Thus, the mechanism by which α-synuclein, a peripheral membrane protein, finds its phospholipid-binding partner(s) in the nerve terminal remains unclear.

Transient Interactions Revealed by Fluorescence Recovery After Photobleaching

In contrast to most biochemical assays, live cell imaging provides real-time information about dynamic cellular processes. For instance, fluorescence recovery after photobleaching (FRAP) can be used to characterize the behavior of a subpopulation of fluorescently tagged protein. After photobleaching a fraction of the fluorescent protein, the movement of unbleached molecules into the bleached area is monitored. The kinetics of fluorescence recovery can be determined and provide quantitative information about behavior of the protein (Axelrod *et al.*, 1976; Phair and Misteli, 2001; Lippincott-Schwartz *et al.*, 2003). Using FRAP, we showed that GFP-α-synuclein is extremely mobile and exchanges rapidly between neighboring synapses, despite its steady-state enrichment at the synapse (Fortin *et al.*, 2005). Nonetheless, the recovery of GFP-α-synuclein is slower than that of A30P-α-synuclein, which behaves essentially the same as the soluble GFP. The recovery of GFP-α-synuclein after photobleaching is not due to the movement of synaptic vesicles, with which α-synuclein is believed to interact, since they exhibit relatively low mobility at rest (Darcy *et al.*, 2006). Indeed, GFP fusions to integral membrane proteins of the synaptic vesicle fail to recover from photobleaching over the time course of these experiments (seconds) (Fortin *et al.*, 2005). Therefore, it appears that α-synuclein experiences interactions at the synapse that impede its movement and result in slower recovery than a "true" soluble protein such as GFP.

To reconcile the synaptic enrichment of GFP-α-synuclein and its high mobility as revealed with FRAP, we proposed that α-synuclein localizes to the synapse by means of transient interactions, exchanging rapidly between bound and unbound states. A low affinity, transient interaction mediating synaptic localization is indeed consistent with previous reports that demonstrated the synaptic localization of α-synuclein using immunofluorescence but failed to isolate significant amounts of the protein from particulate brain fractions (Maroteaux *et al.*,

1988; Jakes *et al.*, 1994; George *et al.*, 1995; Iwai, 2000). Therefore, the use of cell-based assays such as FRAP has allowed us to reconcile two seemingly paradoxical observations: the synaptic enrichment of α-synuclein and the general inability to isolate synuclein/synaptic membrane complexes using traditional biochemical assays. Recently, the use of crosslinking agents to preserve weakly interacting protein complexes prior to biochemical fractionation has provided additional evidence for the short-lived association of α-synuclein with biological membranes (Kim *et al.*, 2006).

Transient interactions between proteins and/or membranes are consistent with the concept of "self-organization" which allows for the stability of cellular structures without compromising their flexibility (Misteli, 2001). For instance, rapid exchange of proteins between bound and unbound forms has previously been demonstrated to underlie the localization of nuclear proteins to stable compartments within the nucleus (Phair and Misteli, 2000). In addition, low affinity interactions between cytoplasmic domains of proteins are sufficient for the formation and maintenance of highly stable ER compartments (Snapp *et al.*, 2003). Rapid exchange between bound and unbound states may serve to ensure that stable cellular structures remain dynamic and amenable to regulation.

Dynamic localization at the presynaptic terminal may be a more general phenomenon than previously appreciated. Recent studies using FRAP and photoactivable GFP fusion proteins have indicated that synapsin also exchanges between adjacent synaptic terminals, as much as 40 μm apart (Tsuriel *et al.*, 2006). The movement of synapsin occurs over a slower time course than α-synuclein, presumably reflecting a higher affinity, and therefore longer-lived, interaction with synaptic vesicles. What is the role of local exchange and redistribution of proteins between synapses? In the case of α-synuclein, rapid exchange may allow for sampling of different synaptic "environments" and is likely to play an important role in the normal function of the protein at the synapse. Factors regulating this dynamic behavior, driving the equilibrium toward association or dissociation, remain to be fully characterized.

Activity-Dependent Dynamics of α-Synuclein

Neuronal activity regulates the localization of several proteins at the nerve terminal. The redistribution

of synaptic proteins may acutely regulate protein composition of the terminal, or even control its long-term stability (De Paola *et al.*, 2003). Alternatively, activity-dependent protein redistribution may reflect basic physiological processes such as the recycling of synaptic vesicles (Sankaranarayanan and Ryan, 2000; Li and Murthy, 2001; Mueller *et al.*, 2004; Fortin *et al.*, 2005). For example, transmembrane proteins of the synaptic vesicle are introduced into the plasma membrane of the terminal during synaptic vesicle exocytosis, then spread laterally into the adjacent axonal plasma membrane (Sankaranarayanan and Ryan, 2000; Li and Murthy, 2001; Fortin *et al.*, 2005). Compensatory endocytosis retrieves both membrane and synaptic vesicle protein, resulting in the formation of new synaptic vesicles that then recluster to the center of the terminal. Thus, the dispersion of synaptic vesicle membrane proteins during neuronal activity simply reflects synaptic vesicle recycling. Synapsin, a peripheral membrane protein of synaptic vesicles thought to regulate the availability of synaptic vesicles by tethering them to the cytoskeleton, also redistributes upon stimulation. However, unlike transmembrane proteins, synapsin dissociates from vesicles prior to their release (dissociation requires only calcium entry, not synaptic vesicle fusion) and accumulates in adjacent axonal regions (Chi *et al.*, 2001). After cessation of activity, synapsin associates with newly formed vesicles and thus reclusters into the terminal. The kinetics of synapsin dispersion vary according to stimulus strength and these kinetics directly control the activity of synapsin in synaptic vesicle mobilization and neurotransmitter release (Chi *et al.*, 2001, 2003).

Like synapsin and transmembrane proteins of the synaptic vesicle, α-synuclein exhibits activity-dependent redistribution from the nerve terminal (Fortin *et al.*, 2005). However, the kinetics of α-synuclein dispersion are unique and therefore define a third type of activity-dependent dynamics for synaptic proteins. Activity-dependent dispersion of α-synuclein requires external calcium and thus occurs after calcium entry and action potential invasion in the terminal. However, the redistribution of α-synuclein during neuronal activity is slower than that of synapsin suggesting that α-synuclein disperses after initial synaptic vesicle mobilization. Tetanus toxin, which blocks synaptic vesicle exocytosis without altering calcium entry into the nerve terminal prevents α-synuclein dispersion, suggesting that exocytosis of synaptic vesicles, not calcium entry *per se*, triggers the movement of α-synuclein. Unlike transmembrane proteins of synaptic vesicles, however, dispersed α-synuclein does not re-accumulate in the peri-synaptic region, suggesting that it does not remain associated with membrane after exocytosis (Fortin *et al.*, 2005). Rather it appears that after exocytosis, α-synuclein dissociates from its binding site and rapidly diffuses away from the synaptic bouton. Recent ultrastructural studies have also provided evidence consistent with the dissociation of α-synuclein from synaptic vesicles that fuse with the plasma membrane during stimulation (Tao-Cheng, 2006). After neuronal activity has ended, the kinetics of α-synuclein binding to newly formed synaptic vesicles will determine the rate at which α-synuclein re-localizes to the nerve terminal. Thus, neuronal activity controls the localization of α-synuclein by regulating its membrane association.

Why does α-synuclein dissociate from synaptic membranes after synaptic vesicle exocytosis? Since α-synuclein has been shown to bind preferentially to small, highly curved liposomes, it may sense the change in membrane curvature that results from the collapse of the synaptic vesicle into the plasma membrane. In addition, it has been estimated that each molecule of α-synuclein binds to a membrane "patch" containing approximately 85 acidic phospholipid molecules (Rhoades *et al.*, 2006). Membrane mixing after synaptic vesicle exocytosis may dilute these acidic phospholipids, greatly reducing the affinity of α-synuclein for the membrane and leading to its dissociation. Recent studies have also shown that cytosolic proteins from the nerve terminal favor the dissociation of α-synuclein from presynaptic membranes (Wislet-Gendebien *et al.*, 2006). Whether these proteins are activated or recruited in an activity-dependent manner remains to be determined.

The role of α-synuclein in the regulation of neurotransmitter release may be influenced by the abundance of the protein at the nerve terminal. For example, it has been shown that striatal dopamine release in α-synuclein knockout mice exhibits immediate facilitation upon strong stimulation (Yavich *et al.*, 2004). In contrast, wild-type neurons require up to two high-frequency stimulus trains before similar facilitation is induced. This delay in the onset of facilitation may simply reflect the activity-dependent loss of α-synuclein from wild-type synaptic terminals. Since the extent of α-synuclein dispersion is graded with respect to increasing stimulus strength, and reclustering is very slow, the amount of α-synuclein at a given terminal and hence its effect on transmitter release may reflect the history of activity at that terminal.

Relationship Between Membrane Binding and Aggregation

Although α-synuclein is natively unfolded in solution (Weinreb *et al.*, 1996), the protein can adopt a variety of distinct conformations depending on its environment (Uversky, 2003; Zhu *et al.*, 2003). The ability of α-synuclein to form fibrillar structures has been proposed to be central in the formation of Lewy bodies and thus the pathogenesis of PD (Spillantini *et al.*, 1998). Although widely believed to be a key pathogenic event in PD, the mechanisms by which aggregated α-synuclein results in neuronal cell death remain unclear. The formation of fibrils by α-synuclein is a multi-step process that begins with the formation of a partially folded β-sheet intermediate able to oligomerize and nucleate protofibril formation (Li *et al.*, 2001; Uversky *et al.*, 2001; Uversky and Fink, 2004; Fink, 2006). Protofibrils, which may themselves be toxic to cells (Volles *et al.*, 2001; Volles and Lansbury, 2002, 2003), eventually convert to amyloid-like fibrils that aggregate to form Lewy bodies (Fink, 2006). Factors regulating the different structural and oligomeric conformations of α-synuclein are therefore likely to play an important role in the pathogenesis of PD.

The ability of α-synuclein to bind membranes has been considered crucial for the physiological role of the protein. Could membrane binding also affect the formation of pathogenic α-synuclein structures? It has been estimated that anchoring a protein to membrane can increase its effective local concentration 1000-fold (Murray *et al.*, 2002). Since the rate of fibril formation *in vitro* is proportional to the concentration of α-synuclein (Wood *et al.*, 1999), membrane association could drive the formation of oligomeric species with a propensity to aggregate (Cole *et al.*, 2002). Membrane binding and subsequent amphipathic α-helix formation could also result in α-synuclein adopting an extended conformation that is more prone to aggregation. Consistent with this, it has been shown that α-synuclein oligomerizes in the presence of fatty acids (Perrin *et al.*, 2001; Sharon *et al.*, 2003) and that membrane-bound α-synuclein can spontaneously aggregate (Lee *et al.*, 2002). However, α-synuclein can also form protofibrils and fibrils in solution without addition of any membranes (Conway *et al.*, 1998). In addition, a strong correlation between the α-helical conformation of α-synuclein and inhibition of fibril formation has been reported (Zhu and Fink, 2003). Membrane binding, and ensuing α-helix formation, could prevent the formation of pathogenic α-synuclein structures, possibly by inhibiting self-association of α-synuclein (Narayanan and Scarlata, 2001), a step required for protofibril and fibril assembly. The actual relationship between membrane binding and aggregation thus remains unclear.

The link between membrane association, structure and aggregation of α-synuclein underscores the importance of protein dynamics for understanding the pathogenesis of PD. Neuronal activity potently regulates the interaction between α-synuclein and synaptic vesicles and thus may contribute to the normal as well as pathogenic role of the protein. For instance, it is currently unknown if α-synuclein remains α-helical, adopts an alternative conformation or unfolds following activity-induced dissociation from synaptic membranes. The fate of the protein after dissociation will not only affect the ability of the protein to re-associate with synaptic vesicles but also the propensity of α-synuclein to fold into pathogenic conformations. Alternatively, if membrane association is a prerequisite for the formation of pathogenic aggregates, activity-dependent dissociation from synaptic vesicles and subsequent dispersion of α-synuclein will be neuroprotective. It is reasonable to expect that other factors, for example protein phosphorylation or protein/protein interactions, will also regulate the dynamics of α-synuclein at the synapse. The identification of these factors will help define pathways that control the dynamics of α-synuclein at the nerve terminal under normal circumstances and may become pathological in PD. Molecular dissection of these pathways will provide novel therapeutic avenues to prevent and/or control the appearance of structural conformations of α-synuclein that are pathogenic.

CONCLUDING REMARKS

The identification of mutations in α-synuclein associated with familial forms of PD provided one of the first molecular entry points to understand the pathogenesis of the disease. However, mutations are rare and do not account for the majority of PD cases, suggesting that dysfunction of the wild-type protein is a more common disease-causing event. Consistent with this, α-synuclein fibrils are found in abundance in Lewy bodies, a cardinal pathological lesion found in most, if not all, cases of PD. α-Synuclein can adopt multiple structural conformations depending on its environment, some of which capable of converting into key intermediates in the assembly of Lewy bodies. Controlling the appearance of these pathogenic intermediates may thus prevent or at least slow down PD. Under normal circumstances, α-synuclein interacts with synaptic vesicle membranes, localizing to the nerve terminal where it has been proposed to regulate neurotransmitter release. Upon membrane binding, α-synuclein undergoes a major conformational change, from relatively unstructured to highly α-helical. Neuronal activity induces the dissociation of the protein from synaptic vesicle membranes and its dispersion from the synapse. How activity-dependent dynamics impact the sampling of different structural conformations by α-synuclein remains unknown. However, the link between membrane association and conformation indicates that the dynamics of α-synuclein probably have an important role in the pathogenesis of PD. Considering that the activity-dependent dynamics of α-synuclein presumably occur in all neurons expressing the protein, what accounts for the unique susceptibility of nigral dopaminergic neurons to degeneration in PD? The distinctive physiology of these neurons, which recently have been shown to rely on unusual Ca^{2+} channels to drive their continuous, pacemaking activity, may render them more susceptible to factors that contribute to disease (Savio Chan *et al.*, 2007). For example, the high activity of these neurons may result in persistent dissociation of α-synuclein from membranes, allowing conformational changes in the protein that favor the formation of pathogenic species. Although α-synuclein has a role in the pathogenesis of PD, the molecular pathways underlying degeneration are unknown. In particular, the relationship between normal and disease-causing functions of α-synuclein remains unclear. In addition, the interaction between α-synuclein and other risk factors for PD remain poorly understood. Rational therapeutic intervention aimed at halting the progression or preventing the onset of degeneration will become possible only once these mechanisms are understood.

REFERENCES

Abeliovich, A., Schmitz, Y., Farinas, I., Choi-Lundberg, D., Ho, W. H., Castillo, P. E., Shinsky, N., Verdugo, J. M., Armanini, M., Ryan, A., Hynes, M., Phillips, H., Sulzer, D., and Rosenthal, A. (2000). Mice lacking alpha-synuclein display functional deficits in the nigrostriatal dopamine system. *Neuron* **25**, 239–252.

Axelrod, D., Koppel, D. E., Schlessinger, J., Elson, E., and Webb, W. W. (1976). Mobility measurement by analysis of fluorescence photobleaching recovery kinetics. *Biophys J* **16**, 1055–1069.

Bisaglia, M., Tessari, I., Pinato, L., Bellanda, M., Giraudo, S., Fasano, M., Bergantino, E., Bubacco, L., and Mammi, S. (2005). A topological model of the interaction between alpha-synuclein and sodium dodecyl sulfate micelles. *Biochemistry* **44**, 329–339.

Bisaglia, M., Schievano, E., Caporale, A., Peggion, E., and Mammi, S. (2006). The 11-mer repeats of human alpha-synuclein in vesicle interactions and lipid composition discrimination: A cooperative role. *Biopolymers* **84**, 310–316.

Bussell, R., Jr., and Eliezer, D. (2001). Residual structure and dynamics in Parkinson's disease-associated mutants of alpha-synuclein. *J Biol Chem* **276**, 45996–46003.

Bussell, R., Jr., and Eliezer, D. (2003). A structural and functional role for 11-mer repeats in alpha-synuclein and other exchangeable lipid binding proteins. *J Mol Biol* **329**, 763–778.

Bussell, R., Jr., and Eliezer, D. (2004). Effects of Parkinson's disease-linked mutations on the structure of lipid-associated alpha-synuclein. *Biochemistry* **43**, 4810–4818.

Cabin, D. E., Shimazu, K., Murphy, D., Cole, N. B., Gottschalk, W., McIlwain, K. L., Orrison, B., Chen, A., Ellis, C. E., Paylor, R., Lu, B., and Nussbaum, R. L. (2002). Synaptic vesicle depletion correlates with attenuated synaptic responses to prolonged repetitive stimulation in mice lacking alpha-synuclein. *J Neurosci* **22**, 8797–8807.

Chandra, S., Chen, X., Rizo, J., Jahn, R., and Sudhof, T. C. (2003). A broken alpha-helix in folded alpha-Synuclein. *J Biol Chem* **278**, 15313–15318.

Chartier-Harlin, M. C., Kachergus, J., Roumier, C., Mouroux, V., Douay, X., Lincoln, S., Levecque, C., Larvor, L., Andrieux, J., Hulihan, M., Waucquier, N., Defebvre, L., Amouyel, P., Farrer, M., and Destee, A. (2004). Alpha-synuclein locus duplication as a cause of familial Parkinson's disease. *Lancet* **364**, 1167–1169.

Chi, P., Greengard, P., and Ryan, T. A. (2001). Synapsin dispersion and reclustering during synaptic activity. *Nat Neurosci* **4**, 1187–1193.

Chi, P., Greengard, P., and Ryan, T. A. (2003). Synaptic vesicle mobilization is regulated by distinct synapsin

I phosphorylation pathways at different frequencies. *Neuron* **38**, 69–78.

Choi, W., Zibaee, S., Jakes, R., Serpell, L. C., Davletov, B., Crowther, R. A., and Goedert, M. (2004). Mutation E46K increases phospholipid binding and assembly into filaments of human alpha-synuclein. *FEBS Lett* **576**, 363–368.

Cole, N. B., Murphy, D. D., Grider, T., Rueter, S., Brasaemle, D., and Nussbaum, R. L. (2002). Lipid droplet binding and oligomerization properties of the Parkinson's disease protein alpha-synuclein. *J Biol Chem* **277**, 6344–6352.

Conway, K. A., Harper, J. D., and Lansbury, P. T. (1998). Accelerated *in vitro* fibril formation by a mutant alpha-synuclein linked to early-onset Parkinson disease. *Nat Med* **4**, 1318–1320.

Darcy, K. J., Staras, K., Collinson, L. M., and Goda, Y. (2006). Constitutive sharing of recycling synaptic vesicles between presynaptic boutons. *Nat Neurosci* **9**, 315–321.

Davidson, W. S., Jonas, A., Clayton, D. F., and George, J. M. (1998). Stabilization of alpha-synuclein secondary structure upon binding to synthetic membranes. *J Biol Chem* **273**, 9443–9449.

De Paola, V., Arber, S., and Caroni, P. (2003). AMPA receptors regulate dynamic equilibrium of presynaptic terminals in mature hippocampal networks. *Nat Neurosci* **6**, 491–500.

Eliezer, D., Kutluay, E., Bussell, R., Jr., and Browne, G. (2001). Conformational properties of alpha-synuclein in its free and lipid-associated states. *J Mol Biol* **307**, 1061–1073.

Fink, A. L. (2006). The aggregation and fibrillation of alpha-synuclein. *Acc Chem Res* **39**, 628–634.

Fortin, D. L., Troyer, M. D., Nakamura, K., Kubo, S., Anthony, M. D., and Edwards, R. H. (2004). Lipid rafts mediate the synaptic localization of alpha-synuclein. *J Neurosci* **24**, 6715–6723.

Fortin, D. L., Nemani, V. M., Voglmaier, S. M., Anthony, M. D., Ryan, T. A., and Edwards, R. H. (2005). Neural activity controls the synaptic accumulation of alpha-synuclein. *J Neurosci* **25**, 10913–10921.

Galvin, J. E., Uryu, K., Lee, V. M., and Trojanowski, J. Q. (1999). Axon pathology in Parkinson's disease and Lewy body dementia hippocampus contains alpha-, beta-, and gamma-synuclein. *Proc Natl Acad Sci USA* **96**, 13450–13455.

George, J. M., Jin, H., Woods, W. S., and Clayton, D. F. (1995). Characterization of a novel protein regulated during the critical period for song learning in the zebra finch. *Neuron* **15**, 361–372.

Ibanez, P., Bonnet, A. M., Debarges, B., Lohmann, E., Tison, F., Pollak, P., Agid, Y., Durr, A., and Brice, A. (2004). Causal relation between alpha-synuclein gene duplication and familial Parkinson's disease. *Lancet* 364, 1169–1171.

Iwai, A. (2000). Properties of NACP/alpha-synuclein and its role in Alzheimer's disease. *Biochim Biophys Acta* 1502, 95–109.

Jakes, R., Spillantini, M. G., and Goedert, M. (1994). Identification of two distinct synucleins from human brain. *FEBS Lett* 345, 27–32.

Jensen, P. H., Nielsen, M. S., Jakes, R., Dotti, C. G., and Goedert, M. (1998). Binding of alpha-synuclein to brain vesicles is abolished by familial Parkinson's disease mutation. *J Biol Chem* 273, 26292–26294.

Jo, E., McLaurin, J., Yip, C. M., St George-Hyslop, P., and Fraser, P. E. (2000). Alpha-Synuclein membrane interactions and lipid specificity. *J Biol Chem* 275, 34328–34334.

Jo, E., Fuller, N., Rand., R. P., St George-Hyslop, P., and Fraser, P. E. (2002). Defective membrane interactions of familial Parkinson's disease mutant A30P alpha-synuclein. *J Mol Biol* 315, 799–807.

Kahle, P. J., Neumann, M., Ozmen, L., Muller, V., Jacobsen, H., Schindzielorz, A., Okochi, M., Leimer, U., van Der Putten, H., Probst, A., Kremmer, E., Kretzschmar, H. A., and Haass, C. (2000). Subcellular localization of wild-type and Parkinson's disease-associated mutant alpha -synuclein in human and transgenic mouse brain. *J Neurosci* 20, 6365–6373.

Kim, Y. S., Laurine, E., Woods, W., and Lee, S. J. (2006). A novel mechanism of interaction between alpha-synuclein and biological membranes. *J Mol Biol* 360, 386–397.

Kruger, R., Kuhn, W., Muller, T., Woitalla, D., Graeber, M., Kosel, S., Przuntek, H., Epplen, J. T., Schols, L., and Riess, O. (1998). Ala30Pro mutation in the gene encoding alpha-synuclein in Parkinson's disease. *Nat Genet* 18, 106–108.

Kubo, S., Nemani, V. M., Chalkley, R. J., Anthony, M. D., Hattori, N., Mizuno, Y., Edwards, R. H., and Fortin, D. L. (2005). A combinatorial code for the interaction of alpha-synuclein with membranes. *J Biol Chem* 280, 31664–31672.

Larsen, K. E., Schmitz, Y., Troyer, M. D., Mosharov, E., Dietrich, P., Quazi, A. Z., Savalle, M., Nemani, V., Chaudhry, F. A., Edwards, R. H., Stefanis, L., and Sulzer, D. (2006). Alpha-synuclein overexpression in PC12 and chromaffin cells impairs catecholamine release by interfering with a late step in exocytosis. *J Neurosci* 26, 11915–11922.

Lee, H. J., Choi, C., and Lee, S. J. (2002). Membrane-bound alpha-synuclein has a high aggregation propensity and the ability to seed the aggregation of the cytosolic form. *J Biol Chem* 277, 671–678.

Li, J., Uversky, V. N., and Fink, A. L. (2001). Effect of familial Parkinson's disease point mutations A30P and A53T on the structural properties, aggregation, and fibrillation of human alpha-synuclein. *Biochemistry* 40, 11604–11613.

Li, Z., and Murthy, V. N. (2001). Visualizing postendocytic traffic of synaptic vesicles at hippocampal synapses. *Neuron* 31, 593–605.

Lippincott-Schwartz, J., Altan-Bonnet, N., and Patterson, G. H. (2003). Photobleaching and photoactivation: Following protein dynamics in living cells. *Nat Cell Biol Suppl*, S7–S14.

Liu, S., Ninan, I., Antonova, I., Battaglia, F., Trinchese, F., Narasanna, A., Kolodilov, N., Dauer, W., Hawkins, R. D., and Arancio, O. (2004). Alpha-synuclein produces a long-lasting increase in neurotransmitter release. *EMBO J* 23, 4506–4516.

Lucke, C., Gantz, D. L., Klimtchuk, E., and Hamilton, J. A. (2006). Interactions between fatty acids and alpha-synuclein. *J Lipid Res* 47, 1714–1724.

Maroteaux, L., Campanelli, J. T., and Scheller, R. H. (1988). Synuclein: A neuron-specific protein localized to the nucleus and presynaptic nerve terminal. *J Neurosci* 8, 2804–2815.

Misteli, T. (2001). The concept of self-organization in cellular architecture. *J Cell Biol* 155, 181–185.

Mueller, V. J., Wienisch, M., Nehring, R. B., and Klingauf, J. (2004). Monitoring clathrin-mediated endocytosis during synaptic activity. *J Neurosci* 24, 2004–2012.

Murphy, D. D., Rueter, S. M., Trojanowski, J. Q., and Lee, V. M. (2000). Synucleins are developmentally expressed, and alpha-synuclein regulates the size of the presynaptic vesicular pool in primary hippocampal neurons. *J Neurosci* 20, 3214–3220.

Murray, D., Arbuzova, A., Honig, B., and McLaughlin, S. (2002). The role of electrostatic and nonpolar interactions in the association of peripheral proteins with membranes. In *Current Topics in Membranes* (S. S. Simon, and T. J. McIntosh, Eds.), Vol. 52, pp. 277–298. Academic Press.

Narayanan, V., and Scarlata, S. (2001). Membrane binding and self-association of alpha-synucleins. *Biochemistry* 40, 9927–9934.

Outeiro, T. F., and Lindquist, S. (2003). Yeast cells provide insight into alpha-synuclein biology and pathobiology. *Science* 302, 1772–1775.

Pals, P., Lincoln, S., Manning, J., Heckman, M., Skipper, L., Hulihan, M., Van den Broeck, M., De Pooter, T., Cras, P., Crook, J., Van Broeckhoven, C., and Farrer, M. J. (2004). Alpha-synuclein promoter confers susceptibility to Parkinson's disease. *Ann Neurol* 56, 591–595.

Perrin, R. J., Woods, W. S., Clayton, D. F., and George, J. M. (2000). Interaction of human alpha-Synuclein and Parkinson's disease variants with phospholipids. Structural analysis using site-directed mutagenesis. *J Biol Chem* 275, 34393–34398.

Perrin, R. J., Woods, W. S., Clayton, D. F., and George, J. M. (2001). Exposure to long chain polyunsaturated fatty acids triggers rapid multimerization of synucleins. *J Biol Chem* **276**, 41958–41962.

Phair, R. D., and Misteli, T. (2000). High mobility of proteins in the mammalian cell nucleus. *Nature* **404**, 604–609.

Phair, R. D., and Misteli, T. (2001). Kinetic modelling approaches to *in vivo* imaging. *Nat Rev Mol Cell Biol* **2**, 898–907.

Polymeropoulos, M. H., Lavedan, C., Leroy, E., Ide, S. E., Dehejia, A., Dutra, A., Pike, B., Root, H., Rubenstein, J., Boyer, R., Stenroos, E. S., Chandrasekharappa, S., Athanassiadou, A., Papapetropoulos, T., Johnson, W. G., Lazzarini, A. M., Duvoisin, R. C., Di Iorio, G., Golbe, L. I., and Nussbaum, R. L. (1997). Mutation in the alpha-synuclein gene identified in families with Parkinson's disease. *Science* **276**, 2045–2047.

Ramakrishnan, M., Jensen, P. H., and Marsh, D. (2003). Alpha-synuclein association with phosphatidylglycerol probed by lipid spin labels. *Biochemistry* **42**, 12919–12926.

Rhoades, E., Ramlall, T. F., Webb, W. W., and Eliezer, D. (2006). Quantification of alpha-synuclein binding to lipid vesicles using fluorescence correlation spectroscopy. *Biophys J* **90**, 4692–4700.

Sankaranarayanan, S., and Ryan, T. A. (2000). Real-time measurements of vesicle-SNARE recycling in synapses of the central nervous system. *Nat Cell Biol* **2**, 197–204.

Sharon, R., Goldberg, M. S., Bar-Josef, I., Betensky, R. A., Shen, J., and Selkoe, D. J. (2001). Alpha-synuclein occurs in lipid-rich high molecular weight complexes, binds fatty acids, and shows homology to the fatty acid-binding proteins. *Proc Natl Acad Sci USA* **98**, 9110–9115.

Sharon, R., Bar-Joseph, I., Frosch, M. P., Walsh, D. M., Hamilton, J. A., and Selkoe, D. J. (2003). The formation of highly soluble oligomers of alpha-synuclein is regulated by fatty acids and enhanced in Parkinson's disease. *Neuron* **37**, 583–595.

Singleton, A. B., Farrer, M., Johnson, J., Singleton, A., Hague, S., Kachergus, J., Hulihan, M., Peuralinna, T., Dutra, A., Nussbaum, R., Lincoln, S., Crawley, A., Hanson, M., Maraganore, D., Adler, C., Cookson, M. R., Muenter, M., Baptista, M., Miller, D., Blancato, J., Hardy, J., and Gwinn-Hardy, K. (2003). Alpha-synuclein locus triplication causes Parkinson's disease. *Science* **302**, 841.

Snapp, E. L., Hegde, R. S., Francolini, M., Lombardo, F., Colombo, S., Pedrazzini, E., Borgese, N., and Lippincott-Schwartz, J. (2003). Formation of stacked ER cisternae by low affinity protein interactions. *J Cell Biol* **163**, 257–269.

Specht, C. G., Tigaret, C. M., Rast, G. F., Thalhammer, A., Rudhard, Y., and Schoepfer, R. (2005). Subcellular localisation of recombinant alpha- and gamma-synuclein. *Mol Cell Neurosci* **28**, 326–334.

Spillantini, M. G., Schmidt, M. L., Lee, V. M., Trojanowski, J. Q., Jakes, R., and Goedert, M. (1997). Alpha-synuclein in Lewy bodies. *Nature* **388**, 839–840.

Spillantini, M. G., Crowther, R. A., Jakes, R., Hasegawa, M., and Goedert, M. (1998). Alpha-synuclein in filamentous inclusions of Lewy bodies from Parkinson's disease and dementia with Lewy bodies. *Proc Natl Acad Sci USA* **95**, 6469–6473.

Takamori, S., Holt, M., Stenius, K., Lemke, E. A., Gronborg, M., Riedel, D., Urlaub, H., Schenck, S., Brugger, B., Ringler, P., Muller, S. A., Rammner, B., Grater, F., Hub, J. S., De Groot, B. L., Mieskes, G., Moriyama, Y., Klingauf, J., Grubmuller, H., Heuser, J., Wieland, F., and Jahn, R. (2006). Molecular anatomy of a trafficking organelle. *Cell* **127**, 831–846.

Tao-Cheng, J. H. (2006). Activity-related redistribution of presynaptic proteins at the active zone. *Neuroscience* **141**, 1217–1224.

Tsuriel, S., Geva, R., Zamorano, P., Dresbach, T., Boeckers, T., Gundelfinger, E. D., Garner, C. C., and Ziv, N. E. (2006). Local sharing as a predominant determinant of synaptic matrix molecular dynamics. *PLoS Biol* **4**, 1572–1587.

Uversky, V. N. (2003). A protein-chameleon: Conformational plasticity of alpha-synuclein, a disordered protein involved in neurodegenerative disorders. *J Biomol Struct Dyn* **21**, 211–234.

Uversky, V. N., and Fink, A. L. (2004). Conformational constraints for amyloid fibrillation: The importance of being unfolded. *Biochim Biophys Acta* **1698**, 131–153.

Uversky, V. N., Li, J., and Fink, A. L. (2001). Evidence for a partially folded intermediate in alpha-synuclein fibril formation. *J Biol Chem* **276**, 10737–10744.

Volles, M. J., and Lansbury, P. T., Jr. (2002). Vesicle permeabilization by protofibrillar alpha-synuclein is sensitive to Parkinson's disease-linked mutations and occurs by a pore-like mechanism. *Biochemistry* **41**, 4595–4602.

Volles, M. J., and Lansbury, P. T., Jr. (2003). Zeroing in on the pathogenic form of alpha-synuclein and its mechanism of neurotoxicity in Parkinson's disease. *Biochemistry* **42**, 7871–7878.

Volles, M. J., Lee, S. J., Rochet, J. C., Shtilerman, M. D., Ding, T. T., Kessler, J. C., and Lansbury, P. T., Jr. (2001). Vesicle permeabilization by protofibrillar alpha-synuclein: Implications for the pathogenesis and treatment of Parkinson's disease. *Biochemistry* **40**, 7812–7819.

Weinreb, P. H., Zhen, W., Poon, A. W., Conway, K. A., and Lansbury, P. T., Jr. (1996). NACP, a protein implicated in Alzheimer's disease and learning, is natively unfolded. *Biochemistry* **35**, 13709–13715.

Wislet-Gendebien, S., D'Souza, C., Kawarai, T., St George-Hyslop, P., Westaway, D., Fraser, P., and Tandon, A. (2006). Cytosolic proteins regulate alpha-synuclein dissociation from presynaptic membranes. *J Biol Chem* **281**, 32148–32155.

Withers, G. S., George, J. M., Banker, G. A., and Clayton, D. F. (1997). Delayed localization of synelfin (synuclein, NACP) to presynaptic terminals in cultured rat hippocampal neurons. *Brain Res Dev Brain Res* **99**, 87–94.

Wood, S. J., Wypych, J., Steavenson, S., Louis, J. C., Citron, M., and Biere, A. L. (1999). Alpha-synuclein fibrillogenesis is nucleation-dependent. Implications for the pathogenesis of Parkinson's disease. *J Biol Chem* **274**, 19509–19512.

Yavich, L., Tanila, H., Vepsalainen, S., and Jakala, P. (2004). Role of alpha-synuclein in presynaptic dopamine recruitment. *J Neurosci* **24**, 11165–11170.

Zarranz, J. J., Alegre, J., Gomez-Esteban, J. C., Lezcano, E., Ros, R., Ampuero, I., Vidal, L., Hoenicka, J., Rodriguez, O., Atares, B., Llorens, V., Gomez Tortosa, E., del Ser, T., Munoz, D. G., and de Yebenes, J. G. (2004). The new mutation, E46K, of alpha-synuclein causes Parkinson and Lewy body dementia. *Ann Neurol* **55**, 164–173.

Zhu, M., and Fink, A. L. (2003). Lipid binding inhibits alpha-synuclein fibril formation. *J Biol Chem* **278**, 16873–16877.

Zhu, M., Li, J., and Fink, A. L. (2003). The association of alpha-synuclein with membranes affects bilayer structure, stability, and fibril formation. *J Biol Chem* **278**, 40186–40197.

19

THE BAC TRANSGENIC APPROACH TO STUDY PARKINSON'S DISEASE IN MICE

X. WILLIAM YANG[1,2,3] AND XIAO-HONG LU[1,2,3]

[1]Center for Neurobehavioral Genetics, Semel Institute for Neuroscience and Human Behavior
[2]Department of Psychiatry and Biobehavioral Sciences
[3]Brain Research Institute, David Geffen School of Medicine at UCLA, Los Angeles, CA 90095, USA

INTRODUCTION

Parkinson's disease (PD) is the second most common form of neurodegenerative disorder with cardinal clinical features of resting tremor, rigidity, and brady-kinesia (Fahn, 2003). The pathological hallmark of PD is the progressive loss of dopaminergic (DA) neurons in the substantia nigra pars compacta (SNc) (McGeer et al., 1977; Trojanowski and Lee, 1998). At present, therapeutic approaches to PD, including dopamine replacement (i.e., with levodopa or other DA agents) and/or surgical approaches, such as deep-brain stimulation (DBS), are exclusively symptomatic treatments. Such treatment options have limitations, since they cannot modify the course of the disease, and are also associated with significant side effects such as dyskinesia (Nutt and Wooten, 2005; DeLong and Wichmann, 2007). Therapeutic strategies that can halt or retard this devastating neurodegenerative process are currently unavailable, but remain an important goal in PD research.

A major challenge for translational research on neurodegenerative disorders including PD is to estab-lish a rational strategy that goes from the disease-causing mutations to the understanding of critical disease mechanisms, and the eventual development of effective therapeutics. As a result of the enor-mous molecular, cellular, and circuitry complexities of the mammalian brain, an important objective in the study of disease pathogenesis and in the testing of therapeutics is the development of animal models that recapitulate genetic and/or phenotypic aspects of the disease.

In PD research, for example, the development of the 1-methyl-4-phenyl-1,2,3,6-tetrahydropyridine (MPTP) model of PD, with acute depletion of DA neurons in non-human primates (Di Monte et al., 2000) and rodents (Ricaurte et al., 1986), has been critical in understanding neuronal circuitry dysfunc-tion as a consequence of the loss of DA neurons in PD, and in developing the current symptomatic treatment of PD (i.e., levodopa and DBS). However, since acute chemical models of PD do not recapitu-late the slowly progressive neurodegenerative proc-ess that occurs in the sporadic and rare familial PD patients (Bloem et al., 1990; Dauer and Przedborski, 2003), they may not be very helpful for studying the mechanisms of late-onset DA neuron loss in PD, and for testing treatment strategies that can prevent the onset of DA neuron degeneration in PD. With the cloning of genes underlying rare familial forms of PD, including α-synuclein, parkin, DJ-1, PINK1, and LRRK2 (Dawson and Dawson, 2003; Moore et al., 2005; Hardy et al., 2006), novel opportunities exist now to model PD pathogenesis in animal models based on these familial mutations.

In developing genetic models of human brain disorders, the mouse has emerged as a leading mammalian model organism not only because it is easy to raise in laboratory conditions (small size, short generation time, etc.) and has many well characterized inbred strains (with defined phenotypes), but also because the mouse genome is amenable to sophisticated genetic manipulations. Over the past two decades, powerful genetic technologies have been developed in mice to stably but randomly insert an exogenous gene into the mouse genome (transgenic mice) or to delete any endogenous gene (or a contiguous stretch of genomic sequence) from the mouse genome (knockout mice). Further refined "conditional" mouse genetic technologies also allow the manipulation of gene expression/gene deletion in mice in a spatially and/or temporally defined manner. These new mouse genetic technologies not only permit modeling of brain disorders such as PD in mice, but also the use of the mouse as a model to dissect critical circuitry and molecular mechanisms underlying the disorder.

In this chapter, we will focus our discussion on a recently developed and powerful mouse genetic technology that uses bacterial artificial chromosomes (BACs) to generate transgenic mice. We will highlight the advantages of the BAC transgenic approach as compared to the alternative genetic approaches, and how the BAC transgenic approach can be applied to model neurodegenerative disorders such as PD.

GENERATION OF BAC TRANSGENIC MICE

Advantages of BAC Mice

Transgenic mice are generated by the direct injection of a DNA fragment into a fertilized one-cell mouse embryo, which results in stable but random integration of multiple copies of the transgene (Palmiter, and Brinster 1985). Unlike the gene-targeting approach in embryonic stem (ES) cells, which is aimed to manipulate the *endogenous* murine genes, the transgenic approach is often utilized to overexpress an *exogenous* gene in order to assess the phenotypic consequences. A basic transgenic construct (Figure 19.1a) consists of a eukaryotic promoter

with associated regulatory elements (i.e., enhancers, suppressors, locus control regions), a coding region such as a cDNA with an open reading frame (ORF) encoding a protein of interest (i.e., marker proteins such as β-galactosidase, green fluorescent protein [GFP], Cre recombinase) which should begin with an efficient mammalian translation initiation sequence or Kozak sequence (i.e., GCC ACC *ATG*) (Kozak, 1986), and finally, a mammalian polyadenylation signal (PA). In conventional transgenic studies prior to the use of large insert genomic clones as transgenes, the average size of the construct, including the promoter and regulatory elements, is less than 20 kilobases (kb). Such conventional transgenes are often plagued by inconsistency in expression due to *positional effects*, which refer to the phenomena of variable, ectopic, or complete lack of transgene expression due to influence from the genomic locus at which the transgene integrated (Giraldo and Montoliu, 2001). Many empirical studies reveal that positional effects are due mostly to the use of smaller genomic DNA fragments that lack critical regulatory elements such as enhancers, suppressors, and locus control regions. An average mammalian gene locus spans about 30 kb of genomic DNA, but these critical transcription regulatory gene elements could be located upstream of or downstream to the coding region, or may also be located in the introns. Thus, large genomic inserts of 100–200 kb are often needed to encompass the entire repertoire of regulatory elements for a given gene to confer accurate, dosage-dependent, integration-site-independent transgene expression *in vivo* (Heintz, 2000). Moreover, large transgenes with an intact genomic locus may also preserve the endogenous gene structures at the locus (i.e., introns and exons), thus for a gene with a complex set of alternatively spliced transcripts, the large genomic insert transgene may express multiple alternatively spliced transcripts while a conventional transgene can only express one pre-selected transcript.

BACs are circular F-element-based circular bacterial genomic clones containing an average of about 180–200 kb genomic DNA (Shizuya et al., 1992). The BAC system has multiple advantages over the older large insert cloning system, yeast artificial chromosome (YAC), which includes clonal stability, low chimerism, and easy manipulation (getting DNA, sequencing, etc). In various genome projects, BACs have largely replaced YACs as the clones of choice for the mapping and sequencing of genomes, and dense databases of BAC contigs have been

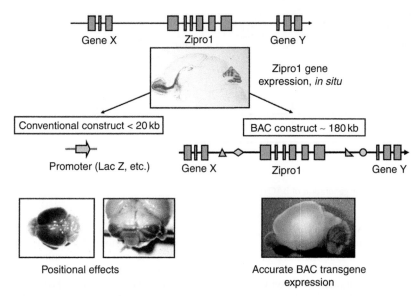

FIGURE 19.1 Advantages of BAC transgenic approach. Zinc-finger transcription factor Zipro1 is used to illustrate the advantages of BAC transgenic approach. Zipro1 is mainly expressed in cerebellum, hippocampus, and olfactory granule cell precursors as shown by *in situ* hybridization. Conventional transgenic approach with 10 kb genomic construct containing the Zipro1 promoter led to ectopic expression of lacZ transgene due to the positional effects (Yang and Heintz, unpublished data). BAC transgenic approach with 131 kb Zipro1 BAC construct results in the accurate lacZ transgene expression (Yang et al., 1997).

mapped to the sequenced model organism genomes of both humans and mice.

The initial use of large genomic inserts for transgenesis was done with YACs, which demonstrated that large genomic inserts are able to confer endogenous-like, dosage-dependent, and integration-site-independent transgene expression (Giraldo and Montolin, 2001). However, since it is relatively difficult to purify intact DNA from YAC clones, a large portion of the YAC transgenic founders generated by direct pronuclear injections may contain fragmented YAC DNA. To overcome such a problem, techniques such as the spheroplast fusion between YAC-harboring yeast clones and murine ES cells are required prior to generating transgenic mice with the ES cells. In order to improve the efficiency, reliability, and simplicity of large genomic DNA transgenesis, bacterial-based genomic clones, including the widely used BACs, and the less commonly used P1-artificial chromosomes (or PACs), have become more widely used instead of the YACs in transgenic experiments (Yang *et al.*, 1997; Heintz, 2001; Yang and Gong, 2005).

The first successful example of using genetically engineered BAC constructs for the generation of transgenic mice was reported in 1997 (Yang *et al.*, 1997). Over the past decade, hundreds of different types of BAC transgenic mice have been generated to study all facets of mammalian biology (Heintz, 2001; Hatten and Heintz, 2005). The collective experience with BAC transgenesis thus far clearly demonstrates that, for the vast majority of the mammalian genes in the genome, an appropriately selected BAC (see next section) is highly likely to confer accurate, dosage-dependent, integration-independent transgene expression *in vivo*. The largest study of this kind to demonstrate the advantage of the BAC transgenic system is the GENSAT BAC transgenic project (Gong *et al.*, 2003). In the GENSAT project (www.gensat.org), which aims to map the expression of mammalian genes in the brain, hundreds of different BAC transgenic mice were generated, expressing the enhanced green fluorescent protein (EGFP) transgene, which was driven by the regulator elements of a gene on the BAC (Gong *et al.*, 2003). The initial report from

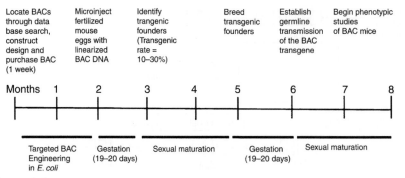

FIGURE 19.2 Timeline for generation of BAC transgenic mice.

the study of 300 such BAC transgenic mice showed that about 90% of the BAC constructs drive accurate transgene expression, and only about 10% of the BACs have poor expression levels and patterns. In most cases, the latter group contained genes with genomic span greater than 100 kb of the genomic DNA. In summary, for the vast majority of mammalian genes with genomic span around 100 kb or less, a carefully chosen BAC transgenic construct has a very high probability of conferring accurate transgene expression, and therefore has a very high probability of overcoming the positional effects that are a severe limitation of the conventional transgenic approach that uses relatively small genomic constructs (<20 kb), as shown in Figure 19.1.

Generation of BAC Transgenic Mice

Selection of BACs for Transgenic Studies

The first step in a BAC transgenic experiment is selecting appropriate BACs for the study via public database search. The first factor to be considered here is BACs from which mammalian species should be used. As a general rule, to model a dominant human genetic disease, human BACs are usually used, since the precise human protein context (in addition to the disease mutation) and any potential human-specific expression patterns could be relevant to disease pathogenesis. On the other hand, murine BACs are usually preferred in the study of the biology of a protein in mice (i.e., overexpression), or to express a marker protein in specific cell types.

Next, in order to ensure that a BAC transgene can recapitulate the expression pattern of a gene of interest on the BAC (termed *target gene*), one should

select BACs that contain at least 50 kb of both the 5′ and 3′ flanking genomic sequences, in addition to the genomic sequence spanning the coding region of the target gene (Heintz, 2001; Gong *et al.*, 2003; Yang and Gong, 2005). For mammalian genes that cover less than 100 kb, such BACs are relatively easy to locate in existing public databases such as the Genome Browser at the University of California at Santa Cruz (http://www.genome.ucsc.edu) and the Ensembl Genome Browser (http://www.ensembl.org/index.html).

For mammalian genes that are 100–200 kb in size, BAC transgenes are less likely but still possible to drive accurate expression of the transgenes (about 10% success rate) (Gong *et al.*, 2003). For example, our laboratory has generated BAC transgenic mice expressing human mutant huntingtin (mhtt) that spans a 170 kb genomic region (from a 240 kb BAC). This BAC transgene expresses full-length human huntingtin which is functional and can rescue the lethality of the murine knockout mice (Gray *et al.*, 2008). After an appropriate BAC is chosen *in silico*, one can readily order them from public repositories such as the BACPAC Resource Center (Oakland Children's Hospital, http://www.bacpac.chori.org/) or a commercial venue (i.e., Invitrogen). The process of locating and obtaining a BAC for *in vivo* experimentation usually takes only a few days (Figure 19.2).

Basic BAC Transgenic Construct Design

The design of a BAC transgenic construct is pivotal to the success of the overall transgenic experiment. Any BAC transgenic construct should have the following basic components: a gene of interest to be overexpressed, which could be an exogenous

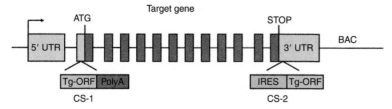

FIGURE 19.3 Basic BAC transgenic construct designs. Construct 1 (CS-1) is designed to overexpress a transgenic open reading frame (Tg-ORF), such as the GFP, lacZ, Cre, and others, but the *target gene* (a cognate gene on the BAC) itself is not overexpressed. CS-2 uses an internal ribosomal entry sequence (IRES) which allows the expression of both the target gene and a Tg-ORF (i.e., GFP, lacZ, or Cre) from the same promoter.

gene (i.e., EGFP or Cre), or a cognate gene on the BAC (i.e., mutant disease gene); a promoter and the regulatory elements from a *target gene* on the BAC, which will be used to drive the expression of the exogenous or the endogenous transgenes; and finally, a mammalian *polyadenylation signal* (i.e., the polyA signal could be an endogenous one from the *target gene*, or an exogenous one from the human growth hormone or phosphoglycerate kinase). Examples of two basic BAC transgenic constructs are shown in Figure 19.3, and a more in-depth discussion on the subject can be found elsewhere (Yang and Gong, 2005).

Engineering BACs by Homologous Recombination in *Escherichia coli*

Prior to generating BAC transgenic mice, appropriate mutations are introduced into BACs. Any types of mutations, including insertion of marker genes (i.e., GFP, Cre recombinase), deletion, or a point mutation, can be readily introduced into a BAC by homologous recombination in *E. coli*. There are two widely used homologous recombination-based methods to engineer BACs in *E. coli*: the RecA-based method and the recombineering method (Figure 19.4a and b).

The first method to engineer a BAC in *E. coli* is the RecA-based method (Yang *et al.*, 1997). The original RecA-based method, as well as its subsequent improved version (Gong *et al.*, 2002), utilizes a shuttle vector that temporarily introduces the *E. coli* RecA gene into the BAC-host bacteria to restore its recombination competence. In the shuttle vector, a recombination cassette is constructed so that two small stretches of homologous DNA sequences are subcloned by polymerase chain reactions (PCRs) from the BAC (each about 200–500 base pairs or bp). In between the two homologous sequences are the modifications to be introduced into

the BAC (i.e., marker gene insertion, point mutation, or deletion). Once the shuttle vector is introduced into the BAC bacteria, two steps of homologous recombination can be sequentially selected to result in the precise placement of the desired modification in a chosen site on the BAC (i.e., between the two homology sequences). The detailed protocol for engineering the BAC using the RecA method can be found elsewhere (Gong and Yang, 2005).

The second widely used method for BAC engineering is the recombineering method, in which the recombination is supported by the bacteriophage-encoded recombination enzymes, *RecE* and *RecT*, from a prophage (Zhang *et al.*, 1998), or the *Red* genes (*gam*, *bet*, and *exo*) from λ phage (Copeland *et al.*, 2001; Lee *et al.*, 2001). The latest and widely used version for recombineering will be illustrated here (Warming *et al.*, 2005) (Figure 19.4b). First, the BAC is transformed into a specialized *E. coli* strain that can inducibly express the γ-*Red* recombination enzymes and is deficient in the metabolic enzyme gene, *galK*. Second, a linear recombination cassette is generated using PCR, with two homology arms as short as 50 bp, each flanking the *galK* gene. This can serve as both a positive and a negative selection marker in an *E. coli* host that is *galK*-deficient. The cassette is then transformed into the specialized recombination-competent bacteria containing the BAC, and subsequently, induction of recombination leads to precise integration of *galK* into the BAC. In the last step, a second recombination cassette containing the intended modifications (i.e., marker gene insertion, point mutation, or deletion), flanked by the same short homology arms as the previous step, is introduced into the BAC. Induction of recombination allows the replacement of *galK* by the intended modifications, and the resulting *galK negative* bacteria can be readily selected.

In summary, both the RecA-based and the recombineering methods are highly efficient and

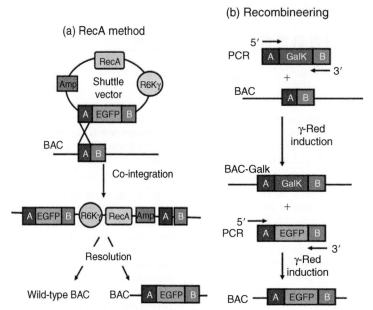

FIGURE 19.4 Engineering BACs by homologous recombination in bacteria. (a) The RecA method. The shuttle vector used for modification contains a *R6Kγ* origin of replication, a RecA gene to support homologous recombination, and a recombination cassette containing two small homology sequences A and B (~500 bp each) flanking the modification to be introduced (e.g., marker insertion, deletion, or point mutation). The shuttle vector plasmid is electroporated into the BAC-host bacteria. Some of the shuttle vectors can undergo homologous recombination with the BAC through either the A or B box, resulting in a complete integration of the shuttle vector into the BAC to form a co-integrate BAC. Bacteria containing co-integrate BACs can be selected by growth on chloramphenicol (chlor) and ampicillin (Amp). If co-integration and resolution occur through different homology boxes, the result is a correctly modified BAC with precise placement of the EGFP marker gene on a chosen locus on the BAC. (b) Recombineering is another BAC modification method based on the bacteriophage-encoded recombination enzymes. The BAC is transformed into an *E. coli* strain that can inducibly express γ-Red recombination enzymes and is deficient in *galK*. A recombination cassette is generated to flank the *galK* gene with two short homology arms. The cassette is then transformed into the BAC in the specialized recombination-competent bacteria, and subsequent induction of homologous recombination leads to precise integration of *galK* into the BAC. A second recombination step, using a cassette containing the intended modifications (i.e., marker gene insertion, deletion, or point mutation) flanked by the same short homology arms, can introduce the modification into a precise location on the BAC.

reliable ways to introduce mutations or marker genes into a BAC. These methods have been widely used in various large-scale BAC engineering projects (Gong *et al.*, 2003; Valenzuela *et al.*, 2003).

Characterization and Preparation of BAC DNA for Transgenic Microinjections

Once the BAC is correctly engineered, appropriate steps should be taken to ensure that the recombination procedures did not introduce any

unwanted rearrangement or deletions. This, however, should be a rather rare occurrence for the BAC modification methods described above. Analyses of modified BACS are often done by restriction enzyme fingerprinting using conventional or pulsed field gel electrophoresis (Yang *et al.*, 1997; Gong and Yang, 2005).

The final critical step for generation of BAC transgenic mice is the purification of intact BAC DNA for pronuclear microinjection into fertilized mouse oocytes. There are three important issues here. First, one needs to purify intact BAC DNA away from fragmented BAC DNA up to 100 kb, which is often present in the DNA preparations. Transgenic integration of such fragmented DNA may introduce positional effects, or the expression of truncated proteins that may confound the studies. Second, one must reduce impurity in the BAC DNA preparations since it may reduce the viability of the injected oocytes, and reduce the rate for generating the BAC transgenic founders. The most successful method, in terms of consistency in generating transgenic founders, is the cesium chloride ultracentrifugation preparation method, which is described in detail elsewhere (Gong and Yang, 2005). Finally, although both circular and linearized BAC DNA (i.e., digested by a restriction enzyme) can be used for generating transgenic mice (Yang et al., 1999), the use of circular BAC DNA runs the risk of integrating BACs that are randomly broken in the middle of the BAC. Thus, a common practice now is to linearize the BAC by digestion with PI-SceI (a rare cutting enzyme which can cut at a single cutting site in the BAC vector) prior to microinjection into the fertilized mouse embryos to generate the transgenic mice (Gong et al., 2003; Gong and Yang, 2005).

Applications of BAC Transgenesis to Study Gene Expression and Gene Function in the Brain

Since the first demonstration of the BAC engineering and BAC transgenic approach (Yang et al., 1997), tremendous strides have been made toward comprehensive and systematic applications of this powerful mouse genetic tool in studying the function and diseases of the mammalian brain (Heintz, 2001; Yang and Gong, 2005). All of these applications are based on the assertion that, for the majority of the genes in a mammalian genome, an appropriately designed BAC transgene has a high probability to confer accurate transgene expression in vivo. The reliability and robustness of the BAC transgenic approach have inspired a myriad of in vivo applications impacting many facets of neuroscience, such as large-scale mapping of brain gene expression (Gong et al., 2003; Zoghbi, 2003), dissecting critical transcriptional regulatory elements for genes (Pfeffer et al., 2002; Mortlock et al., 2003), studying gene function by overexpression (Yang et al., 1999), tracing neuronal circuits (DeFalco et al., 2001), performing cell-type-specific

gene expression profiling (Lobo et al., 2006), and electrophysiology (Day et al., 2006; Kreitzer and Malenka, 2007). In this chapter, we will specifically focus on the application of the BAC transgenic approach to study neurodegenerative disorders, particularly PD.

BAC TRANSGENIC MOUSE MODELS OF NEURODEGENERATION

An important application of BAC transgenesis is to develop mouse models of familial human disorders due to *dominant* genetic mechanisms (Heintz, 2001; Watase and Zoghbi, 2003; Yang and Gong, 2005), as compared to the knockout/knockin technology, which can be used for both *dominant* and *recessive* mechanisms. For a large number of inherited neurodegenerative disorders such as familial Alzheimer's disease (AD), familial PD, polyglutamine disease (i.e., Huntington's disease or HD), and amyotrophic lateral sclerosis (ALS), evidence for dominant genetic mechanisms exist (i.e., dominant inheritance patterns, age-dependent accumulation of toxic proteins). In studying these disorders, conventional transgenic mouse models, overexpressing the toxic mutant proteins have been instrumental in unraveling the dominant disease mechanisms. Moreover, such animal models have been valuable reagents in preclinical studies of candidate therapeutics. In this section, we will discuss the latest advances in applying the BAC transgenic technology to model dominant human neurodegenerative disorders, including PD.

The Advantages of BACs to Model Dominant Neurodegenerative Disorders

A large number of familial neurodegenerative disorders are due to dominant inherited mutations. Examples of such are familial AD due to mutations in amyloid precursor protein (APP) (Price et al., 1998; Tanzi and Bertram, 2005), familial PD due to mutations in α-synuclein or Lrrk2 (Moore et al., 2005; Hardy et al., 2006), all the polyglutamine disorders (Orr and Zoghbi, 2007), and familial ALS mutations in superoxide dismutase 1 (SOD1) (Boillee et al., 2006).

To develop mouse genetic models for these familial neurodegenerative disorders, a set of criteria are needed to evaluate the validity of a given model (Watase and Zoghbi, 2003; Fleming et al., 2005).

First, the mouse model should have *construct valid-ity*, that is to say, the model should have a genetic context that closely resembles that of the patient. Factors that could affect the *construct validity* of a transgenic mouse model include the types of mutations, the expression patterns (including tran-scription and splicing or alternative splicing pat-terns), the expression levels, and whether human or non-human disease genes are used. Second, the mouse model should also have *face validity*, which refers to a model that has recapitulated key behav-ioral and pathological phenotypes (i.e., selective neuronal degeneration) or key molecular pheno-types of a disorder (i.e., biomarkers). Ultimately, a model will be tested for its *predictive validity* to determine whether therapeutic agents that are effec-tive in the mouse model are also effective in the patients or *vice versa*.

Since the late-onset neurodegeneration in humans does not occur until 40–70 years of age, a major challenge in modeling these disorders in mice is the relative short life span of a mouse (about 3 years). Thus, to model human neurodegenera-tive disorders in mice, one would often need to accelerate the disease process by several means in order to "stress" the system, while maintaining a good *construct validity*. The methods to acceler-ate the disease course in mice include: enhancing the expression level of the toxic disease protein by using a stronger promoter, increasing transgene dos-ages (high copy numbers or using homozygotes), or by using more toxic variants of the disease protein (i.e., longer glutamine repeat in polyglutamine dis-orders) (Zoghbi and Botas, 2002). Since enhancing the *face validity* of a mouse model could greatly facilitate its utility in preclinical studies (Beal and Ferrante, 2004), the judicious use of various disease accelerating methods could facilitate the develop-ment of robust and accurate mouse genetic models for both mechanistic and preclinical studies.

Using BAC transgenes to model dominant neu-rodegenerative disorders is advantageous because such models often exhibit excellent *face validity* (i.e., disease-like phenotypes) while still maintain-ing a relatively good *construct validity*. The BAC transgenes have good *construct validity* since they often contain the *entire genomic locus* for the dis-ease gene; therefore, the expression of the mutant gene, including transcription, splicing, translation, and trafficking, is entirely under the control of the endogenous regulatory elements. Moreover, the copy number of the stably integrated BAC transgene usually ranges from 1 to 10 copies, thus the level of disease protein expression in a BAC model is rela-tively physiological, from 0.5- to 5-fold the endog-enous level, as compared to an exogenous promoter, which may drive the mutant protein expression at a level that is much more deviated from the endog-enous level. Additionally, as illustrated before, the BAC transgene is less likely than a conventional transgene to be subject to positional effects. Finally, another significant advantage of the BAC transgenic approach, as compared to the knockin approach to model a dominant human disorder (by introduc-ing human mutations into the endogenous murine homolog of a disease gene), is the fact that the BAC approach allows the usage of the entire human genomic locus for disease modeling. Thus, if any subtle differences between the human and mouse disease gene context (i.e., gene expression regula-tory elements and/or the subtle coding region dif-ferences) may influence the disease process, the use of human BAC transgenic approach would be more advantageous.

Examples supporting the robustness of using large human genomic insert clones (BACs and YACs) to model human neurodegenerative disor-ders include: the YAC transgenic mouse models for AD, HD, and spinal bulbar muscular atrophy; and BAC transgenic mouse models for HD (Gray *et al.*, 2008) and spinocerebellar ataxia 8 (SCA8) (Moseley *et al.*, 2006). Common features of these successful large genomic insert mouse models include the use of human genomic clones containing the entire disease gene loci, modest levels of overex-pression (0.5- to 3-fold the endogenous level), and excellent disease-like phenotypes of these models.

A BAC Transgenic Mouse Model of HD Recapitulates Phenotypic Features of Adult-Onset HD

In further support of the notion that the BAC trans-genic system is a powerful tool to model dominant neurodegenerative disorders in mice; our laboratory has recently developed and characterized a robust, novel BAC transgenic model for HD (Gray *et al.*, 2008). We used a 240 kb human BAC encompassing the entire human huntingtin genomic locus (about 170 kb in size) which expresses full-length mhtt with a 97 polyglutamine repeat. BACHD mice have integrated five copies of the BAC transgene, and overexpress mhtt at about 2-fold the endogenous murine huntingtin level. Full-length mhtt is expressed under the endogenous human regulatory elements on the BAC, and can rescue the embryonic lethality phenotype of the murine huntingtin disease homolog (Hdh) knockout mice (Zeitlin et al., 1995). Phenotypic analyses demonstrate that BACHD mice exhibit robust HD-like phenotypes. At 6 months, these mice begin to exhibit significant Rotarod deficits and electrophysiological alterations, which precede the onset of atrophy and dark neuron degeneration restricted to the cortex and striatum, a pattern that recapitulates that of the HD patients. Finally, unlike the other full-length mhtt mouse models, BACHD mice exhibit an mhtt aggregation pattern reminiscent of that in adult-onset HD: a predominant accumulation of large mhtt aggregates in the neuropil, and only a small amount of mhtt aggregated the nucleus in the cortex. Thus, despite the widespread expression of mhtt, BACHD mice have selective mhtt aggregation and neuropathology that is similar to that found in adult-onset HD patients. In summary, the BACHD model provides an excellent proof-of-principle that BAC transgenesis is a robust technology by which to model neurodegenerative disorders elicited by dominant toxic proteins.

APPLICATIONS OF BAC TRANSGENESIS TO STUDY PD

PD is the second most prevalent neurodegenerative disease, and is characterized pathologically by selective degeneration of DA neurons in the SNc, and by the formation of intraneuronal inclusions known as Lewy bodies (LBs) and Lewy neurites (LNs). Although DA neurons are clearly affected in PD, pathological changes beyond SNc in other brain regions, such as the brain stem and cortex, are also documented (Braak et al., 2003). Five familial genetic mutations are found to cause rare inherited forms of PD. These genes include dominant mutations in α-synuclein and LRRK2, and recessive mutations in Parkin, Pink1, and DJ-1 (Moore et al., 2005; Hardy et al., 2006). Several mouse genetic models based on mutant PD genes (i.e., Parkin, DJ-1, and α-synuclein) have been generated (Fernagut and Chesselet, 2004; Fleming and Chesselet, 2005a). Since review of these existing mouse models of PD is the focus of other chapters in the current book, as well as in recent reviews

elsewhere (Fleming et al., 2005b), it is beyond the scope of this chapter to review the details of the existing PD mice. In brief, however, the existing mouse models of PD have recapitulated nigrostriatal neuronal dysfunction phenotypes, and are thus useful to study aspects of disease pathogenic mechanisms (Masliah et al., 2000; Goldberg et al., 2003, 2005; Itier et al., 2003; Thiruchelvam et al., 2004; Von Coelln et al., 2004). However, most of the existing mouse genetic models do not exhibit the late-onset DA neuron degeneration phenotype of PD (Fernagut and Chesselet, 2004; Perez and Palmiter, 2005). Because the BAC technology has many advantages in modeling neurodegenerative disorders, it would be important to test whether the BAC technology would also be useful in developing robust and accurate PD mouse models. In this section, we will discuss various applications of BAC transgenesis in PD, from modeling PD in mice, to studying PD pathogenic mechanisms, and testing therapeutics.

A BAC Model Reveals Dominant Toxicity of Truncated Parkin Mutant in DA Neurons

Parkin mutations were initially identified in patients with autosomal recessive juvenile parkinsonism (ARJP) (Kitada et al., 1998; Hattori and Mizuno, 2004). Parkin is an E3 ubiquitin ligase (Shimura et al., 2000) which is thought to play a critical function in proteosomal-mediated protein degradation (Chung et al., 2001), receptor trafficking (Fallon et al., 2006), and in maintaining mitochondrial integrity (Greene et al., 2003; Clark et al., 2006a; Park et al., 2006).

Although parkin loss of function is an established etiology in the pathogenesis of ARJP, recent converging lines of evidence also suggest certain parkin mutants may exert dominant toxicity in vivo. PET imaging studies have demonstrated that heterozygous parkin mutants are risk factors for nigrostriatal dysfunction (Hilker et al., 2001). Human genetic studies have demonstrated that certain heterozygous mutants pose an increased risk of sporadic early-onset PD (Clark et al., 2006b) or a lowered age of onset in familial PD (Sun et al., 2006). Moreover, in tissue culture studies, certain parkin mutants, though not wild-type parkin, exert dominant effects including decreased solubility, altered subcellular localization, increased substrate binding affinity, and enhanced cellular susceptibility to mitochondria toxins and oxidative stress (Hyun *et al.*, 2002, 2005; Cookson *et al.*, 2003; Henn *et al.*, 2005; Sriram *et al.*, 2005; Wang *et al.*, 2005a, b). And finally, a recent *Drosophila* genetic study demonstrates that overexpression of two human parkin mutants (point mutation T240R and a C-terminal truncated mutation with a stop codon instead of glutamine at position 311 [Q311X]), though not wild-type parkin, can elicit selective DA neuron loss in the fly (Sang *et al.*, 2007). These converging lines of evidence lead to an important hypothesis that, besides loss of function, certain parkin mutants may also exert dominant toxicity that could lead to dysfunction and degeneration of the mammalian DA neurons.

To test this hypothesis, our laboratory has developed a BAC transgenic mouse model expressing a C-terminal truncated mutant parkin (Q311X, lacking the C-terminal of parkin) in DA neurons. Expression of mutant protein is driven by the dopamine transporter (DAT) promoter on a murine BAC (DAT itself is not overexpressed). The Parkin Q311X mutation was originally reported as a recessive mutation in a Japanese family (Hattori *et al.*, 1998), and the heterozygous phenotype has not been reported to link to PD. We chose to use a parkin-Q311X mutant for our study based on its demonstrated dominant toxicity in cell culture (Hyun *et al.*, 2002, 2005; Sriram *et al.*, 2005) and in *Drosophila* (Sang *et al.*, 2007).

In collaboration with Sheila Fleming and Marie-Francoise Chesselet (UCLA), we have performed longitudinal behavioral studies of these mice using a battery of assays that are sensitive to nigrostriatal dopamine dysfunction (Fleming and Chesselet, 2005; Hwang *et al.*, 2005). Parkin-Q311X mice exhibit slowly progressive motor deficits, (Lu and Yang, unpublished data). Importantly, control tyrosine hydroxylase (TH)-GFP mice overexpressing GFP in the DA neurons (Gong *et al.*, 2003), in the same inbred strain background (FvB) as the Parkin-Q311X mice, do not exhibit any late-onset behavioral deficits. Since the pathological hallmark of PD is degeneration of DA neurons in SNc, we are currently performing stereological analyses on the Parkin-Q311X and wild-type mice to determine whether the mutant mice exhibit late-onset DA neuron degeneration. In summary, our preliminary study demonstrates that the dominant toxicity from a C-terminal truncated mutant parkin can elicit progressive hypokinetic motor deficits in mice. Our study supports the notion that the BAC transgenic approach is suitable to model the dominant toxicity of PD mutant genes in mice.

BAC Transgenic Approach to Model Dominant PD Mutant Genes

Another potential advantage of BAC transgenesis is to model the dominantly inherited familial PD in mice. Currently there are two such mutant genes, α-synuclein and Lrrk2. α-Synuclein was the first gene identified with mutations associated with dominantly inherited PD kindreds (Polymeropoulos *et al.*, 1997). The mutations in α-synuclein include A53T, A30P, E46K, and most recently, multiplications of the wild-type α-synuclein locus (Hardy *et al.*, 2006). The latter mutation is particularly interesting since human patients in families with duplication of α-synuclein (50% increase in the dosage) manifest PD symptoms and/or dementia in their late 40s and 50s, while those with triplication of α-synuclein locus (100% increase in the dosage) resulted in PD-like symptoms and/or dementia in their 30s (Singleton *et al.*, 2003; Nishioka *et al.*, 2006; Fuchs *et al.*, 2007). This inverse relationship between α-synuclein gene dosage and the age of onset clearly indicates that the level of α-synuclein in the brain is a critical factor governing the age of onset for PD and related dementia in patients.

To model α-synuclein dominant toxicity in mice, a variety of conventional transgenic mice have been generated using exogenous promoters (i.e., prion promoter, PDGF promoter, and TH promoter) (Fernagut and Chesselet, 2004). Some of these mouse lines have recapitulated progressive motor deficits and pathological changes such as α-synuclein aggregation and loss of DA neuron terminals. Two of the mouse lines generated using the prion promoter exhibit degeneration of motor neurons in the brain stem and spinal cord, but only one mouse line using the TH promoter (out of several such lines generated) exhibits late-onset loss of TH (+) DA neurons in SNc (Thiruchelvam *et al.*, 2003; Fernagut and Chesselet, 2004). Since the size of the α-synuclein genomic locus in humans is about 100 kb, and α-synuclein multiplication is an established etiology for PD, the use of BAC transgenesis to increase wild-type (or mutant) human α-synuclein gene dosage would have ideal *construct validity* in modeling α-synuclein multiplications in mice. Thus far, one such transgenic mouse model has been generated based on a PAC (Gispert *et al.*, 2003), a type of large insert bacteria clone similar to BAC. Additionally, our laboratory has generated a series of BAC transgenic mouse models of a synuclein multiplication (Murphy and Yang, unpublished data). Analyses of the PAC-α-synuclein mice thus far reveal widespread expression of human α-synuclein, but no behavioral and pathological phenotypes have been reported for these mice.

Lrrk2 is the second dominantly inherited PD mutant gene. At least nine pathogenic mutations have been identified thus far, one of them, G2019S, is found in 5% of familial PD patients and 1.5% of sporadic PD cases (Paisan-Ruiz et al., 2004; Zimprich et al., 2004; Clark et al., 2006b). Thus, the Lrrk2 allele appears to be a relatively common dominant risk allele for sporadic PD. Pathologically Lrrk2 patients closely resemble that of the sporadic PD, exhibiting not only DA neuron loss, but also accumulation of LBs and LNs in substantia nigra, as well as in other brain regions. Some Lrrk2 patients also have Tau pathology (Zimprich et al., 2004; Galpern and Lang, 2006). Several Lrrk2 point mutations have been shown to exhibit dominant toxicity in cultured neuronal cell lines and in primary neurons (Smith et al., 2006; West et al., 2007). Since human Lrrk2 locus is about 140 kb in size and well within the capacity of a BAC, the BAC transgenic approach can also be applied to model Lrrk2 mutant toxicity in vivo. Indeed, two groups have already generated BAC transgenic mice overexpressing wild-type or mutant Lrrk2 (Chenjian Li, Weill Medical College of Cornell University, personal communications; Melrose et al., 2007). Phenotypic analyses of these BAC Lrrk2 mice are underway and may be crucial to address whether dominant toxicity of Lrrk2 can elicit key features of PD in mice.

BAC Transgenic Approach to Study Pathogenic Mechanisms in PD

In addition to generating novel transgenic mouse models of PD, the BAC transgenic approach can also be applied to study pathogenic mechanisms in PD. In this section, we will highlight applications of BAC mice to study pathological PD neural circuits, to identify cell-type-specific gene expression changes, and to test genetic modifiers of PD in vivo.

BAC-GFP Reporter Mice to Study Pathological Neuronal Circuits in PD

It has long been recognized that the loss of DA neurons in SNc results in a series of compensatory changes in the basal ganglia (BG) neuronal circuits, some of which may be maladaptive, and that lead to the clinical symptoms in PD (Albin et al., 1989; DeLong and Wichmann, 2007). The strongest evidence that supports this notion is the demonstration, using non-human primate MPTP models, of the profound circuitry changes in globus pallidus (GP) and subthalamic nucleus (DeLong and Wichmann, 2007). Based on such mechanistic understanding, DBS at these latter sites is clinically beneficial for alleviating motor symptoms in PD patients. In the striatum, the major population of neurons that receive DA neuron input from SNc are medium spiny neurons (MSNs), which constitute about 90% of the neurons in the striatum. MSNs are subdivided into two intermixed subpopulations: the striatonigral neurons (the direct pathway), which project to substantia nigra, and striatopallidal neurons (the indirect pathway), which project to GP externa. A classical model of BG function suggests that loss of DA neurons in the SNc may exert differential effects on the two MSN subtypes: it may activate striatopallidal neurons that normally inhibit movement, and it may inhibit striatonigral neurons that normally activate movement (Albin et al., 1989; Graybiel et al., 2000). Direct evidence for such a model has been difficult to obtain since the two MSN subpopulations are mosaically distributed and are morphologically indistinguishable while they are still alive.

With the advance of BAC transgenesis, particularly the availability of the GENSAT BAC-GFP mice that selectively label the striatonigral neurons (i.e., Drd1-GFP and Chrm4-GFP) and striatopallidal neurons (i.e., Drd2-GFP and AdorA2a-GFP) (Gong et al., 2003), rapid advances have been made to examine the electrophysiological properties of these MSN subtypes under normal physiological conditions, and in models of PD. In a chemical model of PD (i.e., reserpine administration), Surmeier and colleagues discovered that striatopallidal neurons, but not striatonigral neurons, exhibit a rapid and profound loss of glutamatergic synapses due to the dysregulation of a specific calcium channel, Cav1.3 L-type (Day et al., 2006). In a second study, Kreitzer and Malenka (2007) found that striatopallidal neurons selectively express a form of plasticity called endocannabinoid-mediated long-term depression (eCB-LTD), which is dependent on dopamine D2 receptor activation. In the reserpine model of PD, eCB-LTD is absent in the striatopallidal neurons. Interestingly, motor deficits in these models can be rescued by

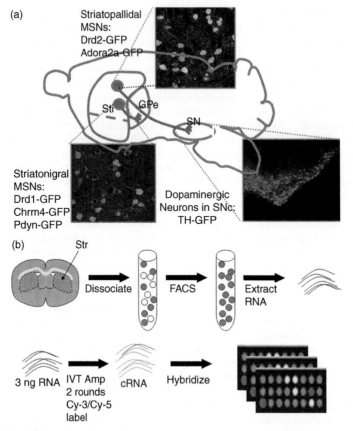

FIGURE 19.5 BAC-GFP mice to study basal ganglia neuronal circuits and FACS array. (a) Schematic drawing of a sagittal mouse brain section illustrating projection of striatopallidal MSNs to the GPe and of striatonigral MSNs to the substantia nigra (SN). GENSAT-GFP mice (i.e., D1-GFP, D2-GFP, and TH-GFP) have been generated to label the neuronal types that are relevant to the pathological PD circuit (Gong et al., 2003). These mice can be obtained from Mutant Mouse Regional Resource Centers (www.mmrrc.org). (b) Schematic drawing of the FACS-array procedure to dissect live striata from BAC-GFP transgenic mice, enzymatically dissociate the striatal neurons and use FACS to purify live (GFP positive and propidium iodide negative) striatonigral or striatopallidal neurons. RNA isolated from the sorted neurons are amplified before being used in microarray studies (Lobo et al., 2006).

either the D2 agonists or inhibitors of endocannabinoid degradation. These elegant studies demonstrate the great potential of using the library of fluorescent protein reporter mice (i.e., GENSAT mice; Figure 19.5a) for striatal, cell-type-specific electrophysiological studies to investigate the pathological neuronal circuits in PD mice.

An important aspect of study in the pathological neuronal circuit in PD is the analysis of cell-type-specific gene expression changes in distinct neuronal types in the brain (i.e., MSNs and DA neurons).

Profiling cell-type-specific gene expression can be done using transcripts from single neurons (Tietjen et al., 2003) obtained by laser-captured microdissection (Kamme et al., 2003). Since certain neuronal types of interest, such as MSN subtypes, are completely intermixed with others and difficult to isolate, several novel technologies have been developed to purify genetically labeled neuronal types for gene expression profiling (Lobo et al., 2006; Nelson et al., 2006).

As shown in Figure 19.5b, we have developed one such approach, termed FACS-array (fluorescent activated cell sorting array), which allows purification of genetically labeled MSN subtypes by FACS. We applied FACS-array to purify striatonigral and striatopallidal MSNs and isolated RNA from about 2000 to 5000 FACS-sorted GFP(+) live neurons (as indicated by propidium iodide negative) for microarray analyses (Lobo *et al.*, 2006). Dozens of novel striatal MSN-subtype-specific genes were identified and validated in the study. One such gene, striatonigral-specific transcription factor Ebf1, was shown to be an essential regulator of striatonigral neuron differentiation. Since FACS-array technology can be used in young adult mice (up to 2 months old), it may also be useful to study early pathological, cell-type-specific gene expression changes in mouse models of PD.

Cre/LoxP Conditional Mouse Model of PD

Another important application of the BAC approach is to develop conditional mouse models to study whether pathological cell–cell interactions could play a role in PD pathogenesis. Many neurodegenerative disorders, including familial PD (i.e., α-synuclein), are characterized by widely expressed toxic proteins that can elicit selective degeneration of a subset of neurons in the brain. The cellular and molecular mechanisms underlying such selective neuronal toxicity remain elusive, but could formerly include a non-cell autonomous component in which neighboring cells (i.e., glia) could exert toxic effects on the vulnerable neurons. Recent genetic analyses, primarily using mouse genetic models, have demonstrated evidence of pathological cell–cell interactions in mouse models of polyglutamine disorders such as HD (Gu *et al.*, 2005), SCA7 (Custer *et al.*, 2006), ALS (Boillee *et al.*, 2006), frontotemporal dementia (Forman *et al.*, 2004), and synucleinopathy (Yazawa *et al.*, 2005). In the latter case, expression of α-synuclein in oligodendrocytes was shown to elicit progressive neuronal toxicity in a mouse model of multi-system atrophy (*ibid*).

In PD, dominantly inherited mutant genes (i.e., α-synuclein and Lrrk2) are also broadly expressed in the brain. Theoretically, expression of these toxic mutant proteins could induce selective DA neuron toxicity via either a cell autonomous mechanism or a non-cell autonomous mechanism. To experimentally address the question *in vivo*, one can develop a conditional mouse model to selectively switch on or off the expression of the toxic disease protein

in distinct neuronal cell types in the brain (Figure 19.6). If switching on the toxic PD protein expression in DA neurons is sufficient to elicit PD-like pheno-types, or switching off the toxic PD protein expression in DA neurons greatly reduces such phenotypes, it would strongly suggest a cell autonomous mechanism. Alternatively, if switching on the mutant PD protein expression in another cell type (i.e., glia or microglia) can elicit DA neuron degeneration and/or switching off the mutant PD protein expression in these cells alleviate the DA neuron phenotype, it would argue for a non-cell autonomous mechanism.

To develop such a conditional mouse model of PD, one would require the use of a site-specific recombinase Cre, which can bind to two properly oriented LoxP sites and efficiently delete the sequence flanked by the two LoxP sites (floxed) (Novak *et al.*, 2000; Branda and Dymecki, 2004). The conditional activation model can be generated by targeting into a ubiquitously expressed murine locus (i.e., Rosa26) (Soriano, 1999), a floxed transcription stop sequence preceding the mutant PD gene. In this case, Cre recombinase can then selectively switch on mutant PD gene expression in different cell types. Our laboratory has used this approach to generate a conditional activation model of HD that permits us to selectively turn on mhtt expression in different neurons in the brain (Gu *et al.*, 2005). This model has clearly demonstrated that cell–cell interactions contribute to cortical pathogenesis in HD (*ibid*). The second type of conditional mouse model is built to switch off expression of a PD gene in specific cell types (conditional inactivation model). BAC models are suitable to generate such a model since BAC transgenes usually have multiple copies integrated, and thus the Cre recombinase can efficiently remove the BAC transgene *in vivo* (Figure 19.6). In such a model, an important exon of the PD gene on the BAC (i.e., the one containing the translation initiation codon) can be floxed. Thus, at base line, this model will be a dominant BAC model of PD, and upon crossing to a Cre mouse line, the expression of the PD gene is selectively switched off in the neurons that express Cre. Our laboratory has used this BAC inactivation strategy to study cell–cell interaction in HD. We have developed BACHD mice in which mhtt-exon 1 is floxed; hence the expression of full-length mhtt can be efficiently removed by Cre (M. Gray and X. W. Yang, unpublished data). This model will be valuable to study how pathological cell–cell

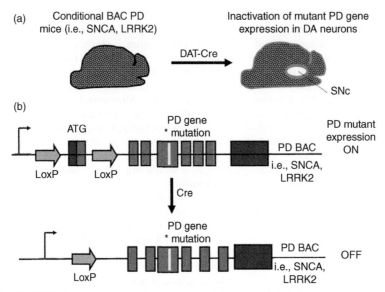

FIGURE 19.6 Conditional BAC mouse model to study pathological cell–cell interactions in PD. (a) Schematic drawing of a conditional BAC PD mice in which expression of the toxic PD gene can be switched off by Cre recombinase in the DA neurons in SNc by Cre (i.e., Dat-Cre). (b) A schematic drawing of a BAC construct in which a critical exon of a mutant PD gene (i.e., the one containing the translation initiation codon) can be flanked by two LoxP sites (floxed). When crossed to a cell-type-specific Cre mouse lines, the expression of the mutant PD gene from the conditional BAC PD model can be selectively switched off in the specific cell types. Using such a model, one may address the pathogenic role of mutant PD gene expression in specific neuronal and non-neuronal cell types in the brain.

interactions elicited by full-length mhtt can contribute to HD pathogenesis.

Similarly combined BAC transgenesis and Cre/LoxP strategies can be applied to study pathological cell–cell interactions in PD. For example, Cre/LoxP conditional BAC transgenic mouse models for mutant α-synuclein or Lrrk2 can be generated. These conditional BAC transgenic models of PD should express the toxic proteins broadly in the brain under their endogenous promoter and regulatory elements. If such models have PD-like motor deficits and neuropathology, one can cross them with Cre mice to selectively remove PD gene expression in distinct cell types such as DA neurons, or non-neuronal cells (i.e., astrocytes, oligodendrocytes, and microglia), to assess the pathogenic roles of disease protein expression in these cells. Such studies may help to uncover cell types in which toxic protein expression is essential to PD pathogenesis, which in turn may help to develop novel therapeutic strategies to modify the disease processes.

Testing Genetic Modifiers of PD in Mice

Besides developing mouse models, BAC transgenesis is also promising in the study of disease mechanisms and in testing therapeutic strategies. One such potential application would be to test genetic modifiers of the disease. Potential genetic modifiers of disease pathogenesis can be readily identified in genetic model organisms suitable for a large-scale genetic screen, such as yeast, *Drosophila*, *C. elegans*, and cultured mammalian cells. A substantial number of these modifiers act through a dominant genetic mechanism which can be tested in mice through BAC transgenesis. Breeding modifier mice to the disease mouse models may reveal the ability of these modifiers to suppress disease pathogenesis in the mammalian brain. Since BAC transgenesis is a highly efficient system and can be readily scaled up (Gong *et al.*, 2003), the BAC transgenic approach may be applied to rapidly test a large number of genetic modifiers as potential therapeutic targets for PD.

POTENTIAL PITFALLS AND SOLUTIONS IN GENERATING BAC MODELS OF NEURODEGENERATION

We have discussed thus far the advantages of the BAC transgenic approach to develop novel mouse models of PD and to study the PD pathogenic mechanism. Like any other technique, the BAC transgenic approach also has its potential pitfalls. In this section, we will discuss some of these pitfalls and ways to mitigate them.

One BAC May Contain Multiple Genes

In generating a BAC transgenic mouse model of a neurodegenerative disorder such as PD, one may encounter the problem that a given BAC contains a genomic locus containing not only the gene of interest (i.e., PD gene) but also one or more other genes. A PD mouse model made from this BAC may overexpress multiple proteins, and thus may confound the interpretation of results. Several methods can be used to address this issue. First, one may generate control BAC mice overexpressing the other genes on the BAC, but not the PD gene (which is deleted). An example of such a BAC control mouse can be found in the study of the overexpression phenotype of a transcription factor, Zipro1 (Yang et al., 1999). Thus, judicious use of control BAC transgenic mice may help to interpret the phenotypes. A second approach, one that is favored by us now, is to delete the coding regions of all the other non-relevant genes on the BAC prior to the generation of BAC transgenic mice. This approach is better now since all the genes in the genome are annotated, thus one can readily locate and delete the other genes on the BAC using homologous recombination in bacteria (Yang et al., 1997; Gong and Yang, 2005). Mice generated by the latter approach should only express the desired mutant proteins, but not other genes on the BAC.

Transgene Integration Effects

BAC transgenes are similar to conventional transgenes in that they may induce rearrangement and/or deletion at the genomic locus that they integrate. Since these transgenes are randomly integrated into the genome in certain genomic loci, such integration effects may induce phenotypes of their own. To avoid such confounding phenotypes unrelated to the expression of the transgene, as a general rule, one should at least analyze two independent transgenic mouse lines derived from different founders. If both mouse lines exhibit dosage-dependent phenotypes that are similar, it is highly unlikely that these phenotypes are due to random transgene integration effects.

Integration of Fragmented BAC Transgene

BACs are large genomic insert clones and are more prone to fragmentation during DNA preparation compared to the smaller conventional DNA inserts. Integration of fragmented BAC inserts could lead to expression of unintended truncated proteins that may complicate the interpretation of the results. Transgene fragmentation is a relatively common problem with YAC transgenes (Giraldo and Montoliu, 2001), but could also occur in BAC transgenes. To avoid the fragmentation of BAC transgene, two measures could be taken. First, one needs to prepare intact BAC DNA for pronuclear injection to generate BAC transgenic mice. Experiences of GENSAT as well as from our laboratory demonstrate that the best method to prepare intact and clean BAC DNA for pronuclear microinjection is the use of cesium chloride ultracentrifugation preparation. Prior to injection, analyses of the BAC DNA preparation using pulsed field gel electrophoresis are also critical to assess the integrity of the BAC DNA (Yang et al., 1997; Gong et al., 2003; Gong and Yang, 2005). With these measures, one can maximize the rate of obtaining transgenic mice with intact BAC inserts.

Inbred Mouse Strain Consideration

Inbred mouse strains have important differences in their behavioral and neuroanatomical characteristics, and in their susceptibility to neurotoxic insults (Schauwecker, 2002). Two commonly used inbred strains in transgenic studies are FvB and C57/BL6 (Watase and Zoghbi, 2003). Oocytes from FvB mice are relatively easy for pronuclear injection of DNA due to their large pronucleus size. FvB mice are also very good breeders with relatively large litter sizes. However, FvB mice have poor visual acuity and are not suitable for visual-based cognitive studies (Voikar et al., 2001). They are also inherently hyperactive (ibid). One known characteristic of FvB mice relevant to modeling neurodegenerative diseases is that they are highly susceptible to neurodegeneration caused by excitotoxins, such as kainate, compared to other mouse strains (Reddy et al., 1998; Hodgson et al., 1999). Several polyglutamine disease transgenic models (i.e., HD and SCA1) generated in FvB background have produced robust neurodegeneration phenotypes (Holmes et al., 2002). Our DAT-ParkinQ311X model (and BACHD model) is generated and maintained in the FvB background, and we are currently determining whether these mice have reproduced the pathological features of PD.

A second inbred mouse strain frequently used in transgenic studies is the C57BL/6 strain. This strain is excellent for certain cognitive behavior tests (Voikar et al., 2001). However, since the C57BL/6 strain is inherently less exploratory and more anxious (ibid), this strain may be difficult for detecting hypokinesia. That the DA neurons are fewer in number in the C57BL/6 strain compared to other strains (such as FvB) is highly relevant to modeling PD (Nelson et al., 1996). Furthermore, several groups have shown that C57BL/6 mice are particularly sensitive to MPTP-induced degeneration of DA neurons (Muthane et al., 1994; German et al., 1996; Sedelis et al., 2000). Thus, C57BL/6 mice may be particularly susceptible to DA neuron damage, at least in the case of MPTP. Mouse models of α-synuclein thus far are generated in either C57BL/6 background or in a F1 background between C57BL/6 and DBA, CBA, or C3H. In summary, both C57BL/6 and FvB strains are both being used to model PD, and additional studies are needed to reveal which inbred strain(s) are appropriate for modeling PD and for testing PD therapeutics.

Species Consideration

Although the main theme of this chapter concerns the generation of BAC transgenic mouse models of PD, one should also keep in mind that there may be other mammalian or vertebrate species that are suitable to study PD, since species-specific differences may determine their vulnerability to a particular PD-related insult (i.e., genetic mutation or environmental toxin). For example, the non-human primate model appears to be much more sensitive to MPTP

than the rodent model (Sherer et al., 2003; Bove et al., 2005). Furthermore, for α-synuclein, one of the pathogenic alleles in humans (A53T) is the normal allele in rodents. Thus, species-specific factors may modulate the phenotypic severity of a PD model. One advantage of the BAC transgenic approach, as compared to the gene-targeting approach, is the possibility that it be used in other vertebrate species such as zebrafish, rats, and sheep (Houdebine, 2000; Yang et al., 2006). Developing BAC models of PD in vertebrate species other than mice may be used to explore these species-specific differences, and help in developing a potentially more robust model of PD, or in taking the advantages of these other models for use in various aspects of drug development (i.e., drug screening in zebrafish models). Finally, a collection of such BAC models in different species may help to elucidate both common pathogenic mechanisms in different models, and species-specific effects that may provide clues to disease pathogenesis and treatment.

SUMMARY AND FUTURE PERSPECTIVE

A major challenge in the study of neurodegenerative disorders such as PD is paving a rational path from the discovery of disease genes to the development of novel, mechanistic-based therapeutics for the patients. A critical step in such a rational approach to study brain diseases is the development of novel animal models that are based on disease mutations and that recapitulate key disease-like phenotypes. The recent advances in mouse molecular genetics, including BAC transgenesis, provide unprecedented opportunities to develop robust and accurate animal models for the study of disease mechanisms, and ultimately may lead to the preclinical testing of potential therapeutics. Applications of BAC transgenic technology, although still in its infancy stage in the PD field, may soon become an essential genetic tool in our quest to understand the disease mechanisms and to search for disease-modifying treatments for PD.

ACKNOWLEDGEMENTS

We would like to thank our colleagues at UCLA and other institutions for their original contributions cited in this chapter; Nat Heintz and NINDS GENSAT project at

Rockefeller University for providing the BAC-GFP mice used in our study and in the illustration; Chenjian Li at Weill Medical College of Cornell University for valuable input on the manuscript. We regret any unintentional omission of any references. Aspects of the work were supported by PHS grants P50 NS38367 and U54 ES12078, an American Parkinson Disease Association Center grant to UCLA; PHS (NINDS) grants 5R01NS049501 and 1R21NS047391, grants from the Hereditary Disease Foundation, and a 2003 Stein-Oppenheimer endowment award (UCLA) to XWY.

REFERENCES

Albin, R. L., Young, A. B., and Penney, J. B. (1989). The functional anatomy of basal ganglia disorders. *Trends Neurosci* 12, 366–375.

Antoch, M. P., Song, E. J., Chang, A. M., Vitaterna, M. H., Zhao, Y., Wilsbacher, L. D., Sangoram, A. M., King, D. P., Pinto, L. H., and Takahashi, J. S. (1997). Functional identification of the mouse circadian Clock gene by transgenic BAC rescue. *Cell* 89, 655–667.

Beal, M. F., and Ferrante, R. J. (2004). Experimental therapeutics in transgenic mouse models of Huntington's disease. *Nat Rev Neurosci* 5, 373–384.

Bloem, B. R., Irwin, I., Buruma, O. J., Haan, J., Roos, R. A., Tetrud, J. W., and Langston, J. W. (1990). The MPTP model: Versatile contributions to the treatment of idiopathic Parkinson's disease. *J Neurol Sci* 97, 273–293.

Boillee, S., Vande Velde, C., and Cleveland, D. W. (2006). ALS: A disease of motor neurons and their nonneuronal neighbors. *Neuron* 52, 39–59.

Bove, J., Prou, D., Perier, C., and Przedborski, S. (2005). Toxin-induced models of Parkinson's disease. *NeuroRx* 2, 484–494.

Braak, H., Del Tredici, K., Rub, U., de Vos, R. A., Jansen Steur, E. N., and Braak, E. (2003). Staging of brain pathology related to sporadic Parkinson's disease. *Neurobiol Aging* 24, 197–211.

Branda, C. S., and Dymecki, S. M. (2004). Talking about a revolution: The impact of site-specific recombinases on genetic analyses in mice. *Dev Cell* 6, 7–28.

Chung, K. K., Dawson, V. L., and Dawson, T. M. (2001). The role of the ubiquitin–proteasomal pathway in Parkinson's disease and other neurodegenerative disorders. *Trends Neurosci* 24, S7–S14.

Clark, I. E., Dodson, M. W., Jiang, C., Cao, J. H., Huh, J. R., Seol, J. H., Yoo, S. J., Hay, B. A., and Guo, M. (2006a). *Drosophila* pink1 is required for mitochondrial function and interacts genetically with parkin. *Nature* 441, 1162–1166.

Clark, L. N., Afridi, S., Karlins, E., Wang, Y., Mejia-Santana, H., Harris, J., Louis, E. D., Cote, L. J., Andrews, H., Fahn, S., Waters, C., Ford, B., Frucht, S., Ottman, R., and Marder, K. (2006b). Case–control study of the parkin gene in early-onset Parkinson disease. *Arch Neurol* 63, 548–552.

Cookson, M. R., Lockhart, P. J., McLendon, C., O'Farrell, C., Schlossmacher, M., and Farrer, M. J. (2003). RING finger 1 mutations in Parkin produce altered localization of the protein. *Hum Mol Genet* 12, 2957–2965.

Copeland, N. G., Jenkins, N. A., and Court, D. L. (2001). Recombineering: A powerful new tool for mouse functional genomics. *Nat Rev Genet* 2, 769–779.

Custer, S. K., Garden, G. A., Gill, N., Rueb, U., Libby, R. T., Schultz, C., Guyenet, S. J., Deller, T., Westrum, L. E., Sopher, B. L., and La Spada, A. R. (2006). Bergmann glia expression of polyglutamine-expanded ataxin-7 produces neurodegeneration by impairing glutamate transport. *Nat Neurosci* 9, 1302–1311.

Dauer, W., and Przedborski, S. (2003). Parkinson's disease: Mechanisms and models. *Neuron* 39, 889–909.

Dawson, T. M., and Dawson, V. L. (2003). Molecular pathways of neurodegeneration in Parkinson's disease. *Science* 302, 819–822.

Day, M., Wang, Z., Ding, J., An, X., Ingham, C. A., Shering, A. F., Wokosin, D., Ilijic, E., Sun, Z., Sampson, A. R., Mugnaini, E., Deutch, A. Y., Sesack, S. R., Arbuthnott, G. W., and Surmeier, D. J. (2006). Selective elimination of glutamatergic synapses on striatopallidal neurons in Parkinson disease models. *Nat Neurosci* 9, 251–259.

DeFalco, J., Tomishima, M., Liu, H., Zhao, C., Cai, X., Marth, J. D., Enquist, L., and Friedman, J. M. (2001). Virus-assisted mapping of neural inputs to a feeding center in the hypothalamus. *Science* 291, 2608–2613.

DeLong, M. R., and Wichmann, T. (2007). Circuits and circuit disorders of the basal ganglia. *Arch Neurol* 64, 20–24.

Di Monte, D. A., McCormack, A., Petzinger, G., Janson, A. M., Quik, M., and Langston, W. J. (2000). Relationship among nigrostriatal denervation, parkinsonism, and dyskinesias in the MPTP primate model. *Mov Disord* 15, 459–466.

Fahn, S. (2003). Description of Parkinson's disease as a clinical syndrome. *Ann NY Acad Sci* 991, 1–14.

Fallon, L., Belanger, C. M., Corera, A. T., Kontogiannea, M., Regan-Klapisz, E., Moreau, F., Voortman, J., Haber, M., Rouleau, G., Thorarinsdottir, T., Brice, A., van Bergen En Henegouwen, P. M., and Fon, E. A. (2006). A regulated interaction with the UIM protein Eps15 implicates parkin in EGF receptor trafficking and PI(3)K-Akt signalling. *Nat Cell Biol* 8, 834–842.

Fernagut, P. O., and Chesselet, M. F. (2004). Alpha-synuclein and transgenic mouse models. *Neurobiol Dis* **17**, 123–130.

Fleming, S. M., and Chesselet, M. F. (2005a). Phenotypical characterization of genetic mouse models of Parkinson's disease. In *Animal Models of Movement Disorders* (M. LeDoux, Ed.), pp. 183–192. Elsevier Press.

Fleming, S. M., Fernagut, P. O., and Chesselet, M. F. (2005b). Genetic mouse models of parkinsonism: Strengths and limitations. *NeuroRx* **2**, 495–503.

Forman, M. S., Trojanowski, J. Q., and Lee, V. M. (2004). Neurodegenerative diseases: A decade of discoveries paves the way for therapeutic breakthroughs. *Nat Med* **10**, 1055–1063.

Fuchs, J., Nilsson, C., Kachergus, J., Munz, M., Larsson, E. M., Schule, B., Langston, J. W., Middleton, F. A., Ross, O. A., Hulihan, M., Gasser, T., and Farrer, M. J. (2007). Phenotypic variation in a large Swedish pedigree due to SNCA duplication and triplication. *Neurology* **68**, 916–922.

Galpern, W. R., and Lang, A. E. (2006). Interface between tauopathies and synucleinopathies: A tale of two proteins. *Ann Neurol* **59**, 449–458.

German, D. C., Nelson, E. L., Liang, C. L., Speciale, S. G., Sinton, C. M., and Sonsalla, P. K. (1996). The neurotoxin MPTP causes degeneration of specific nucleus A8, A9 and A10 dopaminergic neurons in the mouse. *Neurodegeneration* **5**, 299–312.

Giraldo, P., and Montoliu, L. (2001). Size matters: Use of YACs, BACs and PACs in transgenic animals. *Transgenic Res* **10**, 83–103.

Gispert, S., Del Turco, D., Garrett, L., Chen, A., Bernard, D. J., Hamm-Clement, J., Korf, H. W., Deller, T., Braak, H., Auburger, G., and Nussbaum, R. L. (2003). Transgenic mice expressing mutant A53T human alpha-synuclein show neuronal dysfunction in the absence of aggregate formation. *Mol Cell Neurosci* **24**, 419–429.

Goldberg, M. S., Fleming, S. M., Palacino, J. J., Cepeda, C., Lam, H. A., Bhatnagar, A., Meloni, E. G., Wu, N., Ackerson, L. C., Klapstein, G. J., Gajendiran, M., Roth, B. L., Chesselet, M. F., Maidment, N. T., Levine, M. S., and Shen, J. (2003). Parkin-deficient mice exhibit nigrostriatal deficits but not loss of dopaminergic neurons. *J Biol Chem* **278**, 43628–43635.

Goldberg, M. S., Pisani, A., Haburcak, M., Vortherms, T. A., Kitada, T., Costa, C., Tong, Y., Martella, G., Tscherter, A., Martins, A., Bernardi, G., Roth, B. L., Pothos, E. N., Calabresi, P., and Shen, J. (2005). Nigrostriatal dopaminergic deficits and hypokinesia caused by inactivation of the familial Parkinsonism-linked gene DJ-1. *Neuron* **45**, 489–496.

Gong, S., and Yang, X. W. (2005). Modification of bacterial artificial chromosomes (BACs) and preparation of intact BAC DNA for generation of transgenic mice. *Curr Protoc Neurosci*, Chapter 5 Molecular Neuroscience. 5.21.1–5.21.13.

Gong, S., Yang, X. W., Li, C., and Heintz, N. (2002). Highly efficient modification of bacterial artificial chromosomes (BACs) using novel shuttle vectors containing the R6Kgamma origin of replication. *Genome Res* **12**, 1992–1998.

Gong, S., Zheng, C., Doughty, M. L., Losos, K., Didkovsky, N., Schambra, U. B., Nowak, N. J., Joyner, A., Leblanc, G., Hatten, M. E., and Heintz, N. (2003). A gene expression atlas of the central nervous system based on bacterial artificial chromosomes. *Nature* **425**, 917–925.

Graybiel, A. M., Canales, J. J., and Capper-Loup, C. (2000). Levodopa-induced dyskinesias and dopamine-dependent stereotypies: A new hypothesis. *Trends Neurosci* **23**, S71–S77.

Gray, M., Shirasaki, D. I., Cepeda, C., André, V. M., Wilburn, B., Lu, X.-H., Tao, J., Yamazaki, I., Li, S.-H., Sun, Y. E., Li, X.-J., Levine, M.S., and Yang, X. W. (2008). Full-Length Human Mutant Huntingtin with a Stable Polyglutamine Repeat Can Elicit Progressive and Selective Neuropathogenesis in BACHD Mice. *J. Neurosci.* **28**, 6182–6195.

Greene, J. C., Whitworth, A. J., Kuo, I., Andrews, L. A., Feany, M. B., and Pallanck, L. J. (2003). Mitochondrial pathology and apoptotic muscle degeneration in *Drosophila* parkin mutants. *Proc Natl Acad Sci USA* **100**, 4078–4083.

Gu, X., Li, C., Wei, W., Lo, V., Gong, S., Li, S. H., Iwasato, T., Itohara, S., Li, X. J., Mody, I., Heintz, N., and Yang, X. W. (2005). Pathological cell–cell interactions elicited by a neuropathogenic form of mutant huntingtin contribute to cortical pathogenesis in HD mice. *Neuron* **46**, 433–444.

Hardy, J., Cai, H., Cookson, M. R., Gwinn-Hardy, K., and Singleton, A. (2006). Genetics of Parkinson's disease and parkinsonism. *Ann Neurol* **60**, 389–398.

Hatten, M. E., and Heintz, N. (2005). Large-scale genomic approaches to brain development and circuitry. *Annu Rev Neurosci* **28**, 89–108.

Hattori, N., Matsumine, H., Asakawa, S., Kitada, T., Yoshino, H., Elibol, B., Brookes, A. J., Yamamura, Y., Kobayashi, T., Wang, M., Yoritaka, A., Minoshima, S., Shimizu, N., and Mizuno, Y. (1998). Point mutations (Thr240Arg and Gln311Stop) [correction of Thr240Arg and Ala311Stop] in the Parkin gene. *Biochem Biophys Res Commun* **249**, 754–758.

Heintz, N. (2000). Analysis of mammalian central nervous system gene expression and function using bacterial artificial chromosome-mediated transgenesis. *Hum Mol Genet* 9, 937–943.

Hattori, N., and Mizuno, Y. (2004). Pathogenetic mechanisms of parkin in Parkinson's disease. *Lancet* 364, 722–724.

Heintz, N. (2001). BAC to the future: The use of bac transgenic mice for neuroscience research. *Nat Rev Neurosci* 2, 861–870.

Henn, I. H., Gostner, J. M., Lackner, P., Tatzelt, J., and Winklhofer, K. F. (2005). Pathogenic mutations inactivate parkin by distinct mechanisms. *J Neurochem* 92, 114–122.

Hilker, R., Klein, C., Ghaemi, M., Kis, B., Strotmann, T., Ozelius, L. J., Lenz, O., Vieregge, P., Herholz, K., Heiss, W. D., and Pramstaller, P. P. (2001). Positron emission tomographic analysis of the nigrostriatal dopaminergic system in familial parkinsonism associated with mutations in the parkin gene. *Ann Neurol* 49, 367–376.

Hodgson, J. G., Agopyan, N., Gutekunst, C. A., Leavitt, B. R., LePiane, F., Singaraja, R., Smith, D. J., Bissada, N., McCutcheon, K., Nasir, J., Jamot, L., Li, X. J., Stevens, M. E., Rosemond, E., Roder, J. C., Phillips, A. G., Rubin, E. M., Hersch, S. M., and Hayden, M. R. (1999). A YAC mouse model for Huntington's disease with full-length mutant huntingtin, cytoplasmic toxicity, and selective striatal neurodegeneration. *Neuron* 23, 181–192.

Holmes, A., Wrenn, C. C., Harris, A. P., Thayer, K. E., and Crawley, J. N. (2002). Behavioral profiles of inbred strains on novel olfactory, spatial and emotional tests for reference memory in mice. *Genes Brain Behav* 1, 55–69.

Houdebine, L. M. (2000). Transgenic animal bioreactors. *Transgenic Res* 9, 305–320.

Hwang, D. Y., Fleming, S. M., Ardayfio, P., Moran-Gates, T., Kim, H., Tarazi, F. I., Chesselet, M. F., and Kim, K. S. (2005). 3,4-Dihydroxyphenylalanine reverses the motor deficits in Pitx3-deficient aphakia mice: Behavioral characterization of a novel genetic model of Parkinson's disease. *J Neurosci* 25, 2132–2137.

Hyun, D. H., Lee, M., Hattori, N., Kubo, S., Mizuno, Y., Halliwell, B., and Jenner, P. (2002). Effect of wild-type or mutant Parkin on oxidative damage, nitric oxide, antioxidant defenses, and the proteasome. *J Biol Chem* 277, 28572–28577.

Hyun, D. H., Lee, M., Halliwell, B., and Jenner, P. (2005). Effect of overexpression of wild-type or mutant parkin on the cellular response induced by toxic insults. *J Neurosci Res* 82, 232–244.

Itier, J. M., Ibanez, P., Mena, M. A., Abbas, N., Cohen-Salmon, C., Bohme, G. A., Laville, M., Pratt, J., Corti, O., Pradier, L., Ret, G., Joubert, C., Periquet, M., Araujo, F., Negroni, J., Casarejos, M. J., Canals, S.,

Solano, R., Serrano, A., Gallego, E., Sanchez, M., Denefle, P., Benavides, J., Tremp, G., Rooney, T. A., Brice, A., and Garcia de Yebenes, J. (2003). Parkin gene inactivation alters behaviour and dopamine neurotransmission in the mouse. *Hum Mol Genet* 12, 2277–2291.

Kamme, F., Salunga, R., Yu, J., Tran, D. T., Zhu, J., Luo, L., Bittner, A., Guo, H. Q., Miller, N., Wan, J., and Erlander, M. (2003). Single-cell microarray analysis in hippocampus CA1: Demonstration and validation of cellular heterogeneity. *J Neurosci* 23, 3607–3615.

Kitada, T., Asakawa, S., Hattori, N., Matsumine, H., Yamamura, Y., Minoshima, S., Yokochi, M., Mizuno, Y., and Shimizu, N. (1998). Mutations in the parkin gene cause autosomal recessive juvenile parkinsonism. *Nature* 392, 605–608.

Kozak, M. (1986). Point mutations define a sequence flanking the AUG initiator codon that modulates translation by eukaryotic ribosomes. *Cell* 44, 283–292.

Kreitzer, A. C., and Malenka, R. C. (2007). Endocannabinoid-mediated rescue of striatal LTD and motor deficits in Parkinson's disease models. *Nature* 445, 643–647.

Lee, E. C., Yu, D., Martinez de Velasco, J., Tessarollo, L., Swing, D. A., Court, D. L., Jenkins, N. A., and Copeland, N. G. (2001). A highly efficient *Escherichia coli*-based chromosome engineering system adapted for recombinogenic targeting and subcloning of BAC DNA. *Genomics* 73, 56–65.

Lobo, M. K., Karsten, S. L., Gray, M., Geschwind, D. H., and Yang, X. W. (2006). FACS-array profiling of striatal projection neuron subtypes in juvenile and adult mouse brains. *Nat Neurosci* 9, 443–452.

Masliah, E., Rockenstein, E., Veinbergs, I., Mallory, M., Hashimoto, M., Takeda, A., Sagara, Y., Sisk, A., and Mucke, L. (2000). Dopaminergic loss and inclusion body formation in alpha-synuclein mice: Implications for neurodegenerative disorders. *Science* 287, 1265–1269.

McGeer, P. L., McGeer, E. G., and Suzuki, J. S. (1977). Aging and extrapyramidal function. *Arch Neurol* 34, 33–35.

Melrose, H. L., Kent, C. B., Taylor, J. P., Dachsel, J. C., Hinkle, K. M., Lincoln, S. J., Mok, S. S., Culvenor, J. G., Masters, C. L., Tyndall, G. M., Bass, D. I., Ahmed, Z., Andorfer, C. A., Ross, O. A., Wszolek, Z. K., Delldonne, A., Dickson, D. W., and Farrer, M. J. (2007). A comparative analysis of leucine-rich repeat kinase 2 (Lrrk2) expression in mouse brain and Lewy body disease. *Neuroscience* 147, 1047–1058.

Moore, D. J., West, A. B., Dawson, V. L., and Dawson, T. M. (2005). Molecular pathophysiology of Parkinson's disease. *Annu Rev Neurosci* 28, 57–87.

Mortlock, D. P., Guenther, C., and Kingsley, D. M. (2003). A general approach for identifying distant regulatory elements applied to the Gdf6 gene. *Genome Res* 13, 2069–2081.

Moseley, M. L., Zu, T., Ikeda, Y., Gao, W., Mosemiller, A. K., Daughters, R. S., Chen, G., Weatherspoon, M. R., Clark, H. B., Ebner, T. J., Day, J. W., and Ranum, L. P. (2006). Bidirectional expression of CUG and CAG expansion transcripts and intranuclear polyglutamine inclusions in spinocerebellar ataxia type 8. *Nat Genet* 38, 758–769.

Muthane, U., Ramsay, K. A., Jiang, H., Jackson-Lewis, V., Donaldson, D., Fernando, S., Ferreira, M., and Przedborski, S. (1994). Differences in nigral neuron number and sensitivity to 1-methyl-4-phenyl-1,2,3,6-tetrahydropyridine in C57/bl and CD-1 mice. *Exp Neurol* 126, 195–204.

Nelson, E. L., Liang, C. L., Sinton, C. M., and German, D. C. (1996). Midbrain dopaminergic neurons in the mouse: Computer-assisted mapping. *J Comp Neurol* 369, 361–371.

Nelson, S. B., Sugino, K., and Hempel, C. M. (2006). The problem of neuronal cell types: A physiological genomics approach. *Trends Neurosci* 29, 339–345.

Nishioka, K., Hayashi, S., Farrer, M. J., Singleton, A. B., Yoshino, H., Imai, H., Kitami, T., Sato, K., Kuroda, R., Tomiyama, H., Mizoguchi, K., Murata, M., Toda, T., Imoto, I., Inazawa, J., Mizuno, Y., and Hattori, N. (2006). Clinical heterogeneity of alpha-synuclein gene duplication in Parkinson's disease. *Ann Neurol* 59, 298–309.

Novak, A., Guo, C., Yang, W., Nagy, A., and Lobe, C. G. (2000). Z/EG, a double reporter mouse line that expresses enhanced green fluorescent protein upon Cre-mediated excision. *Genesis* 28, 147–155.

Nutt, J. G., and Wooten, G. F. (2005). Clinical practice. Diagnosis and initial management of Parkinson's disease. *N Engl J Med* 353, 1021–1027.

Orr, H. T., and Zoghbi, H. Y. (2007). Trinucleotide repeat disorders. *Annu Rev Neurosci* 30, 575–621.

Paisan-Ruiz, C., Jain, S., Evans, E. W., Gilks, W. P., Simon, J., van der Brug, M., Lopez de Munain, A., Aparicio, S., Gil, A. M., Khan, N., Johnson, J., Martinez, J. R., Nicholl, D., Carrera, I. M., Pena, A. S., de Silva, R., Lees, A., Marti-Masso, J. F., Perez-Tur, J., Wood, N. W., and Singleton, A. B. (2004). Cloning of the gene containing mutations that cause PARK8-linked Parkinson's disease. *Neuron* 44, 595–600.

Palmiter, R. D., and Brinster, R. L. (1985). Transgenic mice. *Cell* 41, 343–345.

Park, J., Lee, S. B., Lee, S., Kim, Y., Song, S., Kim, S., Bae, E., Kim, J., Shong, M., Kim, J. M., and Chung, J. (2006). Mitochondrial dysfunction in *Drosophila* PINK1 mutants is complemented by parkin. *Nature* 441, 1157–1161.

Perez, F. A., and Palmiter, R. D. (2005). Parkin-deficient mice are not a robust model of parkinsonism. *Proc Natl Acad Sci USA* 102, 2174–2179.

Pfeffer, P. L., Payer, B., Reim, G., di Magliano, M. P., and Busslinger, M. (2002). The activation and maintenance of Pax2 expression at the mid–hindbrain boundary is controlled by separate enhancers. *Development* 129, 307–318.

Polymeropoulos, M. H., Lavedan, C., Leroy, E., Ide, S. E., Dehejia, A., Dutra, A., Pike, B., Root, H., Rubenstein, J., Boyer, R., Stenroos, E. S., Chandrasekharappa, S., Athanassiadou, A., Papapetropoulos, T., Johnson, W. G., Lazzarini, A. M., Duvoisin, R. C., Di Iorio, G., Golbe, L. I., and Nussbaum, R. L. (1997). Mutation in the alpha-synuclein gene identified in families with Parkinson's disease. *Science* 276, 2045–2047.

Price, D. L., Tanzi, R. E., Borchelt, D. R., and Sisodia, S. S. (1998). Alzheimer's disease: Genetic studies and transgenic models. *Annu Rev Genet* 32, 461–493.

Reddy, P. H., Williams, M., Charles, V., Garrett, L., Pike-Buchanan, L., Whetsell, W. O., Miller, G., Jr., and Tagle, D. A. (1998). Behavioural abnormalities and selective neuronal loss in HD transgenic mice expressing mutated full-length HD cDNA. *Nat Genet* 20, 198–202.

Ricaurte, G. A., Langston, J. W., Delanney, L. E., Irwin, I., Peroutka, S. J., and Forno, L. S. (1986). Fate of nigrostriatal neurons in young mature mice given 1-methyl-4-phenyl-1,2,3,6-tetrahydropyridine: A neuro-chemical and morphological reassessment. *Brain Res* 376, 117–124.

Sang, T. K., Chang, H. Y., Lawless, G. M., Ratnaparkhi, A., Mee, L., Ackerson, L. C., Maidment, N. T., Krantz, D. E., and Jackson, G. R. (2007). A *Drosophila* model of mutant human parkin-induced toxicity demonstrates selective loss of dopaminergic neurons and dependence on cellular dopamine. *J Neurosci* 27, 981–992.

Schauwecker, P. E. (2002). Modulation of cell death by mouse genotype: Differential vulnerability to excitatory amino acid-induced lesions. *Exp Neurol* 178, 219–235.

Sedelis, M., Hofele, K., Auburger, G. W., Morgan, S., Huston, J. P., and Schwarting, R. K. (2000). MPTP susceptibility in the mouse: Behavioral, neurochemical, and histological analysis of gender and strain differences. *Behav Genet* 30, 171–182.

Sherer, T. B., Kim, J. H., Betarbet, R., and Greenamyre, J. T. (2003). Subcutaneous rotenone exposure causes highly selective dopaminergic degeneration and alpha-synuclein aggregation. *Exp Neurol* 179, 9–16.

Shimura, H., Hattori, N., Kubo, S., Mizuno, Y., Asakawa, S., Minoshima, S., Shimizu, N., Iwai, K., Chiba, T., Tanaka, K., and Suzuki, T. (2000). Familial

Parkinson disease gene product, parkin, is a ubiquitin–protein ligase. *Nat Genet* **25**, 302–305.

Shizuya, H., Birren, B., Kim, U. J., Mancino, V., Slepak, T., Tachiiri, Y., and Simon, M. (1992). Cloning and stable maintenance of 300-kilobase-pair fragments of human DNA in Escherichia coli using an F-factor-based vector. *Proc Natl Acad Sci U S A.* **89**, 8794–8797.

Singleton, A. B., Farrer, M., Johnson, J., Singleton, A., Hague, S., Kachergus, J., Hulihan, M., Peuralinna, T., Dutra, A., Nussbaum, R., Lincoln, S., Crawley, A., Hanson, M., Maraganore, D., Adler, C., Cookson, M. R., Muenter, M., Baptista, M., Miller, D., Blancato, J., Hardy, J., and Gwinn-Hardy, K. (2003). Alpha-synuclein locus triplication causes Parkinson's disease. *Science* **302**, 841.

Smith, W. W., Pei, Z., Jiang, H., Dawson, V. L., Dawson, T. M., and Ross, C. A. (2006). Kinase activity of mutant LRRK2 mediates neuronal toxicity. *Nat Neurosci* **9**, 1231–1233.

Soriano, P. (1999). Generalized lacZ expression with the ROSA26 Cre reporter strain. *Nat Genet* **21**, 70–71.

Sriram, S. R., Li, X., Ko, H. S., Chung, K. K., Wong, E., Lim, K. L., Dawson, V. L., and Dawson, T. M. (2005). Familial-associated mutations differentially disrupt the solubility, localization, binding and ubiquitination properties of parkin. *Hum Mol Genet* **14**, 2571–2586.

Sun, M., Latourelle, J. C., Wooten, G. F., Lew, M. F., Klein, C., Shill, H. A., Golbe, L. I., Mark, M. H., Racette, B. A., Perlmutter, J. S., Parsian, A., Guttman, M., Nicholson, G., Xu, G., Wilk, J. B., Saint-Hilaire, M. H., DeStefano, A. L., Prakash, R., Williamson, S., Suchowersky, O., Labelle, N., Growdon, J. H., Singer, C., Watts, R. L., Goldwurm, S., Pezzoli, G., Baker, K. B., Pramstaller, P. P., Burn, D. J., Chinnery, P. F., Sherman, S., Vieregge, P., Litvan, I., Gillis, T., MacDonald, M. E., Myers, R. H., and Gusella, J. F. (2006). Influence of heterozygosity for parkin mutation on onset age in familial Parkinson disease: The GenePD study. *Arch Neurol* **63**, 826–832.

Tanzi, R. E., and Bertram, L. (2005). Twenty years of the Alzheimer's disease amyloid hypothesis: A genetic perspective. *Cell* **120**, 545–555.

Thiruchelvam, M., McCormack, A., Richfield, E. K., Baggs, R. B., Tank, A. W., Di Monte, D. A., and Cory-Slechta, D. A. (2003). Age-related irreversible progressive nigrostriatal dopaminergic neurotoxicity in the paraquat and maneb model of the Parkinson's disease phenotype. *Eur J Neurosci* **18**, 589–600.

Thiruchelvam, M. J., Powers, J. M., Cory-Slechta, D. A., and Richfield, E. K. (2004). Risk factors for dopaminergic neuron loss in human alpha-synuclein transgenic mice. *Eur J Neurosci* **19**, 845–854.

Tietjen, I., Rihel, J. M., Cao, Y., Koentges, G., Zakhary, L., and Dulac, C. (2003). Single-cell transcriptional analysis of neuronal progenitors. *Neuron* **38**, 161–175.

Trojanowski, J. Q., and Lee, V. M. (1998). Aggregation of neurofilament and alpha-synuclein proteins in Lewy bodies: Implications for the pathogenesis of Parkinson disease and Lewy body dementia. *Arch Neurol* **55**, 151–152.

Valenzuela, D. M., Murphy, A. J., Frendewey, D., Gale, N. W., Economides, A. N., Auerbach, W., Poueymirou, W. T., Adams, N. C., Rojas, J., Yasenchak, J., Chernomorsky, R., Boucher, M., Elsasser, A. L., Esau, L., Zheng, J., Griffiths, J. A., Wang, X., Su, H., Xue, Y., Dominguez, M. G., Noguera, I., Torres, R., Macdonald, L. E., Stewart, A. F., DeChiara, T. M., and Yancopoulos, G. D. (2003). High-throughput engineering of the mouse genome coupled with high-resolution expression analysis. *Nat Biotechnol* **21**, 652–659.

Voikar, V., Koks, S., Vasar, E., and Rauvala, H. (2001). Strain and gender differences in the behavior of mouse lines commonly used in transgenic studies. *Physiol Behav* **72**, 271–281.

Von Coelln, R., Thomas, B., Savitt, J. M., Lim, K. L., Sasaki, M., Hess, E. J., Dawson, V. L., and Dawson, T. M. (2004). Loss of locus coeruleus neurons and reduced startle in parkin null mice. *Proc Natl Acad Sci USA* **101**, 10744–10749.

Wang, C., Ko, H. S., Thomas, B., Tsang, F., Chew, K. C., Tay, S. P., Ho, M. W., Lim, T. M., Soong, T. W., Pletnikova, O., Troncoso, J., Dawson, V. L., Dawson, T. M., and Lim, K. L. (2005a). Stress-induced alterations in parkin solubility promote parkin aggregation and compromise parkin's protective function. *Hum Mol Genet* **14**, 3885–3897.

Wang, C., Tan, J. M., Ho, M. W., Zaiden, N., Wong, S. H., Chew, C. L., Eng, P. W., Lim, T. M., Dawson, T. M., and Lim, K. L. (2005b). Alterations in the solubility and intracellular localization of parkin by several familial Parkinson's disease-linked point mutations. *J Neurochem* **93**, 422–431.

Warming, S., Costantino, N., Court, D. L., Jenkins, N. A., and Copeland, N. G. (2005). Simple and highly efficient BAC recombineering using galK selection. *Nucleic Acids Res* **33**, e36.

Watase, K., and Zoghbi, H. Y. (2003). Modelling brain diseases in mice: The challenges of design and analysis. *Nat Rev Genet* **4**, 296–307.

West, A. B., Moore, D. J., Choi, C., Andrabi, S. A., Li, X., Dikeman, D., Biskup, S., Zhang, Z., Lim, K. L., Dawson, V. L., and Dawson, T. M. (2007). Parkinson's disease-associated mutations in LRRK2 link enhanced GTP-binding and kinase activities to neuronal toxicity. *Hum Mol Genet* **16**, 223–232.

Yang, X. W., and Gong, S. (2005). An overview on the generation of BAC transgenic mice for neuroscience research. *Curr Protoc Neurosci.* 5.20.1–5.20.11

Yang, X. W., Model, P., and Heintz, N. (1997). Homologous recombination based modification in *Escherichia coli* and germline transmission in transgenic mice of a bacterial artificial chromosome. *Nat Biotechnol* 15, 859–865.

Yang, X. W., Wynder, C., Doughty, M. L., and Heintz, N. (1999). BAC-mediated gene-dosage analysis reveals a role for Zipro1 (Ru49/Zfp38) in progenitor cell proliferation in cerebellum and skin. *Nat Genet* 22, 327–335.

Yang, Z., Jiang, H., Chachainasakul, T., Gong, S., Yang, X. W., Heintz, N., and Lin, S. (2006). Modified bacterial artificial chromosomes for zebrafish transgenesis. *Methods* 39, 183–188.

Yazawa, I., Giasson, B. I., Sasaki, R., Zhang, B., Joyce, S., Uryu, K., Trojanowski, J. Q., and Lee, V. M. (2005). Mouse model of multiple system atrophy alpha-synuclein expression in oligodendrocytes causes glial and neuronal degeneration. *Neuron* 45, 847–859.

Zhang, Y., Buchholz, F., Muyrers, J. P., and Stewart, A. F. (1998). A new logic for DNA engineering using recombination in *Escherichia coli*. *Nat Genet* 20, 123–128.

Zimprich, A., Biskup, S., Leitner, P., Lichtner, P., Farrer, M., Lincoln, S., Kachergus, J., Hulihan, M., Uitti, R. J., Calne, D. B., Stoessl, A. J., Pfeiffer, R. F., Patenge, N., Carbajal, I. C., Vieregge, P., Asmus, F., Muller-Myhsok, B., Dickson, D. W., Meitinger, T., Strom, T. M., Wszolek, Z. K., and Gasser, T. (2004). Mutations in LRRK2 cause autosomal-dominant parkinsonism with pleomorphic pathology. *Neuron* 44, 601–607.

Zoghbi, H. Y. (2003). Molecular neuroscience: BAC-to-BAC images of the brain. *Nature* 425, 907–908.

Zoghbi, H. Y., and Botas, J. (2002). Mouse and fly models of neurodegeneration. *Trends Genet* 18, 463–471.

20

VIRAL VECTORS: A POTENT APPROACH TO GENERATE GENETIC MODELS OF PARKINSON'S DISEASE

BERNARD L. SCHNEIDER, MERET N. GAUGLER AND PATRICK AEBISCHER

Brain and Mind Institute, Ecole Polytechnique Fédérale de Lausanne (EPFL), Switzerland

INTRODUCTION

Genetic Modeling of Parkinson's Disease

Generating animal models mimicking the pathogenesis of Parkinson's disease (PD) is essential for both our understanding of the disease and the screening of novel therapeutics. These models should reflect cardinal features of the disease, such as the progressive loss of vulnerable neuronal populations, the formation of Lewy body-like protein aggregates, and the typical motor symptoms.

Neurotoxin-induced experimental animal models were historically the first to be available to the investigation of idiopathic PD. Toxins such as 6-hydroxydopamine (6OHDA), 1-methyl-4-phenyl-1,2,3,6-tetrahydropyridine (MPTP), rotenone, paraquat, reserpine, and methamphetamine have been used to develop parkinsonian models in a wide variety of species (Bove *et al.*, 2005). Both 6-OHDA and MPTP can be used to study the pathophysiology of the degenerating nigrostriatal system, as they replicate some of the neurochemical, morphological, and behavioral changes observed in human disease. However, it is unlikely that toxin-based models reproduce all the primary molecular dysfunctions in idiopathic PD.

In recent years, linkage analysis and screening for mutations have identified nine loci (PARK1–9) associated with genetic forms of PD. Three genes have been found mutated in autosomal dominant PD: α-synuclein (*PARK1*) (Polymeropoulos *et al.*, 1997; Kruger *et al.*, 1998; Singleton *et al.*, 2003), LRRK2 (leucine-rich repeat kinase 2, *PARK8*) (Paisan-Ruiz *et al.*, 2004; Zimprich *et al.*, 2004), and UCHL1 (*PARK5*) (Leroy *et al.*, 1998). Four genes are related to recessive forms: Parkin (*PARK2*) (Kitada *et al.*, 1998), DJ-1 (*PARK7*) (Bonifati *et al.*, 2003), PINK1 (*PARK6*) (Valente *et al.*, 2004), and ATP13A2 (*PARK9*) (Ramirez *et al.*, 2006).

Some of these Mendelian forms are rare, including those caused by α-synuclein, DJ1, and PINK1 mutations. Expectedly, variations are observed in the pathogenesis of genetic forms of PD that frequently present atypical features, such as early onset, dementia, and unusual disease progression when compared to sporadic PD. Still, they have the potential to provide insight into the molecular pathways that lead to the differential death of dopaminergic neurons, as these same toxic mechanisms or loss of maintenance functions might be implicated in the pathogenesis of the idiopathic disease (Hofer and Gasser, 2004; Hardy *et al.*, 2006). Thus, introducing a mutated human transgene into an animal is one possible approach to generate models of PD.

Transgenic Animal Models

Specific knockout mice or mice overexpressing the mutated forms of the aforementioned genes can be valuable for gaining insight into the pathogenic mechanisms involved. However, knockout mice for genes involved in recessive parkinsonism, such as DJ-1 (Chen *et al.*, 2005; Kim *et al.*, 2005) and parkin (Itier *et al.*, 2003; Perez and Palmiter, 2005), have failed to display any loss of dopaminergic neurons. Similarly, attempts at generating classical transgenic mouse models to elucidate the role of α-synuclein and its mutant forms resulted in motor symptoms mainly due to motor neuron dysfunction, while the dopaminergic neurons in the *substantia nigra pars compacta* (SNpc) were mostly spared (Fleming *et al.*, 2005).

These findings suggest distinct species-innate differences in the susceptibility of the SNpc that might reach beyond possible developmental compensatory mechanisms encountered in classical transgenesis.

In order to develop a rapid and efficient method to screen for transgene effects across various species, many groups have reverted to the use of viral vectors to promote the selective overexpression of α-synuclein in the SNpc and thus induce a PD-like pathology. In contrast to transgenic mice, the use of viral vectors to overexpress α-synuclein in the rat and primate SNpc has repeatedly been shown to produce a pronounced and differential loss of dopaminergic neurons (Kirik *et al.*, 2002; Lo Bianco *et al.*, 2002; Eslamboli *et al.*, 2007) (Table 20.1).

Viral Vectors for Targeted and Differential Expression in the SN

As opposed to transgenesis, where targeted and differential expression of the desired transgene in restricted cell populations is still proving difficult, the use of viral vectors and their direct administration to certain specific brain areas puts this within technically manageable reach (Figure 20.1). Targeted expression in a high percentage of the neurons in the SNpc is achievable with a single injection site. In addition, the use of viral vectors allows transgene expression at high doses, which may be needed to induce loss of neurons within a short time period. Expression levels might be particularly important for transgenes such as α-synuclein, whose toxic role has been shown to depend directly on gene dosage and protein abundance (Eriksen *et al.*, 2005).

Alpha-synuclein was linked to PD when three missense mutations (A30P, A53T, and later the E46K) were discovered in affected kindreds (Polymeropoulos *et al.*, 1997; Kruger *et al.*, 1998; Zarranz *et al.*, 2004). This has led to the identification of α-synuclein as the one of the major components of Lewy bodies and glial cytoplasmic inclusions in sporadic PD (Spillantini *et al.*, 1997, 1998), and the protein is now associated with a broad class of neurodegenerative disorders characterized by the presence of protein aggregates containing α-synuclein and referred to as "synucleinopathies." The subsequent discovery of genomic multiplications (duplication or triplication) of a locus containing the α-synuclein gene in several PD families further highlights the importance of the endogenous protein dosage in the disease (Singleton *et al.*, 2003; Chartier-Harlin *et al.*, 2004). Even though the mechanistic relationship between α-synuclein expression level and sporadic forms of PD has not to date been firmly established, the existence of several polymorphisms in its promoter region suggests that protein abundance might present a direct risk factor (Farrer *et al.*, 2001). As the use of protein-overexpressing viral vectors greatly accelerates pathogenesis, careful dosage is a crucial factor in defining the time course of the induced degeneration, and thus in finding a model system displaying the dynamic properties most suited to study the mechanistic questions at hand.

As mentioned above, genes implicated in familial forms of PD can be pathogenic both by a gain of toxic function or the loss of a protective mechanism. While familial forms constitute extreme cases with a monogenic cause, sporadic cases are likely to involve complex interactions between these multiple genetic factors. The use of viral vectors offers a flexible means to rapidly generate combinations of genetic modifications in order to weigh the effects of genes and mutations against each other, and study their interaction with environmental triggers. Such an approach should accelerate the development of appropriate genetic models accurately replicating complex polygenic diseases such as PD.

METHODOLOGY

The Choice of Viral Vector

Viral vectors targeting the central nervous system (CNS) have been widely developed in the past

TABLE 20.1 Comparison of viral models based on α-synuclein overexpression in the substantia nigra of rodents and primates

Vector promoter	Transgene	Targeted species	Time to lesion peak	Infectivity in the SNpc	DA neuron loss in the SNpc	Aggregation	Striatal DA fiber loss	Behavior	References
Lenti PGK	Hu α-Syn WT, A30P, A53T	Rat	6 weeks	≤50%	35% (WT) 33% (A30P) 24% (A53T)	Aggregation in neuronal perikarya and neuritis	20% (A30P) 10% (A53T) 15% (WT)	N/S	Lo Bianco et al. (2002)
AAV2 CBA	Hu α-Syn WT, A53T	Rat	8 weeks	>90% overall	30–80%	Cytoplasmic, granular deposits	ca. 50%	Apomorphine rotation, paw reaching test (ca. 25% of rats)	Kirik et al. (2002)
AAV2 CBA	Hu α-Syn A30P-IRES-GFP	Rat	12 months	N/S	53%	Clumps/ring-like aggregates in neuronal perikarya	N/S	N/S	Klein et al. (2002)
AAV2 CMV	Hu α-Syn WT	Rat	13 weeks	80% at the most infected level	50%	Aggregation in perikarya: membrane-bound and nuclear	N/S	N/S	Yamada et al. (2004)
Lenti CMV	Hu α-Syn WT, A30P	Mouse	10–12 months	N/S	10–25%	Cytoplasmic ubiquitin-positive inclusions	N/S	N/S	Lauwers et al. (2003)
AAV2 CBA	Hu α-Syn WT-IRES-GFP	Mouse	24 weeks	10–80% in a given section	25%	No obvious cytoplasmic inclusions	N/D	N/S	St Martin et al. (2007)
AAV2 CBA	Hu α-Syn WT, A53T	Marmoset	16 weeks	90–95%	30–60%	Cytoplasmic inclusions and granular deposits in the SNpc	40–50%	Head-position test	Kirik et al. (2003)
AAV2/5 CBA	Hu α-Syn WT, A53T	Marmoset	12 months	N/S	>40% overall (A53T)	Inclusions in dendrites and axons of DA-positive neurons	N/S	Full body rotation, head bias, motor coordination errors	Eslamboli et al. (2007)

Hu α-Syn = human α-synuclein; N/S = not specified; N/D = not detectable

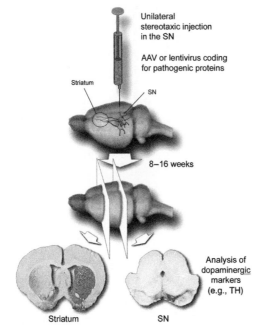

FIGURE 20.1 Schematic description of the procedure. Viral vectors (in most cases lentiviral or AAV vectors) are unilaterally injected in the SN to overexpress pathogenic proteins in neurons. After virus injection, an 8- to 16-week period is usually needed for the lesion to fully develop. The loss of the nigrostriatal neuronal function is analyzed by histological examination to quantify both the number of dopaminergic neurons at the level of the SNpc and the remaining innervation in the striatum, with respect of the non-injected side. Typically, dopaminergic markers including TH, vesicular monoamine transporter 2 (VMAT2), and dopamine transporter (DAT) are used to label the neurons of interest.

decade. While the quiescent nature of post-mitotic neurons constitutes a major obstacle for many vectors, lentiviruses and adeno-associated viruses (AAV) are capable of infecting neurons and/or glial cells at high efficacy. Most remarkably, expression is stable for very long time periods, up to years in the rodent brain. Therefore, they provide means to induce pathology in the SNpc by either overexpressing pathogenic proteins or downregulating factors contributing to the maintenance of dopaminergic neuronal function.

In contrast to transgenic animal models, whose interpretation has to take into consideration the effects of genetic modifications during development and possible adaptive mechanisms masking pathogenesis, the use of viral vectors induces a sudden perturbation in homeostasis of the adult brain, accelerating the development of a pathologic syndrome. Although this could simplify the analysis, particular attention has to be paid to virus handling, in order to carefully control the dosage of transgene and the possible confounding effects of virus injection, such as an inflammatory reaction. The absence of transgenic sequences coding for viral proteins therefore constitutes a crucial factor in the development of viral systems for these applications. Usually, viral vectors induce robust overexpression. In α-synuclein models, degeneration starts to occur within weeks after infection and coincides with the accumulation of the overexpressed protein (Table 20.1). A delay of 2–4 weeks is observed between the acute exposure to the viral vector and the progressive loss of dopaminergic markers in the nigrostriatal pathway. In most cases, loss of dopaminergic neurons is observed over the course of several months. However, the relationship between the level of expression and the time course of the lesion has not been carefully established yet.

Another crucial advantage of viral vector systems is the possibility to use the contralateral hemisphere as an internal control. This is especially useful in motor deficit assessments that rely on functional testing for behavioral asymmetries. A careful optimization of injection volume and viral load is needed to target the SNpc unilaterally and without vector spilling.

Lentivirus

Replication-defective lentiviruses (*Retroviridae* family) derived from human immunodeficiency virus 1 (HIV-1) present both neuronal and glial tropism in the CNS. With a large cloning capacity (up to 10 kb), lentiviral vectors lead to transgene integration into the host genome and long-term expression (>1 year) in both neuronal and non-neuronal cells (Blomer *et al.*, 1997). The recent development of non-integrative lentiviral vectors offers an alternative for long-term trangene expression in postmitotic cell types, avoiding insertional mutagenesis (Philippe *et al.*, 2006; Yanez-Munoz *et al.*, 2006). Lentiviral vectors can be titrated precisely *in vitro* (Delenda and Gaillard, 2005) and lead to stable expression that can be predicted and controlled using inducible systems (Szulc *et al.*, 2006). As they are not retrogradely transported in neuronal cells, HIV-based lentiviruses are injected directly into the SNpc. Additionally, the following limitations need to be carefully addressed: (1) Although viral proteins are not expressed in target cells, impurities and defective viral particles present in concentrated viral preparations (obtained by centrifugation) can induce inflammatory reactions. This is particularly the case following injection into the SN, which inherently appears more susceptible than other brain regions, such as the striatum. (2) Lentiviral vectors show limited diffusion capacities, due to their size and affinity for the extracellular matrix. Multiple injection sites can circumvent this issue, but the procedure remains technically challenging in the rodent SNpc. (3) Lentiviral vectors pseudo-typed with the commonly used VSV-G envelope have a limited tropism for nigral dopaminergic neurons, with infection rates usually below 50% (Lo Bianco *et al.*, 2002).

Adeno-Associated Virus

Recombinant adeno-associated vectors (*Parvoviridae* family) can be produced at high titers and pseudo-typed using capsids derived from several serotypes. Various purification procedures are available, such as ultracentrifugation on cesium chloride gradients or liquid chromatography, allowing for a more stringent quality control of viral batches prior to injection (Grieger *et al.*, 2006). With their small size, AAV vectors can efficiently diffuse through brain parenchyma around the site of injection and, depending on the serotype, can be retrogradely transported along the axons (Kaspar *et al.*, 2002; Burger *et al.*, 2004). They present an excellent tropism for neurons *in vivo*, infecting a high percentage of dopaminergic neurons in the rodent SN (AAV serotypes 1, 2, 5, and 6) (Burger *et al.*, 2004; Paterna *et al.*, 2004). In non-dividing cells, AAV vectors remain mostly episomal and drive long-term expression. Although the assessment of AAV viral infectivity is technically challenging, AAV vectors are promising tools for the development of viral *in vivo* models of PD. Due to high titers and efficient purification, they provide an effective means to target the SNpc with minimal mechanical stress (low injection volume and minimal number of injection sites) and minimal inflammatory reaction. No viral proteins are expressed following transduction of target cells. However, an immune response to capsid proteins can reduce the infectivity of subsequent injections in the case of repeated vector administration into the brain (Mastakov *et al.*, 2002; Peden *et al.*, 2004), a technical constraint avoided by changing capsid serotype between injections. The relatively small packaging capacity of AAV vectors (4.5–5 kb) limits their use to DNA coding sequences that can be efficiently inserted into the viral particles (usually <3 kb).

AAV Serotypes

An increasing focus has recently been put on the seven other naturally occurring AAV serotypes (AAV1 and AAV3–8), which are structurally and functionally different from AAV2. The AAV2 virus genome can be packaged into the capsids of the other AAV serotypes, resulting in a new generation of pseudotyped AAV vectors. Recent studies using vectors derived from alternative AAV serotypes such as AAV1, 4–6 have shown broad tropisms, suggesting that pseudotyping of AAV vectors offers a means to manipulate the pattern of infected cell populations (Gao *et al.*, 2005).

The host range of the AAV6 serotype and its specific properties in regard to the CNS have to date been only poorly investigated. It has been postulated that AAV6 has initially arisen from homologous recombination between AAV1 and AAV2 and would be highly similar to AAV1 in its infection profile (Xiao *et al.*, 1999; Blankinship *et al.*, 2004). AAV1, in turn, has been shown repeatedly to have a broader distribution and higher number of neurons transduced in the CNS than AAV2 (Passini *et al.*, 2003; Wang *et al.*, 2003). Our preliminary data using a control AAV6 virus expressing only green fluorescent protein (GFP), and vectors coding for several forms of α-synuclein under the control of a CMV promoter indeed show a promising infection profile in the SNpc. The significantly higher rate of infection compared to the lentiviral vectors makes it possible to observe more important lesions, and a robust behavioral as well as a cellular phenotype. Also, the vectors derived from the AAV6 offer the advantage of being efficiently retrogradely transported, opening possible alternative ways of administration.

Both lentiviral and AAV vectors present a high tropism for neuronal cells *in vivo*, while glial cells can be infected as well, particularly with the lentiviral vectors. Although PD pathogenesis is likely to be primarily initiated in neuronal cells, a significant role for the glia cannot be excluded. However, this question should be addressed using viral vectors with cell type-specific expression systems.

Adenovirus

Due to inflammatory and innate immune reactions caused by vector administration into the brain parenchyma, adenovirus seems less attractive to model the pathogenesis of PD (Thomas *et al.*, 2001). In the "gutless" version, all adenoviral proteins are supplemented in *trans*, and production depends on a helper adenovirus. Although technically challenging, gutless adenovirus appears less prone to induce immune responses in the host tissues.

A recent study took advantage of the efficient retrograde transport of adenovirus to target the SNpc by infecting nerve terminals in the *striatum* (Kitao *et al.*, 2007). Using genetic recombination, the transgene was overexpressed in dopaminergic neurons of the SNpc projecting to the *striatum*, thereby avoiding virus-induced inflammatory reactions at the site of degeneration. In addition, adenovirus might find applications in the overexpression of large proteins, such as LRRK2, whose coding sequence exceeds the size that can be efficiently packaged in AAV or even lentiviral vectors.

Genes Used for Viral Modeling

Genes associated with familial forms provide a rationale base to understand how intrinsic cell susceptibilities develop into PD, possibly in response to aging and environmental triggers such as toxins. Dominant forms of genetic PD (α-synuclein, LRRK2) are likely to be due to a gained toxic function and therefore might provide causal clues about the pathogenesis of idiopathic PD. Alpha-synuclein is hypothesized to be a triggering event in PD, due to a propensity to change its conformation into amyloidogenic β-sheets as a function of gene dosage and mutations. Thus, α-synuclein has initially been used to generate viral models of PD and is still the most adequate target gene to screen for potential neuroprotective molecules (Table 20.1).

LRRK2 is mutated in some dominant forms of PD, and is therefore another interesting candidate to elucidate mechanisms related to PD pathology (Paisan-Ruiz *et al.*, 2004; Zimprich *et al.*, 2004). LRRK2 mutations appear to be the most com-

mon cause of genetic parkinsonism, accounting for 1–7% of European PD cases and up to 20–40% of PD cases among Ashkenazi Jews and North African Arabs (Zabetian *et al.*, 2006). Its pathogenic role appears to depend on its kinase activity. Increased activity in response to pathogenic mutations suggests a gain of toxic function that remains to be elucidated (West *et al.*, 2005; Smith *et al.*, 2006). However, the large size of the protein and the difficulty to get significant levels of overexpression complicate the implementation of efficient viral transduction for *in vivo* modeling. Dissection of functional domains prior to introduction into viral vectors constitutes a possible approach to address these issues.

Recessive genes involved in PD pathology (parkin, DJ-1, and PINK1), on the other hand, provide valuable insight into protective mechanisms maintaining cellular functions in the SNpc. Clearly, PD-related mutations induce loss of protein or protein function, possibly increasing neuronal susceptibility to insults such as oxidative stress, or decreasing cell capacity to control protein quality and turnover. Parkin has been found to have an E3 ubiquitin ligase activity, targeting a number of substrates including parkin-associated endothelin-receptor like receptor (Pael-R) (Imai *et al.*, 2001), CDCrel-1 (Zhang *et al.*, 2000), and a glycosylated form of α-synuclein (Shimura *et al.*, 2001). Proposed roles for DJ-1 and PINK1 are related to mitochondrial and chaperone functions (Shendelman *et al.*, 2004; Clark *et al.*, 2006; Park *et al.*, 2006).

Although induced loss of gene activity using virus-mediated siRNA expression might be a possible approach to model recessive parkinsonism, no such study has been reported so far. As mentioned above, null mice for parkin and DJ-1 do not display any loss of dopaminergic neurons (Itier *et al.*, 2003; Chen *et al.*, 2005; Kim *et al.*, 2005; Perez and Palmiter,2005). An alternative approach consists in overexpressing known substrates of the enzyme in order to saturate its activity and induce toxic effects by substrate accumulation. For instance, viral gene transfer for CDCrel-1 and Pael-R overexpression, two established substrates of the parkin E3 ubiquitin ligase, has proved deleterious to midbrain dopaminergic function (Dong *et al.*, 2003; Kitao *et al.*, 2007).

Targeted Species

Rats, primates, and to a minor extent mice, have been extensively used for the generation of virus-based PD models (Table 20.1). The rodent SN can be reproducibly targeted by stereotaxic injection, allowing for topical viral infection. Although the mouse SN often appears less vulnerable to genetic insults, it provides a flexible system easily accessible to precise modifications using classical methods of genetic engineering. In contrast, because there are only few genetically engineered rat strains available, protocols involving combined genetic modifications in the rat species usually rely on multiple viral infections.

More specifically, models based on α-synuclein overexpression have been generated in multiple animal species. Although no comparison across species has been reported on yet, data suggest that there might be intrinsic variations in neuronal susceptibilities to α-synuclein toxicity. Interestingly, invertebrate models, such as *Drosophila melanogaster* (Feany and Bender, 2000) and *Caenorhabditis elegans* (Lakso et al., 2003; Kuwahara et al., 2006), show age-related loss of dopaminergic neurons even in the absence of an ortholog to the α-synuclein gene. However, while neuronal loss is observed in the adult rat SN (Kirik et al., 2002; Klein et al., 2002; Lo Bianco et al., 2002), virus-induced overexpression of α-synuclein causes only mild effects in the mouse midbrain (Lauwers et al., 2003; St Martin et al., 2007). Among mammals, an α-synuclein-based virus model in the marmoset (*Callithrix jacchus*) also leads to neuronal loss (Kirik et al., 2003).

However, assessing inter-species differences in vulnerability to α-synuclein is difficult, as the dose of protein expressed needs to be tightly controlled and the relationship between dose, time course, and extent of neuronal lesion to be precisely established. In addition, species specificities in molecular processing of human α-synuclein, such as post-translational modifications, still need to be determined in order to understand to what extent animal models reflect human pathogenesis.

STATE OF THE ART

Rat Models

Alpha-Synuclein Lentivirus

Lentiviruses have been used to overexpress wild-type and mutated forms of human α-synuclein in the adult rat SN. In a study described by Lo Bianco et al. (2002), lentiviral vectors driving transgene expression by the PGK promoter coupled with a downstream post-transcriptional regulatory element of the woodchuck hepatitis virus (WPRE) were injected at two sites above the SN. An equivalent construct expressing β-galactosidase was used as a control vector in order to distinguish the specific toxicity of the pathogenic transgene variants from possible non-specific effects of viral infection and protein overexpression.

The obtained model was able to reproduce the two most salient hallmarks of the human disorder, namely an accumulation of the overexpressed protein in aggregates within neuronal perikarya and neurites, and a selective loss of dopaminergic neurons in the SNpc. With two sites of injection, the virus was shown to diffuse over the entire SN, inducing a loss in tyrosine hydroxylase (TH) reactivity mirroring the differential degeneration of dopaminergic neurons. Importantly, and although the virus also infected non-dopaminergic neurons within the SNpc, the toxicity of the overexpressed α-synuclein was specific for the dopaminergic neuronal population and did not affect the number of GABAergic neurons of the SN *pars reticulata*. While some of the surviving transduced dopaminergic neurons were still found to express the transgene as late as 5 months post-injection, progressive cell loss started 3 weeks after virus administration and reached a plateau after only 6 weeks. All three α-synuclein variants led to about 30% loss of dopaminergic neurons in the SNpc. This degree of lesion did not allow for the monitoring of asymmetrical dopamine release using rotational behavior. Nevertheless, when comparing the 30% of neuronal loss to an overall infection level of 50%, this constituted a major effect.

However, lentiviruses are prone to induce inflammatory reactions and show a limited diffusion capacity within the brain parenchyma. Also, as they are not retrogradely transported in neuronal cells, lentiviruses offer no other option than to be directly injected in the SNpc. Mainly in order to address these limitations of lentiviral vectors, several groups have attempted to develop a similar model using AAV vectors instead.

AAV2 α-Synuclein

AAV2 vectors have been shown to effectively transfer foreign genes in the brain and, depending on the serotype, show a mostly neuronal pattern of infection. They have already been used extensively in preclinical and clinical gene therapy for neurodegenerative diseases.

In the initial AAV2 model described by Kirik *et al.* (2002), the wild-type and A53T α-synuclein variants were expressed on a construct derived from AAV2 and driven by a cytomegalovirus/chicken β-actin (CBA) promoter. A similar construct expressing GFP was used as a control vector. With a single unilateral injection site in the adult rat ventral midbrain, they observed expression of the transgene in over 90% of TH-positive neurons in the SNpc. Additionally, they showed high numbers of transduced cells in the SN *pars reticulata*, the mesencephalic reticular formation, and the ventral tegmental area (VTA). The infection pattern remained restricted to the injected hemisphere, allowing for the contralateral side to be used as an internal control. Alpha-synuclein immunoreactivity was observed at 1 week, while the maximum of transgene expression was reached between 3 and 8 weeks after virus administration and remained robust for more than 6 months. Signs of neuronal pathology developed progressively over the first 2 months following injection, including α-synuclein-positive cytoplasmic inclusions as well as dystrophic and fragmented neurites in both the SNpc and the *striatum*. Reportedly, the effect of the α-synuclein overexpression was twofold: first, a suppression of TH enzyme activity and striatal DA levels by 50% appearing early post virus injection, followed by a loss of nigral dopaminergic neurons. The level of degeneration, however, displayed notable variation, ranging from 30% in some of the animals to 80% in others. This was most probably attributable to variability in targeting, spread and dosage of gene delivery, which are inherent to the administration method. This disparity also accounted for the observed rotational behavior, only marginally impaired in the most affected animals. The model explicitly demonstrated both selective vulnerability of nigral dopaminergic neurons, and transgene-specific toxicity with respect to a control virus coding for GFP.

Surprisingly, this study showed a recovery effect at a later time point. Animals analyzed 27 weeks after virus administration showed a marked recovery of striatal TH-positive fibers, not matched by a recovery of TH-positive nigral cells. These findings suggest that axonal sprouting may have contributed to this effect in long-term surviving animals.

In the model described by Klein *et al.* in 2002, the A30P mutant form of α-synuclein was expressed under the control of the CBA promoter, in a construct

driving co-transcription of GFP from a bicistronic cassette. A construct expressing only the GFP under the control of the same expression cassette was used as a control vector. They were able to demonstrate neuron-specific expression for both vectors. With a single injection site in the adult rat ventral midbrain, the virus diffused over the entire SNpc. Over a 1-year time course, they observed robust expression of GFP in the control-vector injected animals, and of both GFP (although at attenuated levels compared to the control vector) and the primary transgene in animals injected with the α-synuclein/GFP vector. This experiment succeeded in reproducing three hallmark features of the human disease, namely an accumulation of the overexpressed protein in SNpc neuron perikarya, Lewy-like dystrophic neurites in both the SNpc and the *striatum*, and a 53% loss in TH-positve nigral neurons (compared to the contralateral, uninjected hemisphere). The authors reported a specific toxicity of the α-synuclein vector, as the control vector did not induce significant loss of SNpc neurons. However, the interpretation was complicated by the fact that GFP was always co-expressed with the pathogenic transgene. They did not further investigate the specificity of the toxic effect for different neuronal populations.

In 2004, Yamada *et al.* completed another study using an AAV2-derived vector to drive the expression of wild-type human α-synuclein in the adult rat SN under the control of the cytomegalovirus (CMV) promoter. An analog vector expressing GFP was used as a control. They showed successful transduction of more than 80% of the targeted cells at the most infected levels of the SNpc. Due to low expression levels of the transgene, the maximal extent of cell loss in the SNpc appeared at a later time point than in the former studies. Even at the most infected levels of the SNpc, 13 weeks were required to induce a 50% loss of dopaminergic neurons. At this level of lesion, they were not able to establish a robust behavioral phenotype.

On the other hand, the relatively slow time course of their model allowed them to identify several additional cellular features considered crucial to the pathogenesis of idiopathic PD, namely the consistent phosphorylation of α-synuclein at Ser129 and the activation of caspase-9.

Mouse α-Synuclein Model

Compared to the relative consistency of the effects of viral overexpression of α-synuclein in rats, the same approach has shown only limited success in the mouse species.

Using lentiviral vectors coding for human wild-type and A30P α-synuclein under the control of a CMV promoter, Lauwers et al. (2003) performed single stereotaxic injections into different regions of the adult mouse brain. They reported robust overexpression of the transgene for over 10 months in the SN, the *striatum* and the *amygdala*. Similar somal and neuritic accumulations of α-synuclein seemed to occur in all of these brain regions, while there were no differences between the wild-type and the A30P form. Twelve months after virus administration, they observed a 10–25% loss of dopaminergic neurons in the SNpc. However, α-synuclein-induced neurodegeneration was not restricted to the SNpc. Following injection into the *striatum*, there was a 45% decrease in the number of cells expressing the α-synuclein transgene over time. Although Fluoro-Jade B and silver stainings suggested the presence of degenerating cells, this could be partly due to loss of transgene expression.

Recently, St Martin et al. injected AAV2 vectors unilaterally into the SNpc of adult mice. A bicistronic construct under the control of the CBA promoter was used to express either wild-type α-synuclein together with GFP or GFP alone as a control (St Martin et al., 2007). The validity of the control vector, though, is questionable because the GFP is expressed under the attenuating control of the IRES element and thus not directly comparable to the primary transgene in the α-synuclein vector.

Animals were monitored for 24 weeks after virus administration. The AAV vector infected between 10% and 80% of TH-positive neurons in a given section of the SNpc. This variability in the infection rate was attributed in part to the challenge of successfully targeting this structure in the mouse. Reportedly, viral expression of both transgenes was abundant 4 weeks after injection, peaking by 12 weeks, and persistent throughout the whole duration of the experiment. At both 4 and 12 weeks, no significant cell loss was observed in the SNpc. At 24 weeks, a significant 25% loss of TH-positive nigral neurons could be detected with the α-synuclein/GFP bicistronic vector. However, they did not observe significant loss in dopaminergic terminal density in the *striatum*. While there were dystrophic neuritic changes in transduced TH-positive neurons, no specific α-synuclein-positive inclusions were apparent at any time point. In addition, they observed an upregulation of endogenous HSP27 at 4 weeks and

both Hsp70 and Hsp40 at 24 weeks in response to overexpression of the pathogenic transgene. This suggests that the relatively small effects observed in the mouse brain might be due to a potent array of defense mechanisms, which might delay the onset of cell death and lead to the aforementioned species-specific lack of vulnerability.

A previous study using AAV2 vectors driving α-synuclein A53T expression under the control of the PDGF-β promoter reported an absence of lesion in the nigrostriatal pathway at 7 weeks post-injection, despite a high degree on infection (Dong et al., 2002). Following an additional insult using a mild dose of the MPTP toxin, no increase in neuronal vulnerability was observed in response to A53T α-synuclein overexpression.

Altogether, these data suggest that the mouse SNpc displays a relatively high resistance to α-synuclein-mediated toxicity, similar to what had been observed in transgenic mouse lines. Despite these limitations, one must bare in mind the obvious advantages of being able to apply these viral strategies to the mouse species, namely the potential combination of a virus injection with the existing transgenic and knockout lines.

Primate Model: AAV2 α-Synuclein

As mentioned, one of the main advantages of the viral approach for PD modeling is their applicability to a wide variety of species, including primates. In line with the rat model described above, Kirik et al. (2003) went on to inject the same viruses into the SNpc of adult marmosets. They used AAV2 vectors encoding for the α-synuclein wild-type and A53T mutated forms under the control of the CBA promoter. Again, a similar vector expressing GFP was used as a control. With a single site of injection into the adult ventral midbrain, 90–95% of neurons were successfully transduced in the SNpc. The observed overall effects were comparable to those in rats, with the exception of a markedly longer time course. Sixteen weeks post-transduction, they observed a 30–60% loss of the TH-positive neurons in the SNpc, accompanied by a 40–50% reduction of TH-positive innervation in the *striatum*. Overexpression of both wild-type and mutated α-synuclein was sufficient to induce the neuropathological signs of dystrophic, fragmented neurites and granular protein inclusions. They demonstrated specific toxicity of the α-synuclein viruses, as none of these pathological changes were observed with the control virus. Some dopaminergic neuronal loss was observed in the ventral tegmental area (A10 region), suggesting that, in the primate at least this neuronal population presents some susceptibility to α-synuclein. Finally, they were able to establish a consistent behavioral phenotype for the head position test, while the assessment of amphetamine-induced rotation failed to produce a significant effect at any time point.

Recently, the same group completed a long-term follow-up study on the detailed characterization of behavioral and pathological consequences of α-synuclein overexpression in marmoset monkeys (Eslamboli *et al.*, 2007). In contrast to their previous α-synuclein models, Eslamboli *et al.* used AAV2/5 vectors (AAV2 vectors packaged in an AAV5 capsid) to overexpress human wild-type α-synuclein, A53T α-synuclein, or a GFP marker under the control of the CBA promoter. Pseudotyped AAV2/5 vectors transduced dopaminergic nigral and GABAergic neurons as well as glial cells, such as astrocytes ventrally in the SN *pars reticulata* and dorsally in the mesencephalic tegmentum, and oligodendrocytes in the cerebral peduncle. The monkeys were monitored for 1 year after virus administration. On the whole, it appears that the development of α-synuclein-mediated pathology is a relatively slow process in the primate brain when compared to the rat model.

They reported α-synuclein-specific neuronal degeneration in the SNpc, with distinct effects of wild-type and mutated α-synuclein. Overexpression of the wild-type form induced significant cell death in only two out of eight animals (showing 35% and 45% loss of TH-positive neurons in the SNpc), while A53T αsynuclein led to a consistent level of neuronal loss (six out of seven animals displayed lesions above 40%). Also, 1 year after virus administration, the TH-positive fiber terminals in the *striatum* were partly lost, and those remaining were dystrophic in both α-synuclein groups. All animals across both α-synuclein groups showed aggregates of the pathogenic protein in the axonal and dendritic projections of surviving nigral dopaminergic neurons. Moreover, ubiquitin-containing aggregates were observed in the midbrain after 1 year of α-synuclein overexpression. In addition, α-synuclein-induced pathological alterations could be detected by magnetic resonance imaging (MRI), which provides an attractive system to monitor the development of the lesion *in vivo*.

With regard to a behavioral phenotype, the authors defined three distinct phases. Phase one (1–9 weeks after vector surgery) remained pre-symptomatic. Phase two (15–27 weeks after vector surgery) marked a period during which overt behavioral deficits started to appear in both α-synuclein-overexpressing groups. During this time, animals in the α-synuclein wild-type group displayed ipsilesional full body rotations, head turn bias, and some motor coordination errors, while mainly motor coordination errors were apparent in the mutated α-synuclein group. Phase three (33–52 weeks after vector surgery) was defined by a progressive worsening in motor coordination errors in the mutated α-synuclein group.

Alternative Genes

Although α-synuclein overexpression provides an experimental basis for virus-induced PD models in the rodent and primate brain, it is crucial to test a variety of genes directly or indirectly related to genetic cause of the disease in similar conditions. In the end, only a comparison between pathogenic genes will define common cellular and molecular mechanisms, providing a rationale base to devise therapeutics for sporadic PD.

The microtubule-associated protein tau (tau, MAPT) is a protein abundant in the human brain that can aggregate into neurofibrillary tangles considered a hallmark of Alzheimer's disease. There is clear evidence for tau pathology in PD, and both tau and α-synuclein can mutually and synergistically enhance their fibrillization rate (Giasson *et al.*, 2003). Tau mutations are causing autosomal dominant fronto-temporal dementia and parkinsonism linked to chromosome 17 (FTDP-17) (Hutton *et al.*, 1998). Based on these findings, the P301L-mutated form of tau was overexpressed using an AAV viral vector both in the basal forebrain region and in the SNpc of rats. Overexpression consistently led to the formation of neurofibrillary tangles, a loss of dopaminergic neurons, and rotational behavior apparent soon after virus injection (Klein *et al.*, 2006a). A dose–response relation was established using AAV2, 5 and 8 vectors, and the toxic effects of tau expression were compared to viral gene transfer for the GFP reporter protein (Klein *et al.*, 2006b). Despite some neuronal loss observed at high viral doses of the GFP AAV vectors, there remained a marked difference to the levels of loss of TH-positive neurons elicited by the AAV8-driven overexpression of tau (>70% loss). These studies convincingly demonstrate a differential vulnerability of the SNpc to tau overexpression, suggesting that common molecular effectors might be shared by a broad spectrum of neurodegenerative disorders, including parkinsonism, dementia, and Alzheimer's disease.

LRRK2 is currently under intense scrutiny, as elucidating the role of this protein kinase in PD pathogenesis is likely to provide novel insight into mechanisms potentially amenable to pharmacological treatment. A possible gain of toxic function associated with increased kinase activity makes LRRK2 an attractive candidate for *in vivo* modeling based on viral overexpression systems. However, this approach is facing the technical limits of the viral vector systems, as even the lentiviral packaging capacity might not be sufficient to incorporate the entire LRRK2 coding sequence. One alternative consists in using only LRRK2 domains participating in a putative gain of toxic function. In a recent study, lentivirus-driven overexpression of the LRRK2 kinase domain with the G2019S mutation has been reported to induce morphological alterations, increased apoptosis and inclusion formation in rat dopaminergic neurons (MacLeod *et al.*, 2006). Further investigation will be needed to elucidate how LRRK2 affects neuronal function. In particular, it remains to be determined what role the other functional domains of the protein play, and whether these domains are essential to toxic enzyme activity, as already suggested by the presence of pathogenic mutations.

While parkin has proved essential for the survival and cellular function of dopaminergic neurons in the SNpc, no rodent model has been established up to now that would directly target parkin expression. As parkin has been found to be an E3 ubiquitin ligase, one can speculate that loss of its normal function will lead to a toxic accumulation of its protein substrates. Therefore, overexpression of these substrates using viral vectors might mimic some of these loss-of-function effects. A successful attempt in this direction was made with AAV2-mediated overexpression of CDCrel-1, a presynaptic member of the septin protein family that can be ubiquitinated by parkin (Dong *et al.*, 2003). The authors reported a progressive loss of >70% dopaminergic neurons in the rat SNpc, and demonstrated that CDCrel-1 toxicity depends on dopamine synthesis and specifically affects TH-positive neurons.

More recently, the effect of Pael-R overexpression was investigated in the mouse SNpc (Kitao

et al., 2007). Pael-R is a well-described substrate for the parkin ubiquitin ligase that promotes its degradation (Imai *et al.*, 2001). When accumulating in cells, Pael-R induces endoplasmic reticulum stress. Using retrograde adenoviral infection combined with a Cre/Lox recombination, Pael-R overexpression was induced in the adult mouse SNpc. Reportedly, Pael-R activated an unfolded protein response in normal mice. In parkin null mice, this response was enhanced, leading to cell death rescuable by the inhibition of dopamine synthesis.

It cannot be excluded that CDCrel-1 and Pael-R cause cell stress independently of parkin function. In addition, parkin is likely to play a role in the maintenance of dopaminergic neuron integrity that goes beyond the proteasome-mediated degradation of protein substrates. Nevertheless, CDCrel-1 and Pael-R show specific effects on the dopaminergic system that highlight novel insights into the mechanism of neuronal stress in PD, and should broaden the arsenal of relevant genetic models.

Cell Specificity

The question why some neuronal subpopulations such as nigral dopaminergic neurons in PD are selectively degenerating remains poorly understood. The anatomical pattern of the disease provides clues as to the possible mechanisms leading up to the loss of nigral neurons. One possible cause to the demise of these cells might be dopamine-dependent oxidative stress (Lotharius and Brundin, 2002), as dopamine metabolism can give rise to various reactive oxygen species.

Alpha-synuclein might promote the accumulation of dopamine in the cytoplasm. It is possible that α-synuclein protofibrils, that have been proposed to be the toxic species, increase the levels of cytoplasmic dopamine by forming pores akin to those generated by bacterial toxins and permeabilizing synaptic vesicles (Lashuel *et al.*, 2002). Dopamine and dopamine-quinones that arise from the auto-oxidation of dopamine can modify α-synuclein by forming covalent adducts with this protein, inhibiting the transition from protofibrils to fibrils (Conway *et al.*, 2001). In addition to its effects on cellular integrity, oxidative stress resulting from vesicular dopamine leakage might also promote protein misfolding and exacerbate α-synuclein aggregation in the cytoplasm (Lotharius and Brundin, 2002). The possible role of α-synuclein in synaptic-vesicle recycling and neurotransmitter homeostasis also suggests that improper dopamine sequestration might be a consequence of the misfolding of this protein, leading to the specific pattern of neuron loss. However, it remains unclear why some populations of dopaminergic neurons are less affected, for example in the VTA, and why some non-catecholaminergic neurons can display susceptibility along disease progression.

Although injection of viral vectors typically targets vulnerable regions such as the SNpc, some GABAergic neurons and dopaminergic neurons in the VTA usually get transduced as well. Overexpression of α-synuclein has been reported to spare GABAergic neurons following lentiviral injections into the SNpc (Lo Bianco et al., 2002). Recently, in a study by Maingay et al. (2006) AAV2 vectors coding for A53T α-synuclein were directly injected into the VTA. Interestingly, this left the VTA dopaminergic neurons unaffected, although similar injections of 6-OHDA were shown to induce significant cell loss.

These data suggest that viral systems driving overexpression of pathogenic proteins lead to a pattern of cell loss and dysfunction that replicates selective vulnerabilities observed in the human pathology. Thus, it would seem possible to obtain specific toxicity using this mode of gene delivery in adult tissues.

Testing for Neuroprotective Approaches

Although virus-based systems have proved very efficient at generating animal models of PD, they have led to only few demonstrations of successful neuroprotective approaches. Even though neurotrophic factors such as glial cell-derived neurotrophic factor (GDNF) and anti-apoptotic molecules show clear neuroprotection in toxin-based models of PD, their efficacy in genetic models has not been proven yet (Lo Bianco et al., 2004a). Still, most data suggest that genetic causes of neurodegeneration are likely to provide more stringent systems for the testing of therapeutics.

Viral models of genetic PD are currently being implemented for testing both gene therapies and pharmacological compounds. Protective effects of parkin have been reported in the lentiviral α-synuclein model (Lo Bianco et al., 2004b), where viruses coding for both α-synuclein A30P and parkin were co-injected into the adult rat SNpc. While 31% of dopaminergic neurons were lost 6 weeks after co-injection of α-synuclein A30P and control viruses, neuronal loss appeared to be only 9% when parkin

was co-expressed with α-synuclein. Although functional rescue could not be assessed in this model, striatal dopaminergic projections appeared to be protected as well.

In another study, these results were confirmed by injection of mixtures of AAV vectors coding for α-synuclein together with either GFP or parkin into the adult rat SNpc (Yamada et al., 2005). Preliminary data from the same group indicate a similar protective effect in the primate brain (Yasuda et al., 2007). Interestingly, parkin overexpression also shows neuroprotective effects in the AAV model based on mutated tau overexpression in the rat SNpc, highlighting the importance of parkin as a therapeutic target for proteotoxic cellular stress (Klein et al., 2006a).

CONCLUSION AND PERSPECTIVES

The discovery of genetic mutations underlying familial forms of PD has dramatically modified our understanding of the disorder. Alpha-synuclein and particularly its level of expression and propensity to adopt an abnormal conformation are likely to play a causative role in the pathogenic process. As a flexible and rapid method to generate mammalian models of neurodegenerative disorders, viral injection inducing overexpression of pathogenic proteins such as α-synuclein has successfully been used to model PD. This approach replicates some aspects of the pathology by abruptly perturbing the homeostasis of young adult neurons. Nigral cell loss, protein aggregation, and progressive neuronal deficits offer opportunities to assess therapeutic strategies that may slow down disease progression.

Although these results are encouraging, they are unlikely to reflect the multiple genetic and environmental causes that affect dopaminergic neurons along the aging process. As PD usually develops during the last decades of the human life span, any treatment should be tested in conditions as close as possible to the physiology of aged neurons. Among the multiple factors that participate in the aging of the human brain, there appear to be genetic determinants of organismal longevity, possibly implicated in controlling cell resistance to stress. By modulating these complex genetic networks of cellular metabolism, it might be possible to specifically model the cellular conditions of the aged brain and provide a novel basis to assess the effect of virus-induced proteotoxic stress.

There is a rapidly expanding list of genetic factors associated with PD. This multiplicity of determinants highlights the polygenic character of the disease and complicates its analysis. In addition, environmental factors have to be taken into account as a trigger and modulator of the disease process. Nevertheless, there are ways to investigate how pathogenic and protective genes interact, and where environmental causes are likely to intervene. Viral vectors, as a flexible system to genetically engineer the adult animal brain, will be the useful tools to elucidate these questions. Considering the complexity of the disease, there is a clear need to establish an array of model systems. Screening in simple systems, such as yeast and *Drosophila*, will likely continue to pioneer target and lead discovery. However, validation in more complex models, *in vivo* in the mammalian brain and *in vitro* using human neuronal cultures (for instance derived from human stem cells [Schneider *et al.*, 2007]), should provide a more stringent assessment of the efficacy of therapeutics for this devastating disorder.

ABBREVIATIONS

AAV	adeno-associated virus
CBA	cytomegalovirus/chicken β-actin promoter
CNS	central nervous system
DAT	dopamine transporter
FTDP-17	fronto-temporal dementia and parkinsonism linked to chromosome 17
GABA	gamma-aminobutyric acid
GFP	green fluorescent protein
LRRK2	leucine-rich repeat kinase 2
MAPT	microtubule-associated protein tau
MRI	magnetic resonance imaging
Pael-R	parkin-associated endothelin-receptor like receptor
SNpc	*substantia nigra pars compacta*
TH	tyrosine hydroxylase
VMAT	vesicular monoamine transporter
VSV-G	vesicular stomatitis virus glycoprotein
VTA	ventral tegmental area
WPRE	woodchuck hepatitis virus post-transcriptional regulatory element.

REFERENCES

Blankinship, M. J., Gregorevic, P., Allen, J. M. *et al.* (2004). Efficient transduction of skeletal muscle using vectors based on adeno-associated virus serotype 6. *Mol Ther* 10, 671–678.

Blomer, U., Naldini, L., Kafri, T. *et al.* (1997). Highly efficient and sustained gene transfer in adult neurons with a lentivirus vector. *J Virol* 71, 6641–6649.

Bonifati, V., Rizzu, P., van Baren, M. J. *et al.* (2003). Mutations in the DJ-1 gene associated with autosomal recessive early-onset parkinsonism. *Science* 299, 256–259.

Bove, J., Prou, D., Perier, C., and Przedborski, S. (2005). Toxin-induced models of Parkinson's disease. *NeuroRx* 2, 484–494.

Burger, C., Gorbatyuk, O. S., Velardo, M. J. *et al.* (2004). Recombinant AAV viral vectors pseudotyped with viral capsids from serotypes 1, 2, and 5 display differential efficiency and cell tropism after delivery to different regions of the central nervous system. *Mol Ther* 10, 302–317.

Chartier-Harlin, M. C., Kachergus, J., Roumier, C. *et al.* (2004). Alpha-synuclein locus duplication as a cause of familial Parkinson's disease. *Lancet* 364, 1167–1169.

Chen, L., Cagniard, B., Mathews, T. *et al.* (2005). Age-dependent motor deficits and dopaminergic dysfunction in DJ-1 null mice. *J Biol Chem* 280, 21418–21426.

Clark, I. E., Dodson, M. W., Jiang, C. *et al.* (2006). *Drosophila* pink1 is required for mitochondrial function and interacts genetically with parkin. *Nature* 441, 1162–1166.

Conway, K. A., Rochet, J. C., Bieganski, R. M., and Lansbury, P. T., Jr (2001). Kinetic stabilization of the alpha-synuclein protofibril by a dopamine-alpha-synuclein adduct. *Science* 294, 1346–1349.

Delenda, C., and Gaillard, C. (2005). Real-time quantitative PCR for the design of lentiviral vector analytical assays. *Gene Ther* 12(**Suppl 1**), S36–S50.

Dong, Z., Ferger, B., Feldon, J., and Bueler, H. (2002). Overexpression of Parkinson's disease-associated alpha-synucleinA53T by recombinant adeno-associated virus in mice does not increase the vulnerability of dopaminergic neurons to MPTP. *J Neurobiol* 53, 1–10.

Dong, Z., Ferger, B., Paterna, J. C. *et al.* (2003). Dopamine-dependent neurodegeneration in rats induced by viral vector-mediated overexpression of the parkin target protein, CDCrel-1. *Proc Natl Acad Sci USA* 100, 12438–12443.

Eriksen, J. L., Przedborski, S., and Petrucelli, L. (2005). Gene dosage and pathogenesis of Parkinson's disease. *Trends Mol Med* 11, 91–96.

Eslamboli, A., Romero-Ramos, M., Burger, C. *et al.* (2007). Long-term consequences of human alpha-synuclein overexpression in the primate ventral midbrain. *Brain* 130, 799–815.

Farrer, M., Maraganore, D. M., Lockhart, P. *et al.* (2001). Alpha-synuclein gene haplotypes are associated with Parkinson's disease. *Hum Mol Genet* 10, 1847–1851.

Feany, M. B., and Bender, W. W. (2000). A *Drosophila* model of Parkinson's disease. *Nature* **404**, 394–398.

Fleming, S. M., Fernagut, P. O., and Chesselet, M. F. (2005). Genetic mouse models of parkinsonism: Strengths and limitations. *NeuroRx* **2**, 495–503.

Gao, G., Vandenberghe, L. H., and Wilson, J. M. (2005). New recombinant serotypes of AAV vectors. *Curr Gene Ther* **5**, 285–297.

Giasson, B. I., Forman, M. S., Higuchi, M. *et al.* (2003). Initiation and synergistic fibrillization of tau and alpha-synuclein. *Science* **300**, 636–640.

Grieger, J. C., Choi, V. W., and Samulski, R. J. (2006). Production and characterization of adeno-associated viral vectors. *Nat Protoc* **1**, 1412–1428.

Hardy, J., Cai, H., Cookson, M. R., Gwinn-Hardy, K., and Singleton, A. (2006). Genetics of Parkinson's disease and parkinsonism. *Ann Neurol* **60**, 389–398.

Hofer, A., and Gasser, T. (2004). New aspects of genetic contributions to Parkinson's disease. *J Mol Neurosci* **24**, 417–424.

Hutton, M., Lendon, C. L., Rizzu, P. *et al.* (1998). Association of missense and 5'-splice-site mutations in tau with the inherited dementia FTDP-17. *Nature* **393**, 702–705.

Imai, Y., Soda, M., Inoue, H. et al. (2001). An unfolded putative transmembrane polypeptide, which can lead to endoplasmic reticulum stress, is a substrate of Parkin. *Cell* **105**, 891–902.

Itier, J. M., Ibanez, P., Mena, M. A. *et al.* (2003). Parkin gene inactivation alters behaviour and dopamine neurotransmission in the mouse. *Hum Mol Genet* **12**, 2277–2291.

Kaspar, B. K., Erickson, D., Schaffer, D. *et al.* (2002). Targeted retrograde gene delivery for neuronal protection. *Mol Ther* **5**, 50–56.

Kim, R. H., Smith, P. D., Aleyasin, H. *et al.* (2005). Hypersensitivity of DJ-1-deficient mice to 1-methyl-4-phenyl-1,2,3,6-tetrahydropyrindine (MPTP) and oxidative stress. *Proc Natl Acad Sci USA* **102**, 5215–5220.

Kirik, D., Rosenblad, C., Burger, C. et al. (2002). Parkinson-like neurodegeneration induced by targeted overexpression of alpha-synuclein in the nigrostriatal system. *J Neurosci* **22**, 2780–2791.

Kirik, D., Annett, L. E., Burger, C. *et al.* (2003). Nigrostriatal alpha-synucleinopathy induced by viral vector-mediated overexpression of human alpha-synuclein: A new primate model of Parkinson's disease. *Proc Natl Acad Sci USA* **100**, 2884–2889.

Kitada, T., Asakawa, S., Hattori, N. *et al.* (1998). Mutations in the parkin gene cause autosomal recessive juvenile parkinsonism. *Nature* **392**, 605–608.

Kitao, Y., Imai, Y., Ozawa, K. *et al.* (2007). Pael receptor induces death of dopaminergic neurons in the substantia nigra via endoplasmic reticulum stress and

dopamine toxicity, which is enhanced under condition of parkin inactivation. *Hum Mol Genet* **16**, 50–60.

Klein, R. L., King, M. A., Hamby, M. E., and Meyer, E. M. (2002). Dopaminergic cell loss induced by human A30P alpha-synuclein gene transfer to the rat substantia nigra. *Hum Gene Ther* **13**, 605–612.

Klein, R. L., Dayton, R. D., Henderson, K. M., and Petrucelli, L. (2006a). Parkin is protective for substantia nigra dopamine neurons in a tau gene transfer neurodegeneration model. *Neurosci Lett* **401**, 130–135.

Klein, R. L., Dayton, R. D., Leidenheimer, N. J. *et al.* (2006b). Efficient neuronal gene transfer with AAV8 leads to neurotoxic levels of tau or green fluorescent proteins. *Mol Ther* **13**, 517–527.

Kruger, R., Kuhn, W., Muller, T. *et al.* (1998). Ala30Pro mutation in the gene encoding alpha-synuclein in Parkinson's disease. *Nat Genet* **18**, 106–108.

Kuwahara, T., Koyama, A., Gengyo-Ando, K. *et al.* (2006). Familial Parkinson mutant alpha-synuclein causes dopamine neuron dysfunction in transgenic *Caenorhabditis elegans*. *J Biol Chem* **281**, 334–340.

Lakso, M., Vartiainen, S., Moilanen, A. M. *et al.* (2003). Dopaminergic neuronal loss and motor deficits in *Caenorhabditis elegans* overexpressing human alpha-synuclein. *J Neurochem* **86**, 165–172.

Lashuel, H. A., Hartley, D., Petre, B. M., Walz, T., and Lansbury, P. T., Jr (2002). Neurodegenerative disease: Amyloid pores from pathogenic mutations. *Nature* **418**, 291.

Lauwers, E., Debyser, Z., Van Dorpe, J. *et al.* (2003). Neuropathology and neurodegeneration in rodent brain induced by lentiviral vector-mediated overexpression of alpha-synuclein. *Brain Pathol* **13**, 364–372.

Leroy, E., Boyer, R., Auburger, G. et al. (1998). The ubiquitin pathway in Parkinson's disease. *Nature* **395**, 451–452.

Lo Bianco, C., Ridet, J. L., Schneider, B. L., Deglon, N., and Aebischer, P. (2002). Alpha-synucleinopathy and selective dopaminergic neuron loss in a rat lentiviral-based model of Parkinson's disease. *Proc Natl Acad Sci USA* **99**, 10813–10818.

Lo Bianco, C., Deglon, N., Pralong, W., and Aebischer, P. (2004a). Lentiviral nigral delivery of GDNF does not prevent neurodegeneration in a genetic rat model of Parkinson's disease. *Neurobiol Dis* **17**, 283–289.

Lo Bianco, C., Schneider, B. L., Bauer, M. *et al.* (2004b). Lentiviral vector delivery of parkin prevents dopaminergic degeneration in an alpha-synuclein rat model of Parkinson's disease. *Proc Natl Acad Sci USA* **101**, 17510–17515.

Lotharius, J., and Brundin, P. (2002). Pathogenesis of Parkinson's disease: Dopamine, vesicles and alpha-synuclein. *Nat Rev Neurosci* **3**, 932–942.

MacLeod, D., Dowman, J., Hammond, R. *et al.* (2006). The familial parkinsonism gene LRRK2 regulates neurite process morphology. *Neuron* **52**, 587–593.

Maingay, M., Romero-Ramos, M., Carta, M., and Kirik, D. (2006). Ventral tegmental area dopamine neurons are resistant to human mutant alpha-synuclein overexpression. *Neurobiol Dis* 23, 522–532.

Mastakov, M. Y., Baer, K., Symes, C. W. *et al.* (2002). Immunological aspects of recombinant adeno-associated virus delivery to the mammalian brain. *J Virol* 76, 8446–8454.

Paisan-Ruiz, C., Jain, S., Evans, E. W. *et al.* (2004). Cloning of the gene containing mutations that cause PARK8-linked Parkinson's disease. *Neuron* 44, 595–600.

Park, J., Lee, S. B., Lee, S. *et al.* (2006). Mitochondrial dysfunction in *Drosophila* PINK1 mutants is complemented by parkin. *Nature* 441, 1157–1161.

Passini, M. A., Watson, D. J., Vite, C. H. *et al.* (2003). Intraventricular brain injection of adeno-associated virus type 1 (AAV1) in neonatal mice results in complementary patterns of neuronal transduction to AAV2 and total long-term correction of storage lesions in the brains of beta-glucuronidase-deficient mice. *J Virol* 77, 7034–7040.

Paterna, J. C., Feldon, J., and Bueler, H. (2004). Transduction profiles of recombinant adeno-associated virus vectors derived from serotypes 2 and 5 in the nigrostriatal system of rats. *J Virol* 78, 6808–6817.

Peden, C. S., Burger, C., Muzyczka, N., and Mandel, R. J. (2004). Circulating anti-wild-type adeno-associated virus type 2 (AAV2) antibodies inhibit recombinant AAV2 (rAAV2)-mediated, but not rAAV5-mediated, gene transfer in the brain. *J Virol* 78, 6344–6359.

Perez, F. A., and Palmiter, R. D. (2005). Parkin-deficient mice are not a robust model of parkinsonism. *Proc Natl Acad Sci USA* 102, 2174–2179.

Philippe, S., Sarkis, C., Barkats, M. *et al.* (2006). Lentiviral vectors with a defective integrase allow efficient and sustained transgene expression *in vitro* and *in vivo*. *Proc Natl Acad Sci USA* 103, 17684–17689.

Polymeropoulos, M. H., Lavedan, C., Leroy, E. *et al.* (1997). Mutation in the alpha-synuclein gene identified in families with Parkinson's disease. *Science* 276, 2045–2047.

Ramirez, A., Heimbach, A., Grundemann, J. *et al.* (2006). Hereditary parkinsonism with dementia is caused by mutations in ATP13A2, encoding a lysosomal type 5 P-type ATPase. *Nat Genet* 38, 1184–1191.

Schneider, B. L., Seehus, C. R., Capowski, E. E. *et al.* (2007). Over-expression of alpha-synuclein in human neural progenitors leads to specific changes in fate and differentiation. *Hum Mol Genet* 16, 50–51.

Shendelman, S., Jonason, A., Martinat, C., Leete, T., and Abeliovich, A. (2004). DJ-1 is a redox-dependent molecular chaperone that inhibits alpha-synuclein aggregate formation. *PLoS Biol* 2, e362.

Shimura, H., Schlossmacher, M. G., Hattori, N. *et al.* (2001). Ubiquitination of a new form of alpha-synuclein by parkin from human brain: Implications for Parkinson's disease. *Science* 293, 263–269.

Singleton, A. B., Farrer, M., Johnson, J. *et al.* (2003). Alpha-synuclein locus triplication causes Parkinson's disease. *Science* 302, 841.

Smith, W. W., Pei, Z., Jiang, H. *et al.* (2006). Kinase activity of mutant LRRK2 mediates neuronal toxicity. *Nat Neurosci* 9, 1231–1233.

Spillantini, M. G., Schmidt, M. L., Lee, V. M. *et al.* (1997). Alpha-synuclein in Lewy bodies. *Nature* 388, 839–840.

Spillantini, M. G., Crowther, R. A., Jakes, R., Hasegawa, M., and Goedert, M. (1998). Alpha-synuclein in filamentous inclusions of Lewy bodies from Parkinson's disease and dementia with Lewy bodies. *Proc Natl Acad Sci USA* 95, 6469–6473.

St Martin, J. L., Klucken, J., Outeiro, T. F. *et al.* (2007). Dopaminergic neuron loss and up-regulation of chaperone protein mRNA induced by targeted overexpression of alpha-synuclein in mouse substantia nigra. *J Neurochem* 100, 1449–1457.

Szulc, J., Wiznerowicz, M., Sauvain, M. O., Trono, D., and Aebischer, P. (2006). A versatile tool for conditional gene expression and knockdown. *Nat Methods* 3, 109–116.

Thomas, C. E., Birkett, D., Anozie, I., Castro, M. G., and Lowenstein, P. R. (2001). Acute direct adenoviral vector cytotoxicity and chronic, but not acute, inflammatory responses correlate with decreased vector-mediated transgene expression in the brain. *Mol Ther* 3, 36–46.

Valente, E. M., Abou-Sleiman, P. M., Caputo, V. *et al.* (2004). Hereditary early-onset Parkinson's disease caused by mutations in PINK1. *Science* 304, 1158–1160.

Wang, C., Wang, C. M., Clark, K. R., and Sferra, T. J. (2003). Recombinant AAV serotype 1 transduction efficiency and tropism in the murine brain. *Gene Ther* 10, 1528–1534.

West, A. B., Moore, D. J., Biskup, S. *et al.* (2005). Parkinson's disease-associated mutations in leucine-rich repeat kinase 2 augment kinase activity. *Proc Natl Acad Sci USA* 102, 16842–16847.

Xiao, W., Chirmule, N., Berta, S. C. *et al.* (1999). Gene therapy vectors based on adeno-associated virus type 1. *J Virol* 73, 3994–4003.

Yamada, M., Iwatsubo, T., Mizuno, Y., and Mochizuki, H. (2004). Overexpression of alpha-synuclein in rat substantia nigra results in loss of dopaminergic neurons, phosphorylation of alpha-synuclein and activation of caspase-9: Resemblance to pathogenetic changes in Parkinson's disease. *J Neurochem* 91, 451–461.

Yamada, M., Mizuno, Y., and Mochizuki, H. (2005). Parkin gene therapy for alpha-synucleinopathy: A rat model of Parkinson's disease. *Hum Gene Ther* 16, 262–270.

Yanez-Munoz, R. J., Balaggan, K. S., MacNeil, A. *et al.* (2006). Effective gene therapy with nonintegrating lentiviral vectors. *Nat Med* **12**, 348–353.

Yasuda, T., Miyachi, S., Kitagawa, R. *et al.* (2007). Neuronal specificity of alpha-synuclein toxicity and effect of Parkin co-expression in primates. *Neuroscience* **144**, 743–753.

Zabetian, C. P., Hutter, C. M., Yearout, D. *et al.* (2006). LRRK2 G2019S in families with Parkinson disease who originated from Europe and the Middle East: Evidence of two distinct founding events beginning two millennia ago. *Am J Hum Genet* **79**, 752–758.

Zarranz, J. J., Alegre, J., Gomez-Esteban, J. C. *et al.* (2004). The new mutation, E46K, of alpha-synuclein causes Parkinson and Lewy body dementia. *Ann Neurol* **55**, 164–173.

Zhang, Y., Gao, J., Chung, K. K. *et al.* (2000). Parkin functions as an E2-dependent ubiquitin–protein ligase and promotes the degradation of the synaptic vesicle-associated protein, CDCrel-1. *Proc Natl Acad Sci USA* **97**, 13354–13359.

Zimprich, A., Biskup, S., Leitner, P. *et al.* (2004). Mutations in LRRK2 cause autosomal-dominant parkinsonism with pleomorphic pathology. *Neuron* **44**, 601–607.

21

ENVIRONMENTAL EXPLORATIONS OF PARKINSON'S DISEASE USING RODENT GENETIC MODELS

MARIE-FRANCOISE CHESSELET AND PIERRE-OLIVIER FERNAGUT

Departments of Neurology and Neurobiology, David Geffen School of Medicine, UCLA, CA, USA

INTRODUCTION

Rationale for Using Genetic Rodent Models

Parkinson's disease (PD) is primarily a sporadic disease. Indeed, familial forms of PD account for approximately 5% of the cases and usually give rise to young onset forms (i.e. starting before the age of 50). The majority of cases occurs after the age of 65 and appears without family history. However, having a relative with PD increases risk and some of the newly discovered mutation may represent risk factors, as they are more frequent in apparently sporadic cases than in the general population (Litvan *et al.*, 2007a, b). Although age is a clear risk factor, the causes of sporadic PD remain unknown.

A driving force in modern PD research is that studying disease-causing mutations will provide cues for the mechanisms that operate in sporadic PD. The validity of this approach received strong support from the very first mutation identified in the vesicular protein alpha-synuclein. Within weeks of the identification of an A53T point mutation associated with PD in one large family, the Contursi kindred (Polymeropoulos *et al.*, 1997), Spillantini *et al.* (1997) found that wild-type alpha-synuclein is a major component of the Lewy bodies, the cytoplasmic inclusions present in surviving neurons of

the substantia nigra pars compacta, and many other neuronal populations, in patients with sporadic PD. A direct link was thus established between a rare genetic mutation and the much more common form of the disease. This raised hope that more insights would be gained from other mutations, no matter how few patients were affected by the corresponding familial forms of PD.

Unfortunately, although mutations in parkin, UCHL1, DJ1, and PINK1 point to mechanisms that may play a role in PD such as decreases in proteasomal and mitochondrial functions, a direct link between these proteins and sporadic PD has not yet been firmly established (Douglas *et al.*, 2007). Nevertheless all these mutations provide a way to generate new rodent models of PD with construct validity to identify pathophysiological mechanisms leading to at least some forms of the disease (Fleming *et al.*, 2005).

Rationale for Exploring Environmental Factors in PD

In the absence of a strong genetic component at the root of the large majority of PD cases, a role for the environment as a contributing factor has received much attention (Ritz and Yu, 2000; Chade *et al.*, 2006; Kamel *et al.*, 2007). A case for environmental factors linked to industrialized agriculture was made since the 1970s with the work of Andre Barbeau, who showed increased incidence of PD in rural areas (Barbeau *et al.*, 1987). Several factors are associated with modern rural environment, including not only high exposure to pesticides and fertilizers, but also to diesel engines from agricultural machinery. A role for pesticides in PD received further support from structural similarities between the pesticide DDT and the neurotoxin 1-methyl-4-phenyl-1,2,3,6-tetrahydropyrindine (MPTP) when it was discovered that the latter could selectively kill nigrostriatal dopaminergic neurons and cause acute parkinsonism in humans (Langston *et al.*, 1983; Bove *et al.*, 2005).

Since then, several studies have linked environmental exposure to pesticides and an increased risk for PD (Ritz and Yu, 2000; Chade *et al.*, 2006; Kamel *et al.*, 2007). A main problem is that in most cases, exposure assessment is based on recall, and classes of pesticides involved are difficult to identify with certainty. Furthermore, exposed individuals usually have been in contact with more than one agent over long periods of time. Nevertheless, newer studies with improved methodology keep confirming an association between high and prolonged exposure to pesticides and an increased risk of PD. A recent study headed by Dr. Beate Ritz at UCLA based not on recall but on agricultural records mandated by the State law, strongly supports this association in a well-characterized cohort of over 350 newly diagnosed patients and population-based controls. The study detected associations between PD and a number of different classes of pesticides, including some, like paraquat, that have been incriminated in other populations, thus lending further support to the data (Ritz and Costello, 2006; Ritz *et al.*, in preparation). Overall, the case for examining a role for environmental factors in PD remains very strong. Although other environmental factors, such as heavy metal for example, may also increase PD risk, these have not been systematically examined in genetic rodent models of PD. Therefore this review will concentrate on pesticides and related compounds.

Rationale for Combining Exploration of Environmental and Genetic Factors

Despite their high construct value, most lines of mice engineered to express genetic mutations that cause PD in humans do not show a loss of nigrostriatal dopaminergic neurons, the hallmark of PD (Fernagut and Chesselet, 2004; Fleming *et al.*, 2005). However, a loss of tyrosine hydroxylase (TH) positive neuronal cell bodies has been reported in the substantia nigra of mice overexpressing doubly mutated alpha-synuclein under the TH promoter but not wild-type or singly mutated protein (Thiruchelvam *et al.*, 2004). Mice expressing truncated alpha-synuclein have either only a decrease in striatal dopamine or a cell loss that occurs early in development (Chesselet, 2008). Similarly, mice expressing parkin, DJ-1 or PINK1 mutation express behavioral, cellular, and/or molecular anomalies clearly indicating mutation-induced neuronal dysfunction, but no nigrostriatal cell loss (Fleming and Chesselet, 2006). In a few instances, cell loss was observed in alpha-synuclein overexpressing mice but in cell types not normally affected in PD, such as the motor neurons (Fernagut and Chesselet, 2004). This is likely due to the promoters used to drive the transgene.

The general lack of nigrostriatal degeneration in mice that expressed a mutation linked without ambiguity to PD in humans could be due to several factors. Aging is a main risk factor for PD, and although old mice do show behavioral and molecular signs of aging reminiscent to those seen in humans, they may not live long enough to develop the full spectrum of anomalies linked to PD in humans. Indeed, recent research points to the length of the "pre-manifest" phase of the disease (Lang and Obeso, 2004; Langston, 2006). Pathological studies indicate that alpha-synuclein accumulation occurs in other brain regions before the onset of nigral pathology and retrospective studies reveal that a number of symptoms such as sleep disorders, olfactory loss, and affective disturbances can precede the onset of motor signs which lead to the diagnosis of PD (Langston, 2006; Wolters and Braak, 2006). Thus, the available mouse models may model this extended "pre-manifest" phase but not reach the age of nigral cell loss.

Another factor to consider is that experimental animals are not raised in conditions similar to those surrounding the development and adulthood of humans. They are bred in pathogen-free conditions, as free of stress as possible, and receive exquisitely well-balanced diets and nutritional supplements. In addition, the background strains usually used for these experiments have been bred for their longevity and resistance, and they may have a particularly beneficial genetic make up. Finally, the mice may lack the triggering events that precipitate dopaminergic cell loss in humans.

Although patients with sporadic PD do not have mutations in the genes causing familial forms of PD, they seem to express alterations in cellular pathways that are disrupted by these mutations. For example, a decrease in proteasomal function, a mechanism believed to underscore the effects of parkin mutations, has been observed in the substantia nigra of patients with sporadic PD (Olanow and McNaught, 2006). Similarly, mitochondrial dysfunction, long suspected to occur in PD, is probably related to the mechanism of action of DJ-1 and PINK1, two mitochondrial proteins (Greenamyre and Hastings, 2004). Therefore, mice with genetic alterations in genes causing familial forms of PD may provide a sensitized background for the effects of environmental agents.

Most Frequently Used Genetic Rodent Models of PD for Environmental Explorations

Because of the reasoning described in the preceding section, almost all genetic models of PD have been subjected to environmental stressors, in particular MPTP, with the hope to precipitate an elusive dopaminergic cell loss. Because it was the first mutation discovered and the only one with a direct link to sporadic PD (see 1.1), alpha-synuclein overexpressing mice have been more extensively used than the other models. We will describe the data obtained in these mice in more detail, with a brief description of published work in other rodent models at the end of the following section.

RESULTS

MPTP in Genetic Rodent Models of PD

Of course, MPTP is not an environmental toxin. It is a byproduct of heroin synthesis that caused acute parkinsonism in a group of drug addicts in California in the early 1980s (Langston et al., 1983). Its place in this chapter comes from the similarities between its structure and mechanisms of action with those of some environmental pesticides. One link between MPTP and environmental pesticides is the similarity of its structure to that of the now banned insecticide DDT. In addition, MPTP

induces oxidative stress like the pesticide paraquat, which has been linked to an increased risk of PD (Dinis-Oliveira et al., 2006). Furthermore, it is a complex 1 inhibitor like rotenone (Richardson et al., 2007). Rotenone is used as an herbicide and to kill fish in reservoirs. Its environmental impact, however, is probably minimal because it is very unstable and does not accumulate in soil or in tissues. There is no epidemiological data-linking exposure to rotenone and an increased risk of PD but peripheral administration of rotenone in rats can kill nigrostriatal dopaminergic neurons (Betarbet et al., 2000). However, the specificity of this effect is highly variable, not only under different experimental conditions but also among animals treated in the same way (Zhu et al., 2004). Despite its environmental relevance, there is only one published record, to our knowledge, of rotenone administration to genetic mouse models of PD. In this study, the sensitivity of an A30P alpha-synuclein overexpressing mice was not affected by rotenone (Nieto et al., 2006). One reason for the paucity of reports with rotenone may be the variability of its effect. An even more compelling reason is that it does not reliably kill nigrostriatal dopaminergic neurons in mice, probably because of differences in metabolisms of the compound in rats and mice (Richter et al., 2007).

Because it is the toxin of choice to kill nigrostriatal dopaminergic neurons in mice, MPTP has been the most often used compound to mimic environmental insults in genetic mouse models of PD. The results have been greatly variable depending on the type of mouse examined and the experimental conditions. Rathke-Hartlieb et al. (2001) administered a chronic regimen of MPTP to mice expressing mutated alpha-synuclein under the TH promoter. Despite the presence of alpha-synuclein in TH positive neurons of the substantia nigra in these mice, neither the decrease in TH-positive neurons induced by MPTP in the substantia nigra nor the concentrations of catecholamines in the striatum differ in the mutant mice compared with wild-type controls. Similarly, mice overexpressing A53T alpha-synuclein by way of an adeno-associated virus injected into the substantia nigra did not show an increased sensitivity to MPTP (Dong et al., 2002). This result is particularly interesting because this model is relatively acute, and less prone to the development of defense mechanisms than transgenic mice that express mutations from conception.

In contrast to these results, the group of E. Richfield found an increased sensitivity to

MPTP in mice expressing a doubly mutated alpha-synuclein, also under the TH promoter (Richfield et al. 2002). Importantly, the authors attribute this effect to an increased accumulation of the toxin into the dopaminergic neurons resulting from increased expression of the cytoplasmic dopamine transporter (DAT) (Richfield et al., 2002).

This result points to a well-known caveat when examining differences in MPTP toxicity. Because the toxin is metabolized into the active compound, MPP$^+$ by monoamine oxidase and captured by dopaminergic neurons through DAT, alterations of these mechanisms by the mutation could dramatically change the apparent toxicity of MPTP. A more recent report also found an increased loss of TH positive neurons in the substantia nigra and decreased striatal dopamine level in mice expressing the A30P mutated alpha-synuclein that were treated chronically with 80 or 150 mg/kg MPTP (Nieto et al., 2006), but the mechanism of this effect was not investigated further.

Despite the absence of increased neuronal loss after MPTP administration in alpha-synuclein over-expressing mice, there is evidence for a synergistic effect of MPTP and high levels of alpha-synuclein on mitochondrial and axonal morphology in nigrostriatal neurons. In mice expressing wild-type human alpha-synuclein under the Thy1 promoter, Song et al. (2004) found that administration of MPTP twice a week for 2 weeks caused extensive mitochondrial alterations. These alterations included increases in mitochondrial size and filamentous neuritic aggregations. Furthermore they observed evidence of axonal degeneration in the substantia nigra of the transgenic mice. These anomalies were restricted to the substantia nigra (despite the broad expression of alpha-synuclein in this model) and were not observed in non-transgenics, or transgenics not treated with MPTP. Thus, inhibition of complex 1 by a subthreshold dose of MPTP has detrimental effects in mice overexpressing alpha-synuclein. The use of wild-type alpha-synuclein in this model is particularly interesting because it is closer to the sporadic form of the human disease, in which the protein accumulates abnormally but is not mutated at the level of the DNA.

Paraquat Exposure in Mice Overexpressing Alpha-Synuclein

In contrast to MPTP, paraquat is directly relevant to environmental exposure in humans. This widely used pesticide has been repeatedly associated with an increased risk of PD (Hertzman et al., 1990;

Liou et al., 1997). Paraquat causes oxidative stress and kills a small subset (25%) of nigrostriatal dopaminergic neurons in wild-type mice (McCormack et al., 2002, 2005; Fernagut et al., 2007). Several studies have investigated the effects of paraquat in mice overexpressing alpha-synuclein (Manning-Bog et al., 2003; Thiruchelvam et al., 2004; Norris et al., 2007). Surprisingly, mice overexpressing human wild-type or A53T alpha-synuclein under the control of the rat TH promoter were resistant to the toxic effect of paraquat in nigrostriatal dopaminergic neurons (Manning-Bog et al., 2003). The authors attributed this protection to an upregulation of the chaperone hsp70 (Manning-Bog et al., 2003). However, the evidence remains correlative and other mechanisms could play a role as well.

In a more recent study (Norris et al., 2007) a similar regimen of paraquat did not exacerbate the pathology in mice overexpressing A53T alpha-synuclein under the prion promoter, even when older (8–12 month) mice were used. A similar result was obtained in mice overexpressing wild-type alpha synuclein under the Thy1 promoter (Fernagut et al., 2007). In this study, proteinase K-alpha-synuclein aggregates markedly increased in the substantia nigra of transgenic mice (but not wild-type littermate) treated with paraquat. Despite this pathological effect the mild cell loss induced by paraquat in wild-type mice did not increase in transgenics and the behavioral deficits observed in transgenic mice were not modified by paraquat administration. Furthermore, although microglial activation was observed shortly after one administration of paraquat (Purisai et al., 2007), no increase in microglia was observed 5 days after the final treatment of a 3 weeks regimen of the drug, indicating that the massive increase in alpha-synuclein aggregates did not produce a sustained inflammatory response in this model (Fernagut et al., 2007). Overall, it appears that despite increases in the presence of its abnormal forms, alpha-synuclein overexpression does not increase the cell loss induced by short-term (3 weeks) exposure to paraquat alone. Furthermore, in some cases, the upregulation of defense mechanisms induced by overexpression of alpha-synuclein was able to counteract the effect of paraquat on nigrostriatal neurons.

Combined Paraquat and Maneb Exposure in Mice Overexpressing Alpha-Synuclein

Except for rare accidents, most exposures to high levels of pesticides involve a combination of agents either applied together or successively during the life of an individual. A well-characterized model of environmental exposure in rats and mice combines exposure to maneb, a fungicide, and paraquat (Thiruchelvam *et al.*, 2000; Cicchetti *et al.*, 2005; Cory-Slechta *et al.*, 2005). Interestingly, in contrast to paraquat alone, the effect of this regimen was exacerbated both in mice overexpressing a doubly mutated human alpha-synuclein (A30P + A53T) under the control of the rat TH promoter, and in mice overexpressing A53T alpha-synuclein under the control of the murine prion promoter (Thiruchelvam *et al.*, 2004 ; Norris *et al.*, 2007).

Combined maneb and paraquat administration to old rats induces microglial activation (Saint-Pierre *et al.*, 2006) and several lines of evidence indicate that alpha-synuclein promotes microglial activation *in vivo* and *in vitro* (Su *et al.*, 2007). Thus, increased inflammatory processes could be involved in the increased toxicity of this combination of environmental factors in alpha-synuclein overexpressing mice. The combination of paraquat and maneb alters a number of other cellular mechanisms, for example antioxidant defenses and mitochondrial function, which could also be exacerbated by high levels of alpha-synuclein (Zhang *et al.*, 2003; Patel *et al.*, 2006).

Other Genetic Rodent Models

Parkin mutations were the second genetic cause of PD to be identified and they account for a substantial number of recessive juvenile onset cases (Douglas *et al.*, 2007). Despite the presence of synaptic alterations and progressive behavioral deficits in some lines of mice (Goldberg *et al.*, 2003; Itier *et al.*, 2003; Von Coellen *et al.*, 2004), available parkin knock out mice do not show spontaneous loss of dopaminergic neurons. Furthermore, parkin knock out mice did not show increased sensitivity to neurotoxins such as 6-hydroxydopamine and methamphetamine (Perez *et al.*, 2005). However, mice lacking a putative parkin substrate, the PD-associated GPR37/parkin-associated endothelin-like receptor (Pael), are more resistant to MPTP than wild-type mice (Marazziti *et al.*, 2004). The significance of this effect is unclear as it has been difficult to detect increases in Pael in parkin-deficient mice.

The lack of greater sensitivity of parkin-deficient mice to neurotoxins was surprising because parkin overexpression through viral transduction protects wild-type mice from mild MPTP-induced lesions (Paterna *et al.*, 2007). This suggests that functional parkin is neuroprotective against cellular insults, including those that could be caused by environmental toxins. Interestingly, nitrosylation, an effect of oxidative stress, can decrease parkin protective function (Chung *et al.*, 2004). Thus, a link may exist between PD-causing mutations in parkin and mechanisms by which exposure to pesticides increase the risk of sporadic PD (Dinis-Oliveira *et al.*, 2006).

An even more compelling case can be made for a potential interaction between DJ-1 mutations and exposure to environmental toxins because of the protective role of DJ-1 against oxidative stress (Menzies *et al.*, 2005; Meulener *et al.*, 2006). Kim *et al.* (2005) and Manning-Bog *et al.* (2007) found an increased sensitivity to MPTP in mice lacking DJ-1. However, several different lines of DJ-1 deficient mice have an increased level of the dopamine cytoplasmic transporter (Chen *et al.*, 2005; Manning-Bog *et al.*, 2007), which is likely to mediate the increased toxicity of MPTP in these mice (Manning-Bog *et al.*, 2007). It is interesting that, as described earlier, a similar mechanism has been observed in some lines of alpha-synuclein mice (Richfield *et al.*, 2002), suggesting that changes in DAT function may be a common effect of several PD-causing mutations.

As for parkin, DJ-1 overexpression through viral transduction protects wild-type mice from mild MPTP-induced lesions (Kim *et al.*, 2005; Paterna *et al.*, 2007), suggesting that loss of DJ-1 may lead to PD by conferring hypersensitivity to dopaminergic insults (Kim *et al.*, 2005). Here again, inactivation of DJ-1 by environmental exposure (paraquat) could link sporadic and genetic forms of the disease (Mitsumoto *et al.*, 2001).

Little information exists on the sensitivity of other mouse models of PD such as PINK1-deficient mice, which have been much more recently generated (Kitada *et al.*, 2007). Similarly it will be interesting to understand the interactions between environmental toxins and mutations in LRRK2, which are present in a subset of patients with late onset PD, very similar to the classical sporadic forms of the disease (Ozelius *et al.*, 2006).

CONCLUSIONS

Source of Conflicting Results in the Exploration of Environmental Factors in Genetic Models of PD

Despite their high construct validity, genetic rodent models of PD do not reproduce all aspects of the human disease. This should not come as a surprise but it limits the type of questions that can be addressed in these models. The absence of nigrostriatal cell loss despite the presence of a mutation that kills these neurons in humans suggests that efficient defense mechanisms are triggered in the transgenic or knock out mice. Identifying these defense mechanisms may lead to useful insights into ways to counteract the pathophysiological process in PD. A variation in these defense mechanisms in different lines of mice may also explain why some transgenics showed an unexpected decreased sensitivity to the environmental toxin paraquat (Manning-Bog et al., 2003), while others did not (Fernagut et al., 2007).

It is important to remember that the mechanism of actions of some toxins depends on cellular mechanisms that may be affected by the mutation. For example, we have seen that the active metabolite of MPTP, MPP$^+$ enters dopaminergic neurons by way of the cytoplasmic DAT. Increased DAT expression in some lines of alpha-synuclein overexpressors (Richfield et al., 2002) and two lines of DJ-1 knock out mice (Chen et al., 2005; Manning-Bog et al., 2007) seem to explain the apparent increased sensitivity of these mice to MPTP. In these cases, the greater effect of the toxin was due to an increased intracellular level rather than an increased vulnerability of the neurons. It should be noted that MPP$^+$ is sequestered into vesicles in dopaminergic terminals by the vesicular dopamine transporter (VMAT2); therefore, differences in VMAT activity could also change the apparent toxicity of MPTP (German et al., 2000), and it would be important to assess VMAT function in transgenic mice exposed to the toxin.

Future Studies of Environmental Factors in Genetic Rodent Models of PD

The main rationale for exposing genetic rodent models of PD to environmental toxicants has been to generate a more complete model of the disease by inducing nigrostriatal cell loss in mice that failed to reproduce this pathological hallmark of PD. As we have seen, this has met with variable success. A more compelling approach would be to provide a testable model of a "multiple hits" hypothesis for PD, in which pathophysiological mechanisms could be analyzed and therapeutic targets identified (Manning-Bog and Langston, 2007). This is of

course complicated by potential differences not only in mouse physiology but also in the mode of exposure to environmental toxicants. It is extremely difficult to reproduce the duration and complexity of human exposure. Furthermore, naturalistic routes of administration are rarely used and they could influence the mechanisms of toxicity (Rojo et al., 2006).

Valuable information can be obtained, however, even from an imperfect model. The models described in this chapter could be used to dissect out the most effective defense mechanisms that protect nigrostriatal dopaminergic neurons from environmental stressors. There is also a growing need for molecular studies of identified neuronal populations in animals exposed to environmental toxicants (Meurers et al., 2006; Mortazavi et al., 2007). In addition, it will be important to expand the list of toxicants used in experimental studies in rodents to include a broader range of compounds for which evidence of involvement in PD is strengthening (Kamel et al., 2007).

Future Studies of Environmental Factors in Genetic Rodent Models of Risk Factors for PD

So far, studies have focused on genetic models expressing mutations that cause familial forms of PD. These may mimic alterations in cellular pathways that are also involved in the sporadic forms of the disease. However, susceptibility to environmental toxins is more likely due to genetic risk factors. For example, polymorphisms in the VMAT2 promoter leading to the increased levels of expression of the transporter have been associated with a lower risk of PD in women (Glatt et al., 2006), whereas polymorphisms in TNFalpha and IL1beta that increase inflammatory responses are associated with an increased risk of PD (Wahner et al., 2007). So far, genetic associations and epidemiological studies have lacked the power to perform rigorous gene–environment interaction studies. Potential interactions could be tested experimentally by exposing mice expressing mutations that reproduce genetic risk factors for PD to environmental toxins.

Therapeutic and Public Health Implications

The lack of systematic worsening of nigrostriatal cell loss in mice expressing PD-causing mutations when exposed to environmental pesticides should not obscure the compelling epidemiological evidence suggesting a link between prolonged pesticide exposure and PD in humans (Ritz and Yu, 2000; Chade et al., 2006; Kamel et al., 2007). Clearly, risk factors play a role, and using new generations of genetically modified rodents to further elucidate their role may inform genetic studies in humans. In addition, animal work can help identify the mechanisms by which environmental factors may contribute to PD. Evidence for shared mechanisms of action with agents, even banned for years, that emerge as risk factors in individuals that have reached the age at which PD manifests should call for increased scrutiny of newly introduced compounds. Finally the efficient defense mechanisms that protect nigrostriatal dopaminergic neurons in mice expressing PD-causing mutations, and sometimes make them resistant to the effects of environmental toxins, may point to new neuroprotective strategies in humans, the ultimate goal of PD research.

REFERENCES

Barbeau, A., Roy, M., Cloutier, T., Plasse, L., and Paris, S. (1987). Environmental and genetic factors in the etiology of Parkinson's disease. Adv Neurol 45, 299–306.

Betarbet, R., Sherer, T. B., MacKenzie, G., Garcia-Osuna, M., Panov, A. V., and Greenamyre, J. T. (2000). Chronic systemic pesticide exposure reproduces features of Parkinson's disease. Nat Neurosci 3, 1301–1306.

Bove, J., Prou, D., Perier, C., and Przedborski, S. (2005). Toxin-induced models of Parkinson's disease. NeuroRx 2, 484–494.

Chade, A. R., Kasten, M., and Tanner, C. M. (2006). Nongenetic causes of Parkinson's disease. J Neural Transm Suppl 147–151.

Chen, L., Cagniard, B., Mathews, T., Jones, S., Koh, H. C., Ding, Y., Carvey, P. M., Ling, Z., Kang, U. J., and Zhuang, X. (2005). Age-dependent motor deficits and dopaminergic dysfunction in DJ-1 null mice. J Biol Chem 280, 21418–21426.

Chesselet, M.-F. (2008). In vivo alpha-synuclein overexpression in rodents: A useful model of Parkinson's disease? Exp Neurol 209, 22–27.

Chung, K. K., Thomas, B., Li, X., Pletnikova, O., Troncoso, J. C., Marsh, L., Dawson, V. L., and Dawson, T. M. (2004). S-nitrosylation of parkin regulates ubiquitination and compromises parkin's protective function. Science 304, 1328–1331.

Cicchetti, F., Lapointe, N., Roberge-Tremblay, A., Saint-Pierre, M., Jimenez, L., Ficke, B. W., and Gross, R. E. (2005). Systemic exposure to paraquat and maneb models early Parkinson's disease in young adult rats. Neurobiol Dis 20, 360–371.

Cory-Slechta, D. A., Thiruchelvam, M., Barlow, B. K., and Richfield, E. K. (2005). Developmental pesticide models of the Parkinson disease phenotype. Environ Health Perspect 113, 1263–1270.

Dinis-Oliveira, R. J., Remiao, F., Carmo, H., Duarte, J. A., Navarro, A. S., Bastos, M. L., and Carvalho, F. (2006). Paraquat exposure as an etiological factor of Parkinson's disease. Neurotoxicology 27, 1110–1122.

Dong, Z., Ferger, B., Feldon, J., and Bueler, H. (2002). Overexpression of Parkinson's disease-associated alpha-synuclein A53T by recombinant adeno-associated virus in mice does not increase the vulnerability of dopaminergic neurons to MPTP. J Neurobiol 53, 1–10.

Douglas, M. R., Lewthwaite, A. J., and Nicholl, D. J. (2007). Genetics of Parkinson's disease and parkinsonism. Expert Rev Neurother 7, 657–666.

Fernagut, P.-O., and Chesselet, M.-F. (2004). Alpha-synuclein and transgenic mouse models. Neurobiol Dis 17, 123–130.

Fernagut, P. O., Hutson, C. B., Fleming, S. M., Tetreaut, N., Salcedo, J., Masliah, E., and Chesselet, M. F. (2007). Behavioral and histopathological consequences of paraquat intoxication in mice: Effects of alpha-synuclein overexpression. Synapse 61, 991–1001.

Fleming, S. M., and Chesselet, M. F. (2006). Behavioral phenotypes and pharmacology in genetic mouse models of Parkinsonism. Behav Pharmacol 17, 383–391.

Fleming, S. M., Fernagut, P. O., and Chesselet, M. F. (2005). Genetic mouse models of parkinsonism: Strengths and limitations. NeuroRx 2, 495–503.

German, D. C., Liang, C. L., Manaye, K. F., Lane, K., and Sonsalla, P. K. (2000). Pharmacological inactivation of the vesicular monoamine transporter can enhance 1-methyl-4-phenyl-1,2,3,6-tetrahydropyridine-induced neurodegeneration of midbrain dopaminergic neurons, but not locus coeruleus noradrenergic neurons. Neuroscience 101, 1063–1069.

Glatt, C. E., Wahner, A. D., White, D. J., Ruiz-Linares, A., and Ritz, B. (2006). Gain-of-function haplotypes in the vesicular monoamine transporter promoter are protective for Parkinson disease in women. Hum Mol Genet 15, 299–305.

Goldberg, M. S., Fleming, S. M., Palacino, J. J., Cepeda, C., Lam, H. A., Bhatnagar, A., Meloni, E. G., Wu, N., Ackerson, L. C., Klapstein, G. J., Gajendiran, M., Roth, B. L., Chesselet, M. F., Maidment, N. T., Levine, M. S., and Shen, J. (2003). Parkin-deficient mice exhibit nigrostriatal deficits but not loss of dopaminergic neurons. *J Biol Chem* **278**, 43628–43635.

Greenamyre, J. T., and Hastings, T. G. (2004). Biomedicine. Parkinson's–divergent causes, convergent mechanisms. *Science* **304**, 1120–1122.

Hertzman, C., Wiens, M., Bowering, D., Snow, B., and Calne, D. (1990). Parkinson's disease: A case–control study of occupational and environmental risk factors. *Am J Ind Med* **17**, 349–355.

Itier, J. M., Ibanez, P., Mena, M. A., Abbas, N., Cohen-Salmon, C., Bohme, G. A., Laville, M., Pratt, J., Corti, O., Pradier, L., Ret, G., Joubert, C., Periquet, M., Araujo, F., Negroni, J., Casarejos, M. J., Canals, S., Solano, R., Serrano, A., Gallego, E., Sanchez, M., Denefle, P., Benavides, J., Tremp, G., Rooney, T. A., Brice, A., and Garcia de Yebenes, J. (2003). Parkin gene inactivation alters behaviour and dopamine neurotransmission in the mouse. *Hum Mol Genet* **12**, 2277–2291.

Kamel, F., Tanner, C., Umbach, D., Hoppin, J., Alavanja, M., Blair, A., Comyns, K., Goldman, S., Korell, M., Langston, J., Ross, G., and Sandler, D. (2007). Pesticide exposure and self-reported Parkinson's disease in the agricultural health study. *Am J Epidemiol* **165**, 364–374.

Kim, R. H., Smith, P. D., Aleyasin, H., Hayley, S., Mount, M. P., Pownall, S., Wakeham, A., You-Ten, A. J., Kalia, S. K., Horne, P., Westaway, D., Lozano, A. M., Anisman, H., Park, D. S., and Mak, T. W. (2005). Hypersensitivity of DJ-1-deficient mice to 1-methyl-4-phenyl-1,2,3,6-tetrahydropyridine (MPTP) and oxidative stress. *Proc Natl Acad Sci USA* **102**, 5215–5220.

Kitada, T., Pisani, A., Porter, D. R., Yamaguchi, H., Tscherter, A., Martella, G., Bonsi, P., Zhang, C., Pothos, E. N., and Shen, J. (2007). From the Cover: Impaired dopamine release and synaptic plasticity in the striatum of PINK1-deficient mice. *Proc Natl Acad Sci USA* **104**, 11441–11446.

Lang, A. E., and Obeso, J. A. (2004). Time to move beyond nigrostriatal dopamine deficiency in Parkinson's disease. *Ann Neurol* **55**, 761–765.

Langston, J. W. (2006). The Parkinson's complex: Parkinsonism is just the tip of the iceberg. *Ann Neurol* **59**, 591–596.

Langston, J. W., Ballard, P., Tetrud, J. W., and Irwin, I. (1983). Chronic Parkinsonism in humans due to a product of meperidine-analog synthesis. *Science* **219**, 979–980.

Liou, H. H., Tsai, M. C., Chen, C. J., Jeng, J. S., Chang, Y. C., Chen, S. Y., and Chen, R. C. (1997). Environmental risk factors and Parkinson's disease: A case–control study in Taiwan. *Neurology* **48**, 1583–1588.

Litvan, I., Halliday, G., Hallett, M., Goetz, C. G., Rocca, W., Duyckaerts, C., Ben-Shlomo, Y., Dickson, D. W., Lang, A. E., Chesselet, M. F., Langston, W. J., Di Monte, D. A., Gasser, T., Hagg, T., Hardy, J., Jenner, P., Melamed, E., Myers, R. H., Parker, D., Jr., and Price, D. L. (2007a). The etiopathogenesis of Parkinson disease and suggestions for future research. I. *J Neuropathol Exp Neurol* **66**, 251–257.

Litvan, I., Chesselet, M. F., Gasser, T., Di Monte, D. A., Parker, D., Jr., Hagg, T., Hardy, J., Jenner, P., Myers, R. H., Price, D., Hallett, M., Langston, W. J., Lang, A. E., Halliday, G., Rocca, W., Duyckaerts, C., Dickson, D. W., Ben-Shlomo, Y., Goetz, C. G., and Melamed, E. (2007b). The etiopathogenesis of Parkinson disease and suggestions for future research. Part II. *J Neuropathol Exp Neurol* **66**, 329–336.

Manning-Bog, A. B., and Langston, J. W. (2007). Model fusion, the next phase in developing animal models for Parkinson's disease. *Neurotox Res* **11**, 219–240.

Manning-Bog, A. B., McCormack, A. L., Purisai, M. G., Bolin, L. M., and Di Monte, D. A. (2003). Alpha-synuclein overexpression protects against paraquat-induced neurodegeneration. *J Neurosci* **23**, 3095–3099.

Manning-Bog, A. B., Caudle, W. M., Perez, X. A., Reaney, S. H., Paletzki, R., Isla, M. Z., Chou, V. P., McCormack, A. L., Miller, G. W., Langston, J. W., Gerfen, C. R., and Dimonte, D. A. (2007). Increased vulnerability of nigrostriatal terminals in DJ-1-deficient mice is mediated by the dopamine transporter. *Neurobiol Dis* **27**, 141–150.

Marazziti, D., Golini, E., Mandillo, S., Magrelli, A., Witke, W., Matteoni, R., and Tocchini-Valentini, G. P. (2004). Altered dopamine signaling and MPTP resistance in mice lacking the Parkinson's disease-associated GPR37/parkin-associated endothelin-like receptor. *Proc Natl Acad Sci USA* **101**, 10189–10194.

McCormack, A. L., Thiruchelvam, M., Manning-Bog, A. B., Thiffault, C., Langston, J. W., Cory-Slechta, D. A., and Di Monte, D. A. (2002). Environmental risk factors and Parkinson's disease: Selective degeneration of nigral dopaminergic neurons caused by the herbicide paraquat. *Neurobiol Dis* **10**, 119–127.

McCormack, A. L., Atienza, J. G., Johnston, L. C., Andersen, J. K., Vu, S., and Di Monte, D. A. (2005). Role of oxidative stress in paraquat-induced dopaminergic cell degeneration. *J Neurochem* **93**, 1030–1037.

Menzies, F. M., Yenisetti, S. C., and Min, K. T. (2005). Roles of Drosophila DJ-1 in survival of dopaminergic neurons and oxidative stress. *Curr Biol* **15**, 1578–1582.

Meulener, M. C., Xu, K., Thomson, L., Ischiropoulos, H., and Bonini, N. M. (2006). Mutational analysis of DJ-1 in Drosophila implicates functional inactivation by oxidative damage and aging. *Proc Natl Acad Sci USA* **103**, 12517–12522.

Meurers, B. H., Zhu, C., Fernagut, P. O., Mortazavi, F., Fleming, S. M., Oh, M. S., Elashoff, D., DiCarlo, C. D., Seaman, R. L., and Chesselet, M. F. (2006). In between life and death: Functional alterations of dopaminergic neurons in rotenone treated rats that do not exhibit cell loss. *Neurosci Abst* 755, 21.

Mitsumoto, A., Nakagawa, Y., Takeuchi, A., Okawa, K., Iwamatsu, A., and Takanezawa, Y. (2001). Oxidized forms of peroxiredoxins and DJ-1 on two-dimensional gels increased in response to sublethal levels of paraquat. *Free Radic Res* 35, 301–310.

Mortazavi, F., Meurers, B., Oh, M. S., Elashoff, D., and Chesselet, M. F. (2007). Transcriptome analysis of laser-captured dopaminergic neurons in Mice overexpressing human wildtype alpha-synuclein. *Neurosci Abst* 151, 21.

Nieto, M., Gil-Bea, F. J., Dalfo, E., Cuadrado, M., Cabodevilla, F., Sanchez, B., Catena, S., Sesma, T., Ribe, E., Ferrer, I., Ramirez, M. J., and Gomez-Isla, T. (2006). Increased sensitivity to MPTP in human alpha-synuclein A30P transgenic mice. *Neurobiol Aging* 27, 848–856.

Norris, E. H., Uryu, K., Leight, S., Giasson, B. I., Trojanowski, J. Q., and Lee, V. M. (2007). Pesticide exposure exacerbates alpha-synucleinopathy in an A53T transgenic mouse model. *Am J Pathol* 170, 658–666.

Olanow, C. W., and McNaught, K. S. (2006). Ubiquitin–proteasome system and Parkinson's disease. *Mov Disord* 21, 1806–1823.

Ozelius, L. J., Senthil, G., Saunders-Pullman, R., Ohmann, E., Deligtisch, A., Tagliati, M., Hunt, A. L., Klein, C., Henick, B., Hailpern, S. M., Lipton, R. B., Soto-Valencia, J., Risch, N., and Bressman, S. B. (2006). LRRK2 G2019S as a cause of Parkinson's disease in Ashkenazi Jews. *N Engl J Med* 354, 424–425.

Patel, S., Singh, V., Kumar, A., Gupta, Y. K., and Singh, M. P. (2006). Status of antioxidant defense system and expression of toxicant responsive genes in striatum of maneb- and paraquat-induced Parkinson's disease phenotype in mouse: Mechanism of neurodegeneration. *Brain Res* 1081, 9–18.

Paterna, J. C., Leng, A., Weber, E., Feldon, J., and Bueler, H. (2007). DJ-1 and Parkin modulate dopamine-dependent behavior and inhibit MPTP-induced nigral dopamine neuron loss in mice. *Mol Ther* 15, 698–704.

Perez, F. A., Curtis, W. R., and Palmiter, R. D. (2005). Parkin-deficient mice are not more sensitive to 6-hydroxydopamine or methamphetamine neurotoxicity. *BMC Neurosci* 6, 71.

Polymeropoulos, M. H., Lavedan, C., Leroy, E., Ide, S. E., Dehejia, A., Dutra, A., Pike, B., Root, H., Rubenstein, J., Boyer, R., Stenroos, E. S., Chandrasekharappa, S., Athanassiadou, A., Papapetropoulos, T., Johnson, W. G.,

Lazzarini, A. M., Duvoisin, R. C., Di Iorio, G., Golbe, L. I., and Nussbaum, R. L. (1997). Mutation in the alpha-synuclein gene identified in families with Parkinson's disease. *Science* 276, 2045–2047.

Purisai, M. G., McCormack, A. L., Cumine, S., Li, J., Isla, M. Z., and Di Monte, D. A. (2007). Microglial activation as a priming event leading to paraquat-induced dopaminergic cell degeneration. *Neurobiol Dis* 25, 392–400.

Rathke-Hartlieb, S., Kahle, P. J., Neumann, M., Ozmen, L., Haid, S., Okochi, M., Haass, C., and Schulz, J. B. (2001). Sensitivity to MPTP is not increased in Parkinson's disease-associated mutant alpha-synuclein transgenic mice. *J Neurochem* 77, 1181–1184.

Richardson, J. R., Caudle, W. M., Guillot, T. S., Watson, J. L., Nakamaru-Ogiso, E., Seo, B. B., Sherer, T. B., Greenamyre, J. T., Yagi, T., Matsuno-Yagi, A., and Miller, G. W. (2007). Obligatory role for complex I inhibition in the dopaminergic neurotoxicity of 1-methyl-4-phenyl-1,2,3,6-tetrahydropyridine (MPTP). *Toxicol Sci* 95, 196–204.

Richfield, E. K., Thiruchelvam, M. J., Cory-Slechta, D. A., Wuertzer, C., Gainetdinov, R. R., Caron, M. G., Di Monte, D. A., and Federoff, H. J. (2002). Behavioral and neurochemical effects of wild-type and mutated human alpha-synuclein in transgenic mice. *Exp Neurol* 175, 35–48.

Richter, F., Hamann, M., and Richter, A. (2007). Chronic rotenone treatment induces behavioral effects but no pathological signs of parkinsonism in mice. *J Neurosci Res* 85, 681–691.

Ritz, B., and Yu, F. (2000). Parkinson's disease mortality and pesticide exposure in California 1984–1994. *Int J Epidemiol* 29, 323–329.

Ritz, B., and Costello, S. (2006). Geographic model and biomarker-derived measures of pesticide exposure and Parkinson's disease. *Ann NY Acad Sci* 1076, 378–387.

Rojo, A. I., Montero, C., Salazar, M., Close, R. M., Fernandez-Ruiz, J., Sanchez-Gonzalez, M. A., de Sagarra, M. R., Jackson-Lewis, V., Cavada, C., and Cuadrado, A. (2006). Persistent penetration of MPTP through the nasal route induces Parkinson's disease in mice. *Eur J Neurosci* 24, 1874–1884.

Saint-Pierre, M., Tremblay, M. E., Sik, A., Gross, R. E., and Cicchetti, F. (2006). Temporal effects of paraquat/maneb on microglial activation and dopamine neuronal loss in older rats. *J Neurochem* 98, 760–772.

Song, D. D., Shults, C. W., Sisk, A., Rockenstein, E., and Masliah, E. (2004). Enhanced substantia nigra mitochondrial pathology in human alpha-synuclein transgenic mice after treatment with MPTP. *Exp Neurol* 186, 158–172.

Spillantini, M. G., Schmidt, M. L., Lee, V. M., Trojanowski, J. Q., Jakes, R., and Goedert, M.

(1997). Alpha-synuclein in Lewy bodies. *Nature* **388**, 839–840.

Su, X., Maguire-Zeiss, K. A., Giuliano, R., Prifti, L., Venkatesh, K., and Federoff, H. J. (2007). Synuclein activates microglia in a model of Parkinson's disease. *Neurobiol Aging*, Epub ahead of print.

Thiruchelvam, M., Richfield, E. K., Baggs, R. B., Tank, A. W., and Cory-Slechta, D. A. (2000). The nigrostriatal dopaminergic system as a preferential target of repeated exposures to combined paraquat and maneb: Implications for Parkinson's disease. *J Neurosci* **20**, 9207–9214.

Thiruchelvam, M. J., Powers, J. M., Cory-Slechta, D. A., and Richfield, E. K. (2004). Risk factors for dopaminergic neuron loss in human alpha-synuclein transgenic mice. *Eur J Neurosci* **19**, 845–854.

Von Coellen, R., Thomas, B., Savitt, J. M., Lim, K. L., Sasaki, M., Hess, E. J., Dawson, V. L., and Dawson, T. M. (2004). Loss of locus coeruleus neurons and reduced startle in parkin null mice. *Proc Natl Acad Sci USA* **101**, 10744–10749.

Wahner, A. D., Sinsheimer, J. S., Bronstein, J. M., and Ritz, B. (2007). Inflammatory cytokine gene polymorphisms and increased risk of Parkinson disease. *Arch Neurol* **64**, 836–840.

Wolters, E., and Braak, H. (2006). Parkinson's disease: Premotor clinico-pathological correlations. *J Neural Transm Suppl*, 309–319.

Zhang, J., Fitsanakis, V. A., Gu, G., Jing, D., Ao, M., Amarnath, V., and Montine, T. J. (2003). Manganese ethylene-bis-dithiocarbamate and selective dopaminergic neurodegeneration in rat: A link through mitochondrial dysfunction. *J Neurochem* **84**, 336–346.

Zhu, C., Vourc'h, P., Fernagut, P. O., Fleming, S. M., Lacan, S., Dicarlo, C. D., Seaman, R. L., and Chesselet, M. F. (2004). Variable effects of chronic subcutaneous administration of rotenone on striatal histology. *J Comp Neurol* **478**, 418–426.

22

α-SYNUCLEIN, CSPα, SNARES AND NEUROPROTECTION *IN VIVO*

SREEGANGA CHANDRA[1,2] **AND THOMAS C. SÜDHOF**[2,3]

[1]*Program in Cellular Neuroscience, Neurodegeneration and Repair, Yale University, New Haven, CT, USA*
[2]*Department of Neuroscience,* [3]*Howard Hughes Medical Institute, UT Southwestern Medical Center, Dallas, TX,USA*

INTRODUCTION

α-Synuclein and cysteine-string protein-α (CSPα) are presynaptic proteins that are independently linked to neurodegeneration. Dominantly inherited mutations in α-synuclein cause familial Parkinson's disease (PD), and in most cases of PD, α-synuclein forms pathological aggregates known as Lewy bodies. CSPα is a co-chaperone, which when deleted in mice produces progressive neurodegeneration. This chapter focuses on recent studies that established a clear and intriguing genetic link between these two proteins.

α-SYNUCLEIN AND PD

In the last decade, great strides have been made in identifying genes that cause inherited forms of PD (Dauer and Przedborski, 2003; Tan and Skipper, 2007). There are 12 identified loci (PARK 1 – PARK 13; note that PARK 1 and PARK 4 are the same gene) known to cause familial PD. The first familial PD gene to be identified was α-synuclein (PARK 1; SNCA). To date, three dominant α-synuclein mutations – A30P, A53T and E46K – have been identified in separate families with PD (Polymeropoulos *et al.*, 1997; Kruger *et al.*, 1998; Zarranz *et al.*, 2004) (Figure 22.1). More recently duplication and triplication of the wild type α-synuclein locus (PARK 4) (Singleton *et al.*,

2003; Chartier-Harlin *et al.*, 2004; Ibanez *et al.*, 2004) were also shown to cause familial PD. A comparison of families with such α-synuclein gene multiplications indicates that gene dosage is directly related to age of onset and severity of the disease (Singleton, 2005). Furthermore, certain α-synuclein promoter polymorphisms confer susceptibility to PD consistent with the idea that alterations in α-synuclein levels are associated with PD (Pals *et al.*, 2004; Mueller *et al.*, 2005). Finally, Lewy bodies, the pathological hallmark of both sporadic and familial PD, are mainly composed of aggregated α-synuclein, usually in an ubiquitinated form (Spillantini *et al.*, 1997). Thus, α-synuclein is a key player in the pathophysiology of PD. Interestingly, α-synuclein pathology is seen in a variety of other neurodegenerative diseases, including Alzheimer's disease and Lewy body dementia (Galvin *et al.*, 2001). These disorders are classified as "synucleinopathies" and point to the significant role that α-synuclein plays in many neurodegenerative diseases (Galvin *et al.*, 2001).

SYNUCLEINS

α-Synuclein is the founding member of a family of vertebrate-specific proteins (Clayton and George, 1998; Lavedan, 1998) (Figure 22.1). Synucleins consist of three closely related members, namely α-, β-, and γ-synuclein. Synucleins are ~14 kDa

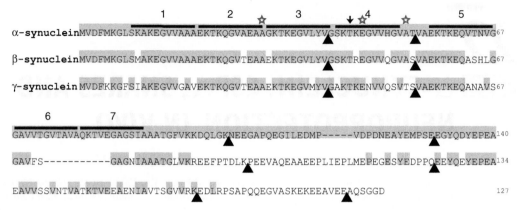

FIGURE 22.1 Sequence alignment of human α-synuclein, β-synuclein and γ-synuclein protein sequences. Residues are colored gray to indicate identity with human α-synuclein. The horizontal lines outline the seven 11 amino acid repeats, the arrowheads the exon boundaries. The stars indicate the position of the three known PD mutations, A30P, A53T and E46K. Arrow indicates location of break in α-helical conformation. (Chandra *et al.*, 2003; Ulmer *et al.*, 2005)

soluble proteins. All synucleins are abundantly expressed in the nervous system and are presynaptically localized (Maroteaux *et al.*, 1988; Clayton and George, 1998). Their expression begins in late embryonic development and peaks around postnatal day 15 similar to other synaptic proteins. α- and β-synuclein are mainly expressed in the central nervous system, while γ-synuclein is enriched in the peripheral nervous synapses (Abeliovich *et al.*, 2000; Ninkina *et al.*, 2003). The physiological/synaptic functions of synucleins remain to be elucidated.

Synucleins have unusual biophysical properties. They are natively unfolded proteins and do not have any secondary or tertiary structure in solution (Weinreb *et al.*, 1996; Eliezer *et al.*, 2001). The atypical folding properties of synucleins are rooted in their characteristic primary sequence (Clayton and George, 1998) (Figure 22.1). All synuclein proteins have a highly conserved N-terminal domain with an 11-residue repeat and a less conserved highly acidic C-terminal domain. The seven imperfect 11-residue repeats have the consensus sequence XKTKEGVXXXX, and are highly reminiscent of amphipathic helices of apolipoproteins. Remarkably, synucleins can bind acidic phospholipid surfaces via their N-terminal domain, and acquire secondary structure (Davidson *et al.*, 1998; Chandra *et al.*, 2003). This folding can be replicated in anionic detergents such as SDS (Eliezer *et al.*, 2001; Chandra *et al.*, 2003). The structure of the

folded conformation of human α-synuclein has been determined using protein NMR (Chandra *et al.*, 2003; Ulmer *et al.*, 2005; Woods *et al.*, 2007) and ESR methods (Jao *et al.*, 2004). On SDS micelles, the folded conformation consists of two amphipathic α-helices with an unfolded C-terminus (Chandra *et al.*, 2003; Ulmer *et al.*, 2005) (Figure 22.2). The two α-helices are oriented in an antiparallel arrangement (Ulmer *et al.*, 2005). Recently, human β and γ-synuclein were also show to adopt this conformation, suggesting it is physiologically relevant (Sung and Eliezer, 2006).

DUALITY OF α-SYNUCLEIN FUNCTION

Human α-synuclein is unique in the synuclein family in that it has a great propensity to aggregate (Beyer, 2006; Fink, 2006; Uversky, 2007). This property has been attributed to a 12 hydrophobic amino acid stretch unique to α-synuclein (Giasson *et al.*, 2001) (aa 71–82; Figure 22.1). α-Synuclein readily forms dimers, then oligomers and subsequently, large insoluble β-sheeted fibrils or amyloids as those seen in neuropathological aggregates (Volles and Lansbury, 2003; Fink, 2006; Uversky, 2007). Conditions that promote the folding of α-synuclein into an α-helical conformation also promote its aggregation (Cole *et al.*, 2002). Post-translational modifications such as

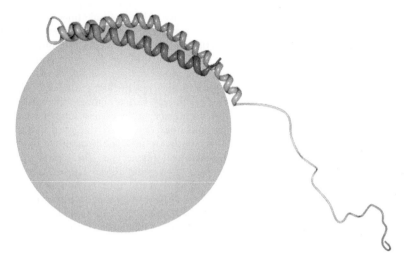

FIGURE 22.2 Folded conformation of α-synuclein on SDS micelle. Ribbon diagram of SDS micelle-bound α-synuclein structure (PDB entry 1XQ8, Chandra *et al.*, 2003, Ulmer *et al.*, 2005). The N-terminal portion of the molecule folds on the micelle surface, with the C-terminal remaining unbound and unfolded. The break in the helix occurs at amino acid 43–45.

phosphorylation and C-terminal truncations, incubation with metal ions such as iron, and oxidative/nitrative challenges also accelerate its aggregation (Fujiwara *et al.*, 2002; Cole *et al.*, 2005; Beyer, 2006; Fink, 2006). Protofibrils are transient β sheet-containing soluble oligomers, formed during fibrillization (Uversky *et al.*, 2001; Volles and Lansbury, 2003). Protofibrillar α-synuclein, in contrast to the monomeric and the fibrillar forms can bind liposomes via a β-sheet-rich structure, and may produce pores in membranes (Volles *et al.*, 2001; Volles and Lansbury, 2003). It is this finding that has lead to the hypothesis that α-synuclein protofibrils are neurotoxic. Interestingly, dopamine reacts with α-synuclein to form covalent adducts that stabilize protofibrils (Conway *et al.*, 2001; Norris *et al.*, 2005). It has been hypothesized that catecholamine protofibril stabilization is the basis for the dopaminergic selectivity of cell death in PD.

The discovery of dominant α-synuclein Parkinson mutations and α-synuclein's tendency to aggregate has lead to the hypothesis that α-synuclein causes PD by a toxic gain-of-function, that is, a pathological function. There is much debate in the field as to which species in the α-synuclein aggregation profile is causal to disease – the soluble protofibrils versus the insoluble aggregates such as Lewy bodies (Volles and Lansbury, 2003). While the

jury is still out on this question, a growing body of evidence favors the protofibrils as the culprit. One of the strongest pieces of evidence is that the A30P mutation actually slows macroscopic aggregation, while the A53T mutation accelerates aggregation *in vitro*. However, both the A53T and A30P mutants promote protofibril formation as compared to wild type α-synuclein (Conway *et al.*, 2000). Further support comes from the finding that overexpression of chaperones in PD models ameliorates neurodegeneration while leaving α-synuclein pathology intact (Auluck *et al.*, 2002), suggesting that the chaperones act on soluble α-synuclein species. Yet, the details of the mechanisms by which α-synuclein protofibrils are toxic to neurons remain to be elucidated.

In contrast, to our insights regarding the pathological functions of α-synuclein, the normal physiological function of synucleins remains largely unknown. However, the discovery that the PARK4 locus is a triplication of a wild type α-synuclein gene has sparked renewed interest in identification of the physiological function of α-synuclein (Singleton *et al.*, 2003; Chartier-Harlin *et al.*, 2004; Ibanez *et al.*, 2004). Furthermore, the awareness that synapses are the first structures to be lost in PD, prior to the loss of neurons, has also fuelled this interest (Dauer and Przedborski, 2003). Several independent lines of research suggest that the physiological function of α-synuclein is neuroprotective (see below) and may play an important role in ameliorating neurodegeneration.

In PARK4 cases, triplication of the wild type locus and increasing synuclein protein levels by approximately twofold is sufficient to cause dominantly inherited PD (Singleton et al., 2003; Miller et al., 2004). This raises the question what is the interplay between the pathological and physiological functions in the pathophysiology of PD? Is PD in these families due to an aberrant conformation of α-synuclein or due to its physiological functions augmented to a deleterious degree or even a loss of synuclein's normal functions by large-scale aggregation? A first step toward dissecting this problem is to define the physiological function of synucleins. Only then can one clearly differentiate the contributions of the physiological and pathological functions of α-synuclein to the pathogenesis of PD.

α-SYNUCLEIN TRANSGENIC MICE

Transgenic mice are invaluable genetic tools to model age-related neurodegenerative diseases as well as to understand the normal functions of disease-associated genes. Transgenic mice are an excellent way to model the functional duality of α-synuclein. Many laboratories have created α-synuclein transgenic mice under the control of different promoters ([Fernagut and Chesselet, 2004]). To a degree, most α-synuclein transgenics show a late–onset phenotype with some measure of cell death and α-synuclein pathology. However, none of the mice available to date show nigrostriatal selectivity of neuronal loss.

In our laboratory, we generated α-synuclein transgenics under the control of the Thy1 promoter (Chandra et al., 2005). Specifically, we generated transgenic mice that overexpress human wild type α-synuclein (analogous to the PARK4 locus), or two of the PD mutants – human A30P and A53T α-synuclein (analogous to the PARK1 locus). We also made transgenic lines that overexpress mouse wild type α-synuclein, which differs from the human ortholog by seven amino acids. All lines have pan-neuronal expression and show 5- to 10-fold overexpression of α-synuclein in the brain and 10- to 20- fold in the spinal cord (Chandra et al., 2005). In spite of numerous papers detailing various α-synuclein transgenics, a side-by-side comparison of these Thy-1 α-synuclein transgenics was

very instructive. All transgenic lines that expressed human α-synuclein developed an overt hind-limb paralytic phenotype. This paralytic phenotype has some similarities to PD, and is accompanied by resting tremor, progressive neuronal loss, insolubility of α-synuclein and α-synuclein pathology. The onset of this phenotype varied amongst the various lines, with 6 months being the earliest time of onset. Most transgenic lines developed the phenotype around 12–14 months of age. The penetrance of this neurodegenerative phenotype also varied between the Thy-1 transgenic lines. Those lines expressing A53T and A30P mutant human α-synuclein showed 50–80% penetrance, while only 20% of transgenics expressing wild type human α-synuclein developed a phenotype. Strikingly, transgenic lines that overexpress mouse wild type α-synuclein never developed this phenotype (tested for >24 months). In spite of having similar expression levels and patterns, these mice remained healthy (six independent lines were tested). This unexpected finding suggests that the seven amino acid changes between human and mouse α-synuclein are important for the pathogenesis of this neurodegenerative phenotype. This finding also indicates that it is possible to tease apart the physiological and pathological functions of α-synuclein using transgenic approaches.

Since α-synuclein transgenics develop both neurodegeneration and α-synuclein pathology, they are particularly useful to test neuroprotective regimes as well as to study genetic modifiers of disease progression. We were especially interested in modifiers that could alter the onset and severity of the neurodegenerative phenotype seen in the α-synuclein transgenics. One of the genes tested for its ability as a modifier is CSPα.

CSPα

CSPα is a synaptic vesicle protein that is linked to neurodegeneration. CSPα contains a DNA-J domain typical for Hsp40-type co-chaperones (Zinsmaier et al., 1990; Gundersen and Umbach, 1992; Chamberlain and Burgoyne, 2000). CSPα binds Hsc70 and the small glutamine-rich tetratricopeptide repeat protein SGT to form an active chaperone complex on the synaptic vesicle (Tobaben et al., 2001). In keeping with its function as a co-chaperone, CSPα can accelerate the intrinsic ATPase activity of Hsc70 (Braun et al., 1996; Chamberlain and Burgoyne, 1997; Tobaben et al., 2001). The

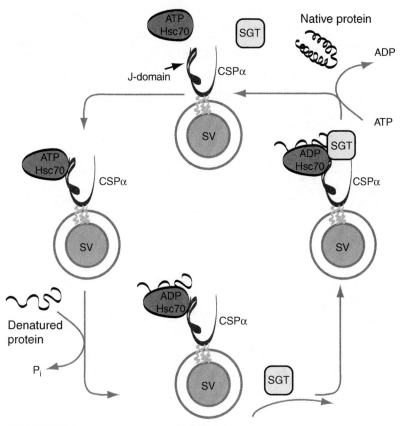

FIGURE 22.3 Model of CSPα as a synaptic vesicle chaperone. CSPα binds to Hsc70 in the presence of ATP and accelerates its intrinsic ATPase activity. The CSPα-Hsc70-ADP complex then binds denatured or misfolded substrates along with SGT, a small guanine-rich tetratricopeptide protein. This chaperone complex uses ATP hydrolysis to catalyze folding of substrates. The full repertoire of presynaptic substrates for this chaperone complex is unknown. Our evidence suggests that the membrane SNARE SNAP-25 is one substrate of the Hsc70/CSPα/SGT complex (Tobaben *et al.*, 2001).

synaptic vesicle localization and co-chaperone activity of CSPα suggest that it may function to prevent the accumulation of misfolded proteins during the incessant operation of a nerve terminal (Figure 22.3). Support for this hypothesis comes from analysis of CSP mutants in flies and mice. In flies, deletion of CSP results in a temperature-sensitive phenotype that includes defects in synaptic transmission and neurodegeneration (Umbach *et al.*, 1994; Zinsmaier *et al.*, 1994). In mice, CSPα knockout (KO) animals are relatively normal at birth, but develop a progressive neurodegeneration that results initially in loss of synapses and eventually neurons. This phenotype manifests as defects in synaptic transmission after

2–3 weeks of age, and lethality after 1–4 months (Fernandez-Chacon *et al.*, 2004). These phenotypes can be readily monitored by vital parameters such as body weight and survival (Figure 22.4, filled squares). In addition to a co-chaperone function, other roles have been proposed for CSP. In flies, CSP may regulate Ca^{2+} influx into nerve terminals (Ranjan *et al.*, 1998; Umbach *et al.*, 1998; Magga *et al.*, 2000), or mediate Ca^{2+}-triggered exocytosis (Chamberlain and Burgoyne, 1998; Dawson-Scully *et al.*, 2000). Fly and vertebrate CSP interact with several proteins *in vitro*, including SNARE proteins and Ca^{2+} channels (Leveque *et al.*, 1998; Nie *et al.*, 1999; Evans and Morgan, 2002).

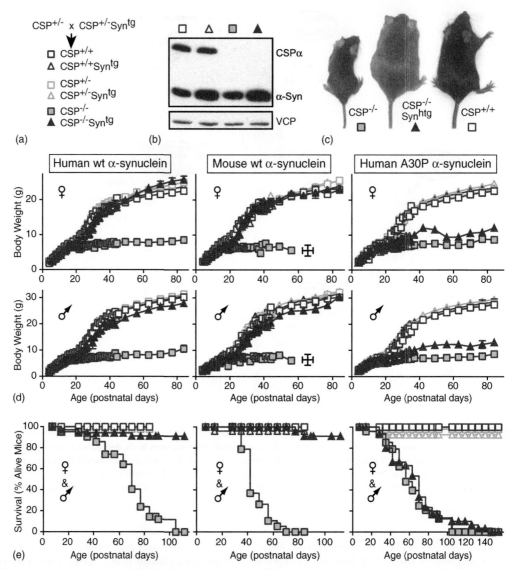

FIGURE 22.4 Effect of transgenic α-synuclein on the growth and survival CSPα KO mice. (a) Breeding strategy. Heterozygous CSPα KO mice (CSP$^{+/-}$) were mated with heterozygous CSPα KO mice containing an α-synuclein transgene (CSP$^{+/-}$Syn^{tg}). (b) Immunoblot analysis of brain proteins from wild type and CSPα KO mice that express or lack transgenic human wild type α-synuclein. Blots were probed with antibodies to CSPα, α-synuclein and VCP (as a loading control). (c) Photograph of CSPα KO mice lacking (CSP$^{-/-}$) or containing the wild type human α-synuclein transgene (CSP$^{-/-}$Syn^{htg}), and of a wild type littermate control mouse (CSP$^{+/+}$) at 10 weeks. (d) Body weights of littermate female (♀; top panels) and male mice (♂; bottom panels) as a function of age. (e) Survival of female and male mice as a function of age. In (d) and (e), the left panels describe animals that express wild type human α-synuclein, central panels wild type mouse α-synuclein and right panels A30P mutant human α-synuclein (Chandra *et al.*, 2005); reproduced with permission.

CSPα/α-SYNUCLEIN TRANSGENIC CROSSES

We initially hypothesized that decreasing synaptic chaperone complex levels would increase α-synuclein aggregation and thereby accelerate the onset and severity of the transgenic PD-like phenotype. To test this experimentally, we used CSPα heterozygous KO animals that have 50% decreased CSPα levels (Fernandez-Chacon et al., 2004) as a mouse model for diminished synaptic chaperone activity. We crossed these animals, which are phenotypically normal, to the human wild type α-synuclein transgenic line and found that heterozygous CSPα levels had no overt effect on the late-onset α-synuclein transgenic phenotypes described above. However, we discovered instead that human wild type α-synuclein transgenics could rescue the lethal neurodegeneration of CSPα homozygous KO animals (Chandra et al., 2005). As shown in Figure 22.4, expression of human wild type α-synuclein completely restored the body weight deficit of CSPα KO mice. More importantly, transgenic α-synuclein expression rescued the lethality of CSPα-deficient mice, with rescued animals living as long as 2 years. This rescue was due to an amelioration of the gliosis and neurodegeneration observed in the CSPα KO mice (Chandra et al., 2005).

Similar to the human transgenic, mouse wild type α-synuclein could also completely rescue the CSPα null mice (Chandra et al., 2005). Surprisingly, the two PD mutants were strikingly divergent in their ability to ameliorate the CSPα KO phenotypes (Chandra et al., 2005). The A53T transgenic mice were able to restore the body weight and survival deficits of CSPα KO mice, while the A30P transgenics were unable to do so (two independent A30P lines tested); Fig 22.4. These opposite results for the A53T and A30P transgenics were unexpected, as all lines have similar expression levels and patterns. In contrast to A53T transgenic mice, the A30P transgenic also could not rescue the gliosis, neurodegeneration and behavioral phenotypes of CSPα KO mice (Chandra et al., 2005). To ensure that the rescue effects were not a consequence of non-specific heat shock protein or chaperone expression, we quantified levels of several ubiquitously expressed constitutive and inducible chaperones in the various lines and found no significant changes. Our finding that the A53T and A30P mutants – 2 protofibril stabilizing mutants – have distinct abilities to rescue the CSPα KO neurodegenerative phenotype further supports the idea that the rescue effect is not mediated by a generic chaperone upregulation.

Is the synuclein-mediated rescue of CSPα KO deficits cell-autonomous? This important question could be successfully addressed in the retina (Chandra et al., 2005). Here, CSPα is normally expressed in both the outer and inner plexiform synapse layers, and deletion of CSPα in mice results in photoreceptor degeneration in the outer plexiform layer (OPL) and in blindness (Schmitz et al., 2006). In α-synuclein transgenics, the transgene has a restrictive expression pattern and is only expressed in the inner plexiform layer. Analysis of CSPα KO mice that express human wild type α-synuclein, revealed that lack of expression of α-synuclein in the OPL causes progressive degeneration of this layer and blindness (Chandra et al., 2005). Synuclein-mediated rescue therefore is cell-autonomous, consistent with both α-synuclein and CSPα acting in the same neuron.

The serendipitous findings described above clearly demonstrate that CSPα and α-synuclein genetically interact. These data also provide unmistakable evidence that the physiological and pathological functions of α-synuclein are mechanistically distinct. First, transgenic lines that overexpressed endogenous mouse α-synuclein did not develop a late-onset PD-like phenotype but were able to abolish the lethality of CSPα KO. Second, transgenic mice that expressed the A53T and A30P mutations both developed the PD-like phenotype reproducibly. Yet, the two mutants diverged diametrically in their ability to rescue the CSPα KO. This is most likely due to the fact that the A53T substitution is a modest change and is the naturally occurring sequence in mouse (Clayton and George, 1998), while the A30P substitution is a drastic change and disrupts both lipid binding and the α-helical conformation of α-synuclein (Jensen et al., 1998; Jo et al., 2002; Fortin et al., 2004; Ulmer and Bax, 2005). In fact, we found no correlation between the transgenic late-onset phenotype and the ability of the same transgenes to rescue the CSPα KO, clearly demonstrating that the roles of α-synuclein in physiology and pathology are distinct and dependent on different protein properties. These observations also imply that the physiological function of α-synuclein prevents neurodegeneration.

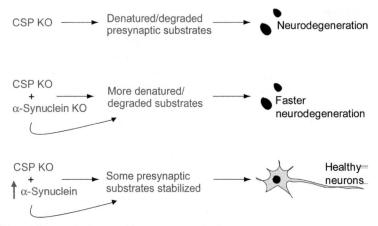

FIGURE 22.5 Model for the genetic interactions between α-synuclein and CSPα in preventing neurodegeneration. Based on the genetic crosses, we hypothesize that α-synuclein acts on one or more CSPα substrates to rescue the neurodegeneration of CSPα KO mice. These presynaptic substrates are critical for maintaining synapses and neurons.

Work from other groups supports our hypothesis that the physiological function of α-synuclein is neuroprotective. First, α-synuclein transgenic mice wherein expression is driven by the tyrosine hydroxylase promoter were unexpectedly found to be resistant to nigrostriatal neurodegeneration induced by the herbicide paraquat (Manning-Bog et al., 2003). Second, induction of α-synuclein expression by histone deacetylase inhibitors was protective against glutamate-induced excitotoxicity (Leng and Chuang, 2006). And finally, in vitro studies that either overexpress α-synuclein or introduce membrane permeable TAT-fusions were protective to oxidative stress and parkin knockdowns in a dose-dependent manner (Albani et al., 2004; Machida et al., 2005). Together, these data suggest that α-synuclein is protective to a variety of neuronal insults.

CSPα/SYNUCLEIN KO CROSSES

To gain further insight into to the physiological functions of synucleins, we generated α-, β-synuclein single and double knockouts. The α-/β-synuclein double KOs are viable, fertile and do not have any overt phenotype, especially with regard to body weight and survival (Chandra et al., 2004). In fact, synuclein null animals (αβγ-synuclein triple KOs) are also viable, fertile and have no overt phenotype (unpublished data). We analyzed excitatory

electrophysiological parameters, short- and long-term plasticity in the hippocampus in these double-KO animals and found no difference from control groups. The only deficit we could detect was a decrease in dopamine levels in the dorsal striatum of α-/β-synuclein double KOs. This decrease was not accompanied by a change in any catabolic metabolites of dopamine, such as DOPAC or HVA. The metabolic basis for this change remains to be elucidated. Overall, the synuclein KO animals are viable and have a mild phenotype (Chandra et al., 2004). These data suggest that synucleins are not essential and have redundant functions. This is in keeping with the fact that synucleins are only expressed in vertebrates and are not evolutionary conserved. It is also important to note that synuclein KOs do not phenocopy the neurodegeneration and lethality of CSPα KO mice. This strongly indicates that α-synuclein and CSPα are not in the same linear pathway but act via convergent pathways.

To further dissect the relationship between α-synuclein and CSPα, we crossed the α-/β-synuclein double KOs to CSPα KO animals and analyzed their body weight and survival (Chandra et al., 2005). Noticeably, CSPα KO animals on an α-/β-synuclein null background had a more severe body weight and survival deficit (Chandra et al., 2005). These triple KO animals started exhibiting a lower body weight at P8 and were half the weight of CSPα KOs by P20. Their lifespan was also reduced to 23.8 days as opposed to 62.3 days for CSPα KO animals. In fact, no CSPα/α-/β-synuclein triple KO animals lived past weaning. This faster mortality was accompanied by accelerated gliosis and neurodegeneration (Chandra *et al.*, 2005). Interestingly, double-KO mice that lacked both CSPα and α-synuclein but contained β-synuclein exhibited an intermediate mortality curve suggesting that α- and β-synuclein are redundant in this effect. These findings provide irrefutable evidence that CSPα and α-synuclein genetically interact.

SUMMARY OF CSPα/α-SYNUCLEIN GENETIC INTERACTIONS

Our studies demonstrate that overexpression of α-synuclein prevents the lethal neurodegeneration of CSPα null animals. Conversely, deletion of α-synuclein exacerbated the CSPα KO phenotypes. Thus, α-synuclein can regulate the CSPα KO phenotypes in both directions. This effect was cell-autonomous. Based on these genetic crosses, it can be concluded that (1) in a synapse, α-synuclein levels are limiting, as endogenous levels were insufficient to rescue the CSPα KO; (2) on a CSPα KO background, α-synuclein is an essential gene; (3) the A30P mutant is a loss-of-physiological-function mutant in addition to being a dominant protofibril-prone mutant and; (4) α-synuclein acts via a pathway that converges at a site downstream of CSPα (Figure 22.5).

BIOCHEMICAL RELATIONSHIP BETWEEN α-SYNUCLEIN AND CSPα

While the genetic evidence that α-synuclein and CSPα interact is strong, the molecular basis for this genetic interaction is unclear. To better understand this interaction biochemically, we first tested evident premises such as do α-synuclein and CSPα physically bind each other or does α-synuclein substitute for CSPα in the Hsc70/CSPα/SGT chaperone complex? In spite of many attempts and approaches, these experiments were unfruitful. Presently, our data does not support the hypothesis that α-synuclein is a chaperone or co-chaperone. This is consistent with the genetic data which suggests that α-synuclein acts at a site downstream of CSPα, possibly at the level of its substrates.

To identify putative CSPα substrates we have utilized quantitative protein profiling. The rationale for this approach is that substrates of CSPα will be misfolded in its absence and targeted for degradation resulting in decreased levels. We compared the levels of ~40 presynaptic proteins in wild type and CSPα KO brains at 6 weeks of age (Chandra *et al.*, 2005). Of these proteins, there were only four prominent changes. Hsc70, Hsp70 and α-synuclein levels were decreased by 20%, corroborating their genetic interactions with CSPα. However, the largest change was observed for the membrane SNARE SNAP-25, which was decreased by 50%. This reduction in SNAP-25 levels preceded the onset of neurodegeneration of CSPα KO mice and can be seen as early as the fifth postnatal day. The magnitude of change in SNAP-25 levels is large and was not even seen in KOs of its interacting partner synaptobrevin 2 (Schoch *et al.*, 2001). This suggests that SNAP-25 is a genuine substrate of the Hsc70/CSPα/SGT complex. To confirm this, we carried out *in vitro* binding experiments in the presence of non-hydrolyzable analogs of ADP and ATP. Hsc70-dependent chaperone complexes bind substrates in the presence of ADP and release them upon folding in the presence of ATP. We could show that a small percentage of SNAP-25, but not the other SNAREs (syntaxin 1a and synaptobrevin 2), bound to Hsc70 in the presence of ADP-β-S (Chandra *et al.*, 2005), lending credence to our premise that SNAP-25 is a substrate of the Hsc70/CSPα/SGT complex. It is presently unclear why SNAP-25, a natively unfolded protein, needs a chaperone. Perhaps, the Hsc70/CSPα/SGT complex acts to dissociate inappropriate SNAP-25 complexes that accumulate over time at synapses, thus allowing for continuous vesicle fusion.

If SNAP-25 is a bona fide substrate of the Hsc70/CSPα/SGT complex, does the reduction of its levels explain the CSPα KO phenotype? The answer is no, SNAP-25 KO heterozygous animals in effect have 50% SNAP-25 levels and are phenotypically normal (Washbourne *et al.*, 2002). This strongly suggests that there are additional substrates for this chaperone complex whose loss or reduction in the CSPα KO is linked to neurodegeneration. Recently, a compendium of synaptic vesicle proteins was published (Takamori *et al.*, 2006). It lists 410 synaptic vesicle proteins and synaptic membrane proteins (Takamori *et al.*, 2006). We have presently only sampled 10% of all synaptic proteins and it is very likely that there are other substrates of the Hsc70/CSPα/SGT chaperone complex.

SNARES AND NEURODEGENERATION

During membrane fusion, SNAP-25 forms a complex with the plasma membrane SNARE protein syntaxin 1 and the synaptic vesicle SNARE protein synaptobrevin/VAMP (Sollner et al., 1993; Jahn and Scheller, 2006). This four helix bundle is known as the SNARE or core complex and is essential for vesicle fusion. Genetic deletions of SNAREs (Schoch et al., 2001; Washbourne et al., 2002) or treatment with clostridal toxins that selectively cleave the SNAREs results in severely impaired or attenuated synaptic transmission (Link et al., 1992; Blasi et al., 1993; Rossetto et al., 2001).

If SNAP-25 is a substrate of the Hsc70/CSPα/SGT chaperone complex, does α-synuclein act via SNAP-25 to rescue the CSPα KO phenotype? In an attempt to answer this question, we first checked whether SNAP-25 levels were restored in CSPα KO animals expressing human wild type α-synuclein transgenics. Surprisingly, the SNAP-25 levels remained decreased in these rescued animals (Chandra et al., 2005). Next, we examined SNARE complex assembly in CSPα KO brains. We found a significant decrease in SNARE complexes in CSPα KO brains that was reversed by transgenic wild type α-synuclein but not A30P mutant α-synuclein (Chandra et al., 2005). This suggests that one way α-synuclein rescues CSPα KO phenotypes is by ameliorating core complex function. This also implies that SNAREs are an important link between α-synuclein, CSPα and neurodegeneration.

The above findings prompt the question is there any connection between the SNARE machinery and neurodegeneration? The best known link is Munc18-1, a SNARE-binding protein that is essential for synaptic vesicle fusion (Rizo and Sudhof, 2002). Munc18-1 KOs develop normally, but die shortly after birth due to profound neurodegeneration (Verhage et al., 2000). In fact, it is not possible to culture munc18-1 null neurons and these neurons undergo apoptosis even in dissociated cultures by 6–7 d.i.v. However, the mechanism by which deletion of munc18-1 causes neurons to undergo neurodegeneration is presently unclear. A second link between SNAREs and neurodegeneration is the finding that prolonged exposure of certain botulinum toxins causes neuronal cell death (Berliocchi et al., 2005). Together, these results emphasize the importance of SNAREs and SNARE-binding proteins in neurodegeneration and the need to further investigate this link.

NEUROPROTECTIVE FUNCTIONS OF SYNUCLEINS AND PD

Our data suggests that the normal physiological function of α-synuclein is neuroprotective to CSPα deletion. Specifically, α-synuclein overexpression could rescue the neurodegeneration and lethality of CSPα KO mice. The mechanism by which α-synuclein prevents the neurodegeneration of CSPα KOs needs to be urgently elucidated. This will help us define the contribution of this physiological function to the pathogenesis of PD.

As reviewed above, in vitro data from other labs further suggests that α-synuclein is neuroprotective to a variety of neuronal insults. Here, a critical question is whether α-synuclein function is neuroprotective to insults known to aggravate nigrostriatal death such as oxidative stress and mitochondrial dysfunction. In fact, there are confounding results showing that the α-synuclein KO is resistant to MPTP, a mitochondrial complex I inhibitor that causes PD symptoms both in rodents and humans (Dauer et al., 2002; Schluter et al., 2003; Robertson et al., 2004). It is therefore imperative to define the neurotoxic agents against which α-synuclein is protective. Synuclein null animals will be extremely useful to address this problem and to define the molecular pathways/mechanisms involved. Answering this question will also help us to determine whether Lewy bodies are toxic aggregates or a loss/sequestration of α-synuclein neuroprotective function.

Another angle to view the physiological role of α-synuclein in PD is through the prism of synapse maintenance. It is becoming increasingly clear from in vivo imaging data that dopaminergic terminals in the striatum are the first structures to be lost in PD (Marek et al., 2001). In most instances, loss of synapses occurs long before the loss of cell bodies. Indeed, synapse loss is a pivotal event in the progression of PD. Preservation of synaptic connections is also critical for PD treatments as the efficacy of dopamine replacement therapy closely depends on the remaining striatal terminals. Since CSPα and α-synuclein play significant roles in synapse maintenance, their functions can thought to be neuroprotective for PD. The pathway by which CSPα and α-synuclein cooperate to maintain synapses, however, still needs to be worked out. It will be interesting to see as other proteins involved in this synapse maintenance pathway are identified, if they also play a role in the neuroprotection/pathogenesis of PD.

OPEN QUESTIONS

There are many questions still unanswered with regards to α-synuclein, CSPα and SNAREs in neurodegeneration and PD. Beginning with α-synuclein, its functional duality needs to be better appreciated and greater effort needs to be made to understand both its physiological and pathological functions. With regards to its pathological function, the chief question is which α-synuclein species is neurotoxic. Is it the protofibrils? If so, how are these species neurotoxic? Do they block mitochondrial electron transport? Or do they impair the cell's ability to deal with its refuse by blocking autophagy or the ubiquitin proteasome system? Presently, there are many labs addressing aspects of these questions, however, a cohesive picture of the pathological functions of α-synuclein is lacking. Obtaining a clearer depiction will be crucial to design therapeutic strategies that halt PD early in its progression.

Our own work shows that the physiological functions of α-synuclein are intricately linked to CSPα, but are far from being complete. First, we have to identify the whole repertoire of CSPα substrates. Only then can we understand the CSPα KO phenotypes of loss of synapses and neurodegeneration as well as the means by which α-synuclein can rescue these deficits. We also need to obtain a detailed understanding of α-synuclein function. This will clarify the contribution of the physiological function of α-synuclein to neurodegeneration. CSPα KO, CSPα KO/ α-synuclein transgenics crosses and αβγ-synuclein KOs will be invaluable tools in addressing these questions. With regards to SNAREs and Munc18-1, what are their links to neurodegeneration? Are SNARE, munc18-1 and CSPα genes mutated in sporadic PD patients?

Finally, all currently available PD treatments are symptomatic and do not slow or halt the progression of the disease. Therefore, there is a strong interest in the development of neuroprotective treatments for PD. *In vitro* data suggests that protection of striatal dopaminergic terminals can offer some protection to the neurons (Dauer and Przedborski, 2003). So it will be important to see if augmenting synapse maintenance pathways, perhaps by increasing CSPα levels or the levels of its substrates will ameliorate the onset and severity of PD. This avenue offers the promise of truly neuroprotective treatments rather than mere symptomatic ones.

REFERENCES

Abeliovich, A., Schmitz, Y., Farinas, I., Choi-Lundberg, D., Ho, W. H., Castillo, P. E., Shinsky, N., Verdugo, J. M., Armanini, M., Ryan, A. *et al.* (2000). Mice lacking alpha-synuclein display functional deficits in the nigrostriatal dopamine system. *Neuron* 25, 239–252.

Albani, D., Peverelli, E., Rametta, R., Batelli, S., Veschini, L., Negro, A., and Forloni, G. (2004). Protective effect of TAT-delivered alpha-synuclein: Relevance of the C-terminal domain and involvement of HSP70. *FASEB J* 18, 1713–1715.

Auluck, P. K., Chan, H. Y., Trojanowski, J. Q., Lee, V. M., and Bonini, N. M. (2002). Chaperone suppression of alpha-synuclein toxicity in a Drosophila model for Parkinson's disease. *Science* 295, 865–868.

Berliocchi, L., Fava, E., Leist, M., Horvat, V., Dinsdale, D., Read, D., and Nicotera, P. (2005). Botulinum neurotoxin C initiates two different programs for neurite degeneration and neuronal apoptosis. *J Cell Biol* 168, 607–618.

Beyer, K. (2006). Alpha-synuclein structure, posttranslational modification and alternative splicing as aggregation enhancers. *Acta Neuropathol (Berl)* 112, 237–251.

Blasi, J., Chapman, E. R., Link, E., Binz, T., Yamasaki, S., De Camilli, P., Sudhof, T. C., Niemann, H., and Jahn, R. (1993). Botulinum neurotoxin A selectively cleaves the synaptic protein SNAP-25. *Nature* 365, 160–163.

Braun, J. E., Wilbanks, S. M., and Scheller, R. H. (1996). The cysteine string secretory vesicle protein activates Hsc70 ATPase. *J Biol Chem* 271, 25989–25993.

Chamberlain, L. H., and Burgoyne, R. D. (1997). Activation of the ATPase activity of heat-shock proteins Hsc70/Hsp70 by cysteine-string protein. *Biochem J* 322(Pt 3), 853–858.

Chamberlain, L. H., and Burgoyne, R. D. (1998). Cysteine string protein functions directly in regulated exocytosis. *Mol Biol Cell* 9, 2259–2267.

Chamberlain, L. H., and Burgoyne, R. D. (2000). Cysteine-string protein: the chaperone at the synapse. *J Neurochem* 74, 1781–1789.

Chandra, S., Chen, X., Rizo, J., Jahn, R., and Sudhof, T. C. (2003). A broken alpha-helix in folded alpha-Synuclein. *J Biol Chem* 278, 15313–15318.

Chandra, S., Fornai, F., Kwon, H. B., Yazdani, U., Atasoy, D., Liu, X., Hammer, R. E., Battaglia, G., German, D. C., Castillo, P. E. *et al.* (2004). Double-knockout mice for alpha- and beta-synucleins: effect on synaptic functions. *Proc Natl Acad Sci USA* 101, 14966–14971.

Chandra, S., Gallardo, G., Fernandez-Chacon, R., Schluter, O. M., and Sudhof, T. C. (2005). Alpha-synuclein cooperates with CSP-alpha in preventing neurodegeneration. *Cell* **123**, 383–396.

Chartier-Harlin, M. C., Kachergus, J., Roumier, C., Mouroux, V., Douay, X., Lincoln, S., Levecque, C., Larvor, L., Andrieux, J., Hulihan, M. *et al.* (2004). Alpha-synuclein locus duplication as a cause of familial Parkinson's disease. *Lancet* **364**, 1167–1169.

Clayton, D. F., and George, J. M. (1998). The synucleins: a family of proteins involved in synaptic function, plasticity, neurodegeneration and disease. *Trends Neurosci* **21**, 249–254.

Cole, N. B., Murphy, D. D., Grider, T., Rueter, S., Brasaemle, D., and Nussbaum, R. L. (2002). Lipid droplet binding and oligomerization properties of the Parkinson's disease protein alpha-synuclein. *J Biol Chem* **277**, 6344–6352.

Cole, N. B., Murphy, D. D., Lebowitz, J., Di Noto, L., Levine, R. L., and Nussbaum, R. L. (2005). Metal-catalyzed oxidation of alpha-synuclein: helping to define the relationship between oligomers, protofibrils, and filaments. *J Biol Chem* **280**, 9678–9690.

Conway, K. A., Lee, S. J., Rochet, J. C., Ding, T. T., Williamson, R. E., and Lansbury, P. T., Jr. (2000). Acceleration of oligomerization, not fibrillization, is a shared property of both alpha-synuclein mutations linked to early-onset Parkinson's disease: Implications for pathogenesis and therapy. *Proc Natl Acad Sci USA* **97**, 571–576.

Conway, K. A., Rochet, J. C., Bieganski, R. M., and Lansbury, P. T., Jr. (2001). Kinetic stabilization of the alpha-synuclein protofibril by a dopamine-alpha-synuclein adduct. *Science* **294**, 1346–1349.

Dauer, W., and Przedborski, S. (2003). Parkinson's disease: Mechanisms and models. *Neuron* **39**, 889–909.

Dauer, W., Kholodilov, N., Vila, M., Trillat, A. C., Goodchild, R., Larsen, K. E., Staal, R., Tieu, K., Schmitz, Y., Yuan, C. A. *et al.* (2002). Resistance of alpha -synuclein null mice to the parkinsonian neurotoxin MPTP. *Proc Natl Acad Sci USA* **99**, 14524–14529.

Davidson, W. S., Jonas, A., Clayton, D. F., and George, J. M. (1998). Stabilization of alpha-synuclein secondary structure upon binding to synthetic membranes. *J Biol Chem* **273**, 9443–9449.

Dawson-Scully, K., Bronk, P., Atwood, H. L., and Zinsmaier, K. E. (2000). Cysteine-string protein increases the calcium sensitivity of neurotransmitter exocytosis in Drosophila. *J Neurosci* **20**, 6039–6047.

Eliezer, D., Kutluay, E., Bussell, R., Jr., and Browne, G. (2001). Conformational properties of alpha-synuclein in its free and lipid-associated states. *J Mol Biol* **307**, 1061–1073.

Evans, G. J., and Morgan, A. (2002). Phosphorylation-dependent interaction of the synaptic vesicle proteins cysteine string protein and synaptotagmin I. *Biochem J* **364**, 343–347.

Fernagut, P. O., and Chesselet, M. F. (2004). Alpha-synuclein and transgenic mouse models. *Neurobiol Dis* **17**, 123–130.

Fernandez-Chacon, R., Wolfel, M., Nishimune, H., Tabares, L., Schmitz, F., Castellano-Munoz, M., Rosenmund, C., Montesinos, M. L., Sanes, J. R., Schneggenburger, R. *et al.* (2004). The synaptic vesicle protein CSP alpha prevents presynaptic degeneration. *Neuron* **42**, 237–251.

Fink, A. L. (2006). The aggregation and fibrillation of alpha-synuclein. *Acc Chem Res* **39**, 628–634.

Fortin, D. L., Troyer, M. D., Nakamura, K., Kubo, S., Anthony, M. D., and Edwards, R. H. (2004). Lipid rafts mediate the synaptic localization of alpha-synuclein. *J Neurosci* **24**, 6715–6723.

Fujiwara, H., Hasegawa, M., Dohmae, N., Kawashima, A., Masliah, E., Goldberg, M. S., Shen, J., Takio, K., and Iwatsubo, T. (2002). Alpha-synuclein is phosphorylated in synucleinopathy lesions. *Nat Cell Biol* **4**, 160–164.

Galvin, J. E., Lee, V. M., and Trojanowski, J. Q. (2001). Synucleinopathies: Clinical and pathological implications. *Arch Neurol* **58**, 186–190.

Giasson, B. I., Murray, I. V., Trojanowski, J. Q., and Lee, V. M. (2001). A hydrophobic stretch of 12 amino acid residues in the middle of alpha-synuclein is essential for filament assembly. *J Biol Chem* **276**, 2380–2386.

Gundersen, C. B., and Umbach, J. A. (1992). Suppression cloning of the cDNA for a candidate subunit of a presynaptic calcium channel. *Neuron* **9**, 527–537.

Ibanez, P., Bonnet, A. M., Debarges, B., Lohmann, E., Tison, F., Pollak, P., Agid, Y., Durr, A., and Brice, A. (2004). Causal relation between alpha-synuclein gene duplication and familial Parkinson's disease. *Lancet* **364**, 1169–1171.

Jahn, R., and Scheller, R. H. (2006). SNAREs – Engines for membrane fusion. *Nat Rev Mol Cell Biol* **7**, 631–643.

Jao, C. C., Der-Sarkissian, A., Chen, J., and Langen, R. (2004). Structure of membrane-bound alpha-synuclein studied by site-directed spin labeling. *Proc Natl Acad Sci USA* **101**, 8331–8336.

Jensen, P. H., Nielsen, M. S., Jakes, R., Dotti, C. G., and Goedert, M. (1998). Binding of alpha-synuclein to brain vesicles is abolished by familial Parkinson's disease mutation. *J Biol Chem* **273**, 26292–26294.

Jo, E., Fuller, N., Rand, R. P., St George-Hyslop, P., and Fraser, P. E. (2002). Defective membrane interactions of familial Parkinson's disease mutant A30P alpha-synuclein. *J Mol Biol* **315**, 799–807.

Kruger, R., Kuhn, W., Muller, T., Woitalla, D., Graeber, M., Kosel, S., Przuntek, H., Epplen, J. T., Schols, L., and Riess, O. (1998). Ala30Pro mutation in the gene encoding alpha-synuclein in Parkinson's disease. *Nat Genet* 18, 106–108.

Lavedan, C. (1998). The synuclein family. *Genome Res* 8, 871–880.

Leng, Y., and Chuang, D. M. (2006). Endogenous alpha-synuclein is induced by valproic acid through histone deacetylase inhibition and participates in neuroprotection against glutamate-induced excitotoxicity. *J Neurosci* 26, 7502–7512.

Leveque, C., Pupier, S., Marqueze, B., Geslin, L., Kataoka, M., Takahashi, M., De Waard, M., and Seagar, M. (1998). Interaction of cysteine string proteins with the alpha1A subunit of the P/Q-type calcium channel. *J Biol Chem* 273, 13488–13492.

Link, E., Edelmann, L., Chou, J. H., Binz, T., Yamasaki, S., Eisel, U., Baumert, M., Sudhof, T. C., Niemann, H., and Jahn, R. (1992). Tetanus toxin action: inhibition of neurotransmitter release linked to synaptobrevin proteolysis. *Biochem Biophys Res Commun* 189, 1017–1023.

Machida, Y., Chiba, T., Takayanagi, A., Tanaka, Y., Asanuma, M., Ogawa, N., Koyama, A., Iwatsubo, T., Ito, S., Jansen, P. H. *et al.* (2005). Common anti-apoptotic roles of parkin and alpha-synuclein in human dopaminergic cells. *Biochem Biophys Res Commun* 332, 233–240.

Magga, J. M., Jarvis, S. E., Arnot, M. I., Zamponi, G. W., and Braun, J. E. (2000). Cysteine string protein regulates G protein modulation of N-type calcium channels. *Neuron* 28, 195–204.

Manning-Bog, A. B., McCormack, A. L., Purisai, M. G., Bolin, L. M., and Di Monte, D. A. (2003). Alpha-synuclein overexpression protects against paraquat-induced neurodegeneration. *J Neurosci* 23, 3095–3099.

Marek, K., Innis, R., van Dyck, C., Fussell, B., Early, M., Eberly, S., Oakes, D., and Seibyl, J. (2001). [123I]beta-CIT SPECT imaging assessment of the rate of Parkinson's disease progression. *Neurology* 57, 2089–2094.

Maroteaux, L., Campanelli, J. T., and Scheller, R. H. (1988). Synuclein: a neuron-specific protein localized to the nucleus and presynaptic nerve terminal. *J Neurosci* 8, 2804–2815.

Miller, D. W., Hague, S. M., Clarimon, J., Baptista, M., Gwinn-Hardy, K., Cookson, M. R., and Singleton, A. B. (2004). Alpha-synuclein in blood and brain from familial Parkinson disease with SNCA locus triplication. *Neurology* 62, 1835–1838.

Mueller, J. C., Fuchs, J., Hofer, A., Zimprich, A., Lichtner, P., Illig, T., Berg, D., Wullner, U., Meitinger, T., and Gasser, T. (2005). Multiple regions of alpha-synuclein are associated with Parkinson's disease. *Ann Neurol* 57, 535–541.

Nie, Z., Ranjan, R., Wenniger, J. J., Hong, S. N., Bronk, P., and Zinsmaier, K. E. (1999). Overexpression of cysteine-string proteins in Drosophila reveals interactions with syntaxin. *J Neurosci* 19, 10270–10279.

Ninkina, N., Papachroni, K., Robertson, D. C., Schmidt, O., Delaney, L., O'Neill, F., Court, F., Rosenthal, A., Fleetwood-Walker, S. M., Davies, A. M. *et al.* (2003). Neurons expressing the highest levels of gamma-synuclein are unaffected by targeted inactivation of the gene. *Mol Cell Biol* 23, 8233–8245.

Norris, E. H., Giasson, B. I., Hodara, R., Xu, S., Trojanowski, J. Q., Ischiropoulos, H., and Lee, V. M. (2005). Reversible inhibition of alpha-synuclein fibrillization by dopaminochrome-mediated conformational alterations. *J Biol Chem* 280, 21212–21219.

Pals, P., Lincoln, S., Manning, J., Heckman, M., Skipper, L., Hulihan, M., Van den Broeck, M., De Pooter, T., Cras, P., Crook, J. *et al.* (2004). Alpha-synuclein promoter confers susceptibility to Parkinson's disease. *Ann Neurol* 56, 591–595.

Polymeropoulos, M. H., Lavedan, C., Leroy, E., Ide, S. E., Dehejia, A., Dutra, A., Pike, B., Root, H., Rubenstein, J., Boyer, R. *et al.* (1997). Mutation in the alpha-synuclein gene identified in families with Parkinson's disease. *Science* 276, 2045–2047.

Ranjan, R., Bronk, P., and Zinsmaier, K. E. (1998). Cysteine string protein is required for calcium secretion coupling of evoked neurotransmission in drosophila but not for vesicle recycling. *J Neurosci* 18, 956–964.

Rizo, J., and Sudhof, T. C. (2002). Snares and Munc18 in synaptic vesicle fusion. *Nat Rev Neurosci* 3, 641–653.

Robertson, D. C., Schmidt, O., Ninkina, N., Jones, P. A., Sharkey, J., and Buchman, V. L. (2004). Developmental loss and resistance to MPTP toxicity of dopaminergic neurones in substantia nigra pars compacta of gamma-synuclein, alpha-synuclein and double alpha/gamma-synuclein null mutant mice. *J Neurochem* 89, 1126–1136.

Rossetto, O., Seveso, M., Caccin, P., Schiavo, G., and Montecucco, C. (2001). Tetanus and botulinum neurotoxins: Turning bad guys into good by research. *Toxicon* 39, 27–41.

Schluter, O. M., Fornai, F., Alessandri, M. G., Takamori, S., Geppert, M., Jahn, R., and Sudhof, T. C. (2003). Role of alpha-synuclein in 1-methyl-4-phenyl-1,2,3,6-tetrahydropyridine-induced parkinsonism in mice. *Neuroscience* 118, 985–1002.

Schmitz, F., Tabares, L., Khimich, D., Strenzke, N., de la Villa-Polo, P., Castellano-Munoz, M., Bulankina, A., Moser, T., Fernandez-Chacon, R., and Sudhof, T. C.

(2006). CSPalpha-deficiency causes massive and rapid photoreceptor degeneration. *Proc Natl Acad Sci USA* **103**, 2926–2931.

Schoch, S., Deak, F., Konigstorfer, A., Mozhayeva, M., Sara, Y., Sudhof, T. C., and Kavalali, E. T. (2001). SNARE function analyzed in synaptobrevin/VAMP knockout mice. *Science* **294**, 1117–1122.

Singleton, A. B. (2005). Altered alpha-synuclein homeostasis causing Parkinson's disease: the potential roles of dardarin. *Trends Neurosci* **28**, 416–421.

Singleton, A. B., Farrer, M., Johnson, J., Singleton, A., Hague, S., Kachergus, J., Hulihan, M., Peuralinna, T., Dutra, A., Nussbaum, R. *et al.* (2003). Alpha-synuclein locus triplication causes Parkinson's disease. *Science* **302**, 841.

Sollner, T., Whiteheart, S. W., Brunner, M., Erdjument-Bromage, H., Geromanos, S., Tempst, P., and Rothman, J. E. (1993). SNAP receptors implicated in vesicle targeting and fusion. *Nature* **362**, 318–324.

Spillantini, M. G., Schmidt, M. L., Lee, V. M., Trojanowski, J. Q., Jakes, R., and Goedert, M. (1997). Alpha-synuclein in Lewy bodies. *Nature* **388**, 839–840.

Sung, Y. H., and Eliezer, D. (2006). Secondary structure and dynamics of micelle bound beta- and gamma-synuclein. *Protein Sci* **15**, 1162–1174.

Takamori, S., Holt, M., Stenius, K., Lemke, E. A., Gronborg, M., Riedel, D., Urlaub, H., Schenck, S., Brugger, B., Ringler, P. *et al.* (2006). Molecular anatomy of a trafficking organelle. *Cell* **127**, 831–846.

Tan, E. K., and Skipper, L. M. (2007). Pathogenic mutations in Parkinson disease. *Hum Mutat* **28**, 641–653.

Tobaben, S., Thakur, P., Fernandez-Chacon, R., Sudhof, T. C., Rettig, J., and Stahl, B. (2001). A trimeric protein complex functions as a synaptic chaperone machine. *Neuron* **31**, 987–999.

Ulmer, T. S., and Bax, A. (2005). Comparison of structure and dynamics of micelle-bound human alpha-synuclein and Parkinson disease variants. *J Biol Chem* **280**, 43179–43187.

Ulmer, T. S., Bax, A., Cole, N. B., and Nussbaum, R. L. (2005). Structure and dynamics of micelle-bound human alpha-synuclein. *J Biol Chem* **280**, 9595–9603.

Umbach, J. A., Zinsmaier, K. E., Eberle, K. K., Buchner, E., Benzer, S., and Gundersen, C. B. (1994). Presynaptic dysfunction in Drosophila csp mutants. *Neuron* **13**, 899–907.

Umbach, J. A., Saitoe, M., Kidokoro, Y., and Gundersen, C. B. (1998). Attenuated influx of calcium ions at nerve endings of csp and shibire mutant Drosophila. *J Neurosci* **18**, 3233–3240.

Uversky, V. N. (2007). Neuropathology, biochemistry, and biophysics of alpha-synuclein aggregation. *J Neurochem* **103**, 17–37.

Uversky, V. N., Lee, H. J., Li, J., Fink, A. L., and Lee, S. J. (2001). Stabilization of partially folded conformation during alpha-synuclein oligomerization in both purified and cytosolic preparations. *J Biol Chem* **276**, 43495–43498.

Verhage, M., Maia, A. S., Plomp, J. J., Brussaard, A. B., Heeroma, J. H., Vermeer, H., Toonen, R. F., Hammer, R. E., van den Berg, T. K., Missler, M. *et al.* (2000). Synaptic assembly of the brain in the absence of neurotransmitter secretion. *Science* **287**, 864–869.

Volles, M. J., and Lansbury, P. T., Jr. (2003). Zeroing in on the pathogenic form of alpha-synuclein and its mechanism of neurotoxicity in Parkinson's disease. *Biochemistry* **42**, 7871–7878.

Volles, M. J., Lee, S. J., Rochet, J. C., Shtilerman, M. D., Ding, T. T., Kessler, J. C., and Lansbury, P. T. (2001). Vesicle permeabilization by protofibrillar alpha-synuclein: Implications for the pathogenesis and treatment of Parkinson's disease. *Biochemistry* **40**, 7812–7819.

Washbourne, P., Thompson, P. M., Carta, M., Costa, E. T., Mathews, J. R., Lopez-Bendito, G., Molnar, Z., Becher, M. W., Valenzuela, C. F., Partridge, L. D. *et al.* (2002). Genetic ablation of the t-SNARE SNAP-25 distinguishes mechanisms of neuroexocytosis. *Nat Neurosci* **5**, 19–26.

Weinreb, P. H., Zhen, W., Poon, A. W., Conway, K. A., and Lansbury, P. T., Jr. (1996). NACP, a protein implicated in Alzheimer's disease and learning, is natively unfolded. *Biochemistry* **35**, 13709–13715.

Woods, W. S., Boettcher, J. M., Zhou, D. H., Kloepper, K. D., Hartman, K. L., Ladror, D. T., Qi, Z., Rienstra, C. M., and George, J. M. (2007). Conformation-specific binding of alpha-synuclein to novel protein partners detected by phage display and NMR spectroscopy. *J Biol Chem*, e-publication.

Zarranz, J. J., Alegre, J., Gomez-Esteban, J. C., Lezcano, E., Ros, R., Ampuero, I., Vidal, L., Hoenicka, J., Rodriguez, O., Atares, B. *et al.* (2004). The new mutation, E46K, of alpha-synuclein causes Parkinson and Lewy body dementia. *Ann Neurol* **55**, 164–173.

Zinsmaier, K. E., Hofbauer, A., Heimbeck, G., Pflugfelder, G. O., Buchner, S., and Buchner, E. (1990). A cysteine-string protein is expressed in retina and brain of Drosophila. *J Neurogenet* **7**, 15–29.

Zinsmaier, K. E., Eberle, K. K., Buchner, E., Walter, N., and Benzer, S. (1994). Paralysis and early death in cysteine string protein mutants of Drosophila. *Science* **263**, 977–980.

23

INSIGHTS FROM ZEBRAFISH PD MODELS AND THEIR POTENTIALS FOR IDENTIFYING NOVEL DRUG TARGETS AND THERAPEUTIC COMPOUNDS

SU GUO

Department of Biopharmaceutical Sciences, Programs in Biological Sciences and Human Genetics, University of California, San Francisco, CA, USA

INTRODUCTION

The zebrafish, *Danio rerio*, a vertebrate model organism for genetics that has emerged in the past decade or so, is a relatively uncharted system for modeling Parkinson's disease (PD). Despite the fact that only a handful of papers have been published so far describing the efforts toward modeling PD in zebrafish, many salient features of this model organism warrant particular attention, especially in the consideration of therapeutic drug discovery. In this review, I will: (1) introduce the salient features of zebrafish as well as describe available molecular and chemical genetic tools in zebrafish, (2) discuss available PD models in zebrafish, highlighting their strengths as well as limitations toward understanding the etiology and pathogenesis of PD, and toward facilitating therapeutic drug discovery, (3) Discuss the promising progress of using zebrafish to study the development and regeneration of dopamine (DA) neurons, the degeneration of which is the primary cause of movement defects in PD.

THE HISTORY OF ZEBRAFISH AS A VERTEBRATE GENETIC MODEL ORGANISM

A new kid on the block will have to prove his/her worth of existence. As a new model organism that is emerged in the era when quite a few sophisticated genetic systems are available, the zebrafish *D. rerio*, is faced with the same question: why zebrafish? Here I give a brief overview on the rise and shine of zebrafish, in order to highlight the salient features that this model organism has (for a detailed historical perspective, see [Grunwald and Eisen, 2002]).

Although zebrafish swam to light about a decade ago, pioneering work on this system started in the 1970s, by George Streisinger, an accomplished phage geneticist with a desire to apply the same powerful forward genetic approach used in phages to a vertebrate system. Zebrafish were chosen, because they are small-sized and diploid; they can be easily maintained in large quantities in laboratory conditions and bred year around on a weekly basis to produce hundreds of progeny at each

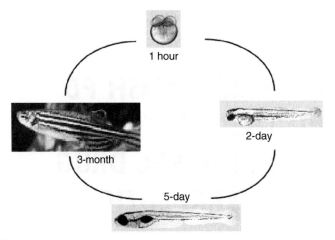

FIGURE 23.1   The life cycle of zebrafish. Hundreds of eggs can be produced from a single mating. A typical verte-brate body plan is laid out by 2 days post-fertilization. By 5 days post-fertilization, the larval zebrafish are free-living, hunt for food, and escape from predators. By 3 months post-fertilization, zebrafish are ready for reproduction. Adapted from Guo (2004).

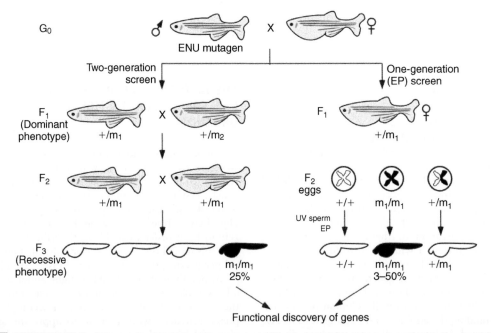

FIGURE 23.2   Schemes for forward genetic screens. Adult male zebrafish are mutagenized with the chemical muta-gen ENU (ethyl nitrosourea). ENU induces mutations in the spermatogonia. Mutagenized males are mated with wild-type females to produce the FI generation. The FI generation is heterozygous for any induced mutations, thus, only dominant mutations can be recovered if genetic screens are conducted on the FI generation. To identify recessive muta-tions, FI fish need to be crossed to obtain F3 generations, in which homozygous mutations are present (two-generation screen). Alternatively, FI female fish can be induced to undergo gynogenesis by applying early pressure (EP); therefore, homozygous mutations can be recovered in the F2 generation. Adapted from Guo (2004).

mating (Figure 23.1). Moreover, external development and a complete optical transparency of zebrafish embryos allow the entire process of vertebrate development to unfold in a Petri dish in front of curious and observing eyes of a researcher. These salient features have given zebrafish its particular strength toward studying vertebrate-related biology by the means of forward genetics (Figure 23.2) and through imaging cellular processes in an intact animal. Two large-scale forward genetic screens, carried out a decade ago, have proved the point, and led to the isolation of a variety of mutants which affect many processes of vertebrate development, that are visible under a simple dissecting microscope (Driever *et al.*, 1996; Haffter *et al.*, 1996).

MOLECULAR GENETICS IN ZEBRAFISH

The ultimate fruition of forward genetic analyses is to identify the gene whose function is critical for the phenotype of interest. Toward this aim, a battery of molecular tools have been developed for zebrafish, including the genetic linkage map, radiation hybrid map, synteny map, physical map, and ultimately, a nearly to be completed genomic blueprint (Knapik *et al.*, 1998; Geisler *et al.*, 1999; Hukriede *et al.*, 1999; Barbazuk *et al.*, 2000; *Woods et al., 2000*). These molecular tools have greatly facilitated the discovery of genes from mutations. In addition, tools for insertional mutagenesis have been established, which allow a rapid cloning of genes from mutations (Amsterdam *et al.*, 1999).

Complementing forward genetic analyses, reverse genetic methods have also been developed. These include the morpholino antisense knockdown (Nasevicius and Ekker, 2000) and TILLING (Wienholds *et al.*, 2002). While the morpholino antisense-based approach is fast, easy, and effective in transiently inactivating genes during early stages of development, the TILLING method allows the identification of permanent lesions in any gene of interest by employing the same strengths that have made zebrafish a favorable vertebrate system for forward genetic analysis (Figure 23.3). Taken together, these molecular genetic tools make zebrafish an attractive vertebrate system for studying the molecular mechanisms of PD pathogenesis.

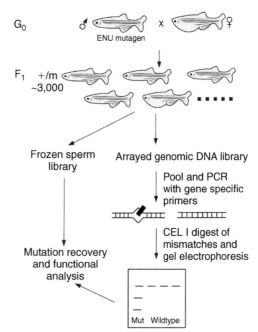

FIGURE 23.3 The scheme of the reverse genetic method, TILLING. Similar to a forward genetic screen, the F1 generation that is heterozygous for any induced mutations are raised to adulthood. A frozen sperm library is from F1 fish. In addition, genomic DNA are prepared from individual F1 fish to establish a corresponding genomic DNA library. To identify a mutation in a particular gene of interest, gene-specific primers are used to amplify PCR products from the genomic DNA library. PCR products are digested with CEL1 enzyme, which cuts at mismatches. Gel electrophoresis is carried out to identify which F1 carries a mutation in the gene of interest. Once the F1 carrying a mutation in the gene of interest is identified, the line can be recovered from the frozen sperm library. Adapted from Guo (2004).

CHEMICAL GENETICS IN ZEBRAFISH

The ease of maintenance and availability in large quantities has also made zebrafish an attractive system for small molecule-based drug discovery efforts (Figure 23.4). Larval zebrafish are highly permeable to water-soluble chemicals, which can also gain readily access to adult zebrafish blood circulation systems, through rich blood vessels underneath the gills. Compounds dissolved in the tank water

FIGURE 23.4 Small molecule screen in zebrafish. View of 96-well plates containing 2-day-old zebrafish embryos treated with small molecule compounds directly dissolved in the medium. Note the chemical in the top left well disrupts embryonic development.

have access to the adult zebrafish brain (Lau *et al.*, 2006; Ninkovic *et al.*, 2006). Compared to *in vitro* biochemical or cell culture-based small molecule screens, a chemical screen in a whole organism has the obvious advantage, because of the relevance of *in vivo* contexts. Zebrafish-based small molecule screens have been carried out successfully. For example, a single point mutation in the zebrafish *gridlock/hey2* gene disrupts aortic blood flow akin to a condition known as aortic coarctation in humans. A small molecule screen of ~5000 compounds has led to the identification of a chemical that can suppress the mutant condition (Peterson *et al.*, 2004).

PD models in zebrafish, as discussed in the following paragraphs, may be used, not only to study the molecular genetic mechanisms underlying PD, but also to screen for compounds that may alleviate, halt, or even reverse the disease phenotypes.

PD MODELS IN ZEBRAFISH

Why Zebrafish?

As one can appreciate from reading other chapters in this book, a large body of work has been carried out to model PD in organisms ranging from yeast, worms, flies, rodents, to primates. Despite the tremendous advancements in our knowledge of PD pathogenesis that have been gained from these studies, a clear picture of mechanisms underlying

DA neuron-selective degeneration is yet to emerge. Moreover, the development of treatments for PD is still at its infant stage.

With a typical vertebrate neuroanatomy and the amenability to large-scale genetic and chemical screens (Guo, 2004), zebrafish could potentially offer significant insights not only into the disease mechanisms, but also into the development of treatments for PD.

Strength and Weakness of Using Zebrafish to Model PD

The strengths of zebrafish for modeling PD include: (1) the zebrafish neuroanatomy is typical of that of vertebrates, with a central nervous system (CNS) composed of forebrain, midbrain, hindbrain, and spinal cord, (2) zebrafish genes in general show high degree of homology with their counterparts across vertebrate species including humans. For example, at the protein level, ~60–80% identity can be expected between zebrafish and the human counterparts, (3) compared to other vertebrate model organisms such as mice, zebrafish is much more amenable for forward genetic as well as chemical screens.

However, the sophisticated homologous recombination-based gene targeting technology is currently not available in zebrafish, making the generation of genetic recessive loss-of-function PD models a challenging task. The recently developed reverse genetic method TILLING (see Figure 23.4) shall lead to the production of these PD models in the near future. Compared to flies and worms, the life cycle of zebrafish (~2–3 months) is considerably longer, making any type of genetic and modifier screens a more time-consuming endeavor.

As with any animal models for human diseases, cautions shall be taken when extrapolating findings from animal systems to humans. Such considerations also make it a good argument to have a variety of animal models in hand, from which conserved features may be observed and universal conclusions can be drawn.

Chemically Induced Models of PD in Zebrafish

PD as the second most common neurodegenerative disorder affects approximately 1% of the population over the age of 65, and ~4–5% over the age of 85 (Lang and Lozano, 1998a, b). While genetic analyses of familial cases have identified several genes that, when mutated, lead to PD (Dawson and Dawson, 2003), it is worth noting that a majority of PD cases are sporadic. Therefore, environmental factors interacting with genetic components are likely to play an important role in the development of PD. Neurotoxins that have been strongly implicated in the pathogenesis of sporadic PD include l-methyl-4-phenyl-1,2,3,6-tetiahydropyridine (MPTP), which was initially identified as a contaminant of meperidine (a synthetic heroin) (Langston et al., 1983), and pesticides such as rotenone and paraquat (Le Couteur et al., 1999).

Given the ease of delivering chemical compounds to zebrafish (by simply adding them to the tank water, which enables access to the CNS), the potential PD-inducing effects of MPTP, its metabolite MPP$^+$, and the pesticides including rotenone and paraquat, have been evaluated in both larval and adult zebrafish (Anichtchik et al., 2004; Bretaud et al., 2004; Lam et al., 2005; McKinley et al., 2005). The reason that larval zebrafish, which are ~5 days post-fertilization to 2–3 weeks old, are tested alongside with adult zebrafish, is the following: First, larval zebrafish are particularly amenable to large-scale genetic or chemical screening due to their small size (~1–2 mm in length). This feature allows the use of larval zebrafish for screening in 96- or 384-well formats. Such ease of screening demands an inquiry into the possibility of observing PD-related phenotypes in larval zebrafish. Second, although PD mostly manifests as an old-age disease, juvenile onset of PD has also been reported (Kitada et al., 1998). Hence, the age dependence of PD may be due to the fact that neurotoxic agents take time to build up and exert their effects in the CNS. A larger dose of neurotoxic agents will likely shorten the incubation time required for the appearance of disease symptoms. When neurotoxins such as MPTP, MPP$^+$, rotenone, or paraquat are given to larval zebrafish, PD relevant effects have been observed. Three independent studies have shown that treatment of larval zebrafish with MPTP or MPP$^+$ leads to a reduced appearance of brain dopaminergic neurons (Bretaud et al., 2004; Lam et al., 2005; McKinley et al., 2005) (Figure 23.5). The concentration of MPTP administered in the tank water ranges from 25 μM to 1.6 mM, whereas the exposure time all begin when the embryos are ~24 h post-fertilization,

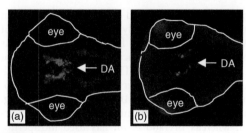

FIGURE 23.5 MPTP effects on DA neurons in larval zebrafish. (a) Control 3-day-old zebrafish embryo showing DA neurons labeled in red (with TH antibody). (b) MPP$^+$-treated 3-day-old zebrafish embryo showing a significant reduction in the appearance of DA neurons.

which is the onset of dopaminergic neuron development in zebrafish. Several lines of evidence suggest that MPTP (or MPP$^+$) treatment affected the survival rather than the development of DA neurons: First, deprenyl, which acts to inhibit the activity of monoamine oxidase B (MAOB) that converts MPTP to its active metabolite MPP$^+$, can protect dopaminergic neurons in the presence of MPTP. This result suggests that the loss of dopaminergic neurons in zebrafish is due to the same toxic metabolite MPP$^+$. Second, morpholino knockdown of the dopamine transporter (DAT), which is necessary for the uptake of MPP$^+$ into dopaminergic neurons, has shown some degree of protection against MPTP toxicity. However, the efficacy of the DAT morpholino in knocking down endogenous DAT activity was not examined in the study (McKinley et al., 2005). The other issue relating to the validity of this MPTP neurotoxin-induced PD model is whether the loss of DA neurons represents a selective event as in humans, that is, other groups of neurons are largely spared. Such selective effect of MPTP has indeed been observed in larval zebrafish. It will be of interest in the future to treat larval zebrafish of a later stage and see if DA neuronal loss can be similarly observed, in order to further determine whether MPTP affects the development or the survival of DA neurons. Moreover, it will be of great interest to know whether a recovery of DA neuronal loss can be observed after the removal of MPTP.

The acute effects of MPTP (and MPP$^+$) on adult zebrafish have also been examined (Anichtchik et al., 2004; Bretaud et al., 2004). While reduced locomotor activity has been observed in treated animals from both studies, neither study has observed a loss of DA neurons, despite the fact that one study has reported a significant reduction of the brain level of catecholamines upon MPTP treatment (Anichtchik et al., 2004). Anichtchik et al. has observed a locomotor deficit with an MPTP dose of 25 mg/kg, which was delivered through intramuscular injection, whereas Bretaud et al. has used intraperitoneal injection and did not observe any locomotor deficit until the dose of 225 mg/kg was used. This difference in dose sensitivity is likely due to the difference in the method of drug delivery, although the potential difference in the genetic background of the fish used in these studies may also need to be considered.

It is unclear why DA neuronal loss has not been observed in adult zebrafish. It is worth noting that higher doses of MPTP as compared to those used in humans or primates, together with prolonged exposures, are also required to elicit DA neuronal loss in rodents (Heikkila et al., 1984). It is also of interest to note that DA neuronal loss has been observed in larval zebrafish, which are bathed continuously in MPTP (or MPP$^+$). Thus, it will be important to determine whether prolonged and multiple exposures to high doses of MPTP might ultimately lead to DA neuronal death in adult zebrafish.

In addition to locomotor deficits, MPTP has also elicited remarkable peripheral phenotypes in zebrafish, including a difficulty in respiration and darkened skin pigmentation. It should be further validated whether the change in skin pigmentation might be related to a loss of norepinephrine contents or neuronal terminals in MPTP-treated fish. If so, this easy-to-observe phenotype might serve as a convenient indicator of MPTP toxicity on catecholamine terminals, and could be used as a simple assay for high throughput screens of chemicals that can counteract the neurotoxic effects of MPTP.

In recent years, exposure to pesticides have been strongly linked to the development of PD (Le Couteur et al., 1999). It has been reported that rotenone and paraquat exposure lead to DA neuronal loss in rodents (Betarbet et al., 2000; McCormack et al., 2002), although variable results have also been obtained (Perier et al., 2003).

These pesticides have been experimented in both larval and adult zebrafish, individually or in combination (Bretaud et al., 2004). They were delivered to zebrafish by direct administration in the tank water. Such delivery method is analogous to the possible routes of exposure that may incur in humans.

Larval zebrafish were exposed to either rotenone or paraquat from 24 h post-fertilization to 5 days post-fertilization, at concentrations ranging from 5 µg/L to 30 µg/L (rotenone), and 5 mg/L to 600 mg/L (paraquat), respectively. These exposed larval zebrafish did not exhibit discernible defects, including both locomotor activity and DA neuronal appearance. But interestingly, exposure to a combination of both drugs (at much lower doses, 10 µg/L rotenone + 50 mg/L paraquat) caused pronounced cardiovascular defects in developing embryos. It is unclear why individual pesticides had no effect on embryos, but a combination elicited strong defects in particular in the cardiovascular system. Several possibilities exist. Rotenone and paraquat may target different parts of the mitochondrial complex I, and the combined use may have a synergistic effect. The embryonic cardiovascular system may be more vulnerable to the toxic insults of these pesticides than DA neurons. It remains to be determined whether these pesticides elicit an oxidative stress in these embryos.

Exposure of adult zebrafish to either rotenone or paraquat did not produce any discernible effects either. Rotenone is known to be very toxic to fishes and has been used as a fish-killing agent. A concentration as low as 10 µg/L is lethal to adult zebrafish after a few days of exposure. An exposure regimen of 2 µg/L rotenone was employed for 4 weeks (with a daily change of fresh pesticide-containing solution). After this period of exposure, adult zebrafish exhibited normal locomotor activity. Likewise, exposure to paraquat at 5 mg/L for 4 weeks did not alter locomotor activity in adult zebrafish. In addition, these pesticides did not elicit the peripheral effects (including altered pigmentation and increased rates of breathing) that were observed after exposure to MPTP. Several possibilities exist as to why exposure to these pesticides did not induce PD-like phenotypes. First, it is formally possible that the appropriate concentration × time of exposure has yet to be established. In the case of rotenone, the extreme peripheral toxicity limits a possible exposure to higher concentrations of the pesticide. In the case of paraquat, higher concentration could possibly be tried. Prolonged exposure may also be necessary. Second, zebrafish may have excellent detoxifying mechanisms peripherally to break down these compounds before they have the opportunity to reach the CNS. Third, the adult fish brain is known to have very good regenerative abilities. For example, MPTP-damaged catecholamine systems in goldfish brain areas undergo spontaneous recovery (Poli *et al.*, 1992). It is possible that any potential damage to DA neurons caused by these pesticides might be efficiently and timely repaired through incorporation of new neurons generated by adult neural stem cells. At present, little is understood about the regenerative capacity of DA system in adult zebrafish. Future studies will shed important lights on this interesting possibility.

In addition to neurotoxin-induced loss of DA neurons, it has been well established that both antipsychotics and antidepressants can have extrapyramidal side effects (EPS) leading to movement disorders in individuals who are treated with these medications (Kelly and Miller, 1975; Errea-Abad *et al.*, 1998). This so-called drug-induced parkinsonism usually develops within 1 month of the initiation of the offending medication in approximately 60% patients and in approximately 90% within 3 months (Van Gerpen, 2002). The likely risk factors include prior history of movement defects, age, gender, and genetically determined differences in drug metabolism and possibly drug action. The conditions are generally reversible, once the medications are removed, suggesting that the drugs produce an interference with neuronal function rather than killing the neurons. A report describes that antipsychotics fluphenazine and haloperidol also impair locomotor activity in larval zebrafish (Giacomini *et al.*, 2006). This assay may be used in future genetic screens to identify genes that mediate dopamine's effect on locomotor coordination.

Genetic Models of PD in Zebrafish

Studies of rare, familial cases have identified genes, when mutated, lead to early onset of PD. These genes include α-synuclein, Parkin, UCHL-1, DJ-1, Pink-1, and Park-8 (also known as LRRK2) (Farrer, 2006). While no stable transgenic or mutational lines in these genes have been reported in zebrafish, one recent study has examined the effects of transient knockdown of DJ-1 using morpholino antisense oligonucleotides (Bretaud *et al.*, 2007). Loss-of-function of DJ-1 leads to early-onset PD (Bonifati *et al.*, 2003). Despite intensive studies, the physiological role of DJ-1 remains poorly understood. Functions ranging from protection against oxidative stress or proteasome inhibition-induced cell death, to transcriptional activation, and chaperone

activity have been reported (Lev *et al.*, 2006). One highly conserved orthologue of DJ-1 was identified in zebrafish, which shares ~83% identity in amino acid sequences with the human counterpart. Zebrafish DJ-1 is expressed in early embryos as well as in many different tissues in adult zebrafish. The number of dopaminergic neurons in embryonic and larval zebrafish was assessed upon reduction of DJ-1 activity. No difference in DA neuronal number was found between control and DJ-1 morphants. But interestingly, DJ-1 morphants are more sensitive to oxidative stress, as evidenced by a significant reduction of DA neurons in DJ-1 morphants as compared to control embryos upon H_2O_2 exposure. DJ-1 morphants also have increased levels of superoxide dismutase (SOD1) mRNA.

While the link between DJ-1 and oxidative stress is quite intriguing, it remains unclear whether the loss of DA neurons was selective, and whether it is due to neuronal degeneration or abnormal development.

The recent development of TILLING method to selectively inactivate genes of interest provides an opportunity to generate stable inactivation of PD-related genes in zebrafish. The availability of these mutant zebrafish shall provide new tools to further evaluate whether genetic models of PD in zebrafish may recapitulate key features of the disease, in particular, the selective degeneration of DA neurons in specific regions of the brain.

STUDYING THE DEVELOPMENT AND REGENERATION OF DA NEURONS IN ZEBRAFISH

Most of the PD symptoms are caused by the degeneration of DA neurons. Replenishment of brain DA with the L-dopa or DA agonists remains the pre-dominant symptomatic treatment for PD. Transplantation of DA neurons into the patient's brain could potentially be a cure for PD. However, two major obstacles severely hinder its use: First, there is a limited availability of pure DA neurons to serve as a source for transplantation. Second, the survivability of transplanted DA neurons is poor. In order to overcome these obstacles, a clear understanding of the mechanisms underlying the development and connectivity of DA neurons is important. Moreover, an ideal treatment for PD would be to somehow stimulate the endogenous adult neural

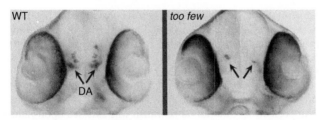

FIGURE 23.6 Developmental loss of DA neurons in the *too few* mutant zebrafish. Ventral view of 2-day-old zebrafish embryos showing the loss of DA neurons in the too few mutant (labeled with TH antibody).

stem cells to re-initiate the developmental program toward DA neurons. Whether this is a possibility or not, we still know very little. In the following paragraphs, I will discuss the current progress and future potential of using zebrafish to elucidate the development, connectivity, and possible regeneration of DA neurons.

DA Neurons in Zebrafish

DA neurons are uniquely defined by their expression of tyrosine hydroxylase (TH), the rate-limiting enzyme involved in converting tyrosine to L-dopa, and their lack of expression of dopamine-beta-hydroxylase (DBH), which converts DA to noradrenaline. In the vertebrate CNS, DA neurons are found in stereotypic locations including the olfactory bulb, retina, ventral forebrain, and ventral midbrain (Smeets and Reiner, 1994; Goridis and Rohrer, 2002). In zebrafish, DA neurons are detected in the olfactory bulb, retina, and ventral forebrain (Guo et al., 1999b; Ma, 2003). Dye-tracing experiments suggest that subsets of DA neurons in the ventral forebrain make ascending projections to the telencephalon, resembling the mammalian midbrain DA neurons (Rink and Wullimann, 2001, 2002). Pharmacological and behavioral analyses provide further evidence that the DA systems in zebrafish play critical roles in regulating movement coordination and reward-associated behaviors (Bretaud et al., 2004; Rink and Guo, 2004; Lau et al., 2006).

Development of DA Neurons in Zebrafish

The development of DA neurons has been studied in mice, mainly through targeted inactivation of candidate genes that are expressed in the vicinity of developing DA neurons. These analyses reveal that the ventral midbrain DA neurons, whose projections to the telencephalon constitute the mesostriatal, mesolimbic, and mesocortical pathways, are critically dependent on the secreted factors Sonic Hedgehog (SHH) and FGF8 (Ye et al., 1998), and the transcription regulators Nurr1 (Zetterström et al., 1997; Saucedo-Cardenas et al., 1998), Lmx1a, Msx1 (Andersson et al., 2006), Lmx1b (Smidt et al., 2000), Pitx3 (Smits et al., 2005), and Neurogenin 2 (Andersson et al., 2006; Kele et al., 2006) for their development. Similarly, the ventral forebrain DA neurons, which play important roles in regulating hormonal homeostasis, require the secreted factors SHH and BMP7 (Ohyama et al., 2005), and the transcription regulators Dlx and Pax6 for their proper development (Vitalis et al., 2000; Andrews et al., 2003). Taken together, these studies reveal the involvement of regionally expressed signaling molecules as well as a complex array of DNA-binding transcription regulators in DA neuron development.

In recent years, an unbiased forward genetic approach has been taken to study the development of DA neurons in zebrafish (Guo et al., 1999a). Molecular genetic characterizations of the zebrafish mutant *foggy* have revealed an unexpected role of regulated transcription elongation during DA neuron development (Guo et al., 2000). Analyses of the mutants *too few* (*tof/fezl*) (Figure 23.6) and *neurogenin 1* (*ngn1*) have uncovered the importance of these region-specific DNA-binding transcription regulators in the early commitment of progenitor cells to a ventral forebrain DA neuronal lineage (Levkowitz et al., 2003 Jeong et al., 2006;). In particular, the study of *ngn1* reveals that the roles of neurogenins in DA neuron development are conserved between zebrafish (Jeong et al., 2006) and mammals (Andersson et al., 2006). Recently, our positional cloning of the *motionless* mutation has shown that a subunit of the mediator complex, which has been associated with schizophrenia in humans (Philibert et al., 2001), is essential for, and when overexpressed can increase, the production of DA neurons (Wang *et al.*, 2006). Taken together, the unbiased forward genetic studies in zebrafish have revealed conserved features, as well as uncovered novel regulatory mechanisms in DA neuron development.

CONCLUSIONS

In this chapter, we discussed the salient features of zebrafish, including a typical vertebrate neuroanatomy, conserved orthologous gene sequences, and the amenability to forward genetic and small molecule screening. We also discussed the use of zebrafish for PD-related studies, including the establishment of chemical-induced PD models, genetic PD models, as well as the study of DA neuron development. These studies are still at early stages. Future work in refining these models, establishing new models, as well as looking into the connectivity, function, and regeneration of DA systems in zebrafish shall make further significant contributions to understanding PD as well as developing therapeutics for treating the disease.

REFERENCES

Amsterdam, A., Burgess, S., Golling, G., Chen, W., Sun, Z., Townsend, K., Farrington, S., Haldi, M., and Hopkins, N. (1999). A large-scale insertional mutagenesis screen in zebrafish. *Genes Dev* 13, 2713–2724.

Andersson, E., Jensen, J. B., Parmar, M., Guillemot, F., and Bjorklund, A. (2006). Development of the mesencephalic dopaminergic neuron system is compromised in the absence of neurogenin 2. *Development epress.*

Andrews, G. L., Yun, K., Rubenstein, J., and Mastick, G. S. (2003). Dlx transcription factors regulate differentiation of dopaminergic neurons of the ventral thalamus. *Mol Cell Neurosci* 23, 107–120.

Anichtchik, O. V., Kaslin, J., Peitsaro, N., Scheinin, M., and Panula, P. (2004). Neurochemical and behavioral changes in zebrafish *Danio rerio* after systemic administration of 6-hydroxydopamine and 1-methyl-4-phenyl-1,2,3,6-tetrahydropyridine. *J Neurochem* 88, 443–453.

Barbazuk, W. B., Korf, I., Kadavi, C., Heyen, J., Tate, S., Wun, E., Bedell, J. A., McPherson, J. D., and Johnson, S. L. (2000). The syntenic relationship of the zebrafish and human genomes. *Genome Res* 10, 1351–1358.

Betarbet, R., Sherer, T. B., MacKenzie, G., Garcia-Osuna, M., Panov, A. V., and Greenamyre, J. T. (2000). Chronic systemic pesticide exposure reproduces features of Parkinson's disease. *Nat Neurosci* 3, 1301–1306.

Bonifati, V., Rizzu, P., van Baren, M., Schaap, O., Breedveld, G. J., Krieger, E., Dekker, M. C., Squitieri, F., Ibanez, P., Joosse, M. *et al.* (2003). Mutations in the DJ-1 gene associated with autosomal recessive early-onset parkinsonism. *Science* 299, 256–259.

Bretaud, S., Lee, S., and Guo, S. (2004). Sensitivity of zebrafish to environmental toxins implicated in Parkinson's disease. *Neurotoxicol Teratol* 26, 857–864.

Bretaud, S., Allen, C., Ingham, P. W., and Bandmann, O. (2007). p53-dependent neuronal cell death in a DJ-1-deficient zebrafish model of Parkinson's disease. *J Neurochem* 100, 1626–1635.

Dawson, T. M., and Dawson, V. L. (2003). Rare genetic mutations shed light on the pathogenesis of Parkinson's disease. *J Clin Invest* 111, 145–151.

Driever, W., Solnica-Krezel, L., Schier, A. F., Neuhauss, S. C. F., Malicki, J., Stemple, D. L., Stainier, D. Y. R., Zwartkruis, F., Abdelilah, S., Rangini, Z. *et al.* (1996). A genetic screen for mutations affecting embryogenesis in zebrafish. *Development* 123, 37–46.

Errea-Abad, J. M., Ara-Callizo, J. R., and Aibar-Remón, C. (1998). Drug-induced parkisnonism: Clinical aspects compared with Parkinson disease. *Rev Neurol* 27, 35–39.

Farrer, M. J. (2006). Genetics of Parkinson disease: Paradigm shifts and future prospects. *Nat Rev Genet* 7, 306–318.

Geisler, R., Rauch, G. J., Baier, H., Van Bebber, F., Brobeta, L., Dekens, M. P., Finger, K., Fricke, C., Gates, M. A., Geiger, H. *et al.* (1999). A radiation hybrid map of the zebrafish genome. *Nat Genet* 23, 86–89.

Giacomini, N. J., Rose, B., Kobayashi, K., and Guo, S. (2006). Antipsychotics produce locomotor impairment in larval zebrafish. *Neurotoxicol Teratol* 28, 245–250.

Goridis, C., and Rohrer, H. (2002). Specification of catecholaminergic and serotonergic neurons. *Nat Rev Neurosci* 3, 531–541.

Grunwald, D. J., and Eisen, J. S. (2002). Headwaters of the zebrafish-emergence of a new model vertebrate. *Nat Rev Genet* 3, 717–724.

Guo, S. (2004). Linking genes to brain, behavior, and neurological diseases: What can we learn from zebrafish? *Genes Brain Behav* 3, 63–74.

Guo, S., Driever, W., and Rosenthal, A. (1999a). Mutagenesis in Zebrafish: Studying the Brain Dopamine Systems. *Handbook of Molecular-Genetic Techniques for Brain and Behavior Research*, 166–176, Chapter 2.1.8.

Guo, S., Wilson, S. W., Cooke, S., Chitnis, A. B., Driever, W., and Rosenthal, A. (1999b). Mutations in the zebrafish unmask shared regulatory pathways controlling the development of catecholaminergic neurons. *Dev Biol* 208, 473–487.

Guo, S., Yamaguchi, Y., Schilbach, S., Wada, T., Goddard, A., Lee, J., French, D., Handa, H., and Rosenthal, A. (2000). A regulator of transcriptional elongation, which is required for vertebrate neuronal development. *Nature* 408, 366–369.

Haffter, P., Granato, M., Brand, M., Mullins, M. C., Hammerschmidt, M., Kane, D. A., Odenthal, J., Van Eeden, F. J. M., Jiang, Y. J., Heisenberg, C. P. *et al.* (1996). The identification of genes with unique and essential function in the development of the zebrafish, *Danio rerio. Development* 123, 1–36.

Heikkila, R. E., Hess, A., and Duvoisin, R. (1984). Dopaminergic neurotoxicity of 1-methyl-4-phenyl-1,2,3,6-tetrahydropyridine in mice. *Science* 224, 1451–1453.

Hukriede, N. A., Joly, L., Tsang, M., Miles, J., Tellis, P., Epstein, J. A., Barbazuk, W. B., Li, F. N., Paw, B., Postlewait, J. H. *et al.* (1999). Radiation hybrid mapping of the zebrafish genome. *Proc Natl Acad USA* 96, 9745–9750.

Jeong, J., Einhorn, Z., Mercurio, S., Lee, S., Lau, B., Mione, M., Wilson, S. W., and Guo, S. (2006). Neurogenin1 is a determinant of zebrafish basal forebrain dopaminergic neurons and is regulated by the conserved zinc finger protein Tof/Fezl. *Proc Natl Acad Sci* 103, 5143–5148.

Kele, J., Simplicio, N., Ferri, A. L. M., Mira, H., Guillemot, F., Arenas, E., and Abg, S. L. (2006). Neurogenin 2 is required for the development of ventral midbrain dopaminergic neurons. *Development* 133, 495–505.

Kelly, P. H., and Miller, R. J. (1975). The interaction of neuroleptic and muscarinic agents with central dopaminergic systems. *Br J Pharmacol* 54, 115–121.

Kitada, T. *et al.* (1998). Mutations in the parkin gene cause autosomal recessive juvenile parkinsonism. *Nature* 392, 605–608.

Knapik, E. W., Goodman, A., Ekker, M., Chevrette, M., Delgado, J., Neuhauss, S. C. F., Shimoda, N., Driever, W., Fishman, M. C., and Jacob, H. (1998). A microsatellite genetic linkage map for zebrafish. *Nat Genet* 18, 338–343.

Lam, C. S., Korzh, V., and Strahle, U. (2005). Zebrafish embryos are susceptible to the dopaminergic neurotoxin MPTP. *Eur J Neurosci* 21, 1758–1762.

Lang, A. E., and Lozano, A. M. (1998a). Parkinson's disease. First of two parts. *N Engl J Med* 339, 1044–1053.

Lang, A. E., and Lozano, A. M. (1998b). Parkinson's disease. Second of two parts. *N Engl J Med* 339, 1130–1143.

Langston, J. W., Ballard, P., Tetrud, J. W., and Irwin, I. (1983). Chronic parkinsonism in humans due to a product of meperidine analogue synthesis. *Science* 219, 979–980.

Lau, B., Bretaud, S., Huang, Y., Lin, E., and Guo, S. (2006). Dissociation of food and opiate preference by a genetic mutation in zebrafish. *Genes Brain Behav* 5, 497–505.

Le Couteur, D. G., McLean, A. J., Taylor, M. C., Woodham, B. L., and Board, P. G. (1999). Pesticides and parkinson's disease. *Biomed Pharmacother* 53, 122–130.

Lev, N., Roncevich, D., Ickowicz, D., Melamed, E., and Offen, D. (2006). Role of DJ-1 in Parkinson's disease. *J Mol Neurosci* 29, 215–225.

Levkowitz, G., Zeller, J., Sirotkin, H. I., French, D., Schilbach, S., Hashimoto, H., Hibi, M., Talbot, W. S., and Rosenthal, A. (2003). Zinc finger protein too few controls the development of monoaminergic neurons. *Nat Neurosci* 6, 28–33.

Ma, P. M. (2003). Catecholaminergic systems in the zebrafish IV. Organization and projection pattern of dopaminergic neurons in the diencephalon. *J Comp Neurol* 460, 13–37.

McCormack, A. L., Thiruchelvam, M., Manning-Bog, A. B., Thiffault, C., Langston, J. W., Cory-Slechta, D. A., and Di Monte, D. A. (2002). Environmental risk factors and Parkinson's disease: Selective degeneration of nigral dopaminergic neurons caused by the herbicide paraquat. *Neurobiol Dis* 10, 119–127.

McKinley, E. T., Baranowski, T. C., Blavo, D. O., Cato, C., Doan, T. N., and Rubinstein, A. L. (2005). Neuroprotection of MPTP-induced toxicity in zebrafish dopaminergic neurons. *Brain Res Mol Brain Res* 141, 128–137.

Nasevicius, A., and Ekker, S. C. (2000). Effective targeted gene "knockdown" in zebrafish. *Nat Genet* 26, 216–220.

Ninkovic, J., Folchert, A., Makhankov, Y. V., Neuhauss, S. C., Sillaber, I., Straehle, U., and Bally-Cuif, L. (2006). Genetic identification of AChE as a positive modulator of addiction to the psychostimulant D-amphetamine in zebrafish. *J Neurobiol* 66, 463–475.

Ohyama, K., Ellis, P., Kimura, S., and Placzek, M. (2005). Directed differentiation of neural cells to hypothalamic dopaminergic neurons. *Development* 132, 5185–5197.

Perier, C., Bove, J., Vila, M., and Przedborski, S. (2003). The rotenone model of Parkinson's disease. *Trends Neurosci* 26, 345–346.

Peterson, R. T., Shaw, S. Y., Peterson, T. A., Milan, D. J., Zhong, T. P., Schreiber, S. L., MacRae, C. A., and Fishman, M. C. (2004). Chemical suppression of a genetic mutation in a zebrafish model of aortic coarctation. *Nat Biotech* 22, 595–599.

Philibert, R. A., Sandhu, H. K., Hutton, A. M., Wang, Z., Arndt, S., Andreasen, N. C., Crowe, R., and Wassink, T. H. (2001). Population-based association analyses of the HOPA12bp polymorphism for schizophrenia and hypothyroidism. *Am J Med Genet* 105, 130–134.

Poli, A., Gandolfi, O., Lucchi, R., and Barnabei, O. (1992). Sponatenous recovery of MPTP-damaged catecholamine systems in goldfish brain areas. *Brain Res* 585, 128–134.

Rink, E., and Guo, S. (2004). The *too few* mutant selectively affects subgroups of monoaminergic neurons in the zebrafish forebrain. *Neuroscience* 127, 147–154.

Rink, E., and Wullimann, M. F. (2001). The teleostean (zebrafish) dopaminergic system ascending to the subpallium (striatum) is located in the basal diencephalon (posterior tuberculum). *Brain Res* 889, 316–330.

Rink, E., and Wullimann, M. F. (2002). Connections of the ventral telencephalon and tyrosine hydroxylase distribution in the zebrafish brain (*Danio rerio*) lead to identification of an ascending dopaminergic system in a teleost. *Brain Res Bull* 57, 385–387.

Saucedo-Cardenas, O., Quintanahau, J. D., Le, W. D., Smidt, M. P., Cox, J. J., Demayo, F., Burbach, J. P. H., and Conneely, O. M. (1998). Nurr1 is essential for the induction of the dopaminergic phenotype and the survival of ventral mesencephalic late dopaminergic precursor neurons. *Proc Natl Acad Sci USA* 95, 4013–4018.

Smeets, W. J. A. J., and Reiner, A. (1994). *Phylogeny and Development of Catecholamine Systems in the CNS of Vertebrates.* Cambridge University Press, Cambridge, England.

Smidt, M. P., Asbreuk, C. H. J., Cox, J. J., Chen, H., Johnson, R. L., and Burbach, J. P. H. (2000). A second independent pathway for development of mesencephalic dopaminergic neurons require *Lmx1b. Nat Neurosci* 3, 337–341.

Smits, S. M., Mathon, D. S., Burbach, J. P., Ramakers, G. M., and Smidt, M. P. (2005). Molecular and cellular alterations in the Pitx3-deficient midbrain dopaminergic system. *Mol Cell Neurosci* 30, 352–363.

Streisinger, G., Walker, C., Dower, N., Knauber, D., and Singer, F. (1981). Production of clones of homozygous diploid zebra fish (*Brachydanio rerio*). *Nature* 291, 293–296.

Van Gerpen, J. A. (2002). Drug-induced parkinsonism. *Neurologist* 8, 363–370.

Vitalis, T., Cases, O., Engelkamp, D., Verney, C., and Price, D. J. (2000). Defects of tyrosine hydroxylase-immunoreactive neurons in the brains of mice lacking the transcription factor Pax6. *J Neurosci* 20, 6501–6516.

Wang, X., Yang, N., Uno, E., Roeder, R. G., and Guo, S. (2006). A subunit of the mediator complex regulates vertebrate neuronal development. *Proc Natl Acad Sci USA* 103, 17284–17289.

Wienholds, E., Schulte-Merker, S., Walderich, B., and Plasterk, R. H. A. (2002). Target-selected inactivation of the zebrafish rag1 gene. *Science* 297, 99–102.

Woods, I. G., Kelly, P. D., Chu, F., Ngo-Hazelett, P., Yan, Y. L., Huang, H., Postlewait, J. H., and Talbot, W. S. (2000). A comparative map of the zebrafish genome. *Genome Res* 10, 1903–1914.

Ye, W., Shimamura, K., Rubenstein, J. L. R., Hynes, M. A., and Rosenthal, A. (1998). FGF8 and Shh signals control dopaminergic and serotonergic cell fate in the anterior neural plate. *Cell* 93, 755–766.

Zetterström, R. H., Solomin, L., Jansson, L., Hoffer, B. J., Olson, L., and Perlmann, T. (1997). Dopamine neuron agenesis in Nurr-1-deficient mice. *Science* 276, 248–250.

V

MULTICELLULAR INVERTEBRATE MODELS

24
PARKINSON'S DISEASE: INSIGHTS FROM INVERTEBRATES

ELLEN B. PENNEY AND BRIAN D. MCCABE

Department of Physiology and Cellular Biophysics and Department of Neuroscience, Center for Motor Neuron Biology and Disease, Columbia University, New York, NY, USA

For you, ye little Race, the impartial Powers
Subject to Chances, and to Pains like ours,
Misfortunes shock you, and Disorders maim,
And dire Diseases rack your tender Frame.

De Bombyce (the Silk-worm) by
Marco Girolamo Vida (1519)

Parkinson's disease (PD) is the most common neurodegenerative movement disorder, affecting 1% of the population over the age of 60 (de Lau and Breteler, 2006). Over the past two decades, significant progress has been made in our understanding of the genetic and environmental factors that predispose some people to develop PD; however, the exact molecular and cellular mechanisms that underlie the disease remain unclear and current treatments are primarily palliative. An arsenal of scientific approaches is being applied to study PD and the following chapters detail a relatively recent addition to the armory of human disease biology – modeling disease in invertebrate genetic organisms. In this overview, we will outline the rationale for these studies and highlight both the promise and the pitfalls that invertebrate model systems offer to elucidate PD pathogenic pathways and identify therapeutic targets for human PD.

ADVANTAGES OF INVERTEBRATE MODELS

Like any other model system, invertebrates have a distinct set of advantages and disadvantages. In this chapter, we will concern ourselves with the two most prominent invertebrate genetic models organisms – the nematode *Caenorhabditis elegans* and the fruit fly *Drosophila melanogaster*. Both have the advantage that they are small, inexpensive to cultures and have short life spans. The rapid developments of these models is particularly advantageous for the study of neurodegenerative disease as nervous system maturation and subsequent neuronal degeneration occur over a matter of days. As genetic tools, both models have the advantage of fully sequenced genomes, which have revealed remarkable genetic conservation to humans; though commonly a single gene in a fly or a worm suffices to perform functions for which humans may utilize many genes (Consortium, 1998; Adams *et al.*, 2000). Importantly, the intrinsic advantages of these model organisms have been exploited over nearly 100 years of manipulation (in the case of *Drosophila*), to generate incredibly tractable and powerful genetic model systems. Molecular and genetic techniques have been developed which allow rapid, cell-specific labeling, gene knockdown, and gene over-expression. The combination of these techniques with the short life spans of these models enables extremely fast, high-throughput *in vivo* screening. This expedites identification of involved molecular pathways.

At first glance the neuroanatomy of invertebrates seems very different to that of mammals; however, upon closer inspection, many similarities do exist at the molecular and cellular levels between vertebrate and invertebrate neurons (Consortium, 1998; Adams *et al.*, 2000). For example, ion channels, receptors, neurotransmitters (NT), and the machinery of NT release are similar in both structure and function between vertebrates and invertebrates (Bargmann, 1998; Yoshihara *et al.*, 2001). Many components of cell death pathways, such as apoptosis, were first described in invertebrates and since confirmed to function in an analogous manner in vertebrates (Liu and Hengartner, 1999; Richardson and Kumar, 2002).The use of invertebrate model organisms for the study of human disease-associated genes has increased significantly in the last decade, and this trend seems set to continue as the remarkable and unexpected parallels between invertebrate and human physiology continue to be revealed.

INVERTEBRATE NEURODEGENERATION

The relevance of invertebrate biology to human neurodegenerative disease is further emphasized by the finding that mutations in some invertebrate homologs of human disease-associated genes produce neurodegeneration in these model systems. For example, the lipid storage disease mucolipidosis type IV (MLT-IV) is associated with impaired neuronal development and retinal degeneration in humans. Mutants of the *C. elegans* homolog of the MLT-IV gene have defective endocytosis and dysfunctional lipid storage (Fares and Greenwald, 2001; Hersh *et al.*, 2002). In *Drosophila,* the crumbs mutation causes light-dependent retinal degeneration. Mutations in the human homolog of crumbs produce severe retinitis pigmentosa and Leber's congenital amaurosis (den Hollander *et al.*, 2001; Izaddoost *et al.*, 2002; Johnson *et al.*, 2002; Pellikka *et al.*, 2002).

Intensive familial genetic studies of PD have revealed a growing number of genes associated with both sporadic and familial forms of the disease including alpha-synuclein, LRRK2, DJ-1, parkin, and pink. All of these genes have homologs in *C. elegans* and *Drosophila* with the exception of alpha-synuclein (Meulener *et al.*, 2005; Ved *et al.*, 2005b; Clark *et al.*, 2006; Park *et al.*, 2006; Sakaguchi-Nakashima *et al.*, 2007; Sang *et al.*, 2007; Schmidt *et al.*, 2007; Liu *et al.*, 2008). As our understanding of the genetics underlying human PD increases, the number of invertebrate homologs of PD-associated genes will undoubtedly rise, provding an opportunity to exploit the advantages offered by

these model systems to understand the roles of these genes.

MODEL SYSTEM: *C. ELEGANS*

C. elegans are small (~1mm long), soil dwelling, unsegmented, bilaterally symmetrical nematodes (roundworms) that have been used as a model organism for the last four decades (Brenner, 1974). The life span of *C. elegans* is ~3 weeks and their life cycle consists of an embryonic phase followed by four larval stages, in the last stage of which the animals are sexually mature (Barry Wood, 1988; Riddle, 1997). The majority of *C. elegans* are hermaphrodites. In the wild, males are rare accounting for only 0.05% of the population. The generation time of *C. elegans* is ~3 days and each hermaphrodite can bear several hundred progeny, so large numbers of animals can be rapidly produced (Barry Wood, 1988; Riddle, 1997).

Every one of the ~1000 somatic cells in *C. elegans* has been identified and its lineage mapped (Barry Wood, 1988; Riddle, 1997). *C. elegans* do not have a distinct, confined central nervous system (CNS) but do contain a structure, the central nerve ring, where processes from many neurons connect. The *C. elegans* hermaphrodite nervous system contains 302 neurons that make ~10,000 connections and many of these connections have been identified (Barry Wood, 1988; Riddle, 1997). The dopaminergic system in hermaphroditic *C. elegans* consists of eight neurons involved in mechanosensation – four symmetrically arranged cephalic cells and two pairs of bilateral anterior and posterior deirids (Nass and Blakely, 2003). The male animals contain three additional pairs of dopaminergic neurons in the tail and four male-specific dopaminergic spicule cells that are important for mating behavior (Nass and Blakely, 2003).

C. elegans are diploid and have five pairs of autosomes plus one pair of sex chromosomes (Barry Wood, 1988). The genome of *C. elegans* has been completely sequenced and contains ~20,000 genes (Consortium, 1998). The genome of *Caenorhabditis briggsae*, a related nematode, has also been sequenced, which allows comparative genomics (Stein *et al.*, 2003). A high density of identified polymorphisms in *C. elegans* enables mutations to be rapidly mapped. Transgenic animals containing rescue constructs can be generated in a matter of days

(Barry Wood, 1988; Riddle, 1997). The animals' small size, short life span, and large brood size facilitate large-scale forward mutagenesis screens involving hundreds of thousands of animals (Jorgensen and Mango, 2002). New automated technology assists the screening process by enabling *C. elegans* to be sorted based on macroscopic or microscopic phenotypes (Burns *et al.*, 2006; Rohde *et al.*, 2007). The self-fertilizing capacity of the hermaphrodites allows mutations to be easily bred to homozygosity and male *C. elegans* can be used to breed animals that contain multiple mutations or genetic constructs. Many gene mutants are available through centralized stock centers (Haag, 2007). Genes can also be manipulated with RNA interference (RNAi), a powerful method to decrease expression of specific proteins that was first identified in *C. elegans* (Fire *et al.*, 1998).

Anatomical changes due to developmental or degenerative processes can easily be followed *in vivo* through the transparent animal. This can be facilitated by labeling specific cells or proteins with transgenic green fluorescent protein (GFP) reporter constructs (Chalfie *et al.*, 1994). Finally, the function of specific proteins can now be studied *in vitro* through the use of primary embryonic cell cultures, which differentiate into many other cell types(Bianchi and Driscoll, 2006). These cultures also allow electrophysiology to be performed on differentiated neurons, which is otherwise difficult in *C. elegans* because of its hard, pressurized cuticle (Christensen *et al.*, 2002).

The characteristics of *C. elegans* and the techniques that have been developed for working with the animal offer a system in which genetics, anatomy, and cellular function can be both easily studied and quickly manipulated.

MODEL SYSTEM: *D. MELANOGASTER*

Drosophila melanogaster, a common kitchen pest better known as fruit flies, are two-winged insects of the order Diptera that have been used as a model organism for almost a century (Morgan, 1910). The developmental speed of *Drosophila* varies with temperature but when raised at 25°C, *Drosophila* have a generation time of ~10 days and a life span of ~40 days. The life cycle of *Drosophila* is complex consisting of an embryonic phase, three larval stages, pupation, and finally adult life. Male and female flies become sexually productive within a few hours after eclosion (emergence from the pupal case) and each female fly can lay hundreds of eggs over the course of several days, leading to large brood numbers.

The *Drosophila* nervous system is far more complex than that of *C. elegans*. The *Drosophila* CNS is bilaterally symmetrical and includes a bi-lobed brain and a ventral nerve cord, which is a fusion of the subesophageal ganglion, three thoracic ganglia, and several abdominal ganglia (Bolenstein, D. 1994). The ganglia within the ventral nerve cord contain motor neurons and interneurons necessary to control the body segment that they innervate, whereas the brain contains neurons involved in sensory processing, learning, and memory (Margulies *et al.*, 2005; Keene and Waddell, 2007). During pupation, the body plan of the animal is drastically altered. In the nervous system, existing neurons are re-wired and new neurons develop (Truman, 1990). The larval CNS consists of ~125,000 neurons whereas the more complex adult CNS contains ~250,000 neurons that make millions of connections. Dopaminergic neurons are widely distributed in the *Drosophila* CNS and they are known to be important for learning and memory, mating behavior, and motor control. (Tempel *et al.*, 1984; Neckameyer, 1998; Monastirioti, 1999; Pendleton *et al.*, 2002; Zhang *et al.*, 2007).

The *Drosophila* genome consists of one pair of sex chromosomes (X,Y) and three pairs of autosomes (2,3, and 4) (Cooper, K.W. 1994). The genome has been completely sequenced and although it has ~50% more base pairs than the *C. elegans* genome, there are only ~14,000 protein coding genes (Adams *et al.*, 2000).

As in *C. elegans,* many tools have been developed that allow *Drosophila* to be easily and rapidly genetically manipulated. In *Drosophila*, after mutagenesis screens, mapping can quickly be done either by recombination mapping or through the use of a deficiency kit, a collection of strains that carry small, overlapping deletions of the genome (Ryder *et al.*, 2007). Unwanted recombination can be prevented and genes of interest can be tracked through the use of 'balancer chromosomes'. These chromosomes carry numerous inversions, making recombination unlikely, and genes that cause both a dominant visible phenotype and recessive lethality (Greenspan, 1997). The use of balancer chromosomes therefore allows mutations of interest to be easily maintained and followed over many generations without loss or repair.

Specific cells and proteins can be transgenically fluorescently labeled and followed *in vivo* over time, which is facilitated in the larval stages by the animals' transparent cuticle. Development of the yeast GAL4/UAS expression system for use in *Drosophila* allows spatial and temporal control of gene expression (Fischer *et al.*, 1988). GAL4 is a yeast transcription factor that will drive the expression of genes downstream of a specific DNA sequence known as upstream activating sequences (UAS) (Fischer *et al.*, 1988; Brand and Perrimon, 1993). This system is frequently used to over-express proteins and drive rescue constructs in a cell-specific manner. The newly developed LexA/LexOp system, which works on the same principles as GAL4/UAS, will allow the expression of multiple genes to be manipulated in the same animal (Lai and Lee, 2006). The 'mosaic analysis with a repressible cell marker' (MARCM) system is another powerful tool and enables the cell autonomous effects of mutations to be studied through the generation and labeling of individual cellular clones, which carry a homozygous mutation in an otherwise heterogenous animal (Lee and Luo, 2001; Wu and Luo, 2006).

Electrophysiology is frequently performed in *Drosophila* and this technique may allow neuronal function to be assessed throughout both development and neurodegenerative processes. This facilitates the study of experience-dependent functional change and may enhance the study of mechanisms that affect neuronal function early in a neurodegenerative process, before cell death is evident.

Drosophila, like *C. elegans*, have many features that make them an exceptional model organism. The *Drosophila* tools that have been developed over the last century allow every aspect of a gene's function to be studied including the affects it may have on anatomy, cellular morphology, and neuronal function.

FLIES VERSUS WORMS

Invertebrates are incredibly tractable systems in which to study biological processes and both *C. elegans* and *Drosophila* offer many advantages that vertebrates cannot provide. These models organisms, however, are not created equal.

C. elegans offer a more rapid system for *in vivo* screening and high-throughput drug discovery than *Drosophila*. Screens in *Drosophila* are usually performed on the scale of tens of thousands, whereas *C. elegans* screens are commonly done on the scale of hundreds of thousands (Jorgensen and Mango, 2002). *C. elegans* are hermaphrodites so in a mutagenesis screen each mutant animal will produce its own homozygous line without requiring mating, which increases the speed of the high-throughput experiments and enables screens to completely saturate the genome more rapidly. RNAi is also an extremely powerful tool in *C. elegans* and can be used to knockdown the expression of specific genes with efficiency and speed that cannot be matched with any current *Drosophila* tools. Since the discovery of RNAi, it has been developed for use in many model organisms including *Drosophila* but in no other system is it as convenient as in *C. elegans*, which can simply be raised on food containing the double-stranded DNA that initiates the RNAi pathway (Mello and Fire, 1995; Kamath and Ahringer, 2003). Finally, the anatomy of the worm is less complex than that of the fly and every individual neuron has been characterized, including both their connections and lineage (Barry Wood, 1988; Riddle, 1997). Therefore, the study of how genes and cells affect circuitry and how circuitry controls behavior is possible.

Drosophila also innately provide several distinct advantages for the study of human disease-associated mechanisms. The *Drosophila* retina is frequently used in studies of disease-associated toxicity (Ghosh and Feany, 2004). Cell death in the fly eye is easy to quantify and as an externally visible portion of the nervous system, it enables rapid screening for genetic modifiers and therapeutic agents (Min and Benzer, 1999; Georgiev *et al.*, 2005; Muraro and Moffat, 2006). The MARCM technique is also a uniquely powerful tool that allows the study of a mutations' cell autonomous effects. Whereas the simplicity of *C. elegans* can be an advantage, the relative complexity of the *Drosophila* CNS and the resulting learned behaviors enable molecular changes to be correlated with higher brain function (Keene and Waddell, 2007). *In vivo* electrophysiology and high-resolution imaging of synapses are easier in *Drosophila* than in *C. elegans*, which may be important for determining the morphological and functional changes that occur during a disease-associated process. Finally, the neuroanatomical sophistication of *Drosophila* versus *C. elegans* may be important for studying the non-cell autonomous aspects of PD disease, such as circuit disregulation.

GENERATING INVERTEBRATE DISEASE MODELS

Invertebrate disease models can be generated in three general ways; toxin exposures, forward genetic screens, and reverse genetic manipulations. Toxin models are created by exposing animals to substances that are either known as disease producing causal factors or can produce disease-like symptoms. In recent years, as disease-associated genes have been discovered, toxin models have become less common; however, one notable exception to this general trend is the use of mitochondrial toxins as a model for PD. Environmental toxins, such as the pesticide rotenone, are associated with PD (Priyadarshi *et al.*, 2000) as is the mitochondrial toxin, methyl-4-phenyl-1,2,3,6-tetrahydropyridine (MPTP) which produces parkinsonism in humans (Langston *et al.*, 1983). The oxidative stress compound 6-hydroxydopamine (6-OHDA), also a suspected mitochondrial toxin, is selectively taken up by dopaminergic neurons (Sachs and Jonsson, 1975; Andrew *et al.*, 1993; Betarbet *et al.*, 2002). Animal models based on exposure to all these compounds can produce Parkinson-like cell death (Ungerstedt, 1968; Burns *et al.*, 1983; Betarbet *et al.*, 2000). Mitochondrial toxins may be an important causal factor in the development of sporadic PD and are widely studied (Bové *et al.*, 2005).

Disease models can also be generated through forward genetic screens, in which animals are mutagenized and then selected for analysis based on behavioral (movement disorders) or cellular (degeneration) phenotypes produced. This method allows phenotype producing genes to be selected in an unbiased manner so that unknown genes can be identified. As an increasing number of disease-associated genes have been described in affected kindreds, models that target identified human disease causing genes have become favored and forward genetic screens are primarily used to identify other members of involved pathways by searching for genetic enhancers or suppressors of the disease phenotype.

Reverse genetics, the final method by which disease models are produced, is currently the most favored method. In these models, genes already known to be associated with human disease are specifically targeted so that their function and the pathways through which they act can be studied. When a homolog of the disease-associated gene exists, it can be mutated or if the mutant gene is believed to have a toxic "gain-of-function" effect, the human mutant gene can be introduced into the organism. Invertebrate models allow not only the study of a gene's function but also the elucidation of the involved molecular pathways, identification of enhancers or suppressors through genetic screens.

IMPORTANT *C. ELEGANS* MODELS

Toxin models of PD in *C. elegans* have been effectively used to identify pathogenic pathways and neuroprotective compounds. These models are produced by exposing animals to compounds that are associated with human PD and cause nigrostriatal degeneration in mammalian models including 6-OHDA, MPTP, and rotenone (Ungerstedt, 1968; Sachs and Jonsson, 1975; Burns *et al.*, 1983; Langston *et al.*, 1983; Betarbet *et al.*, 2000; Priyadarshi *et al.*, 2000; Betarbet *et al.*, 2002). In the *C. elegans* 6-OHDA model, dopaminergic neurons die in a manner dependent on the dopamine transporter (DAT), which suggests that this transporter may control the selectivity of the degeneration by allowing toxin entry into dopaminergic neurons (Nass *et al.*, 2002). This model has been used for a small forward genetic screen which identified that DAT mutations are important suppressors of 6-OHDA toxicity and thereby validated the use of this model to screen for suppressors of toxicity (Nass *et al.*, 2005). The tractability of *C. elegans* for use in rapid high-throughput screens suggests great promise for future drug discovery research.

Several reverse genetic models of PD have also been made in *C. elegans*. Models have been made for multiple genes associated with the human disease including alpha-synuclein, parkin, and DJ-1. For the alpha-synuclein models, human wild-type or mutant alpha-synuclein is over-expressed in dopaminergic neurons where the mutant re-produces many of the cellular changes seen in PD including Lewy body-like aggregates (Kuwahara *et al.*, 2006). No cell death occurs, but the level of DA is decreased and animals have behavior phenotypes, such as abnormal modulation of locomotion. Interestingly, although the mutant protein caused a more pronounced and statistically significant phenotype, over-expression of the wild-type human protein also trends toward similar changes (Kuwahara *et al.*, 2006).

Other genetic models have been produced by knocking out the parkin gene or knocking down the levels of DJ-1 (Ved *et al.*, 2005a). Both of these manipulations made the animals more susceptible to mitochondrial complex I inhibitors, whereas their susceptibility to other toxins was not enhanced, and they were partially rescued by anti-oxidants or mitochondrial complex II activators (Ved *et al.*, 2005a). This research therefore supports the theory that mitochondrial damage is an essential component of PD and that PD-associated pathways may converge on the mitochondria (Schapira, 2008).

Finally, other genes potentially regulated by PD pathways have been identified in genome-wide microarray experiments on an alpha-synuclein over-expression model (Vartiainen et al., 2006). In this research, the gene expression profile of *C. elegans* over-expressing human mutant alpha-synuclein was compared to that of animals over-expressing normal human alpha-synuclein. Five hundred genes with more than a twofold change in expression were identified including many genes important for mitochondrial function and the ubiquitin-proteasome pathway.

IMPORTANT *DROSOPHILA* MODELS

The first *Drosophila* model of PD was created in 2000 by Feany and Bender who found that over-expression of wild-type and mutant human alpha-synuclein produced a neurodegenerative phenotype. The phenotype induced by alpha-synuclein was reminiscent of human PD in that there was adult onset of dopaminergic cell death, retinal degeneration, lewy body formation, and locomotor defects (Feany and Bender, 2000). Feany went on to explore the mechanism of these noxious effects and discovered that phosphorylation of specific alpha-synuclein amino acids is essential for toxicity and that the inclusion formation that occurs without phosphorylation is correlated with damage protection (Chen and Feany, 2005).

Several reverse genetic *Drosophila* models of PD have now been created including mutants of DJ-1, parkin, and pink (Greene *et al.*, 2003; Meulener *et al.*, 2005; Clark *et al.*, 2006). Parkin mutations cause muscle degeneration, sterility. DJ-1 appears to protect against oxidative stress and DJ-1 mutant flies are more susceptible to a number of environmental toxins (Meulener *et al.*, 2005). DJ-1 was shown to not only respond to oxidative stress but

also be biochemically modified by oxidative toxins, which may indicate that the interplay between environmental toxins and DJ-1 increases susceptibility to PD.

Finally, the first model of mutant *pink*-associated degeneration was generated in *Drosophila*. In this model, the *Drosophila* homolog of Pink was mutated and although the phenotype produced was distinctly different from PD (Clark *et al.*, 2006), it was surprisingly similar to the phenotype observed in parkin mutants (Greene *et al.*, 2003). Importantly, expression of human Pink can rescue the Pink loss-of-function (LOF) spermatid phenotype, indicating that *Drosophila* Pink and human Pink can function redundantly. Over-expression of Parkin can also partially suppress the Pink LOF phenotype and double mutants of Pink and Parkin are not significantly different than single gene mutants. This suggests that Parkin and Pink act in the same molecular pathway with Parkin functioning downstream of Pink. This work is therefore an elegant illustration of how *Drosophila* genetics can be used to elucidate human disease-associated molecular pathways.

MODIFIER SCREENS

One of the advantages of invertebrate models, as previously mentioned, is the ease with which they can be used to identify other members of involved molecular pathways. This is frequently done through the use of modifier screens that identify involved molecules in either a biased or an unbiased manner. For modifier screens to be performed, ideally the phenotype should be easy to score and externally visible. Degeneration of the fly eye can be used for these types of screens (Min and Benzer, 1999; Georgiev *et al.*, 2005; Muraro and Moffat, 2006). New automated screening and sorting of C. *elegans* has improved the speed and size of screens that can be performed and similar techniques are being developed for flies (Furlong *et al.*, 2001; Braungart *et al.*, 2004; Pulak, 2006).

Modifiers screens have been used to find suppressors of several other neurodegenerative disorders. Polyglutamine toxicity, for example, was found to be suppressed by regulators of the ubiquitin-proteasome pathway, chaperone proteins, and histone acetylation proteins (Warrick *et al.*, 1999; Fernandez-Funez *et al.*, 2000). Chaperone proteins, such as heat shock protein 40 (HSP40), were identified in a polyglutamine suppression screen and may also modify PD toxicity (Warrick *et al.*, 1999; Kazemi-Esfarjani and Benzer, 2000; Auluck *et al.*, 2002). Transcriptional regulation has long been implicated in polyglutamine disorders and histone acetylation regulators were identified as suppressors of ataxin-1 toxicity in a modifier screen performed by Fernandez-Funez and colleagues. Since then regulation of histone acetylation has been implicated in many polyglutamine disorders and drugs that modify histone acetylation have been taken to clinical trials (Fernandez-Funez *et al.*, 2000; Steffan *et al.*, 2001; Hockly *et al.*, 2003; Minamiyama *et al.*, 2004).

Modifiers of a *Drosophila* PD model have been directly studied by Greene *et al.* (2005) who identified several oxidative stress components as enhancers of the pathology of parkin LOF mutants. As previously mentioned, modifier screens have also been validated in the 6-OHDA toxicity model in *C. elegans* (Nass *et al.*, 2005). A hypothesis-based RNAi screen has also been performed in an alpha-synuclein *C. elegans* model and several potential modifiers were identified, including trafficking proteins, regulators of G-protein signaling, and transcription factors (Hamamichi *et al.*, 2008). These screens offer immense potential for identifying other components of PD-associated pathways and as modifier screens are performed on more invertebrate models, our understanding of PD will become more complete.

DRUG DISCOVERY

One of the most exciting advantages that invertebrate models offer is the ability to rapidly screen large numbers of drug compounds. No high-throughput drug screens of invertebrate PD models have yet been performed, but proof of principle experiments have been done showing that a *C. elegans* MPTP toxin model is suitable for high-throughput drug screens (Braungart *et al.*, 2004).

Several drugs have been found to be effective in *Drosophila* models of neurodegenerative disorders. Histone deacetylase inhibitors are protective in models of polyglutamine disease and alpha-synuclein over-expression models of PD (Steffan *et al.*, 2001; Kontopoulos *et al.*, 2006). Geldanamycin, an anti-tumor agent, has also been shown to be neuroprotective in an alpha-synuclein *Drosophila*

model of PD in which it acts to mobilize the stress response and increase levels of chaperone proteins such as HSP70 (Kontopoulos *et al.*, 2006). Once invertebrate models of PD begin to be widely used in drug screens, large libraries of compounds could be screened in a fast, cheap, and effective manner.

POTENTIAL LIMITATIONS WITH INVERTEBRATE MODELS

There are both conceptual and practical drawbacks to invertebrate models. First, as described, neuroanatomy of invertebrates is vastly different than that of humans and invertebrate models seem unlikely to contribute to the study of PD at a "system function" level because invertebrates lack basal ganglia and corresponding inputs and outputs.

Even when invertebrates are solely used to study the molecular mechanisms underlying human disease, specific invertebrate models may not always be relevant to human disease. For example, it is unclear why expression of wild-type alpha-synuclein is toxic in *Drosophila*. It may be due to expression levels, location of expression, or lack of necessary protective systems. Whatever the cause, the system may no longer correctly mimic the human disease.

Invertebrate homologs may not always be structural or functionally related to the corresponding human protein. For example, the human LRRK2 has N-terminal domains that are not conserved in the invertebrate LRRK2 (Marín, 2006). The phenotype of the *Drosophila* LOF Parkin mutant is not rescued by the human Parkin protein, which raises questions about whether the human and the *Drosophila* Parkin protein serve the same cellular function (Sang *et al.*, 2007). The fact that not all animal models may be relevant to human disease must be firmly kept in mind and each model must be individually evaluated.

SUMMARY

Invertebrates are tractable model systems in which to study PD. Numerous invertebrate models of PD have been generated using toxin exposures, over-expression of disease proteins, and disruption of disease genes. These models recapitulate some of the clinical features of PD including dysfunction or death of dopaminergic neurons, locomotor defects, and decreased life span. Significant contributions to our understanding of PD have already been made with these models including the identification of novel genetic interactions and phenotype modifying small molecules. The full power of invertebrate genetics has not yet been brought to bear on the study of PD in high-throughput modifier and drug screens. Once this begins, invertebrate systems will be able to rapidly advance both our understanding of the mechanisms underlying PD and the identification of therapeutic drugs.

REFERENCES

Adams, M. D., Celniker, S. E., Holt, R. A., Evans, C. A., Gocayne, J. D., Amanatides, P. G., Scherer, S. E., Li, P. W., Hoskins, R. A., Galle, R. F. et al. (2000). The genome sequence of *Drosophila melanogaster*. *Science* 287, 2185–2195.

Andrew, R., Watson, D. G., Best, S. A., Midgley, J. M., Wenlong, H., and Petty, R. K. (1993). The determination of hydroxydopamines and other trace amines in the urine of parkinsonian patients and normal controls. *Neurochem Res* 18, 1175–1177.

Auluck, P. K., Chan, H. Y., Trojanowski, J. Q., Lee, V. M., and Bonini, N. M. (2002). Chaperone suppression of alpha-synuclein toxicity in a *Drosophila* model for Parkinson's disease. *Science* 295, 865–868.

Bargmann, C. I. (1998). Neurobiology of the *Caenorhabditis elegans* genome. *Science* 282, 2028–2033.

Betarbet, R., Sherer, T. B., MacKenzie, G., Garcia-Osuna, M., Panov, A. V., and Greenamyre, J. T. (2000). Chronic systemic pesticide exposure reproduces features of Parkinson's disease. *Nat Neurosci* 3, 1301–1306.

Betarbet, R., Sherer, T. B., and Greenamyre, J. T. (2002). Animal models of Parkinson's disease. *Bioessays* 24, 308–318.

Bianchi, L., and Driscoll, M. (2006). Culture of embryonic *C. elegans* cells for electrophysiological and pharmacological analyses. *WormBook*, 1–15.

Bodenstein, D. (1994). In "Biology of Drosophila" (M. Demerec ed.) Cold Spring Harbor Laboratory Press, New York. pp 319–325.

Bové, J., Prou, D., Perier, C., and Przedborski, S. (2005). Toxin-induced models of Parkinson's disease. *NeuroRx* 2, 484–494.

Braungart, E., Gerlach, M., Riederer, P., Baumeister, R., and Hoener, M. C. (2004). *Caenorhabditis elegans* MPP+ model of Parkinson's disease for high-throughput drug screenings. *Neurodegener Dis* 1, 175–183.

Brand, A. H., and Perrimon, N. (1993). Targeted gene expression as a means of altering cell fates and generating dominant phenotypes. *Development* 118, 401–415.

Brenner, S. (1974). The genetics of *Caenorhabditis elegans*. *Genetics* 77, 71–94.

Burns, A. R., Kwok, T. C., Howard, A., Houston, E., Johanson, K., Chan, A., Cutler, S. R., McCourt, P., and Roy, P. J. (2006). High-throughput screening of small molecules for bioactivity and target identification in *Caenorhabditis elegans*. *Nat Protoc* 1, 1906–1914.

Burns, R. S., Chiueh, C. C., Markey, S. P., Ebert, M. H., Jacobowitz, D. M., and Kopin, I. J. (1983). A primate model of parkinsonism: Selective destruction of dopaminergic neurons in the pars compacta of the substantia nigra by N-methyl-4-phenyl-1,2,3,6-tetrahydropyridine. *Proc Natl Acad Sci USA* 80, 4546–4550.

C. elegans Sequencing Consortium. (1998). Genome sequence of the nematode *C. elegans*: A platform for investigating biology. *Science* 282, 2012–2018.

Chalfie, M., Tu, Y., Euskirchen, G., Ward, W. W., and Prasher, D. C. (1994). Green fluorescent protein as a marker for gene expression. *Science* 263, 802–805.

Chen, L., and Feany, M. B. (2005). Alpha-synuclein phosphorylation controls neurotoxicity and inclusion formation in a *Drosophila* model of Parkinson disease. *Nat Neurosci* 8, 657–663.

Christensen, M., Estevez, A., Yin, X., Fox, R., Morrison, R., McDonnell, M., Gleason, C., Miller, D. M., and Strange, K. (2002). A primary culture system for functional analysis of *C. elegans* neurons and muscle cells. *Neuron* 33, 503–514.

Clark, I. E., Dodson, M. W., Jiang, C., Cao, J. H., Huh, J. R., Seol, J. H., Yoo, S. J., Hay, B. A., and Guo, M. (2006). *Drosophila* pink1 is required for mitochondrial function and interacts genetically with parkin. *Nature* 441, 1162–1166.

Cooper, K. W. (1994). *Biology of Drosophila*. (M. Demerec ed.) pp 10–16. Cold Spring Harbor Laboratory Press, New York.

de Lau, L. M., and Breteler, M. M. (2006). Epidemiology of Parkinson's disease. *Lancet neurol* 5, 525–535.

den Hollander, A. I., Heckenlively, J. R., van den Born, L. I., de Kok, Y. J., van der Velde-Visser, S. D., Kellner, U., Jurklies, B., van Schooneveld, M. J., Blankenagel, A., Rohrschneider, K. et al. (2001). Leber congenital amaurosis and retinitis pigmentosa with Coats-like exudative vasculopathy are associated with mutations in the crumbs homologue 1 (CRB1) gene. *Am J Hum Genet* 69, 198–203.

Fares, H., and Greenwald, I. (2001). Regulation of endocytosis by CUP-5, the *Caenorhabditis elegans* mucolipin-1 homolog. *Nat Genet* 28, 64–68.

Feany, M. B., and Bender, W. W. (2000). A *Drosophila* model of Parkinson's disease. *Nature* **404**, 394–398.

Fernandez-Funez, P., Nino-Rosales, M. L., de Gouyon, B., She, W. C., Luchak, J. M., Martinez, P., Turiegano, E., Benito, J., Capovilla, M., Skinner, P. J. et al. (2000). Identification of genes that modify ataxin-1-induced neurodegeneration. *Nature* **408**, 101–106.

Fire, A., Xu, S., Montgomery, M. K., Kostas, S. A., Driver, S. E., and Mello, C. C. (1998). Potent and specific genetic interference by double-stranded RNA in *Caenorhabditis elegans*. *Nature* **391**, 806–811.

Fischer, J. A., Giniger, E., Maniatis, T., and Ptashne, M. (1988). GAL4 activates transcription in *Drosophila*. *Nature* **332**, 853–856.

Furlong, E. E., Profitt, D., and Scott, M. P. (2001). Automated sorting of live transgenic embryos. *Nat Biotechnol* **19**, 153–156.

Georgiev, P., Garcia-Murillas, I., Ulahannan, D., Hardie, R. C., and Raghu, P. (2005). Functional INAD complexes are required to mediate degeneration in photoreceptors of the *Drosophila* rdgA mutant. *J Cell Sci* **118**, 1373–1384.

Ghosh, S., and Feany, M. B. (2004). Comparison of pathways controlling toxicity in the eye and brain in *Drosophila* models of human neurodegenerative diseases. *Hum Mol Genet* **13**, 2011–2018.

Greenspan, R. J. (1997). Fly Pushing: The Theory and Practice of Drosophila Genetics. New York: Cold Spring Harbor Laboratory Press.

Greene, J. C., Whitworth, A. J., Kuo, I., Andrews, L. A., Feany, M. B., and Pallanck, L. J. (2003). Mitochondrial pathology and apoptotic muscle degeneration in *Drosophila* parkin mutants. *Proc Natl Acad Sci USA* **100**, 4078–4083.

Greene, J. C., Whitworth, A. J., Andrews, L. A., Parker, T. J., and Pallanck, L. J. (2005). Genetic and genomic studies of *Drosophila* parkin mutants implicate oxidative stress and innate immune responses in pathogenesis. *Hum Mol Genet* **14**, 799–811.

Haag, E. S. (2007). Dial-a-mutant: Web-based knockout collections for model organisms. *Biol Cell* **99**, 343–347.

Hamamichi, S., Rivas, R. N., Knight, A. L., Cao, S., Caldwell, K. A., and Caldwell, G. A. (2008). Hypothesis-based RNAi screening identifies neuroprotective genes in a Parkinson's disease model. *Proc Natl Acad Sci USA* **105**, 728–733.

Hersh, B. M., Hartwieg, E., and Horvitz, H. R. (2002). The *Caenorhabditis elegans* mucolipin-like gene cup-5 is essential for viability and regulates lysosomes in multiple cell types. *Proc Natl Acad Sci USA* **99**, 4355–4360.

Hockly, E., Richon, V. M., Woodman, B., Smith, D. L., Zhou, X., Rosa, E., Sathasivam, K., Ghazi-Noori, S., Mahal, A., Lowden, P. A. et al. (2003). Suberoylanilide hydroxamic acid, a histone deacetylase inhibitor, ameliorates motor deficits in a mouse model of Huntington's disease. *Proc Natl Acad Sci USA* **100**, 2041–2046.

Izaddoost, S., Nam, S. C., Bhat, M. A., Bellen, H. J., and Choi, K. W. (2002). *Drosophila* crumbs is a positional cue in photoreceptor adherens junctions and rhabdomeres. *Nature* **416**, 178–183.

Johnson, K., Grawe, F., Grzeschik, N., and Knust, E. (2002). *Drosophila* crumbs is required to inhibit light-induced photoreceptor degeneration. *Curr Biol* **12**, 1675–1680.

Jorgensen, E. M., and Mango, S. E. (2002). The art and design of genetic screens: *Caenorhabditis elegans*. *Nat Rev Genet* **3**, 356–369.

Kamath, R. S., and Ahringer, J. (2003). Genome-wide RNAi screening in *Caenorhabditis elegans*. *Methods* **30**, 313–321.

Kazemi-Esfarjani, P., and Benzer, S. (2000). Genetic suppression of polyglutamine toxicity in *Drosophila*. *Science* **287**, 1837–1840.

Keene, A. C., and Waddell, S. (2007). *Drosophila* olfactory memory: Single genes to complex neural circuits. *Nat Rev Neurosci* **8**, 341–354.

Kontopoulos, E., Parvin, J. D., and Feany, M. B. (2006). Alpha-synuclein acts in the nucleus to inhibit histone acetylation and promote neurotoxicity. *Hum Mol Genet* **15**, 3012–3023.

Kuwahara, T., Koyama, A., Gengyo-Ando, K., Masuda, M., Kowa, H., Tsunoda, M., Mitani, S., and Iwatsubo, T. (2006). Familial Parkinson mutant alpha-synuclein causes dopamine neuron dysfunction in transgenic *Caenorhabditis elegans*. *J Biol Chem* **281**, 334–340.

Lai, S. L., and Lee, T. (2006). Genetic mosaic with dual binary transcriptional systems in *Drosophila*. *Nat Neurosci* **9**, 703–709.

Langston, J. W., Ballard, P., Tetrud, J. W., and Irwin, I. (1983). Chronic parkinsonism in humans due to a product of meperidine-analog synthesis. *Science* **219**, 979–980.

Lee, T., and Luo, L. (2001). Mosaic analysis with a repressible cell marker (MARCM) for *Drosophila* neural development. *Trends Neurosci* **24**, 251–254.

Liu, Q. A., and Hengartner, M. O. (1999). The molecular mechanism of programmed cell death in *C. elegans*. *Ann N Y Acad Sci* **887**, 92–104.

Liu, Z., Wang, X., Yu, Y., Li, X., Wang, T., Jiang, H., Ren, Q., Jiao, Y., Sawa, A., Moran, T. et al. (2008). A *Drosophila* model for LRRK2-linked parkinsonism. *Proc Natl Acad Sci USA* **105**, 2693–2698.

Margulies, C., Tully, T., and Dubnau, J. (2005). Deconstructing memory in *Drosophila*. *Curr Biol* **15**, R700–R713.

Marín, I. (2006). The Parkinson disease gene LRRK2: Evolutionary and structural insights. *Mol Biol Evol* **23**, 2423–2433.

Mello, C., and Fire, A. (1995). DNA transformation. *Methods Cell Biol* **48**, 451–482.

Meulener, M., Whitworth, A. J., Armstrong-Gold, C. E., Rizzu, P., Heutink, P., Wes, P. D., Pallanck, L. J., and Bonini, N. M. (2005). *Drosophila* DJ-1 mutants are selectively sensitive to environmental toxins associated with Parkinson's disease. *Curr Biol* **15**, 1572–1577.

Min, K. T., and Benzer, S. (1999). Preventing neurodegeneration in the *Drosophila* mutant bubblegum. *Science* **284**, 1985–1988.

Minamiyama, M., Katsuno, M., Adachi, H., Waza, M., Sang, C., Kobayashi, Y., Tanaka, F., Doyu, M., Inukai, A., and Sobue, G. (2004). Sodium butyrate ameliorates phenotypic expression in a transgenic mouse model of spinal and bulbar muscular atrophy. *Hum Mol Genet* **13**, 1183–1192.

Monastirioti, M. (1999). Biogenic amine systems in the fruit fly *Drosophila melanogaster*. *Microsc Res Tech* **45**, 106–121.

Morgan, T. H. (1910). Sex limited inheritance in *Drosophila*. *Science* **32**, 120–122.

Muraro, N. I., and Moffat, K. G. (2006). Down-regulation of torp4a, encoding the *Drosophila* homologue of torsinA, results in increased neuronal degeneration. *J Neurobiol* **66**, 1338–1353.

Nass, R., and Blakely, R. D. (2003). The *Caenorhabditis elegans* dopaminergic system: Opportunities for insights into dopamine transport and neurodegeneration. *Annu Rev Pharmacol Toxicol* **43**, 521–544.

Nass, R., Hall, D. H., Miller, D. M., 3rd, and Blakely, R. D. (2002). Neurotoxin-induced degeneration of dopamine neurons in *Caenorhabditis elegans*. *Proc Natl Acad Sci USA* **99**, 3264–3269.

Nass, R., Hahn, M. K., Jessen, T., McDonald, P. W., Carvelli, L., and Blakely, R. D. (2005). A genetic screen in *Caenorhabditis elegans* for dopamine neuron insensitivity to 6-hydroxydopamine identifies dopamine transporter mutants impacting transporter biosynthesis and trafficking. *J Neurochem* **94**, 774–785.

Neckameyer, W. S. (1998). Dopamine and mushroom bodies in *Drosophila*: Experience-dependent and -independent aspects of sexual behavior. *Learn Mem* **5**, 157–165.

Park, J., Lee, S. B., Lee, S., Kim, Y., Song, S., Kim, S., Bae, E., Kim, J., Shong, M., Kim, J. M., and Chung, J. (2006). Mitochondrial dysfunction in *Drosophila* PINK1 mutants is complemented by parkin. *Nature* **441**, 1157–1161.

Pellikka, M., Tanentzapf, G., Pinto, M., Smith, C., McGlade, C. J., Ready, D. F., and Tepass, U. (2002). Crumbs, the *Drosophila* homologue of human CRB1/RP12, is essential for photoreceptor morphogenesis. *Nature* **416**, 143–149.

Pendleton, R. G., Rasheed, A., Sardina, T., Tully, T., and Hillman, R. (2002). Effects of tyrosine hydroxylase mutants on locomotor activity in *Drosophila*: A study in functional genomics. *Behav Genet* **32**, 89–94.

Priyadarshi, A., Khuder, S. A., Schaub, E. A., and Shrivastava, S. (2000). A meta-analysis of Parkinson's disease and exposure to pesticides. *Neurotoxicology* **21**, 435–440.

Pulak, R. (2006). Techniques for analysis, sorting, and dispensing of *C. elegans* on the COPAS flow-sorting system. *Methods Mol Biol* **351**, 275–286.

Richardson, H., and Kumar, S. (2002). Death to flies: *Drosophila* as a model system to study programmed cell death. *J Immunol Methods* **265**, 21–38.

Riddle, D. L., Blumenthal, T., Meyer, B. J., and Priess, J. R. (1998). In *C. elegans* 11 (D.L. Riddle ed.) pp 1–22. Cold Spring Harbor Laboratory Press, New York.

Rohde, C. B., Zeng, F., Gonzalez-Rubio, R., Angel, M., and Yanik, M. F. (2007). Microfluidic system for on-chip high-throughput whole-animal sorting and screening at subcellular resolution. *Proc Natl Acad Sci USA* **104**, 13891–13895.

Ryder, E., Ashburner, M., Bautista-Llacer, R., Drummond, J., Webster, J., Johnson, G., Morley, T., Chan, Y. S., Blows, F., Coulson, D. et al. (2007). The DrosDel deletion collection: A *Drosophila* genomewide chromosomal deficiency resource. *Genetics* **177**, 615–629.

Sachs, C., and Jonsson, G. (1975). Mechanisms of action of 6-hydroxydopamine. *Biochem Pharmacol* **24**, 1–8.

Sakaguchi-Nakashima, A., Meir, J. Y., Jin, Y., Matsumoto, K., and Hisamoto, N. (2007). LRK-1, a *C. elegans* PARK8-related kinase, regulates axonal-dendritic polarity of SV proteins. *Curr Biol* **17**, 592–598.

Sang, T. K., Chang, H. Y., Lawless, G. M., Ratnaparkhi, A., Mee, L., Ackerson, L. C., Maidment, N. T., Krantz, D. E., and Jackson, G. R. (2007). A *Drosophila* model of mutant human parkin-induced toxicity demonstrates selective loss of dopaminergic neurons and dependence on cellular dopamine. *J Neurosci* **27**, 981–992.

Schapira, A. H. (2008). Mitochondria in the aetiology and pathogenesis of Parkinson's disease. *Lancet neurol* **7**, 97–109.

Schmidt, E., Seifert, M., and Baumeister, R. (2007). *Caenorhabditis elegans* as a model system for Parkinson's disease. *Neurodegener Dis* **4**, 199–217.

Steffan, J. S., Bodai, L., Pallos, J., Poelman, M., McCampbell, A., Apostol, B. L., Kazantsev, A., Schmidt, E., Zhu, Y. Z., Greenwald, M. et al. (2001). Histone deacetylase inhibitors arrest polyglutamine-dependent neurodegeneration in *Drosophila*. *Nature* **413**, 739–743.

Stein, L. D., Bao, Z., Blasiar, D., Blumenthal, T., Brent, M. R., Chen, N., Chinwalla, A., Clarke, L.,

Clee, C., Coghlan, A. et al. (2003). The genome sequence of *Caenorhabditis briggsae*: A platform for comparative genomics. *PLoS Biol* **1**, E45.

Tempel, B. L., Livingstone, M. S., and Quinn, W. G. (1984). Mutations in the dopa decarboxylase gene affect learning in *Drosophila*. *Proc Natl Acad Sci USA* **81**, 3577–3581.

Truman, J. W. (1990). Metamorphosis of the central nervous system of *Drosophila*. *J Neurobiol* **21**, 1072–1084.

Ungerstedt, U. (1968). 6-Hydroxy-dopamine induced degeneration of central monoamine neurons. *Eur J Pharmacol* **5**, 107–110.

Vartiainen, S. Pehkonen, P., Lakso, M. Nass, R., and Wong, G. (2006). Identification of gene expression changes overexpressing human A-synuclein. *Neurobiol. Dis.* **22**, 477–486.

Ved, R., Saha, S., Westlund, B., Perier, C., Burnam, L., Sluder, A., Hoener, M., Rodrigues, C. M., Alfonso, A., Steer, C. et al. (2005). Similar patterns of mitochondrial vulnerability and rescue induced by genetic modification of alpha-synuclein, parkin, and DJ-1 in *Caenorhabditis elegans*. *J Biol Chem* **280**, 42655–42668.

Ved, R., Saha, S., Westlund, B., Perier, C., Burnam, L., Sluder, A., Hoener, M. C., Rodrigues, C. M., Alfonso, A.,

Steer, C. et al. (2005a). Similar patterns of mitochondrial vulnerability and rescue induced by genetic modification of alpha-synuclein, parkin, and DJ-1 in *Caenorhabditis elegans*. *J Biol Chem* **280**, 42655–42668.

Warrick, J. M., Chan, H. Y., Gray-Board, G. L., Chai, Y., Paulson, H. L., and Bonini, N. M. (1999b). Suppression of polyglutamine-mediated neurodegeneration in *Drosophila* by the molecular chaperone HSP70. *Nat Genet* **23**, 425–428.

Wood, W. B., (1988). In The nematode *C. elegans* (W.B. Wood ed.) pp 1–16. Cold Spring Harbor Laboratory Press, New York.

Wu, J. S., and Luo, L. (2006). A protocol for mosaic analysis with a repressible cell marker (MARCM) in *Drosophila*. *Nat Protoc* **1**, 2583–2589.

Yoshihara, M., Ensminger, A. W., and Littleton, J. T. (2001). Neurobiology and the *Drosophila* genome. *Funct Integr Genomics* **1**, 235–240.

Zhang, K., Guo, J. Z., Peng, Y., Xi, W., and Guo, A. (2007). Dopamine-mushroom body circuit regulates saliency-based decision-making in *Drosophila*. *Science* **316**, 1901–1904.

25

DROSOPHILA MODELS FOR PARKINSON'S DISEASE RESEARCH

NANCY M. BONINI

Department of Biology, Howard Hughes Medical Institute, University of Pennsylvania, Philadelphia, PA, USA

INTRODUCTION

The simple fruit fly, *Drosophila melanogaster*, has many advantages for providing new insight into human neurodegenerative disease mechanisms. The fly is among those systems, like *C. elegans*, that have many molecular genetic tools for manipulating genes, discovering modifiers and assessing gene activity. With a relatively rapid reproductive cycle – from egg to adult fly in 10 days – and short lifespan – about 80–90 days maximum lifespan, the fly has the additional advantages of a complex brain and nervous system for the study of neurological disease mechanisms. Such studies can be approached in the fly at many levels. For example, as the result of manipulating gene activity, anatomical and histological effects as well as functional effects in select behaviors can be examined in parallel.

Among the many approaches, one of the most powerful available in the fly is the ready ability to target the expression of a gene to particular cell types. This includes dopaminergic neurons, of special importance to Parkinson's disease, but also includes targeting expression to select anatomical structures, like the eye. Through single generation F1 screens, one can find dominant modifiers of a phenotype. In addition, one has the advantage to not only select in a single generation dominant modifiers, but one can also perform mosaic analysis selecting for homozygous mutant tissue in the background of a heterozygous animal. This has the advantage of allowing the isolation of modifiers that are organismal lethal mutations. The general strengths of *Drosophila* for genomic and molecular genetic approaches in a variety of contexts are covered in various reviews (Rubin *et al.*, 2000; Driscoll and Gerstbrein, 2003; Bier, 2005; Echeverri *et al.*, 2006).

A key to the use of any genetic organism for human neurological or other disease mechanisms is the extent to which the system recapitulates fundamental aspects of the "gold standard," meaning the human disease situation. Genomic analysis of the fly has revealed remarkable conservation of genes and gene pathways, indicating that the fundamental framework of the human genome is reflected in the fly. This has proven particularly powerful for many developmental mechanisms. For the special situation of modeling a human disease situation in the fly, it is of paramount importance, as this indicates that genes that influence the process in the fly are highly likely to then also function in the human process. Note that the use of simple organisms as well allows focus on those mechanisms that may be most fundamental to the disease process, and, although the manifestation in the fly might not be identical to the human, the fly can still reveal basic insight. In this regard, the fly has proven its worth as a powerful complementary tool for the study of human disease mechanisms. This review is intended to be illustrative of the types of approaches and findings that can be gleaned from studies of parkinsonism genes and mechanisms in the fly. For other reviews of Parkinson's gene activities in the fly, see (Muqit and

Feany, 2002; Bilen and Bonini, 2005; Whitworth et al., 2006).

MODELS OF PARKINSON'S DISEASE IN THE FLY

α-Synuclein

Abnormal accumulation of α-synuclein is associated with sporadic Parkinson's disease, and dominant point mutations in α-synuclein as well as increases in gene copy number cause familial Parkinson's disease (reviewed in Farrer, 2006; Hardy et al., 2006). These data indicate that increased expression of α-synuclein causes parkinsonism, and provide the foundation for models of α-synuclein toxicity in lower organisms with the directed expression of the human gene.

The initial fly model for Parkinson's disease was generated with expression of human wild-type, A30P or A53T mutant α-synuclein in the fly brain generally, or selectively to dopaminergic neurons (Feany and Bender, 2000). Expression of any of these forms of the protein leads to age-dependent loss of dopaminergic neuron integrity: flies are born with the normal complement of tyrosine hydroxylase (TH) positive neurons, which drops by ~50% in select clusters by 30d (Feany and Bender, 2000; Auluck et al., 2002).

These flies also show other hallmarks of α-synuclein toxicity and parkinsonism. They display abnormal accumulation of α-synuclein into structures that by electron microscopy resemble Lewy bodies, with a homogenous core surrounded by a halo of filaments (Feany and Bender, 2000). The accumulations also immunostain for ubiquitin and antibodies that detect aggregated forms of α-synuclein in Lewy bodies of human tissue (Auluck et al., 2002). Most strikingly, the flies display a loss of normal locomotor climbing activity with age, with A30P showing more severe effects (Feany and Bender, 2000). Although data suggest that the dopaminergic neurons may be compromised for TH levels rather than lost entirely (Auluck et al., 2005), expression of α-synuclein appears to have a deleterious effect to dopaminergic neurons in the fly brain – reduction in TH immunostaining or loss of the cells, effects on locomotor activity – causing features that strikingly resemble Parkinson's disease. The loss of TH immunostaining

is blocked by anti-cell death genes, suggesting it reflects loss or compromise of the cells (Periquet et al., 2007). The effects of α-synuclein in the fly appear specific to dopaminergic neurons within the brain, although eye phenotypes have been reported (Feany and Bender, 2000).

Microarray studies on heads of pre-symptomatic α-synuclein flies reveal that select transcripts are affected, enriched in those that affect lipid processing, membrane transport, and, interesting in light of the role of other parkinsonism genes (see below) mitochondrial function (Scherzer et al., 2003). Proteomic studies confirm just under half of the transcriptional changes, and indicate early changes in actin cytoskeletal and mitochondrial components (Xun et al., 2007).

Using this basic model, a number of studies have revealed modifiers of the toxicity of α-synuclein to dopaminergic neurons. Abnormal folding or misfolding of the protein is likely involved, based on the abnormal accumulation of the protein into Lewy-body like structures. Consistent with this, upregulation of the molecular chaperone Hsp70 has been shown to mitigate toxicity of α-synuclein to dopaminergic neurons (Auluck et al., 2002). Moreover, compromise of molecular chaperone activity accelerates α-synuclein toxicity, indicating that the toxicity of the protein to dopaminergic neuron integrity is highly dependent on molecular chaperone activity (Auluck et al., 2002).

These findings are an example of how fly findings can be extended to the clinical human situation (Figure 25.1). As in the fly, abnormal accumulations of α-synuclein immunostain for chaperones in human synucleinopathies, including Parkinson's disease (Auluck et al., 2002), suggest that the activity of molecular chaperones may modulate human disease like they modulate α-synuclein toxicity in the fly. Strikingly, polymorphisms in one of the human Hsp70 genes which compromises the stress-induced activation of the gene with heat shock, has been linked to Parkinson's disease (Wu et al., 2004). This finding illustrates the power of studies in the fly, providing a striking example by which a finding in a fly model for the disease is extended, with critical therapeutic implications, to the human situation.

In the fly, such studies have also been extended to pharmacological protection from α-synuclein toxicity (Auluck and Bonini, 2002). An advantage of Drosophila is the relatively modest blood–brain barrier. This means that drugs or compounds can be tested for effects in flies to determine whether, in

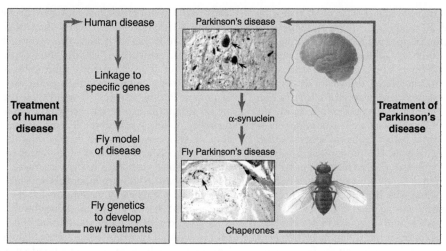

FIGURE 25.1   The parallel approach of the fly in defining new mechanistic insight and the foundation for new therapeutics, for human Parkinson's disease. The figure highlights how the fly was used to develop a Parkinson's disease model (Feany and Bender, 2000), and then the model used to define a role of molecular chaperones in modulating of disease phenotype in fly, then the findings extended to human synucleinopathies including Parkinson's disease (Auluck et al., 2002). From Helfand (2002). Reprinted with permission from AAAS.

principle, they might work effectively against a deleterious neurological condition in an animal *in vivo* with an intact nervous system. Thus, by simply feeding flies a compound, one can determine whether there is a modulatory effect on the neurological condition. In such a manner, feeding flies geldanamycin has been shown to protect against α-synuclein toxicity to dopaminergic neurons. Geldanamycin is an inhibitor of Hsp90 activity (Whitesell *et al.*, 1994), which is involved in the regulation of many proteins, including heat shock factor (Zou *et al.*, 1998). Target validation studies show that the effect of geldanamycin is mediated through its ability to activate heat shock factor, rather than through other target proteins in the fly (Auluck *et al.*, 2005).

Phosphorylation of α-synuclein on S129 has been implicated in human disease, as α-synuclein isolated from Lewy bodies in Parkinson's disease and other synucleinopathies is phosphorylated at this site (Fujiwara *et al.*, 2002). α-Synuclein is also phosphorylated at this site in the fly brain with age (Takahashi *et al.*, 2003). Studies in the fly have extended these findings to reveal a functional role of S129 phosphorylation in accumulation and toxicity of α-synuclein (Chen and Feany, 2005). Mutating S129 to an alanine prevents phosphorylation at this site, and also fully prevents toxicity of α-synuclein

to dopaminergic neurons. The complementary mutation to an aspartate, which mimics constitutive phosphorylation of S129, enhances the toxicity of α-synuclein, such that loss of dopaminergic neuron integrity proceeds more rapidly; enhanced effects on the eye are also reported (Chen and Feany, 2005). These studies further implicate G protein-coupled receptor kinase 2 as a kinase that can, *in vivo*, enhance toxicity of wild-type α-synuclein through S129 phosphorylation.

Interestingly, studies on inclusion formation of α-synuclein in this situation suggested the possibility that accumulations of α-synuclein may be protective (Chen and Feany, 2005). Accumulation of α-synuclein to protease K-resistant forms was analyzed, revealing an inverse relationship between protease K sensitivity and S129 phosphorylation: the non-toxic S129A form readily accumulates into protease K-resistant inclusions, whereas the toxic forms of α-synuclein (S129D or wild-type α-synuclein with upregulated kinase activity) fail to accumulate to protease K-resistant forms. Parallel studies on other pathogenic disease proteins suggest that gross inclusions may be protective or innocuous rather than toxic, and rather, that it may be smaller, oligomeric forms that are the toxic species (Caughey and Lansbury, 2003; Arrasate *et al.*, 2004). C-terminal truncation of α-synuclein (amino acid 1–120 only, removing the C-terminal 20 amino acids), enhances toxicity and accumulation of protease K-resistant forms, but also enhances levels of soluble oligomeric species of the protein (Periquet *et al.*, 2007). Further studies with the phosphorylated forms are important, in order to assess whether enhanced toxicity in that situation – where there are opposite effects on the large inclusions that are protease K resistant – show a similar correlation with oligomeric species. Other studies directing α-synuclein protein selectively to the nucleus or cytoplasm further raise the possibility that α-synuclein may have, at least in part, a nuclear toxic role interacting with histones to affect acetylation levels (Kontopoulos *et al.*, 2006).

Additional insight into α-synuclein toxicity has been found from studies in yeast that have then been extended to other models including the fly. In yeast, human α-synuclein causes toxicity due to a block in trafficking from the endoplasmic reticulum (ER) to the Golgi (Cooper *et al.*, 2006). Upregulation of Rab1 ortholog Ypt1, a GTPase that functions at this step, can suppress the α-synuclein-dependent trafficking defect. As yeast cells are very different from dopaminergic neurons, it was perhaps surprising that these results were confirmed in the fly and other genetic systems. In *Drosophila*, expression of Rab1 was shown to indeed mitigate α-synuclein toxicity to dopaminergic neurons in the brain *in vivo* (Cooper *et al.*, 2006). A finding like this in yeast that extends to *C. elegans*, *Drosophila* and mammalian cells clearly shows the benefit of modeling Parkinson's disease in multiple systems. The striking effect on trafficking in yeast may translate to effects in neurons on synaptic function, components of which are highly dependent upon proper ER–Golgi transport.

These varied and distinct components of the toxicity as revealed in flies – chaperone modulation, oligomers versus inclusions, nuclear versus cytoplasmic localization, role of phosphorylation, trafficking and synaptic function – will require integration to obtain a fuller picture of what comprises α-synuclein toxicity. Further, studies of interactions between α-synuclein and other Parkinson's disease genes are also underway in flies (see below), and should reveal how α-synuclein biology fits into the more complete picture of Parkinson's disease.

Parkin

The human *PARK2* gene is a recessive parkinsonism gene encoding Parkin, whose loss of function leads to disease in autosomal recessive juvenile parkinsonism (AR-JP). This indicates that, in contrast to the above model with overexpression of human α-synuclein, generation of fly models requires knock-out of the orthologous gene (although see below for some dominant effects upon expression of mutant Parkin proteins). The protein, Parkin, is an E3 ubiquitin ligase that probably functions in turnover of select substrates (Farrer, 2006; Hardy *et al.*, 2006); one idea would be that loss of Parkin leads to excess accumulation of its target substrate proteins, which then leads to degeneration.

In flies, *parkin* knock-out animals are viable, have a shortened lifespan and locomotor defects (Greene *et al.*, 2003). However, the most insightful analysis came from detailed studies of additional phenotypes of infertility and wing posture. These shared problems immediately point to a mitochondrial issue: in the fly, mitochondrial fusion is involved in sperm maturation, and loss of mitochondrial function leads to loss of integrity of the flight muscles, resulting in problems with wing posture. Indeed detailed analysis revealed dramatic loss of mitochondrial structural integrity in *parkin* fly mutants. The effects are quite severe in flight muscle tissue, with fragmentation of the mitochondria and gross loss of the normally highly regular cristae intermembranous structure.

Independently generated *parkin* mutants confirm the initial findings, and further reveal sensitivity of the animals to a variety of stressors, including cold shock and the oxidative toxin paraquat (Pesah *et al.*, 2004) which has been implicated in parkinsonism (Dauer and Przedborski, 2003). Effects on dopaminergic neuronal morphology or integrity have been detected, as well as effects on head dopamine levels, although not all models show a similarly severe effect (Greene *et al.*, 2003; Cha *et al.*, 2005; Whitworth *et al.*, 2005). Additional results suggest that c-Jun N-terminal kinase (JNK) may become inappropriately activated in *parkin* mutant dopaminergic neurons, leading to their dysfunction and/or loss (Cha *et al.*, 2005).

Although mitochondrial dysfunction has long been implicated in Parkinson's disease, it was perhaps unexpected to find that loss of Parkin in a simpler system like *Drosophila* leads to widespread effects on mitochondrial integrity, rather than a phenotype that is limited to dopaminergic neurons. One interpretation is that the fly reflects a fundamental/non-redundant function of the gene, which in the context of the more complex human organism, translates to a specific effect of parkinsonism.

Further studies on *parkin* mutants for analysis of gene expression changes and pathways affected have highlighted that upregulation of oxidative stress components and genes of innate immunity occurs in the mutants (Greene *et al.*, 2005). The upregulation of oxidative stress genes presumably reflects the effect of the animal to cope with severe disruption of mitochondrial function. Effects on innate immunity genes may either indicate that *parkin* has a role in immunity; alternatively, it is conceivable that the morphological disruption of the tissue upon loss of mitochondrial integrity, as with the flight muscle, may be inducing an immunity response, or that an immunity response is coupled with loss of mitochondrial integrity to oxidative stress.

These studies also identified *glutathione-S-transferase* as an upregulated gene, suggesting a response of the animal to help deal with detoxifying oxidatively damaged proteins. Upregulation and loss of *glutathione-S-transferase* activity modulate the *parkin* dopaminergic phenotype (Greene *et al.*, 2005), indicating that effects of loss of *parkin* are critically sensitive to the glutathione detoxification pathway. These effects of the glutathione pathway may be revealing special properties about *parkin* function, especially given genetic interactions, but it is also possible that upregulation of this pathway may function in parallel to generally promote the health of the animal. Overall, one wonders where *parkin* fits into energy and metabolic pathways; does *parkin* gene activity affect a global energetic process, or is there a very specific pathway or set of proteins being affected with mitochondrial dysfunction being a symptom but not a cause.

These studies have highlighted the loss-of-function role of *parkin* as reflected in familial AR-JP. However, additional studies in the fly indicate that expression of mutant Parkin proteins with point mutations or truncations, reflecting familial disease situations where only a single copy of the gene appears mutated, can exert a dominant toxic effect on age-dependent integrity and function of dopaminergic neurons (Sang *et al.*, 2007). The effect of such mutant forms of Parkin is specific to dopaminergic neurons, in having strong effects on the excitation of dopaminergic neurons and locomotor activity of the fly – as revealed by analysis using a tiny fly rotarod apparatus – and morphology of dopaminergic neurons. With similar expression levels, serotonergic neurons as well as retinal neurons are unaffected.

This model for Parkin toxicity and loss of dopaminergic neuron integrity associated with a parkinsonism situation, was further used to address the potential role of the neurotransmitter dopamine in loss of integrity of the neurons (Sang *et al.*, 2007). Numerous data suggest the possibility that dopamine itself, due to the oxidative byproducts of its biosynthetic pathways, may contribute to the demise of dopaminergic neurons in Parkinson's disease. By manipulating levels of the vesicular monoamine transporter (*Drosophila* VMAT, or DVMAT), which transports dopamine into synaptic vesicles, dopamine levels were raised or lowered, and effects on Parkin toxicity assessed.

These studies showed that increasing dopamine levels by DVMAT knock-down accelerated Parkin degenerative effects, whereas decreasing dopamine levels by DVMAT upregulation mitigated Parkin effects. Manipulating DVMAT alone had no effect on dopaminergic cell number or locomotor activity. This finding may reflect a general principle of the role of dopamine in degeneration of dopaminergic neurons in the face of any toxic insult and/or those that induce parkinsonism and Parkinson's disease. It will be of interest to determine whether manipulations of DVMAT affect the toxicity of other parkinsonism situations in the fly. Another possibility these studies raise, however, is that, given that dopamine covalently modulates and inactivates normal Parkin protein (LaVoie *et al.*, 2005), dopamine may enhance deleterious consequences of misfolded and toxic conformations of such mutant forms of Parkin, and by this means enhance toxicity of the protein.

Pink1

The human *PARK6* gene encodes PTEN-induced kinase 1, or Pink1, with recessive loss-of-function mutation of *Pink1*, like with the human *parkin* gene *PARK2*, causing parkinsonism (Farrer, 2006; Hardy *et al.*, 2006). Although both genes cause parkinsonism in humans, studies of this gene in the fly revealed an unexpected and dramatic similarity in effects of their mutation: loss of *Pink1* in the fly looks exactly like loss of *parkin* (Clark *et al.*, 2006; Park *et al.*, 2006; Yang *et al.*, 2006). Indeed, every aspect of the phenotype is identical: *Pink1* mutants are viable with a shortened lifespan, infertile and have a wing postural defect. Detailed anatomical analysis confirmed an identical loss of mitochondrial integrity. The mutant animals also show sensitivity to stresses like paraquat and rotenone, but also to more general stress situations such as salt osmotic stress and protein misfolding induced by dietary dithiothreitol. Decreases in dopaminergic neuron number and abnormal mitochondrial morphology in dopaminergic neurons of *Pink1* mutants have also been detected (Park *et al.*, 2006; Yang *et al.*, 2006).

The fact that loss of *Pink1* so strikingly resembles loss of *parkin* indicated that the two genes may be in the same genetic pathway. In flies, this can be readily tested. Therefore, it was determined whether upregulation of *parkin* gene activity can rescue *Pink1* mutants, as well as the reciprocal situation – whether upregulation of *Pink1* gene activity can rescue *parkin* mutants. These experiments revealed that upregulation of *parkin* rescues *Pink1* mutants, but not the reverse; *Pink1* upregulation cannot rescue *parkin* mutation (Clark *et al.*, 2006; Park *et al.*, 2006; Yang *et al.*, 2006). Moreover, double mutants of flies lacking both *parkin* and *Pink1* are not worse than either mutant alone. These findings suggest a model whereby Parkin functions downstream of Pink1, with an ultimate effect on mitochondrial integrity.

Several observations are worth noting. Although mutants in both Pink1 and Parkin protein affect mitochondrial integrity and function, Pink1 localizes largely to mitochondria (Clark *et al.*, 2006), although Parkin largely does not (although some Parkin protein does [Zhang *et al.*, 2005]). So a "simple" model whereby Pink1 and Parkin both function within mitochondria may not be so straightforward, although Pink1 and Parkin may both modulate the same substrate or set of substrates which influence mitochondrial function and integrity. Alternatively, the genes may affect another process, the symptoms of which are loss of mitochondrial integrity, but whose primary target(s) is other energy or protective pathways.

The studies by Park *et al.* (2006) indicate that the *Pink1* mutant phenotype is not only mitigated by *parkin*, but also by upregulation of a mitochondrial BCL2 anti-cell death ortholog protein Buffy. This finding indicates that one does not have to reverse the effects of loss of Pink1 (or perhaps Parkin); one needs to only keep the mitochondria from undergoing structural disintegration to protect against the mutant phenotype. Underscoring this is the finding that upregulation of superoxide dismutase to protect against oxidative damage can protect against *Pink1* mutation in flies (Wang *et al.*, 2006) – this finding is fundamentally similar to those on *parkin* mutants with

the glutathione pathway as noted above. Indeed, for "therapeutic" rescue, one does not have to reverse the initial insult (in this case, providing back the missing gene), one may only have to provide the ability to maintain mitochondrial integrity or promote health/protection of the cells against toxic insults. This has significant therapeutic implications. These findings also highlight the critical importance of testing for interactions among the different Parkinson's disease familial genes in such organisms like the fly where this is a simple task, and highlight the mitochondria as an organelle critical in the disease.

DJ-1

Like mutations in *Pink1* and *parkin*, mutations in *DJ-1* (the *PARK7* gene) are associated with recessive parkinsonism (Farrer, 2006; Hardy *et al.*, 2006). Although humans have a single *DJ-1* gene, flies have two, *DJ-1a* (or *DJ-1alpha*) and *DJ-1b* (or *DJ-1beta*). The *DJ-1a* gene is expressed in testes, whereas *DJ-1b* is ubiquitous, including in the brain (Menzies *et al.*, 2005; Meulener *et al.*, 2005; Park *et al.*, 2005). A number of loss-of-function mutants for DJ-1 protein in the fly have been generated. Loss of both genes or loss of *DJ-1b* alone is viable, but causes selective sensitivity to oxidative toxins paraquat and rotenone which are associated with parkinsonism in humans (Menzies *et al.*, 2005; Meulener *et al.*, 2005; Park *et al.*, 2005). In one study an effect on locomotor activity was noted that becomes worse upon paraquat exposure (Park *et al.*, 2005). Although some studies show no effect on dopaminergic neuron integrity (Meulener *et al.*, 2005; Park *et al.*, 2005), another study suggests increased survival of dopaminergic neurons in *DJ-1b* mutants with prolonged age, potentially due to an increase in *DJ-1a* brain expression (Menzies *et al.*, 2005). One study shows a striking rough eye phenotype with RNAi knock-down of *DJ-1a*, a loss of dopaminergic neurons and interactions with Akt signaling pathways (Yang *et al.*, 2005), although eye and that type of dopaminergic neuron phenotype are not seen in genetic deficiency mutants. Dopaminergic neuron effects can be modest, relatively hard to detect and may be sensitive to genetic background; however, that an RNAi line shows an eye phenotype not reflected in genetic deficiencies is of concern as it may be due to off-target or non-specific effects of RNAi (see below).

An interesting feature of the DJ-1 protein is that it undergoes oxidative modification on cysteine residues, which may play a role in regulating its activity (Wilson *et al.*, 2003; Kinumi *et al.*, 2004; Shendelman *et al.*, 2004). In flies, DJ-1b protein is also seen to undergo oxidative modification (Meulener *et al.*, 2005). Study of this modification in flies has revealed that oxidation occurs on C104 (analogous to C106 in humans), and occurs with both exposure of animals to paraquat and normally with aging (Meulener *et al.*, 2006) – an interesting observation since age is the greatest risk factor for neurodegenerative diseases including Parkinson's disease. Meulener *et al.* (2006) found that C104 mutations that either prevent or mimic oxidative modification yield a functionally inactive protein. The finding that mutant forms prevent oxidative modification indicates that this amino acid is critical for DJ-1b function; that mutations that mimic oxidative modification also inactive the protein suggests the possibility that over-oxidation of the protein functionally inactivates the protein. Moreover, DJ-1 oxidation with age is a feature not only of flies, but also of mouse and of human brain tissue. In flies, old flies that have ~50% of their DJ-1 oxidized are also strikingly sensitive to paraquat. In addition, with oxidative stress, a dramatically greater extent of DJ-1b becomes oxidized compared to the situation in young flies. Taken together, these studies raise the possibility that oxidation that occurs with normal aging and in response to oxidative agents may inactivate DJ-1 function (Meulener *et al.*, 2006). This inactivation of DJ-1 function could then potentially contribute to sporadic Parkinson's disease. Additional testing of this model requires the identification of additional modifiers of DJ-1, as well as greater knowledge of the protein's function and how it is affected by oxidative modification of the protein.

Toxin Models for Environmentally Triggered Parkinsonism

Epidemiological and mammalian toxin studies suggest links between toxins and parkinsonism (Dauer and Przedborski, 2003). Thus, beyond genetic mutation or over-expression models, a complete repertoire of Parkinson's disease models in a genetic system like the fly should take into account environmental effects. Relevant toxins are being incorporated into genetic models for familial genes, noted with paraquat and other drug exposure above. Nevertheless, models in the fly that can show effects of toxins on their own on dopaminergic neuron integrity and/or dopaminergic neuron function are

important to develop. As with any other phenotype in the fly, sensitivity to environmental toxins can itself be amenable to genetic analysis (e.g., modifier screens of rotenone sensitivity, not in a Parkinson's disease gene background). In this light, models have been developed using either rotenone or paraquat.

Exposure of *Drosophila* to rotenone causes striking loss of locomotor activity and dopaminergic neurons (Coulom and Birman, 2004). This appears to be a specific effect, despite the systemic exposure of the flies being fed rotenone, as serotonergic neurons are not affected. Feeding flies l-DOPA – to increase dopamine levels – protects against locomotor deficits, suggesting the effects of rotenone on climbing can be counteracted by raising dopamine levels. No effect of l-DOPA on dopaminergic neuron number is seen, indicating l-DOPA treatment does not protect against rotenone effects on the cells, but it is possible that raising dopamine promotes function of the surviving cells. In addition, co-exposure of flies to rotenone with melatonin, an antioxidant, protects against rotenone effects at both the locomotor and dopaminergic neuronal number level (Coulom and Birman, 2004). One general complication in such studies is the extent to which co-exposure of such counteracting compounds may react outside of the cells (for instance, inactivating the toxic drug, or inactivating its effects in the digestive system such that less drug reaches the brain), rather than acting at the level of the cells in the brain.

Paraquat exposure also targets the dopaminergic system: feeding flies with this drug results in morphological effects on dopaminergic neurons in a manner mitigated by feeding flies with dopamine or l-DOPA (Chaudhuri *et al.*, 2007). In a counter-intuitive manner, genetic mutations that lower dopamine pathway function reduce toxicity of paraquat, whereas genetic situations that raise dopamine pathway function enhance toxicity. This finding is reminiscent of the effects of expression of toxic Parkin mutant forms with manipulation of dopaminergic pathway function (above; [Sang *et al.*, 2007]), and endorses the idea that dopamine pathways may contribute to the toxicity of genes and toxins to this class (dopaminergic) of neurons.

PERSPECTIVES

These studies provide an overview, illustrating the extent of studies in *Drosophila* of models for Parkinson's disease using both familial gene forms and mutations, as well as integrating these with the environmental component that is so critical to the human situation. These studies have provided insight in the human disease, and established the foundation for new therapeutic approaches (see Figure 25.1). Future directions are many and varied, although there are also some areas that require resolution or focus.

The Precision of the Models for Providing Insight into Genes and the Integration of Genes and Toxins

One overall issue that underlies these studies concerns the models. Despite the fact that generally phenotypes of the same model generated or studies by different laboratories gives similar results, there are differences or inabilities to see select phenotypes in different situations (noted above, as well as [Shendelman et al., 2004]). Issues that may affect the penetrance of an effect include genetic background; some of the assays being used are likely very sensitive to this. Many behaviors, which include sensitivity to toxins and locomotor activity, are notoriously sensitive to genetic background and subtle environmental conditions (time of day, distractions in the surroundings, room temperature, humidity, among others). The dopaminergic neuron assays used in studies of Parkinson's disease genes and conditions can be difficult to reproduce in different situations. Isogenizing lines to obtain a uniform genetic background for studies, such that the same result with the identical reagents can be rigorous reproduced, would be helpful for the field.

A more and more common approach that will be incorporated into these models is siRNA approaches to knock-down gene function. This proves to be a powerful and fruitful approach for generating models, as well as for defining modifier genes. However, there are overall issues that are emerging in a number of contexts regarding differences between genetic mutations and gene knock-down by siRNA. Off-target effects of siRNAs are now well-established and the criteria to accept a result by siRNA as truly reflecting the disruption of the targeted gene need to become and have become more rigorous (Echeverri et al., 2006). In this manner, the field can assess more rigorously the extent to which different results truly reflect the gene activity, as one can imagine that partial knock-down versus null activity of a gene may yield different results that could be

important and interesting to define, if real. Indeed, whereas a precise as possible reflection of the human disease situation in the fly is important, key examples where the phenotype in the fly has provided astonishing insight regarding fundamental function of parkinsonism genes – parkin and Pink1 – also exist. Thus, it is important to assess both extents of the fly strengths – how precise is the model, but also what do the properties of the model indicate about fundamental function of the gene(s) derived from those features of the fly model that do not necessarily fit an exact replica of the human disease.

Beyond these issues is the overall point that we are only just beginning to glean insight from the fly in our studies of Parkinson's disease phenotypes, both of genes and toxins. The parkin and Pink1 studies highlight the importance of mitochondrial function to Parkinson's disease gene activity. To what extent are mitochondria hit in the other parkinsonism situations? There are data that implicate mitochondrial function in the fly (noted above), and DJ-1 protein may localize to mitochondria in flies (Park et al., 2005); genetic pathway interactions between DJ-1, parkin and Pink1 may reveal insight. Although DJ-1 mutants show a specificity to oxidative toxins that is not seen in parkin and Pink1, the partial overlap of phenotypes may reflect partial overlap of function. Questions of how the genes identified in familial parkinsonism interact with each other and the environment are only beginning to be addressed.

What are the substrates of these genes? Effects of α-synuclein on dopaminergic neuron integrity can be protected against by upregulation of Parkin (Haywood and Staveley, 2006), suggesting an involvement of Parkin in α-synuclein toxicity. Will upregulation of Pink1 also protect? Will reduction of function of parkin, Pink1 or DJ-1 enhance? Overall, an advantage of the fly will be to push potential genetic interactions between the different genes in a manner not possible in mammalian systems, in order to provide insight for focused testing of interactions in higher systems. Interestingly, upregulation of Hsp70, which protects against α-synuclein toxicity to dopaminergic neurons, also protects against paraquat toxicity (Bilen and Bonini, 2005). This illustrates that, despite the effects of a chaperone like Hsp70 on folding or accumulation of misfolded α-synuclein, Hsp70 may also promote general robustness of the animal in response to a variety of toxic situations and function in this manner as a protective agent.

Genes to Therapeutic Targets and Drugs

As highlighted by a number of examples, a great strength of the fly is the power of performing genetic screens for modifiers of the human disease phenotype. Such genes provide insight not only into basic mechanisms, but also provide potential therapeutic targets. Thus, *Drosophila* holds the promise of not only defining genes, but also drugs and compounds that affect degeneration (e.g., Auluck and Bonini, 2002; Marsh and Thompson, 2006). In particular, the fly allows identification of those compounds that may, in principle, work in the context of an intact organism with an intact brain. The lack of a significant blood–brain barrier means that compounds can be screened in the fly without the need to specifically accommodate the problem of exposure to neurons in the brain. This greatly streamlines such screens.

It is also possible to test and define the target or targets of compounds or drugs by combining the genetics together with therapeutic compounds (e.g., Auluck *et al.*, 2005). Typically such an approach requires much knowledge of the drug and defining potential targets through other types of studies as well, because defining targets typically requires consolidation of findings from many different lines of investigation. It is also clear that, if a drug or compound is deemed safe, target validation, although ultimately important for further understanding and development of that or other therapeutic agents, is not necessarily critical for practical application. There are a large number of well-used compounds for which we are still defining targets through multiple types of approaches (e.g., lithium [Phiel and Klein, 2001; Quiroz *et al.*, 2004]).

In addition, however, there is always the issue of whether a compound defined in a lower organism like the fly will translate to mammals, including humans. Despite some remarkable success (e.g., histone deacetylase inhibitors for polyglutamine disease situations, [Butler and Bates, 2006]), this clearly always remains a question. To date, most approaches have used general compounds for which research has established some conservation of mechanism. But how powerful would such a lower organism be for screening new or more specific compounds? Another underlying idea that may hold is that for a compound to work effectively in humans, with their vastly more complex biology and issues of access to the brain, the compound should work fantastically well in a lower system

like the fly (e.g., Hsp70, as one example). The idea of combinatorial treatment, typically for many types of disease treatment, may prove highly effective and can be tested in principle in models like the fly (Agrawal *et al.*, 2005).

A final note concerns the nature of screens to define the foundation for therapeutic targets. Typically, the goals of defining gene mechanism in disease and defining gene targets for therapeutics can be at odds. That is, when using a disease model to define modifiers – which can provide insight into both biology and therapeutics – those modifiers that can be more readily understood often have effects on their own in the absence of the disease situation. The reason is that if they have effects on their own, they can be studied for their function and biological activity to define what their role is. However, genes or targets that when manipulated have effects are not ideal therapeutic targets. Rather, an ideal target would be manipulated with no side effects or other effects. Along this same line, for defining genes that may serve best as therapeutic targets, it is ideal if the gene modifies in a loss-of-function manner, as it is easier to knock a gene activity out than to enhance it. This does not mean that defining genes that function upon upregulation are not useful; often there are loss-of-function approaches to lead to upregulation of that target gene, although at least one step removed. In the end, the ideal targets that may serve the most use therapeutically would be loss-of-function targets that have no effect other than effective treatment of the disease situation when their activity is knocked down.

CLOSING COMMENTS

Finally, these studies highlight the still-to-be-reached potential of *Drosophila* in its contribution to the problems of human disease. Studies like these have dramatically stimulated the use of the fly (as well as other simple genetic systems) for studying neurodegenerative and other disease situations (Driscoll and Gerstbrein, 2003; Bier, 2005). Despite this, such studies are arguably still largely in the beginning phases, with new models of various aspects of human disease situations, and new types of gene modifiers including microRNAs, continually being developed to further the extraordinary use of the fly as a key to revealing new insight and therapeutics for disease (e.g., Pagliarini and

Xu, 2003; Bilen *et al.*, 2006; Koh *et al.*, 2006; Jung and Bonini, 2007).

ACKNOWLEDGEMENTS

Thanks to Derek Lessing and Nan Liu for critical reading; NMB receives funding from the David and Lucile Packard Foundation, the NIA and the NINDS. NMB is an Investigator of the Howard Hughes Medical Institute.

REFERENCES

Agrawal, N., Pallos, J., Slepko, N., Apostol, B. L., Bodai, L., Chang, L. W., Chiang, A. S., Thompson, L. M., and Marsh, J. L. (2005). Identification of combinatorial drug regimens for treatment of Huntington's disease using *Drosophila*. *Proc Natl Acad Sci USA*. **102**, 3777–3781.

Arrasate, M., Mitra, S., Schweitzer, E. S., Segal, M. R., and Finkbeiner, S. (2004). Inclusion body formation reduces levels of mutant Huntingtin and the risk of neuronal death. *Nature* **431**, 805–810.

Auluck, P. K., and Bonini, N. M. (2002). Pharmacological prevention of Parkinson disease in *Drosophila*. *Nat Med* **8**, 1185–1186.

Auluck, P. K., Chan, H. Y., Trojanowski, J. Q., Lee, V. M., and Bonini, N. M. (2002). Chaperone suppression of alpha-synuclein toxicity in a *Drosophila* model for Parkinson's disease. *Science* **295**, 865–868.

Auluck, P. K., Meulener, M. C., and Bonini, N. M. (2005). Mechanisms of suppression of {alpha}-synuclein neurotoxicity by geldanamycin in *Drosophila*. *J Biol Chem* **280**, 2873–2878.

Bier, E. (2005). *Drosophila*, the golden bug, emerges as a tool for human genetics. *Nat Rev Genet* **6**, 9–23.

Bilen, J., and Bonini, N. M. (2005). *Drosophila* as a model for human neurodegenerative disease. *Annu Rev Genet* **39**, 153–171.

Bilen, J., Liu, N., Burnett, B. G., Pittman, R. N., and Bonini, N. M. (2006). MicroRNA pathways modulate polyglutamine-induced neurodegeneration. *Mol Cell* **24**, 157–163.

Butler, R., and Bates, G. P. (2006). Histone deacetylase inhibitors as therapeutics for polyglutamine disorders. *Nat Rev Neurosci* **7**, 784–796.

Caughey, B., and Lansbury, P. T. (2003). Protofibrils, pores, fibrils, and neurodegeneration: Separating the responsible protein aggregates from the innocent bystanders. *Annu Rev Neurosci* **26**, 267–298.

Cha, G. H., Kim, S., Park, J., Lee, E., Kim, M., Lee, S. B., Kim, J. M., Chung, J., and Cho, K. S. (2005). Parkin negatively regulates JNK pathway in the dopaminergic neurons of *Drosophila*. *Proc Natl Acad Sci USA* **102**, 10345–10350.

Chaudhuri, A., Bowling, K., Funderburk, C., Lawal, H., Inamdar, A., Wang, Z., and O'Donnell, J. M. (2007). Interaction of genetic and environmental factors in a *Drosophila* parkinsonism model. *J Neurosci* **27**, 2457–2467.

Chen, L., and Feany, M. B. (2005). Alpha-synuclein phosphorylation controls neurotoxicity and inclusion formation in a *Drosophila* model of Parkinson disease. *Nat Neurosci* **8**, 657–663.

Clark, I. E., Dodson, M. W., Jiang, C., Cao, J. H., Huh, J. R., Seol, J. H., Yoo, S. J., Hay, B. A., and Guo, M. (2006). *Drosophila* pink1 is required for mitochondrial function and interacts genetically with parkin. *Nature* **441**, 1162–1166.

Cooper, A. A., Gitler, A. D., Cashikar, A., Haynes, C. M., Hill, K. J., Bhullar, B., Liu, K., Xu, K., Strathearn, K. E., Liu, F., Cao, S., Caldwell, K. A., Caldwell, G. A., Marsischky, G., Kolodner, R. D., Labaer, J., Rochet, J. C., Bonini, N. M., and Lindquist, S. (2006). Alpha-synuclein blocks ER–Golgi traffic and Rab1 rescues neuron loss in Parkinson's models. *Science* **313**, 324–328.

Coulom, H., and Birman, S. (2004). Chronic exposure to rotenone models sporadic Parkinson's disease in *Drosophila melanogaster*. *J Neurosci* **24**, 10993–10998.

Dauer, W., and Przedborski, S. (2003). Parkinson's disease: Mechanisms and models. *Neuron* **39**, 889–909.

Driscoll, M., and Gerstbrein, B. (2003). Dying for a cause: Invertebrate genetics takes on human neurodegeneration. *Nat Rev Genet* **4**, 181–194.

Echeverri, C. J., Beachy, P. A., Baum, B., Boutros, M., Buchholz, F., Chanda, S. K., Downward, J., Ellenberg, J., Fraser, A. G., Hacohen, N., Hahn, W. C., Jackson, A. L., Kiger, A., Linsley, P. S., Lum, L., Ma, Y., Mathey-Prevot, B., Root, D. E., Sabatini, D. M., Taipale, J., Perrimon, N., and Bernards, R. (2006). Minimizing the risk of reporting false positives in large-scale RNAi screens. *Nat Methods* **3**, 777–779.

Farrer, M. J. (2006). Genetics of Parkinson disease: Paradigm shifts and future prospects. *Nat Rev Genet* **7**, 306–318.

Feany, M. B., and Bender, W. W. (2000). A *Drosophila* model of Parkinson's disease. *Nature* **404**, 394–398.

Fujiwara, H., Hasegawa, M., Dohmae, N., Kawashima, A., Masliah, E., Goldberg, M. S., Shen, J., Takio, K., and Iwatsubo, T. (2002). Alpha-synuclein is phosphorylated in synucleinopathy lesions. *Nat Cell Biol* **4**, 160–164.

Greene, J. C., Whitworth, A. J., Kuo, I., Andrews, L. A., Feany, M. B., and Pallanck, L. J. (2003). Mitochondrial pathology and apoptotic muscle degeneration in

Drosophila parkin mutants. *Proc Natl Acad Sci USA.* 100, 4078–4083.

Greene, J. C., Whitworth, A. J., Andrews, L. A., Parker, T. J., and Pallanck, L. J. (2005). Genetic and genomic studies of *Drosophila* parkin mutants implicate oxidative stress and innate immune responses in pathogenesis. *Hum Mol Genet* 14, 799–811.

Hardy, J., Cai, H., Cookson, M. R., Gwinn-Hardy, K., and Singleton, A. (2006). Genetics of Parkinson's disease and parkinsonism. *Ann Neurol* 60, 389–398.

Haywood, A. F., and Staveley, B. E. (2006). Mutant alpha-synuclein-induced degeneration is reduced by parkin in a fly model of Parkinson's disease. *Genome* 49, 505–510.

Helfand, S. L. (2002). Neurobiology. Chaperones take flight. *Science* 295, 809–810.

Jung, J., and Bonini, N. (2007). CREB-binding protein modulates repeat instability in a *Drosophila* model for polyQ disease. *Science*.

Kinumi, T., Kimata, J., Taira, T., Ariga, H., and Niki, E. (2004). Cysteine-106 of DJ-1 is the most sensitive cysteine residue to hydrogen peroxide-mediated oxidation *in vivo* in human umbilical vein endothelial cells. *Biochem Biophys Res Commun* 317, 722–728.

Koh, K., Evans, J. M., Hendricks, J. C., and Sehgal, A. (2006). A *Drosophila* model for age-associated changes in sleep: Wake cycles. *Proc Natl Acad Sci USA* 103, 13843–13847.

Kontopoulos, E., Parvin, J. D., and Feany, M. B. (2006). Alpha-synuclein acts in the nucleus to inhibit histone acetylation and promote neurotoxicity. *Hum Mol Genet* 15, 3012–3023.

LaVoie, M. J., Ostaszewski, B. L., Weihofen, A., Schlossmacher, M. G., and Selkoe, D. J. (2005). Dopamine covalently modifies and functionally inactivates parkin. *Nat Med* 11, 1214–1221.

Marsh, J. L., and Thompson, L. M. (2006). *Drosophila* in the study of neurodegenerative disease. *Neuron* 52, 169–178.

Menzies, F. M., Yenisetti, S. C., and Min, K. T. (2005). Roles of *Drosophila* DJ-1 in survival of dopaminergic neurons and oxidative stress. *Curr Biol* 15, 1578–1582.

Meulener, M., Whitworth, A. J., Armstrong-Gold, C. E., Rizzu, P., Heutink, P., Wes, P. D., Pallanck, L. J., and Bonini, N. M. (2005). *Drosophila* DJ-1 mutants are selectively sensitive to environmental toxins associated with Parkinson's disease. *Curr Biol* 15, 1572–1577.

Meulener, M. C., Xu, K., Thomson, L., Ischiropoulos, H., and Bonini, N. M. (2006). Mutational analysis of DJ-1 in *Drosophila* implicates functional inactivation by oxidative damage and aging. *Proc Natl Acad Sci USA* 103, 12517–12522.

Muqit, M. M., and Feany, M. B. (2002). Modelling neurodegenerative diseases in *Drosophila*: A fruitful approach? *Nat Rev Neurosci* 3, 237–243.

Pagliarini, R. A., and Xu, T. (2003). A genetic screen in *Drosophila* for metastatic behavior. *Science* 302, 1227–1231.

Park, J., Kim, S. Y., Cha, G. H., Lee, S. B., Kim, S., and Chung, J. (2005). *Drosophila* DJ-1 mutants show oxidative stress-sensitive locomotive dysfunction. *Gene* 361, 133–139.

Park, J., Lee, S. B., Lee, S., Kim, Y., Song, S., Kim, S., Bae, E., Kim, J., Shong, M., Kim, J. M., and Chung, J. (2006). Mitochondrial dysfunction in *Drosophila* PINK1 mutants is complemented by parkin. *Nature* 441, 1157–1161.

Periquet, M., Fulga, T., Myllykangas, L., Schlossmacher, M. G., and Feany, M. B. (2007). Aggregated {alpha}-synuclein mediates dopaminergic neurotoxicity *in vivo*. *J Neurosci* 27, 3338–3346.

Pesah, Y., Pham, T., Burgess, H., Middlebrooks, B., Verstreken, P., Zhou, Y., Harding, M., Bellen, H., and Mardon, G. (2004). *Drosophila* parkin mutants have decreased mass and cell size and increased sensitivity to oxygen radical stress. *Development* 131, 2183–2194.

Phiel, C. J., and Klein, P. S. (2001). Molecular targets of lithium action. *Annu Rev Pharmacol Toxicol* 41, 789–813.

Quiroz, J. A., Gould, T. D., and Manji, H. K. (2004). Molecular effects of lithium. *Mol Interv* 4, 259–272.

Rubin, G. M., Yandell, M. D., Wortman, J. R., Gabor Miklos, G. L., Nelson, C. R., Hariharan, I. K., Fortini, M. E., Li, P. W., Apweiler, R., Fleischmann, W., Cherry, J. M., Henikoff, S., Skupski, M. P., Misra, S., Ashburner, M., Birney, E., Boguski, M. S., Brody, T., Brokstein, P., Celniker, S. E., Chervitz, S. A., Coates, D., Cravchik, A., Gabrielian, A., Galle, R. F., Gelbart, W. M., George, R. A., Goldstein, L. S., Gong, F., Guan, P., Harris, N. L., Hay, B. A., Hoskins, R. A., Li, J., Li, Z., Hynes, R. O., Jones, S. J., Kuehl, P. M., Lemaitre, B., Littleton, J. T., Morrison, D. K., Mungall, C., O'Farrell, P. H., Pickeral, O. K., Shue, C., Vosshall, L. B., Zhang, J., Zhao, Q., Zheng, X. H., and Lewis, S. (2000). Comparative genomics of the eukaryotes. *Science* 287, 2204–2215.

Sang, T. K., Chang, H. Y., Lawless, G. M., Ratnaparkhi, A., Mee, L., Ackerson, L. C., Maidment, N. T., Krantz, D. E., and Jackson, G. R. (2007). A *Drosophila* model of mutant human parkin-induced toxicity demonstrates selective loss of dopaminergic neurons and dependence on cellular dopamine. *J Neurosci* 27, 981–992.

Scherzer, C. R., Jensen, R. V., Gullans, S. R., and Feany, M. B. (2003). Gene expression changes presage neurodegeneration in a *Drosophila* model of Parkinson's disease. *Hum Mol Genet* 12, 2457–2466.

Shendelman, S., Jonason, A., Martinat, C., Leete, T., and Abeliovich, A. (2004). DJ-1 is a redox-dependent molecular chaperone that inhibits alpha-synuclein aggregate formation. *PLoS Biol* **2**, e362.

Takahashi, M., Kanuka, H., Fujiwara, H., Koyama, A., Hasegawa, M., Miura, M., and Iwatsubo, T. (2003). Phosphorylation of alpha-synuclein characteristic of synucleinopathy lesions is recapitulated in alpha-synuclein transgenic *Drosophila*. *Neurosci Lett* **336**, 155–158.

Wang, D., Qian, L., Xiong, H., Liu, J., Neckameyer, W. S., Oldham, S., Xia, K., Wang, J., Bodmer, R., and Zhang, Z. (2006). Antioxidants protect PINK1-dependent dopaminergic neurons in *Drosophila*. *Proc Natl Acad Sci USA* **103**, 13520–13525.

Whitesell, L., Mimnaugh, E. G., De Costa, B., Myers, C. E., and Neckers, L. M. (1994). Inhibition of heat shock protein HSP90-pp60v-src heteroprotein complex formation by benzoquinone ansamycins: Essential role for stress proteins in oncogenic transformation. *Proc Natl Acad Sci USA* **91**, 8324–8328.

Whitworth, A. J., Theodore, D. A., Greene, J. C., Benes, H., Wes, P. D., and Pallanck, L. J. (2005). Increased glutathione S-transferase activity rescues dopaminergic neuron loss in a *Drosophila* model of Parkinson's disease. *Proc Natl Acad Sci USA* **102**, 8024–8029.

Whitworth, A. J., Wes, P. D., and Pallanck, L. J. (2006). *Drosophila* models pioneer a new approach to drug discovery for Parkinson's disease. *Drug Discov Today* **11**, 119–126.

Wilson, M. A., Collins, J. L., Hod, Y., Ringe, D., and Petsko, G. A. (2003). The 1.1-A resolution crystal structure of DJ-1, the protein mutated in autosomal recessive early onset Parkinson's disease. *Proc Natl Acad Sci USA* **100**, 9256–9261.

Wu, Y. R., Wang, C. K., Chen, C. M., Hsu, Y., Lin, S. J., Lin, Y. Y., Fung, H. C., Chang, K. H., and Lee-Chen, G. J. (2004). Analysis of heat-shock protein 70 gene polymorphisms and the risk of Parkinson's disease. *Hum Genet* **114**, 236–241.

Xun, Z., Sowell, R. A., Kaufman, T. C., and Clemmer, D. E. (2007). Protein expression in a *Drosophila* model of Parkinson's disease. *J Proteome Res* **6**, 348–357.

Yang, Y., Gehrke, S., Haque, M. E., Imai, Y., Kosek, J., Yang, L., Beal, M. F., Nishimura, I., Wakamatsu, K., Ito, S., Takahashi, R., and Lu, B. (2005). Inactivation of *Drosophila* DJ-1 leads to impairments of oxidative stress response and phosphatidylinositol 3-kinase/Akt signaling. *Proc Natl Acad Sci USA* **102**, 13670–13675.

Yang, Y., Gehrke, S., Imai, Y., Huang, Z., Ouyang, Y., Wang, J. W., Yang, L., Beal, M. F., Vogel, H., and Lu, B. (2006). Mitochondrial pathology and muscle and dopaminergic neuron degeneration caused by inactivation of *Drosophila* Pink1 is rescued by Parkin. *Proc Natl Acad Sci USA* **103**, 10793–10798.

Zhang, L., Shimoji, M., Thomas, B., Moore, D. J., Yu, S. W., Marupudi, N. I., Torp, R., Torgner, I. A., Ottersen, O. P., Dawson, T. M., and Dawson, V. L. (2005). Mitochondrial localization of the Parkinson's disease related protein DJ-1: Implications for pathogenesis. *Hum Mol Genet* **14**, 2063–2073.

Zou, J., Guo, Y., Guettouche, T., Smith, D. F., and Voellmy, R. (1998). Repression of heat shock transcription factor HSF1 activation by HSP90 (HSP90 complex) that forms a stress-sensitive complex with HSF1. *Cell* **94**, 471–480.

26

CAENORHABDITIS ELEGANS MODELS OF PARKINSON'S DISEASE: A ROBUST GENETIC SYSTEM TO IDENTIFY AND CHARACTERIZE ENDOGENOUS AND ENVIRONMENTAL COMPONENTS INVOLVED IN DOPAMINE NEURON DEGENERATION

RICHARD NASS AND RAJA S. SETTIVARI

Department of Pharmacology and Toxicology, Center for Environmental Health, and Stark Neuroscience Research Institute, Indiana University School of Medicine, Indianapolis, IN, USA

INTRODUCTION

In the early 1960s Sydney Brenner suggested that the little known free-living nematode *Caenorhabditis elegans* (*C. elegans*) could be utilized in dissecting the genetic pathways involved in organ development and behavior. Almost 40 years later, he along with Robert Horvitz and John Sulston were awarded the 2002 Nobel Prize in Physiology or Medicine for their seminal studies using the worm to explore the molecular basis of development and programmed cell death. The utility of *C. elegans* in elucidating the molecular basis of gene regulation and development was further highlighted by the award of the 2006 Nobel Prize in Physiology or Medicine to *C. elegans* researchers Andrew Fire and Craig Mello for their discovery of RNA interference (RNAi), the evolutionarily conserved process of sequence-specific degradation of mRNAs. Their discoveries

and proofs of fundamental mechanisms involved in gene regulation across phyla has provided powerful applications in functional genomics, and holds significant promise in understanding and treating a number of human diseases and diverse conditions ranging from immunological disorders and aging, to carcinogenesis and neurodegeneration.

The high genetic and neurobiochemical conservation between *C. elegans* and humans, and facile laboratory methodologies developed by Sydney Brenner, Andy Fire, and others is a key reason that the worm has significant potential to contribute to our understanding of the molecular basis of neurodegenerative diseases. The human genome contains less than 20% more genes than found in the worm genome (The *C. elegans* Sequencing Consortium, 1998; Hiller *et al.*, 2006; Nass and Chen, 2008). Molecular pathways involved in animal and cell development, including programmed cell death, are highly conserved between both organisms. Also, most of the known signaling and neurotransmitter systems, including enzyme and molecular pathways involved in the production of neurotransmitters such as acetylcholine, glutamate, α-aminobutyric acid, serotonin, and dopamine (DA) are present in this nematode (Rand and Nonet, 1997; Bargmann, 1998; Thomas and Lockery, 1999; Nass and Blakely, 2003). Furthermore, ion channels and other components involved in mammalian synaptic neurotransmission and neurotransmitter reuptake are highly conserved between worms and humans. This strong molecular conservation between the worm and humans suggests that paradigms developed using C. elegans in neurodegenerative studies, will have significant relevance to human disease. In this review, we describe the utility of the nematode C. elegans to model the debilitating human neurodegenerative disease Parkinson's disease (PD), and discuss its potential for the identification of molecular and environmental components that could contribute to this devastating disorder.

C. ELEGANS AS A GENETIC MODEL

C. elegans is a powerful genetic model system for exploring the molecular mechanisms of neuronal function and disease (Rand and Nonet, 1997; Riddle et al., 1997; Nass and Chen, 2008). It is an anatomically simple organism containing approximately a thousand cells, in which 302 are neurons, and whose wiring diagram has been determined. Its small size (1 mm long), large brood size (approximately 350 progeny from a single hermaphrodite), short generation time (3.5 days at 20°C), and ease of maintenance in the laboratory (tens of thousands of animals can be grown on a 100 mm agar plate overlayed with bacteria) facilitate rapid and inexpensive production of animal's experimental analysis (Riddle et al., 1997; Wood, 1988). The transparency of the animal allows for the visualization in vivo of individual cells and developmental processes such as organ development and programmed cell death. Transgenic animals containing green fluorescent reporter fusions or rescue constructs can be generated and identified within 4 days, and the worms transparency greatly facilitates neuronal spatial and structure–function analysis (White et al., 1986; Chalfie et al., 1994; Mello and Fire, 1995; Miller et al., 1999). Worm strains can also be stored almost

indefinitely in a –80°C low temperature freezer or in liquid nitrogen, allowing for analysis at later dates.

Large-scale primary cultures, first developed by Laird Bloom in 1993, also provide the opportunity to investigate neuronal function on a cellular level (Bloom, 1993). The cells differentiate and appear to retain many of their in vivo cellular properties (Bloom, 1993; Carvelli et al., 2004; Bianchi and Driscoll, 2006). Electrical properties can be determined in many cell types, and are amendable to cell sorting with specific phenotypes that allow for the development of tissue-specific cDNA libraries (Christensen et al., 2002; Zang et al., 2002; Bianchi and Driscoll, 2006; Fox et al., 2005; Touroutine et al., 2005). The nematode can also easily be grown in liquid medium in standard 384- or 96-well microtiter plates, allowing for the high or medium throughput screening (HTS) of animals with particular behavioral phenotypes or optical properties (Link et al., 2000; Kaletta and Hengartner, 2006; Nass and Hamza, 2007). Furthermore the lack of a blood–brain barrier that allows for the permeability of exogenous compounds into its nervous systems, facilitates the identification of neurotherapeutic leads that inhibit cellular dysfunctions and death (Chapter 27; Link et al., 2000; Westlund et al., 2004; Nass and Chen, 2008).

One of the more significant advantages of C. elegans is the ease of generating genetic knockdown or deletion mutants to identify gene function. Forward genetic screens incorporating chemical or physical mutagens can rapidly identify genes involved in a particular cellular process, and have a significant advantage over many other model systems in that no a priori knowledge is needed about its function in order to determine whether the gene plays a role in the particular behavioral phenotype or cellular process (Riddle et al., 1997; Jorgensen and Mango, 2002). The identification of the mutant gene can now be identified within as little as a week, and since the animals can easily be mated with each other, generation of strains with several genetic mutations can be generated within days (Wicks et al., 2001). Over 8,000 mutants contributed from different laboratories around the world are also available at a federally funded C. elegans gene knockout consortium, as well as the Japanese National Bioresource Project, and are available on request at a nominal cost from the C. elegans Genetic Stock Center (see Caenorhabditis elegans WWW Server: http://elegans.swmed.edu/).

C. elegans also provides opportunities for facile reverse genetics screens that can determine the function of specific genes. Popular and powerful approaches using reverse genetics include mutagenesis to induce and identify genetic nulls and, or as mentioned above, incorporating RNAi screens (Fire *et al.*, 1998; Montgomery *et al.*, 1998; Timmmons and Fire, 1998). *RNAi* can be used as a convenient way to decrease protein expression, and in *C. elegans* can be obtained by introducing gene-specific double-stranded RNA (dsRNA) into the animals either by microinjection, soaking in solutions containing dsRNA, or feeding the animals bacteria that express dsRNA. RNAi screens utilizing bacteria expressing dsRNA is an especially facile and rapid approach to identifying genes involved in a particular cellular process or pathway. Bacteria feeding libraries are available containing *E. coli* strains each containing dsRNA for a specific gene within the *C. elegans* genome, and combinatorially represent almost all of the known *C. elegans* genes, facilitating the completion of a whole genome screen in as little as a month (Maeda *et al.*, 2001; Kamath and Ahringer, 2003; Kamath, *et al.*, 2003; Buckingham *et al.*, 2004).

The similarities between the worm and the human nervous system and the ease and strength of performing forward and reverse genetic screens suggest that *C. elegans* will be a powerful tool to elucidate the molecular components and pathways involved in PD. As discussed below and in Chapter 27, these same attributes and the ease of incorporating *C. elegans* into high throughput genetic and chemical screens should also provide opportunities to identify novel therapeutic targets and compounds that can inhibit and or protect against DA neuron dysfunction and cell death.

TOXIN-ASSOCIATED *C. ELEGANS* MODELS OF PD

PD was systematically described by the British neurologist James Parkinson in a 1817 publication in which he called the disorder the Shaking Palsy (Parkinson, 1817). In the early 1960s a fundamental hallmark of the disorder was identified as the loss of DA neurons in the substantia nigra pars compacta in the ventral midbrain (see Chapters 1–3). As described in detail in several chapters within this book, the etiology of the disease is unknown but it is believed to be multifactorial, with significant contributions involving both genes and the environment that results in oxidative damage, and

proteasome and mitochondrial dysfunction (see Chapters 15, 17, 21; Mitchell *et al.*, 1996; Jenner, 1998; Masliah *et al.*, 2000).

There is not a single experimental paradigm or animal model that fully recapitulates the human disease. One of the more common experimental approaches is to expose vertebrates to the relatively specific DA neuron neurotoxins 6-hydroxydopamine (6-OHDA), 1-methyl-4-phenyl-1,2,3,6-tetrahydropyridine (MPTP), or 1-methyl-4-phenylpyridinium ion (MPP$^+$ [the active metabolite of MPTP]) (see Chapters 11–13). These compounds confer DA neuronal death, and within a few weeks, the animals display parkinsonian behaviors. Both 6-OHDA and MPP$^+$ are transported into the cell by the high affinity Na$^+$- and Cl$^-$-dependent dopamine transporter (DAT), which is also the target of drugs of abuse that include cocaine, amphetamine, and methyphenidate (Gainetdinov *et al.*, 1998). More recently insecticides and fungicides such as paraquat, rotenone, and Maneb have also been utilized to generate PD-associated models, (see Chapters 14–15; Thiruchelvam *et al.*, 2003; Hoglinger *et al.*, 2006).

The *C. elegans* toxin-induced PD models recapitulate many aspects of the vertebrate models and the human disease. On our way to develop the first *C. elegans* PD models, we generated transgenic strains that expressed the green fluorescent protein (GFP) in the eight DA neurons in the hermaphrodite (Nass *et al.*, 2001). These neurons are clearly visible in the live animals as they move under a simple dissecting fluorescent microscope, and provided the opportunity for the first time to visualize DA neurons clearly *in vivo* (Figure 26.1). Brief exposures of the animals to 6-OHDA result in both a time and concentration-dependent loss of DA neurons within the head (Nass *et al.*, 2002; Nass and Hamza, 2007). Consistent with the vertebrate 6-OHDA PD models, the toxin-induced effect requires the expression of DAT, and can be inhibited with DAT agonists (e.g., amphetamine) or antagonists (e.g., cocaine) (Figure 26.2) (Nass *et al.*, 2002). The dependence on the DAT for the 6-OHDA-induced cell death is remarkable; exposure of wild-type (WT) and DAT knockout animals to as high as 50 mM 6-OHDA for 2 h results in all the WT worms showing significant DA neuronal death, while the DAT deletion animals are completely void of any apparent degeneration (Nass *et al.*, 2002). The toxin also appears to affect only the DA neurons since abnormal cellular morphologies are not seen in other cell types. Furthermore, the 6-OHDA-induced cell death does not appear to

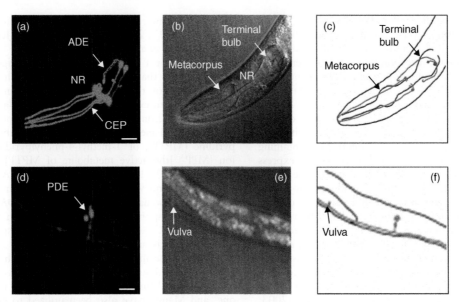

FIGURE 26.1   Visualization of all eight DA neurons in living, adult *C. elegans* hermaphrodites using DAT::GFP transcriptional fusions. (a) 3D reconstruction of confocal epifluorescence from head DA neurons in a $P_{dat:1}$::GFP transgenic line. Arrows identify CEP and ADE processes. NR refers to the nerve ring. (b) DIC image of animal in panel (a). (c) Schematic drawing showing location of DA neurons in the head relative to the pharynx. In this top view two pairs of CEP neurons project dendritic endings to the tip of the nose and one pair of ADE neurons extend ciliated processes to amphids adjacent to the terminal bulb of the pharynx. (d) 3D reconstruction of confocal epifluorescence of the PDE neurons. Both PDE cell bodies are apparent. (e) DIC image of animal in panel (d). (f) Schematic drawing showing left-hand member of pair of PDE neurons in lateral location posterior to vulva. All scale bars = 25 μm. Anterior is to the left. See Reference Nass *et al.*, 2002 for details. Reproduced with permission from *PNAS* (Nass *et al.*, 2002).

be through necrotic or a classical apoptotic mechanisms, since swollen cell bodies or membrane whirls are absent, and the caspases CED-3 and CED-4 are not required for the neurodegeneration. These data are consistent with more recent vertebrate studies that suggest that the DA neurons may degenerate by a caspase-independent mechanism.

Low concentrations of the complex I inhibitor MPP$^+$ also causes the apparent loss of DA neurons, as well as movement defects reminiscent of PD. Braungart and colleagues treated *C. elegans* with various concentrations of MPP$^+$ in liquid for 2 days and found the IC$_{50}$'s at 590 μM for the mobility defects, which include slow and uneven movements (Braungart *et al.*, 2004). DA neuronal loss, as determined by GFP fluorescence, was also correlated with the movement deficits, but other cell types are likely affected since the loss of DA neurons have previously been shown to cause only mild phenotypes (Sawin, 1996; Sawin *et al.*, 2000; Braungart *et al.*, 2004).

C. elegans is also sensitive to the mitochondria complex I inhibitor rotenone, the insecticide that has been weakly associated with the propensity to develop PD (see Chapters 14, 21; Hoglinger *et al.*, 2006). The worm shows a positive dose-dependent relationship with rotenone exposure and increased mortality, and the effect appears to be through inhibition of respiration (Ved *et al.*, 2005). Unlike in the PD-associated worm mutants (see below), the DA neurons do not show adverse morphological changes following chronic exposure to rotenone, although thioflavin-positive inclusions are found in a significant number of cells. Anti-oxidants and complex II stimulation also protects against the cellular toxicity (Ved *et al.*, 2005). Overall the toxin-associated PD worm models have significant overlap with vertebrates, and suggest that the worm will be a powerful *in vivo* genetic model to dissect the contributions that neurotoxins may play in DA neuron vulnerability and PD-associated cell death.

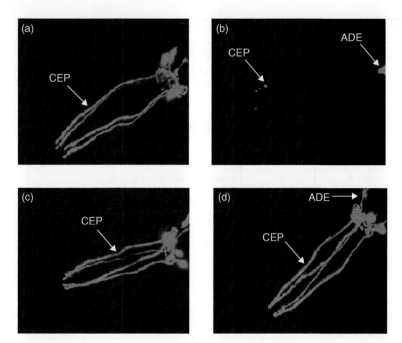

FIGURE 26.2 Genetic suppression of 6-OHDA sensitivity of DA neurons in *C. elegans*. (a) P$_{dat-1}$::GFP animals exposed to vehicle; (b) P$_{dat-1}$::GFP worms exposed to 6-OHDA; (c) P$_{dat-1}$::GFP, *dat-1*($\triangle$dat-1 worms) exposed to vehicle; (d) P$_{dat-1}$::GFP, $\triangle$dat-1 worms exposed to 6-OHDA. All scale bars = 25 μm. See Reference Nass *et al.*, 2002 for details. Reproduced with permission from *PNAS* (Nass *et al.*, 2002).

GENE-ASSOCIATED *C. ELEGANS* MODELS OF PD

As discussed above and within this book, genetic and epidemiological studies suggest that idiopathic PD likely involves molecular pathways affecting oxidative stress responses and affect both proteasomal and mitochondrial function. Mutations within genes such as α-synuclein, parkin, DJ-1, PINK-1, and LRRK2 have been associated with the rare, familial forms of PD, and have been ascribed to less than 5% of all known cases of PD (see Chapter 2). Although there are not clear linkages between these genes and the propensity to develop idiopathic PD, the similarities between the familial and idiopathic cases on the cellular and molecular level suggests significant overlap in the pathogenesis, and that they likely involve cellular dysfunction in protein degradation, aggregation, and oxidative stress.

As described in several other chapters in this book, the first genetic mutation associated with sporadic PD was identified as the pre-synaptic protein α-synuclein (see Chapter 2). Although the precise function of α-synuclein is not known, the protein may interact with synaptic vesicles and could be involved in the regulation of both DA biosynthesis and DAT function (see Chapters 18, 22; Lee *et al.*, 2001). Mutations with α-synuclein alter the structure of α-synuclein, and affect DA neuron integrity and viability, and likely interfere with the proteasomal degradation pathway (see Chapters 18, 22, 33, 40, 42; Masliah *et al.*, 2000).

Of the genes identified to date that have been associated with PD, only α-synuclein does not have a clear ortholog in *C. elegans* (Figure 26.3). One advantage of characterizing a protein that is not normally expressed in a biological system is the lack of functional interference from the endogenous molecule. We generated the first gene-associated *C. elegans* PD model in collaboration with Dr. Garry Wong at Kuopio University in Finland by expressing pan-neuronally or selectively in the DA or cholinergic neurons in human WT or mutant A53T α-synuclein (Lakso *et al.*, 2003). Expression of either

Human gene	*C. elegans* gene	% Similarity	References
Parkin	*pdr-1*	41	Springer *et al.*, 2005
UCHL1	*ubh-1*	57	Ha *et al.*, 2006
	ubh-2	56	
	ubh-3	58	
	ubh-4	46	
PINK1	*pink-1*	50	*C. elegans* Sequencing Consortium, 2003
DJ1	*djr-1.1*	66	Ved *et al.*, 2005
	djr-1.2	62	-
LRRK2	*lrk-1*	48	Sakaguchi-Nakashima *et al.*, 2007
Nurr1	*nhr-6*	46	Gissendanner *et al.*, 2004
α-Synuclein	?	?	Lakso *et al.*, 2003
DAT	*dat-1*	64	Jayanthi *et al.*, 1998; Nass *et al.*, 2002, 2003
VMAT	*cat-1*	65	Duerr *et al.*, 1999
TH	*cat-2*	70	Sulston, 1975; Lints and Emmons, 1999
MAO	*amx-1*	36	Sulston *et al.*, 1992; Nass *et al.*, 2001, 2003
	amx-2	40	
	amx-3	46	

FIGURE 26.3   PD and DA neuron-associated orthologs in *C. elegans*. BLAST was performed at NCBI website: http://www.ncbi.nlm.nih.gov/blast/bl2seq/wblast2.cgi.

protein confers motor deficits when expressed behind the pan-neuronal promotor, and cause DA neuron cell death when expressed behind either the pan-neuronal or DAT promotor. α-Synuclein containing aggregates were also detected in some of the DA neurons (Lakso *et al.*, 2003). Expression studies of human A30P α-synuclein also cause locomotor defects and DA neuronal loss, and the addition of exogenous DA inhibits the behavioral abnormalities, which is consistent with vertebrate studies (Kuwahara *et al.*, 2006). α-Synuclein may confer cellular toxicity by disrupting vesicle docking and transport; a yeast genetic screen by Cooper and colleagues (2006) identified a Rab GTPase that suppresses α-synuclein toxicity in yeast, *Drosophila*, *C. elegans*, and rat midbrain cultures. These results are also consistent with mice α-synuclein deletion studies showing vesicle trafficking disruption, and provide further evidence of utility of the invertebrate PD models to recapitulate and contribute to our understanding of DA neuron cell death.

Microarray gene expression studies in *C. elegans* show that α-synuclein causes significant changes in molecular pathways associated with the mitochondria, proteasome, cellular development, and a number of histones (Vartianainen *et al.*, 2006a). The expression changes associated with nuclear histones are consistent with vertebrate studies indicating α-synuclein inhibits histone acetylation *in vitro* and promotes toxicity (Kontopoulos *et al.*, 2006). The PD-associated environmental toxin paraquat also increases α-synuclein expression in the nucleus, consistent with the putative contribution that herbicides may play in PD (see Chapter 15; Manning-Bo *et al.*, 2003).

C. elegans overexpression of human α-synuclein may also increase overall fitness. Worms expressing the human WT or A53T α-synuclein genes live on average 25% longer than controls, while α-synuclein expression protects against some forms of oxidative stress (Vartianainen *et al.*, 2006b; Nass, unpublished). These studies are intriguing considering recent studies suggesting that in vertebrates α-synuclein may also function as a chaperone and to protect against pro-apoptotic stimuli, and suggests that *C. elegans* could be a useful genetic model to begin to identify the molecular components involved in the α-synuclein neuroprotection (see Chapters 22 and 36; Sidhu *et al.*, 2004).

C. elegans contains a single ortholog for the parkin protein that is 41% similar to the mammalian homolog (Figure 26.3). Two other genes associated with recessive PD have also been explored in *C. elegans*. As reviewed in prior chapters in this book, Parkin encodes an E3 ubiquitin ligase that ubiquinates proteins that are marked for degradation by the proteasome. In the worm, the endogenous parkin is expressed in both muscles and neurons, and interacts with E2 enzymes and an E4 ubiquitin ligase (Springer *et al.*, 2005). Mutations in parkin can inhibit *C. elegans* larva development, confer hypersensitivity to ER stress, and can cause significant cellular aggregation (Springer *et al.*, 2005; Ved *et al.*, 2005). Loss-of-function parkin mutants are also almost twofold more sensitive to 6-OHDA, suggesting that Parkin plays a role in toxin-induced cellular stress (Nass and Chen 2008). A53T α-synuclein expression in a parkin mutant background results in lethality, suggesting that the cellular stress and protein aggregation generated by α-synuclein mutations may significantly compromise the proteasome degradation system (Springer *et al.*, 2005; Ved *et al.*, 2005).

Genetic knockdown of the worm ortholog to DJ-1, as with several other PD-associated gene and toxin models, leads to increases in sensitivity to rotenone (Figure 26.3) (Ved *et al.*, 2005). Compounds that increase cellular energy and bypass the complex I inhibition at least partially rescues the phenotype. Recently the *C. elegans* otholog for LRRK2, LRK-1, was identified as being required for polarization of synaptic vesicle proteins along the axons and dendrites (Sakaguchi-Nakashima *et al.*, 2007). Wolozin and colleagues report that overexpression of the mammalian LRRK2 in *C. elegans* also protects against rotenone toxicity (see Chapter 32; Wolozin *et al.*, 2008). Overall these results support a strong role of the PD-associated proteins affecting proteasomal and mitochondrial function, and suggest that *C. elegans* is a viable model for PD-associated DA neuron dysfunction and cell death.

GENETIC SCREENS TO IDENTIFY NOVEL PD-ASSOCIATED GENES AND PATHWAYS

As described above, *C. elegans* provides a remarkable opportunity to identify novel molecular pathways involved in DA neuron cell death. A simple forward genetic screen could identify novel genes in as little time as 2 weeks or less (Wicks *et al.*, 2001). A power of forward genetics is that no *a prior* knowledge of the gene function or molecular pathway is necessary in order to identify whether a protein plays a role in a cellular process; the requirement is that the molecules are essential for the phenotype assayed. A typical screen would involve exposing the *C. elegans* hermaphrodites to the mutagen ethylmethanesulphonate (EMS) for several hours (Jorgensen *et al.*, 2002). Following mutagenesis, the second generation of the self-fertilizing animals would be examined for changes in the phenotype. The mutated genes conferring the phenotype would then be identified using genetic map and common molecular technologies.

We have performed similar screens as described above to identify novel regulators of toxin-induced DA neuron cell death (Figure 26.4). Animals expressing GFP in the DA neurons were exposed to the mutagen, and those animals in the second generation in which the DA neurons were resistant to 6-OHDA were selected for further analysis. Our initial screen identified a number of mutants that showed significant neuroprotection against the toxin (Nass *et al.*, 2005). Three of the mutants were novel *dat-1* alleles conferring complete tolerance to 6-OHDA and display movement defects (Nass *et al.*, 2005; McDonald *et al.*, 2007). As expected these mutations confer changes in DAT expression, localization, and function. Our other mutants have mutations in pathways independent of the transporter, and involve molecules likely involved in cellular stress and viability. These studies provide proof-of-concept that we should be able to identify proteins involved in DA neuron cell death.

Reverse genetic screens have been fruitful in identifying genes involved in PD-associated phenotypes. Genome-wide RNAi screening utilizing a *C. elegans* library has recently identified a number of genes involved in α-synuclein aggregation. van Ham and colleagues identified 80 genes that following a decrease in protein expression conferred greater α-synuclein aggregation in muscle cells (van Ham *et al.*, 2008). 49 of the genes have human orthologs, and a number of these have previously been found to play a role in PD-associated phenotypes in other cell and animal models. The study identified genes involved in vesicle transport, lipid metabolism, protein quality control, and herbicide and insecticide detoxification (Nollen *et al.*, 2008; see Chapters 2, 15, 18, 21, 22, 25, 43). Others have also used a targeted gene approach with RNAi in the worm to identify putative modifiers of α-synuclein toxicity and identified

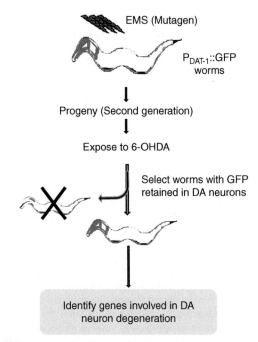

EMS (Mutagen)

P_{DAT-1}::GFP worms

↓

Progeny (Second generation)

↓

Expose to 6-OHDA

↓

Select worms with GFP retained in DA neurons

Identify genes involved in DA neuron degeneration

FIGURE 26.4 A genetic screen to identify proteins that protect against DA neuron cell death in *C. elegans*. Animals that retain GFP in the DA neurons following mutagenesis and 6-OHDA exposure has a mutation in a gene that protects against toxin-induced neurodegeneration.

proteins involved in protein trafficking and in transport (Hamamichi *et al.*, 2008). Overall these studies further showcase the combinatorial use of *C. elegans* and RNAi methodologies to identify novel proteins that may play a role in idiopathic PD.

TOOLS TO IDENTIFY EXOGENOUS COMPOUNDS THAT MAY CONTRIBUTE TO DA NEURON DEGENERATION

As described throughout this book, the cause of idiopathic PD is likely multifactorial and includes both genetic and environmental contributions. Although this past decade has identified a number of genes that may be involved in the development of PD, the role of environmental compounds in the initiation and progression of the disease is not well understood (see Chapter 21). For example, several

epidemiological studies have established an increased risk for PD with occupational exposure to heavy metals, including Fe^{2+}, Al^{3+}, Cu^{2+}, and Mn^{2+} (Zayed *et al.*, 1990; Kienzl *et al.*, 1995; Montgomery, 1995; Gorell *et al.*, 1999; Hudnell, 1999). Individuals exposed in the workplace to Fe^{2+} or Al^{3+} for 30 years or more have a greater propensity to develop PD, and there is a positive relationship in developing PD and the ingestion of foods containing higher levels of Fe^{2+} or Al^{3+}. Interestingly, PD patients may also have higher concentrations of Fe^{2+} or Al^{3+} in the substantia nigra (Oestreicher *et al.*, 1994).

The strongest association with heavy metal exposure and the susceptibility to develop PD is with the trace element Mn^{2+} (Hudnell, 1999; Gorell *et al.*, 1999). Mn^{2+} is an essential metal that is required for normal growth and development, and is a cofactor in a number of enzymes in energy production and oxidative stress. Acute Mn^{2+} toxicity results in symptoms similar to those seen in patients with PD, including rigidity, tremors, and bradykinesia (Chia *et al.*, 1993; Calne *et al.*, 1994; Stredrick *et al.*, 2004). Overexposure to Mn^{2+} causes a reduction in TH activity, inhibition of complex-I in the mitochondria, increase in the production of ROS, and loss of DA neurons in the substantia nigra (Parenti *et al.*, 1988; Migheli *et al.*, 1999; Tomas-Camardiel *et al.*, 2002). Mn^{2+} also causes an increase in aggregation and fibrillation of α-synuclein and parkin *in vitro*, and decreases the viability of α-synuclein expressing dopaminergic cells (Uversky *et al.*, 2001; Pifl *et al.*, 2004). Overexpression of parkin also protects against Mn^{2+}-induced DA neuron cell death *in vitro* (Higashi *et al.*, 2004) Mn^{2+}-induced Parkinsonism, also called manganism, has been associated with Mn^{2+} mining and welding, and these occupations may be particularly susceptible to developing early-onset PD (Hudnell, 1999; Gorell *et al.*, 1999; Witholt *et al.*, 2000; Racette *et al.*, 2001). Currently there are active lawsuits in US courts addressing some of these issues.

A significant hurdle in many vertebrate studies in identifying toxins that may increase DA neuron vulnerability to cell death is the considerable amount of time and effort involved in performing full-scale toxicological screening. These types of studies in rodents can often last a year or more, and the lack of rapid genetic tools and HTS technologies can make it difficult to subsequently identify the molecular pathways involved in the toxicological responses. *C. elegans* provides an opportunity to relatively rapidly identify or validate environmental neurotoxins that may

Heavy metal	Normal DA neuron morphology (% worms)
Control	100
Cu^{+2}	98
Fe^{+2}	97
Al^{+3}	86
Mn^{+2}	81

FIGURE 26.5 Exposure to heavy metals confers DA neuron degeneration in *C. elegans*. Animals were briefly exposed to heavy metals, and DA neuron integrity was evaluated 48 hrs later.

contribute to the DA neuron degeneration *in vivo* (Jones *et al.*, 1996; Nass and Hamza, 2007).

Since exposures of heavy metals have been associated with the propensity to develop PD and parkinsonism, we asked whether brief exposures to heavy metals can cause DA neuron cell death in *C. elegans*. Our studies suggest that several metals confer significant DA neuron degeneration *in vivo* (Figure 26.5; manuscript in preparation). Morphologically the heavy metal-induced neurodegeneration has similarities to those of neurons following exposure to PD-associated toxins. Mn^{2+} appears to be an especially potent DA neurotoxin, and its toxicity may be dependent on the worm ortholog of the mammalian divalent metal transporter (DMT-1), SMF (Figure 26.5; manuscript in preparation; Garrick *et al.*, 2003). Our studies are consistent with a number of cellular, vertebrate, and epidemiological studies suggesting that Mn^{2+} exposures may be an environmental factor that contributes to idiopathic PD.

The development of this *C. elegans* model should also allow one to quickly explore the molecular basis of the Mn^{2+}-induced cell death. Genetic screens can be utilized as described above in which following mutagenesis, animals can be selected for those that still maintain DA neuron integrity. Subsequent genetic mapping will identify genes involved in Mn^{2+}-induced neurotoxicity. In concert, microarray studies could identify genes that may be involved in the toxicity, and subsequent RNAi inhibition experiments could quickly confirm the role the genes may have in the DA neurodegeneration.

The crossing of the GFP expressing DA neuronal strains into animals that lack the DA, DAT, or PD-associated proteins can also assist in elucidating the role these proteins may play in DA neuron vulnerability. And unlike most analogous vertebrate studies that can take upward of many months or longer, the *C. elegans* toxicity studies can be completed in a matter of days to weeks, and can provide significant insight into the molecular pathways involved in the cell death.

LIMITATIONS OF *C. ELEGANS* MODELS OF PD-ASSOCIATED DA NEURON CELL DEATH

C. elegans is a powerful tool to identify and characterize molecules involved in DA neurodegeneration, but as in many models there can be limitations. For example, because of the relatively small number of neurons and synaptic connections in the worm, the roles of cell–cell interactions in affecting DA neuron vulnerability may not be completely recapitulated. As discussed in the clinical and non-human primate section of this book, PD is a complex disorder that can involve several neurotransmitter systems, including DAergic, acetylcholinergic, GABAergic, and serotonergic (see Chapters 1–3, 5). Neurons often have thousands of synaptic connections that can interact with many subtypes. The *C. elegans* DA neurons also have synaptic connections, but each may generate 50 connections or less, and many of these are *en passant* (White *et al.*, 1986; Riddle *et al.*, 1997). Although on the molecular level the conservation with vertebrate systems is high, cell–cell interactions and the contributions they could play in PD may not he completely recapitulated in the worm.

Another potential limitation of using *C. elegans* for PD-associated cell death studies is that the natural overabundance or absence, or overexpression of a specific protein in *C. elegans* could affect the apparent role the protein may be playing in neurodegeneration. For example, *C. elegans* does not have a clear ortholog to the α-synuclein gene. Although this does not preclude the worm from robust studies involving α-synuclein, as clearly reviewed above, its role in the cell death or putative interactions with other PD-associated proteins may be difficult to define. If for example, α-synuclein interacts with DAT to affect DA neuron vulnerability to endogenous or exogenous toxins in mammals, then the lack of α-synuclein in the worm model may differentially affect the interpretation of the role that DAT may play in PD. Conversely, if the worm has a greater number of orthologs for a particular gene, assigning a precise role that the analogous homologue may play in PD may be erroneous. Two examples of greater number of genes in a particular class found in the worm relative to humans are the ABC transporters and cytochrome P-450s. Both proteins are involved in xenobiotic detoxification, so it may not be too surprising that C. elegans has a greater number of homologs that may possibly protect against toxic compounds found in its natural soil habitat.

The solubility of some environmental chemicals may also limit opportunities characterize them in C. elegans. Although the sensitivity of the cells within C. elegans to pharmacological and toxicological agents, and the active sites and binding constants are remarkably similar with those found in mammals, the higher exposure concentrations that is necessary to penetrate the cuticle (often at low millimolar or high micromolar concentration) could preclude some compounds from being tested to determine whether they may contribute to DA neurodegeneration. For example there is correlative evidence that the pesticide Maneb may contribute to the development of PD (see Chapter 21). The solubility of the Mn-based fungicide Maneb in aqueous solution is very low, approximately $0.5\,\mu g/mL$ water or just below $2\,\mu M$, which could limit the utility of the worm in exploring the potential role Maneb may have in inducing DA neuron cell death. Fortunately C. elegans can also be cultured in up to 2% DMSO, and are remarkably resistant to a number of organic molecules, so alternative culture or exposure conditions, may make this more feasible (Rand and Johnson, 1995; Nass and Hamza, 2007). Furthermore, there are a number of C. elegans strains that have mutations in the outer cuticle, and have greater permeability to exogenous molecules. These strains may also be useful in evaluating the role that more hydrophobic compounds may have on DA neuron vulnerability and cell death (Rand and Johnson, 1995).

C. elegans is a powerful system to identify or characterize novel endogenous molecules, molecular pathways, or exogenous compounds that may contribute to DA neuron vulnerability, and on the molecular level, this model system can provide significant insight into the components and mechanisms involved in cell death. The worm though is not a mammal, and there are likely protein or cellular interactions that are not completely recapitulated in this model system. Furthermore, PD-associated genetic screens, although very rapid and robust, may not completely define or characterize the mammalian protein of interest. It is important therefore, as with any model, to further evaluate and test in other model systems.

PERSPECTIVES

The high degree of similarity on the molecular and genetic level between humans and C. elegans suggest that the molecules and pathways that are involved in DA neuron degeneration in the worm will likely as a result have a correlates in humans. The array of genetic tools available with C. elegans, including the strength and ease of performing forward and reverse genetics, provide remarkable opportunities to rapidly identify and characterize novel genes and endogenous molecules that may be involved in PD. The sensitivity of this invertebrate to exogenous neurotoxins and its ability to grow in multi-well formats suggest that C. elegans can be utilized to screen environmental agents or xenobiotics to determine whether these compounds may contribute to DA neuron vulnerability and PD. As described in the following chapter, the worm is also easily amendable to chemical genetic screens that can be used to identify novel therapeutic compounds and targets that may inhibit the progression of DA neurodegeneration associated with PD (see Chapter 27). The screens should also be able to identify molecular pathways involved in the neuroprotection.

As described in this book, idiopathic PD is the second most prevalent neurodegenerative disease, and although described almost 200 years ago, and rigorously explored on the molecular level within the past 40 years, the causative agents and biochemical pathways involved in the disorder have yet to be elucidated. This tiny nematode called C. elegans may finally provide us with the molecular tools necessary to combat this devastating disease.

REFERENCES

Bargmann, C. I. (1998). Neurobiology of the Caenorhabditis elegans genome. Science 282, 2028–2033.

Bianchi, L., and Driscoll, M. (2006). Culture of embryonic C. elegans cells for electrophysiological and pharmacological analyses, *Wormbook*, ed. The *C. elegans* Research Community. http://www.wormbook.org.

Bloom, L. (1993). Genetic and molecular analysis of genes required for axon outgrowth in *Caenorhabditis elegans*. PhD Thesis, Massachusetts Institute of Technology, Cambridge, MA.

Braungart, E., Gerlach, M., Riederer, P., Baumeister, R., and Hoener, M. C. (2004). *Caenorhabditis elegans* MPP+ model of Parkinson's disease for high-throughput drug screening. *Neurodegener Dis* 1, 175–183.

Buckingham, S. D., Esmaeili, B., Wood, M., and Sattelle, D. B. (2004). RNA interference: From model organisms towards therapy for neural and neuromuscular disorders. *Hum Mol Genet* 45, R275–R288.

Calne, D. B., Chu, N. S., Huang, C. C., Lu, C. S., and Olanow, W. (1994). Manganism and idiopathic parkinsonism: Similarities and differences. *Neurology* 44, 1583–1586.

Carvelli, L., McDonald, P. W., Blakely, R. D., and Defelice, L. J. (2004). Dopamine transporters depolarize neurons by a channel mechanism. *Proc Natl Acad Sci USA* 101, 16046–16051.

Chalfie, M., Tu, Y., Euskirchen, G., Ward, W. W., and Prasher, D. C. (1994). Green fluorescent protein as a marker for gene expression. *Science* 263, 802–805.

Chia, S. E., Foo, S. C., Gan, S. L., Jeyaratnam, J., and Tian, C. S. (1993). Neurobehavioral functions among workers exposed to manganese ore. *Scand J Work Environ Health* 19, 264–270.

Christensen, M., Estevez, A., Yin, X., Fox, R., Morrison, R., McDonnell, M., Gleason, C., Miller, D. M., and Strange, K. (2002). A primary culture system for functional analysis of *C. elegans* neurons and muscle cells. *Neuron* 33, 503–514.

Cooper, A. A., Gitler, A. D., Casikar, A., Haynes, C. M., Hill, K. J., Bhullar, B., Liu, K., Xu, K., Strathearn, K. E., Liu, F. *et al.* (2006). α-Synuclein blocks ER–Golgi traffic and Rab1 rescues neuron loss in Parkinson's models. *Science* 313, 324–328.

Duerr, J. S., Frisby, D. L., Gaskin, J., Duke, A., Asermely, K., Huddleston, D., Eiden, L. E., and Rand, J. B. (1999). The cat-1 gene of Caenorhabditis elegans encodes a vesicular monoamine transporter required for specific monoamine-dependent behaviors. *J. Neurosci.* 19, 72–84.

Fire, A., Xu, S., Montgomery, M. K., Kostas, S. A., Driver, S. E., and Mello, C. C. (1998). Potent and specific genetic interference by double-stranded RNA in *Caenorhabditis elegans*. *Nature* 391, 806–811.

Fox, R. M., Von Stetina, S. E., Barlow, S. J., Shaffer, C., Olszewski, K. L., Moore, J. H., Dupuy, D., Vidal, M., and Miller, D. M., III (2005). A gene expression fingerprint of *C. elegans* embryonic motor neurons. *BMC Genomics* 6, 42.

Gainetdinov, R. R., Jones, S. R., Fumagalli, F., Wightman, R. M., and Caron, M. G. (1998). Re-evaluation of the role of the dopamine transporter in dopamine system homeostasis. *Brain Res Brain Res Rev* 26, 148–153.

Garrick, M. D., Dolan, K. G., Horbinski, C., Ghio, A. J., Higgins, D., Porubcin, M., Moore, E. G., Hainsworth, L. N., Umbreit, J. N., Conrad, M. E. *et al.* (2003). DMT1: A mammalian transporter for multiple metals. *Biometals* 16, 41–54.

Gissendanner, C. R., Crossgrove, K., Kraus, K. A., Maina, C. V., and Sluder, A. E. (2004). Expression and function of conserved nuclear receptor genes in *C. elegans*. *Dev Biol* 266, 399–416.

Gorell, J. M., Johnson, C. C., Rybicki, B. A., Peterson, E. L., Kortsha, G. X., Brown, G. G., and Richardson, R. J. (1999). Occupational exposure to manganese, copper, lead, iron, mercury, and zinc and the risk of Parkinson's disease. *Neurotoxicology* 20, 239–247.

Ha, M. K., Cho, J. S., Baik, O., Lee, K. H., Koo, H., and Chung, K. Y. (2006). *Caenorhabditis elegans* as a screening tool for the endothelial cell-derived putative aging-related proteins. *Proteomics* 6, 3339–3351.

Hamamichi, S., Rivas, R. N., Knight, A. L., Cao, S., Caldwell, K. A., Caldwell, G. A. (2008). Hypothesis-based RNAi screening identifies neuroprotective genes in a Parkinson's disease model. *PNAS* 105, 728–732.

Higashi, Y., Asanuma, M., Miyazaki, I., Hattori, N., Mizuno, Y., and Ogawa, N. (2004). Parkin attenuates manganese-induced dopaminergic cell death. *J Neurochem* 89, 1490–1497.

Hiller, L. W., Coulson, A., Murray, J. I., Bao, Z., Sulston, J. E., and Waterson, R. H. (2006). Genomics in *C. elegans*: So many genes, such a little worm. *Genome Res* 15, 1651–1660.

Hoglinger, G. U., Oerte, W. H., and Hirsch, E. C. (2006). The rotenone model of parkinsonism – The five year inspection. *J. Neural Transm Suppl* 70, 269–270.

Hudnell, H. K. (1999). Effects from environmental Mn exposures: A review of the evidence from non-occupational exposure studies. *Neurotoxicology* 20, 379–397.

Jayanthi, L. D., Apparsundaram, S., Malone, M. D., Ward, E., Miller, D. M., Eppler, M., and Blakely, R. D. (1998). The Caenorhabditis elegans gene T23G5.5 encodes an antidepressant- and cocaine sensitive dopamine transporter. *Mol. Pharmacol.* 54, 601–609.

Jenner, P. (1998). Oxidative mechanisms in nigral cell death in Parkinson's disease. *Mov Disord* 13, 24–34.

Jones, D., Strngham, E. G., Babich, S. L., and Candido, E. P. M. (1996). Transgenetic strains of the nematode *C. elegans* in biomonitoring and toxicology: Effects of captan and related compounds on the stress response. *Toxicology* 109, 119–127.

Jorgensen, E. M., and Mango, S. E. (2002). The art and design of genetic screens: *Caenorhabditis elegans*. *Nature Rev Genet* **3**, 356–369.

Kaletta, T., and Hengartner, M. O. (2006). Finding function in novel targets: *C. elegans* as a model organism. *Nat Rev Drug Discov* **5**, 387–398.

Kamath, R. S., and Ahringer, J. (2003). Genome-wide RNAi screening in *Caenorhabditis elegans*. *Methods* **30**, 313–321.

Kamath, R. S., Fraser, A. G., Dong, Y., Poulin, G., Durbin, R., Gotta, M., Kanapin, A., Le Bot, N., Moreno, S., Sohrmann, M. *et al.* (2003). Systematic functional analysis of the *Caenorhabditis elegans* genome using RNAi. *Nature* **421**, 231–237.

Kienzl, E., Puchinger, L., Jellinger, K., Linert, W., Stachelberger, H., and Jameson, R. F. (1995). The role of transition metals in the pathogenesis of Parkinson's disease. *J Neurol Sci* **134 Suppl**, 69–78.

Kontopoulos, E., Parvin, J. D., and Feany, M. B. (2006). α-Synuclein acts in the nucleus to inhibit histone acetylation and promote neurotoxicity. *Hum Mol Genet* **15**, 3012–3023.

Kuwahara, T., Koyama, A., Gengyo-Ando, K., Masuda, M., Kowa, H., Tsunoda, M., Mitani, S., and Iwatsubo, T. (2006). Familial Parkinson mutant α-synuclein causes dopamine neuron dysfunction in transgenic *Caenorhabditis elegans*. *J Biol Chem* **281**, 334–340.

Lakso, M., Vartianen, S., Moilanen, A. M., Sirvio, J., Thomaas, J. H., Nass, R., Blakely, R. D., and Wong, G. (2003). Dopaminergic neuronal loss and motor deficits in *Caenorhabditis elegans* overexpressing human α-synuclein. *J Neurochem* **86**, 165–172.

Lee, F. J., Liu, F., Pristupa, Z. B., and Niznik, H. B. (2001). Direct binding and functional coupling of alpha-synuclein to the dopamine transporters accelerate dopamine-induced apoptosis. *FASEB J* **15**, 916–926.

Link, E. M., Hardiman, G., Sluder, A. E., Johnson, C. D., and Liu, L. X. (2000). Therapeutic target discovery using *Caenorhabditis elegans*. *Pharmacogenomics* **1**, 203–217.

Lints, R., and Emmons, S. W. (1999). Patterning of dopaminergic neurotransmitter identity among *Caenorhabditis elegans* ray sensory neurons by a TGFbeta family signaling pathway and Hox gene. *Development* **126**, 5819–5831.

Maeda, I., Kohara, Y., Yamamoto, M., and Sugimoto, A. (2001). Large-scale analysis of gene function in *Caenorhabditis elegans* by high-throughput RNAi. *Curr Biol* **11**, 171–176.

Manning-Bo, A. B., McCormack, A. L., Purisai, M. G., Bolin, L. M., and DiMonte, D. A. (2003). α-Synuclein overexpression protects against paraquat-induced neurodegeneration. *J Neurosci* **23**, 3095.

Masliah, E., Rockenstein, E., Veinbergs, I., Mallory, M., Hashimoto, M., Takeda, A., Sagara, Y., Sisk, A., and Mucke, L. (2000). Dopaminergic loss and inclusion body formation in alpha-synuclein mice: Implications for neurodegenerative disorders. *Science* **287**, 1265–1269.

Matthews, D. J., and Kopczynski, J. (2001). Using model-system genetics for drug-based target discovery. *Drug Discov Today* **6**, 141–149.

McDonald, P. W., Hardie, S. L., Jessen, T. N., Carvelli, L., Matthies, D. S., and Blakely, R. D. (2007). Vigorous activity in *Caenorhabditis elegans* requires efficient clearance of dopamine mediated by synaptic localization of the dopamine transporter DAT-1. *J Neurosci* **27**, 14216–14227.

Mello, C., and Fire, A. (1995). DNA transformation. *Methods in Cell Biol* **48**, 451–482.

Miller, D. M, III, Desai, N. S., Hardin, D. C., Piston, D. W., Patterson, G. H. *et al.* (1999). Two color GFP expression for *C. elegans*. *Biotechniques* **26**, 914–921.

Migheli, R., Godani, C., Sciola, L., Delogu, M. R., Serra, P. A., Zangani, D., De Natale, G., Miele, E., and Desole, M. S. (1999). Enhancing effect of manganese on L-DOPA-induced apoptosis in PC12 cells: Role of oxidative stress. *J Neurochem* **73**, 1155–1163.

Mitchell, S. L., Kiely, D. K., Kiel, D. P., and Lipsitz, L. A. (1996). The epidemiology, clinical characteristics, and natural history of older nursing home residents with a diagnosis of Parkinson's disease. *J Am Geriatr Soc* **44**, 394–399.

Montgomery, E. B., Jr. (1995). Heavy metals and the etiology of Parkinson's disease and other movement disorders. *Toxicology* **97**, 3–9.

Montgomery, MK., Xu, S., and Fire, A. (1998). RNA as a target of double-stranded RNA-mediated genetic interference in *Caenorhabditis elegans*. *Proc Natl Acad Sci USA* **95**, 15502–15507.

Nass, R., and Blakely, R. B. (2003). The *Caenorhabditis elegans* dopaminergic system: Opportunities for insights into dopamine transport and neurodegeneration. *Annu Rev Pharmacol Toxicol* **43**, 521–544.

Nass, R., and Hamza, I. (2007). The nematode *C. elegans* as a model to explore toxicology *in vivo*: Solid and axenic growth culture conditions and compound exposure parameters. *Curr Protocols Toxicol*, 1.9.1–1.9.18.

Nass, R., Chen, L. (2008). *Caenorhabditis elegans* models of human neurodegenerative diseases: A powerful tool to identify molecular mechanisms and novel therapeutic targets (P. Michael Conn, Ed.), pp. 91–101. Humana Press, Totowa, NJ.

Nass, R., Miller, D. M., and Blakely, R. D. (2001). *C. elegans*: A novel pharmacogenetic model to study Parkinson's disease. *Parkinsonism Relat Disord* **7**, 185–191.

Nass, R., Hall, D. H., Miller, D. M., III, and Blakely, R. D. (2002). Neurotoxin-induced degeneration of dopamine neurons in *Caenorhabditis elegans*. *Proc Natl Acad Sci USA* **99**, 3264–3269.

Nass, R., Hahn, M. K., Jessen, T., McDonald, P. W., Carvelli, L., and Blakely, R. D. (2005). A genetic screen in *Caenorhabditis elegans* for dopamine neuron insensitivity to 6-hydroxydopamine identifies dopamine transporter mutants impacting transporter biosynthesis and trafficking. *J Neurochem* **94**, 774–785.

Oestreicher, E., Sengstock, G. J., Riederer, P., Olanow, C. W., Dunn, A. J., and Arendash, G. W. (1994). Degeneration of nigrostriatal dopaminergic neurons increases iron within the substantia nigra: A histochemical and neurochemical study. *Brain Res* **660**, 8–18.

Parenti, M., Rusconi, L., Cappabianca, V., Parati, E. A., and Groppetti, A. (1988). Role of dopamine in manganese neurotoxicity. *Brain Res* **473**, 236–240.

Parkinson, J. (1817). *An Essay on the Shaking Palsy.* Whittingham and Rowland, London.

Pifl, C., Khorchide, M., Kattinger, A., Reither, H., Hardy, J., and Hornykiewicz, O. (2004). Alpha-synuclein selectively increases manganese-induced viability loss in SK-N-MC neuroblastoma cells expressing the human dopamine transporter. *Neurosci Lett* **354**, 34–37.

Racette, B. A., McGee-Minnich, L., Moerlein, S. M., Mink, J. W., Videen, T. O., and Perlmutter, J. S. (2001). Welding-related parkinsonism: Clinical features, treatment, and pathophysiology. *Neurology* **56**, 8–13.

Rand, J. B, Johnson, C. D. (1995). Genetic pharmacology: Interactions between drugs and gene products in *Caenorhabditis elegans*. In Methods in Cell Biology, pp. 187–204. New York.

Rand, J. B., and Nonet, M. L. (1997). Synaptic transmission. In *C. elegans* II (D. L. Riddle, T. Blumenthal, B. J. Meyer, and J. R. Priess, Eds.), pp. 611–643. Cold Spring Harbor Laboratory Press, New York.

Richmond, J. E, and Jorgensen, E. M. (1999). One GABA and two acerylcholine receptors function at the *C. elegans* neuromuscular junction. *Nat Neurosci* **2**, 791–797.

Riddle, D. L., Blumenthal, T., Meyer, B. J., and Priess, J. R. (Eds.) (1997). *C. elegans* II. Cold Spring Harbor Laboratory Press, New York.

Sakaguchi-Nakashima, A., Meir, JY., Jin, Y., Matsumoto, K., and Hisamoto, N. (2007). LRK-1, a *C. elegans* PARK8-related kinase, regulates axonal-dendritic polarity of SV proteins. *Curr Biol* **17**, 592–598.

Sawin, E. R. (1996) Genetic and cellular analysis of modulated behaviors in *Caenorhabditis elegans*. Massachusetts Institute of Technology-Biology Dept. Ph.D thesis.

Sawin, E. R., Ranganathan, R., and Horvitz, H. R. (2000). *C. elegans* locomotory rate is modulated by the environment through a dopaminergic pathway and experience through a serotonergic pathway. *Neuron* **26**, 619–631.

Sidhu, A., Wersinger, C., Moussa, C. E., and Vernier, P. (2004). The role of α-synuclein in both neuroprotection and neurodegeneration. *Ann NY Acad Sci* **1035**, 250–270.

Springer, W., Hoppe, T., Schmidt, E., and Baumeister, R. (2005). A *Caenorhabditis elegans* parkin mutant with altered solubility couples α-synuclein aggregation to proteotoxic stress. *Hum Mol Genet* **14**, 3407.

Stredrick, D. L., Stokes, A. H., Worst, T. J., Freeman, W. M., Johnson, E. A., Lash, L. H., Aschner, M., and Vrana, K. E. (2004). Manganese-induced cytotoxicity in dopamine-producing cells. *Neurotoxicology* **25**, 543–553.

Sulston, J., Dew, M., and Brenner, S. (1975). Dopaminergic neurons in the nematode Caenorhabditis elegans. *J. Comp. Neurol.* **15**, 215–226.

Sultson, J., Du, Z., Thomas, K., Wilson, R., Hillier, L. et al. (1992). The *C. elegans* genome sequencing project: a beginning. *Nature* **356**, 37–41.

The *C. elegans* Sequencing Consortium (1998). Genome sequence of the nematode *C. elegans*: A platform for investigating biology. *Science* **282**, 2012–2018.

The *C. elegans* Sequencing Consortium (2003). *Nematode Sequencing Project*. Wellcome Trust Sanger Institute, Hinxton, Cambridge UK.

Thiruchelvam, M., McCormack, A., Richfield, E. K., Baggs, R. B., Tank, A. W., Di Monte, D. A., and Cory-Slechta, D. A. (2003). Age-related irreversible progressive nigrostriatal dopaminergic neurotoxicity in the paraquat and maneb model of the Parkinson's disease phenotype. *Eur J Neruosci* **18**, 589–600.

Thomas, J. H., and Lockery, S. H. (1999). Neurobiology. In *C. elegans: A Practical Approach* (I. A. Hope, Ed.), pp. 143–179. Oxford University Press, New York.

Timmons, L., and Fire, A. (1998). Specific interference by ingested dsRNA. *Nature* **395**, 854.

Tomas-Camardiel, M., Herrera, A. J., Venero, J. L., Cruz Sanchez-Hidalgo, Cano, J., and Machado, A. (2002). Differential regulation of glutamic acid decarboxylase mRNA and tyrosine hydroxylase mRNA expression in the aged manganese-treated rats. *Brain Res Mol Brain Res* **103**, 116–129.

Touroutine, D., Fox, R. M., Von Stetina, S. E., Burdina, A., Miller, D. M., III, and Richmond, J. E. (2005). acr-16 encodes an essential subunit of the levamisole-resistant nicotinic receptor at the *Caenorhabditis elegans* neuromuscular junction. *J Biol Chem* **280**, 27013–27021.

Uversky, V. N., Li, J., and Fink, A. L. (2001). Metal-triggered structural transformations, aggregation, and fibrillation of human alpha-synuclein. A possible molecular NK between Parkinson's disease and heavy metal exposure. *J Biol Chem* **276**, 44284–44296.

van Ham, T. J., Thijssen, K. L., Breitling, R., Hofstra, R. M. W., Plasterk, R. H. A., and Nollen, E. A. A. (2008). *C. elegans* model identifies genetic modifiers of α-synuclein inclusion formation during aging. *PLoS Genet* **4**(3), e10000027.

Vartiainen, S., Aarnio, V., Lakso, M., and Wong, G. (2006a). Increased lifespan in transgenic *Caenorhabditis elegans* overexpressing human α-synuclein. *Exp Gerontol* **41**, 871–876.

Vartiainen, S., Pehkonen, P., Lakso, M., Nass, R., and Wong, G. (2006b). Identification of gene expression changes in transgenic *C. elegans* overexpressing human α-synuclein. *Neurobiol Dis* **22**, 477–486.

Ved, R., Saha, S., Westlund, B., Perier, C., Burnam, L., Sluder, A., Hoener, M., Rodrigues, CM., Alfonso, A., Steer, C., Liu, L., Przedborski, S., and Wolozin, B. (2005). Similar patterns of mitochondrial vulnerability and rescue induced by genetic modification of alpha-synuclein, parkin, and DJ-1 in *Caenorhabditis elegans*. *J Biol Chem* **280**, 42655–42668.

Westlund, B., Stilwell, G., and Sluder, A. (2004). Invertebrate disease models in neurotherapeutic discovery. *Curr Opin Drug Discov* **7**, 169–178.

Wicks, S. R., Yeh, R. T., Gish, W. R., Waterston, R. H., and Plasterk, R. H. (2001). Rapid gene mapping in *Caenorhabditis elegans* using a high density polymorphism map. *Nat Genet* **28**, 160–164.

Witholt, R., Gwiazda, R. H., and Smith, D. R. (2000). The neurobehavioral effects of subchronic manganese exposure in the presence and absence of pre-parkinsonism. *Neurotoxicol Teratol* **22**, 851–861.

White, J. G., Southgate, E., Thompson, J. N., and Brenner, S. (1986). The structure of the nervous system of the nematode *Caenorhabditis elegans*. *Philos Trans Roy Soc London Ser* **B314**, 1–340.

Wolozin, B., Saha, S., Guillily, M., Ferree, A., and Riley, M. (2008). Investigating convergent actions of genes linked to familiar Parkinson's disease. *Neurodegener Dis* **5**, 3–4.

Wood, W. B. (1988). Introduction to *C. elegans*. In *The Nematode Caenorhabditis elegans* (W. B. Wood, Ed.), pp. 1–16. Cold Spring Harbor Laboratory Press, New York.

Zayed, J., Ducic, S., Campanella, G., Panisset, J. C., Andre, P., Masson, H., and Roy, M. (1990). Environmental factors in the etiology of Parkinson's disease. *Can J Neurol Sci* **17**, 286–291.

Zhang, Y., Ma, C., Delohery, T., Nasipak, B., Foat, B. C., Bounoutas, A., Bussemaker, H. J., Kim, S. K., and Chalfie, M. (2002). Identification of genes expressed in *C. elegans* touch receptor neurons. *Nature* **418**, 331–335.

27

C. *ELEGANS* GENETIC STRATEGIES TO IDENTIFY NOVEL PARKINSON'S DISEASE-ASSOCIATED THERAPEUTIC TARGETS AND LEADS

RICHARD NASS

*Department of Pharmacology and Toxicology, Center for Environmental Health, and
Stark Neuroscience Research Institute, Indiana University School of Medicine,
Indianapolis, IN, USA*

INTRODUCTION

As described within this book, Parkinson's disease (PD) results in the loss of dopamine neurons in the *substantia nigra pars compacta*. It is a chronic and progressive disease, and is the second most prevalent neurodegenerative disease (following Alzheimer's disease) that affects roughly 10% of individuals over 65 years old, and over 4 million people worldwide. Idiopathic PD likely involves both genetic and environmental components, and is characterized by at least three major symptoms that include resting tremors, bradykinesia, and rigidity (Chapter 1). The most effective and commonly prescribed medication is Levadopa (L-dopa). L-dopa is metabolized in the brain to dopamine and is used in treating the motor deficits, although its efficacy declines within a few years and can be associated with significant side effects including dyskinesias (Chapter 1). Other therapeutics that can provide temporary symptomatic relief include dopamine agonists, monoamine oxidase (MAO) inhibitors, and anti-cholinergics; currently there is not a cure for PD.

Despite the recognition of PD as a medical condition for almost 200 years, the high incidence in society, and intensive research within the past decades, a therapeutic has not been identified that can slow the DA neuronal death or the progression of the disease (Chapters 1 and 6; Parkinson, 1817). Why is this? Certainly one answer is because of the complexity of the dopaminergic pathways in the vertebrate brain and the difficulty in modeling PD in experimental systems. These issues are addressed in many chapters within this book. The human brain contains over 100 billion neurons and tens of thousands of DA neurons, each of these cells capable of making many thousands of synaptic connections. This complexity makes it very difficult to identify novel compounds or therapeutic targets when it is not yet clear the primary insults or molecular mechanisms leading to the DA neuron death. Most experimental vertebrate models are also similarly complex, and rodents and *in vitro* models such as cell and tissue culture models often do not fully reflect the *in vivo* state of PD-affected DA neurons that could rapidly facilitate the identification of promising therapeutic leads. These last points are especially evident in the recent failures of the apoptosis inhibitors CEP-1347 and TCH346 in clinical trials; remarkably they each demonstrated significant success in inhibiting cell death *in vitro* and *in vivo*, but were not effective in

PD patients (Waldmeier *et al.*, 2006). Finally, genetic manipulations in vertebrate systems are either not possible or are difficult and very time consuming, and can often take many months to a year or more.

Off-target pharmacology and toxicology may also contribute to the delay of PD therapeutics reaching the market place (Carroll and Fitzgerald, 2003; Liebler and Guengerich, 2005; Guengerich and MacDonald, 2007). During the past 30 years, drug discovery has largely been based on compound screens *in vitro* that involve either the identification of therapeutic leads that alter a known protein's structure or function that is associated with a particular disease, or on cell-based assays that modify a disease phenotype. Compounds identified in these screens are further developed *in vitro* and *in vivo* to increase specificity and efficacy, and decrease toxicity. These screening paradigms, although very effective in identifying and validating therapeutic targets, may not identify unexpected pharmacological interactions or toxicities until late in drug development or in pre-clinical trials, delaying product release due to drug reformulations or clinical trial termination.

A *C. elegans* screening platform for therapeutic target and lead development can be a powerful and rapid alternative to compound screens in *in vitro* and *cell culture* models (Link *et al.*, 2000; Kaletta *et al.*, 2003; Artal-Sanz *et al.*, 2006; Kaletta and Hengartner, 2006). In this chapter we briefly describe the utility of *C. elegans* to identify and characterize novel PD-associated therapeutic targets and lead compounds, and their associated molecular pathways.

C. ELEGANS: ADDITIONAL ATTRIBUTES AMENDABLE TO DRUG DISCOVERY AND TARGET VALIDATION FOR PD

As described in Chapters 24 and 26, *C. elegans* is a powerful tool for genetic analysis and determination of molecular mechanisms involved PD-associated DA neuron cell death. These characteristics will not be recapitulated in detail here, except to reiterate that many of these same attributes are why this organism is favorable as a drug and target discovery platform (for excellent reviews, see Kaletta *et al.*, 2003; Artal-Sanz *et al.*, 2006; Kaletta and Hengartner, 2006). As discussed earlier, *C. elegans* genes and biochemical pathways are strongly conserved with humans, and the ease and rapidity of

generating transgenic animals, deletion mutants, genetic crosses and mapping, and protein expression knockdowns, allows for quick analysis of the molecular targets and pathways involved in a therapeutic effect (Riddle *et al.*, 1997; Chapters 24 and 26). The animal's small size, large brood size, short generation time, transparency, and ease of growth in agar or liquid culture in 96- or 384-well microtiter plates also facilitates rapid compound screening and phenotypic analysis of neurons and other cell populations (Kaletta and Hengartner, 2006; Chapter 26; Nass and Hamza, 2007).

C. elegans has provided key insight into the molecular mechanisms involved in several human neurodegenerative disorders, including PD, it is therefore likely that this conservation will extend to therapeutic target identification and validation (Nass and Chen, 2008). As reviewed in Chapters 24 and 26, *C. elegans* toxin and genetic PD models have been developed that recapitulates many aspects of the human disorder (Chapter 24 and 26). *C. elegans* Alzheimer's disease models have demonstrated that human presenilin can substitute for the worm presenilin gene, and expression of the human B-amyloid (AB) peptide can induce worm amyloid deposits, suggesting strong conservation within these disease pathways (Levitan and Greenwald, 1995; Sherrington *et al.*, 1995; Link, 2001). Overexpression of a human Huntingtin fragment with an expanded polyglutamine in the worm causes specific polyglutamine length-dependent neuronal dysfunction and aggregation, also recapitulating significant aspects of the human disorder (Faber *et al.*, 1999). Furthermore, regulators of the polyglutamine aggregation have been successfully identified using a florescence marker in a genome-wide RNAi screen, further supporting the utility of the worm in human neurodegenerative studies (Nollen *et al.*, 2004).

C. elegans is a powerful and facile system for drug discovery and target validation. The worm is sensitive to a number of human neuroreactive compounds including acetylcholine receptor agonist, anesthetics, cholinesterase inhibitors, immunosuppressants, and serotonin, GABA, and dopamine-related compounds, and these drugs appear to interact with the worm homologs and at the orthologous active sites (Rand and Johnson, 1995; Nass and Blakely, 2003; Artal-Sanz *et al.*, 2006; Chapter 26). A significant advantage that a *C. elegans* compound HTS platform has relative to an *in vitro* or tissue culture platform is that the therapeutic target is in its native biological environment, and is therefore more likely to mimic the *in vivo* state (Horrobin, 2001). This screening protocol would also more likely identify novel therapeutic targets and molecular pathways, as well as identify off-target toxicological effects, since the whole organism screening approach provides an opportunity for multiple types of drug interactions, and is not bias for a specific therapeutic target. Furthermore, the *C. elegans* RNAi feeding library allows for the rapid identification of drug targets and the associated biochemical pathways without the necessity of cloning or genetic mapping (Fire *et al.*, 1998; Montgomery *et al.*, 1998; Kamath *et al.*, 2003; Artal-Sanz *et al.*, 2006). Finally, the lack of a blood brain barrier that could inhibit some compounds from reaching the worm's nervous system, and the ease of growth of the nematode in up to 2% DMSO, the solvent and antioxidant commonly used in drug screening libraries, greatly facilitates *C. elegans* HTS (Chapter 26).

C. ELEGANS FORWARD AND REVERSE GENETIC SCREENS TO IDENTIFY PD-ASSOCIATED THERAPEUTIC TARGETS

C. elegans provides opportunities to identify novel drug targets through forward and reverse genetics (Artal-Sanz *et al.*, 2006; Chapters 24 and 26). The 6-OHDA screen described in Chapter 26 is an example of a forward genetic screen. The strength of this screen is that no *a priori* knowledge about the molecular pathways involved in the DA neurodegeneration is needed to identify novel therapeutic targets. In this particular screen, worms that express the green fluorescent protein (GFP) reporter in the DA neurons are then exposed to a mutagen, and allowing for homozygosity of the mutation in two generations, the progeny are exposed to the PD-associated DA neurotoxin 6-OHDA (Chapter 26; Nass *et al.*, 2005). After 3 days of growth on bacteria plates, the worm's DA neurons are examined under a fluorescent dissecting scope, and those that retain the GFP in the DA neurons likely contain mutations within a biochemical pathway that protects against the neuronal death. The genes involved in conferring the 6-OHDA resistant phenotype would then be identified by using standard worm methods involving genetic crosses, genetic and physical maps, and/or single-nucleotide polymorphism analysis (snip-SNP) (Wicks, 2001). A similar screen was used to identify a number of animals with mutations in the dopamine transporter (DAT) and other proteins, and provides proof-of-concept that this screening protocol could be a useful tool to identify proteins involved in PD-associated toxin-induced DA cell death (Nass *et al.*, 2005; Chapter 26).

A reverse genetic screen to identify modifiers of a PD-associated phenotype of α-synuclein aggregation in *C. elegans* was recently described by Nollen and colleagues (van Ham *et al.*, 2008; Chapter 26). In this assay, the investigators screened human α-synuclein expressing worms against a RNA interference (RNAi) bacteria library to identify animals in which α-synuclein would prematurely form internal inclusions (see Chapter 26; van Ham *et al.*, 2008). As described in Chapter 26, each bacteria clone in a RNAi library can be induced to express dsRNA to a single *C. elegans* gene, and when the bacteria is engulfed by the worm, causes a decrease in expression of that particular protein. One strength of this type of protocol is that animals can rapidly be screened in 96-well titer plates following growth in wells containing the different bacteria clones. The whole genome can be assayed in about a month, and is cost effective since it involves only about 200 microtiter plates (Artal-Sanz *et al.*, 2006; Kaletta and Hengartner, 2006). Another significant advantage of this screen is that the identification of the gene involved in the modulation of the PD-associated phenotype does not depend on cloning or genetic mapping.

Nollen and colleagues followed a similar protocol and screened almost 17,000 bacteria clones (representing approximately 85% of the genome) (van Ham *et al.*, 2008). The investigators identified 80 genes that are involved in premature inclusion formation. Remarkably, even though α-synuclein is not an endogenous worm protein, and the protein was expressed in the muscle and not the DA neurons (i.e., where the α-synuclein containing inclusions are often observed in PD), the investigators were able to identify a number of novel proteins as well as those that have previously been implicated in aging and other PD models (van Ham *et al.*, 2008). Each of these proteins is a potential therapeutic target, and its affect on other PD-associated proteins or mutations can rapidly be accessed to determine where it may enhance or suppress inclusion formation or neurodegeneration. This study also reinforces the utility of the *C. elegans* screening platform in identifying genes that may be involved in a human disorder, even though the gene conferring the pathology is not evolutionarily conserved.

C. ELEGANS HTS TO IDENTIFY DA NEUROPROTECTIVE COMPOUNDS

Most of the *C. elegans* PD models rely on the over-expression or deletion of a PD-related protein or the exposure of the animals to a neurotoxin that would cause DA neuronal death (Chapter 26). A relatively simple screen to identify compounds that could protect against the cell death could follow an algorithm as in Figure 27.1a. Young animals (L1 larvae) co-expressing the GFP and human A53T α-synuclein in the DA neurons, which upon maturing would normally cause DA neurodegeneration (Figure 27.2), would be distributed in 96-well microtiter plates, with each well containing a single compound from a drug library (Lakso *et al.*, 2003). The animals would be allowed to grow into adulthood, and following the several day incubation, fluorescence in each well would be determined in a fluorescent plate reader. Those wells emitting a strong fluorescent signal would likely contain a compound that protects against DA neurodegeneration. The compounds would then be tested in secondary assays, such as in primary cell, heterologous expression systems, or rodents to confirm and/or characterize the compound's neuroprotective properties.

A *C. elegans* toxin-based therapeutic lead screen can follow a similar algorithm, yet provides opportunities to select compounds that protect against different steps of the neurodegeneration process. For example, we could pre-expose the animals to each compound in the library, wash the animals, and then expose to a DAT-specific toxin to select for compounds that may induce neuroprotective proteins involved in an early response to insult. Conversely, exposing the animals to a drug library after toxin exposure may select for compounds that specifically inhibit or decrease further cell death only after the neurodegenerative cascade has been initiated. The latter protocol will also likely avoid false positives that may inhibit uptake of the toxin through DAT.

In a project largely supported by the Michael J. Fox Foundation for Parkinson's Research, we co-exposed young worms expressing GFP in the DA neurons to 6-OHDA and to a 2000 chemical library and other select compounds. Our exposure protocol allowed for acute 30 min co-exposures of 3–5 mM 6-OHDA/1% DMSO with the tested compound, as well as longer exposures up to 72 hours in liquid media (Figure 27.1b), or on agar plates.

This paradigm allowed us to recover candidate neuroprotectants targeting all steps of the neurodegeneration process. DAT antagonists were also expected to be one class of compounds recovered in the screen, which would assist in validating the screen (Nass *et al.*, 2002).

Our screen resulted in the identification of a number of compounds that significantly protected against 6-OHDA-induced DA neuron cell death. As expected, some of these included DAT antagonists, antidepressants, and neuroleptics, which could interfere with DAT to inhibit the uptake of 6-OHDA (Tatsumi, *et al.*, 1997; Tatsumi *et al.*, 1999; Nass *et al.*, 2002). We also identified a number of compounds that are unlikely to interact with DAT, including the human NMDA receptor antagonist dextromethorphan, the antimicrobial agent colistimethate, the flower component trichilenone, and several human D_2 agonists (Figure 27.3). Dextromethorphan also protects against MPTP toxicity in a rodent PD model, providing further proof of concept that the assay is capable of identifying compounds that protect against vertebrate DA neuron cell death (Zhang *et al.*, 2004).

The identification of the D_2 agonist quinpirole (QPR), bromocriptine (BMC), and the mixed D_1/D_2 agonist dihydrexidine as neuroprotective compounds is also notable. Both D_1 and D_2 receptor families have previously been implicated in playing roles in neuroprotection, and D_2 agonist can prevent the loss of substantial nigral neurons and DA in the striatum typically observed in 6-OHDA-lesioned rats (McLaughlin, 2001). Remarkably, quinpirole (at low [mM]) can also provide significant protection against 6-OHDA-induced DA neuron cell death in the worm if the worm is exposed to the agonist *prior* to toxin exposure, suggesting that activation of a signaling cascade could be involved in the neuroprotection. Secondary assays show that the neuroprotection is dose dependent, and consistent with mammalian studies, quinpirole and bromocriptine did not interfere with DA transport when DAT-1 is expressed in HEK293 cells. Finally, our preliminary studies suggest that the absence of D_1, D_2, and D_3 receptors in the worm do not alter the neuroprotective properties of the agonist, suggesting that the D_2 agonists are not directly dependent on these receptors to confer neuroprotection. It is possible that the agonist may interact with other unidentified DA receptors or a completely independent signaling pathway.

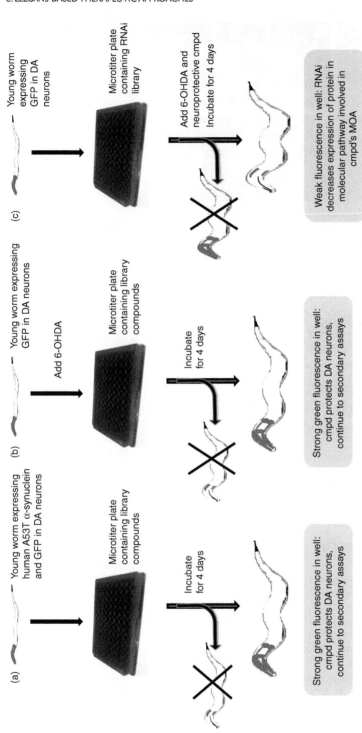

FIGURE 27.1 Chemical screens to identify therapeutic leads and MOA. (a) A putative screen to identify compounds that protect against α-synuclein-induced DA neurodegeneration. Several hundred embryos or L1 larvae expressing human α-synuclein and GFP in the DA neurons are placed in wells in a microtiter plate in which each well contains a potential therapeutic lead. Following growth in liquid or on agar media, the total fluorescence in each well is determined by a fluorescent plate reader. Wells that emit a strong fluorescent signal contain a compound that likely protects against α-synuclein-induced DA neuron degeneration, and should be tested in secondary assays. (b) A putative screen to identify compounds that protect against 6-OHDA-induced DA neuron degeneration. Assay performed similar as (a) except worms do not express α-synuclein, and 6-OHDA is added to assay. (c) A putative MOA screen to identify drug targets and their associated molecular pathways. Young animals are added to each well each containing a bacteria clone expressing dsRNA to a single C. elegans gene. Wells in which animals show little or no GFP fluorescence likely have a reduction in expression of a protein that is a necessary for the drugs efficacy.

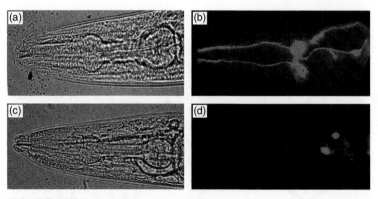

FIGURE 27.2 Expression of human A53T α-synuclein causes *C. elegans* DA neurons to degenerate. Anterior of adult animal in WT (a) and (b) shows normal compliment of DA neurons, while A53T expressing animals (c) and (d) often show loss of cell bodies and processes. In (b) only four of the six neurons are clearly visible because of the focal plane; in (d) only the two ADE cell bodies and part of the processes are present. Reproduced with permission from Nass and Nichols (2007).

Compound Name	Therapeutic Class
Dextromethorphan	NMDA receptor antagonist
Alfaxalone	GABA receptor agonist
Quinpirole	D_2 receptor agonist
Bromocriptine	D_2 receptor agonist
Dihydrexidine	D_1/D_2 receptor agonist

FIGURE 27.3 Mammalian receptor agonist and antagonist protect against 6-OHDA-induced DA neurodegeneration. The scoring assay is similar to that described in Nass *et al.* (2002).

C. ELEGANS AND MOA GENETIC SCREENS

A significant hurdle in moving a promising therapeutic lead to pre-clinical trials is the identification of its molecular targets and its interacting molecular pathways. The MOA (mode of action) is especially important for target validation and optimization of lead compounds. Ideally the mechanisms of serious adverse off-target effects should be identified early in the drug discovery process to reduce misplaced effort and financial resources (Artal-Sanz *et al.*, 2006).

Forward and reverse chemical genetic screens in *C. elegans* provide opportunities to accelerate the identification of therapeutic targets and MOA of compounds for human neurodegenerative diseases (Chapter 24; Artal-Sanz *et al.*, 2006; Kaletta and Hengartner, 2006). The strong conservation of biochemical and disease pathways between *C. elegans* and humans suggest that the therapeutic targets and their MOAs in the worm will likely have significant correlates in man. Furthermore, since the screen occurs within a whole animal, novel genes or pathways that are not expected to play a role in the phenotype can rapidly be identified. Prior examples of successful drug target and MAO studies in *C. elegans* include human anticancer, anticonvulsant, and anitdepressant compounds (Artal-Sanz *et al.*, 2006).

An example of a reverse genetic screen that may identify the therapeutic target and MOA of quinpirole from our screen is outlined in Figure 27.1c. In this chemical reverse genetics screen, young animals expressing GFP in the DA neurons are incubated in 96-well microtitre plates in which each well contains a bacteria culture from a single clone from the RNAi library. 6-OHDA and quinpirole are added, and following an appropriate incubation period, total fluorescence is determined in each well. Wells that contains a **weak or absent** fluorescent signal contain bacteria expressing dsRNA to a *C. elegans* gene that is involved in quinpirole's neuroprotective effects. This screen should identify quinpirole's molecular target and its affected molecular pathways. A strength of this approach is that gene involved in suppression of the drug effects is immediately known. Another strength of this *in vivo* screen is that genes responsible for synergistic or off-target effects could be identified quickly if other phenotypes are observed.

An alternative to a reverse genetic screen using the RNAi library described above would be a forward chemical genetic screen. In this screen, larvae expressing GFP in the DA neurons would be mutagenized as described in Chapter 26, and the F2 progeny exposed to 6-OHDA and quinpirole. Animals in which quinpirole did *not* protect against 6-OHDA-induced DA neuron cell death would be selected, and using classical worm crosses and mapping or SNP analysis, the gene carrying the mutation would be identified. An advantage of this approach is that it is a relatively unbiased approach to identifying suppressors or enhancers of genes affected by the compound, and concerns of RNAi sensitivities in neurons will not need to be addressed.

PERSPECTIVES

C. elegans is a powerful genetic model for elucidating and characterizing the molecular components involved in human cellular processes and disease. This recognition by the scientific community has culminated in the awards of two Nobel prizes in Physiology or Medicine this past decade. The highly conserved genome, facile genetic analysis and manipulations, and the ease and rapidity of genetic and chemical screens coupled with HTS methodologies, suggest that *C. elegans* will play an integral role in the identification in the future of novel human therapeutic targets and leads. The needs of 21st century medicine for the acceleration of human drug discovery, optimization, and target validation may finally be realized by this tiny, unobtrusive, and ancient nematode called *C. elegans*.

REFERENCES

Artal-Sanz, M., de Jong, L., and Tavernarakis, N. (2006). *Caenorhabditis elegans*: A versatile platform for drug discovery. *Biotechnol J*, 1405–1418.

Bargmann, C. I. (1998). Neurobiology of the *Caenorhabditis elegans* genome. *Science* 282, 2028–2033.

Braungart, E., Gerlach, M., Riederer, P., Baumeister, R., and Hoener, M. C. (2004). Caenorhabditis elegans MPP+ model of Parkinson's disease for high-throughput drug screening. *Neurodegener Dis* 1, 175–183.

Carroll, P. M., and Fitzgerald, K. (ed.) (2003). *Model Organisms in Drug Discovery*. John Wiley and Sons, Holbroken, NY.

Faber, P. W., Alter, J. R., MacDonald, M. E., and Hard, A. C. (1999). Polyglutamine-mediated dysfunction and apoptotic death of a *Caenorhabditis elegans* sensory neuron. *Proc Natl Acad Sci USA* 101, 6403–6408.

Fire, A., Xu, S., Montgomery, M. K., Kostas, S. A., Driver, S. E., and Mello, C. C. (1998). Potent and specific genetic interference by double-stranded RNA in *Caenorhabditis elegans*. *Nature* 391, 806–811.

Guengerich, F. P., and MacDonald, J. S. (2007). Applying mechanisms of chemical toxicity to predict drug safety. *Chem Res Toxicol* 20, 344–369.

Horrobin, D. F. (2001). Realism in drug discovery – Cassandra be right? *Nat Biotechnol* 19, 1099–1100.

Kaletta, T., and Hengartner, M. O. (2006). Finding function in novel targets: C. *elegans* as a model organism. *Nat Rev Drug Discov* 5, 387–398.

Kaletta, T., Butler, L., and Bogaert, T. (2003). *Caenorhabditis elegans*: functional genomics in drug discovery: expanding paradigms. In *Model Organisms in Drug Discovery* (P. M. Carroll, and K. Fitzgerald, (Eds.). John Wiley and Sons.

Kamath, R. S., and Ahringer, J. (2003). Genome-wide RNAi screening in Caenorhabditis elegans. *Methods* 30, 313–321.

Kamath, R. S., Fraser, A. G., Dong, Y., Poulin, G., Durbin, R., Gotta, M., Kanapin, A., Le Bot, N., Moreno, S., Sohrmann, M. *et al.* (2003). Systematic functional analysis of the *Caenorhabditis elegans* genome using RNAi. *Nature* 421, 231–237.

Lakso, M., Vartianen, S., Moilanen, A. M., Sirvio, J., Thomoas, J. H., Nass, R., Blakely, R. D., and Wong, G. (2003). Dopaminergic neuronal loss and motor deficits in *Caenorhabditis elegans* overexpressing human α-synuclein. *J Neurochem* 86, 165–172.

Liebler, D. C., and Guengerich, F. P. (2005). Elucidating mechanisms of drug-induced toxicity. *Nat Rev* 4, 410–420.

Link, C. D. (2001). Transgenic invertebrates models of age-associated neurodegenerative diseases. *Mech Ageing Dev* 122, 1639–1649.

Link, E. M., Hardiman, G., Sluder, A. E., Johnson, C. D., and Liu, L. X. (2000). Therapeutic target discovery using *Caenorhabditis elegans*. *Pharmacogenomics* 1, 203–217.

McLaughlin, B. (2001). *Dopamine Neurotoxicity and Neurodegeneration*. Humana Press, Totowa, NJ, pp. 195–231.

Montgomery, M. K., Xu, S., and Fire, A. (1998). RNA as a target of double-stranded RNA-mediated genetic interference in *Caenorhabditis elegans*. *Proc Natl Acad Sci USA* 95, 15502–15507.

Nass, R., Hahn, M. K., Jessen, T., McDonald, P. W., Carvelli, L., and Blakely, R. D. (2005). A genetic screen in *Caenorhabditis elegans* for dopamine neuron insensitivity to 6-hydroxydopamine identifies dopamine transporter mutants impacting transporter biosynthesis and trafficking. *J Neurochem* **94**, 774–785.

Nass, R., Hall, D. H., Miller, D. M., and Blakely, R. D., III (2002). Neurotoxin-induced degeneration of dopamine neurons in *Caenorgabditis elegans*. *Proc Nat Acade Sci USA* **99**, 3264–3269.

Nass, R., and Blakely, R. B. (2003). The *Caenorhabditis elegans* dopaminergic system: Opportunities for insights into dopamine transport and neurodegeneration. *Annu Rev Pharmacol Toxicol* **43**, 521–544.

Nass, R., and Hamza, I. (2007). The nematode *C. elegans* as a model to explore toxicology in vivo: solid and axenic growth culture conditions and compound exposure parameters. *Curr Protocols Toxicol* **15**, 1.9.1–1.9.18.

Nass, R., Nichols, S. (2007). Invertebrates as powerful genetic models for human neurodegenerative disease. D.R. Sibley, I. Hanin, M. Kuhar, P. Skolnick, (Eds.), pp. 567–588. John Wiley and Sons, Inc, Holboken, New Jersey

Nollen, E. A. *et al.* (2004). Genome-wide RNA interference screen identifies previously undescribed regulators of polyglutamine aggregation. *Proc Natl Acad Sci USA* **101**, 6403–6408.

Parkinson, J. (1817). *An Essay on the Shaking Palsy.* Whittingham and Rowland, London.

Rand, J.B., Johnson, C.D. (1995). Genetic pharmacology: Interactions between drugs and gene products in *Caenorhabditis elegans*. In Methods in Cell Biology, pp. 187–204. New York.

Riddle, D. L., Blumenthal, T., Meyer, B. J., and Priess, J. R. (Eds.) (1997). *C. elegans* II. Cold Spring Harbor Laboratory Press, New York.

Sluder, A., and Baumeister, R. (2004). *From genes to drugs: target validation in Caenorhabditis elegans.* Drug Discovery Today: Technologies, Elsevier, 171–177.

Tatsumi, M., Groshan, K., Blakely, R. D., and Richelson, E. (1997). Pharmacological profile of antidepressants and related compounds at human monoamine transporters. *Eur J Pharmacol* **340**, 249–258.

Tatsumi, M., Jansen, K., Blakely, R. D., and Richelson, E. (1999). Pharmacological profile of neuroleptics at human monoamine transporters. *Eur J Pharmacol* **368**, 277–283.

Timmons, L., and Fire, A. (1998). Specific interference by ingested dsRNA. *Nature* **395**, 854.

van Ham, T. J., Thijssen, K. L., Breitling, R., Hofstra, R. M. W., Plasterk, R. H. A., and Nollen, E. A. A. (2008). *C. elegans* model identifies genetic modifiers of α-synuclein inclusion formation during aging. *PLoS Genet* **4**(3), e10000027.

Vartiainen, S., Pehkonen, P., Lakso, M., Nass, R., and Wong, G. (2006). Identification of gene expression changes in transgenic *C. elegans* overexpressing human α-synuclein. *Neurobio. Dis* **22**, 477–486.

Waldmeier, P., Bozyczko-Coyne, D., Williams, M., and Vaught, J. L. (2006). Recent clinical failures in Parkinson's disease with apoptosis inhibitors underline the need for a paradigm shift in drug discovery for neurodegenerative diseases. *Biochem Pharmacol* **72**, 1197–1206.

Westlund, B., Stilwell, G., and Sluder, A. (2004). Invertebrate disease models in neurotherapeutic discovery. *Curr. Opin. Drug Discov* **7**, 169–178.

Wicks, S. R., Yeh, R. T., Gish, W. R., Waterston, R. H., and Plasterk, R. H. (2001). Rapid gene mapping in *Caenorhabditis elegans* using a high density polymorphism map. *Nat Genet* **28**, 160–164.

Zhang, W., Wang, T., Qin, L., Gao, H. M., Wilson, B., Ali, S. F., Zhang, W., Hong, J., and Liu, B. (2004). Neuroprotective effect of dextromethorphan in the MPTP Parkinson's disease model: role of NADPH oxidase. *FASEB J* **18**, 589–591.

VI

CELL-BASED MODELS

IV

CELL-BASED MODELS

28

OVERVIEW: CELL-BASED MODELS OF PARKINSON'S DISEASE

D. JAMES SURMEIER, Ph.D.

Department of Physiology, Feinberg School of Medicine Northwestern University
303 E. Chicago Ave, Chicago, IL 60611

THE CHALLENGE OF UNRAVELING THE ETIOLOGY OF PARKINSON'S DISEASE

In the last decade, there have been major strides made toward understanding the etiology of Parkinson's disease (PD). But, much remains. For decades, the assault on the roots of PD has been based upon two key assumptions. The first assumption is that PD is first and foremost a disease of dopaminergic neurons in the substantia nigra pars compacta (SNc). This assumption is based in large measure on the tight correlation between the severity of the disease (at least as measured by motor dysfunction) and the loss of these neurons, together with the success of levodopa therapy in alleviating the motor symptoms of the disease (Dauer and Przedborski, 2003; Fahn and Sulzer, 2004). Although this assumption has been challenged in recent years (Braak *et al.*, 2004), there remain compelling reasons to believe that there is something special about SNc dopaminergic neurons, and understanding why they die will lead to a fundamental insight into the pathogenesis in PD.

The second operating assumption is that toxins capable of producing a near selective loss of SNc dopaminergic neurons tell us something important about disease mechanisms. There is no doubt that toxins like 1-methyl-4-phenyl-1, 2, 3, 6-tetrahydropyridine (MPTP) have provided us with an extremely valuable window into cellular function and the consequences of oxidative stress on neurons. The correlation between aging, PD and loss of mitochondrial oxidative phosphorylation capacity in the mesencephalon has buttressed the scientific rationale for pursuing this line of study. Cell-based models, particularly cell lines with a dopaminergic phenotype, like PC-12 (see Chapter 29), N27 cells (see Chapter 36) and primary cultures of embryonic dopaminergic neurons (see Chapter 30) have proven to be excellent tools for dissecting the cascade of molecular events initiated by these toxins and the identification of potentially neuroprotective compounds, some of which are now in clinical trials. Yet, there are legitimate concerns about toxins as a window into the disease process in PD. Certainly, the vast majority of PD cases are not linked to ingestion of MPTP. Moreover, the cellular specificity of MPTP and 6-OHDA turn upon their "accidental" structural similarity to dopamine (DA) and consequent uptake by the plasma membrane DA transporter. The recent discovery that structurally dissimilar pesticides, like rotenone, have a similar mode of action and produce a similar pattern of cell loss has certainly given new impetus to this field of study (Betarbet *et al.*, 2000).

The discovery of genetic mutations that increase the probability of developing PD (Moore *et al.*, 2005) has had a fundamental impact on the pursuit of pathogenic mechanisms in PD, redirecting much of the field away from the conventional approaches. Unlike the situation with toxins, there is no doubt that these mutations are linked to pathogenesis. The problem is that for the most part, we don't know the function of the proteins coded for by these genes. Cell-based models are proving extremely valuable tools for figuring out what these proteins do and how PD-linked mutations alter their function. Studies at this stage are not limited to cells with a dopaminergic phenotype. All of the genes linked to PD thus far are expressed broadly, not just in dopaminergic neurons (Moore *et al.*, 2005). This suggests the pathology in at least familial cases of PD is sensitive to modifiers that elevate the negative consequences of these mutations (or replication in the case of α-synuclein) in SNc dopaminergic neurons. Two genes linked to autosomal dominant forms of PD – *SNCA* and *LRRK2* – are the subject of chapters here. Chapters 29, 31 and 36 review what is know about the function of α-synuclein (the protein encoded by SNCA) and the consequences of PD-associated mutations; Chapter 31 also discusses the potential pathogenic effect of α-synuclein on autophagy. Chapter 32 focuses on insights gained into the function of LRRK2 from cell-based assays.

CELL-BASED MODELS AND PATHOGENIC MECHANISMS IN PD

At the outset, it is worth noting that cell-based models have both advantages and disadvantages. These models are particularly valuable in tracking the consequences of toxin exposure and genetic mutations in a cellular environment. Studies of 6-OHDA, MPP$^+$ and pesticide toxins have relied heavily upon cell lines, like PC-12 (Chapter 29) and N27 cells (Chapter 36), with phenotypic features resembling those of post-mitotic SNc dopaminergic neurons. The advantage of cell lines is ease of generation, phenotypic homogeneity (although see Malagelda and Greene), transfectability and accessibility to pharmacological manipulation as well as physiological and anatomical analysis. The obvious disadvantage is their uncertain expression of genes that might turn out to be causally linked to the susceptibility of SNc DA neurons in PD. This concern is much less prominent with dopaminergic neurons cultured from embryonic brains (Chapter 30). The use of these cultures has provided key information about how the dopaminergic phenotype might

contribute to toxin vulnerability. That said, their heterogeneity limits their utility for biochemical and molecular assays that require a substantial cell numbers. The development of transgenic mouse lines in which dopaminergic neurons report their expression of tyrosine hydroxylase (Gong *et al.*, 2003) could not only enhance the utility of this preparation for physiological analysis, but also enable the creation of cultures enriched in dopaminergic neurons by the use of fluorescence-activated cell sorting. The evolution of strategies to specify the dopaminergic phenotype in embryonic stem (ES) cells (see Chapter 34) promises to provide another important model system. These studies suggest that a potentially significant limitation of both cell lines and primary cultures is their uncertain relationship to fully differentiated, adult dopaminergic neurons and, as a consequence, their relevance for the study of an aging-linked disease like PD. This issue is developed in Chapter 34 where they present fascinating new data showing that loss of one fox2a gene leads to an aging-related loss of SNc dopaminergic neurons and parkinsonian symptoms in mice.

At this stage of investigation, the utility of cell-based models in the pursuit of genetic mechanisms in PD is perhaps stronger, since even the basic function of most of the proteins encoded by these genes is still being explored. Chapter 32 does an excellent job of discussing the strengths and weaknesses of the various neuronal and non-neuronal cell-based models in the pursuit of LRRK2 function. Similarly, Chapters 29, 31 and 36 also do an excellent job in discussing these issues within the context of α-synuclein function. Most of these models are amenable to a variety of pharmacological and genetic manipulations.

An important issue raised in these chapters is the role of interacting proteins and aging in determining the functional consequences of mutations linked to PD. None of the mouse models of PD recapitulate the pattern of pathology seen in PD, particularly the degeneration of SNc DA neurons. The obvious interpretation of this result is that allelic variation and environmental factors, like pesticides, must be important factors in diseases progression.

INSIGHTS FROM THE STUDIES OF CELL-BASED MODELS

In addition to visiting methodological issues related to cell-based experimental approaches, the chapters in this section do an excellent job of summarizing recent insights into the potential pathogenic mechanisms in PD.

- Escobar-Khondiker *et al.* provides an insightful summary of work with primary cultures of mesencephalic dopaminergic neurons examining the cascade of events triggered by mitochondrial toxins like MPP+ that lead to the production of reactive oxygen species and eventual cell death, as well as how this death spiral might be stopped by trophic factors.
- Sun *et al.* summarizes several lines of study implicating PKCδ in the response of dopaminergic N27 neurons to toxins and oxidative stress. The high expression level of PKCδ in vulnerable SNc DA neurons is a particularly intriguing observation and identifies it as a potential therapeutic target. Their chapter also summarizes new work suggesting that α-synuclein has both neuroprotective and deleterious effects in response to toxin exposure.
- Malagelada and Greene summarize a vast range of studies using PC12 cells as models in PD. One the most fascinating studies they describe that takes full advantage of the PC12 model is the description of Serial Analysis of Gene Expression (SAGE) strategy to characterize the transcriptome before and after exposure to 6-OHDA. Mining this data set has identified old friends, but a host of new ones that could provide novel insights into the cellular response to stress underlying PD.
- Kaushik *et al.* summarize the evidence linking defects in non-proteasomal handling of misfolded or damaged proteins by autophagy – microautophagy, macroautophagy and chaperone-mediated autophagy – in the pathogenesis of PD. In addition to providing a scholarly summary of the features of each form of autophagy, they discuss the potential role of DA adducts of α-synuclein and mutations of α-synuclein in "clogging up" this critical pathway for clearance of pathogenic proteins. They also discuss the complications associated with therapeutic interventions to correct PD-associated defects.
- Cookson *et al.* provides a very thoughtful overview of the current understanding of LRRK2 function. Mutations in this gene are associated with an autosomal dominant form of PD with a relatively high penetrance in some ethnic groupings (e.g., Ishihara *et al.*, 2007). With a keen awareness of experimental pitfalls, Cookson *et al.* carefully map out what we can safely conclude about the functional domains of LRRK2 and how these functions are affected by mutations linked to PD. The authors point out the limitations posed by not knowing the natural substrates of LRRK2 and how the interaction between the GTPase and kinase domains might regulate activity. Insightfully, they note that since the toxic effects of mutant LRRK2 appear to be dependent upon its kinase activity and directly targeting kinase activity is not a viable therapeutic strategy, this regulatory interaction between the GTPase and kinase domains could be a target to shoot for.
- The chapter by Kittappa *et al.* has a distinctive focus. The authors summarize efforts to determine the lineage of vulnerable mesencephalic dopaminergic neurons and the factors that control their differentiation and survival. They detail an elegant series of experiments pointing to the conclusion that SNc dopaminergic neurons are derived from floor plate cells that are distinct from the precursors that generate the majority of neurons in the brain. This result has fundamental implications for how we approach selective vulnerability in PD. Their review also discusses evidence that the transcription factor fox2a, a member of the forkhead family, is necessary and sufficient for specification of mesencephalic dopaminergic neurons. Not only does this have implications for the generation of dopaminergic neurons from ES cells, it has implications for the etiology of PD as well since fox2a haploinsufficiency leads to a late onset degeneration of mesencephalic dopaminergic neurons (that is not accompanied by Lewy body formation) and a parkinsonian phenotype.

REFERENCES

Betarbet, R., Sherer, T. B., MacKenzie, G., Garcia-Osuna, M., Panov, A. V., and Greenamyre, J. T. (2000). Chronic systemic pesticide exposure reproduces features of Parkinson's disease. *Nat Neurosci* 3, 1301–1306.

Braak, H., Ghebremedhin, E., Rub, U., Bratzke, H., and Del Tredici, K. (2004). Stages in the development of Parkinson's disease-related pathology. *Cell Tissue Res* 318, 121–134.

Dauer, W., and Przedborski, S. (2003). Parkinson's disease: Mechanisms and models. *Neuron* **39**, 889–909.

Fahn, S., and Sulzer, D. (2004). Neurodegeneration and neuroprotection in Parkinson disease. *NeuroRx* **1**, 139–154.

Gong, S., Zheng, C., Doughty, M. L., Losos, K., Didkovsky, N., Schambra, U. B., Nowak, N. J., Joyner, A., Leblanc, G., Hatten, M. E., and Heintz, N. (2003). A gene expression atlas of the central nervous system based on bacterial artificial chromosomes. *Nature* **425**, 917–925.

Ishihara, L., Gibson, R. A., Warren, L., Amouri, R., Lyons, K., Wielinski, C., Hunter, C., Swartz, J. E., Elango, R., Akkari, P. A., Leppert, D., Surh, L., Reeves, K. H., Thomas, S., Ragone, L., Hattori, N., Pahwa, R., Jankovic, J., Nance, M., Freeman, A., Gouider-Khouja, N., Kefi, M., Zouari, M., Ben Sassi, S., Ben Yahmed, S., El Euch-Fayeche, G., Middleton, L., Burn, D. J., Watts, R. L., and Hentati, F. (2007). Screening for Lrrk2 G2019S and clinical comparison of Tunisian and North American Caucasian Parkinson's disease families. *Mov Disord* **22**, 55–61.

Moore, D. J., West, A. B., Dawson, V. L., and Dawson, T. M. (2005). Molecular pathophysiology of Parkinson's disease. *Annu Rev Neurosci* **28**, 57–87.

29

PC12 CELLS AS A MODEL FOR PARKINSON'S DISEASE RESEARCH

CRISTINA MALAGELADA AND LLOYD A. GREENE

Department of Pathology and Neurobiology
Columbia University, New York, NY, USA

INTRODUCTION

Established cell lines represent potentially useful models for studying the causes as well as for finding means to treat a variety of neurodegenerative disorders including Parkinson's disease (PD). Our purpose here will be to focus on one such cell line – the rat PC12 pheochromocytoma cell – that has been widely applied (as reported in over 300 publications at the time this chapter was prepared in March, 2007) for studies related to PD. We will discuss the potential advantages and weaknesses of this model for PD research as well as provide specific examples of how it has been employed. We will also consider approaches that might be used to further enhance the utility of the cell line for PD research. As the reader is likely to be aware, PC12 cells are by no means the only line with potential use for studying PD. Additional cell lines derived from tumors, immortalized neuroblasts and from neural stem have been used for this purpose (Choi *et al.*, 1992; Spina *et al.*, 1992; Pifl *et al.*, 1996; Martinat *et al.*, 2004; Leak *et al.*, 2006). Although a detailed review of such lines is beyond the practical scope of what we wish to accomplish in this chapter, it should be considered that many of the strengths and weaknesses of PC12 cells as models for PD research will pertain to other lines as well.

WHAT ARE PC12 CELLS?

PC12 cells were established as a clonal cell line from an induced, transplantable rat pheochromocytoma

(Greene and Tischler, 1976). When maintained in growth medium, the cells proliferate and like their chromaffin cell counterparts, they synthesize, store and can be stimulated to release catecholamines (Greene and Tischler, 1976; Greene and Rein, 1977). Although they possess dopamine beta hydroxylase, the principle monoamine present in the cells is dopamine (Greene and Tischler, 1976). The catecholamines are stored in chromaffin granules and synaptic vesicles and the cells possess active transporters for uptake of catecholamines across plasma membrane and into granules/vesicles. One of the most salient features of the line is its capacity to respond to nerve growth factor (NGF) by exiting the cell cycle and undergoing neuronal differentiation (Greene and Tischler, 1976). The neuronally differentiated cells highly resemble sympathetic neurons in phenotype; they extend branched axons, are electrically excitable and responsive to neurotransmitters, express a variety of neuronal markers and retain their capacity for dopamine synthesis, storage, uptake and release (Greene and Tischler, 1976; Greene and Rein, 1977; Greene *et al.*, 1998; Teng *et al.*, 1998).

WHAT ARE THE POTENTIAL ADVANTAGES OF PC12 CELLS FOR PD RESEARCH?

As an established cell line, PC12 cells share the general advantages conferred by the capacity for unlimited proliferation. They can be produced in large numbers with a minimum of effort and, since

Parkinson's Disease: molecular and therapeutic insights from model systems

they are derived from a single progenitor, represent a relatively homogeneous population that can be used to provide replicate cultures. For these reasons, the line has been widely used for the purposes of screening and in this way has been used to identify both environmental agents that may cause PD (Collins *et al.*, 1992; Cobuzzi *et al.*, 1994; Mattammal *et al.*, 1995; Offen *et al.*, 1997; Gassen *et al.*, 1998; Seaton *et al.*, 1998; Yang and Sun, 1998; Lamensdorf *et al.*, 2000; Offen *et al.*, 2000; Bharat *et al.*, 2002; Nie *et al.*, 2002; Sheng *et al.*, 2002; Chuenkova and Pereira, 2003; Gelinas *et al.*, 2004; Barlow *et al.*, 2005; Feng *et al.*, 2005; Sousa and Castilho, 2005; Wang *et al.*, 2005; Zheng *et al.*, 2005; An *et al.*, 2006; Chai *et al.*, 2006; Chen *et al.*, 2006; Guan *et al.*, 2006; Hirata *et al.*, 2006; Weinreb *et al.*, 2006; Xu *et al.*, 2007) as well as compounds that may be used to treat the disease (Gassen *et al.*, 1998; Mayo *et al.*, 1998; Seaton *et al.*, 1998; Alves Da Costa *et al.*, 2002; Bharat *et al.*, 2002; Nie *et al.*, 2002; Sheng *et al.*, 2002; Chuenkova and Pereira, 2003; Am *et al.*, 2004; Gelinas *et al.*, 2004; Iwashita *et al.*, 2004; Masutani *et al.*, 2004; Feng *et al.*, 2005; Sousa and Castilho, 2005; Wang *et al.*, 2005; Zheng *et al.*, 2005; An *et al.*, 2006; Chai *et al.*, 2006; Chen *et al.*, 2006; Guan *et al.*, 2006; Hirata *et al.*, 2006; Lipman *et al.*, 2006; Meuer *et al.*, 2006; Weinreb *et al.*, 2006; Xu *et al.*, 2007). In addition, the cells can be genetically modified by transfection or viral infection and stably altered sublines can be selected and propagated (Teng *et al.*, 1998). These properties have been valuable both for probing the molecular properties of cell death in PD models and for creating and characterizing genetic models of PD (Forloni *et al.*, 2000; Panet *et al.*, 2001; Stefanis *et al.*, 2001b; Tanaka *et al.*, 2001; Seo *et al.*, 2002; Biswas *et al.*, 2005; Gorman *et al.*, 2005; Peng *et al.*, 2005; Smith *et al.*, 2005; Jiang *et al.*, 2006; Malagelada *et al.*, 2006; Tyurina *et al.*, 2006).

Another strength (as well as potential weakness) of cell lines and of PC12 cells for PD studies is that the model permits very rapid experimentation. Most uses of the line employ PD mimetics that cause neuronal degeneration and death within 1–2 days (Hartley *et al.*, 1994; Walkinshaw and Waters, 1994; Sawada *et al.*, 1996). Genetic manipulation by gene overexpression or knockdown also occurs within this interval making it possible to evaluate results within a short time frame. Likewise, effects of stable gene manipulations (such as overexpression of synuclein mutants) are also manifested within a short period of time (Tanaka *et al.*, 2001).

An additional relevant feature of the line is that it has been the subject of numerous studies (over 10,000 to date) and is therefore very well characterized with respect to its properties and gene expression patterns. Also, the line has been highly studied as a model for how neurons die in response to a variety of stresses and these provide an important background for specific work on PD (Rukenstein *et al.*, 1991; Valavanis *et al.*, 2001; Greene *et al.*, 2006; Xu and Greene, 2006). Finally, as the line has been increasingly exploited for studies of PD, this in turn increases its potential attractiveness for future work in the field.

The capacity to undergo neuronal differentiation in response to NGF is an especially useful and important feature of PC12 cells for PD studies. Thus, on one hand it is possible to generate large numbers of cells, and on the other, to convert the population to a post-mitotic state with many of the features of sympathetic and dopaminergic neurons. Among potential advantages of this for PD research is that the neuronally differentiated cells directly model sympathetic neurons which are one of the neuron types affected by the disease (Koike and Takahashi, 1997) and that they share many relevant properties with CNS dopaminergic neurons. In addition, growth factors have been considered as possible treatments for PD (Date *et al.*, 1990; Hyman *et al.*, 1991; Lin *et al.*, 1993) and the capacity of PC12 cells to respond to NGF and other such neurotrophic agents (Shimoke and Chiba, 2001; Chuenkova and Pereira, 2003; Salinas *et al.*, 2003; Meuer *et al.*, 2006) provides the opportunity to understand the mechanisms by which growth factors promote neuroprotection in the disease. This property also provides the possibility to assess additional growth factors for protective capacity and to screen reagents that mimic growth factor actions in PD.

HOW HAVE PC12 CELLS BEEN USED TO MODEL PD?

Toxin Models

The majority of PC12 cell–based studies of PD to date have utilized toxin models that are also widely employed *in vivo*. The major PD mimetics used for this purpose have been 6-OHDA and MPP$^+$ (1-methyl-4-phenylpyridinium ion) (Hartley *et al.*, 1994; Walkinshaw and Waters, 1994; Blum *et al.*,

2001). The complex I inhibitors rotenone, dieldrin, paraquat and CCCP have also been used to model PD in PC12 cell studies (Fleming *et al.*, 1994; Yang and Sun, 1998; Sherer *et al.*, 2003; Hirata and Nagatsu, 2005). Finally, several groups have used PC12 cells to demonstrate the toxicity of additional agents with the potential to be present in the environment or to be synthesized *in vivo* (Cobuzzi *et al.*, 1994; Offen *et al.*, 1997). Where characterized, the dose of toxin employed dictates whether death is apoptotic or necrotic in nature. At lower doses, the majority of cells die by an apoptotic mechanism and at high doses, death is non-apoptotic or necrotic. Most studies aim for doses that evoke apoptotic death and under such conditions, cell death generally appears to commence within 16–20 h and to reach a maximum level (generally about 40–60%) by 24–48 h (Blum *et al.*, 1997; Blum *et al.*, 2001; Ryu *et al.*, 2002; Biswas *et al.*, 2005).

Oxidative Stress Models

Based on the notion that oxidative stress is a relevant aspect of PD, several studies have used hydrogen peroxide (Drukarch *et al.*, 1996; Gassen *et al.*, 1998; Saito *et al.*, 2007) exposure or glutathione depletion to model the disease in PC12 cells (Jha *et al.*, 2000; Bharat *et al.*, 2002; Chinta and Andersen, 2006).

Catecholamines

Dopamine and DOPA have been observed to promote death of PC12 cells (Walkinshaw and Waters, 1995; Kang *et al.*, 1998; Panet *et al.*, 2001; Chen *et al.*, 2003) and since these compounds are synthesized in SNpc neurons, it has been argued that application of such compounds is a potentially valid model for PD (Walkinshaw and Waters, 1995; Blum *et al.*, 2001).

Metal Ions

Bivalent cations such as Zn^{++} and Mn^{++} have been suggested to play roles in the etiology of PD and several studies have shown that these elevate the death-promoting actions of the above treatments on PC12 cells (Galvani *et al.*, 1995; Roth *et al.*, 2002; Reaney and Smith, 2005).

Overexpression of Wild-Type and Mutant α-Synucleins

Several groups have shown that acute or constitutive expression of disease-related mutants of α-synuclein negatively affects the survival and functional properties of PC12 cells (Stefanis *et al.*, 2001b; Tanaka *et al.*, 2001; Peng *et al.*, 2005; Smith *et al.*, 2005; Jiang *et al.*, 2006; Larsen *et al.*, 2006). Such findings have supported the use of this approach to model the role of α-synuclein both in familial and sporadic PD.

Additional Genetic Models

One study has used PC12 cells to characterize the functional actions of the PD-associated gene Parkin (Tanaka *et al.*, 2001).

WHAT ARE THE POTENTIAL DRAWBACKS OF USING PC12 CELLS TO MODEL AND STUDY PD?

Just as there are potential advantages of using PC12 cells as models for PD research, there are also several potential shortcomings and deficits of the system. We discuss these below along with possible ameliorative solutions.

Relevance

The foremost issue is that there is no assurance that any of the findings made in the *in vitro* PC12 cell model will be directly relevant or applicable to PD. This arises from two major sources. The first, which is shared with most other model systems, is that the significance of the data produced with PC12 cells can be no better than the aptness of the means that are used to simulate PD. The vast majority of PD-directed work with PC12 cells has thus far employed acute toxin or oxidative stress models. The general strengths and weaknesses of such models are reviewed elsewhere in this volume and it is on these that such PC12 cell studies will or will not yield relevant findings. The second potential relevance issue is that irrespective of the means to mimic PD, findings with PC12 cells in particular may not be relevant to the disease. That is, means by which PC12 cells degenerate and die in PD models may be very different from what happens in PD patients.

Given these issues, what can be done to minimize the problem and to assure relevancy of the model? One important approach has been to utilize and compare results with multiple PD mimetics. If, for instance, toxin/oxidative stress models are relevant to PD because they involve mitochondrial and metabolic dysfunction, then one should expect to have the same findings across a variety of PC12 cell studies using different means to promote such dysfunction. A second key approach is to verify to the extent possible, whether findings with the PC12 cell system are replicable in other cellular and animal PD models and in PD itself. The latter can be achieved by immunohistochemical or biochemical comparison of tissue from postmortem PD and non-PD tissue. As reviewed below, one example has been to verify that gene products such as RTP801 which are induced in the PC12 cell model, are also selectively upregulated in neurons of the substantia nigra (SN) of PD patients as well as in SN neurons of animals treated with 1-methyl-4-phenyl-1,2,3,6-tetrahydropyridine (MPTP) (Malagelada *et al.*, 2006). A third potentially valuable approach is to create and exploit PC12 cell models of PD based on known genetic defects related to familial PD. The major example here thus far has been the generation and characterization of PC12 cell lines that express mutant synuclein. In this light, it will be valuable to extend this approach to additional gene defects known to be associated with parkinsonism.

Cell–Cell Interactions

Although the homogeneity of cell cultures can provide important advantages for research, this also has the major potential drawback that the cells do not interact with other cell types such as glia and do not form synaptic connections as they would *in vivo*. Such interactions can significantly influence neuronal responses in PD models and affect experimental outcome and interpretation (Falkenburger and Schulz, 2006). There is no easy solution for this potential issue other than, as noted above, to compare findings made with PC12 cells and other *in vitro* models with animal models and with PD itself.

Metabolic Properties of PC12 Cells

Like many neoplastic cell lines, as pointed out by Basma *et al.* (1992), PC12 cells (at least in the NGF-untreated state) have a high rate of glycolysis and are hence relatively insensitive to the complex I inhibitors that are used to model PD. This might account, for example, for the difference in levels of 6-OHDA that are required to kill PC12 cells versus cultured sympathetic neurons (Malagelada *et al.*, 2006). While this point poses a potential drawback for use of PC12 cells and other neoplastic lines in PD studies, it does not appear to be insurmountable. For one, suitable adjustment of complex I inhibitor concentrations appears to provoke PC12 cell death that is similar to that achieved in catecholaminergic neurons *in vitro* as well as *in vivo*. Moreover, a number of the biochemical responses evoked in PC12 cells by complex I inhibitors have also been observed in primary neurons as well as in PD (cf. Ryu *et al.*, 2005; Malagelada *et al.*, 2006). An alternative solution is to lower the concentration of glucose in the culture medium used for PD-related experiments or to include suitable inhibitors of glycolysis such as 2-deoxyglucose (Basma *et al.*, 1992). Such manipulations enhance the toxic effects of MPTP on PC12 cells and these findings have reinforced the idea that complex I inhibitors do indeed kill PC12 cells by a mechanism reliant on blockade of mitochondrial respiration (Basma *et al.*, 1992).

Death Versus Dysfunction

A fourth potential weakness of PD-related PC12 cell studies to date is that these have mainly focused on cell death rather than on cellular dysfunction. Although neuron death is an irreversible endpoint in PD, dysfunction is also key in the pathophysiology of the disease and its prevention could be as or more important than that of blocking neuron death itself. This issue may be exacerbated by the acute nature of most PD studies with PC12 cells. Possible approaches to deal with this question would be to use levels of PD mimetics that cause dysfunction rather than death and to focus more on the issue of dysfunction rather than on death. For instance, a clearly seen response of neuronal PC12 cells to PD mimetics is degeneration of neurites. This often occurs before or without cell death. However, there have been few studies on how this occurs or how to prevent it. Genetic models may again be important here as indicated by studies with cells overexpressing mutant synuclein. In addition to excess cell death, such cells showed a variety of defects including impaired neurite outgrowth (Stefanis *et al.*,

2001b). It should be noted that placing further emphasis on dysfunction may not necessarily be at odds with studies of death. It may well be that the same cellular responses lead to both and that understanding the molecular basis of each endpoint will provide insight to preventing the other.

Genetic Drift and Reproducibility

Like all cell lines, a potential liability of working with PC12 cells is the effect of genetic drift and differences in handling conditions on reproducibility of findings from one laboratory to another or even within the same laboratory. Solutions to this overall issue have been addressed in detail elsewhere and include strict adherence to culture techniques that minimize selection of substrains and the maintenance and use of replicate frozen cell stocks (Greene et al., 1998).

APPLICATIONS OF THE PC12 CELL MODEL TO PD

As discussed above, PC12 cells possess a number of potential advantages as cellular models for PD-related studies. In the following sections, we provide several examples of their application for this purpose.

Transcriptional Changes Associated with PD

Irrespective of the initiating events, loss of neurons generally requires the transcriptional induction of death-associated genes. In an effort to identify such genes, our laboratory used Serial Analysis of Gene Expression (SAGE) to compare the transcriptomes of neuronal PC12 cells before and after 8 h of exposure to 6-OHDA (Angelastro et al., 2000; Ryu et al., 2002). The rationales were that 6-OHDA may evoke changes in gene expression similar to those that occur in PD and that because cell death commences at about 16–24 h in this model, an 8-h time point would detect transcriptional changes with the potential to be causally involved in promoting death. This analysis revealed induction (by sixfold or more) of approximately 1,200 of the 14,000 transcripts detected in neuronal PC12 cells. Of these, nearly 600 have been linked to known genes

(Ryu et al., 2005). Mining of these findings have in turn engendered the following studies to date.

6-OHDA Induces Both Pro- and Anti-apoptotic Genes

Functional consideration of the identified genes regulated by 6-OHDA in the neuronal PC12 cell model revealed not only that a number could be associated with cell death, but also that many had potential pro-survival and neuroprotective activities (Ryu et al., 2005). This led to two major suggestions which have implications for strategies to treat PD. One is that neuron degeneration in PD may be the result of the activation of multiple pro-apoptotic genes, each of which may contribute to bringing the neurons over a critical threshold required for death. The implication of such a model is that interference with the expression or activity of any one of these genes may be sufficient to provide neuroprotection by bringing the level of pro-apoptotic stimuli below the threshold required for neuron death. The other suggestion was that PD-related stresses may also induce compensatory pro-survival genes by which the neurons attempt to stave off death. From this point of view, the balance between neuron life and death in PD may reflect the relative level of induction of pro- and anti-apoptotic genes. This led to the suggestion that it may be important to expend further effort to identify those pro-survival genes that are induced in PD and to seek means to augment their expression.

Unfolded Protein Response and Endoplasmic Reticulum Stress

An unanticipated finding of the SAGE study was the observation that a large number (approximately 70) of the induced identified transcripts encoded proteins involved in the cellular responses to unfolded proteins and endoplasmic reticulum (ER) stress (Ryu et al., 2002). Under normal conditions, proteins synthesized in lumen of the ER are correctly folded with the help of chaperones and translocated to other cellular compartments. In response to certain cellular stresses, misfolded proteins can accumulate in the ER, setting off an unfolded protein response (UPR) and ER stress response in which a number of proteins are upregulated that augment the function of the ER and of the cellular protein degradation machinery (Marciniak and Ron, 2006). These responses can lead either to successful clearance of unfolded proteins and cell protection or to cell demise. Stimulation of the ER stress response in 6-OHDA-treated neuronal PC12 cells was confirmed by observations that the key stress sensor kinases PERK and IRE1alpha were activated in the model. There was also upregulation at the protein level of ER-stress-associated chaperone HSP70 and of the transcription factors CHOP (Gadd153) and ATF4. Such findings were replicated with the PD mimetics MPP$^+$ and rotenone and were confirmed in cultures of primary sympathetic neurons. Based on experiments with sympathetic neurons null for the ER stress sensor PERK, it was suggested that the UPR and ER stress-response machinery acted to protect neurons from stress and that if this system failed or were overwhelmed, then it could also play an active role in triggering degeneration and death (Ryu et al., 2002).

Such findings have naturally raised the issue of whether they are limited to PC12 cells and whether they may be relevant to PD. One positive indication came from prior studies that have suggested that a form of juvenile PD associated with mutation of the Parkin gene is caused by ER stress due to the abnormal ER accumulation of a Parkin substrate (Imai et al., 2001). In addition, studies in other cellular models have also found that toxin PD mimetics induce UPR and ER stress responses (Holtz and O'Malley, 2003; Kheradpezhouh et al., 2003; Yamamuro et al., 2006) as does overexpression of mutant synuclein in PC12 cells (Smith et al., 2005). In the latter studies, evidence was presented that ER stress played a causal role in cell death. The ER-stress-associated transcription factor CHOP is also induced in SN neurons in 6-OHDA and MPTP mouse models of PD (Silva et al., 2005). In addition, animals null for CHOP expression showed reduced death in response to 6-OHDA (Silva et al., 2005). Finally, a recent study has documented the elevation of markers for ER stress in SN neurons in postmortem brains of PD patients (Hoozemans et al., 2007). Taken together, these findings support the idea that the UPR and ER stress responses may play causal roles in neuron death and degeneration in PD and they affirm that findings made in the PC12 cellular model have the potential to be highly relevant to the disease.

Nix and the JNK Pathway

Among the genes that were induced by 6-OHDA in the PC12 cell SAGE study was Nix, a BH3-only pro-apoptotic member of the Bcl2 family. Wilhelm et al. (2007) confirmed that 6-OHDA induces Nix protein as well as transcripts in neuronal PC12 cells and that its overexpression is sufficient to promote their apoptotic death. Moreover, it was found that Nix appears to be required for death in the 6-OHDA model in that overexpression of s-Nix, a dominant-negative form of the protein, was protective. Evidence was presented that Nix promotes death by activation of the JNK/c-Jun signaling pathway and does so by interacting with POSH (plenty of SH3 domains), a scaffold protein that acts upstream of JNKs. Interestingly, Nix did not require its BH3 domain to induce death of neuronal PC12 cells, but did require the presence of a transmembrane domain that permits its association with mitochondria. Past work has implicated activation of the JNK/c-Jun pathway in neuron death in a variety of disorders including PD (Kang et al., 1998; Luo et al., 1998; Soldner et al., 1999; Blum et al., 2001; Tieu et al., 2001). The present findings with PC12 cells now provide a potential mechanism by which this pathway is activated in the disease. However, it remains to be seen whether or not such findings can be extended to PD.

p53 and Puma

A variety of findings have implicated the transcription factor p53 in neuron death in several PD models, including PC12 cells (Soldner et al., 1999; Blum et al., 2001; Tieu et al., 2001). However, it was important to define the effectors of p53 that may mediate death in PD. Puma is a BH3-only pro-apoptotic Bcl2 family member and p53 target that was also found to be significantly elevated in our SAGE study of 6-OHDA-treated PC12 cells. Western blotting confirmed that Puma protein is also induced in this system (Biswas et al., 2005). In our study and a subsequent study (Nair, 2006) upregulation of Puma in 6-OHDA-exposed PC12 cells was found to be dependent on p53. Significantly, Puma was required for death in response to 6-OHDA in that short hairpin RNAs (shRNAs) specifically targeted to Puma were highly protective (Biswas et al., 2005). Such findings raise the intriguing possibilities, which remain to be explored in depth, that p53 and Puma may play significant roles in neuron degeneration and death in PD.

RTP801

The most highly induced transcript (97-fold) to emerge from the SAGE study was that encoding RTP801 (also known as REDD1 or Dig 2) (Ryu et al.,

2005). RTP801 is a stress-response protein that was described in PC12 cells and that can be induced by stimuli including DNA damage, oxidative stress, hypoxia and energy depletion (Ellisen *et al.*, 2002; Shoshani *et al.*, 2002; Wang *et al.*, 2003). The function of RTP801 with respect to survival depends on the cell type and state. In proliferating PC12 cells, RTP801 overexpression is protective from oxidative stress. In contrast, in neuronal PC12 cells, overexpression of the protein is sufficient to promote apoptotic death (Shoshani *et al.*, 2002). Such findings led us to further explore the potential involvement of RTP801 in PD models and in PD (Malagelada *et al.*, 2006). Induction of RTP801 at the protein level was confirmed not only in the 6-OHDA model, but also in neuronal PC12 cells exposed to MPP$^+$ or rotentone. Short hairpin RNAs directed against RTP801 protected neuronal PC12 cells against all three PD mimetics and suppressed death of 6-OHDA-treated cultured sympathetic neurons. In contrast, there was no protection against apoptosis triggered by NGF deprivation. Thus, RTP801 appears to be required for death in our PD models, but not for death evoked by loss of trophic support.

We next explored the potential relevance of RTP801 induction to neuron death in animal models and in PD itself. In the mouse acute MPTP model, RTP801 expression was significantly upregulated in SN neurons within 24 h of treatment. We also observed that RTP801 expression was highly elevated in neuromelanin-positive neurons within the SN of human postmortem PD brains. In contrast, the same brains showed no expression of RTP801 in cerebellar granule neurons. These encouraging findings thus support the hypothesis that RTP801 may play an important role in the pathophysiology of PD and confirm that a gene recognized in studies with 6-OHDA-treated PC12 cells can indeed be relevant to PD.

The mechanisms by which RTP801 may induce neuron death in PD are not fully understood. Such knowledge is potentially important since it may lead to the development of strategies to provide neuroprotection in the disease. Because of their properties as discussed above, neuronal PC12 cells represent a very useful tool for dissecting the means by which RTP801 is regulated and how its induction leads to neuron death. In several systems, RTP801 has been described as a negative regulator of the mammalian target of rapamycin (mTOR) protein (Brugarolas *et al.*, 2004; Corradetti *et al.*, 2005). mTOR is a protein kinase that plays important roles in a variety of cellular behaviors including protein synthesis, proliferation, autophagy and survival. We observed that in 6-OHDA-treated neuronal PC12 cells, mTOR activity is greatly decreased and that this is temporally correlated with RTP801 induction. Moreover, we found that repression of mTOR activity or promotion of death by RTP801 required the presence of the tuberous sclerosis protein TSC2. The latter has been found previously to participate in mTOR regulation. These findings represent initial steps in understanding how RTP801 may participate in PD, but serve to illustrate the potential utility of neuronal PC12 cells for this purpose.

α-Synuclein

α-Synuclein is a presynaptic protein that has been associated with both familial and sporadic PD. PC12 cell–based studies have been aimed at understanding the regulation of neuronal levels of α-synuclein and at exploring the mechanism by which it (and its A53T and A30P mutant forms in particular) cause neuronal degeneration and death.

Regulation of α-Synuclein Expression and Degradation

Studies with PC12 cells revealed that α-synuclein expression is very low in the cells when in a nonneuronal state and that its expression is greatly increased in the presence of NGF by a transcription-dependent mechanism (Stefanis *et al.*, 2001a). Similar responses to neurotrophic factors have been confirmed in neurons. Recent experiments with PC12 cells reveal that such regulation occurs via both the ras/ERK and PI3 kinase pathways and have identified several regulatory regions for this within the α-synuclein gene (Clough and Stefanis, 2007). Such findings raise the possibility of designing means to manipulate α-synuclein levels in neurons. Additional studies have used PC12 cells that constitutively express exogenous wild-type and mutant synuclein (A53T and A30P) to explore the mechanism of α-synuclein turnover in neuronal cells. These have revealed degradation by the chaperone-mediated autophagy pathway and that mutant α-synucleins are both poorly degraded by this pathway and that their presence also impairs the lysosomal degradation of other cellular proteins (Cuervo *et al.*, 2004). An additional study with PC12 cells reported that α-synuclein is degraded by both the lysosomal and proteasomal pathways (Webb *et al.*, 2003). Consistent with this, blockade of proteasomal activity in PC12 cells produced the formation of α-synuclein-immunoreactive inclusions (Rideout *et al.*, 2001). Taken together, such studies have illuminated potential mechanisms by which the levels of α-synuclein may be regulated in neurons and the means by which this protein may be turned over. They also suggest novel mechanisms by which mutant α-synuclein may affect its own turnover as well as that of other cellular proteins and how this in turn may contribute to the pathophysiology of PD.

α-Synuclein and PD

To explore the potential mechanisms by which α-synuclein and its mutant forms may affect neuronal function and survival in PD, several groups have generated and exploited PC12 cell lines that overexpress these proteins either in a constitutive or regulated manner. In one study, cells overexpressing mutant α-synuclein were reported to show impaired proteasomal activity and enhanced sensitivity (in terms of death) to proteasomal inhibitors (Tanaka *et al.*, 2001). Subsequent work indicated elevated levels of basal cell death in such lines and that this was at least in part due to ER stress and mitochondrial dysfunction (Smith *et al.*, 2005). In another study, independent PC12 cell lines expressing mutant α-synuclein were reported to exhibit impairment in both proteasomal and lysosomal function, elevated autophagic cell death, loss of chromaffin granules and reduced capacity to secrete catecholamines (Stefanis *et al.*, 2001b). In contrast, work from a third group failed to find effects of overexpressed wild-type or mutant α-synucleins on proteasomal function (Martin-Clemente *et al.*, 2004). The reason for this apparent discrepancy is unclear, but could arise from a difference in PC12 cell strains. Relevant to this, Stefanis *et al.* (2001b) reported variation in response of various passages of PC12 cells to overexpression of mutant α-synuclein. This point highlights one of the potential weaknesses of the system and reinforces the importance of verifying studies with PC12 cells with additional models and, as far as possible, with material from PD patients.

As noted above, the work by Cuervo *et al.* (2004) delved into the mechanism by which mutant α-synuclein impairs cellular lysosomal activity which can in turn affect neuronal cell function and survival. Also of potential relevance to PD, several studies employing PC12 cells (as well as cultured chromaffin cells) demonstrated that overexpression of wild-type as well as mutant α-synucleins increases cytosolic concentrations of catecholamines and interferes with catecholamine release by affecting a late stage of exocytosis (Larsen *et al.*, 2006; Mosharov *et al.*, 2006). All-in-all, such studies with PC12 cells have provided novel insights to potential roles of α-synuclein and of its mutants in the pathophysiology of PD. If such roles can be verified, then PC12 cells that overexpress wild-type and mutant α-synucleins may serve as models for screening drugs that interfere with α-synuclein expression, aggregation or impairment of cellular function.

SUMMARY AND PERSPECTIVES

We have reviewed here the utility and application of the PC12 cell model for studying PD. As any model system, the line possesses both potential advantages and weaknesses for this purpose. Findings to date with the line are on the whole encouraging and indicate that it will continue to be profitably exploited for uses such as identifying novel environmental agents with potential to cause PD; unraveling the mechanisms by which genetic and environmental factors associated with PD promote neuron degeneration and death; discovering gene products and their regulatory pathways that may be involved in the pathophysiology of PD; and screening for molecules with the capacity to prevent or suppress the progression of PD. Ultimately, the successful and most effective use of data derived from PC12 cells will depend on verifying their relevance in animal models of PD and in PD itself. Conversely, findings from animal models and PD will inform our future exploitation of PC12 cells.

ACKNOWLEDGEMENTS

A portion of the work described here was supported by grants from the NIH-NINDS, Parkinson's Disease Foundation and American Parkinson's Disease Association.

REFERENCES

Alves Da Costa, C., Paitel, E., Vincent, B., and Checler, F. (2002). Alpha-synuclein lowers p53-dependent apoptotic response of neuronal cells. Abolishment by 6-hydroxydopamine and implication for Parkinson's disease. *J Biol Chem* 277, 50980–50984.

Am, O. B., Amit, T., and Youdim, M. B. (2004). Contrasting neuroprotective and neurotoxic actions of respective metabolites of anti-Parkinson drugs rasagiline and selegiline. *Neurosci Lett* 355, 169–172.

An, L. J., Guan, S., Shi, G. F., Bao, Y. M., Duan, Y. L., and Jiang, B. (2006). Protocatechuic acid from *Alpinia oxyphylla* against MPP+-induced neurotoxicity in PC12 cells. *Food Chem Toxicol* 44, 436–443.

Angelastro, J. M., Klimaschewski, L., Tang, S., Vitolo, O. V., Weissman, T. A., Donlin, L. T., Shelanski, M. L., and Greene, L. A. (2000). Identification of diverse nerve growth factor-regulated genes by serial analysis of gene expression (SAGE) profiling. *Proc Natl Acad Sci USA* 97, 10424–10429.

Barlow, B. K., Lee, D. W., Cory-Slechta, D. A., and Opanashuk, L. A. (2005). Modulation of antioxidant defense systems by the environmental pesticide maneb in dopaminergic cells. *Neurotoxicology* 26, 63–75.

Basma, A. N., Heikkila, R. E., Saporito, M. S., Philbert, M., Geller, H. M., and Nicklas, W. J. (1992). 1-Methyl-4-(2′-ethylphenyl)-1,2,3,6-tetrahydropyridine-induced toxicity in PC12 cells is enhanced by preventing glycolysis. *J Neurochem* 58, 1052–1059.

Bharat, S., Cochran, B. C., Hsu, M., Liu, J., Ames, B. N., and Andersen, J. K. (2002). Pre-treatment with R-lipoic acid alleviates the effects of GSH depletion in PC12 cells: Implications for Parkinson's disease therapy. *Neurotoxicology* 23, 479–486.

Biswas, S. C., Ryu, E., Park, C., Malagelada, C., and Greene, L. A. (2005). Puma and p53 play required roles in death evoked in a cellular model of Parkinson disease. *Neurochem Res* 30, 839–845.

Blum, D., Wu, Y., Nissou, M. F., Arnaud, S., Alim Louis, B., and Verna, J. M. (1997). p53 and Bax activation in 6-hydroxydopamine-induced apoptosis in PC12 cells. *Brain Res* 751, 139–142.

Blum, D., Torch, S., Lambeng, N., Nissou, M., Benabid, A. L., Sadoul, R., and Verna, J. M. (2001). Molecular pathways involved in the neurotoxicity of 6-OHDA, dopamine and MPTP: Contribution to the apoptotic theory in Parkinson's disease. *Prog Neurobiol* 65, 135–172.

Brugarolas, J., Lei, K., Hurley, R. L., Manning, B. D., Reiling, J. H., Hafen, E., Witters, L. A., Ellisen, L. W., and Kaelin, W. G., Jr. (2004). Regulation of mTOR function in response to hypoxia by REDD1 and the TSC1/TSC2 tumor suppressor complex. *Genes Dev* 18, 2893–2904.

Chai, Y., Niu, L., Sun, X. L., Ding, J. H., and Hu, G. (2006). Iptakalim protects PC12 cell against H₂O₂-induced oxidative injury via opening mitochondrial ATP-sensitive potassium channel. *Biochem Biophys Res Commun* 350, 307–314.

Chen, J., Tang, X. Q., Zhi, J. L., Cui, Y., Yu, H. M., Tang, E. H., Sun, S. N., Feng, J. Q., and Chen, P. X. (2006). Curcumin protects PC12 cells against 1-methyl-4-phenylpyridinium ion-induced apoptosis by bcl-2–mitochondria–ROS–iNOS pathway. *Apoptosis* 11, 943–953.

Chen, X. C., Zhu, Y. G., Zhu, L. A., Huang, C., Chen, Y., Chen, L. M., Fang, F., Zhou, Y. C., and Zhao, C. H. (2003). Ginsenoside Rg1 attenuates dopamine-induced apoptosis in PC12 cells by suppressing oxidative stress. *Eur J Pharmacol* 473, 1–7.

Chinta, S. J., and Andersen, J. K. (2006). Reversible inhibition of mitochondrial complex I activity following chronic dopaminergic glutathione depletion *in vitro*: Implications for Parkinson's disease. *Free Radic Biol Med* 41, 1442–1448.

Choi, H. K., Won, L., Roback, J. D., Wainer, B. H., and Heller, A. (1992). Specific modulation of dopamine expression in neuronal hybrid cells by primary cells from different brain regions. *Proc Natl Acad Sci USA* 89, 8943–8947.

Chuenkova, M. V., and Pereira, M. A. (2003). PDNF, a human parasite-derived mimic of neurotrophic factors, prevents caspase activation, free radical formation, and death of dopaminergic cells exposed to the Parkinsonism-inducing neurotoxin MPP+. *Brain Res Mol Brain Res* 119, 50–61.

Clough, R. L., and Stefanis, L. (2007). A novel pathway for transcriptional regulation of alpha-synuclein. *FASEB J* 21, 596–607.

Cobuzzi, R. J., Jr., Neafsey, E. J., and Collins, M. A. (1994). Differential cytotoxicities of N-methyl-beta-carbolinium analogues of MPP+ in PC12 cells: Insights into potential neurotoxicants in Parkinson's disease. *J Neurochem* 62, 1503–1510.

Collins, M. A., Neafsey, E. J., Matsubara, K., Cobuzzi, R. J., Jr., and Rollema, H. (1992). Indole-N-methylated beta-carbolinium ions as potential brain-bioactivated neurotoxins. *Brain Res* 570, 154–160.

Corradetti, M. N., Inoki, K., and Guan, K. L. (2005). The stress-inducted proteins RTP801 and RTP801L are negative regulators of the mammalian target of rapamycin pathway. *J Biol Chem* 280, 9769–9772.

Cuervo, A. M., Stefanis, L., Fredenburg, R., Lansbury, P. T., and Sulzer, D. (2004). Impaired degradation of mutant alpha-synuclein by chaperone-mediated autophagy. *Science* 305, 1292–1295.

Date, I., Notter, M. F., Felten, S. Y., and Felten, D. L. (1990). MPTP-treated young mice but not aging mice show partial recovery of the nigrostriatal dopaminergic system by stereotaxic injection of acidic fibroblast growth factor (aFGF). *Brain Res* **526**, 156–160.

Drukarch, B., Jongenelen, C. A., Schepens, E., Langeveld, C. H., and Stoof, J. C. (1996). Glutathione is involved in the granular storage of dopamine in rat PC 12 pheochromocytoma cells: Implications for the pathogenesis of Parkinson's disease. *J Neurosci* **16**, 6038–6045.

Ellisen, L. W., Ramsayer, K. D., Johannessen, C. M., Yang, A., Beppu, H., Minda, K., Oliner, J. D., McKeon, F., and Haber, D. A. (2002). REDD1, a developmentally regulated transcriptional target of p63 and p53, links p63 to regulation of reactive oxygen species. *Mol Cell* **10**, 995–1005.

Falkenburger, B. H., and Schulz, J. B. (2006). Limitations of cellular models in Parkinson's disease research. *J Neural Transm Suppl*, 261–268.

Feng, W., Wei, H., and Liu, G. T. (2005). Pharmacological study of the novel compound FLZ against experimental Parkinson's models and its active mechanism. *Mol Neurobiol* **31**, 295–300.

Fleming, L., Mann, J. B., Bean, J., Briggle, T., and Sanchez-Ramos, J. R. (1994). Parkinson's disease and brain levels of organochlorine pesticides. *Ann Neurol* **36**, 100–103.

Forloni, G., Bertani, I., Calella, A. M., Thaler, F., and Invernizzi, R. (2000). Alpha-synuclein and Parkinson's disease: Selective neurodegenerative effect of alpha-synuclein fragment on dopaminergic neurons *in vitro* and *in vivo*. *Ann Neurol* **47**, 632–640.

Galvani, P., Fumagalli, P., and Santagostino, A. (1995). Vulnerability of mitochondrial complex I in PC12 cells exposed to manganese. *Eur J Pharmacol* **293**, 377–383.

Gassen, M., Gross, A., and Youdim, M. B. (1998). Apomorphine enantiomers protect cultured pheochromocytoma (PC12) cells from oxidative stress induced by H_2O_2 and 6-hydroxydopamine. *Mov Disord* **13**, 661–667.

Gelinas, S., Bureau, G., Valastro, B., Massicotte, G., Cicchetti, F., Chiasson, K., Gagne, B., Blanchet, J., and Martinoli, M. G. (2004). Alpha and beta estradiol protect neuronal but not native PC12 cells from paraquat-induced oxidative stress. *Neurotox Res* **6**, 141–148.

Gorman, A. M., Szegezdi, E., Quigney, D. J., and Samali, A. (2005). Hsp27 inhibits 6-hydroxydopamine-induced cytochrome c release and apoptosis in PC12 cells. *Biochem Biophys Res Commun* **327**, 801–810.

Greene, L. A., and Tischler, A. S. (1976). Establishment of a noradrenergic clonal line of rat adrenal pheochromocytoma cells which respond to nerve growth factor. *Proc Natl Acad Sci USA* **73**, 2424–2428.

Greene, L. A., and Rein, G. (1977). Release of (3H)norepinephrine from a clonal line of pheochromocytoma cells (PC12) by nicotinic cholinergic stimulation. *Brain Res* **138**, 521–528.

Greene, L. A., Farinelli. S. E., Cunningham, M. E., and Park, D. S. (1998). Culturing Nerve Cells, 2nd edn. MIT press, Cambridge.

Greene, L. A., Liu, D. X., Troy, C. M., and Biswas, S. C. (2006). Cell cycle molecules define a pathway required for neuron death in development and disease. *Biochim Biophys Acta*.

Guan, S., Jiang, B., Bao, Y. M., and An, L. J. (2006). Protocatechuic acid suppresses MPP^+-induced mitochondrial dysfunction and apoptotic cell death in PC12 cells. *Food Chem Toxicol* **44**, 1659–1666.

Hartley, A., Stone, J. M., Heron, C., Cooper, J. M., and Schapira, A. H. (1994). Complex I inhibitors induce dose-dependent apoptosis in PC12 cells: Relevance to Parkinson's disease. *J Neurochem* **63**, 1987–1990.

Hirata, Y., and Nagatsu, T. (2005). Rotenone and CCCP inhibit tyrosine hydroxylation in rat striatal tissue slices. *Toxicology* **216**, 9–14.

Hirata, Y., Meguro, T., and Kiuchi, K. (2006). Differential effect of nerve growth factor on dopaminergic neurotoxin-induced apoptosis. *J Neurochem* **99**, 416–425.

Holtz, W. A., and O'Malley, K. L. (2003). Parkinsonian mimetics induce aspects of unfolded protein response in death of dopaminergic neurons. *J Biol Chem* **278**, 19367–19377.

Hoozemans, J. J., van Haastert, E. S., Eikelenboom, P., de Vos, R. A., Rozemuller, J. M., and Scheper, W. (2007). Activation of the unfolded protein response in Parkinson's disease. *Biochem Biophys Res Commun* **354**, 707–711.

Hyman, C., Hofer, M., Barde, Y. A., Juhasz, M., Yancopoulos, G. D., Squinto, S. P., and Lindsay, R. M. (1991). BDNF is a neurotrophic factor for dopaminergic neurons of the substantia nigra. *Nature* **350**, 230–232.

Imai, Y., Soda, M., Inoue, H., Hattori, N., Mizuno, Y., and Takahashi, R. (2001). An unfolded putative transmembrane polypeptide, which can lead to endoplasmic reticulum stress, is a substrate of Parkin. *Cell* **105**, 891–902.

Iwashita, A., Yamazaki, S., Mihara, K., Hattori, K., Yamamoto, H., Ishida, J., Matsuoka, N., and Mutoh, S. (2004). Neuroprotective effects of a novel poly (ADP-ribose) polymerase-1 inhibitor, 2-[3-[4-(4-chlorophenyl)-1-piperazinyl] propyl]-4(3H)-quinazolinone (FR255595), in an *in vitro* model of cell death and in mouse 1-methyl-4-phenyl-1,2,3,6-tetrahydropyridine model of Parkinson's disease. *J Pharmacol Exp Ther* **309**, 1067–1078.

Jha, N., Jurma, O., Lalli, G., Liu, Y., Pettus, E. H., Greenamyre, J. T., Liu, R. M., Forman, H. J., and Andersen, J. K. (2000). Glutathione depletion in PC12 results in selective inhibition of mitochondrial complex

I activity. Implications for Parkinson's disease. *J Biol Chem* 275, 26096–26101.

Jiang, H., Wu, Y. C., Nakamura, M., Liang, Y., Tanaka, Y., Holmes, S., Dawson, V. L., Dawson, T. M., Ross, C. A., and Smith, W. W. (2006). Parkinson's disease genetic mutations increase cell susceptibility to stress: Mutant alpha-synuclein enhances $H(2)O(2)$- and Sin-1-induced cell death. *Neurobiol Aging*.

Kang, C. D., Jang, J. H., Kim, K. W., Lee, H. J., Jeong, C. S., Kim, C. M., Kim, S. H., and Chung, B. S. (1998). Activation of c-jun N-terminal kinase/stress-activated protein kinase and the decreased ratio of Bcl-2 to Bax are associated with the auto-oxidized dopamine-induced apoptosis in PC12 cells. *Neurosci Lett* 256, 37–40.

Kheradpezhouh, M., Shavali, S., and Ebadi, M. (2003). Salsolinol causing parkinsonism activates endoplasmic reticulum-stress signaling pathways in human dopaminergic SK-N-SH cells. *Neurosignals* 12, 315–324.

Koike, Y., and Takahashi, A. (1997). Autonomic dysfunction in Parkinson's disease. *Eur Neurol* 38(Suppl 2), 8–12.

Lamensdorf, I., Eisenhofer, G., Harvey-White, J., Hayakawa, Y., Kirk, K., and Kopin, I. J. (2000). Metabolic stress in PC12 cells induces the formation of the endogenous dopaminergic neurotoxin, 3,4-dihydroxyphenylacetaldehyde. *J Neurosci Res* 60, 552–558.

Larsen, K. E., Schmitz, Y., Troyer, M. D., Mosharov, E., Dietrich, P., Quazi, A. Z., Savalle, M., Nemani, V., Chaudhry, F. A., Edwards, R. H., Stefanis, L., and Sulzer, D. (2006). Alpha-synuclein overexpression in PC12 and chromaffin cells impairs catecholamine release by interfering with a late step in exocytosis. *J Neurosci* 26, 11915–11922.

Leak, R. K., Liou, A. K., and Zigmond, M. J. (2006). Effect of sublethal 6-hydroxydopamine on the response to subsequent oxidative stress in dopaminergic cells: Evidence for preconditioning. *J Neurochem* 99, 1151–1163.

Lin, L. F., Doherty, D. H., Lile, J. D., Bektesh, S., and Collins, F. (1993). GDNF: A glial cell line-derived neurotrophic factor for midbrain dopaminergic neurons. *Science* 260, 1130–1132.

Lipman, T., Tabakman, R., and Lazarovici, P. (2006). Neuroprotective effects of the stable nitroxide compound Tempol on 1-methyl-4-phenylpyridinium ion-induced neurotoxicity in the Nerve Growth Factor-differentiated model of pheochromocytoma PC12 cells. *Eur J Pharmacol* 549, 50–57.

Luo, Y., Umegaki, H., Wang, X., Abe, R., and Roth, G. S. (1998). Dopamine induces apoptosis through an oxidation-involved SAPK/JNK activation pathway. *J Biol Chem* 273, 3756–3764.

Malagelada, C., Ryu, E. J., Biswas, S. C., Jackson-Lewis, V., and Greene, L. A. (2006). RTP801 is elevated in Parkinson brain substantia nigral neurons and mediates death in cellular models of Parkinson's disease by a mechanism involving mammalian target of rapamycin inactivation. *J Neurosci* 26, 9996–10005.

Marciniak, S. J., and Ron, D. (2006). Endoplasmic reticulum stress signaling in disease. *Physiol Rev* 86, 1133–1149.

Martin-Clemente, B., Alvarez-Castelao, B., Mayo, I., Sierra, A. B., Diaz, V., Milan, M., Farinas, I., Gomez-Isla, T., Ferrer, I., and Castano, J. G. (2004). Alpha-synuclein expression levels do not significantly affect proteasome function and expression in mice and stably transfected PC12 cell lines. *J Biol Chem* 279, 52984–52990.

Martinat, C., Shendelman, S., Jonason, A., Leete, T., Beal, M. F., Yang, L., Floss, T., and Abeliovich, A. (2004). Sensitivity to oxidative stress in DJ-1-deficient dopamine neurons: An ES-derived cell model of primary Parkinsonism. *PLoS Biol* 2, e327.

Masutani, H., Bai, J., Kim, Y. C., and Yodoi, J. (2004). Thioredoxin as a neurotrophic cofactor and an important regulator of neuroprotection. *Mol Neurobiol* 29, 229–242.

Mattammal, M. B., Haring, J. H., Chung, H. D., Raghu, G., and Strong, R. (1995). An endogenous dopaminergic neurotoxin: Implication for Parkinson's disease. *Neurodegeneration* 4, 271–281.

Mayo, J. C., Sainz, R. M., Uria, H., Antolin, I., Esteban, M. M., and Rodriguez, C. (1998). Melatonin prevents apoptosis induced by 6-hydroxydopamine in neuronal cells: Implications for Parkinson's disease. *J Pineal Res* 24, 179–192.

Meuer, K., Pitzer, C., Teismann, P., Kruger, C., Goricke, B., Laage, R., Lingor, P., Peters, K., Schlachetzki, J. C., Kobayashi, K., Dietz, G. P., Weber, D., Ferger, B., Schabitz, W. R., Bach, A., Schulz, J. B., Bahr, M., Schneider, A., and Weishaupt, J. H. (2006). Granulocyte-colony stimulating factor is neuroprotective in a model of Parkinson's disease. *J Neurochem* 97, 675–686.

Mosharov, E. V., Staal, R. G., Bove, J., Prou, D., Hananiya, A., Markov, D., Poulsen, N., Larsen, K. E., Moore, C. M., Troyer, M. D., Edwards, R. H., Przedborski, S., and Sulzer, D. (2006). Alpha-synuclein overexpression increases cytosolic catecholamine concentration. *J Neurosci* 26, 9304–9311.

Nair, V. D. (2006). Activation of p53 signaling initiates apoptotic death in a cellular model of Parkinson's disease. *Apoptosis* 11, 955–966.

Nie, G., Cao, Y., and Zhao, B. (2002). Protective effects of green tea polyphenols and their major component, (−)-epigallocatechin-3-gallate (EGCG), on 6-hydroxydopamine-induced apoptosis in PC12 cells. *Redox Rep* 7, 171–177.

Offen, D., Ziv, I., Barzilai, A., Gorodin, S., Glater, E., Hochman, A., and Melamed, E. (1997). Dopamine-melanin induces apoptosis in PC12 cells; possible implications for the etiology of Parkinson's disease. *Neurochem Int* **31**, 207–216.

Offen, D., Sherki, Y., Melamed, E., Fridkin, M., Brenneman, D. E., and Gozes, I. (2000). Vasoactive intestinal peptide (VIP) prevents neurotoxicity in neuronal cultures: Relevance to neuroprotection in Parkinson's disease. *Brain Res* **854**, 257–262.

Panet, H., Barzilai, A., Daily, D., Melamed, E., and Offen, D. (2001). Activation of nuclear transcription factor kappa B (NF-kappaB) is essential for dopamine-induced apoptosis in PC12 cells. *J Neurochem* **77**, 391–398.

Peng, X., Tehranian, R., Dietrich, P., Stefanis, L., and Perez, R. G. (2005). Alpha-synuclein activation of protein phosphatase 2A reduces tyrosine hydroxylase phosphorylation in dopaminergic cells. *J Cell Sci* **118**, 3523–3530.

Pifl, C., Hornykiewicz, O., Giros, B., and Caron, M. G. (1996). Catecholamine transporters and 1-methyl-4-phenyl-1,2,3,6-tetrahydropyridine neurotoxicity: Studies comparing the cloned human noradrenaline and human dopamine transporter. *J Pharmacol Exp Ther* **277**, 1437–1443.

Reaney, S. H., and Smith, D. R. (2005). Manganese oxidation state mediates toxicity in PC12 cells. *Toxicol Appl Pharmacol* **205**, 271–281.

Rideout, H. J., Larsen, K. E., Sulzer, D., and Stefanis, L. (2001). Proteasomal inhibition leads to formation of ubiquitin/alpha-synuclein-immunoreactive inclusions in PC12 cells. *J Neurochem* **78**, 899–908.

Roth, J. A., Horbinski, C., Higgins, D., Lein, P., and Garrick, M. D. (2002). Mechanisms of manganese-induced rat pheochromocytoma (PC12) cell death and cell differentiation. *Neurotoxicology* **23**, 147–157.

Rukenstein, A., Rydel, R. E., and Greene, L. A. (1991). Multiple agents rescue PC12 cells from serum-free cell death by translation- and transcription-independent mechanisms. *J Neurosci* **11**, 2552–2563.

Ryu, E. J., Harding, H. P., Angelastro, J. M., Vitolo, O. V., Ron, D., and Greene, L. A. (2002). Endoplasmic reticulum stress and the unfolded protein response in cellular models of Parkinson's disease. *J Neurosci* **22**, 10690–10698.

Ryu, E. J., Angelastro, J. M., and Greene, L. A. (2005). Analysis of gene expression changes in a cellular model of Parkinson disease. *Neurobiol Dis* **18**, 54–74.

Saito, Y., Nishio, K., Ogawa, Y., Kinumi, T., Yoshida, Y., Masuo, Y., and Niki, E. (2007). Molecular mechanisms of 6-hydroxydopamine-induced cytotoxicity in PC12 cells: Involvement of hydrogen peroxide-dependent and -independent action. *Free Radic Biol Med* **42**, 675–685.

Salinas, M., Diaz, R., Abraham, N. G., Ruiz de Galarreta, C. M., and Cuadrado, A. (2003). Nerve growth factor protects against 6-hydroxydopamine-induced oxidative stress by increasing expression of heme oxygenase-1 in a phosphatidylinositol 3-kinase-dependent manner. *J Biol Chem* **278**, 13898–13904.

Sawada, H., Shimohama, S., Tamura, Y., Kawamura, T., Akaike, A., and Kimura, J. (1996). Methylphenylpyridium ion (MPP$^+$) enhances glutamate-induced cytotoxicity against dopaminergic neurons in cultured rat mesencephalon. *J Neurosci Res* **43**, 55–62.

Seaton, T. A., Cooper, J. M., and Schapira, A. H. (1998). Cyclosporin inhibition of apoptosis induced by mitochondrial complex I toxins. *Brain Res* **809**, 12–17.

Seo, B. B., Nakamaru-Ogiso, E., Flotte, T. R., Yagi, T., and Matsuno-Yagi, A. (2002). A single-subunit NADH-quinone oxidoreductase renders resistance to mammalian nerve cells against complex I inhibition. *Mol Ther* **6**, 336–341.

Sheng, G. Q., Zhang, J. R., Pu, X. P., Ma, J., and Li, C. L. (2002). Protective effect of verbascoside on 1-methyl-4-phenylpyridinium ion-induced neurotoxicity in PC12 cells. *Eur J Pharmacol* **451**, 119–124.

Sherer, T. B., Betarbet, R., Testa, C. M., Seo, B. B., Richardson, J. R., Kim, J. H., Miller, G. W., Yagi, T., Matsuno-Yagi, A., and Greenamyre, J. T. (2003). Mechanism of toxicity in rotenone models of Parkinson's disease. *J Neurosci* **23**, 10756–10764.

Shimoke, K., and Chiba, H. (2001). Nerve growth factor prevents 1-methyl-4-phenyl-1,2,3,6-tetrahydropyridine-induced cell death via the Akt pathway by suppressing caspase-3-like activity using PC12 cells: Relevance to therapeutical application for Parkinson's disease. *J Neurosci Res* **63**, 402–409.

Shoshani, T., Faerman, A., Mett, I., Zelin, E., Tenne, T., Gorodin, S., Moshel, Y., Elbaz, S., Budanov, A., Chajut, A., Kalinski, H., Kamer, I., Rozen, A., Mor, O., Keshet, E., Leshkowitz, D., Einat, P., Skaliter, R., and Feinstein, E. (2002). Identification of a novel hypoxia-inducible factor 1-responsive gene, RTP801, involved in apoptosis. *Mol Cell Biol* **22**, 2283–2293.

Silva, R. M., Ries, V., Oo, T. F., Yarygina, O., Jackson-Lewis, V., Ryu, E. J., Lu, P. D., Marciniak, S. J., Ron, D., Przedborski, S., Kholodilov, N., Greene, L. A., and Burke, R. E. (2005). CHOP/GADD153 is a mediator of apoptotic death in substantia nigra dopamine neurons in an *in vivo* neurotoxin model of parkinsonism. *J Neurochem* **95**, 974–986.

Smith, W. W., Jiang, H., Pei, Z., Tanaka, Y., Morita, H., Sawa, A., Dawson, V. L., Dawson, T. M., and Ross, C. A. (2005). Endoplasmic reticulum stress and mitochondrial cell death pathways mediate A53T mutant alpha-synuclein-induced toxicity. *Hum Mol Genet* **14**, 3801–3811.

Soldner, F., Weller, M., Haid, S., Beinroth, S., Miller, S. W., Wullner, U., Davis, R. E., Dichgans, J., Klockgether, T., and Schulz, J. B. (1999). MPP$^+$ inhibits proliferation of PC12 cells by a p21(WAF1/Cip1)-dependent pathway and induces cell death in cells lacking p21(WAF1/Cip1). *Exp Cell Res* 250, 75–85.

Sousa, S. C., and Castilho, R. F. (2005). Protective effect of melatonin on rotenone plus Ca^{2+}-induced mitochondrial oxidative stress and PC12 cell death. *Antioxid Redox Signal* 7, 1110–1116.

Spina, M. B., Squinto, S. P., Miller, J., Lindsay, R. M., and Hyman, C. (1992). Brain-derived neurotrophic factor protects dopamine neurons against 6-hydroxydopamine and N-methyl-4-phenylpyridinium ion toxicity: Involvement of the glutathione system. *J Neurochem* 59, 99–106.

Stefanis, L., Kholodilov, N., Rideout, H. J., Burke, R. E., and Greene, L. A. (2001a). Synuclein-1 is selectively up-regulated in response to nerve growth factor treatment in PC12 cells. *J Neurochem* 76, 1165–1176.

Stefanis, L., Larsen, K. E., Rideout, H. J., Sulzer, D., and Greene, L. A. (2001b). Expression of A53T mutant but not wild-type alpha-synuclein in PC12 cells induces alterations of the ubiquitin-dependent degradation system, loss of dopamine release, and autophagic cell death. *J Neurosci* 21, 9549–9560.

Tanaka, Y., Engelender, S., Igarashi, S., Rao, R. K., Wanner, T., Tanzi, R. E., Sawa, A. V. L. D., Dawson, T. M., and Ross, C. A. (2001). Inducible expression of mutant alpha-synuclein decreases proteasome activity and increases sensitivity to mitochondria-dependent apoptosis. *Hum Mol Genet* 10, 919–926.

Teng, K. K., Angelastro, J. M., Cunningham, M. E., Farinelli, S. E., and Greene, L. A. (1998). *Cultured PC12 cells: A Model for Neuronal Function, Differentiation and Survival.* Academic Press, Orlando, FL.

Tieu, K., Ashe, P. C., Zuo, D. M., and Yu, P. H. (2001). Inhibition of 6-hydroxydopamine-induced p53 expression and survival of neuroblastoma cells following interaction with astrocytes. *Neuroscience* 103, 125–132.

Tyurina, Y. Y., Kapralov, A. A., Jiang, J., Borisenko, G. G., Potapovich, A. I., Sorokin, A., Kochanek, P. M., Graham, S. H., Schor, N. F., and Kagan, V. E. (2006). Oxidation and cytotoxicity of 6-OHDA are mediated by reactive intermediates of COX-2 overexpressed in PC12 cells. *Brain Res* 1093, 71–82.

Valavanis, C., Hu, Y., Yang, Y., Osborne, B. A., Chouaib, S., Greene, L., Ashwell, J. D., and Schwartz, L. M. (2001). Model cell lines for the study of apoptosis *in vitro. Methods Cell Biol* 66, 417–436.

Walkinshaw, G., and Waters, C. M. (1994). Neurotoxin-induced cell death in neuronal PC12 cells is mediated by induction of apoptosis. *Neuroscience* 63, 975–987.

Walkinshaw, G., and Waters, C. M. (1995). Induction of apoptosis in catecholaminergic PC12 cells by l-DOPA. Implications for the treatment of Parkinson's disease. *J Clin Invest* 95, 2458–2464.

Wang, G., Qi, C., Fan, G. H., Zhou, H. Y., and Chen, S. D. (2005). PACAP protects neuronal differentiated PC12 cells against the neurotoxicity induced by a mitochondrial complex I inhibitor, rotenone. *FEBS Lett* 579, 4005–4011.

Wang, Z., Malone, M. H., Thomenius, M. J., Zhong, F., Xu, F., and Distelhorst, C. W. (2003). Dexamethasone-induced gene 2 (dig2) is a novel pro-survival stress gene induced rapidly by diverse apoptotic signals. *J Biol Chem* 278, 27053–27058.

Webb, J. L., Ravikumar, B., Atkins, J., Skepper, J. N., and Rubinsztein, D. C. (2003). Alpha-synuclein is degraded by both autophagy and the proteasome. *J Biol Chem* 278, 25009–25013.

Weinreb, O., Amit, T., Bar-Am, O., Sagi, Y., Mandel, S., and Youdim, M. B. (2006). Involvement of multiple survival signal transduction pathways in the neuroprotective, neurorescue and APP processing activity of rasagiline and its propargyl moiety. *J Neural Transm Suppl*, 457–465.

Wilhelm, M., Xu, Z., Kukekov, N. V., Gire, S., and Greene, L. A. (2007). Proapoptotic Nix activates the JNK pathway by interacting with POSH and mediates death in a Parkinson disease model. *J Biol Chem* 282, 1288–1295.

Xu, J., Wei, C., Xu, C., Bennett, M. C., Zhang, G., Li, F., and Tao, E. (2007). Rifampicin protects PC12 cells against MPP(+)-induced apoptosis and inhibits the expression of an alpha-synuclein multimer. *Brain Res* 1139, 220–225.

Xu, Z., and Greene, L. A. (2006). Activation of the apoptotic JNK pathway through the Rac1-binding scaffold protein POSH. *Methods Enzymol* 406, 479–489.

Yamamuro, A., Yoshioka, Y., Ogita, K., and Maeda, S. (2006). Involvement of endoplasmic reticulum stress on the cell death induced by 6-hydroxydopamine in human neuroblastoma SH-SY5Y cells. *Neurochem Res* 31, 657–664.

Yang, W. L., and Sun, A. Y. (1998). Paraquat-induced cell death in PC12 cells. *Neurochem Res* 23, 1387–1394.

Zheng, H., Gal, S., Weiner, L. M., Bar-Am, O., Warshawsky, A., Fridkin, M., and Youdim, M. B. (2005). Novel multifunctional neuroprotective iron chelator-monoamine oxidase inhibitor drugs for neurodegenerative diseases: *In vitro* studies on antioxidant activity, prevention of lipid peroxide formation and monoamine oxidase inhibition. *J Neurochem* 95, 68–78.

30

DISSOCIATED MESENCEPHALIC CULTURES: A RESEARCH TOOL TO MODEL DOPAMINERGIC CELL DEATH IN PARKINSON'S DISEASE

MYRIAM ESCOBAR-KHONDIKER[1,2,3], DAMIEN TOULORGE[1,2], SERGE GUERREIRO[1,2], ETIENNE C. HIRSCH[1,2] AND PATRICK P. MICHEL[1,2,4]

[1]*INSERM Unité Mixte de Recherche S679, Experimental Neurology and Therapeutics, Paris, France*
[2]*Université Pierre et Marie Curie-Paris 6, Paris, France*
[3]*Experimental Neurology, Philipps University, Marburg, Germany*
[4]*Centre de Recherche Pierre Fabre, Castres, France*

Nigrostriatal dopaminergic (DA) neurons are critically involved in the control of voluntary movements (DeLong, 1990; Grillner and Mercuri, 2002). As a result, their loss leads to profoundly disabling motor symptoms in Parkinson's disease (PD) (Agid, 1991). The identification and characterization of the signals and factors that control the survival and function of these neurons is therefore of interest as it may not only provide key insights to the mechanisms leading to PD, but may also help in the development of new neuroprotective/neurorestorative treatments for the disease.

Dissociated primary midbrain cultures provide a simplified, but nevertheless attractive model system to analyze the cellular and molecular mechanisms that control DA cell survival in a tightly controlled environment. These cultures are usually obtained from embryonic and less often from early post-natal mice or rats. Embryonic mesencephalic cultures are prepared typically from mouse or rat embryos at gestational ages 13 and 14–15.5, respectively (Henze *et al.*, 2005). The main advantage of this type of cultures is the presence of authentic DA neurons in the context of their physiological neighbors, that is, other mesencephalic neurons and various populations of glial cells including astrocytes and microglia (Collier *et al.*, 2003). DA neurons are characterized phenotypically by the presence of tyrosine hydroxylase (TH), the rate-limiting enzyme in the synthesis of dopamine, and the absence of dopamine-β-hydroxylase, the enzyme that converts dopamine into noradrenaline (Traver *et al.*, 2006). DA neurons represent 2–5% (up to 50% in the case of post-natal cultures) of the entire population of cultured mesencephalic neurons (Michel and Agid, 1996; Larsen *et al.*, 2002) which makes this model system inevitably more challenging to manipulate than immortalized DA cell lines which contain a homogenous population of cells. This heterogeneity constitutes an obstacle for biochemical assays of proteins which are not specifically expressed by TH⁺ DA neurons (Falkenburger and Schulz, 2006). This problem may be partially overcome, however, by studying the expression of these proteins, at the cellular level, using double or triple immunofluorescence staining (Troadec *et al.*, 2002). It may also turn

389

out to be an advantage since the presence of non-DA neurons may allow to demonstrate the selectivity of the degenerative process affecting TH^+ neurons (Michel et al., 1990). Another possible drawback is that DA neurons in dissociated cultures are in an isolated state that precludes any excitatory inputs from other brain nuclei. This may represents, however, an interesting feature if these inputs influence the survival of DA neurons during development and possibly in the adult brain as suggested previously (Michel et al., 2007), and if the culture conditions reproduce the destabilizing effect resulting from their absence (Salthun-Lassalle et al., 2004, 2005).

Here, we describe a number of experimental paradigms in which mesencephalic cultures proved to be valuable to explore the pathomechanisms of PD. More specifically, we will focus in this chapter on experimental settings used to study the mechanisms of DA cell death resulting from deficits in mitochondrial complex I activity, increased and uncontrolled production of reactive oxygen species (ROS), dysfunction of the ubiquitin–proteasome system, aberrant activation of the cell cycle machinery, lack of appropriate trophic factors, alterations in neuronal excitability in relation with abnormal calcium homeostasis and inflammatory-related processes. When relevant, we will also describe how and in what conditions these models of DA cell degeneration have been used to test molecules of potential therapeutical interest.

DA CELL DEATH CAUSED BY MITOCHONDRIAL COMPLEX I INHIBITORS

Experimental and clinical studies suggest that mitochondria are key effectors of the cell death process in PD (Swerdlow et al., 1996; Orth and Schapira et al., 2002). More specifically, it has been suggested that DA neurons die as a consequence of a chronic deficit in mitochondrial complex I activity. This explains why dissociated embryonic mesencephalic cultures have been used extensively to address the question of the mechanisms that link mitochondrial complex I dysfunction to DA neuronal death in PD (Mytilineou et al., 1985; Sanchez-Ramos et al., 1986; Michel et al., 1989, 1990; Michel and Agid, 1992; Spina et al., 1992; Koutsilieri et al., 1995; Lannuzel et al., 2003; Ren et al., 2005; Kanthasamy et al., 2006). Here, we will describe experimental

settings in which mitochondrial complex I inhibitors, 1-methyl-4-phenylpyridinium (MPP^+), the active metabolite of the xenobiotic chemical 1-methyl-4-phenyl-1,2,3,6-tetrahydropyridine (MPTP) and two plant toxins, rotenone and annonacin, have been used to trigger DA cell death in mesencephalic cultures.

The MPP^+ Model of DA Cell Death

Several studies have shown that midbrain DA neurons are highly and selectively sensitive to MPP^+ in a range of concentrations comprised between 0.1 and $10\,\mu M$, either in rat or in mouse mesencephalic cultures (Mytilineou et al., 1985; Michel et al., 1990; Koutsilieri et al., 1995; Kanthasamy et al., 2006). The high degree of selectivity of MPP^+ for DA neurons was established on the basis of two observations: (1) non-DA neurons from mesencephalic cultures were resistant to the toxin up to concentrations that kill more than 90% of the DA neurons (Figure 30.1) (Sanchez-Ramos et al., 1986; Michel et al., 1989, 1990; Saporito et al., 1992); (2) cholinergic neurons from septal cultures were totally insensitive to the pyridinium in the same experimental setting (Michel et al., 1990). Interestingly, blocking the dopamine transporter with mazindol substantially prevented MPP^+-induced DA cell death in mesencephalic cultures (Sanchez-Ramos et al., 1986; Lotharius and O'Malley 2000) confirming that the pyridinium gained access to the intracellular compartment of DA neurons via the neurotransmitter uptake system (Storch et al., 2004). Similar to what was observed after systemic administration of MPTP (Herkenham et al., 1991), MPP^+ initially affected the morphology and function of DA nerve endings in mesencephalic cultures before that of corresponding cell bodies (Michel et al., 1990; Koutsilieri et al., 1995): (1) neuritic extensions appeared already distorted and fragmented even when the loss of TH^+ cell bodies remained limited; (2) dopamine uptake which occurs preferentially at the level of DA nerve endings was reduced at concentrations of the toxin that did not affect the number TH^+ cell bodies (Michel et al., 1990). Glial cell line-derived neurotrophic factor (GDNF), a prototypical trophic factor for DA neurons (Lin et al., 1993), was not directly protective against MPP^+ but it stimulated the regrowth of injured DA fibers after withdrawal of the toxin (Hou et al., 1996). Other neurotrophic

(a)

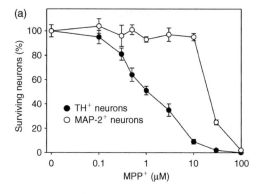

(b)

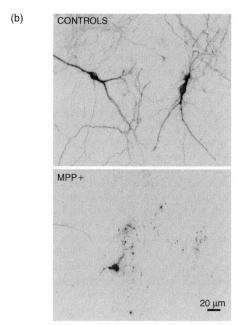

FIGURE 30.1 Selective toxicity of MPP$^+$ for DA neurons in rat primary mesencephalic cultures. (a) MPP$^+$ kills DA (TH$^+$) neurons without affecting non-DA (MAP-2$^+$) neurons in a range of concentrations comprised between 0.3 and 10 μM. At higher concentrations, all populations of neurons are affected regardless of their neurotransmitter's phenotype. (b) Immunodetection of TH$^+$ neurons in control and MPP$^+$-treated cultures. MPP$^+$ was applied at 3 μM for 48 h between days 4 and 6 *in vitro*.

When mesencephalic cultures were exposed to low concentration of MPP$^+$ (<3 μM), a high proportion of DA neurons in culture presented a fragmented nuclear chromatin and expressed the activated form of caspase-3 indicating that cell demise occurred via apoptosis (Hartmann *et al.*, 2000). It is worth noting that caspase-3 was also detected under its activated form in the population of DA neurons that are committed to degeneration in PD (Hartmann *et al.*, 2000). Caspase inhibition produced by either broad range or specific inhibitors was reported to reduce MPP$^+$-induced DA cell death, either partially (Dodel *et al.*, 1998; Bilsland *et al.*, 2002) or totally (Kanthasamy *et al.*, 2006). Other studies have demonstrated, however, that caspase inactivation caused a switch from apoptosis to necrosis without providing protection (Hartmann *et al.*, 2001). Note that other mechanisms described to participate in MPP$^+$-induced DA cell death will be addressed subsequently in the chapter.

The toxic effects of MPTP/MPP$^+$ *in vivo* were not always recapitulated in mesencephalic cultures. For instance, DA neurons in culture were resistant to MPTP (Sanchez-Ramos *et al.*, 1986; our own unpublished observation) probably for reasons inherent to this model system: (1) astrocytes which convert MPTP into MPP$^+$ via their monoamine oxidase (MAO)-B are present at a relatively low density in this type of cultures; (2) once produced in astrocytes, MPP$^+$ is not released in a natural three-dimensional brain environment and as a consequence is too diluted in the culture medium to generate toxic concentrations. In addition, whereas DA neurons from rat and mouse embryonic mesencephalic cultures were equally sensitive to MPP$^+$, only adult DA neurons in the mouse brain were selectively affected by systemic administration of MPTP (Chiueh *et al.*, 1984) probably because the bioavailability of the tetrahydropyridine and/or its conversion by MAOB is limited in the rat brain (Yazdani *et al.*, 2006). Finally, MPP$^+$-induced cell death *in vitro* appeared irreversible and even progressed further after stopping the exposure to the toxin (Michel *et al.*, 1990; Hou *et al.*, 1996), whereas adult DA neurons in the mouse brain partially recover in acute and subacute regimens of MPTP intoxication (Mitsumoto *et al.*, 1998; Petroske *et al.*, 2001).

The model of MPP$^+$-induced DA cell death was also used to define the structural requirements of other pyridiniums that could potentially operate as potential environmental neurotoxins for DA neurons (Michel *et al.*, 1989; Saporito *et al.*, 1992).

peptides such as transforming growth factors (TGFs)-β and fibroblast growth factor (FGF)-2 protected DA cell bodies against the effects of the pyridinium (Krieglstein *et al.*, 1995).

These studies revealed that these requirements were highly stringent so that only a few structural changes on the original structure of MPP$^+$ allowed to preserve the selective toxicity of this class of compounds for DA neurons (Michel *et al.*, 1989; Saporito *et al.*, 1992). Among these compounds, 2'CH$_3$-MPP$^+$ and 4'-NH$_2$-MPP$^+$ appeared the most potent (Michel *et al.*, 1989; Saporito *et al.*, 1992). Interestingly, systemic administration of the corresponding tetrahydropyridines to mice resulted in selective DA toxicity (Saporito *et al.*, 1992).

DA Cell Death Induced by Complex I Inhibitors Unrelated Structurally to MPP$^+$: Rotenone and Annonacin

The high susceptibility of DA neurons to MPP$^+$ and to some of its close analogs has raised the question as to whether other complex I inhibitors structurally unrelated to MPP$^+$ could also operate as DA neurotoxins in humans (Michel *et al.*, 1989; Betarbet *et al.*, 2000; Langston 2002; Lannuzel *et al.*, 2003, 2007). Two plant toxins, the pesticide rotenone (Betarbet *et al.*, 2000) and the acetogenin annonacin (Lannuzel *et al.*, 2003, 2007), are two prime candidates: (1) both rotenone and annonacin kill DA neurons when administrated systemically to rats using osmotic minipumps (Betarbet *et al.*, 2000; Höglinger *et al.*, 2003a; Champy *et al.*, 2004); (2) case–control studies have revealed that the consumption of soursop, a plant that contains high amounts of annonacin, may be responsible for the abnormally high frequency of atypical parkinsonism in the island of Guadeloupe in the French West Indies and in New Caledonia (Caparros-Lefebvre and Elbaz, 1999; Angibaud *et al.*, 2004; Lannuzel *et al.*, 2007); (3) epidemiological studies have implicated exposure to pesticides in particular rotenone, as a significant risk factor for PD (Betarbet *et al.*, 2002; Langston, 2002).

Several reports have shown that rotenone and annonacin kill DA neurons in mesencephalic cultures with a higher potency than MPP$^+$. The EC50s for rotenone and annonacin were on the order of 20–40 nM after 24 h of treatment as compared to 1–2 μM for the pyridinium (Ahmadi *et al.*, 2003; Lannuzel *et al.*, 2003; Ren *et al.*, 2005). However, rotenone and annonacin also reduced the survival of non-DA mesencephalic neurons probably due to the fact that these two molecules are highly lipophilic and not taken up selectively by DA neurons (Lannuzel *et al.*, 2003; Sakka *et al.*, 2003).

Toxic effects were seen at even lower concentrations with both compounds when the incubation time was extended by several days (Lannuzel *et al.*, 2003; Sakka *et al.*, 2003). At low concentrations, the toxic effects of rotenone were more specific to DA neurons since neuronal demise resulted in part from the production of toxic catabolites of the neurotransmitter dopamine (Ren *et al.*, 2005) (see next chapter for details). Similar to what was observed with MPP$^+$, the effect of rotenone required the activation of caspase-3 and was prevented by caspase inhibitors (Ahmadi *et al.*, 2003; Casarejos *et al.*, 2006).

Annonacin caused the death of DA neurons in mesencephalic cultures via a mechanism that mostly resulted from impairment of energy metabolism. Indeed, annonacin-induced DA cell death was prevented by two hexoses, glucose and its glycolyzable isomer mannose, which both operated by partially restoring intracellular ATP levels which were decreased as a consequence of mitochondrial complex I inhibition (Lannuzel *et al.*, 2003). Deoxyglucose, a non-metabolizable glucose/mannose analog, reversed these neuroprotective effects probably by competition, at the glucose transporter sites. Other hexoses such as galactose and fructose were not protective because they were poorly taken up by DA neurons (Lannuzel *et al.*, 2003). Attempts to restore oxidative phosphorylation with substrates of the citric acid cycle, lactate or pyruvate, failed to provide protection to DA neurons whereas idoacetate, an inhibitor of glycolysis, inhibited survival promotion by glucose and mannose indicating that both hexoses acted upstream of the mitochondria by stimulating the glycolytic flux in these neurons (Lannuzel *et al.*, 2003).

IMPLICATION OF ROS IN DA CELL DEATH

ROS are supposed to play a key role in neurodegenerative processes in PD (Hirsch, 1992; Michel *et al.*, 2002; Dauer and Przedborski, 2003; Moore *et al.*, 2005). Therefore, efforts have been made to better understand the nature of the mechanisms by which oxidative stress may trigger DA cell neurodegeneration.

Using free radical sensitive fluorochromes, Lotharius and O'Malley (2000) demonstrated that MPP$^+$ causes an early and sustained rise of ROS in cultured DA neurons. It was suggested that ROS production may occur as a mere consequence of mitochondrial complex I blockade (Hasegawa *et al.*, 1990; Koopman *et al.*, 2005). Alternatively, it was also proposed that when used at low concentrations, MPP^+ by redistributing vesicular dopamine to the cytoplasm caused the production of toxic catabolites of dopamine by autoxidation (Lotharius and O'Malley, 2000). This was supported by the fact that depleting cells of newly synthesized and/ or stored dopamine with α-methyl-*p*-tyrosine or reserpine, respectively, significantly attenuated both ROS production and cell death, whereas enhancing intracellular dopamine content exacerbated sensitivity to MPP^+. This observation supported the concept that dopamine may possibly become an endogenous neurotoxin in PD (Michel and Hefti, 1990). It is worth mentioning that neuroprotection resulting from dopamine depletion was improved by energy supplementation with the mitochondrial complex II substrate succinate suggesting that a mean lethal dose of the toxin killed DA cells by two concurrent mechanisms: dopamine-dependent ROS production and ATP depletion (Lotharius *et al.*, 2000). Still consistent with a role of ROS in MPP^+-induced cell death, two antioxidants, the C3 carboxyfullerene derivative and EUK-134, a catalase and superoxide dismutase (SOD) mimetic, protected DA neurons from the toxin (Pong *et al.*, 2000). The D_1/D_2 preferential agonist, lisuride, was also reported to have neuroprotective effects against MPP^+ due to its antioxidant activity (Gille *et al.*, 2002). Other classical antioxidants such as l-acetyl-carnitine, β-carotene, and α-tocopherol were, however, totally ineffective in the same experimental setting (Sanchez-Ramos *et al.*, 1988).

Rotenone intoxication was also found to stimulate ROS production in DA neurons (Radad *et al.*, 2006). It was suggested that ROS were generated as a consequence of the depolymerizing action of the alkaloid on microtubules (Ren *et al.*, 2005). Indeed, microtubule disorganization caused disruption in the axonal transport and thus the accumulation of DA vesicles in the soma leading to increased ROS production due to oxidation of cytosolic dopamine leaked from vesicles (Ren *et al.*, 2005; Feng, 2006). This explains why the toxicity of rotenone was significantly reduced by the microtubule stabilizing drug taxol and partially reproduced by microtubule-depolymerizing agents such as colchicine or nocodazole. Interestingly, Parkin, one of the most frequently mutated genes in PD, was reported to stabilize microtubules (Feng, 2006) which may account for the fact that cultured DA neurons from

Parkin null mice were more susceptible to rotenone, although another explanation has been proposed (Casarejos *et al.*, 2006). Importantly, the depolymerizing activity of rotenone on microtubules and the subsequent autoxidation of dopamine were apparently not related to the inhibitory action of the toxin on the mitochondrial respiratory chain (Ren *et al.*, 2005). Nevertheless, the mitochondrial deficit still participated to neurodegeneration and it was not restricted to DA neurons. As a consequence, the selectivity of rotenone for DA neurons remained only partial. Similar to what was observed *in vitro*, the selectivity of rotenone for DA neurons in the substantia nigra was only partial when the toxin was administered systemically to rats (Höglinger *et al.*, 2003a; Lapointe *et al.*, 2004). The group III metabotropic glutamate receptor agonist l-(+)-2-amino-4-phosphonobutyric acid (l-AP4) and neurotrophic factors, such as nerve growth factor, brain-derived neurotrophic factor (BDNF), and GDNF attenuated rotenone toxicity on DA neurons by activating an $ERK_{1/2}$-dependent mechanism that caused the stabilization of microtubules (Jiang *et al.*, 2006a, b).

The complex I inhibitor, annonacin, shares the same site of inhibition at the complex I level with rotenone (Miyoshi *et al.*, 1998). Annonacin also caused ATP depletion and increased ROS production but by a mechanism that was apparently not linked to dopamine autoxidation (Lannuzel *et al.*, 2003). It was shown that ROS emitted as the consequence of the blockade of the mitochondrial complex I by annonacin were not crucial for DA cell death (Lannuzel *et al.*, 2003) since restoration of ROS to control levels by antioxidants such as N-acetylcysteine or trolox, a vitamin E soluble analog, failed to protect DA neurons against the acetogenin (Lannuzel *et al.*, 2003).

Besides MPP^+ and rotenone, tetrahydrobiopterin (BH_4), an endogenous cofactor for dopamine synthesis, and methamphetamine, an illicit psychostimulant, were found to induce DA cell degeneration by a mechanism that involves dopamine autoxidation. BH_4-treated DA neurons showed increased level of oxidized proteins and most interestingly were protected by antioxidants. DA cell death induced by BH_4 was apoptotic as it required the activation of caspase-3 and the release of cytochrome-C from the mitochondria (Lee *et al.*, 2007). The effect of methamphetamine, appeared restricted to neuritic processes since DA cell bodies were spared by the treatment (Larsen *et al.*, 2002).

The mechanisms of ROS-mediated DA cell death have also been addressed by exposing mesencephalic cultures to the synthetic catechol compound 6-hydroxydopamine (6-OHDA), a prototypic DA neurotoxin used *in vivo* to generate lesions of the nigrostriatal system. Like MPP^+, rotenone, and BH_4, 6-OHDA was found to kill DA neurons via the production of autoxidative degradation products (Michel and Hefti, 1990). The autoxidation process occurred initially in the culture medium due to the instability of 6-OHDA at physiological pH (Michel and Hefti, 1990). Because the degradation products were not substrates of the dopamine transporter, 6-OHDA had none or a very limited selectivity for DA neurons *in vitro* (Michel and Hefti, 1990; Kramer *et al.*, 1999). Despite this limitation, this model system was proved useful to study the mechanisms of ROS-mediated apoptotic DA cell death. In particular, several groups reported that the effects of 6-OHDA were mediated through the activation of caspase-8 and caspase-3 (Von Coelln *et al.*, 2001; Han *et al.*, 2003; Choi *et al.*, 2004; Kanthasamy *et al.*, 2006). More specifically, it was shown that the inhibition of caspase-3 protected DA neurons by preventing the cleavage of PKCδ, an enzyme which may be crucially involved in DA cell death in PD (Kanthasamy *et al.*, 2006). It should be mentioned, however, that while caspase inhibitors provided robust protection against the deleterious effect of 6-OHDA on DA cell bodies, they remained ineffective to maintain functional DA cell nerve endings (Von Coelln *et al.*, 2001). This finding suggested that caspase inhibitors may have a limited therapeutic use in PD, if used alone. In contrast, GDNF favored exclusively the regrowth of DA nerve terminals damaged by 6-OHDA treatment (Kramer *et al.*, 1999).

It should be mentioned that dopamine and also its immediate precursor l-DOPA were found to produce detrimental effects which were similar to those produced by 6-OHDA (Michel and Hefti, 1990; Mena *et al.*, 1997), suggesting that the motor complications observed in PD patients after a long-term treatment with l-DOPA may result from neurotoxic effects (Agid *et al.*, 1999; Olanow *et al.*, 2004). These results remain, however, controversial because (1) the toxic concentrations of dopamine and l-DOPA were relatively high and (2) the density of glial cells which was low in the test experimental conditions did not reflect the physiological environment of the brain (Agid *et al.*, 1999; Olanow *et al.*, 2004). Indeed, when the density of glial cells was higher or when a glia-conditioned culture medium was applied to the cultures, l-DOPA exerted trophic effects for DA neurons (Mena *et al.*, 1997). Interestingly, l-DOPA was also found to promote the recovery of striatal innervation in rats with partial lesions of the nigrostriatal pathway (Murer *et al.*, 1998).

Iron, a transition metal which acts as a catalyst for the production of hydroxyl radicals through the Fenton reaction, was found to be increased in the brain of PD patients (Dexter *et al.*, 1989; Gerlach *et al.*, 2006) suggesting that it may operate as a trigger for oxidative stress-mediated neurodegeneration in PD. To explore this possibility, models of iron-mediated oxidative stress have also been developed using mesencephalic cultures (Michel *et al.*, 1992; Troadec *et al.* 2001, 2002). In a model developed by Troadec *et al.* (2001), trace amounts of iron ($\sim 1\,\mu M$) present in the culture medium were found to be sufficient to induce DA cell demise. The death process was slow and favored when the density of glial cells was reduced. The outer side of the plasma membrane was probably the initial target of hydroxyl radicals produced by an iron-catalyzed Fenton-type reaction since the membrane-impermeable enzyme catalase (Beckman *et al.*, 1988), which prevents this reaction, afforded robust protection in this experimental setting. Intracellular oxidative stress which occurred secondarily as a consequence of the ROS-mediated attack of the plasma membrane was controlled by caspases and was instrumental for cell death (Troadec *et al.*, 2001, 2002). Consistent with a crucial role of ROS in DA cell death in this model system, various antioxidants, including the lazaroid U-74389, dipyridamole, and the vitamin E analog, trolox, were strongly neuroprotective. More surprisingly, low levels of the neurotransmitter noradrenaline also prevented DA cell demise. This effect occurred via an antioxidant mechanism that bypassed adrenoceptors. The presence of a catechol moiety in the chemical structure of noradrenaline appeared to be instrumental for this effect (Troadec *et al.*, 2001). Interestingly, stimulation of the $ERK_{1/2}$-dependent signaling pathway by cAMP elevating agents (Troadec *et al.*, 2002) or by depolarizing concentrations of K^+ (Michel *et al.*, 2003) strongly potentiated the survival promoting effect of antioxidants and that of noradrenaline. In line with these observations, lesions of the locus coeruleus noradrenergic system have been suspected to play a critical role in the progression of DA cell death in PD (Marien *et al.*, 2004).

A different strategy to study ROS-induced DA cell death was to reduce the endogenous antioxidant glutathione (GSH) using l-buthionine sulfoximine (BSO). GSH depletion to levels, that cause total cell loss in cultures containing neurons and glial cells, had no effect on cell viability in enriched neuronal cultures suggesting that cell death in this model system was the consequence of events mediated by glial cells. The lazaroid U-83836E (Grasbon-Frodl *et al.*, 1996), the antioxidant ascorbic acid, and the lipoxygenase (LOX) inhibitor nordihydroguaiaretic acid (Mytilineou *et al.*, 1999) also provided protection against BSO toxicity, indicating that arachidonic acid metabolism through the LOX pathway and the generation of ROS played a role in the loss of DA cell viability. Depletion in GSH by BSO treatment also increased the susceptibility of DA neurons to MPP$^+$, which in itself did not have any effect on GSH metabolism (Nakamura *et al.*, 1997). Interestingly, *N*-acetyl-l-cysteine which stimulates GSH synthesis was found protective in another model system of post-natal ventral midbrain neuron-cortical astrocyte cocultures in which DA cell death was spontaneous (Mena *et al.*, 1997). Unexpectedly, however, the effects of *N*-acetyl-l-cysteine were mimicked by l-DOPA in this experimental setting (Mena *et al.*, 1997) indicating that this catechol derivative had the potential like noradrenaline to operate as an antioxidant for DA neurons probably due to its catechol structure (Troadec *et al.*, 2001).

ROLE OF THE UBIQUITIN–PROTEASOME SYSTEM IN DA CELL DEATH

The ubiquitin–proteasome degradation pathway, a proteolytic system of degradation, is suspected to be dysfunctional in familial and also in sporadic forms of PD (Vila and Przedborski, 2004; Moore *et al.*, 2005). The role of the proteasome in DA cell death was investigated by treating mesencephalic cultures with two specific inhibitors, epoxomycin and lactacystin. Both compounds caused neuronal cell death by apoptosis rather specifically within the population of TH$^+$ cells (Rideout *et al.*, 2005). Proteasomal inhibition also resulted in the formation of ubiquitin and α-synuclein-positive cytoplasmic inclusions in the surviving mesencephalic DA neurons (Rideout *et al.*, 2005). These inclusions were reminiscent of aggregates observed in the brains of PD patients known as Lewy bodies. BH$_4$

and rotenone, which both cause DA cell death via a ROS-dependent mechanism involving the autoxidation of dopamine (see previous paragraph), also caused an increase in ubiquitin immunoreactivity in cultured DA neurons, suggesting an alteration in the pattern of protein degradation in these experimental paradigms (Zeevalk and Bernard, 2005; Lee *et al.*, 2007).

It was reported that subtoxic concentrations of rotenone and MPP$^+$ not only increased free radical production, but also reduced proteasomal activity as a consequence of ATP depletion in mesencephalic cultures (Höglinger *et al.*, 2003b). The question was therefore to determine if a moderate impairment of the proteasomal function could exacerbate the effects of non-toxic or partially toxic concentrations of mitochondrial complex I inhibitors (Höglinger *et al.*, 2003b). Proteasome inhibitors caused ATP depletion but had no effect on ROS production. They operated in synergy with MPP$^+$ or rotenone to reduce DA cell survival in the situation where ROS produced by complex I inhibitors did not exceed 40% above baseline. The accumulation of oxidized proteins was also observed in the presence of the combined treatments indicating that the capacity of the neurons for detoxification was probably impaired in this paradigm. Preventing the emission of ROS with antioxidants or stimulating the glycolytic flux with high levels of glucose protected DA neurons from the synergistic toxic effects of the two treatments (Höglinger *et al.*, 2003b) suggesting that deficits caused by impairment of the mitochondrial and proteasomal functions may operate in concert to promote neurodegeneration in PD.

INVOLVEMENT OF CELL CYCLE-DEPENDENT MECHANISMS IN DA CELL DEATH

A substantial body of evidence suggests that cell division and cell death are intimately related and use many of the same mechanisms for their execution (Herrup *et al.*, 2004; Copani *et al.*, 2007). This may be particularly true in PD: (1) phosphorylation of the retinoblastoma protein (Rb), a molecular trigger of cell cycle progression, is observed in the DA neurons committed to death; (2) several genes responsible for the familial forms of PD have been implicated in cancer or in cell cycle regulation (West *et al.*, 2005). Mesencephalic cultures were exposed to low concentrations of MPP$^+$ to better understand the molecular mechanisms that lead to activation of cell cycle components in post-mitotic DA neurons. MPP$^+$-treated DA neurons were immunoreactive for the G1-phase-associated cyclin D1, S-phase-associated cyclin E, G2-phase-associated cyclin A, and M-phase-associated cyclin B in decreasing proportions (Höglinger et al., 2007). Cyclin D1 DA neurons still had rather well preserved neuritic processes, whereas in those expressing the late-phase-associated cyclin B only dying cell bodies were detectable. This provided indirect evidence that each stage of the neurodegenerative process may be associated to a specific phase of the cell cycle. The S phase is typically initiated by phosphorylation of Rb, leading to the release of E2F transcription factors. Unbound E2F proteins autoinduce E2F gene expression and transactivate E2F-target genes, such as PCNA, which is required for DNA synthesis (Höglinger et al., 2007). After MPP$^+$ intoxication, but not in control conditions, numerous TH$^+$ cells became immunoreactive for phosphorylated Rb, E2F-1, PCNA, or BrdU further comforting the idea that DNA replication takes place in degenerating TH$^+$ neurons. The activation of E2F-1 was probably instrumental for the progression of the cell death process since blocking E2F-1 transcription with an antisense oligonucleotide protected cultured DA neurons from MPP$^+$. The presence of polyploid DA neurons in the substantia nigra of PD patients confirmed that these neurons have the potential to duplicate their DNA in a pathological context (Höglinger et al., 2007).

These results are also reminiscent of earlier studies showing that several antimitotics, the synthetic deoxynucleosides ara-C and fluorodeoxyuridine (Michel et al., 1997), and the purine analog cyclin-dependent kinase (cdks) inhibitors, olomoucine and roscovitine (Mourlevat et al., 2003), were highly efficient in protecting DA neurons from their spontaneous demise in mesencephalic cultures. A similar observation was reported with other purine derivatives adenosine and cAMP (Michel and Agid, 1996, 1999; Mourlevat et al., 2003), which can also act as cdk inhibitors although it is not their primary function. The data from Höglinger et al. (2007) suggest that the rescuing effects provided by these compounds may result from blockade of the cell cycle machinery in the DA neurons committed to death. Neuroprotection was suggested to occur, however, via an indirect mechanism that required the repression of a subpopulation of immature astrocytes which was specifically deleterious for DA neurons (Michel et al., 1999; Mourlevat et al., 2003).

PREVENTION OF DA CELL DEATH BY TROPHIC PEPTIDES

We, and others, have made the observation that DA neurons from midbrain embryonic cultures degenerate selectively and spontaneously, at a relatively slow rate (Lin et al., 1993; Michel and Agid, 1996; Michel et al., 1997) whereas non-DA neurons (mostly GABA-ergic and serotoninergic neurons) remain rather unaffected in the course of this process. This experimental paradigm has been used initially for the detection of peptides that exert a trophic/neuroprotective activity on DA neurons (Knusel et al., 1990; Lin et al., 1993) and, later on, as a model system to study the mechanisms underlying DA cell death (Douhou et al., 2001; Salthun-Lassalle et al., 2004, 2005). Several members of the TGF-β superfamily including GDNF, its close homologs neurturin, artemin and persephin, bone morphogenetic factors, and TGFs-β rescued DA neurons from their spontaneous demise (Lin et al., 1993; Poulsen et al., 1994; Krieglstein et al., 1995; Roussa et al., 2004; Zihlmann et al., 2005). The neurotrophins, BDNF and NT-4, insulin, the insulin growth factors I and II, and FGF-2 and -8 also promoted DA cell survival in this model system (Knusel et al., 1990; Hyman et al., 1991; Roussa et al., 2004). Note that the effect of FGF-2 was mediated by astroglial cells (Knusel et al.., 1990) via the production of TGFs-β (Krieglstein et al., 1998). Priming of the cultures with serum proteins was also essential to reveal the effects of BDNF and NT-4 on DA neurons for a reason that has not been clearly elucidated (Krieglstein et al., 1996). Cultured DA neurons obtained from transgenic mice overexpressing the human free radical scavenging enzyme Cu/Zn–SOD were protected against spontaneous DA cell death (Sanchez-Ramos et al., 1997) indicating that trophic factor deficiency may possibly cause DA cell death by oxidative stress-related mechanisms.

Stimulating the secretion of GDNF, BDNF, or FGF-2 in the culture medium by application of the D$_1$/D$_2$ dopamine receptor agonist apomorphine or the D$_3$ dopamine receptor preferential agonist pramipexole afforded protection too to DA neurons (Guo et al., 2002; Du et al., 2005; Li et al., 2006). The effect of apomorphine and pramipexole were mediated via DA receptors. The effect of pramipexole was also reported to occur partly via an antioxidant mechanism that bypassed dopamine receptors and did not implicate trophic factors (Ling *et al.*, 1999).

The trophic peptides mentioned previously play most likely a key role in the development of DA neurons (Knusel *et al.*, 1990; Oo *et al.*, 2003; Zhang *et al.*, 2007). Some of them, in particular GDNF and TGFs-β, were found to retain their trophic/neuroprotective activity in the adult brain (Love *et al.*, 2005; Schober *et al.*, 2007) and GDNF was reported to have restorative and trophic effects when injected directly in the putamen of PD patients (Love *et al.*, 2005).

ROLE OF ACTIVITY-DEPENDENT MECHANISMS IN DA CELL DEATH

Control of DA Cell Survival by Voltage-Gated Ca^{2+} Channels

That the survival of DA neurons can be controlled by their electrical activity was established using the model of spontaneous DA cell death mentioned previously (Michel and Agid, 1996; Murer *et al.*, 1999; Douhou *et al.*, 2001; Salthun-Lassalle *et al.*, 2004, 2005). In particular, low-level activation of voltage-gated sodium (Nav) channels by the alkaloid veratridine or the α-scorpion toxin (Catterall, 1980) proved to be highly effective in preventing the DA cell loss (Salthun-Lassalle *et al.*, 2004). Whether the sodium channel agonists acted directly on DA neurons or indirectly via a network effect that recruited other neurons has not been totally established. As expected, tetrodotoxin (TTX), a selective blocker of Nav channels, prevented the rescue by both veratridine and the α-scorpion toxin. However, TTX did not reduce the survival of other neurons, mostly GABA-ergic, contained in these cultures, suggesting that only DA neurons needed electrical stimulation to survive. We may assume that this requirement was probably revealed by the isolated state of these neurons in dissociated cultures, one that precluded any excitatory input (Salthun-Lassalle *et al.*, 2004). Interestingly, there is some evidence that sick DA neurons may become progressively hyperpolarized and reduced to silence long before they die in PD (reviewed in Michel *et al.*, 2006, 2007).

The effects on DA neurons of chronic depolarization by high concentrations of extracellular

K^+ were also tested. Andreeva *et al.* (1996) initially reported that this treatment did not increase the survival of DA neurons in dissociated mesencephalic cultures. Subsequent studies showed, however, that it had a robust effect in this experimental setting, but only when ionotropic glutamate receptors, that is, N-methyl-d-aspartate (NMDA) or α-amino-3-hydroxy-5-methyl-4-isoxazolepropionate (AMPA)/kainate receptors, were blocked concurrently by specific antagonists to prevent secondary excitotoxic stress caused by K^+-mediated release of glutamate (Murer *et al.*, 1999; Douhou *et al.*, 2001) through the reverse operation of glutamate transporters (Rossi *et al.*, 2000). The effect of high K^+ concentrations in the presence of glutamate receptors antagonists was resistant to TTX, however, which indicates that the survival of DA neurons was controlled in this paradigm by mechanisms that bypass Nav channels.

It was proposed on the basis of the following observations that intracellular Ca^{2+} levels play a key role in the survival of DA neurons: (1) concentrations of veratridine which were optimally neuroprotective in dissociated mesencephalic cultures caused a moderate but sustained elevation of intracellular calcium levels; (2) TTX which abolished the rescuing effect of veratridine for DA neurons also prevented the calcium rise induced by this treatment (Salthun-Lassalle *et al.*, 2004). The calcium influx generated by neuroprotective concentrations of veratridine was mediated by T-type voltage-gated calcium (Cav) channels, since it was abolished by two known blockers of this calcium channel subtype, the neuroleptic flunarizine, and nickel used at low concentrations (Santi *et al.*, 2002). Similar to TTX, flunarizine abolished the protective effects of the alkaloid on DA neurons (Salthun-Lassalle *et al.*, 2004). These observations led to the conclusion that DA neurons treated with veratridine were dependent for their survival on a mechanism that generates T-type Cav currents downstream of Nav channel activation. The bee venom toxin, apamin, a selective blocker of Ca^{2+} activated K^+ (SK) channels and the Na^+/K^+-ATPase blocker ouabain exerted a neuroprotective effect for DA neurons which also required the activation of T-type Cav channels (Salthun-Lassalle *et al.*, 2004). Interestingly, missense mutations in the gene coding for the Na^+/K^+-ATPase-3 cause rapid-onset dystonia–parkinsonism (de Carvalho Aguia *et al.*, 2004), a curious disorder in which dystonia and parkinsonian signs develop rapidly and irreversibly over hours to weeks.

A moderate rise in intracellular calcium levels was also detected when dissociated mesencephalic cultures were exposed to neuroprotective concentrations of K^+ in the presence of NMDA or AMPA/kainate receptor blockers to avoid excitotoxic stress inherent to this treatment (Douhou *et al.*, 2001). Both the calcium elevation and the increase in survival associated to it were resistant to a blockade of T-type Cav channels with flunarizine (Salthun-Lassalle *et al.*, 2004). They were prevented, however, by nifedipine, a selective blocker of high-voltage activated L-type calcium channels (Douhou *et al.*, 2001; Salthun-Lassalle *et al.*, 2004). A protective function has also been ascribed to calcium currents generated through N-type (ω-conotoxin MVIIA-sensitive) Cav channels (Salthun-Lassalle *et al.*, 2005) following the activation of neurokinin receptors by endogenous peptide agonists, including substance P, NKA, and NKB (Salthun-Lassalle *et al.*, 2005). The recruitment of N-type calcium channels by stimulation of tachykinin receptors required the activation of sodium currents by a mechanism that has not been yet elucidated (Salthun-Lassalle *et al.*, 2005). A schematic representation of the mechanisms that link DA neuron survival to Cav channel activation in mesencephalic cultures is given in Figure 30.2.

Given that the effects of the various depolarizing signals were reproduced by caspase inactivation (Salthun-Lassalle *et al.*, 2004), one may assume that the effect of the calcium elevation was to stimulate a prosurvival signaling pathway or to prevent the activation of a cell death program that was intrinsic to DA neurons (Salthun-Lassalle *et al.*, 2005). Several candidate mechanisms have received experimental support. In particular, it was suggested that K^+-induced depolarization could operate via the upregulation of Nurr1 (Volpicelli *et al.*, 2004), an orphan nuclear receptor involved in the development and survival of DA neurons (Vitalis *et al.*, 2005) or via the recruitment of TrkB receptors to the plasma membrane (Meyer-Franke *et al.*, 1998) making DA neurons possibly more responsive to the cognate ligand for these receptors, BDNF in the presence of depolarizing concentrations of K^+. This last possibility was not likely since the effect of high K^+ concentrations on DA neurons was not reduced by an antibody that neutralizes the biological activity of BDNF (Murer *et al.*, 1999). That neuroprotection due to calcium elevation could result indirectly from a mechanism involving other putative trophic factors secreted in the culture medium was also not supported experimentally (Salthun-Lassalle *et al.*, 2004, 2005).

Control of DA Cell Survival by Ligand-Gated Ion Channels

In addition, other studies have shown that DA neurons in cultures can be protected from death by low-level stimulation of several types of ligand-gated ion channels (Michel *et al.*, 1999; Jeyarasasingam *et al.*, 2002). More specifically, low-level activation of NMDA receptors was found to enhance the survival promoting effects of BDNF for DA neurons (Franke *et al.*, 2000). The activation of Purinergic P_{2X} receptors by ATP or by its non-hydrolyzable analog α,β-methylene-ATP also rescued DA of purinergic neurons from death (Michel *et al.*, 1999). Interestingly, Parkin and UCHL-1, two proteins that are implicated in familial PD and play key roles in the ubiquitin–proteasome protein degradation pathway (Vila and Przedborski, 2004), were also reported recently to stimulate neuronal excitability by potentiating inward currents generated through ATP-gated P_{2X} receptors (Sato *et al.*, 2006). Finally, the depolarizing agent nicotine was reported to reduce the death of cultured DA neurons exposed to MPP^+ (Jeyarasasingam *et al.*, 2002). In line with this observation, nicotine was found to be neuroprotective in animal models of PD and is also suspected to reduce the risk of developing the disease in the population of smokers (Parain *et al.*, 2003; Quik *et al.*, 2006).

Conversely, it has been proposed that overexcitation of DA neurons by excessive activation of NMDA receptors may trigger DA cell degeneration in PD and in animal models of the disease through an excitotoxic process (Rodriguez *et al.*, 1998). Mesencephalic neurons were indeed highly sensitive to excitotoxic stress generated, either directly, by high concentrations of glutamate, or indirectly, by depolarizing concentrations of K^+. The cell death process was, however, not specific to DA neurons (Andreeva *et al.*, 1996; Douhou *et al.*, 2001). Furthermore, in paradigms of MPP^+- or annonacin-induced DA cell death, which are more relevant to PD, blockade of NMDA receptors proved to be ineffective against neuronal demise (Michel and Agid, 1992; Lannuzel *et al.*, 2003).

ROLE OF INFLAMMATORY PROCESSES IN DA CELL DEATH

It is becoming increasingly evident that inflammation plays an important role in the pathogenesis of PD (McGeer *et al.*, 2001; Hirsch *et al.*, 2003). In particular, it has been suggested that microglial cells, the resident immune cells of the brain, may be involved in an inflammatory response that perpetuates and accentuates the effects of the initial insult that causes DA neurons to degenerate. Mesencephalic cultures have been used to explore the mechanisms that underlie microglia-mediated facilitation of DA cell death and contribute to microglial cell proliferation in the course of neurodegeneration.

Mechanisms Underlying Microglial Cell Proliferation

The mitogenic action that MPP$^+$ exerts on microglial cells was studied using neuron/glia mesencephalic cultures (Henze *et al.*, 2005). Microglial cell proliferation was apparently not the mere consequence of DA cell demise caused by MPP$^+$ since the pyridinium produced the same effect in a model system of neuronal/glial cortical cultures where DA neurons were absent. Consistent with this observation, the proliferative effect of MPP$^+$ was also detectable in neuron-free microglia/astrocyte mesencephalic cultures (Henze *et al.*, 2005). It disappeared, however, when the toxin was added to

(a)

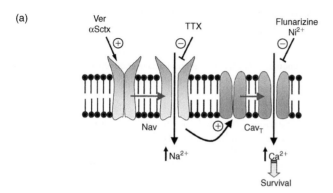

(b)

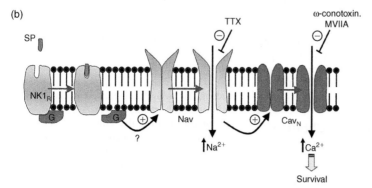

FIGURE 30.2 Schematic representation of the mechanisms that link DA cell survival to Cav channel activation in mesencephalic cultures. (a) Both veratridine and the α-scorpion toxin prevent the spontaneous and selective loss of DA neurons through sequential activation of Nav and T-type Cav channels. (b) Substance P is neuroprotective in the same experimental setting through the activation of neurokinin-1 receptors and a mechanism that recruits sequentially Nav and N-type calcium (ω-conotoxin MVIIA-sensitive) channels. Note that other tachykinins NKA and NKB are protective via activation of Nav and N-type Cav channels through a mechanism involving both neurokinin-1 and -3 receptors. The mechanism that couple neurokinin receptor activation to the opening of Nav channels has not been yet elucidated but it may possibly involve a G protein regulated pathway.

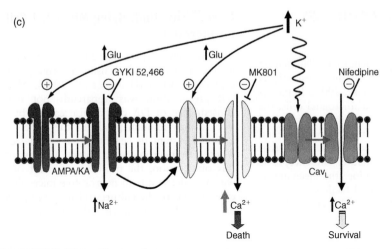

FIGURE 30.2 (*Continued*) (c) K^+-induced depolarization also protects DA neurons, but only when ionotropic glutamate receptors, that is, NMDA or AMPA/kainate receptors, are blocked concurrently by specific antagonists to prevent secondary excitotoxic stress caused by K^+-mediated release of glutamate. Note that the depolarizing signals are maximally effective in protecting DA neurons when their application results in a calcium elevation that is between 35% and 80% of control values. Adapted from Douhou *et al.* (2001) and Salthun-Lassalle *et al.* (2004, 2005).

pure microglial cell cultures suggesting that astrocytes played a key role in the mitogenic mechanism. Accordingly, the proliferation of microglial cells in response to MPP$^+$ treatment was mimicked by granulocyte macrophage colony-stimulating factor (GM-CSF), a proinflammatory cytokine produced by astrocytes and was blocked by a neutralizing antibody to GM-CSF (Henze *et al.*, 2005). Based on these results, it was proposed that the microglial reaction observed following MPP$^+$ exposure depends on astrocytic factors, for example GM-CSF, a finding that may have therapeutic implications.

Mechanisms of DA Cell Death Facilitation by Microglial Cells

It was proposed that the extracellular matrix molecule laminin may participate to the activation of microglial cells and to neurodegeneration following MPP$^+$ treatment (Wang *et al.*, 2006). Supporting this view, an antibody against a non-integrin laminin receptor (LR) or a synthetic pentapeptide inhibitor for LR, YIGSR, significantly attenuated DA neurotoxicity induced by MPP$^+$ treatment in neuron/glia mesencephalic cultures. Interestingly, the phagocytic activity of MPP$^+$-treated microglial cells

revealed by the incorporation of fluorescent microspheres and their potential to produce ROS was also reduced by treatment with LRAb. Conversely, the presence of a soluble form of laminin, a putative ligand for LR, in the culture medium, dose-dependently stimulated microglial cell activation and DA cell toxicity comforting the idea that cell–extracellular matrix interactions were crucially involved in MPP$^+$-induced microgliosis (Wang *et al.*, 2006).

The mechanism by which microglial cells facilitated DA cell death caused by MPP$^+$ or rotenone resulted from the production of superoxide by the microglial NADPH oxidase also called phagocytic oxidase (Gao *et al.*, 2003a, b): (1) in mixed mesencephalic neuron/glia cultures, DA neurons from NADPH oxidase deficient (gp91phox −/−) mice were more resistant to rotenone and MPP$^+$ neurotoxicity than DA neurons from gp91phox +/+ mice; (2) in neuron-enriched cultures, the addition of microglia prepared from gp91phox +/+ mice but not from gp91phox −/− mice markedly increased the loss of DA neurons; (3) inhibition of NADPH oxidase by apocynin attenuated rotenone or MPP$^+$-induced neurotoxicity only in the presence of microglia from gp91phox +/+ mice. Blocking laminin/LR interactions prevented superoxide production induced by MPP$^+$ suggesting that extracellular matrix interactions may participate indirectly to NADPH oxidase activation (Wang *et al.*, 2006). Suggesting that gene mutations may possibly exacerbate inflammatory processes in familial PD, cultured DA neurons from Parkin null mice appeared more susceptible to microglial-induced DA cell death caused via NADPH oxidase activation (Casarejos *et al.*, 2006). Note that substance P, which was found strongly neuroprotective in a paradigm of spontaneous DA cell death (Salthun-Lassalle *et al.*, 2005), was also reported by others to favor DA cell death via the activation of microglial NADPH oxidase (Block *et al.*, 2006). The origin of this discrepancy is currently unknown. Consistent with these different observations, genetic invalidation of NADPH oxidase was reported to be protective in the MPTP mouse model of PD (Wu *et al.*, 2003). Cycloxygenase-2 (COX-2), one of the targets of non-steroidal anti-inflammatory drugs, was induced by MPP^+ in both microglial cells and neurons in mesencephalic cultures. Yet, pharmacological inhibition of COX-2 by a specific inhibitor DuP697 afforded only marginal protection against neurodegeneration suggesting that this enzyme was not crucially involved in DA cell death in this experimental setting (Wang *et al.*, 2005).

The bacterial toxin lipopolysaccharide (LPS) was also used to model in mesencephalic cultures, inflammatory changes associated to neurodegeneration in PD (Gayle *et al.*, 2002, Wang *et al.*, 2004). Several papers demonstrated that LPS dose-dependently reduced the number of TH^+ neurons via a mechanism that required the activation of microglial cells. At the lowest concentrations of LPS tested, cell death was selective for DA neurons (Gayle *et al.*, 2002; Wang *et al.*, 2004). LPS-induced DA cell death was mediated by microglial cells through the activation of NADPH oxidase and the production of superoxide as in the models of MPP^+ or rotenone intoxication. Together with this, the SOD/catalase mimetic MnTMPyP was neuroprotective against LPS-induced toxicity. Interleukin-10, an immunosuppressive cytokine, the neuropeptide pituitary adenylate cyclase activating polypeptide (PACAP) 38, and the internal PACAP4–6 tripeptide Gly-Ile-Phe (GIF), the opioid receptor antagonist naloxone, and the codeine analog dextromethorphan were also protective in these conditions by inhibition of the NADPH oxidase (Wang *et al.*, 2004; Li *et al.*, 2005; Qian *et al.*, 2006; Yang *et al.*, 2006). PACAP and GIF exerted their effect by a mechanism that was independent of conventional PACAP receptors. Pharmacophore modeling revealed that naloxone, dextromorphan, and GIF possess common physicochemical features that confer them anti-inflammatory properties.

Note that the proinflammatory cytokine tumor necrosis factor (TNF) was released in substantial amounts by microglial cells in LPS-treated mesencephalic cultures. Its sequestration with an engineered dominant negative TNF compound (XENP-345) prevented microglial cell activation and provided partial protection against selective DA cell death (McCoy *et al.*, 2006). This indicated that TNF may possibly amplify neurodegeneration either by a direct mechanism on DA neurons and/or by perpetuating the production of microglial-derived extracellular ROS or reactive nitrogen species (McCoy *et al.*, 2006). Interestingly, neuroprotection was also observed with XENP-345 in an animal model of PD (McCoy *et al.*, 2006). Genetic invalidation of TNF receptors has also been found to be protective in animal models of PD (Sriram *et al.*, 2002), but these results were not substantiated by others (Rousselet *et al.*, 2002).

In conclusion, we described a number of experimental settings in which dissociated mesencephalic cultures proved to be a valuable tool to explore the mechanisms of neuronal cell death in PD. These observations which have only a predictive value of what could occur in a pathophysiological situation may become, however, of key interest when validated by studies performed in animal models of PD and comforted by postmortem analysis of brain samples. Mesencephalic culture models may therefore be useful for the identification of hit compounds and their transformation to lead molecules for the treatment of PD. Medium-throughput screening performed with automated cell culture workstations may facilitate this search.

ACKNOWLEDGEMENTS

The work by the authors mentioned in this chapter received support from Institut de la Santé et de la Recherche Médicale (INSERM) and Université Pierre et Marie Curie-Paris 6. PPM was supported by Centre de Recherche Pierre Fabre.

REFERENCES

Agid, Y. (1991). Parkinson's disease: Pathophysiology. *Lancet* 337, 1321–1324.

Ahmadi, F. A., Linseman, D. A., Grammatopoulos, T. N., Jones, S. M., Bouchard, R. J., Freed, C. R., Heidenreich, K. A., and Zawada, W. M. (2003). The pesticide rotenone induces caspase-3-mediated apoptosis in ventral mesencephalic dopaminergic neurons. *J Neurochem* 87, 914–921.

Andreeva, N., Ungethum, U., Heldt, J., Marschhausen, G., Altmann, T., Andersson, K., and Gross, J. (1996). Elevated potassium enhances glutamate vulnerability of dopaminergic neurons developing in mesencephalic cell cultures. *Exp Neurol* 137, 255–262.

Angibaud, G., Gaultier, C., and Rascol, O. (2004). Atypical parkinsonism and Annonaceae consumption in New Caledonia. *Mov Disord* 19, 603–604.

Beckman, J. S., Minor, R. L., Jr., White, C. W., Repine, J. E., Rosen, G. M., and Freeman, B. A. (1988). Superoxide dismutase and catalase conjugated to polyethylene glycol increases endothelial enzyme activity and oxidant resistance. *J Biol Chem* 263, 6884–6892.

Betarbet, R., Sherer, T. B., MacKenzie, G., Garcia-Osuna, M., Panov, A. V., and Greenamyre, J. T. (2000). Chronic systemic pesticide exposure reproduces features of Parkinson's disease. *Nat Neurosci* 3, 1301–1306.

Betarbet, R., Sherer, T. B., Di Monte, D. A., and Greenamyre, J. T. (2002). Mechanistic approaches to Parkinson's disease pathogenesis. *Brain Pathol* 12, 499–510.

Bilsland, J., Roy, S., Xanthoudakis, S., Nicholson, D. W., Han, Y., Grimm, E., Hefti, F., and Harper, S. J. (2002). Caspase inhibitors attenuate 1-methyl-4-phenylpyridinium toxicity in primary cultures of mesencephalic dopaminergic neurons. *J Neurosci* 22, 2637–2649.

Block, M. L., Li, G., Qin, L., Wu, X., Pei, Z., Wang, T., Wilson, B., Yang, J., and Hong, J. S. (2006). Potent regulation of microglia-derived oxidative stress and dopaminergic neuron survival: Substance P vs. dynorphin. *FASEB J* 20, 251–258.

Caparros-Lefebvre, D., and Elbaz, A. (1999). Possible relation of atypical parkinsonism in the French West Indies with consumption of tropical plants: A case–control study. Caribbean Parkinsonism Study Group. *Lancet* 354, 281–286.

Casarejos, M. J., Menendez, J., Solano, R. M., Rodriguez-Navarro, J. A., Garcia de Yebenes, J., and Mena, M. A. (2006). Susceptibility to rotenone is increased in neurons from parkin null mice and is reduced by minocycline. *J Neurochem* 97, 934–946.

Catterall, W. A. (1980). Neurotoxins that act on voltage-sensitive sodium channels in excitable membranes. *Annu Rev Pharmacol Toxicol* 20, 15–43.

Champy, P., Höglinger, G. U., Feger, J., Gleye, C., Hocquemiller, R., Laurens, A., Guerineau, V., Laprevote, O., Medja, F., Lombes, A., Michel, P. P., Lannuzel, A., Hirsch, E. C., and Ruberg, M. (2004).

Annonacin, a lipophilic inhibitor of mitochondrial complex I, induces nigral and striatal neurodegeneration in rats: Possible relevance for atypical parkinsonism in Guadeloupe. *J Neurochem* 88, 63–69.

Chiueh, C. C., Markey, S. P., Burns, R. S., Johannessen, J. N., Pert, A., and Kopin, I. J. (1984). Neurochemical and behavioral effects of systemic and intranigral administration of N-methyl-4-phenyl-1,2,3,6-tetrahydropyridine in the rat. *Eur J Pharmacol* 100, 189–194.

Choi, W. S., Eom, D. S., Han, B. S., Kim, W. K., Han, B. H., Choi, E. J., Oh, T. H., Markelonis, G. J., Cho, J. W., and Oh, Y. J. (2004). Phosphorylation of p38 MAPK induced by oxidative stress is linked to activation of both caspase-8- and -9-mediated apoptotic pathways in dopaminergic neurons. *J Biol Chem* 279, 20451–20460.

Collier, T. J., Steece-Collier, K., McGuire, S., and Sortwell, C. E. (2003). Cellular models to study dopaminergic injury responses. *Ann NY Acad Sci* 991, 140–151.

Copani, A., Caraci, F., Hoozemans, J. J., Calafiore, M., Angela Sortino, M., and Nicoletti, F. (2007). The nature of the cell cycle in neurons: Focus on a "non-canonical" pathway of DNA replication causally related to death. *Biochim Biophys Acta* 1772, 409–412.

Dauer, W., and Przedborski, S. (2003). Parkinson's disease: Mechanisms and models. *Neuron* 39, 889–909.

de Carvalho Aguiar, P., Sweadner, K. J., Penniston, J. T., Zaremba, J., Liu, L., Caton, M., Linazasoro, G., Borg, M., Tijssen, M. A., Bressman, S. B., Dobyns, W. B., Brashear, A., and Ozelius, L. J. (2004). Mutations in the Na$^+$/K$^+$-ATPase alpha3 gene ATP1A3 are associated with rapid-onset dystonia parkinsonism. *Neuron* 43, 169–175.

DeLong, M. R. (1990). Primate models of movement disorders of basal ganglia origin. *Trends Neurosci* 13, 281–285.

Dexter, D. T., Wells, F. R., Lees, A. J., Agid, F., Agid, Y., Jenner, P., and Marsden, C. D. (1989). Increased nigral iron content and alterations in other metal ions occurring in brain in Parkinson's disease. *J Neurochem* 52, 1830–1836.

Dodel, R. C., Du, Y., Bales, K. R., Ling, Z. D., Carvey, P. M., and Paul, S. M. (1998). Peptide inhibitors of caspase-3-like proteases attenuate 1-methyl-4-phenylpyridinum-induced toxicity of cultured fetal rat mesencephalic dopamine neurons. *Neuroscience* 86, 701–707.

Douhou, A., Troadec, J. D., Ruberg, M., Raisman-Vozari, R., and Michel, P. P. (2001). Survival promotion of mesencephalic dopaminergic neurons by depolarizing concentrations of K$^+$ requires concurrent inactivation of NMDA or AMPA/kainate receptors. *J Neurochem* 78, 163–174.

Du, F., Li, R., Huang, Y., Li, X., and Le, W. (2005). Dopamine D3 receptor-preferring agonists induce neurotrophic effects on mesencephalic dopamine neurons. *Eur J Neurosci* 22, 2422–2430.

Falkenburger, B. H., and Schulz, J. B. (2006). Limitations of cellular models in Parkinson's disease research. *J Neural Transm*(**Suppl**), 261–268.

Feng, J. (2006). Microtubule: A common target for parkin and Parkinson's disease. *Neuroscientist* 12, 469–476.

Franke, B., Bayatti, N., and Engele, J. (2000). Neurotrophins require distinct extracellular signals to promote the survival of CNS neurons *in vitro*. *Exp Neurol* 165, 125–135.

Gao, H. M., Liu, B., and Hong, J. S. (2003a). Critical role for microglial NADPH oxidase in rotenone-induced degeneration of dopaminergic neurons. *J Neurosci* 23, 6181–6187.

Gao, H. M., Liu, B., Zhang, W., and Hong, J. S. (2003b). Critical role of microglial NADPH oxidase-derived free radicals in the *in vitro* MPTP model of Parkinson's disease. *FASEB J* 17, 1954–1956.

Gayle, D. A., Ling, Z., Tong, C., Landers, T., Lipton, J. W., and Carvey, P. M. (2002). Lipopolysaccharide (LPS)-induced dopamine cell loss in culture: Roles of tumor necrosis factor-alpha, interleukin-1beta, and nitric oxide. *Dev Brain Res* 133, 27–35.

Gerlach, M., Double, K. L., Youdim, M. B., and Riederer, P. (2006). Potential sources of increased iron in the substantia nigra of parkinsonian patients. *J Neural Transm*(**Suppl**), 133–142.

Gille, G., Rausch, W. D., Hung, S. T., Moldzio, R., Janetzky, B., Hundemer, H. P., Kolter, T., and Reichmann, H. (2002). Pergolide protects dopaminergic neurons in primary culture under stress conditions. *J Neural Transm* 109, 633–643.

Grasbon-Frodl, E. M., Andersson, A., and Brundin, P. (1996). Lazaroid treatment prevents death of cultured rat embryonic mesencephalic neurons following glutathione depletion. *J Neurochem* 67, 1653–1660.

Grillner, P., and Mercuri, N. B. (2002). Intrinsic membrane properties and synaptic inputs regulating the firing activity of the dopamine neurons. *Behav Brain Res* 130, 149–169.

Guo, H., Tang, Z., Yu, Y., Xu, L., Jin, G., and Zhou, J. (2002). Apomorphine induces trophic factors that support fetal rat mesencephalic dopaminergic neurons in cultures. *Eur J Neurosci* 16, 1861–1870.

Han, B. S., Hong, H. S., Choi, W. S., Markelonis, G. J., Oh, T. H., and Oh, Y. J. (2003). Caspase-dependent and -independent cell death pathways in primary cultures of mesencephalic dopaminergic neurons after neurotoxin treatment. *J Neurosci* 23, 5069–5078.

Hartmann, A., Hunot, S., Michel, P. P., Muriel, M. P., Vyas, S., Faucheux, B. A., Mouatt-Prigent, A., Turmel, H., Srinivasan, A., Ruberg, M., Evan, G. I., Agid, Y., and Hirsch, E. C. (2000). Caspase-3: A vulnerability factor and final effector in apoptotic death of dopaminergic neurons in Parkinson's disease. *Proc Natl Acad Sci USA* 97, 2875–2880.

Hartmann, A., Troadec, J. D., Hunot, S., Kikly, K., Faucheux, B. A., Mouatt-Prigent, A., Ruberg, M., Agid, Y., and Hirsch, E. C. (2001). Caspase-8 is an effector in apoptotic death of dopaminergic neurons in Parkinson's disease, but pathway inhibition results in neuronal necrosis. *J Neurosci* 21, 2247–2255.

Hasegawa, E., Takeshige, K., Oishi, T., Murai, Y., and Minakami, S. (1990). 1-Methyl-4-phenylpyridinium (MPP^+) induces NADH-dependent superoxide formation and enhances NADH-dependent lipid peroxidation in bovine heart submitochondrial particles. *Biochem Biophys Res Commun* 170, 1049–1055.

Henze, C., Hartmann, A., Lescot, T., Hirsch, E. C., and Michel, P. P. (2005). Proliferation of microglial cells induced by 1-methyl-4-phenylpyridinium in mesencephalic cultures results from an astrocyte-dependent mechanism: Role of granulocyte macrophage colony-stimulating factor. *J Neurochem* 95, 1069–1077.

Herkenham, M., Little, M. D., Bankiewicz, K., Yang, S. C., Markey, S. P., and Johannessen, J. N. (1991). Selective retention of MPP^+ within the monoaminergic systems of the primate brain following MPTP administration: An *in vivo* autoradiographic study. *Neuroscience* 40, 133–158.

Herrup, K., Neve, R., Ackerman, S. L., and Copani, A. (2004). Divide and die: Cell cycle events as triggers of nerve cell death. *J Neurosci* 24, 9232–9239.

Hirsch, E. C. (1992). Why are nigral catecholaminergic neurons more vulnerable than other cells in Parkinson's disease? *Ann Neurol* 32, S88–S93.

Hirsch, E. C., Breidert, T., Rousselet, E., Hunot, S., Hartmann, A., and Michel, P. P. (2003). The role of glial reaction and inflammation in Parkinson's disease. *Ann NY Acad Sci* 991, 214–228.

Höglinger, G. U., Feger, J., Prigent, A., Michel, P. P., Parain, K., Champy, P., Ruberg, M., Oertel, W. H., and Hirsch, E. C. (2003a). Chronic systemic complex I inhibition induces a hypokinetic multisystem degeneration in rats. *J Neurochem* 84, 491–502.

Höglinger, G. U., Carrard, G., Michel, P. P., Medja, F., Lombes, A., Ruberg, M., Friguet, B., and Hirsch, E. C. (2003b). Dysfunction of mitochondrial complex I and the proteasome: Interactions between two biochemical deficits in a cellular model of Parkinson's disease. *J Neurochem* 86, 1297–1307.

Höglinger, G. U., Breunig, J. J., Depboylu, C., Rouaux, C., Michel, P. P., Alvarez-Fischer, D., Boutillier, A. L., Degregori, J., Oertel, W. H., Rakic, P., Hirsch, E. C., and Hunot, S. (2007). The pRb/E2F cell-cycle pathway mediates cell death in Parkinson's disease. *Proc Natl Acad Sci USA* 104, 3585–3590.

Hou, J. G., Lin, L. F., and Mytilineou, C. (1996). Glial cell line-derived neurotrophic factor exerts neurotrophic effects on dopaminergic neurons *in vitro* and promotes their survival and regrowth after damage by 1-methyl-4-phenylpyridinium. *J Neurochem* **66**, 74–82.

Hyman, C., Hofer, M., Barde, Y. A., Juhasz, M., Yancopoulos, G. D., Squinto, S. P., and Lindsay, R. M. (1991). BDNF is a neurotrophic factor for dopaminergic neurons of the substantia nigra. *Nature* **350**, 230–232.

Jeyarasasingam, G., Tompkins, L., and Quik, M. (2002). Stimulation of non-α7 nicotinic receptors partially protects dopaminergic neurons from 1-methyl-4-phenylpyridinium-induced toxicity in culture. *Neuroscience* **109**, 275–285.

Jiang, Q., Yan, Z., and Feng, J. (2006a). Neurotrophic factors stabilize microtubules and protect against rotenone toxicity on dopaminergic neurons. *J Biol Chem* **29(281)**, 29391–29400.

Jiang, Q., Yan, Z., and Feng, J. (2006b). Activation of group III metabotropic glutamate receptors attenuates rotenone toxicity on dopaminergic neurons through a microtubule-dependent mechanism. *J Neurosci* **26**, 4318–4328.

Kanthasamy, A. G., Anantharam, V., Zhang, D., Latchoumycandane, C., Jin, H., Kaul, S., and Kanthasamy, A. (2006). A novel peptide inhibitor targeted to caspase-3 cleavage site of a proapoptotic kinase protein kinase C delta (PKCdelta) protects against dopaminergic neuronal degeneration in Parkinson's disease models. *Free Radic Biol Med* **41**, 1578–1589.

Knusel, B., Michel, P. P., Schwaber, J. S., and Hefti, F. (1990). Selective and nonselective stimulation of central cholinergic and dopaminergic development *in vitro* by nerve growth factor, basic fibroblast growth factor, epidermal growth factor, insulin and the insulin-like growth factors I and II. *J Neurosci* **10**, 558–570.

Koopman, W. J., Verkaart, S., Visch, H. J., van der Westhuizen, F. H., Murphy, M. P., van den Heuvel, L. W., Smeitink, J. A., and Willems, P. H. (2005). Inhibition of complex I of the electron transport chain causes O_2 mediated mitochondrial outgrowth. *Am J Physiol Cell Physiol* **288**, C1440–C1450.

Koutsilieri, E., Chen, T. S., Kruzik, P., and Rausch, W. D. (1995). A morphometric analysis of bipolar and multipolar TH-IR neurons treated with the neurotoxin MPP^+ in co-cultures from mesencephalon and striatum of embryonic C57BL/6 mice. *J Neurosci Res* **41**, 197–205.

Kramer, B. C., Goldman, A. D., and Mytilineou, C. (1999). Glial cell line derived neurotrophic factor promotes the recovery of dopamine neurons damaged by 6-hydroxydopamine *in vitro*. *Brain Res* **851**, 221–227.

Krieglstein, K., Suter-Crazzolara, C., Fischer, W. H., and Unsicker, K. (1995). TGF-β superfamily members promote survival of midbrain dopaminergic neurons and protect them against MPP^+ toxicity. *EMBO J* **14**, 736–742.

Krieglstein, K., Maysinger, D., and Unsicker, K. (1996). The survival response of mesencephalic dopaminergic neurons to the neurotrophins BDNF and NT-4 requires priming with serum: Comparison with members of the TGF-beta superfamily and characterization of the serum-free culture system. *J Neural Transm Suppl* **47**, 247–258.

Krieglstein, K., Reuss, B., Maysinger, D., and Unsicker, K. (1998). Transforming growth factor-β mediates the neurotrophic effect of fibroblast growth factor-2 on midbrain dopaminergic neurons. *Eur J Neurosci* **10**, 2746–2750.

Langston, J. W. (2002). Parkinson's disease: Current and future challenges. *Neurotoxicology* **23**, 443–450.

Lannuzel, A., Michel, P. P., Höglinger, G. U., Champy, P., Jousset, A., Medja, F., Lombes, A., Darios, F., Gleye, C., Laurens, A., Hocquemiller, R., Hirsch, E. C., and Ruberg, M. (2003). The mitochondrial complex I inhibitor annonacin is toxic to mesencephalic dopaminergic neurons by impairment of energy metabolism. *Neuroscience* **121**, 287–296.

Lannuzel, A., Höglinger, G. U., Verhaeghe, S., Gire, L., Belson, S., Escobar-Khondiker, M., Poullain, P., Oertel, W. H., Hirsch, E. C., Dubois, B., and Ruberg, M. (2007). Atypical parkinsonism in Guadeloupe: A common risk factor for two closely related phenotypes? *Brain* **130**, 816–827.

Lapointe, N., St-Hilaire, M., Martinoli, M. G., Blanchet, J., Gould, P., Rouillard, C., and Cicchetti, F. (2004). Rotenone induces non-specific central nervous system and systemic toxicity. *FASEB J* **18**, 717–719.

Larsen, K. E., Fon, E. A., Hastings, T. G., Edwards, R. H., and Sulzer, D. (2002). Methamphetamine-induced degeneration of dopaminergic neurons involves autophagy and upregulation of dopamine synthesis. *J Neurosci* **22**, 8951–8960.

Lee, S. Y., Moon, Y., Hee Choi, D., Jin Choi, H., and Hwang, O. (2007). Particular vulnerability of rat mesencephalic dopaminergic neurons to tetrahydrobiopterin: Relevance to Parkinson's disease. *Neurobiol Dis* **25**, 112–120.

Li, A., Guo, H., Luo, X., Sheng, J., Yang, S., Yin, Y., Zhou, J., and Zhou, J. (2006). Apomorphine-induced activation of dopamine receptors modulates FGF-2 expression in astrocytic cultures and promotes survival of dopaminergic neurons. *FASEB J* **20**, 1263–1265.

Li, G., Cui, G., Tzeng, N. S., Wei, S. J., Wang, T., Block, M. L., and Hong, J. S. (2005). Femtomolar concentrations of dextromethorphan protect mesencephalic dopaminergic neurons from inflammatory damage. *FASEB J* **19**, 489–496.

Lin, L. F., Doherty, D. H., Lile, J. D., Bektesh, S., and Collins, F. (1993). GDNF: A glial cell line-derived neurotrophic factor for midbrain dopaminergic neurons. *Science* 260, 1130–1132.

Ling, Z. D., Robie, H. C., Tong, C. W., and Carvey, P. M. (1999). Both the antioxidant and D3 agonist actions of pramipexole mediate its neuroprotective actions in mesencephalic cultures. *J Pharmacol Exp Ther* 289, 202–210.

Lotharius, J., and O'Malley, K. L. (2000). The parkinsonism-inducing drug 1-methyl-4-phenylpyridinium triggers intracellular dopamine oxidation. A novel mechanism of toxicity. *J Biol Chem* 275, 38581–38588.

Love, S., Plaha, P., Patel, N. K., Hotton, G. R., Brooks, D. J., and Gill, S. S. (2005). Glial cell line-derived neurotrophic factor induces neuronal sprouting in human brain. *Nat Med* 11, 703–704.

Marien, M. R., Colpaert, F. C., and Rosenquist, A. C. (2004). Noradrenergic mechanisms in neurodegenerative diseases: A theory. *Brain Res Rev* 45, 38–78.

McCoy, M. K., Martinez, T. N., Ruhn, K. A., Szymkowski, D. E., Smith, C. G., Botterman, B. R., Tansey, K. E., and Tansey, M. G. (2006). Blocking soluble tumor necrosis factor signaling with dominant-negative tumor necrosis factor inhibitor attenuates loss of dopaminergic neurons in models of Parkinson's disease. *J Neurosci* 26, 9365–9375.

McGeer, P. L., Yasojima, K., and McGeer, E. G. (2001). Inflammation in Parkinson's disease. *Adv Neurol* 86, 83–89.

Mena, M. A., Khan, U., Togasaki, D. M., Sulzer, D., Epstein, C. J., and Przedborski, S. (1997). Effects of wild-type and mutated copper/zinc superoxide dismutase on neuronal survival and L-DOPA-induced toxicity in postnatal midbrain culture. *J Neurochem* 69, 21–33.

Meyer-Franke, A., Wilkinson, G. A., Kruttgen, A., Hu, M., Munro, E., Hanson, M. G., Jr., Reichardt, L. F., and Barres, B. A. (1998). Depolarization and cAMP elevation rapidly recruit TrkB to the plasma membrane of CNS neurons. *Neuron* 21, 681–693.

Michel, P. P., and Agid, Y. (1992). The glutamate antagonist, MK-801, does not prevent dopaminergic cell death induced by the 1-methyl-4-phenylpyridinium ion (MPP+) in rat dissociated mesencephalic cultures. *Brain Res* 597, 233–240.

Michel, P. P., and Agid, Y. (1996). Chronic activation of the cyclic AMP signaling pathway promotes development and long-term survival of mesencephalic dopaminergic neurons. *J Neurochem* 67, 1633–1642.

Michel, P. P., and Hefti, F. (1990). Toxicity of 6-hydroxydopamine and dopamine for dopaminergic neurons in culture. *J Neurosci Res* 26, 428–435.

Michel, P. P., Dandapani, B. K., Sanchez-Ramos, J., Efange, S., Pressman, B. C., and Hefti, F. (1989). Toxic effects of potential environmental neurotoxins related to 1-methyl-4-phenylpyridinium on cultured rat dopaminergic neurons. *J Pharmacol Exp Ther* 248, 842–850.

Michel, P. P., Dandapani, B. K., Knusel, B., Sanchez-Ramos, J., and Hefti, F. (1990). Toxicity of 1-methyl-4-phenylpyridinium for rat dopaminergic neurons in culture: Selectivity and irreversibility. *J Neurochem* 54, 1102–1109.

Michel, P. P., Vyas, S., and Agid, Y. (1992). Toxic effects of iron for cultured mesencephalic dopaminergic neurons derived from rat embryonic brains. *J Neurochem* 59, 118–127.

Michel, P. P., Ruberg, M., and Agid, Y. (1997). Rescue of mesencephalic dopamine neurons by anticancer drug cytosine arabinoside. *J Neurochem* 69, 1499–1507.

Michel, P. P., Marien, M., Ruberg, M., Colpaert, F., and Agid, Y. (1999). Adenosine prevents the death of mesencephalic dopaminergic neurons by a mechanism that involves astrocytes. *J Neurochem* 72, 2074–2082.

Michel, P. P., Hirsch, E. C., and Agid, Y. (2002). Parkinson disease: Mechanisms of cell death. *Rev Neurol* 158, S24–S32.

Michel, P. P., Salthun-Lassalle, B., Henaff, M., Höglinger, G., Ruberg, M., Colpaert, F., and Marien, M. (2003). Neurotrophic effects of noradrenaline for dopaminergic neurons are potentiated by chronic depolarization and GDNF. *Abstr Soc Neurosci* 784, 26.

Michel, P. P., Ruberg, M., and Hirsch, E. C. (2006). Dopaminergic neurons reduced to silence by oxidative stress: An early step in the death cascade in Parkinson's disease? *Sci STKE* 2006(332), pe19.

Michel, P. P., Alvarez-Fischer, D., Guerreiro, S., Hild, A., Hartmann, A., and Hirsch, E. C. (2007). Role of activity-dependent mechanisms in the control of dopaminergic neuron survival. *J Neurochem* 101, 289–297.

Mitsumoto, Y., Watanabe, A., Mori, A., and Koga, N. (1998). Spontaneous regeneration of nigrostriatal dopaminergic neurons in MPTP-treated C57BL/6 mice. *Biochem Biophys Res Commun* 248, 660–663.

Miyoshi, H., Ohshima, M., Shimada, H., Akagi, T., Iwamura, H., and McLaughlin, J. L. (1998). Essential structural factors of annonaceous acetogenins as potent inhibitors of mitochondrial complex. *Biochim Biophys Acta* 1365, 443–452.

Moore, D. J., West, A. B., Dawson, V. L., and Dawson, T. M. (2005). Molecular pathophysiology of Parkinson's disease. *Annu Rev Neurosci* 28, 57–87.

Mourlevat, S., Troadec, J. D., Ruberg, M., and Michel, P. P. (2003). Prevention of dopaminergic neuronal death by cyclic AMP in mixed neuronal/glial mesencephalic cultures requires the repression of presumptive astrocytes. *Mol Pharmacol* 64, 578–586.

Murer, M. G., Dziewczapolski, G., Menalled, L. B., Garcia, M. C., Agid, Y., Gershanik, O., and Raisman-Vozari, R. (1998). Chronic levodopa is not toxic for remaining dopamine neurons, but instead promotes their recovery, in rats with moderate nigrostriatal lesions. *Ann Neurol* 43, 561–575.

Murer, M. G., Raisman-Vozari, R., Yan, Q., Ruberg, M., Agid, Y., and Michel, P. P. (1999). Survival factors promote BDNF protein expression in mesencephalic dopaminergic neurons. *Neuroreport* 10, 801–805.

Mytilineou, C., Cohen, G., and Heikkila, R. E. (1985). 1-Methyl-4-phenylpyridine (MPP$^+$) is toxic to mesencephalic dopamine neurons in culture. *Neurosci Lett* 57, 19–24.

Mytilineou, C., Kokotos Leonardi, E. T., Kramer, B. C., Jamindar, T., and Olanow, C. W. (1999). Glial cells mediate toxicity in glutathione-depleted mesencephalic cultures. *J Neurochem* 73, 112–119.

Nakamura, K., Wang, W., and Kang, U. J. (1997). The role of glutathione in dopaminergic neuronal survival. *J Neurochem* 69, 1850–1858.

Olanow, C. W., Agid, Y., Mizuno, Y., Albanese, A., Bonuccelli, U., Damier, P., De Yebenes, J., Gershanik, O., Guttman, M., Grandas, F., Hallett, M., Hornykiewicz, O., Jenner, P., Katzenschlager, R., Langston, W. J., LeWitt, P., Melamed, E., Mena, M. A., Michel, P. P., Mytilineou, C., Obeso, J. A., Poewe, W., Quinn, N., Raisman-Vozari, R., Rajput, A. H., Rascol, O., Sampaio, C., and Stocchi, F. (2004). Levodopa in the treatment of Parkinson's disease: Current controversies. *Mov Disord* 19, 997–1005.

Oo, T. F., Kholodilov, N., and Burke, R. E. (2003). Regulation of natural cell death in dopaminergic neurons of the substantia nigra by striatal glial cell line-derived neurotrophic factor *in vivo*. *J Neurosci* 23, 5141–5148.

Orth, M., and Schapira, A. H. (2002). Mitochondrial involvement in Parkinson's disease. *Neurochem Int* 40, 533–541.

Parain, K., Hapdey, C., Rousselet, E., Marchand, V., Dumery, B., and Hirsch, E. C. (2003). Cigarette smoke and nicotine protect dopaminergic neurons against the 1-methyl-4-phenyl-1,2,3,6 tetrahydropyridine parkinsonian toxin. *Brain Res* 984, 224–232.

Petroske, E., Meredith, G. E., Callen, S., Totterdell, S., and Lau, Y. S. (2001). Mouse model of Parkinsonism: A comparison between subacute MPTP and chronic MPTP/probenecid treatment. *Neuroscience* 106, 589–601.

Pong, K., Doctrow, S. R., and Baudry, M. (2000). Prevention of 1-methyl-4-phenylpyridinium- and 6-hydroxydopamine-induced nitration of tyrosine hydroxylase and neurotoxicity by EUK-134, a superoxide dismutase and catalase mimetic, in cultured dopaminergic neurons. *Brain Res* 881, 182–189.

Poulsen, K. T., Armanini, M. P., Klein, R. D., Hynes, M. A., Phillips, H. S., and Rosenthal, A. (1994). TGF beta 2 and TGF beta 3 are potent survival factors for midbrain dopaminergic neurons. *Neuron* 13, 1245–1252.

Qian, L., Block, M. L., Wei, S. J., Lin, C. F., Reece, J., Pang, H., Wilson, B., Hong, J. S., and Flood, P. M. (2006). Interleukin-10 protects lipopolysaccharide-induced neurotoxicity in primary midbrain cultures by inhibiting the function of NADPH oxidase. *J Pharmacol Exp Ther* 319, 44–52.

Quik, M., Parameswaran, N., McCallum, S. E., Bordia, T., Bao, S., McCormack, A., Kim, A., Tyndale, R. F., Langston, J. W., and Di Monte, D. A. (2006). Chronic oral nicotine treatment protects against striatal degeneration in MPTP-treated primates. *J Neurochem* 98, 1866–1875.

Radad, K., Rausch, W. D., and Gille, G. (2006). Rotenone induces cell death in primary dopaminergic culture by increasing ROS production and inhibiting mitochondrial respiration. *Neurochem Int* 49, 379–386.

Ren, Y., Liu, W., Jiang, H., Jiang, Q., and Feng, J. (2005). Selective vulnerability of dopaminergic neurons to microtubule depolymerization. *J Biol Chem* 280, 34105–34112.

Rideout, H. J., Lang-Rollin, I. C., Savalle, M., and Stefanis, L. (2005). Dopaminergic neurons in rat ventral midbrain cultures undergo selective apoptosis and form inclusions, but do not up-regulate iHSP70, following proteasomal inhibition. *J Neurochem* 93, 1304–1313.

Rodriguez, M. C., Obeso, J. A., and Olanow, C. W. (1998). Subthalamic nucleus-mediated excitotoxicity in Parkinson's disease: A target for neuroprotection. *Ann Neurol* 44, S175–S188.

Rossi, D. J., Oshima, T., and Attwell, D. (2000). Glutamate release in severe brain ischaemia is mainly by reversed uptake. *Nature* 403, 316–321.

Roussa, E., Farkas, L. M., and Krieglstein, K. (2004). TGF-β promotes survival on mesencephalic dopaminergic neurons in cooperation with Shh and FGF-8. *Neurobiol Dis* 16, 300–310.

Rousselet, E., Callebert, J., Parain, K., Joubert, C., Hunot, S., Hartmann, A., Jacque, C., Perez-Diaz, F., Cohen-Salmon, C., Launay, J. M., and Hirsch, E. C. (2002). Role of TNF-alpha receptors in mice intoxicated with the parkinsonian toxin MPTP. *Exp Neurol* 177, 183–192.

Sakka, N., Sawada, H., Izumi, Y., Kume, T., Katsuki, H., Kaneko, S., Shimohama, S., and Akaike, A. (2003). Dopamine is involved in selectivity of dopaminergic neuronal death by rotenone. *Neuroreport* 14, 2425–2428.

Salthun-Lassalle, B., Hirsch, E. C., Wolfart, J., Ruberg, M., and Michel, P. P. (2004). Rescue of mesencephalic dopaminergic neurons in culture by low-level stimulation of voltage-gated sodium channels. *J Neurosci* **24**, 5922–5930.

Salthun-Lassalle, B., Traver, S., Hirsch, E. C., and Michel, P. P. (2005). Substance P, neurokinins A and B, and synthetic tachykinin peptides protect mesencephalic dopaminergic neurons in culture via an activity-dependent mechanism. *Mol Pharmacol* **68**, 1214–1224.

Sanchez-Ramos, J., Barrett, J. N., Goldstein, M., Weiner, W. J., and Hefti, F. (1986). 1-Methyl-4-phenylpyridinium (MPP$^+$) but not 1-methyl-4-phenyl-1,2,3,6-tetrahydropyridine (MPTP) selectively destroys dopaminergic neurons in cultures of dissociated rat mesencephalic neurons. *Neurosci Lett* **72**, 215–220.

Sanchez-Ramos, J. R., Michel, P., Weiner, W. J., and Hefti, F. (1988). Selective destruction of cultured dopaminergic neurons from fetal rat mesencephalon by 1-methyl-4-phenylpyridinium: Cytochemical and morphological evidence. *J Neurochem* **50**, 1934–1944.

Sanchez-Ramos, J. R., Song, S., Facca, A., Basit, A., and Epstein, C. J. (1997). Transgenic murine dopaminergic neurons expressing human Cu/Zn superoxide dismutase exhibit increased density in culture, but no resistance to methylphenylpyridinium-induced degeneration. *J Neurochem* **68**, 58–67.

Santi, C. M., Cayabyab, F. S., Sutton, K. G., McRory, J. E., Mezeyova, J., Hamming, K. S., Parker, D., Stea, A., and Snutch, T. P. (2002). Differential inhibition of T-type calcium channels by neuroleptics. *J Neurosci* **22**, 340–396.

Saporito, M. S., Heikkila, R. E., Youngster, S. K., Nicklas, W. J., and Geller, H. M. (1992). Dopaminergic neurotoxicity of 1-methyl-4-phenylpyridinium analogs in cultured neurons: Relationship to the dopamine uptake system and inhibition of mitochondrial respiration. *J Pharmacol Exp Ther* **260**, 1400–1409.

Sato, A., Arimura, Y., Manago, Y., Nishikawa, K., Aoki, K., Wada, E., Suzuki, Y., Osaka, H., Setsuie, R., Sakurai, M., Amano, T., Aoki, S., Wada, K., and Noda, M. (2006). Parkin potentiates ATP-induced currents due to activation of P2X receptors in PC12 cells. *J Cell Physiol* **209**, 172–182.

Schober, A., Peterziel, H., von Bartheld, C. S., Simon, H., Krieglstein, K., and Unsicker, K. (2007). GDNF applied to the MPTP-lesioned nigrostriatal system requires TGF-beta for its neuroprotective action. *Neurobiol Dis* **25**, 378–391.

Spina, M. B., Squinto, S. P., Miller, J., Lindsay, R. M., and Hyman, C. (1992). Brain-derived neurotrophic factor protects dopamine neurons against 6-hydroxydopamine and N-methyl-4-phenylpyridinium ion toxicity: Involvement of the glutathione system. *J Neurochem* **59**, 99–106.

Sriram, K., Matheson, J. M., Benkovic, S. A., Miller, D. B., Luster, M. I., and O'Callaghan, J. P. (2002). Mice

deficient in TNF receptors are protected against dopaminergic neurotoxicity: Implications for Parkinson's disease. *FASEB J* **16**, 1474–1476.

Storch, A., Ludolph, A. C., and Schwarz, J. (2004). Dopamine transporter: Involvement in selective dopaminergic neurotoxicity and degeneration. *J Neural Transm* **111**, 1267–1286.

Swerdlow, R. H., Parks, J. K., Miller, S. W., Tuttle, J. B., Trimmer, P. A., Sheehan, J. P., Bennett, J. P., Jr., Davis, R. E., and Parker, W. D., Jr. (1996). Origin and functional consequences of the complex I defect in Parkinson's disease. *Ann Neurol* **40**, 663–671.

Traver, S., Marien, M., Martin, E., Hirsch, E. C., and Michel, P. P. (2006). The phenotypic differentiation of locus ceruleus noradrenergic neurons mediated by brain-derived neurotrophic factor is enhanced by corticotropin releasing factor through the activation of a cAMP-dependent signaling pathway. *Mol Pharmacol* **70**, 30–40.

Troadec, J. D., Marien, M., Darios, F., Hartmann, A., Ruberg, M., Colpaert, F., and Michel, P. P. (2001). Noradrenaline provides long-term protection to dopaminergic neurons by reducing oxidative stress. *J Neurochem* **79**, 200–210.

Troadec, J. D., Marien, M., Mourlevat, S., Debeir, T., Ruberg, M., Colpaert, F., and Michel, P. P. (2002). Activation of the mitogen-activated protein kinase (ERK$_{1/2}$) signaling pathway by cyclic AMP potentiates the neuroprotective effect of the neurotransmitter noradrenaline on dopaminergic neurons. *Mol Pharmacol* **62**, 1043–1052.

Vila, M., and Przedborski, S. (2004). Genetic clues to the pathogenesis of Parkinson's disease. *Nat Med* **1(Suppl)**, S58–S62.

Vitalis, T., Cases, O., and Parnavelas, J. G. (2005). Development of the dopaminergic neurons in the rodent brainstem. *Exp Neurol* **191**, S104–S112.

Volpicelli, F., Perrone-Capano, C., Da Pozzo, P., Colucci-D'Amato, L., and di Porzio, U. (2004). Modulation of nurr1 gene expression in mesencephalic dopaminergic neurones. *J Neurochem* **88**, 1283–1294.

von Coelln, R., Kugler, S., Bahr, M., Weller, M., Dichgans, J., and Schulz, J. B. (2001). Rescue from death but not from functional impairment: Caspase inhibition protects dopaminergic cells against 6-hydroxydopamine-induced apoptosis but not against the loss of their terminals. *J Neurochem* **77**, 263–273.

Wang, T., Liu, B., Qin, L., Wilson, B., and Hong, J. S. (2004). Protective effect of the SOD/catalase mimetic MnTMPyP on inflammation-mediated dopaminergic neurodegeneration in mesencephalic neuronal-glial cultures. *J Neuroimmunol* **147**, 68–72.

Wang, T., Pei, Z., Zhang, W., Liu, B., Langenbach, R., Lee, C., Wilson, B., Reece, J. M., Miller, D. S., and Hong, J. S. (2005). MPP$^+$-induced COX-2 activation and subsequent dopaminergic neurodegeneration. *FASEB J* 19, 1134–1136.

Wang, T., Zhang, W., Pei, Z., Block, M., Wilson, B., Reece, J. M., Miller, D. S., and Hong, J. S. (2006). Reactive microgliosis participates in MPP$^+$-induced dopaminergic neurodegeneration: Role of 67 kDa laminin receptor. *FASEB J* 20, 906–915.

West, A. B., Dawson, V. L., and Dawson, T. M. (2005). To die or grow: Parkinson's disease and cancer. *Trends Neurosci* 28, 348–352.

Wu, D. C., Teismann, P., Tieu, K., Vila, M., Jackson-Lewis, V., Ischiropoulos, H., and Przedborski, S. (2003). NADPH oxidase mediates oxidative stress in the 1-methyl-4-phenyl-1,2,3, 6-tetrahydropyridine model of Parkinson's disease. *Proc Natl Acad Sci USA* 100, 6145–6150.

Yang, S., Yang, J., Yang, Z., Chen, P., Fraser, A., Zhang, W., Pang, H., Gao, X., Wilson, B., Hong, J. S., and Block, M. L. (2006). Pituitary adenylate cyclase-activating polypeptide (PACAP) 38 and PACAP4–6 are neuroprotective through inhibition of NADPH oxidase: Potent regulators of microglia-mediated oxidative stress. *J Pharmacol Exp Ther* 319, 595–603.

Yazdani, U., German, D. C., Liang, C. L., Manzino, L., Sonsalla, P. K., and Zeevalk, G. D. (2006). Rat model of Parkinson's disease: Chronic central delivery of 1-methyl-4-phenylpyridinium (MPP$^+$). *Exp Neurol* 200, 172–183.

Zeevalk, G. D., and Bernard, L. P. (2005). Energy status, ubiquitin proteasomal function, and oxidative stress during chronic and acute complex I inhibition with rotenone in mesencephalic cultures. *Antioxid Redox Signal* 7, 662–672.

Zhang, J., Pho, V., Bonasera, S. J., Holtzman, J., Tang, A. T., Hellmuth, J., Tang, S., Janak, P. H., Tecott, L. H., and Huang, E. J. (2007). Essential function of HIPK2 in TGFβ-dependent survival of midbrain dopamine neurons. *Nat Neurosci* 10, 77–86.

Zihlmann, K. B., Ducray, A. D., Schaller, B., Huber, A. W., Krebs, S. H., Andres, R. H., Seiler, R. W., Meyer, M., and Widmer, H. R. (2005). The GDNF family members neurturin, artemin and persephin promote the morphological differentiation of cultured ventral mesencephalic dopaminergic neurons. *Brain Res Bull* 68, 42–53.

31

SELECTIVE AUTOPHAGY IN THE PATHOGENESIS OF PARKINSON'S DISEASE

SUSMITA KAUSHIK, ESTHER WONG AND
ANA MARIA CUERVO

*Department of Anatomy and Structural Biology, Marion Bessin Liver Research Center,
Institute for Aging Research, Albert Einstein College of Medicine, Bronx, NY, USA*

INTRODUCTION: AUTOPHAGY, TYPES AND FUNCTIONS

To maintain homeostasis and general "well-being" cells rely on constant turnover of intracellular proteins and organelles. Two major proteolytic systems contribute to this continuous turnover, the ubiquitin/proteasome system (UPS) and the lysosomal system (Lecker *et al.*, 2006). The degradation of intracellular constituents – soluble proteins and organelles – in lysosomes is generically known as *autophagy* (Klionsky, 2005). Over the past decade, it has been increasingly recognized that autophagy is not only involved in cellular "housekeeping", but also plays a key role in immune surveillance of pathogens, cellular response to stressors, development and as an essential component of the quality control systems in cells, by promoting the clearance of abnormal or altered components (Cuervo, 2004a; Levine and Klionsky, 2004; Klionsky, 2005; Mizushima, 2005). This last function is the main focus of this chapter in the context of protein conformational disorders, specially, with regard to neurodegenerative disorders.

Three types of autophagy have been described in mammals: *microautophagy*, *macroautophagy* and *chaperone-mediated autophagy* (Figure 31.1) (Cuervo, 2004a). Although all three pathways mediate lysosomal degradation of intracellular components,

they differ in (i) the nature of the substrates degraded, (ii) the mechanisms by which the substrates are delivered to the lysosomes and (iii) the conditions that result in the activation of the individual pathways.

Microautophagy is characterized by the invagination or tubulation of the lysosomal membrane, thus trapping regions of cytosol in vesicles that pinch off from the membrane and get degraded in the lysosomal lumen (Mortimore *et al.*, 1988; Muller *et al.*, 2000) (Figure 31.1a). Although this pathway is the least characterized form of autophagy, it is generally accepted that microautophagy is constitutively active and it is responsible for the turnover of both whole organelles and proteins (Roberts *et al.*, 2003; Yokota, 2003; Dubouloz *et al.*, 2005). In contrast, macroautophagy is considered an inducible form of autophagy, even though most cell types have considerable basal macroautophagic activity which is essential for the maintenance of cellular homeostasis ([Hara *et al.*, 2006; Komatsu *et al.*, 2006] and reviewed in [Cuervo, 2006]). Macroautophagy involves the sequestration of cytosol by *de novo* formation of a double membrane, which seals on itself to form a vesicle known as an autophagosome (Mizushima *et al.*, 2002; Noda *et al.*, 2002) (Figure 31.1b). The engulfed cytosolic contents are degraded by lysosomal hydrolases following the fusion of the autophagosome with mature lysosomes. Genetic

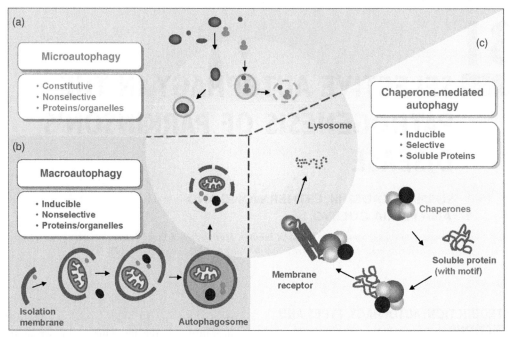

FIGURE 31.1 Types of autophagy in mammalian cells. The degradation of intracellular components in lysosomes can be attained through three different autophagic pathways in mammalian cells. In microautophagy (a), whole regions of the cytosol are sequestered by lysosomes through invaginations or tubulations of the lysosomal membrane. In macroautophagy (b), cytosolic cargo is first enclosed in a double membrane vesicle or autophagosome, which then fuses directly with lysosomes to facilitate degradation of the substrates by the lysosomal proteases. Soluble cytosolic proteins can also be degraded by CMA (c). In this pathway, the critical steps are the interaction of the substrate protein with a particular cytosolic chaperone and its translocation across the lysosomal membrane through a protein complex.

screens in yeast allowed the identification of a family of genes, the autophagy-related genes or ATGs, which codify for proteins involved in different steps of the macroautophagic process – from formation of the autophagosome to its fusion with lysosomes – and in the regulation of this type of autophagy (Klionsky et al., 2003). Knocking down ATG genes has allowed to establish a direct connection between alterations in this form of autophagy and several human disorders including among others cancer, muscle disorders, immune dysfunctions, metabolic disorders and neurodegeneration (Cuervo, 2004b). Both macro- and microautophagy are autophagic pathways with large capability, as whole regions of the cytosol are engulfed at once and delivered to lysosomes. However, because of this characteristic feature, they lack selectivity when it concerns degradation of soluble cytosolic proteins.

Chaperone-mediated autophagy (CMA), on the other hand, is a highly selective type of autophagy in which only soluble cytosolic proteins that contain the broad consensus motif (KFERQ) in their amino acid sequence can be targeted for lysosomal degradation (Majeski and Dice, 2004; Massey et al., 2006a) (Figure 31.1c). This motif is recognized by the cytosolic chaperone, heat shock cognate protein of 70 kDa (hsc70) (Chiang et al., 1989), which in complex with other co-chaperones delivers the substrate to the lysosomal membrane. Once at the membrane, the complex binds to the receptor for this pathway, the lysosome-associated membrane protein type-2A (LAMP-2A) (Cuervo and Dice, 1996) (Figure 31.2). The substrate protein is then unfolded and translocates through a multimeric translocation complex, assisted by a resident lysosomal chaperone (lys-hsc70) (Agarraberes et al., 1997). Similar to macroautophagy, some basal CMA activity is detectable in almost all cell types, but maximal activation of this pathway is attained under stress conditions such as during prolonged nutrient deprivation (Cuervo et al., 1995), mild oxidative stress (Kiffin et al., 2004) and exposure to toxins

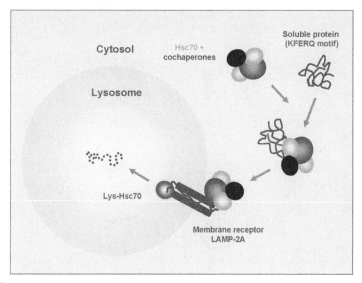

FIGURE 31.2 CMA. Soluble cytosolic proteins bearing a particular targeting motif in their sequence are selectively recognized by a cytosolic chaperone, hsc70, and its associated co-chaperones. The chaperones/substrate complex is targeted to the lysosomal membrane where it interacts with a receptor protein, LAMP-2A. The substrate is then disassembled and unfolded before it can be translocated through the translocation complex into the lysosomal lumen, assisted by a resident chaperone on the luminal side of the membrane (lys-hsc70).

(Cuervo *et al.*, 1999). As reviewed in detail in the following sections, although the capability of CMA is considerably lower than that of the other two autophagic pathways, the selectivity of this pathway offers cells particular advantages. A number of proteins of neuropathological importance (α-synuclein, huntingtin, tau, amyloid precursor protein) contain the CMA-targeting motif, and are thus putative CMA substrates. In this chapter, we review recent findings on the role of CMA in the degradation of α-synuclein, a major component of Lewy bodies (the protein inclusions detected in the affected neurons of Parkinson's disease), and the possible role that α-synuclein-mediated alterations in CMA could have on the pathogenesis of the group of neurodegenerative disorders broadly known as synucleopathies.

AUTOPHAGY IN NEURODEGENERATION

Intracellular Proteolytic Systems and Neurodegeneration

Aberration in protein homeostasis caused by protein "mishandling" is a key event in many neurodegenerative diseases. This is marked by the almost invariant occurrence of protein inclusions which includes Lewy bodies (LBs) in Parkinson's disease (PD), neurofibrillary tangles in Alzheimer's disease (AD), Bunina bodies in amyotrophic lateral sclerosis (ALS), and nuclear inclusions in Huntington's disease (HD) and in spinocerebellar ataxia (SCA) (Chung *et al.*, 2001; Ross and Pickart, 2004). Dysfunction in the intracellular quality control systems – chaperones and proteases – have been proposed to underlie such protein accumulation in inclusions. In these instances, disease-causing mutations and/or posttranslational modifications of specific proteins promote their misfolding or organization into complex structures (oligomers, fibrils, aggregates) that are often toxic for cells. These proteins are usually detected as abnormal by intracellular chaperones which promote their removal by degradation through the UPS or by autophagy (Goldberg, 2003). The UPS is essential for the regulatory degradation of short-lived proteins and subsets of altered proteins (i.e., newly synthesized proteins that fail to acquire their final folded conformation, oxidized proteins, and so on) (Goldberg, 2003) (Figure 31.3a). However, pathological misfolded or altered proteins often assume complex conformations that prevent their entry into

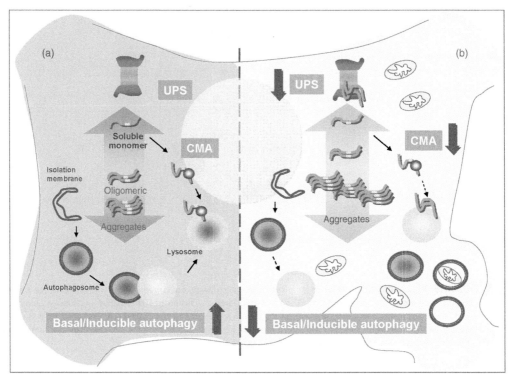

FIGURE 31.3 Dynamic changes in the intracellular proteolytic systems in response to protein conformational changes. (a) The UPS is the favored and dominant route of clearance for intracellular soluble proteins. Some soluble proteins are also selectively degraded by CMA. However, when a cytosolic protein is aggregate prone, it becomes a poor proteasome substrate and may accumulate. Under such circumstances, macroautophagy becomes the main route of clearance for the aggregated proteins. (b) In late stages of the disorder, inadequate level of macroautophagy activity often becomes evident. This may be caused by decline in lysosomal protease activities and/or inefficient vesicle transport and fusion between autophagosomes and lysosomes. This may account for the accumulation of autophagic vacuoles and protein inclusions observed in the pathological states of many neurodegenerative diseases.

the narrow catalytic axial pore of the barrel-shaped proteasome complex (Bennett *et al.*, 2005). It is also common that the interaction of these altered proteins with the proteasome core causes steric occlusions, thereby precluding efficient proteasomal clearance of other proteins (Stefanis *et al.*, 2001; Verhoef *et al.*, 2002; Bennett *et al.*, 2005) (Figure 31.3b). Conceivably, this proteasomal impairment further drives the aggregation process and exacerbates the toxicity of the misfolded proteins (Figure 31.3b). As discussed in more detail below, similar principles apply to the degradation of these proteins via CMA. Unfolding of the substrate protein is required before its translocation into the lysosomal lumen via CMA (Salvador *et al.*, 2000), thus,

making the degradation of protein complexes or aggregates through this pathway not possible. In the same way as for the proteasome, interactions of some of these abnormal proteins with the CMA translocation complex at the lysosomal membrane results in blockage of this system, precluding the degradation of other CMA substrates (Salvador *et al.*, 2000; Cuervo *et al.*, 2004) (Figure 31.3b). The "bulk removal mechanism" characteristic of the macroautophagic pathway makes it better suited to handle aggregates and inclusions (Figure 31.3).

Compensatory activation of the autophagosomal/lysosomal system in general has been well reported in many protein conformational disorders (Cataldo *et al.*, 1995; Kegel *et al.*, 2000;

Bendiske and Bahr, 2003). Expression of huntingtin, the protein found in protein inclusions associated with Huntington's disease, results in upregulation of macroautophagy along with the general stimulation of endosomal–lysosomal activity (Kegel et al., 2000). Enhanced macroautophagy and lysosomal protease levels were also observed in different AD models (Cataldo et al., 1995; Bendiske and Bahr, 2003; Butler et al., 2005). Interventions aimed to enhance macroautophagy activity have been successful in facilitating the removal of some of these "misbehaving" proteins (i.e., mutant peripheral myelin protein, mutant proteins with expanded polyglutamine tracts and α-synuclein) in different cellular models (Ravikumar et al., 2002; Fortun et al., 2003; Webb et al., 2003; Iwata et al., 2005). Furthermore, pharmacological stimulation of macroautophagy with analogs of rapamycin, an inhibitor of the negative regulator of autophagy, mTOR, reduced huntingtin aggregates, decreased toxicity and improved disease-associated symptoms in animal models for HD (Ravikumar et al., 2004). These promising studies thus support that the compensatory upregulation of macroautophagy may help to slow the pathogenic progression and could explain the extended disease duration characteristic of many age-related neurodegenerative disorders.

In addition to this compensatory activation of macroautophagy in response to aggregating proteins, recent studies have provided conclusive evidence supporting a critical role for macroautophagy in neuronal homeostasis, even in the absence of neurodegenerative disease-associated mutant proteins. Two groups have independently shown that the selective suppression of basal macroautophagy in mouse brain causes diffuse accumulation of ubiquitinated abnormal proteins and ubiquitin-positive inclusions, thereby leading to neuronal toxicity and neurodegeneration (Hara et al., 2006; Komatsu et al., 2006). These studies emphasize that constitutive macroautophagy is essential for neuronal survival.

Autophagy and Protein Inclusions

Many pathogenic proteins are susceptible to degradation by both the proteasomal and the autophagic pathways (Webb et al., 2003; Cuervo et al., 2004). Although what controls the choice of one degradation system over another for a protein is not fully understood, factors such as the intrinsic propensity of the protein to aggregate seem crucial in this decision (Figure 31.3). Indeed, protein inclusions have been postulated to be transitory entities, plausibly linking the proteasomal and the autophagic pathways (Fortun et al., 2003; Lim et al., 2006). Proteins which are poor substrates for the proteasome may be channeled to protein inclusions which serve as a staging ground for their alternative removal by macroautophagy. Indeed, many neurodegenerative disease-associated protein inclusions are enriched in p62, a novel protein that binds ubiquitinated proteins in inclusions and one of the Atg proteins, the Light-chain 3 protein (LC3), crucial for autophagosome formation (Bjorkoy et al., 2005). Indeed, p62 and LC3 proteins were found to form a rim around huntingtin inclusions in cultured cells. The proposed model foresees p62 interacting with the polyubiquitin chains in the aggregated proteins, and bringing along LC3 with the rest of the autophagy machinery required for the formation of an autophagosome around the protein aggregate. Some controversy has recently arisen about this model, because different electron micropy studies have failed to detect autophagosomes surrounding the aggregates (Jia et al., 2007). Very rapid clearance of the autophagosomes once formed or their rapid trafficking to other cellular regions to promote fusion with lysosomes could explain these discrepancies.

Inclusion formation is an active process relying on the transport of microaggregates along the microtubules to form perinuclear inclusions (Iwata et al., 2005). Hence it is tempting to suggest that inclusion formation may be an active process utilized by the cells to divert proteasomal load toward the autophagic pathways. Alternatively, it has been suggested that macroautophagy does not directly clear aggregates but instead it clears aggregate precursors (Rubinsztein, 2006). The removal of such precursors may shift the equilibrium away from aggregate formation, thereby reducing the size and number of protein inclusions.

Autophagic Failure in Neurodegenerative Disorders?

Upregulation of macroautophagy prevents intracellular accumulation of protein aggregates. How then do we account for the persistence of intraneuronal protein inclusions and cell toxicity in the pathological states? A possible answer to this question could be inferred from the accumulation of autophagic vacuoles reported as pathological findings in certain neurodegenerative disorders, such as AD (Cataldo *et al.*, 1995; Kegel *et al.*, 2000; Yu *et al.*, 2005) (Figure 31.3b). This proliferation of autophagic vacuoles in degenerating cells has often been interpreted as an indication that over-activation of macroautophagy mediates neurodegeneration. However, emerging evidence supports that a failure of the macroautophagic process in advanced states of the disease is responsible for such accumulation of autophagosomes. Macroautophagy is initially induced in most of these diseases to protect neurons against abnormal toxic proteins, stress and damaged organelles that may cause cell toxicity and apoptosis. However, overloading of this system or impairment at different steps due to other etiologic factors in the disease could lead to the failure of macroautophagy and the resultant accumulation of autophagic vacuoles. At this stage, an inadequate level of macroautophagic activity contributes toward the degeneration process and neuronal cell death (Boland and Nixon, 2006) (Figure 31.3b).

Different factors could account for the failure of the autophagic systems under these circumstances. Both CMA and macroautophagy have been reported to decline with age (Cuervo *et al.*, 2005). The fusion of autophagosome to lysosome is decreased in aging cells and a defect in the ability of lysosomes to degrade cargo delivered by the reduced number of autophagosomes that succeed in the fusion has also been reported (Terman *et al.*, 1999; Brunk and Terman, 2002; Butler *et al.*, 2005). This age-related impairment in autophagy may directly contribute to the accumulation of altered proteins. Furthermore, the decreased turnover rates of intracellular components may explain the accumulation of dysfunctional mitochondria in aging neurons, and the consequent increase in oxidative stress (Brunk and Terman, 2002) (Figure 31.3b). As the impairment persists, destabilization of autophagosome and autophagolysosome membranes triggers release of undigested products and hydrolytic enzymes into the cytosol (Boland and Nixon, 2006).

Although the particular sequence of alterations in the intracellular proteolytic systems in neurodegenerative disorders could vary depending on the cell types and the intrinsic properties of the pathogenic protein, there are some common principles that likely apply to all these conditions and may set the basis for generic treatments applicable to a whole subset of these disorders. According to our proposed model (Figure 31.3), therapeutic interventions based on manipulations of the proteolytic

pathways should be customized depending on the stage. Thus, for example, upregulation of macroautophagy could be beneficial to prolong the duration of the compensation stage, but it may have detrimental effects during the failure stage when the clearance of autophagosomes is already compromised. Better characterization of the molecular defects underlying the failure of each of the different proteolytic systems is essential for any future restorative effort.

α-SYNUCLEIN AND AUTOPHAGY

Most of the general principles described in the previous section apply to PD and other synucleopathies, in which α-synuclein is the main component of the protein inclusions detected in the affected neurons. However, α-synuclein has become a particularly interesting molecule for the study of the alterations of the proteolytic systems in these disorders, because it is the first pathogenic protein for which connections with the UPS, macroautophagy and CMA have been established (Webb *et al.*, 2003; Cuervo *et al.*, 2004).

Interactions of pathogenic mutants of α-synuclein with the catalytic component of the UPS result in blockage of this pathway (Bennett *et al.*, 2005). Although macroautophagy does not seem to contribute significantly to the degradation of wild-type α-synuclein (Webb *et al.*, 2003), the mutant forms of this protein are preferentially degraded through this pathway (Webb *et al.*, 2003). Blockage of macroautophagy in cultured cells expressing different pathogenic mutant forms of α-synuclein favors accumulation and aggregation of these proteins, which have been detected inside autophagosomes by immunogold electron microscopy (Webb *et al.*, 2003; Wilson *et al.*, 2004). Although the mechanisms that activate macroautophagy in cells expressing the mutant forms of α-synuclein remain unknown, it is interesting to note that pharmacological blockage of the catalytic activities of the proteasome results in upregulation of macroautophagy (Iwata *et al.*, 2005). It is thus possible that the inhibitory effect of mutant α-synuclein on the proteasome system is behind the activation of macroautophagy in those cells.

Finally, degradation of wild-type α-synuclein via CMA and blockage of this autophagic pathway by mutant and posttranslationally modified forms of α-synuclein have also been reported (Cuervo *et al.*, 2004; Martinez-Vicente et al., 2008). This particular relationship between α-synuclein and CMA is the main focus of this section.

Chaperone-Mediated Autophagy: What Sets It Apart from the Other Types of Autophagy?

As described in the introduction, one of the unique features of CMA when compared to the other lysosomal pathways of protein degradation is the selectivity of this pathway. Selectivity of CMA is conferred by the interaction of the cytosolic chaperone involved in this pathway with a particular pentapeptide motif in the sequence of the substrate proteins (Figure 31.2) (Dice, 1990). The advantages of the selective nature of CMA can be inferred by considering the context in which CMA is maximally activated. Thus, activation of CMA during nutrient starvation follows that of macroautophagy (Cuervo *et al.*, 1995). During the first hours of starvation, the high capability of the "in-bulk" degradation mediated by macroautophagy is enough to guarantee a first wave of amino acids and other essential macromolecules required to maintain protein synthesis in these conditions when nutrients are scarce (Mizushima, 2005). However, if starvation persists, the random nature of macroautophagy could compromise cell viability by degrading components essential for survival under these stress conditions. It is then when CMA is activated to promote selective degradation of non-essential proteins, such as enzymes involved in metabolic pathways that shut down during starvation (Cuervo *et al.*, 1995). Selectivity may also be behind the reason for CMA activation during mild oxidative stress or after exposure to particular toxic compounds that alter protein conformation (Cuervo *et al.*, 1999; Kiffin *et al.*, 2004). In this case, the selectivity of CMA would allow removal of the affected proteins (i.e., oxidized or misfolded) without compromising the normal functional proteins in the vicinity.

The other property that sets CMA apart from the other types of autophagy is the mechanism utilized for delivery of cargo into the lysosomal lumen. In contrast to macro- and microautophagy where the sequestered proteins included in a whole region of the cytosol are delivered all at once, soluble cytosolic proteins are translocated in a molecule-by-molecule manner into lysosomes. A subset of cytosolic chaperones associated to the cytosolic side of the lysosomal membrane, an integral membrane protein (LAMP-2A) and a chaperone resident in the lysosomal lumen are the minimal components of the translocation system (Majeski and Dice, 2004; Massey *et al.*, 2006a). Because of the nature of the translocation process, substrate disassembly, if delivered by the chaperones as a protein complex, and unfolding are required before the substrate can reach the lysosomal lumen (Salvador *et al.*, 2000). Although the role of the different lysosomal membrane-associated chaperones in unfolding and translocation of the substrate proteins remains to be elucidated, modifications of the substrate proteins that prevent them from unfolding completely block their translocation into lysosomes (Salvador *et al.*, 2000). The requirement for unfolding and the presence of a selective receptor for the CMA substrates at the lysosomal membrane make this a saturable form of autophagy.

CMA of α-Synuclein

Analysis of the sequence of α-synuclein revealed the presence of a CMA consensus motif (^{95}vkkdq^{99}). Mutations in this motif increased the half-life of α-synuclein, thus supporting the degradation of a percentage of intracellular α-synuclein by CMA (Cuervo *et al.*, 2004). Furthermore, α-synuclein interacts with the chaperones and the membrane receptor involved in CMA, and it is possible to reproduce direct translocation across the lysosome membrane of wild-type α-synuclein in a well-established *in vitro* system using isolated lysosomes (Cuervo *et al.*, 2004). Thus, α-synuclein fulfills all the standard criteria that catalog a particular protein as a *bona fide* CMA substrate (Figure 31.4).

In contrast, pathogenic mutant forms of α-synuclein failed to be degraded by CMA (Cuervo *et al.*, 2004). None of the mutations described till date are in the CMA-targeting motif, and consequently do not interfere with the ability of the chaperone to recognize mutant α-synuclein and deliver it to lysosomes. Once at the lysosomal membrane, although mutant proteins bind with abnormal high affinity to the lysosomal receptor, these, however, fail to translocate into the lysosomal lumen (Cuervo *et al.*, 2004) (Figure 31.4b). Although mutations in α-synuclein are present only in a small percentage of PD patients, altered CMA of α-synuclein may still be important for the pathogenesis of idiopathic forms of PD. We have found that certain posttranslational

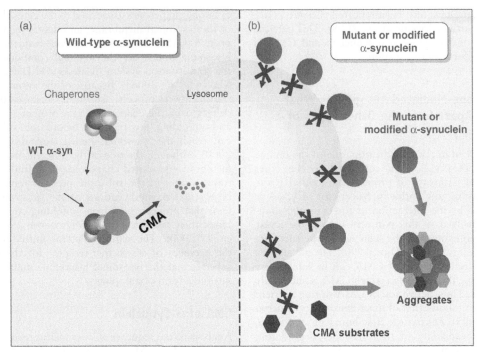

FIGURE 31.4 CMA and α-synuclein. (a) A percentage of intracellular α-synuclein can be selectively degraded by lysosomes via CMA. (b) Pathogenic mutations in α-synuclein and particular posttranslational modifications (see text) alter the lysosomal translocation of α-synuclein via CMA. Because the abnormal proteins bind to the lysosomal membrane with unusual high affinity, the consequences of impaired CMA of α-synuclein are twofold: on one hand, decreased degradation of α-synuclein contributes to increase the cytosolic levels of this protein and to facilitate its aggregation; on the other hand, the blockage of CMA by the altered α-synucleins, prevents the degradation of other CMA substrates, which could then accumulate in the protein inclusions, characteristic of the disease state.

modifications of α-synuclein affect its ability to be degraded by CMA differently (Martinez-Vicente et al., 2008). Phosphorylation and nitration of α-synuclein results in reduced lysosomal uptake. However, because these modified forms of α-synuclein bind loosely to the CMA receptor, they can be easily displaced by other CMA substrate proteins, and do not affect CMA (Martinez-Vicente et al., 2008). More interesting are the modifications that result from exposure of α-synuclein to dopamine or its oxidized product dopaminochrome. Dopamine-modified forms of α-synuclein are unable to form fibrils, which makes them particularly toxic for cells (Norris et al., 2005). From the CMA point of view, the behavior of these modified forms of α-synuclein mimics what has been described for the mutant proteins (Martinez-Vicente et al., 2008). Dopamine-modified α-synuclein cannot be translocated into

lysosomes via CMA, but its high affinity binding to the CMA receptor prevents the degradation of other CMA substrates through this pathway. Studies in cultured ventromedial neurons exposed to dopamine have revealed that the CMA blockage in these cells is directly dependent on the presence of α-synuclein, as similar treatment in neurons from α-synuclein (−/−) mice did not have a significant effect on CMA activity. The resemblance between the pathological mutants and the dopamine-modified α-synuclein, among all the other posttranslational modifications analyzed, is particularly interesting, as dopaminergic neurons are one of the first neuronal types affected in PD. Although the specific mechanisms by which mutant and dopamine-modified α-synuclein block CMA remain to be elucidated, we have found that both forms of the protein organize in high molecular weight oligomeric complexes at the lysosomal membrane (Martinez-Vicente et al., 2008). Direct steric blockage of the translocation complex by the oligomers, changes in the lateral mobility of membrane proteins or changes in essential lysosomal properties due to oligomer formation at the lysosomal membrane are all possible mechanisms that require future investigation.

Efforts to elucidate the cellular consequences of the blockage of CMA in PD neurons have confirmed a critical role of CMA as part of the cellular response to stress. RNA interference against LAMP-2A in different types of cultured cells renders these cells more susceptible to different stressors (Massey et al., 2006b). Interestingly, cells with blocked CMA display higher rates of macroautophagy, even under basal conditions. This upregulation of macroautophagy does not seem to be responsible for the decrease in cell viability after exposure to stressors, because inhibition of this pathway results in even higher rates of cell death (Massey et al., 2006b).

Based on these findings, we propose that blockage of CMA by pathological mutants and dopamine-modified α-synuclein may contribute to the loss of neurons in PD by diminishing their defense against common stressors, such as oxidative stress. The reported reduction in CMA activity with age (Cuervo and Dice, 2000) brings aging, once again, as an aggravating factor in the course of the disease.

EXPERIMENTAL MODELS FOR THE STUDY OF THE INTERPLAY BETWEEN α-SYNUCLEIN AND AUTOPHAGY

The critical role that the different intracellular proteolytic systems play in the cellular quality control and the distinctive effect of α-synuclein, and probably other pathogenic proteins, on these systems justifies the need for detailed molecular studies of this interplay. Following, we described some of the in vitro and in vivo models that could be used for these studies.

Autophagy in Cultured Cells

The study of autophagy in cultured cells has benefited enormously from the molecular dissection of macroautophagy that the genetic studies have offered and from the novel critical components identified for CMA. In the context of PD, both the degradation of the pathogenic proteins and the

effect of these proteins on the autophagic pathways can now be analyzed in cultured cells. For a detailed description of the methods to measure autophagy, readers are referred to Klionsky et al. (2007).

Macroautophagy can be tracked in cultured cells using recombinant Atg proteins fused to fluorescent proteins which allows identification of autophagosomes, the intracellular compartments directly linked to macroautophagy (Mizushima et al., 2004). Morphometric electron microscopy studies of the cells of interest also offer valuable information on macroautophagy. Complementation of these image-based assays with functional assays is strongly encouraged, because as pointed out earlier, an increase in the number and size of autophagic vacuoles is not necessarily indicative of macroautophagy upregulation, but it could reflect possible problems in their clearance by lysosomes. Measurement of the degradation of components known to be processed in lysosomes, such as the autophagosome marker LC3 or the pool of long-lived intracellular proteins, allows distinction between both possibilities. Enhanced degradation of such components will support macroautophagy upregulation, whereas reduced degradation rates will be observed in conditions resulting in poor autophagosome clearance (Klionsky et al., 2007). Different options are now available for the blockage of macroautophagy in cultured cells which includes both inhibition by pharmacological agents, such as inhibitors of phosphoinositol-3-kinase type I, required for macroautophagy induction (Seglen and Gordon, 1982), or siRNA against essential Atg proteins, such as Atg 5 or 7 (Mizushima et al., 2002). The choice of activators of macroautophagy is certainly more limited and the cell-type specificity of some of these activators complicates their use. Rapamycin has been one of the more extensively used enhancers of macroautophagy, based on the fact that mTOR, principal target of rapamycin, is a repressor of macroautophagy (Ueno et al., 1999). However, recent evidence supports the existence of an mTOR-independent form of macroautophagy which remains poorly characterized. Agents such as lithium (Sarkar et al., 2005) and trehalose (Sarkar et al., 2007), activate macroautophagy in an mTOR-independent manner in some cell types, but their applicability to other cell types has been questioned.

Study of CMA in cultured cells is more challenging because CMA activity does not cause morphological changes in the lysosomal compartment. Changes in the major components of this pathway, namely the lysosomal chaperone and LAMP-2A, and in the number of lysosomes active for CMA (those containing both markers), are useful to assess CMA activity in cultured cells (Massey *et al.*, 2006b). A more accurate measurement is to examine the degradation rates of long-lived proteins by CMA, which can be easily differentiated from the degradation through macroautophagy because of the insensitivity of CMA to phosphoinositol-3-kinase inhibitors. Lastly, the definitive assay for the accurate measurement of CMA requires the isolation of the pool of lysosomes engaged in this process and the analysis of their ability to take up and degrade well-characterized CMA substrates (Klionsky *et al.*, 2007). CMA can be inhibited in cultured cells by siRNA against LAMP-2A (Massey *et al.*, 2006b). Note that interference of the cytosolic chaperone is not advisable as this chaperone is involved in many other cellular processes not related to CMA (Klionsky *et al.*, 2007). Upregulation of CMA is challenging, but can be attained by increasing the lysosomal levels of LAMP-2A (Cuervo and Dice, 1996). Mistargeting of LAMP-2A to the plasma membrane limits its exogenous expression to a maximum of fourfold, but these levels seem enough to significantly increase CMA rates (Cuervo and Dice, 1996).

Selective blockage of CMA or macroautophagy in different neuronal types in culture could be useful models to study early and late cellular changes in PD-affected neurons, respectively.

In Vivo Models with Disregulated Autophagy

Models based on the expression of different α-synuclein mutants in different non-mammalian eukaryotes, such as yeast, worms or flies, are particularly attractive for the study of PD-related changes in macroautophagy, because genetic manipulations of this autophagic pathway in these systems are considerably easier than in mammals, where existence of several homologs for each ATG gene is not unusual (Klionsky *et al.*, 2007). However, the applicability of these models for the study of the relationship between α-synuclein and CMA is questionable, as CMA has so far only been described in mammals, and LAMP-2A, the critical component of this pathway, is not conserved in those non-mammalian species.

Mice knock-out for essential macroautophagy components die few hours after birth, thus, limiting

their applicability for the study of neurodegeneration (Kuma *et al.*, 2004). Tissue-specific knock-outs have been very valuable for linking defective macroautophagy activity to neurodegeneration (Hara *et al.*, 2006; Komatsu *et al.*, 2006). Future development of neuronal type-specific and of conditional knock-outs models, in which blockage can be applied later in life will be necessary to mimic the changes in the autophagic system described in PD. Although overexpression of certain Atgs is enough to upregulate macroautophagy in yeast (Klionsky *et al.*, 2007), this approach is not effective in mammalian cells and consequently, transgenic mouse models with upregulated macroautophagy are still missing.

A mouse model with selective impairment of CMA has not been generated yet. LAMP-2A, the limiting component for CMA, has been eliminated in mice but by knocking down the complete LAMP-2 gene (Tanaka *et al.*, 2000). Because this gene undergoes alternative splicing and the two other variants, LAMP-2B and -2C, have been proposed to participate in macroautophagy and lysosomal biogenesis, respectively, the phenotype of the LAMP-2 (−/−) mice goes beyond that expected from CMA blockage alone. The LAMP-2 (−/−) mice, however, could be a good model to mimic advanced stages of neurodegenerative disorders, wherein both CMA and macroautophagy are altered. Decreased rates of protein degradation and massive accumulation of autophagic vacuoles in different tissues are common features of the LAMP-2 (−/−) mice (Tanaka *et al.*, 2000). However, because LAMP-2 participates in other intracellular processes, such as cholesterol trafficking and metabolism (Eskelinen *et al.*, 2002; Eskelinen *et al.*, 2004), these mice also present phenotypic manifestations independent of autophagy. Selective blockage of only the LAMP-2A isoform would be more desirable to reproduce the changes in the cellular environment in early stages of PD, resulting from CMA impaired activity.

Our group has generated a bitransgenic LAMP-2A mouse model in which expression of an extra copy of LAMP-2A can be modulated in different tissues. This model has become very useful to correct the age-related decline in CMA by maintaining LAMP-2A in old rodent at levels comparable to the ones in young mice (Zhang, C. and Cuervo, A. M., submitted). Crosses between mice overexpressing LAMP-2A in specific neuronal groups with some of the α-synuclein mutant or overexpressing mice could be attractive models to evaluate the contribution of CMA blockage to the progression of the disease. However, because of the lack of overt phenotype in most α-synuclein transgenic mice, added challenges (oxidation or other stresses) may be needed to test the beneficial effects of restoring or upregulating CMA in those models.

MANIPULATIONS OF AUTOPHAGY AS FUTURE THERAPEUTICS FOR PD

The dynamic nature of the changes that the autophagic system undergoes in response to the presence of pathogenic α-synuclein should be taken into account when considering manipulations of autophagy as possible therapeutic alternatives for PD and other α-synucleopathies. For example, upregulation of macroautophagy could be beneficial during the compensatory stage, as it would contribute to the removal of the aggregates and other toxic forms of the pathogenic protein, slowing down the progression of the disease. However, at advanced stages, when clearance of autophagic vacuoles is already compromised, upregulation of this system would contribute to further clogging of the affected neurons.

The ideal interventions would be those aimed at maintaining an adequate autophagic activity in neurons through enhancement of both catabolic activity and delivery mechanisms. The reasons for the failure of macroautophagy are not completely understood. Defects in autophagosome/lysosome fusion or in the degradation of the autophagic content once delivered to lysosomes have been proposed. To date, there are no available methods to enhance fusion, as the mechanisms mediating this autophagic step are, for the most part, unknown. However, degradation of the cargo once inside lysosomes could be improved by increasing the enzymatic content, and/or decreasing the lysosomal pH. This could help lysosomes to handle the increased amount of cargo delivered by the autophagosomes.

The initial studies proposing upregulation of macroautophagy in order to promote removal of aggregates or toxic conformations of the pathogenic proteins were received with some level of concern, as maintained activation of macroautophagy was thought to be responsible for cellular death in most cellular settings. However, the recent reports that constitutive macroautophagy is essential for neuronal homeostasis could explain the favorable effect observed with rapamycin in animal models of Huntington's disease. Rapamycin treatment is not, however, a good choice because it targets one of the major cellular kinase complexes involved in the regulation of several other intracellular processes. Future identification of more selective activators of macroautophagy should help overcome these limitations.

Prevention of the blockage of the UPS and of CMA by the pathogenic proteins may require combined interventions on the chaperones and the proteolytic systems, because once the proteins are organized in complex structures they cannot be degraded by any of these two mechanisms. An alternative option would be to prevent blockage of these systems by facilitating the formation of aggregates, which could be removed via macroautophagy.

Validation of these different types of interventions and of their possible cellular consequences will require the development of novel cellular and animal models to reproduce different stages of the disease.

CONCLUDING REMARKS

The intracellular quality control systems play a critical role in the maintenance of cellular homeostasis and in the defense against alterations in the cellular proteome. Efficient removal of aggregate-prone proteins is required to prevent their toxicity to the cells. The ability of each of the different intracellular proteolytic systems to eliminate such "misbehaving" proteins is directly dependent on the protein's conformation. The UPS and CMA are the main routes of degradation for soluble monomeric proteins, although they could also be degraded by macroautophagy. However, once organized in complexes, macroautophagy is the only option for their efficient removal. Understanding what determines degradation through one or the other proteolytic pathway and the consequences of the interactions of these toxic protein structures with the different proteolytic systems, is crucial before treatments based on modifying different proteolytic systems could be implemented. The recent advances in the molecular characterization of macroautophagy and CMA now offer the possibility of using genetic manipulations to mimic the changes in these two autophagic systems in different pathologies, including neurodegenerative disorders. The current challenge is however to reproduce the temporal sequence of events and the cell-type specificity described in the disease.

REFERENCES

Agarraberes, F., Terlecky, S., and Dice, J. (1997). An intralysosomal hsp70 is required for a selective pathway of lysosomal protein degradation. *J Cell Biol* **137**, 825–834.

Bendiske, J., and Bahr, B. (2003). Lysosomal activation is a compensatory response against protein accumulation and associated synaptopathogenesis – an approach for slowing Alzheimer's disease? *J Neuropathol Exp Neurol* **62**, 451–463.

Bennett, E., Bence, N., Jayakumar, R., and Kopito, R. (2005). Global impairment of the ubiquitin–proteasome system by nuclear or cytoplasmic protein aggregates precedes inclusion body formation. *Mol Cell* **17**, 351–365.

Bjorkoy, G., Lamark, T., Brech, A., Outzen, H., Perander, M., Overvatn, A., Stenmark, H., and Johansen, T. (2005). p62/SQSTM1 forms protein aggregates degraded by autophagy and has a protective effect on huntingtin-induced cell death. *J Cell Biol* **171**, 603–614.

Boland, B., and Nixon, R. (2006). Neuronal macroautophagy: From development to degeneration. *Mol Aspects Med* **27**, 503–519, Mol. Aspects of Med. 27, 503–519.

Brunk, U. T., and Terman, A. (2002). The mitochondrial–lysosomal axis theory of aging: Accumulation of damaged mitochondria as a result of imperfect autophagocytosis. *Eur J Biochem* **269**, 1996–2002.

Butler, D., Brown, Q., Chin, D., Batey, L., Karim, S., Mutneja, M., Karanian, D., and Bahr, B. (2005). Cellular responses to protein accumulation involve autophagy and lysosomal enzymes activation. *Rejuvenation Res* **8**, 227–237.

Cataldo, A., Barnett, J. L., Berman, S. A., Li, J., Quarless, S., Bursztajn, S., Lippa, C., and Nixon, R. A. (1995). Gene expression and cellular content of cathepsin D in human neocortex: Evidence for early upregulation of the endosomal–lysosomal system in pyramidal neurons in Alzheimer's disease. *Neuron* **14**, 1–20.

Chiang, H., Terlecky, S., Plant, C., and Dice, J. (1989). A role for a 70 kDa heat shock protein in lysosomal degradation of intracellular protein. *Science* **246**, 382–385.

Chung, K., Dawson, V., and Dawson, T. (2001). The role of the ubiquitin–proteasomal pathway in Parkinson's disease and other neurodegenerative disorders. *Trends Neurosci* **24**, S7–S14.

Cuervo, A. M. (2006). Autophagy in neurons: It is not all about food. *Trends Mol Med* **12**, 461–464.

Cuervo, A. M. (2004a). Autophagy: In sickness and in health. *Trends Cell Biol* **14**, 70–77.

Cuervo, A. M. (20004b). Autophagy: Many paths to the same end. *Mol Cell Biochem* **263**, 55–72.

Cuervo, A. M., and Dice, J. F. (1996). A receptor for the selective uptake and degradation of proteins by lysosomes. *Science* **273**, 501–503.

Cuervo, A. M., and Dice, J. F. (2000). Age-related decline in chaperone-mediated autophagy. *J Biol Chem* **275**, 31505–31513.

Cuervo, A. M., Knecht, E., Terlecky, S. R., and Dice, J. F. (1995). Activation of a selective pathway of lysosomal proteolysis in rat liver by prolonged starvation. *Am J Physiol* **269**, C1200–C1208.

Cuervo, A. M., Hildebrand, H., Bomhard, E. M., and Dice, J. F. (1999). Direct lysosomal uptake of alpha 2-microglobulin contributes to chemically induced nephropathy. *Kidney Int* **55**, 529–545.

Cuervo, A. M., Stefanis, L., Fredenburg, R., Lansbury, P. T., and Sulzer, D. (2004). Impaired degradation of mutant alpha-synuclein by chaperone-mediated autophagy. *Science* **305**, 1292–1295.

Cuervo, A. M., Bergamini, E., Brunk, U. T., Droge, W., Ffrench, M., and Terman, A. (2005). Autophagy and aging: The importance of maintaining "clean" cells. *Autophagy* **1**, 131–140.

Dice, J. (1990). Peptide sequences that target cytosolic proteins for lysosomal proteolysis. *Trends Biochem Sci* **15**, 305–309.

Dubouloz, F., Deloche, O., Wanke, V., Cameroni, E., and De Virgilio, C. (2005). The TOR and EGO protein complexes orchestrate microautophagy in yeast. *Mol Cell* **19**, 15–26.

Eskelinen, E., Illert, A., Tanaka, Y., Schwarzmann, G., Blanz, J., Von Figura, K., and Saftig, P. (2002). Role of LAMP-2 in lysosome biogenesis and autophagy. *Mol Biol Cell* **13**, 3355–3368.

Eskelinen, E., Schmidt, C. *et al.* (2004). Disturbed cholesterol traffic but normal proteolytic function in LAMP-1/LAMP-2 double-deficient fibroblasts. *Mol Biol Cell* **15**, 3132–3145.

Fortun, J., Dunn, W. A., Jr., Joy, S., Li, J., and Notterpek, L. (2003). Emerging role for autophagy in the removal of aggresomes in Schwann cells. *J Neurosci* **23**, 10672–10680.

Goldberg, A. (2003). Protein degradation and protection against misfolded or damaged proteins. *Nature* **18**, 895–899.

Hara, T., Nakamura, K., Matsui, M., Yamamoto, A., Nakahara, Y., Suzuki-Migishima, R., Yokoyama, M., Mishima, K., Saito, I., Okano, H., and Mizushima, N. (2006). Suppression of basal autophagy in neural cells causes neurodegenerative disease in mice. *Nature* **441**, 885–889.

Iwata, A., Riley, B. E., Johnston, J. A., and Kopito, R. R. (2005). HDAC6 and microtubules are required for autophagic degradation of aggregated huntingtin. *J Biol Chem* **280**, 40282–40292.

Jia, K., Hart, A., and Levine, B. (2007). Autophagy genes protect against disease caused by polyglutamine expansion proteins in *Caenorhabditis elegans*. *Autophagy* **3**, 21–25.

Kegel, K., Kim, M., Sapp, E., McIntyre, C., Castano, J., Aronin, N., and Difiglia, M. (2000). Huntingtin expression stimulates endosomal–lysosomal activity, endosome tubulation, and autophagy. *J Neurosci* **20**, 7268–7278.

Kiffin, R., Christian, C., Knecht, E., and Cuervo, A. M. (2004). Activation of chaperone-mediated autophagy during oxidative stress. *Mol Biol Cell* **15**, 4829–4840.

Klionsky, D., Cuervo, A., and Seglen, P. (2007). Methods for monitoring autophagy from yeast to human. *Autophagy* **3**, [Epub ahead of print].

Klionsky, D. J. (2005). Autophagy. *Curr Biol* **15**, R282–R283.

Klionsky, D. J., Cregg, J. M., Dunn, W. A., Jr., Emr, S. D., Sakai, Y., Sandoval, I. V., Sibirny, A., Subramani, S., Thumm, M., Veenhuis, M., and Ohsumi, Y. (2003). A unified nomenclature for yeast autophagy-related genes. *Dev Cell* **5**, 539–545.

Komatsu, M., Waguri, S., Chiba, T., Murata, S., Iwata, J., Tanida, I., Ueno, T., Koike, M., Uchiyama, Y., Kominami, E., and Tanaka, K. (2006). Loss of autophagy in the central nervous system causes neurodegeneration in mice. *Nature* **441**, 880–884.

Kuma, A., Hatano, M., Matsui, M., Yamamoto, A., Nakaya, H., Yoshimori, T., Ohsumi, Y., Tokuhisa, T., and Mizushima, N. (2004). The role of autophagy during the early neonatal starvation period. *Nature* **432**, 032–036.

Lecker, S., Goldberg, A., and Mitch, W. (2006). Protein degradation by the ubiquitin–proteasome pathway in normal and disease states. *J Am Soc Nephrol* **17**, 1807–1819.

Levine, B., and Klionsky, D. J. (2004). Development by self-digestion: Molecular mechanisms and biological functions of autophagy. *Dev Cell* **6**, 463–477.

Lim, K., Dawson, V., and Dawson, T. (2006). Parkin-mediated lysine 63-linked polyubiquitination: A link to protein inclusions formation in Parkinson's and other conformational diseases? *Neurobiol. Aging* **27**, 524–529.

Majeski, A., and Dice, J. (2004). Mechanisms of chaperone-mediated autophagy. *Int J Biochem Cell Biol* **36**, 2435–2444.

Martinez-Vicente, M., Talloczy, Z., Kaushik, S., Massey, A. C., Mazzulli, J., Mosharov, E. V., Hodara, R., Fredenburg, R., Wu, D. C., Follenzi, A., Dauer, W., Przedborski, S., Ischiropoulos, H., Lansbury, P. T., Sulzer, D., Cuervo, A. M., (2008). Dopamine-modified alpha-synuclein blocks chaperone-mediated autophagy. *J Clin Invest* **118**, 777–788.

Massey, A., Zhang, C., and Cuervo, A. (2006a). Chaperone-mediated autophagy in aging and disease. *Curr Top Dev Biol* **73**, 205–235.

Massey, A. C., Kaushik, S., Sovak, G., Kiffin, R., and Cuervo, A. M. (2006b). Consequences of the selective blockage of chaperone-mediated autophagy. *Proc Natl Acad Sci USA* **103**, 5905–5910.

Mizushima, N. (2005). The pleiotropic role of autophagy: From protein metabolism to bactericide. *Cell Death Differ* **12**, 1535–1541.

Mizushima, N., Ohsumi, Y., and Yoshimori, T. (2002). Autophagosome formation in mammalian cells. *Cell Struct Funct* **27**, 421–429.

Mizushima, N., Yamamoto, A., Matsui, M., Yoshimori, T., and Ohsumi, Y. (2004). *In vivo* analysis of autophagy in response to nutrient starvation using transgenic mice expressing a fluorescent autophagosome marker. *Mol Biol Cell* **15**, 1101–1111.

Mortimore, G. E., Lardeux, B. R., and Adams, C. E. (1988). Regulation of microautophagy and basal protein turnover in rat liver. Effects of short-term starvation. *J Biol Chem* **263**, 2506–2512.

Muller, O., Sattler, T., Flotenmeyer, M., Schwarz, H., Plattner, H., and Mayer, A. (2000). Autophagic tubes: Vacuolar invaginations involved in lateral membrane sorting and inverse vesicle budding. *J Cell Biol* **151**, 519–528.

Noda, T., Suzuki, K., and Ohsumi, Y. (2002). Yeast autophagosomes: *De novo* formation of a membrane structure. *Trends Cell Biol* **12**, 231–235.

Norris, E., Giasson, B., Hodara, R., Xu, S., Trojanowski, J., Ischiropoulos, H., and Lee, V. (2005). Reversible inhibition of alpha-synuclein fibrillization by dopaminochrome-mediated conformational alterations. *J Biol Chem* **280**, 21212–21219.

Ravikumar, B., Duden, R., and Rubinsztein, D. (2002). Aggregate-prone proteins with polyglutamine and polyalanine expansions are degraded by autophagy. *Hum Mol Genet* **11**, 1107–1117.

Ravikumar, B., Vacher, C., Berger, Z., Davies, J. E., Luo, S., Oroz, L. G., Scaravilli, F., Easton, D. F., Duden, R., O'Kane, C. J., and Rubinsztein, D. C. (2004). Inhibition of mTOR induces autophagy and reduces toxicity of polyglutamine expansions in fly and mouse models of Huntington disease. *Nat Genet* **36**, 585–595.

Roberts, P., Moshitch-Moshkovitz, S., Kvam, E., O'Toole, E., Winey, M., and Goldfarb, D. (2003). Piecemeal microautophagy of nucleus in *Saccharomyces cerevisiae*. *Mol Biol Cell* **14**, 129–141.

Ross, C., and Pickart, C. (2004). The ubiquitin–proteasome pathway in Parkinson's disease and other neurodegenerative diseases. *Trends Cell Biol* **14**, 703–711.

Rubinsztein, D. (2006). The roles of intracellular protein-degradation pathways in neurodegeneration. *Nature* **443**, 780–786.

Salvador, N., Aguado, C., Horst, M., and Knecht, E. (2000). Import of a cytosolic protein into lysosomes by chaperone-mediated autophagy depends on its folding state. *J Biol Chem* **275**, 27447–27456.

Sarkar, S., Davies, J., Huang, Z., Tunnacliffe, A., and Rubinsztein, D. (2007). Trehalose, a novel mTOR-independent autophagy enhancer, accelerates the clearance of mutant Huntingtin and α-synuclein. *J Biol Chem* **282**, 5641–5652.

Sarkar, S., Flot, R. A., Berger, Z., Imarisio, S., Cordenier, A., Pasco, M., Cook, L., and Rubinsztein, D. (2005). Lithium induces autophagy by inhibiting inositol monophosphatase. *J Cell Biol* **170**, 1101–1111.

Seglen, P., and Gordon, P. (1982). 3-Methyladenine: Specific inhibitor of autophagic/lysosomal protein degradation in isolated hepatocytes. *Proc Natl Acad Sci USA* **79**, 1889–1892.

Stefanis, L., Larsen, K., Rideout, H., Sulzer, D., and Greene, L. (2001). Expression of A53T mutant but not wild-type alpha-synuclein in PC12 cells induces alterations of the ubiquitin-dependent degradation system, loss of dopamine release, and autophagic cell death. *J Neurosci* **21**, 9549–9560.

Tanaka, Y., Guhde, G., Suter, A., Eskelinen, E.-L., Hartmann, D., Lullmann-Rauch, R., Janssen, P., Blanz, J., von Figura, K., and Saftig, P. (2000). Accumulation of autophagic vacuoles and cardiomyopathy in Lamp-2-deficient mice. *Nature* **406**, 902–906.

Terman, A., Dalen, H. *et al.* (1999). Ceroid/lipofuscin-loaded human fibroblasts show decreased survival time and diminished autophagocytosis during amino acid starvation. *Exp Gerontol* **34**, 943–957.

Ueno, T., Ishidoh, K., Mineki, R., Tanida, I., Murayama, K., Kadowaki, M., and Kominami, E. (1999). Autolysosomal membrane-associated betaine homocysteine methyltransferase, limited degradation fragment of a sequestered cytosolic enzyme monitoring autophagy. *J Biol Chem* **274**, 15222–15229.

Verhoef, L., Lindsten, K., Masucci, M., and Dantuma, N. (2002). Aggregate formation inhibits proteasomal degradation of polyglutamine proteins. *Hum Mol Genet* **11**, 2689–2700.

Webb, J., Ravikumar, B., Atkins, J., Skepper, J., and Rubinsztein, D. (2003). Alpha-synuclein is degraded by both autophagy and the proteasome. *J Biol Chem* **278**, 25009–25013.

Wilson, C., Murphy, D., Giasson, B., Zhang, B., Trojanowski, J., and Lee, V. (2004). Degradative organelles containing mislocalized α- and β-synuclein proliferate in presenilin-1 null neurons. *J Cell Biol* **165**, 335–346.

Yokota, S. (2003). Degradation of normal and proliferated peroxisomes in rat hepatocytes: Regulation of peroxisomes quantity in cells. *Microsc Res Tech* **61**, 151–160.

Yu, W. H., Cuervo, A. M. *et al.* (2005). Macroautophagy – a novel β-amyloid peptide-generating pathway activated in Alzheimer's disease. *J Cell Biol* **171**, 87–98.

32

THE ROLE OF LRRK2 KINASE ACTIVITY IN CELLULAR PD MODELS

MARK R. COOKSON, ELISA GREGGIO AND PATRICK LEWIS

Cell Biology and Gene Expression Unit, Laboratory of Neurogenetics,
National Institute on Aging, Bethesda, MD, USA

CELL-BASED MODELS FOR PARKINSON DISEASE RESEARCH

In this chapter, we will discuss some of the advantages and disadvantages of cell-based models related to inherited Parkinson disease (PD). A disclaimer that should be placed at the front of any such review is that, by necessity, cell models will not capture the range of events that occur in the intact CNS. Where cell models are most useful is in understanding the behavior of proteins in a relatively intact cellular context. Specifically for the study of genetic diseases, it is possible to examine a range of mutations, both pathogenic and artificial, hypothesis testing, variants, in a consistent background.

In the general class of studies referred to as cell models, one would take a neuronal preparation and express the protein of interest in it, then measure a suitable phenotype. There are a variety of important flavors of this approach; one might use dopaminergic neurons from the midbrain and express α-synuclein from a viral vector. In some of the experiments discussed below, we have used primary cortical neurons and plasmid transfections of LRRK2. In many studies, neuroblastoma cell lines are also used; for some of the studies discussed here we have used human dopaminergic lines such as SH-SY5Y or M17. The output could be electrophysiological, imaging based or, something that we have focused on, cell death assays. Non-neuronal

cells or cell lines are also used in some of the assays. For example, cancer cell lines such as HEK293 or COS7 can be used for larger-scale transfection experiments for enzyme activity measurements or protein–protein interaction studies. There are legitimate concerns about how readily one might be able to extrapolate from simple assay systems where the timescale is compressed into days compared to the decades of a neurological illness. But, this is also often an advantage, not only for those of us who have limited patience, but also in being able to perform more assays in a limited amount of time. One might also be concerned about whether protein–protein interactions are maintained in heterologous systems and at expression levels that often exceed those seen for the endogenous proteins. These concerns mandate the confirmation of putative interactors in additional contexts, such as endogenous proteins in the brain, which is sometimes extremely hard to do.

Here, we will focus on the study of a kinase, LRRK2, mutations in which are associated with inherited PD. This is a suitable test system for many of the concepts listed above. As we will discuss, LRRK2 is a large complex kinase with multiple mutations that are associated with familial PD. Understanding the biology of LRRK2 is complicated by the presence of multiple domains with different (and often unclear) functions. Cell-based models may allow us to dissect out which domains are critical, by designing point and truncating mutations to test specific hypothesis. We will illustrate this by discussing recent results that have implicated the kinase domain in toxic effects of this protein. At the time of writing no reports of how the protein behaves *in vivo*, we have an interesting opportunity to review a manageable amount of literature without knowing which observations will stand the more rigorous test of animal models. We will label the clearest predictions, with appropriate *caveats*, from cell models that the reader may be able to review at a future date to see which have retained support. Where such systems might be used as testing grounds for future therapeutics will also be explored. Before doing this, it is important to briefly review the genetics of *LRRK2* as it has relevance for interpretation of some of the studies.

LRRK2: GENETIC AND SEQUENCE ANALYSES

LRRK2 Mutations Cause Dominant PD

In 2004, two groups described mutations in the *LRRK2* gene that are associated with autosomal dominant inherited PD (Paisan-Ruiz *et al.*, 2004; Zimprich *et al.*, 2004). Multiple mutations were found in several families that had been collected independently from different parts of the world. Subsequently, a mutation was found in the same gene (Funayama *et al.*, 2005) in the family from Japan originally used to generate the linkage region on chromosome 12 that had been labeled PARK8 (Funayama *et al.*, 2002). Since that time, several more mutations have been described including one, G2019S, which has a high prevalence in specific populations (reviewed in Cookson *et al.*, 2005). Not all mutations are convincingly pathogenic: some may be polymorphic variants that are not causal for PD but have been found in one or two sporadic cases.

Before discussing the protein product of *LRRK2*, it is worth reiterating what is known about the phenotype of cases with mutations. The majority of cases are remarkably similar to sporadic PD, with loss of dopaminergic neurons that project from the substantia nigra to the striatum and Lewy bodies, the intracellular protein inclusions typical of PD (e.g., Khan *et al.*, 2005). However, some variation in the disease is seen in postmortem studies, with some cases having nigral loss but no Lewy bodies

(Funayama *et al.*, 2002) and other having protein inclusions that are ubiquitin or tau positive (Zimprich *et al.*, 2004). Some additional clinical features not normally seen in sporadic PD have also been reported, including amyotrophy resulting from loss of anterior horn neurons in the spinal cord (Zimprich *et al.*, 2004). This variation in disease presentation is of interest because it implies that the relationship between mutation status and phenotype is complex, especially given that there is no clear relationship between specific mutations and specific phenotypes, meaning that *LRRK2* likely plays roles relevant to several neurodegenerative conditions. Furthermore, mutations are not fully penetrant and so there are individual mutation carriers who do not develop disease (e.g., Kay *et al.*, 2005). This might suggest that either the mutations have rather subtle effects or that the disease process requires additional factors for full expression. Despite these subtleties, the overall message is that *LRRK2* disease overlaps substantially with typical PD and therefore its study may be informative for the more common sporadic form.

Sequence and Predicted Domain Structure of LRRK2

Both groups who initially cloned *LRRK2* mutations examined the sequence and predicted the presence of two catalytic domains and at least two protein–protein interaction domains within the 2500 amino acid LRRK2 protein sequence. The protein is named for a series of leucine-rich repeats (LRR) toward the N-terminus and a kinase domain located toward the C-terminus. In between these two is a GTPase domain that has been labeled as a ROC domain for Ras in complex proteins and an intervening sequence identified as a COR (C-terminal of Roc) domain that is characteristic of a family of complex kinases (Bosgraaf and Van Haastert, 2003). Finally, a WD40 domain that, like the LRRs, may be important in protein–protein interactions is present at the extreme C-terminus.

Four of these regions, which are assumed to be independently folded and would thus fulfill the definition of a protein domain, contain one or more mutations whose pathogenicity can be definitively assigned. The GTPase domain contains several mutations, including R1441C and R1441G, that were identified in the first cloning papers (Paisan-Ruiz *et al.*, 2004; Zimprich *et al.*, 2004). The COR region contains Y1699C, present in a family from a region of the United Kingdom (Lincolnshire) who have Lewy body-positive PD (Khan *et al.*, 2005). The kinase domain contains the very common G2019S mutations and the I2020T mutation from the original Japanese family (Funayama *et al.*, 2005). There are mutations described in both the LRRs and the WD40 domains, although the pathogenicity of these is less certain as segregation of the disease with mutation in large families remains to be established. For example, the G2385R mutation in the WD40 domain is common in some Asian populations (Fung *et al.*, 2006) but has low penetrance and is not clearly pathogenic.

The N-terminal region of LRRK2 is not currently well annotated. There may be further repeat sequences that are related to ankyrin repeats, another protein–protein interaction motif (Marin, 2006). An interesting aspect of this region is that it is what makes LRRK2 different from the otherwise closely related LRRK1. LRRK1 is smaller than LRRK2 and most of the difference is that LRRK1 has a shorter N-terminal region. There are no reported mutations in LRRK1, suggesting that this protein does not have a relationship to PD, and there are no pathogenic mutations in LRRK2 in this unique region. It has been suggested that LRRK2 is closely related to the MAPKKK family of kinases (West *et al.*, 2005). However, this logic is false because alignment of all protein kinases separates LRRK1 and LRRK2 into their own grouping (Manning *et al.*, 2002), most closely related to the RIPK (receptor interacting protein kinases) family that play critical roles in cellular stress responses (Meylan and Tschopp, 2005). Therefore, LRRK1 and LRRK2 have a unique place within the kinase superfamily, but only LRRK2 is associated with PD.

ENZYMATIC ACTIVITIES AND THE NORMAL FUNCTION OF LRRK2

Of the predicted domains of LRRK2, the putative kinase and GTPase enzymatic domain have attracted most attention because of the presence of many mutations in these regions, and because they have potential therapeutic implications. Using an *in vitro* kinase assay, Gloeckner and colleagues demonstrated that LRRK2 does indeed contain an active kinase domain. By contrasting a wild-type epitope-tagged LRRK2 construct immunoprecipitated from heterologous cells with a construct containing an artificial mutation (K1906M) at a critical residue within the kinase domain, and with autophosphorylation used as a read-out, they were able to detect kinase activity in wild type above that seen in the kinase-dead form (Gloeckner *et al.*, 2006). Similar results have now been reported by multiple groups using a range of constructs and either autophosphorylation or generic kinase substrates such as myelin basic protein as a measure of kinase activity (Table 32.1). Thus far, however, there have been no reports of detectable kinase activity from purified, recombinant protein from bacterial sources. This is relevant to the interpretation of data from *in vitro* studies, which are plagued by both high background phosphorylation as witnessed by detectable autophosphorylation in "kinase-dead" forms, presumably from either contaminating kinases or incomplete eradication of activity, and by high variability. Another *caveat* to these data is the absence of a substrate of physiological relevance. Whilst it is possible that an autophosphorylation activity may play a part in the normal cellular role of LRRK2, this has not yet been thoroughly examined and the detected *in vitro* autophosphorylation

TABLE 32.1   Kinase-dead variants used to confirm LRRK2 activity in recent studies

Study	Assay	Kinase dead	Kinase activity
Gloeckner *et al.*	Autophosphorylation	K1906M	25% WT
Greggio *et al.*	Autophosphorylation	Triple mutant	10% WT
Macleod *et al.*	MBP	K1906M	Decreased
Smith *et al.*	Autophosphorylation	D1994N/Double mutant	Decreased
West *et al.*	Autophosphorylation	K1906M	Decreased
Ito *et al.*	Autophosphorylation/MBP	K1906M/T2035A	Decreased

activity may be an artifact of the technique. It is, therefore, important to note that detectable alterations in autophosphorylation and phosphorylation of a generic substrate may not correlate with activity directed toward an *in vivo* substrate; this will be discussed below in the context of how pathogenic mutations affect activity. The key point relevant to the discussion of cellular models is that one can use artificial mutations like the kinase-dead variants to establish fairly clearly that the enzyme activity reported is dependent on the protein expressed rather than, for example, a co-precipitated kinase. Although COS7 or HEK293 cells used in all of the above studies are cancerous cell lines and not neurons, they are used here as factories for producing usable amounts of protein with a number of variations.

Alignment of the LRRK2 GTPase domain with LRRK1, and representative members of the Ras and Rho families of small GTPases reveals significant homology between these domains, particularly with regard to residues involved in the co-ordination of the γ-phosphate of GTP. Intriguingly, several residues that have been shown to be critical for GTPase activity (specifically within the γ-phosphate co-ordination pocket and at residue 1398) differ from those found in the Ras and Rho families, suggesting that both LRRK2 and LRRK1 have low intrinsic GTPase activities. LRRK2 (Smith *et al.*, 2006; Greggio *et al.*, 2007; West *et al.*, 2007) and LRRK1 (Korr *et al.*, 2005; Greggio *et al.*, 2007) can bind GTP, as demonstrated by both precipitation of epitope-tagged LRRK2 from whole cell lysates via GTP immobilized on sepharose beads and by binding of radioactively labeled GTP to the GTPase domain. These properties are disrupted by the alteration of residues predicted to be critical for GTP binding (e.g., K1347 and T1348), or by competition with a molar excess of unlabeled GTP. Recent results have suggested that, using assays on proteins expressed in HEK293 and mouse N2a neuroblastoma cells, wild-type LRRK2 possesses only a low GTPase activity (Ito *et al.*, 2007). This may be due to the absence of essential protein co-factors, such as GTPase activating proteins (GAPs) and GTP exchange factors (GEFs) that could be required for cleavage of the β/γ phosphate bond of GTP by LRRK2 within the cell. How much GTPase activity is present in complexes containing LRRK2 from the brain requires therefore further clarification, but in cellular models at least there is an opportunity to measure activity.

In terms of how LRRK2 acts within the cell, the interplay between these two enzymatic functions is of great interest. Small GTPases such as Ras operate as molecular switches for associated kinases, and an attractive hypothesis is that the GTPase domain of LRRK2 acts to regulate its kinase domain. This could occur through cycling from an "on" state (with GTP bound) to an "off" state (with GDP bound), a process perhaps controlled by associated GEFs and GAPs. There is some experimental evidence that this may be the case: mutations within the GTPase domain that have been shown to result in an increase in the interaction between LRRK2 and GTP may also cause an increase in kinase activity. In addition to this, disruption of the GTP binding properties of LRRK2 results in a concomitant decrease in kinase activity (Smith *et al.*, 2006; Ito *et al.*, 2007), suggesting that GTP binding is indeed important for this function as has been shown for LRRK1 (Korr *et al.*, 2005). There is also some evidence suggesting that incubating LRRK2 with GTP containing a sulfide bond rather than a phosphate bond between its β and γ phosphate groups (GTP-γS, which can be bound, but not cleaved by GTPases) leads to an increase in kinase activity – presumably due to the GTPase domain being irreversibly activated. Again, however, there are conflicting reports regarding this, and clarification of the role of the GTPase domain of LRRK2 in the regulation of its kinase activity awaits further investigation. The converse process, that is, the regulation of the GTPase domain by the kinase domain, has been examined by investigating the GTP binding properties of kinase-dead forms of LRRK2. Kinase-dead LRRK2 mutants retain GTP binding activity (West *et al.*, 2007), making it unlikely that there is a feedback from the kinase to the GTPase domain.

These considerations demonstrate that the two predicted catalytic activities are robust, at least in the types of models we are discussing here of transfecting plasmids into heterologous cell lines for recombinant protein production. However, they do not predict a specific function for LRRK2. Recently, MacLeod *et al.* (2006) reported that mutant LRRK2 interferes with neurite outgrowth in cellular models and this process seems to be dependent upon kinase activity. They demonstrated that dopaminergic neurons expressing mutant LRRK2 but not wild type have shorter neuronal processes and unaltered soma diameter. Suppression of LRRK2 with shRNAs or the K1906M, kinase dead, mutant led to an opposite phenotype with increased neurite process length and complexity, suggesting that LRRK2 is involved in the regulation of neurite process morphology and maintenance. Therefore, cell-based models are beginning to be used to predict normal function of LRRK2. In fact, MacLeod and colleagues moved this through to an *in vivo* setting by measuring neuronal morphology in the brains of rodents where the LRRK2 kinase domain had been expressed using viral vectors. With the *caveat* that this was the same laboratory and not an independent replication, the prediction from cell-based models held for the intact brain, with the predicted effects of LRRK2 kinase on neuronal morphology.

PATHOGENIC LRRK2 MUTATIONS

Toxic Effects of Mutant LRRK2 and Kinase Activity

The dominant mechanism of inheritance of *LRRK2* mutations suggests that they are associated with a gain of function, which might be predicted to have detrimental effects. For example, and hypothetically, an increase in kinase activity could alter the phosphorylation state of substrates involved in a pro-apoptotic pathway. A major limitation with current studies is that authentic, physiological substrates of LRRK2 remain to be identified, as discussed above. In the absence of an authentic *in vivo* substrate, one method to investigate whether LRRK2 pathological mutations interfere with the kinase activity is to examine autophosphorylation *in vitro* as a read-out of enzymatic activity. To date, several groups have reported that the G2019S and I2020T mutations, located in the magnesium binding loop at the beginning of the activation segment of the kinase domain, significantly increase autophosphorylation activity compared to wild-type protein (West *et al.*, 2005; Gloeckner *et al.*, 2006; Greggio *et al.*, 2006, 2007; MacLeod *et al.*, 2006; Smith *et al.*, 2006). Conflicting results have been reported regarding the kinase activity of LRRK2 containing mutations outside the kinase domain, for example, the Y1699C mutation in the COR domain and the R1441C/G mutations in the ROC domain. West *et al.* (2005, 2007) observed an increase of 1.5- to 2-fold autophosphorylation activity for these mutations, whilst we were not able to detect any significant difference between these mutants

and wild-type LRRK2 (Greggio *et al.*, 2006, 2007). MacLeod *et al.* (2006) also reported increased autophosphorylation for G2019S and I2020T, but no increase for R1441G. The disparity between these findings suggests that LRRK2 autophosphorylation assays may produce different results under different experimental conditions. It is likely that LRRK2 requires a set of interactors, adaptor proteins and/ or co-factors to carry out its normal function, and that the levels of autophosphorylation observed *in vitro* may not reflect activity in terms of a protein complex phosphorylating heterologous substrates. This may be one situation where heterologous cell lines (variously HEK293 and derivatives or COS7) may not capture the full function of a protein in the brain, which is a limitation of these results. But, it is also unlikely that expression of multiple variants with appropriate negative controls such as kinase-dead versions of the same will ever be attempted in mouse brain due to the complexity and expense of those experiments, limiting comparisons to those between studies.

Smith *et al.* (2005) showed that overexpression of mutant LRRK2 (R1441C, Y1699C and G2019S mutants) increased cellular toxicity when expressed for 2 days in either SH-SY5Y neuroblastoma cells or primary cortical neurons. Overexpression of mutant LRRK2 caused condensed and fragmented nuclei, suggesting that mutated proteins can drive toxicity by triggering apoptosis, although further details of the mechanism(s) involved remain to be elucidated. Therefore, although their effects on kinase activity are variable, all convincingly pathogenic mutations examined to date induce toxicity. We addressed this apparent discrepancy by comparing the toxic effect of kinase-dead versions of mutant LRRK2 with kinase active forms (Greggio *et al.*, 2006); similar results have been reported by others (Smith *et al.*, 2006). In our experiments, we transfected GFP-tagged LRRK2 plasmid constructs into either primary cortical neurons or SH-SY5Y cells and measured cell death by morphological indices 2–4 days later. In our studies and in others (Smith *et al.*, 2006), the advantage of cell models to test multiple variants of a protein to test a specific hypothesis was used. In these cases kinase-dead variants, for example, the triple kinase-dead K1906A/ D1994A/D2017A or the single mutants D1994N and K1906A were used in comparison to versions with the kinase activity intact but also containing pathogenic mutations. Inactivation of the kinase domain significantly ameliorates the toxic effects of mutant LRRK2, even if a pathological mutation was present in the inactive kinase. A small *caveat* is that in some of our experiments, 96 h post-transfection, the kinase-dead proteins showed increased toxicity, suggesting that the effects of kinase-dead mutations might be to delay rather than absolutely prevent toxicity. This has some implications for therapeutic design, but also highlights that all of these experiments should be repeated using *in vivo* approaches where a more chronic set of experiments could be performed. It would also be helpful to calibrate the relative level of activity of LRRK2 in the cultured cells compared to the endogenous activity. To date, all experiments have been driven by strong promoters (such as CMV) in culture and we therefore do not have a sense of how much LRRK2 activity is required for toxic effects.

However, and despite these *caveats*, the key message is that the cellular toxicity of mutant LRRK2 can be prevented or ameliorated by inactivation of the kinase domain and that this can be explored using cell models. Because of this, the kinase domain is an appealing therapeutic target for LRRK2-associated PD. The presence of pathological mutations outside the kinase domain that are not consistently associated with increased activity, however, suggests that a more complex mechanism for the observed toxicity may exist. An analogy can be made between mutant LRRK2 and mutant SOD1, a genetic cause of autosomal dominant amyotrophic lateral sclerosis (ALS). Similar to LRRK2, pathological mutations in SOD1 occur throughout the protein sequence. Some mutations alter SOD1 enzymatic activity, but others do not. All mutations that have been examined, however, possess an enhanced propensity to aggregate, both *in vivo* and *in vitro*, suggesting that alteration of enzymatic activity only partially contributes to the toxic gain of function.

Mirroring the SOD1 phenotype, the overexpression of LRRK2 in cells results in an increased accumulation of the protein into spheroid-like inclusion bodies resembling aggresomes (Greggio *et al.*, 2006; MacLeod *et al.*, 2006). These LRRK2-positive aggregates are perinuclear ubiquitin-positive bodies surrounded by a cage of vimentin filaments that can be efficiently dispersed by addition of nocodazole, a microtubule polymerization inhibitor and thus are similar to aggresomes formed by misfolded proteins in many cell types. We were able to correlate the formation of inclusion bodies with the presence of an active LRRK2 kinase, suggesting

that kinase activity and the inclusion formation phenotype may share a common mechanism. Sequestering LRRK2 into inclusion bodies may represent a way of scavenging misfolded or abnormally active proteins from cells. However, there are strong *caveats* about these experiments. First, and as discussed above, we have not calibrated expression levels compared to the endogenous gene, which means that we may be expressing a protein at very high levels. The inclusion body phenotype might be a result of high levels of expression as much as the innate tendency to misfolding of this large protein. Second, the tendency to form inclusion bodies is variable between different mutations; for example, the effect is quite strong with mutations that are associated with unusual pathologies in humans, such as Y1699C, but weaker with mutations associated with typical Lewy body parkinsonism such as G2019S. Some studies failed to see inclusion bodies at all, unless mutant LRRK2 was co-expressed with other aggregation prone proteins (Smith *et al.*, 2005). Taken together, these *in vitro* observations suggest that LRRK2 has an intrinsic ability to misfolding but do not demonstrate that this process is part of the pathogenic process *in vivo*. There have been contradictory results reported for whether LRRK2 is a component of Lewy bodies, which might be expected if the protein tends to misfold and/or aggregate in the sporadic disease (Giasson *et al.*, 2006; Miklossy *et al.*, 2006; Zhu *et al.*, 2006a, b). In our studies, staining with a C-terminal antibody showed that about 10% of brain stem Lewy bodies contain LRRK2 (Greggio *et al.*, 2006), suggesting that this protein is not a major or required component of inclusion bodies in sporadic PD. This is perhaps an example of observations in heterologous cell lines that do not readily extrapolate to the *in vivo* setting and are a limitation of some of these models.

Recently, we compared the R144C and Y1699C mutations located in the ROC and COR domains of LRRK2 with the corresponding K745C and F1021C mutations in the homologous kinase LRRK1, previously characterized by Korr *et al.* (2005). Although the two kinases share the same domain structure and a high degree of similarity (~26% identity and ~45% similarity at the amino acid level), LRRK2 is intrinsically more toxic than LRRK1 in cell culture models and is more prone to misfolding, measured indirectly by aggresome formation. One possible explanation for this is that the larger size of LRRK2 (2527 amino acids versus 2038 amino acids) may increase its propensity to undergo misfolding. LRRK2 is also a more active kinase than LRRK1, based on autophosphorylation assays, and this may have repercussions on its cellular toxicity (Greggio *et al.*, 2007). Again, this is the kind of study where cell models are an advantage. Rather than making transgenic mouse lines for six protein variants where, at least with hindsight, only the two pathogenic LRRK2 lines might show phenotypes, we were able to provide a first estimate of the relative tendency of the two proteins to damage neurons.

LRRK2 as a Therapeutic Target

Several studies suggest that the kinase activity of LRRK2 is an attractive and tractable therapeutic target for PD. The experiments supporting this idea come from cell models where it has been shown that whereas pathogenic mutations are toxic to neurons, the equivalent kinase-dead versions are not. The prediction from these experiments is that a small molecule inhibitor of the kinase would be able to prevent the toxic effects of LRRK2. As no small molecule inhibitors have yet been developed, it seems likely that cell-based models and *in vitro* kinase activity measurements will provide an important basis for developing and testing any such compounds. For example, heterologous cell lines can be used to produce recombinant LRRK2, which is currently inactive when produced in *E. coli*. Eventually, promising inhibitors would need to be tested in whole animal (transgenic) models to ensure that the same mechanisms hold in more intact systems, but cell assays with their higher throughput would be helpful in determining the most potent inhibitors. Comparing LRRK2 and LRRK1, or other related kinases, might be used to assess the selectivity of any candidate inhibitors.

There are other kinase inhibitors that have been successfully applied as therapeutic agents in diseases where hyperactive kinases contribute to pathogenesis. A known limitation of many kinase inhibitor derives from their mode of action. As competitive inhibitors that sit in the ATP binding pocket of kinases, many inhibitors suffer from lack of specificity because the ATP binding pocket of kinases is relatively conserved structurally. However, additional tactics might be helpful to generate small molecule inhibitors of LRRK2. A few studies have examined the GTP binding and potential hydrolytic activity of LRRK2. West *et al.*

(2007) recently reported that mutations outside of the kinase domain result in an increase in precipitation of LRRK2 by GTP-sepharose as measured by immunoblot. Because the GTPase region appears to regulate kinase activity, it may be possible to generate compounds that bind to this (or other) regulatory regions and thus inhibit the kinase in a non-competitive manner, which may lead to better specificity. An important, but tricky, set of experiments would be to try and separate out the function of the GTPase and kinase domains. Our current model is that the GTP binding region exists mainly to modulate kinase activity, but it is possible that there may be distinguishable and important roles of the GTPase alone. Cellular models may be helpful in being able to test multiple, and quite complex, variants. For example, how would a kinase-dead LRRK2 compare to one where both kinase and GTPase activities had been removed?

Finally, while there has been a great deal of emphasis on the catalytic domains of LRRK2, rather little attention has been given to the protein–protein interaction domains. Cell-based models have provided parkin (Smith *et al.*, 2005), cdc37 and Hsp90 (Gloeckner *et al.*, 2006) as potential interactors for LRRK2. It seems likely that there are additional interactors, given the large size of LRRK2 and the presence of multiple protein–protein interaction domains, and hence searching for additional components of the LRRK2 complex(es) will be valuable. The current proposed interactors also require further validation, as none have been shown to interact with specific regions of LRRK2. *In vivo* proofs of interactions are also needed. We would also argue that understanding what LRRK2 does that differs from LRRK1 is important in understanding why the former is associated with PD but the latter is not. Protein interaction partners may be important in the differing functions of the two kinases.

SUMMARY AND THE PROS AND CONS OF CELL MODELS

To summarize the studies published so far on LRRK2 from cellular systems: (i) some pathological mutations increase autophosphorylation activity and enhance cellular toxicity *in vitro*; (ii) kinase activity mediates neuronal toxicity which has important therapeutic implications; (iii) neuronal toxicity is more dramatic in mutant LRRK2 compared to the closely related kinase LRRK1, which is less active and (iv) not all mutations increase activity using autophosphorylation measures, implying that there are subtleties of their mechanisms that remain to be established.

Using this series of studies as an instructive tool, what do we know about cell-based models, and their relative disadvantages and advantages? The major limitation is that as cell culture preparations are not *in vivo* models, there are inevitable concerns about whether results can be extrapolated to the intact animal. We have perhaps limited this concern in the present discussion of LRRK2 because there are few opportunities to make the comparisons as there are rather limited animal models available. But, in one study where deliberate comparisons were made to pros are that many variants can be examined in the same background, allowing for the assessment of whether mutations cause consistent effects and allows for multiple artificial variants to be constructed dissecting out functional contributions of specific activities. One facet that is both advantageous and difficult is the time course – cell experiments can be completed in a few days but the accelerated process means that we often cannot be sure that mechanisms (e.g., for toxicity) remain the same.

ACKNOWLEDGEMENT

This research was supported by the Intramural Research Program of the NIH, National Institute on Aging.

REFERENCES

Bosgraaf, L., and Van Haastert, P. J. (2003). Roc, a Ras/GTPase domain in complex proteins. *Biochim Biophys Acta* **1643**, 5–10.

Cookson, M. R., Xiromerisiou, G., and Singleton, A. (2005). How genetics research in Parkinson's disease is enhancing understanding of the common idiopathic forms of the disease. *Curr Opin Neurol* **18**, 706–711.

Funayama, M., Hasegawa, K., Kowa, H., Saito, M., Tsuji, S., and Obata, F. (2002). A new locus for Parkinson's disease (PARK8) maps to chromosome 12p11.2-q13.1. *Ann Neurol* **51**, 296–301.

Funayama, M., Hasegawa, K., Ohta, E., Kawashima, N., Komiyama, M., Kowa, H., Tsuji, S., and Obata, F. (2005). An LRRK2 mutation as a cause for the parkinsonism in the original PARK8 family. *Ann Neurol* **57**, 918–921.

Fung, H. C., Chen, C. M., Hardy, J., Singleton, A. B., and Wu, Y. R. (2006). A common genetic factor for Parkinson disease in ethnic Chinese population in Taiwan. *BMC Neurol* **6**, 47.

Giasson, B. I., Covy, J. P., Bonini, N. M., Hurtig, H. I., Farrer, M. J., Trojanowski, J. Q., and Van Deerlin, V. M. (2006). Biochemical and pathological characterization of Lrrk2. *Ann Neurol* **59**, 315–322.

Gloeckner, C. J., Kinkl, N., Schumacher, A., Braun, R. J., O'Neill, E., Meitinger, T., Kolch, W., Prokisch, H., and Ueffing, M. (2006). The Parkinson disease causing LRRK2 mutation I2020T is associated with increased kinase activity. *Hum Mol Genet* **15**, 223–232.

Greggio, E., Jain, S., Kingsbury, A., Bandopadhyay, R., Lewis, P., Kaganovich, A., van der Brug, M. P., Beilina, A., Blackinton, J., Thomas, K. J., Ahmad, R., Miller, D. W., Kesavapany, S., Singleton, A., Lees, A., Harvey, R. J., Harvey, K., and Cookson, M. R. (2006). Kinase activity is required for the toxic effects of mutant LRRK2/dardarin. *Neurobiol Dis* **23**, 329–341.

Greggio, E., Lewis, P. A., van der Brug, M. P., Ahmad, R., Kaganovich, A., Ding, J., Beilina, A., Baker, A. K., Cookson, M. R. (2007). Mutations in LRRK2/dardarin associated with Parkinson disease are more toxic than equivalent mutations in the homologous kinase LRRK1. *J Neurochem* **102**, 93–102.

Ito, G., Okai, T., Fujino, G., Takeda, K., Ichijo, H., Katada, T., and Iwatsubo, T. (2007). GTP binding is essential to the protein kinase activity of LRRK2, a causative gene product for familial Parkinson's disease. *Biochemistry* **46**, 1380–1388.

Kay, D. M., Kramer, P., Higgins, D., Zabetian, C. P., and Payami, H. (2005). Escaping Parkinson's disease: A neurologically healthy octogenarian with the LRRK2 G2019S mutation. *Mov Disord* **20**, 1077–1078.

Khan, N. L., Jain, S., Lynch, J. M., Pavese, N., Abou-Sleiman, P., Holton, J. L., Healy, D. G., Gilks, W. P., Sweeney, M. G., Ganguly, M., Gibbons, V., Gandhi, S., Vaughan, J., Eunson, L. H., Katzenschlager, R., Gayton, J., Lennox, G., Revesz, T., Nicholl, D., Bhatia, K. P., Quinn, N., Brooks, D., Lees, A. J., Davis, M. B., Piccini, P., Singleton, A. B., and Wood, N. W. (2005). Mutations in the gene LRRK2 encoding dardarin (PARK8) cause familial Parkinson's disease: Clinical, pathological, olfactory and functional imaging and genetic data. *Brain* **128**, 2786–2796.

Korr, D., Toschi, L., Donner, P., Pohlenz, H. D., Kreft, B., and Weiss, B. (2005). LRRK1 protein kinase activity is stimulated upon binding of GTP to its Roc domain. *Cell Signal* **18**, 910–920.

MacLeod, D., Dowman, J., Hammond, R., Leete, T., Inoue, K., and Abeliovich, A. (2006). The familial Parkinsonism gene LRRK2 regulates neurite process morphology. *Neuron* **52**, 587–593.

Manning, G., Whyte, D. B., Martinez, R., Hunter, T., and Sudarsanam, S. (2002). The protein kinase complement of the human genome. *Science* **298**, 1912–1934.

Marin, I. (2006). The Parkinson disease gene LRRK2: evolutionary and structural insights. *Mol Biol Evol* **23**, 2423–2433.

Meylan, E., and Tschopp, J. (2005). The RIP kinases: Crucial integrators of cellular stress. *Trends Biochem Sci* **30**, 151–159.

Miklossy, J., Arai, T., Guo, J. P., Klegeris, A., Yu, S., McGeer, E. G., and McGeer, P. L. (2006). LRRK2 expression in normal and pathologic human brain and in human cell lines. *J Neuropathol Exp Neurol* **65**, 953–963.

Paisan-Ruiz, C., Jain, S., Evans, E. W., Gilks, W. P., Simon, J., van der Brug, M., Lopez de Munain, A., Aparicio, S., Gil, A. M., Khan, N., Johnson, J., Martinez, J. R., Nicholl, D., Carrera, I. M., Pena, A. S., de Silva, R., Lees, A., Marti-Masso, J. F., Perez-Tur, J., Wood, N. W., and Singleton, A. B. (2004). Cloning of the gene containing mutations that cause PARK8-linked Parkinson's disease. *Neuron* **44**, 595–600.

Smith, W. W., Pei, Z., Jiang, H., Moore, D. J., Liang, Y., West, A. B., Dawson, V. L., Dawson, T. M., and Ross, C. A. (2005). Leucine-rich repeat kinase 2 (LRRK2) interacts with parkin, and mutant LRRK2 induces neuronal degeneration. *Proc Nat l Acad Sci USA* **102**, 18676–18681.

Smith, W. W., Pei, Z., Jiang, H., Dawson, V. L., Dawson, T. M., and Ross, C. A. (2006). Kinase activity of mutant LRRK2 mediates neuronal toxicity. *Nat Neurosci* **9**, 1231–1233.

West, A. B., Moore, D. J., Biskup, S., Bugayenko, A., Smith, W. W., Ross, C. A., Dawson, V. L., and Dawson, T. M. (2005). Parkinson's disease-associated mutations in leucine-rich repeat kinase 2 augment kinase activity. *Proc Natl Acad Sci USA* **102**, 16842–16847.

West, A. B., Moore, D. J., Choi, C., Andrabi, S. A., Li, X., Dikeman, D., Biskup, S., Zhang, Z., Lim, K. L., Dawson, V. L., and Dawson, T. M. (2007). Parkinson's disease-associated mutations in LRRK2 link enhanced GTP-binding and kinase activities to neuronal toxicity. *Hum Mol Genet* **16**, 223–232.

Zhu, X., Babar, A., Siedlak, S. L., Yang, Q., Ito, G., Iwatsubo, T., Smith, M. A., Perry, G., and Chen, S. G. (2006a). LRRK2 in Parkinson's disease and dementia with Lewy bodies. *Mol Neurodegen* **1**, 17.

Zhu, X., Siedlak, S. L., Smith, M. A., Perry, G., and Chen, S. G. (2006b). LRRK2 protein is a component of Lewy bodies. *Ann Neurol* **60**, 617–618, author reply 618–619.

Zimprich, A., Biskup, S., Leitner, P., Lichtner, P., Farrer, M., Lincoln, S., Kachergus, J., Hulihan, M., Uitti, R. J., Calne, D. B., Stoessl, A. J., Pfeiffer, R. F., Patenge, N., Carbajal, I. C., Vieregge, P., Asmus, F., Muller-Myhsok, B., Dickson, D. W., Meitinger, T., Strom, T. M., Wszolek, Z. K., and Gasser, T. (2004). Mutations in LRRK2 cause autosomal-dominant parkinsonism with pleomorphic pathology. *Neuron* **44**, 601–607.

33

USING YEAST AS A MODEL SYSTEM FOR THE GENETIC DISSECTION OF α-SYNUCLEIN TOXICITY

VICENTE SANCENON,[1] **SUE-ANN LEE**[1,2] **AND PAUL J. MUCHOWSKI**[1,2,3,4]

[1]*Gladstone Institute of Neurological Disease, University of California, San Francisco, CA, USA*
[2]*Biomedical Sciences Program, University of California, San Francisco, CA, USA*
[3]*Department of Biochemistry and Biophysics, University of California, San Francisco, CA, USA*
[4]*Department of Neurology, University of California, San Francisco, CA, USA*

α-SYNUCLEIN: A KEY PLAYER IN PARKINSON'S DISEASE AND SYNUCLEINOPATHIES

Nearly two centuries after Dr. James Parkinson reported the cardinal symptoms of *"shaking palsy"*, Parkinson's disease (PD) remains a predominantly idiopathic disorder. Over the past decade, the identification of causal mutations that co-segregate with rare, hereditary forms of PD is beginning to increase our understanding of the etiopathology of the disease. For example, missense mutations (A30P, E46K, and A53T) in the human *SNCA* gene, which encodes the small presynaptic protein α-synuclein, cause rare, familial forms of PD (Kruger *et al.*, 1998; Polymeropoulos *et al.*, 1997; Zarranz *et al.*, 2004). Nevertheless, ubiquitinated inclusions of α-synuclein, or Lewy bodies, are a pathologic hallmark of both hereditary and sporadic PD and other neurological disorders collectively referred to as α-synucleinopathies (Mezey *et al.*, 1998). The mechanisms of α-synuclein toxicity have only started to be unraveled. *In vitro*, α-synuclein aggregation correlates with mitochondrial deficits, cellular trafficking defects, vesicle permeabilization, and proteasomal impairment (Hsu *et al.*, 2000; Gosavi *et al.*, 2002; Lashuel *et al.*, 2002; Petrucelli *et al.*,

2002). Despite the multiplicity of genetic and environmental factors that lead to PD, the presence of α-synuclein pathology in virtually all PD patients indicates that this protein represents a juncture between multiple pathogenic pathways. Therefore, *in vitro* and *in vivo* models of α-synuclein aggregation and toxicity have become the subject of basic and applied studies to understand the etiology of PD and identify effective therapies.

WHY USE YEAST TO MODEL CELL-AUTONOMOUS ASPECTS OF NEUROLOGICAL DISEASES ASSOCIATED WITH PROTEIN MISFOLDING?

The budding yeast *Saccharomyces cerevisiae*, commonly known as baker's yeast, is widely used to study human diseases, including cancer, neurological syndromes, and mitochondrial disorders (Barrientos, 2003; McMurray and Gottschling, 2003; Sherman and Muchowski, 2003). Since yeast lack neuron-specialized functions, it might seem counterintuitive to use yeast models to study neurological diseases. However, nearly one-third of the 6,217 predicted protein-encoding genes of yeast exhibit statistically significant similarity to at least one human gene, and two-thirds have at least one domain with homology to human genes (Botstein *et al.*, 1997). Importantly, many key cellular pathways are conserved from yeast to humans, including protein folding, degradation, trafficking, signal transduction, cell-cycle control, DNA replication, transcription, translation, chromatin remodeling, and even processes that underlie neurological diseases in humans, such as prion propagation (Wickner, 1994) and expansion of trinucleotide repeats (Richard and Dujon, 1996).

Further, yeast has several distinct advantages over many other experimental systems. Its main advantage is its genetic tractability. Unlike other model organisms, genes can be introduced easily into yeast via episomal plasmids or integrative vectors. Yeast genes can also be deleted, mutated, overexpressed, and tagged at a small fraction of the time and cost for other model organisms. Indeed, a veritable toolbox of techniques is available for genetic manipulations on a genome-wide level in yeast, including libraries for functional genomic and proteomic analyses. Thus, unicellular eukaryotic models have potential utility for unraveling the intracellular mechanisms of pathogenicity and identifying the key cellular processes that may be relevant to neurological disorders.

MODELING α-SYNUCLEINOPATHY IN YEAST

Although the yeast genome does not contain a homolog of the α-synuclein gene, heterologous expression of α-synuclein in yeast exerts profound physiological consequences in this organism. For example, α-synuclein binds to membranes, inhibits endogenous phospholipase D (PLD) activity, and is cleared by proteasomal degradation and autophagy, recapitulating physiologically relevant properties of this protein in mammalian cells (Outeiro and Lindquist, 2003; Zabrocki *et al.*, 2005). Pioneering studies by Outeiro and Lindquist (2003) elegantly demonstrated that α-synuclein becomes cytotoxic in yeast, as in humans, in a concentration-dependent manner. Taking advantage of well-established techniques to manipulate gene dosage in yeast, these authors showed that, while a single copy of the α-synuclein gene has no effect on cell growth, two copies dramatically impairs it. The familial mutation A53T does not modify the toxic properties of α-synuclein in yeast, whereas the A30P mutation is

only partially toxic when expressed from an extrachromosomal high-copy plasmid. Interestingly, this dose-dependent toxic phenotype correlates with a particular localization pattern. In the single-copy strains, both WT and A53T α-synuclein localize almost exclusively to the plasma membrane, whereas in the two-copy strains they form cytoplasmic inclusions. While original studies believed these structures to be classical inclusion bodies, studies from our laboratory and others indicate that they are composed of intracellular accumulations of vesicular structures that arise due to intracellular trafficking defects (Gitler *et al.*, 2008; Soper *et al.*, 2008; Sancenon *et al.*, 2008). Consistent with these observations Lindquist and colleagues demonstrated that expression of α-synuclein in yeast causes an ER-Golgi block (Cooper *et al.*, 2006).

The α-synuclein-induced growth inhibition in yeast is accompanied by cellular consequences that are reminiscent of those believed to be important in PD, such as proteasome impairment, heat-shock and oxidative stress, formation of ubiquitin-positive α-synuclein inclusion bodies, and emergence of apoptotic markers (Outeiro and Lindquist, 2003; Dixon *et al.*, 2005; Flower *et al.*, 2005). While it may sound surprising that a unicellular organism such as yeast can undergo apoptosis, certain conditions can induce yeast cell death with typical hallmarks of mammalian apoptosis, including DNA degradation and chromatin condensation, externalization of phosphatidyl serine to the outer leaflet of the plasma membrane, and release of cytochrome c (reviewed by Frohlich *et al.*, 2007). Some of these events are in part mediated by the yeast metacaspase Yca1p. In agreement with these observations, proteasome mutations and oxidative agents promote inclusion formation and increase the vulnerability to α-synuclein (Dixon *et al.*, 2005; Zabrocki *et al.*, 2005; Sharma *et al.*, 2006), whereas heat-shock induction and compounds with antioxidant and metal-chelating properties protect against α-synuclein-induced cellular damage (Flower *et al.*, 2005; Griffioen *et al.*, 2006). During stationary phase conditions, which are characterized by increased oxidative stress and proteasome inhibition (Chen *et al.*, 2005a, b), α-synuclein reduces the ability of yeast cells to withstand aging. Although cells overexpressing α-synuclein exhibit signs of apoptosis, the potential role of the pro-apoptotic metacaspase Yca1p in α-synucleinopathy in yeast remains unclear since the deletion of the *YCA1* gene (*yca1*△)

was reported to both abrogate (Flower *et al.*, 2005) and augment (Griffioen *et al.*, 2006) α-synuclein toxicity. Finally, yeast cells overexpressing α-synuclein accumulate lipid droplets (Outeiro and Lindquist, 2003), an outcome that mimics the inhibitory effect of this protein on the hydrolysis of stored triglycerides in cultured cells (Cole *et al.*, 2002). Therefore, expression of α-synuclein in yeast recapitulates a subset of features observed in neurons and thus provides a tractable platform to dissect molecular mechanisms that underlie α-synuclein toxicity.

GENETIC AND PHARMACOLOGIC DISSECTION OF α-SYNUCLEIN TOXICITY

Despite the low incidence of α-synuclein mutations in humans affected by PD, the presence of α-synuclein-positive Lewy bodies in sporadic PD suggests that other genetic factors may predispose dopaminergic neurons to become vulnerable to α-synuclein. From a clinical perspective, evidence suggests the existence of genetic modifiers of PD. Thus, it will be important to identify susceptibility factors that predispose humans to a genetically complex disorder such as PD, because they may affect a significant proportion of the population. Furthermore, the products of these genes may be attractive targets to explore new therapeutic approaches, or may be targets of currently approved drugs that could be moved quickly to the clinic. Unfortunately, genetic risk factors are very difficult to identify because of their often modest phenotypic effects and low penetrance, interactions with genetic and non-genetic factors, and limited availability of human samples. Moreover, mapping single-nucleotide polymorphisms (SNPs) linked to PD in human populations can be technically challenging and time-consuming. These limitations make model organisms such as yeast an especially attractive means to identify therapeutic targets and lead candidates to help guide genetic and clinical studies in human populations with PD.

Yeast as Platform for the Discovery of Therapeutic Drugs and Drug Targets

Provided that a particular phenotype can be easily scrutinized, yeast provides an invaluable tool to identify genetic and chemical modifiers of the phenotype. As described in section "Modeling α-Synucleinopathy in Yeast," α-synuclein toxicity is monitored in yeast as a growth or viability defect, a phenotype that is easily assayed in yeast by any of several techniques, including spotting assays, growth curves or vital dyes. Image analysis or spectroscopic measurements enable the collection of quantitative data on cell growth or viability. Finally, before starting to screen for toxicity modifiers, any type of assay needs to be optimized and adapted to a high-throughput format.

Once a toxicity model and assay have been chosen and optimized, two types of approaches can be applied to identify genes and compounds that modulate the toxicity of α-synuclein. Whereas the molecules that inhibit α-synuclein toxicity are potential therapeutic leads, mutant alleles that alter this phenotype are putative therapeutic targets. In this case, two types of therapeutic strategies can be envisioned. Genes that suppress the toxicity of α-synuclein when silenced or increase toxicity when overexpressed are attractive targets for inhibitory drugs, whereas genes that increase toxicity when silenced or suppress toxicity when overexpressed should be considered as targets for enhancing drugs.

So far, two genetic screens have been successfully applied to identify genetic modifiers of α-synuclein toxicity in yeast (Willingham *et al.*, 2003; Cooper *et al.*, 2006). The first screen (Willingham *et al.*, 2003) took advantage of the Yeast Genome Deletion Set, a commercially available collection of arrayed yeast strains carrying loss-of-function alleles in approximately 85% of all predicted open reading frames (ORFs) (Giaever *et al.*, 2002). The second screen (Cooper *et al.*, 2006) took advantage of the Yeast FLEXGene Collection comprising approximately 3,000 yeast ORFs in a galactose-inducible expression system. The therapeutic implications of these studies are extensively discussed in section "Genetic Analyses in Yeast Implicate Two Major Pathways Mediating α-Synuclein Toxicity."

Epistasis studies are also useful for guiding the design of therapeutic strategies. These studies can establish the hierarchy of the identified genetic modifiers and their sequential involvement in specific pathways. By analyzing the simultaneous effect of two modifiers on toxicity, such studies can determine if two genes act in the same pathway, in parallel and independent pathways, or in interdependent pathways. This type of analysis can help determining the preferred target or combination of targets that would maximize therapeutic specificity and minimize off-side effects based on the predicted hierarchy of the genetic modifiers.

While genetic approaches are useful to identify therapeutic targets, small molecule screens can directly identify leads with potential therapeutic activity, circumventing the requirement of having information on the targets. Yeast constitutes an excellent platform to conduct small molecule screens for the discovery of lead compounds because of its simple growth conditions and cost-effectiveness. Libraries of small molecules can be systematically screened in yeast to identify compounds that rescue the growth defect induced by α-synuclein. However, to increase the bioavailability and effectiveness of the compounds, the strain of interest has to be genetically modified. Typically, two strategies are used to accomplish this requirement: inactivating members of the ATP-binding cassette (ABC) family of pleiotropic drug resistance (PDR) efflux pumps or their transcriptional regulators, which results in enhanced drug sensitivity, or eliminating genes of the ergosterol biosynthetic pathway, which results in increased membrane permeability.

Small molecule screens in yeast have already identified novel compounds of potential therapeutic relevance, such as inhibitors of the pro-inflammatory MAP kinase p38α, inhibitors of polyglutamine aggregation, and stimulators of autophagic clearance of misfolded proteins, including α-synuclein (Zhang et al., 2005; Friedmann et al., 2006; Sarkar et al., 2007). Interestingly, antioxidants and metal-chelating compounds were retrieved in a screen of chemicals that alleviate α-synuclein toxicity, hinting a potential therapeutic venue for the treatment of α-synucleinopathies (Griffioen et al., 2006).

An alternative strategy to identify pathways and compounds in yeast with potential therapeutic properties is based on the so-called Phenotype MicroArray™ (PM) technology (Biolog). This platform offers a wide variety of phenotypical tests utilizing respiration as a quantitative reporter of the physiological state of the cells. PM allows to evaluate multiple cellular functions of a particular yeast model, or to identify compounds that have beneficial or detrimental effects on that model.

Genetic and chemical screens provide complementary information. For example, the therapeutic targets identified by genetic approaches can improve drug discovery by providing candidate genes for target-based drug screening. The advantage of these screens is their expected higher hit ratio and the availability of target-specific collections of molecules. On the other hand, the bioactive leads identified by chemical approaches can help identify therapeutic targets and cellular pathways, which reveal the sites and mechanisms of action of the leads. So far, yeast is the only model organism that enables rapid and facile identification of drug targets employing genome-wide approaches. In the next section we review three of the most common techniques used to identify drug targets in yeast.

The first technique takes advantage of varying gene dosage in yeast. When yeast strains are transformed with a genomic or cDNA library of genes and treated with a drug, yeast clones that harbor multiple copies of the drug target gene will be more resistant to the effects of the drug. Previously unknown targets of the drug can be identified in this way and studied to determine the mechanism of drug action. These screens are not limited to libraries of yeast genes; as long as the gene products can be functionally expressed in yeast, libraries derived from human DNA (e.g., brain lysate) are usable. For example, this technique has been successfully used to identify the molecular targets for known antifungals, such as tunicamycin (Rine et al., 1983), ketoconazole (Launhardt et al., 1998) and soraphen (Vahlensieck et al., 1994), therapeutic molecules (Lum et al., 2004), growth inhibitors (Luesch et al., 2005), and components of the target of rapamycin (TOR) pathway (Butcher et al., 2005).

A second technique complements the overexpression approach described above by reducing the gene dosage to identify drug targets. This technique takes advantage of the Yeast Genome Deletion Set, which contains diploid mutant yeast strains that are heterozygous for single-gene deletions. Reducing the number of copies of a drug target gene in yeast from two to one sensitizes the diploid strain to the drug, a phenomenon known as drug-induced haploinsufficiency (Giaever et al., 1999). The systematic, unbiased nature of this approach makes it particularly attractive to identify the cellular pathways that are affected by a given drug, as demonstrated by Baetz et al. (2004), who used a yeast genome-wide haploinsufficiency screen to show that dihydromotuporamine C targets the sphingolipid metabolism pathway.

Another technique is based on the theory that inhibition of a drug target results in genome-wide changes in gene expression that match the changes when the drug target gene is deleted (Hughes et al. 2000). This technique involves DNA microarray and computational analyses. For example, RNA from a yeast strain treated with a drug that has an unknown target can be hybridized to a chip bearing an array of all yeast genes. The expression profile can then be compared to a reference database of expression profiles from mutations and chemical treatments. Alternatively, if a specific drug target is of interest, RNA from a mutant yeast strain with a deletion of the gene encoding that target can be hybridized to yeast gene chip arrays to display the relative abundance of all mRNAs in the absence of the drug target. Libraries of small molecules can then be screened in wild-type yeast to identify compounds that recapitulate the expression pattern of the mutant yeast strain.

Advantages and Disadvantages of Using Yeast Models for the Discovery of Drugs and Drug targets

Using yeast as a living test tube for the discovery of genes and chemicals that modulate the toxicity of α-synuclein has several advantages over other model systems. The short life cycle and low cost of manipulation of yeast enable one to conduct screens in a timescale and throughput level nearly comparable to *in vitro* systems. Yet yeast provides a cellular context to study chemical and genetic interactions that is an *in vivo* model. These properties have two major implications. First, yeast confers the ability to screen an enormous diversity of molecules and virtually entire genomes in a relatively short time period. And second, yeast screens can identify genes in regulatory pathways or molecules that target those pathways.

The relative ease and efficiency with which one can recombine DNA in yeast has lead in the past decade to the development of a complete set of libraries for functional genomic and proteomic studies that have significantly benefited many areas of biomedicine. These libraries, which are commercially available, constitute an unsurpassed tool to characterize the mode of action of drugs (e.g., the targets of bioactive compounds), and to dissect the pathways implicated in a particular phenotype (e.g., the toxicity of α-synuclein).

However, the information obtained by these studies has to be interpreted cautiously. Despite the significant degree of conservation between yeast and mammalian genomes, yeast lacks gene families specialized in neuronal differentiation and function. This fact has two major consequences. First, neuron-specific genes that might be relevant to

therapeutic targets cannot be detected by genetic screens in yeast. Second, compounds that are effective in yeast may have unpredictable side-effects in mammalian systems. In addition, as a unicellular organism, yeast is not suitable to model the neuronal circuitry dysfunction that underlies neurodegenerative disorders. For these reasons, yeast has to be kept in mind simply as a primary platform for the rapid discovery and preliminary development of putative leads and targets. Ultimately, their relevance and efficacy for the treatment of human disease has obligatorily to be validated in cellular and animal models.

Genetic Analyses in Yeast Implicate Two Major Pathways Mediating α-Synuclein Toxicity

To identify extragenic modifiers of α-synuclein toxicity, two genome-wide screens have been conducted in yeast models of α-synucleinopathy. A loss-of-function genetic screen using the Yeast Genome Deletion Set identified 86 null alleles in non-essential genes that augment the growth defect caused by overexpression of α-synuclein from a multicopy plasmid (Willingham *et al.*, 2003). A functional classification of the hits revealed two categories of genes largely represented in the screen: those involved in lipid metabolism and vesicle-mediated transport (Table 33.1). Independently, an overexpression genetic screen using the Yeast FLEXGene collection identified 34 genes that suppress and 20 genes that enhance the growth defect caused by overexpression of α-synuclein from two genomic loci plus a centromeric extrachromosomal plasmid (Cooper *et al.*, 2006). Consistent with previous results, modifiers involved in vesicle-mediated transport were enriched (Table 33.1). Remarkably, putative human orthologs could be assigned to at least 20 (87%) of these genes (Table 33.1), suggesting that the identified interactions may be conserved in neurons. Consistent with these results, emerging evidence from various experimental paradigms implicates α-synuclein in lipid metabolism and vesicle trafficking, as discussed below.

α-Synuclein: The Lipid Connection

Although its physiological function has not been well established, α-synuclein is structurally similar to fatty acid binding proteins, suggesting a role in lipid metabolism (Sharon *et al.*, 2001). In fact, most of the first 100 N-terminal residues of α-synuclein comprise a lipid-binding domain that forms an amphipathic alpha helix required for interaction with membranes (Davidson et al., 1998; Chandra et al., 2002). Although α-synuclein localizes to pre-synaptic terminals under physiological conditions, the widespread distribution of α-synuclein inclusions in neuronal bodies and neurites in PD suggest that neurotoxicity may arise from an abnormal protein/lipid interaction in multiple subcellular compartments under pathological conditions.

The analysis of brains of PD patients, α-synuclein-null transgenic mice, and neurons overexpressing α-synuclein indicates that α-synuclein facilitates the uptake and incorporation of saturated and unsaturated fatty acyl chains to brain phospholipids, including the mitochondria-specific lipid cardiolipin (Sharon et al., 2003b; Castagnet et al., 2005; Golovko et al., 2005, 2006; Barcelo-Coblijn et al., 2007). In this way, α-synuclein could modulate membrane fluidity and facilitate the coupling between electron transfer complexes in the mitochondrial inner membrane (Ellis et al., 2005). Conceivably, moderate levels of α-synuclein may alter mitochondrial membrane fluidity in yeast, leading to mitochondrial dysfunction and growth defects. Moreover, deletion of the carnitine acetyltransferase gene (yap1△), which is involved in fatty acid transport into mitochondria, synergistically increases this defect (Table 33.1). This finding supports the hypothesis that energy imbalance and oxidative stress underlie cellular dysfunction in PD. Indeed, some neurotoxins that alter mitochondrial electron transport and induce oxidative stress in dopaminergic neurons are recognized risk factors for PD and have been used to model this disorder in mammals (reviewed in chapters 9–19; Przedborski and Ischiropoulos, 2005).

Apart from being a substrate of α-synuclein, lipids are a major constituent of Lewy bodies (Gai et al., 2000) and may play an important role in α-synuclein-dependent pathogenesis. In cultured neurons and mouse brains, polyunsaturated fatty acids promote the oligomerization of α-synuclein (Sharon et al., 2003a), a process that may be initiated on the surfaces of membranes (Cole et al., 2002). This scenario has been reconstructed in yeast, where the nucleation of α-synuclein inclusions is initiated in the plasma membrane and increased by stimulating phospholipid biosynthesis (Zabrocki et al., 2005). Interestingly, inactivating genes required for peroxisome biogenesis (pex2△, pex8△ and pox1△) or for intravacuolar breakdown of lipids (cvt17△) sensitizes yeast cells to α-synuclein (Table 33.1).

In eukaryotes, lipids are degraded in vacuoles/lysosomes, and fatty acids are oxidized in peroxisomes. Thus, it is reasonable to envision a model in which α-synuclein, vacuoles/lysosomes, and peroxisomes coordinately regulate the cellular content of lipids and fatty acids, which in turn may influence the aggregation state of α-synuclein. Therefore, mutations that compromise vacuolar/lysosomal and peroxisomal structure or function likely disrupt this homeostatic balance, resulting in α-synuclein accumulation and subsequent cytotoxicity. Concomitantly, these processes may further damage the integrity of these organelles, leading to a pernicious cycle of α-synuclein inclusion formation organelle dysfunction, and imbalance in lipid metabolism. Indeed, lysosomal dysfunction has been implicated in several neurodegenerative diseases (Bahr and Bendiske, 2002). Interestingly, α-synuclein accumulation causes lysosomal pathology in transgenic mice (Rockenstein et al., 2005), raising the possibility that enhancing lysosomal function might be a potential therapeutic strategy for PD (Lee et al., 2004). Although a link between peroxisomes and PD has not been established, peroxisome biogenesis disorders underlie severe neurological conditions, such as the Zellweger syndrome, characterized by defects in brain development and growth of the myelin sheath owing to the inability of the central nervous system to carry out the β-oxidation of long chain fatty acids (reviewed by Faust et al., 2005). A mouse model of this disorder, generated by targeted disruption of the murine PEX2 gene, exhibits extended neuronal lipidosis in the brain (Faust et al., 2001). The exacerbation of α-synuclein toxicity by pex null alleles in yeast raises the interesting possibility that peroxisomal defects may contribute in some manner to PD.

Genetic screens in yeast have also linked α-synuclein toxicity to alterations in lipid signaling. In mammalian cells, α-synuclein may participate in signal transduction cascades by inhibiting PLD, a process that has been modeled in yeast cells (Jenco et al., 1998; Ahn et al., 2002; Outeiro and Lindquist, 2003). PLD converts phosphatidylcholine (PC) into phosphatidic acid (PA), a phospholipid that regulates numerous cellular processes, including vesicle trafficking (reviewed by Wang et al., 2006). Interestingly, loss of function of OPI3 (opi3△), a gene required for PC biosynthesis, increases

■ **TABLE 33.1** **Genetic modifiers of α-synuclein toxicity that are involved in lipid metabolism and vesicle trafficking**

Yeast[a]	Human[b]	Biological function[c,d]
Loss-of-function enhancers (Willingham *et al.*, 2003)		
arl3Δ	*ARP1*	*Yeast*: Ras GTPase involved in recruiting tethering complexes to the *trans*-Golgi network *Human*: ARF-related GTPase implicated in the Golgi-to-plasma membrane and endosome-to-Golgi transport
Cog6Δ	*COG6*	*Yeast and human*: Component of the COG tethering complex that mediates fusion of transport vesicles to Golgi compartments
cvt17Δ	*None*	*Yeast*: Lipase required for vacuolar lysis of lipid vesicles
dpp1Δ	*PPAP2B*	*Yeast*: DGPP phosphatase involved in lipid metabolism and signaling *Human*: PA phosphatase involved in lipid metabolism and signaling
msb3Δ	*USP6NL*	*Yeast*: GAP of Ypt/Rab GTPases involved in exocytosis (Sec4p), ER-to-Golgi (Ypt1p), endosome-to-Golgi, intra-Golgi, and Golgi-to-ER (Ypt6p), post-Golgi (Ypt31p, Ypt32p), or endocytic and vacuolar (Ypt51p) transport *Human*: GAP of the GTPase Rab5 involved in neurotransmitter receptor endocytosis
opi3Δ	*PEMT*	*Yeast and human*: PL methyltransferase involved in the bisoynthesis of PC
pex2Δ	*PXMP3*	*Yeast and human*: Component of peroxisomal import machinery required for peroxisome biogenesis
pex8Δ	*None*	*Yeast*: Intraperoxisomal organizer of the peroxisomal import machinery required for peroxisome biogenesis
pox1Δ	*ACOX1*	*Yeast*: Peroxisomal fatty-acyl coenzyme A oxidase involved in the beta-oxidation of fatty acids *Human*: Acyl-coenzyme A oxidase required for peroxisome biogenesis
tlg2Δ	*STX16*	*Yeast*: Syntaxin-like SNARE protein that mediates fusion of endosome-derived vesicles with the trans Golgi network *Human*: SNARE protein involved in vesicle transport within the Golgi
vps52Δ	*VPS52*	*Yeast and human*: Component of the GARP complex required for recycling of proteins from endosomes to the late Golgi
vps24Δ	*CHMP3*	*Yeast and human*: Component of the ESCRT-III complex involved in the multivesicular body sorting pathway
vps28Δ	*VPS28*	*Yeast and human*: Component of the ESCRT-I complex involved in the multivesicular body sorting pathway
vps60Δ	*CHMP5*	*Yeast and human*: Component of the ESCRT-III complex involved in the multivesicular body sorting pathway
yat1Δ	*CRAT*	*Yeast and human*: Carnitine acetyltransferase involved in transport of fatty acids into the mitochondria
Overexpression suppressors (Cooper *et al.*, 2006)		
BRE5	None	*Yeast*: Ubiquitin protease cofactor of Ubp3p
ERV29	SURF4	*Yeast*: Transmembrane protein involved in packaging soluble secretory proteins into ER-derived transport vesicles *Human*: Integral membrane protein of unknown function
UBP3	USP10	*Yeast*: Ubiquitin protease that interacts with Bre5p to co-regulate anterograde and retrograde transport between the ER and cis-Golgi compartments *Human*: Ubiquitin protease involved in the degradation of the androgen receptor
YKT6	YKT6	*Yeast*: SNARE protein involved in ER-to-Golgi and retrograde transport to *cis*-Golgi compartments, post-Golgi and endocytic traffic to the vacuole, and homotypic vacuole fusion *Human*: Neuronal-specific SNARE protein of unknown function
YPT1	RAB1A	*Yeast and human*: Rab GTPase involved in the ER-to-Golgi transport and endocytosis

(Continued)

▆ **TABLE 33.1** **(Continued)**

Yeast[a]	Human[b]	Biological function[c,d]
Overexpression enhancers (Cooper *et al.*, 2006)		
GYP8	*TBC20*	*Yeast*: GTPase-activating protein (GAP) of Ypt/Rab GTPases involved in exocytosis (Sec4p), ER-to-Golgi (Ypt1p), or post Golgi (Ypt31p, Ypt32p) transport
		Human: Putative GAP of Rab GTPases
PMR1	*ATP2C1*	*Yeast and human*: High affinity Ca^{2+}/Mn^{2+} P-type ATPase of the Golgi involved in Ca^{2+} dependent protein sorting and processing
SLY41	*SLC35E1*	*Yeast*: Protein involved in the ER-to-Golgi transport
		Human: Nucleotide sugar transporter of the ER and Golgi involved in protein glycosylation

[a]Yeast modifier allele.
[b]Putative human ortholog.
[c]ARF = ADP-ribosylation factor; CHMP = charged multivesicular body protein; COG = conserved oligomeric Golgi; DGPP = diacylglycerol pyrophosphate; ESCRT = endosomal sorting complex required for transport; GAP = GTPase-activating protein; GARP = Golgi-associated retrograde protein; PA = phosphatidic acid; PC = phosphatidylcholine; PL = phospholipid; SNARE = soluble N-ethylmaleimide-sensitive factor attachment protein receptor.
[d]Information obtained from the *Saccharomyces* Genome Database (http://www.yeastgenome.org/) and the UCSC genome browser (http://genome.ucsc.edu/).

cellular sensitivity to α-synuclein (Table 33.1), presumably by depleting the substrate of PLD. Similarly, a null allele of the yeast diacylglycerol pyrophosphate (DGPP) phosphatase (*dpp1△*), which independently generates PA from DGPP, has similar effects (Table 33.1). These interactions suggest that α-synuclein toxicity may result from an attenuation of PA signaling and predict that the activity of downstream effectors of this pathway may be downregulated in models of α-synucleinopathy. If this hypothesis is validated, agonizing PA function or upregulating the activity of PA targets might constitute potential strategies for treating α-synucleinopathies. However, these strategies remain speculative at this time.

Crosstalk between α-Synuclein and Cellular Trafficking

Ever since its implication in the pathogenesis of PD and other neurological disorders, considerable effort has been made to elucidate the biological function of α-synuclein. Several findings suggest a role for α-synuclein in synaptic plasticity and neurotransmission. However, the precise contribution of α-synuclein to these processes is controversial. For example, α-synuclein appears to stimulate synaptic vesicle formation and potentiate synaptic transmission in rat hippocampal neurons (Murphy *et al.*, 2000;

Liu *et al.*, 2004). These findings are consistent with the depletion of resting synaptic vesicles and synaptic deficits in α-synuclein-null mice (Cabin *et al.*, 2002). In contrast, an independent phenotypical analysis of α-synuclein-knockout animals endorsed an inhibitory role for α-synuclein in dopamine neurotransmission (Abeliovich *et al.*, 2000). A third investigation failed to detect any measurable alteration in synaptic plasticity or in the ultrastructure of synaptic terminals of α-synuclein-null mice (Chandra *et al.*, 2004). These apparently contradictory analyses, which may simply reflect strain-specific effects of the three mice lines, indicate that α-synuclein is a non-essential protein that may facilitate the fine-tuning of vesicle dynamics at the synapse.

The molecular mechanisms by which α-synuclein modulates synaptic transmission have begun to be elucidated in the last few years. *In vitro* and *in vivo* studies have determined that a particular combination of acidic phospholipids directs α-synuclein to synaptic vesicles (Fortin *et al.*, 2004; Kubo *et al.*, 2005). Upon reaching its destination, α-synuclein regulates synaptic transmission through effects on the recycling or exocytosis of synaptic vesicles. For example, α-synuclein might enhance synaptic strength by attenuating dopamine re-uptake from the synapse through the endocytosis of dopamine transporters (Wersinger and Sidhu, 2003). Since this process is dependent on the integrity of the tubulin cytoskeleton, α-synuclein may be part of an adaptor complex that tethers cargo proteins to the microtubular network (Wersinger and Sidhu, 2005).

An alternative hypothesis supports the idea that α-synuclein modulates endocytic trafficking through the PLD pathway. Based on the observation that α-synuclein-mediated inhibition of PLD activity is abrogated by phosphorylation (Pronin et al., 2000), it has been proposed that neuronal activity may stimulate vesicle recycling and receptor endocytosis by inducing phosphorylation of α-synuclein, and consequently, activation of PLD (Leng et al., 2001; Lotharius and Brundin, 2002). In contrast, more recent studies on α-synuclein-regulated secretion in catecholaminergic cells showed that α-synuclein hampers neurotransmitter release by blocking a step of exocytosis that precedes the fusion of docked vesicles (Larsen et al., 2006). Intriguingly, evidence from transgenic mice suggests that α-synuclein favors exocytosis of neurotransmitters by assisting in the assembly of soluble N-ethylmaleimide-sensitive factor attachment protein receptors (SNAREs) involved in the fusion of synaptic vesicles to the plasma membrane (Chandra et al., 2005). These observations indicate that α-synuclein might be a multifaceted regulator of synaptic processes, orchestrating numerous steps of vesicle recycling exocytosis, and endocytosis.

How does α-synuclein become neurotoxic? Singleton et al. (2003) and Chartier-Harlin et al. (2004) showed that increased expression of wild-type α-synuclein causes early-onset PD. This observation has important clinical and molecular implications. First, it suggests that the clinical onset of sporadic PD may be initiated when the progressive accumulation of misfolded protein exceeds a critical threshold. Second, it validates transgenic cellular and animal models based on the overexpression of α-synuclein. Third, it indicates that α-synuclein toxicity may simply result from an imbalance in the physiological processes regulated by this protein, rather than a simple gain-of-toxic function mechanism. For example, although α-synuclein associates with microtubular proteins to regulate synaptic vesicle transport, its accumulation impairs microtubule-dependent trafficking (Lee et al., 2006). In some cellular models, α-synuclein strengthens synaptic transmission (Liu et al., 2004); in others, it impedes neurotransmitter release (Larsen et al., 2006). These studies indicate that subtle variations in α-synuclein

expression levels may account for differential effects in diverse systems, perhaps due to changes in the stoichiometry of α-synuclein-containing complexes. Indeed, physiological partners of α-synuclein have been detected in Lewy bodies, suggesting that these proteins are sequestered under pathological conditions. Thus, the physiological and pathological effects of α-synuclein may be more closely linked than previously imagined.

What new contributions may yeast models add to the existing knowledge of α-synuclein biology? As outlined above, relevant aspects of the pathophysiology of α-synuclein can be reconstituted by heterologous expression in yeast. These features, when combined with the readily available arsenal of powerful genetic techniques in yeast, have enabled the establishment of yeast-based platforms to identify components of several trafficking pathways that buffer α-synuclein toxicity. The human orthologs of these proteins represent pharmacologic targets to correct or attenuate the pathological imbalance induced by α-synuclein accumulation.

α-Synuclein and Endosomal Sorting

Genetic analyses in yeast suggest that α-synuclein disrupts the endosomal trafficking of transmembrane proteins destined for intravacuolar degradation. This hypothetical mechanism of toxicity is based on the observation that loss-of-function alleles (vps24Δ, vps28Δ and vps60Δ) of genes encoding distinct subunits of endosomal sorting complexes required for transport (ESCRT) exacerbate α-synuclein-induced growth defects in yeast (Table 33.1). These complexes participate in the multivesicular body pathway and are widely conserved in eukaryotes (reviewed by Slagsvold et al., 2006). In humans, ESCRTs may play a role in neurodegeneration, as they transduce apoptotic signals in neuronal cells (Mahul-Mellier et al., 2006). Interestingly, mutations in the charged multivesicular body protein 2B (CHMP2B), a subunit of human ESCRT-III, are linked to autosomal dominant forms of frontotemporal dementia (Skibinski et al., 2005). Further, Almeida et al. (2006) documented defects in multivesicular body sorting in a neuronal model of β-amyloidopathy. Although crosstalk between the endosomal sorting machinery of neurons and α-synuclein has not been established, yeast studies provide evidence that genetically determined endosomal deficits may predispose to α-synucleinopathies. This possibility has implications for the treatment of PD and merits further investigation.

α-Synuclein and Rab-Mediated Membrane Trafficking

Yeast screens have also implicated small Rab GTPases and other components of Rab-dependent pathways in the pathogenicity of α-synuclein. To understand the relevance of this finding, we need to briefly review the mechanisms of Rab function. Rab GTPases constitute a conserved family of proteins that govern membrane traffic in eukaryotic cells by cycling between an inactive GDP-bound state and an active GTP-bound state (reviewed by Grosshans *et al.*, 2006). This nucleotide-dependent switch mechanism is controlled by a collection of guanidine nucleotide exchange factors (GEFs) and GTPase-activating proteins (GAPs). Upon switching to the GTP-bound state, Rabs are recognized by a specific subset of downstream effectors that convert the signal of the activated Rab into essential steps of membrane traffic. Rab effectors include the Golgi-associated retrograde protein (GARP) and the conserved oligomeric Golgi (COG) tethering complexes, and the SNARE fusion complexes (Table 33.1).

In yeast, an early cytotoxic effect of α-synuclein accumulation is an apparent blockage of ER-to-Golgi transport (Cooper *et al.*, 2006). Interestingly, toxicity is enhanced by loss of function or overexpression of members of the GTPase (*arl3△*), GAP (*GYP8*, *msb3△*), tethering (*cog6△*, *vps52△*), and SNARE (*tlg2△*) families that facilitate vesicle transport to, within, or from the Golgi apparatus (Table 33.1). In contrast, α-synuclein toxicity can be reversed by overexpressing proteins that regulate anterograde and retrograde transport between the ER and Golgi, including the Rab GTPase Ypt1p and the SNARE Ykt6p (Table 33.1). These observations indicate that α-synuclein may compromise Golgi-dependent functions in yeast by disrupting multiple rather than single trafficking pathways. Indeed, α-synuclein produces Golgi fragmentation in cultured cells and nigral neurons in patients with PD (Gosavi *et al.*, 2002; Fujita *et al.*, 2006). Although most of the modifiers identified in these studies are highly conserved in eukaryotes, their mammalian counterparts have diversified to acquire cell-type and species-specific functions. Therefore, detailed investigation of the neuron-specific functions of the human orthologs of α-synuclein toxicity modifiers may provide information on potential

mechanisms of toxicity in neurons, as illustrated below by the SNARE and the Rab families.

The role of SNAREs in the nervous system is exemplified by well-characterized members of the family that carry out neuron-specific functions, including synaptic transmission, endosomal recycling of receptors, and neurite outgrowth (reviewed by Wang and Tang, 2006). Therefore, impairing the activity of SNARE proteins may severely damage neuronal functions. In fact, mutations in SNARE proteins have been implicated in neurological disorders, such as schizophrenia and the CEDNIK syndrome (Saito *et al.*, 2001; Sprecher *et al.*, 2005). In yeast, deletion of the Golgi-specific SNARE gene *TLG2* (*tlg2△*) enhances α-synuclein toxicity, whereas overexpression of the more promiscuous *YKT6* reverses this effect, suggesting that α-synuclein triggers toxicity by reducing the availability of these proteins. In humans, syntaxin 16 and Ykt6, the putative orthologs of these proteins, are moderately and highly enriched in brain tissue, respectively (Tang *et al.*, 1998; Hasegawa *et al.*, 2004). Although the neuron-specific functions of these proteins are unclear, it is tempting to speculate that impairment of normal SNARE functions by α-synuclein may contribute in some manner to neuronal demise in α-synucleinopathies. Interestingly, one of the proposed physiological functions of α-synuclein is to ensure the proper assembly of synaptic SNARE proteins (Chandra *et al.*, 2005). Thus, compounds that stabilize or promote the formation of SNARE complexes may have potential beneficial effects in the treatment of PD and other α-synucleinopathies. The mode of action of certain antipsychotic drugs, such as haloperidol and chlorpromazine, which increase the levels of the presynaptic SNARE SNAP-25 in the hippocampus (Barr *et al.*, 2006), or antidepressant drugs, such as fluoxetine, reboxetine, and desipramine, which reduce SNARE complex formation (Bonanno *et al.*, 2005) illustrates the potential feasibility of intervening at this level for treating neurological disorders.

Another example of the potential implications of yeast studies for human neurological disorders is illustrated by the increase in α-synuclein toxicity upon elimination of the GAP Msb3, which is involved in several transport pathways. The closest human ortholog of Msb3, USP6NL, is a GAP of Rab5, which is involved in neurotransmitter receptor endocytosis (Haas *et al.*, 2005). Alterations in the Rab5 guanidine exchange cycle have been linked to juvenile-onset forms of neurodegenerative diseases, including amyotrophic lateral sclerosis, primary lateral sclerosis, and spastic paraplegia (reviewed by Carney *et al.*, 2006). Consistent with these observations, α-synuclein interacts with Rab GTPases that are involved in docking and fusion of synaptic vesicles. These interactions are enhanced in patients with Lewy body disease, suggesting Rab sequestration and defective Rab signaling under pathological conditions (Dalfo *et al.*, 2004). Thus, an impairment in neuronal trafficking by synuclein, either in the cell body or at the synapse, may contribute to distinct neuropathies, including PD. Remarkably, Rab1, the mammalian ortholog of yeast Ypt1p, protects against loss of dopaminergic neurons in animal models of α-synucleinopathy (Cooper *et al.*, 2006), providing an experimental validation of the yeast studies and a potential pharmacologic target for early therapeutic intervention.

CONCLUSIONS

Over the past few decades, significant interest in using yeast as a tool to dissect fundamental cellular processes resulted in the generation of a set of remarkably valuable tools for genetic manipulation on a genome-wide scale that confer to yeast multiple advantages that are unparalleled in other model organisms. First, inducible systems, techniques to modulate gene dosage, and well-established methods of biochemical characterization can be used to examine early molecular events triggered by overexpression of the pathogenic protein. Detailed examination of these events may assist in discriminating between causal events, coping responses, and unlinked epiphenomena. Second, the availability of commercial collections of mutants and expression systems, the short life cycle of yeast cells, and the affordability of these techniques allow genome-wide genetic screens to be conducted more rapidly and at lower cost in yeast than in other eukaryotic systems. Third, the ability to systematically manipulate individual genes in a homogeneous genetic background allows one to assess the contribution of single genes or alleles to a complex phenotype. Therefore, genetic screens in yeast can identify putative susceptibility factors of genetically complex disorders that are otherwise difficult to discern by classical linkage analyses in heterogeneous human populations. Fourth, the same capability allows the relationships between multiple genetic factors to be dissected by epistatic analyses. Finally, yeast models can provide a drug discovery platform for the identification of lead compounds that counteract the toxicity of α-synuclein in living cells.

Like all experimental paradigms, however, yeast models have several notable limitations for the study of human disease and should be thought of simply as a toolbox or "living test tube" for future analyses in more physiologically relevant systems. The yeast genome lacks gene families expanded in higher eukaryotes specialized in nervous system structure, function, and development. Therefore, genetic studies in yeast cannot detect relevant disease-processes modulated by genes that are exclusively expressed in neurons. Another limitation is the absence of multicellular organization. As a result, cellular dysfunction resulting from the decline of neuronal networks cannot be modeled in yeast, and genetic and molecular abnormalities cannot be correlated with electrophysiological or behavioral phenotypes.

Despite these limitations, yeast are well suited for modeling cell autonomous effects of α-synucleinopathies and identifying genes and pathways that modulate the detrimental effects of elevated α-synuclein levels, providing insights into the molecular mechanisms of α-synuclein toxicity. Remarkably, many of these genes are conserved in humans. Such orthologs may contain SNPs that increase or decrease susceptibility to PD, and may also be pharmacologic targets for disease-modifying therapies. In some cases, the relevance of these modifier genes to the neuropathology of PD has been validated in cellular and animal models. For example, the sensitization of yeast cells to α-synuclein upon removal of glutathione S-transferases (Willingham *et al.*, 2003) has been validated by genetic approaches in two independent *Drosophila* models of PD (Whitworth *et al.*, 2005; Trinh *et al.*, 2008). These studies have shown that glutathione S-transferase activity prevents the loss of dopaminergic neurons in flies and suggest a neuroprotective role for drugs that stimulate glutathione metabolism. Similarly, the neuroprotective role of Rab1 has been validated in neuronal and animal models of α-synucleinopathy (Cooper *et al.*, 2006).

In summary, yeast studies may be used to infer the effects of α-synuclein accumulation at the synapse. Genetic interactions in yeast are consistent with existing models of mitochondrial defects and lysosomal pathology contributing to neuronal dysfunction in PD. Further, studies in yeast have implicated specific cellular functions that might be defective in degenerating neurons, including peroxisomal-dependent functions, PA signaling, endosomal sorting, SNARE-mediated fusion, and Rab-controlled vesicle trafficking. This multiplicity of potential sources of cellular damage, even in an organism as simple as yeast, may serve as a sobering corollary of the intricate etiology of PD. On a more promising note, it discloses several levels for prospective therapeutic intervention. Although further analyses are needed to unravel the interactions betweens these pathways and to validate their significance in higher eukaryotic organisms, these studies highlight the necessity of exploring therapeutic strategies aimed at restoring neuronal homeostasis at different levels, including peroxisomal and lysosomal function, phospholipid balance, and synaptic trafficking, for the treatment of α-synucleinopathies.

REFERENCES

Abeliovich, A., Schmitz, Y., Farinas, I., Choi-Lundberg, D., Ho, W. H., Castillo, P. E., Shinsky, N., Verdugo, J. M., Armanini, M., Ryan, A., Hynes, M., Phillips, H., Sulzer, D., and Rosenthal, A. (2000). Mice lacking α-synuclein display functional deficits in the nigrostriatal dopamine system. *Neuron* 25, 239–252.

Ahn, B. H., Rhim, H., Kim, S. Y., Sung, Y. M., Lee, M. Y., Choi, J. Y., Wolozin, B., Chang, J. S., Lee, Y. H., Kwon, T. K., Chung, K. C., Yoon, S. H., Hahn, S. J., Kim, M. S., Jo, Y. H., and Min, D. S. (2002). α-Synuclein interacts with phospholipase D isozymes and inhibits pervanadate-induced phospholipase D activation in human embryonic kidney-293 cells. *J Biol Chem* 277, 12334–12342.

Almeida, C. G., Takahashi, R. H., and Gouras, G. K. (2006). β-Amyloid accumulation impairs multivesicular body sorting by inhibiting the ubiquitin–proteasome system. *J Neurosci* 26, 4277–4288.

Baetz, K., McHardy, L., Gable, K., Tarling, T., Rebierioux, D., Bryan, J., Anderson, R. J., Dunn, T., Hieter, P., and Roberge, M. (2004). Yeast genome-wide drug-induced haploinsufficiency screen to determine drug mode of action. *Proc Natl Acad Sci USA* 101, 4525–4530.

Bahr, B. A., and Bendiske, J. (2002). The neuropathogenic contributions of lysosomal dysfunction. *J Neurochem* 83, 481–489.

Barcelo-Coblijn, G., Golovko, M. Y., Weinhofer, I., Berger, J., and Murphy, E. J. (2007). Brain neutral lipids mass is increased in α-synuclein gene-ablated mice. *J Neurochem* 101, 132–141.

Barr, A. M., Young, C. E., Phillips, A. G., and Honer, W. G. (2006). Selective effects of typical antipsychotic drugs on SNAP-25 and synaptophysin in the hippocampal trisynaptic pathway. *Int J Neuropsychopharmacol* 9, 457–463.

Barrientos, A. (2003). Yeast models of human mitochondrial diseases. *IUBMB Life* 55, 83–95.

Bonanno, G., Giambelli, R., Raiteri, L., Tiraboschi, E., Zappettini, S., Musazzi, L., Raiteri, M., Racagni, G., and Popoli, M. (2005). Chronic antidepressants reduce depolarization-evoked glutamate release and protein interactions favoring formation of SNARE complex in hippocampus. *J Neurosci* 25, 3270–3279.

Botstein, D., Chervitz, S. A., and Cherry, J. M. (1997). Yeast as a model organism. *Science* 277, 1259–1260.

Butcher, R. A., Bhullar, B. S., Perlstein, E. O., Marsischky, G., LaBaer, J., and Schreiber, S. L. (2005). Micro-array based method for monitoring yeast overexpression strains reveals small-molecule targets in TOR pathway. *Nature Chem Biol* 2, 103–109.

Cabin, D. E., Shimazu, K., Murphy, D., Cole, N. B., Gottschalk, W., McIlwain, K. L., Orrison, B., Chen, A., Ellis, C. E., Paylor, R., Lu, B., and Nussbaum, R. L. (2002). Synaptic vesicle depletion correlates with attenuated synaptic responses to prolonged repetitive stimulation in mice lacking α-synuclein. *J Neurosci* 22, 8797–8807.

Carney, D. S., Davies, B. A., and Horazdovsky, B. F. (2006). Vps9 domain-containing proteins: Activators of Rab5 GTPases from yeast to neurons. *Trends Cell Biol* 16, 27–35.

Castagnet, P. I., Golovko, M. Y., Barcelo-Coblijn, G. C., Nussbaum, R. L., and Murphy, E. J. (2005). Fatty acid incorporation is decreased in astrocytes cultured from α-synuclein gene-ablated mice. *J Neurochem* 94, 839–849.

Chandra, S., Chen, X., Rizo, J., Jahn, R., and Sudhof, T. C. (2002). A broken alpha-helix in folded α-synuclein. *J Biol Chem* 278, 15313–15318.

Chandra, S., Fornai, F., Kwon, H. B., Yazdani, U., Atasoy, D., Liu, X., Hammer, R. E., Battaglia, G., German, D. C., Castillo, P. E., and Südhof, T. C. (2004). Double-knockout mice for α- and β-synucleins: Effect on synaptic functions. *Proc Natl Acad Sci USA* 101, 14966–14971.

Chandra, S., Galardo, G., Fernández.Chacón, R., Schlüter, O. M., and Südhof, T. C. (2005). α-Synuclein cooperates with CSPα in preventing neurodegeneration. *Cell* 123, 383–396.

Chartier-Harlin, M. C., Kachergus, J., Roumier, C., Mouroux, V., Douay, X., Lincoln, S., Levecque, C., Larvor, L., Andrieux, J., Hulihan, M., Waucquier, N., Defebvre, L., Amouyel, P., Farrer, M., and Destee, A. (2004). α-Synuclein locus duplication as a cause of familial Parkinson's disease. *Lancet* 364, 1167–1169.

Chen, Q., Ding, Q., and Keller, J. N. (2005a). The stationary phase model of aging in yeast for the study of oxidative stress and age-related neurodegeneration. *Biogerontoloy* **6**, 1–13.

Chen, Q., Thorpe, J., and Keller, J. N. (2005b). α-Synuclein alters proteasome function, protein synthesis, and stationary phase viability. *J Biol Chem* **280**, 30009–30017.

Cole, N. B., Murphy, D. D., Grider, T., Rueter, S., Brasaemle, D., and Nussbaum, R. L. (2002). Lipid droplet binding and oligomerization properties of the Parkinson's disease protein α-synuclein. *J Biol Chem* **277**, 6344–6352.

Cooper, A. A., Gitler, A. D., Cashikar, A., Haynes, C. M., Hill, K. J., Bhullar, B., Liu, K., Xu, K., Strathearn, K. E., Liu, F., Cao, S., Caldwell, K. A., Caldwell, G. A., Marsischky, G., Kolodner, R. D., Labaer, J., Rochet, J. C., Bonini, N. M., and Lindquist, S. (2006). α-Synuclein blocks ER-Golgi traffic and Rab1 rescues neuron loss in Parkinson's models. *Science* **313**, 324–328.

Dalfo, E., Borrachina, M., Rosa, J. L., Ambrosio, S., and Ferrer, I. (2004). Abnormal α-synuclein interactions with rab3a and rabphilin in diffuse Lewy body disease. *Neurobiol Dis* **16**, 92–97.

Davidson, W. S., Jonas, A., Clayton, D. F., and George, J. M. (1998). Stabilization of α-synuclein secondary structure upon binding to synthetic membranes. *J Biol Chem* **273**, 9443–9449.

Dixon, C., Mathias, N., Zweig, R. M., Davis, D. A., and Gross, D. S. (2005). α-Synuclein targets the plasma membrane via the secretory pathway and induces toxicity in yeast. *Genetics* **170**, 47–59.

Ellis, C. E., Murphy, E. J., Mitchell, D. C., Golovko, M. Y., Scaglia, F., Barcelo-Coblijn, G. C., and Nussbaum, R. L. (2005). Mitochondrial lipid abnormality and electron transport chain impairment in mice lacking α-synuclein. *Mol Cell Biol* **25**, 10190–10201.

Faust, P. L., Su, H. M., Moser, A., and Moser, H. W. (2001). The peroxisome deficient PEX2 Zellweger mouse: Pathologic and biochemical correlates of lipid dysfunction. *J Mol Neurosci* **16**, 289–297.

Faust, P. L., Banka, D., Siriratsivawong, R., Ng, V. G., and Wikander, T. M. (2005). Peroxisome biogenesis disorders: The role of peroxisomes and metabolic dysfunction in developing brain. *J Inherit Metab Dis* **28**, 369–383.

Flower, T. R., Chesnokova, L. S., Froelich, C. A., Dixon, C., and Witt, S. N. (2005). Heat shock prevents α-synuclein-induced apoptosis in a yeast model of Parkinson's disease. *J Mol Biol* **351**, 1081–1100.

Fortin, D. L., Troyer, M. D., Nakamura, K., Kubo, S., Anthony, M. D., and Edwards, R. H. (2004). Lipid rafts mediate the synaptic localization of α-synuclein. *J Neurosci* **24**, 6715–6723.

Frohlich, K. U., Fussi, H., and Ruckenstuhl, (2007). Yeast apoptosis – from genes to pathways. *Semin Cancer Biol* **17**, 112–121.

Friedmann, Y., Shriki, A., Bennett, E. R., Golos, S., Diskin, R., Marbach, I., Bengal, E., and Engelberg, D. (2006). JX401, a p38α inhibitor containing a 4-benzylpiperidine motif, identified via a novel screening system in yeast. *Mol Pharmacol* **70**, 1395–1405.

Fujita, Y., Ohama, E., Takatama, M., Al-Sarraj, S., and Okamoto, K. (2006). Fragmentation of Golgi apparatus of nigral neurons with α-synuclein-positive inclusions in patients with Parkinson's disease. *Acta Neuropathol* **112**, 261–265.

Gai, W. P., Yuan, H. X., Li, X. Q., Power, J. T., Blumbergs, P. C., and Jensen, P. H. (2000). *In situ* and *in vitro* study of colocalization and segregation of α-synuclein, ubiquitin, and lipids in Lewy bodies. *Exp Neurol* **166**, 324–333.

Giaever, G., Shoemaker, D. D., Jones, T. W., Liang, H., Winzeler, E. A., Astromoff, A., and Davis, R. W. (1999). Genomic profiling of drug sensitivities via induced haploinsufficiency. *Nat Genet* **21**, 278–283.

Giaever, G., Chu, A. M., Ni, L., Connelly, C., Riles, L., Veronneau, S., Dow, S., Lucau-Danila, A., Anderson, K., Andre, B., Arkin, A. P., Astromoff, A., El-Bakkoury, M., Bangham, R., Benito, R., Brachat, S., Campanaro, S., Curtiss, M., Davis, K., Deutschbauer, A., Entian, K. D., Flaherty, P., Foury, F., Garfinkel, D. J., Gerstein, M., Gotte, D., Guldener, U., Hegemann, J. H., Hempel, S., Herman, Z., Jaramillo, D. F., Kelly, D. E., Kelly, S. L., Kotter, P., LaBonte, D., Lamb, D. C., Lan, N., Liang, H., Liao, H., Liu, L., Luo, C., Lussier, M., Mao, R., Menard, P., Ooi, S. L., Revuelta, J. L., Roberts, C. J., Rose, M., Ross-Macdonald, P., Scherens, B., Schimmack, G., Shafer, B., Shoemaker, D. D., Sookhai-Mahadeo, S., Storms, R. K., Strathern, J. N., Valle, G., Voet, M., Volckaert, G., Wang, C. Y., Ward, T. R., Wilhelmy, J., Winzeler, E. A., Yang, Y., Yen, G., Youngman, E., Yu, K., Bussey, H., Boeke, J. D., Snyder, M., Philippsen, P., Davis, R. W., and Johnston, M. (2002). Functional profiling of the *Saccharomyces cerevisiae* genome. *Nature* **418**, 387–391.

Gitler, A. D., Bevis, B. J., Shorter, J., Strathearn, K. E., Hamamichi, S., Su, L. J., Caldwell, K. A., Caldwell, G. A., Rochet, J. C., McCaffery, J. M., Barlowe, C., and Lindquist, S. (2008). The Parkinson's disease protein α-synuclein disrupts cellular Rab homeostasis. *Proc. Natl. Acad. Sci. USA* **105**, 145–150.

Golovko, M. Y., Faergeman, N. J., Cole, N. B., Castagnet, P. I., Nussbaum, R. L., and Murphy, E. J. (2005). α-Synuclein gene deletion decreases brain palmitate uptake and alters the palmitate metabolism in the absence of α-synuclein palmitate binding. *Biochemistry* **44**, 8251–8259.

Golovko, M. Y., Rosenberger, T. A., Faergeman, N. J., Feddersen, S., Cole, N. B., Pribill, I., Berger, J., Nussbaum, R. L., and Murphy, E. J. (2006). Acyl-CoA synthetase activity links wild-type but not mutant α-synuclein to brain arachidonate metabolism. *Biochemistry* **45**, 6956–6966.

Gosavi, N., Lee, H. J., Lee, J. S., Patel, S., and Lee, S. J. (2002). Golgi fragmentation occurs in the cells with prefibrillar α-synuclein aggregates and precedes the formation of fibrillar inclusion. *J Biol Chem* **277**, 48984–48992.

Griffioen, G., Duhamel, H., Van Damme, N., Pellens, K., Zabrocki, P., Pannecouque, C., van Leuven, F., Winderickx, J., and Wera, S. (2006). A yeast-based model of α-synucleinopathy identifies compounds with therapeutic potential. *Biochim. Biophys. Acta* **1762**, 312–318.

Grosshans, B. L., Ortiz, D., and Novick, P. (2006). Rabs and their effectors: Achieving specificity in membrane traffic. *Proc Natl Acad Sci USA* **103**, 11821–11827.

Haas, A. K., Fuchs, E., Kopajtich, R., and Barr, F. A. (2005). A GTPase-activating protein controls Rab5 function in endocytic trafficking. *Nat Cell Biol* **7**, 887–893.

Hasegawa, H., Yang, Z., Oltedal, L., Davanger, S., and Hay, J. C. (2004). Intramolecular protein-protein and protein-lipid interactions control the conformation and subcellular targeting of neuronal Ykt6. *J Cell Sci* **117**, 4495–4508.

Hsu, L. J., Sagara, Y., Arroyo, A., Rockenstain, E., Sisk, A., Mallory, M., Wong, J., Takenouchi, T., Hashimoto, M., and Masliah, E. (2000). α-Synuclein promotes mitochondrial deficit and oxidative stress. *Ann J Pathol* **157**, 401–410.

Hughes, T. R., Marton, M. J., Jones, A. R., Roberts, C. J., Stoughton, R., Armour, C. D., Bennett, H. A., Coffrey, E., Dai, H., and He, Y. D. (2000). Functional discovery via a compendium of expression profiles. *Cell* **102**, 109–126.

Jenco, J. M., Rawlingson, A., Daniels, B., and Morris, A. J. (1998). Regulation of phospholipase D2: Selective inhibition of mammalian phospholipase D isoenzymes by α- and β-synucleins. *Biochemistry* **37**, 4901–4909.

Kruger, R., Kuhn, W., Muller, T., Woitalla, D., Graeber, M., Kosel, S., Przuntek, H., Epplen, J. T., Schols, L., and Riess, O. (1998). Ala30Pro mutation in the gene encoding α-synuclein in Parkinson's disease. *Nat Genet* **18**, 106–108.

Kubo, S., Nemani, V. M., Chalkley, R. J., Anthony, M. D., Hattori, N., Mizuno, Y., Edwards, R. H., and Fortin, D. L. (2005). A combinatorial code for the interaction of alpha-synuclein with membranes. *J Biol Chem* **280**, 31664–31672.

Larsen, K. E., Schmitz, Y., Troyer, M. D., Mosharov, E., Dietrich, P., Quazi, A. Z., Savalle, M., Nemani, V.,

Chaudhry, F. A., Edwards, R. H., Stefanis, L., and Sulzer, D. (2006). α-Synuclein overexpression in PC12 and chromaffin cells impairs catecholamine release by interfering with a late step in exocytosis. *J Neurosci* **26**, 11915–11922.

Lashuel, H. A., Hartley, D., Petre, B. M., Walz, T., and Lansbury, P. T. (2002). Neurodegenerative disease: Amyloid pores from pathogenic mutations. *Nature* **418**, 291.

Launhardt, H., Hinnen, A., and Munden, T. (1998). Drug-induced phenotypes provide a tool for the functional analysis of yeast genes. *Yeast* **14**, 935–942.

Lee, H. J., Khoshaghideh, F., Patel, S., and Lee, S. J. (2004). Clearance of α-synuclein oligomeric intermediates via the lysosomal degradation pathway. *J Neurosci* **24**, 1888–1896.

Lee, H. J., Khoshaghideh, F., Lee, S., and Lee, S. J. (2006). Impairment of microtubule-dependent trafficking by overexpression of α-synuclein. *Eur J Neurosci* **24**, 3153–3162.

Leng, Y., Chase, T. N., and Bennett, M. C. (2001). Muscarinic receptor stimulation induces translocation of an α-synuclein oligomer from plasma membrane to a light vesicle fraction in cytoplasm. *J Biol Chem* **276**, 28212–28218.

Liu, S., Ninan, I., Antonova, I., Battaglia, F., Trinchese, F., Narasanna, A., Kolodilov, N., Dauer, W., Hawkins, R. D., and Arancio, O. (2004). α-Synuclein produces a long-lasting increase in neurotransmitter release. *EMBO J* **23**, 4506–4516.

Lotharius, J., and Brundin, P. (2002). Impaired dopamine storage resulting from α-synuclein mutations may contribute to the pathogenesis of Parkinson's disease. *Hum Mol Genet* **11**, 2395–2407.

Luesch, H., Wu, T. Y., Ren, P., Gray, N. S., Schultz, P. G., and Supek, F. (2005). A genome-wide overexpression screen in yeast for small molecule target identification. *Chem Biol* **12**, 55–63.

Lum, P. Y., Armour, C. D., Stepaniants, S. B., Cavet, G., Wolf, M. K., Butler, J. S., Hinshaw, J. C., Garnier, P., Prestwich, G. D., Leonardson, A., Garett-Engele, P., Rush, C. M., Bard, M., Schimmack, G., Phillips, J. W., Roberts, C. J., and Shoemaker, D. D. (2004). Discovering modes of action for therapeutic compounds using a genome-wide screen of yeast heterozygotes. *Cell* **116**, 121–137.

Mahul-Mellier, A. L., Hemming, F. J., Blot, B., Fraboulet, S., and Sadoul, R. (2006). Alix, making a link between apoptosis-linked gene-2, the endosomal sorting complexes required for transport, and neuronal death *in vivo*. *J Neurosci* **26**, 542–549.

McMurray, M., and Gottschling, D. E. (2003). An age-induced switch to a hyper-combinational state. *Science* **301**, 1859–1860.

Mezey, E., Dehejia, A. M., Harta, G., Tresser, N., Suchy, S. F., Nussbaum, R. L., Brownstein, M. J., and Polymeropoulos, M. H. (1998). α-Synuclein is present in Lewy bodies in sporadic Parkinson's disease. *Mol Psychiatry* 3, 493–499.

Murphy, D. D., Rueter, S. M., Trojanowski, J. Q., and Lee, V. M. (2000). Synucleins are developmentally expressed, and, α-synuclein regulates the size of the presynaptic vesicular pool in primary hippocampal neurons. *J Neurosci* 20, 3214–3220.

Outeiro, T. F., and Lindquist, S. (2003). Yeast cells provide insight into α-synuclein biology and pathobiology. *Science* 302, 1772–1775.

Petrucelli, L., O'Farrell, C., Lockhart, P. J., Baptista, M., Kehoe, K., Vink, L., Choi, P., Wolozin, B., Farrer, M., Hardy, J., and Cookson, M. R. (2002). Parkin protects against the toxicity associated with mutant α-synuclein. Proteasome dysfunction selectively affects catecholaminergic neurons. *Neuron* 36, 1007–1019.

Polymeropoulos, M. H., Lavedan, C., Leroy, E., Ide, S. E., Dehejia, A., Dutra, A., Pike, B., Root, H., Rubenstein, J., Boyer, R., Stenroos, E. S., Chandrasekharappa, S., Athanassiadou, A., Papapetropoulos, T., Johnson, W. G., Lazzarini, A. M., Duvoisin, R. C., Di Iorio, G., Golbe, L. I., and Nussbaum, R. L. (1997). Mutation in the α-synuclein gene identified in families with Parkinson's disease. *Science* 276, 2045–2047.

Pronin, A. N., Morris, A. J., Surguchov, A., and Benovic, J. L. (2000). Synucleins are a novel class of substrates for G protein-coupled receptor kinases. *J Biol Chem* 275, 26515–26522.

Przedborski, S., and Ischiropoulos., H. (2005). Reactive oxygen and nitrogen species: Weapons of neuronal destruction in models of Parkinson's disease. *Antioxid Redox Signal* 7, 685–693.

Richard, G. F., and Dujon, B. (1996). Distribution and variability of trinucleotide repeats in the genome of the yeast *Saccharomyces cerevisiae*. *Gene* 174, 165–174.

Rine, J., Hansen, W., Hardeman, E., and Davis, R. W. (1983). Targeted selection of recombinant clones through gene dosage effects. *Proc Natl Acad Sci USA* 80, 6750–6754.

Rockenstein, E., Schwach, G., Ingolic, E., Adame, A., Crews, L., Mante, M., Pfragner, R., Schreiner, E., Windisch, M., and Masliah, E. (2005). Lysosomal pathology associated with α-synuclein accumulation in transgenic models using an eGFP fusion protein. *J Neurosci Res* 80, 247–259.

Saito, T., Guan, F., Papolos, D. F., Rajouria, N., Fann, C. S., and Lachman, H. M. (2001). Polymorphism in SNAP29 gene promoter region associated with schizophrenia. *Mol Psychiatry* 6, 193–201.

Sancenon, V., Lee, S., Griffith, J., Outeiro, T. F., Masliah, E., Reggiori, F., Muchowski, P. J. (2008).

Phosphorylation by casein kinase I at Ser-129 attenuates endocytosis defects caused by α-synuclein. Submitted Manuscript.

Sarkar, S., Perlstein, E. O., Imarisio, S., Pineau, S., Cordenier, A., Maglathlin, R. L., Webster, J. A., Lewis, T. A., O'kane, C. J., Schreiber, S. L., and Rubinsztein, D. C. (2007). Small molecules enhance autophagy and reduce toxicity in Huntington's disease models. *Nat Chem Biol* 3, 331–338.

Sharma, N., Brandis, K. A., Herrera, S. K., Johnson, B. E., Vaidya, T., Shrestha, R., and Debburman, S. K. (2006). α-Synuclein budding yeast model: Toxicity enhanced by impaired proteasome and oxidative stress. *J Mol Neurosci* 28, 161–178.

Sharon, R., Goldberg, M. S., Bar-Joseph, I., Betensky, R. A., Shen, J., and Selkoe, D. J. (2001). α-Synuclein occurs in lipid-rich high molecular weight complexes, binds fatty acids, and shows homology to the fatty acid-binding proteins. *Proc Natl Acad Sci USA* 98, 9110–9115.

Sharon, R., Bar-Joseph, I., Frosch, M. P., Walsh, D. M., Hamilton, J. A., and Selkoe, D. J. (2003a). The formation of highly soluble oligomers of α-synuclein is regulated by fatty acids and enhanced in Parkinson's disease. *Neuron* 37, 583–595.

Sharon, R., Bar-Joseph, I., Mirick, G. E., Serhan, C. N., and Selkoe, D. J. (2003b). Altered fatty acid composition of dopaminergic neurons expressing α-synuclein and human brains with α-synucleinopathies. *J Biol Chem* 278, 49874–49881.

Sherman, M. Y., and Muchowski, P. J. (2003). Making yeast tremble: Yeast models as tools to study neurodegenerative disorders. *Neuromolecular Med* 4, 133–146.

Singleton, A. B., Farrer, M., Johnson, J., Singleton, A., Hague, S., Kachergus, J., Hulihan, M., Peuralinna, T., Dutra, A., Nussbaum, R., Lincoln, S., Crawley, A., Hanson, M., Maraganore, D., Adler, C., Cookson, M. R., Muenter, M., Baptista, M., Miller, D., Blancato, J., Hardy, J., and Gwinn-Hardy, K. (2003). α-Synuclein locus triplication causes Parkinson's disease. *Science* 302, 841.

Skibinski, G., Parkinson, N. J., Brown, J. M., Chakrabarti, L., Lloyd, S. L., Hummerich, H., Nielsen, J. E., Hodges, J. R., Spillantini, M. G., Thusgaard, T., Brandner, S., Brun, A., Rossor, M. N., Gade, A., Johannsen, P., Sorensen, S. A., Gydesen, S., Fisher, E. M., and Collinge, J. (2005). Mutations in the endosomal ESCRTIII-complex subunit CHMP2B in frontotemporal dementia. *Nat Genet* 37, 806–808.

Slagsvold, T., Pattni, K., Malerod, L., and Stenmark, H. (2006). Endosomal and non-endosomal functions of ESCRT proteins. *Trends Cell Biol* 16, 317–326.

Soper, J. H., Roy, S., Stieber, A., Lee, E., Wilson, R. B., Trojanowski, J. Q., Burd, C. G., and Lee, V. M.

(2008). α-Synuclein-induced aggregation of cytoplasmic vesicles in *Saccharomyces cerevisiae*. *Mol. Biol. Cell* 19, 1093–1103.

Sprecher, E., Ishida-Yamamoto, A., Mizrahi-Koren, M., Rapaport, D., Goldsher, D., Indelman, M., Topaz, O., Chefetz, I., Keren, H., O'brien, T. J., Bercovich, D., Shalev, S., Geiger, D., Bergman, R., Horowitz, M., and Mandel, H. (2005). A mutation in SNAP29, coding for a SNARE protein involved in intracellular trafficking, causes a novel neurocutaneous syndrome characterized by cerebral dysgenesis, neuropathy, ichthyosis, and palmoplantar keratoderma. *Am J Hum Genet* 77, 242–251.

Tang, B. L., Low, D. Y., Lee, S. S., Tan, A. E., and Hong, W. (1998). Molecular cloning and localization of human syntaxin 16, a member of the syntaxin family of SNARE proteins. *Biochem Biophys Res Commun* 242, 673–679.

Trinh, K., Moore, K., Wes, P., Muchowski, P. J., Dey, J., Andrews, L., and Pallanck, L. (2008). Induction of the phase II detoxification pathway suppresses neuron loss in Drosophila models of Parkinson's disease. *J Neurosci* 28, 465–472.

Vahlensieck, H. F., Pridzun, L., Reichenbach, H., and Hinnen, A. (1994). Identification of the yeast *ACC1* gene product (acetyl-coA-carboxylase) as the target of the polyketide fungicide soraphen A. *Curr Genet* 25, 95–100.

Wang, Y., and Tang, B. L. (2006). SNAREs in neurons-beyond synaptic vesicle exocytosis. *Mol Membr Biol* 23, 377–384.

Wang, X., Devaiah, S. P., Zhang, W., and Welti, R. (2006). Signaling functions of phosphatidic acid. *Prog Lipid Res* 45, 250–278.

Wersinger, C., and Sidhu, A. (2005). Disruption of the interaction of α-synuclein with microtubules enhances cell surface recruitment of the dopamine transporter. *Biochemistry* 44, 11362–13624.

Wersinger, C., and Sidhu, A. (2003). Attenuation of dopamine transporter activity by α-synuclein. *Neurosci Lett* 340, 189–192.

Whitworth, A. J., Theodore, D. A., Greene, J. C., Benes, H., Wes, P. D., and Pallanck, L. J. (2005). Increased glutathione S-transferase activity rescues dopaminergic neuron loss in a Drosophila model of Parkinson's disease. *Proc Natl Acad Sci USA* 102, 8024–8029.

Wickner, R. B. (1994). [URE3] as an altered URE2 protein: Evidence for a prion analog in *Saccharomyces cerevisiae*. *Science* 264, 566–569.

Willingham, S., Outeiro, T. F., DeVit, M. J., Lindquist, S. L., and Muchowski, P. J. (2003). Yeast genes that enhance the toxicity of a mutant huntingtin fragment or α-synuclein. *Science* 302, 1769–1772.

Zabrocki, P., Pellens, K., Vanhelmont, T., Vandebroek, T., Griffioen, G., Wera, S., Van Leuven, F., and Winderickx, J. (2005). Characterization of α-synuclein aggregation and synergistic toxicity with protein tau in yeast. *FEBS J* 272, 1386–1400.

Zarranz, J. J., Alegre, J., Gomez-Esteban, J. C., Lezcano, E., Ros, R., Ampuero, I., Vidal, L., Hoenicka, J., Rodríguez, O., Atares, B., Llorens, V., Gomez Tortosa, E., Munoz, D. G., and de Yebenes, J. G. (2004). The new mutation, E46K, of α-synuclein causes Parkinson and Lewy body dementia. *Ann Neurol* 55, 164–173.

Zhang, X., Smith, D. L., Meriin, A. B., Engemann, S., Russel, D. E., Roark, M., Washington, S. L., Maxwell, M. M., Marsh, J. L., Thompson, L. M., Wanker, E. E., Young, A. B., Housman, D. E., Bates, G. P., Sherman, M. Y., and Kazantsev, A. G. (2005). A potent small molecule inhibits polyglutamine aggregation in Huntington's disease neurons and suppresses neurodegeneration *in vivo*. *Proc Natl Acad Sci USA* 102, 892–897.

34

THE ROLE OF THE FOXA2 GENE IN THE BIRTH AND DEATH OF DOPAMINE NEURONS

RAJA KITTAPPA, WENDY CHANG AND RONALD McKAY

Laboratory of Molecular Biology, National Institutes of Neurological Disorders, Stroke, Bethesda, MD, USA

INTRODUCTION

The transplantation of fetal midbrain tissue may relieve many of the symptoms of Parkinson's disease (PD), although this remains controversial (Laguna Goya *et al.*, 2007). Unfortunately, human fetal tissue is in short supply making it difficult to control the quality of the donor cells and develop transplantation technologies. Embryonic stem (ES) cells are a promising alternative to primary human fetal tissue and dopamine neurons derived from ES cells are one of the best examples of stem cell-based cell replacement therapies. The transplantation of undifferentiated mouse and human ES cells provides some limited relief for motor problems in an animal model of PD but also causes a high frequency of lethal teratomas (Bjorklund *et al.*, 2002; Roy *et al.*, 2006). However, when their differentiation is controlled, transplanted neurons synthesize dopamine and reverse motor deficits for long periods without tumor formation (Kim *et al.*, 2002; Rodríguez-Gómez *et al.*, 2007). These neurons derived in the laboratory express genes characteristic of the DA neurons in the substantia nigra that are at high risk in PD. A clearer understanding of the embryonic development of dopamine neurons would make their *ex vivo* derivation more efficient and reduce the tumor risk.

Although some of the genes involved in the differentiation of these neurons are known, a comprehensive view of the specification of dopamine neurons has remained elusive. In this chapter, we review new findings defining the precursor domains in the ventral midbrain showing that DA neurons are derived the floor plate. The floor plate is specified by the action of a transcription factor, foxa2, that also plays a high level role in the birth of dopamine neurons in the laboratory (Sasaki and Hogan, 1994; Ferri *et al.*, 2007; Kittappa *et al.*, 2007).

The floor plate origin of DA neurons is consistent with new data in the developing limb challenging the long-standing postulate of embryology that organizers do not contribute significantly to the adult tissue (Harfe *et al.*, 2004). Like the limb, two interacting organizers regulate the development of DA neurons. Morphogenic signals secreted by these organizers influence the differentiation of DA neurons from ES cells. The promise of ES-cell-based technologies requires that we understand how interactions between transcription factors and external signals restricted precursors to a DA fate. This information will improve the methods for the production of dopamine neurons from ES cells and reduce the risk of tumors.

Routine access to dopamine neurons is not simply a logistical achievement, as it also promises new insights into the mechanisms that control their function and death. Fox genes, which encode members of the forkhead transcription factor family, regulate cell survival and play important roles in the development of many tissues. There is growing evidence that compromised fox gene function leads to age-dependent phenotypes (Panowski *et al.*, 2007; van der Horst and Burgering, 2007). Mice with only a single copy of a fox gene acquire a late-onset degeneration of dopamine neurons associated with motor deficits (Kittappa *et al.*, 2007). This spontaneous cell death preferentially affects neurons that are at most risk in PD and suggests that this transcription factor is a target of the age-dependent changes that cause PD. In spite of the breadth of advances in the genetics of PD, we still lack a comprehensive explanation for the sensitivity of DA neurons in the substantia nigra to aging. The central survival role of fox proteins may allow a unified view of genes that contribute to PD and focus new therapeutic approaches on the common effector pathway that controls the birth and death of these cells.

THE FLOOR PLATE ORIGIN OF MIDBRAIN DOPAMINE NEURONS

In the fetal mouse, dopamine neurons are born in the ventral midbrain from embryonic day 11.5 during gestation (Zetterström *et al.*, 1997). In the human fetus, dopamine neurons are born between 4.5 and 6 weeks (Puelles and Verney, 1998). Specific types of neurons are born from distinct domains of neural progenitors which are individually specified in a precise region along the rostro-caudal and dorsoventral axes of the embryonic central nervous system (CNS). Each of these progenitor domains is demarcated and, often, specified by a particular transcription factor or factors (Treacy and Rosenfeld, 1992; Chalepakis *et al.*, 1993). The identification of the location of the progenitor domain which gives rise to dopamine neurons in the ventral midbrain and the transcription factors which regulate the identity of this domain are important starting points in the understanding of the development of dopamine neurons.

The ventral part of the embryonic CNS has been shown to be patterned by the secreted morphogen, sonic hedgehog (shh). In the embryonic spinal cord, different cell fates, including motoneurons, are induced at different concentrations of shh protein (Jessell, 2000). Normally, shh is produced precisely at the ventral midline within a special organizer called the floor plate and, in transplantation studies,

the floor plate, itself, has been shown to induce ventral fates in dorsal neural tissue (Hirano *et al.*, 1991; Placzek *et al.*, 1991).

In the midbrain, tissue recombination experiments with explants that putatively contain DA neuron precursors and the floor plate lead to the generation of increased numbers of dopamine neurons (Hynes *et al.*, 1995; Wang *et al.*, 1995; Ye *et al.*, 1998). This result was interpreted as an induction or specification of the DA neuron fate by morphogenic signals from the adjacent organizer regions, the floor plate and the isthmus (the cells at the boundary between the midbrain and the hindbrain, the isthmic organizer). Previous explant and grafting studies in the chick showed that these cells had a strong patterning effect that could also be achieved by application of the growth factors, FGF8 and Shh (Ye *et al.*, 1998). Transgenic expression of a constitutively active form of the shh receptor in the embryonic mouse midbrain leads to the induction of ectopic dopamine neurons in the dorsal midbrain (Hynes *et al.*, 2000).

These results suggested that shh from the floor plate and FGF8 from the midbrain–hindbrain boundary morphogenically determine the dopamine fate in the midbrain. In addition, Shh and FGF8 increase the numbers of dopamine neurons that can be derived from ES cells (Lee *et al.*, 2000; Kim *et al.*, 2002). In a modified ES cell protocol, shh similarly increases the number of motor neurons (Wichterle *et al.*, 2002). These results are consistent with shh acting to morphogenically induce the DA fate. Closer examination of the mesencephalic floor plate during embryonic development of dopamine neurons reveals inconsistencies with this model. In the embryonic spinal cord, specific types of neurons are born in two separate groups, one on the left side and one on the right side of the ventral midline. The production of neurons in two separate groups reflects the presence of the floor plate which induces them. In contrast, dopamine progenitors are generated in a single group that spans the ventral midline and floor plate. These cells co-express foxa2, a gene that is responsible for the generation of the floor plate (Sasaki *et al.*, 1994). The location and expression of foxa2 suggests that dopamine neurons are not induced by the floor plate but, instead, are directly derived from the floor plate.

The distinction between neurons derived from the floor plate and the immediately adjacent tissue is important because organizers, like the floor plate and the isthmus, are generally considered to have a transient role that precludes a direct lineage relation with adult cells. Given the potential differences between these different precursor cells, it was necessary to directly assess the lineage link between the floor plate and dopamine neurons. An Shh is expressed in the floor plate and, in a functional sense, shh expression defines the floor plate. A transgenic mouse expressing the cre recombinase gene under the control of shh transcriptional regulatory elements was bred to a "floxed" reporter in which beta-galactosidase expression is irreversibly turned on in cells containing cre activity (Harfe et al., 2004). This type of genetic lineage tracing has been used to determine lineage products of Shh-expressing cells in the limb (Harfe et al., 2004) and elsewhere in the CNS (Zervas et al., 2004; Louvi et al., 2007). Consistent with earlier results in the spinal cord, the descendents of the floor plate exist in a narrow medial stripe and no neurons were labeled in the hindbrain using this lineage tracing approach. In contrast, in the midbrain, a much larger number of cells around the midline expressed beta-galactosidase. The majority of DA neurons become post-mitotic before embryonic day E14.5, and on this day all of the tyrosine hydroxylase (TH) expressing cells co-express beta-galactosidase. This result definitively establishes that dopamine neurons are generated from the floor plate. Other neurons including brn-3a-expressing cells destined for the red nucleus are also labeled in these animals. (Kittappa et al., 2007).

A separate study has shown that fetal midbrain progenitors sorted for the floor-plate-specific cell surface protein, corin, generate dopamine neurons much more efficiently than corin-negative cells. In this manner, the properties of the floor plate lineage can be further exploited to produce purer populations of dopamine neurons for analysis or clinical use (Ono et al., 2007).

Their floor plate origin has consequences for our understanding of the development of DA neurons. One issue that arises immediately is whether the dopamine fate is determined by the actions of shh as a morphogen. The treatment of embryonic spinal cord explants and ES-derived neural progenitors with increasing concentrations of shh induces ventral progenitors at the expense of dorsal progenitors (Marti et al., 1995; Wichterle et al., 2002). Analogous experiments carried out on dissociated midbrain progenitors similarly show that some ventral fates are induced at the expense of dorsal fates.

However, floor plate, the most ventral fate, is not induced in these experiments suggesting shh does not act as a morphogen for midbrain dopamine precursors (Kittappa et al., 2007).

It is important to point out that these data do not exclude other roles for shh in the development of dopamine neurons. Conditional inactivation of shh receptor using nestin-cre disrupts the morphogenic induction of ventral markers in the midbrain, but dopamine neurons remain intact. When the shh receptor is ablated at an earlier time, using engrailed-cre, dopamine neurons fail to develop (Blaess et al., 2006). This is consistent with data showing that floor plate induction by the notochord is restricted to a short temporal window during early development (Placzek et al., 1993). These data suggest that the midbrain floor plate is induced by an early interaction between the notochord and the ventral midbrain. The floor plate then directly generates dopamine neurons from precursors at the midline. In a second step, the shh-expressing floor plate determines the fate of other neuron types in the ventrolateral midbrain by a morphogenic graded signaling mechanism. This mechanism has implications for the use of Shh in protocols for the in vitro generation of DA neurons from ES cells as it suggests that the precise period when shh induces the dopaminergic fate must still be defined.

The growth factor FGF8 is a second important signaling molecule in the ventral midbrain. FGF8 is expressed in the isthmus, the boundary of the midbrain and hindbrain, and can induce ectopic midbrain or hindbrain fates (Crossley et al., 1996; Irving and Mason, 1999). A zebrafish mutation in fgf8 affects formation of the isthmus but has no effect of the development of dopamine neurons (Holzschuh et al., 2003). The Shh also has little effect on DA neuron generation in zebrafish. However, the value of this model is complicated by the fact that DA neurons are found in the diencephalon and striatum and not in the mesencephalon of the teleost fish (Guo et al., 1999; Jeong et al., 2006). In mice, FGF8 interacts with Otx2, Gbx2, Lmx1b, and Wnt1 to control the activity of the isthmic organizer (Adams et al., 2000; Matsunaga et al., 2002).

A number of studies have suggested that wnt signaling plays an important role in the differentiation of dopamine neurons. Wnt-1 is expressed in a subset of ventral midbrain cells and lineage studies have shown that some of these cells become dopamine neurons (Zervas et al., 2004). In vitro studies suggest that wnt-1, wnt-3a, and wnt-5a have non-overlapping roles in dopamine neuron development but, so far, genetic evidence for a role in DA neuron specification only exists for wnt-1 (Castelo-Branco *et al.*, 2003; Prakash *et al.*, 2006). In explants of the rat midbrain, FGF8 treatment generates increased numbers of dopamine neurons (Ye *et al.*, 1998) but more recent work suggests that fgf effects are mediated by the induction of wnt-1 and that fgfs do not affect dopamine neuron development in the absence of wnt signaling (Prakash *et al.*, 2006). The Shh and FGF8 co-operate to generate increased numbers of dopamine neurons from ES cells (Lee *et al.*, 2000). Genetic experiments in the mouse have shown that FGF receptors in the ventral midbrain may regulate the proliferation of dopamine precursors but they are not required for the induction of the dopamine fate (Saarimaki-Vire *et al.*, 2007). A role for fgf in the control of cell division and, therefore, the number of dopamine neurons may reconcile the disparate outcomes of these many studies.

Why the midbrain floor plate is neurogenic while the floor plates of the hindbrain and spinal cord are non-neurogenic is unresolved. The floor plate is initially induced by the notochord, but, as the embryo gets older, the notochord is displaced such that it terminates rostrally near the midbrain–hindbrain boundary. At later stages, after floor plate induction, signals from the notochord would influence the ventral hindbrain and spinal cord, but not the ventral midbrain or more rostral parts of the brain. For example, the gene, nkx6.1, is expressed widely in the embryonic ventral CNS but is not expressed in the developing, ventral telencephalon. Small pieces of notochord grafted into the embryonic forebrain induce nkx6.1 expression (Qiu *et al.*, 1998). A notochord-derived signal would be an attractive candidate for the neurogenesis competence factor in the floor plate. Downstream of such a signal, the transcription factors, otx2 and mash1, may be involved in establishing neurogenesis in the midbrain floor plate. In overexpression studies in the chick embryo, otx2, and mash1 are able to induce some of the genes expressed in the embryonic midbrain floor plate (Ono *et al.*, 2007).

Is it possible to reconcile a floor plate origin for dopamine neurons with the earlier work suggesting that shh induces the dopamine fate? An important early study foreshadowed the true origin of dopamine neurons. The gene, gli2, acts downstream of shh and is required for the early development of the floor plate. In the spinal cord of gli2 −/− embryos, the most ventral neurons, including motoneurons, are compromised but not completely lost. Within the midbrain of gli2 −/− embryos, in contrast, dopamine neurons are absent (Matise *et al.*, 1998). The authors of this study concluded that the midbrain must contain a special class of cells, which they term the "ventral intermediate region," and this region has a special dependence on the floor plate, not previously observed in the midbrain or the spinal cord. Despite this interesting, earlier interpretation of this effect, these data are consistent and more clearly understood as a consequence of the floor plate origin of dopamine neurons. The newer lineage data clearly suggest an alternative interpretation for these results, that the dopamine neurons are derived from the floor plate.

As mentioned previously, the transgenic misexpression of a constitutively active shh receptor induced dopamine neurons in the embryonic midbrain. Importantly, the ectopic dopamine neurons were restricted to the dorsal midline in a "mirror image domain," relative to the endogenous dopamine neurons born at the ventral midline (Hynes *et al.*, 2000). One possible interpretation is that the roof plate, an organizer expressing BMPs at the dorsal midline, was converted to a floor plate and, subsequently, to dopamine neurons in these transgenic embryos. While experimental evidence for the conversion of a roof plate to a floor plate is lacking in this study, the conversion of a floor plate to a roof plate by dorsalizing factors has been observed (Tremblay *et al.*, 1996). In experiments parallel to the mouse transgenic studies, chick embryos were electroporated with the same constitutively active shh receptors, at later developmental stages than the transgenic study. While ventral markers were upregulated, the ectopic induction of dopamine neurons was not reported (Hynes *et al.*, 2000), consistent with the more recent studies showing that shh induction of dopamine neurons is not coupled to its late effects as a morphogen (Blaess *et al.*, 2006; Kittappa *et al.*, 2007).

These experiments suggest that dopamine neurons are derived from cells that are distinct from the precursors that generate the majority of neurons in the brain. The floor plate origin of dopamine neurons suggests conditions for further analysis of the mechanisms drive dopamine neuron fate from stem cells (see section "The Interaction of foxa2 and Other Transcription Factors in The Development of Dopamine Neurons") and regulate survival of dopamine neurons in neurodegenerative disease (see section "Spontaneous Degeneration of Dopamine Neurons in foxa2 Mutant Mice").

THE INTERACTION OF FOXA2 AND OTHER TRANSCRIPTION FACTORS IN THE DEVELOPMENT OF DOPAMINE NEURONS

Foxa2 is a forkhead transcription factor, formerly known as hepatocyte nuclear factor-3B (HNF-3B), which is expressed in the floor plate and important for floor plate development (Ang and Rossant, 1994; Sasaki et al., 1994; Weinstein et al., 1994). Forkhead transcription factors regulate cell fate, patterning, and survival (Wijchers et al., 2006). In the nematode, foxo, which is related to foxa2 is a critical regulator of lifespan (Brunet et al., 1999). A number of groups have shown that foxa2 and the related transcription factor, foxa1, are expressed in dopamine progenitors and dopamine neurons (Thuret et al., 2004; Ferri et al., 2007; Kittappa et al., 2007). Foxa2 nullizygous embryos die in utero between E8.5 and E10.5 but neural progenitors can be cultured in vitro, from these early embryos. Foxa2 null progenitors are unable to differentiate into dopamine neurons but they are able to differentiate into ventrolateral neurons which are born outside the floor plate (Kittappa et al., 2007). After the induction and initial differentiation of dopamine neurons, foxa1 and foxa2 are both required for the continued differentiation of dopamine neurons (Ferri et al., 2007). Overexpression of foxa2 induces dopamine neurons from cultured midbrain progenitors (Kittappa et al., 2007). A mouse ES cell line was generated, which inducibly expresses foxa2 when treated with doxycycline. Doxycycline treatment during neural differentiation of these cells leads to a greater than sevenfold increase in the number of dopamine neurons generated (Kittappa et al., 2007). In a much earlier study, transgenic expression of foxa2 in the mouse embryonic CNS led to the generation of ectopic dopamine neurons. At that time, the authors concluded that the new dopamine neurons were induced by an ectopic floor plate induced by foxa2. Newer lineage data and the loss and gain-of-function experiments, described above, suggest that foxa2 directly and autonomously drove the generation of dopamine neurons, in this experiment (Hynes et al., 1995). Taken together, these data suggest that foxa2 is

both necessary and sufficient for the specification of midbrain dopamine neurons.

While many other transcription factors have been implicated in the development of dopamine neurons, no other single transcription factor is, both, necessary and sufficient for dopamine neuron development. Nurr1 is important for the proper differentiation of dopamine neurons (Zetterström et al., 1997), but in the absence of Nurr1, dopamine neurons are born and project to the striatum, normally, but do not express TH and other genes critical for dopamine neuron function (Witta et al., 2000). In addition, cultured Nurr1 null midbrain progenitors differentiate into TH-expressing dopamine neurons, suggesting that the effects of Nurr1 are actually cell non-autonomous (Eells et al., 2001). Pitx3 is important for the development of the substantia nigra and, more specifically, the ventral tier neurons of the substantia nigra (Smidt et al., 1997; van den Munckhof et al., 2003; Smidt et al., 2004; Jacobs et al., 2007). Pitx3 overexpression does not promote differentiation of dopamine neurons (Sakurada et al., 1999). The Lmx1 proteins are very likely critical for differentiation of midbrain dopamine neurons (Andersson et al., 2006) but neither lmx1a nor lmx1b mutants demonstrate a significant loss of dopamine neurons (Smidt et al., 2000; Kittappa, unpublished). The effect of a loss of both lmx1a and lmx1b on the development of dopamine neurons has not yet been shown. The overexpression of lmx1a, but not lmx1b, in ES cells differentiating to neurons, improves the percentage of TH-expressing neurons which express other markers of midbrain dopamine neurons but the overexpression of lmx1a has not been shown to increase the number of dopamine neurons in culture (Andersson et al., 2006). Considering the widespread expression of lmx1a in the cortex, hippocampus, hypothalamus, ventral midbrain, cerebellum, and spinal cord, it is little surprise that lmx1a cannot specifically determine midbrain dopamine fate in naïve neural progenitors.

Finally, engrailed transcription factors have been shown to be important for the survival of dopamine neurons during development (Simon et al., 2001; Alberi et al., 2004). Overexpression of engrailed in ES cells, neural progenitors, or the embryonic CNS has not been shown but, as engrailed transcription factors are so widely expressed in the embryonic CNS, like lmx1 transcription factors, that it is unlikely that they specifically determine the fate of dopamine neurons. Still, it is worth pointing out that foxa2 and engrailed have been shown to specifically interact and to co-ordinately regulate gene expression (Foucher *et al.*, 2003). It will be interesting, in the future, to cross foxa2 mutants with mice with mutations in these other transcription factors to look for genetic interactions. The ES cell line which inducibly expresses foxa2 and, perhaps, similar cell lines, generated in the future, may be well suited for transplantation into animal models of PD or into PD patients, themselves. Foxa2 is known to bind to regulatory sequences in the AADC (Raynal *et al.*, 1998) and TH (Kessler *et al.*, 2003) promoters but it will be interesting to determine what targets of foxa2 are critical for dopamine differentiation. Foxa2 is a critical transcriptional regulator of the secreted, signaling molecules, shh (Epstein *et al.*, 1999), and netrin (Rastegar *et al.*, 2002). It will be interesting to assess whether the functional properties of differentiated dopamine neurons are regulated by these pathways.

Although foxa2 appears to have a central role in the differentiation of dopamine neurons, it is almost certain that it does so in concert with known or yet-to-be-discovered transcription factors. For example, foxa2 and engrailed-1 have been shown to specifically interact and to co-ordinately regulate gene expression (Foucher *et al.*, 2003). In the embryonic midbrain, foxa2 is initially expressed in a broad domain of neural progenitors. The medial cells co-express lmx1 transcription factors and foxa2, and become dopamine neurons. The more lateral cells in the foxa2 domain initially differentiate to motoneurons, expressing phox2 and islet proteins, of the oculomotor nucleus. In a second stage of differentiation, lateral foxa2 progenitors differentiate to brn-3a-expressing neurons which will populate the red nucleus. Unlike dopamine neurons, neither oculomotor neurons nor neurons of the red nucleus continue to express foxa2 so it is unlikely that foxa2 has a role in the postmitotic life of these neurons. It is now necessary to identify the signals which pattern and subdivide the foxa2-expressing floor plate, itself. Also, it will be interesting, in the future, to cross foxa2 mutants with mice with mutations in these other transcription factors to look for additional and stronger CNS phenotypes.

SPONTANEOUS DEGENERATION OF DOPAMINE NEURONS IN FOXA2 MUTANT MICE

Foxa2, which plays such an important role in the development of midbrain dopamine neurons,

prenatally, continues to be expressed in all of the dopamine neurons of the adult midbrain. Forkhead transcription factors are important mediators of cell survival (Wijchers *et al.*, 2006) and foxa2 would be an appropriate candidate for a regulator of the survival of adult midbrain dopamine neurons. The foxa2 gene plays a critical early role in the development of the extraembryonic and definitive endoderm and as a consequence foxa2 null animals die at an early stage of development. Animals that carry only a single foxa2 gene are fertile and viable, although but they do have a defect in stride alternation. Instead of alternating their left and right limbs when they walk, they hop with both forelimbs moving together, simultaneously, and both hindlimbs moving together, simultaneously. This stride alternation defect is reminiscent of netrin-1 null neonates (netrin-1 is a target gene of foxa2) and also of dopamine transporter (DAT) null mice (Cyr *et al.*, 2003). In addition, approximately 10–15% of foxa2 mutants have a swaying posture, poor leg strength, and an inability to right themselves when flipped on their back. In many respects, this phenotype is similar to the swaying mouse which has a mutation in wnt1, a secreted factor which influences the development of dopamine neurons during embryonic development (see above).

Much later in life, foxa2 mutant mice spontaneously develop major defects in posture and movement (Kittappa *et al.*, 2007). The most obvious late-onset phenotype is a significantly kinked posture caused by extreme muscle flexion and rigidity on one side of the animal. Rigidity is also observed in the limbs and, in severe cases, mice lose use of their hindlimbs. Horizontal movement is slowed and vertical movement, or rearing, is absent in these mice. These late-onset phenotypes occur almost exclusively in mice older than 18 months. Approximately one-third of the mutant mice older than 18 months develops these late-onset motor problems. Additionally, there is a degree of strain specificity. These defects have been observed in a C57BL/6 background but never in a CD-1 background. It is noteworthy that C57BL/6 mice, when compared to other strains, demonstrate an enhanced sensitivity to the dopamine neuron-selective neurotoxin, 1-methyl-4-phenyl-1,2,3,6-tetrahydropyridine (MPTP) (Cook *et al.*, 2003).

6-Hydroxydopamine (6-OHDA) is a potent neurotoxin which selectively kills dopamine neurons. Mice and rats which are unilaterally lesioned with 6-OHDA demonstrate considerable rotational movement when challenged with amphetamine.

This assay is easily quantitated and the extent of rotational movement is directly related to the severity of the lesion. Older foxa2 mutants which demonstrate these spontaneous, late-onset motor phenotypes similarly show significant rotational behavior when challenged with 6-OHDA. Younger mutant mice and older mutants which have not yet developed the late-onset phenotypes do not demonstrate this amphetamine-induced rotational behavior. These data suggest that the late-onset defects observed in the foxa2 mice are related to a loss of dopaminergic function in the nigro-striatal system. Immunohistochemistry for TH and Nissl staining reveals a significant loss of midbrain dopamine neurons strikingly reminiscent of PD. This loss of dopamine neurons is only observed in older foxa2 mutants with the late-onset phenotypes – it has not been seen, at all, in younger foxa2 mutants and older mutants which have not developed the late-onset phenotypes (Kittappa et al., 2007).

Just as the posture of these animals is asymmetric, the loss of dopamine neurons – in all but the most severe cases in which almost all of the dopamine neurons is lost – is also asymmetric. Motor deficits in PD almost always occur unilaterally, at the first signs of the disease, and PET scanning, similarly, shows an asymmetric loss of dopamine in most patients (Cheesman et al., 2005). Ventral tier neurons of the substantia nigra have been shown to be at greater risk in PD and these neurons are protected by specific growth factors (Damier et al., 1999; Murase and Mckay, 2006). In foxa2 mutants, ventral tier neurons also show an increased sensitivity. Gliosis is often observed in neurodegenerative disease (Maragakis and Rothstein, 2006) and is implicated in PD (Hirsch et al., 2003). Increased gliosis is observed in foxa2 mutants with late-onset deficits, not only in the substantia nigra but throughout the ventral midbrain. Lewy bodies are cellular inclusions, easily observed by staining with hematoxylin and eosin, that are often seen throughout the postmortem brain of PD patients. Lewy bodies are immunoreactive for a number of proteins including ubiquitin, alpha-synuclein, neurofilament, and torsin A (Olanow et al., 2004). In foxa2 heterozygous mice, Lewy bodies are not a major feature of the pathology.

Other mutations in mice that cause dopamine neuron degeneration include mice with mutations in GIRK2 (weaver; Roffler-Tarlov and Graybiel, 1984; Verina et al., 1997; Martí et al., 2007), TGF-alpha (Blum, 1998), ATM (Eilam et al., 1998; Eilam et al., 2003), the engrailed genes (Alberi et al.,

2004; Sgado et al., 2006; Sonnier et al., 2007), and HIPK2 (Zhang et al., 2007). The weaver mutation was one of the first mouse behavioral mutations to be studied. The story of the subsequent study of weaver mice reminds us that it is often difficult to trace a clear path between genotype and phenotype (Herrup, 1996). There are many features of these mice pertinent to the general question of the death of dopamine neurons but, in the context of this paper, we note that the time course of cell loss is different from foxa2 heterozygous mice. These previously described mice demonstrate a loss of neurons at early stages of their lives instead of in old age.

Recently, a mouse with a hypomorphic allele of the vesicular monoamine transporter 2 (VMAT2) has been shown to develop mild degeneration of dopamine neurons, late in life (Caudle et al., 2007). Both the extent of degeneration of dopamine neurons and accompanying motor deficits are much less severe than foxa2 mutant mice. It will be interesting to investigate the relationship of this novel hypomorphic allele of VMAT2 with mice which are heterozygous for VMAT2 which have recently been shown to demonstrate a reduction in anxiety behaviors, similar to depression (Fukui et al., 2007). Targeting the mitochondria in dopamine neurons also leads to the postnatal death of dopamine neurons (Ekstrand et al., 2007). These mice all have something to teach us but the late-onset of degeneration of dopamine neurons in foxa2 mutant mice may offer a unique insight. The late-onset, asymmetry, and cellular sensitivity parallel features of the onset and progression of PD not seen in other models. The phenotype of these mice suggests that limited function of the foxa2 gene may contribute to PD.

TOWARDS A UNIFIED VIEW OF GENETIC AND SPONTANEOUS RISK FACTORS

The dopamine projections from the ventral midbrain reach out to large areas of the nervous system. Loss of these projections in PD or their dysfunction in degenerative and psychiatric disease constitutes a major medical burden. For several years our group has developed new *ex vivo* methods to generate dopamine neurons that are now an every day reality in laboratories around the world. The successful differentiation of dopamine neurons from ES cells suggests a comprehensive understanding of the biochemistry of dopamine neurons is possible. In the United States approximately 15% of patients with

PD have a family history of the disease and mutations in a number of genes contribute to hereditary forms of PD. These genes have been divided into two classes, the PARK genes that are thought to cause disease and genes that are associated with PD. The PARK genes include alpha-synuclein (SNCA), parkin, PINK1, and LRRK2 (Sulzer, 2007; http://www.pdgene.org/linkage.asp). To date, mice generated with mutations in these genes do not demonstrate degeneration of dopamine neurons, either during fetal or juvenile development or, spontaneously, in old age. The simplest explanation for this is that the PARK mutations interact with alleles at other loci. The different frequencies of LRRK2 mutations in North African and European populations are a striking example of this point (Lesage et al., 2005; Clark et al., 2006; Ishihara et al., 2007; Lesage et al., 2007).

The phenotype of the foxa2 mice and the central role of fox genes in survival signaling suggest that the effects of PARK and other mutations on dopamine neurons can be assessed by making the appropriate compound mutations. It will also be instructive to assess whether environmental toxins suspected to cause sporadic PD compromise the foxa2 survival pathway. This can be achieved in animal models or in human dopamine neurons that will soon be routinely available from ES cells. In short, when combined with advances in stem cell biology and genetics, the phenotype of foxa2 heterozygous mice will provide important insight into the mechanisms that cause the age-dependent cellular pathology seen in the midbrain of patients with PD.

REFERENCES

Adams, K. A., Maida, J. M., Golden, J. A., and Riddle, R. D. (2000). The transcription factor Lmx1b maintains Wnt1 expression within the isthmic organizer. *Development* **127**(9), 1857–1867.

Alberi, L., Sgado, P., and Simon, H. H. (2004). Engrailed genes are cell-autonomously required to prevent apoptosis in mesencephalic dopaminergic neurons. *Development* **131**(13), 3229–3236.

Andersson, E., Tryggvason, U., Deng, Q., Friling, S., Alekseenko, Z. *et al.* (2006). Identification of intrinsic determinants of midbrain dopamine neurons. *Cell* **124**, 393–405.

Ang, S. L., and Rossant, J. (1994). HNF-3 beta is essential for node and notochord formation in mouse development. *Cell* **78**, 561–574.

Bjorklund, L. M., Sánchez-Pernaute, R., Chung, S., Andersson, T., Chen, I. Y., McNaught, K. S., Brownell, A. L., Jenkins, B. G., Wahlestedt, C., Kim, K. S., and Isacson, O. (2002). Embryonic stem cells develop into functional dopaminergic neurons after transplantation in a Parkinson rat model. *Proc Natl Acad Sci USA* **99**(4), 2344–2349.

Blaess, S., Corrales, J. D., and Joyner, A. L. (2006). Sonic hedgehog regulates Gli activator and repressor functions with spatial and temporal precision in the mid/hindbrain region. *Development* **133**(9), 1799–1809.

Blum, M. (1998). A null mutation in TGF-alpha leads to a reduction in midbrain dopaminergic neurons in the substantia nigra. *Nat Neurosci* **1**(5), 374–377.

Brunet, A., Bonni, A., Zigmond, M. J., Lin, M. Z., Juo, P. *et al.* (1999). Akt promotes cell survival by phosphorylating and inhibiting a Forkhead transcription factor. *Cell* **96**, 857–868.

Castelo-Branco, G., Wagner, J., Rodriguez, F. J., Kele, J., Sousa, K., Rawal, N., Pasolli, H. A., Fuchs, E., Kitajewski, J., and Arenas, E. (2003). Differential regulation of midbrain dopaminergic neuron development by Wnt-1, Wnt-3a, and Wnt-5a. *Proc Natl Acad Sci USA* **100**(22), 12747–12752.

Caudle, W. M., Richardson, J. R., Wang, M. Z., Taylor, T. N., Guillot, T. S., McCormack, A. L., Colebrooke, R. E., Di Monte, D. A., Emson, P. C., and Miller, G. W. (2007). Reduced vesicular storage of dopamine causes progressive nigrostriatal neurodegeneration. *J Neurosci* **27**(30), 8138–8148.

Chalepakis, G., Stoykova, A., Wijnholds, J., Tremblay, P., and Gruss, P. (1993). Pax: Gene regulators in the developing nervous system. *J Neurobiol* Oct; **24**(10), 1367–1384.

Cheesman, A. L., Barker, R. A., Lewis, S. J., Robbins, T. W., Owen, A. M. *et al.* (2005). Lateralisation of striatal function: Evidence from 18F-dopa PET in Parkinson's Disease. *J Neurol Neurosurg Psychiatry* **76**, 1204–1210.

Clark, L. N., Wang, Y., Karlins, E., Saito, L., Mejia-Santana, H., Harris, J., Louis, E. D., Cote, L. J., Andrews, H., Fahn, S., Waters, C., Ford, B., Frucht, S., Ottman, R., and Marder, K. (2006). Frequency of LRRK2 mutations in early- and late-onset Parkinson disease. *Neurology* **67**(10), 1786–1791.

Cook, R., Lu, L., Gu, J., Williams, R. W., and Smeyne, R. J. (2003). Identification of a single QTL, Mptp1, for susceptibility to MPTP-induced substantia nigra pars compacta neuron loss in mice. *Brain Res Mol Brain Res* **110**(2), 279–288.

Crossley, P. H., Martinez, S., and Martin, G. R. (1996). Midbrain development induced by FGF8 in the chick embryo. *Nature* **380**(6569), 66–68.

Cyr, M., Beaulieu, J. M., Laakso, A., Sotnikova, T. D., Yao, W. D., Bohn, L. M., Gainetdinov, R. R., and Caron, M. G. (2003). Sustained elevation of extracellular dopamine causes motor dysfunction and selective degeneration of striatal GABAergic neurons. *Proc Natl Acad Sci USA* 100(19), 11035–11040.

Damier, P., Hirsch, E. C., Agid, Y., and Graybiel, A. M. (1999). The substantia nigra of the human brain II. Patterns of loss of dopamine-containing neurons in Parkinson's disease. *Brain* 122(Pt 8), 1437–1448.

Eells, J. B., Rives, J. E., Yeung, S. K., and Nikodem, V. M. (2001). *In vitro* regulated expression of tyrosine hydroxylase in ventral midbrain neurons from Nurr1-null mouse pups. *J Neurosci Res* **May 15;** 64(4), 322–330.

Eilam, R., Peter, Y., Elson, A., Rotman, G., Shiloh, Y., Groner, Y., and Segal, M. (1998). Selective loss of dopaminergic nigro-striatal neurons in brains of Atm-deficient mice. *Proc Natl Acad Sci USA* 95(21), 12653–12656.

Eilam, R., Peter, Y., Groner, Y., and Segal, M. (2003). Late degeneration of nigro-striatal neurons in ATM-/- mice. *Neuroscience* 121(1), 83–98.

Ekstrand, M. I., Terzioglu, M., Galter, D., Zhu, S., Hofstetter, C., Lindqvist, E., Thams, S., Bergstrand, A., Hansson, F. S., Trifunovic, A., Hoffer, B., Cullheim, S., Mohammed, A. H., Olson, L., and Larsson, N. G. (2007). Progressive parkinsonism in mice with respiratory-chain-deficient dopamine neurons. *Proc Natl Acad Sci USA* 104(4), 1325–1330.

Epstein, D. J., McMahon, A. P., and Joyner, A. L. (1999). Regionalization of Sonic hedgehog transcription along the anteroposterior axis of the mouse central nervous system is regulated by Hnf3-dependent and -independent mechanisms. *Development* 126, 281–292.

Ferri, A. L., Lin, W., Mavromatakis, Y. E., Wang, J. C., Sasaki, H., Whitsett, J. A., and Ang, S. L. (2007). Foxa1 and Foxa2 regulate multiple phases of midbrain dopaminergic neuron development in a dosage-dependent manner. *Development* 134(15), 2761–2769.

Foucher, I., Montesinos, M. L., Volovitch, M., Prochiantz, A., and Trembleau, A. (2003). Joint regulation of the MAP1B promoter by HNF3beta/Foxa2 and Engrailed is the result of a highly conserved mechanism for direct interaction of homeoproteins and Fox transcription factors. *Development* 130(9), 1867–1876.

Fukui, M., Rodriguiz, R. M., Zhou, J., Jiang, S. X., Phillips, L. E., Caron, M. G., and Wetsel, W. C. (2007). Vmat2 heterozygous mutant mice display a depressive-like phenotype. *J Neurosci* 27(39), 10520–10529.

Guo, S., Wilson, S. W., Cooke, S., Chitnis, A. B., Driever, W., and Rosenthal, A. (1999). Mutations in the zebrafish unmask shared regulatory pathways controlling the development of catecholaminergic neurons. *Dev Biol* 208(2), 473–487.

Harfe, B. D., Scherz, P. J., Nissim, S., Tian, H., McMahon, A. P., and Tabin, C. J. (2004). Evidence for an expansion-based temporal Shh gradient in specifying vertebrate digit identities. *Cell* 118(4), 517–528.

Herrup, K. (1996). The weaver mouse: A most cantankerous rodent. *Proc Natl Acad Sci USA* 93, 10541–10542.

Hirano, S., Fuse, S., and Sohal, G. S. (1991). The effect of the floor plate on pattern and polarity in the developing central nervous system. *Science* 251(4991), 310–313.

Hirsch, E. C., Breidert, T., Rousselet, E., Hunot, S., Hartmann, A., and Michel, P. P. (2003). The role of glial reaction and inflammation in Parkinson's disease. *Ann NY Acad Sci* 991, 214–228.

Holzschuh, J., Hauptmann, G., and Driever, W. (2003). Genetic analysis of the roles of Hh, FGF8, and nodal signaling during catecholaminergic system development in the zebrafish brain. *J Neurosci* 23(13), 5507–5519.

Hynes, M., Poulsen, K., Tessier-Lavigne, M., and Rosenthal, A. (1995). Control of neuronal diversity by the floor plate: Contact-mediated induction of midbrain dopaminergic neurons. *Cell* 80, 95–101.

Hynes, M., Ye, W., Wang, K., Stone, D., Murone, M., Sauvage, F., and Rosenthal, A. (2000). The seven-transmembrane receptor smoothened cell-autonomously induces multiple ventral cell types. *Nat Neurosci* 3(1), 41–46.

Irving, C., and Mason, I. (1999). Regeneration of isthmic tissue is the result of a specific and direct interaction between rhombomere 1 and midbrain. *Development* 126(18), 3981–3989.

Ishihara, L., Gibson, R. A., Warren, L., Amouri, R., Lyons, K., Wielinski, C., Hunter, C., Swartz, J. E., Elango, R., Akkari, P. A., Leppert, D., Surh, L., Reeves, K. H., Thomas, S., Ragone, L., Hattori, N., Pahwa, R., Jankovic, J., Nance, M., Freeman, A., Gouider-Khouja, N., Kefi, M., Zouari, M., Ben Sassi, S., Ben Yahmed, S., El Euch-Fayeche, G., Middleton, L., Burn, D. J., Watts, R. L., and Hentati, F. (2007). Screening for Lrrk2 G2019S and clinical comparison of Tunisian and North American Caucasian Parkinson's disease families. *Mov Disord* 22(1), 55–61.

Jacobs, F. M., Smits, S. M., Noorlander, C. W., von Oerthel, L., van der Linden, A. J., Burbach, J. P., and Smidt, M. P. (2007). Retinoic acid counteracts developmental defects in the substantia nigra caused by Pitx3 deficiency. *Development* 134(14), 2673–2684.

Jeong, J. Y., Einhorn, Z., Mercurio, S., Lee, S., Lau, B., Mione, M., Wilson, S. W., and Guo, S. (2006). Neurogenin1 is a determinant of zebrafish basal forebrain dopaminergic neurons and is regulated by the conserved zinc finger protein Tof/Fezl. *Proc Natl Acad Sci USA* **103**(13), 5143–5148.

Jessell, T. M. (2000). Neuronal specification in the spinal cord: Inductive signals and transcriptional codes. *Nat Rev Genet* **1**(1), 20–29.

Kessler, M. A., Yang, M., Gollomp, K. L., Jin, H., and Iacovitti, L. (2003). The human tyrosine hydroxylase gene promoter. *Brain Res Mol Brain Res* **112**(1-2), 8–23.

Kim, J.-H., Auerbach, J. M., Rodriguez-Gomez, J. A., Velasco, I., Gavin, D. *et al.* (2002). Dopamine neurons derived from embryonic stem cells function in an animal model of Parkinson's disease. *Nature* **418**, 50–56.

Kittappa, R., Chang, W. W., Awatramani, R. B., and McKay, R. D. (2007). The foxa2 gene controls the birth and spontaneous degeneration of dopamine neurons in old age. *PLoS Biol* **5**(12), e325.

Laguna Goya, R., Tyers, P., and Barker, R. A. (2008). The search for a curative cell therapy in Parkinson's disease. *J Neurol Sci* **265**, 32–42.

Lee, S. H., Lumelsky, N., Studer, L., Auerbach, J. M., and McKay, R. D. (2000). Efficient generation of midbrain and hindbrain neurons from mouse embryonic stem cells. *Nat Biotechnol* **18**, 675–679.

Lesage, S., Ibanez, P., Lohmann, E., Pollak, P., Tison, F., Tazir, M., Leutenegger, A. L., Guimaraes, J., Bonnet, A. M., Agid, Y., Durr, A., and Brice, A.French Parkinson's Disease Genetics Study Group (2005). G2019S LRRK2 mutation in French and North African families with Parkinson's disease. *Ann Neurol* **58**(5), 784–787.

Lesage, S., Janin, S., Lohmann, E., Leutenegger, A. L., Leclere, L., Viallet, F., Pollak, P., Durif, F., Thobois, S., Layet, V., Vidailhet, M., Agid, Y., Dürr, A., Brice, A., Bonnet, A. M., Borg, M., Broussolle, E., Damier, P., Destée, A., Martinez, M., Penet, C., Rasco, O., Tison, F., Tranchan, C., and Vérin, M.French Parkinson's Disease Genetics Study Group (2007). LRRK2 exon 41 mutations in sporadic Parkinson disease in Europeans. *Arch Neurol* **64**(3), 425–430.

Louvi, A., Yoshida, M., and Grove, E. A. (2007). The derivatives of the Wnt3a lineage in the central nervous system. *J Comp Neurol* **504**(5), 550–569.

Maragakis, N. J., and Rothstein, J. D. (2006). Mechanisms of Disease: Astrocytes in neurodegenerative disease. *Nat Clin Pract Neurol* **2**, 679–689.

Marti, E., Bumcrot, D. A., Takada, R., and McMahon, A. P. (1995). Requirement of 19K form of Sonic hedgehog for induction of distinct ventral cell types in CNS explants. *Nature* **375**(6529), 322–325.

Martí, J., Santa-Cruz, M. C., Bayer, S. A., Ghetti, B., and Hervás, J. P. (2007). Generation and survival of midbrain dopaminergic neurons in weaver mice. *Int J Dev Neurosci* **25**(5), 299–307.

Matise, M. P., Epstein, D. J., Park, H. L., Platt, K. A., and Joyner, A. L. (1998). Gli2 is required for induction of floor plate and adjacent cells, but not most ventral neurons in the mouse central nervous system. *Development* **125**(15), 2759–2770.

Matsunaga, E., Katahira, T., and Nakamura, H. (2002). Role of Lmx1b and Wnt1 in mesencephalon and metencephalon development. *Development* Nov;**129**(22), 5269–5277.

Murase, S., and McKay, R. D. (2006). A specific survival response in dopamine neurons at most risk in Parkinson's disease. *J Neurosci* **26**(38), 9750–9760.

Olanow, C. W., Perl, D. P., DeMartino, G. N., and McNaught, K. S. (2004). Lewy-body formation is an aggresome-related process. *Lancet Neurol* **3**, 496–503.

Ono, Y., Nakatani, T., Sakamoto, Y., Eri, Mizuhara., Minaki, Y., Kumai, M., Hamaguchi, A., Nishimura, M., Inoue, Y., Hayashi, H., Takahashi, J., and Imai, T. (2007). Differences in neurogenic potential in floor plate cells along an anteroposterior location: Midbrain dopaminergic neurons originate from mesencephalic floor plate cells. *Development* **134**, 3213–3225.

Panowski, S. H., Wolff, S., Aguilaniu, H., Durieux, J., and Dillin, A. (2007). PHA-4/Foxa mediates diet-restriction-induced longevity of *C. elegans*. *Nature* **447**(7144), 550–555.

Placzek, M., Yamada, T., Tessier-Lavigne, M., Jessell, T., and Dodd, J. (1991). Control of dorsoventral pattern in vertebrate neural development: Induction and polarizing properties of the floor plate. *Development* Suppl 2, 105–122.

Placzek, M., Jessell, T. M., and Dodd, J. (1993). Induction of floor plate differentiation by contact-dependent, homeogenetic signals. *Development* **117**(1), 205–218.

Prakash, N., Brodski, C., Naserke, T., Puelles, E., Gogoi, R., Hall, A., Panhuysen, M., Echevarria, D., Sussel, L., Weisenhorn, D. M., Martinez, S., Arenas, E., Simeone, A., and Wurst, W. (2006). A Wnt1-regulated genetic network controls the identity and fate of midbrain-dopaminergic progenitors *in vivo*. *Development* **133**(1), 89–98.

Puelles, L., and Verney, C. (1998). Early neuromeric distribution of tyrosine-hydroxylase-immunoreactive neurons in human embryos. *J Comp Neurol* **394**(3), 283–308.

Qiu, M., Shimamura, K., Sussel, L., Chen, S., and Rubenstein, J. (1998). Control of anteroposterior and dorsoventral domains of Nkx-6.1 gene expression relative to other Nkx genes during vertebrate CNS development. *Mech Dev* **72**(1-2), 77–88.

Rastegar, S., Albert, S., Le Roux, I., Fischer, N., Blader, P., Muller, F., and Strahle, U. (2002). A floor plate enhancer of the zebrafish netrin1 gene requires Cyclops (Nodal) signalling and the winged helix transcription factor FoxA2. *Dev Biol* **252**(1), 1–14.

Raynal, J. F., Dugast, C., Le Van Thai, A., and Weber, M. J. (1998). Winged helix hepatocyte nuclear factor 3 and POU-domain protein brn-2/N-oct-3 bind overlapping sites on the neuronal promoter of human aromatic L-amino acid decarboxylase gene. *Brain Res Mol Brain Res* 56, 227–237.

Rodríguez-Gómez, J. A., Lu, J. Q., Velasco, I., Rivera, S., Zoghbi, S. S., Liow, J. S., Musachio, J. L., Chin, F. T., Toyama, H., Seidel, J., Green, M. V., Thanos, P. K., Ichise, M., Pike, V. W., Innis, R. B., and McKay, R. D. (2007). Persistent dopamine functions of neurons derived from embryonic stem cells in a rodent model of Parkinson disease. *Stem Cells* 25(4), 918–928.

Roffler-Tarlov, S., and Graybiel, A. M. (1984). Weaver mutation has differential effects on the dopamine-containing innervation of the limbic and nonlimbic striatum. *Nature* 307(5946), 62–66.

Roy, N. S., Cleren, C., Singh, S. K., Yang, L., Beal, M. F., and Goldman, S. A. (2006). Functional engraftment of human ES cell-derived dopaminergic neurons enriched by coculture with telomerase-immortalized midbrain astrocytes. *Nat Med* 12(11), 1259–1268.

Saarimaki-Vire, J., Peltopuro, P., Lahti, L., Naserke, T., Blak, A. A., Vogt Weisenhorn, D. M., Yu, K., Ornitz, D. M., Wurst, W., and Partanen, J. (2007). Fibroblast growth factor receptors cooperate to regulate neural progenitor properties in the developing midbrain and hindbrain. *J Neurosci* 27(32), 8581–8592.

Sakurada, K., Ohshima-Sakurada, M., Palmer, T. D., and Gage, F. H. (1999). Nurr1, an orphan nuclear receptor, is a transcriptional activator of endogenous tyrosine hydroxylase in neural progenitor cells derived from the adult brain. *Development* 126(18), 4017–4026.

Sasaki, H., and Hogan, B. L. (1994). HNF-3 beta as a regulator of floor plate development. *Cell* 76, 103–115.

Serafini, T., Colamarino, S. A., Leonardo, E. D., Wang, H., Beddington, R., Skarnes, W. C., and Tessier-Lavigne, M. (1996). Netrin-1 is required for commissural axon guidance in the developing vertebrate nervous system. *Cell* 87(6), 1001–1014.

Sgado, P., Alberi, L., Gherbassi, D., Galasso, S. L., Ramakers, G. M., Alavian, K. N., Smidt, M. P., Dyck, R. H., and Simon, H. H. (2006). Slow progressive degeneration of nigral dopaminergic neurons in postnatal Engrailed mutant mice. *Proc Natl Acad Sci USA* 103(41), 15242–15247.

Simon, H. H., Saueressig, H., Wurst, W., Goulding, M. D., and O'Leary, D. D. (2001). Fate of midbrain dopaminergic neurons controlled by the engrailed genes. *J Neurosci* 21(9), 3126–3134.

Smidt, M. P., van Schaick, H. S., Lanctot, C., Tremblay, J. J., Cox, J. J., van der Kleij, A. A., Wolterink, G., Drouin, J., and Burbach, J. P. (1997). A homeodomain gene Ptx3 has highly restricted brain expression in mesencephalic dopaminergic neurons. *Proc Natl Acad Sci USA* 94(24), 13305–13310.

Smidt, M. P., Asbreuk, C. H., Cox, J. J., Chen, H., Johnson, R. L., and Burbach, J. P. (2000). A second independent pathway for development of mesencephalic dopaminergic neurons requires Lmx1b. *Nat Neurosci* 3(4), 337–341.

Smidt, M. P., Smits, S. M., Bouwmeester, H., Hamers, F. P., van der Linden, A. J., Hellemons, A. J., Graw, J., and Burbach, J. P. (2004). Early developmental failure of substantia nigra dopamine neurons in mice lacking the homeodomain gene Pitx3. *Development* 131(5), 1145–1155.

Sonnier, L., Le Pen, G., Hartmann, A., Bizot, J. C., Trovero, F., Krebs, M. O., and Prochiantz, A. (2007). Progressive loss of dopaminergic neurons in the ventral midbrain of adult mice heterozygote for Engrailed1. *J Neurosci* 27(5), 1063–1071.

Sulzer, D. (2007). Multiple hit hypotheses for dopamine neuron loss in Parkinson's disease. *Trends Neurosci* 30(5), 244–250.

Thuret, S., Bhatt, L., O'Leary, D. D., and Simon, H. H. (2004). Identification and developmental analysis of genes expressed by dopaminergic neurons of the substantia nigra pars compacta. *Mol Cell Neurosci* 25(3), 394–405.

Treacy, M. N., and Rosenfeld, M. G. (1992). Expression of a family of POU-domain protein regulatory genes during development of the central nervous system. *Annu Rev Neurosci* 15, 139–165.

Tremblay, P., Pituello, F., and Gruss, P. (1996). Inhibition of floor plate differentiation by Pax3: Evidence from ectopic expression in transgenic mice. *Development* 122(8), 2555–2567.

van den Munckhof, P., Luk, K. C., Ste-Marie, L., Montgomery, J., Blanchet, P. J., Sadikot, A. F., and Drouin, J. (2003). Pitx3 is required for motor activity and for survival of a subset of midbrain dopaminergic neurons. *Development* 130(11), 2535–2542.

van der Horst, A., and Burgering, B. M. (2007). Stressing the role of FoxO proteins in lifespan and disease. *Nat Rev Mol Cell Biol* 8(6), 440–450.

Verina, T., Norton, J. A., Sorbel, J. J., Triarhou, L. C., Laferty, D., Richter, J. A., Simon, J. R., and Ghetti, B. (1997). Atrophy and loss of dopaminergic mesencephalic neurons in heterozygous weaver mice. *Exp Brain Res* 113(1), 5–12.

Wang, M. Z., Jin, P., Bumcrot, D. A., Marigo, V., McMahon, A. P. *et al.* (1995). Induction of dopaminergic neuron phenotype in the midbrain by Sonic hedgehog protein. *Nat Med* 1, 1184–1188.

Weinstein, D. C., Ruiz i Altaba, A., Chen, W. S., Hoodless, P., Prezioso, V. R. *et al.* (1994). The winged-helix transcription factor HNF-3 beta is required for notochord development in the mouse embryo. *Cell* 78, 575–588.

Wichterle, H., Lieberam, I., Porter, J. A., and Jessell, T. M. (2002). Directed differentiation of embryonic stem cells into motor neurons. *Cell* 110(3), 385–397.

Wijchers, P. J., Burbach, J. P., and Smidt, M. P. (2006). In control of biology: Of mice, men and Foxes. *Biochem J* 397(2), 233–246.

Witta, J., Baffi, J. S., Palkovits, M., Mezey, E., Castillo, S. O., and Nikodem, V. M. (2000). Nigrostriatal innervation is preserved in Nurr1-null mice, although dopaminergic neuron precursors are arrested from terminal differentiation. *Brain Res Mol Brain Res* 84(1-2), 67–78.

Ye, W., Shimamura, K., Rubenstein, J. L., Hynes, M. A., and Rosenthal, A. (1998). FGF and Shh signals control dopaminergic and serotonergic cell fate in the anterior neural plate. *Cell* 93(5), 755–766.

Zervas, M., Millet, S., Ahn, S., and Joyner, A. L. (2004). Cell behaviors and genetic lineages of the mesencephalon and rhombomere 1. *Neuron* 43(3), 345–357.

Zetterström, R. H., Solomin, L., Jansson, L., Hoffer, B. J., Olson, L., and Perlmann, T. (1997). Dopamine neuron agenesis in Nurr1-deficient mice. *Science* 276(5310), 248–250.

Zhang, J., Pho, V., Bonasera, S. J., Holtzman, J., Tang, A. T., Hellmuth, J., Tang, S., Janak, P. H., Tecott, L. H., and Huang, E. J. (2007). Essential function of HIPK2 in TGFbeta-dependent survival of midbrain dopamine neurons. *Nat Neurosci* 10(1), 77–86.

35

EMBRYONIC STEM CELL-BASED MODELS OF PARKINSON'S DISEASE

MARK J. TOMISHIMA[1] AND LORENZ STUDER[2]

[1]SKI Stem Cell Research Facility, Developmental Biology Program, Sloan-Kettering Institute, New York, NY, USA
[2]Developmental Biology Program, Sloan-Kettering Institute, New York, NY, USA

INTRODUCTION

Parkinson's disease (PD) is primarily caused by an age-related, progressive degeneration of midbrain dopamine (mDA) neurons. A major goal of stem cell research is the development of cell-based therapies aimed at replacing the cells lost in PD patients. Traditionally, such therapies have used fetal midbrain tissue, but more recently embryonic stem cells (ESCs) have been intensively studied as a cell source for transplantation. The development of protocols that direct ESCs to mDA is important for transplantation biology, but has also created a unique opportunity to study PD pathogenesis at the cellular level. In this chapter, we will explore such alternative uses for ESCs aimed at modeling PD *in vitro*.

ADVANTAGES OF ESCs

ES-derived mDA neurons provide a number of advantages over embryo-derived neurons. One of the main advantages is the *unlimited supply* of neurons: few biochemical, or "ChIP on Chip" experiments have been performed on mDA neurons since the amount of embryo-derived tissue required is prohibitive. ES-based differentiations can be scaled up to provide the required number of mDA neurons. These mDA neurons are thought to be *authentic*: they are derived from cells with

a normal karyotype and express genes characteristic of mDA neurons. Furthermore, at least in the case of mouse ESCs, there is evidence that ES-derived mDA neurons restore neural function *in vivo* in a model of PD (Kim *et al.*, 2002; Barberi *et al.*, 2003; Rodriguez-Gomez *et al.*, 2007). ESCs are *pluripotent*, meaning that they can differentiate into a wide range of cell types. This means that a PD-related genotype can be differentiated into many different cell types *in vitro*. For example, Yamashita *et al.* (2006) differentiated ESCs from a parkinsonian genotype into mDA neurons and oligodendrocytes and found that only the mDA neurons were sensitive to oxidative stress. Genetically modified ESCs can also be introduced into a developing mouse embryo to subject the modified cells into a partial (chimera) or complete mouse (tetraploid aggregation). Besides pluripotency, the ease with which the ESC *genome* can be *modified* has enabled the revolution in mouse genetics. Genetic modification can create ESCs that mimic natural PD mutations or fluoresce when they express a gene of interest. Instead of engineered mutations, disease-specific ESCs can be isolated from embryos discarded after a positive preimplantation genetic diagnosis (PGD embryos) since some forms of PD are genetic. Another exciting avenue to explore is to induce ES-like cells (iPS cells) from fibroblasts (Takahashi *et al.*, 2007; Yu *et al.*, 2007). This revolutionary new technology should make the production of pluripotent cells from PD patients relatively simple. Directing the differentiation of ESCs or iPS cell lines carrying disease-causing mutations might help clarify the cellular events that cause PD, and provide a powerful high-throughput screening platform for drug compounds that slow or block the destruction of human mDA neurons.

It is thought that genetics is only partially responsible for PD since most cases are sporadic and therefore likely unrelated to genotype (although see [Pamphlett, 2004; Ahn et al., 2008]). Epidemiological studies have correlated many environmental factors with PD. A more thorough understanding of PD will likely emerge by screening the thousands of chemicals that we come into contact with in our daily lives. This is difficult to comprehensively test in a relevant system. One approach to identifying compounds that might contribute to PD is to use epidemiology to identify populations with higher than expected levels of PD. Such an approach led to the finding that rotenone, a frequently used pesticide, causes a disease that resembles PD in rats (Betarbet et al., 2000). A more aggressive screening of compounds initially identified through epidemiological studies could be achieved with the unlimited numbers of ESC-derived mDA neurons. Interestingly, a recent epidemiological study found that the highest risk factor for PD was a family history of the disease (Dick et al., 2007b). There were no strong correlations to currently known genetic mutations, suggesting that many PD-related genes remain to be discovered (Dick et al., 2007a). However, it is also possible that something unique to the families' environment causes the high incidence of disease.

ESC GENETIC MODIFICATION

One advantage of using ESCs is the ability to precisely alter the genome of karotypically normal cells. To create knock-out or knock-in mice, the genetic modification is first constructed in vitro before site-specific, homologous recombination into the ESC genome. Genetically engineered ESCs are then used to create mice carrying the desired mutation. While not as refined as mouse techniques, progress has been made in modifying the human ESC genome as well. Homologous recombination has been achieved in HPRT, Pou5F1, and Rosa26, although the procedure is difficult and therefore not yet in common use (Zwaka and Thomson, 2003; Urbach et al., 2004; Irion et al., 2007). Because

of the difficulty of making human knock-out and knock-in lines, RNAi has been used to knock-down expression of specific genes in human ESCs (Liu et al., 2005; Zaehres et al., 2005). The creation of RNAi hESC lines is usually faster and easier than generating knock-out hESC lines. However, incomplete knock-down of target gene and off-target effects can complicate data interpretation.

Another approach to genetic modification is the use of conventional, plasmid-based transgenes whereby a promoter–gene–polyadenylation cassette is randomly integrated into the ESC genome. The gene in question could be mutated and might behave as a dominant negative allele that induces a toxic gain of function in the target cell, providing an opportunity to study the effects of the mutation in question. In other cases, the transgene might read out the activity of a promoter. Such transgenes facilitate developmental biology studies, permit cell isolation for transplantation, and can aid high-throughput screens designed to identify genes or compounds that act on the cell type or pathway in question. One variation on the transgenic strategy is the use of bacterial artificial chromosomes (BACs). BACs are large pieces of genomic DNA (~100–300 kb) that are maintained and amplified in bacteria. BACs are thought to more faithfully mimic gene expression because their large size isolates the transgene from the surrounding chromatin, and because more promoter/enhancer sequences are present compared to small, conventional transgenes (Lee et al., 2001). BACs have been used to model human disease by expressing mutated alleles that cause disease symptoms in transgenic mice (Gu et al., 2005), and have been used in mouse ESCs as a mechanism to visualize gene expression and isolate specific cell populations (Tomishima et al., 2007) or as a means to study promoter activity (Xian et al., 2005). A second transgene variation is the use of viral vectors to deliver a transgene. Such vectors are more difficult to create but easier to deliver to the ESCs. It should be noted that a nearly infinite combination of these techniques can be used. For example, RNAi and BACs can be delivered via viral vectors (Wade-Martins et al., 2001; Barton and Medzhitov, 2002; Brummelkamp et al., 2002; Devroe and Silver, 2002).

Engineering human ESCs that express disease-associated alleles will provide an opportunity to study these cells at a cellular level. Using this approach, mutant and wild type human mDA neurons could be produced in unlimited quantities. One major unknown factor is the kinetics of an ESC-based model system: PD is a disease common in aged populations and is rarely observed in patients <40 years of age. Therefore it could be possible that under normal conditions cellular phenotypes of the disease would take just as long to develop *in vitro*. If true, it might still be possible to uncover phenotypes in a more reasonable timeframe by sensitizing mutant neurons using low concentrations of chemicals known to cause the selective destruction of mDA neurons such as MPTP or rotenone. Alternatively, such studies might reveal cellular changes that precede clinical symptoms serving as surrogate markers of the disease.

MAKING AUTHENTIC MDA NEURONS

ESCs can develop into any cell of the adult organism, at least in the case of mouse ESCs. This remarkable plasticity poses an interesting problem for using ESCs: how can we direct differentiation into a single specific cell type? Work in ESCs has shown that, in the simplest terms, the derivation of mDA neurons can be defined in four interdependent stages: *neural induction*, *patterning*, dopamine neuron *specification*, and *survival*.

Neural Induction

ESCs are defined by their ability to yield any of the three primary germ layers, and ESCs must first be directed to become embryonic ectoderm prior to adopting a neural fate (*neural induction*). The specified ectodermal cells default into neural cells unless alternative signals convert them to an epidermal fate. In practical terms, a number of techniques have been developed to direct ESCs to a neural fate (reviewed in [Tomishima and Studer, 2003]). These techniques include co-culture with specific bone marrow stromal cells such as the PA-6 or MS5 cell lines, embryoid body-based neural induction methods, neural overgrowth, and default induction strategies. One of the simplest mechanisms to coax ESCs to become neural is called default induction. Default induction produces neural cells in the absence of aggregation or co-culture with feeder cells. Instead, it relies on a combination of cell density and media components to induce neural differentiation (Ying *et al.*, 2003). This initial study demonstrated that neurons exposed to Sonic Hedgehog (SHH) and FGF8 expressed tyrosine hydroxylase (*TH*), the

rate-limiting enzyme in the production of dopamine (Ying *et al.*, 2003), suggesting that ESCs could be *patterned* (see below) toward an mDA neuron state using default induction. However, more extensive analysis of the *TH*-expressing neuron populations derived via default induction suggested lack of expression of many key mDA neuron markers such as *FoxA2*, *Lmx1a*, and *En* (Andersson *et al.*, 2006; Parmar and Li, 2007). Based on these findings it was suggested that ESC differentiation toward authentic mDA neurons requires the transgenic activation of intrinsic cell fate determinants such as *Lmx1A* (Andersson *et al.*, 2006). A later study compared default induction to stromal feeder cell-based induction protocols and found that the method of neural induction was critical for the subsequent expression of mDA neuron markers. Parmar and Li (2007) found that default induction led to the expression of *TH* and *Nurr1* but not *En*, *FoxA2*, and *Lmx1a*. However, parallel studies using co-culture with bone marrow stromal feeder cells for neural induction yielded neurons that expressed *TH*, *Nurr1* as well as the complete panel of markers *En*, *FoxA2*, and *Lmx1a*. Together, these results suggest that the production of genuine mDA neurons from ESCs via extrinsic factors requires the collaboration of SHH and FGF8 and a set of undefined factors that can be supplied – at least in part – via co-culture with bone marrow stromal cells.

Early work using bone marrow stromal cells suggested that this neural induction technique made a high proportion of mDA neurons without the need to add extrinsic factors such as SHH and FGF8 (Kawasaki *et al.*, 2000). Our work has shown that the addition of extrinsic factors to bone marrow stromal cell-based protocol could direct ESCs to primarily make forebrain, midbrain, hindbrain, or spinal cord neurons depending on the combination and timing of factor addition (Barberi *et al.*, 2003). Likewise, the serum-free EB (SFEB) method was shown to make a higher proportion of forebrain neurons (Watanabe *et al.*, 2005). However, at least for human ESCs, it has been clearly shown that the addition of extrinsic factors to SFEB-like culture conditions can direct fate toward mDA neurons and motor neurons (Li *et al.*, 2005b; Yan *et al.*, 2005; Roy *et al.*, 2006). The efficiency of mDA neuron generation from human ESCs using SFEB-based neural induction approaches are comparable to that observed using stromal feeder-based neural induction (Perrier *et al.*, 2004). One way to reconcile all of the above-mentioned results is to posit that the method of neural induction biases the neural patterning state of ESC progeny but that this state can be further modified via addition of extrinsic patterning factors. It is also important to note that the addition of extrinsic factors is exquisitely sensitive to appropriate timing and concentration. Further work is needed to define the effect of neural induction methods on neural patterning at the molecular level and to characterize epigenetic states during neural induction that convey competence for further specification into authentic mDA neurons.

Patterning

At the time of neural induction, newly specified cells receive signals that determine positional identity in a time-dependent manner. This four-dimensional information grid is known as *patterning*, and it ultimately defines the specific cell types that emerge during neural development. For mDA neuron development *in vivo*, two main signals are thought to provide patterning information: FGF8 provides the anterior–posterior information, and SHH provides the dorsal–ventral cue. *Fgf8* is the central player in establishing the isthmic organizer: the structure responsible for directing transcriptional cascades necessary for midbrain and anterior hindbrain development (Wurst and Bally-Cuif, 2001; Sato *et al.*, 2004; Zervas *et al.*, 2005). Ectopic FGF8 application can direct developing forebrain to adopt midbrain–hindbrain patterns of gene expression, altering anterior–posterior positional information (Liu *et al.*, 1999). While *Fgf8* controls downstream transcriptional cascades, the continued expression of *Fgfs* requires feedback from downstream genes such as *En* and *Wnt1* discussed below. An appropriate anterior–posterior pattern is only part of the positional information needed to make an mDA neuron: developing progenitors still require SHH that provides a ventralizing effect. SHH is first produced in the notochord, and its expression later induces the floor plate to produce SHH as well. SHH is an extracellular protein released from the notochord and floor plate, forming a ventral-to-dorsal gradient in the neural tube. It is often called a "morphogen" since small differences in the SHH concentrations cause changes in cell fate.

DA Neuron Specification

After the ventral midbrain is patterned, additional secreted factors collaborate to produce mDA neurons.

Midbrain astrocytes secrete many factors that aid in mDA neuron specification and in survival (Prochiantz *et al.*, 1979; Engele, 1998a,b). One recent example is the finding that ventral midbrain glia secrete WNT5a, which in turn promotes the differentiation of mDA neurons from their progenitors (Castelo-Branco *et al.*, 2006). *Wnt1* is also expressed in the midbrain and has important effects on mDA progenitor cells, although, in contrast to *Wnt5A*, it causes enhanced proliferation of midbrain progenitors (Arenas, 2005). Based on these data, it is likely that the *Wnts* have multiple effects (in addition to specification) at each stage of development of mDA neurons. Another family of secreted proteins that promotes DA neuron specification is the TGFβ family of molecules: these proteins are released from the ventral midline and are important for the production of authentic mDA neurons particularly the expression of *En1* in mDA neurons (Farkas *et al.*, 2003). Similar to the effects reported for Wnt molecules, there is evidence that TGFβs act on many stages of mDA neuron development (Farkas *et al.*, 2003; Roussa *et al.*, 2004; Roussa *et al.*, 2006).

DA Neuron Survival

Once mDA neurons have differentiated, they require trophic support to remain alive. Many different molecules support mDA neuron survival from fetal tissue although only a few have been tested on ESC-derived mDA neurons. As mentioned above, early work showed that astrocytes markedly improved mDA neuron survival *in vitro*. Recent studies using human ESCs suggest that there are still astrocyte-derived factors to be discovered (Roy *et al.*, 2006). DA neuron survival promoted by astrocytes provided an *in vitro* system to isolate a number of factors now used in ESC-derived mDA culture systems (Prochiantz *et al.*, 1979; Engele, 1998a, b). One of the first factors identified was GDNF: the glial cell line-derived neurotrophic factor (Lin *et al.*, 1993). A few members of the TGFβs family were also isolated from astrocytes (Poulsen *et al.*, 1994; Krieglstein *et al.*, 1995). BDNF was isolated biochemically from the striatum, and was shown to increase survival of mDA neurons (Hyman *et al.*, 1991). Ascorbic acid was initially shown to increase the numbers of TH neurons from primary culture (Kalir and Mytilineou, 1991) and has since been included in many ES-derived culture systems (Lee *et al.*, 2000). More recently, FGF20

TABLE 35.1 Commonly used patterning and survival factors for making ESC-derived mDA neurons

	Species	FGF8	SHH	Ascorbic acid	cAMP	BDNF	GDNF	TGFβ	NT-3	WNT3a	Astrocytes	Stromal feeders
Perrier et al., 2004	human	✓	✓	✓	✓	✓	✓	✓				✓
Yan et al., 2005	human	✓	✓	✓	✓	✓	✓					
Yang et al., 2008	human	✓	✓	✓	✓	✓	✓	✓		✓		
Park et al., 2005	human		✓	✓								✓
Zeng et al., 2006	human						✓					✓
Roy et al., 2006	human	✓	✓								✓	
Sanchez-Pernaute et al., 2005	monkey	✓	✓	✓	✓	✓	✓	✓				
Takagi et al., 2006	monkey					✓			✓			
Lee et al., 2000	mouse	✓	✓	✓	✓							
Barberi et al., 2003	mouse	✓	✓	✓		✓						✓
Andersson et al., 2006	mouse	✓	✓									
Parmar and Li, 2007	mouse	✓	✓									✓

has been shown to enhance the survival of fetal mDA neurons in culture (Ohmachi, 2000; Murase, 2006) and from monkey ES-derived mDA neurons (Takagi *et al.*, 2005). Table 35.1 shows many of the commonly used patterning and survival factors used in making mDA neurons from ESCs.

EXAMPLES OF ESC-BASED PD APPROACHES

The ability to direct ESCs to mDA neurons has provided a new way to model PD pathogenesis. Using the methods described above, we can now produce mDA neurons *in vitro* to test hypotheses about PD pathogenesis, or to screen for genes or compounds that inhibit the disease process(es). This type of approach is still in its infancy, but there are proof-of-principle experiments that demonstrate the power of ESC-derived neurons.

The first example of such an approach was the study of *DJ-1* (PARK7), first identified as being relevant to PD through human genetic studies (PARK7 mutation) (van Duijn *et al.*, 2001; Bonifati *et al.*,

2003; Klein and Schlossmacher, 2006). To test how *DJ-1* affects mDA neurons, *DJ-1* knock-out ESCs were made and differentiated into mDA neurons. *DJ-1* null ESC-derived mDA neurons were found to be more sensitive to oxidative stress when compared to normal controls (Martinat *et al.*, 2004; Shendelman *et al.*, 2004).

A second example of this approach examined the role of α-synuclein. The function of α-synuclein is not known, but the protein is largely expressed in presynaptic boutons in the nervous system. Mutations in α-synucleins were identified in families with PD (Polymeropoulos *et al.*, 1997). ESCs were engineered to overexpress *Nurr1*, a nuclear orphan receptor known to enhance the differentiation of ESC-derived mDA neurons (Kim *et al.*, 2002). These *Nurr1* transgenic ESCs were subsequently used to express wild type, A30P, A53T α-synuclein during differentiation to mDA neurons (Yamashita *et al.*, 2006). Using this strategy, the authors found that the disease-associated alleles caused increased susceptibility of oxidative stress, proteasome inhibition, and mitochondrial inhibition. These phenotypes are not universal since ESC-derived oligodendrocytes did not show such differences. All genotypes had the same viability until around d28 after differentiation, when reduced viability and cytoplasmic aggregates formed in the mutant mDA neurons (Yamashita et al., 2006).

To date, there have been at least 12 genes that correlate (at least in some studies) with PD (Parkinson's disease mutational database-www. thepi.org). So far, only mouse ESCs have been used to model PD, but human ESCs will likely provide a better model of human disease. This will require the refinement of methods to genetically modify human ESCs. Only around 10% of the cases of PD appear to be genetic, and it is thought that understanding these rare cases will provide insight into the more common sporadic cases of the disease. Recently, it has been shown that some forms of sporadic PD are due to genetic mutations not passed through the germline (Ahn et al., 2008). This observation could increase the importance of understanding the pathogenesis of genetically inherited PD. The production of preimplantation genetic diagnostic (PGD) human ESCs could be a complementary method to discover how such mutations cause PD. Genetic testing is now available for the *parkin* and *PINK1* genes (www.genome.gov/10001217#4). The recent discovery that adult human fibroblasts can be reverted to an ES-like cell is likely to cause a rapid increase in such research since derivation of cell lines no longer requires a human blastocyst (Takahashi et al., 2007; Yu et al., 2007).

FROM DEVELOPMENT TO HUMAN DISEASE

A reciprocal approach to understanding PD is to learn how mDA neurons develop *in vivo*, how the neural circuitry is established during development, and how such cells function over the life of an organism. A comprehensive understanding of mDA neurons could lead to insights about how PD starts and progresses. Developmental biology has made enormous progress in elucidating the molecular cues that make and maintain mDA neurons yet there is much to learn. In many ways, this basic science approach is diametrically opposed to the clinical methods described above, and yet has arguably provided a wealth of information relative to human disease. For example, ESC-based methods for making mDA neurons could not have progressed without an understanding of midbrain development. One limitation of this approach is the difficulty of studying human development. Therefore nearly all of the data available to date were derived from mouse development or other model organisms. Human ESC biology might provide a novel alternative route modeling human development *in vitro*.

Engrailed

There are two *Engrailed* (*En*) genes in mice and humans, and their function and expression patterns partially overlap (Joyner, 1996; Liu and Joyner, 2001; Simon et al., 2004). Since little is known about *En* and human development, the data reviewed below focuses on mouse development. Expression of both *En* genes is controlled by *Fgf8* before E13, and both *En1* and *Fgf8* are expressed in the mesencephalon and rostral cerebellum (r1) at E8.5 – about 12 h before *En2* expression begins (Zetterstrom et al., 1997; Simon et al., 2004; Zervas et al., 2005). *En1* continues to be expressed in cells that do not express *En2* later in development. *En1* mutants are not viable and have an almost complete loss of the midbrain/r1 region by E9.5 (Wurst et al., 1994). The loss of a midbrain and r1 is caused at least partially by cell death, suggesting that the region is initially established but cannot be maintained without *En1* expression (Chi et al., 2003). *En2* mutants, on the other hand, have a much milder phenotype limited to the patterning and growth of the cerebellar folia (Joyner et al., 1991; Millen et al., 1994). While the different phenotypes and expression patterns underscore the differences between the two *En* genes, some experiments highlight their similarities: *En1/En2* double mutants completely lack the tectum and cerebellum (Liu and Joyner, 2001; Simon et al., 2004) and *En2* can substitute for *En1* when targeted to the *En1* locus. In fact, Drosophila *En* can reverse the brain phenotypes when expressed in mutant mice, although it does not substitute for *En* functions outside of the nervous system (Hanks et al., 1998). More recent *En* dose-response experiments have revealed that *En1* plays a larger role in the development of the midbrain, whereas *En2* is more important in the cerebellum with certain folia being completely dependent on *En2* (Sgaier et al., 2007). One interesting finding is that continued expression of *En* is necessary to maintain mDA neurons throughout life. Heterozygous adult *En1–/+* mice progressively lose mDA neurons in a Swiss background (Sonnier et al., 2007); the same phenotype is found in *En1–/+; En2–/–* mutants in a C57BL6 background

(Sgado *et al.*, 2006). To the best of our knowledge, *En* has not yet been associated with clinical PD, and *En* loss of function mutations have not yet been systematically explored in ESC-based model systems.

FoxA1 and A2

The forkhead transcription factors *FoxA1/2* (*HNF3α/β*) have been known to play a critical role in nervous system development of the notochord and floor plate for over a decade (Lai *et al.*, 1991; Ang and Rossant, 1994; Sasaki and Hogan, 1994; Weinstein *et al.*, 1994). While it had been clear that these genes were important to ventralize the neural tube, it has recently emerged that mDA neurons are directly derived from floor plate cells (Ferri *et al.*, 2007; Kittappa *et al.*, 2007). Furthermore, much like the situation with *En*, continuous *FoxA1/2* expression is required for the development and the maintenance of mDA neurons. One third of the heterozygous *FoxA2* animals over 18 months of age spontaneously develop motor abnormalities (Kittappa *et al.*, 2007). These abnormalities occur asymmetrically in the SNc, causing preferential degeneration of TH cells on one side of the brain. *FoxA2* overexpression in mouse ESCs markedly increases the number of TH neurons after differentiation, even in the presence of cyclopamine (Kittappa *et al.*, 2007). This suggests that the main role of *Shh* is to cause *FoxA2* expression, and *FoxA2* is known to regulate *En1/2* (Ferri *et al.*, 2007). This would place *FoxA2* at or near the beginning of the transcriptional cascade needed to make and maintain mDA neurons.

Pitx3

The bicoid-related homeobox gene *Pitx3* is expressed in all mDA neurons although the absence of *Pitx3* affects mDA neurons differently. *Pitx3* is required for the survival of some SNc mDA neurons in adult mice (Hwang *et al.*, 2003; Nunes *et al.*, 2003; van den Munckhof *et al.*, 2003; Smidt *et al.*, 2004; Maxwell *et al.*, 2005). At e12.5, a variety of phenotypes have been reported: no change in TH+ numbers (van den Munckhof *et al.*, 2003), altered distribution but no change in numbers (Smidt *et al.*, 2004), or a striking reduction in number (Zhao *et al.*, 2004). Zhao *et al.* used an engineered null mutant whereas the other groups used the naturally occurring but hypomorphic *aphakia* mutant (Rieger *et al.*, 2001). The surviving *Pitx3*–/– SNc neurons do not express TH (Zhao

et al., 2004). It is not clear why the loss of *Pitx3* affects SNc dopamine neurons differently than VTA or RA mDA neurons, but it has been hypothesized that the unique gene expression pattern requires *Pitx3* expression for the proper functioning of the SNc neuron. *Pitx3* controls the expression of *Ahd2*, an aldehyde dehydrogenase that oxidizes retinaldehyde to retinoic acid. Some of the developmental defects in the SNc caused by the absence of *Pitx3* can be reversed by the application of retinoic acid (Jacobs *et al.*, 2007). One explanation for this result is that retinoic acid restores the mDA pattern that was lost without *Pitx3*.

Pitx3 overexpression in ESCs enhances their differentiation into mDA neurons (Chung *et al.*, 2005; Maxwell *et al.*, 2005), whereas another study found that only some mDA markers were expressed after differentiation of *Pitx3*-overexpressing ESCs (Martinat *et al.*, 2004). In Martinat *et al.*, *Nurr1* and *Pitx3* co-expression was shown to enhance maturation of ESC-derived mDA neurons. Clinically, a family with a *Pitx3* mutation has been identified, and a few family members were homozygous and had significant neurological problems consistent with the phenotype of *Pitx3* mutant mice. More recently, it has been reported that a single nucleotide polymorphism (SNP) in the *Pitx3* promoter correlates with sporadic PD (Fuchs *et al.*, 2007).

Nurr1

Nurr1 is an orphan nuclear receptor expressed more widely in the CNS, unlike *Pitx3* whose CNS expression is limited to mDA neurons. In *Nurr1*–/– mice, mDA neurons are initially established, but there is conflicting evidence regarding the fate of *Nurr1*–/– cells perinatally. These discrepancies might depend on the genetic background of the mice or the marker used to visualize the mDA population (Saucedo-Cardenas *et al.*, 1998; Le *et al.*, 1999; Wallen *et al.*, 1999). Continued expression of *Nurr1* is required for the expression of *TH*, *DAT*, *VMAT2*, and *AADC* in mDA neurons (Smits *et al.*, 2003; Wang *et al.*, 2003), and old *Nurr1* +/– animals displayed motor impairments consistent with PD (Jiang *et al.*, 2005). Neuropathological studies have suggested loss of NURR1 protein in some PD patients (Chu *et al.*, 2006). As noted above, Martinat *et al.* (2004) argue that *Nurr1* and *Pitx3* must be co-expressed to induce ESC-derived neurons to become mDA neurons, whereas others reported that *Nurr1* alone was sufficient to facilitate conversion into mDA neurons (Chung *et al.*, 2002; Kim *et al.*, 2002, p. 223). A direct comparison between ESCs overexpressng *Nurr1* and *Pitx3* found that excess *Nurr1* alone globally enhanced mDA differentiation whereas excess *Pitx3* specifically enhanced mDA neurons from the A9 region, the neurons that project to the striatum (Chung *et al.*, 2005). An interesting side note is that *Nurr1* can induce a non-neuronal dopaminergic cell type during ESC differentiation. This indicates that some of the transcriptional cascades that induce biochemical maturation of dopamine neuron phenotype may be independent of cellular context (Sonntag *et al.*, 2004).

WNTs

The WNTs are a family of lipoproteins that are known to control many stages of nervous system development. As noted above, a key role for *Wnt1* is stabilization of signals in the isthmic organizer and the specification of midbrain versus hindbrain areas (Li *et al.*, 2005a). WNT5a is secreted from ventral midbrain glia (Schulte *et al.*, 2005; Castelo-Branco *et al.*, 2006), and both *Wnt1* and *Wnt5a* were expressed in the *Nurr1*+ domain as mDA neurons are born *in vivo* (Castelo-Branco *et al.*, 2003). *In vitro* experiments with WNT-conditioned media have shown that WNT1 increases the overall number of TuJ1+ cells in ventral mesencephalic cultures and enhanced their differentiation into mDA neurons. This result suggests that WNT1 has a sequential effect on VM precursors, initially expanding the progenitor but later enhancing its differentiation. WNT5a does not display mitogenic activity but rather promotes differentiation of midbrain progenitors into postmitotic dopamine neurons (Castelo-Branco *et al.*, 2003).

Lmx1a and b

Lmx1a

Lmx1a plays an early role in the development of mDA neurons by suppressing alternative fates within the ventral midbrain. This effect is mediated in part via direct suppression of *Nkx6.1* critical in establishing motor neuron identity within the midbrain. *Lmx1a* is also critical in orchestrating a transcriptional cascade needed for terminal differentiation. *Lmx1a* activates another transcription factor *Msx1* that drives mDA progenitors toward

a neuronal fate (Andersson *et al.*, 2006). Mouse ESCs engineered to express *Lmx1a* differentiate efficiently into authentic mDA neurons when cultured in the presence of SHH and FGF8 (Andersson *et al.*, 2006).

Lmx1b

Lmx1b is early in the cascade of the transcriptional network controlling mDA neuron production: it is essential for *Fgf8* and *Wnt1* expression in the isthmic organizer (Adams *et al.*, 2000). Aside from this patterning role, *Lmx1b* is also required to make properly differentiated mDA neurons. *Lmx1b* knockouts yield TH-expressing cells that lack expression of *Pitx3*. Interestingly, an ectopic population of cells was observed that expresses *Pitx3* but lacking TH (Smidt *et al.*, 2000).

FUTURE APPLICATIONS FOR ESC-BASED DA NEURONS

ESC-based DA neurons provide an ideal substrate for large-scale, high-throughput screens. The derivation of unlimited numbers of karotypically normal and authentic human mDA neurons should soon become routine technology. In parallel, there has been considerable progress in our ability to genetically manipulate human ESCs following the foot steps of mouse genetics. Genetic modification can be used to engineer ESC reporter lines that read out mDA neuron specification, survival, or metabolic state. Such reporter lines will be powerful tools in the development of ESC-based throughput screening assays.

Genetic modifications can also be harnessed for direct genetic modeling of the disease state via introduction of disease-associated alleles. In addition to the candidate gene approaches discussed above, there are a number of large-scale ESC-based tools that may be attractive for future disease modeling efforts. For example libraries of ESC lines have be generated that trap most known mouse genes (Nord *et al.*, 2006) (Lexicon, BayGenomics Consortium) and similar efforts are under development for human ESCs (Dhara and Benvenisty, 2004). Such libraries could be adapted for the identification of candidate genes involved in DA neuron specification and death. The recent breakthroughs in DNA sequencing and SNP-based association studies (for review [Christensen and Murray, 2007, p. 250]) will require the use of a relevant functional readout in order to screen for functionally relevant genetic changes. ESC-based genetic modeling approaches may provide a versatile platform to this end. Finally the iPS technology should enable rapid progress in generating sets of PD-specific human ESC lines from hundreds of patients. Such libraries of PD-specific iPS cell lines will provide a unique experimental tool to dissect genetic contribution to disease state, defining PD subtypes and ultimately for drug development.

With the emergence of these novel ESC-based tools for PD modeling it should become possible to screen very large libraries of small molecules, drugs, or genes. Such studies should contribute to our understanding of PD pathogenesis and help unraveling the signaling cascades underlying the death of these neurons in culture. ESC-based genetic labeling techniques should permit the direct isolation of mDA neurons at any stage of development. The use of purified cell populations will facilitate global gene and protein expression studies or studies addressing the importance of epigenetic changes during DA neuron development and cell death. Studies with purified ESC progeny will also be critical in defining the contribution of cell autonomous versus non-autonomous factors in the development of disease as illustrated recently in ESC-based models of ALS (Di Giorgio *et al.*, 2007; Nagai *et al.*, 2007). Finally, ESC-based assays may also be critical in screening for compounds that cause the selective death of mDA neurons as a novel tool in the armamentarium for toxicological and environmental studies related to PD. While the future for cell therapy in PD remains uncertain, the use of human ES-derived mDA neurons offers vast opportunities to use these cells now for learning about PD pathogenesis and for the identification of novel drug targets.

REFERENCES

Adams, K. A., Maida, J. M., Golden, J. A., and Riddle, R. D. (2000). The transcription factor Lmx1b maintains Wnt1 expression within the isthmic organizer. *Development* **127**, 1857–1867.

Ahn, T. B., Kim, S. Y., Kim, J. Y., Park, S. S., Lee, D. S., Min, H. J., Kim, Y. K., Kim, S. E., Kim, J. M., Kim, H. J., Cho, J., and Jeon, B. S. (2007). Alpha-synuclein gene duplication is present in sporadic Parkinson disease. *Neurology* 2008. Jan 1; 70(1) 43–49.

Andersson, E., Tryggvason, U., Deng, Q., Friling, S., Alekseenko, Z., Robert, B., Perlmann, T., and Ericson, J. (2006 Jan 27). Identification of intrinsic determinants of midbrain dopamine neurons. *Cell* **124(2)**, 393–405.

Ang, S. L., and Rossant, J. (1994). HNF-3 beta is essential for node and notochord formation in mouse development. *Cell* **78**, 561–574.

Arenas, E. (2005). Engineering a dopaminergic phenotype in stem/precursor cells: Role of Nurr1, glia-derived signals, and Wnts. *Ann NY Acad Sci* **1049**, 51–66.

Barberi, T., Klivenyi, P., Calingasan, N. Y., Lee, H., Kawamata, H., Loonam, K., Perrier, A. L., Bruses, J., Rubio, M. E., Topf, N., Tabar, V., Harrison, N. L., Beal, M. F., Moore, M. A., and Studer, L. (2003). Neural subtype specification of fertilization and nuclear transfer embryonic stem cells and application in parkinsonian mice. *Nat Biotechnol* **21**, 1200–1207.

Barton, G. M., and Medzhitov, R. (2002). Retroviral delivery of small interfering RNA into primary cells. *Proc Natl Acad Sci USA* **99**, 14943–14945.

Betarbet, R., Sherer, T. B., MacKenzie, G., Garcia-Osuna, M., Panov, A. V., and Greenamyre, J. T. (2000). Chronic systemic pesticide exposure reproduces features of Parkinson's disease. *Nat Neurosci* **3**, 1301–1306.

Bonifati, V., Rizzu, P., van Baren, M. J., Schaap, O., Breedveld, G. J., Krieger, E., Dekker, M. C., Squitieri, F., Ibanez, P., Joosse, M., van Dongen, J. W., Vanacore, N., van Swieten, J. C., Brice, A., Meco, G., van Duijn, C. M., Oostra, B. A., and Heutink, P. (2003). Mutations in the DJ-1 gene associated with autosomal recessive early-onset parkinsonism. *Science* **299**, 256–259.

Brummelkamp, T. R., Bernards, R., and Agami, R. (2002). Stable suppression of tumorigenicity by virus-mediated RNA interference. *Cancer Cell* **2**, 243–247.

Castelo-Branco, G., Wagner, J., Rodriguez, F. J., Kele, J., Sousa, K., Rawal, N., Pasolli, H. A., Fuchs, E., Kitajewski, J., and Arenas, E. (2003). Differential regulation of midbrain dopaminergic neuron development by Wnt-1, Wnt-3a, and Wnt-5a. *Proc Natl Acad Sci USA* **100**, 12747–12752.

Castelo-Branco, G., Sousa, K. M., Bryja, V., Pinto, L., Wagner, J., and Arenas, E. (2006). Ventral midbrain glia express region-specific transcription factors and regulate dopaminergic neurogenesis through Wnt-5a secretion. *Mol Cell Neurosci* **31**, 251–262.

Chi, C. L., Martinez, S., Wurst, W., and Martin, G. R. (2003). The isthmic organizer signal FGF8 is required for cell survival in the prospective midbrain and cerebellum. *Development* **130**, 2633–2644.

Christensen, K., and Murray, J. C. (2007). What genome-wide association studies can do for medicine. *N Engl J Med.* **356**, 1094–1097.

Chu, Y., Le, W., Kompoliti, K., Jankovic, J., Mufson, E. J., and Kordower, J. H. (2006). Nurr1 in Parkinson's disease and related disorders. *J Comp Neurol* **494**, 495–514.

Chung, S., Sonntag, K. C., Andersson, T., Bjorklund, L. M., Park, J. J., Kim, D. W., Kang, U. J., Isacson, O., and Kim, K. S. (2002). Genetic engineering of mouse embryonic stem cells by Nurr1 enhances differentiation and maturation into dopaminergic neurons. *Eur J Neurosci* **16**, 1829–1838.

Chung, S., Hedlund, E., Hwang, M., Kim, D. W., Shin, B. S., Hwang, D. Y., Jung Kang, U., Isacson, O., and Kim, K. S. (2005). The homeodomain transcription factor Pitx3 facilitates differentiation of mouse embryonic stem cells into AHD2-expressing dopaminergic neurons. *Mol Cell Neurosci* **28**, 241–252.

Devroe, E., and Silver, P. A. (2002). Retrovirus-delivered siRNA. *BMC Biotechnol* **2**, 15.

Dhara, S. K., and Benvenisty, N. (2004). Gene trap as a tool for genome annotation and analysis of X chromosome inactivation in human embryonic stem cells. *Nucleic Acids Res* **32**, 3995–4002.

Di Giorgio, F. P., Carrasco, M. A., Siao, M. C., Maniatis, T., and Eggan, K. (2007). Non-cell autonomous effect of glia on motor neurons in an embryonic stem cell-based ALS model. *Nat Neurosci* **10**, 608–614.

Dick, F. D., De Palma, G., Ahmadi, A., Osborne, A., Scott, N. W., Prescott, G. J., Bennett, J., Semple, S., Dick, S., Mozzoni, P., Haites, N., Wettinger, S. B., Mutti, A., Otelea, M., Seaton, A., Soderkvist, P., and Felice, A. (2007a). Gene–environment interactions in parkinsonism and Parkinson's disease: The Geoparkinson study. *Occup Environ Med* **64**, 673–680.

Dick, F. D., De Palma, G., Ahmadi, A., Scott, N. W., Prescott, G. J., Bennett, J., Semple, S., Dick, S., Counsell, C., Mozzoni, P., Haites, N., Wettinger, S. B., Mutti, A., Otelea, M., Seaton, A., Soderkvist, P., and Felice, A. (2007b). Environmental risk factors for Parkinson's disease and parkinsonism: The Geoparkinson study. *Occup Environ Med* **64**, 666–672.

Engele, J. (1998a). Changing responsiveness of developing midbrain dopaminergic neurons for extracellular growth factors. *J Neurosci Res* **51**, 508–516.

Engele, J. (1998b). Spatial and temporal growth factor influences on developing midbrain dopaminergic neurons. *J Neurosci Res* **53**, 405–414.

Farkas, L. M., Dunker, N., Roussa, E., Unsicker, K., and Krieglstein, K. (2003). Transforming growth factor-beta(s) are essential for the development of midbrain dopaminergic neurons *in vitro* and *in vivo*. *J Neurosci* **23**, 5178–5186.

Ferri, A. L., Lin, W., Mavromatakis, Y. E., Wang, J. C., Sasaki, H., Whitsett, J. A., and Ang, S. L. (2007). Foxa1 and Foxa2 regulate multiple phases of midbrain dopaminergic neuron development in a dosage-dependent manner. *Development* **134**, 2761–2769.

Fuchs, J., Mueller, J. C., Lichtner, P., Schulte, C., Munz, M., Berg, D., Wullner, U., Illig, T., Sharma, M., and Gasser, T. (2007). The transcription factor PITX3 is associated with sporadic Parkinson's disease. *Neurobiol Aging* 2007. Sep 28.

Gu, X., Li, C., Wei, W., Lo, V., Gong, S., Li, S. H., Iwasato, T., Itohara, S., Li, X. J., Mody, I., Heintz, N., and Yang, X. W. (2005). Pathological cell–cell interactions elicited by a neuropathogenic form of mutant Huntingtin contribute to cortical pathogenesis in HD mice. *Neuron* **46**, 433–444.

Hanks, M. C., Loomis, C. A., Harris, E., Tong, C. X., Anson-Cartwright, L., Auerbach, A., and Joyner, A. (1998). Drosophila engrailed can substitute for mouse Engrailed1 function in mid–hindbrain, but not limb development. *Development* **125**, 4521–4530.

Hwang, D. Y., Ardayfio, P., Kang, U. J., Semina, E. V., and Kim, K. S. (2003). Selective loss of dopaminergic neurons in the substantia nigra of Pitx3-deficient aphakia mice. *Brain Res Mol Brain Res* **114**, 123–131.

Hyman, C., Hofer, M., Barde, Y. A., Juhasz, M., Yancopoulos, G. D., Squinto, S. P., and Lindsay, R. M. (1991). BDNF is a neurotrophic factor for dopaminergic neurons of the substantia nigra. *Nature* **350**, 230–232.

Irion, S., Luche, H., Gadue, P., Fehling, H. J., Kennedy, M., and Keller, G. (2007). Identification and targeting of the ROSA26 locus in human embryonic stem cells. *Nat Biotechnol* **25**, 1477–1482.

Jacobs, F. M., Smits, S. M., Noorlander, C. W., von Oerthel, L., van der Linden, A. J., Burbach, J. P., and Smidt, M. P. (2007). Retinoic acid counteracts developmental defects in the substantia nigra caused by Pitx3 deficiency. *Development* **134**, 2673–2684.

Jiang, C., Wan, X., He, Y., Pan, T., Jankovic, J., and Le, W. (2005). Age-dependent dopaminergic dysfunction in Nurr1 knockout mice. *Exp Neurol* **191**, 154–162.

Joyner, A. L. (1996). Engrailed, Wnt and Pax genes regulate midbrain–hindbrain development. *Trends Genet* **12**, 15–20.

Joyner, A. L., Herrup, K., Auerbach, B. A., Davis, C. A., and Rossant, J. (1991). Subtle cerebellar phenotype in mice homozygous for a targeted deletion of the En-2 homeobox. *Science* **251**, 1239–1243.

Kalir, H. H., and Mytilineou, C. (1991). Ascorbic acid in mesencephalic cultures: Effects on dopaminergic neuron development. *J Neurochem* **57**, 458–464.

Kawasaki, H., Mizuseki, K., Nishikawa, S., Kaneko, S., Kuwana, Y., Nakanishi, S., Nishikawa, S. I., and Sasai, Y. (2000). Induction of midbrain dopaminergic neurons from ES cells by stromal cell-derived inducing activity. *Neuron* **28**, 31–40.

Kim, J. H., Auerbach, J. M., Rodriguez-Gomez, J. A., Velasco, I., Gavin, D., Lumelsky, N., Lee, S. H.,

Nguyen, J., Sanchez-Pernaute, R., Bankiewicz, K., and McKay, R. (2002). Dopamine neurons derived from embryonic stem cells function in an animal model of Parkinson's disease. *Nature* **418**, 50–56.

Kittappa, R., Chang, W. W., Awatramani, R. B., and McKay, R. D. (2007). The foxa2 gene controls the birth and spontaneous degeneration of dopamine neurons in old age. *PLoS Biol* **5**, e325.

Klein, C., and Schlossmacher, M. G. (2006). The genetics of Parkinson disease: Implications for neurological care. *Nat Clin Pract Neurol* **2**, 136–146.

Krieglstein, K., Suter-Crazzolara, C., Hotten, G., Pohl, J., and Unsicker, K. (1995). Trophic and protective effects of growth/differentiation factor 5, a member of the transforming growth factor-beta superfamily, on midbrain dopaminergic neurons. *J Neurosci Res* **42**, 724–732.

Lai, E., Prezioso, V. R., Tao, W. F., Chen, W. S., and Darnell, J. E., Jr. (1991). Hepatocyte nuclear factor 3 alpha belongs to a gene family in mammals that is homologous to the Drosophila homeotic gene fork head. *Genes Dev* **5**, 416–427.

Le, W., Conneely, O. M., Zou, L., He, Y., Saucedo-Cardenas, O., Jankovic, J., Mosier, D. R., and Appel, S. H. (1999). Selective agenesis of mesencephalic dopaminergic neurons in Nurr1-deficient mice. *Exp Neurol* **159**, 451–458.

Lee, E. C., Yu, D., Martinez de Velasco, J., Tessarollo, L., Swing, D. A., Court, D. L., Jenkins, N. A., and Copeland, N. G. (2001). A highly efficient *Escherichia coli*-based chromosome engineering system adapted for recombinogenic targeting and subcloning of BAC DNA. *Genomics* **73**, 56–65.

Lee, S. H., Lumelsky, N., Studer, L., Auerbach, J. M., and McKay, R. D. (2000). Efficient generation of midbrain and hindbrain neurons from mouse embryonic stem cells. *Nat Biotechnol* **18**, 675–679.

Li, J. Y., Lao, Z., and Joyner, A. L. (2005a). New regulatory interactions and cellular responses in the isthmic organizer region revealed by altering Gbx2 expression. *Development* **132**, 1971–1981.

Li, X. J., Du, Z. W., Zarnowska, E. D., Pankratz, M., Hansen, L. O., Pearce, R. A., and Zhang, S. C. (2005b). Specification of motoneurons from human embryonic stem cells. *Nat Biotechnol* **23**, 215–221.

Lin, L. F., Doherty, D. H., Lile, J. D., Bektesh, S., and Collins, F. (1993). GDNF: A glial cell line-derived neurotrophic factor for midbrain dopaminergic neurons. *Science* **260**, 1130–1132.

Liu, A., and Joyner, A. L. (2001). EN and GBX2 play essential roles downstream of FGF8 in patterning the mouse mid/hindbrain region. *Development* **128**, 181–191.

Liu, A., Losos, K., and Joyner, A. L. (1999). FGF8 can activate Gbx2 and transform regions of the rostral mouse brain into a hindbrain fate. *Development* **126**, 4827–4838.

Liu, Y. P., Dambaeva, S. V., Dovzhenko, O. V., Garthwaite, M. A., and Golos, T. G. (2005). Stable plasmid-based siRNA silencing of gene expression in human embryonic stem cells. *Stem Cells Dev* **14**, 487–492.

Martinat, C., Shendelman, S., Jonason, A., Leete, T., Beal, M. F., Yang, L., Floss, T., and Abeliovich, A. (2004). Sensitivity to oxidative stress in DJ-1-deficient dopamine neurons: An ES-derived cell model of primary Parkinsonism. *PLoS Biol* **2**, e327.

Maxwell, S. L., Ho, H. Y., Kuehner, E., Zhao, S., and Li, M. (2005). Pitx3 regulates tyrosine hydroxylase expression in the substantia nigra and identifies a subgroup of mesencephalic dopaminergic progenitor neurons during mouse development. *Dev Biol* **282**, 467–479.

Millen, K. J., Wurst, W., Herrup, K., and Joyner, A. L. (1994). Abnormal embryonic cerebellar development and patterning of postnatal foliation in two mouse Engrailed-2 mutants. *Development* **120**, 695–706.

Nagai, M., Re, D. B., Nagata, T., Chalazonitis, A., Jessell, T. M., Wichterle, H., and Przedborski, S. (2007). Astrocytes expressing ALS-linked mutated SOD1 release factors selectively toxic to motor neurons. *Nat Neurosci* **10**, 615–622.

Nord, A. S., Chang, P. J., Conklin, B. R., Cox, A. V., Harper, C. A., Hicks, G. G., Huang, C. C., Johns, S. J., Kawamoto, M., Liu, S., Meng, E. C., Morris, J. H., Rossant, J., Ruiz, P., Skarnes, W. C., Soriano, P., Stanford, W. L., Stryke, D., von Melchner, H., Wurst, W., Yamamura, K., Young, S. G., Babbitt, P. C., and Ferrin, T. E. (2006). The International Gene Trap Consortium Website: A portal to all publicly available gene trap cell lines in mouse. *Nucleic Acids Res* **34**, D642–D648.

Nunes, I., Tovmasian, L. T., Silva, R. M., Burke, R. E., and Goff, S. P. (2003). Pitx3 is required for development of substantia nigra dopaminergic neurons. *Proc Natl Acad Sci USA* **100**, 4245–4250.

Pamphlett, R. (2004). Somatic mutation: A cause of sporadic neurodegenerative diseases? *Med Hypotheses* **62**, 679–682.

Parmar, M., and Li, M. (2007). Early specification of dopaminergic phenotype during ES cell differentiation. *BMC Dev Biol* **7**, 86.

Park, C. H., Minn, Y. K., Lee, J. Y., Choi, D. H., Chang, M. Y., Shim, J. W., Ko, J. Y., Koh, H. C., Kang, M. J., Kang, J. S., Rhie, D. J., Lee, Y. S., Son, H., Moon, S. Y., Kim, K. S., and Lee, S. H. (2005). In vitro and in vivo analyses of human embryonic stem cell-derived dopamine neurons. *J Neurochem.* **92(5)**, 1265–1276.

Perrier, A. L., Tabar, V., Barberi, T., Rubio, M. E., Bruses, J., Topf, N., Harrison, N. L., and Studer, L. (2004). Derivation of midbrain dopamine neurons from human embryonic stem cells. *Proc Natl Acad Sci USA* **101**, 12543–12548.

Polymeropoulos, M. H., Lavedan, C., Leroy, E., Ide, S. E., Dehejia, A., Dutra, A., Pike, B., Root, H., Rubenstein, J., Boyer, R., Stenroos, E. S., Chandrasekharappa, S., Athanassiadou, A., Papapetropoulos, T., Johnson, W. G., Lazzarini, A. M., Duvoisin, R. C., Di Iorio, G., Golbe, L. I., and Nussbaum, R. L. (1997). Mutation in the alpha-synuclein gene identified in families with Parkinson's disease. *Science* **276**, 2045–2047.

Poulsen, K. T., Armanini, M. P., Klein, R. D., Hynes, M. A., Phillips, H. S., and Rosenthal, A. (1994). TGF beta 2 and TGF beta 3 are potent survival factors for midbrain dopaminergic neurons. *Neuron* **13**, 1245–1252.

Prochiantz, A., di Porzio, U., Kato, A., Berger, B., and Glowinski, J. (1979). *In vitro* maturation of mesencephalic dopaminergic neurons from mouse embryos is enhanced in presence of their striatal target cells. *Proc Natl Acad Sci USA* **76**, 5387–5391.

Rieger, D. K., Reichenberger, E., McLean, W., Sidow, A., and Olsen, B. R. (2001). A double-deletion mutation in the Pitx3 gene causes arrested lens development in aphakia mice. *Genomics* **72**, 61–72.

Rodriguez-Gomez, J. A., Lu, J. Q., Velasco, I., Rivera, S., Zoghbi, S. S., Liow, J. S., Musachio, J. L., Chin, F. T., Toyama, H., Seidel, J., Green, M. V., Thanos, P. K., Ichise, M., Pike, V. W., Innis, R. B., and McKay, R. D. (2007). Persistent dopamine functions of neurons derived from embryonic stem cells in a rodent model of Parkinson disease. *Stem Cells* **25**, 918–928.

Roussa, E., Farkas, L. M., and Krieglstein, K. (2004). TGF-beta promotes survival on mesencephalic dopaminergic neurons in cooperation with Shh and FGF-8. *Neurobiol Dis* **16**, 300–310.

Roussa, E., Wiehle, M., Dunker, N., Becker-Katins, S., Oehlke, O., and Krieglstein, K. (2006). Transforming growth factor beta is required for differentiation of mouse mesencephalic progenitors into dopaminergic neurons *in vitro* and *in vivo*: Ectopic induction in dorsal mesencephalon. *Stem Cells* **24**, 2120–2129.

Roy, N. S., Cleren, C., Singh, S. K., Yang, L., Beal, M. F., and Goldman, S. A. (2006). Functional engraftment of human ES cell-derived dopaminergic neurons enriched by coculture with telomerase-immortalized midbrain astrocytes. *Nat Med* **12**, 1259–1268.

Sanchez-Pernaute, R., Studer, L., Ferrari, D., Perrier, A., Lee, H., Vinuela, A., and Isacson, O. (2005). Long-term survival of dopamine neurons derived from parthenogenetic primate embryonic stem cells (cyno-1) after transplantation. *Stem Cells* **128**, 914–922.

Sasaki, H., and Hogan, B. L. (1994). HNF-3 beta as a regulator of floor plate development. *Cell* **76**, 103–115.

Sato, T., Joyner, A. L., and Nakamura, H. (2004). How does Fgf signaling from the isthmic organizer induce midbrain and cerebellum development? *Dev Growth Differ* **46**, 487–494.

Saucedo-Cardenas, O., Quintana-Hau, J. D., Le, W. D., Smidt, M. P., Cox, J. J., De Mayo, F., Burbach, J. P., and Conneely, O. M. (1998). Nurr1 is essential for the induction of the dopaminergic phenotype and the survival of ventral mesencephalic late dopaminergic precursor neurons. *Proc Natl Acad Sci USA* **95**, 4013–4018.

Schulte, G., Bryja, V., Rawal, N., Castelo-Branco, G., Sousa, K. M., and Arenas, E. (2005). Purified Wnt-5a increases differentiation of midbrain dopaminergic cells and dishevelled phosphorylation. *J Neurochem* **92**, 1550–1553.

Sgado, P., Alberi, L., Gherbassi, D., Galasso, S. L., Ramakers, G. M., Alavian, K. N., Smidt, M. P., Dyck, R. H., and Simon, H. H. (2006). Slow progressive degeneration of nigral dopaminergic neurons in postnatal Engrailed mutant mice. *Proc Natl Acad Sci USA* **103**, 15242–15247.

Sgaier, S. K., Lao, Z., Villanueva, M. P., Berenshteyn, F., Stephen, D., Turnbull, R. K., and Joyner, A. L. (2007). Genetic subdivision of the tectum and cerebellum into functionally related regions based on differential sensitivity to engrailed proteins. *Development* **134**, 2325–2335.

Shendelman, S., Jonason, A., Martinat, C., Leete, T., and Abeliovich, A. (2004). DJ-1 is a redox-dependent molecular chaperone that inhibits alpha-synuclein aggregate formation. *PLoS Biol* **2**, e362.

Simon, H. H., Thuret, S., and Alberi, L. (2004). Midbrain dopaminergic neurons: Control of their cell fate by the engrailed transcription factors. *Cell Tissue Res* **318**, 53–61.

Smidt, M. P., Asbreuk, C. H., Cox, J. J., Chen, H., Johnson, R. L., and Burbach, J. P. (2000). A second independent pathway for development of mesencephalic dopaminergic neurons requires Lmx1b. *Nat Neurosci* **3**, 337–341.

Smidt, M. P., Smits, S. M., Bouwmeester, H., Hamers, F. P., van der Linden, A. J., Hellemons, A. J., Graw, J., and Burbach, J. P. (2004). Early developmental failure of substantia nigra dopamine neurons in mice lacking the homeodomain gene Pitx3. *Development* **131**, 1145–1155.

Smits, S. M., Ponnio, T., Conneely, O. M., Burbach, J. P., and Smidt, M. P. (2003). Involvement of Nurr1 in specifying the neurotransmitter identity of ventral midbrain dopaminergic neurons. *Eur J Neurosci* **18**, 1731–1738.

Sonnier, L., Le Pen, G., Hartmann, A., Bizot, J. C., Trovero, F., Krebs, M. O., and Prochiantz, A. (2007). Progressive loss of dopaminergic neurons in the ventral midbrain of adult mice heterozygote for Engrailed1. *J Neurosci* **27**, 1063–1071.

Sonntag, K. C., Simantov, R., Kim, K. S., and Isacson, O. (2004). Temporally induced Nurr1 can induce a nonneuronal dopaminergic cell type in embryonic stem cell differentiation. *Eur J Neurosci* 19, 1141–1152.

Takagi, Y., Takahashi, J., Saiki, H., Morizane, A., Hayashi, T., Kishi, Y., Fukuda, H., Okamoto, Y., Koyanagi, M., Ideguchi, M., Hayashi, H., Imazato, T., Kawasaki, H., Suemori, H., Omachi, S., Iida, H., Itoh, N., Nakatsuji, N., Sasai, Y., and Hashimoto, N. (2005). Dopaminergic neurons generated from monkey embryonic stem cells function in a Parkinson primate model. *J Clin Invest*. 115(1), 102–109.

Takahashi, K., Tanabe, K., Ohnuki, M., Narita, M., Ichisaka, T., Tomoda, K., and Yamanaka, S. (2007). Induction of pluripotent stem cells from adult human fibroblasts by defined factors. *Cell* 2007. Nov 30; 131(5): 861–72.

Tomishima, M. J., Hadjantonakis, A. K., Gong, S., and Studer, L. (2007). Production of green fluorescent protein transgenic embryonic stem cells using the GENSAT bacterial artificial chromosome library. *Stem Cells* 25, 39–45.

Tomishima, M. J., and Studer, L. (2003). Embryonic stem cell differentiation into neural cells. In *Neural Stem Cells: Development and Transplantation* (J. E. Bottenstein, Ed.). Kluwer Academic Publishers, Norwell, MA.

Urbach, A., Schuldiner, M., and Benvenisty, N. (2004). Modeling for Lesch-Nyhan disease by gene targeting in human embryonic stem cells. *Stem Cells* 22, 635–641.

van den Munckhof, P., Luk, K. C., Ste-Marie, L., Montgomery, J., Blanchet, P. J., Sadikot, A. F., and Drouin, J. (2003). Pitx3 is required for motor activity and for survival of a subset of midbrain dopaminergic neurons. *Development* 130, 2535–2542.

van Duijn, C. M., Dekker, M. C., Bonifati, V., Galjaard, R. J., Houwing-Duistermaat, J. J., Snijders, P. J., Testers, L., Breedveld, G. J., Horstink, M., Sandkuijl, L. A., van Swieten, J. C., Oostra, B. A., and Heutink, P. (2001). Park7, a novel locus for autosomal recessive early-onset parkinsonism, on chromosome 1p36. *Am J Hum Genet* 69, 629–634.

Wade-Martins, R., Smith, E. R., Tyminski, E., Chiocca, E. A., and Saeki, Y. (2001). An infectious transfer and expression system for genomic DNA loci in human and mouse cells. *Nat Biotechnol* 19, 1067–1070.

Wallen, A., Zetterstrom, R. H., Solomin, L., Arvidsson, M., Olson, L., and Perlmann, T. (1999). Fate of mesencephalic AHD2-expressing dopamine progenitor cells in NURR1 mutant mice. *Exp Cell Res* 253, 737–746.

Wang, Z., Benoit, G., Liu, J., Prasad, S., Aarnisalo, P., Liu, X., Xu, H., Walker, N. P., and Perlmann, T. (2003). Structure and function of Nurr1 identifies a class of ligand-independent nuclear receptors. *Nature* 423, 555–560.

Watanabe, K., Kamiya, D., Nishiyama, A., Katayama, T., Nozaki, S., Kawasaki, H., Watanabe, Y., Mizuseki, K., and Sasai, Y. (2005). Directed differentiation of telencephalic precursors from embryonic stem cells. *Nat Neurosci* 8, 288–296.

Weinstein, D. C., Ruiz i Altaba, A., Chen, W. S., Hoodless, P., Prezioso, V. R., Jessell, T. M., and Darnell, J. E., Jr. (1994). The winged-helix transcription factor HNF-3 beta is required for notochord development in the mouse embryo. *Cell* 78, 575–588.

Wurst, W., and Bally-Cuif, L. (2001). Neural plate patterning: Upstream and downstream of the isthmic organizer. *Nat Rev Neurosci* 2, 99–108.

Wurst, W., Auerbach, A. B., and Joyner, A. L. (1994). Multiple developmental defects in Engrailed-1 mutant mice: An early mid–hindbrain deletion and patterning defects in forelimbs and sternum. *Development* 120, 2065–2075.

Xian, H. Q., Werth, K., and Gottlieb, D. I. (2005). Promoter analysis in ES cell-derived neural cells. *Biochem Biophys Res Commun* 327, 155–162.

Yamashita, H., Nakamura, T., Takahashi, T., Nagano, Y., Hiji, M., Hirabayashi, T., Amano, T., Yagi, T., Sakai, N., Kohriyama, T., and Matsumoto, M. (2006). Embryonic stem cell-derived neuron models of Parkinson's disease exhibit delayed neuronal death. *J Neurochem* 98, 45–56.

Yan, Y., Yang, D., Zarnowska, E. D., Du, Z., Werbel, B., Valliere, C., Pearce, R. A., Thomson, J. A., and Zhang, S. C. (2005). Directed differentiation of dopaminergic neuronal subtypes from human embryonic stem cells. *Stem Cells* 23, 781–790.

Yang, D., Zhang, Z. J., Oldenburg, M., Ayala, M., and Zhang, S. C. (2008). Human embryonic stem cell-derived dopaminergic neurons reverse functional deficit in parkinsonian rats . *Stem Cells* 26, 55–63.

Ying, Q. L., Stavridis, M., Griffiths, D., Li, M., and Smith, A. (2003). Conversion of embryonic stem cells into neuroectodermal precursors in adherent monoculture. *Nat Biotechnol* 21, 183–186.

Yu, J., Vodyanik, M. A., Smuga-Otto, K., Antosiewicz-Bourget, J., Frane, J. L., Tian, S., Nie, J., Jonsdottir, G. A., Ruotti, V., Stewart, R., Slukvin, , II, and Thomson, J. A. (2007). Induced pluripotent stem cell lines derived from human somatic cells. *Science* 2007 Dec 21, 318(5858), 1917–1920.

Zaehres, H., Lensch, M. W., Daheron, L., Stewart, S. A., Itskovitz-Eldor, J., and Daley, G. Q. (2005). High-efficiency RNA interference in human embryonic stem cells. *Stem Cells* 23, 299–305.

Zeng, X., Chen, J., Deng, X., Liu, Y., Rao, M. S., Cadet, J.L., and Freed, W. J. (2006). An in vitro model of human dopaminergic neurons derived from embryonic stem cells: MPP$^+$ toxicity and GDNF neuroprotection. *Neuropsychopharmacology* **31**, 2708–2715.

Zervas, M., Blaess, S., and Joyner, A. L. (2005). Classical embryological studies and modern genetic analysis of midbrain and cerebellum development. *Curr Top Dev Biol* **69**, 101–138.

Zetterstrom, R. H., Solomin, L., Jansson, L., Hoffer, B. J., Olson, L., and Perlmann, T. (1997). Dopamine neuron agenesis in Nurr1-deficient mice. *Science* **276**, 248–250.

Zhao, S., Maxwell, S., Jimenez-Beristain, A., Vives, J., Kuehner, E., Zhao, J., O'Brien, C., de Felipe, C., Semina, E., and Li, M. (2004). Generation of embryonic stem cells and transgenic mice expressing green fluorescence protein in midbrain dopaminergic neurons. *Eur J Neurosci* **19**, 1133–1140.

Zwaka, T. P., and Thomson, J. A. (2003). Homologous recombination in human embryonic stem cells. *Nat Biotechnol* **21**, 319–321.

36

NEUROPROTECTIVE AND NEUROTOXIC PROPERTIES OF α-SYNUCLEIN IN CELL CULTURE MODELS OF DOPAMINERGIC DEGENERATION

FANENG SUN, VELLAREDDY ANANTHARAM, HUAJUN JIN, DANHUI ZHANG, ARTHI KANTHASAMY AND ANUMANTHA G. KANTHASAMY

Parkinson's Disorder Research Laboratory, Iowa Center for Advanced Neurotoxicology, Department of Biomedical Sciences, Iowa State University, Ames, IA, USA

α-Synuclein was originally identified in cholinergic vesicles of *Torpedo californica,* the Pacific torpedo ray (Maroteaux *et al.,* 1988). The protein was named as α-synuclein, because of its predominant cellular localization at synapse and the nuclear envelope of neurons. Subsequently, the mammalian homolog of the *Torpedo* synuclein was isolated and named as γ-synuclein. α-Synuclein is highly expressed in the central nervous system (CNS), especially in the substantia nigra, caudate nucleus, amygdala, and hippocampus. It is currently known that both α- and γ-synuclein belong to the same gene family which also includes β-synuclein and synoretin (Suh and Checler, 2002). All the synuclein family proteins contain the KTKEGV consensus domains. Human α-synuclein, a natively unfolded protein with 140 amino acids, consists of three structurally distinct motifs: an N-terminal amphipathic region, a central non-Aβ component (NAC) domain, and a C-terminal acidic tail (Recchia *et al.,* 2004). As shown in Figure 36.1, the N-terminal amphipathic region, containing a majority of the repeats of the KTKEGV consensus sequence, has the capacity to associate with negatively charged phospholipids (Cookson, 2005). Upon binding to lipid, native unfolded α-synuclein changes to an α-helix configuration. All three point mutations of α-synuclein: A30P, E46K, and A53T, are exclusively located in this region. The E46K and A53T mutations of α-synuclein potentiate its lipid binding and accelerate filament formation, while A30P reduces the binding capacity and slows down the formation of fibrillar species. This suggests that the amphipathic helix conformation favors the formation of α-synuclein aggregation (Choi *et al.,* 2004).

The highly negatively charged C-terminal tail of α-synuclein has several phosphorylation sites: Tyr 125, 133, and 136, and Ser 129. Approximately 90% of α-synuclein in the urea-insoluble fraction prepared from brain samples of synucleinopathy is phosphorylated on Ser 129 (Fujiwara *et al.,* 2002). The posttranslational modifications in the C-terminus include the nitration on Tyr 125, 133, and 136, and possible glycosylation at an unidentified position. The inhibitory effect of the acidic C-terminus on aggregation is based on the observation that the C-terminal truncated form of α-synuclein more readily forms fibrillar filaments (Murray *et al.,* 2003). NAC-region (66–95) initially was identified as a major component secondary to Aβ in Alzhemier's plaques. The NAC domain of α-synuclein, which is absent from β- and γ-synuclein, is

FIGURE 36.1   Structural features of human α-synuclein proteins. The three structural domains of α-synuclein include an N-terminal amphipathic region (1–65), a central NAC domain (66–90), and a C-terminal acidic tail (91–140). The majority of the signature consensus domains (imperfect KTKEGV sequence, black solid box) of synuclein family proteins are located at the N-terminus. A30P, E46K, and A53T are the three human mutations associated with familial PD. The acidic C-terminus contains several amino acids (Y125, S-129, Y133, and Y136), that could bear posttranslational modifications such as phosphorylation and nitration.

hydrophobic, and amyloidgenicity of NAC is crucial for the formation of the β-sheet structure of α-synuclein. Likely, β-sheet structure promotes oligomerization of protein to form the so-called protofibril and the subsequent filament, which eventually lead to the protein aggregation in Lewy bodies (Bodles *et al.*, 2001; Giasson *et al.*, 2001).

PHYSIOLOGICAL FUNCTION OF α-SYNUCLEIN

Currently, the physiological function of α-synuclein is under intense investigation. Consistent with its high expression at the presynaptic terminals, α-synuclein has been thought to play a role in synaptic transmission. It is postulated that α-synuclein may play a role in dopamine (DA) synthesis, vesicle transport, and release of DA including DA uptake by dopamine transporter (DAT) and vesicular monoamine transporter (VMAT) in the dopaminergic system. Neurochemical studies revealed impairment in paired stimuli-triggered DA release at the nigrostriatal terminals, and reduced striatal DA levels in α-synuclein knockout mice (Abeliovich *et al.*, 2000). Similarly, suppression of α-synuclein expression reduces the number of synaptic vesicles, especially the vesicles of reserve pool in hippocampal neurons, suggesting the important regulatory roles of α-synuclein in presynaptic vesicle formation and maintenance (Cabin *et al.*, 2002). A recent study by Larsen and colleagues (2006) indicated that α-synuclein interferes with secretory exocytosis of transmitter release. In line with α-synuclein's role in neurotransmission, knockout or mutation of α-synuclein has been shown to lower the capacity of the DA storage pool (Yavich *et al.*, 2004). Modulation of phospholipase D2 (PLD2) by α-synuclein in clathrin-mediated endocytosis for the presynaptic

vesicle recycling has been suggested to be the underlying regulatory role of α-synuclein in the process (Lotharius and Brundin, 2002a).

CHAPERONE ACTIVITY OF α-SYNUCLEIN

Because of its homology and interaction with 14-3-3, α-synuclein has also been postulated to function as a chaperone protein (Ostrerova *et al.*, 1999). Protein–protein interactions often dictate much of the cellular signaling influenced by α-synuclein within various cellular systems. Indeed α-synuclein has been shown to regulate the activity and function of several proteins associated with DA homeostasis and cellular signaling (Table 36.1).

TYROSINE HYDROXYLASE AND DOPAMINE SYNTHESIS

Tyrosine hydroxylase (TH) is the rate limiting enzyme in the synthesis of DA, and is activated when phosphorylated on any of its serine residues: Ser 19, Ser 31, and Ser 40. Various protein kinases like MAPK and ERK1/2 regulate this reversible phosphorylation of TH on residue Ser 31 (Royo *et al.*, 2004). Protein kinase A phosphorylates TH on Ser 31 and Ser 40 and Ca^{++}-dependent protein kinase C (PKC) modulates activity of TH in other models (Albert *et al.*, 1984; Cahill *et al.*, 1989; Kobori *et al.*, 2004; Sura *et al.*, 2004). Recently, we showed PKCδ negatively regulates TH–Ser 40 phosphorylation and DA synthesis via phosphatase 2B (Zhang *et al.*, 2007). We noted that α-synuclein overexpression suppresses PKCδ levels (unpublished observation), but it is yet to be determined whether PKCδ has any role in modulating the dopaminergic neurotransmission.

■ **TABLE 36.1 Summary of proteins that interact with α-synuclein**

Proteins	References
PLD2 (Phospholipase D2)	(Jenco et al., 1998; Payton et al., 2004)
UCH-L1 (Ubiquitin ligase 1)	(Liu et al., 2002)
Parkin	(Choi et al., 2001; Shimura et al., 2001; Oluwatosin-Chigbu et al., 2003)
Synphilin	(Engelender et al., 1999; Ribeiro et al., 2002)
14-3-3	(Ostrerova et al., 1999)
PKC, BAD, ERK	(Ostrerova et al., 1999)
Elk-1/Erk-2 complex	(Iwata et al., 2001b)
MAPK	(Iwata et al., 2001a)
Tubulin	(Alim et al., 2002)
Cytochrome oxidase IV (COX IV)	(Elkon et al., 2002)
Dopamine transporter (DAT)	(Kobayashi et al., 2004)
Aβ, Tau	(Yoshimoto et al., 1995; Jensen et al., 1997)
Calmodulin	(Lee et al., 2002a; Martinez et al., 2003)
Protein kinase Cδ	(Kaul et al., 2005a)
PP2A	(Peng et al., 2005)

Chaperone proteins 14-3-3 and α-synuclein have been suggested to interact with TH, and exert opposite regulatory effects on TH activity and DA synthesis (Sidhu et al., 2004a). Preferential interaction of 14-3-3 with phosphorylated TH stabilize the protein to its active conformation, and thus maximizes enzymatic activity for DA synthesis; whereas association of α-synuclein with dephosphorylated TH likely suppresses TH activation by stabilizing TH in its inactive form (Sidhu et al., 2004a). Additionally, Peng and coworkers (2005) showed that activation of PP2A by α-synuclein results in the dephosphorylation of TH, and thus decreased DA synthesis. Due to its chaperone activity, α-synuclein has also been postulated to directly interact with kinases associated with DA homeostasis such as MAPK and PKCs (Ostrerova et al., 1999; Iwata et al., 2001a; Baptista et al., 2003). Furthermore, overexpression of familial α-synuclein mutants have been shown to affect DA homeostasis in both cell culture and animal models of dopaminergic degeneration (Lotharius et al., 2002; Orth et al., 2004). This effect of α-synuclein on DA metabolism can lead to excessive release or production of DA. Excessive DA production can result in the formation of free radicals due to auto-oxidation of DA, which can be deleterious to the neurons (Luo et al., 1998; Lotharius and O'Malley, 2001; Jenner, 2003).

VESICULAR TRANSPORT AND TRAFFICKING

In addition to its regulatory role of TH activity, α-synuclein also participates in the other cellular events maintaining DA homeostasis, such as modulation of the plasma membrane DAT. DA, once synthesized, is stored in lipid-bound synaptic vesicles that protect the dopaminergic neurons from its auto-oxidative effects. In DA neurons, sequestration of cytosolic DA into synaptic vesicles by VMAT2 is essential for the neurons to avoid the neurotoxicity of DA. The uptake and storage of DA in the synaptic vesicles is regulated by the VMAT2 (Weihe and Eiden, 2000). VMAT2 also attenuates the neurotoxicity of MPP^+, a known dopaminergic toxin, by sequestering it safely in synaptic vesicles (Gainetdinov et al., 1998a). Human positron emission tomography (PET) studies have revealed an enhanced loss of VMAT2 in dopaminergic neurons, indicating that vesicular dysfunction might be an important contributing factor in Parkinson's disease (PD) (Lee et al., 2000). α-Synuclein negatively controls DA release by acting on VMAT2 activity through its inhibitory action on PLD2 (Lotharius and Brundin, 2002b; Sidhu et al., 2004b). Overexpression of human A53T α-synuclein mutant has been shown to downregulate the expression of VMAT, thus impairing the vesicular storage and cytosolic accumulation of DA (Lotharius et al., 2002). Also, disruption of the integrity of the vesicular membrane, presumably as the result of the formation of α-synuclein protofibrils, has been suggested to account for the cell type-specific neurotoxicity in DA neurons, since α-synuclein overexpression appears to dissipate the proton gradient across the vesicle membrane and remarkably elevates the cytosolic DA level (Mosharov et al., 2006).

DA TRANSPORTER FUNCTION

The DA receptor (DAT) belongs to the Na^+/Cl^--dependent transporter family of monoamine transporters involved in DA homeostasis through clearance of excess neurotransmitter from the synaptic clefts (Gallant et al., 2003; Mortensen and Amara, 2003). Similar to α-synuclein, DAT is expressed in the pre-synaptic terminals and is crucial for effective maintenance of DA neurotransmission in dopaminergic nerve terminals (Gainetdinov et al., 1998b; Chen and Reith, 2004). DAT function involves the phosphorylation of certain N-terminal residues by various protein kinases including PKC, resulting in redistribution of the transporter between the plasma membrane and the cytoplasm (Pristupa et al., 1998; Daniels and Amara, 1999; Melikian and Buckley, 1999; Foster et al., 2002). Recent studies have shown that overexpression of the human wild-type α-synuclein led to a reduction in DAT activity due to reduced DA uptake, but not due to DAT trafficking or transcriptional regulation (Wersinger et al., 2003a). The opposite effect of α-synuclein on DAT-mediated DA uptake has been reported; α-synuclein attenuates the activity of coexpressed DAT, and suppresses the DA-related oxidative stress in the neurons (Wersinger and Sidhu, 2003; Wersinger et al., 2003a). However, other studies have shown that α-synuclein positively regulates DAT activity and enhances the neurotoxicity of DA and MPP$^+$ (Lee et al., 2001).

α-SYNUCLEIN MUTATIONS IN PD

Several lines of evidence suggest that, in both sporadic and familial forms of PD, protein aggregates within dopaminergic neurons of the substantia nigra are a common feature. Although several proteins have been found in the Lewy bodies, fibrillar α-synuclein is the major component of the intracellular protein inclusions (Choi et al., 2001). Familial PD has been linked to missense and genomic multiplication mutations of the α-synuclein gene. Autosomal dominant mutations in the α-synuclein gene have been shown to be associated with familial PD. Three different missense mutations, A53T, A30P, and E46R have been found in patients of familial PD (Polymeropoulos et al., 1997; Kruger et al., 1998; Zarranz et al., 2004). Triplication and duplication of the α-synuclein locus has also been found in several families with PD (Singleton et al., 2003; Chartier-Harlin et al., 2004). Though the gene triplication/duplication occurs in rare cases of PD, gene multiplication apparently may result in the elevation of the α-synuclein protein level and insoluble protein aggregates, which may mediate

PD pathogenesis (Johnson et al., 2004; Miller et al., 2004; Hofer et al., 2005). A recent study also linked α-synuclein promoter's susceptibility to sporadic PD (Pals et al., 2004). These studies clearly suggest that overproduction of α-synuclein can be a risk factor for PD.

α-SYNUCLEIN PHOSPHORYLATION

The Ser 129 phosphorylation appears to be a very important posttranslational modification associated with Lewy bodies (Anderson et al., 2006), and the pathological relevance of the modification is manifested by its role in promoting α-synuclein fibrillation or ubiquitination (Hasegawa et al., 2002; et al., 2006). Ser 129 is constitutively phosphorylated in transfected HEK293 and PC12 cells; this may be mediated by kinases such as casein kinase (CK I and CK II) and downstream kinases of G-protein coupled receptors (GPCRs) (Okochi et al., 2000; Pronin et al., 2000). α-Synuclein interaction with PLD2 is important for regulation of vesicle release of DA into the synaptic environment, and Ser 129 phosphorylation can attenuate this interaction, thus altering DA homeostasis (Lotharius et al., 2002). Recently, however, investigators have also suggested that vesicle trafficking could be PLD2 independent (Abeliovich et al., 2000). Y39, Y125, Y133, and Y136 tyrosine residues are well conserved in all the α-synuclein homologs, as well as in the β-synuclein paralogs, indicating that these residues are important in synuclein functioning. Activation of Pyk2/RAFTK in COS7 cells can phosphorylate α-synuclein via the src kinase family of enzymes, and this tyrosine phosphorylation can serve as a neuroprotective mechanism in the case of deleterious nitrosylation of synuclein at Y125 (Nakamura et al., 2002; Takahashi et al., 2003). Tyrosine phosphorylation has also been suggested to have an effect on the regulation of synaptic vesicles in lieu of the fact that tau, synuclein, and src-PTK members interact with each other at various levels of cellular signaling (Lee et al., 1998; Trojanowski and Lee, 2000). Hypothetical suggestions regarding the importance of these interactions are based on the fact that tau might help bring src-PTKs in close proximity to α-synuclein and then lead to tyrosine phosphorylation, which plays an important role in the development of synaptic plasticity (Clayton and George, 1999). Recently, Wakamatsu and colleagues (2007) reported accumulation of phosphorylated α-synuclein in dopaminergic neurons of transgenic mice that express human α-synuclein. It has been reported that substitution of Ser 129 of α-synuclein with alanine (S129A) reduces the formation of intracellular protein aggregation (Smith et al., 2005b). Phosphorylation of α-synuclein at Ser 129 leads to an increase in formation of its insoluble aggregated oligomers. Further, serine hyperphosphorylated forms have been isolated from human brain tissues, transgenic mice, and fly neurons (Fujiwara et al., 2002; Kahle et al., 2002; Neumann et al., 2002; Takahashi et al., 2002).

α-SYNUCLEIN AGGREGATION

Several studies have linked ubiquitin proteasomal dysfunction to α-synuclein aggregation in primary mesencephalic neurons, dopaminergic neuronal cells, and in animal models (Rideout et al., 2001; McNaught et al., 2002a, b). Extensive studies also suggest various factors that could promote α-synuclein aggregation. First, as the major component, α-synuclein tends to self-aggregate; cross-linking of nitrated tyrosine by dinitrated bond can form urea/detergent-insoluble dimers or trimers of α-synuclein (El-Agnaf et al., 1998b; Giasson et al., 2000; Souza et al., 2000; Takahashi et al., 2002). Transglutaminase, found in Lewy bodies, has been shown to induce intramolecular cross-linking of α-synuclein (Junn et al., 2003; Andringa et al., 2004). Mitochondrial inhibition has also been shown to result in the formation of α-synuclein aggregation in cell culture and animal models (Lee et al., 2002b; Sherer et al., 2003; Fornai et al., 2005).

NEUROPROTECTIVE EFFECT OF α-SYNUCLEIN IN DOPAMINERGIC NEURONS

To understand the role of α-synuclein in dopaminergic degeneration in PD, numerous groups have examined the effect of overexpression of either wild-type or mutant α-synuclein on dopaminergic neurons in cell culture, as well as in transgenic and knockout animals. In cell culture studies, overexpression of wild type, but not A53T or A30P α-synuclein mutants, protected against caspase-3 activation and apoptotic cell death induced by

several neurotoxicants in the neocortical cell line (Alves da Costa et al., 2006). We showed that overexpression of wild-type α-synuclein but not mutant α-synuclein attenuated PKCδ-dependent apoptotic cascade in dopaminergic cells (Kaul et al., 2005a). Similarly, α-synuclein overexpression was also shown to protect a human dopaminergic cell line (SH-SY5Y cells) from cytotoxicity from Parkin knockdown and DA treatment (Machida et al., 2005; Colapinto et al., 2006). Studies have shown α-synuclein exerts its neuroprotective effect via inactivation of Jun kinase or inhibition of caspase-3 activation (Hashimoto et al., 2002; Li and Lee, 2005). Others have shown that nanomolar concentrations of α-synuclein can activate the PI3/Akt cell survival signal pathway, which renders neurons more resistant to serum deprivation, oxidative stress, and excitotoxicity.

The effect of α-synuclein on neuronal viability has been suggested to be dependent on several other factors, such as its intracellular abundance, cell types, or the types of stimuli (Seo et al., 2002; Xu et al., 2002; Zourlidou et al., 2003). In addition, α-synuclein has been shown to play a protective role in animal PD models (Hashimoto et al., 2002; Manning-Bog et al., 2003; Lee et al., 2006). Animals treated with the herbicide paraquat showed increased α-synuclein expression and aggregation in the brains. This increased expression and aggregation of α-synuclein results in neuroprotection. In neuronal cells overexpressing α-synuclein, the intracellular retrograde transport system has been shown to play a crucial role in aggregate formation, and that these aggregates are thought to represent a neuroprotective response (Hasegawa et al., 2004). Increased α-synuclein expression in response to dopaminergic toxins, for example, 1-methyl-4-phenyl-1,2,3,4-tetrahydropyridine (MPTP), rotenone, paraquat, suggests that an increase of α-synuclein represents an adaptive response to toxic stimuli and α-synuclein overexpression in transgenic mice does not consistently result in neuronal damage, nor does it exacerbate neurodegeneration caused by MPTP or other dopaminergic toxins (Masliah et al., 2000; Matsuoka et al., 2001; Lee et al., 2006). Therefore, the neuroprotective property of α-synuclein may be used for cell survival strategies (Lee et al., 2006). α-Synuclein is a major synaptic protein and therefore it is not surprising that the normal level of α-synuclein may have some neuroprotective functions in CNS. Table 36.2 summarizes key studies on the neuroprotective effects of α-synuclein.

TABLE 36.2 Neuroprotection by α-synuclein

Key findings	Models	References
Wild-type α-synuclein, not A53T mutant, protects cells from apoptosis	TSM1 neuronal cells	(Alves da Costa *et al.*, 2006)
Wild-type α-synuclein protects SH-SY5 cells from apoptosis and dopamine metabolite accumulation as the result of Parkin loss	SH-SY5Y	(Machida *et al.*, 2005)
Wild-type α-synuclein alleviates cytotoxicity of dopamine in SH-SY5 cells, and upregulates DJ-1 expression	SH-SY5Y	(Colapinto *et al.*, 2006)
Wild-type α-synuclein inactivates JNK by upregulating JNK-interacting protein JIP-1b/IB1 during oxidative stress	GT1-7 murine hypothalamic tumor cell line	(Hashimoto *et al.*, 2002)
Wild-type, A30P human α-synuclein, but not A53T α-synuclein, β-synuclein or mouse α-synuclein, protect cells from apoptosis via inhibiting cas pase-3	SH-SY5Y	(Li and Lee, 2005)
Wild-type α-synuclein, not A53T mutant, suppressed MPP$^+$-induced activation of apoptosis via interaction with PKCδ and BAD	N27 cells	(Kaul *et al.*, 2005a)

NEUROTOXIC EFFECT OF α-SYNUCLEIN

Overproduction and/or accumulation of α-synuclein in cultured neuronal cells causes selective degeneration in dopaminergic neurons but not in non-dopaminergic neurons, suggesting selective toxicity (Xu *et al.*, 2002). Also, in mice expressing the A53T human α-synuclein mutation there is an early onset of neurodegeneration and α-synuclein aggregation in the brain (Lee *et al.*, 2002c). In cell culture models, direct neurotoxicity of α-synuclein was manifested by the increased cell death of SH-SY5Y cells following exposure to mutant, aggregated α-synuclein or NAC fragment (El-Agnaf *et al.*, 1998a; Sung *et al.*, 2001). Endocytotic uptake of α-synuclein involving Rab5A was hypothesized to be crucial for its observed neurotoxicity (Sung *et al.*, 2001). Expression of wild type, A30P, or A53T mutant human α-synuclein induces apoptosis in the mouse nodose ganglion neurons (Saha *et al.*, 2000). Studies with neuronal cell lines indicated that different types of cell death including mitochondria-related, endoplasmic reticulum stress cell death or autophagic cell death are involved in the neurotoxicity of either wild-type or mutant α-synuclein (Hsu *et al.*, 2000; Stefanis *et al.*, 2001; Smith *et al.*, 2005a). α-Synuclein also appears to enhance the vulnerability of cells to a variety of neurotoxins. Overexpression of human α-synuclein in human SH-SY5Y neuroblastoma cells, especially the C-terminal truncated form, A30P and A53T mutants, significantly potentiates the oxidative damage and cell death triggered by MPP$^+$ or H_2O_2 (Kanda *et al.*, 2000). Consistently, expression of mutant α-synuclein in human BE-M17 neuroblastoma cells results in more profound neuronal death following exposure to iron, which promotes free radical generation (Ostrerova-Golts *et al.*, 2000). Inducible expression of mutant α-synuclein inhibits proteasome activity and renders PC12 cells more susceptible to proteasome inhibitor-induced apoptosis (Tanaka *et al.*, 2001). Coexpression of wild-type or A30P α-synuclein with a DAT in SH-SY5Y cells revealed DA-dependent cell death accompanied by collapse of cellular membrane potential, oxidative stress, and mitochondrial abnormalities (Moussa *et al.*, 2004). Table 36.3 summarizes some key studies that describe the neurotoxic properties of α-synuclein.

In essence, the neuroprotective form of α-synuclein can be readily converted to toxic gain-of-function forms, under certain conditions such as overproduction, oxidative modification, and oligomerization, thus demonstrating the observed opposing roles of α-synuclein.

███ **TABLE 36.3   Neurotoxicity of α-synuclein**

Key findings	Models	References
α-Synuclein, especially aggregated form, directly provokes neurotoxicity	SH-SY5Y cells	(El-Agnaf et al., 1998a)
Rab5A-specific endocytosis of α-synuclein mediates the neurotoxicity of exogenous α-synuclein	Rat hippocampal neuronal cells H19-7	(Sung et al., 2001)
Wild-type, A30P or A53T mutant human α-synuclein, not γ-synuclein, induces apoptosis	Mouse nodose ganglion neurons	(Saha et al., 2000)
Formation of α-synuclein aggregates, compromised mitochondria activity, increased ROS generation in α-synuclein overexpressing cells	Hypothalamic neuronal cell line (GT1-7)	(Hsu et al., 2000)
A53T mutant, but not wild-type α-synuclein, impairs ubiquitin proteasome and lysosomal degradation and enhances autophagy cell death	PC12 cells	(Stefanis et al., 2001)
A53T mutant α-synuclein induces mitochondria-mediated and ER stress-mediated apoptosis	PC12 cells	(Smith et al., 2005b)
C-terminal truncated, A30P, and A53T mutant α-synuclein potentiates oxidative damage and cell death triggered by MPP^+ or H_2O_2	SH-SY5Y	(Kanda et al., 2000)
Synuclein increases the vulnerability of cells to neurotoxicity of iron (A53T > A30P > wild type), and iron induces α-synuclein aggregation in the same order	Human BE-M17 neuroblastoma cells	(Ostrerova-Golts et al., 2000)
A30P α-synuclein inhibits proteasomal activity and sensitizes cells to mitochondria apoptosis	PC12 cells	(Tanaka et al., 2001)
Wild-type α-synuclein inhibits proteasomal activity and enhances dieldrin-induced apoptosis	N27 cells	(Sun et al., 2005)
Selective neurotoxicity of A53T α-synuclein, but not wild type, to dopamine neurons in mesencephalic primary culture	Human mesencephalic primary culture	(Zhou et al., 2002)
Dopamine-dependence of the neurotoxicity of wild-type α-synuclein	Human mesencephalic primary culture	(Xu et al., 2002)
Coexpression of wild-type or A30P α-synuclein with DAT causes mitochondria pathologies, oxidative stress and dopamine-dependent neuron death	SH-SY5Y	(Moussa et al., 2004)

IMMORTALIZED MESENCEPHALIC CELL LINE (N27) AS A MODEL SYSTEM FOR ELUCIDATING α-SYNUCLEIN FUNCTION

We have recently established an immortalized rat mesencephalic dopaminergic neuronal cell line as an excellent model system for studying dopaminergic degeneration. The N27 ($1RB_3AN_{27}$) cell line was initially developed by Dr. Prasad and his coworkers; these cells exhibited and retained most of the key features of dopaminergic neurons such as

the expression of neuron-specific enolase, nestin, TH, and DAT, and the production of homovanillic acid, a DA metabolite (Prasad et al., 1994). Upon differentiation, N27 cells acquire morphologic and functional features of the post-mitotic DA neurons, such as enlargement of the cell body, growth of neuronal processes, upregulation of TH and DAT, and increased DA production (Figure 36.2). A study by Clarkson and colleagues (1999) demonstrated that differentiated N27 cells are even more vulnerable to MPP^+ and 6-OHDA-induced neurotoxicity than

Undifferentiated   Differentiated

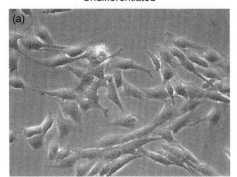 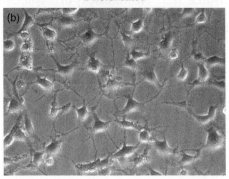

FIGURE 36.2   Rat mesencephalic dopaminergic neuronal cells (N27 cells). The phase contrast image (20×) shows undifferentiated N27 cells (panel a) and differentiated N27 cells (panel b). Dibutyryl 3,5-cyclic adenosine monophosphate (dbcAMP, 2.0 mM) was supplemented in the growth medium to induce morphological and biochemical alterations characteristic of differentiation of N27 cells.

undifferentiated cells. We subsequently established that undifferentiated N27 cells are also highly susceptible to apoptotic cell death induced by dopaminergic toxins similar to primary neuronal cultures.

In N27 cells, we recently showed that oxidative stress, multiple caspases, and PKCδ mediate apoptotic cell death induced by several dopaminergic toxins including MMT, dieldrin, MPP⁺, and manganese (Anantharam *et al.*, 2002; Kaul *et al.*, 2003; Kitazawa *et al.*, 2003; Latchoumycandane *et al.*, 2005). We clearly established that PKCδ is an oxidative stress-sensitive kinase in this cell culture PD model (Kanthasamy *et al.*, 2003). Oxidative stress activates PKCδ by proteolysis in which caspase-3 cleaves the native kinase (72–74 kDa) resulting in 41-kDa catalytically active and 38-kDa regulatory fragments, to persistently activate the kinase. Phosphorylation of PKCδ at tyrosine residue 311 is essential for the proteolytic cleavage of the kinase during oxidative stress (Kaul *et al.*, 2005b). The proteolytic activation of PKCδ plays a key role in promoting apoptotic cell death in various cell types including neuronal cells (Kikkawa *et al.*, 2002; Brodie and Blumberg, 2003; Kanthasamy *et al.*, 2003). Studies from this cell line are consistent with reduced cellular antioxidant capacity, increased oxidative stress, and impaired mitochondrial function as observed during dopaminergic degeneration. Overexpression of loss-of-function dominant-negative mutant PKCδ^{D327A} (caspase-cleavage resistant), PKCδ^{K376R} (kinase inactive), and PKCδ^{Y311F} (phosphorylation defective) proteins also attenuates dopaminergic neurons from MPP⁺- and oxidative stress-induced apoptotic cell death. Suppression of caspase-3-dependent proteolytic activation of PKCδ by small interfering RNA (siRNA) also prevented MPP⁺-induced dopaminergic degeneration (Yang *et al.*, 2004). In addition to the proapoptotic role, PKCδ may also amplify apoptotic signaling via positive feedback activation of the caspase cascade (Kanthasamy *et al.*, 2003). Thus, the dual role of PKCδ as a mediator and amplifier of apoptosis was established in this cell culture model and may be important in the pathogenesis of PD. PKCδ is also highly expressed in these cells, compared to several other non-dopaminergic neuronal cells (unpublished observations) and also colocalizes with TH in these cells. These results were subsequently confirmed in the mouse nigral tissue, where PKCδ is also highly expressed and colocalizes with TH (Zhang *et al.*, 2007). Further investigation revealed PKCδ negatively regulates TH activity and DA synthesis by enhancing protein phosphatase-2A activity in N27 cells (Zhang *et al.*, 2007). Many results obtained in N27 cells were readily reproducible in animal models, indicating that N27 cells are very reliable cell culture model of PD.

After we established the N27 cell line as a model system for studying dopaminergic degeneration, we generated stable N27 cell lines overexpressing wild-type and A53T mutant α-synuclein (Kaul *et al.*, 2005a). N27 cells overexpressing wild-type α-synuclein were highly resistant to MPP⁺-induced cytotoxicity, mitochondrial cytochrome c release, and subsequent caspase-3 activation, without affecting reactive oxygen species (ROS) generation (Kaul *et al.*, 2005a). Co-immunoprecipitation studies revealed MPP$^+$ treatment induced the physical association of α-synuclein with pro-apoptotic proteins PKCδ and BCl-2 associated death protein (BAD), but not with the anti-apoptotic protein Bcl-2. The physical association between PKCδ and α-synuclein did not involve direct phosphorylation. On the contrary, in A53T α-synuclein mutant expressing cells, MPP$^+$-induced apoptotic cell death signaling including activation of caspase-3, PKCδ, and fragmentation, was exacerbated. These results suggested that normal level of wild-type α-synuclein is neuroprotective whereas A53T α-synuclein is neurotoxic and may mediate the effects via interaction with pro-apoptotic molecules BAD and PKCδ. Unlike MPP$^+$, human α-synuclein exacerbated dieldrin-induced increases in caspase-3 activity and fragmentation compared to vector expressing cells (Sun *et al.*, 2005). In the N27 model system we demonstrated that human α-synuclein can be neuroprotective or neurotoxic, depending on the duration and type of neurotoxins exposed (Kaul *et al.*, 2005a; Sun *et al.*, 2005).

Since recent evidence indicate that abnormal accumulation and aggregation of α-synuclein and ubiquitin–proteasome system dysfunction can contribute to the degenerative processes of PD, we examined the effect of human α-synuclein on dieldrin-induced impairment in dysfunction in N27 cell line (Sun *et al.*, 2005). Baseline proteasomal activity in human α-synuclein overexpressing cells was 50% less than vector expressing N27 cells, suggesting that α-synuclein overexpression significantly attenuated baseline proteasomal activity. Further, overexpression of human α-synuclein also exacerbated dieldrin-induced decreases in proteasomal activity by more than 60% compared to vector expressing N27 cells. Confocal microscopic analysis revealed that α-synuclein-positive protein aggregates colocalized with ubiquitin protein in dieldrin-treated cells, and these aggregates were distinct from autophagosomes and lysosomes. The dieldrin-induced proteasomal dysfunction in α-synuclein cells also resulted in significant accumulation of ubiquitin protein conjugates; proteasomal inhibition preceded cell death. In these studies human α-synuclein overexpression predisposed N27 cells to proteasomal dysfunction, which can be further exacerbated by the pesticide dieldrin (Sun *et al.*, 2005).

Collectively, N27 cells appear to be an ideal *in vitro* model system, revealing the molecular and cellular mechanisms leading to dopaminergic degeneration. Unlike PC12 cells (pheochromocytoma

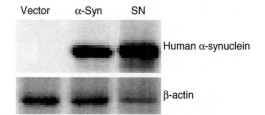

FIGURE 36.3 Expression of human wild-type α-synuclein in N27 dopaminergic cells. pCEP4 expression vector containing coding sequence for human α-synuclein was transfected into N27 cells using Lipofectamine Plus reagent. Stable expression was achieved with prolonged hygromycin screening. Cell lysates were obtained from vector transfected cells (Vec, lane 1), or α-synuclein expressing cells (α-Syn, lane 2) for Western blot analysis with α-synuclein antibody. Lysate of the rat substantia nigra (SN) was included to show that exogenously introduced α-synuclein is expressed at physiological levels comparable to the substantia nigra region (Lysate, lane 3). Protein amount: 20 μg.

of rat adrenal medulla), which have also been extensively used in PD study as a model system, N27 cells are derived from the mesencephalic region and maintain most of the key features of the dopaminergic phenotype, such as TH and DAT expression, DA synthesis, and high susceptibility to dopaminergic neurotoxins. Additionally, N27 cells express high levels of PKCδ, similar to those found in the substantia nigra of animals (unpublished observations). Therefore, N27 cells appear to be an excellent cell culture model system for studying dopaminergic-specific neuronal events prior to *in vivo* rodent models. Recently, we demonstrated that in N27 cells and in the mouse substantia nigra, PKCδ participates in the regulation of TH activity and DA synthesis via modulation of PP2A activity. However, N27 cells, like any other cell used in PD studies, are different from the post-mitotic neurons present in adult animal brains; therefore, it is conceivable that results acquired from N27 cells may not be extrapolated to the *in vivo* situation. We believe that any finding derived from studies using any cell model, including N27 cells, requires further validation using *in vivo* models.

Recently we observed that α-synuclein levels in α-synuclein overexpressing N27 cells was comparable to the level of α-synuclein in rat substantia nigra (Figure 36.3). Overexpression of human

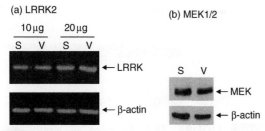

FIGURE 36.4 Lack of effect of α-synuclein on LRRK2 (a) and MEK1/2 (b) protein expression. Western blot was performed on whole cell lysates obtained from α-synuclein expressing (S) and vector control (V) N27 cells. β-Actin was used as a loading control. Overexpression of human α-synuclein did not have an effect on protein expression levels of LRRK2 and MEK1/2. Both proteins have been implicated in the pathogenesis of PD.

α-synuclein also did not have any effect on protein expression levels of LRRK2, a kinase implicated in familial PD (Figure 36.4a and MEK1/2, an ERK kinase activator (Figure 36.4b), both of which have been implicated in the pathogenesis of PD.

In conclusion, α-synuclein expressing cells appear to be slightly resistant to cytotoxicity induced by acute exposure to dopaminergic toxins, probably by sequestering proapoptotic molecules PKCδ and BAD (Kaul *et al.*, 2005a). However, the α-synuclein cells can also be more susceptible to apoptotic cell death induced by chronic exposure to dopaminergic toxins presumably because of impairment in UPS dysfunction, protein misfolding, and aggregation (Sun *et al.*, 2005). However, α-synuclein appears to be detrimental to DA neurons, considering the self-aggregation propensity of the protein during toxic insults and the possibility of chronic exposure to environmental factors, which is a dominant risk factor for PD.

ACKNOWLEDGEMENTS

The authors acknowledge the support from National Institute of Health (NIH) grants NS45133, NS 38644, ES10586 and Eugene and Linda Lloyd Professorship. The authors also thank Ms. Keri Hendersen and Dr. Siddharth Kaul for their assistance in the preparation of this manuscript.

ABBREVIATIONS

BAD	BCl-2 associated death protein
DA	dopamine
DAT	dopamine transporter
MMT	methylcyclopentadienyl manganese tricarbonyl
MPP$^+$	1-methyl-4-phenylpyridinium ion
MPTP	1-methyl-4-phenyl-1,2,3,4-tetrahydropyridine
NAC	non-Aβ component
6-OHDA	6-hydroxyl dopamine
PD	Parkinson's disease
PKCδ	protein kinase C delta
PLD2	Phospholipase D2
TH	tyrosine hydroxylase
Uch-L1	ubiquitin C-terminal hydrolase L1
UPS	ubiquitin–proteasome system
VMAT-2	vesicular monoamine transporter 2.

REFERENCES

Abeliovich, A., Schmitz, Y., Farinas, I., Choi-Lundberg, D., Ho, W. H., Castillo, P. E., Shinsky, N., Verdugo, J. M., Armanini, M., Ryan, A., Hynes, M., Phillips, H., Sulzer, D., and Rosenthal, A. (2000). Mice lacking alpha-synuclein display functional deficits in the nigrostriatal dopamine system. *Neuron* 25, 239–252.

Albert, K. A., Helmer-Matyjek, E., Nairn, A. C., Muller, T. H., Haycock, J. W., Greene, , Goldstein, M., and Greengard, P. (1984). Calcium/phospholipid-dependent protein kinase (protein kinase C) phosphorylates and activates tyrosine hydroxylase. *Proc Natl Acad Sci* 81, 7713–7717.

Alim, M. A., Hossain, M. S., Arima, K., Takeda, K., Izumiyama, Y., Nakamura, M., Kaji, H., Shinoda, T., Hisanaga, S., and Ueda, K. (2002). Tubulin seeds alpha-synuclein fibril formation. *J Biol Chem* 277, 2112–2117.

Alves da Costa, C., Dunys, J., Brau, F., Wilk, S., Cappai, R., and Checler, F. (2006). 6-Hydroxydopamine but not 1-methyl-4-phenylpyridinium abolishes alpha-synuclein anti-apoptotic phenotype by inhibiting its proteasomal degradation and by promoting its aggregation. *J Biol Chem* 281, 9824–9831.

Anantharam, V., Kitazawa, M., Wagner, J., Kaul, S., and Kanthasamy, A. G. (2002). Caspase-3-dependent proteolytic cleavage of protein kinase C delta is essential for oxidative stress-mediated dopaminergic cell death after exposure to methylcyclopentadienyl manganese tricarbonyl. *J Neurosci* 22, 1738–1751.

Anderson, J. P., Walker, D. E., Goldstein, J. M., de Laat, R., Banducci, K., Caccavello, R. J., Barbour, R., Huang, J.,

Kling, K., Lee, M., Diep, L., Keim, P. S., Shen, X., Chataway, T., Schlossmacher, M. G., Seubert, P., Schenk, D., Sinha, S., Gai, W. P., and Chilcote, T. J. (2006). Phosphorylation of Ser-129 is the dominant pathological modification of alpha-synuclein in familial and sporadic Lewy body disease. *J Biol Chem* **281**, 29739–29752.

Andringa, G., Lam, K. Y., Chegary, M., Wang, X., Chase, T. N., and Bennett, M. C. (2004). Tissue transglutaminase catalyzes the formation of alpha-synuclein crosslinks in Parkinson's disease. *FASEB J* **18**, 932–934.

Baptista, M. J., O'Farrell, C., Daya, S., Ahmad, R., Miller, D. W., Hardy, J., Farrer, M. J., and Cookson, M. R. (2003). Co-ordinate transcriptional regulation of dopamine synthesis genes by alpha-synuclein in human neuroblastoma cell lines. *J Neurochem* **85**, 957–968.

Bodles, A. M., Guthrie, D. J., Greer, B., and Irvine, G. B. (2001). Identification of the region of non-Abeta component (NAC) of Alzheimer's disease amyloid responsible for its aggregation and toxicity. *J Neurochem* **78**, 384–395.

Brodie, C., and Blumberg, P. M. (2003). Regulation of cell apoptosis by protein kinase C delta. *Apoptosis* **8**, 19–27.

Cabin, D. E., Shimazu, K., Murphy, D., Cole, N. B., Gottschalk, W., McIlwain, K. L., Orrison, B., Chen, A., Ellis, C. E., Paylor, R., Lu, B., and Nussbaum, R. L. (2002). Synaptic vesicle depletion correlates with attenuated synaptic responses to prolonged repetitive stimulation in mice lacking alpha-synuclein. *J Neurosci* **22**, 8797–8807.

Cahill, A. L., Horwitz, J., and Perlman, R. L. (1989). Phosphorylation of tyrosine hydroxylase in protein kinase C-deficient PC12 cells. *Neuroscience* **30**, 811–818.

Chartier-Harlin, M. C., Kachergus, J., Roumier, C., Mouroux, V., Douay, X., Lincoln, S., Levecque, C., Larvor, L., Andrieux, J., Hulihan, M., Waucquier, N., Defebvre, L., Amouyel, P., Farrer, M., and Destee, A. (2004). Alpha-synuclein locus duplication as a cause of familial Parkinson's disease. *Lancet* **364**, 1167–1169.

Chen, N., and Reith, M. E. (2004). Interaction between dopamine and its transporter: Role of intracellular sodium ions and membrane potential. *J Neurochem* **89**, 750–765.

Choi, P., Golts, N., Snyder, H., Chong, M., Petrucelli, L., Hardy, J., Sparkman, D., Cochran, E., Lee, J. M., and Wolozin, B. (2001). Co-association of parkin and alpha-synuclein. *Neuroreport* **12**, 2839–2843.

Choi, W., Zibaee, S., Jakes, R., Serpell, L. C., Davletov, B., Crowther, R. A., and Goedert, M. (2004). Mutation E46K increases phospholipid binding and assembly into filaments of human alpha-synuclein. *FEBS Lett* **576**, 363–368.

Clarkson, E. D., Edwards-Prasad, J., Freed, C. R., and Prasad, K. N. (1999). Immortalized dopamine neurons: A model to study neurotoxicity and neuroprotection. *Proc Soc Exp Biol Med* **222**, 157–163.

Clayton, D. F., and George, J. M. (1999). Synucleins in synaptic plasticity and neurodegenerative disorders. *J Neurosci Res* **58**, 120–129.

Colapinto, M., Mila, S., Giraudo, S., Stefanazzi, P., Molteni, M., Rossetti, C., Bergamasco, B., Lopiano, L., and Fasano, M. (2006). Alpha-synuclein protects SH-SY5Y cells from dopamine toxicity. *Biochem Biophys Res Commun* **349**, 1294–1300.

Cookson, M. R. (2005). The biochemistry of Parkinson's disease. *Annu Rev Biochem* **74**, 29–52.

Daniels, G. M., and Amara, S. G. (1999). Regulated trafficking of the human dopamine transporter. Clathrin-mediated internalization and lysosomal degradation in response to phorbol esters. *J Biol Chem* **274**, 35794–35801.

El-Agnaf, O. M., Jakes, R., Curran, M. D., Middleton, D., Ingenito, R., Bianchi, E., Pessi, A., Neill, D., and Wallace, A. (1998a). Aggregates from mutant and wild-type alpha-synuclein proteins and NAC peptide induce apoptotic cell death in human neuroblastoma cells by formation of beta-sheet and amyloid-like filaments. *FEBS Lett* **440**, 71–75.

El-Agnaf, O. M., Jakes, R., Curran, M. D., and Wallace, A. (1998b). Effects of the mutations Ala30 to Pro and Ala53 to Thr on the physical and morphological properties of alpha-synuclein protein implicated in Parkinson's disease. *FEBS Lett* **440**, 67–70.

Elkon, H., Don, J., Melamed, E., Ziv, I., Shirvan, A., and Offen, D. (2002). Mutant and wild-type alpha-synuclein interact with mitochondrial cytochrome c oxidase. *J Mol Neurosci* **18**, 229–238.

Engelender, S., Kaminsky, Z., Guo, X., Sharp, A. H., Amaravi, R. K., Kleiderlein, J. J., Margolis, R. L., Troncoso, J. C., Lanahan, A. A., Worley, P. F., Dawson, V. L., Dawson, T. M., and Ross, C. A. (1999). Synphilin-1 associates with alpha-synuclein and promotes the formation of cytosolic inclusions. *Nat Genet* **22**, 110–114.

Fornai, F., Schluter, O. M., Lenzi, P., Gesi, M., Ruffoli, R., Ferrucci, M., Lazzeri, G., Busceti, C. L., Pontarelli, F., Battaglia, G., Pellegrini, A., Nicoletti, F., Ruggieri, S., Paparelli, A., and Sudhof, T. C. (2005). Parkinson-like syndrome induced by continuous MPTP infusion: Convergent roles of the ubiquitin–proteasome system and alpha-synuclein. *Proc Natl Acad Sci* **102**, 3413–3418.

Foster, J. D., Pananusorn, B., and Vaughan, R. A. (2002). Dopamine transporters are phosphorylated on

N-terminal serines in rat striatum. *J Biol Chem* **277**, 25178–25186.

Fujiwara, H., Hasegawa, M., Dohmae, N., Kawashima, A., Masliah, E., Goldberg, M. S., Shen, J., Takio, K., and Iwatsubo, T. (2002). Alpha-synuclein is phosphorylated in synucleinopathy lesions. *Nat Cell Biol* **4**, 160–164.

Gainetdinov, R. R., Fumagalli, F., Wang, Y. M., Jones, S. R., Levey, A. I., Miller, G. W., and Caron, M. G. (1998a). Increased MPTP neurotoxicity in vesicular monoamine transporter 2 heterozygote knockout mice. *J Neurochem* **70**, 1973–1978.

Gainetdinov, R. R., Jones, S. R., Fumagalli, F., Wightman, R. M., and Caron, M. G. (1998b). Re-evaluation of the role of the dopamine transporter in dopamine system homeostasis. *Brain Res Brain Res Rev* **26**, 148–153.

Gallant, P., Malutan, T., McLean, H., Verellen, L., Caveney, S., and Donly, C. (2003). Functionally distinct dopamine and octopamine transporters in the CNS of the cabbage looper moth. *Eur J Biochem* **270**, 664–674.

Giasson, B. I., Duda, J. E., Murray, I. V., Chen, Q., Souza, J. M., Hurtig, H. I., Ischiropoulos, H., Trojanowski, J. Q., and Lee, V. M. (2000). Oxidative damage linked to neurodegeneration by selective alpha-synuclein nitration in synucleinopathy lesions. *Science* **290**, 985–989.

Giasson, B. I., Murray, I. V., Trojanowski, J. Q., and Lee, V. M. (2001). A hydrophobic stretch of 12 amino acid residues in the middle of alpha-synuclein is essential for filament assembly. *J Biol Chem* **276**, 2380–2386.

Hasegawa, M., Fujiwara, H., Nonaka, T., Wakabayashi, K., Takahashi, H., Lee, V. M., Trojanowski, J. Q., Mann, D., and Iwatsubo, T. (2002). Phosphorylated alpha-synuclein is ubiquitinated in alpha-synucleinopathy lesions. *J Biol Chem* **277**, 49071–49076.

Hasegawa, T., Matsuzaki, M., Takeda, A., Kikuchi, A., Akita, H., Perry, G., Smith, M. A., and Itoyama, Y. (2004). Accelerated alpha-synuclein aggregation after differentiation of SH-SY5Y neuroblastoma cells. *Brain Res* **1013**, 51–59.

Hashimoto, M., Hsu, L. J., Rockenstein, E., Takenouchi, T., Mallory, M., and Masliah, E. (2002). Alpha-synuclein protects against oxidative stress via inactivation of the c-Jun N-terminal kinase stress-signaling pathway in neuronal cells. *J Biol Chem* **277**, 11465–11472.

Hofer, A., Berg, D., Asmus, F., Niwar, M., Ransmayr, G., Riemenschneider, M., Bonelli, S. B., Steffelbauer, M., Ceballos-Baumann, A., Haussermann, P., Behnke, S., Kruger, R., Prestel, J., Sharma, M., Zimprich, A., Riess, O., and Gasser, T. (2005). The role of alpha-synuclein gene multiplications in early-onset Parkinson's disease and dementia with Lewy bodies. *J Neural Transm* **112**, 1249–1254.

Hsu, L. J., Sagara, Y., Arroyo, A., Rockenstein, E., Sisk, A., Mallory, M., Wong, J., Takenouchi, T., Hashimoto, M., and Masliah, E. (2000). Alpha-synuclein promotes mitochondrial deficit and oxidative stress. *Am J Pathol* **157**, 401–410.

Iwata, A., Maruyama, M., Kanazawa, I., and Nukina, N. (2001a). Alpha-synuclein affects the MAPK pathway and accelerates cell death. *J Biol Chem* **276**, 45320–45329.

Iwata, A., Miura, S., Kanazawa, I., Sawada, M., and Nukina, N. (2001b). Alpha-synuclein forms a complex with transcription factor Elk-1. *J Neurochem* **77**, 239–252.

Jenco, J. M., Rawlingson, A., Daniels, B., and Morris, A. J. (1998). Regulation of phospholipase D2: Selective inhibition of mammalian phospholipase D isoenzymes by alpha- and beta-synucleins. *Biochemistry* **37**, 4901–4909.

Jenner, P. (2003). Oxidative stress in Parkinson's disease. *Ann Neurol* **53**(Suppl 3), S26–36, Discussion S36–38.

Jensen, P. H., Hojrup, P., Hager, H., Nielsen, M. S., Jacobsen, L., Olesen, O. F., Gliemann, J., and Jakes, R. (1997). Binding of Abeta to alpha- and beta-synucleins: identification of segments in alpha-synuclein/NAC precursor that bind Abeta and NAC. *Biochem J* **323**(Pt 2), 539–546.

Johnson, J., Hague, S. M., Hanson, M., Gibson, A., Wilson, K. E., Evans, E. W., Singleton, A. A., McInerney-Leo, A., Nussbaum, R. L., Hernandez, D. G., Gallardo, M., McKeith, I. G., Burn, D. J., Ryu, M., Hellstrom, O., Ravina, B., Eerola, J., Perry, R. H., Jaros, E., Tienari, P., Weiser, R., Gwinn-Hardy, K., Morris, C. M., Hardy, J., and Singleton, A. B. (2004). SNCA multiplication is not a common cause of Parkinson disease or dementia with Lewy bodies. *Neurology* **63**, 554–556.

Junn, E., Ronchetti, R. D., Quezado, M. M., Kim, S. Y., and Mouradian, M. M. (2003). Tissue transglutaminase-induced aggregation of alpha-synuclein: Implications for Lewy body formation in Parkinson's disease and dementia with Lewy bodies. *Proc Natl Acad Sci* **100**, 2047–2052.

Kahle, P. J., Neumann, M., Ozmen, L., Muller, V., Jacobsen, H., Spooren, W., Fuss, B., Mallon, B., Macklin, W. B., Fujiwara, H., Hasegawa, M., Iwatsubo, T., Kretzschmar, H. A., and Haass, C. (2002). Hyperphosphorylation and insolubility of alpha-synuclein in transgenic mouse oligodendrocytes. *EMBO Rep* **3**, 583–588.

Kanda, S., Bishop, J. F., Eglitis, M. A., Yang, Y., and Mouradian, M. M. (2000). Enhanced vulnerability to oxidative stress by alpha-synuclein mutations and C-terminal truncation. *Neuroscience* **97**, 279–284.

Kanthasamy, A. G., Kitazawa, M., Kanthasamy, A., and Anantharam, V. (2003). Role of proteolytic activation of protein kinase C delta in oxidative stress-induced apoptosis. *Antioxid Redox Signal* 5, 609–620.

Kaul, S., Kanthasamy, A., Kitazawa, M., Anantharam, V., and Kanthasamy, A. G. (2003). Caspase-3 dependent proteolytic activation of protein kinase C delta mediates and regulates 1-methyl-4-phenylpyridinium (MPP+)-induced apoptotic cell death in dopaminergic cells: Relevance to oxidative stress in dopaminergic degeneration. *Eur J Neurosci* 18, 1387–1401.

Kaul, S., Anantharam, V., Kanthasamy, A., and Kanthasamy, A. G. (2005a). Wild-type alpha-synuclein interacts with pro-apoptotic proteins PKC delta and BAD to protect dopaminergic neuronal cells against MPP + -induced apoptotic cell death. *Brain Res Mol Brain Res* 139, 137–152.

Kaul, S., Anantharam, V., Yang, Y., Choi, C. J., Kanthasamy, A., and Kanthasamy, A. G. (2005b). Tyrosine phosphorylation regulates the proteolytic activation of protein kinase C delta in dopaminergic neuronal cells. *J Biol Chem* 280, 28721–28730.

Kikkawa, U., Matsuzaki, H., and Yamamoto, T. (2002). Protein kinase C delta (PKC delta): Activation mechanisms and functions. *J Biochem (Tokyo)* 132, 831–839.

Kitazawa, M., Anantharam, V., and Kanthasamy, A. G. (2003). Dieldrin induces apoptosis by promoting caspase-3-dependent proteolytic cleavage of protein kinase C delta in dopaminergic cells: Relevance to oxidative stress and dopaminergic degeneration. *Neuroscience* 119, 945–964.

Kobayashi, H., Ide, S., Hasegawa, J., Ujike, H., Sekine, Y., Ozaki, N., Inada, T., Harano, M., Komiyama, T., Yamada, M., Iyo, M., Shen, H. W., Ikeda, K., and Soraa, I. (2004). Study of association between {alpha}synuclein gene polymorphism and methamphetamine psychosis/dependence. *Ann NY Acad Sci* 1025, 325–334.

Kobori, N., Waymire, J. C., Haycock, J. W., Clifton, G. L., and Dash, P. K. (2004). Enhancement of tyrosine hydroxylase phosphorylation and activity by glial cell line-derived neurotrophic factor. *J Biol Chem* 279, 2182–2191.

Kruger, R., Kuhn, W., Muller, T., Woitalla, D., Graeber, M., Kosel, S., Przuntek, H., Epplen, J. T., Schols, L., and Riess, O. (1998). Ala30Pro mutation in the gene encoding alpha-synuclein in Parkinson's disease. *Nat Genet* 18, 106–108.

Larsen, K. E., Schmitz, Y., Troyer, M. D., Mosharov, E., Dietrich, P., Quazi, A. Z., Savalle, M., Nemani, V., Chaudhry, F. A., Edwards, R. H., Stefanis, L., and Sulzer, D. (2006). Alpha-synuclein overexpression in PC12 and chromaffin cells impairs catecholamine release by interfering with a late step in exocytosis. *J Neurosci* 26, 11915–11922.

Latchoumycandane, C., Anantharam, V., Kitazawa, M., Yang, Y., Kanthasamy, A., and Kanthasamy, A. G.

(2005). Protein kinase C delta is a key downstream mediator of manganese-induced apoptosis in dopaminergic neuronal cells. *J Pharmacol Exp Ther* 313, 46–55.

Lee, C. S., Samii, A., Sossi, V., Ruth, T. J., Schulzer, M., Holden, J. E., Wudel, J., Pal, P. K., de la Fuente-Fernandez, R., Calne, D. B., and Stoessl, A. J. (2000). *In vivo* positron emission tomographic evidence for compensatory changes in presynaptic dopaminergic nerve terminals in Parkinson's disease. *Ann Neurol* 47, 493–503.

Lee, D., Lee, S. Y., Lee, E. N., Chang, C. S., and Paik, S. R. (2002a). Alpha-synuclein exhibits competitive interaction between calmodulin and synthetic membranes. *J Neurochem* 82, 1007–1017.

Lee, F. J., Liu, F., Pristupa, Z. B., and Niznik, H. B. (2001). Direct binding and functional coupling of alpha-synuclein to the dopamine transporters accelerate dopamine-induced apoptosis. *FASEB J* 15, 916–926.

Lee, G., Newman, S. T., Gard, D. L., Band, H., and Panchamoorthy, G. (1998). Tau interacts with src-family non-receptor tyrosine kinases. *J Cell Sci* 111 (Pt 21), 3167–3177.

Lee, H. G., Zhu, X., Takeda, A., Perry, G., and Smith, M. A. (2006). Emerging evidence for the neuroprotective role of alpha-synuclein. *Exp Neurol* 200, 1–7.

Lee, H. J., Shin, S. Y., Choi, C., Lee, Y. H., and Lee, S. J. (2002b). Formation and removal of alpha-synuclein aggregates in cells exposed to mitochondrial inhibitors. *J Biol Chem* 277, 5411–5417.

Lee, M. K., Stirling, W., Xu, Y., Xu, X., Qui, D., Mandir, A. S., Dawson, T. M., Copeland, N. G., Jenkins, N. A., and Price, D. L. (2002c). Human alpha-synuclein-harboring familial Parkinson's disease-linked Ala-53 --> Thr mutation causes neurodegenerative disease with alpha-synuclein aggregation in transgenic mice. *Proc Natl Acad Sci USA* 99, 8968–8973.

Li, W., and Lee, M. K. (2005). Antiapoptotic property of human alpha-synuclein in neuronal cell lines is associated with the inhibition of caspase-3 but not caspase-9 activity. *J Neurochem* 93, 1542–1550.

Liu, Y., Fallon, L., Lashuel, H. A., Liu, Z., and Lansbury, P. T., Jr. (2002). The UCH-L1 gene encodes two opposing enzymatic activities that affect alpha-synuclein degradation and Parkinson's disease susceptibility. *Cell* 111, 209–218.

Lotharius, J., and O'Malley, K. L. (2001). Role of mitochondrial dysfunction and dopamine-dependent oxidative stress in amphetamine-induced toxicity. *Ann Neurol* 49, 79–89.

Lotharius, J., and Brundin, P. (2002a). Impaired dopamine storage resulting from alpha-synuclein mutations may contribute to the pathogenesis of Parkinson's disease. *Hum Mol Genet* 11, 2395–2407.

Lotharius, J., and Brundin, P. (2002b). Pathogenesis of Parkinson's disease: Dopamine, vesicles and alpha-synuclein. *Nat Rev Neurosci* 3, 932–942.

Lotharius, J., Barg, S., Wiekop, P., Lundberg, C., Raymon, H. K., and Brundin, P. (2002). Effect of mutant alpha-synuclein on dopamine homeostasis in a new human mesencephalic cell line. *J Biol Chem* 277, 38884–38894.

Luo, Y., Umegaki, H., Wang, X., Abe, R., and Roth, G. S. (1998). Dopamine induces apoptosis through an oxidation-involved SAPK/JNK activation pathway. *J Biol Chem* 273, 3756–3764.

Machida, Y., Chiba, T., Takayanagi, A., Tanaka, Y., Asanuma, M., Ogawa, N., Koyama, A., Iwatsubo, T., Ito, S., Jansen, P. H., Shimizu, N., Tanaka, K., Mizuno, Y., and Hattori, N. (2005). Common antiapoptotic roles of parkin and alpha-synuclein in human dopaminergic cells. *Biochem Biophys Res Commun* 332, 233–240.

Manning-Bog, A. B., McCormack, A. L., Purisai, M. G., Bolin, L. M., and Di Monte, D. A. (2003). Alphasynuclein overexpression protects against paraquatinduced neurodegeneration. *J Neurosci* 23, 3095–3099.

Maroteaux, L., Campanelli, J. T., and Scheller, R. H. (1988). Synuclein: A neuron-specific protein localized to the nucleus and presynaptic nerve terminal. *J Neurosci* 8, 2804–2815.

Martinez, J., Moeller, I., Erdjument-Bromage, H., Tempst, P., and Lauring, B. (2003). Parkinson's disease-associated alpha-synuclein is a calmodulin substrate. *J Biol Chem* 278, 17379–17387.

Masliah, E., Rockenstein, E., Veinbergs, I., Mallory, M., Hashimoto, M., Takeda, A., Sagara, Y., Sisk, A., and Mucke, L. (2000). Dopaminergic loss and inclusion body formation in alpha-synuclein mice: Implications for neurodegenerative disorders. *Science* 287, 1265–1269.

Matsuoka, Y., Vila, M., Lincoln, S., McCormack, A., Picciano, M., LaFrancois, J., Yu, X., Dickson, D., Langston, W. J., McGowan, E., Farrer, M., Hardy, J., Duff, K., Przedborski, S., and Di Monte, D. A. (2001). Lack of nigral pathology in transgenic mice expressing human alpha-synuclein driven by the tyrosine hydroxylase promoter. *Neurobiol Dis* 8, 535–539.

McNaught, K. S., Bjorklund, L. M., Belizaire, R., Isacson, O., Jenner, P., and Olanow, C. W. (2002a). Proteasome inhibition causes nigral degeneration with inclusion bodies in rats. *Neuroreport* 13, 1437–1441.

McNaught, K. S., Mytilineou, C., Jnobaptiste, R., Yabut, J., Shashidharan, P., Jennert, P., and Olanow, C. W. (2002b). Impairment of the ubiquitin–proteasome system causes dopaminergic cell death and inclusion body formation in ventral mesencephalic cultures. *J Neurochem* 81, 301–306.

Melikian, H. E., and Buckley, K. M. (1999). Membrane trafficking regulates the activity of the human dopamine transporter. *J Neurosci* 19, 7699–7710.

Miller, D. W., Hague, S. M., Clarimon, J., Baptista, M., Gwinn-Hardy, K., Cookson, M. R., and Singleton, A. B. (2004). Alpha-synuclein in blood and brain from familial Parkinson disease with SNCA locus triplication. *Neurology* 62, 1835–1838.

Mortensen, O. V., and Amara, S. G. (2003). Dynamic regulation of the dopamine transporter. *Eur J Pharmacol* 479, 159–170.

Mosharov, E. V., Staal, R. G., Bove, J., Prou, D., Hananiya, A., Markov, D., Poulsen, N., Larsen, K. E., Moore, C. M., Troyer, M. D., Edwards, R. H., Przedborski, S., and Sulzer, D. (2006). Alpha-synuclein overexpression increases cytosolic catecholamine concentration. *J Neurosci* 26, 9304–9311.

Moussa, C. E., Wersinger, C., Tomita, Y., and Sidhu, A. (2004). Differential cytotoxicity of human wild type and mutant alpha-synuclein in human neuroblastoma SH-SY5Y cells in the presence of dopamine. *Biochemistry* 43, 5539–5550.

Murray, I. V., Giasson, B. I., Quinn, S. M., Koppaka, V., Axelsen, P. H., Ischiropoulos, H., Trojanowski, J. Q., and Lee, V. M. (2003). Role of alpha-synuclein carboxy-terminus on fibril formation *in vitro*. *Biochemistry* 42, 8530–8540.

Nakamura, T., Yamashita, H., Nagano, Y., Takahashi, T., Avraham, S., Avraham, H., Matsumoto, M., and Nakamura, S. (2002). Activation of Pyk2/RAFTK induces tyrosine phosphorylation of alpha-synuclein via Src-family kinases. *FEBS Lett* 521, 190–194.

Neumann, M., Kahle, P. J., Giasson, B. I., Ozmen, L., Borroni, E., Spooren, W., Muller, V., Odoy, S., Fujiwara, H., Hasegawa, M., Iwatsubo, T., Trojanowski, J. Q., Kretzschmar, H. A., and Haass, C. (2002). Misfolded proteinase K-resistant hyperphosphorylated alpha-synuclein in aged transgenic mice with locomotor deterioration and in human alphasynucleinopathies. *J Clin Invest* 110, 1429–1439.

Okochi, M., Walter, J., Koyama, A., Nakajo, S., Baba, M., Iwatsubo, T., Meijer, L., Kahle, P. J., and Haass, C. (2000). Constitutive phosphorylation of the Parkinson's disease associated alpha-synuclein. *J Biol Chem* 275, 390–397.

Oluwatosin-Chigbu, Y., Robbins, A., Scott, C. W., Arriza, J. L., Reid, J. D., and Zysk, J. R. (2003). Parkin suppresses wild-type alpha-synuclein-induced toxicity in SHSY-5Y cells. *Biochem Biophys Res Commun* 309, 679–684.

Orth, M., Tabrizi, S. J., Tomlinson, C., Messmer, K., Korlipara, L. V., Schapira, A. H., and Cooper, J. M. (2004). G209A mutant alpha synuclein expression specifically enhances dopamine induced oxidative damage. *Neurochem Int* 45, 669–676.

Ostrerova, N., Petrucelli, L., Farrer, M., Mehta, N., Choi, P., Hardy, J., and Wolozin, B. (1999). Alpha-synuclein shares physical and functional homology with 14-3-3 proteins. *J Neurosci* **19**, 5782–5791.

Ostrerova-Golts, N., Petrucelli, L., Hardy, J., Lee, J. M., Farer, M., and Wolozin, B. (2000). The A53T alpha-synuclein mutation increases iron-dependent aggregation and toxicity. *J Neurosci* **20**, 6048–6054.

Pals, P., Lincoln, S., Manning, J., Heckman, M., Skipper, L., Hulihan, M., Van den Broeck, M., De Pooter, T., Cras, P., Crook, J., Van Broeckhoven, C., and Farrer, M. J. (2004). Alpha-synuclein promoter confers susceptibility to Parkinson's disease. *Ann Neurol* **56**, 591–595.

Payton, J. E., Perrin, R. J., Woods, W. S., and George, J. M. (2004). Structural determinants of PLD2 inhibition by alpha-synuclein. *J Mol Biol* **337**, 1001–1009.

Peng, X., Tehranian, R., Dietrich, P., Stefanis, L., and Perez, R. G. (2005). Alpha-synuclein activation of protein phosphatase 2A reduces tyrosine hydroxylase phosphorylation in dopaminergic cells. *J Cell Sci* **118**, 3523–3530.

Polymeropoulos, M. H., Lavedan, C., Leroy, E., Ide, S. E., Dehejia, A., Dutra, A., Pike, B., Root, H., Rubenstein, J., Boyer, R., Stenroos, E. S., Chandrasekharappa, S., Athanassiadou, A., Papapetropoulos, T., Johnson, W. G., Lazzarini, A. M., Duvoisin, R. C., Di Iorio, G., Golbe, L. I., and Nussbaum, R. L. (1997). Mutation in the alpha-synuclein gene identified in families with Parkinson's disease. *Science* **276**, 2045–2047.

Prasad, K. N., Carvalho, E., Kentroti, S., Edwards-Prasad, J., Freed, C., and Vernadakis, A. (1994). Establishment and characterization of immortalized clonal cell lines from fetal rat mesencephalic tissue. *In Vitro Cell Dev Biol Anim* **30A**, 596–603.

Pristupa, Z. B., McConkey, F., Liu, F., Man, H. Y., Lee, F. J., Wang, Y. T., and Niznik, H. B. (1998). Protein kinase-mediated bidirectional trafficking and functional regulation of the human dopamine transporter. *Synapse* **30**, 79–87.

Pronin, A. N., Morris, A. J., Surguchov, A., and Benovic, J. L. (2000). Synucleins are a novel class of substrates for G protein-coupled receptor kinases. *J Biol Chem* **275**, 26515–26522.

Recchia, A., Debetto, P., Negro, A., Guidolin, D., Skaper, S. D., and Giusti, P. (2004). Alpha-synuclein and Parkinson's disease. *FASEB J* **18**, 617–626.

Ribeiro, C. S., Carneiro, K., Ross, C. A., Menezes, J. R., and Engelender, S. (2002). Synphilin-1 is developmentally localized to synaptic terminals, and its association with synaptic vesicles is modulated by alpha-synuclein. *J Biol Chem* **277**, 23927–23933.

Rideout, H. J., Larsen, K. E., Sulzer, D., and Stefanis, L. (2001). Proteasomal inhibition leads to formation of ubiquitin/alpha-synuclein-immunoreactive inclusions in PC12 cells. *J Neurochem* **78**, 899–908.

Royo, M., Daubner, S. C., and Fitzpatrick, P. F. (2004). Specificity of the MAP kinase ERK2 for phosphorylation of tyrosine hydroxylase. *Arch Biochem Biophys* **423**, 247–252.

Saha, A. R., Ninkina, N. N., Hanger, D. P., Anderton, B. H., Davies, A. M., and Buchman, V. L. (2000). Induction of neuronal death by alpha-synuclein. *Eur J Neurosci* **12**, 3073–3077.

Seo, J. H., Rah, J. C., Choi, S. H., Shin, J. K., Min, K., Kim, H. S., Park, C. H., Kim, S., Kim, E. M., Lee, S. H., Lee, S., Suh, S. W., and Suh, Y. H. (2002). Alpha-synuclein regulates neuronal survival via Bcl-2 family expression and PI3/Akt kinase pathway. *FASEB J* **16**, 1826–1828.

Sherer, T. B., Kim, J. H., Betarbet, R., and Greenamyre, J. T. (2003). Subcutaneous rotenone exposure causes highly selective dopaminergic degeneration and alpha-synuclein aggregation. *Exp Neurol* **179**, 9–16.

Shimura, H., Schlossmacher, M. G., Hattori, N., Frosch, M. P., Trockenbacher, A., Schneider, R., Mizuno, Y., Kosik, K. S., and Selkoe, D. J. (2001). Ubiquitination of a new form of alpha-synuclein by parkin from human brain: Implications for Parkinson's disease. *Science* **293**, 263–269.

Sidhu, A., Wersinger, C., Moussa, C. E., and Vernier, P. (2004a). The role of alpha-synuclein in both neuroprotection and neurodegeneration. *Ann NY Acad Sci* **1035**, 250–270.

Sidhu, A., Wersinger, C., and Vernier, P. (2004b). Does alpha-synuclein modulate dopaminergic synaptic content and tone at the synapse? *FASEB J* **18**, 637–647.

Singleton, A. B., Farrer, M., Johnson, J., Singleton, A., Hague, S., Kachergus, J., Hulihan, M., Peuralinna, T., Dutra, A., Nussbaum, R., Lincoln, S., Crawley, A., Hanson, M., Maraganore, D., Adler, C., Cookson, M. R., Muenter, M., Baptista, M., Miller, D., Blancato, J., Hardy, J., and Gwinn-Hardy, K. (2003). Alpha-synuclein locus triplication causes Parkinson's disease. *Science* **302**, 841.

Smith, W. W., Jiang, H., Pei, Z., Tanaka, Y., Morita, H., Sawa, A., Dawson, V. L., Dawson, T. M., and Ross, C. A. (2005a). Endoplasmic reticulum stress and mitochondrial cell death pathways mediate A53T mutant alpha-synuclein-induced toxicity. *Hum Mol Genet* **14**, 3801–3811.

Smith, W. W., Margolis, R. L., Li, X., Troncoso, J. C., Lee, M. K., Dawson, V. L., Dawson, T. M., Iwatsubo, T., and Ross, C. A. (2005b). Alpha-synuclein phosphorylation enhances eosinophilic cytoplasmic inclusion formation in SH-SY5Y cells. *J Neurosci* **25**, 5544–5552.

Souza, J. M., Giasson, B. I., Chen, Q., Lee, V. M., and Ischiropoulos, H. (2000). Dityrosine cross-linking promotes formation of stable alpha-synuclein polymers.

Implication of nitrative and oxidative stress in the pathogenesis of neurodegenerative synucleinopathies. *J Biol Chem* **275**, 18344–18349.

Stefanis, L., Larsen, K. E., Rideout, H. J., Sulzer, D., and Greene, (2001). Expression of A53T mutant but not wild-type alpha-synuclein in PC12 cells induces alterations of the ubiquitin-dependent degradation system, loss of dopamine release, and autophagic cell death. *J Neurosci* **21**, 9549–9560.

Suh, Y. H., and Checler, F. (2002). Amyloid precursor protein, presenilins, and alpha-synuclein: Molecular pathogenesis and pharmacological applications in Alzheimer's disease. *Pharmacol Rev* **54**, 469–525.

Sun, F., Anantharam, V., Latchoumycandane, C., Kanthasamy, A., and Kanthasamy, A. G. (2005). Dieldrin induces ubiquitin–proteasome dysfunction in alpha-synuclein overexpressing dopaminergic neuronal cells and enhances susceptibility to apoptotic cell death. *J Pharmacol Exp Ther* **315**, 69–79.

Sung, J. Y., Kim, J., Paik, S. R., Park, J. H., Ahn, Y. S., and Chung, K. C. (2001). Induction of neuronal cell death by Rab5A-dependent endocytosis of alpha-synuclein. *J Biol Chem* **276**, 27441–27448.

Sura, G. R., Daubner, S. C., and Fitzpatrick, P. F. (2004). Effects of phosphorylation by protein kinase A on binding of catecholamines to the human tyrosine hydroxylase isoforms. *J Neurochem* **90**, 970–978.

Takahashi, T., Yamashita, H., Nakamura, T., Nagano, Y., and Nakamura, S. (2002). Tyrosine 125 of alpha-synuclein plays a critical role for dimerization following nitrative stress. *Brain Res* **938**, 73–80.

Takahashi, T., Yamashita, H., Nagano, Y., Nakamura, T., Ohmori, H., Avraham, H., Avraham, S., Yasuda, M., and Matsumoto, M. (2003). Identification and characterization of a novel Pyk2/related adhesion focal tyrosine kinase-associated protein that inhibits alpha-synuclein phosphorylation. *J Biol Chem* **278**, 42225–42233.

Tanaka, Y., Engelender, S., Igarashi, S., Rao, R. K., Wanner, T., Tanzi, R. E., Sawa, A., V, L. D., Dawson, T. M., and Ross, C. A. (2001). Inducible expression of mutant alpha-synuclein decreases proteasome activity and increases sensitivity to mitochondria-dependent apoptosis. *Hum Mol Genet* **10**, 919–926.

Trojanowski, J. Q., and Lee, V. M. (2000). "Fatal attractions" of proteins. A comprehensive hypothetical mechanism underlying Alzheimer's disease and other neurodegenerative disorders. *Ann NY Acad Sci* **924**, 62–67.

Wakamatsu, M., Ishii, A., Ukai, Y., Sakagami, J., Iwata, S., Ono, M., Matsumoto, K., Nakamura, A., Tada, N., Kobayashi, K., Iwatsubo, T., and Yoshimoto, M. (2007). Accumulation of phosphorylated alpha-synuclein in dopaminergic neurons of transgenic mice that express human alpha-synuclein. *J Neurosci Res* **85**, 1819–1825.

Weihe, E., and Eiden, L. E. (2000). Chemical neuroanatomy of the vesicular amine transporters. *FASEB J* **14**, 2435–2449.

Wersinger, C., and Sidhu, A. (2003). Attenuation of dopamine transporter activity by alpha-synuclein. *Neurosci Lett* **340**, 189–192.

Wersinger, C., Prou, D., Vernier, P., and Sidhu, A. (2003a). Modulation of dopamine transporter function by alpha-synuclein is altered by impairment of cell adhesion and by induction of oxidative stress. *FASEB J* **17**, 2151–2153.

Xu, J., Kao, S. Y., Lee, F. J., Song, W., Jin, L. W., and Yankner, B. A. (2002). Dopamine-dependent neurotoxicity of alpha-synuclein: A mechanism for selective neurodegeneration in Parkinson disease. *Nat Med* **8**, 600–606.

Yang, Y., Kaul, S., Zhang, D., Anantharam, V., and Kanthasamy, A. G. (2004). Suppression of caspase-3-dependent proteolytic activation of protein kinase C delta by small interfering RNA prevents MPP+ -induced dopaminergic degeneration. *Mol Cell Neurosci* **25**, 406–421.

Yavich, L., Tanila, H., Vepsalainen, S., and Jakala, P. (2004). Role of alpha-synuclein in presynaptic dopamine recruitment. *J Neurosci* **24**, 11165–11170.

Yoshimoto, M., Iwai, A., Kang, D., Otero, D. A., Xia, Y., and Saitoh, T. (1995). NACP, the precursor protein of the non-amyloid beta/A4 protein (A beta) component of Alzheimer disease amyloid, binds A beta and stimulates A beta aggregation. *Proc Natl Acad Sci* **92**, 9141–9145.

Zarranz, J. J., Alegre, J., Gomez-Esteban, J. C., Lezcano, E., Ros, R., Ampuero, I., Vidal, L., Hoenicka, J., Rodriguez, O., Atares, B., Llorens, V., Gomez Tortosa, E., del Ser, T., Munoz, D. G., and de Yebenes, J. G. (2004). The new mutation, E46K, of alpha-synuclein causes Parkinson and Lewy body dementia. *Ann Neurol* **55**, 164–173.

Zhang, D., Kanthasamy, A., Yang, Y., and Anantharam, V. (2007). Protein kinase C delta negatively regulates tyrosine hydroxylase activity and dopamine synthesis by enhancing protein phosphatase-2A activity in dopaminergic neurons. *J Neurosci* **27**, 5349–5362.

Zhou, W., Schaack, J., Zawada, W. M., and Freed, C. R. (2002). Overexpression of human alpha-synuclein causes dopamine neuron death in primary human mesencephalic culture. *Brain Res* **926**, 42–50.

Zourlidou, A., Payne Smith, M. D., and Latchman, D. S. (2003). Modulation of cell death by alpha-synuclein is stimulus-dependent in mammalian cells. *Neurosci Lett* **340**, 234–238.

37

POSTNATALLY DERIVED VENTRAL MIDBRAIN DOPAMINE NEURON CULTURES AS A MODEL SYSTEM FOR STUDYING NEUROTOXICITY AND PARKINSON'S DISEASE

DAVID SULZER[1], LOUIS-ERIC TRUDEAU[2] AND
STEPHEN RAYPORT[3]

[1]Department of Neurology, Psychiatry, & Pharmacology, Columbia University,
Department of Molecular Therapeutics, NYS Psychiatric Institute, New York, NY, USA
[2]Departments of Pharmacology, Psychiatry and Physiology, CNS Research Group,
Faculty of Medicine, Université de Montréal, Montréal, Québec, Canada
[3]Department of Psychiatry, Columbia University, Department of Molecular Therapeutics,
NYS Psychiatric Institute, New York, NY, USA

INTRODUCTION

Neuronal cultures were introduced in 1907 by Ross Granville Harrison, who showed that amphibian spinal cord explants can grow on a protein matrix. Harrison's culture studies are acknowledged as providing proof of the "neuron theory" of the brain, that is, that the nervous system is composed of discrete neurons, along with the discovery that axonal growth cones extend toward targets (Keshishian, 2004). Despite his role as a founder of neuroscience and the vast fundamental insights on synaptic function, development, and disease provided by neuronal culture, Harrison, who lived to the age of 89, was never awarded a Nobel prize, apparently as many felt that results from the tissue culture approach were suspect.

Embryonically derived ("embryonic") ventral midbrain (VM) cultures for studying isolated dopamine (DA) neurons were introduced in 1979 by Alain Prochiantz and collaborators, who developed a method that provided for neuronal survival of over 2 weeks (Prochiantz et al., 1979). Prochiantz's approach has proved invaluable for studies of neuronal outgrowth, synaptic properties, and cell toxicity and death, and the majority of neuronal culture studies on Parkinson's disease (PD) neurodegeneration continue to use it.

There are practical advantages in using more mature neurons for synaptic studies and diseases associated with aging. Techniques to maintain postnatally derived ("postnatal") VM cultures, however, require additional effort, due in part to the toxicity of DA itself, as well as requirements for media factors and astrocyte contact for long-term survival. Successful techniques for long-term culture of postnatal VM DA neurons were introduced in 1992 by two groups, the team of the Nakajimas at Purdue University (Masuko et al., 1992), and by Stephen Rayport and David Sulzer at Columbia

(Rayport *et al.*, 1992), with a third study soon following by David Cardozo in David Potter's group at Harvard (Cardozo, 1993). These systems have now been used in scores of studies by many labs, particularly using the Rayport approach, in which the neurons are plated on a pre-established bed of astrocytes, following a technique introduced for cortical neuronal culture (Huettner and Baughman, 1986). There are many steps and "tricks" in these protocols, and detailed protocol chapters are available from Trudeau's (Fasano *et al.*, 2007) and Sulzer's (Staal *et al.*, 2007) labs, and a highly detailed booklet can be downloaded from http://www.sulzerlab.org/download.html.

While most studies using postnatal VM cultures use neurons derived from mice or rats within 3 days following birth, cultured DA neurons derived from 45-day-old mice have survived in culture for over 2 weeks (Trudeau lab, unpublished), and a report by Gregory Brewer suggests that even adult VM neurons could be maintained in long-term culture (Brewer, 1995).

Another postnatal culture method not described here is to use "organotypic" cultures in which brain slices are cultured (Gahwiler *et al.*, 1997). These are particularly useful for studying the effects of synaptic connections between different brain regions, and have been used successfully to study postnatal DA neurons (Plenz and Kitai, 1998; Day *et al.*, 2006).

The purpose of this article is to examine the use of postnatal cultures for studies in dopaminergic neuronal degeneration and to provide a basis for comparison with the other model systems discussed in this volume.

CHARACTERISTICS OF POSTNATAL VM CULTURES

Neurochemical Identification

The fraction of DA neurons in postnatal cultures, identified by immunostaining for the DA synthetic enzyme tyrosine hydroxylase (TH), depends chiefly on the dissection. For the entire VM, we tend to observe ~20% DA neurons, while smaller dissections encompassing only the ventral tegmental area (VTA) can provide ~70% DA neurons (Rayport and Sulzer, unpublished). Cultures can be prepared from fluorescently labeled DA neurons (see below) using fluorescence-activated cell sorting (FACS), reaching 95–97% DA neurons (Mendez et al., 2008).

These procedures offer substantial advantages for analysis by biochemical and analytical techniques, such as Western blotting and HPLC.

The high probability of recording from a DA neuron in postnatal culture facilitates electrophysiological and optical analysis, and recent attempts to identify living DA neurons by fluorescent labeling have further assisted these efforts. The first such approach used a fluorescent serotonin analog, 5,7-dihydroxytryptamine, which is apparently a substrate for DA transporter (DAT) and makes DA neurons fluorescent (Silva *et al.*, 1988). This technique is challenging; there can be severe phototoxic damage following observation, although minimizing fluorescent excitation can limit the damage; it may also label non-DA neurons (Trudeau, unpublished observations) including serotonergic neurons (Franke *et al.*, 2002). Another fluorescent cationic compound that is also used for mitochondrial labeling, 4-(4-(dimethylamino)styrl)-N-methylpyridinium, has been identified as a DAT substrate and has been used to identify cultured DA neurons (Schwartz *et al.*, 2003); the compound, however, is accumulated at a slower rate by other neurons and we have found it difficult to robustly identify DA neurons from its fluorescent signal.

A subsequent attempt at producing fluorescent living DA neurons in culture was to inject fluorescent microspheres into the axon terminal regions in the ventral striatum *in vivo*: these were taken up by endocytosis and retrogradely transported to lysosomes in the cell bodies, so that cultured neurons that had originally projected from the VTA to the nucleus accumbens possessed fluorescent cell bodies in culture: 86% of cells so labeled were TH+ (Rayport *et al.*, 1992), exactly matching the original anatomical data (Swanson, 1982) (Figure 37.1).

The introduction of mutant mouse lines that express fluorescent DA neurons provides a particularly useful means of neuronal identification with less effort in preparation. The first such approach was introduced by Elio Raviola's lab, using an alkaline phosphatase that is transported to the plasma membrane linked to a TH promoter; individual cultured retinal DA neurons could be recognized for electrophysiological recording by fluorescently conjugated antibodies or histochemical reaction (Gustincich *et al.*, 1997).

A simpler fluorescent mouse line to use for this purpose was developed by Kazuto Kobayashi and colleagues to express green fluorescent protein (GFP) under control of a TH promoter (Yoshizaki *et al.*,

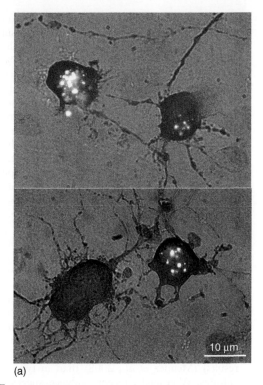

(a)

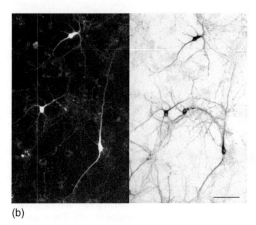

(b)

FIGURE 37.1 Identification of living DA neurons in VTA culture. (a) Retrograde labeling. Neonatal pups were stereotaxically injected with fluorescent latex microspheres (LumaFluor) 2 days prior to the preparation of cultures. A fine dissection of the VTA was done for the preparation of cultures. Cultures were fixed after 3 days *in vitro* (DIV) and immunostained for TH using diaminobenzidine to produce a dark reaction product. Many fields contained only DA neurons: two fields are shown, each with two neurons. Of the four TH+ neurons, three were retrogradely labeled, as seen by the bright spots overlaying the TH staining, in the double exposure (bright field for TH and fluorescence for the microspheres). (Credit: Stephen Rayport, David Sulzer, Geetha Rajendran). (b) Postnatal VM neurons derived from TH-GFP mice. The left panel shows GFP fluorescence in the living culture, and the left panel shows TH immunoreactivity using diaminobenzidine. Note that all three GFP+ neurons are TH+, while one TH+ neuron lacks appreciable levels of GFP. (Credit: Yvonne Schmitz.)

2004). In Sulzer's lab, at 2 weeks in postnatal culture, GFP labels ~50% of the neurons that express TH, while 97% of GFP-labeled neurons are TH+ (Figure 37.1; Yvonne Schmitz *et al.*, unpublished). A higher fraction of TH− GFP+ neurons was reported by Trudeau's group (Jomphe *et al.*, 2005), although they noted that colocalization improved with time in culture. Due to the high cytoplasmic fluorescent signal in this line, it is well adapted for video microscopy of neurite development (Yvonne Schmitz, manuscript in preparation).

Another mouse line that provides fluorescent DA neurons was developed by Rayport and colleagues to express "floxed" CRE-responsive GFP with a cre "knocked in" to the DAT site (Zhuang *et al.*, 2005). The resulting mice require additional breeding effort and maintain a lower than normal expression of DAT, but have the striking advantage that they can be used to examine the effects of a variety of other floxed mutations. Recently, DAT *cre* mice have been developed that do not interfere with normal DAT expression (Backman *et al.*, 2006). A red fluorescent protein (RFP) under control of the TH promoter has been introduced by Douglas McMahon's group (Zhang *et al.*, 2004), although in our hands the number of RFP+ neurons in postnatal VM culture is low, and tends to decrease with time in culture.

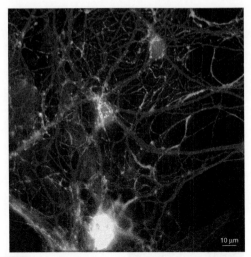

FIGURE 37.2 Immunocytochemical characterization of VTA culture. A VTA culture was first immunostained for GABA (Sigma), using diaminobenzidine to produce an opaque reaction product obscuring further immunostaining (displayed in blue). In a second round of immunostaining, DA neurons were visualized by immunostaining for tyrosine hydroxylase (TH; Chemicon; displayed in green) and for glutamate (Glu2B, Diasorin; displayed in red). There are three neurons in the field, two GABAergic (in upper half of field), and one that double stains for TH and glutamate (near bottom of field; merged colors appear yellow). Within the neuropil, many fine processes, presumably axonal, are studded with varicosities. Varicosities show staining for all three neurotransmitters, as well as well as double staining for TH and glutamate. (Credit: Myra P. Joyce, Stephen Rayport.)

single cells has not yet been observed *in vitro* using single-cell RT-PCR (Trudeau, unpublished results).

There are few neurons that are not labeled for either GABA or TH. A very few neurons, <1%, can be labeled for tryptophan hydroxylase, and presumably produce serotonin; these may have resided in the raphe nucleus and result from imperfect dissection, or derive from a small serotonergic population in the VM (Hui Zhang, unpublished results), or could reflect altered protein expression in culture. The neurochemical identity of the relatively small fraction of remaining neurons is unknown, but electrophysiological recordings from DA neurons indicate ongoing synaptic glutamate input. These inputs may be from "genuine" glutamatergic neurons that release glutamate as their primary transmitter, or reflect glutamate release from nominally DA and GABA neurons (Sulzer *et al.*, 1998), which in DA neurons is due to expression of the vesicular glutamate transporter, VGLUT2 (Dal Bo *et al.*, 2004). DA neuronal release of glutamate, and certainly VGLUT2 expression, appears more commonly in single neuron "microcultures" (Figure 37.3), in which DA neurons tend to make glutamatergic synapses (autapses) on themselves (Sulzer *et al.*, 1998) (Hui Zhang, unpublished results) (Mendez *et al.*, 2008) than in typical VM cultures. Alternately, recent evidence suggests that there may be a contingent of non-DAergic VGLUT2 expressing neurons in the VTA (Yamaguchi *et al.*, 2007), and some of these may provide ongoing glutamatergic input. Neuropeptides are also present within and are presumably released from cultured DA neurons, including cholecystokinin, which is present in 55% of cultured retrogradely labeled mesoaccumbens DA neurons (Rayport *et al.*, 1992).

Role of Astrocytes

Astrocytes provide soluble factors such as glutathione (Mena *et al.*, 1997a), remove toxins (Rosenberg *et al.*, 1992), and establish various forms of glial–neuronal interactions via the accumulation, synthesis, release, and metabolism of transmitters such as glutamate and ATP. As discussed, the most widely used VM postnatal culture system involves plating neurons onto pre-established astrocyte monolayers. Despite numerous attempts, we have been unable to maintain healthy postnatal DA neurons for more than about 3 days in the absence of physical contact between the neurons and astrocytes; simply exposing neurons to nearby astrocytes, as the "Banker" technique

The vast majority of the non-DA neurons in postnatal VM cultures are GABAergic on the basis of γ-aminobutyric acid (GABA) or glutamic acid decarboxylase (GAD) immunostaining (Figure 37.2), providing the basis for several studies (Michel and Trudeau, 2000; Bergevin *et al.*, 2002; Michel *et al.*, 2004a, b). We typically find neurons that are labeled for TH or GABA combined represent >90% of the cultured neurons, although immunostaining for GABA antigens can be less robust than for TH, and often require significant "tweaking". We have also consistently observed a small population of neurons that label for both TH and GABA (Rayport and Sulzer, unpublished results), consistent with a recent anatomical study (Olson and Nestler, 2007), although GAD and TH colocalization within

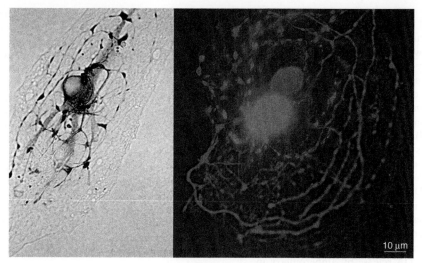

FIGURE 37.3 Microculture of single DA neuron. In the left panel, a single DA neuron, identified by TH immunostaining, is shown growing in a microisland culture. The agarose used to create an inhospitable substrate is seen around the island, with its characteristic mottled pattern. The island is formed by an aerosolized drop of collagen, populated by a single astrocyte, which grows to cover the collagen dot. A subsequently plated neuron then grows on the astrocyte. Its processes flow radially, as they are constrained to the island. In addition to forcing neurons to form autaptic interactions that reveal the neurotransmitter profile of the neuron, immunocytochemical studies can be conducted on single cells, and their processes. In the case of DA neurons, this revealed that the intensity of TH labeling could be strikingly different from process to process of the same cell, as is evident in the pseudocolor transformation shown in the right panel. This and other observations indicate that DA neurons show considerable functional heterogeneity extending beyond gradations in TH trafficking to synaptic varicosities releasing different mixes of transmitters. (Credit: Ling Lin, Stephen Rayport.)

(Banker, 1980) has done successfully for hippocampal cultures, has been unsuccessful, suggesting that DA neurons require an unidentified contact-dependent factor provided by astrocytes.

We typically use cortically derived astrocytes, and select the source of species and genotype based on the experimental aims. Often we use rat-derived astrocytes for mouse neuronal cultures, as it can obviate concerns about genetic background when comparing mutant lines or whether a particular effect is due to glial or neuronal mechanisms. We typically add fluorodeoxyuridine a few days after plating to inhibit astrocyte and microglial mitosis. Microglia in particular can cause neuronal damage, and studies of neurodegeneration in culture require very healthy controls if they are to be meaningful. It is quite straightforward to plate microglia from any genetic background several days after the mitotic inhibitors, and these can be activated and divide, providing the means to observe neuroinflammatory effects.

An important point in examining growth factors and other potential neurotrophic compounds is that the medium must generally be "conditioned" by astrocytes (Takeshima *et al.*, 1994; Burke *et al.*, 1998) or the medium is toxic to both the experimental and control groups (it is important in culture systems to add the same vehicle to both groups). This is particularly important when using serum-free medium. The requirement for astrocyte conditioning is at least in part due to the trace glutamate content of commercial sources of tissue culture media that can lead to excitotoxicity, and this is buffered by conversion for glutamate to glutamine in astrocytes (Rosenberg and Aizenman, 1989). Astrocytes also provide glutathione or a glutathione precursor during the conditioning, which is highly protective against L-DOPA and DA-mediated neurotoxicity in both embryonic and postnatal VM cultures (Mena *et al.*, 1997a, b), and may certainly be providing additional factors.

Longevity

Postnatal VM cultures exhibit a high rate of DA neuron death following plating, which is apoptotic; while we originally believed that this was due to trauma, it follows a time course identical to apoptotic DA neuronal death *in vivo*, and it is possible that some of the same cell death mechanisms occur both during normal development and in culture (Burke *et al.*, 1998). The rate of DA neuron death levels off by 4 days in culture, which is also when we begin to observe significant excitatory synaptic input during whole cell recording. Synaptic connections onto the DA neurons continue to develop over time, and DA neurons can be clearly observed to form more axonal varicosities and longer, highly branched axons, over the first 2 weeks in culture. In postnatal VM rat cultures, we have never been able to measure quantal DA release by amperometry in cultures younger than 3 weeks (Sulzer lab, unpublished results), presumably because the density of release sites is low. Stimulation-evoked release is increased after 2 weeks in culture (Fortin *et al.*, 2006), although lower levels of evoked DA release can be measured in the media by HPLC of much younger cultures.

Due to the reduced rate of cell death and ongoing increase in evoked DA release, we tend to perform experiments on cultures from 2 to 4 weeks *in vitro*. We have sometimes maintained postnatal VM cultures for as long as 8 months. Typically, however, we find neuronal loss to begin again after 4 weeks *in vitro*. This we believe is mostly due to increased osmolarity as the medium evaporates: it may be that careful control of osmolarity will result in much longer lived cultures. Steve Potter has introduced culture dish lids that are permeable to oxygen and carbon dioxide, and relatively impermeable to water, resulting in neuronal cultures that survive over a year (Potter and DeMarse, 2001).

Response to Neurotrophic Factors

An important goal of cell culture toxicity studies in general has been to identify trophic factors necessary for neuron survival, following Rita Levi-Montalcini's study of peripheral neuron cultures in 1953 that provided the original characterization of nerve growth factor response (Levi-Montalcini and Angeletti, 1968). As noted, postnatal VM cultures can be maintained in serum-free, highly defined media, and are good systems for studying potential neuroprotective factors. Postnatal VM cultures appear to be far more restricted in their response to growth factors than embryonic cultures. Neuronal death of SN, but not VTA DA neurons in postnatal culture is specifically inhibited by glial-derived neurotrophic factor (GDNF), which also increases the size of the cell bodies (Burke *et al.*, 1998), neurite outgrowth (Bourque and Trudeau, 2000), and DA release (Pothos *et al.*, 1998). Other growth factors that rescue cultured embryonic DA neurons from cell death, including basic fibroblast growth factor, epidermal growth factor, and transforming growth factor beta, were each ineffective in preventing DA neuronal death in postnatal cultures (Burke *et al.*, 1998), although one study indicates that basic fibroblast growth factor may inhibit synapse formation (Forget *et al.*, 2006). We suspect that the lack of response of these growth factors is because postnatal cultures are in a more mature state that resembles neurons in postnatal animals.

Neurotransmitter Receptors in Postnatal DA Neurons

Ionotropic receptors

DA neurons in postnatal culture possess robust excitatory responses to glutamate with both *N*-methyl-D-aspartate (NMDA) and α-amino-3-hydroxy-5-methyl-4-isoxazolepropionate (AMPA)/kainate components (Sulzer *et al.*, 1998). We have observed excitatory responses to nicotine and inhibitory hyperpolarizing responses to GABA-A (D.S. unpublished results).

Metabotropic receptors

A variety of G-protein-coupled receptors are present on cultured VM DA neurons. The neurons hyperpolarize in response to D2 DA receptor agonists (Rayport *et al.*, 1992; Congar *et al.*, 2002), a property that can be used to differentiate between DA and GABAergic neurons (Rayport *et al.*, 1992). The neurons likewise hyperpolarize in response to GABAb agonists (Cardozo and Bean, 1995). In both cases this is due to activation of G-protein-coupled inwardly rectifying channels, with a particularly pronounced expression of GIRK2 subunits, especially in SN DA neurons (Davila *et al.*, 2003). They also respond with excitation to neurotensin via an NST1 receptor (St-Gelais *et al.*, 2004) and inhibition via an orphanin F/Q receptor (Prasad and Amara, 2001). Some subcellular localization of receptors on the cultured neurons has been accomplished using fluorescent ligands and antireceptor antibodies; D2 receptors are found in "hot spots" on the soma, proximal dendrites (Rayport and Sulzer, 1995; Rayport, 1998), and putative axonal varicosities (Joyce and Rayport, 2000).

Calcium Channels, Calbindin, and Spontaneous Activity

Due to its role in neurotoxicity and PD (Chan *et al.*, 2007), calcium handling is particularly important in model systems of neurodegeneration. Cardozo and Bean (Cardozo and Bean, 1995) reported that components of the calcium currents in cultured postnatal VM DA neurons were inhibited by nimodipine (L-type channels), omega-conotoxin GVIA (N-type channels), and by omega-agatoxin-IVA (P/Q channels), with a significant amount of unidentified remaining current not inhibited by the blockers (Cardozo and Bean, 1995).

The presence of the calcium-binding cytosolic protein calbindin is potentially very important for PD studies, as human VM DA neurons that express calbindin predominantly in the VTA show preferential survival in PD (Damier *et al.*, 1999) and the mouse MPTP model (Liang *et al.*, 1996). We find that ~80% of cultured VTA neurons and ~25% of SN neurons express calbindin by immunocytochemistry (Eugene Mosharov, unpublished results).

Spontaneous activity in DA neurons *in vivo* occurs with a mean frequency of ~4 Hz, and appears to be driven either by sodium or L-type calcium channels (Chan *et al.*, 2007); DA neurons in postnatal culture can also be driven by soidum of L-type calcium channels (Chan et al., 2007); DA neurons is postnatal culture can show similiar spontaneous activity (Prasad and Amara, 2001; St.-Gelais et al., 2004).

Synaptic Morphology

Electron microscopy indicates that postnatal presynaptic DA terminals contain numerous 40 nm diameter small synaptic vesicles and a small fraction (generally <1%) of small dense core vesicles of 100–150 nm diameter (Sulzer and Rayport, 1990; Rayport *et al.*, 1992), both of which accumulate the osmophilic DA analog, 5-hydroxy-DA, and thus are likely to release DA (Figure 37.4). The presynaptic terminals tend to contain small asymmetric

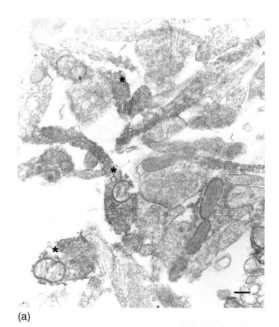

(a)

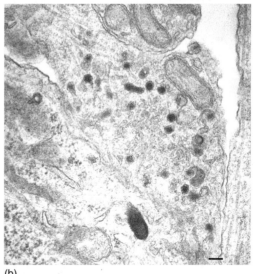

(b)

FIGURE 37.4 Presynaptic DA terminals in VM culture. The upper figure shows TH+ terminals (*) in the midst of TH-neurites. The poor membrane preservation is due to detergent treatment during the immunolabel protocol. (Scale bar = 200 nm). The lower figure shows 5OHDA labeled synaptic vesicles in a DA terminal, a procedure for which detergent treatment is not required. (Scale bar = 100 nm). (Credit: David Sulzer, Roland Staal, Stephen Rayport.)

synapses similar to those *in situ*, while some terminals at least in microculture have larger asymmetric specializations that may correspond to glutamate release sites (Sulzer *et al.*, 1998). An important difference from the *in vivo* situation, however, appears to be that the cultured terminals have a lower density of small synaptic vesicles, although this impression has not been addressed carefully. If so, the kinetics of some parameters of neurotransmitter release, such as the "refilling of the releasable pool" could be different from terminals *in situ*.

In vitro, DA release occurs at presynaptic varicosities on axonal processes. It should be noted that "varicosities" observed after fixation may arise due to osmotic stress during fixation and may not correspond to genuine presynaptic sites. To avoid these artifacts the osmolarity of the fixation solution must be identical to that of the growth medium (Yvonne Schmitz, unpublished observations).

ASSAY METHODS

In addition to standard electrophysiological and immunocytochemical methods, postnatal VM cultures have provided unique opportunities for analysis.

DA Pools

The packaging of DA is widely suspected to play an important role in PD pathogenesis (Sulzer, 2007). Healthy DA neurons in postnatal culture release DA by synaptic vesicle exocytosis (Pothos *et al.*, 1998) and via non-exocytic mechanisms with amphetamine (Sulzer and Rayport, 1990) by reverse transport through the DAT (Sulzer *et al.*, 1993). An advantage to the system is that it is straightforward to use HPLC to measure DA and its metabolites in the media, and to remove the media, lyse the cells, and measure intracellular DA. The ability to measure the entire content of DA has revealed some surprises, for example, that under some conditions, amphetamine can release more DA than was present in the culture when it was added (Larsen *et al.*, 2002).

The ability to directly approach DAergic synaptic terminals by an amperometric electrode has provided the only means to date by which quantal neurotransmitter release has been measured from central synapses. This technique has elucidated

effects on quantal release by GDNF, DA synthesis, VMAT2, and second messenger systems (Pothos *et al.*, 1998; Pothos *et al.*, 2000; Staal *et al.*, 2004). Fluorescent styryl dyes of the "FM" series have been used to observe fusion of recycling small synaptic vesicles in postnatal DA terminals (Pothos *et al.*, 1998; Jomphe *et al.*, 2005; Fortin *et al.*, 2006).

The system has an avid DA reuptake system (Prasad and Amara, 2001) with sufficient DAT currents that in some cases they can drive activity (Ingram *et al.*, 2002). DAT expression in these cultures provides sufficient activity to measure effects of blockers including cocaine and nomifensine and releasers including amphetamine (Sulzer and Rayport, 1990; Sulzer *et al.*, 1993; Sulzer *et al.*, 1996; Fon *et al.*, 1997); our impression is that DAT is more highly expressed than in embryonic cultures, although this has not been rigorously examined.

While intracellular DA can be measured by HPLC as above, nearly all of the transmitter is in synaptic vesicles, while the toxic component may be due to DA free in the cytosol. This has recently become measurable in the cell bodies using intracellular patch electrochemistry (Mosharov *et al.*, 2003; Eugene Mosharov, manuscript unpublished), indicating multiple forms of regulation.

Midbrain DA neurons are well established to release DA from dendrites, and nearly all evidence points to release during fusion of unusually shaped vesicles that presumably do not recycle (Pickel *et al.*, 2002). Dendritic DA release *in vivo* requires comparatively lower extracellular calcium levels (Chen and Rice, 2001). In postnatal VM cultures, a portion of DA release that can be measured by HPLC persists in relatively low levels of calcium ($500\,\mu M$), and may predominantly reflect dendritic release (Fortin *et al.*, 2006).

Toxicity Studies

The postnatal VM system has been used for numerous studies to model aspects of PD including studies using DA neurotoxins such as methamphetamine (Cubells *et al.*, 1994; Larsen *et al.*, 2002), and MPP^+ and rotenone (Dauer *et al.*, 2002). Toxic effects of L-DOPA and DA have been examined in the system (Mena *et al.*, 1997a, c), as has the related role of these compounds in producing neuromelanin (Sulzer *et al.*, 2000) and additional DA derivatives (Greggio *et al.*, 2005). Various protective mechanisms against toxic effects have been examined including comparative effects of growth factor (Burke *et al.*, 1998; Kholodilov *et al.*, 2004) and antioxidant enzymes and peptides (Przedborski *et al.*, 1996; Mena *et al.*, 1997a, b, c) and stress-induced autophagic pathways (Sulzer *et al.*, 2000; Larsen *et al.*, 2002; Cuervo *et al.*, 2004). Relatively recently, the system has been used to examine the comparative response of neurons from mouse lines with mutations involved in PD, including alpha-synuclein (Kholodilov *et al.*, 1999; Dauer *et al.*, 2002; Petrucelli *et al.*, 2002; Cuervo *et al.*, 2004) and parkin (Petrucelli *et al.*, 2002).

Several assays can be used to gather data on neurotoxicity. The most common is to count surviving neurons after a particular treatment. We prefer to immunostain for TH and a marker for GABAergic neurons, either the synthetic enzyme GAD, or GABA itself, or for MAP2 for total neurons (Przedborski *et al.*, 1996; Przedborski *et al.*, 2002). This provides information on whether a particular regimen is selective for one population, and also addresses the issue of whether loss of TH is due to actual neuronal death or a loss of TH signal. We have also used the DNA dye Hoechst 33342 to observe apoptotic profiles (Larsen *et al.*, 2002), as well as histochemical means to detect the presence of activated caspase-3, although the time that the profiles are present is relatively short, and cells can die of non-apoptotic mechanisms.

Typically, we count all immunostained neurons in at least five cultures per experimental group, and conduct ANOVA statistical comparisons, assuming that each culture is independent and is a member of a normally distributed population. Comparisons should be between cultures that were prepared on the same day, and if a particular mutation is examined, it is advisable to compare cultures prepared from wild-type littermates. The counts should be performed by an observer blind to the experimental treatments.

We have recently observed that most of the inter-culture variability results from loss of astrocytes and neurons during tissue processing, and that there is a lower variability in the periphery of the cultures, where the neurons are more protected from the turbulence or solution changes. We therefore now assess neuronal density by counting the number of immunoreactive cells in adjacent fields-of-view of the periphery at 200× magnification (~0.8 mm^2 viewing field) and taking the average as a representative for each dish (Eugene Mosharov, unpublished methods), an approach that reduces the variability between cultures.

Neurite outgrowth can be an important parameter in toxicity studies, although the extensive and complex patterns make quantitation challenging. Neurite measurements have been performed using several approaches, all of which have drawbacks. Simply tracing neurites is tedious, and immunostaining does not indicate the neuron to which a neurite belongs. For many experiments, we place the cell body within the middle of a field, and count the number of primary neurites that extend pass a 120 μm circular perimeter, and analyze the non-Gaussian data by chi-square test (Larsen *et al.*, 2002). The fluorescent DA neurons from various mouse lines (see section "DA Pools") provide a means to measure neurite outgrowth and retraction with video microscopy, although the experimenter needs to be cautious to limit phototoxic damage (Yvonne Schmitz, unpublished results). Alternatively, microcultures of single neurons can be used to confirm that neurites belong to the cell in question, although this limits the number of neurons that can be observed and introduces additional variables including the size and condition of the island.

We should emphasize that control cultures must be outstandingly healthy or it will be nearly impossible to separate the parameter in question from the neurotoxicity seen in unhealthy cultures. In this regard, we routinely observe cultures under phase or differential interference contrast (DIC) optics prior to the experiment to ensure that they are healthy and to ruthlessly exclude cultures that show visible signs of damage. There should be no floating debris. Healthy neurons are phase-bright (or show strong pseudo-3D appearance under DIC), have smooth somatic membranes, prominent nuclei, and processes that can be followed for at least a couple of cell body diameters before they are submerged in the astrocytic monolayer. Healthy neurons depend on healthy glia, which best support neurons when they are confluent but spread out with a fine "ground glass" appearance (Grierson *et al.*, 1990).

Optical Probes

A variety of optical probes have been used in living culture to examine the state of the cells, particularly in PD models. This preparation provided the first evidence of intraneuronal "hot spots" of oxygen radical stress, using the fluorogenic probe dichlorofluorescein (Cubells *et al.*, 1994) (Figure 37.5). Vital optical dyes have also been used for endosomes (Cubells *et al.*, 1994), lysosomes (Sulzer and

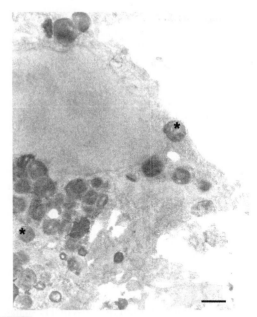

FIGURE 37.6 Methamphetamine-induced autophagic vacuole induction. Methamphetamine treatment destroys DA neurites and induces the formation of large autophagic vacuoles (asterisks) that fill large portions of the cell body. Scale bar = 1 μm. (Credit: Kristin Larsen, David Sulzer.)

FIGURE 37.5 Methamphetamine-induced oxidative stress in VM culture. Under DIC optics, a cell body and neuronal process of a single neuron in a VM culture exposed to 10 μM methamphetamine for 48 h appears swollen, particularly at varicosities, and has several blebs. Fluorescence imaging with 2,7-dichlorofluorescin diacetate (DCF, Molecular Probes) reveals discrete sites of oxidative stress in the cell body and along the process (DCF fluorescence is shown superimposed on a flame scale). This demonstrates localized oxidative stress within the process. In the cell body, DCF fluorescence follows a punctate distribution similar in pattern to the distribution of endocytic organelles. (Credit: Joseph Cubells, Stephen Rayport, Geetha Rajendran, David Sulzer.) (NB: Black and white version in Cubells *et al.*, *J Neurosci*, 1994, Figure 8.)

Rayport, 1990), autophagic vacuoles (Larsen *et al.*, 2002), apoptosis (Burke *et al.*, 1998), and mitochondria and endoplasmic reticulum, particularly using fluorescent probes available from Molecular Probes (Eugene, OR). Similar studies have been conducted in postnatal striatal cultures (Petersen *et al.*, 2001).

Electron Microscopy

Electron microscope preparations of the cultures are greatly facilitated by use of Aclar, a plastic film that can be used as a culture substrate in place of glass coverslips (Masurovsky and Bunge, 1968). The fixation, dehydration, and embedding steps can all be conducted with the culture adherent to the Aclar. The Aclar is then peeled away to expose the underside of the culture for sectioning. This approach has provided significant insight to presynaptic structure of the cultured neurons (Rayport *et al.*, 1988; Sulzer and Rayport, 1990) (see section "Synaptic Morphology") as well as changes in organelles and morphology due to stress responses following neurotoxic interventions (Figure 37.6).

Single-Cell RT-PCR

Dissociated neurons, including those in long-term culture provide a cleaner preparation for single-cell RT-PCR than cells in slice, where it can be

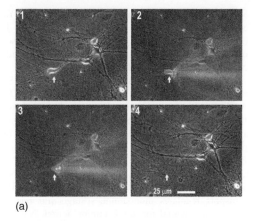

(a)

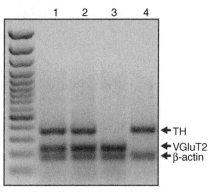

(b)

FIGURE 37.7 mRNA profiling of cultured DA neurons by single-cell RT-PCR. (a) Image set illustrating the collection of individual DA neurons in culture. Postnatal mesencephalic cultures were prepared using P0–P2 pups of the transgenic mouse line TH-EGFP/21–31 that expresses the GFP gene under the control of the TH promoter. DA neurons were thus identified by the expression of GFP (green). Note in images 2 and 3 the shadow of the macro patch pipette as it approaches the neuron of interest and starts to aspirate the complete cell body. When the pipette is retired (image 4), the cell body has been completely removed, leaving behind the neuron's major dendrites and the underlying astrocyte monolayer. (b) RNA expression profile of four single GFP-expressing neurons collected as shown (a). Neurons 1, 2 and 4 contain TH mRNA and are thus dopaminergic, but only the first two express mRNA for the vesicular glutamate transporter type 2 (VGluT2). Neuron 3 is a pure glutamatergic neuron. (Credit: Jose Alfredo Mendez, Marie-Josée Bourque, University of Montreal.)

difficult to exclude contamination from nearby cells (Eberwine *et al.*, 1992). An example of single-cell RT-PCR in postnatal VM cultures is shown in Figure 37.7. (Dalbo *et al.*, 2004; St-Gelais *et al.*, 2008).

An interesting use of these cultures that has not been pursued past an initial paper is that DA neurons can be fixed and identified by TH immunocytochemistry, then removed from the supporting astrocytes with a glass pipette, and the mRNA amplified by PCR, an approach that has been used to confirm expression of alpha-synuclein by DA neurons (Kholodilov *et al.*, 1999).

ACKNOWLEDGEMENTS

We thank the members of the Sulzer lab who have worked to characterize this system, including Viviana Davila, Yelena Kanter, Eugene Mosharov, Asa Petersen, Kester Phillips, Emmanuel Pothos, Yvonne Schmitz, Roland Staal, and Hui Zhang; and collaborators including those in Robert Burke, Serge Przedborski, and Maria Mena's laboratories. We also thank Marie-Josée Bourque for her work in optimizing cultures in the Trudeau lab. We are grateful to Sansana Sawasdikosol, Deirdre Batson, and Geetha Rajendran for their help in developing the culture approach, and Robert Baughman, Paul A. Rosenberg, and Michael M. Segal for their advice.

REFERENCES

Backman, C. M., Malik, N., Zhang, Y., Shan, L., Grinberg, A., Hoffer, B. J., Westphal, H., and Tomac, A. C. (2006). Characterization of a mouse strain expressing Cre recombinase from the 3' untranslated region of the dopamine transporter locus. *Genesis* **44**, 383–390.

Banker, G. A. (1980). Trophic interactions between astroglial cells and hippocampal neurons in culture. *Science* **209**, 809–810.

Bergevin, A., Girardot, D., Bourque, M. J., and Trudeau, L. E. (2002). Presynaptic mu-opioid receptors regulate a late step of the secretory process in rat ventral tegmental area GABAergic neurons. *Neuropharmacology* **42**, 1065–1078.

Bourque, M. J., and Trudeau, L. E. (2000). GDNF enhances the synaptic efficacy of dopaminergic neurons in culture. *Eur J Neurosci* **12**, 3172–3180.

Brewer, G. J. (1995). Serum-free B27/neurobasal medium supports differentiated growth of neurons from the striatum, substantia nigra, septum, cerebral cortex, cerebellum, and dentate gyrus. *J Neurosci Res* **42**, 674–683.

Burke, R. E., Antonelli, M., and Sulzer, D. (1998). Glial cell line-derived neurotrophic growth factor inhibits apoptotic death of postnatal substantia nigra dopamine neurons in primary culture. *J Neurochem* **71**, 517–525.

Cardozo, D. L. (1993). Midbrain dopaminergic neurons from postnatal rat in long-term primary culture. *Neuroscience* **56**, 409–421.

Cardozo, D. L., and Bean, B. P. (1995). Voltage-dependent calcium channels in rat midbrain dopamine neurons: Modulation by dopamine and GABAB receptors. *J Neurophysiol* **74**, 1137–1148.

Chan, C. S., Guzman, J. N., Ilijic, E., Mercer, J. N., Rick, C., Tkatch, T., Meredith, G. E., and Surmeier, D. J. (2007). "Rejuvenation" protects neurons in mouse models of Parkinson's disease. *Nature*.

Chen, B. T., and Rice, M. E. (2001). Novel Ca2+ dependence and time course of somatodendritic dopamine release: Substantia nigra versus striatum. *J Neurosci* **21**, 7841–7847.

Congar, P., Bergevin, A., and Trudeau, L. E. (2002). D2 receptors inhibit the secretory process downstream from calcium influx in dopaminergic neurons: Implication of k(+) channels. *J Neurophysiol* **87**, 1046–1056.

Cubells, J. F., Rayport, S., Rajendran, G., and Sulzer, D. (1994). Methamphetamine neurotoxicity involves vacuolation of endocytic organelles and dopamine-dependent intracellular oxidative stress. *J Neurosci* **14**, 2260–2271.

Cuervo, A. M., Stefanis, L., Fredenburg, R., Lansbury, P. T., and Sulzer, D. (2004). Impaired degradation of mutant alpha-synuclein by chaperone-mediated autophagy. *Science* **305**, 1292–1295.

Dal Bo, G., St-Gelais, F., Danik, M., Williams, S., Cotton, M., and Trudeau, L. E. (2004). Dopamine neurons in culture express VGLUT2 explaining their capacity to release glutamate at synapses in addition to dopamine. *J Neurochem* **88**, 1398–1405.

Damier, P., Hirsch, E. C., Agid, Y., and Graybiel, A. M. (1999). The substantia nigra of the human brain: I. Nigrosomes and the nigral matrix, a compartmental organization based on calbindin D(28K) immunohistochemistry. *Brain* **122**, 1421–1436.

Dauer, W., Kholodilov, N., Vila, M., Trillat, A. C., Goodchild, R., Larsen, K. E., Staal, R., Tieu, K., Schmitz, Y., Yuan, C. A., Rocha, M., Jackson-Lewis, V., Hersch, S., Sulzer, D., Przedborski, S., Burke, R., and Hen, R. (2002). Resistance of alpha -synuclein null mice to the parkinsonian neurotoxin MPTP. *Proc Natl Acad Sci USA* **99**, 14524–14529.

Davila, V., Yan, Z., Craciun, L. C., Logothetis, D., and Sulzer, D. (2003). D3 dopamine autoreceptors do not activate G-protein-gated inwardly rectifying potassium channel currents in substantia nigra dopamine neurons. *J Neurosci* **23**, 5693–5697.

Day, M., Wang, Z., Ding, J., An, X., Ingham, C. A., Shering, A. F., Wokosin, D., Ilijic, E., Sun, Z.,

Sampson, A. R., Mugnaini, E., Deutch, A. Y., Sesack, S. R., Arbuthnott, G. W., and Surmeier, D. J. (2006). Selective elimination of glutamatergic synapses on striatopallidal neurons in Parkinson disease models. *Nat Neurosci* **9**, 251–259.

Eberwine, J., Yeh, H., Miyashiro, K., Cao, Y., Nair, S., Finnell, R., Zettel, M., and Coleman, P. (1992). Analysis of gene expression in single live neurons. *Proc Natl Acad Sci USA* **89**, 3010–3014.

Fasano C, Thibault D, Trudeau L. -E (2008) Culture of postnatal mesencephalic dopamine neurons. *Curr Protoc Neurosci*, in press.

Fon, E. A., Pothos, E. N., Sun, B. C., Killeen, N., Sulzer, D., and Edwards, R. H. (1997). Vesicular transport regulates monoamine storage and release but is not essential for amphetamine action. *Neuron* **19**, 1271–1283.

Forget, C., Stewart, J., and Trudeau, L. E. (2006). Impact of basic FGF expression in astrocytes on dopamine neuron synaptic function and development. *Eur J Neurosci* **23**, 608–616.

Fortin, G. D., Desrosiers, C. C., Yamaguchi, N., and Trudeau, L. E. (2006). Basal somatodendritic dopamine release requires snare proteins. *J Neurochem* **96**, 1740–1749.

Franke, H., Grosche, J., Illes, P., and Allgaier, C. (2002). 5,7-Dihydroxytryptamine – a selective marker of dopaminergic or serotonergic neurons? *Naunyn Schmiedebergs Arch Pharmacol* **366**, 315–318.

Gahwiler, B. H., Capogna, M., Debanne, D., McKinney, R. A., and Thompson, S. M. (1997). Organotypic slice cultures: A technique has come of age. *Trends Neurosci* **20**, 471–477.

Greggio, E., Bergantino, E., Carter, D., Ahmad, R., Costin, G. E., Hearing, V. J., Clarimon, J., Singleton, A., Eerola, J., Hellstrom, O., Tienari, P. J., Miller, D. W., Beilina, A., Bubacco, L., and Cookson, M. R. (2005). Tyrosinase exacerbates dopamine toxicity but is not genetically associated with Parkinson's disease. *J Neurochem* **93**, 246–256.

Grierson, J. P., Petroski, R. E., Ling, D. S. F., and Geller, H. M. (1990). Astrocyte topography and tenascin/cytotactin expression: Correlation with the ability to support neuritic outgrowth. *Dev Brain Res* **55**, 11–19.

Gustincich, S., Feigenspan, A., Wu, D. K., Koopman, L. J., and Raviola, E. (1997). Control of dopamine release in the retina: A transgenic approach to neural networks. *Neuron* **18**, 723–736.

Huettner, J. E., and Baughman, R. W. (1986). Primary culture of identified neurons from the visual cortex of postnatal rats. *J Neurosci* **6**, 3044–3061.

Ingram, S. L., Prasad, B. M., and Amara, S. G. (2002). Dopamine transporter-mediated conductances increase excitability of midbrain dopamine neurons. *Nat Neurosci* **5**, 971–978.

Jomphe, C., Bourque, M. J., Fortin, G. D., St-Gelais, F., Okano, H., Kobayashi, K., and Trudeau, L. E. (2005). Use of TH-EGFP transgenic mice as a source of identified dopaminergic neurons for physiological studies in postnatal cell culture. *J Neurosci Methods* **146,** 1–12.

Joyce, M. P., and Rayport, S. (2000). Mesoaccumbens dopamine neuron synapses reconstructed *in vitro* are glutamatergic. *Neuroscience* **99,** 445–456.

Keshishian, H. (2004). Ross Harrison's "The outgrowth of the nerve fiber as a mode of protoplasmic movement". *J Exp Zoolog A Comp Exp Biol* **301,** 201–203.

Kholodilov, N. G., Neystat, M., Oo, T. F., Lo, S. E., Larsen, K. E., Sulzer, D., and Burke, R. E. (1999). Increased expression of rat synuclein in the substantia nigra pars compacta identified by mRNA differential display in a model of developmental target injury. *J Neurochem* **73,** 2586–2599.

Kholodilov, N., Yarygina, O., Oo, T. F., Zhang, H., Sulzer, D., Dauer, W., and Burke, R. E. (2004). Regulation of the development of mesencephalic dopaminergic systems by the selective expression of glial cell line-derived neurotrophic factor in their targets. *J Neurosci* **24,** 3136–3146.

Larsen, K. E., Fon, E. A., Hastings, T. G., Edwards, R. H., and Sulzer, D. (2002). Methamphetamine-induced degeneration of dopaminergic neurons involves autophagy and upregulation of dopamine synthesis. *J Neurosci* **22,** 8951–8960.

Levi-Montalcini, R., and Angeletti, P. U. (1968). Nerve growth factor. *Physiol Rev* **48,** 534–569.

Liang, C. L., Sinton, C. M., Sonsalla, P. K., and German, D. C. (1996). Midbrain dopaminergic neurons in the mouse that contain calbindin-D28k exhibit reduced vulnerability to MPTP-induced neurodegeneration. *Neurodegeneration* **5,** 313–318.

Masuko, S., Nakajima, S., and Nakajima, Y. (1992). Dissociated high-purity dopaminergic neuron cultures from the substantia nigra and the ventral tegmental area of the postnatal rat. *Neuroscience* **49,** 347–364.

Masurovsky, E. B., and Bunge, R. P. (1968). Fluoroplastic coverslips for long-term nerve tissue culture. *Stain Technol* **43,** 161–165.

Mena, M. A., Davila, V., and Sulzer, D. (1997a). Neurotrophic effects of L-DOPA in postnatal midbrain dopamine neuron/cortical astrocyte cocultures. *J Neurochem* **69,** 1398–1408.

Mena, M. A., Casarejos, M. J., Carazo, A., Paino, C. L., and Garcia de Yebenes, J. (1997b). Glia protect fetal midbrain dopamine neurones in culture from L-DOPA toxicity through multiple mechanisms. *J Neural Transm* **104,** 317–328.

Mena, M. A., Khan, U., Togasaki, D. M., Sulzer, D., Epstein, C. J., and Przedborski, S. (1997c). Effects of wild-type and mutated copper/zinc superoxide dismutase

on neuronal survival and L-DOPA-induced toxicity in postnatal midbrain culture. *J Neurochem* **69,** 21–33.

Mendez, J. A., Bourque, M.-J., Bourdeau, M. L., Danik, M., Williams, S., Lacaille, J.-C., and Trudeau, L.-E. (2008). Developmental and contact-dependent regulation of vesicular glutamate transporter expression in dopamine neurons. *Journal of Neuroscience*, in press.

Michel, F. J., and Trudeau, L. E. (2000). Clozapine inhibits synaptic transmission at GABAergic synapses established by ventral tegmental area neurones in culture. *Neuropharmacology* **39,** 1536–1543.

Michel, F. J., Robillard, J. M., and Trudeau, L. E. (2004a). Regulation of rat mesencephalic GABAergic neurones through muscarinic receptors. *J Physiol* **556,** 429–445.

Michel, F. J., Robillard, J., and Trudeau, L. E. (2004b). Regulation of rat mesencephalic GABAergic neurones through muscarinic receptors. *J Physiol.*

Mosharov, E. V., Gong, L. W., Khanna, B., Sulzer, D., and Lindau, M. (2003). Intracellular patch electrochemistry: Regulation of cytosolic catecholamines in chromaffin cells. *J Neurosci* **23,** 5835–5845.

Olson, V. G., and Nestler, E. J. (2007). Topographical organization of GABAergic neurons within the ventral tegmental area of the rat. *Synapse* **61,** 87–95.

Petersen, A., Larsen, K. E., Behr, G. G., Romero, N., Przedborski, S., Brundin, P., and Sulzer, D. (2001). Expanded CAG repeats in exon 1 of the Huntington's disease gene stimulate dopamine-mediated striatal neuron autophagy and degeneration. *Hum Mol Genet* **10,** 1243–1254.

Petrucelli, L., O'Farrell, C., Lockhart, P. J., Baptista, M., Kehoe, K., Vink, L., Choi, P., Wolozin, B., Farrer, M., Hardy, J., and Cookson, M. R. (2002). Parkin protects against the toxicity associated with mutant alpha-synuclein. *Neuron* **36,** 1007–1019.

Pickel, V. M., Chan, J., and Nirenberg, M. J. (2002). Region-specific targeting of dopamine D2-receptors and somatodendritic vesicular monoamine transporter 2 (VMAT2) within ventral tegmental area subdivisions. *Synapse* **45,** 113–124.

Plenz, D., and Kitai, S. T. (1998). Regulation of the nigrostriatal pathway by metabotropic glutamate receptors during development. *J Neurosci* **18,** 4133–4144.

Pothos, E. N., Davila, V., and Sulzer, D. (1998). Presynaptic recording of quanta from midbrain dopamine neurons and modulation of the quantal size. *J Neurosci* **18,** 4106–4118.

Pothos, E. N., Larsen, K. E., Krantz, D. E., Liu, Y., Haycock, J. W., Setlik, W., Gershon, M. D., Edwards, R. H., and Sulzer, D. (2000). Synaptic vesicle transporter expression regulates vesicle phenotype and quantal size. *J Neurosci* **20,** 7297–7306.

Potter, S. M., and DeMarse, T. B. (2001). A new approach to neural cell culture for long-term studies. *J Neurosci Methods* **110,** 17–24.

Prasad, B. M., and Amara, S. G. (2001). The dopamine transporter in mesencephalic cultures is refractory to physiological changes in membrane voltage. *J Neurosci* **21**, 7561–7567.

Prochiantz, A., di Porzio, U., Kato, A., Berger, B., and Glowinski, J. (1979). *In vitro* maturation of mesencephalic dopaminergic neurons from mouse embryos is enhanced in presence of their striatal target cells. *Proc Natl Acad Sci USA* **76**, 5387–5391.

Przedborski, S., Khan, U., Kostic, V., Carlson, E., Epstein, C. J., and Sulzer, D. (1996). Increased superoxide dismutase activity improves survival of cultured postnatal midbrain neurons. *J Neurochem* **67**, 1383–1392.

Przedborski, S., Jackson-Lewis, V., Sulzer, D., Naini, A., Romero, N., Chen, C., and Arias, J. (2002). Transgenic superoxide dismutase overproducer: Murine. *Methods Enzymol* **349**, 180–190.

Rayport, S. G. (1998). Imaging dopamine receptors on living neurons in culture. In *Receptor Localization: Laboratory Methods and Procedures* (M. A. Ariano, Ed.), pp. 197–219. John Wiley & Sons, New York.

Rayport, S., and Sulzer, D. (1995). Visualization of antipsychotic binding to living mesolimbic neurons reveals D2 receptor mediated, acidotropic and lipophilic components. *J Neurochem* **65**, 691–703.

Rayport, S., Monaco, J., and Sawasdikosol, S. (1988). Identified mesoaccumbens dopamine neurons *in vitro*. *Soc Neurosci Abstr* **14**, 932.

Rayport, S., Sulzer, D., Shi, W. X., Sawasdikosol, S., Monaco, J., Batson, D., and Rajendran, G. (1992). Identified postnatal mesolimbic dopamine neurons in culture: Morphology and electrophysiology. *J Neurosci* **12**, 4264–4280.

Rosenberg, P. A., and Aizenman, E. (1989). Hundred-fold increase in neuronal vulnerability to glutamate toxicity in astrocyte-poor cultures of rat cerebral cortex. *Neurosci Lett* **103**, 162–168.

Rosenberg, P. A., Amin, S., and Leitner, M. (1992). Glutamate uptake disguises neurotoxic potency of glutamate agonists in cerebral cortex in dissociated cell culture. *J Neurosci* **12**, 56–61.

Schwartz, J. W., Blakely, R. D., and DeFelice, L. J. (2003). Binding and transport in norepinephrine transporters. Real-time, spatially resolved analysis in single cells using a fluorescent substrate. *J Biol Chem* **278**, 9768–9777.

Silva, N. L., Mariani, A. P., Harrison, N. L., and Barker, J. L. (1988). 5,7-Dihydroxytryptamine identifies living dopaminergic neurons in mesencephalic cultures. *Proc Natl Acad Sci USA* **85**, 7346–7350.

St-Gelais, F., Legault, M., Bourque, M. J., Rompre, P. P., and Trudeau, L. E. (2004). Role of calcium in neurotensin-evoked enhancement in firing in mesencephalic dopamine neurons. *J Neurosci* **24**, 2566–2574.

Staal, R. G., Mosharov, E. V., and Sulzer, D. (2004). Dopamine neurons release transmitter via a flickering fusion pore. *Nat Neurosci* **7**, 341–346.

Staal, R. G. W., Rayport, S., and Sulzer, D. (2007). Amperometric detection of dopamine exocytosis from synaptic termminals. In *Electrochemical Methods in Neuroscience* (A. C. Michael, and L. M. Borland, Eds.), pp. 337–352. CRC Press, Boca Raton, Florida.

Sulzer, D. (2007). Multiple hit hypotheses for dopamine neuron loss in Parkinson's disease. *Trends Neurosci* **30**, 244–250.

Sulzer, D., and Rayport, S. (1990). Amphetamine and other psychostimulants reduce pH gradients in midbrain dopaminergic neurons and chromaffin granules: A mechanism of action. *Neuron* **5**, 797–808.

Sulzer, D., Maidment, N. T., and Rayport, S. (1993). Amphetamine and other weak bases act to promote reverse transport of dopamine in ventral midbrain neurons. *J Neurochem* **60**, 527–535.

Sulzer, D., St Remy, C., and Rayport, S. (1996). Reserpine inhibits amphetamine action in ventral midbrain culture. *Mol Pharmacol* **49**, 338–342.

Sulzer, D., Joyce, M. P., Lin, L., Geldwert, D., Haber, S. N., Hattori, T., and Rayport, S. (1998). Dopamine neurons make glutamatergic synapses *in vitro*. *J Neurosci* **18**, 4588–4602.

Sulzer, D., Bogulavsky, J., Larsen, K. E., Behr, G., Karatekin, E., Kleinman, M. H., Turro, N., Krantz, D., Edwards, R. H., Greene, L. A., and Zecca, L. (2000). Neuromelanin biosynthesis is driven by excess cytosolic catecholamines not accumulated by synaptic vesicles. *Proc Natl Acad Sci USA* **97**, 11869–11874.

Swanson, L. W. (1982). The projections of the ventral tegmental area and adjacent regions: A combined fluorescent retrograde tracer and immunofluorescence study in the rat. *Brain Res Bull* **9**, 321–353.

Takeshima, T., Johnston, J. M., and Commissiong, J. W. (1994). Mesencephalic type 1 astrocytes rescue dopaminergic neurons from death induced by serum deprivation. *J Neurosci* **14**, 4769–4779.

Yamaguchi, T., Sheen, W., and Morales, M. (2007). Glutamatergic neurons are present in the rat ventral tegmental area. *Eur J Neurosci* **25**, 106–118.

Yoshizaki, T., Inaji, M., Kouike, H., Shimazaki, T., Sawamoto, K., Ando, K., Date, I., Kobayashi, K., Suhara, T., Uchiyama, Y., and Okano, H. (2004). Isolation and transplantation of dopaminergic neurons generated from mouse embryonic stem cells. *Neurosci Lett* **363**, 33–37.

Zhang, D. Q., Stone, J. F., Zhou, T., Ohta, H., and McMahon, D. G. (2004). Characterization of genetically labeled catecholamine neurons in the mouse retina. *Neuroreport* **15**, 1761–1765.

Zhuang, X., Masson, J., Gingrich, J. A., Rayport, S., and Hen, R. (2005). Targeted gene expression in dopamine and serotonin neurons of the mouse brain. *J Neurosci Methods* **143**, 27–32.

38

YEAST CELLS AS A DISCOVERY PLATFORM FOR PARKINSON'S DISEASE AND OTHER PROTEIN MISFOLDING DISEASES

KAREN L. ALLENDOERFER, LINHUI JULIE SU AND SUSAN LINDQUIST

Whitehead Institute for Biomedical Research, and Howard Hughes Medical Institute, 9 Cambridge Center, Cambridge, MA

INTRODUCTION: YEAST AS A MODEL SYSTEM FOR UNDERSTANDING HUMAN DISEASE

One reason we lack effective therapeutics and a deep biological understanding for many of the most common and devastating human neurodegenerative diseases is the absence of experimental disease models that can be used in high-throughput screening and genetic analysis. Mouse models can provide invaluable information about disease processes in a whole organism, but knockouts and transgenics are costly, have long generation times and complicated husbandry, and usually develop symptoms similar to the human disease quite slowly. Fruit fly and worm models combine the advantages of well-established genetics and relative ease of handling with an intact nervous system. But practically, they cannot be employed in screens involving hundreds of thousands of assays using the very small quantities of compounds available from large chemical libraries. Neuronal cells are delicate, difficult to manipulate, and immortalization reprograms their normal apoptotic and senescence programs, thus hindering the development of lines with normal sensitivities to proteotoxic stress that are amenable to high-throughput experiments.

In contrast, the yeast *Saccharomyces cerevisiae* has a short generation time (1.5–3h) and grows in a highly reproducible way on a variety of carbon sources, in simple defined media (Sherman, 2002; see also "Getting Started with Yeast," the Saccharomyces Genome Database, SGD, and the Yeast Resource Center, Table 38.1). These properties make it readily amenable to high-throughput screens in 96-, 384-, and even 1536-well plates. It is also, by far, the best-characterized and most readily manipulable eukaryotic cell, and a staggering variety of genetic tools have been developed to facilitate its analysis. In cases where disease processes impinge on basic cellular functions yeast can provide an invaluable starting point for later analysis in more complex organisms.

Humans have enjoyed association with *S. cerevisiae* and its close relatives for thousands of years. Its metabolism, geared to rapid conversion of sugars into ethanol and CO_2, has been employed for the production of alcoholic beverages and the leavening of bread since early historical times. It was the desire of brewers to produce more uniform, higher quality beers that led to the early genetic analysis and manipulation of yeast cells (reviewed in Polaina, 2002). By simple chance, the organism happened to

TABLE 38.1 Yeast Internet resources

Getting Started with Yeast by Fred Sherman	http://www.dbb.urmc. rochester.edu/labs/sherman_ f/StartedYeast.html
Saccharomyces Genome Database (SGD)	http://www.yeastgenome. org/
Some of the most useful of the SGD's "External Links" for researchers new to the field include:	
The Comprehensive Yeast Genome Database (from MIPS)	http://www.mips.gsf.de/ genre/proj/yeast/index.jsp
Yeast Global Microarray Viewer	http://www.transcriptome. ens.fr/ymgv/
Yeast Resource Center	http://www.depts. washington.edu/yeastrc/ pages/overview.html
Yeast GFP Fusion Localization Database	http://www.yeastgfp.ucsf. edu/
Yeast FLEXgene collection (Harvard Institute of Proteomics)	http://www.hip.harvard. edu/research/yeast_flexgene/
Information Hyperlinked over Proteins	http://www.ihop-net. org/UniPub/iHOP/
BioGRID General Repository for Interaction Datasets	http://www.thebiogrid.org

have several critical properties that led to it becoming the extraordinary model it is today. Though it has a diploid life cycle, early genetic manipulations allowed stable vigorous propagation in the haploid state (Sherman, 2002; Mell and Burgess, 2003) and strains with distinct phenotypes could be mated. Moreover, meiosis can be readily induced in the diploid organism, and the four haploid meiotic products remain together in a single ascus that is readily dissected. These properties allowed the recovery and phenotypic characterization of recessive mutants, tests of allelic complementation and recombination, immediate means to distinguish between simple and complex traits, the determination of epistatic genetic relationships, and an early and extensive genetic map (Sherman, 2002; Mell and Burgess, 2003).

The organism is readily transformed with extra-chromosomal plasmids or by genomic insertion, providing a wealth of information on diverse gene functions. Promoters with different and readily manipulated expression patterns are available, and this led to the development of a wide range of vectors and expression cassettes. Indeed there are now libraries of plasmids for manipulating the expression of every open reading frame (ORF) in the yeast genome (Gelperin *et al.*, 2005; Cooper *et al.*, 2006; Sopko *et al.*, 2006).

Yeast cells also have unusually high rates of homologous recombination (Sherman, 2002). This led to the development of techniques for site-directed modification of the genome with rapid and simple methods (e.g., Orr-Weaver *et al.*, 1983; Struhl, 1983; Roitgrund *et al.*, 1993). Any wild-type (WT) gene can be readily replaced with any desired mutation in its normal chromosomal location.

Using such techniques, yeast researchers – an unusually collaborative group – have developed genome-wide libraries of isogenic strains, each of which contains a single mutated allele: these include deletion mutations (Giaever *et al.*, 2002) and reduced expression alleles (Mnaimneh *et al.*, 2004) for genetic analysis, green fluorescent protein (GFP) fusions for protein localization (GFP Fusion Localization Database, Table 38.1; Huh *et al.*, 2003), and purification-tag fusions for the recovery and analysis of individual proteins and protein complexes (Ghaemmaghami *et al.*, 2003). These have all been made readily available to the entire scientific community. Robust and detailed databases have been developed and are available online. The SGD not only serves as a central repository for genome, protein, and microarray data, it includes sequence data from *Saccharomyces* sibling species, makes available a variety of bioinformatics tools, points to a variety of other resources, and provides links to any yeast paper published, among many other services (see Table 38.1).

Yeast has one of the smallest genomes of any free-living organism. It was therefore the first eukaryotic genome to be sequenced (Goffeau *et al.*, 1996). There are about ~6,000 genes in yeast. The majority of these have motifs in common with human proteins and nearly half of them have clear human counterparts. It is common to swap yeast genes for their human relatives while retaining full functionality.

Over the years this combination of features (and others that must be omitted to save space) have led to the development of extraordinary methods of genetic and cell biological analysis that can be performed at a fraction of the cost in time and materials required in other systems, with one success rapidly breeding another (Guthrie and Fink, 1991;

Johnston, 1994; Forsburg, 2001; Boone *et al.*, 2007; Madhani, 2007). For example, screens for synthetic lethal genetic interactions (combinations of two mutations which themselves have little or no phenotype, but which together are severely toxic or inviable) have been used extensively to identify genes whose products buffer one another's functions or impinge on the same essential pathway. In the current era, this process can be automated, allowing any query mutation to be crossed to an array of approximately 4,700 viable deletion mutants to determine networks of interactions among genes on an unprecedented scale (Tong *et al.*, 2001; Tong and Boone, 2006). Out of the millions of such crosses that have now been done, more than 35,000 genetic interactions have been mapped (Schuldiner *et al.*, 2005; Collins *et al.*, 2007a; Collins *et al.*, 2007b; see the BioGRID General Repository for Interaction Datasets, Table 38.1). Several large-scale proteomics projects have been completed (Kolkman *et al.*, 2005), both by two-hybrid analysis and by high-throughput mass spectroscopy (Causier, 2004), yielding more than 30,000 protein–protein interactions. Furthermore, thousands of expression profiles have been done with diverse genetic, chemical, environmental, and nutritional perturbations (e.g., Gasch *et al.*, 2000). Because these experiments tend to be carried out with a few common laboratory strains under standard culture conditions, data can readily be pooled for enhanced computational power (Irizarry *et al.*, 2005). Thus, in the past decade yeast has progressed from being a good model, to a super model. There is simply no other organism we understand so well and can manipulate so readily to learn about fundamental cellular processes.

Scientific Achievements Using Yeast Models

Because the fundamental underpinnings of cell biology are conserved in all eukaryotes, the use of yeast as a model system has yielded stunning insights into important problems in human biology and medicine. Here we will touch on just a very few, to allow the reader who is new to yeast to gain just a brief notion of these accomplishments.

In the 1970s, Leland Hartwell recognized the potential of using yeast to study the cell cycle. In an elegant series of genetic experiments, his group identified more than one hundred genes specifically involved in cell cycle control, called the CDC

(cell division cycle) genes (Hartwell *et al.*, 1974). Hartwell and colleagues also studied the sensitivity of yeast cells to irradiation, and on the basis of these findings, introduced the concept of the checkpoint. Checkpoints are highly regulated pauses or breaks in one portion of the cell cycle, to allow time for DNA repair, proper spindle function, etc. before the cell continues to the next phase. This insures a correct execution and ordering of each cell cycle phase (Hartwell and Weinert, 1989). Meanwhile, other yeast researchers provided remarkable insights on the essential processes of DNA replication, recombination, repair, transcription, chromatin remodeling, and chromosome segregation (discussed in Snustad and Simmons, 2005; Griffiths *et al.*, 2007; Madhani, 2007). While yeast cells do not get cancer, essential features of cell biology are conserved in all eukaryotic cells, and studies in yeast have helped to transform our understanding of the processes that go awry in oncogenic transformation.

These and many other discoveries point to the value of yeast models for the assignment of functions to genes that are linked to diseases, even if the functions of these genes are currently unclear. For many such genes, the first clue to their function, once discovered in other organisms, was homology to a yeast gene. Human disease genes whose functions have been assigned by identifying their yeast homologs include those that cause neurofibromatosis type 1, ataxia telangiectasia, Werner's syndrome, and Friedreich's ataxia (Ballester *et al.*, 1990; Greenwell *et al.*, 1995; Morrow *et al.*, 1995; Watt *et al.*, 1996; Koenig and Mandel, 1997). In the last example, yeast cells lacking the homolog of frataxin, a human mitochondrial gene (FRDA) (*yfh1*, yeast frataxin homolog 1), developed abnormal mitochondrial iron metabolism and oxidative stress (reviewed in Pandolfo, 2001; Scherzer and Feany, 2004). The observations in yeast were confirmed in human tissues and have led to clinical trials of antioxidants, which have shown promise in ameliorating FRDA-associated cardiomyopathy (Pandolfo, 2001; Hausse *et al.*, 2002).

Other genes linked to human neurodegenerative diseases that have yeast homologs or orthologs include: *ATP7B* (linked to Wilson's disease), *CACNA1A* (a calcium channel subunit linked to spinocerebellar ataxia type 6), *CLN3* (linked to ceroid lipofuscinosis), *PARK2* (parkin; a ubiquitin E3 ligase linked to Parkinson's disease, PD), *SCA2* (spinocerebellar ataxia type 2), *SOD1* (superoxide dismutase 1, linked to amyotrophic lateral sclerosis,

ALS), and *UCHL1* (ubiquitin carboxyl-terminal este-rase L1, linked to PD) (Rubin *et al.*, 2000; Scherzer and Feany, 2004). These offer promising avenues for further research but there are still complexities to be solved. For example, there is a large and grow-ing number of known human ubiquitin ligases (at least 110 genes), and at least 15 of these share simi-larity to parkin (*PARK2*). The closest yeast ortholog to parkin is YKR017c, which joins a family of ~20 yeast ubiquitin ligase genes. In yeast, the overall number of ubiquitin ligases is small, but the diversity of function (i.e., the conditions under which the dif-ferent ligases function) is high. In humans, the diver-sity of function is further multiplied by the diversity of cell types. How to relate yeast and human data in these cases is not trivial.

PROTEIN HOMEOSTASIS AND HUMAN DISEASE

The Protein Folding Problem

Several decades ago, Anfinsen proposed that the primary amino acid sequence of a protein contains all the information required for it to fold, and that a protein molecule will spontaneously assume the conformation of greatest thermodynamic stabil-ity in a given environment (Anfinsen, 1973). This principle transformed our understanding of protein form and function. However, as a practical matter, it only holds true for small, single-domain proteins at very dilute concentrations *in vitro*. It does not capture the problem that proteins have when they fold in the dynamic, crowded environment of the living cell, where protein concentrations are on the order of 200–300 mg/mL. It is now generally accepted that a substantial fraction, perhaps even as much as 30%, of all proteins made in cells never fold properly and are degraded (Schubert *et al.*, 2000; reviewed in Goldberg, 2003). How could this be? The high price of protein misfolding is offset by the fact that once proteins are properly folded, localized, and assembled tremendous efficiencies are gained from having them so close together.

This protein folding problem is as old as life itself, and a large number of quality control (QC) mechanisms and protein compartmentalization processes have evolved to handle it. These have generally been well conserved from prokaryotes to eukaryotes, and include the following.

Chaperone Proteins, Each Binding to Specific Protein Folding Intermediates

For example, the chaperone Hsp70 binds short, extended hydrophobic polypeptide chains that would normally be buried in the hydrophobic core of a folded protein. Hsp60 recognizes molten glob-ules, proteins that have acquired secondary struc-ture but have not yet locked into the closely packed excluded volumes of a folded protein (Young *et al.*, 2004). In contrast, Hsp90 recognizes a more select group of substrates, proteins that have achieved a considerable degree of structure in one domain but remain highly flexible in others. This is typical of signal transducers that regulate many key biologi-cal pathways (Whitesell and Lindquist, 2005).

Chaperones are found in virtually every compart-ment of the cell. They work in partnership with a complex array of co-chaperones and other cofactors with various functions. They help chaperones recog-nize substrates in a combinatorial fashion, promote ATP hydrolysis, conformational change, or substrate release. Chaperones generally maintain proteins in a soluble state and simply release them for refold-ing. But chaperones can also promote degradation by releasing substrates in association with the pro-teolytic machinery. Chaperones are associated with many of the protein aggregates found in neurode-generative diseases, and it is thought that they might be aggregated there due to being overwhelmed by the number of substrates. In keeping with this idea, the overexpression of chaperones is neuroprotective in at least some cases (Barral *et al.*, 2004; Auluck *et al.*, 2005; Kiaei *et al.*, 2005; Brown, 2007).

Protein Remodeling Factors, Such As hsp104 (in yeast) and p97 (in Higher Eukaryotes)

Certain large, multi-domain, multi-subunit proteins interact with substrates that are already improperly folded and actively rearrange them using the energy of ATP to do so. They also decon-struct certain large protein complexes that need to be disassembled as part of their normal functional cycle. Protein remodeling factors belong to a larger class of proteins that share a particular type of ATPase domain known as AAA+. Each monomer has two such domains. Protein remodeling factors often interact with a variety of chaperones and cofactors to determine the fate of their substrates. The roles of these remodeling factors in combating the misfolding of neurodegenerative disease proteins is just beginning to be uncovered (Cashikar *et al.*, 2005; Vacher *et al.*, 2005).

Osmolytes Such As Trehalose, Betaine and Triethylamine-*N*-Oxide (Sometimes Called "Chemical Chaperones")

Osmolytes change the hydrodynamic interactions between proteins and the aqueous milieu and promote maintenance of the folded state (Somero, 1986; Bolen and Baskakov, 2001). Osmolytes also affect the stability of DNA and biological membranes. Different organisms have repeatedly evolved the use of different osmolytes to stabilize proteins. For example, trehalose is a disaccharide that has been found in a wide variety of species in addition to *S. cerevisiae*, including the 'resurrection plant' *Selaginella lepidophylla*, certain brine shrimp, and nematode worms, and plays a role in the ability of these species to withstand almost complete desiccation for long periods of time, and in thermotolerance (Singer and Lindquist, 1998a). Mammalian kidney cells use a variety of organic osmolytes such as betaine and methylamines to cope with the high extracellular osmolarity during normal operation of the urinary concentrating mechanism (Kempson and Montrose, 2004; Yancey, 2005). Taurine is also an important osmolyte in the response of mammalian cells to osmotic swelling (Lambert, 2004). Though different organisms and cell types employ different osmolytes, their general effects on protein folding seem universal. Indeed, although mammals do not make trehalose themselves, orally administered trehalose can prevent the development of brain pathology and delay the progress of symptoms such as motor dysfunction in mouse models of Huntington's disease (HD) (Tanaka *et al.*, 2004; Tanaka *et al.*, 2005).

Elaborate Proteolytic Machineries for the Selective Degradation of Misfolded Proteins

Cells use highly regulated energy-dependent proteases to turn over naturally short-lived proteins as well as misfolded and damaged proteins. In eukaryotic cells the ubiquitin–proteasome system degrades cytosolic and nuclear proteins of the cytosol and, through retrograde transport, the proteins of the endoplasmic reticulum (ER). An important step in this process is specific recognition of the substrate by one of many ubiquitin ligases (E3s), followed by generation and conjugation of a polyubiquitin degradation signal. Degradation is then carried out by the 26S proteasome (Ciechanover, 1994).

When the folding process fails, molecular chaperones can increase the susceptibility of certain substrates to proteolysis by releasing their substrates to particular proteolytic pathways (Hayes and Dice, 1996). The exquisite specificity of the degradation machinery provides cells with the capacity to destroy abnormal or damaged proteins rapidly while avoiding destruction of normal proteins. Defects in protein degradation have been repeatedly implicated in neurodegenerative disease (Kopito, 2000; Sakamoto, 2002; Jiang and Beaudet, 2004; Vila and Przedborski, 2004).

The Sequestration of Proteins into Diverse Membrane-Bound Compartments

Another layer of regulation of protein homeostasis that has evolved in eukaryotes to manage the increased complexity of their proteome includes elaborate internal membrane-bound compartments, such as the nuclear envelope (NE), mitochondria, Golgi apparatus, ER, lysosomes, and endosomes, all of which segregate distinct functional groups of proteins. Many proteins traffic between these compartments, in many cases via transport vesicles, which are also used for secretion from the cell (Devos *et al.*, 2004).

Specialized protein complexes localized to particular compartment membranes play vital roles in the dynamics and specificity of vesicle and protein trafficking, allowing each organelle to perform its distinct functions. The SNARE (soluble *N*-ethylmaleimide-sensitive factor attachment protein receptor) superfamily of proteins, present in both yeast and mammalian cells, mediates fusion of transport vesicles with the cell membrane or with target compartments (reviewed in Rothman and Warren, 1994; Jahn and Scheller, 2006). SNARE protein complexes can function in multiple transport steps with varying degrees of specificity. Clathrin/adaptin complexes are responsible for endocytosis and vesicular trafficking between the Golgi, lysosomes, and endosomes; the COPI complex is involved in intra-Golgi and retrograde Golgi-to-ER trafficking; and the COPII complex supports vesicle movement from the ER to the Golgi. These are also conserved from yeast to humans (reviewed in Kirschhausen, 2000; Bonifacino and Lippincott-Schwartz, 2003).

Problems in protein trafficking have been implicated in many human diseases (Shields and Arvan, 1999), including Alzheimer's disease (AD) (Shields and Arvan, 1999; Kins *et al.*, 2006), peripheral neuropathies such as Charcot-Marie Tooth (Sanders *et al.*, 2001), and cystic fibrosis (Ameen *et al.*, 2007). The extraordinary degree of overlap of cellular protein quality systems between diverse eukaryotes enables the investigation of the misfolding and mistrafficking of proteins linked to human diseases in simpler systems, from fruit flies to nematodes to yeast.

Neurodegenerative Protein Folding Diseases

The seemingly disparate diseases of Parkinson (PD), Alzheimer (AD), Huntington (HD), amyotrophic lateral sclerosis (ALS), several prion diseases, frontal temporal dementias, various spinocerebellar ataxias, dentatorubro pallidoluysian atrophy (DRPLA), and others, some of which themselves represent collections of diverse pathologies (Ellison *et al.*, 2003), appear superficially to have little in common with each other. Yet one feature they do share is the occurrence of brain lesions consisting of an accumulation of misfolded, aggregated proteins that are intimately associated with neurodegeneration (Muchowski, 2002; Trojanowski, 2004). The aggregates may be nuclear, cytoplasmic, or extracellular. Proteins in these complexes are often ubiquitinated and associated with chaperones (Muchowski, 2002), suggesting that protein QC has gone awry or was overwhelmed. In each disease, a different protein is the major constituent of the aggregate. For example, in AD there are two: AB, a peptide derived from amyloid precursor protein (APP) primarily present in extracellular plaques, and tau, a primary constituent of neurofibrillary tangles (reviewed in LaFerla *et al.*, 2007). In ALS the major aggregating protein is SOD. In the prion diseases, aggregates are largely composed of the prion protein PrP, and in HD the culprit is huntingtin (htt) (reviewed by Zoghbi and Orr, 2000). Lesion-associated proteins in neurodegenerative disease furthermore share a propensity to form an ordered fibrillar structure called amyloid (reviewed by Rochet and Lansbury, 2000). However, recent studies support the hypothesis that prefibrillar intermediates (protofibrils) or earlier oligomeric assemblages may be the key toxic species (Lansbury and Brice, 2002;

Kayed *et al.*, 2003; Glabe, 2004; Lacor *et al.*, 2007) rather than the mature amyloid fibrils.

In PD, these lesions are called Lewy bodies (LBs) and are prominent in the cytoplasm of dopaminergic (DA) neurons in the *substantia nigra pars compacta*. Though found in other neurons as well, it is the *substantia nigra* dysfunction that leads to the characteristic tremors of PD. LBs are composed primarily of full-length and truncated forms of the protein alpha-synuclein (α-syn) (Spillantini *et al.*, 1997). They also include various molecular chaperones, including hsp70 and hsp40, ubiquitin, and other components of the ubiquitin–proteasome degradation machinery that are commonly associated with other disease-associated aggregates (Goedert, 2001; Sherman and Goldberg, 2001). LBs also accumulate in a disease termed dementia with Lewy bodies (DLB) (Sherman and Goldberg, 2001) where there is degeneration of cortical areas in addition to the *substantia nigra*. They may even be associated with the etiology of AD (Kotzbauer *et al.*, 2001), although their presence in AD plaques of some brains might simply reflect the co-occurrence of two relatively common diseases. Interestingly, in a separate condition, multiple system atrophy (MSA), α-syn accumulates within glial cells as glial cytoplasmic inclusions (GCIs; Morris *et al.*, 2000). GCIs are generally only seen in the context of MSA, and have rounded contours rather than the target-like shape of LBs. However, as in LBs, they are ubiquitinated and contain hsp70 protein (Kawamoto *et al.*, 2007).

Neurodegenerative disorders may initiate when age-related declines in cell function combine with underlying genetic susceptibilities and environmental insults to raise protein misfolding problems above the capacity of cellular protein QC systems to cope with them. For α-syn and PD, not only are misfolded α-syn aggregates a hallmark of the disease (Spillantini *et al.*, 1997), but mutations in α-syn that are associated with its misfolding cause early onset of PD (Polymeropoulos *et al.*, 1997; Kruger *et al.*, 1998). Moreover, mutations in genes that are involved in protein degradation are associated with other early onset forms of PD. These include mutations in the genes *PARK2*, which encodes parkin, a ubiquitin ligase (Kitada *et al.*, 1998); and *PARK5*, encoding the ubiquitin C-terminal hydrolase-L1 (Leroy *et al.*, 1998). Moreover, chaperones and remodeling factors that promote normal protein folding ameliorate α-syn toxicity in *Drosophila melanogaster* neurodegenerative models and mammalian cell culture (Auluck *et al.*, 2002; Outeiro *et al.*, 2006). Finally, activity of the 20S proteasome produces truncated, amyloidogenic, α-syn fragments similar to those found in PD patients and animal models (Liu *et al.*, 2005; Kim *et al.*, 2006). Incomplete degradation of α-syn, especially coupled with an overloaded proteasome capacity, might produce truncated α-syn fragments that induce and enhance the aggregation of full-length protein. These aggregates could in turn further reduce proteasome activity, leading to further accumulation and aggregation of α-syn in a vicious cytotoxic cycle (Liu *et al.*, 2005).

Thus PD and many common neurodegenerative diseases can be thought of as diseases of protein folding and protein trafficking homeostasis. We have taken advantage of the similarities in protein homeostasis in all eukaryotic cells, and the ease of genetic manipulation in yeast, to develop yeast models of protein misfolding diseases. We have then used these models to investigate mechanisms of cytotoxicity and to screen for novel therapeutic targets. We will focus on a yeast model developed in our laboratory, independently in other laboratories, that provides an important model of PD. We and others have also developed a yeast model of HD by expressing a disease-associated fragment of human htt exon I. The modifiers of α-syn and htt toxicity in yeast also modify the toxicity of these proteins in neuronal cells. Remarkably, the modifiers that affect α-syn and htt toxicity are very different, showing that these disease models are not simply revealing general effects of protein misfolding but are specifically probing their unique pathobiologies in a disease-relevant manner. We hope that these models, and other yeast models of neurodegenerative disease, will help the field to move rapidly from basic biological insights to defining therapeutic strategies and screening for compounds that ameliorate toxicity. We realize that many seminal aspects of neuronal pathobiology cannot be approached with these models. However, we are convinced that yeast models will provide a valuable adjunct to studies in other model systems and provide insights relevant to human pathology.

CURRENT PD GENETICS AND CELLULAR MECHANISMS OF TOXICITY

Various aspects of PD are extensively reviewed in other chapters of this book, and the reader is strongly encouraged to consult them. Here we briefly review general features of PD and associated genes to place them in context relevant to yeast models. PD is the most common neurodegenerative movement disorder. Neurologically its symptoms include muscle rigidity and tremor, as well as the slowing of physical movements. One pathological hallmark of PD is the selective loss of the DA neurons comprising the *pars compacta* of the *substantia nigra* (Fearnley and Lees, 1991) and the presence of proteinaceous inclusion bodies (LBs), and Lewy neurites, in the affected cells (Dawson and Dawson, 2003). While most cases of PD (~95%) are sporadic, understanding the genetic and cellular mechanism(s) of familial forms of PD should provide critical information about the root causes and ultimate treatment strategies for both types of PD (Vila and Przedborski, 2004; Olanow and McNaught, 2006; Raichur *et al.*, 2006).

A number of gene mutations have been shown to be associated with PD, and these are inherited in a Mendelian manner (Table 38.2; adapted from Hardy *et al.*, 2006). The identification of these mutated genes shows that several cellular pathways are important in the etiology of PD, including the proteasomal degradation system (discussed above), protein oxidative stress, and mitochondrial dysfunction (Dawson and Dawson, 2003). Exciting evidence is beginning to accumulate that the genes associated with PD can affect each other, suggesting that several PD genes may be affecting the same cellular pathway. For example, recent work in *Drosophila melanogaster* showed that the knockout phenotype of *Drosophila Pink1* (*PARK6*) could be rescued by expression of the *Drosophila parkin* (*PARK2*) gene. Yet the converse was not true, suggesting that *parkin* may act downstream in the same pathway as *Pink1* (Park *et al.*, 2006). The power of high-throughput genetic analysis in yeast offers particular promise for this type of analysis (see below).

The finding that α-syn, an intrinsically unfolded, abundant protein normally associated with presynaptic terminals, is a major component of LBs has made α-syn a major focus of PD research (Spillantini *et al.*, 1998). The normal function of α-syn within neuronal cells is not well understood. It is a small 14-kDa protein principally associated with phospholipids in membranes and presynaptic vesicles in neurons. Within biological membranes, α-syn associates with GM1 gangliosides, membrane-associated glycosphingolipids that are concentrated in the CNS and interact with growth factors and their receptors. α-Syn is recruited by GM1 to

Human gene	Protein	Function
PARK1,4 (SNCA)	α-Synuclein	Associated with presynaptic terminals; possible role in ER, Golgi trafficking
PARK2	Parkin	Ubiquitin E3 ligase
PARK3	Unknown	Unknown
PARK5	UCH-L1	Ubiquitin C-terminal hydrolase
PARK6	Pink1	Mitochondrial kinase
PARK7	DJ-1	Oxidative stress signaling molecule
PARK8 (LRRK2)	Dardarin	Cytosolic (leucine-rich repeat) kinase
PARK9	ATP13A2	P-type ATPase

caveolae and lipid raft regions (Martinez *et al.*, 2007). In neurons it appears to act in cooperation with CSP (cysteine string protein)-α and may have a role in the proper folding of SNARE proteins at the presynaptic membrane (Chandra *et al.*, 2005). In our yeast model it has been implicated in vesicular trafficking between the ER and Golgi (Cooper *et al.*, 2006) and at other steps as well (Gitler *et al.*, 2008).

Familial PD has been linked with rare missense mutations in the gene coding for human α-syn (*SNCA*): A53T ([Polymeropoulos *et al.*, 1997]; nucleotide mutation G209A [Papapetropoulos *et al.*, 2003]), A30P (Kruger *et al.*, 1998), and E46K (Zarranz *et al.*, 2004). And importantly, it is linked with genetic variants expected to yield moderate increases in expression of WT α-syn, that is duplications and triplications of the *SNCA* locus on one chromosome (Singleton *et al.*, 2003; Chartier-Harlin *et al.*, 2004). These are inherited in an autosomal dominant manner. Certain haplotypes in the *SNCA* promoter are also overrepresented in PD patients (Izumi *et al.*, 2001; Holzmann *et al.*, 2003; Pals *et al.*, 2004). Both WT and A53T α-syn are toxic to DA neurons when expressed in *Drosophila* and transgenic mice; A30P shows variable toxicity. Patients with different α-syn mutations share many, but not all, clinical features with each other and with those with idiopathic PD. Interestingly, brains from patients with the E46K mutation exhibit LBs

in neurons of the *substantia nigra* as well as in the cortex, resembling a sporadic neurodegenerative disorder known as DLB (Jellinger, 2003). These patients also have clinical symptoms similar to that of DLB (Zarranz *et al.*, 2004).

Oxidative stress is of particular importance in PD. The protein folding mechanisms of the cell are in constant and dynamic states of flux, and because both protein folding and degradation require energy, they are strongly affected by changes in the energy balance of the cell. Several animal and cellular models for PD have been generated by exposure to drugs, such as rotenone, paraquat, or MPTP (1-methyl-4-phenyl 1,2,3,6-tetrahydropyridine), which are known to induce reactive oxygen species (ROS) and cause an imbalance in the redox conditions of the cells. An *SNCA* mutation associated with PD has been linked to enhanced dopamine-induced oxidative damage (Orth *et al.*, 2004), and α-syn itself has been shown to be upregulated in subpopulations of cultured neurons and to protect these cells from apoptosis following exposure to media lacking antioxidants (Quilty *et al.*, 2006). DJ-1 (*PARK7*) (Bonifati *et al.*, 2003) is a redox-activated chaperone (Martinat *et al.*, 2004; Shendelman *et al.*, 2004) that includes α-syn among its client proteins and inhibits its aggregation (Shendelman *et al.*, 2004; Zhou *et al.*, 2006). A crystal structure has been obtained for the yeast homolog of DJ-1, YDR533Cp, revealing its structural similarities with both human DJ-1 and the bacterial chaperone Hsp31 (Wilson *et al.*, 2004). Recent work has revealed a protective role for PINK1 against oxidative-stress-induced cell death via its ability to phosphorylate the mitochondrial chaperone TRAP1 and suppress cytochrome c release from mitochondria (Pridgeon *et al.*, 2007).

Individuals with mutations in the *LRRK2* (leucine-rich repeat kinase; *PARK8*) gene can also develop late-onset PD or Parkinson-like symptoms with diverse histopathologies, sometimes with LBs, sometimes without (Zimprich *et al.*, 2004). The protein encoded by *LRRK2*, dardarin, is a large, complex, multi-domain protein containing both Ras GTPase-like (Roc) and kinase (MAPKKK) domains (Guo *et al.*, 2007). Mutations in LRRK2/dardarin (Bialecka *et al.*, 2005; Hernandez *et al.*, 2005) lead to an enhancement in kinase activity by dysregulation of the kinase domain (Guo *et al.*, 2007; Luzon-Toro *et al.*, 2007) and are inherited in an autosomal dominant pattern. *LRRK2* mutations are the most common genetic cause of PD (Hernandez *et al.*, 2005;

Khan *et al.*, 2005) and also account for up to 40% of patients with sporadic disease in some ethnic groups (reviewed in Morris, 2007). Despite the relative variability in the clinical presentation and histopathology of patients with *LRRK2* mutations, these mutations are not found in patients with AD, MSA, progressive supranuclear palsy, or frontotemporal dementia (Hernandez *et al.*, 2005).

Mutations in *ATP13A2* (*PARK9*), a P-type ATPase with unknown function that likely couples the hydrolysis of ATP to the transport of metal cations across membranes, were recently identified as being associated with PD (Ramirez *et al.*, 2006). The specificity of ATP13A2 metal transport is not known. However, exposure to certain metals, especially inhaled forms of manganese, can lead to PD-like symptoms (Olanow, 2004). In addition, iron has been found to be overrepresented in the *substantia nigra* of patients with PD (Double *et al.*, 2000; Zecca *et al.*, 2004). The presence of Cu^{2+} enhances the self-oligomerization of WT, monomeric α-syn *in vitro* (Paik *et al.*, 1999), and the oligomerization of the A53 mutant is enhanced relative to that of the WT by the activity of Cu, Zn-SOD (Kang and Kim, 2003). These observations suggest that the regulation of metal ions, in particular Mn^{2+}, Fe^{2+}, and Cu^{2+}, may play a critical role in the toxicity of α-syn, whether these are related to ATP13A2 function or not is an exciting subject for further study.

YEAST PD MODELS BASED ON α-SYN EXPRESSION

Yeast does not have a clear homolog for *SNCA*, the gene encoding human α-syn. However, small lipid-binding proteins that peripherally associate with membranes such as α-syn does evolve rapidly and would be hard to detect by sequence identity. In 2003 we reported our yeast α-syn model (Outeiro and Lindquist, 2003). Other groups working independently (Chen *et al.*, 2005; Dixon *et al.*, 2005; Flower *et al.*, 2005; Sharma *et al.*, 2006) have also constructed models of PD by expressing or overexpressing different forms of human α-syn in yeast. The finding that α-syn affects yeast biology in many ways that are similar to its effects on neurons suggests that at least the pathobiology and also the normal biology of the protein is recapitulated in yeast. Indeed they suggest that the pathobiology is related to its normal biology (see below).

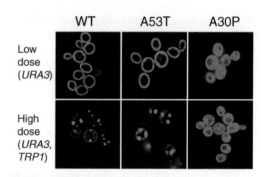

FIGURE 38.1 Fluorescence microscopy of yeast cells expressing three different fusions of α-syn-GFP: WT, A53T, and A30P mutants. Low does strain has α-syn insertes at the URA3 locus, high dc strain has insertion at the URA3 an TRPI 10ci. Adapted from Outeiro and Lindquist, 2003.

In our model (Outeiro and Lindquist, 2003), human α-syn (WT, A53T, or A30P) protein was fused at its carboxy terminus to GFP with a four-amino-acid linker. These fusion proteins exhibited no evidence of proteolysis in yeast cells, in contrast to a previous report for α-syn-GFP fusions in other cell types (McLean *et al.*, 2001). To insure that results were not an artifact of the large GFP tag we also examined cells expressing WT, A53T, and A30P α-syn alone, without GFP tags. Using immunofluorescence with an α-syn antibody, the inclusions detected by this method were similar in number, size, and distribution to those obtained with the α-syn-GFP fusion proteins (see Figure 38.1). However, membrane localization was reduced, probably as a result of the fixation and permeabilization procedures required for immunofluorescence (data not shown). Most importantly the general biological effects that the α-syn-GFP fusion created in yeast cells (see below) were also observed when α-syn was expressed without the fusion. Thus, in this particular cell type with this particular fusion, GFP provides a powerful tool for studying protein distributions and, most importantly, dynamic changes in distribution, in real time in living cells, without compromising the biological integrity of the protein. They also allow us to circumvent the experimental difficulties and potential inconsistencies of indirect immunohistochemical methods that are particularly problematic with membrane proteins.

In creating our model we took particular advantage of two features of yeast biology. First, we employed a galactose-regulated promoter, *GAL1*, one of the most tightly regulated promoters known. Its expression is controlled by the combinatorial action of a repressor, which binds the promoter in the presence of glucose, and an activator, which binds in the presence of galactose. For routine growth and maintenance of stocks we passage cells in glucose media, where expression is completely off. This allows us to avoid any selective pressures from leaky α-syn expression, which could lead to the accumulation of hidden genetic variation. For experimental analysis we pre-grow cells in raffinose and then shift them to galactose. In raffinose medium glucose repression is alleviated, but the leaky α-syn expression that occurs under these conditions has no detectable effects on the cell for short periods. After allowing time for the repressor to disappear, a switch from raffinose to galactose medium allows us to rapidly and synchronously induce α-syn expression in all cells in the culture.

Second, we took advantage of the unusually high rates of homologous recombination in yeast cells to integrate the α-syn-GFP constructs directly into the yeast genome. We chose sites of known auxotrophic mutations (mutations that render an organism unable to synthesize a given nutrient). Integration simultaneously restored prototrophy for that nutrient and insured that the insertion was not producing any harmful effects on its own, a common problem with random integrations. Further, integration insured that the gene was inherited faithfully by every cell, avoiding the cell-to-cell variation typical of extra-chromosomal plasmids. In various experiments we have employed cells with single or double integrations at the *URA3*, *HIS3*, or *TRP1* loci.

Expression of a Low Dose of α-Syn May Provide a Model of Normal α-Syn Function

When cells expressing a low dose (integrated at *TRP1*) of WT α-syn-GFP are shifted from raffinose to galactose, the protein rapidly becomes localized to the plasma membrane, with a much smaller quantity found in the cytoplasm (Figure 38.1; Outeiro and Lindquist, 2003). As controls, we used mitochondrial- and nuclear-localized GFPs and confirmed that α-syn did not localize to the membranes of these organelles (not shown). The high concentration of α-syn at the plasma membrane *in vivo* is in good agreement with its selective binding to phospholipid vesicles with specific lipid compositions *in vitro* (Jo *et al.*, 2000). It is also, in its own way, consistent with the known concentration of α-syn at the synapse in mammalian neurons. Using their own model of α-syn expression in yeast, the laboratory of David Gross has reported that α-syn first associates with membranes in the ER and then traffics via the Golgi out to the plasma membrane (Dixon *et al.*, 2005), where the final trafficking step involves the fusion of vesicles to the plasma membrane. This final fusion step is highly regulated in neurons but is constitutive in yeast. Accordingly, if α-syn initially follows a common trafficking pathway in yeast and neurons, it would be expected to end up concentrated at the plasma membrane in yeast but to accumulate at the synapse in neurons.

Cells expressing a low dose of WT, A30P, or A53T α-syn (inserted at *URA3*) grew at about the same rate as cells carrying a vector with the same auxotrophic marker (Outeiro and Lindquist, 2003), with cells expressing WT or A53T α-syn exhibiting only a subtle reduction in final culture density in liquid medium (Figure 38.2). Recently, electron microscopy of cells expressing WT α-syn established that these cells do have a subtle defect in vesicle trafficking. Specifically, small vesicles accumulate, juxtaposed to the peripheral ER that is localized near the plasma membrane. This indicates that α-syn reduces vesicle docking or fusion (Gitler *et al.*, 2008; see Figure 38.3), an observation that was confirmed in cell-free vesicle trafficking assays (Gitler *et al.*, 2008).

This observation is strikingly reminiscent of findings in mammalian cells and mice. Mice in which the endogenous α-syn gene has been deleted are normal in appearance, but have reduced pools of synaptic vesicles (Cabin *et al.*, 2002) and exhibit an enhancement in the activity-dependent release of dopamine (Abeliovich *et al.*, 2000). In separate experiments, neuronal cells that overexpress α-syn were found to exhibit a reduction in stimulation-dependent neurotransmitter release. This was not due to a reduction in the dopamine present in individual vesicles but to a reduction in their fusion with the plasma membrane (Larsen *et al.*, 2006; Gitler *et al.*, 2008). Thus, data from yeast and mammalian cells both suggest that α-syn acts to reduce the fusion of vesicles with acceptor membranes.

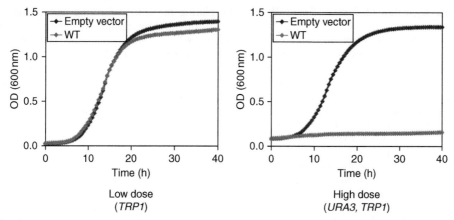

FIGURE 38.2 α-Syn overexpression inhibits growth of yeast cells in liquid galactose medium. A low dose of α-syn integrated at URA3 had little-to-no effect on growth whereas a high dose of α-syn integrated at URA3 and TRPI completely inhibited growth (gray lines).

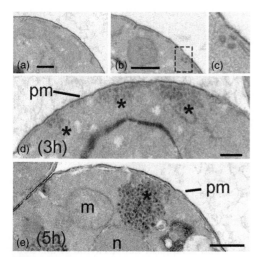

FIGURE 38.3 Electron microscopy of α-syn-overexpressing yeast cells reveals a time- and dosage-dependent accumulation of vesicles. Control strain (a) and NoTox α-syn strain (b) at 3h. (c) Higher magnification of vesicles accumulated near peripheral ER at 3h in the strain. Accumulation of vesicles (marked by *) in HiTox strain at 3h (d) and 5h (e). Scale bars = 0.5 μm. Excerpted from Gitler et al., 2008.

Expression of a High Dose of α-Syn Provides a Model of α-Syn Toxicity

In attempting to reconstitute the pathology of α-syn in yeast we took several factors into consideration: (1)

cellular QC and protein trafficking systems are highly conserved, (2) these systems are compromised with age, and (3) age is the single greatest risk factor for PD. We reasoned that simply increasing the level of α-syn expression in yeast might exceed the QC system of the cell and mimic a cellular aging scenario associated with α-syn toxicity. Indeed, we found that integrating additional copies of WT α-syn-GFP into the yeast genome (at the URA3 and TRP1 loci) increased its accumulation and caused a remarkable change in its localization. A few hours after induction, the vast majority of the α-syn protein appears in large cytoplasmic inclusions (Figure 38.1, two copies). These inclusions were not formed simply by excess protein unable to find docking sites at the plasma membrane; in time course experiments using the high dose (URA TRP1) strain, α-syn first accumulated at the membrane, formed small foci there, and was later found away from the membrane in larger cytoplasmic inclusions (Gitler et al., 2007). Moreover, much less protein was present at the membrane in high dose strains than in low dose strains (Figure 38.1). The results for WT α-syn were similar whether cells were grown on plates or in liquid medium.

The A53T mutant behaved much like WT α-syn in both strains, except that it moved from membrane association to inclusion more rapidly than WT protein in the high dose strain, and small inclusions could be observed in some cells even in cells carrying only a single copy of α-syn. In contrast, A30P α-syn maintained its cytoplasmic distribution and did not form inclusions, regardless of dosage (Figure 38.1, high dose).

To assess the effects of α-syn on cell growth, the optical density of the culture was continuously monitored after cells were transferred from raffinose to galactose (Figure 38.2). A low dose of WT or A53T α-syn had little or no inhibitory effect on growth, while high dose (inserted at *URA3*, *TRP1*) completely inhibited it. In contrast, A30P had no impact on growth, even when expressed from the 2μ plasmid (Outeiro and Lindquist, 2003). The lack of toxicity with A30P was initially perplexing. Our recent work suggests that this might relate to the fact that the protein requires co-expression of WT α-syn to be toxic (Rubio-Texeira, M., Outeiro, T. F., Su, L. J. and Lindquist, S., unpublished observations).

Taking advantage of the fact that chromosomal locations can exert subtle effects on gene expression, even with genes driven by the powerful *GAL1* promoter, we created yet another strain, that has α-syn integrated at *HIS3* and *TRP1* (Cooper *et al.*, 2006). The level of α-syn expression was ∼30–40% less than in the strain carrying insertions at *URA3* and *TRP1* (Gitler *et al.*, manuscript in preparation). This had a large effect on toxicity, creating a strain that can still grow after galactose induction, albeit slowly. We have designated these strains HiTox (*URA3* and *TRP1*) and IntTox (*HIS3* and *TRP1*) (Gitler *et al.*, 2008). As discussed below, in the sections on chemical and genetic screens, having such carefully regulated levels of toxicity has been extremely useful for particular purposes. This illustrates another advantage of the yeast system, as achieving such precise regulation of α-syn expression is difficult in cultured mammalian cells.

The extreme dosage dependence of α-syn in yeast is highly unusual. Indeed, we have never observed such extreme dosage sensitivity before. It is, however, strikingly reminiscent of α-syn dosage sensitivity in humans. Certain familial forms of early onset PD in humans are caused by duplication or triplication of the α-syn gene (Singleton *et al.*, 2003; Chartier-Harlin *et al.*, 2004). Furthermore, mutations in the promoter and in the 3′ end of the gene are associated with PD susceptibility, likely because they increase its level of its expression (Holzmann *et al.*, 2003; Pals *et al.*, 2004; Winkler *et al.*, 2007).

To further characterize the toxicity of HiTox, we monitored the loss of cellular viability following induction of WT α-syn (Cooper *et al.*, 2006, and Su *et al.*, unpublished observations) in two ways,

by plating for colony forming units remaining in the culture and by staining with the dye propidium iodide, which stains dead cells. The lag between α-syn induction and cell death (with just a few percent of cells dying 4 h post-induction, reaching ∼30% cell death at 6 h and ∼50% at 8 h) afforded an opportunity to assay the biological effects of the protein before irreversible toxicity confounded results. At 6 h the IntTox strain exhibits ∼10% cell death, but we have not yet characterized it as well as we have the HiTox strain (Su *et al.*, unpublished observations).

As described in more detail in Outeiro and Lindquist (2003) and below, overexpression of α-syn in the HiTox model recapitulates many aspects of α-syn toxicity in mammalian cells and PD patients, including ubiquitination of α-syn inclusions, impairment of the ubiquitin–proteasome system (Iwatsubo *et al.*, 1996; Dauer and Przedborski, 2003), A53T-mediated enhancement of toxicity, accumulation of lipid droplets (den Jager, 1969; Gai *et al.*, 2000), the production of ROS (Giasson *et al.*, 2000; Dauer and Przedborski, 2003), and mitochondrial pathology (Ramsey and Giasson, 2007). α-Syn has biochemical properties and a structural motif that suggests it is a member of the fatty acid-binding protein family (Sharon *et al.*, 2001). In mammalian cells α-syn binds to lipid droplets; overexpression of WT and A53T α-syn, but not A30P, in HeLa cells results in lipid droplet accumulation in the cytoplasm (Cole *et al.*, 2002). Remarkably, the same pattern is observed in yeast cells (Figure 38.4).

One reported similarity between the effects of α-syn on yeast and mammalian cells, however, must be considered with caution. α-Syn has been reported by three different laboratories to inhibit phospholipase D (PLD) (Jenco *et al.*, 1998; Ahn *et al.*, 2002; Outeiro and Lindquist, 2003; Payton *et al.*, 2004). PLD is not essential in yeast, but becomes essential in cells carrying mutations in Sec14, a phosphatidylinositol phophatidylcholine transfer protein, and Cki1, a choline kinase. Levels of α-syn expression that did not inhibit WT cells killed cells carrying these mutations, implying that α-syn inhibits PLD in yeast. However, others have found it difficult to reproduce the effects of α-syn on mammalian PLD *in vitro* (Rappley, I., personal communication). Further, moderate levels of α-syn expression do not inhibit sporulation in yeast, the stage in the cell cycle when PLD becomes essential. Adding to the mystery, PLD mutations in sporulating cells have a phenotype that is strikingly reminiscent

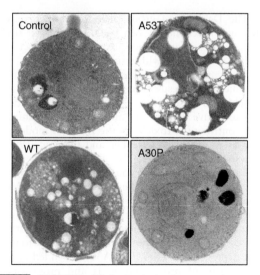

FIGURE 38.4 Lipid droplet accumulation in yeast with integrated into URA3 and TRPI α-syn. Electron micrographs showing lipid droplet accumulation in the cytoplasm of yeast cells expressing WT or A53T, but not A30P mutant α-syn. Excerpted from Outeiro and Lindquist, 2003.

of the reduced vesicle fusion observed with low-dose α-syn expression in yeast and mammalian cells. Specifically, mutations in PLD block the fusion of precursor vesicles to the prospore membrane, resulting in the accumulation of docked but unfused vesicles (Nakanishi *et al.*, 2006). Thus relationship between α-syn and the activity of this enzyme warrant further investigation.

To identify gene expression changes in response to α-syn toxicity during the early period before cell death, we performed microarray analysis comparing the HiTox strain α-syn, the single copy *URA3* strain, and the empty vector control strains at 0, 2, 4, and 6 h after the induction of α-syn. These time points delineate a progressive increase in α-syn expression and specific alterations in α-syn localization, from its presence at the plasma membrane (at 2 h) to its formation of inclusions (at 4 h), and culminating in cell death (~30% at 6 h; Cooper *et al.*, 2006; Gitler *et al.*, 2008; Yeger-Lotem, E., Su, L. J. and Lindquist, S., unpublished observations). We compared these changes in gene expression with the pathobiology undergone by α-syn-expressing yeast cells in the same time frame. We found that genes involved in mitochondrial processes and ER stress are perturbed even at 2 h post-induction. Consistent

with these array data, the earliest biological defects observed in HiTox α-syn cells are vesicle accumulations and vesicular trafficking defects, at 3 h post-induction (Cooper *et al.*, 2006; Gitler *et al.*, 2008; and below, genetic screens section). By 4 h post-induction, HiTox cells expressing α-syn differentially express numerous genes in classes highly enriched for ER stress, vesicle trafficking, sterol biosynthesis, and mitochondrial function. These cells exhibit defective mitochondrial morphology and abnormal mitochondrial DNA, consistent with altered mitochondrial gene expression (Su, L. J. and Lindquist, S., unpublished observation). They also accumulate damaging ROS and undergo significant oxidative and nitrosative stress, as measured by reactivity to fluorescent probes (e.g., CM-H2DCFDA and DAF-FM) and antibodies against nitrosylated proteins (Su, L. J. and Lindquist, S., unpublished observation). These results, too, are consistent with reports in mammalian PD models and in humans, wherein there has been a long-standing realization that there is a relationship between PD and mitochondrial stress, ROS, and nitrosative damage (Hara *et al.*, 2005; Abou-Sleiman *et al.*, 2006; Uehara *et al.*, 2006). Further evidence that the toxicity of α-syn in yeast is directly relevant to neurons is provided below, in results from our high-throughput screens.

Other Models of α-Syn Toxicity

Other groups (Chen *et al.*, 2005; Dixon *et al.*, 2005; Sharma *et al.*, 2006) have also used *S. cerevisiae* GFP-α-syn models to shed light on α-syn biology and pathobiology. In all of the models discussed here, low doses of WT and A53T α-syn were localized to the cell membrane, whereas A30P α-syn exhibited a more diffuse cytoplasmic localization. Flower *et al.* (2005) also observed the accumulation of toxic ROS in their model, and an accompanying sensitivity of the α-syn-expressing cells to hydrogen peroxide (H_2O_2). They propose that α-syn-induced ROS accumulation triggers apoptotic cell death mediated by the yeast pro-apoptotic caspase *YCA1* (Flower *et al.*, 2005). Surprisingly, the A30P α-syn mutant inhibited cell growth and exhibited H_2O_2 sensitivity in their model, perhaps due to extremely high levels of expression in their strain background (Flower *et al.*, 2005).

The laboratory of David Gross (Dixon *et al.*, 2005) took advantage of the extensive previous analysis of protein trafficking in yeast, to discover that WT and A53T α-syn are specifically targeted to the plasma membrane via the classical secretory pathway, whereas A30P mutants do not enter that pathway. When trafficking was blocked at different stages in the vesicle trafficking pathway by temperature-sensitive mutants that act at specific steps in that pathway, WT and A53T α-syn trafficking was also blocked at that step. In contrast, A30P synuclein remained dispersed throughout the cytosol in all cases. The authors' N-terminal GFP construct exhibited a lower level of toxicity than our C-terminal fusion. In order to produce significant toxicity in the WT background, the authors made use of a clever trick that is commonly employed by yeast geneticists to increase protein expression to very high levels. The N-terminal GFP fusion construct was expressed from a 2μ plasmid that contained two selectable markers, one that allows the production of uracil, the other, the production of leucine. However, the leucine marker on this plasmid carried a promoter deletion; cells can only survive under leucine selection when the plasmid is driven to very high copy number. Under these circumstances WT and A53T α-syn were toxic but not A30P. The fact that most of the results in this study correlate well with others using untagged synuclein and C-terminally tagged synuclein (Outeiro and Lindquist, 2003; Cooper et al., 2006), which do not require high levels of expression for toxicity, are reassuring. Moreover another study (Sharma et al., 2006) found that N-terminally tagged α-syn is degraded more rapidly than C-terminally tagged protein, and this may be the sole reason that much higher expression levels were required with this construct. However, it is also possible that N-terminally tagged synuclein behaves differently from untagged and C-terminally tagged proteins (which are indistinguishable from each other). Given the importance of the observations with the secretory pathway mutants, it would be helpful if they were repeated with untagged synuclein or C-terminally tagged synuclein.

Dixon et al. (2005) also found that at moderate levels of expression the N-terminally tagged WT and A53T forms of α-syn induced a heat-shock response. Further, these forms became toxic in yeast in conjunction with mutations in the 20S proteasome. Sharma et al. (2006) also found enhanced toxicity when α-syn expression was coupled with different proteasomal mutants. Taken together these data suggest that the protein QC system normally renders membrane-targeted α-syn non-toxic, but it can be overwhelmed when α-syn is genetically compromised or overexpressed.

The DebBurman group also found that yeast lacking mitochondrial Mn-SOD were more susceptible to oxidants when expressing α-syn. This finding is consistent with results of Flower et al., an earlier synthetic lethal screen (see below) and with our microarray results (Su, L. J. and Lindquist, S., unpublished observations), and supports the idea that oxidative damage and mitochondrial dysfunction contribute to toxicity in the yeast PD model as well as in human PD (Jenner and Olanow, 1996; Dawson and Dawson, 2003).

Lastly, Chen et al. (2005a, b) employed stationary phase cells to study the effect of α-syn expression on aging cells. Numerous studies have shown that stationary phase yeast cells recapitulate many features of aging in mammalian cells (e.g., Ashrafi et al., 1999). Using these cells, the group found that expression of 1X WT or A30P α-syn led to an impairment of protein synthesis and proteasome complex assembly. Moreover, expression of 1X WT or A30P α-syn had a dramatic effect on viability, whereas stationary phase cells expressing tau, a protein involved in the pathobiology of AD, were virtually indistinguishable from cells expressing an empty vector. Taken together, these results suggest a susceptibility of aged cells to α-syn, and that yeast might provide important insight into the relationship between aging and α-syn toxicity.

SUCCESS OF YEAST PD MODELS IN HIGH-THROUGHPUT SCREENING

A major advantage of the yeast system is the tractability of high-throughput screens. We and others have used and continue to use a variety of different screens to identify factors that influence α-syn toxicity. (Figure 38.6).

Yeast Two-Hybrid System

The first yeast high-throughput screen with α-syn was not designed to investigate α-syn biology or pathobiology. Rather it simply employed yeast as a tool to investigate protein–protein interactions (Engelender et al., 1999). Typically in the yeast two-hybrid system, one protein (bait) is fused to the binding domain of a transcription factor, and the other protein (prey) is fused to the activation domain of the transcription factor. When these proteins associate and reconstitute the activity of the split transcriptional factor in the yeast nucleus, the interaction drives the activation of a reporter gene, such as beta-galactosidase. The interaction is assayed via the readout of the reporter activity (Stelzl and Wanker, 2006). Here, Engelender and coworkers screened the human brain cDNA libraries using various constructs of α-syn as "bait," and found only one interacting protein, synphilin-1. Although the function of synphilin-1 is not known, the protein is found in LBs (Wakabayashi et al., 2000), it is a substrate of parkin (Chung et al., 2001), and its ubiquitination promotes the formation of both α-syn and synphilin-1 into LB-like inclusions in cell culture models (Lim et al., 2005).

While the example mentioned above demonstrates the value of yeast two-hybrid system in identifying an α-syn interacting partner, it is rather striking that no other proteins were identified using this approach. It is possible that α-syn, which preferentially interacts with the plasma membrane, is unattractive bait in a system that relies on both the bait protein and its interacting partner to enter and form an association in the nucleus to provide a positive readout. Alternative yeast two-hybrid systems have been developed, such as the reverse Ras recruitment system and the G-protein fusion, which facilitate identification of interacting partners of membrane proteins (Stagljar and Fields, 2002). These new alternative systems that employ soluble or membrane-bound bait proteins might be more useful in identifying additional α-syn interacting proteins, thereby providing further insight into α-syn function or toxicity.

Synthetic Lethal Screen

In collaboration with us, the laboratory of Paul Muchowski (Willingham et al., 2003) took advantage of a collection of 4,850 haploid gene deletion mutant strains (Winzeler et al., 1999; Giaever et al., 2002) to identify genes that modulate the toxicity associated with α-syn expression. These strains were transformed with a 2μ α-syn plasmid, which reduces but does not completely block growth. Although 2μ plasmids usually exist in high copy numbers, when they carry toxic expression constructs, cells with low copy numbers are selected. Indeed, the accumulation of α-syn in these cells on average is lower than that of the HiTox two copy strain. Mutants that exacerbated growth defects when expressing α-syn were isolated. From the library, 86 mutants exhibited enhanced toxicity with α-syn expression. Sixty-five

percent (56/86) of these corresponded to genes for which a function has been determined experimentally or can be predicted; thirty-two percent were in categories functionally related to lipid metabolism and vesicular transport. Our results showed a large degree of overlap in the functional categories affected in the recent transcriptional profile experiment in a *Drosophila* PD model (Scherzer et al., 2003), and in our microarray analysis (Yeger-Lotem, E., Su, L.J. and Lindquist, S., unpublished observations), where the major categories identified were lipid metabolism, membrane transport, and energy genes. Most importantly, they strongly correlated with what we have learned from the cell biology of our model system, which shows signs of oxidative stress.

While screening a deletion series has been a fabulous tool and generated many important advances, in the time since our synthetic lethal screen was first reported, we and other groups have found that this type of screen can be subject to technical difficulties resulting in hits that are not reproducible. Spontaneous mutations can arise during routine growth and maintenance of the deletion strains that may have unknown effects on toxicity. If a and α types are then produced by a simple mating type switch and those used to generate a diploid strain for further screening, these spontaneous mutations will not be complemented during the screen and their effects will persist, possibly masking the intended effects of the original deletion. Of the original 86 hits we reported, 12 have been confirmed as α-syn toxicity enhancers (ARL3, COD2/COG6, NBP2, OPI3, PTK2, SAC2/VPS52, SOD2, TLG2, THR1, VPS24, VPS28, YGL226w). Several additional original hits have not been reproduced (STP2, MSB3, UBC8, YKL100c, Muchowksi, P., personal communication; and Paul and DebBurman, 2006). The remaining hits await further testing in a fresh genetic background.

Overexpression Screens

In a more recent collaborative study with the Harvard Institute of Proteomics (HIP), we completed a screen for genes whose overexpression would enhance or suppress α-syn toxicity (Cooper et al., 2006; Gitler et al., manuscript in preparation). To address some of the technical issues from the deletion screen, in designing our overexpression screen we expressed both α-syn and the genes from the library using the same galactose-inducible promoter, *GAL1*. We were therefore able to perform all maintenance and manipulations prior to the actual screen in non-inducing conditions, in the absence of galactose, reducing the likelihood that spontaneous, unwanted changes in the genome would accumulate during routine growth and maintenance of the cells.

To find modifiers of α-syn toxicity, a library of cells of the IntTox strain, each harboring one of ~5,000 galactose-regulated, sequence-verified ORFs (the Yeast FLEXGene collection, Table 38.1) was arrayed in 96-well plates and stamped onto galactose medium (Figure 38.5). The plasmid carried a *CEN* (centromeric element) so that it would be present at one copy per cell and the ORFs carried no fusions or tags.

The intermediate level of toxicity in this strain allowed us to screen simultaneously for genes that would enhance or suppress the toxicity of α-syn. Given the extreme dosage sensitivity of α-syn toxicity, we expected to find genes such as *MIG1* and *MIG3*, which downregulate expression from the galactose promoter, and we did. This established that the screen was working properly, but of course these genes were not themselves of further interest, except as controls. This overexpression screen identified more than 50 genes that modify α-syn toxicity. In keeping with the complexity of the genetic and environmental factors associated with PD, these genes fell into several different functional classes, including vesicle trafficking, metal ion transport, osmolyte synthesis, protein phosphorylation,

and nitrosative stress (Cooper *et al.*, 2006; Gitler *et al.*, manuscript in preparation).

The largest class of genetic modifiers encompassed genes involved in vesicular trafficking, specifically at the ER/Golgi step (Cooper *et al.*, 2006). There was excellent inherent logic in the hits, with the suppressors in this class predicted to function in promoting anterograde transport (from ER to Golgi) while the enhancers were predicted to inhibit it. For example, *RAB1*, a GTPase that promotes the movement of vesicles from ER to Golgi suppressed toxicity. *GYP8*, the GTPase activating protein that converts Rab1 from its active GTP-bound state to its inactive GDP-bound state, enhanced toxicity. Furthermore, the general functions of the genes suggest that α-syn is likely to be inhibiting the docking or fusion of vesicles to the Golgi rather than to inhibit vesicle budding from the ER (Cooper *et al.*, 2006).

This finding was in keeping with observations from the Cooper laboratory. They found that α-syn expression induced ER stress and asked if it might do so by blocking the endoplasmic reticulum-associated degradation (ERAD) of misfolded ER proteins. Interestingly, the degradation of one substrate, a mutation in Sec61 (Sec61–2p) was not affected, but the degradation of another, a mutated form of carboxy peptidase Y (CPY*) was. The difference between these two ERAD substrates is that CPY*, but not Sec61–2p, requires trafficking from the ER to the Golgi prior to degradation. Taken together these results and those of our genetic

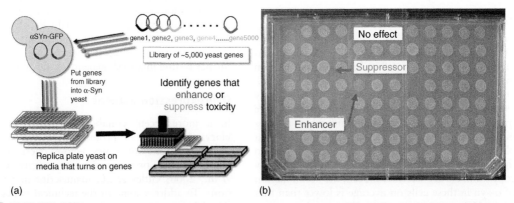

(a)   (b)

FIGURE 38.5 Yeast screen for α-syn toxicity enhancers and suppressors. (a) A library of 5,000 individual yeast ORFs, and a moderately toxic α-syn construct, were both expressed under the control of the galactose-inducible GAL1 promoter. (b) The strains were plated on galactose media and monitored for changes in growth revealing suppression or enhancement of toxicity.

screen indicated that α-syn causes a block in ER-to-Golgi trafficking. To test this directly, we monitored the trafficking of two proteins, CPY WT and alkaline phosphatase, which acquire various modifications as they traffic from the ER, through the Golgi, to the vacuole. Indeed in the HiTox strain trafficking was completely halted before the protein reached the Golgi, in the IntTox strain it was reduced (Cooper et al., 2006 and Cooper, A.A. and Lindquist, S., unpublished data).

In separate experiments (Gitler et al., 2008) we used a variety of approaches to investigate the specific nature of the ER-Golgi trafficking defect caused by overexpression of α-syn. First, using purified α-syn in reconstituted cell-free assays, we found that addition of WT, but not A30P, α-syn protein to semi-intact membranes inhibited ER to Golgi transport in a dose-dependent manner, and it specifically inhibited the tethering and/or fusion of vesicles to the ER, not the budding of vesicles from the ER. (As mentioned earlier, A30P does not interact with membranes in yeast, making it a perfect control.) Ultrastructural analysis established that α-syn caused a time- and dose-dependent block in trafficking. Initially, vesicles accumulated proximal to the ER and plasma membranes. At later time points, vesicles globally mislocalized in the cell interior. With the HiTox strain, the vesicles were more heterogeneous and included Golgi markers indicating that other Rab-regulated trafficking steps were disrupted. Synuclein was directly involved in the formation of these vesicle clusters and co-localized with them by immunostaining with gold particles (Gitler et al., 2008). In keeping with the more diverse mislocalization of vesicles in the HiTox strain, it could not be rescued by Rab1 alone.

The key question was then to determine whether the genes that rescue yeast cells from α-syn toxicity are relevant to neurons. Indeed, the human homolog of Rab1/Ypt1 suppressed α-syn toxicity in DA neurons of three other model systems including fruit flies (in collaboration with Bonini, N., University of Pennsylvania), in nematodes (in collaboration with Caldwell, G., University of Alabama), and in mixed primary rat neuronal cultures (in collaboration with Rochet, J.-C., Purdue University; Cooper et al., 2006).

Reasoning that that the distinct biology of DA neurons might make them more sensitive to perturbations in other trafficking steps we tested two other genes in nematodes and rat primary cultures (Caldwell, G., Rochet, J.-C. and Lindquist,

S., unpublished results; Gitler et al., 2008; and Gitler et al., manuscript in preparation). We chose Rab8a and Rab3a. Rab8a is the most closely related to Rab1 by sequence homology (Gitler et al., 2008) but functions at a post-Golgi step (Huber et al., 1993). Rab3a is highly expressed in neurons (Stettler et al., 1994; Gurkan et al., 2005), concentrates at presynaptic sites (Stettler et al., 1994), and plays an important role in active transport and docking of neurotransmitter vesicles prior to their regulated fusion and release (Geppert et al., 1994; Leenders et al., 2001).

In addition to proteins involved with vesicle trafficking, this study also found that genes for three metal ion transporters modify the toxicity of α-syn. Pmr1p, a protein that transports Mn^{2+} and Ca^{2+} ions from the cytoplasm into the Golgi (Durr et al., 1998), was found to enhance toxicity when co-expressed with α-syn (Cooper et al., 2006). The enhancement of α-syn toxicity by Pmr1p has since been confirmed in DA neurons of nematodes (Caldwell, G. and Lindquist, S., unpublished results), and provides an attractive potential therapeutic target. An additional yeast metal ion transporter, Ccc1p, was found to suppress toxicity of α-syn. This transporter sequesters Mn^{2+} and Fe^{2+} ions into the vacuole of yeast cells (Lapinskas et al., 1996). Notably, exposure to inhaled forms of manganese (e.g., in miners of manganese) has long been associated with increased risk of developing a Parkinson-like syndrome (Weiss, 2006). It is therefore particularly striking that of the dozens of metal transporters that were in the original library, two of the three are involved in the transport of manganese.

The third transporter, YOR291w, is a transmembrane ATPase that is a member of the metal transport family, but its specificity is unknown. The human homolog of this highly conserved protein, ATP13A2, is believed to couple the hydrolysis of ATP to the transport of cations across various cellular membranes. Remarkably, mutants in ATP13A2, also known as PARK9, were recently shown to cause early onset PD (Ramirez et al., 2006), establishing again that the yeast model captures relevant aspects of the human disease.

Another important class of genes isolated from this screen involved trehalose biosynthesis and metabolism. Trehalose, discussed above, is a chemical chaperone found in yeast that promotes correct protein folding, inhibits aggregation, and allows organisms, even where it is not endogenously synthesized, to survive extreme conditions of environmental stress (Singer and Lindquist, 1998a). We isolated three genes involved in trehalose metabolism as suppressors of α-syn toxicity: Ugp1p, an enzyme that produces the precursor trehalose-6-phosphate (Elbein *et al.*, 2003); Tps3p, a regulatory subunit of the trehalose synthase complex; and Nth1p, an enzyme that degrades trehalose. High concentrations of osmolytes can also interfere with protein folding (Singer and Lindquist, 1998b) suggesting that the beneficial effects of trehalose on some proteotoxic stresses might be countered by detrimental effects on others. Remarkably, however, oral administration of trehalose reduces protein aggregation and extends lifespan in a mouse model of HD (Tanaka *et al.*, 2005) and reduces protein aggregation and muscle weakness in a mouse model of oculopharyngeal muscular dystrophy (Davies *et al.*, 2006). Moreover, a recent report indicates trehalose has an additional role to play in protein homeostasis by inducing autophagy and promoting the clearance of misfolded mutant htt and α-syn in mammalian cells (Sarkar *et al.*, 2007). Because osmolytes such as trehalose are small, can cross the blood–brain barrier, and are generally not toxic they might be potential therapeutic modulators of proteotoxic stress.

These and a host of other genes isolated in our overexpression screen suggest that yeast cells expressing α-syn will provide a powerful model for finding genetic modifiers that may play an important role in PD diagnostics and in the discovery of new therapeutic strategies (see below).

In a separate overexpression screen, Flower *et al.* (2007) exploited the sensitivity of α-syn, particularly the A30P form, to H_2O_2 in their own model of α-syn toxicity. They used a high copy yeast genomic library to screen for suppressors of the toxicity of A30P combined with H_2O_2 and found one protein, Ypp1, which specifically antagonized ROS accumulation and toxicity when overexpressed. Ypp1 bound to A30P α-syn and altered the localization of that protein in these cells, directing it towards degradation in the vacuole. To define the pathway by which Ypp1 suppressed A30P toxicity, the authors assessed deletions in 116 nonessential genes involved in autophagy, cytosol-to-vacuole transport, endocytosis, and vacuole sorting pathways for the ability to antagonize Ypp1 rescue. The hits from this screen were enriched for genes involved in the class E vacuolar sorting pathway and endocytosis, suggesting that Ypp1 mediates trafficking of A30P α-syn to the vacuole via the endocytic pathway. This hypothesis was confirmed by the analysis of GFP-A30P α-syn localization in various genetic backgrounds that showed that deletion of endocytic pathway genes prevented localization of A30P α-syn to the vacuole, whereas deletion of Sec12, a control gene involved in the secretory pathway, had no effect on A30P α-syn localization. Furthermore, Ypp1 was suggested to have a role in receptor-mediated endocytosis, because addition of mating factor α drives Ypp1 into the vacuole and is thereby sufficient to protect cells expressing A30P α-syn from toxicity (Flower *et al.*, 2007).

The authors postulate that Ypp1 is likely to interact with WT or A53T α-syn, but that the unique conformation of A30P α-syn results in a stronger association between Ypp1 and A30P α-syn than between Ypp1 and WT or A53T α-syn. This unique interaction is sufficient to target A30P towards vacuolar degradation and protect cells from A30P toxicity. Previously, it has been shown that WT, but not mutant, α-syn is preferentially degraded in the lysosome via chaperone-mediated autophagy (Cuervo *et al.*, 2004). Therefore, results from this genomic screen provided insight into how cells utilize alternate routes to degrade misfolded mutant α-syn.

Chemical Screens

We have also used our yeast PD model to perform a high-throughput chemical screen for small molecules that ameliorate α-syn toxicity (Su *et al.*, manuscript in preparation). Over 115,000 compounds from various collections, including commercial libraries, natural products, and NCI collections, were screened for their ability to restore growth in yeast cells expressing toxic levels of α-syn. After retesting, four structurally related compounds were found to antagonize α-syn-mediated inclusion formation and α-syn toxicity at low micromolar concentrations. This suggests that these compounds may have a highly specific cellular target. These four compounds did not rescue growth of yeast cells expressing galactose-inducible polyglutamine (polyQ) huntingtin, indicating specificity for α-syn toxicity.

Further analysis revealed that the four compounds rescued the ER-to-Golgi vesicular trafficking defect in yeast. They also greatly reduced the formation of α-syn foci and restored membrane binding. The target of these compounds, however, is unlikely to be α-syn itself: the compounds have no effect on the conformational states of purified α-syn (data not shown). The two most potent compounds also rescued DA neurons in two established models of α-syn toxicity where genetic rescue was also demonstrated: *C. elegans* DA neurons and primary cultures of rat midbrain transduced with α-syn (Caldwell, G., Rochet, J.-C. and Lindquist, S., unpublished, and Su *et al.*, manuscript in preparation). Microarray analysis indicated that in addition to ER stress, α-syn causes mitochondrial toxicity. Further analysis indicated that the compounds partially correct this mitochondrial toxicity. Importantly, we also tested the effects of the compounds on DA neurons killed by rotenone (a mitochondrial poison that can produce toxicities similar to those seen in PD). Unlike Rab1, our lead compounds antagonized both rotenone and A53T α-syn toxicity (data not shown). Due to the extensive connections between mitochondrial stress and PD, these results suggest that the mechanism of action for these compounds is central to PD pathobiology.

A Proposed Mechanism for the Sensitivity of DA Neurons to α-Syn

It is now clear that the pathology of PD is much broader than toxicity to DA neurons (reviewed in Ahlskog, 2007), although DA neurons certainly are amongst the most vulnerable population (Naoi and Maruyama 1999; Jackson-Lewis and Smeyne, 2005). Yet, α-syn is expressed throughout the brain and our work suggests that α-syn misfolding and accumulation are likely to inhibit vesicle trafficking in many cell types. The special susceptibility of DA neurons may be due to the fact that particular features of their biology make them unusually vulnerable to the general toxicity of α-syn. DA neurons face unusually high oxidative stresses, particularly from high iron levels and dopamine oxidation. Dopamine synthesis, transport, and storage all provide unique opportunities for creating oxidative stress in the cell. Enzymatic metabolism of dopamine by cellular monoamine oxidase produces H_2O_2 (Lotharius *et al.*, 2002), and the catechol ring of dopamine is inherently unstable, easily oxidizing to dopamine-o-quinone and ROS (Perez and Hastings, 2004). Dopamine is synthesized in the cytosol, but it is normally rapidly pumped into synaptic vesicles by the vesicular monoamine transporter (VMAT2). Once sequestered, dopamine's potentially toxic breakdown is limited by low vesicular pH and the absence of monoamine oxidase. However, defects in the early secretory pathway caused by α-syn accumulation could result in a shortage of synaptic vesicles and a reduced level of VMAT2 at the synapse. This would impede dopamine sequestration and produce a rise in cytosolic dopamine. The inhibition of vesicular trafficking by α-syn would more severely affect dopamine-producing neurons, because neurotransmitters produced by other neurons are less likely to produce ROS and be toxic when unsequestered. Indeed, reduced VMAT expression in mice results in altered dopamine homeostasis and age-associated DA neuron dysfunction and degeneration, lending support to this hypothesis (Caudle *et al.*, 2007).

As discussed above, experiments in both mice and yeast suggest that α-syn inhibits the fusion of trafficking vesicles with their acceptor membranes. At higher concentrations, or when α-syn is misregulated, this reduction in trafficking would reduce the function of those dopamine-containing vesicles that are produced, causing a further deficit in dopamine signaling. Finally, the vesicle clustering that results from increased α-syn activity would create a high local concentration of α-syn, increasing the opportunity for the protein to rearrange into oligomeric species, which the work of other laboratories has indicated to be an important contribution to toxicity (Lansbury and Brice, 2002; Kayed *et al.*, 2003; Glabe, 2004; Lacor *et al.*, 2007). Thus, results from a number of laboratories, combined with data obtained from the yeast model, strongly suggest that α-syn's pathological function is related to its normal biology.

Schizosaccharomyces pombe: A Fission Yeast Model?

The fission yeast, *Schizosaccharomyces pombe*, is also used as a genetic model and in high-throughput screening (Forsburg, 2001). Like *S. cerevisiae*, *S. pombe* shares with humans a conservation of protein folding and protein QC pathways (Wood *et al.*, 2002; Brandis *et al.*, 2006), but the two yeasts are genetically as different from each other as either one is from animals (Sipiczki, 2000). The DebBurman group, who produced an *S. cerevisiae* model (Sharma *et al.*, 2006), also produced an *S. pombe* model overexpressing different GFP-labeled α-syn mutants (Brandis *et al.*, 2006). Similar to the behavior of α-syn in *S. cerevisiae*, WT and A53T α-syn formed prominent protein aggregates in *S. pombe*, whereas A30P did not. However, unlike in

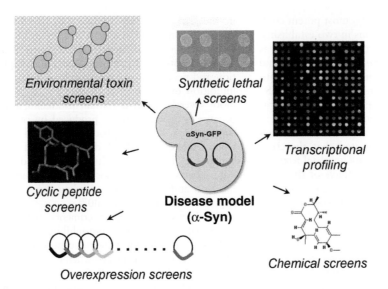

FIGURE 38.6 Diverse high-throughput screens made possible using yeast.

S. cerevisiae, WT and A53T α-syn did not target to the plasma membrane in this system, not even at low concentrations of α-syn or as a precursor to aggregation. Furthermore, even WT and A53T α-syn were not toxic to *S. pombe*, despite the extensive aggregation, even at high concentrations. The *S. pombe* model may thereby provide insight into mechanisms by which cells could be protected from α-syn toxicity.

YEAST MODELS OF HD

This book focuses on PD. Given that protein misfolding is such a deeply rooted problem in biology, it is not surprising that yeast cells can provide useful models for other protein misfolding diseases as well. Indeed it is useful to compare and contrast findings obtained with PD models with those obtained from other disease models. Due to limitations on space we will consider only yeast models of HD. HD and other diseases caused by polyQ expansions are also dauntingly complex. However, they provide a particularly interesting case for dissecting the relationships between protein misfolding and toxicity. The disease-causing proteins that contain polyQ expansions seem completely unrelated to each other

and their toxicities are first manifested in, and most severely affect, different specific cell types. Perplexingly, many of these proteins are highly expressed in cell types where they appear to produce no toxicity at all (Schilling *et al.*, 1995; Sharp *et al.*, 1995; Nishiyama *et al.*, 1996; Nishiyama *et al.*, 1997). Yet, for virtually all of these proteins disease occurs when the normal polyQ tract expands beyond 35 to 40 residues and the longer the polyQ expansion is, the earlier the disease onset and the more severe the phenotype. Further, when a polyQ expansion is placed in a protein that normally does not contain one, it is sufficient to cause neurodegenerative disease in mice (Ordway *et al.*, 1997; Tallaksen-Greene *et al.*, 2003). These observations suggest that common proteotoxic features overlie the specificities inherent in the different diseases. It is this reasoning that has led many investigators to produce models of polyQ expansion disease in diverse organisms and cell types (Sathasivam *et al.*, 1999; Zoghbi and Orr, 2000; Di Prospero and Fischbeck, 2005; Schiffer *et al.*, 2007).

We and others have developed yeast strains that express fragments of the human htt exon I, containing 16 amino acids at its amino terminus followed by polyQ tracts of different lengths (Krobitsch and Lindquist, 2000; Muchowski *et al.*, 2000;

Hughes and Olson, 2001; Hughes *et al.*, 2001; Meriin *et al.*, 2002; Duennwald *et al.*, 2006a; Duennwald *et al.*, 2006b). This exon I fragment is found in the htt aggregates that occur in the neurons of HD patients (Mende-Mueller *et al.*, 2001) and it is itself sufficient to produce neurodegeneration in mouse models (Mangiarini *et al.*, 1996). The critical length for polyQ toxicity in the yeast models (Duennwald *et al.*, 2006b) is similar to the threshold value for disease symptoms in humans, and the increasing toxicity with length in yeast is reminiscent of the earlier onset and increasing severity of polyQ diseases with increasing polyQ length in humans.

The baffling commonalities and differences between the polyQ diseases suggest that a variety of factors might influence the toxicity of proteins containing polyQ expansions: sequences flanking the Q tract, differences in the proteomes of the cells in which they are expressed, the age and metabolism of the cell, etc. These possibilities are easier to test in than in most other cell types because large numbers of genetically identical cells can be queried, with simple genetic manipulations and changes in growth condition. Indeed, we and others have found all of these factors, even small alterations in regions flanking the polyQ tract, can have a profound influence on the toxicity of htt exon I. For example, a small proline-rich region that is adjacent to the polyQ tract in the normal human htt protein is strongly protective, and removing it can transform a completely benign protein to a deadly one (Duennwald *et al.*, 2006b). Changes in the expression of just a single other protein in the cell can also shift a htt exon I from deadly to benign and *vice versa*. In some cases this appears to be due to direct interaction with the polyQ protein that can titrate out an essential protein (Schaffar *et al.*, 2004). In other cases, interaction with other nonessential Q-rich proteins can influence the conformational state of the htt fragment (Duennwald *et al.*, 2006a). Synthetic lethality screens have indicated that a large number of other cellular proteins can shift a mildly toxic htt exon I protein to a deadly species (Willingham *et al.*, 2003). And, not surprisingly, given that the toxicity of htt is related to protein misfolding, changes in the concentrations of chaperones and protein remodeling factors such as, TRiC, Hsp104, and the small heat-shock proteins can alleviate toxicity in yeast cells (reviewed in Barral *et al.*, 2004; Cashikar *et al.*, 2005; Duennwald *et al.*, 2006a; Tam *et al.*, 2006).

Remarkably, htt exon I toxicity also depends on another factor that would have been very difficult to establish in any other cell type: the conformational status of another glutamine-rich protein in the yeast cell, Rnq1. Rnq (named for being rich in asparagine and glutamine residues) can exist in two different self-perpetuating states, one soluble and the other a self-perpetuating amyloid that constitutes the prion factor known as [RNQ^+] (Sondheimer and Lindquist, 2000). The normal biological function of Rnq1 is unknown, but it influences the ability of other glutamine-rich proteins to adopt altered conformations (Derkatch *et al.*, 2001; Osherovich and Weissman, 2001). It is only when Rnq1 is in the prion state that the htt exon I fragment is toxic (Meriin *et al.*, 2002; Duennwald *et al.*, 2006a). Because Rnq is not conserved in mammals, this observation might seem to be a bizarre oddity of yeast cells, and perhaps it is. However, the human proteome contains a large number of glutamine- and asparagine-rich proteins that are likely to have a capacity to change their conformational status in a similar way (Michelitsch and Weissman, 2000; Si *et al.*, 2003; Shorter and Lindquist, 2005) and it therefore seem reasonable to suggest that changes in the conformation of such proteins in neurons might have a profound effect on toxicity.

In any case, experiments in several laboratories have begun to cross-validate results obtained in yeast cells with whole mice or neuronal cells in culture, and *vice versa*. To cite just a few examples: the chaperones Hsp70, Hsp40, and TRiC, and the protein remodeling factor, Hsp104, reduce the aggregation of htt exon I fragments *in vitro*, reduce aggregation and toxicity in yeast, and prolong survival in fly, mouse, and mammalian cell models of HD (Muchowski *et al.*, 2000; Vacher *et al.*, 2005; Marsh and Thompson, 2006; Tam *et al.*, 2006). A natural product present in green tea, initially found to inhibit the aggregation of pure polyQ proteins *in vitro*, EGCG [(-)-epigallocatechin gallate] alleviates htt toxicity both in our yeast HD model and in a fly model of HD (Ehrnhoefer *et al.*, 2006). These and other data suggest the potential of chemical screens with our yeast HD model to find compounds that are not toxic and that alleviate htt toxicity by mechanisms that also work in higher organisms.

Specificity of the Models

The huntingtin protein and α-syn appear to produce their toxic effects when they misfold. However, in keeping with the results of many biological assays, in which the two proteins had very different biological effects, all of the genetic screens that have been performed identified disparate sets of genes for the two models. Moreover, chemical compounds that alleviate toxicity of htt do not relieve the toxicity of α-syn, and vice versa. This establishes that the toxicities are not due to non-specific effects of misfolded protein, but to specific features of their misfunctioning that interface with cell biology in diverse ways. It seems likely that differently therapeutic strategies will likely be required to combat these diseases.

CONCLUSIONS

We present yeast cells as a discovery platform to find genetic factors and chemical compounds that modify the toxicity of proteins that are prone to misfoldings and producing toxic gain-of-function phenotypes in man. The introduction of human proteins into yeast can be used to model important cellular elements of at least two neurodegenerative diseases and this simple approach will likely be useful for other neurodegenerative diseases as well. The yeast models do not simply show non-specific toxicity due to the accumulation of misfolded proteins. On the contrary, the toxicity in the different models is highly specific and distinct. Because, the diseases modeled in this way affect specific types of neurons, it was, until recently, a commonly held belief that they could only be fruitfully studied in animal models and cell cultures of those particular neurons. However, in at least some cases the specificities of the diseases may simply be the result of particular neurons simply being more vulnerable to general cellular defects.

An additional advantage of these models in drug screening is that they have a capacity to focus compound searches on the initiating events of the diseases, particularly those involved with the misfolding and aggregation of the toxic protein. Compounds that prevent the pathological process from ever beginning would clearly be the most useful. In addition, yeast cells provide a whole collection of technical advantages for high-throughput screening – speed, robustness, low cost, genetic manipulability – making for a compelling combination. Of course, there are entire realms of neuronal pathology that will never be modeled in yeast.

But at those levels where pathology can be modeled, it is our hope that yeast cells will provide valuable assistance to the heroic efforts that are being made in more complex systems to find cures for these devastating diseases.

ACKNOWLEDGEMENTS

We thank Andrew Steele, Jessica Brown, Drs. Brooke Bevis, Melissa Geddie, Pavan Auluck, Martin Duennwald, and other members of the Lindquist lab for their thoughtful comments and suggestions. This work was funded in part by Udall program grant NS038372 from the NIH, a Target Validation grant from the Michael J. Fox Foundation for Parkinson's Research, and support from the Whitehead Institute Regenerative Biology Initiative to Susan Lindquist; and a postdoctoral fellowship from the American Cancer Society to Linhui Julie Su. Susan Lindquist is an Investigator of the Howard Hughes Medical Institute and a Fellow of the Radcliffe Institute for Advanced Study at Harvard University.

REFERENCES

Abeliovich, A., Schmitz, Y., Farinas, I., Choi-Lundberg, D., Ho, W. H., Castillo, P. E., Shinsky, N., Verdugo, J. M., Armanini, M., Ryan, A., Hynes, M., Phillips, H., Sulzer, D., and Rosenthal, A. (2000). Mice lacking α-synuclein display functional deficits in the nigrostriatal dopamine system. *Neuron* **25**, 239–252.

Abou-Sleiman, P. M., Muqit, M. M., and Wood, N. W. (2006). Expanding insights of mitochondrial dysfunction in Parkinson's disease. *Nat Rev Neurosci* **7**, 207–219.

Ahlskog, J. E. (2007). Beating a dead horse: Dopamine and Parkinson disease. *Neurology* **69**, 1701–1711.

Ahn, B. H., Rhim, H., Kim, S. Y., Sung, Y. M., Lee, M. Y., Choi, J. Y., Wolozin, B., Chang, J. S., Lee, Y. H., Kwon, T. K., Chung, K. C., Yoon, S. H., Hahn, S. J., Kim, M. S., Jo, Y. H., and Min do, S. (2002). Alpha-synuclein interacts with phospholipase D isozymes and inhibits pervanadate-induced phospholipase D activation in human embryonic kidney-293 cells. *J Biol Chem* **277**, 12334–12342.

Ameen, N., Silvis, M., and Bradbury, N. A. (2007). Endocytic trafficking of CFTR in health and disease. *J Cyst Fibros* **6**, 1–14.

Anfinsen, C. B. (1973). Principles that govern the folding of protein chains. *Science* **181**, 223–230.

Ashrafi, K., Sinclair, D., Gordon, J. I., and Guarente, L. (1999). Passage through stationary phase advances replicative aging in *Saccharomyces cerevisiae*. *Proc Natl Acad Sci USA* **96**, 9100–9105.

Auluck, P. K., Chan, H. Y. E., Trojanowski, J. Q., Lee, V. M. Y., and Bonini, N. M. (2002). Chaperone suppression of alpha-synuclein toxicity in a Drosophila model for Parkinson's disease. *Science* **295**, 865–868.

Auluck, P. K., Meulener, M. C., and Bonini, N. M. (2005). Mechanisms of suppression of alpha-synuclein neurotoxicity by geldanamycin in Drosophila. *J Biol Chem* **280**, 2873–2878.

Ballester, R., Marchuk, D., Boguski, M., Saulino, A., Letcher, R., Wigler, M., and Collins, F. (1990). The NF1 locus encodes a protein functionally related to mammalian GAP and yeast IRA proteins. *Cell* **63**, 851–859.

Barral, J. M., Broadley, S. A., Schaffar, G., and Hartl, F. U. (2004). Roles of molecular chaperones in protein misfolding diseases. *Semin Cell Dev Biol* **15**, 17–29.

Bialecka, M., Hui, S., Klodowska-Duda, G., Opala, G., Tan, E. K., and Drozdzik, M. (2005). Analysis of LRRK 2 G 2019 S and I 2020 T mutations in Parkinson's disease. *Neurosci Lett* **390**, 1–3.

Bolen, D. W., and Baskakov, I. V. (2001). The osmophobic effect: Natural selection of a thermodynamic force in protein folding. *J Mol Biol* **310**, 955–963.

Bonifacino, J. S., and Lippincott-Schwartz, J. (2003). Coat proteins: Shaping membrane transport. *Nat Rev Mol Cell Biol* **4**, 409–414.

Bonifati, V., Rizzu, P., van Baren, M. J., Schaap, O., Breedveld, G. J., Krieger, E., Dekker, M. C., Squitieri, F., Ibanez, P., Joosse, M., van Dongen, J. W., Vanacore, N., van Swieten, J. C., Brice, A., Meco, G., van Duijn, C. M., Oostra, B. A., and Heutink, P. (2003). Mutations in the *DJ-1* gene associated with autosomal recessive early-onset parkinsonism. *Science* **299**, 256–259.

Boone, C., Bussey, H., and Andrews, B. J. (2007). Exploring genetic interactions and networks with yeast. *Nat Rev Genet* **8**, 437–449.

Brandis, K. A., Holmes, I. F., England, S. J., Sharma, N., Kukreja, L., and DebBurman, S. K. (2006). Alpha-synuclein fission yeast model: Concentration-dependent aggregation without plasma membrane localization or toxicity. *J Mol Neurosci* **28**, 179–191.

Brown, I. R. (2007). Heat shock proteins and protection of the nervous system. *Ann NY Acad Sci.* **1113**, 147–158.

Cabin, D. E., Shimazu, K., Murphy, D., Cole, N. B., Gottschalk, W., McIlwain, K. L., Orrison, B., Chen, A., Ellis, C. E., Paylor, R., Lu, B., and Nussbaum, R. L. (2002). Synaptic vesicle depletion correlates with attenuated synaptic responses to prolonged repetitive stimulation in mice lacking α-synuclein. *J Neurosci* **22**, 8797–8807.

Cashikar, A. G., Duennwald, M., and Lindquist, S. L. (2005). A chaperone pathway in protein disaggregation, Hsp26 alters the nature of protein aggregates to facilitate reactivation by Hsp104. *J Biol Chem* **280**, 23869–23875.

Caudle, W. M., Richardson, J. R., Wang, M. Z., Taylor, T. N., Guillot, T. S., McCormack, A. L., Colebrooke, R. E., Di Monte, D. A., Emson, P. C., and Miller, G. W. (2007). Reduced vesicular storage of dopamine causes progressive nigrostriatal neurodegeneration. *J Neurosci* **27**, 8138–8148.

Causier, B. (2004). Studying the interactome with the yeast two-hybrid system and mass spectrometry. *Mass Spectrom Rev* **23**, 350–367.

Chandra, S., Gallardo, G., Fernandez-Chacon, R., Schluter, O. M., and Sudhof, T. C. (2005). Alpha-synuclein cooperates with CSPalpha in preventing neurodegeneration. *Cell* **123**, 383–396.

Chartier-Harlin, M. C., Kachergus, J., Roumier, C., Mouroux, V., Douay, X., Lincoln, S., Levecque, C., Larvor, L., Andrieux, J., Hulihan, M., Waucquier, N., Defebvre, L., Amouyel, P., Farrer, M., and Destee, A. (2004). Alpha-synuclein locus duplication as a cause of familial Parkinson's disease. *Lancet* **364**, 1167–1169.

Chen, Q., Thorpe, J., and Keller, J. N. (2005). Alpha-synuclein alters proteasome function, protein synthesis, and stationary phase viability. *J Biol Chem* **280**, 30009–30017.

Chung, K. K., Zhang, Y., Lim, K. L., Tanaka, Y., Huang, H., Gao, J., Ross, C. A., Dawson, V. L., and Dawson, T. M. (2001). Parkin ubiquitinates the alpha-synuclein-interacting protein, synphilin-1: implications for Lewy-body formation in Parkinson disease. *Nat Med* **7**, 1144–1150.

Ciechanover, A. (1994). The ubiquitin–proteasome proteolytic pathway. *Cell* **79**, 13–21.

Cole, N. B., Murphy, D. D., Grider, T., Rueter, S., Brasaemle, D., and Nussbaum, R. L. (2002). Lipid droplet binding and oligomerization properties of the Parkinson's disease protein alpha-synuclein. *J Biol Chem* **277**, 6344–6352.

Collins, S. R., Kemmeren, P., Zhao, X. C., Greenblatt, J. F., Spencer, F., Holstege, F. C., Weissman, J. S., and Krogan, N. J. (2007a). Toward a comprehensive atlas of the physical interactome of *Saccharomyces cerevisiae*. *Mol Cell Proteomics* **6**, 439–450.

Collins, S. R., Miller, K. M., Maas, N. L., Roguev, A., Fillingham, J., Chu, C. S., Schuldiner, M., Gebbia, M., Recht, J., Shales, M., Ding, H., Xu, H., Han, J., Ingvarsdottir, K., Cheng, B., Andrews, B., Boone, C., Berger, S. L., Hieter, P., Zhang, Z., Brown, G. W., Ingles, C. J., Emili, A., Allis, C. D., Toczyski, D. P., Weissman, J. S., Greenblatt, J. F., and Krogan, N. J. (2007b). Functional dissection of protein complexes involved in yeast chromosome biology using a genetic interaction map. *Nature* **446**, 806–810.

Cooper, A. A., Gitler, A. D., Cashikar, A., Haynes, C. M., Hill, K. J., Bhullar, B., Liu, K., Xu, K., Strathearn, K. E., Liu, F., Cao, S., Caldwell, K. A., Caldwell, G. A., Marsischky, G., Kolodner, R. D., Labaer, J., Rochet, J. C., Bonini, N. M., and Lindquist, S. (2006). Alpha-synuclein blocks ER-Golgi traffic and Rab1 rescues neuron loss in Parkinson's models. *Science* **313**, 324–328.

Cuervo, A. M., Stefanis, L., Fredenburg, R., Lansbury, P. T., and Sulzer, D. (2004). Impaired degradation of mutant alpha-synuclein by chaperone-mediated autophagy. *Science* **305**, 1292–1295.

Dauer, W., and Przedborski, S. (2003). Parkinson's disease: Mechanisms and models. *Neuron* **39**, 889–909.

Davies, J. E., Sarkar, S., and Rubinsztein, D. C. (2006). Trehalose reduces aggregate formation and delays pathology in a transgenic mouse model of oculopharyngeal muscular dystrophy. *Hum Mol Genet* **15**, 23–31.

Dawson, T. M., and Dawson, V. L. (2003). Molecular pathways of neurodegeneration in Parkinson's disease. *Science* **302**, 819–822.

den Jager, W. A. (1969). Sphingomyelin in Lewy inclusion bodies in Parkinson's disease. *Arch Neurol* **21**, 615.

Derkatch, I. L., Bradley, M. E., Hong, J. Y., and Liebman, S. W. (2001). Prions affect the appearance of other prions: The story of [PIN(+)]. *Cell* **106**, 171–182.

Devos, D., Dokudovskaya, S., Alber, F., Williams, R., Chait, B. T., Sali, A., and Rout, M. P. (2004). Components of coated vesicles and nuclear pore complexes share a common molecular architecture. *PLoS Biol* **2**, e380.

Di Prospero, N. A., and Fischbeck, K. H. (2005). Therapeutics development for triplet repeat expansion diseases. *Nat Rev Genet* **6**, 756–765.

Dixon, C., Mathias, N., Zweig, R. M., Davis, D. A., and Gross, D. S. (2005). Alpha-synuclein targets the plasma membrane via the secretory pathway and induces toxicity in yeast. *Genetics* **170**, 47–59.

Double, K. L., Gerlach, M., Youdim, M. B., and Riederer, P. (2000). Impaired iron homeostasis in Parkinson's disease. *J Neural Transm Suppl*, 37–58.

Duennwald, M. L., Jagadish, S., Giorgini, F., Muchowski, P. J., and Lindquist, S. (2006a). A network of protein interactions determines polyglutamine toxicity. *Proc Natl Acad Sci USA* **103**, 11051–11056.

Duennwald, M. L., Jagadish, S., Muchowski, P. J., and Lindquist, S. (2006b). Flanking sequences profoundly alter polyglutamine toxicity in yeast. *Proc Natl Acad Sci USA* **103**, 11045–11050.

Durr, G., Strayle, J., Plemper, R., Elbs, S., Klee, S. K., Catty, P., Wolf, D. H., and Rudolph, H. K. (1998). The medial-Golgi ion pump Pmr1 supplies the yeast secretory pathway with Ca^{2+} and Mn^{2+} required for glycosylation, sorting, and endoplasmic reticulum-associated protein degradation. *Mol Biol Cell* **9**, 1149–1162.

Ehrnhoefer, D. E., Duennwald, M., Markovic, P., Wacker, J. L., Engemann, S., Roark, M., Legleiter, J., Marsh, J. L., Thompson, L. M., Lindquist, S., Muchowski, P. J., and Wanker, E. E. (2006). Green tea (-)-epigallocatechin-gallate modulates early events in huntingtin misfolding and reduces toxicity in Huntington's disease models. *Hum Mol Genet* **15**, 2743–2751.

Elbein, A. D., Pan, Y. T., Pastuszak, I., and Carroll, D. (2003). New insights on trehalose: A multifunctional molecule. *Glycobiology* **13**, 17R–27R.

Ellison, D., Love, S., Chimelli, L., Harding, B., Lowe, J., and Vinters, H. (2003). *Neuropathology: A Reference Text of CNS Pathology*. Mosby. Philadelphia PA.

Engelender, S., Kaminsky, Z., Guo, X., Sharp, A. H., Amaravi, R. K., Kleiderlein, J. J., Margolis, R. L., Troncoso, J. C., Lanahan, A. A., Worley, P. F., Dawson, V. L., Dawson, T. M., and Ross, C. A. (1999). Synphilin-1 associates with alpha-synuclein and promotes the formation of cytosolic inclusions. *Nat Genet* **22**, 110–114.

Fearnley, J. M., and Lees, A. J. (1991). Ageing and Parkinson's disease: Substantia nigra regional selectivity. *Brain* **114**, 2283–2301.

Flower, T. R., Chesnokova, L. S., Froelich, C. A., Dixon, C., and Witt, S. N. (2005). Heat shock prevents alpha-synuclein-induced apoptosis in a yeast model of Parkinson's disease. *J Mol Biol* **351**, 1081–1100.

Flower, T. R., Clark-Dixon, C., Metoyer, C., Yang, H., Shi, R., Zhang, Z., and Witt, S. N. (2007). YGR198w (YPP1) targets A30P alpha-synuclein to the vacuole for degradation. *J Cell Biol* **177**, 1091–1104.

Forsburg, S. L. (2001). The art and design of genetic screens: Yeast. *Nat Rev Genet* **2**, 659–669.

Gai, W. P., Yuan, H. X., Li, X. Q., Power, J. T., Blumbergs, P. C., and Jensen, P. H. (2000). *In situ* and *in vitro* study of colocalization and segregation of alpha-synuclein, ubiquitin, and lipids in Lewy bodies. *Exp Neurol* **166**, 324–333.

Gasch, A. P., Spellman, P. T., Kao, C. M., Carmel-Harel, O., Eisen, M. B., Storz, G., Botstein, D., and Brown, P. O. (2000). Genomic expression programs in the response of yeast cells to environmental changes. *Mol Biol Cell* **11**, 4241–4257.

Gelperin, D. M., White, M. A., Wilkinson, M. L., Kon, Y., Kung, L. A., Wise, K. J., Lopez-Hoyo, N., Jiang, L., Piccirillo, S., Yu, H., Gerstein, M., Dumont, M. E., Phizicky, E. M., Snyder, M., and Grayhack, E. J. (2005). Biochemical and genetic analysis of the yeast proteome with a movable ORF collection. *Genes Dev* **19**, 2816–2826.

Geppert, M., Bolshakov, V. Y., Siegelbaum, S. A., Takei, K., De Camilli, P., Hammer, R. E., and Sudhof, T. C. (1994). The role of Rab3A in neurotransmitter release. *Nature* **369**, 493–497.

Ghaemmaghami, S., Huh, W. K., Bower, K., Howson, R. W., Belle, A., Dephoure, N., O'Shea, E. K., and Weissman, J. S. (2003). Global analysis of protein expression in yeast. *Nature* **425**, 737–741.

Giaever, G., Chu, A. M., Ni, L., Connelly, C., Riles, L., Veronneau, S., Dow, S., Lucau-Danila, A., Anderson, K., Andre, B., Arkin, A. P., Astromoff, A., El-Bakkoury, M., Bangham, R., Benito, R., Brachat, S., Campanaro, S., Curtiss, M., Davis, K., Deutschbauer, A., Entian, K. D., Flaherty, P., Foury, F., Garfinkel, D. J., Gerstein, M., Gotte, D., Guldener, U., Hegemann, J. H., Hempel, S., Herman, Z., Jaramillo, D. F., Kelly, D. E., Kelly, S. L., Kotter, P., LaBonte, D., Lamb, D. C., Lan, N., Liang, H., Liao, H., Liu, L., Luo, C., Lussier, M., Mao, R., Menard, P., Ooi, S. L., Revuelta, J. L., Roberts, C. J., Rose, M., Ross-Macdonald, P., Scherens, B., Schimmack, G., Shafer, B., Shoemaker, D. D., Sookhai-Mahadeo, S., Storms, R. K., Strathern, J. N., Valle, G., Voet, M., Volckaert, G., Wang, C. Y., Ward, T. R., Wilhelmy, J., Winzeler, E. A., Yang, Y., Yen, G., Youngman, E., Yu, K., Bussey, H., Boeke, J. D., Snyder, M., Philippsen, P., Davis, R. W., and Johnston, M. (2002). Functional profiling of the *Saccharomyces cerevisiae* genome. *Nature* **418**, 387–391.

Giasson, B. I., Duda, J. E., Murray, I. V., Chen, Q., Souza, J. M., Hurtig, H. I., Ischiropoulos, H., Trojanowski, J. Q., and Lee, V. M. (2000). Oxidative damage linked to neurodegeneration by selective α-synuclein nitration in synucleinopathy lesions. *Science* **290**, 985–989.

Gitler, A., Bevis, B., Shorter, J., Strathearn, K., Hamamichi, S., Su, L., Caldwell, K., Caldwell, G., Rochet, J., McCaffery, J., Barlowe, C., and Lindquist, S. (2008). The Parkinson's disease protein alpha-synuclein disrupts cellular Rab homeostasis. *Proc Natl Acad Sci USA* **105**, 145–150.

Glabe, C. G. (2004). Conformation-dependent antibodies target diseases of protein misfolding. *Trends Biochem Sci* **29**, 542–547.

Goedert, M. (2001). Alpha-synuclein and neurodegenerative diseases. *Nat Rev Neurosci* **2**, 492–501.

Goffeau, A., Barrell, B. G., Bussey, H., Davis, R. W., Dujon, B., Feldmann, H., Galibert, F., Hoheisel, J. D., Jacq, C., Johnston, M., Louis, E. J., Mewes, H. W., Murakami, Y., Philippsen, P., Tettelin, H., and Oliver, S. G. (1996). Life with 6000 genes. *Science* **274**, 546, 563–546.

Goldberg, A. L. (2003). Protein degradation and protection against misfolded or damaged proteins. *Nature* **426**, 895–899.

Greenwell, P. W., Kronmal, S. L., Porter, S. E., Gassenhuber, J., Obermaier, B., and Petes, T. D. (1995). TEL1, a gene involved in controlling telomere length in *S. cerevisiae*, is homologous to the human ataxia telangiectasia gene. *Cell* **82**, 823–829.

Griffiths, A., Wessler, S., Lewontin, R., and Carroll, S. (2007). *Introduction to Genetic Analysis*. W.H. Freeman. Gordonsville VA.

Guo, L., Gandhi, P. N., Wang, W., Petersen, R. B., Wilson-Delfosse, A. L., and Chen, S. G. (2007). The Parkinson's disease-associated protein, leucine-rich repeat kinase 2 (LRRK2), is an authentic GTPase that stimulates kinase activity. *Exp Cell Res* **313**, 3658–3670.

Gurkan, C., Lapp, H., Alory, C., Su, A. I., Hogenesch, J. B., and Balch, W. E. (2005). Large-scale profiling of Rab GTPase trafficking networks: The membrome. *Mol Biol Cell* **16**, 3847–3864.

Guthrie, C., and Fink, G. R. (1991). *Guide to Yeast Genetics and Molecular Biology*. Academic Press, San Diego, CA.

Hara, M. R., Agrawal, N., Kim, S. F., Cascio, M. B., Fujimuro, M., Ozeki, Y., Takahashi, M., Cheah, J. H., Tankou, S. K., Hester, L. D., Ferris, C. D., Hayward, S. D., Snyder, S. H., and Sawa, A. (2005). S-nitrosylated GAPDH initiates apoptotic cell death by nuclear translocation following siahl binding. *Nat. Cell. Biol* **7**, 665–674.

Hardy, J., Cai, H., Cookson, M. R., Gwinn-Hardy, K., and Singleton, A. (2006). Genetics of Parkinson's disease and parkinsonism. *Ann Neurol* **60**, 389–398.

Hartwell, L. H., and Weinert, T. A. (1989). Checkpoints: Controls that ensure the order of cell cycle events. *Science* **246**, 629–634.

Hartwell, L. H., Culotti, J., Pringle, J. R., and Reid, B. J. (1974). Genetic control of the cell division cycle in yeast. *Science* **183**, 46–51.

Hausse, A. O., Aggoun, Y., Bonnet, D., Sidi, D., Munnich, A., Rotig, A., and Rustin, P. (2002). Idebenone and reduced cardiac hypertrophy in Friedreich's ataxia. *Heart* **87**, 346–349.

Hayes, S. A., and Dice, J. F. (1996). Roles of molecular chaperones in protein degradation. *J Cell Biol* **132**, 255–258.

Hernandez, D., Paisan Ruiz, C., Crawley, A., Malkani, R., Werner, J., Gwinn-Hardy, K., Dickson, D., Wavrant Devrieze, F., Hardy, J., and Singleton, A. (2005). The dardarin G 2019 S mutation is a common cause of Parkinson's disease but not other neurodegenerative diseases. *Neurosci Lett* **389**, 137–139.

Holzmann, C., Krüger, R., Saecker, A. M., Schmitt, I., Schöls, L., Berger, K., and Riess, O. (2003). Polymorphisms of the alpha-synuclein promoter: Expression analyses and association studies in Parkinson's disease. *J Neural Transm* **110**, 67–76.

Huber, L. A., Pimplikar, S., Parton, R. G., Virta, H., Zerial, M., and Simons, K. (1993). Rab8, a small GTPase involved in vesicular traffic between the TGN and the basolateral plasma membrane. *J Cell Biol* **123**, 35–45.

Hughes, R. E., and Olson, J. M. (2001). Therapeutic opportunities in polyglutamine disease. *Nat Med* 7, 419–423.

Hughes, R. E., Lo, R. S., Davis, C., Strand, A. D., Neal, C. L., Olson, J. M., and Fields, S. (2001). Altered transcription in yeast expressing expanded polyglutamine. *Proc Natl Acad Sci USA* 98, 13201–13206.

Huh, W. K., Falvo, J. V., Gerke, L. C., Carroll, A. S., Howson, R. W., Weissman, J. S., and O'Shea, E. K. (2003). Global analysis of protein localization in budding yeast. *Nature* 425, 686–691.

Irizarry, R. A., Warren, D., Spencer, F., Kim, I. F., Biswal, S., Frank, B. C., Gabrielson, E., Garcia, J. G., Geoghegan, J., Germino, G., Griffin, C., Hilmer, S. C., Hoffman, E., Jedlicka, A. E., Kawasaki, E., Martinez-Murillo, F., Morsberger, L., Lee, H., Petersen, D., Quackenbush, J., Scott, A., Wilson, M., Yang, Y., Ye, S. Q., and Yu, W. (2005). Multiple-laboratory comparison of microarray platforms. *Nat Methods* 2, 345–350.

Iwatsubo, T., Yamaguchi, H., Fujimuro, M., Yokosawa, H., Ihara, Y., Trojanowski, J. Q., and Lee, V. M. (1996). Purification and characterization of Lewy bodies from the brains of patients with diffuse Lewy body disease. *Am J Pathol* 148, 1517–1529.

Izumi, Y., Morino, H., Oda, M., Maruyama, H., Udaka, F., Kameyama, M., Nakamura, S., and Kawakami, H. (2001). Genetic studies in Parkinson's disease with an alpha-synuclein/NACP gene polymorphism in Japan. *Neurosci Lett* 300, 125–127.

Jackson-Lewis, V., and Smeyne, R. J. (2005). MPTP and SNpc DA neuronal vulnerability: role of dopamine, superoxide and nitric oxide in neurotoxicity. Minireview. *Neurotox Res* 7, 193–202.

Jahn, R., and Scheller, R. H. (2006). SNAREs – engines for membrane fusion. *Nat Rev Mol Cell Biol* 7, 631–643.

Jellinger, K. A. (2003). Neuropathological spectrum of synucleinopathies. *Mov Disord* 18(Suppl 6), S2–S12.

Jenco, J. M., Rawlinson, A., Daniels, B., and Morris, A. J. (1998). Regulation of phospholipase D2: Selective inhibition of mammalian phospholipase D isoenzymes by alpha- and beta-synucleins. *Biochemistry* 37, 4901–4909.

Jenner, P., and Olanow, C. W. (1996). Oxidative stress and the pathogenesis of Parkinson's disease. *Neurology* 47, S161–S170.

Jiang, Y. H., and Beaudet, A. L. (2004). Human disorders of ubiquitination and proteasomal degradation. *Curr Opin Pediatr* 16, 419–426.

Jo, E., McLaurin, J., Yip, C. M., St George-Hyslop, P., and Fraser, P. E. (2000). Alpha-synuclein membrane interactions and lipid specificity. *J Biol Chem* 275, 34328–34334.

Johnston, J. R. (1994). *Molecular genetics of yeast – a practical approach*. Oxford University Press, Oxford.

Kang, J. H., and Kim, K. S. (2003). Enhanced oligomerization of the alpha-synuclein mutant by the Cu, Zn-superoxide dismutase and hydrogen peroxide system. *Mol Cells* 15, 87–93.

Kawamoto, Y., Akiguchi, I., Shirakashi, Y., Honjo, Y., Tomimoto, H., Takahashi, R., and Budka, H. (2007). Accumulation of Hsc70 and Hsp70 in glial cytoplasmic inclusions in patients with multiple system atrophy. *Brain Res* 1136, 219–227.

Kayed, R., Head, E., Thompson, J. L., McIntire, T. M., Milton, S. C., Cotman, C. W., and Glabe, C. G. (2003). Common structure of soluble amyloid oligomers implies common mechanism of pathogenesis. *Science* 300, 486–489.

Kempson, S. A., and Montrose, M. H. (2004). Osmotic regulation of renal betaine transport: Transcription and beyond. *Pflugers Arch* 449, 227–234.

Khan, N. L., Jain, S., Lynch, J. M., Pavese, N., Abou-Sleiman, P., Holton, J. L., Healy, D. G., Gilks, W. P., Sweeney, M. G., Ganguly, M., Gibbons, V., Gandhi, S., Vaughan, J., Eunson, L. H., Katzenschlager, R., Gayton, J., Lennox, G., Revesz, T., Nicholl, D., Bhatia, K. P., Quinn, N., Brooks, D., Lees, A. J., Davis, M. B., Piccini, P., Singleton, A. B., and Wood, N. W. (2005). Mutations in the gene LRRK2 encoding dardarin (PARK8) cause familial Parkinson's disease: Clinical, pathological, olfactory and functional imaging and genetic data. *Brain* 128, 2786–2796.

Kiaei, M., Kipiani, K., Petri, S., Chen, J., Calingasan, N. Y., and Beal, M. F. (2005). Celastrol blocks neuronal cell death and extends life in transgenic mouse model of amyotrophic lateral sclerosis. *Neurodegener Dis* 2, 246–254.

Kim, H. J., Lee, D., Lee, C. H., Chung, K. C., Kim, J., and Paik, S. R. (2006). Calpain-resistant fragment(s) of alpha-synuclein regulates the synuclein-cleaving activity of 20S proteasome. *Arch Biochem Biophys* 455, 40–47.

Kins, S., Lauther, N., Szodorai, A., and Beyreuther, K. (2006). Subcellular trafficking of the amyloid precursor protein gene family and its pathogenic role in Alzheimer's disease. *Neurodegener Dis* 3, 218–226.

Kirschhausen, T. (2000). Three ways to make a vesicle. *Nat Rev Mol Cell Biol* 1, 187–198.

Kitada, T., Asakawa, S., Hattori, N., Matsumine, H., Yamamura, Y., Minoshima, S., Yokochi, M., Mizuno, Y., and Shimizu, N. (1998). Mutations in the parkin gene cause autosomal recessive juvenile parkinsonism. *Nature* 392, 605–608.

Koenig, M., and Mandel, J. L. (1997). Deciphering the cause of Friedreich ataxia. *Curr Opin Neurobiol* 7, 689–694.

Kolkman, A., Slijper, M., and Heck, A. J. (2005). Development and application of proteomics technologies in *Saccharomyces cerevisiae*. *Trends Biotechnol* 23, 598–604.

Kopito, R. R. (2000). Aggresomes, inclusion bodies and protein aggregation. *Trends Cell Biol* 10, 524–530.

Kotzbauer, P. T., Trojanowsk, J. Q., and Lee, V. M. (2001). Lewy body pathology in Alzheimer's disease. *J Mol Neurosci* 17, 225–232.

Krobitsch, S., and Lindquist, S. (2000). Aggregation of huntingtin in yeast varies with the length of the polyglutamine expansion and the expression of chaperone proteins. *Proc Natl Acad Sci USA* 97, 1589–1594.

Kruger, R., Kuhn, W., Muller, T., Woitalla, D., Graeber, M., Kosel, S., Przuntek, H., Epplen, J. T., Schols, L., and Riess, O. (1998). Ala30Pro mutation in the gene encoding alpha-synuclein in Parkinson's disease. *Nat Genet* 18, 106–108.

Lacor, P. N., Buniel, M. C., Furlow, P. W., Clemente, A. S., Velasco, P. T., Wood, M., Viola, K. L., and Klein, W. L. (2007). Abeta oligomer-induced aberrations in synapse composition, shape, and density provide a molecular basis for loss of connectivity in Alzheimer's disease. *J Neurosci* 27, 796–807.

LaFerla, F. M., Green, K. N., and Oddo, S. (2007). Intracellular amyloid-beta in Alzheimer's disease. *Nat Rev Neurosci* 8, 499–509.

Lambert, I. H. (2004). Regulation of the cellular content of the organic osmolyte taurine in mammalian cells. *Neurochem Res* 29, 27–63.

Lansbury, P. T., and Brice, A. (2002). Genetics of Parkinson's disease and biochemical studies of implicated gene products. *Curr Opin Cell Biol* 14, 653–660.

Lapinskas, P. J., Lin, S. J., and Culotta, V. C. (1996). The role of the *Saccharomyces cerevisiae* CCC1 gene in the homeostasis of manganese ions. *Mol Microbiol* 21, 519–528.

Larsen, K. E., Schmitz, Y., Troyer, M. D., Mosharov, E., Dietrich, P., Quazi, A. Z., Savalle, M., Nemani, V., Chaudhry, F. A., Edwards, R. H., Stefanis, L., and Sulzer, D. (2006). Alpha-synuclein overexpression in PC12 and chromaffin cells impairs catecholamine release by interfering with a late step in exocytosis. *J Neurosci* 26, 11915–11922.

Leenders, A. G., Lopes da Silva, F. H., Ghijsen, W. E., and Verhage, M. (2001). Rab3a is involved in transport of synaptic vesicles to the active zone in mouse brain nerve terminals. *Mol Biol Cell* 12, 3095–3102.

Leroy, E., Boyer, R., Auburger, G., Leube, B., Ulm, G., Mezey, E., Harta, G., Brownstein, M. J., Jonnalagada, S., Chernova, T., Dehejia, A., Lavedan, C., Gasser, T., Steinbach, P. J., Wilkinson, K. D., and Polymeropoulos, M. H. (1998). The ubiquitin pathway in Parkinson's disease. *Nature* 395, 451–452.

Lim, K. L., Chew, K. C., Tan, J. M., Wang, C., Chung, K. K., Zhang, Y., Tanaka, Y., Smith, W., Engelender, S., Ross, C. A., Dawson, V. L., and Dawson, T. M. (2005). Parkin mediates nonclassical, proteasomal-independent ubiquitination of synphilin-1: Implications for Lewy body formation. *J Neurosci* 25, 2002–2009.

Liu, C. W., Giasson, B. I., Lewis, K. A., Lee, V. M., Demartino, G. N., and Thomas, P. J. (2005). A precipitating role for truncated alpha-synuclein and the proteasome in alpha-synuclein aggregation: Implications for pathogenesis of Parkinson disease. *J Biol Chem* 280, 22670–22678.

Lotharius, J., Barg, S., Wiekop, P., Lundberg, C., Raymon, H. K., and Brundin, P. (2002). Effect of mutant alpha-synuclein on dopamine homeostasis in a new human mesencephalic cell line. *J Biol Chem* 277, 38884–38894.

Luzon-Toro, B., de la Torre, E. R., Delgado, A., Perez-Tur, J., and Hilfiker, S. (2007). Mechanistic insight into the dominant mode of the Parkinson's disease-associated G2019S LRRK2 mutation. *Hum Mol Genet* 16, 2031–2039.

Madhani, H. D. (2007). *From a to Alpha: Yeast as a Model for Cellular Differentiation.* Cold Spring Harbor Laboratory Press, Woodbury, NY.

Mangiarini, L., Sathasivam, K., Seller, M., Cozens, B., Harper, A., Hetherington, C., Lawton, M., Trottier, Y., Lehrach, H., Davies, S. W., and Bates, G. P. (1996). Exon 1 of the HD gene with an expanded CAG repeat is sufficient to cause a progressive neurological phenotype in transgenic mice. *Cell* 87, 493–506.

Marsh, J. L., and Thompson, L. M. (2006). Drosophila in the study of neurodegenerative disease. *Neuron* 52, 169–178.

Martinat, C., Shendelman, S., Jonason, A., Leete, T., Beal, M. F., Yang, L., Floss, T., and Abeliovich, A. (2004). Sensitivity to oxidative stress in DJ-1-deficient dopamine neurons: An ES-derived cell model of primary parkinsonism. *PLoS Biol* 2, e327.

Martinez, Z., Zhu, M., Han, S., and Fink, A. L. (2007). GM1 specifically interacts with alpha-synuclein and inhibits fibrillation. *Biochemistry* 46, 1868–1877.

McLean, P. J., Kawamata, H., and Hyman, B. T. (2001). Alpha-synuclein-enhanced green fluorescent protein fusion proteins form proteasome sensitive inclusions in primary neurons. *Neuroscience* 104, 901–912.

Mell, J. C., and Burgess, S. M. (2003). Yeast as a model genetic organism. *Encyclopedia of Life Sciences*, http://www.els.net/. John Wiley & Sons, Ltd., Chichester, UK.

Mende-Mueller, L. M., Toneff, T., Hwang, S. R., Chesselet, M. F., and Hook, V. Y. (2001). Tissue-specific proteolysis of Huntingtin (htt) in human brain: Evidence of enhanced levels of N- and C-terminal htt fragments in Huntington's disease striatum. *J Neurosci* 21, 1830–1837.

Meriin, A. B., Zhang, X., He, X., Newnam, G. P., Chernoff, Y. O., and Sherman, M. Y. (2002).

Huntington toxicity in yeast model depends on poly-glutamine aggregation mediated by a prion-like protein Rnq1. *J Cell Biol* **157**, 997–1004.

Michelitsch, M. D., and Weissman, J. S. (2000). A census of glutamine/asparagine-rich regions: Implications for their conserved function and the prediction of novel prions. *Proc Natl Acad Sci USA* **97**, 11910–11915.

Mnaimneh, S., Davierwala, A. P., Haynes, J., Moffat, J., Peng, W. T., Zhang, W., Yang, X., Pootoolal, J., Chua, G., Lopez, A., Trochesset, M., Morse, D., Krogan, N. J., Hiley, S. L., Li, Z., Morris, Q., Grigull, J., Mitsakakis, N., Roberts, C. J., Greenblatt, J. F., Boone, C., Kaiser, C. A., Andrews, B. J., and Hughes, T. R. (2004). Exploration of essential gene functions via titratable promoter alleles. *Cell* **118**, 31–44.

Morris, H. (2007). Autosomal dominant Parkinson's disease and the route to new therapies. *Expert Rev Neurother* **7**, 649–656.

Morris, H. R., Vaughan, J. R., Datta, S. R., Bandopadhyay, R., Rohan De Silva, H. A., Schrag, A., Cairns, N. J., Burn, D., Nath, U., Lantos, P. L., Daniel, S., Lees, A. J., Quinn, N. P., and Wood, N. W. (2000). Multiple system atrophy/progressive supranuclear palsy: Alpha-synuclein, synphilin, tau, and APOE. *Neurology* **12**, 1918–1920.

Morrow, D. M., Tagle, D. A., Shiloh, Y., Collins, F. S., and Hieter, P. (1995). TEL1, an *S. cerevisiae* homolog of the human gene mutated in ataxia telangiectasia, is functionally related to the yeast checkpoint gene MEC1. *Cell* **82**, 831–840.

Muchowski, P. J. (2002). Protein misfolding, amyloid formation, and neurodegeneration: A critical role for molecular chaperones? *Neuron* **35**, 9–12.

Muchowski, P. J., Schaffar, G., Sittler, A., Wanker, E. E., Hayer-Hartl, M. K., and Hartl, F. U. (2000). Hsp70 and hsp40 chaperones can inhibit self-assembly of polyglutamine proteins into amyloid-like fibrils. *Proc Natl Acad Sci USA* **97**, 7841–7846.

Nakanishi, H., Morishita, M., Schwartz, C. L., Coluccio, A., Engebrecht, J., and Neiman, A. M. (2006). Phospholipase D and the SNARE Sso1p are necessary for vesicle fusion during sporulation in yeast. *J Cell Sci* **119**, 1406–1415.

Naoi, M., and Maruyama, W. (1999). Cell death of dopamine neurons in aging and Parkinson's disease. *Mech Ageing Dev* **111**, 175–188.

Nishiyama, K., Murayama, S., Goto, J., Watanabe, M., Hashida, H., Katayama, S., Nomura, Y., Nakamura, S., and Kanazawa, I. (1996). Regional and cellular expression of the Machado-Joseph disease gene in brains of normal and affected individuals. *Ann Neurol* **40**, 776–781.

Nishiyama, K., Nakamura, K., Murayama, S., Yamada, M., and Kanazawa, I. (1997). Regional and cellular expression of the dentatorubral-pallidoluysian atrophy gene in brains of normal and affected individuals. *Ann Neurol* **41**, 599–605.

Olanow, C. W. (2004). Manganese-induced parkinsonism and Parkinson's disease. *Ann NY Acad Sci* **1012**, 209–223.

Olanow, C. W., and McNaught, K. S. (2006). Ubiquitin–proteasome system and Parkinson's disease. *Mov Disord* **21**, 1806–1823.

Ordway, J. M., Tallaksen-Greene, S., Gutekunst, C. A., Bernstein, E. M., Cearley, J. A., Wiener, H. W., Dure, L. S.t, Lindsey, R., Hersch, S. M., Jope, R. S., Albin, R. L., and Detloff, P. J. (1997). Ectopically expressed CAG repeats cause intranuclear inclusions and a progressive late onset neurological phenotype in the mouse. *Cell* **91**, 753–763.

Orr-Weaver, T. L., Szostak, J. W., and Rothstein, R. J. (1983). Genetic applications of yeast transformation with linear and gapped plasmids. *Methods Enzymol* **101**, 228–245.

Orth, M., Tabrizi, S. J., Tomlinson, C., Messmer, K., Korlipara, L. V., Schapira, A. H., and Cooper, J. M. (2004). G209A mutant alpha synuclein expression specifically enhances dopamine induced oxidative damage. *Neurochem Int* **45**, 669–676.

Osherovich, L. Z., and Weissman, J. S. (2001). Multiple Gln/Asn-rich prion domains confer susceptibility to induction of the yeast [PSI(+)] prion. *Cell* **106**, 183–194.

Outeiro, T. F., and Lindquist, S. (2003). Yeast cells provide insight into alpha-synuclein biology and pathobiology. *Science* **302**, 1772–1775.

Outeiro, T. F., Klucken, J., Strathearn, K. E., Liu, F., Nguyen, P., Rochet, J. C., Hyman, B. T., and McLean, P. J. (2006). Small heat shock proteins protect against alpha-synuclein-induced toxicity and aggregation. *Biochem Biophys Res Commun* **351**, 631–638.

Paik, S. R., Shin, H. J., Lee, J. H., Chang, C. S., and Kim, J. (1999). Copper(II)-induced self-oligomerization of alpha-synuclein. *Biochem J* **340**(Pt 3), 821–828.

Pals, P., Lincoln, S., Manning, J., Heckman, M., Skipper, L., Hulihan, M., Van den Broeck, M., De Pooter, T., Cras, P., Crook, J., Van Broeckhoven, C., and Farrer, M. J. (2004). Alpha-synuclein promoter confers susceptibility to Parkinson's disease. *Ann Neurol* **56**, 591–595.

Pandolfo, M. (2001). Molecular basis of Friedreich ataxia. *Mov Disord* **16**, 815–821.

Papapetropoulos, S., Ellul, J., Paschalis, C., Athanassiadou, A., Papadimitriou, A., and Papapetropoulos, T. (2003). Clinical characteristics of the alpha-synuclein mutation (G209A)-associated Parkinson's disease in comparison with other forms of familial Parkinson's disease in Greece. *Eur J Neurol* **10**, 281–286.

Park, J., Lee, S. B., Lee, S., Kim, Y., Song, S., Kim, S., Bae, E., Kim, J., Shong, M., Kim, J. M., and Chung, J. (2006). Mitochondrial dysfunction in Drosophila PINK1 mutants is complemented by parkin. *Nature* **441**, 1157–1161.

Paul, A. G., and DebBurman, S. K. (2006). Role, Evaluation of Stp2p-dependent α-Synuclein Toxicity in Yeast: Role of GAPDH?. *Impulse, The Premier Journal for Undergraduate Publications in the Neurosciences* **3**, 1–11.

Payton, J. E., Perrin, R. J., Woods, W. S., and George, J. M. (2004). Structural determinants of PLD2 inhibition by alpha-synuclein. *J Mol Biol* **337**, 1001–1009.

Perez, R. G., and Hastings, T. G. (2004). Could a loss of alpha-synuclein function put dopaminergic neurons at risk? *J Neurochem* **89**, 1318–1324.

Polaina, J. (2002). Brewer's yeast: Genetics and biotechnology. *Appl Mycol Biotechnol* **2**, 1–17.

Polymeropoulos, M. H., Lavedan, C., Leroy, E., Ide, S. E., Dehejia, A., Dutra, A., Pike, B., Root, H., Rubenstein, J., Boyer, R., Stenroos, E. S., Chandrasekharappa, S., Athanassiadou, A., Papapetropoulos, T., Johnson, W. G., Lazzarini, A. M., Duvoisin, R. C., Di Iorio, G., Golbe, L. I., and Nussbaum, R. L. (1997). Mutation in the alpha-synuclein gene identified in families with Parkinson's disease. *Science* **276**, 2045–2047.

Pridgeon, J. W., Olzmann, J. A., Chin, L. S., and Li, L. (2007). PINK1 protects against oxidative stress by phosphorylating mitochondrial chaperone TRAP1. *PLoS Biol* **5**, e172.

Quilty, M. C., King, A. E., Gai, W. P., Pountney, D. L., West, A. K., Vickers, J. C., and Dickson, T. C. (2006). Alpha-synuclein is upregulated in neurones in response to chronic oxidative stress and is associated with neuroprotection. *Exp Neurol* **199**, 249–256.

Raichur, A., Vail, S., and Gorin, F. (2006). Dynamic modeling of alpha-synuclein aggregation for the sporadic and genetic forms of Parkinson's disease. *Neuroscience* **142**, 859–870.

Ramirez, A., Heimbach, A., Grundemann, J., Stiller, B., Hampshire, D., Cid, L. P., Goebel, I., Mubaidin, A. F., Wriekat, A. L., Roeper, J., Al-Din, A., Hillmer, A. M., Karsak, M., Liss, B., Woods, C. G., Behrens, M. I., and Kubisch, C. (2006). Hereditary parkinsonism with dementia is caused by mutations in ATP13A2, encoding a lysosomal type 5 P-type ATPase. *Nat Genet* **38**, 1184–1191.

Ramsey, C. P., and Giasson, B. I. (2007). Role of mitochondrial dysfunction in Parkinson's disease: Implications for treatment. *Drugs Aging* **24**, 95–105.

Rochet, J. C., and Lansbury, P. T. (2000). Amyloid fibrillogenesis: Themes and variations. *Curr Opin Struct Biol* **10**, 13–15.

Roitgrund, C., Steinlauf, R., and Kupiec, M. (1993). Donation: A new, facile method of gene replacement in yeast. *Mol Gen Genet* **237**, 306–310.

Rothman, J. E., and Warren, G. (1994). Implications of the SNARE hypothesis for intracellular membrane topology and dynamics. *Curr Biol* **4**, 220–233.

Rubin, G. M., Yandell, M. D., Wortman, J. R., Gabor Miklos, G. L., Nelson, C. R., Hariharan, I. K., Fortini, M. E., Li, P. W., Apweiler, R., Fleischmann, W., Cherry, J. M., Henikoff, S., Skupski, M. P., Misra, S., Ashburner, M., Birney, E., Boguski, M. S., Brody, T., Brokstein, P., Celniker, S. E., Chervitz, S. A., Coates, D., Cravchik, A., Gabrielian, A., Galle, R. F., Gelbart, W. M., George, R. A., Goldstein, L. S. B., Gong, F., Guan, P., Harris, N. L., Hay, B. A., Hoskins, R. A., Li, J., Li, Z., Hynes, R. O., Jones, S. J. M., Kuehl, P. M., Lemaitre, B., Littleton, J. T., Morrison, D. K., Mungall, C., O'Farrell, P. H., Pickeral, O. K., Shue, C., Vosshall, L. B., Zhang, J., Zhao, Q., Zheng, X. H., Zhong, F., Zhong, W., Gibbs, R., Venter, J. C., Adams, M. D., and Lewis, S. (2000). Comparative genomics of the eukaryotes. *Science* **287**, 2204–2215.

Sakamoto, K. M. (2002). Ubiquitin-dependent proteolysis: Its role in human diseases and the design of therapeutic strategies. *Mol Genet Metab* **77**, 44–56.

Sanders, C. R., Ismail-Beigi, F., and McEnery, M. W. (2001). Mutations of peripheral myelin protein 22 result in defective trafficking through mechanisms which may be common to diseases involving tetraspan membrane proteins. *Biochemistry* **40**, 9453–9459.

Sarkar, S., Davies, J. E., Huang, Z., Tunnacliffe, A., and Rubinsztein, D. C. (2007). Trehalose, a novel mTOR-independent autophagy enhancer, accelerates the clearance of mutant huntingtin and alpha-synuclein. *J Biol Chem* **282**, 5641–5652.

Sathasivam, K., Hobbs, C., Mangiarini, L., Mahal, A., Turmaine, M., Doherty, P., Davies, S. W., and Bates, G. P. (1999). Transgenic models of Huntington's disease. *Philos Trans R Soc Lond B Biol Sci* **354**, 963–969.

Schaffar, G., Breuer, P., Boteva, R., Behrends, C., Tzvetkov, N., Strippel, N., Sakahira, H., Siegers, K., Hayer-Hartl, M., and Hartl, F. U. (2004). Cellular toxicity of polyglutamine expansion proteins: Mechanism of transcription factor deactivation. *Mol Cell* **15**, 95–105.

Scherzer, C. R., and Feany, M. B. (2004). Yeast genetics targets lipids in Parkinson's disease. *Trends Genet* **20**, 273–277.

Scherzer, C. R., Jensen, R. V., Gullans, S. R., and Feany, M. B. (2003). Gene expression changes presage neurodegeneration in a Drosophila model of Parkinson's disease. *Hum Mol Genet* **12**, 2457–2466.

Schiffer, N. W., Broadley, S. A., Hirschberger, T., Tavan, P., Kretzschmar, H. A., Giese, A., Haass, C., Hartl, F. U., and Schmid, B. (2007). Identification of anti-prion compounds as efficient inhibitors of polyglutamine protein aggregation in a zebrafish model. *J Biol Chem* **282**, 9195–9203.

Schilling, G., Sharp, A. H., Loev, S. J., Wagster, M. V., Li, S. H., Stine, O. C., and Ross, C. A. (1995). Expression of the Huntington's disease (IT15) protein product in HD patients. *Hum Mol Genet* **4**, 1365–1371.

Schubert, U., Anton, L. C., Gibbs, J., Norbury, C. C., Yewdell, J. W., and Bennink, J. R. (2000). Rapid degradation of a large fraction of newly synthesized proteins by proteasomes. *Nature* **404**, 770–774.

Schuldiner, M., Collins, S. R., Thompson, N. J., Denic, V., Bhamidipati, A., Punna, T., Ihmels, J., Andrews, B., Boone, C., Greenblatt, J. F., Weissman, J. S., and Krogan, N. J. (2005). Exploration of the function and organization of the yeast early secretory pathway through an epistatic miniarray profile. *Cell* **123**, 507–519.

Sharma, N., Brandis, K. A., Herrera, S. K., Johnson, B. E., Vaidya, T., Shrestha, R., and DebBurman, S. K. (2006). Alpha-synuclein budding yeast model. *J Mol Neurosci* **28**, 161–178.

Sharon, R., Goldberg, M. S., Bar-Josef, I., Betensky, R. A., Shen, J., and Selkoe, D. J. (2001). Alpha-synuclein occurs in lipid-rich high molecular weight complexes, binds fatty acids, and shows homology to the fatty acid-binding proteins. *Proc Natl Acad Sci USA* **98**, 9110–9115.

Sharp, A. H., Loev, S. J., Schilling, G., Li, S. H., Li, X. J., Bao, J., Wagster, M. V., Kotzuk, J. A., Steiner, J. P., Lo, A. et al. (1995). Widespread expression of Huntington's disease gene (IT15) protein product. *Neuron* **14**, 1065–1074.

Shendelman, S., Jonason, A., Martinat, C., Leete, T., and Abeliovich, A. (2004). DJ-1 is a redox-dependent molecular chaperone that inhibits alpha-synuclein aggregate formation. *PLoS Biol* **2**, e362.

Sherman, F. (2002). Getting started with yeast. *Methods Enzymol* **350**, 3–41.

Sherman, M. Y., and Goldberg, A. L. (2001). Cellular defenses against unfolded proteins. *Neuron* **29**, 15–32.

Shields, D., and Arvan, P. (1999). Disease models provide insights into post-golgi protein trafficking, localization and processing. *Curr Opin Cell Biol* **11**, 489–494.

Shorter, J., and Lindquist, S. (2005). Prions as adaptive conduits of memory and inheritance. *Nat Rev Genet* **6**, 435–450.

Si, K., Lindquist, S., and Kandel, E. R. (2003). A neuronal isoform of the aplysia CPEB has prion-like properties. *Cell* **115**, 879–891.

Singer, M. A., and Lindquist, S. (1998a). Multiple effects of trehalose on protein folding *in vitro* and *in vivo*. *Mol Cell* **1**, 639–648.

Singer, M. A., and Lindquist, S. (1998b). Thermotolerance in *Saccharomyces cerevisiae*: The Yin and Yang of trehalose. *Trends Biotechnol* **16**, 460–468.

Singleton, A. B., Farrer, M., Johnson, J., Singleton, A., Hague, S., Kachergus, J., Hulihan, M., Peuralinna, T., Dutra, A., Nussbaum, R., Lincoln, S., Crawley, A., Hanson, M., Maraganore, D., Adler, C., Cookson, M. R., Muenter, M., Baptista, M., Miller, D., Blancato, J., Hardy, J., and Gwinn-Hardy, K. (2003). Alpha-synuclein locus triplication causes Parkinson's disease. *Science* **302**, 841.

Sipiczki, M. (2000). Where does fission yeast sit on the tree of life? *Genome Biol* **1**, Reviews1011.

Snustad, D., and Simmons, M. (2005). *Principles of Genetics*. Wiley. Hoboken, NJ.

Somero, G. N. (1986). Protons, osmolytes, and fitness of internal milieu for protein function. *Am J Physiol* **251**, R197–R213.

Sondheimer, N., and Lindquist, S. (2000). Rnq1: An epigenetic modifier of protein function in yeast. *Mol Cell* **5**, 163–172.

Sopko, R., Huang, D., Preston, N., Chua, G., Papp, B., Kafadar, K., Snyder, M., Oliver, S. G., Cyert, M., Hughes, T. R., Boone, C., and Andrews, B. (2006). Mapping pathways and phenotypes by systematic gene overexpression. *Mol Cell* **21**, 319–330.

Spillantini, M. G., Schmidt, M. L., Lee, V. M., Trojanowski, J. Q., Jakes, R., and Goedert, M. (1997). Alpha-synuclein in Lewy bodies. *Nature* **388**, 839–840.

Spillantini, M. G., Crowther, R. A., Jakes, R., Hasegawa, M., and Goedert, M. (1998). Alpha-synuclein in filamentous inclusions of Lewy bodies from Parkinson's disease and dementia with Lewy bodies. *Proc Natl Acad Sci USA* **95**, 6469–6473.

Stagljar, I., and Fields, S. (2002). Analysis of membrane protein interactions using yeast-based technologies. *Trends Biochem Sci* **27**, 559–563.

Stelzl, U., and Wanker, E. E. (2006). The value of high quality protein–protein interaction networks for systems biology. *Curr Opin Chem Biol* **10**, 551–558.

Stettler, O., Moya, K. L., Zahraoui, A., and Tavitian, B. (1994). Developmental changes in the localization of the synaptic vesicle protein rab3A in rat brain. *Neuroscience* **62**, 587–600.

Struhl, K. (1983). Direct selection for gene replacement events in yeast. *Gene* **26**, 231–241.

Tallaksen-Greene, S. J., Ordway, J. M., Crouse, A. B., Jackson, W. S., Detloff, P. J., and Albin, R. L. (2003). Hprt(CAG)146 mice: Age of onset of behavioral abnormalities, time course of neuronal intranuclear inclusion accumulation, neurotransmitter marker alterations, mitochondrial function markers, and susceptibility to 1-methyl-4-phenyl-1,2,3,6-tetrahydropyridine. *J Comp Neurol* **465**, 205–219.

Tam, S., Geller, R., Spiess, C., and Frydman, J. (2006). The chaperonin TRiC controls polyglutamine aggregation and toxicity through subunit-specific interactions. *Nat Cell Biol* 8, 1155–1162.

Tanaka, M., Machida, Y., Niu, S., Ikeda, T., Jana, N. R., Doi, H., Kurosawa, M., Nekooki, M., and Nukina, N. (2004). Trehalose alleviates polyglutamine-mediated pathology in a mouse model of Huntington disease. *Nat Med* 10, 148–154.

Tanaka, M., Machida, Y., and Nukina, N. (2005). A novel therapeutic strategy for polyglutamine diseases by stabilizing aggregation-prone proteins with small molecules. *J Mol Med* 83, 343–352.

Tong, A. H., and Boone, C. (2006). Synthetic genetic array analysis in *Saccharomyces cerevisiae*. *Methods Mol Biol* 313, 171–192.

Tong, A. H., Evangelista, M., Parsons, A. B., Xu, H., Bader, G. D., Page, N., Robinson, M., Raghibizadeh, S., Hogue, C. W., Bussey, H., Andrews, B., Tyers, M., and Boone, C. (2001). Systematic genetic analysis with ordered arrays of yeast deletion mutants. *Science* 294, 2364–2368.

Trojanowski, J. Q. (2004). Protein mis-folding emerges as a "drugable" target for discovery of novel therapies for neuropsychiatric diseases of aging. *Am J Geriatr Psychiatry* 12, 134–135.

Uehara, T., Nakamura, T., Yao, D., Shi, Z. Q., Gu, Z., Ma, Y., Masliah, E., Nomura, Y., and Lipton, S. A. (2006). S-nitrosylated protein-disulphide isomerase links protein misfolding to neurodegeneration. *Nature* 441, 513–517.

Vacher, C., Garcia-Oroz, L., and Rubinsztein, D. C. (2005). Overexpression of yeast hsp104 reduces polyglutamine aggregation and prolongs survival of a transgenic mouse model of Huntington's disease. *Hum Mol Genet* 14, 3425–3433.

Vila, M., and Przedborski, S. (2004). Genetic clues to the pathogenesis of Parkinson's disease. *Nat Med* 10, S58–S62.

Wakabayashi, K., Engelender, S., Yoshimoto, M., Tsuji, S., Ross, C. A., and Takahashi, H. (2000). Synphilin-1 is present in Lewy bodies in Parkinson's disease. *Ann Neurol* 47, 521–523.

Watt, P. M., Hickson, I. D., Borts, R. H., and Louis, E. J. (1996). SGS1, a homologue of the Bloom's and Werner's syndrome genes, is required for maintenance of genome stability in *Saccharomyces cerevisiae*. *Genetics* 144, 935–945.

Weiss, B. (2006). Economic implications of manganese neurotoxicity. *Neurotoxicology* 27, 362–368.

Whitesell, L., and Lindquist, S. L. (2005). HSP90 and the chaperoning of cancer. *Nat Rev Cancer* 5, 761–772.

Willingham, S., Outeiro, T. F., DeVit, M. J., Lindquist, S. L., and Muchowski, P. J. (2003). Yeast genes that enhance the toxicity of a mutant huntingtin fragment or alpha-synuclein. *Science* 302, 1769–1772.

Wilson, M. A., St Amour, C. V., Collins, J. L., Ringe, D., and Petsko, G. A. (2004). The 1.8-A resolution crystal structure of YDR533Cp from *Saccharomyces cerevisiae*: A member of the DJ-1/ThiJ/PfpI superfamily. *Proc Natl Acad Sci USA* 101, 1531–1536.

Winkler, S., Hagenah, J., Lincoln, S., Heckman, M., Haugarvoll, K., Lohmann-Hedrich, K., Kostic, V., Farrer, M., and Klein, C. (2007). Alpha-synuclein and Parkinson disease susceptibility. *Neurology* 69, 1745–1750.

Winzeler, E. A., Shoemaker, D. D., Astromoff, A., Liang, H., Anderson, K., Andre, B., Bangham, R., Benito, R., Boeke, J. D., Bussey, H., Chu, A. M., Connelly, C., Davis, K., Dietrich, F., Dow, S. W., El Bakkoury, M., Foury, F., Friend, S. H., Gentalen, E., Giaever, G., Hegemann, J. H., Jones, T., Laub, M., Liao, H., Davis, R. W. et al. (1999). Functional characterization of the *S. cerevisiae* genome by gene deletion and parallel analysis. *Science* 285, 901–906.

Wood, V., Gwilliam, R., Rajandream, M. A., Lyne, M., Lyne, R., Stewart, A., Sgouros, J., Peat, N., Hayles, J., Baker, S., Basham, D., Bowman, S., Brooks, K., Brown, D., Brown, S., Chillingworth, T., Churcher, C., Collins, M., Connor, R., Cronin, A., Davis, P., Feltwell, T., Fraser, A., Gentles, S., Goble, A., Hamlin, N., Harris, D., Hidalgo, J., Hodgson, G., Holroyd, S., Hornsby, T., Howarth, S., Huckle, E. J., Hunt, S., Jagels, K., James, K., Jones, L., Jones, M., Leather, S., McDonald, S., McLean, J., Mooney, P., Moule, S., Mungall, K., Murphy, L., Niblett, D., Odell, C., Oliver, K., O'Neil, S., Pearson, D., Quail, M. A., Rabbinowitsch, E., Rutherford, K., Rutter, S., Saunders, D., Seeger, K., Sharp, S., Skelton, J., Simmonds, M., Squares, R., Squares, S., Stevens, K., Taylor, K., Taylor, R. G., Tivey, A., Walsh, S., Warren, T., Whitehead, S., Woodward, J., Volckaert, G., Aert, R., Robben, J., Grymonprez, B., Weltjens, I., Vanstreels, E., Rieger, M., Schafer, M., Muller-Auer, S., Gabel, C., Fuchs, M., Dusterhoft, A., Fritzc, C., Holzer, E., Moestl, D., Hilbert, H., Borzym, K., Langer, I., Beck, A., Lehrach, H., Reinhardt, R., Pohl, T. M., Eger, P., Zimmermann, W., Wedler, H., Wambutt, R., Purnelle, B., Goffeau, A., Cadieu, E., Dreano, S., Gloux, S., Lelaure, V., Mottier, S., Galibert, F., Aves, S. J., Xiang, Z., Hunt, C., Moore, K., Hurst, S. M., Lucas, M., Rochet, M., Gaillardin, C., Tallada, V. A., Garzon, A., Thode, G., Daga, R. R., Cruzado, L., Jimenez, J., Sanchez, M., del Rey, F., Benito, J., Dominguez, A., Revuelta, J. L., Moreno, S., Armstrong, J., Forsburg, S. L., Cerutti, L., Lowe, T., McCombie, W. R., Paulsen, I., Potashkin, J., Shpakovski, G. V., Ussery, D., Barrell, B. G., and Nurse, P. (2002). The genome sequence of *Schizosaccharomyces pombe*. *Nature* 415, 871–880.

Yancey, P. H. (2005). Organic osmolytes as compatible, metabolic and counteracting cytoprotectants in high osmolarity and other stresses. *J Exp Biol* **208**, 2819–2830.

Young, J. C., Agashe, V. R., Siegers, K., and Hartl, F. U. (2004). Pathways of chaperone-mediated protein folding in the cytosol. *Nat Rev Mol Cell Biol* **5**, 781–791.

Zarranz, J. J., Alegre, J., Gomez-Esteban, J. C., Lezcano, E., Ros, R., Ampuero, I., Vidal, L., Hoenicka, J., Rodriguez, O., Atares, B., Llorens, V., Gomez Tortosa, E., del Ser, T., Munoz, D. G., and de Yebenes, J. G. (2004). The new mutation, E46K, of alpha-synuclein causes Parkinson and Lewy body dementia. *Ann Neurol* **55**, 164–173.

Zecca, L., Youdim, M. B., Riederer, P., Connor, J. R., and Crichton, R. R. (2004). Iron, brain ageing and neurodegenerative disorders. *Nat Rev Neurosci* **5**, 863–873.

Zhou, W., Zhu, M., Wilson, M. A., Petsko, G. A., and Fink, A. L. (2006). The oxidation state of DJ-1 regulates its chaperone activity toward alpha-synuclein. *J Mol Biol* **356**, 1036–1048.

Zimprich, A., Biskup, S., Leitner, P., Lichtner, P., Farrer, M., Lincoln, S., Kachergus, J., Hulihan, M., Uitti, R. J., Calne, D. B., Stoessl, A. J., Pfeiffer, R. F., Patenge, N., Carbajal, I. C., Vieregge, P., Asmus, F., Muller-Myhsok, B., Dickson, D. W., Meitinger, T., Strom, T. M., Wszolek, Z. K., and Gasser, T. (2004). Mutations in LRRK2 cause autosomal-dominant parkinsonism with pleomorphic pathology. *Neuron* **44**, 601–607.

Zoghbi, H. Y., and Orr, H. T. (2000). Glutamine repeats and neurodegeneration. *Annu Rev Neurosci* **23**, 217–247.

VII

CELL-FREE MODELS

39

OVERVIEW OF CELL-FREE SYSTEMS FOR THE STUDY OF PARKINSON'S DISEASE

CHUN ZHOU

Department of Neurology, Columbia University, New York, NY, USA

INTRODUCTION

As mentioned on several occasions in this book, Parkinson's disease (PD) is the second most common neurodegenerative disorder after Alzheimer's dementia. In the chapter on PD genetics, it is stressed that while PD arises essentially as a sporadic condition, it may occasionally be inherited (Vila and Przedborski, 2004). Over the past decade, there has been an exponential increase in the number of studies dedicated to these rare familial forms of PD (fPD). The rationale for studying these fPD syndromes is based on the phenotypic similarities between the familial and sporadic forms of PD (sPD), implying that the two phenotypes share important pathogenic mechanisms and, consequently, that analysis of fPD will shed light onto key molecular pathways in both disorders. Along this vein, it is incontestable that the discovery of various genetic loci linked to fPD and subsequently of the proteins encoded by these genes has provided us with a unique opportunity to unravel the myths surrounding the molecular pathogenesis of PD.

As shown in this book, so far, scientists have used various experimental models to probe the neurobiology of PD, including cellular models and vertebrate and invertebrate animals and more recently cell-free systems as well. The latter model system is the topic of this section of the book and, as we will see, by zeroing in on a particular physicochemical aspect of the protein under investigation, this simplified approach may provide crucial insights into the intrinsic biochemical functions or properties of PD-related proteins (Table 39.1).

CELL-FREE SYSTEM

But, one may wonder what is a cell-free system? A cell-free system, as discussed here and in the three chapters that compose this section of the book, can be defined as an experimental setting in a test tube aimed at studying protein functions or biochemical properties. Various assays can be done in test tubes to analyze protein functions/properties, such as protein folding, protein cleavage, enzyme activity and so on. The basis of a cell-free system study is the expression of target proteins, which will be analyzed in test tubes under well-defined conditions. In general, there are three ways to generate a target protein. A simple method is *in vitro* translation, in which translation occurs entirely in a test tube. All of the components necessary for translating mRNA can be obtained from cell lysates that are highly efficient in protein synthesis, such as reticulocyte lysates. Under the appropriate conditions, a synthetic or purified RNA added to such a system will be translated efficiently into the target protein. The second method is the production of recombinant proteins. One of the simplest methods here involves the cloning of the cDNA of the target protein into a bacterial plasmid which contains a transcriptional promoter. When introduced into the appropriate bacterial host, large amounts

TABLE 39.1 Identified fPD-related genes/proteins with suggested functions

Locus	Gene/protein	Suggested function
PARK1 and PARK4	α-synuclein	Regulate synaptic vesicles
PARK2	parkin	E3 ubiquitin ligase
PARK5	UCH-L1	Ubiquitin ligase and hydrolase
PARK6	PINK1	Mitochondrial kinase
PARK7	DJ-1	Anti-oxidant protein
PARK8	LRRK2	Protein kinase

of mRNA will be transcribed, which, in turn, will be translated into protein. The recombinant protein can then be purified away from all of the bacterial proteins. However, many eukaryotic proteins require post-translational modifications for maximal activity, whereas bacteria do not have the machinery required to accomplish complex modification. Therefore, when these modifications are required, eukaryotic cells, such as yeast cells and insect cells, can be used for the production of recombinant protein. These eukaryotic cells can execute most of the post-translational modifications required by mammalian proteins. The third way to produce a target protein is to overexpress such a protein in mammalian cells; immunoprecipitation is then performed to enrich the target protein for cell-free system assays. Among these three methods, recombinant proteins are preferred since the purification process is used to purify recombinant proteins. On the other hand, *in vitro* translated and immunoprecipitated proteins may contain possible contaminants of other proteins. However, the latter two methods are often technically more convenient as compared to the production of recombinant proteins.

Once target proteins are produced, researchers usually provide an artificial setting, which contains all of the elements necessary to assist a biochemical reaction or phenomenon, in order to study intrinsic functions of a target protein and further, to seek detailed molecular mechanisms through the modification of this well-controlled artificial setting.

CHAPTER CONTENT

In this section, the target proteins, which are PD-related proteins, can be divided into either

enzymes, such as parkin, PINK1, and LRRK2, or proteins with unique structures, such as native unfolded α-synuclein and dimeric DJ-1. Therefore, enzyme assays and protein structure assays are the main focus of PD-related protein analyses in a cell-free system. In Chapter 43, the author has discussed the findings using cell-free systems on two extensively studied proteins, α-synuclein and DJ-1, and has summarized recent cell-free system studies on UCH-L1, PINK1, and LRRK2. In Chapter 42, the author has provided a biochemical view of the structural dynamics of free α-synuclein and membrane-bound α-synuclein and how such knowledge acquired from cell-free systems can shed a light on deciphering the pathological role of α-synuclein in PD. Lastly, in Chapter 41, novel methods are developed to enrich neural α-synuclein for identifying α-synuclein-associated proteins and to use α-synuclein as a biomarker for PD, with a sensitive ELISA to detect α-synuclein in a variety of biological specimens.

CRITICAL INSIGHT PROVIDED BY CELL-FREE SYSTEM

Here, we will take parkin as an example to discuss how cell-free system studies are critical for establishing the concept of protein intrinsic properties and for shedding light on our understanding of the functions and pathological roles of PD-related proteins.

One of the common properties of PD-related proteins is that most of them are enzymes, such as the extensively studied parkin, an E3 ubiquitin ligase, and the newly identified protein kinases (PINK1 and LRRK2). This property highlights the importance of a cell-free system study since it is a classic well-established system to study intrinsic enzyme activities. To this end, the parkin studies in cell-free systems have provided us with a good example on how the cell-free model can play an important role in deciphering the functions of PD-related proteins. In 1998, after the first identification of the PD-related gene, parkin (Kitada *et al.*, 1998), a link between parkin and ubiquitin was immediately suggested due to its ubiquitin homology (Nussbaum, 1998). However, the first direct evidence showing parkin as an E3 ligase comes from a cell-free system study – *in vitro* ubiquitination assay (Imai *et al.*, 2000; Shimura *et al.*, 2000; Zhang *et al.*, 2000b). As we know, ubiquitin ligases are an essential component of the cellular machinery that covalently modifies protein substrates with ubiquitin. Ubiquitination results from a sequential action of ubiquitin-activating (E1), -conjugating (E2), and -ligase (E3) enzymes, respectively. This *in vitro* ubiquitination assay was elegantly designed to contain parkin either immunoprecipitated from parkin overexpressing cells or made by *in vitro* transcription, ATP regenerating system, recombinant E1, E2, and ubiquitin. Under this well-defined system, parkin was shown to have the intrinsic property of E3 ligase activity, whereas fPD-linked parkin mutants showed disrupted E3 ligase activity, suggesting a loss-of-function pathology in this recessive fPD. Notably, in the same studies, the researchers also showed evidence for the involvement of parkin in the ubiquitination system in cellular models, but in an indirect manner rather than through direct evidence, as compared to the *in vitro* ubiquitination assay. Here, we can appreciate that the cell-free system provided the "definitive" evidence of parkin's E3 ligase activity, which marked a clear direction for the follow-up studies in both cellular and *in vivo* models.

Since then, numerous parkin "substrates" have been suggested, such as CDCrel-1 (Zhang *et al.*, 2000a), O-glycosylated α-synuclein (Shimura *et al.*, 2001), synphilin-1 (Chung *et al.*, 2001), Pael R (Imai *et al.*, 2001), and p38/JTV-1 (Corti *et al.*, 2003). It was hypothesized that the loss of parkin E3 ligase function results in the dysfunction of the polyubiquitin-related ubiquitin–proteasome system (UPS) and the toxic accumulation of its target substrates. Such a hypothesis is attractive because it fits with the following two facts: first, the classic function of an E3 ubiquitin ligase is to polyubiquitinate its substrate, resulting in the degradation of the substrate through the UPS; second, a dysfunction of the UPS is believed to be one of the major pathological mechanisms of PD. However, this hypothesis remains to be further tested since contradictory to the expected results, some of the reported parkin "substrates" failed to accumulate in parkin-deficient mice.

One possibility to explain the above unexpected results is that parkin may perform a UPS-independent role through its E3 ubiquitin ligase activity. In fact, two major types of protein ubiquitination exist in mammalian cells: first, classically, a multiubiquitin chain, with four or more ubiquitin moieties, is assembled through isopeptide bonds via Lys-48 or Lys-29 of ubiquitin, and such polyubiquitination targets proteins for degradation by the 26 S proteasome (Hershko and Ciechanover, 1992); second, in contrast to the classical situation, the conjugation of a single ubiquitin moiety

(monoubiquitination) or the ligation of a polyubiquitin chain through Lys-63 (polyubiquitination) is a post-translational modification, which serves as a cellular signal for the endocytosis of membrane proteins (Nakatsu *et al.*, 2000; Haglund *et al.*, 2003), or for protein sorting and trafficking, which can promote recycling or degradation by the lysosome (Hicke, 2001; Katzmann *et al.*, 2002). Thus, it is possible that parkin performs its non-degradation role through either Lys-63-linked polyubiquitination or monoubiquitination.

Indeed, two independent studies using modified *in vitro* ubiquitination assays addressed the intrinsic ubiquitination property of parkin (Hampe *et al.*, 2006; Matsuda *et al.*, 2006). The improvement in the modified assay is that the researchers used purified recombinant parkin, rather than immunoprecipitated or *in vitro*-translated parkin, in which one cannot exclude the possibilities that contaminating E3 ligases or cofactors of ubiquitination reactions may interfere with or modify parkin activity, making results difficult to interpret. Consistently, both studies revealed an intrinsic property of parkin to mediate monoubiquitination. A subsequent study confirmed the monoubiquitin property of parkin by identifying a new parkin substrate, PICK1 (protein interacting with C-kinase 1) (Joch *et al.*, 2007). In addition, an *in vitro* ubiquitination assay also revealed that parkin can mediate Lys-63-linked polyubiquitination of misfolded DJ-1, serving as a signal for targeting misfolded DJ-1 to aggresomes (Olzmann *et al.*, 2007). Thus, cell-free systems once again have opened a novel direction for parkin study, especially future *in vivo* studies. It is likely that this new direction may lead to the identification of novel molecular pathways which are critical to the pathogenesis of PD.

At this point, it is not clear which ubiquitination type belongs to the physiological role of parkin, or whether parkin performs multi-functional ubiquitination to modulate multiple cellular processes. However, from the studies of parkin in the past decade, we can clearly see that cell-free systems have provided critical insights into the definition of parkin's major biochemical property, an E3 ubiquitin ligase, and the molecular details of the types of ubiquitination, the latter of which is critical to deciphering the physiological functions of parkin.

At the same time, we can also see the major limitation of the cell-free system studies. While the cell-free system provides valuable information about the intrinsic biochemical properties of PD-related proteins, especially enzymes, as we can see from the above example of parkin study, or from the other functional studies of PINK1 and LRRK2 (*in vitro* kinase assay), whether or not such biochemical properties perform important functions *in vivo*, however, remains to be further validated in the more complex models, such as cellular models and animal models. In the case of parkin E3 ubiquitin ligase function, how parkin performs its protective role through what type(s) of ubiquitination on what physiological substrate(s) in PD needs to be further studied.

INTERPRETATION OF CELL-FREE SYSTEM FINDINGS

In the following paragraph, we will discuss how cell-free system studies are critical for another type of PD-related protein, the aggregation-prone protein α-synuclein. But more importantly, we will focus on how we should treat cell-free system studies, how to interpret cell-free system data properly, and why cell-free system models need to be integrated with other models, such as cellular models and *in vivo* models, to study PD as a whole.

α-Synuclein was identified as a presynaptic protein (George *et al.*, 1995) and its mutation causes a rare form of fPD. The fact that α-synuclein is the major component of intracellular inclusions known as Lewy bodies (LB), a hallmark of PD, attracted extensive studies with the hopes of decoding the mechanism(s) of LB formation and its possible pathological roles in PD. Soon after the identification of the link between α-synuclein and neurodegenerative disease, a biochemical study in a cell-free system revealed that α-synuclein exhibits a striking random three-dimensional structure, a so-called natively unfolded structure, in solution (Weinreb *et al.*, 1996). Later, *in vitro* fibrillization assays using recombinant α-synuclein further revealed the intrinsic properties of α-synuclein as developing a more β-sheet structure and forming protein fibrils (Conway *et al.*, 1998; El-Agnaf *et al.*, 1998; Giasson *et al.*, 1999). The above finding stimulated the major hypothesis on the gain-of-toxicity of α-synuclein, that as an aggregation-prone protein, α-synuclein produces its toxicity through the formation of aggregates, such as protofibrils or fibrils.

Due to the lack of an intact biological environment, the finding of α-synuclein fibril formation in a cell-free system is only physiologically or pathologically important if we have evidence to show that *in vitro* generated α-synuclein filaments resemble those produced *in vivo*. Indeed, a later study confirmed that *in vitro* generated α-synuclein filaments resemble the major ultrastructural elements of authentic LB from human patients (Spillantini *et al.*, 1998). Thus, the knowledge acquired from a cell-free system could be interpreted as a biochemical explanation for the finding that α-synuclein forms the major filamentous component of LB from PD patients.

Later, one of the key advantages of cell-free systems, the well-controlled experimental conditions, was utilized in the study of modulators of the formation of α-synuclein aggregates. To this end, the *in vitro* assay showed that oxidative stressors, such as iron and peroxide (Hashimoto *et al.*, 1999) and nitrative stressors (Souza *et al.*, 2000) enhanced α-synuclein aggregation. The effects of nitrative stress on α-synuclein found in a cell-free system were interesting since previous studies showed that oxidative and nitrative stresses are potential pathogenic mediators of PD (Ischiropoulos, 1998). Again, it is after the finding that nitrated α-synuclein is present in the major filamentous building blocks of LB from human patients, that the role of oxidative and nitrative effects on α-synuclein becomes biologically important and is considered as one of the mechanisms underlying PD (Giasson *et al.*, 2000).

Therefore, although cell-free systems may be efficient and accurate in identifying the intrinsic functions/properties of a protein, it is necessary to test the physiological or pathological role in cellular studies or *in vivo* models, in order to apply the knowledge acquired from a cell-free system properly and accurately.

While some results from cell-free systems are successfully validated *in vivo*, other findings in cell-free systems lack clear biological/pathological meaning due to the lack of *in vivo* confirmation. For example, from cell-free system studies, it has been shown that some oligomeric α-synuclein species are annular or pore-like structure and that α-synuclein is a lipid-bound protein at least in some structural forms (Volles and Lansbury, Jr., 2002; Zhu *et al.*, 2003). Therefore, it has been suggested that α-synuclein oligomers might form pores in intracellular membranes such as the plasma membrane. Some indirect evidence of this comes from observations that cells expressing α-synuclein have increased cation permeability (Furukawa *et al.*, 2006), but the pores formed by α-synuclein lack direct confirmation from either cellular studies or *in vivo* studies. Thus, we must be cautious when we interpret such data. That is, although the existence of an α-synuclein pore structure formed in a cell-free system maybe a mechanism of α-synuclein's gain-of-toxicity, this hypothesis needs to be confirmed in cell or animal models. In addition, it remains to be studied further, whether or not, or how α-synuclein binds to cell membranes, and what membranes of which organelles. The latter will clarify the physiological or pathological meaning for such observations in cell-free systems.

CONCLUSION AND PERSPECTIVE

Given the above discussion, it is clear that cell-free models provide us with a powerful tool for defining the intrinsic functions/properties of a given protein, which are crucial for the study of PD-related proteins and to shed light onto the possible roles of these proteins in PD. However, the following two major points need to be taken into consideration in the interpretation of cell-free system studies. First, due to the different advantages or pitfalls of the methods used to produce target proteins, the data from one method need to be interpreted carefully and, if possible, confirmation by another method is necessary. Second, due to various well-established cell-free system assays, cell-free system data can usually open up a direction for *in vivo* studies. However, *in vivo* confirmation is necessary for the extension of knowledge from cell-free systems to the biological roles in PD.

In the future, a more complex cell-free system is needed to study the PD-related protein effects on organelles, such as mitochondria. For example, it has been shown that α-synuclein overexpression causes mitochondrial defects (Smith *et al.*, 2005). Whether or not α-synuclein has a direct effect on mitochondria is not clear, since α-synuclein is predominantly a cytoplasmic protein. Therefore, a well-designed cell-free system containing isolated mitochondria and purified α-synuclein with various structures, from monomer, to protofibril, to filament, would address this issue. Given that multiple organelle abnormalities, such as ER stress, mitochondrial and lysosomal dysfunction, are suggested as underlying the pathology of PD, such complex cell-free models could provide insight into the detailed molecular pathways of PD-related proteins.

With various established cell-free system assays, we can expect that the research on newly identified PD-related proteins, such as PINK1 and LRRK2, will advance quickly. The integration of knowledge acquired from cell-free models, cellular studies, and *in vivo* models will serve the research of PD as a whole, leading to a ultimate understanding of the pathogenesis of PD and possible strategies for the treatment of PD.

REFERENCES

Chung, K. K., Zhang, Y., Lim, K. L., Tanaka, Y., Huang, H., Gao, J., Ross, C. A., Dawson, V. L., and Dawson, T. M. (2001). Parkin ubiquitinates the alpha-synuclein-interacting protein, synphilin-1: Implications for Lewy-body formation in Parkinson disease. *Nat Med* **7**, 1144–1150.

Conway, K. A., Harper, J. D., and Lansbury, P. T. (1998). Accelerated *in vitro* fibril formation by a mutant alpha-synuclein linked to early-onset Parkinson disease. *Nat Med* **4**, 1318–1320.

Corti, O., Hampe, C., Koutnikova, H., Darios, F., Jacquier, S., Prigent, A., Robinson, J. C., Pradier, L., Ruberg, M., Mirande, M., Hirsch, E., Rooney, T., Fournier, A., and Brice, A. (2003). The p38 subunit of the aminoacyl-tRNA synthetase complex is a Parkin substrate: Linking protein biosynthesis and neurodegeneration. *Hum Mol Genet* **12**, 1427–1437.

El-Agnaf, O. M., Jakes, R., Curran, M. D., Middleton, D., Ingenito, R., Bianchi, E., Pessi, A., Neill, D., and Wallace, A. (1998). Aggregates from mutant and wild-type alpha-synuclein proteins and NAC peptide induce apoptotic cell death in human neuroblastoma cells by formation of beta-sheet and amyloid-like filaments. *FEBS Lett* **440**, 71–75.

Furukawa, K., Matsuzaki-Kobayashi, M., Hasegawa, T., Kikuchi, A., Sugeno, N., Itoyama, Y., Wang, Y., Yao, P. J., Bushlin, I., and Takeda, A. (2006). Plasma membrane ion permeability induced by mutant alpha-synuclein contributes to the degeneration of neural cells. *J Neurochem* **97**, 1071–1077.

George, J. M., Jin, H., Woods, W. S., and Clayton, D. F. (1995). Characterization of a novel protein regulated during the critical period for song learning in the zebra finch. *Neuron* **15**, 361–372.

Giasson, B. I., Uryu, K., Trojanowski, J. Q., and Lee, V. M. (1999). Mutant and wild type human alpha-synucleins assemble into elongated filaments with distinct morphologies *in vitro*. *J Biol Chem* **274**, 7619–7622.

Giasson, B. I., Duda, J. E., Murray, I. V., Chen, Q., Souza, J. M., Hurtig, H. I., Ischiropoulos, H., Trojanowski, J. Q., and Lee, V. M. (2000). Oxidative damage linked to neurodegeneration by selective alpha-synuclein nitration in synucleinopathy lesions. *Science* **290**, 985–989.

Haglund, K., Sigismund, S., Polo, S., Szymkiewicz, I., Di Fiore, P. P., and Dikic, I. (2003). Multiple monoubiquitination of RTKs is sufficient for their endocytosis and degradation. *Nat Cell Biol* 5, 461–466.

Hampe, C., rdila-Osorio, H., Fournier, M., Brice, A., and Corti, O. (2006). Biochemical analysis of Parkinson's disease-causing variants of Parkin, an E3 ubiquitin-protein ligase with monoubiquitylation capacity. *Hum Mol Genet* 15, 2059–2075.

Hashimoto, M., Hsu, L. J., Xia, Y., Takeda, A., Sisk, A., Sundsmo, M., and Masliah, E. (1999). Oxidative stress induces amyloid-like aggregate formation of NACP/alpha-synuclein *in vitro* [In Process Citation]. *Neuroreport* 10, 717–721.

Hershko, A., and Ciechanover, A. (1992). The ubiquitin system for protein degradation. *Annu Rev Biochem* 61, 761–807.

Hicke, L. (2001). Protein regulation by monoubiquitin. *Nat Rev Mol Cell Biol* 2, 195–201.

Imai, Y., Soda, M., and Takahashi, R. (2000). Parkin suppresses unfolded protein stress-induced cell death through its E3 ubiquitin-protein ligase activity. *J Biol Chem* 275, 35661–35664.

Imai, Y., Soda, M., Inoue, H., Hattori, N., Mizuno, Y., and Takahashi, R. (2001). An unfolded putative transmembrane polypeptide, which can lead to endoplasmic reticulum stress, is a substrate of Parkin. *Cell* 105, 891–902.

Ischiropoulos, H. (1998). Biological tyrosine nitration: A pathophysiological function of nitric oxide and reactive oxygen species. *Arch Biochem Biophys* 356, 1–11.

Joch, M., Ase, A. R., Chen, C. X., MacDonald, P. A., Kontogiannea, M., Corera, A. T., Brice, A., Seguela, P., and Fon, E. A. (2007). Parkin-mediated monoubiquitination of the PDZ protein PICK1 regulates the activity of acid-sensing ion channels. *Mol Biol Cell* 18, 3105–3118.

Katzmann, D. J., Odorizzi, G., and Emr, S. D. (2002). Receptor downregulation and multivesicular-body sorting. *Nat Rev Mol Cell Biol* 3, 893–905.

Kitada, T., Asakawa, S., Hattori, N., Matsumine, H., Yamamura, Y., Minoshima, S., Yokochi, M., Mizuno, Y., and Shimizu, N. (1998). Mutations in the parkin gene cause autosomal recessive juvenile parkinsonism. *Nature* 392, 605–608.

Matsuda, N., Kitami, T., Suzuki, T., Mizuno, Y., Hattori, N., and Tanaka, K. (2006). Diverse effects of pathogenic mutations of Parkin that catalyze multiple monoubiquitylation *in vitro*. *J Biol Chem* 281, 3204–3209.

Nakatsu, F., Sakuma, M., Matsuo, Y., Arase, H., Yamasaki, S., Nakamura, N., Saito, T., and Ohno, H. (2000). A di-leucine signal in the ubiquitin moiety. Possible involvement in ubiquitination-mediated endocytosis. *J Biol Chem* 275, 26213–26219.

Nussbaum, R. L. (1998). Putting the parkin into Parkinson's. *Nature* 392, 544–545.

Olzmann, J. A., Li, L., Chudaev, M. V., Chen, J., Perez, F. A., Palmiter, R. D., and Chin, L. S. (2007). Parkin-mediated K63-linked polyubiquitination targets

misfolded DJ-1 to aggresomes via binding to HDAC6. *J Cell Biol* 178, 1025–1038.

Shimura, H., Hattori, N., Kubo, S., Mizuno, Y., Asakawa, S., Minoshima, S., Shimizu, N., Iwai, K., Chiba, T., Tanaka, K., and Suzuki, T. (2000). Familial Parkinson disease gene product, parkin, is a ubiquitin-protein ligase. *Nat Genet* 25, 302–305.

Shimura, H., Schlossmacher, M. G., Hattori, N., Frosch, M. P., Trockenbacher, A., Schneider, R., Mizuno, Y., Kosik, K. S., and Selkoe, D. J. (2001). Ubiquitination of a new form of alpha-synuclein by parkin from human brain: Implications for Parkinson's disease. *Science* 293, 263–269.

Smith, W. W., Jiang, H., Pei, Z., Tanaka, Y., Morita, H., Sawa, A., Dawson, V. L., Dawson, T. M., and Ross, C. A. (2005). Endoplasmic reticulum stress and mitochondrial cell death pathways mediate A53T mutant alpha-synuclein-induced toxicity. *Hum Mol Genet* 14, 3801–3811.

Souza, J. M., Giasson, B. I., Chen, Q., Lee, V. M., and Ischiropoulos, H. (2000). Dityrosine cross-linking promotes formation of stable alpha-synuclein polymers. Implication of nitrative and oxidative stress in the pathogenesis of neurodegenerative synucleinopathies. *J Biol Chem* 275, 18344–18349.

Spillantini, M. G., Crowther, R. A., Jakes, R., Hasegawa, M., and Goedert, M. (1998). α-Synuclein in filamentous inclusions of Lewy bodies from Parkinson's disease and dementia with Lewy bodies. *Proc Natl Acad Sci USA* 95, 6469–6473.

Vila, M., and Przedborski, S. (2004). Genetic clues to the pathogenesis of Parkinson's disease. *Nat Med* 10(Suppl), S58–S62.

Volles, M. J., and Lansbury, P. T., Jr. (2002). Vesicle permeabilization by protofibrillar alpha-synuclein is sensitive to Parkinson's disease-linked mutations and occurs by a pore-like mechanism. *Biochemistry* 41, 4595–4602.

Weinreb, P. H., Zhen, W., Poon, A. W., Conway, K. A., and Lansbury, P. T., Jr. (1996). NACP, a protein implicated in Alzheimer's disease and learning, is natively unfolded. *Biochemistry* 35, 13709–13715.

Zhang, Y., Gao, J., Chung, K. K., Huang, H., Dawson, V. L., and Dawson, T. M. (2000a). Parkin functions as an E2-dependent ubiquitin- protein ligase and promotes the degradation of the synaptic vesicle-associated protein, CDCrel-1. *Proc Natl Acad Sci USA* 97, 13354–13359.

Zhang, Y., Gao, J., Chung, K. K., Huang, H., Dawson, V. L., and Dawson, T. M. (2000b). Parkin functions as an E2-dependent ubiquitin-protein ligase and promotes the degradation of the synaptic vesicle-associated protein, CDCrel-1. *Proc Natl Acad Sci USA* 97, 13354–13359.

Zhu, M., Li, J., and Fink, A. L. (2003). The association of alpha-synuclein with membranes affects bilayer structure, stability, and fibril formation. *J Biol Chem* 278, 40186–40197.

40

MICROGLIAL ACTIVATION IN A MOUSE MODEL OF α-SYNUCLEIN OVEREXPRESSION

KATHLEEN A. MAGUIRE-ZEISS[1], XIAOMIN SU[2,*] AND HOWARD J. FEDEROFF[1,3]

[1]*Department of Neuroscience, Georgetown University Medical Center, Washington, DC, USA*
[2]*Department of Microbiology and Immunology, University of Rochester School of Medicine and Dentistry, Rochester, NY, USA*
[3]*Department of Neurosurgery, University of California San Francisco, San Francisco, CA, USA*
Current Address: Department of Neurosurgery, University of California San Francisco, San Francisco, CA, USA

INTRODUCTION

The etiology of Parkinson's disease (PD) is unknown, but increasing evidence suggests that despite disparate initiating mechanisms including both genetic and environmental factors convergent pathogenic mechanisms exist leading to the invariant dopaminergic neuronal cell death which is the hallmark feature of PD (Dauer and Przedborski, 2003; Maguire-Zeiss and Federoff, 2003; Greenamyre and Hastings, 2004; Hodaie *et al.*, 2007). One genetic determinant, α-synuclein, has moved to the forefront of PD research because missense mutations in and overexpression of the wild-type protein as well as SYN's propensity to form oligomers have been linked to human disease. In addition, large intracytoplasmic protein aggregates in surviving substantia nigra (SN) dopaminergic neurons (DANs) the so-called Lewy bodies are replete with SYN and serve as a second hallmark pathological feature of PD (Spillantini *et al.*, 1998). A third common histopathologic feature of PD is the presence of activated microglia. Inflammation is associated with a broad spectrum of neurodegenerative diseases, including PD and other synucleinopathies, Alzheimer's disease, amyotrophic lateral sclerosis, Creutzfeldt–Jakob disease, Huntington's disease, multiple sclerosis, Pick's disease and other tauopathies (McGeer and McGeer, 2004; Boillee *et al.*, 2006; Cagnin *et al.*, 2006; Griffin, 2006; Kim and Joh, 2006; Klegeris *et al.*, 2007). The association of inflammation with PD was reported in 1988 when the presence of activated microglia in the SN of PD cases were revealed (McGeer *et al.*, 1988). Subsequent studies have established the involvement of a spectrum of inflammatory mediators in this disorder (Kim and Joh, 2006; Liu, 2006; Sawada *et al.*, 2006). In this chapter we review evidence supporting a role for SYN, microglia, proinflammatory molecules and the attendant oxidative stress in PD focusing on *early* proinflammatory events in a transgenic mouse model of PD with tyrosine hydroxylase promoter-directed overexpression of human *α-synuclein* ($SYN_{WT+/+}$).

ROLE OF SYNUCLEIN IN PD

Synuclein (SYN) is a 16-kDa cytoplasmic protein whose cellular function has yet to be completely defined. Genetic data demonstrates that both mutations in and overexpression of SYN correlates with the occurrence of PD (Polymeropoulos *et al.*, 1996, 1997; Polymeropoulos, 2000; Singleton *et al.*,

2003). The exact mechanism for SYN's toxic gain of function is not completely understood but thought to involve misfolding of this protein which natively adopts a random coiled conformation. Tissue culture and animal models of SYN overexpression demonstrate the propensity for this protein to form oligomers and higher order aggregates. Misfolded SYN is also present in Lewy bodies one of the hallmark pathological features of PD (Spillantini *et al.*, 1998). These intracytoplasmic proteinaceous inclusions are localized to surviving DANs and contain large amounts of SYN and ubiquitin. Both the formation and the role of Lewy bodies in PD pathogenesis remain elusive. However, recent studies have begun to shed light on the cause of SYN aggregation and its relevance to the pathogenesis of PD.

The accumulation of aggregated SYN is determined by an equilibrium between production and degradation. Several factors have been reported to enhance SYN misfolding and aggregation, including protein overexpression, mutation, chemical modification and molecular interaction (Figure 40.1). SYN

has a high self-aggregation propensity and overexpression and/or mutation are known to accelerate this aggregation process (Conway *et al.*, 2000a,b Greenbaum *et al.*, 2005). Chemical modification of SYN also has a direct effect on the normal conformation of this protein, thus affecting its aggregation propensity. In fact, exposure to oxidative and nitrative species stabilizes SYN filaments presumably through dityrosine cross-linking. Nitrated and/or phosphorylated SYN have been observed in synucleinopathy lesions suggesting a pathogenic role for this protein (Giasson *et al.*, 2000; Souza *et al.*, 2000; Paxinou *et al.*, 2001; Fujiwara *et al.*, 2002; Norris *et al.*, 2003). Furthermore, dopamine (DA) the neurotransmitter that is diminished in PD facilitates formation of toxic soluble oligomers, but inhibits formation of less toxic fibrils suggesting a conspiracy between DA and SYN to facilitate the demise of DANs (Cappai *et al.*, 2005; Maguire-Zeiss *et al.*, 2005, 2006). Alterations in SYN conformation also arise from the physical interaction between cellular SYN and other molecules such as

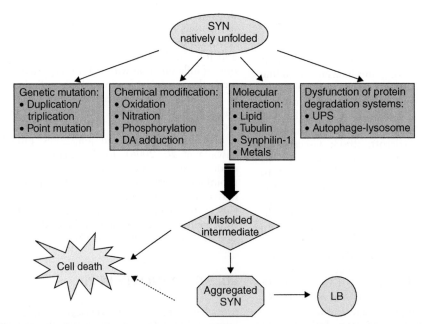

FIGURE 40.1 Multiple pathways lead to SYN aggregation. SYN is natively unfolded under normal physiological conditions. Genetic mutations, chemical modifications, molecular interactions and dysfunction of protein degradation systems induce SYN to form misfolded intermediates including β-sheet rich oligomers, protofibrils, stable amyloid fibrils and finally highly aggregated SYN. Increasing evidence suggests that misfolded intermediates are implicated in PD pathogenesis, leading to neuronal cell death. In PD the few surviving DAN accumulate abnormally aggregated SYN sequestered into Lewy bodies (LB).

lipids, tubulin, synphilin-1 and metals (Chung et al., 2001; Uversky et al., 2001; Lee et al., 2002; Alim et al., 2004). Finally, SYN degradation occurs through both the ubiquitin–proteasome (UPS) and the autophage–lysosomal systems whose functions are impacted by mutated and modified/misfolded SYN conformers which lead to further protein accumulation and toxicity (Stefanis et al., 2001; Meredith et al., 2002; Cuervo et al., 2004; Lee et al., 2004; Martinez-Vicente et al., 2008).

SYN aggregation is also appreciated following exposure to neurotoxicants with an attendant increase in oxidative stress. 1-Methyl-4-phenyl-1,2,3,6-tetrahydropyridine (MPTP) exposure was first linked to nigrostriatal dysfunction following inadvertent ingestion by heroin addicts and has been the most common neurotoxicant animal model of PD (Langston et al., 1983, 1984a, b, 1999; Kowall et al., 2000). Treatment of mice with MPTP results in oxidative modification and aggregation of SYN as well as specific DAN injury (Seniuk et al., 1990; Vila et al., 2000; Przedborski et al., 2001; Schmidt and Ferger, 2001; Gomez-Santos et al., 2002; Dauer and Przedborski, 2003; Song et al., 2004). MPTP also links another feature of PD, glial activation, with SYN such that following toxicant treatment mice display nigrostriatal microglial activation and astrocytosis as well as changes in SYN conformation (Kohutnicka et al., 1998; Thomas et al., 2004, 2005). Together these studies link SYN, glial activation and oxidative stress with the convergent pathogenic pathway that leads to DAN death in PD.

MICROGLIA-DIRECTED PROINFLAMMATORY EVENTS AND PD

Microglia constitute a network of central nervous system (CNS) immunocompetent cells comprising 5–20% of the brain cell population with a region specific distribution (Dobrenis, 1998; Perry, 1998; Alliot et al., 1999). The most widely accepted thesis is that microglial precursors enter the developing CNS from the vascular compartment, ventricles and meninges during in utero development, followed by spread through the parenchyma and differentiation into ramified cells (Cuadros and Navascues, 1998). During adulthood microglia can be replenished slowly but continuously by recently arrived bone marrow derived cells, which enter the brain parenchyma and assume a ramified morphology identical to that of intrinsic microglia (Flugel et al., 2001).

Along with astrocytes, microglia comprise the innate immune response in the brain and are responsible for mediating responses to foreign proteins and cell debris (reviewed in Kim and Joh, 2006). The primary functions of microglia are to provide continuous surveillance of the parenchyma and to protect the CNS during injury and disease. In the non-activated/resting state microglia usually display a ramified morphology. When faced with a foreign protein/molecule or cell debris microglia go through a staged process of activation which includes changes in cell morphology from the resting ramified shape to an amoeboid profile, increased cell surface receptor expression, and stimulated production and release of chemokines and cytokines (Kreutzberg, 1996). In the course of their activities, microglia produce reactive oxygen species (ROS) through activation of membrane-bound NADPH oxidase (Gao et al., 2003a, b; Wu et al., 2003, 2005). Continuous release of inflammatory molecules such as tumor necrosis factor alpha (TNFα) and increased production of ROS is both proinflammatory and neurotoxic.

Microglia-associated inflammation is a pathological hallmark of *end-stage* PD presumably due to the immune surveillance response to dead DANs (McGeer et al., 1988; McGeer and McGeer, 2004; Croisier et al., 2005). Enhanced expression of the proninflammatory molecules, TNFα, IL1β, IL6 and INFγ has been shown in basal ganglia as well as cerebrospinal fluid of PD patients (Hunot et al., 1996; Hirsch et al., 1998; Nagatsu et al., 2000). The PD inflammatory process which includes activation/proliferation of microglia, secretion of proinflammatory cytokines and production of free radicals is multifaceted. First, activation maybe an event secondary to the neurodegenerative process such that as DANs die releasing cellular contents microglia are recruited and activated to remove cell debris (Giasson et al., 2000; Przedborski et al., 2001; Mandel et al., 2005). This "end-stage" activation would be predicted to occur later in the disease process when DANs are dying but prior to diagnosis as patients present clinical symptoms after approximately 60–70% of DANs are already dead. Second and perhaps the most intriguing microglial activation maybe a primary event occurring in response to the release of molecules from damaged/stressed neurons prior to cell death which would engender a feed forward cycle of microglial activation, proinflammatory molecule release and oxidative stress resulting in cycle reinforcement and finally neurodegeneration (Zhang *et al.*, 2005; Su *et al.*, 2007; Thomas *et al.*, 2007; Reynolds et al., 2008). In this case it is anticipated that microglial activation occurs earlier during the course of disease prior to cell death. Third, it is conceivable that microglial activation can occur at the initiation of disease either due to a self-limited injury or subtle genetic vulnerability (i.e., increased SYN) and during this epoch microglia are capable of resolving the initial insult but remain primed for a second "hit." This second "hit" could be a neurotoxicant exposure, traumatic brain injury or an intrinsic alteration in the rates of protein production and degradation. Following reactivation microglia would enter a pathological proinflammatory state leading to neurodegeneration. In all scenarios, activated microglia exert their neurotoxic effects by releasing proinflammatory cytokines such as TNFα, IL1β, IL6 and IFNγ, free radicals including ROS and nitric oxide (NO), as well as inflammatory

mediators such as prostaglandins PGE2, leading to DAN damage and death (Figure 40.2).

SYN$_{WT+/+}$ MICE: A MODEL TO STUDY SYN AND MICROGLIA AS PROINFLAMMATORY CO-CONSPIRATORS IN PD PATHOGENESIS

The mechanism by which SYN promotes neurodegeneration in PD is unknown. However, there is a clear linkage between misfolded SYN, neurotoxicant exposure, oxidative stress and cell death. Furthermore, glial activation has been linked to MPTP treatment where both microglial activation and astrocytosis occur in the nigrostriatum (Kohutnicka *et al.*, 1998; Thomas *et al.*, 2004). MPTP treatment also causes aggregation of SYN, suggesting a relationship between misfolded SYN and glial activation. Taken together these studies support the convergent pathogenesis pathway of PD that includes SYN, oxidative stress, protein aggregation and microglial activation. We sought to determine using a mouse model of human α-synuclein

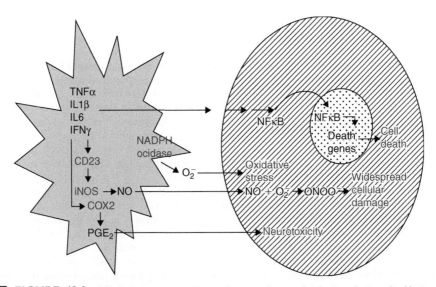

FIGURE 40.2 Potential neurotoxic effects of activated microglia. Activated microglia (depicted in gray) produce proinflammatory cytokines including TNFα, IL1β, IL6 and INFγ, which induce downstream signaling pathways to activate NFκB, leading to neuronal cell death (depicted in stripes). In addition, these cytokines will induce iNOS through CD23, resulting in the release of NO. NO interacts with NADPH oxidase-derived O_2^- to generate $ONOO^-$ and cause widespread cellular damage. Furthermore, the cytokines can activate COX2 to produce PGE2, leading to neurotoxicity.

overexpression in the neurons most vulnerable in PD (SYN$_{WT+/+}$) whether microglial activation is an early pathologic event and then using primary cultured microglia whether SYN *directs* microglial activation (Su *et al.*, 2007).

We have previously reported in heterozygous transgenic mice that overexpress human wild-type α-synuclein driven by a rat tyrosine hydroxylase promoter (hwα-SYN-5; Richfield *et al.*, 2002) increased DA transporter density and dystrophic

neurites. In our current studies the transgene locus has been homozygosed (SYN$_{WT+/+}$). We demonstrate that SYN overexpression results in increased numbers of activated microglia as enumerated by unbiased stereology of ionized calcium-binding adaptor molecule 1 positive (Iba1+) microglia in the SN of SYN$_{WT+/+}$ mice compared to non-transgenic animals (Su *et al.*, 2007; Figure 40.3a). We counted both resting and activated microglia based on morphological characteristics and found an increase in activated microglia but no change in total numbers of microglia; illustrating a change of activation state and not proliferation or recruitment to the SN at this time point. Furthermore, mRNA isolated from the SN and striatum (STR) demonstrated enhanced expression of the prototypical proinflammatory cytokine TNFα in 1-month-old SYN$_{WT+/+}$ mice compared with age-matched non-transgenic animals (Su *et al.*, 2007; Figure 40.3b). Iba1 mRNA was also increased in the SN of mice corroborating the immunohistological data. This SYN-dependent activation occurs at a time well in advance of DAN death, suggesting that SYN overexpression is driving microglial activation and proinflammatory events *early* in the pathogenesis in this model of PD (Thiruchelvam *et al.*, 2004). We posit that early microglial activation plays a role in the later development of PD and in support of this hypothesis studies have demonstrated that intranigral lipopolysaccharide (LPS) injections induce potent early microglial activation (within 2 days) with subsequent degeneration of DANs (up to 1 year after LPS injection; Castano *et al.*, 1998). Similarly, Purisai *et al.* (2006) demonstrated that induction of microglial activation by LPS followed by a single dose of paraquat triggered loss of DAN and that this degeneration was dependent on NADPH oxidase activity.

In our model, if SYN is the central molecule responsible for direct activation of microglia there must be extracellular SYN or a portion of the protein available to activate resident microglia through either phagocytosis or receptor-mediated mechanisms. Release of SYN has been demonstrated in a culture model following neural activity and in cell culture models of SYN overexpression (Fortin *et al.*, 2005; Lee *et al.*, 2005; Su *et al.*, 2007) suggesting that SYN may be available to directly activate microglia. Further work is required to establish that SYN is released from neurons *in vivo* in our model.

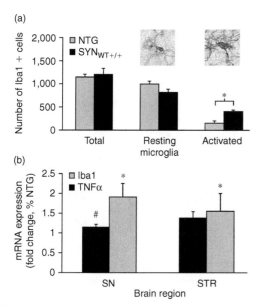

FIGURE 40.3 Early expression of SYN activates microglia in SYN$_{WT+/+}$ mice. Iba1+ cells were enumerated from 1-month-old SYN$_{WT+/+}$ and C57Bl6 (NTG) mice (n = 4/genotype) following immunohistochemistry for the microglia marker, ionized calcium-binding adaptor molecule 1 (Iba1). Representative images of resting and activated microglia are shown (inset; 60× magnification; Olympus AX70, Melville, NY). Here, we demonstrate a statistically significant increase in activated SN microglia (a; *, p = 0.0001). RNA from the SN and STR of 1-month-old SYN$_{WT+/+}$ and NTG was subjected to qRT-PCR to determine changes in Iba1 and TNFα expression levels. There is a significant increase in Iba1 expression in the SN while TNFα expression is increased in both the SN and STR (b; #, p = 0.03; *, p < 0.05; from Su *et al.*, 2007).

Cultured microglia have been widely utilized to examine the effect of SYN on activation of proinflammatory processes. We utilized enriched primary microglial cultures from non-transgenic mice exposed to purified SYN to establish an activation state (Su et al., 2007). Exogenous treatment with SYN (50 nM) resulted in approximately 50% of the cultured microglia to become activated as determined by a change in Iba1+ cell morphology. At very high protein concentrations (250 nM) over 80% of the cultured microglia exhibited an amoeboid/activated morphology. As shown in Figure 40.4 this SYN-directed microglial activation resulted in a proinflammatory state exemplified by a time-dependent increase in mRNA expression levels of TNFα, IL1β, IL6, COX2, NOX2 and iNOS (Figure 40.4a; Su et al., 2007). An elevation in TNFα protein released into the culture media was also demonstrated (Figure 40.4b; Su et al., 2007). Lastly we determined that the cellular ROS production was dependent on SYN (Figure 40.4c; Su et al., 2007). Our studies demonstrate a robust time and SYN dose-dependent increase in the production of cytokines, inflammation-related enzymes and ROS consistent with a microglia induced proinflammatory cascade and reflective of activation in our animal model (SYN$_{WT+/+}$ mice). Although increased release of TNFα was not unexpected the increased production of other cytokines as well as the concomitant upregulation of COX2, NOX2, iNOS and ROS is setting in motion a self-propagating cycle of inflammation that may prime microglia to future insults such as neurotoxicant exposure (Figure 40.5). Zhang et al. (2005) also demonstrated an increase in microglial activation following treatment of primary mesencephalic neuron–glia cultures with aggregated SYN. Furthermore using this system SYN-directed microglial activation resulted in decreased numbers of TH+ neurons and decreased DA uptake which were NADPH oxidase dependent (Zhang et al., 2005).

Rapid activation of microglia in response to external stimuli is mediated by a variety of surface pattern recognition receptors (PRRs) (Block et al., 2007). One PRR CD36 is upregulated in Alzheimer's disease where it is responsible for internalization and degradation of fibrillar β-amyloid (Husemann et al., 2002; Medeiros et al., 2004; Cho et al., 2005). Since fibrillar SYN shares a similar structure with fibrillar β-amyloid, CD36 is proposed to play a role in SYN-dependent microglial activation

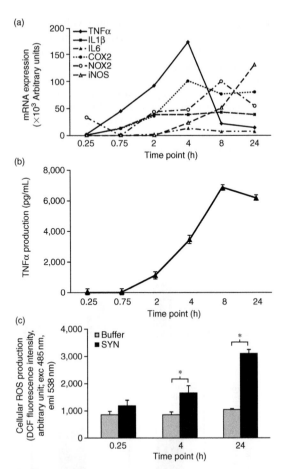

FIGURE 40.4 Exogenous SYN induces the expression of microglial-derived proinflammatory molecules. We have previously demonstrated a SYN dose-dependent increase in activated cultured microglia based on unbiased stereological counts of Iba1+ amoeboid microglia (Su et al., 2007). In this experiment primary microglia-enriched cultures from C57/Bl6 mice were treated with 250 nM SYN or buffer only for 0.25, 0.75, 2, 4, 8 and 24 h. (a) RNA extracted from treated cells was subjected to qRT-PCR to evaluate the temporal expression of TNFα, IL1 β, IL6, COX2, NOX2 and iNOS. (b) TNFα protein levels in the conditioned media were determined using an ELISA (enzyme-linked immunoabsorbent assay). Cellular ROS was determined using 2', 7'-dichlorofluorescein diacetate (c; *, $p < 0.05$; DCF). Here we demonstrate SYN- and time-dependent changes in the expression of proinflammatory molecules and cellular ROS production consistent with increased microglial activation. Adapted from Su et al. (2007).

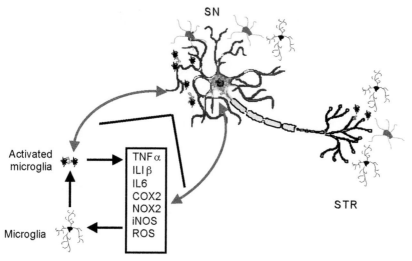

FIGURE 40.5 SYN and activated microglia: a vicious cycle leading to neurodegeneration. Utilizing SYN$_{WT+/+}$ mice we demonstrate increased microglial activation and TNFα release in SN DANs. We posit that released SYN can directly activate microglia initiating a proinflammatory state that includes increased expression of TNFα, ILIβ, IL6, COX2, NOX2, iNOS and ROS. This proinflammatory environment promotes the activation of neighboring microglia as well as direct damage to the DAN resulting in increased SYN release continuing the cycle until neurons are compromised by repeated exposure to ROS and proinflammatory insults eventually leading to DAN death and loss of striatal fibers. (SN = substantia nigra DAN cell body; STR = striatal presynaptic terminals; astrocytes are shown surrounding the SN DAN and STR and contain large cell bodies (light gray); aggregated SYN is present in both the SN and STR; both resting and activated microglia are as indicated in the figure (black cell body; gray processes.)

(Zhang *et al.*, 2005; Klegeris *et al.*, 2007; Su *et al.*, 2007). We have demonstrated that enriched microglial cultures from CD36-deficient mice are partially protected from SYN-mediated activation compared to non-transgenic mice (Su *et al.*, 2007). Furthermore, CD36 knockout microglial cultures display reduced extracellular regulated protein-serine kinase phosphorylation (pERK1/2) and also decreased ROS production. This work is consistent with others that have demonstrated that SYN activates stress signaling protein kinases (Klegeris *et al.*, 2007). Likewise, another PRR, the MAC1 receptor, also known as complement receptor 3 (CR3)/integrin CD11b/CD18, fulfills a dual function as an adhesion molecule and a PRR that binds several diverse ligands, including iCR3b fragment of complement (Ross and Vetvicka, 1993), mutant SYN (A30P and A53T) (Zhang *et al.*, 2007) and LPS (Pei *et al.*, 2007). The expression of MAC1 has been shown to be elevated in the postmortem brains of another neurodegenerative disease,

Alzheimer's disease, but its fate in PD remains to be determined (Akiyama and McGeer, 1990).

CONCLUSION AND FUTURE DIRECTIONS: SYN$_{WT+/+}$ MICE

Our findings of increased microglial activation in young SYN$_{WT+/+}$ mice, well in advance of any evidence of neurodegeneration, suggest that microglia are playing an active role in the progression of PD rather than reacting after substantial cell death has ensued. We conclude that SYN is the central player in the neurotoxicity associated with PD and that through overexpression and misfolding can directly cause neuronal damage; moreover, SYN release through a still obliquely understood mechanism appears to facilitate direct activation of microglia. In our model of SYN-mediated nigrostriatal dysfunction we envision a temporal and spatial

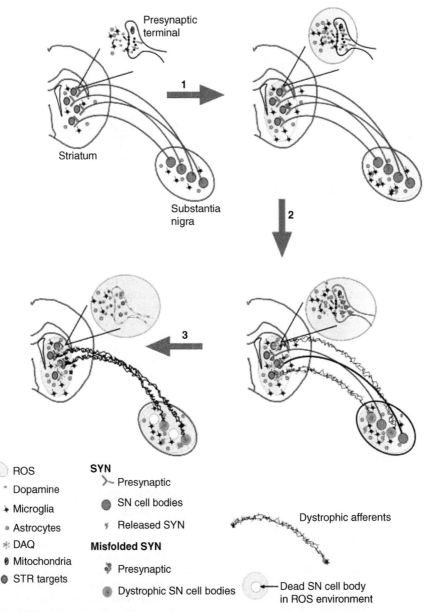

FIGURE 40.6 Model of nigrostriatal dysfunction in $SYN_{WT+/+}$ mice. A schematic of the proposed temporal and spatial progression of PD pathogenesis in $SYN_{WT+/+}$ mice is depicted. We have demonstrated pathologic microglial activation in 1-month-old $SYN_{WT+/+}$ mice. Here we propose that in $SYN_{WT+/+}$ mice between 1 month and 6 months the overexpressed and released SYN acts directly on the presynaptic terminal and DAN cell body to activate microglia. The graded response to continued microglial activation and increased ROS in both brain regions is depicted in gray and represented by arrow 1. As mice continue to age ROS (gray shading) will increase due to a combination of the formation of dopamine quinone (DAQ), DAQ-modified proteins, misfolded SYN and continuous microglial activation leading to dystrophic neurites and the formation of SYN aggregates within DANs (arrow 2). When $SYN_{WT+/+}$ mice are 19–24 months old ROS continues to increase and DAN die resulting in another burst of microglial activation and presynaptic dysfunction (arrow 3).

progression of pathogenesis (Figure 40.6). We posit that early pathology will be evident in both the SN and STR. Specifically, at a young age (1–6 months old) we propose that a small amount of secreted SYN sets into motion a feed forward cycle of microglial activation and increased oxidant stress, resulting in more released and misfolded SYN conformers (Figure 40.6, arrow 1). As mice age, there will be an increase in ROS due to a combination of increased synaptic DA, dopamine quinone (DAQ)-modified proteins, misfolded SYN conformers and continuous activation of microglia (Figure 40.6, arrow 2). In older animals (late, 19+ months), ROS will continue to increase with the appearance of dystrophic afferents and attendant presynaptic dysfunction. At these later time points SN DAN will report large increases in ROS and microglial activation and finally cell death. Once microglia are activated a proinflammatory cascade is set into motion that left unchecked will lead to neurotoxic levels of cytokines and ROS. Although activated microglia can revert back to the resting state through the actions of IL10 and fractalkine we hypothesize that repeated stimulation would readily reactivate this primed population of microglia. It is this action of SYN on microglia and the consequent increase in ROS that facilitates the increased production of misfolded SYN which in turn reactivates the microglia resulting in a vicious cycle resulting in death of DAN (Figure 40.5). This model is supported by our observation of increased microglial activation and proinflammatory molecules in 1-month-old $SYN_{WT+/+}$ mice and further by our demonstration of a direct effect of exogenous SYN on cultured microglia.

In conclusion, the $SYN_{WT+/+}$ mice represent a novel model to further dissect important pathogenic epochs of disease with regard to microglial activation. First, the finding that microglia are activated at 1 month of age suggests that early anti-inflammatory intervention may preclude later pathology. Secondly, our *in vitro* work demonstrating a direct effect of SYN on microglia advocates for prevention of SYN release. We have previously identified single chain antibodies specific for different forms of SYN that may be useful in a gene therapeutic approach to direct SYN for more efficient degradation (Maguire-Zeiss *et al.*, 2006). Thirdly, our observation that CD36 is involved in SYN-directed microglial activation portends the usefulness for small molecule inhibitors of CD36 as anti-inflammatory agents in this model. In conclusion, $SYN_{WT+/+}$ mice represent an important model

for testing novel therapeutic agents to prevent aberrant microglial activation and for dissecting critical stages of disease progression.

REFERENCES

Akiyama, H., and McGeer, P. L. (1990). Brain microglia constitutively express beta-2 integrins. *J Neuroimmunol* 30, 81–93.

Alim, M. A., Ma, Q. L., Takeda, K., Aizawa, T., Matsubara, M., Nakamura, M., Asada, A., Saito, T., Kaji, H., Yoshii, M., Hisanaga, S., and Ueda, K. (2004). Demonstration of a role for alpha-synuclein as a functional microtubule-associated protein. *J Alzheimers Dis* 6, 435–442, discussion 443–449..

Alliot, F., Godin, I., and Pessac, B. (1999). Microglia derive from progenitors, originating from the yolk sac, and which proliferate in the brain. *Brain Res Dev Brain Res* 117, 145–152.

Block, M. L., Zecca, L., and Hong, J. S. (2007). Microglia-mediated neurotoxicity: Uncovering the molecular mechanisms. *Nat Rev Neurosci* 8, 57–69.

Boillee, S., Vande Velde, C., and Cleveland, D. W. (2006). ALS: A disease of motor neurons and their nonneuronal neighbors. *Neuron* 52, 39–59.

Cagnin, A., Kassiou, M., Meikle, S. R., and Banati, R. B. (2006). *In vivo* evidence for microglial activation in neurodegenerative dementia. *Acta Neurol Scand Suppl* 185, 107–114.

Cappai, R., Leck, S. L., Tew, D. J., Williamson, N. A., Smith, D. P., Galatis, D., Sharples, R. A., Curtain, C. C., Ali, F. E., Cherny, R. A., Culvenor, J. G., Bottomley, S. P., Masters, C. L., Barnham, K. J., and Hill, A. F. (2005). Dopamine promotes alpha-synuclein aggregation into SDS-resistant soluble oligomers via a distinct folding pathway. *FASEB J* 19, 1377–1379.

Castano, A., Herrera, A. J., Cano, J., and Machado, A. (1998). Lipopolysaccharide intranigral injection induces inflammatory reaction and damage in nigrostriatal dopaminergic system. *J Neurochem* 70, 1584–1592.

Cho, S., Park, E. M., Febbraio, M., Anrather, J., Park, L., Racchumi, G., Silverstein, R. L., and Iadecola, C. (2005). The class B scavenger receptor CD36 mediates free radical production and tissue injury in cerebral ischemia. *J Neurosci* 25, 2504–2512.

Chung, K. K., Zhang, Y., Lim, K. L., Tanaka, Y., Huang, H., Gao, J., Ross, C. A., Dawson, V. L., and Dawson, T. M. (2001). Parkin ubiquitinates the alpha-synuclein-interacting protein, synphilin-1: Implications for Lewy-body formation in Parkinson disease. *Nat Med* 7, 1144–1150.

Conway, K. A., Harper, J. D., and Lansbury, P. T., Jr (2000a). Fibrils formed *in vitro* from alpha-synuclein and two mutant forms linked to Parkinson's disease are typical amyloid. *Biochemistry* 39, 2552–2563.

Conway, K. A., Lee, S. J., Rochet, J. C., Ding, T. T., Williamson, R. E., and Lansbury, P. T., Jr (2000b). Acceleration of oligomerization, not fibrillization, is a shared property of both alpha-synuclein mutations linked to early-onset Parkinson's disease: Implications for pathogenesis and therapy. *Proc Natl Acad Sci USA* 97, 571–576.

Croisier, E., Moran, L. B., Dexter, D. T., Pearce, R. K., and Graeber, M. B. (2005). Microglial inflammation in the parkinsonian substantia nigra: Relationship to alpha-synuclein deposition. *J Neuroinflammation* 2, 14.

Cuadros, M. A., and Navascues, J. (1998). The origin and differentiation of microglial cells during development. *Prog Neurobiol* 56, 173–189.

Cuervo, A. M., Stefanis, L., Fredenburg, R., Lansbury, P. T., and Sulzer, D. (2004). Impaired degradation of mutant alpha-synuclein by chaperone-mediated autophagy. *Science* 305, 1292–1295.

Dauer, W., and Przedborski, S. (2003). Parkinson's disease: Mechanisms and models. *Neuron* 39, 889–909.

Dobrenis, K. (1998). Microglia in cell culture and in transplantation therapy for central nervous system disease. *Methods* 16, 320–344.

Flugel, A., Bradl, M., Kreutzberg, G. W., and Graeber, M. B. (2001). Transformation of donor-derived bone marrow precursors into host microglia during autoimmune CNS inflammation and during the retrograde response to axotomy. *J Neurosci Res* 66, 74–82.

Fortin, D. L., Nemani, V. M., Voglmaier, S. M., Anthony, M. D., Ryan, T. A., and Edwards, R. H. (2005). Neural activity controls the synaptic accumulation of alpha-synuclein. *J Neurosci* 25, 10913–10921.

Fujiwara, H., Hasegawa, M., Dohmae, N., Kawashima, A., Masliah, E., Goldberg, M. S., Shen, J., Takio, K., and Iwatsubo, T. (2002). Alpha-synuclein is phosphorylated in synucleinopathy lesions. *Nat Cell Biol* 4, 160–164.

Gao, H. M., Liu, B., and Hong, J. S. (2003a). Critical role for microglial NADPH oxidase in rotenone-induced degeneration of dopaminergic neurons. *J Neurosci* 23, 6181–6187.

Gao, H. M., Liu, B., Zhang, W., and Hong, J. S. (2003b). Critical role of microglial NADPH oxidase-derived free radicals in the *in vitro* MPTP model of Parkinson's disease. *FASEB J* 17, 1954–1956.

Giasson, B. I., Duda, J. E., Murray, I. V., Chen, Q., Souza, J. M., Hurtig, H. I., Ischiropoulos, H., Trojanowski, J. Q., and Lee, V. M. (2000). Oxidative damage linked to neurodegeneration by selective alpha-synuclein nitration in synucleinopathy lesions. *Science* 290, 985–989.

Gomez-Santos, C., Ferrer, I., Reiriz, J., Vinals, F., Barrachina, M., and Ambrosio, S. (2002). MPP+ increases alpha-synuclein expression and ERK/MAP-kinase phosphorylation in human neuroblastoma SH-SY5Y cells. *Brain Res* 935, 32–39.

Greenamyre, J. T., and Hastings, T. G. (2004). Biomedicine. Parkinson's – divergent causes, convergent mechanisms. *Science* 304, 1120–1122.

Greenbaum, E. A., Graves, C. L., Mishizen-Eberz, A. J., Lupoli, M. A., Lynch, D. R., Englander, S. W., Axelsen, P. H., and Giasson, B. I. (2005). The E46K mutation in alpha-synuclein increases amyloid fibril formation. *J Biol Chem* 280, 7800–7807.

Griffin, W. S. (2006). Inflammation and neurodegenerative diseases. *Am J Clin Nutr* 83, 470S–474S.

Hirsch, E. C., Hunot, S., Damier, P., and Faucheux, B. (1998). Glial cells and inflammation in Parkinson's disease: A role in neurodegeneration? *Ann Neurol* 44, S115–S120.

Hodaie, M., Neimat, J. S., and Lozano, A. M. (2007). The dopaminergic nigrostriatal system and Parkinson's disease: Molecular events in development, disease, and cell death, and new therapeutic strategies. *Neurosurgery* 60, 17–28, discussion 28–30.

Hunot, S., Boissiere, F., Faucheux, B., Brugg, B., Mouatt-Prigent, A., Agid, Y., and Hirsch, E. C. (1996). Nitric oxide synthase and neuronal vulnerability in Parkinson's disease. *Neuroscience* 72, 355–363.

Husemann, J., Loike, J. D., Anankov, R., Febbraio, M., and Silverstein, S. C. (2002). Scavenger receptors in neurobiology and neuropathology: Their role on microglia and other cells of the nervous system. *Glia* 40, 195–205.

Kim, Y. S., and Joh, T. H. (2006). Microglia, major player in the brain inflammation: Their roles in the pathogenesis of Parkinson's disease. *Exp Mol Med* 38, 333–347.

Klegeris, A., McGeer, E. G., and McGeer, P. L. (2007). Therapeutic approaches to inflammation in neurodegenerative disease. *Curr Opin Neurol* 20, 351–357.

Kohutnicka, M., Lewandowska, E., Kurkowska-Jastrzebska, I., Czlonkowski, A., and Czlonkowska, A. (1998). Microglial and astrocytic involvement in a murine model of Parkinson's disease induced by 1-methyl-4-phenyl-1,2,3,6-tetrahydropyridine (MPTP). *Immunopharmacology* 39, 167–180.

Kowall, N. W., Hantraye, P., Brouillet, E., Beal, M. F., McKee, A. C., and Ferrante, R. J. (2000). MPTP induces alpha-synuclein aggregation in the substantia nigra of baboons. *Neuroreport* 11, 211–213.

Kreutzberg, G. W. (1996). Microglia: A sensor for pathological events in the CNS. *Trends Neurosci* 19, 312–318.

Langston, J. W., Ballard, P., Tetrud, J. W., and Irwin, I. (1983). Chronic parkinsonism in humans due to a product of meperidine-analog synthesis. *Science* 219, 979–980.

Langston, J. W., Irwin, I., Langston, E. B., and Forno, L. S. (1984a). 1-Methyl-4-phenylpyridinium ion (MPP+): Identification of a metabolite of MPTP, a toxin selective to the substantia nigra. *Neurosci Lett* **48**, 87–92.

Langston, J. W., Langston, E. B., and Irwin, I. (1984b). MPTP-induced parkinsonism in human and non-human primates – clinical and experimental aspects. *Acta Neurol Scand Suppl* **100**, 49–54.

Langston, J. W., Forno, L. S., Tetrud, J., Reeves, A. G., Kaplan, J. A., and Karluk, D. (1999). Evidence of active nerve cell degeneration in the substantia nigra of humans years after 1-methyl-4-phenyl-1,2,3,6-tetrahydropyridine exposure. *Ann Neurol* **46**, 598–605.

Lee, H. J., Khoshaghideh, F., Patel, S., and Lee, S. J. (2004). Clearance of alpha-synuclein oligomeric intermediates via the lysosomal degradation pathway. *J Neurosci* **24**, 1888–1896.

Lee, H. J., Patel, S., and Lee, S. J. (2005). Intravesicular localization and exocytosis of alpha-synuclein and its aggregates. *J Neurosci* **25**, 6016–6024.

Lee, H. J., Shin, S. Y., Choi, C., Lee, Y. H., and Lee, S. J. (2002). Formation and removal of alpha-synuclein aggregates in cells exposed to mitochondrial inhibitors. *J Biol Chem* **277**, 5411–5417.

Liu, B. (2006). Modulation of microglial pro-inflammatory and neurotoxic activity for the treatment of Parkinson's disease. *AAPS J* **8**, E606–E621.

Maguire-Zeiss, K. A., and Federoff, H. J. (2003). Convergent pathobiologic model of Parkinson's disease. *Ann NY Acad Sci* **991**, 152–166.

Maguire-Zeiss, K. A., Short, D. W., and Federoff, H. J. (2005). Synuclein, dopamine and oxidative stress: Co-conspirators in Parkinson's disease? *Brain Res Mol Brain Res* **134**, 18–23.

Maguire-Zeiss, K. A., Wang, C. I., Yehling, E., Sullivan, M. A., Short, D. W., Su, X., Gouzer, G., Henricksen, L. A., Wuertzer, C. A., and Federoff, H. J. (2006). Identification of human alpha-synuclein specific single chain antibodies. *Biochem Biophys Res Commun* **349**, 1198–1205.

Mandel, S., Grunblatt, E., Riederer, P., Amariglio, N., Jacob-Hirsch, J., Rechavi, G., and Youdim, M. B. (2005). Gene expression profiling of sporadic Parkinson's disease substantia nigra pars compacta reveals impairment of ubiquitin–proteasome subunits, SKP1A, aldehyde dehydrogenase, and chaperone HSC-70. *Ann NY Acad Sci* **1053**, 356–375.

Martinez-Vicente, M., Talloczy, Z., Kaushik, S., Massey, A. C., Mazzulli, J., Mosharov, E. V., Hodara, R., Fredenburg, R., Wu, D. C., Follenzi, A., Dauer, W., Przedborski, S., Ischiropoulos, H., Lansbury, P. T., Sulzer, D., and Cuervo, A. M. (2008). Dopamine-modified alpha-synuclein blocks chaperone-mediated autophagy. *J Clin Invest* **118**, 777–788.

McGeer, P. L., and McGeer, E. G. (2004). Inflammation and the degenerative diseases of aging. *Ann NY Acad Sci* **1035**, 104–116.

McGeer, P. L., Itagaki, S., Boyes, B. E., and McGeer, E. G. (1988). Reactive microglia are positive for HLA-DR in the substantia nigra of Parkinson's and Alzheimer's disease brains. *Neurology* **38**, 1285–1291.

Medeiros, L. A., Khan, T., El Khoury, J. B., Pham, C. L., Hatters, D. M., Howlett, G. J., Lopez, R., O'Brien, K. D., and Moore, K. J. (2004). Fibrillar amyloid protein present in atheroma activates CD36 signal transduction. *J Biol Chem* **279**, 10643–10648.

Meredith, G. E., Totterdell, S., Petroske, E., Santa Cruz, K., Callison, R. C., Jr, and Lau, Y. S. (2002). Lysosomal malfunction accompanies alpha-synuclein aggregation in a progressive mouse model of Parkinson's disease. *Brain Res* **956**, 156–165.

Nagatsu, T., Mogi, M., Ichinose, H., and Togari, A. (2000). Changes in cytokines and neurotrophins in Parkinson's disease. *J Neural Transm Suppl*, 277–290.

Norris, E. H., Giasson, B. I., Ischiropoulos, H., and Lee, V. M. (2003). Effects of oxidative and nitrative challenges on alpha-synuclein fibrillogenesis involve distinct mechanisms of protein modifications. *J Biol Chem* **278**, 27230–27240.

Paxinou, E., Chen, Q., Weisse, M., Giasson, B. I., Norris, E. H., Rueter, S. M., Trojanowski, J. Q., Lee, V. M., and Ischiropoulos, H. (2001). Induction of alpha-synuclein aggregation by intracellular nitrative insult. *J Neurosci* **21**, 8053–8061.

Pei, Z., Pang, H., Qian, L., Yang, S., Wang, T., Zhang, W., Wu, X., Dallas, S., Wilson, B., Reece, J. M., Miller, D. S., Hong, J. S., and Block, M. L. (2007). MAC1 mediates LPS-induced production of superoxide by microglia: The role of pattern recognition receptors in dopaminergic neurotoxicity. *Glia* **55**, 1362–1373.

Perry, V. H. (1998). A revised view of the central nervous system microenvironment and major histocompatibility complex class II antigen presentation. *J Neuroimmunol* **90**, 113–121.

Polymeropoulos, M. H. (2000). Genetics of Parkinson's disease. *Ann NY Acad Sci* **920**, 28–32.

Polymeropoulos, M. H., Higgins, J. J., Golbe, L. I., Johnson, W. G., Ide, S. E., Di Iorio, G., Sanges, G., Stenroos, E. S., Pho, L. T., Schaffer, A. A., Lazzarini, A. M., Nussbaum, R. L., and Duvoisin, R. C. (1996). Mapping of a gene for Parkinson's disease to chromosome 4q21-q23. *Science* **274**, 1197–1199.

Polymeropoulos, M. H., Lavedan, C., Leroy, E., Ide, S. E., Dehejia, A., Dutra, A., Pike, B., Root, H., Rubenstein, J., Boyer, R., Stenroos, E. S., Chandrasekharappa, S., Athanassiadou, A., Papapetropoulos, T., Johnson, W. G., Lazzarini, A. M., Duvoisin, R. C., Di Iorio, G., Golbe, L. I., and Nussbaum, R. L. (1997). Mutation in the alpha-synuclein gene identified in families with Parkinson's disease. *Science* 276, 2045–2047.

Przedborski, S., Jackson-Lewis, V., Naini, A. B., Jakowec, M., Petzinger, G., Miller, R., and Akram, M. (2001). The parkinsonian toxin 1-methyl-4-phenyl-1,2,3,6-tetrahydropyridine (MPTP): A technical review of its utility and safety. *J Neurochem* 76, 1265–1274.

Purisai, M. G., McCormack, A. L., Cumine, S., Li, J., Isla, M. Z., and Di Monte, D. A. (2006). Microglial activation as a priming event leading to paraquat-induced dopaminergic cell degeneration. *Neurobiol Dis* 25, 392–400.

Reynolds, A. D., Kadiu, I., Garg, S. K., Glanzer, J. G., Nordgren, T., Ciborowski, P., Banerjee, R., and Gendelman, H. E. (2008). Nitrated alpha-synuclein and microglial neuroregulatory activities. *J Neuroimmune Pharmacol* 3, 59–74.

Richfield, E. K., Thiruchelvam, M. J., Cory-Slechta, D. A., Wuertzer, C., Gainetdinov, R. R., Caron, M. G., Di Monte, D. A., and Federoff, H. J. (2002). Behavioral and neurochemical effects of wild-type and mutated human alpha-synuclein in transgenic mice. *Exp Neurol* 175, 35–48.

Ross, G. D., and Vetvicka, V. (1993). CR3 (CD11b, CD18): A phagocyte and NK cell membrane receptor with multiple ligand specificities and functions. *Clin Exp Immunol* 92, 181–184.

Sawada, M., Imamura, K., and Nagatsu, T. (2006). Role of cytokines in inflammatory process in Parkinson's disease. *J Neural Transm Suppl*, 373–381.

Schmidt, N., and Ferger, B. (2001). Neurochemical findings in the MPTP model of Parkinson's disease. *J Neural Transm* 108, 1263–1282.

Seniuk, N. A., Tatton, W. G., and Greenwood, C. E. (1990). Dose-dependent destruction of the coeruleus-cortical and nigral-striatal projections by MPTP. *Brain Res* 527, 7–20.

Singleton, A. B., Farrer, M., Johnson, J., Singleton, A., Hague, S., Kachergus, J., Hulihan, M., Peuralinna, T., Dutra, A., Nussbaum, R., Lincoln, S., Crawley, A., Hanson, M., Maraganore, D., Adler, C., Cookson, M. R., Muenter, M., Baptista, M., Miller, D., Blancato, J., Hardy, J., and Gwinn-Hardy, K. (2003). Alpha-synuclein locus triplication causes Parkinson's disease. *Science* 302, 841.

Song, D. D., Shults, C. W., Sisk, A., Rockenstein, E., and Masliah, E. (2004). Enhanced substantia nigra mitochondrial pathology in human alpha-synuclein transgenic mice after treatment with MPTP. *Exp Neurol* 186, 158–172.

Souza, J. M., Giasson, B. I., Chen, Q., Lee, V. M., and Ischiropoulos, H. (2000). Dityrosine cross-linking promotes formation of stable alpha-synuclein polymers. Implication of nitrative and oxidative stress in the pathogenesis of neurodegenerative synucleinopathies. *J Biol Chem* 275, 18344–18349.

Spillantini, M. G., Crowther, R. A., Jakes, R., Hasegawa, M., and Goedert, M. (1998). Alpha-synuclein in filamentous inclusions of Lewy bodies from Parkinson's disease and dementia with lewy bodies. *Proc Natl Acad Sci USA* 95, 6469–6473.

Stefanis, L., Larsen, K. E., Rideout, H. J., Sulzer, D., and Greene, L. A. (2001). Expression of A53T mutant but not wild-type alpha-synuclein in PC12 cells induces alterations of the ubiquitin-dependent degradation system, loss of dopamine release, and autophagic cell death. *J Neurosci* 21, 9549–9560.

Su, X., Maguire-Zeiss, K. A., Giuliano, R., Prifti, L., Venkatesh, K., and Federoff, H. J. (2007). Synuclein activates microglia in a model of Parkinson's disease. *Neurobiol Aging*.

Thiruchelvam, M. J., Powers, J. M., Cory-Slechta, D. A., and Richfield, E. K. (2004). Risk factors for dopaminergic neuron loss in human alpha-synuclein transgenic mice. *Eur J Neurosci* 19, 845–854.

Thomas, D. M., Walker, P. D., Benjamins, J. A., Geddes, T. J., and Kuhn, D. M. (2004). Methamphetamine neurotoxicity in dopamine nerve endings of the striatum is associated with microglial activation. *J Pharmacol Exp Ther* 311, 1–7.

Thomas, D. M., Francescutti-Verbeem, D. M., and Kuhn, D. M. (2005). Gene expression profile of activated microglia under conditions associated with dopamine neuronal damage. *FASEB J* 20, 515–7.

Thomas, M. P., Chartrand, K., Reynolds, A., Vitvitsky, V., Banerjee, R., and Gendelman, H. E. (2007). Ion channel blockade attenuates aggregated alpha synuclein induction of microglial reactive oxygen species: Relevance for the pathogenesis of Parkinson's disease. *J Neurochem* 100, 503–519.

Uversky, V. N., Li, J., and Fink, A. L. (2001). Metal-triggered structural transformations, aggregation, and fibrillation of human alpha-synuclein. A possible molecular link between Parkinson's disease and heavy metal exposure. *J Biol Chem* 276, 44284–44296.

Vila, M., Vukosavic, S., Jackson-Lewis, V., Neystat, M., Jakowec, M., and Przedborski, S. (2000). Alpha-synuclein up-regulation in substantia nigra dopaminergic neurons following administration of the parkinsonian toxin MPTP. *J Neurochem* 74, 721–729.

Wu, D. C., Teismann, P., Tieu, K., Vila, M., Jackson-Lewis, V., Ischiropoulos, H., and Przedborski, S. (2003). NADPH oxidase mediates oxidative stress in the 1-methyl-4-phenyl-1,2,3,6-tetrahydropyridine model of Parkinson's disease. *Proc Natl Acad Sci USA* 100, 6145–6150.

Wu, X. F., Block, M. L., Zhang, W., Qin, L., Wilson, B., Zhang, W. Q., Veronesi, B., and Hong, J. S. (2005).

The role of microglia in paraquat-induced dopaminergic neurotoxicity. *Antioxid Redox Signal* 7, 654–661.

Zhang, W., Wang, T., Pei, Z., Miller, D. S., Wu, X., Block, M. L., Wilson, B., Zhang, W., Zhou, Y., Hong, J. S., and Zhang, J. (2005). Aggregated alpha-synuclein activates microglia: A process leading to disease progression in Parkinson's disease. *FASEB J* 19, 533–542.

Zhang, W., Dallas, S., Zhang, D., Guo, J. P., Pang, H., Wilson, B., Miller, D. S., Chen, B., McGeer, P. L., Hong, J. S., and Zhang, J. (2007). Microglial PHOX and Mac-1 are essential to the enhanced dopaminergic neurodegeneration elicited by A30P and A53T mutant alpha-synuclein. *Glia* 55, 1178–1188.

41

PURIFICATION AND QUANTIFICATION OF NEURAL α-SYNUCLEIN: RELEVANCE FOR PATHOGENESIS AND BIOMARKER DEVELOPMENT

BRIT MOLLENHAUER AND MICHAEL G. SCHLOSSMACHER

Center for Neurologic Diseases, Brigham and Women's Hospital, Harvard Medical School, Boston, MA, USA

INTRODUCTION

Several human neurodegenerative illnesses result in parkinsonism either without or with accompanying cognitive impairment. These include among others Parkinson disease (PD), dementia with Lewy bodies (DLB), the Lewy body variant of Alzheimer's disease (AD), progressive supranuclear palsy, and prion disease. A subgroup of these disorders is characterized by the deposition of protein aggregates and are therefore referred to as "synucleinopathy diseases;" they share the deposition of insoluble α-synuclein (αS) aggregates in the form of Lewy-type inclusions and glial cytoplasmic inclusions in PD, DLB, and multiple system atrophy (MSA), respectively (Spillantini *et al.*, 1997; Wakabayashi *et al.*, 1998; Jellinger, 2003). Ten years ago, the first heterozygous mutation (Ala53Thr) in the αS-encoding *SNCA* gene was discovered that co-segregated with the phenotype of autosomal-dominant parkinsonism in affected family members (Polymeropoulos *et al.*, 1997) (reviewed in Farrer, 2006). Reports of two additional *SNCA* missense mutations (Ala30Pro and Glu46Lys) linked to heritable PD/DLB helped establish the concept that dysregulation

of human αS as a critically important factor in disease development. In 2003, an autosomal-dominant PD phenotype with cognitive impairment and autonomic dysfunction was linked to a triplication event in the *SNCA* gene (Singleton *et al.*, 2003). Other reports of gene multiplication events followed (Chartier-Harlin *et al.*, 2004; Farrer *et al.*, 2004; Ibanez *et al.*, 2004), which collectively demonstrated that the severity of the disease phenotype could be linked to *SNCA* dosage (Singleton *et al.*, 2003; Chartier-Harlin *et al.*, 2004; Farrer *et al.*, 2004; Ibanez *et al.*, 2004). Therefore, the development of a heritable Parkinson-plus syndrome could be attributed to an increased expression rate of the wild-type protein and suggested a "gain of neurotoxic function" event *in vivo*. According to autopsy studies, >75% of sporadic PD cases are histopathologically linked to αS misprocessing (reviewed in Jellinger, 2003).

It is generally believed that the dynamics of αS homeostasis *in vivo* can be altered by different types of metabolic changes leading to synucleinopathy (Cookson, 2005), such as through: an increased propensity to aggregate (Conway *et al.*, 2000); an enhanced production rate (Outeiro and

Lindquist, 2003); sustained phosphorylation with resulting insolubility (Liu *et al.*, 2002; Chen and Feany, 2005; Anderson *et al.*, 2006); and decreased degradation rates (Liu *et al.*, 2002; Cuervo *et al.*, 2004). Recent work has focused on several pathways of αS degradation involving either proteasomal activity or lysosomal function (Ardley *et al.*, 2004; Cuervo *et al.*, 2004; Zhou and Freed, 2004; Lee *et al.*, 2005; Shin *et al.*, 2005). However, a third mechanism of downregulating the intracellular αS steady state may be the release into extracellular fluids (El-Agnaf *et al.*, 2003). Extracellular, full-length αS has been previously detected in biological fluids, including plasma and conditioned cell media (Lee *et al.*, 2005; Masliah *et al.*, 2005). Other mechanisms of αS processing within neurons may be sequestration, thereby leading to insoluble inclusion formation (Spillantini *et al.*, 1997) and partial proteolysis, thereby generating meta-stable intermediates that may affect cell viability (Tofaris *et al.*, 2006; Periquet *et al.*, 2007).

The community of synucleinopathy researchers is faced with three unmet needs: (1) a better understanding of the pathogenetic steps by which increased expression of wild-type αS promotes neurotoxicity; (2) the development of disease-linked biomarkers that help gauge the risk of PD in healthy persons, ascertain the diagnosis of PD in symptomatic subjects, and determine the rate of disease progression; and (3) the generation of cause-directed therapy that sees the lowering of specific αS variants and/or the total αS steady state *in vivo* (Masliah *et al.*, 2005; Klein and Schlossmacher, 2006; Scherzer *et al.*, 2007). Here, we describe two protocols that are aimed at: affinity enrichment of neural αS from mammalian brain and cerebrospinal fluid (CSF) for the purpose of identifying αS-associated proteins; and development of a sensitive enzyme-linked immunoabsorbent assay (ELISA) protocol for the precise measurement of αS species in a variety of biological specimens.

PURIFICATION OF BRAIN αS AND IDENTIFICATION OF INTERACTING PROTEINS

Affinity Enrichment of αS from Mouse Brain

To purify αS from the adult central nervous system and to identify interacting proteins, we first developed a serial antibody (Ab) column protocol using a non-specific Ig antibody-carrying resing for clearance, followed by one (or two) specific, high-titred polyclonal Ab(s). We then examined the final fractionated eluates by SDS/PAGE/Western blotting (WB) to determine the degree of enrichment and subjected chosen fractions to unbiased mass spectrometry (MS). The protocol, which is graphically summarized in Figure 41.1, entailed for the non-specific αS column the covalent conjugation of "7071FT" (flow-through) immunoglobulins (Ig), and for the αS-specific Ab column(s) the coupling of Ig from affinity purified sera, "mSA-1" (or 7071AP, or mAb syn-1, or two in tandem); these columns were generated using 50 mM sodium borate (pH 8.2 [ImmunoPure, Pierce]) and 1 ml of immobilized protein G beads. Mouse brain homogenates (Schlossmacher and Shimura, 2005) and, subsequently, lumbar drain-derived CSF (LD-CSF, see below) were first applied onto the 7071FT Ig column to remove non-specifically bound proteins. The flow-through of 7071FT Ig was then applied onto the αS-specific Ig column(s). The latter was (were) extensively washed and proteins were eluted with 3.5 ml elution buffer (0.2 M glycine, pH 2.0) into seven 500 μl fractions to collect αS and αS-associated proteins. Column eluates were re-equilibrated with 1 M Tris (pH 7.5). When we processed wild-type mouse brain homogenates using said protocol, the second and third fraction of the final column eluate consistently contained the highest levels of detectable, 16 kDa full-length αS, as detected by routine SDS/PAGE/WB (Figure 41.1); these fractions were then chosen for further processing by trypsin digestion and MS.

Identification of αS-Interacting Proteins from Mouse Brain

Eluates from mouse brain (and later, from human LD-CSF, see below) were reduced in 8 M urea/1% SDS/100 mM ammonium bicarbonate/10 mM DTT (pH 8.6) and then alkylated in 30 mM iodoacetamide. Each eluate was re-electrophoresed on an 8–16% tris/glycine gels. Gels were stained with SimplyBlue SafeStain (Invitrogen Life Technologies) before being imaged and sliced according to molecular weight standards (not shown). The gel slices were destained, washed, rinsed, and dried before being digested with trypsin (Promega) for 36 h. Peptides were extracted, vacuum concentrated, and loaded onto a 75 mm nanospray capillary spraying into LCQ DECA XP plus (Thermofinnigan). Each LCQ MS

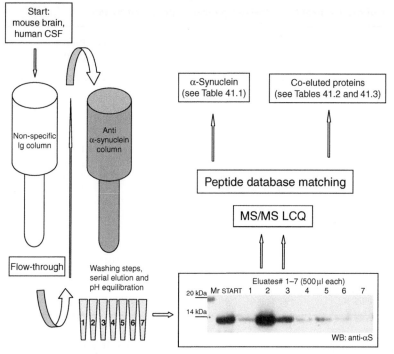

FIGURE 41.1 Protocol for the affinity enrichment of neural α-synuclein. Serial antibody-based affinity column chromatography is followed by SDS/PAGE and WB analysis and the processing of selected eluates by MS for structural identification (see text for details); WB (bottom right panel) of eluting fractions was carried out with monoclonal antibody Syn-1 (1:1000).

scan was followed by three MS/MS scans (performed at the Proteomics Center at Brigham & Women's Hospital, Cambridge, MA). Peptide identifications were made using Sequest (Bioworks Browser version 3.2); sequential database searches were performed using the RefSeqHuman FASTA database from the EMBL European Bioinformatics Institute.

Among the 177 distinct peptides identified by MS in fractions #2 and #3 from mouse brain, 26 corresponded to 4 distinct murine αS-specific sequences (Table 41.1); within the list of αS-co-eluting proteins from adult mouse brain we readily identified 6 groups of potential relevance to PD pathogenesis (Table 41.2). These contained: (i) β-synuclein, which may have been co-eluted because of its high sequence similarity to αS (Spillantini *et al.*, 1995; Hashimoto *et al.*, 2001), and non-synuclein-type presynaptic proteins, that is, syntaxin-binding protein-1 and "similar to vacuolar protein sorting 13D protein;" (ii) actin and three tubulin proteins as abundant cytoskeletal proteins; (iii) subunits of H+-transporting ATP synthase and malate dehydrogenase-2 as peptides from mitochondrial proteins; (iv) three histone proteins, that is, histone 1 isoforms H2b, H3g, and H4c, which provided independent validation of the recent findings by Kontopoulos *et al.* (2006); (v) three adult hemoglobin chain proteins, which is likely to underlie the recent observation of abundant SNCA expression during heme synthesis (Miller *et al.*, 2005; Scherzer *et al.*, Proc Nat Acad Sci USA 2008 in press); and (vi) several intriguing proteins with possible relevance to human synucleinopathies, for example, myelin basic protein (MBP), which is of interest because of the elusive etiology of the oligodendroglial synucleinopathy, MSA, and the α-subunit of Rho GDP dissociation inhibitor that is functionally linked to Rab proteins; the latter are known to modulate αS-mediated toxicity (Cooper *et al.*, 2006). The list of our working inventory of αS-interacting proteins from mouse brain can be found in Table 41.2. To date,

■ TABLE 41.1   List of mouse brain αS-derived peptides identified by MS

Mouse brain	Peptide fragment sequence		Score			MW	Accession	Peptide (hits)	% Area
Trypsin-generated peptide fragment #	Eluate #2 [28/106]	MH+	Sf	XC	Sp	RSp		Ions	Peak area
	αS isoform NACP140		2.84	30.30	14485.1	6678047		15 (15 0 0 0 0)	0.79
T12	K.EQVTNVGGAVVTGVTAVAQK.T	1929.16446	0.85	3.56	683.8	1		22/57	ND
T11–T12	K.TKEQVTNVGGAVVTGVTAVAQK.T	2158.44170	0.95	5.02	914.2	1		25/63	250838560
T13–T14	K.TVEGAGNIAAATGFVK.K	1506.68545	0.81	3.06	868.1	1		19/45	253462128
Trypsin digested peptide fragment #	Eluate #3 [18/71]								
	αS isoform NACP140	1929.16446	2.67	30.24	14485.1	6678047		6 (6 0 0 0 0)	0.19
T12	K.EQVTNVGGAVVTGVTAVAQK.T	2158.44170	0.65	2.70	502.5	1		20/57	280954688
T11–T12	K.TKEQVTNVGGAVVTGVTAVAQK.T	1506.68545	0.83	3.38	940.8	1		26/63	137086128
T13–T14	K.TVEGAGNIAAATGFVK.K	1929.16446	0.90	3.48	1056.9	1		21/45	122072664

we have independently validated the association of soluble, full-length αS with MBP (and *vice versa*) in co-immunoprecipitation experiments of human brain homogenates, and gathered *in vivo* evidence for the synchronized expression of wild-type αS with subunits of human hemoglobin during hematopoiesis. The majority of proteins that co-eluted with full-length, soluble αS in our protocol awaits further studies to discern whether they are true interactors in the living brain, or whether they only bound murine αS during lysis condition and purification. Of note, to date we have not yet identified synphilin as a constituent of these eluates.

PURIFICATION OF CSF αS AND IDENTIFICATION OF CO-ELUTING PROTEINS

Following re-equilibration of our serial Ab column system, which never before had seen a human αS-containing specimen, we loaded 40 ml of cell-free CSF (LD-CSF). We then proceeded with the enrichment of αS through identical washing, elution, SDS/PAGE/WB, and in-gel trypsin digestion steps, as done before for mouse brain (Figure 41.1). When analyzing the digested peptide samples from elution fractions #2 and #3 by MS, we identified

10 of the 16 predicted fragments of the human αS sequence; these 10 peptides encompassed 76 residues of the 140 amino acid long protein (54%). Importantly, with the identification of Ala53 and Ser87 in trypsin fragments #10 and #13, we unequivocally confirmed the identity of the αS protein as the human ortholog (B. Mollenhauer *et al.*, 2008, in press).

Furthermore, eluate fractions #2 and #3 contained 146 additional peptides, corresponding to a total of 58 proteins, which were present in CSF eluates in association with αS. The alphabetically ordered list of our preliminary inventory of αS interactors in CSF is shown in Table 41.3; it includes – among others – four interesting groups of potential relevance: (i) known CSF constituents, such as choroid plexus-derived cystatin C and plasma-derived albumin precursor proteins; (ii) neuronal constituents such as prion protein, the precursor of GABA-A receptor γ-subunit, and calgranulin B (also referred to as S100 calcium-binding protein); (iii) histone protein H2b (see above); and (iv) two ubiquitin–proteasome pathway-related proteins (Table 41.3). Collectively, these await further examination as *bona fide* αS-interacting proteins. Nevertheless, we have now begun to explore quantitative analyses by comparing serum and CSF values for αS and albumin concentrations, and

αS co-eluting proteins from adult mouse brain in alphabetical order

α-synuclein	Malate dehydrogenase 2, NAD (mitochondrial)
Actin, β, cytoplasmic	Monoclonal antibody κ light chain
Actin, γ 2, smooth muscle, enteric	Myelin basic protein
Actin-like	Odd Oz/ten-m homolog 3
Aldolase 1, A isoform	P lysozyme structural
α-Tubulin	Plakophilin 1
ATP synthase, H+ transporting mitochondrial F1 complex, β-subunit	Polo-like kinase 2
ATP synthase, H+ transporting, mitochondrial F1 complex, α-subunit	Procollagen, type I, α 1
ATP synthase, H+ transporting, mitochondrial F1 complex, O-subunit	Protease, serine, 2
β-Synuclein	Protease, serine, 3
BC026645 protein	Rho GDP dissociation inhibitor (GDI) α
Calmodulin 2	Rhomboid family 1
cDNA sequence BC031593	RIKEN cDNA 1810049H19 gene
Cleavage and polyadenylation specific factor 5	RIKEN cDNA 2310001L23
Collapsin response mediator protein 1	RIKEN cDNA 2310058N18
Creatine kinase	RIKEN cDNA 2410039E07 gene
Desmosomal cytoskeletal connector molecule	RIKEN cDNA 2900073G15
Eno1 protein	RIKEN cDNA 4631426H08
Enolase 2, γ neuronal	RIKEN cDNA 6330509G02
Enolase 3, β muscle	Similar to vacuolar protein sorting 13D
g polo-like kinase 2	Syntaxin binding protein 1
Glyceraldehyde-3-phosphate dehydrogenase (phosphorylating)	Triosephosphate isomerase 1
Hemoglobin α, adult chain 1	Try10-like trypsinogen
Hemoglobin β 1 chain (B1) (Major)	Trypsinogen 10
Hemoglobin, β adult minor chain	Trypsinogen 16
Histone 1, H2bb	Tubulin, α 1
Histone 1, H3g	Tubulin, α 2; tubulin α 2
Histone 1, H4c	Tubulin, β 2
Hypothetical protein 4732456N10	Tumor protein D52-like 2
Ig heavy-chain V region precursor	Tyrosine 3-monooxygenase/tryptophan 5-monooxygenase activation protein
Immunoglobulin light chain variable region	Ubiquitin A-52 residue ribosomal protein fusion product 1
Integral membrane protein 2B	
L-NAME induced actin cytoskeletal protein	

Note: Keratin and other skin-derived proteins as known contaminants are not listed.

through comparative MS of CSF collected from a well characterized cohort of PD subjects and normal control persons (Mollenhauer *et al.*, 2006a).

QUANTIFICATION OF αS BY ELISA USING CONCENTRATED CSF

Development of Second Generation Sandwich ELISA

Given the physiological presence of full-length αS in normal adult human CSF, as unequivocally demonstrated by our MS and WB experiments, we focused on CSF αS detection by ELISA as a potential platform for biomarker development. Recently, Tokuda *et al.* (2006) had to employ concentration of human CSF specimen for reliable signal detection in their first-generation sandwich ELISA (ELISA-I). Given the potential for non-uniform protein loss during CSF processing in this approach study, and in light of the fact that potential blood cell (or plasma) contamination of CSF was not specifically addressed, we set out to develop a second generation (ELISA-II) protocol with increased sensitivity for the quantification of CSF αS.

■ **TABLE 41.3   List of human CSF αS-associated proteins identified by MS**

αS co-eluting proteins from adult human CSF in alphabetical order

α-Synuclein isoform NACP140; non-A4 component of amyloid precursor	Palate, lung and nasal epithelium carcinoma associated protein precursor
Adducin 2 isoform a; adducin-2 (β); β adducin	Prion protein; Prion protein (p27-30)
Albumin precursor	Prolactin-induced protein; prolactin-inducible protein
α 1 actin precursor; α skeletal muscle actin	Proline rich 4 (lacrimal); lacrimal proline rich protein
Annexin IV; annexin IV (placental anticoagulant protein II); placental	S100 calcium-binding protein A9; calgranulin B
β Cysteine string protein; dnaJ homolog subfamily C member X	Similar to CYorf16 protein
	Similar to eukaryotic translation initiation factor 4A1
Cadherin 4, type 1 preproprotein; cadherin 4, R-cadherin (retinal)	Similar to hypothetical protein
Calmodulin-like skin protein	Similar to Ig heavy chain precursor V region (VDH26) – human
Clusterin; complement-associated protein SP-40	Similar to Ig κ variable region
Cystatin C precursor; cystatin 3; γ-trace; post-γ-globulin	Similar to IgG heavy chain
Dermcidin precursor; AIDD protein	Similar to IgG heavy chain variable region
Fc fragment of IgG binding protein; IgG Fc binding protein	Similar to leukocyte IgG-like receptor, subfamily B
Fc fragment of IgG binding protein; IgG Fc binding protein	Similar to metalloproteinase inhibitor 4 precursor (TIMP-4)
Fibronectin 1 isoform 1 preproprotein; cold-insoluble globulin	Similar to natural killer cell transcript 4
γ-Aminobutyric acid A receptor, γ 2 precursor	Similar to putative protein
H2b histone family, member Q; H2b histone	Similar to testicular metalloprotease-like, disintegrin-like, cysteine
Hypothetical protein XP_303178	Similar to The KIAA0191 gene is expressed ubiquitously
Hypothetical protein KIAA1434	Similar to zinc finger homeobox protein 2
Hypothetical protein XP_211685	SRp25 nuclear protein; HSVI binding protein
Hypothetical protein XP_303008	Tau-tubulin kinase
Hypothetical protein XP_305010	Tigger transposable element derived 7; jerky (mouse) homolog-like
Hypothetical protein XP_305285	Ubiquitin and ribosomal protein S27a precursor
Hypothetical protein XP_305800	Ubiquitin-activating enzyme E1; A1S9T and BN75 temperature sensitivity
Hypothetical protein XP_305944	Ubiquitously expressed transcript isoform 2
Immunoglobulin lambda-like polypeptide 1 isoform a precursor	v-akt murine thymoma viral oncogene homolog 2; Murine thymoma viral
Lipocalin 2 (oncogene 24p3)	
Lysozyme precursor	

To this end, CSF specimens from living subjects were obtained by routine lumbar puncture, centrifuged at 3200 rpm at 4°C within 30 min after collection and stored at 4°C until diagnostics were completed. Cell-free aliquots were then stored at −80°C prior to ELISA analysis. Total protein and cell counts were recorded for each sample. Upon thawing on ice, NP40 (0.5%) and protease inhibitors were added before re-centrifugation at

15,000×g. For validation steps, we also carried out CSF concentration, where samples were dialyzed against 10 mM ammonium acetate in Slide-A-Lyzer (10,000 MWCO [Pierce]) cassettes and lyophilized to dryness. Samples were re-constituted in 200 μl H₂O.

For ELISA-II, 96- and 384-well MaxiSorp plates were coated with capturing Ab (mSA-1; 7071AP) and diluted in coating buffer (NaHCO₃

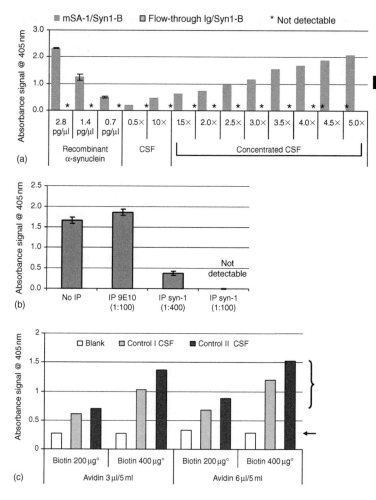

(a)

(b)

(c)

FIGURE 41.2 Characterization of ELISA-II detection of α-synuclein in concentrated CSF. (a) Serial dilutions of recombinant αS and aliquots of cell-free CSF were loaded (200 μl/well) onto a 96-well plate and read by ELISA-II (mSA-1/Syn1-B) and ELISA (flow-through Ig/Syn1-B). CSF was loaded either without concentration (0.5×, 1.0×) or after concentration by dialysis and lyophilization (1.5×–5.0×). Asteriks denote lack of signal above background. (b) Depletion of αS concentration signals from LD-CSF was carried out by immunoprecipitation (IP) of 1 ml aliquots of CSF with two doses of anti-αS Ab, Syn-1, in comparison with anti-myc Ab, 9E10, and a mock IP arm (beads only). The graph shows signals for concentrated CSF supernatants after IP, as read on a 96-well plate by ELISA-II (mSA-1/Syn1-B) at dilutions of 1:200 and 1:2000, respectively. (c) Optimization of αS signal detection by ELISA-II in two control CSF specimens (CSF-I, -II) as a result of increasing concentrations of biotin and avidin present in the assaying arm of the ELISA. Blank bars refer to background signal that does not change (arrow); bracket indicates the rise in specific signal for the two CSF specimens following optimization (see text for details).

with 0.2% NaN_3, pH 9.6) in 200 and 50 μl volumes per well, respectively. Following washes with PBS/0.05% Tween-20 (PBS-T), plates were blocked in 1.125% fish skin gelatine. Biotinylated Syn-1 Ab (as the assaying Ab) was generated using 200 μg Sulfo-NHS-LC Biotin (Pierce). Following four washes, ExtrAvidin phosphatase (Sigma) was applied. Color development was carried out by using Fast-*p*-Nitrophenyl Phosphate (Sigma) and monitored kinetically at OD 405 nm every 2.5 min. Saturation kinetics were examined for identification of time points in which standards and most sample dilutions were in the log phase (Mollenhauer *et al.*, 2008 in press).

Characterization of Second Generation Sandwich ELISA

To first determine the signal range for αS from cell-free CSF, aliquots were dialyzed, lyophilized, and loaded onto an ELISA-II plate together with recombinant αS (Figure 41.2a). There, a concentration-dependent increase in OD absorbance was observed for serial dilutions of the recombinant protein, for the two samples of un-concentrated CSF and for eight aliquots of concentrated LD-CSF. When no CSF (or recombinant protein) was loaded onto wells that were coated with Ab mSA-1, the OD absorbance signal equalled the background signal that was generated by CSF loaded onto non-specific, rabbit Ig (7071FT Ig)-coated wells (Figure 41.2a). From these results we calculated that the low end of our ELISA-II sensitivity approached 10 pg of recombinant human αS per 200 μl well; however to more convincingly separate CSF αS signals from background absorbance rates, we still required >1.5-fold concentration of CSF (Figure 41.2a). To further substantiate the specificity of these results, we next performed immunodepletion experiments. The addition of Ab, syn-1, to CSF for the purpose of removing αS by immunoprecipitation prior to loading the remaining supernatant onto the ELISA-II plate led to a dose-dependent depletion of the OD absorbance signal, as expected. This effect was specific, as signal loss did not occur when either no Ab was added during the protocol or when a myc protein-specific Ab, 9E10, was used instead (Figure 41.2b).

Biochemical Optimization of ELISA to Quantify αS in Native CSF

We next sought to improve the sensitivity of ELISA-II for the following reasons: (1) to better separate our low – but specific – αS CSF signals from the background absorbance reading; (2) to reduce the number of processing steps prior to loading, thereby minimizing the chance of non-uniform protein loss during CSF concentration; (3) to reduce the necessary volume CSF required from each patient; and (4) to thereby facilitate high-throughput (384-well plates) analysis of larger patient cohorts.

For the measurement of αS in un-concentrated CSF, we introduced two key modifications to the assaying arm of our ELISA-II protocol: We first doubled the concentration of biotin for the covalent tagging of the assaying mAb, syn-1, and then increased the concentration of ExtrAvidin–alkaline phosphatase to further amplify the signal (Figure 41.2c). Accordingly, when we now examined un-concentrated aliquots of LD-CSF together with CSF from a second, neurologically healthy donor, we observed a substantial increase in the absorbance rates in both CSF specimens; however, we saw no concomitant rise in background signals (Figure 41.2c).

This improvement in our signal amplification step permitted us from thereon to reproducibly quantify human αS below 1 pg/μl without any CSF concentration. For practical purposes, this modified sandwich ELISA-II developed specifically for CSF αS measurements (and referred to as ELISA mSA-1/Syn1-**BB**) now permitted the direct analysis of native human CSF. Importantly, its specificity remained preserved, as demonstrated by the parallel analysis of brain tissue from *snca*-null and wild-type mice (Mollenhauer *et al.*, 2008, in press).

The ELISA (mSA-1/Syn1-**BB**) also featured low intra-plate signal and day-to-day signal variability on these 384-well plates (Mollenhauer *et al.*, 2008, in press) and led to a >90% recovery rate for CSF αS proteins, as demonstrated in serial spiking experiments using recombinant human protein (Figure 41.3b).

QUANTIFICATION OF CSF αS IN LIVING SUBJECTS AND POSTMORTEM BRAIN

Collection of CSF from Clinically Well Characterized, Living Donors

Following the biochemical optimization of αS quantification in native CSF, we carried out a cross-sectional pilot study using previously collected CSF specimens from 100 subjects (Figure 41.4a). CSF from controls and patients with neurodegenerative diseases were collected at the University of Goettingen and at the Paracelsus Elena-Klinik in Kassel (both in Germany). Samples from Creutzfeldt–Jakob disease (CJD) patients were collected at the "National Surveillance Unit for Spongiform Encephalopathies" in Goettingen, Germany. The study was approved by the ethics committees of all institutions including the institutional review board at Brigham & Women's Hospital. For all CSF donors, neuroimaging was performed to exclude structural causes of dementia and parkinsonism. A group of neurological controls (NCO) encompassed gender-matched subjects without a neurodegenerative illness (e.g., chronic headache, multiple sclerosis; chronic inflammatory demyelinating polyradiculopathy; paraneoplastic syndrome). The clinical diagnosis of "definite PD" was rendered according to the UK Parkinson's Disease Society Brain Bank criteria (Gibb, 1988);

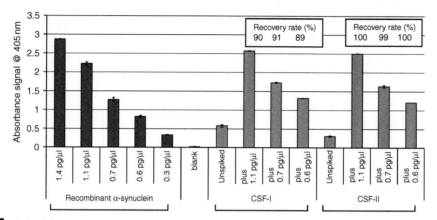

FIGURE 41.3 Characterization of ELISA-II using recombinant α-synuclein and native CSF. Detection of αS in unprocessed CSF (CSF-I and -II) and recovery in CSF spiked with aliquots of recombinant human αS; samples were analyzed in duplicate by ELISA-II (mSA-1/Syn1-*BB*) on a 384-well plate. The recovery rate of spiked recombinant protein was determined and listed, as indicated.

DLB patients had dementia before the onset of parkinsonism and fulfilled DSMIV criteria for dementia and McKeith criteria for the clinical diagnosis of "probable DLB" (McKeith *et al.*, 1996; McKeith *et al.*, 2005); AD patients fulfilled the DSMIV criteria for AD, and NINCDS-ADRDA criteria for the clinical diagnosis of "probable AD" (McKhann *et al.*, 1984; Gilman *et al.*, 1998).

Cross-sectional Measurement of CSF αS in 165 Donors

CSF aliquots from 100 donors were analyzed by ELISA (mSA-1/Syn1-*BB*) using 384 well plates. The demographic details and assay results for total protein content and αS values from these subjects have been described elsewhere (Mollenhauer *et al.*, in press). The corrected OD values for CSF αS (minus the OD value for the background signal) from all donor groups are displayed in box plot format in Figure 41.4a. The mean CSF αS absorbance rate was highest in the CJD group (mean, >1.3 ± SD, 1.2 [read at a 1:40 dilution]; *n* = 8) followed – in decreasing order – by the AD group (0.9 ± 0.8, *n* = 35), neurological controls (NCO, 0.8 ± 0.9, *n* = 38), the DLB group (0.6 ± 0.6, *n* = 54), and PD subjects (0.4 ± 0.2, *n* = 5) (Figure 41.4a). Using OD signals from our

recombinant standards, we also intrapolated the individual and group concentrations of CSF αS in pg/μl in these specimens. There, the spectrum of CSF αS ranged from a mean of 300 ± 248 pg/μl in the CJD group to 4.7 ± 4.8 and 2.3 ± 1.0, and 2.4 ± 0.5 pg/μl in NCO and PD cases, respectively (data not shown Mollenhauer *et al.*, 2008, in press).

Statistical Analyses of CSF αS ELISA Data

We carried out Monte Carlo re-sampling approximation tests, an(c)ovas, follow-up *post hoc* tests, and discriminant and regression analyses. Age and total CSF protein concentrations were significantly positively related to the logarithmic concentrations of CSF αS ($p = 0.0149$ and $p = 0.0051$, respectively). Nevertheless, the CSF αS absorbance values in the PD/DLB group were significantly lower than the mean for the AD/NCO group ($p = 0.0072$), and CJD cases had a significantly higher CSF αS level than all other groups, as shown for example in the comparison of CJD and NCO ($p < 0.001$). These collective results were confirmed at the αS protein concentration level in the same Caucasian cohort of 100 donors (Mollenhauer et al., 2008, in press) and in an independently analyzed Asian Cohort (Tokuda *et al.*, 2006).

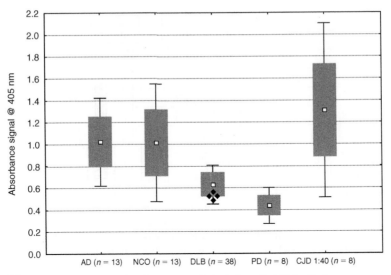

(a)   ❖ indicating the value of one neuropathologically verified DLB patient

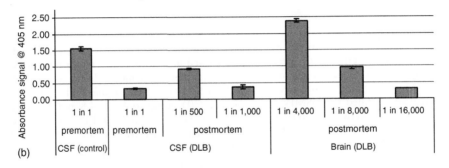

(b)

FIGURE 41.4 ELISA-based αS quantification in a cross-sectional study using CSF from living donors and in post-mortem specimens. (a) Aliquots of un-concentrated CSF (50 μl/well) from 100 living subjects were loaded in duplicates onto 384 well plates and analyzed by ELISA-II (mSA-1/Syn1-BB). A box plot for the absorbance value at 405 nm of CSF αS is shown for five subject groups examined. Note, CSF specimens from CJD cases were diluted (1:40) prior to loading. The value for a single patient with autopsy confirmation is marked in the DLB cohort (see (b)). Statistical group comparisons are indicated by brackets and parentheses (■ mean; □ mean ± standard error of the mean; ⊥ mean ± 0.89 confidence interval). (b) Comparative analysis of CSF αS collected *in vivo* from one control and one DLB patient (as shown in [a]), and of CSF and frontal cortex collected at autopsy from the same DLB subject. The αS reactivity was measured in duplicates using ELISA-II (mSA-1/Syn1-BB) on a 384-well plate. Note, the reduced CSF αS concentration detected before death in the DLB patient (versus control), the ~1000-fold elevation in CSF αS at 12 h after death, and the overall minute amounts of CSF αS seen when compared with the soluble brain αS level from the same case.

Postmortem Analyses of Nine Study Participants

Among the 165 subjects included in our pilot study, we learned of 9 deaths to date. As predicted from the clinical diagnosis, all 8 study subjects with CJD have died. At autopsy, the working diagnosis of prion disease was confirmed in all 8 cases (not shown). Furthermore, one study subject with the clinical diagnosis of probable DLB from our pilot study has expired, and brain autopsy confirmed the clinical diagnosis; routine immunohistochemistry revealed widespread brainstem and cortical synucleinopathy, as detected by anti-αS, LB509 (not shown). In the same DLB case, CSF was also collected by ventricular puncture at the time of autopsy. The direct comparison of CSF collected *intra vitam* versus postmortem demonstrated an almost 1000-fold rise in the CSF αS signal as a result of death (Figure 41.4b). Furthermore, frontal cortex-derived specimens obtained from the same DLB subject also demonstrated the minute amount of αS that is present in CSF when compared to the αS concentration present in brain homogenates (Figure 41.4b).

DISCUSSION

Our data are in support of several findings with relevance to the pathogenesis of PD, to the possible diagnosis of synucleinopathies, and to the development of platforms for possible drug validation in pre-clinical trials.

First, we provide unequivocal evidence that αS is present in cell-free human CSF, which we collected through an indwelling lumbar drain (LD-CSF) from a neurologically healthy adult. LD-CSF was processed for affinity enrichment and eluates were subjected to MS analyses. There, we identified peptides corresponding to 76 out of the predicted 140 amino acid of the human αS sequence (54%) (B. Mollenhauer et al., 2008, in press). The lack of detection of other αS fragments by MS to date may be explained by specific post-translational modifications, such as oxidation of methionine residues, N-terminal acetylation events, and phosphorylation at the C-terminus (Anderson et al., 2006). Although murine β-synuclein co-purified with αS from adult mouse brain homogenates (Table 41.2), neither β-synuclein nor γ-synuclein fragments were detected among the digests of human CSF αS eluates. These collective findings identify in αS a normal constituent of adult human CSF.

Future studies will: (1) determine the concentration gradient between serum and CSF αS versus that of albumin (the latter represents an informative ratio routinely investigated in neurological conditions [Reiber, 2001; Mollenhauer et al., 2007]); (2) identify the actual sources of both plasma and CSF αS; and (3) delineate the metabolic rate of CSF αS turnover *in vivo*, as has been recently shown elegantly for AD-linked amyloid β-protein (Seubert et al., 1992; Bateman et al., 2006) and transthyretin (Redzic et al., 2005). It is possible that several

factors, such as neuronal integrity, the blood–brain barrier, plasma filtration rates, and choroid plexus function, which has a key role in the production of CSF and absorption of CSF proteins (Redzic et al., 2005), may together regulate the concentration of CSF αS *in vivo*.

Second, we provide a first inventory of proteins by MS that co-eluted with endogenous αS from adult mouse brain (Table 41.2) and from human CSF (Table 41.3). It is important to note, that given the low abundance of αS in CSF, we purposefully chose a washing protocol during its enrichment from LD-CSF that minimized potential losses. Therefore, we may have enriched for proteins with relatively low affinity for αS binding *in vivo* (Table 41.3). From a validation perspective, many of the murine αS-interacting proteins and all of the CSF candidates that co-eluted await further validation studies to separate true αS interactors from non-specific binders.

Third, we demonstrate the development, validation, and application of a sandwich-type ELISA method, which permits – for the first time – the direct quantification of neural αS without preceding processing steps. The estimated low end of its sensitivity at the current time is 0.1 pg/μl. Despite our confirmatory MS data, we ensured that the signals generated by ELISA (mSA-1/Syn1-**BB**) (and two other sandwiches) were specific and that the assay detected both total and oligomeric species of soluble αS (not shown). In parallel work, we have now optimized our protocols for the sandwich ELISA-based detection and quantification of both human and rodent αS from the following sources: brain homogenates, native CSF (see above), cell lysates of dopaminergic and cholinergic neurons (Mollenhauer et al., 2008, in press), plasma, serum, and whole blood (Scherzer et al., 2008, in press). We predict that the versatility, specificity, and sensitivity of such assays will further advance research into several aspects of human synucleinopathies (Fjorback et al., 2007) as well as their modeling in cells and animals (Feany and Bender, 2000; Masliah et al., 2000).

Fourth, we present the results of a first, large cross-sectional pilot study of CSF αS quantification in 165 subjects, which was performed using a 384-well plate ELISA format. Our studies demonstrate reduced mean CSF αS signals in patients treated for clinically diagnosed synucleinopathy disorder, that is, PD and DLB, and possibly MSA, when compared to control subjects that included healthy controls (not shown), NCO, patients with AD and those with CJD

(Figure 41.4a (Mollenhauer *et al.*, 2008, in press)). The differences between the synucleinopathy group (PD and DLB) and the non-synucleinopathy group (controls and AD) remained statistically significant in both studies (despite their substantial overlap) even after correction of CSF αS levels for age, gender, and the total CSF protein content. Future studies will compare CSF αS from DLB patients age-matched healthy control donors, de novo PD patients, and subjects with early versus advanced AD subjects, and also examine other disease-linked CSF markers for stratification purposes, for example amyloid β-protein, tau and heart-type fatty acid-binding protein (Hansson *et al.*, 2006; Mollenhauer *et al.*, 2006b; Seubert *et al.*, 1992). As with all biomarker development programs, our cross-sectional data need to be confirmed in longitudinal, prospective studies, and need to include subjects that have not yet been medicated for their illness. Nevertheless, the observed relation between higher HY stage and lower CSF αS concentration highlights the potential of our ELISA as a biochemical PD progression marker (Tokuda *et al.*, 2006).

Fifth, we present evidence for a marked rise in mean CSF αS concentrations and a 55-fold elevation in the CSF αS-to-CSF total protein ratio in living subjects with CJD ($p < 0.001$) (Figure 41.4a). We speculate that the steep rise in CSF αS levels in our definite CJD cases resulted from pan-neural degeneration *in vivo*. Future analyses will determine whether CSF αS may be of value in the earlier diagnosis of prion disease.

Sixth, in direct support of our findings in living CJD patients (Figure 41.4a), we provide evidence for the marked elevation in CSF αS concentrations after brain death. This observation was confirmed in our patient with definite DLB, where we measured an ~1000-fold increase in postmortem CSF αS when compared to the corresponding concentration *in vivo* (Figure 41.4b); it was also supported by the ELISA analysis of 12 additional CSF specimens collected by ventricular puncture after death (not shown).

There are several relevant implications of our collective results: (1) the quantification of αS from human CSF, when collected postmortem, is unlikely to yield informative data as to its concentration *in vivo* and (2) the specific reduction of CSF αS in our 79 living subjects with Lewy body disease is unlikely explained by cell death alone. Future studies will address whether any of the currently four speculative mechanisms underlies this reduction: (i)

a change in *SNCA* gene expression rate or in post-translational modification(s) of αS; (ii) an increased uptake and thus removal of αS from CSF, possibly related to pharmacotherapy with dopaminergic agonists; (iii) a decreased rate in release by an αS of yet unknown mechanism into CSF that may (or may not) be related to increased intracellular accumulation of αS; or (iv) any combination thereof.

In summary, our protocols may support a broad research agenda that relates to the etiology of PD, to the biochemical diagnosis of neurodegenerative diseases, and to the development of drug validation platforms in pre-clinical experiments. All three areas require the precise quantification of αS and elucidation of its *bonafide* binding partners, *in vivo*.

REFERENCES

Anderson, J. P., Walker, D. E., Goldstein, J. M., de Laat, R., Banducci, K., Caccavello, R. J., Barbour, R., Huang, J., Kling, K., Lee, M., Diep, L., Keim, P. S., Shen, X., Chataway, T., Schlossmacher, M. G., Seubert, P., Schenk, D., Sinha, S., Gai, W. P., and Chilcote, T. J. (2006). Phosphorylation of Ser 129 is the dominant pathological modification of alpha-synuclein in familial and sporadic Lewy body disease. *J Biol Chem* **281**, 29739–29752.

Ardley, H. C., Scott, G. B., Rose, S. A., Tan, N. G., and Robinson, P. A. (2004). UCH-L1 aggresome formation in response to proteasome impairment indicates a role in inclusion formation in Parkinson's disease. *J Neurochem* **90**, 379–391.

Bateman, R. J., Munsell, L. Y., Morris, J. C., Swarm, R., Yarasheski, K. E., and Holtzman, D. M. (2006). Human amyloid-beta synthesis and clearance rates as measured in cerebrospinal fluid *in vivo*. *Nat Med* **12**, 856–861.

Chartier-Harlin, M.-C., Kachergus, J., Roumier, C., Mouroux, V., Douay, X., Lincoln, S., Levecque, C., Larvor, L., Andrieux, J., and Hulihan, M. (2004). Alpha-synuclein locus duplication as a cause of familial Parkinson's disease. *Lancet* **364**, 1167–1169.

Chen, L., and Feany, M. B. (2005). Alpha-synuclein phosphorylation controls neurotoxicity and inclusion formation in a *Drosophila* model of Parkinson disease. *Nat Neurosci* **8**, 657–663.

Clemens, R., Scherzerl, Jeffrey A. Grass, Zhixiang Liaol, Bin Zheng, Imelda Pepivani, Aron C. Eklund, Paul A. Ney, Juliana Ng, Meghan McGoldrick, Olaf Riess, Brit Moellenhauer, Emery H. Bresnick, and Michael G. Schlossmacher, GATA-1 transcription factor directly regulates expression of parkinson's disease-linked a-synuclein. *Proc Nat Acad Sci USA* 2008.

Conway, K. A., Lee, S. J., Rochet, J. C., Ding, T. T., Williamson, R. E., and Lansbury, P. T., Jr. (2000). Acceleration of oligomerization, not fibrillization, is a shared property of both alpha-synuclein mutations linked to early-onset Parkinson's disease: Implications for pathogenesis and therapy. *Proc Natl Acad Sci USA* 97, 571–576.

Cookson, M. R. (2005). The biochemistry of Parkinson's disease. *Annu Rev Biochem* 74, 29–52.

Cooper, A. A., Gitler, A. D., Cashikar, A., Haynes, C. M., Hill, K. J., Bhullar, B., Liu, K., Xu, K., Strathearn, K. E., Liu, F., Cao, S., Caldwell, K. A., Caldwell, G. A., Marsischky, G., Kolodner, R. D., Labaer, J., Rochet, J. C., Bonini, N. M., and Lindquist, S. (2006). Alpha-synuclein blocks ER–Golgi traffic and Rab1 rescues neuron loss in Parkinson's models. *Science* 313, 324–328.

Cuervo, A. M., Stefanis, L., Fredenburg, R., Lansbury, P. T., and Sulzer, D. (2004). Impaired degradation of mutant alpha-synuclein by chaperone-mediated autophagy. *Science* 305, 1292–1295.

El-Agnaf, O. M., Salem, S. A., Paleologou, K. E., Cooper, L. J., Fullwood, N. J., Gibson, M. J., Curran, M. D., Court, J. A., Mann, D. M., Ikeda, S., Cookson, M. R., Hardy, J., and Allsop, D. (2003). Alpha-synuclein implicated in Parkinson's disease is present in extracellular biological fluids, including human plasma. *FASEB J* 17, 1945–1947.

Farrer, M., Kachergus, J., Forno, L., Lincoln, S., Wang, D. S., Hulihan, M., Maraganore, D., Gwinn-Hardy, K., Wszolek, Z., Dickson, D., and Langston, J. W. (2004). Comparison of kindreds with parkinsonism and alpha-synuclein genomic multiplications. *Ann Neurol* 55, 174–179.

Farrer, M. J. (2006). Genetics of Parkinson disease: Paradigm shifts and future prospects. *Nat Rev Genet* 7, 306–318.

Feany, M. B., and Bender, W. W. (2000). A *Drosophila* model of Parkinson's disease. *Nature* 404, 394–398.

Fjorback, A. W., Varming, K., and Jensen, P. H. (2007). Determination of alpha-synuclein concentration in human plasma using ELISA. *Scand J Clin Lab Invest* 67, 431–435.

Gibb, W. R. (1988). Accuracy in the clinical diagnosis of parkinsonian syndromes. *Postgrad Med J* 64, 345–351.

Gilman, S., Low, P., Quinn, N., Albanese, A., Ben-Shlomo, Y., Fowler, C., Kaufmann, H., Klockgether, T., Lang, A., Lantos, P., Litvan, I., Mathias, C., Oliver, E., Robertson, D., Schatz, I., and Wenning, G. (1998). Consensus statement on the diagnosis of multiple system atrophy. American Autonomic Society and American Academy of Neurology. *Clin Auton Res* 8, 359–362.

Hansson, O., Zetterberg, H., Buchhave, P., Londos, E., Blennow, K., and Minthon, L. (2006). Association between CSF biomarkers and incipient Alzheimer's disease in patients with mild cognitive impairment: A follow-up study. *Lancet Neurol* 5, 228–234.

Hashimoto, M., Rockenstein, E., Mante, M., Mallory, M., and Masliah, E. (2001). Beta-synuclein inhibits alpha-synuclein aggregation: A possible role as an anti-parkinsonian factor. *Neuron* 32, 213–223.

Ibanez, P., Bonnet, A. M., Debarges, B., Lohmann, E., Tison, F., Pollak, P., Agid, Y., Durr, A., and Brice, A. (2004). Causal relation between alpha-synuclein gene duplication and familial Parkinson's disease. *Lancet* 364, 1169–1171.

Jellinger, K. (2003). Synucleinopathies. In *Neurodegeneration. The Molecular Pathology of Dementia and Movement Disorders* (D. Dickson, Ed.), Vol. 1, pp. 155–225. ISN Neuropath Press, Basel, Switzerland.

Klein, C., and Schlossmacher, M. G. (2006). The genetics of Parkinson disease: Implications for neurological care. *Nat Clin Pract Neurol* 2, 136–146.

Kontopoulos, E., Parvin, J. D., and Feany, M. B. (2006). Alpha-synuclein acts in the nucleus to inhibit histone acetylation and promote neurotoxicity. *Hum Mol Genet* 15, 3012–3023.

Lee, H. J., Patel, S., and Lee, S. J. (2005). Intravesicular localization and exocytosis of alpha-synuclein and its aggregates. *J Neurosci* 25, 6016–6024.

Li, W., West, N., Colla, E., Pletnikova, O., Troncoso, J. C., Marsh, L., Dawson, T. M., Jakala, P., Hartmann, T., Price, D. L., and Lee, M. K. (2005). Aggregation promoting C-terminal truncation of alpha-synuclein is a normal cellular process and is enhanced by the familial Parkinson's disease-linked mutations. *Proc Natl Acad Sci USA* 102, 2162–2167.

Liu, Y., Fallon, L., Lashuel, H. A., Liu, Z., and Lansbury, P. T. (2002). The UCH-L1 gene encodes two opposing enzymatic activities that affect alpha-synuclein degradation and Parkinson's disease susceptibility. *Cell* 111, 209–218.

Masliah, E., Rockenstein, E., Veinbergs, I., Mallory, M., Hashimoto, M., Takeda, A., Sagara, Y., Sisk, A., and Mucke, L. (2000). Dopaminergic loss and inclusion body formation in alpha-synuclein mice: Implications for neurodegenerative disorders. *Science* 287, 1265–1269.

Masliah, E., Rockenstein, E., Adame, A., Alford, M., Crews, L., Hashimoto, M., Seubert, P., Lee, M., Goldstein, J., Chilcote, T., Games, D., and Schenk, D. (2005). Effects of alpha-synuclein immunization in a mouse model of Parkinson's disease. *Neuron* 46, 857–868.

McKeith, I. G., Galasko, D., Kosaka, K., Perry, E. K., Dickson, D. W., Hansen, L. A., Salmon, D. P., Lowe, J., Mirra, S. S., Byrne, E. J., Lennox, G., Quinn, N. P.,

Edwardson, J. A., Ince, P. G., Bergeron, C., Burns, A., Miller, B. L., Lovestone, S., Collerton, D., Jansen, E. N., Ballard, C., de Vos, R. A., Wilcock, G. K., Jellinger, K. A., and Perry, R. H. (1996). Consensus guidelines for the clinical and pathologic diagnosis of dementia with Lewy bodies (DLB): Report of the consortium on DLB international workshop. *Neurology* 47, 1113–1124.

McKeith, I. G., Dickson, D. W., Lowe, J., Emre, M., O'Brien, J. T., Feldman, H., Cummings, J., Duda, J. E., Lippa, C., Perry, E. K., Aarsland, D., Arai, H., Ballard, C. G., Boeve, B., Burn, D. J., Costa, D., Del Ser, T., Dubois, B., Galasko, D., Gauthier, S., Goetz, C. G., Gomez-Tortosa, E., Halliday, G., Hansen, L. A., Hardy, J., Iwatsubo, T., Kalaria, R. N., Kaufer, D., Kenny, R. A., Korczyn, A., Kosaka, K., Lee, V. M. Y., Lees, A., Litvan, I., Londos, E., Lopez, O. L., Minoshima, S., Mizuno, Y., Molina, J. A., Mukaetova-Ladinska, E. B., Pasquier, F., Perry, R. H., Schulz, J. B., Trojanowski, J. Q., and Yamada, M.for the Consortium on DLB (2005). Diagnosis and management of dementia with Lewy bodies: Third report of the DLB consortium. *Neurology* 65, 1863–1872.

McKhann, G., Drachman, D., Folstein, M., Katzman, R., Price, D., and Stadlan, E. M. (1984). Clinical diagnosis of Alzheimer's disease: Report of the NINCDS-ADRDA Work Group under the auspices of Department of Health and Human Services Task Force on Alzheimer's Disease. *Neurology* 34, 939–944.

Miller, D. W., Johnson, J. M., Solano, S. M., Hollingsworth, Z. R., Standaert, D. G., and Young, A. B. (2005). Absence of alpha-synuclein mRNA expression in normal and multiple system atrophy oligodendroglia. *J Neural Transm* 112, 1613–1624.

Mollenhauer, B., Cullen, V., Krastins, B., Trenkwalder, C., Sarracino, D. A., and Schlossmacher, M. G. (2006a). Characterisation and quantification of alpha-synuclein release into cell cultured medium and cerebrospinal fluid. *Mov Disord* 21(Suppl 15), 72.

Mollenhauer, B., Trenkwalder, C., von Ahsen, N., Bibl, M., Steinacker, P., Brechlin, P., Schindehuette, J., Poser, S., Wiltfang, J., and Otto, M. (2006b). Beta-amlyoid 1-42 and tau-protein in cerebrospinal fluid of patients with Parkinson's disease dementia. *Dement Geriatr Cogn Disord* 22, 200–208.

Mollenhauer B., , Steinacker, P, , Bahn, E., Bibl, M., Brechlin, P., Schlossmacher, M. G., Locascio, J. J., Wiltfang, J., Kretzschmar, H. A., Poser, S., Trenkwalder, C., and Otto, M. (2007). Serum heart-type fatty acid-binding protein and cerebrospinal fluid tau: Marker candidates for dementia with Lewy bodies. *Neurodegenerative Diseases* 4, 366–375.

Outeiro, T. F., and Lindquist, S. (2003). Yeast cells provide insight into alpha-synuclein biology and pathobiology. *Science* 302, 1772–1775.

Periquet, M., Fulga, T., Myllykangas, L., Schlossmacher, M. G., and Feany, M. B. (2007). Aggregated alpha-synuclein mediates dopaminergic neurotoxicity *in vivo*. *J Neurosci* 27, 3338–3346.

Polymeropoulos, M. H., Lavedan, C., Leroy, E., Ide, S. E., Dehejia, A., Dutra, A., Pike, B., Root, H., Rubenstein, J., Boyer, R., Stenroos, E. S., Chandrasekharappa, S., Athanassiadou, A., Papapetropoulos, T., Johnson, W. G., Lazzarini, A. M., Duvoisin, R. C., Di Iorio, G., Golbe, L. I., and Nussbaum, R. L. (1997). Mutation in the alpha-synuclein gene identified in families with Parkinson's disease. *Science* 276, 2045–2047.

Redzic, Z. B., Preston, J. E., Duncan, J. A., Chodobski, A., Szmydynger, , and Chodobska, J. (2005). The choroid plexus cerebrospinal fluid system. In *From Development to Aging Current Topics in Developmental Biology* (G. P. Schatten, Ed.), Vol. 71, pp. 1–52. Academic Press, San Diego, CA.

Reiber, H. (2001). Dynamics of brain-derived proteins in cerebrospinal fluid. *Clin Chim Acta* 310, 173–186.

Scherzer, C. R., Eklund, A. C., Morse, L. J., Liao, Z., Locascio, J. J., Fefer, D., Schwarzschild, M. A., Schlossmacher, M. G., Hauser, M. A., Vance, J. M., Sudarsky, L. R., Standaert, D. G., Growdon, J. H., Jensen, R. V., and Gullans, S. R. (2007). Molecular markers of early Parkinson's disease based on gene expression in blood. *Proc Natl Acad Sci USA*.

Schlossmacher, M. G., and Shimura, H. (2005). Parkinson's disease: Assays for the ubiquitin ligase activity of neural Parkin. *Methods Mol Biol* 301, 351–369.

Seubert, P., Vigo-Pelfrey, C., Esch, F., Lee, M., Dovey, H., Davis, D., Sinha, S., Schlossmacher, M., Whaley, J., Swindlehurst, C. et al. (1992). Isolation and quantification of soluble Alzheimer's beta-peptide from biological fluids. *Nature* 359, 325–327.

Shin, Y., Klucken, J., Patterson, C., Hyman, B. T., and McLean, P. J. (2005). The co-chaperone carboxyl terminus of Hsp70-interacting protein (CHIP) mediates α-synuclein degradation decisions between proteasomal and lysosomal pathways. *J Biol Chem* 280, 23727–23734.

Singleton, A. B., Farrer, M., Johnson, J., Singleton, A., Hague, S., Kachergus, J., Hulihan, M., Peuralinna, T., Dutra, A., Nussbaum, R., Lincoln, S., Crawley, A., Hanson, M., Maraganore, D., Adler, C., Cookson, M. R., Muenter, M., Baptista, M., Miller, D., Blancato, J., Hardy, J., and Gwinn-Hardy, K. (2003). Alpha-synuclein locus triplication causes Parkinson's disease. *Science* 302, 841.

Spillantini, M. G., Divane, A., and Goedert, M. (1995). Assignment of human alpha-synuclein (SNCA) and beta-synuclein (SNCB) genes to chromosomes 4q21 and 5q35. *Genomics* 27, 379–381.

Spillantini, M. G., Schmidt, M. L., Lee, V. M., Trojanowski, J. Q., Jakes, R., and Goedert, M.

(1997). Alpha-synuclein in Lewy bodies. *Nature* **388**, 839–840.

Tofaris, G. K., Garcia Reitbock, P., Humby, T., Lambourne, S. L., O'Connell, M., Ghetti, B., Gossage, H., Emson, P. C., Wilkinson, L. S., Goedert, M., and Spillantini, M. G. (2006). Pathological changes in dopaminergic nerve cells of the substantia nigra and olfactory bulb in mice transgenic for truncated human alpha-synuclein(1-120): Implications for Lewy body disorders. *J Neurosci* **26**, 3942–3950.

Tokuda, T., Salem, S. A., Allsop, D., Mizuno, T., Nakagawa, M., Qureshi, M. M., Locascio, J. J., Schlossmacher, M. G., and El-Agnaf, O. M. (2006).

Decreased alpha-synuclein in cerebrospinal fluid of aged individuals and subjects with Parkinson's disease. *Biochem Biophys Res Commun* **349**, 162–166.

Wakabayashi, K., Yoshimoto, M., Tsuji, S., and Takahashi, H. (1998). Alpha-synuclein immunoreactivity in glial cytoplasmic inclusions in multiple system atrophy. *Neurosci Lett* **249**, 180–182.

Zhou, W., and Freed, C. R. (2004). Tyrosine-to-cysteine modification of human alpha-synuclein enhances protein aggregation and cellular toxicity. *J Biol Chem* **279**, 10128–10135.

42

PROTEIN FOLDING AND AGGREGATION IN *IN VITRO* MODELS OF PARKINSON'S DISEASE: STRUCTURE AND FUNCTION OF α-SYNUCLEIN

DAVID ELIEZER

Department of Biochemistry and Program in Structural Biology, Weill Medical College of Cornell University, New York, NY, USA

α-SYNUCLEIN IN PARKINSON'S DISEASE

Rapid advances in genetics have led to the identification of a growing number of genes that are responsible for familial forms of several neurodegenerative diseases, including Parkinson's disease (PD). The first PD-linked gene, eventually designated as PARK1 (OMIM 168600), was recognized in 1996 (Polymeropoulos *et al.*, 1996) and subsequently identified in 1997 to be the gene encoding the protein α-synuclein (Polymeropoulos *et al.*, 1997), previously designated as SNCA (Jakes *et al.*,

1994). An alanine to threonine mutation at position 53 of the 140 residue α-synuclein primary sequence (see Figure 42.1) was the first specific genetic lesion to be associated with PD, causing an autosomal dominant early-onset form of the disease.

Familial forms constitute only a small fraction of the total incidence of PD, and mutations in α-synuclein are an extremely rare cause of even familial PD. Furthermore, familial inheritance has in the past been considered an exclusionary criterion for a diagnosis of PD (Hardy *et al.*, 2006). Nevertheless, a number of reported observations

FIGURE 42.1 Primary sequence of human α-synuclein, highlighting motifs identified through genetics, sequence analysis, and structural studies. The sites of the three mutations, A30P, A53T, and E46K, associated with Parkinsonism are shown in white on grey. The seven imperfect 11-residue repeats are delineated by spaces and within each repeat the imperfect KTKEGV-like motif is underlined. The hydrophobic NAC region, corresponding to a proteolytic fragment originally identified as a component of Alzheimer's disease (AD) plaques, is highlighted in light grey. The residual helical structure in the N-terminal end of the free state of the protein is highlighted in medium. The two helices formed in the micelle-bound state of the protein are indicated by dashed lines beneath the sequence. The acidic C-terminal tail is highlighted in darker grey.

suggest that α-synuclein may play an important role in idiopathic as well as familial PD. The first of these observations was the identification of α-synuclein as a primary protein component of the intraneuronal Lewy body deposits (Spillantini *et al.*, 1997) which are a pathological hallmark of the disease and often considered and required for a diagnosis of (idiopathic) PD at autopsy. Subsequently, increases in the number of wild-type α-synuclein alleles were shown to cause PD (Singleton *et al.*, 2003; Chartier-Harlin *et al.*, 2004). In addition, there has been some success in producing animal models of PD through the transgenic introduction of human α-synuclein (Feany and Bender, 2000; Masliah *et al.*, 2000; Lakso *et al.*, 2003). As a result, considerable effort has been focused on understanding the biology of α-synuclein in order to elucidate pathways that are important in the etiology of PD.

Potential Role for Aggregation of α-Synuclein in PD

One pathway in which α-synuclein is clearly involved and which may play a crucial role in PD is the aggregation of the protein from its highly soluble native form into the amyloid fibril form that is found within Lewy body and other α-synuclein deposits (Spillantini *et al.*, 1998a). The role of protein misfolding and aggregation in human disease has received intense attention because proteins found in disease-related deposits have been increasingly linked to the etiology of these diseases through genetics. In addition to α-synuclein, important examples include (but are not limited to) the amyloid-β precursor protein (APP) (linked to Alzheimer's Disease, AD) (Goate *et al.*, 1991), tau (linked to frontotemporal dementia with Parkinsonism) (Hutton *et al.*, 1998; Poorkaj *et al.*, 1998; Spillantini *et al.*, 1998b), transthyretin (linked to familial amyloidotic polyneuropathy) (Tawara *et al.*, 1983; Saraiva *et al.*, 1984), and lysozyme (linked to hereditary non-neuropathic systemic amyloidosis) (Pepys *et al.*, 1993). In each of these cases, mutations that lead to familial and usually autosomal dominant forms of the disorder are found in the very protein which forms deposits that are characteristic hallmarks of the respective disease. This has led to the so-called amyloid hypothesis, originally developed for AD (Selkoe, 1991; Hardy and Higgins, 1992), which postulates that the aggregation of these proteins plays a causative role in the associated disorders.

The precise manner by which aggregation may lead to cell death and disease remains unclear and this has led to a number of variations of the original hypothesis, including the so-called oligomer hypothesis (Lambert *et al.*, 1998; Walsh *et al.*, 1999; Conway *et al.*, 2000b) and the amyloid ion channel hypothesis (Arispe *et al.*, 1993).

Potential Role for Normal Function of α-Synuclein in PD

In addition to participating in aggregation pathways, α-synuclein has been implicated as a regulatory protein in pathways controlling synaptic vesicle formation, trafficking, and fusion. Because a deficit of the neurotransmitter dopamine is a crucial aspect of PD, the involvement of α-synuclein in the regulation of neurotransmitter-containing vesicles has led to suggestions that this aspect of its function may somehow be involved causally in PD. Supporting this idea, recent work has demonstrated that α-synuclein toxicity is linked to the blockage of transport vesicle fusion pathways (Cooper *et al.*, 2006), and that α-synuclein can act in a neuroprotective manner to compensate for the loss of one component of the chaperone system that regulates the folding of the synaptic vesicle fusion machinery (Chandra *et al.*, 2005). As is the case with aggregation, however, the specific mechanisms by which perturbations in the normal physiological function of α-synuclein might lead to the development of disease remains unclear.

A Role for *In Vitro* Structural Studies of α-Synuclein in PD Models

In vitro studies by definition take place outside the context of a living organism and are therefore inherently limited in their ability to recapitulate or model the complex processes that are associated with disease pathogenesis. *In vivo* or even *in situ* models, however, suffer from this very complexity, which often occludes the detailed processes that underlie the diagnostic clinical or phenotypical signatures. *In vitro* models, which are usually comprised of a relatively small number of well-defined components, offer the advantage of simplicity and clarity, and the opportunity to trace in detail the connection between cause and effect. Their value, therefore, lies in translating observations of the behavior of simple reconstituted systems into testable hypotheses regarding the mechanisms that underlie complex disease processes, which can then be evaluated in more complete cellular or animal models.

In the case of α-synuclein, *in vitro* structural studies can provide models for both of the pathways described above: the aggregation of the protein into amyloid fibrils and the involvement of the protein in synaptic vesicle regulation. The purpose of this chapter is to describe the current state of these models, and where appropriate to point out their value in providing a structural framework, based on *in vitro* studies of purified recombinant α-synuclein, for understanding and interpreting the observations made in other more complex contexts, and for exploring new structure-based hypotheses.

STRUCTURE IN THE FREE STATE OF α-SYNUCLEIN

Wild-Type α-Synuclein Contains Residual Structure

Early efforts to characterize the structure of α-synuclein using primarily optical methods such as circular dichroism and Fourier-transform infrared spectroscopy, which report on secondary structure content, revealed the surprising result that the protein did not have a typical native structure, and did not exhibit pH- or gaunidinium-induced unfolding transitions (Weinreb *et al.*, 1996), as would be expected if it were a well-folded protein. These results indicated that the protein belonged to the newly emerging category of intrinsically unstructured (also referred to as natively unfolded) proteins (Wright and Dyson, 1999; Dyson and Wright, 2005). This was confirmed by the first detailed NMR study of the protein, performed in our lab, which showed that the protein was indeed largely unfolded under physiological conditions (Eliezer *et al.*, 2001). Nevertheless, despite the clear absence of fixed secondary or tertiary structure in free α-synuclein, our NMR study revealed clear departures from a fully random coil-like structure, as indicated primarily through an analysis of NMR chemical shifts, which are highly sensitive to polypeptide backbone conformations. Our data revealed that the N-terminal 100 residues of the protein showed a slight but detectable preference for helical structures. This preference was most notable in the 40 N-terminal residues of the protein, and strongest

for residues 18 through 31. In contrast, the acidic C-terminal 40 residues of α-synuclein exhibited NMR chemical shifts that are consistent with more extended structural preferences. Initially we considered this region to have a random structure, but a potential weak preference for elements of strand-like structure is also consistent with the data.

PD-Linked α-Synuclein Mutations Influence Its Conformation

Since the discovery of the first PD-linked α-synuclein mutation, A53T (Polymeropoulos *et al.*, 1997), two additional mutations have been linked to PD or parkinsonism, A30P (Kruger *et al.*, 1998), and E46K (Zarranz *et al.*, 2004). *In vitro* studies of the aggregation of the A30P and A53T mutants (Conway *et al.*, 1998; Giasson *et al.*, 1999; Narhi *et al.*, 1999), and more recently of the E46K mutant as well (Choi *et al.*, 2004; Greenbaum *et al.*, 2005) revealed that all the three mutations enhance the propensity of α-synuclein to form oligomeric species. While the A53T and E46K mutations also accelerate the formation of mature amyloid fibrils, the A30P mutation does not (Conway *et al.*, 2000b), with the conversion of the rapidly formed A30P oligomeric species to mature fibrils occurring more slowly than for the other variants. The effects of PD-linked mutations on α-synuclein oligomerization rates *in vitro* suggest that these mutations may exert a similar effect *in vivo*, which in the context of the amyloid hypothesis (or, in consideration of the effects of A30P, the oligomer hypothesis) may in turn underlie the ability of the mutations to cause disease.

The effects of single point mutations on the aggregation rates of recombinant purified α-synuclein preparations *in vitro* suggest an underlying mutation-induced change in the physiochemical properties of the protein. To investigate the nature of such a change, efforts have been made to characterize the effects of the PD-linked mutations on the structural properties of the aggregation-competent-free state of α-synuclein. Studies using lower resolution optical techniques revealed that the gross structural properties of the free protein were insensitive to the presence of the three mutations (Conway *et al.*, 1998; Narhi *et al.*, 1999; Li *et al.*, 2001; Greenbaum *et al.*, 2005). We used high-resolution NMR spectroscopy to investigate the effects of the A30P and A53T mutations in greater detail (Bussell and Eliezer, 2001). Our results, based on a chemical shift analysis, revealed that the A30P mutation significantly perturbed the residual helical structure that was observed in the wild-type protein in the vicinity of the mutation. This result is also supported by measurements of fluorescence energy transfer between donor acceptor pairs in the N-terminal region of α-synuclein, which provide evidence for a partial population of compact helix-like structures in the wild-type protein that is perturbed by the A30P mutation (Lee *et al.*, 2004). In contrast the A53T mutation had a much less pronounced effect on residues in the immediate vicinity of the mutation. However, this mutation did lead to an apparent longer range decrease in helical preference for residues ranging approximately from positions 33 to 83. For the E46K mutant, circular dichroism spectra confirm a highly disordered state similar to that of the wild-type protein and the two other PD-linked mutants (Greenbaum *et al.*, 2005), but no high-resolution structural analysis has been reported to date.

Intramolecular Contacts in α-Synuclein may Regulate Aggregation

Based on our discovery of residual structure in free α-synuclein and the effect of the A30P mutant on

this structure we originally proposed the hypothesis that different regions within the same α-synuclein molecule may form transient contacts that might occlude epitopes that mediate intermolecular interactions leading to aggregation, and thereby exert an inhibitory effect on the self-assembly of the protein (Bussell and Eliezer, 2001). Our original model posited that residual structure at the N-terminus of the protein, where a significant signature for helical conformations was observed experimentally, would take the form of an amphipathic helix, providing a hydrophobic surface that could interact favorably with the non-Aβ component (NAC) region, the most hydrophobic region of α-synuclein, extending from residue 61 to 95 and originally identified as a proteolytic α-synuclein fragment associated with amyloid plaques (Ueda *et al.*, 1993). Such intramolecular contacts would interfere with intermolecular contacts involving the NAC region, which are thought to be critical in the early stages of α-synuclein amyloid fibril formation (Giasson *et al.*, 2001). The A30P mutation was experimentally observed to abrogate the residual helical structure in its vicinity, and would therefore be expected to relieve such intramolecular contacts and favor intermolecular contacts leading to oligomerization of the protein (see Figure 42.2).

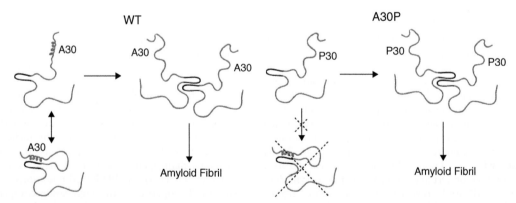

FIGURE 42.2 Schematic model for the effects of transient intramolecular interactions on the aggregation of α-synuclein. The model posits that the residual amphipathic helial structure observed around position 25 using NMR chemical shifts may interact intramolecularly, but transiently, with the hydrophobic NAC region, which is thought to nucleate β-sheet structure formation, and thereby decrease the rate of fibril-nucleating intermolecular interactions involving this region (Bussell and Eliezer, 2001). The A30P mutation destroys the residual helical structure, and is therefore hypothesized to abrogate the intramolecular interaction and thereby increase the rate of intermolecular NAC contacts and consequent oligomer (but not necessarily mature fibril) formation. Similar models involving contacts between the acidic C-terminal tail and the NAC region were later proposed as well (Bertoncini *et al.*, 2005b; Dedmon *et al.*, 2005b).

Our proposal that the free state of α-synuclein may contain transient intramolecular contacts was consistent with existing data showing that the hydrodynamic radius of the protein (Morar *et al.*, 2001; Uversky *et al.*, 2002a), as well as the radius of gyration determined from x-ray scattering measurements (Uversky *et al.*, 2002a), were smaller than would be expected for a fully random coil-like ensemble of structures, suggesting that the molecule was somehow constrained to adopt more compact conformations. Recent experiments monitoring electron transfer between donor–acceptor pairs introduced at different sites in α-synuclein also suggest a somewhat compact structure with dynamic long-range contacts (Lee *et al.*, 2005; Lee *et al.*, 2007). NMR paramagnetic relaxation enhancement (PRE) experiments, which monitor NMR line broadening as a function of distance from spin labels introduced at different sites in the protein, were able to directly detect a close spatial proximity of regions of α-synuclein that were distant from one another in sequence (Bertoncini *et al.*, 2005b; Dedmon *et al.*, 2005b), providing experimental support for the existence of transient intramolecular contacts. In one report, an ensemble of structures calculated based on experimental constraints contained significant contacts between residues 10–40 and residues 70–80 (Dedmon *et al.*, 2005b), consistent with our initially hypothesized intramolecular contact involving residual helical structure in the N-terminal portion of the protein and the NAC region. However, stronger contacts were deduced to occur between the acidic C-terminal tail of the protein and either the NAC region, the N-terminal portion of the protein or both. Such contacts have been proposed to be mediated by hydrophobic interactions between the few hydrophobic residues in the C-terminus and the NAC region (Bertoncini *et al.*, 2005b), or by electrostatic interactions between the negatively charged C-terminus and positive residues in the N-terminal region of the protein (Bernado *et al.*, 2005).

C-terminal interactions with the NAC or N-terminal regions of α-synuclein could act to prevent intermolecular interactions, in the manner already discussed above, and release of such interactions could promote the aggregation of the protein. This model, which is consistent with data showing that C-terminally truncated α-synuclein has a higher propensity to aggregate (Crowther *et al.*, 1998; Serpell *et al.*, 2000; Murray *et al.*, 2003), has been proposed based on the observations that polycations such as spermine or spermidine, which accelerate α-synuclein aggregation (Antony *et al.*, 2003; Goers *et al.*, 2003b) bind to the acidic C-terminus and may release it from any intramolecular contacts (Fernandez *et al.*, 2004). In addition, two PD-linked mutations, A30P and A53T, were proposed to release such contacts (Bertoncini *et al.*, 2005a), although it is unclear how or why these mutations would cause such an effect, as neither mutation occurs in a site of purported contact with the C-terminus and neither significantly alters hydrophobic character or electric charge. An alternative model is that the highly charged C-terminus of α-synuclein acts to retard aggregation through charge repulsion between molecules. The effects of polycations in enhancing aggregation could occur primarily through charge neutralization, and the effects of PD-linked mutations may act by altering intramolecular contacts other than those involving the C-terminus. While the specific details of which intramolecular transient contacts within free α-synuclein are most important remain to be established, our initial model which posits that such interactions act to favor a compact state and to shield regions involved in nucleating aggregation (Bussell and Eliezer, 2001) is now generally accepted.

STRUCTURE IN THE LIPID-BOUND STATE OF α-SYNUCLEIN

α-Synuclein Binds to Lipid Membranes *In Vivo* and *In Vitro*

A functional association of α-synuclein with lipid membranes was proposed in the report describing the original isolation of the protein (Maroteaux *et al.*, 1988), and was subsequently rationalized (George *et al.*, 1995) by noting the resemblance of a series of 11-residue imperfect repeats present within the α-synuclein primary sequence (see Figure 42.1) to repeats that are commonly found in the lipid-binding domains of the exchangeable apolipoproteins (Segrest *et al.*, 1992). Binding of the protein to artificial and natural lipid membranes was confirmed *in vitro* (Davidson *et al.*, 1998; Jensen *et al.*, 1998), and was shown using circular dichroism to be associated with a disorder-to-order transition in which the largely unstructured free protein became highly helical upon binding to membranes (Davidson *et al.*, 1998). The lipid-binding affinity of α-synuclein is increased in the presence of negatively charged phospholipids (Davidson *et al.*, 1998; Jo *et al.*, 2000), but weaker binding to purely zwitterionic membranes can also observed using circular dichroism, NMR, or fluorescence correlation spectroscopy (Jo *et al.*, 2000; Bussell and Eliezer, 2004; Rhoades *et al.*, 2006). α-Synuclein was reported, based on size exclusion chromatography, to bind more readily to more highly curved small unilamellar lipid vesicles, when compared to larger multilamellar vesicles (MLVs) (Davidson *et al.*, 1998), but these results effectively relied on a comparison of total protein-to-lipid ratios and did not take into consideration that much of the lipid content of MLVs is not surface accessible. When the entire lipid content of MLVs is made accessible by introducing α-synuclein prior to vesicle formation, robust binding was observed (Jo *et al.*, 2000). Furthermore, a quantitative comparison of α-synuclein-binding affinities to small versus large unilamellar vesicles using fluorescence correlation spectroscopy did not reveal a significant curvature dependence (Rhoades *et al.*, 2006). α-Synuclein was suggested to have a preference for binding to defects in lipid membranes based on calorimetry studies (Nuscher *et al.*, 2004) and according to ESR studies influences the structure and properties of lipid bilayers to which it binds (Ramakrishnan *et al.*, 2003; Kamp and Beyer, 2006).

Binding of α-synuclein to lipid membranes *in vivo* remains more controversial. Electron microscopy using immunogold labeling shows that α-synuclein is localized to the vicinity of synaptic vesicles in presynaptic termini, but is only occasionally found directly associated with the vesicle exterior (Iwai *et al.*, 1995; Clayton and George, 1999). Similarly, in fractionation studies, synuclein is found predominantly in cytosolic fractions (George *et al.*, 1995), with lesser quantities of the protein found in fractions containing synaptic vesicles (Irizarry *et al.*, 1996; Kahle *et al.*, 2000). Fluorescence resonance energy transfer (FRET) studies between a membrane-partitioned dye and immunofluorescently labeled α-synuclein document a close association of the protein with membranes in primary cortical neurons (McLean *et al.*, 2000a), but are not sensitive to the presence of membrane-free protein. When transfected into HeLa cells (Cole *et al.*, 2002) or yeast (Outeiro and Lindquist, 2003), α-synuclein is found both in the cytoplasm

and localized to membrane structures. When HeLa cells were loaded with fatty acids, the protein was found at the surface of intracellular lipid droplets, which are surrounded by a phospholipids monolayer (Cole *et al.*, 2002), and the protein was also found to promote lipid droplet formation in yeast (Outeiro and Lindquist, 2003).

Recent work suggests that efficient membrane binding by α-synuclein requires a highly precise lipid composition (Kubo *et al.*, 2005), and may also be influenced by other cytosolic components (Kim *et al.*, 2006), providing potential explanations for the disparity between *in vitro* and *in vivo* observations. In addition, careful *in vitro* experiments using more sophisticated spectroscopic techniques show that the affinity of synuclein to small lipid vesicles is not very high (Bussell and Eliezer, 2004; Rhoades *et al.*, 2006), with an estimated dissociation constant, K_D, of around 100 μM for a plasma membrane-like phosphatidylserine content (Rhoades *et al.*, 2006), consistent with the bipartite distribution of the protein *in vivo*. Thus, consistent with its exchangeable lipoprotein-like sequence, α-synuclein appears to naturally bind membranes in an easily reversible fashion, but *in vivo* microenvironments may strongly effect its local partitioning between membrane bound and free forms.

Detergent Micelle-Bound α-Synuclein Adopts a Non-Compact Helical Structure

From optical spectroscopy studies (Davidson *et al.*, 1998; Jo *et al.*, 2000; Perrin *et al.*, 2000), lipid-bound α-synuclein was known to have a highly helical structure. In addition, based on the similarity of its sequence to the alpolipoprotein family (George *et al.*, 1995), the lipid-bound structure was predicted to take the form of 5 distinct amphipathic helices spanning residues 1 through 94 that would bind to the surface of membranes rather than inserting in a transmembrane geometry (Davidson *et al.*, 1998). Our initial NMR studies showed that lipid-binding was mediated by α-synuclein residues 1 through 102, and that the helical structure of the bound protein was restricted to this lipid-bound region (Eliezer *et al.*, 2001). We also showed that the SDS micelle-bound state of α-synuclein mimicked the lipid-bound state insofar as the same residues, 1–102, were responsible for detergent micelle binding in a helical conformation and the behavior of the remaining residues, belonging to the acidic C-terminal tail, was the same in both the lipid- and detergent-bound states (Eliezer *et al.*, 2001).

Establishing the SDS micelle-bound state of α-synuclein as a reasonable mimic of the lipid-bound state led to a number of subsequent studies of the detailed structure of the micelle-bound protein. We and a second group used NMR chemical shifts and short range NOEs (which reflect distances between individual nuclei) to show that the helical structure in the micelle-bound region of the protein took the form of two longer helices with a single break between them, rather than the five predicted shorter helices (Bussell and Eliezer, 2003; Chandra *et al.*, 2003). We suggested that the discrepancy between the predicted and observed number of helices was caused by an alteration in the periodicity of the helical structure, which would allow the amphipathic nature of the helices to remain intact over longer distances (Bussell and Eliezer, 2003), an idea first envisioned in the context of models of apolipoprotein structures (Segrest *et al.*, 1999). Subsequent measurements using ESR spectroscopy (Jao *et al.*, 2004), as well by ourselves using NMR (Bussell *et al.*, 2005) confirmed the presence of this altered periodicity in lipid- and micelle-bound α-synuclein.

The presence of two separate helices raised the possibility of inter-helix interactions within α-synuclein, which we explored using NMR PRE experiments. We showed that the two micelle-bound helices did not approach each other closely enough to permit direct contact between the helices (Bussell *et al.*, 2005). This is consistent with a failure by ourselves, as well as others (Chandra *et al.*, 2003; Ulmer *et al.*, 2005) to find any indication of long-range NOEs between the two helices, even when using selective labeling strategies for side chain methyl groups (Ulmer *et al.*, 2005).

A combination of PRE measurements and NMR residual dipolar coupling (RDC) measurements, combined with a novel technology for calculating NMR structures by empirical fitting of existing structure segments in the Protein Data Bank to RDC data (Delaglio *et al.*, 2000; Kontaxis *et al.*, 2005) finally resulted in a structure calculation for the detergent micelle-bound form of α-synuclein (Ulmer *et al.*, 2005). This structure confirms many of the previous results based on our own and others' studies, including the presence of two long well-formed helices, now defined as extending from residues 3 through 37 (helix 1) and residues

45 through 92 (helix 2). Surprisingly, however, the helices were found to be oriented antiparallel to one another, despite the absence of any tertiary interactions between the helices. Both helices were also shown to be slightly curved, with a curvature conforming to a sphere of diameter 153 Å for helix 1 and 82 Å for helix 2. The somewhat higher curvature of helix 2 was attributed to a slight kink at residue 66–68, consistent with a locally decreased indication for helical structure observed in this region in earlier reports (Bussell and Eliezer, 2003; Chandra *et al.*, 2003). The curvature of both helices is significantly less than would be expected if they conformed to the surface of a spherical micelle (estimated diameter 46 Å) suggesting that in the complex, the micelle is ellipsoidal, rather than spherical.

In addition to the two micelle-bound helices, residues 93–97 were found to be ordered and to adopt an extended conformation. This is consistent with our earlier observations that residues beyond the helix-terminating phenylalanine 94 showed short range NOEs indicative of ordered structure (Bussell and Eliezer, 2003), and with our observation that residues up to lysine 102 were immobilized upon binding to micelles (Eliezer *et al.*, 2001). Furthermore, the linker region between helix 1 and helix 2, consisting of residues 38–44, was found to adopt an extended, irregular, but well-ordered conformation. This is consistent with NMR relaxation parameters for backbone nitrogens in this region (Bisaglia *et al.*, 2005; Bussell *et al.*, 2005), as well as order parameters (Ulmer *et al.*, 2005), which indicate relatively restricted bond librations and are inconsistent with a highly flexible state.

α-Synuclein Remains on the Surface of Detergent Micelles

Based on a sequence analysis, α-synuclein was predicted to bind to the surface of lipid membranes rather than inserting in a transmembrane fashion. For the detergent micelle-bound state, NMR PRE measurements using spin-label doped micelles confirmed that neither of the micelle-bound protein helices traverses the micelle interior, as would be expected for a transmembrane helix (Bisaglia *et al.*, 2005; Bussell *et al.*, 2005). In addition, we showed that the helices do not penetrate below carbon 4 of the detergent acyl chains, placing an upper limit of about 5 Å on how deeply below the headgroup region the helix backbones can penetrate. However, some variations in the depth to which the protein is embedded into the micelle surface were observed. In particular, the linker region between the two helices was observed to be embedded more deeply than surrounding regions (Bisaglia *et al.*, 2005; Bussell *et al.*, 2005), and the sixth of the seven 11-residue repeats, corresponding to residues 68 through 79, is also more deeply buried and protected from solvent (Bussell *et al.*, 2005). Measurements of saturation transfer between detergent protons and protein amide protons suggest that the majority of the protein backbone residues lie at a depth below that of carbon 2 of the detergent acyl chains (Ulmer and Bax, 2005), which resides approximately 2.5 Å beneath the headgroup region.

Does Lipid Vesicle-Bound α-Synuclein Resemble the Micelle-Bound Protein?

A major remaining question regarding the structure of membrane-bound α-synuclein relates to the effect of using detergent micelles as a membrane mimetic in the high-resolution NMR studies. Although our NMR studies showed that the residues that are directly involved in binding lipid vesicles are the same as those responsible for micelle binding (Eliezer *et al.*, 2001), the slow tumbling rate of the vesicles precludes direct observation of NMR signals originating from these same residues (whereas resonances originating in the C-terminal tail, which remains unbound and flexible, are readily observed). Thus, it is clear that the same region of the protein that is observed in the micelle-bound structure binds to vesicles and adopts a highly helical structure, but the details of this structure remain to be fully elucidated. One study that has provided partial information used ESR spectroscopy of α-synuclein spin labeled at different sites (Jao *et al.*, 2004) to show that residues 59–90 adopt a helical structure, confirming that most of helix 2 is intact in the vesicle-bound state, and further showed that the periodicity of this helical structure was indeed matched to the 11-residue repeats present in the protein's sequence, as postulated by us (Bussell and Eliezer, 2003) and subsequently shown for the micelle-bound state as well (Bussell *et al.*, 2005). The line shape of ESR signals originating from residues in helix 2 also indicate that these residues are not involved in well-formed tertiary structure, confirming the absence of contacts between helix 2 and other regions of the protein. Residues that

reside in the linker region between the two helices in the micelle-bound state were also observed to be ordered in the vesicle-bound state, but the secondary structure of the linker region was not evaluated. In addition, the immersion depth of the center of helix 2 was estimated to be approximately 1–4 Å, in good agreement with observations made using spin labels (Bussell *et al.*, 2005) and saturation transfer (Ulmer and Bax, 2005) for the micelle-bound state.

All of the information provided by the ESR study of vesicle-bound α-synuclein is consistent with observations of the micelle-bound state, including the ordered nature of the linker region. At present, however, it remains unclear if the vesicle-bound linker region adopts the extended irregular structure that is observed in the micelle-bound state (Ulmer *et al.*, 2005). Furthermore, it is not known whether the antiparallel orientation of the two α-synuclein helices in the micelle-bound state is preserved in the vesicle-bound state. This latter point is of particular interest, because the dimensions of the two helices (approximate lengths of 52.5 and 72 Å for helix 1 and 2, respectively) are on the same order as the expected diameter of an SDS detergent micelle (46 Å for a spherical topology). Therefore, we postulated that the break observed between the micelle-bound helices may be a result of the small size of the micelle and the lack of room for the propagation of a longer unbroken helix (Bussell and Eliezer, 2003). Furthermore, the antiparallel orientation of the helices may also be a result of the small micelle size, since the second helix is essentially restricted to return in the direction from which the first originated. We recently performed experiments that begin to address this issue by using pulsed ESR to measure long distances between different pairs of spin-labeled sites in α-synuclein bound to two different micelle varieties (Borbat *et al.*, 2006), one with an acyl chain of length 12 (SDS) and one with an acyl chain of length 16 (lyso 1-palmitoylphosphatidylglycerol). These experiments confirmed that the two helices are oriented antiparallel to one another in both micelle types, but showed that when bound to the larger micelle, the two helices splay further apart and the distance between their ends increases significantly, suggesting that the antiparallel geometry may indeed be imposed by a small micelle size (see Figure 42.3). Further experiments will be required to determine if the helices become more collinear on a lipid vesicle surface, and if they do so, whether

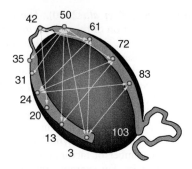

FIGURE 42.3 Schematic model of micelle-bound α-synuclein based on ESR data (Borbat *et al.*, 2006) and consistent with NMR structure calculations (Ulmer *et al.*, 2005). White arrows indicate the distances measured between different pairs of spin-labeled residues using pulsed ESR methods. Upon increasing the size of the micelle, the distances between residue pairs 24/61, 13/72, and 3/83 increase significantly, indicating that the two helices can splay further apart when the protein is bound to a larger particle. (Reproduced with permission from *J Am Chem Soc* 2006, 128, 1004–1005. Copyright 2006 American Chemical Society.)

this is accompanied by a transition of the linker region to a helical conformation, forming one long helix, or whether the linker region retains the irregular conformation observed in the micelle context.

PD-Linked α-Synuclein Mutations Affect Its Lipid-Bound Structure Differently

The affect of PD-linked mutations on the structure of lipid-bound α-synuclein are of natural interest because they may be linked to the etiology of PD by affecting either the normal function or the aggregation of α-synuclein. Circular dichroism measurements of total helix content in the presence of synthetic lipid vesicles of varied compositions demonstrated that there is very little difference between the helical content induced in the wild-type protein and the A53T PD-linked mutant (Jo *et al.*, 2000; Perrin *et al.*, 2000). In the case of the A30P mutant, however, a significantly decreased amount of helix was induced compared to the wild-type protein under identical conditions (Perrin *et al.*, 2000; Jo *et al.*, 2002) and the decrease in helical structure was accompanied by an apparent decrease in lipid-binding affinity. This observation was consistent with the

reduced ability of the A30P mutant to bind to rat brain vesicles (Jensen *et al.*, 1998) (the A53T mutant resembled the wild type in this assay), and with the inability of the A30P (but not A53T) mutant to localize to membranes in yeast (Outeiro and Lindquist, 2003) and to lipid droplets in cultured cell lines (Cole *et al.*, 2002). However, FRET studies in cultured neurons demonstrate that A30P α-synuclein retains some ability to bind membranes within cells (McLean *et al.*, 2000a). For the more recently discovered E46K mutation, an increased lipid affinity based on a liposome pull down assay was reported (Choi *et al.*, 2004), but no information is available regarding structure in the lipid-bound state of this mutant as of yet.

To further investigate potential differences between the lipid-bound structures of wild-type, A30P and A53T synucleins, we characterized the detergent micelle-bound states of all the three variants using solution state NMR, as well as the lipid vesicle-bound states using circular dichroism, NMR and limited proteolysis (Bussell and Eliezer, 2004). Although NMR could not be used to directly monitor vesicle-bound regions of α-synuclein, both unbound protein and the flexible lipid-free C-terminal tail can be directly observed. We were therefore able to show that all three synuclein variants bound to lipid vesicles in a reversible fashion, resulting in an equilibrium between free and bound protein. Under identical conditions, the wild type and A53T proteins bound vesicles to a similar extent, while the A30P mutant bound more weakly, resulting in a greater population of free protein. This increased population of free protein is responsible for the decreased helical content observed for the A30P mutant in the presence of vesicles by ourselves and others using circular dichroism (Perrin *et al.*, 2000; Jo *et al.*, 2002). Importantly, the overall helical content of the bound state of the A30P mutant is nearly identical to that of the wild-type and A53T variants. An analysis of NMR chemical shifts and NOEs for detergent micelle-bound A30P and A53T α-synuclein confirmed that the structure of the latter mutant is essentially identical to that of the wild type. For the A30P mutant, a slight decrease in helical structure was evident locally around the site of the mutation, as indicated primarily by a dramatic decrease in the α-carbon chemical shift of residue 28. Apparently the introduction of the proline residue, as expected, interferes with helix formation, possibly causing a kink in the helix, but this perturbation is local.

Structure calculations for the detergent micelle-bound A30P and A53T α-synuclein mutants (Ulmer and Bax, 2005), carried out using the same protocol developed for the wild-type protein (Ulmer et al., 2005) confirmed that the A53T structure is essentially identical to that of the wild type. For the A30P mutation, the mutation-induced local structural perturbation is accompanied by an increase in local mobility, as indicated by decreased values of backbone amide group order parameters and alignment tensor magnitudes, which precluded a determination of the structure around the site of the mutation even with the use of the RDC-based molecular fragment replacement method. Therefore, the precise nature of the structure around the site of the mutation remains unclear. There is clearly a perturbation of the helical structure of helix 1 at position 28, as we also surmised based on chemical shifts, but the conformation of residues 29 through 44 remains ambiguous. The structure of helix 2 remains unmodified. Interestingly, although the structure of the A30P mutant remains unperturbed at sites distant from the mutation, small chemical shift perturbations are observed as far away as 30 residues from the site of the mutation, indicating a change in the external environment of residues distant from position 30, and suggesting an alteration in the shape of the detergent micelle itself.

Membrane-Bound Structure May Play a Role in α-Synuclein Aggregation

Although it is clear that α-synuclein undergoes an aggregation process in the brains of those suffering from PD, the starting point of that aggregation process in vivo remains unknown. Since α-synuclein is known to populate at least two distinct states in vivo, being found in both cytosolic- and membrane-associated forms, either of these contexts may be considered as a potential initiation point for aggregation in vivo. Importantly, although the hallmark Lewy body deposits of PD are formed in the cell soma, where in general α-synuclein has been observed to be largely cytoplasmic, a recent report indicates that a great deal of α-synuclein aggregation occurs in synapses as well (Kramer and Schulz-Schaeffer, 2007), where the membrane bound form of the protein may be more common. In vitro, the free state of the protein has been shown to lead, under appropriate conditions, to the formation of authentic amyloid fibrils (Conway et al., 2000a; Serpell et al., 2000) resembling those isolated from

PD inclusions (Serpell et al., 2000). The effects of lipids on the aggregation of the protein in vitro have also been investigated, and have resulted in apparently conflicting observations, where in some reports lipids have been shown to inhibit the aggregation of the protein (Narayanan and Scarlata, 2001; Zhu and Fink, 2003), while in others lipids or fatty acids are found to facilitate α-synuclein self-assembly (Lee et al., 2002; Sharon et al., 2003). Some studies report both effects, depending on the synuclein variant (Jo et al., 2004). A number of factors could contribute to these discrepancies. For example, the precise composition of lipid bilayers has been shown to effect both α-synuclein binding (Kubo et al., 2005) and the ability of the membranes to promote α-synuclein aggregation (Perrin et al., 2001). Importantly, however, the effect of lipids or fatty acids on synuclein aggregation has been shown to depend strongly on the lipid-to-protein ratio (Necula et al., 2003; Zhu et al., 2003), with lower ratios favoring, and higher ratios disfavoring aggregation.

A potential explanation for these observations is that when sufficient membrane or micelle surface area is present to allow for adequate segregation of each bound protein monomer, aggregation is prevented. However, when an insufficient surface area is available, individual protein monomers compete for binding to a limited space, resulting in a high local concentration, which in turn drives the aggregation process. In the latter case, it remains unclear whether the interactions leading to amyloid fibril formation occur between membrane-bound protein molecules or membrane proximal, but unbound, molecules. Given the two-dimensional nature of membrane surfaces, it is likely that membrane bound α-synuclein experiences a higher effective local protein concentration, which in turn suggests, but does not prove, that lipid-induced aggregation initiates from the helical membrane-bound conformation. In the case of micelle-bound α-synuclein, it has been shown that reducing the concentration of available micelles leads to changes in NMR spectra of the protein that are consistent with environmental and/or conformational changes in the protein (Ulmer et al., 2005), indicating that a protein-to-micelle stoichiometry higher than 1 directly affects the micelle-bound state.

Membrane-Bound Structure is Linked to α-Synuclein Function

Although the normal function of α-synuclein remains poorly understood, several lines of evidence suggest that the membrane-bound state of the protein is necessary for this function. First, the sequence of synuclein is closely related to that of the exchangeable apolipoproteins, strongly suggesting a lipid-associated function. In addition, we showed that the synuclein sequence is also related to the adipophilin/perilipin family of proteins (Bussell and Eliezer, 2003), which are responsible for coating lipid droplets and regulating lipid storage and metabolism in adipocytes, again suggesting a functional link to lipids. Second, a number of studies of synuclein knockout mice or of synuclein knockdown in cultured cells have implicated the protein in pathways associated with the trafficking and fusion of neurotransmitter-carrying synaptic vesicles (Abeliovich *et al.*, 2000; Murphy *et al.*, 2000; Cabin *et al.*, 2002; Chandra *et al.*, 2004; Liu *et al.*, 2004; Martin *et al.*, 2004; Chandra *et al.*, 2005), to which synuclein is expected to bind in its helical conformation. Finally, inhibition by α-synuclein of its most reliably confirmed target, phospholipase D (PLD) (Jenco *et al.*, 1998), has been shown to require the helical lipid-bound conformation of the protein (Payton *et al.*, 2004). It is likely that many other protein–protein interactions involving α-synuclein will have a similar requirement.

ROLE OF α-SYNUCLEIN STRUCTURE IN PD MODELS

Testing the Importance of α-Synuclein Aggregation

Aggregation of α-synuclein has been proposed to play an important role in PD based largely on the appearance of aggregated protein in Lewy body and Lewy neurite deposits and on the increased rate of *in vitro* oligomerization of PD-linked α-synuclein mutants. One approach for evaluating the importance of α-synuclein aggregation in disease is to study correlations of aggregation propensity with disease-linked phenotypes. The above-described structural studies provide several models for α-synuclein aggregation that can facilitate this process by guiding the rational design of new α-synuclein mutants with different aggregation propensities. This notion was first suggested by us (Bussell and Eliezer, 2001) in the context of modifying the sequence of α-synuclein in order to modulate the stability of the residual helical structure observed around position 25 in the protein. Since this structure is postulated to interact with and

occlude the NAC region, mutants that stabilize this structure should more efficiently segregate the NAC region, leading to a reduced tendency to aggregate, while mutants that destabilize this structure should have the opposite effect. A similar strategy could be employed based on those models that include interactions between the C-terminal tail of the protein and either the NAC region or positively charged N-terminal residues, with mutations that decrease either hydrophobicity or negative charge in the tail expected to lead to more rapid aggregation. To date, however, the generation and characterization of such designed mutants remains in progress, as does evaluation of their effects *in situ* or *in vivo*, which is required to test the correlation between aggregation and toxicity. Notably, however, such correlative studies of several mutated α-synuclein constructs generated without the use of structural models have been reported recently, albeit with somewhat different conclusions regarding the role of α-synuclein aggregation in disease. (Periquet *et al.*, 2007; Volles and Lansbury, 2007).

In principle, structural studies of α-synuclein can also aid in the design of reagents that target the aggregation of the protein. Unfortunately, because of the highly dynamic nature of the free state of α-synuclein, identification of such reagents cannot proceed via strategies commonly employed in structure-based drug design. Thus, most efforts at discovering such reagents have focused either on traditional screening methods or on exploring the use of naturally occurring α-synuclein interacting compounds, peptides, and proteins, including α-synuclein-derived peptide sequences, antibodies, and other potential interaction partners. Examples of naturally occurring binding partners include small peptides derived from the NAC region and fused to delivery sequences (El-Agnaf *et al.*, 2004), phage-display optimized single-chain antibody fragments (Emadi *et al.*, 2004), other synuclein family members (Hashimoto *et al.*, 2001; Uversky *et al.*, 2002a) as well as chaperones (Dedmon *et al.*, 2005a), and reagents such as flavonoids (Zhu *et al.*, 2004). Screening methods have identified other reagents that interact with α-synuclein and interfere with at least some aspects of its self-assembly (Conway *et al.*, 2001; Masuda *et al.*, 2006; Mazzulli *et al.*, 2006). Efforts to determine the detailed mode of interaction of these and other α-synuclein aggregation modulators remain challenging, and are further complicated by the observation that a number of these reagents appear to interact with oligomeric forms of the protein, rather than directly with the free state. Importantly, the potential toxicity of oligomeric species suggests that great care must be used in evaluating the consequences of interrupting α-synuclein aggregation at the later stages of the process.

Unlike its free state, the lipid-bound state of α-synuclein is well ordered, and a high-resolution structural model based on the micelle-bound state of the protein is available (Ulmer *et al.*, 2005). Such a model could be combined with structure-based drug design approaches to identify potential small molecule binding partners. As discussed above, it is possible that the lipid-bound state of the protein serves as a starting point for the aggregation of the protein *in vivo*, and such compounds could inhibit this pathway. To date, however, such design efforts have not been reported, possibly because the unusual topology of the structure presents a greater challenge to structure-based design methodologies than more typical globular protein targets.

Testing the Importance of α-Synuclein Function

Although discerning the normal function of α-synuclein has proven a difficult challenge, continuing progress is being made toward this goal. While most approaches to this problem involve the use of intact cells or living organisms, structural information can also be helpful in trying to understand how α-synuclein may function. Furthermore, to the extent that specific α-synuclein functions can be identified, structural studies can again provide a context for designing tests of the relevance of such functions to the toxic or disease-linked properties of the protein.

α-synuclein was originally identified as a protein enriched both in presynaptic nerve terminals and at the nuclear envelope of cells (Maroteaux *et al.*, 1988). Although some evidence has been presented for the presence of α-synuclein in the nucleus (McLean *et al.*, 2000b; Goers *et al.*, 2003a; Kontopoulos *et al.*, 2006; Yu *et al.*, 2007), and for interactions with nuclear proteins such as histones (Goers *et al.*, 2003a), most studies have focused on a functional role for the protein in the synapse. Because PD is a dopamine deficit disorder, attention has been given to the potential role of α-synuclein in regulating dopamine homeostasis at the synapse. A number of *in vivo* studies have revealed alterations in dopamine levels in mice lacking α-synuclein (Abeliovich *et al.*, 2000; Chandra *et al.*, 2004; Yavich *et al.*, 2004), and α-synuclein has been proposed to regulate dopamine uptake into synaptic vesicles by influencing levels of the vesicular dopamine transporter (Lotharius *et al.*, 2002; Drolet *et al.*, 2004) and to act on dopamine synthesis pathways (Perez *et al.*, 2002). α-Synuclein has also been shown to directly interact with and regulate the activity of the human dopamine transporter (hDAT), possibly by influencing its trafficking (Lee *et al.*, 2001; Wersinger and Sidhu, 2003; Wersinger *et al.*, 2003). An upregulation of hDAT activity by α-synuclein (Lee *et al.*, 2001) would be consistent with experiments performed using cultured cells (Fountaine and Wade-Martins, 2007) and mice (Dauer *et al.*, 2002; Schluter *et al.*, 2003; Drolet *et al.*, 2004; Robertson *et al.*, 2004; Klivenyi *et al.*, 2006), showing that removal of α-synuclein is protective against the effects of the toxin MPP$^+$, since this toxin is taken up into cells by hDAT. Indeed a reduction of transporter activity upon α-synuclein knockdown was reported in cell culture studies (Fountaine and Wade-Martins, 2007), although in knockout mice transporter activity (Dauer *et al.*, 2002) and levels (Dauer *et al.*, 2002; Schluter *et al.*, 2003) were unaffected. Thus, hDAT appears to be a good candidate for a functionally relevant α-synuclein interaction partner.

The specific region of hDAT to which α-synuclein binds had been identified as the terminal 22 residues of the cytoplasmic C-terminal tail of the transporter, and within the α-synuclein sequence, the hDAT interacting region falls within residues 58–107, which includes the hydrophobic NAC region (Lee *et al.*, 2001; Wersinger *et al.*, 2003). At present, however, the structural basis for the interaction between the two proteins is not known. Because hDAT is an integral membrane protein, it seems likely that the membrane-bound form of α-synuclein would mediate the interaction between the two proteins, and that helix 2 in particular, which encompasses the NAC region, would be involved. Destabilization of helix 2 might therefore be expected to interfere with any functional hDAT interactions. Although mutants that destabilize helix 2 of membrane-bound α-synuclein have not been reported, we have recently showed that in β-synuclein, a closely related member of the synuclein family, the absence of 11 residues from the NAC region of the protein leads to a significant destabilization of helix 2 of the membrane-bound conformation (Sung and Eliezer, 2006). Thus, based on structural studies, it might be expected that β-synuclein may not be as effective as α-synuclein in influencing hDAT function. At present, however, no reported data are available to evaluate this hypothesis.

α-Synuclein has also been shown to interact with and inhibit the activity of the enzyme PLD, which hydrolyzes phosphatidylcholine to phosphatidic acid. Phosphatidic acid is believed to act as a signal for the biogenesis of synaptic vesicles, suggesting that α-synuclein may regulate synaptic vesicle formation. α-Synuclein may also play a role in regulating synaptic vesicle storage and trafficking, as suggested by observations of altered vesicle pool sizes in some studies of α-synuclein knockout in mice (Cabin *et al.*, 2002) or in studies of α-synuclein knockdown in cell culture (Murphy *et al.*, 2000). Most recently, evidence has accumulated that α-synuclein may regulate synaptic vesicle fusion. In particular, α-synuclein has been shown to rescue a severe neurodegeneration that results in mice upon removal of CSPα (Chandra *et al.*, 2005), a protein that acts as a co-chaperone for the assembly of SNARE complexes required for vesicle fusion at the synapse. Depletion of both α- and β-synuclein in mice leads to an upregulation of the protein complexin, which binds to SNARE complexes and modulates the release of vesicle contents (Chandra *et al.*, 2004). Furthermore, when transfected into yeast, α-synuclein acts to block traffic between the endoplasmic reticulum (ER) and the Golgi, and this block can be rescued by Rab overexpression (Cooper *et al.*, 2006), suggesting that it may occur at the vesicle fusion step of the transport process. Finally, α-synuclein overexpression in cultured chromaffin cells impairs the release of catecholamines by inhibiting the fusion of so-called "docked" vesicles, apparently by blocking late steps in the exocytosis process (Larsen *et al.*, 2006).

The direct inhibition of PLD by α-synuclein is one of the best characterized functional interactions of the protein. *In vitro* studies with recombinant protein constructs demonstrated that this interaction requires the presence of the acidic C-terminal tail of the protein as well as of exon 4 (residues 56–102), and is dependent on the helical conformation of the membrane-bound protein (Payton *et al.*, 2004). PLD inhibition was also confirmed in studies examining the effect of α-synuclein transfection in yeast (Outeiro and Lindquist, 2003) and the activity was observed to correlate with membrane binding. Thus it is clear that the membrane-bound

form of the protein is responsible for this function, consistent with the fact that PLD localizes to membranes in cells (McDermott *et al.*, 2004) and that its substrate is a component of lipid bilayers. Although helix 2 of the membrane-bound state of α-synuclein is required for PLD inhibition (since exon 4 largely encompasses this helix), it is not sufficient, as deletion of the C-terminal tail completely abrogates activity. The unstructured nature of the acidic C-terminal tail of α-synuclein has lead to suggestions that it may be involved in protein–protein interactions, which is indeed the case for PLD, and has also been reported for interactions with other proteins such as tau (Jensen *et al.*, 1999). Based on the well-structured nature of helix 2 and the highly extended nature of the C-terminal tail, an hypothetical model for α-synuclein–PLD interactions can be constructed by invoking stabilizing interactions between helix 2 and membrane-bound regions of PLD, accompanied by insertion of the acidic C-terminal tail into the enzymatic active site of the protein.

In addition to a role in vesicle budding, PLD has also been implicated in exocytic pathways and may be involved in regulating vesicle fusion, as phosphatidic acid is known to be an activator of the enzyme phosphatidylinositol-4-phosphate 5-kinase, which generates phosphatidylinositol-4,5-bisphosphate, a lipid that interacts with and activates components of the vesicle fusion machinery such as the protein synaptotagmin. In addition, phosphatidic acid may act in both vesicle budding and fusion by altering the local properties of the lipid bilayer in order to facilitate membrane curvature and instability, which may be required for both processes (McDermott *et al.*, 2004). The potential involvement of PLD activity in vesicle fusion, combined with the evidence for the role of α-synuclein in this process suggests that the interaction between these two proteins may take place in the context of fusion events. This hypothesis can also provide an avenue for explaining an additional puzzling feature of α-synuclein, the fact that the normal function of the protein requires membrane binding, but the protein is only loosely associated with membrane vesicles both *in vitro* and *in vivo*. Possibly, efficient membrane binding by α-synuclein depends on the presence of a particular membrane topology, which may only be present during vesicle fusion. Interestingly, the interactions of the protein with small micelles are much tighter than with lipid vesicles (Bussell and Eliezer, 2004), suggesting that the micelle, but not the vesicle surface, may present an optimal topology. The most obvious difference between micelles and vesicles is the high degree of curvature of the former, a property that accommodates the unusual hairpin structure observed for the micelle-bound protein by allowing the two α-synuclein helices to bind to effectively, oppositely oriented membrane surfaces. Intriguingly, the presence of closely apposed membrane surfaces also occurs whenever a synaptic vesicle is in close proximity to the plasma membrane (i.e., when vesicles are budding or fusing).

Putting all of the above observations together results in a novel model for α-synuclein function, illustrated in Figure 42.4, which posits that α-synuclein functions to stabilize and anchor synaptic vesicles at the membrane surface, either after they bud or while they are "docked" and ready to fuse, by using its two membrane-binding helices to span the space between the vesicle and membrane surfaces. According to this model, the transient association of α-synuclein with membrane vesicles *in vitro* and *in vivo* results from their non-optimal geometry for α-synuclein binding. Tight binding is only achieved for those few vesicles in the synapse that are "docked" in close proximity to another membrane. In addition, vesicle fusion may be highly sensitive to

precise α-synuclein levels. Too much protein could over stabilize docked vesicles, possibly blocking late steps in the refilling of the "readily releasable pool," and therefore decrease exocytosis as observed in chromaffin cells and in yeast. Overexpression or upregulation of other vesicle trafficking or fusion proteins such as Rabs or complexins may circumvent this block either by increasing fusion efficiency or by increasing the overall number of "docked" vesicles. Too little α-synuclein, however, could lead to a disregulation of the fusion of docked vesicles, possible accounting for the more rapid recovery of dopamine release upon repeated stimulation (Abeliovich *et al.*, 2000; Yavich *et al.*, 2004). In cells where the normal fusion machinery is compromised, however, such as when CSPα is knocked out, increased α-synuclein levels may help to drive some degree of vesicle fusion by anchoring vesicles in close proximity to the membrane.

Although the above model can, to some extent, explain many disparate results related to normal α-synuclein function, it is highly speculative. Nevertheless, it serves as a good illustration of the role that structural studies can play in trying to understand observations obtained using a wide spectrum of techniques applied to systems ranging from highly purified *in vitro* preparations to intact

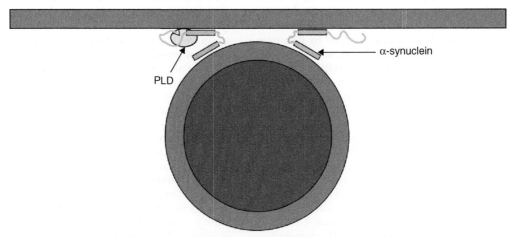

FIGURE 42.4 Model for the hypothetical involvement of α-synuclein in tethering docked synaptic vesicles (shaded darkly for the interior and more lightly for the surrounding membrane bilayer) in the proximity of the plasma membrane (shaded stripe) using its helical hairpin structure. Helix 1 (the N-terminal helix) is proposed to interact with the docked vesicles while helix 2 (the C-terminal helix) binds to the plasma membrane. This positions both helix 2 and the highly unstructured C-terminal tail for interactions with PLD in order to regulate the production of phosphatidic acid, which may alter the local properties of the membrane in such a way as to favor either fusion or budding.

living organisms. The value of such a model lies not so much in its ultimate success as in motivating new experiments which, irrespective of whether they support or invalidate it, will provide valuable new insights and move our understanding forward.

FUTURE DIRECTIONS

Structural studies of α-synuclein, both in its disordered-free state and in its highly helical lipid-bound state have provided important insights into the protein's behavior both *in vitro* and *in vivo*. These studies can be extended in number of important directions in order to improve our understanding of synuclein function and pathology at the molecular level. Specifically, α-synuclein is known to be phosphorylated at both serine/threonine (Okochi *et al.*, 2000; Pronin *et al.*, 2000) and tyrosine (Ellis *et al.*, 2001; Nakamura *et al.*, 2001) sites, and phosphorylation may have important effects on both the function and aggregation of the protein, but the structural consequences of phosphorylation are not presently known. α-Synuclein has also been reported to undergo a number of other covalent modifications, such as ubiquitination (Hasegawa *et al.*, 2002; Tofaris *et al.*, 2003), sumoylation (Dorval and Fraser, 2006), transglutamination (Citron *et al.*, 2002; Junn *et al.*, 2003; Andringa *et al.*, 2004), methionine oxidation (Uversky *et al.*, 2002b), and tyrosine nitration (Giasson *et al.*, 2000), all of which can be expected to influence the structure of the protein, with functional consequences. Finally, genetic and proteomic studies have implicated a large number of proteins in potentially direct interactions with α-synuclein (Payton *et al.*, 2001; Scherzer *et al.*, 2003; Willingham *et al.*, 2003; Zhou *et al.*, 2004; Jin *et al.*, 2007), and the structural changes that accompany those interactions that are verifiable are likely to be highly informative. Pursuing these directions will provide new information that can be used to evaluate the role of different α-synuclein pathways, involving either the aggregation or the normal function of the protein or both, in PD.

REFERENCES

Abeliovich, A., Schmitz, Y., Farinas, I., Choi-Lundberg, D., Ho, W. H., Castillo, P. E., Shinsky, N., Verdugo, J. M., Armanini, M., Ryan, A., Hynes, M., Phillips, H.,

Sulzer, D., and Rosenthal, A. (2000). Mice lacking alpha-synuclein display functional deficits in the nigrostriatal dopamine system. *Neuron* **25**, 239–252.

Andringa, G., Lam, K. Y., Chegary, M., Wang, X., Chase, T. N., and Bennett, M. C. (2004). Tissue transglutaminase catalyzes the formation of alpha-synuclein crosslinks in Parkinson's disease. *Faseb J* **18**, 932–934.

Antony, T., Hoyer, W., Cherny, D., Heim, G., Jovin, T. M., and Subramaniam, V. (2003). Cellular polyamines promote the aggregation of alpha-synuclein. *J Biol Chem* **278**, 3235–3240.

Arispe, N., Rojas, E., and Pollard, H. B. (1993). Alzheimer disease amyloid beta protein forms calcium channels in bilayer membranes: Blockade by tromethamine and aluminum. *Proc Natl Acad Sci USA* **90**, 567–571.

Bernado, P., Bertoncini, C. W., Griesinger, C., Zweckstetter, M., and Blackledge, M. (2005). Defining long-range order and local disorder in native alpha-synuclein using residual dipolar couplings. *J Am Chem Soc* **127**, 17968–17969.

Bertoncini, C. W., Fernandez, C. O., Griesinger, C., Jovin, T. M., and Zweckstetter, M. (2005). Familial mutants of alpha-synuclein with increased neurotoxicity have a destabilized conformation. *J Biol Chem* **280**, 30649–30652.

Bertoncini, C. W., Jung, Y. S., Fernandez, C. O., Hoyer, W., Griesinger, C., Jovin, T. M., and Zweckstetter, M. (2005). Release of long-range tertiary interactions potentiates aggregation of natively unstructured alpha-synuclein. *Proc Natl Acad Sci USA* **102**, 1430–1435.

Bisaglia, M., Tessari, I., Pinato, L., Bellanda, M., Giraudo, S., Fasano, M., Bergantino, E., Bubacco, L., and Mammi, S. (2005). A topological model of the interaction between alpha-synuclein and sodium dodecyl sulfate micelles. *Biochemistry* **44**, 329–339.

Borbat, P., Ramlall, T. F., Freed, J. H., and Eliezer, D. (2006). Inter-helix distances in lysophospholipid micelle-bound alpha-synuclein from pulsed ESR measurements. *J Am Chem Soc* **128**, 10004–10005.

Bussell, R., Jr., and Eliezer, D. (2001). Residual structure and dynamics in Parkinson's disease-associated mutants of alpha-synuclein. *J Biol Chem* **276**, 45996–46003.

Bussell, R., Jr., and Eliezer, D. (2003). A structural and functional role for 11-mer repeats in alpha-synuclein and other exchangeable lipid binding proteins. *J Mol Biol* **329**, 763–778.

Bussell, R., Jr., and Eliezer, D. (2004). Effects of Parkinson's disease-linked mutations on the structure of lipid-associated alpha-synuclein. *Biochemistry* **43**, 4810–4818.

Bussell, R., Jr., Ramlall, T. F., and Eliezer, D. (2005). Helix periodicity, topology, and dynamics of membrane-associated alpha-synuclein. *Protein Sci* **14**, 862–872.

Cabin, D. E., Shimazu, K., Murphy, D., Cole, N. B., Gottschalk, W., McIlwain, K. L., Orrison, B., Chen, A., Ellis, C. E., Paylor, R., Lu, B., and Nussbaum, R. L. (2002). Synaptic vesicle depletion correlates with attenuated synaptic responses to prolonged repetitive stimulation in mice lacking alpha-synuclein. *J Neurosci* **22**, 8797–8807.

Chandra, S., Chen, X., Rizo, J., Jahn, R., and Sudhof, T. C. (2003). A broken alpha-helix in folded alpha-synuclein. *J Biol Chem* **278**, 15313–15318.

Chandra, S., Fornai, F., Kwon, H. B., Yazdani, U., Atasoy, D., Liu, X., Hammer, R. E., Battaglia, G., German, D. C., Castillo, P. E., and Sudhof, T. C. (2004). Double-knockout mice for alpha- and beta-synucleins: Effect on synaptic functions. *Proc Natl Acad Sci USA* **101**, 14966–14971.

Chandra, S., Gallardo, G., Fernandez-Chacon, R., Schluter, O. M., and Sudhof, T. C. (2005). Alpha-synuclein cooperates with CSPalpha in preventing neurodegeneration. *Cell* **123**, 383–396.

Chartier-Harlin, M. C., Kachergus, J., Roumier, C., Mouroux, V., Douay, X., Lincoln, S., Levecque, C., Larvor, L., Andrieux, J., Hulihan, M., Waucquier, N., Defebvre, L., Amouyel, P., Farrer, M., and Destee, A. (2004). Alpha-synuclein locus duplication as a cause of familial Parkinson's disease. *Lancet* **364**, 1167–1169.

Choi, W., Zibaee, S., Jakes, R., Serpell, L. C., Davletov, B., Crowther, R. A., and Goedert, M. (2004). Mutation E46K increases phospholipid binding and assembly into filaments of human alpha-synuclein. *FEBS Lett* **576**, 363–368.

Citron, B. A., Suo, Z., SantaCruz, K., Davies, P. J., Qin, F., and Festoff, B. W. (2002). Protein crosslinking, tissue transglutaminase, alternative splicing and neurodegeneration. *Neurochem Int* **40**, 69–78.

Clayton, D. F., and George, J. M. (1999). Synucleins in synaptic plasticity and neurodegenerative disorders. *J Neurosci Res* **58**, 120–129.

Cole, N. B., Murphy, D. D., Grider, T., Rueter, S., Brasaemle, D., and Nussbaum, R. L. (2002). Lipid droplet binding and oligomerization properties of the Parkinson's disease protein alpha-synuclein. *J Biol Chem* **277**, 6344–6352.

Conway, K. A., Harper, J. D., and Lansbury, P. T. (1998). Accelerated *in vitro* fibril formation by a mutant alpha-synuclein linked to early-onset Parkinson disease. *Nat Med* **4**, 1318–1320.

Conway, K. A., Harper, J. D., and Lansbury, P. T., Jr. (2000). Fibrils formed *in vitro* from alpha-synuclein and two mutant forms linked to Parkinson's disease are typical amyloid. *Biochemistry* **39**, 2552–2563.

Conway, K. A., Lee, S. J., Rochet, J. C., Ding, T. T., Williamson, R. E., and Lansbury, P. T., Jr. (2000). Acceleration of oligomerization, not fibrillization, is a shared property of both alpha-synuclein mutations linked to early-onset Parkinson's disease: Implications for pathogenesis and therapy. *Proc Natl Acad Sci USA* **97**, 571–576.

Conway, K. A., Rochet, J. C., Bieganski, R. M., and Lansbury, P. T., Jr. (2001). Kinetic stabilization of the alpha-synuclein protofibril by a dopamine- alpha-synuclein adduct. *Science* **294**, 1346–1349.

Cooper, A. A., Gitler, A. D., Cashikar, A., Haynes, C. M., Hill, K. J., Bhullar, B., Liu, K., Xu, K., Strathearn, K. E., Liu, F., Cao, S., Caldwell, K. A., Caldwell, G. A., Marsischky, G., Kolodner, R. D., Labaer, J., Rochet, J. C., Bonini, N. M., and Lindquist, S. (2006). Alpha-synuclein blocks ER-Golgi traffic and Rab1 rescues neuron loss in Parkinson's models. *Science* **313**, 324–328.

Crowther, R. A., Jakes, R., Spillantini, M. G., and Goedert, M. (1998). Synthetic filaments assembled from C-terminally truncated alpha-synuclein. *FEBS Lett* **436**, 309–312.

Dauer, W., Kholodilov, N., Vila, M., Trillat, A. C., Goodchild, R., Larsen, K. E., Staal, R., Tieu, K., Schmitz, Y., Yuan, C. A., Rocha, M., Jackson-Lewis, V., Hersch, S., Sulzer, D., Przedborski, S., Burke, R., and Hen, R. (2002). Resistance of alpha-synuclein null mice to the parkinsonian neurotoxin MPTP. *Proc Natl Acad Sci USA* **99**, 14524–14529.

Davidson, W. S., Jonas, A., Clayton, D. F., and George, J. M. (1998). Stabilization of alpha-synuclein secondary structure upon binding to synthetic membranes. *J Biol Chem* **273**, 9443–9449.

Dedmon, M. M., Christodoulou, J., Wilson, M. R., and Dobson, C. M. (2005). Heat shock protein 70 inhibits alpha-synuclein fibril formation via preferential binding to prefibrillar species. *J Biol Chem* **280**, 14733–14740.

Dedmon, M. M., Lindorff-Larsen, K., Christodoulou, J., Vendruscolo, M., and Dobson, C. M. (2005). Mapping long-range interactions in alpha-synuclein using spin-label NMR and ensemble molecular dynamics simulations. *J Am Chem Soc* **127**, 476–477.

Delaglio, F., Kontaxis, G., and Bax, A. (2000). Protein structure determination using molecular fragment replacement and NMR dipolar couplings. *J Am Chem Soc* **122**, 2142–2143.

Dorval, V., and Fraser, P. E. (2006). Small ubiquitin-like modifier (SUMO) modification of natively unfolded proteins tau and alpha-synuclein. *J Biol Chem* **281**, 9919–9924.

Drolet, R. E., Behrouz, B., Lookingland, K. J., and Goudreau, J. L. (2004). Mice lacking alpha-synuclein have an attenuated loss of striatal dopamine following prolonged chronic MPTP administration. *Neurotoxicology* **25**, 761–769.

Dyson, H. J., and Wright, P. E. (2005). Intrinsically unstructured proteins and their functions. *Nat Rev Mol Cell Biol* 6, 197–208.

El-Agnaf, O. M., Paleologou, K. E., Greer, B., Abogrein, A. M., King, J. E., Salem, S. A., Fullwood, N. J., Benson, F. E., Hewitt, R., Ford, K. J., Martin, F. L., Harriott, P., Cookson, M. R., and Allsop, D. (2004). A strategy for designing inhibitors of alpha-synuclein aggregation and toxicity as a novel treatment for Parkinson's disease and related disorders. *Faseb J* 18, 1315–1317.

Eliezer, D., Kutluay, E., Bussell, R., Jr., and Browne, G. (2001). Conformational properties of alpha-synuclein in its free and lipid-associated states. *J Mol Biol* 307, 1061–1073.

Ellis, C. E., Schwartzberg, P. L., Grider, T. L., Fink, D. W., and Nussbaum, R. L. (2001). alpha-synuclein is phosphorylated by members of the Src family of protein-tyrosine kinases. *J Biol Chem* 276, 3879–3884.

Emadi, S., Liu, R., Yuan, B., Schulz, P., McAllister, C., Lyubchenko, Y., Messer, A., and Sierks, M. R. (2004). Inhibiting aggregation of alpha-synuclein with human single chain antibody fragments. *Biochemistry* 43, 2871–2878.

Feany, M. B., and Bender, W. W. (2000). A drosophila model of Parkinson's disease. *Nature* 404, 394–398.

Fernandez, C. O., Hoyer, W., Zweckstetter, M., Jares-Erijman, E. A., Subramaniam, V., Griesinger, C., and Jovin, T. M. (2004). NMR of alpha-synuclein-polyamine complexes elucidates the mechanism and kinetics of induced aggregation. *Embo J* 23, 2039–2046.

Fountaine, T. M., and Wade-Martins, R. (2007). RNA interference-mediated knockdown of alpha-synuclein protects human dopaminergic neuroblastoma cells from MPP(+) toxicity and reduces dopamine transport. *J Neurosci Res* 85, 351–363.

George, J. M., Jin, H., Woods, W. S., and Clayton, D. F. (1995). Characterization of a novel protein regulated during the critical period for song learning in the zebra finch. *Neuron* 15, 361–372.

Giasson, B. I., Uryu, K., Trojanowski, J. Q., and Lee, V. M. (1999). Mutant and wild type human alpha-synucleins assemble into elongated filaments with distinct morphologies *in vitro*. *J Biol Chem* 274, 7619–7622.

Giasson, B. I., Duda, J. E., Murray, I. V., Chen, Q., Souza, J. M., Hurtig, H. I., Ischiropoulos, H., Trojanowski, J. Q., and Lee, V. M. (2000). Oxidative damage linked to neurodegeneration by selective alpha-synuclein nitration in synucleinopathy lesions. *Science* 290, 985–989.

Giasson, B. I., Murray, I. V., Trojanowski, J. Q., and Lee, V. M. (2001). A hydrophobic stretch of 12 amino acid residues in the middle of alpha-synuclein is essential for filament assembly. *J Biol Chem* 276, 2380–2386.

Goate, A., Chartier-Harlin, M. C., Mullan, M., Brown, J., Crawford, F., Fidani, L., Giuffra, L., Haynes, A., Irving, N., James, L. et al. (1991). Segregation of a missense mutation in the amyloid precursor protein gene with familial Alzheimer's disease. *Nature* 349, 704–706.

Goers, J., Manning-Bog, A. B., McCormack, A. L., Millett, I. S., Doniach, S., Di Monte, D. A., Uversky, V. N., and Fink, A. L. (2003a). Nuclear localization of alpha-synuclein and its interaction with histones. *Biochemistry* 42, 8465–8471.

Goers, J., Uversky, V. N., and Fink, A. L. (2003). Polycation-induced oligomerization and accelerated fibrillation of human alpha-synuclein *in vitro*. *Protein Sci* 12, 702–707.

Greenbaum, E. A., Graves, C. L., Mishizen-Eberz, A. J., Lupoli, M. A., Lynch, D. R., Englander, S. W., Axelsen, P. H., and Giasson, B. I. (2005). The E46K mutation in alpha-synuclein increases amyloid fibril formation. *J Biol Chem* 280, 7800–7807.

Hardy, J. A., and Higgins, G. A. (1992). Alzheimer's disease: The amyloid cascade hypothesis. *Science* 256, 184–185.

Hardy, J., Cai, H., Cookson, M. R., Gwinn-Hardy, K., and Singleton, A. (2006). Genetics of Parkinson's disease and parkinsonism. *Ann Neurol* 60, 389–398.

Hasegawa, M., Fujiwara, H., Nonaka, T., Wakabayashi, K., Takahashi, H., Lee, V. M., Trojanowski, J. Q., Mann, D., and Iwatsubo, T. (2002). Phosphorylated alpha-synuclein is ubiquitinated in alpha-synucleinopathy lesions. *J Biol Chem* 277, 49071–49076.

Hashimoto, M., Rockenstein, E., Mante, M., Mallory, M., and Masliah, E. (2001). beta-Synuclein inhibits alpha-synuclein aggregation: A possible role as an anti-parkinsonian factor. *Neuron* 32, 213–223.

Hutton, M., Lendon, C. L., Rizzu, P., Baker, M., Froelich, S., Houlden, H., Pickering-Brown, S., Chakraverty, S., Isaacs, A., Grover, A., Hackett, J., Adamson, J., Lincoln, S., Dickson, D., Davies, P., Petersen, R. C., Stevens, M., de Graaff, E., Wauters, E., van Baren, J., Hillebrand, M., Joosse, M., Kwon, J. M., Nowotny, P., Heutink, P. et al. (1998). Association of missense and 5'-splice-site mutations in tau with the inherited dementia FTDP-17. *Nature* 393, 702–705.

Irizarry, M. C., Kim, T. W., McNamara, M., Tanzi, R. E., George, J. M., Clayton, D. F., and Hyman, B. T. (1996). Characterization of the precursor protein of the non-A beta component of senile plaques (NACP) in the human central nervous system. *J Neuropathol Exp Neurol* 55, 889–895.

Iwai, A., Masliah, E., Yoshimoto, M., Ge, N., Flanagan, L., de Silva, H. A., Kittel, A., and Saitoh, T. (1995).

The precursor protein of non-A beta component of Alzheimer's disease amyloid is a presynaptic protein of the central nervous system. *Neuron* **14**, 467–475.

Jakes, R., Spillantini, M. G., and Goedert, M. (1994). Identification of two distinct synucleins from human brain. *FEBS Lett* **345**, 27–32.

Jao, C. C., Der-Sarkissian, A., Chen, J., and Langen, R. (2004). Structure of membrane-bound alpha-synuclein studied by site-directed spin labeling. *Proc Natl Acad Sci USA* **101**, 8331–8336.

Jenco, J. M., Rawlingson, A., Daniels, B., and Morris, A. J. (1998). Regulation of phospholipase D2: Selective inhibition of mammalian phospholipase D isoenzymes by alpha- and beta-synucleins. *Biochemistry* **37**, 4901–4909.

Jensen, P. H., Nielsen, M. S., Jakes, R., Dotti, C. G., and Goedert, M. (1998). Binding of alpha-synuclein to brain vesicles is abolished by familial Parkinson's disease mutation. *J Biol Chem* **273**, 26292–26294.

Jensen, P. H., Hager, H., Nielsen, M. S., Hojrup, P., Gliemann, J., and Jakes, R. (1999). alpha-Synuclein binds to Tau and stimulates the protein kinase A – catalyzed tau phosphorylation of serine residues 262 and 356. *J Biol Chem* **274**, 25481–25489.

Jin, J., Jane Li, G., Davis, J., Zhu, D., Wang, Y., Pan, C., and Zhang, J. (2007). Identification of novel proteins interacting with both a-synuclein and DJ-1. *Mol Cell Proteomics*.

Jo, E., McLaurin, J., Yip, C. M., St George-Hyslop, P., and Fraser, P. E. (2000). alpha-Synuclein membrane interactions and lipid specificity. *J Biol Chem* **275**, 34328–34334.

Jo, E., Fuller, N., Rand, R. P., St George-Hyslop, P., and Fraser, P. E. (2002). Defective membrane interactions of familial Parkinson's disease mutant A30P alpha-synuclein. *J Mol Biol* **315**, 799–807.

Jo, E., Darabie, A. A., Han, K., Tandon, A., Fraser, P. E., and McLaurin, J. (2004). alpha-Synuclein-synaptosomal membrane interactions: implications for fibrillogenesis. *Eur J Biochem* **271**, 3180–3189.

Junn, E., Ronchetti, R. D., Quezado, M. M., Kim, S. Y., and Mouradian, M. M. (2003). Tissue transglutaminase-induced aggregation of alpha-synuclein: Implications for Lewy body formation in Parkinson's disease and dementia with Lewy bodies. *Proc Natl Acad Sci USA* **100**, 2047–2052.

Kahle, P. J., Neumann, M., Ozmen, L., Muller, V., Jacobsen, H., Schindzielorz, A., Okochi, M., Leimer, U., van Der Putten, H., Probst, A., Kremmer, E., Kretzschmar, H. A., and Haass, C. (2000). Subcellular localization of wild-type and Parkinson's disease-associated mutant alpha-synuclein in human and transgenic mouse brain. *J Neurosci* **20**, 6365–6373.

Kamp, F., and Beyer, K. (2006). Binding of alpha-synuclein affects the lipid packing in bilayers of small vesicles. *J Biol Chem* **281**, 9251–9259.

Kim, Y. S., Laurine, E., Woods, W., and Lee, S. J. (2006). A novel mechanism of interaction between alpha-synuclein and biological membranes. *J Mol Biol* **360**, 386–397.

Klivenyi, P., Siwek, D., Gardian, G., Yang, L., Starkov, A., Cleren, C., Ferrante, R. J., Kowall, N. W., Abeliovich, A., and Beal, M. F. (2006). Mice lacking alpha-synuclein are resistant to mitochondrial toxins. *Neurobiol Dis* **21**, 541–548.

Kontaxis, G., Delaglio, F., and Bax, A. (2005). Molecular fragment replacement approach to protein structure determination by chemical shift and dipolar homology database mining. *Methods Enzymol* **394**, 42–78.

Kontopoulos, E., Parvin, J. D., and Feany, M. B. (2006). Alpha-synuclein acts in the nucleus to inhibit histone acetylation and promote neurotoxicity. *Hum Mol Genet* **15**, 3012–3023.

Kramer, M. L., and Schulz-Schaeffer, W. J. (2007). Presynaptic alpha-synuclein aggregates, not Lewy bodies, cause neurodegeneration in dementia with Lewy bodies. *J Neurosci* **27**, 1405–1410.

Kruger, R., Kuhn, W., Muller, T., Woitalla, D., Graeber, M., Kosel, S., Przuntek, H., Epplen, J. T., Schols, L., and Riess, O. (1998). Ala30Pro mutation in the gene encoding alpha-synuclein in Parkinson's disease. *Nat Genet* **18**, 106–108.

Kubo, S., Nemani, V. M., Chalkley, R. J., Anthony, M. D., Hattori, N., Mizuno, Y., Edwards, R. H., and Fortin, D. L. (2005). A combinatorial code for the interaction of alpha-synuclein with membranes. *J Biol Chem* **280**, 31664–31672.

Lakso, M., Vartiainen, S., Moilanen, A. M., Sirvio, J., Thomas, J. H., Nass, R., Blakely, R. D., and Wong, G. (2003). Dopaminergic neuronal loss and motor deficits in Caenorhabditis elegans overexpressing human alpha-synuclein. *J Neurochem* **86**, 165–172.

Lambert, M. P., Barlow, A. K., Chromy, B. A., Edwards, C., Freed, R., Liosatos, M., Morgan, T. E., Rozovsky, I., Trommer, B., Viola, K. L., Wals, P., Zhang, C., Finch, C. E., Krafft, G. A., and Klein, W. L. (1998). Diffusible, nonfibrillar ligands derived from Abeta1–42 are potent central nervous system neurotoxins. *Proc Natl Acad Sci USA* **95**, 6448–6453.

Larsen, K. E., Schmitz, Y., Troyer, M. D., Mosharov, E., Dietrich, P., Quazi, A. Z., Savalle, M., Nemani, V., Chaudhry, F. A., Edwards, R. H., Stefanis, L., and Sulzer, D. (2006). Alpha-synuclein overexpression in PC12 and chromaffin cells impairs catecholamine release by interfering with a late step in exocytosis. *J Neurosci* **26**, 11915–11922.

Lee, F. J., Liu, F., Pristupa, Z. B., and Niznik, H. B. (2001). Direct binding and functional coupling of alpha-synuclein to the dopamine transporters accelerate dopamine-induced apoptosis. *Faseb J* **15**, 916–926.

Lee, H. J., Choi, C., and Lee, S. J. (2002). Membrane-bound alpha-synuclein has a high aggregation propensity and the ability to seed the aggregation of the cytosolic form. *J Biol Chem* 277, 671–678.

Lee, J. C., Langen, R., Hummel, P. A., Gray, H. B., and Winkler, J. R. (2004). Alpha-synuclein structures from fluorescence energy-transfer kinetics: Implications for the role of the protein in Parkinson's disease. *Proc Natl Acad Sci USA* 101, 16466–16471.

Lee, J. C., Gray, H. B., and Winkler, J. R. (2005). Tertiary contact formation in alpha-synuclein probed by electron transfer. *J Am Chem Soc* 127, 16388–16389.

Lee, J. C., Lai, B. T., Kozak, J. J., Gray, H. B., and Winkler, J. R. (2007). Alpha-synuclein tertiary contact dynamics. *J Phys Chem B* 111, 2107–2112.

Li, J., Uversky, V. N., and Fink, A. L. (2001). Effect of familial Parkinson's disease point mutations A30P and A53T on the structural properties, aggregation, and fibrillation of human alpha-synuclein. *Biochemistry* 40, 11604–11613.

Liu, S., Ninan, I., Antonova, I., Battaglia, F., Trinchese, F., Narasanna, A., Kolodilov, N., Dauer, W., Hawkins, R. D., and Arancio, O. (2004). alpha-Synuclein produces a long-lasting increase in neurotransmitter release. *Embo J* 23, 4506–4516.

Lotharius, J., Barg, S., Wiekop, P., Lundberg, C., Raymon, H. K., and Brundin, P. (2002). Effect of mutant alpha-synuclein on dopamine homeostasis in a new human mesencephalic cell line. *J Biol Chem* 277, 38884–38894.

Maroteaux, L., Campanelli, J. T., and Scheller, R. H. (1988). Synuclein: A neuron-specific protein localized to the nucleus and presynaptic nerve terminal. *J Neurosci* 8, 2804–2815.

Martin, E. D., Gonzalez-Garcia, C., Milan, M., Farinas, I., and Cena, V. (2004). Stressor-related impairment of synaptic transmission in hippocampal slices from alpha-synuclein knockout mice. *Eur J Neurosci* 20, 3085–3091.

Masliah, E., Rockenstein, E., Veinbergs, I., Mallory, M., Hashimoto, M., Takeda, A., Sagara, Y., Sisk, A., and Mucke, L. (2000). Dopaminergic loss and inclusion body formation in alpha-synuclein mice: Implications for neurodegenerative disorders. *Science* 287, 1265–1269.

Masuda, M., Suzuki, N., Taniguchi, S., Oikawa, T., Nonaka, T., Iwatsubo, T., Hisanaga, S., Goedert, M., and Hasegawa, M. (2006). Small molecule inhibitors of alpha-synuclein filament assembly. *Biochemistry* 45, 6085–6094.

Mazzulli, J. R., Mishizen, A. J., Giasson, B. I., Lynch, D. R., Thomas, S. A., Nakashima, A., Nagatsu, T., Ota, A., and Ischiropoulos, H. (2006). Cytosolic catechols inhibit alpha-synuclein aggregation and facilitate the formation of intracellular soluble oligomeric intermediates. *J Neurosci* 26, 10068–10078.

McDermott, M., Wakelam, M. J., and Morris, A. J. (2004). Phospholipase D. *Biochem Cell Biol* 82, 225–253.

McLean, P. J., Kawamata, H., Ribich, S., and Hyman, B. T. (2000). Membrane association and protein conformation of alpha-synuclein in intact neurons, effect of Parkinson's disease-linked mutations. *J Biol Chem* 275, 8812–8816.

McLean, P. J., Ribich, S., and Hyman, B. T. (2000). Subcellular localization of alpha-synuclein in primary neuronal cultures: Effect of missense mutations. *J Neural Transm Suppl* 24, 53–63.

Morar, A. S., Olteanu, A., Young, G. B., and Pielak, G. J. (2001). Solvent-induced collapse of alpha-synuclein and acid-denatured cytochrome c. *Protein Sci* 10, 2195–2199.

Murphy, D. D., Rueter, S. M., Trojanowski, J. Q., and Lee, V. M. (2000). Synucleins are developmentally expressed, and alpha-synuclein regulates the size of the presynaptic vesicular pool in primary hippocampal neurons. *J Neurosci* 20, 3214–3220.

Murray, I. V., Giasson, B. I., Quinn, S. M., Koppaka, V., Axelsen, P. H., Ischiropoulos, H., Trojanowski, J. Q., and Lee, V. M. (2003). Role of alpha-synuclein carboxy-terminus on fibril formation *in vitro*. *Biochemistry* 42, 8530–8540.

Nakamura, T., Yamashita, H., Takahashi, T., and Nakamura, S. (2001). Activated Fyn phosphorylates alpha-synuclein at tyrosine residue 125. *Biochem Biophys Res Commun* 280, 1085–1092.

Narayanan, V., and Scarlata, S. (2001). Membrane binding and self-association of alpha-synucleins. *Biochemistry* 40, 9927–9934.

Narhi, L., Wood, S. J., Steavenson, S., Jiang, Y., Wu, G. M., Anafi, D., Kaufman, S. A., Martin, F., Sitney, K., Denis, P., Louis, J. C., Wypych, J., Biere, A. L., and Citron, M. (1999). Both familial Parkinson's disease mutations accelerate alpha-synuclein aggregation. *J Biol Chem* 274, 9843–9846.

Necula, M., Chirita, C. N., and Kuret, J. (2003). Rapid anionic micelle-mediated alpha-synuclein fibrillization *in vitro*. *J Biol Chem* 278, 46674–46680.

Nuscher, B., Kamp, F., Mehnert, T., Odoy, S., Haass, C., Kahle, P. J., and Beyer, K. (2004). Alpha-synuclein has a high affinity for packing defects in a bilayer membrane: A thermodynamics study. *J Biol Chem* 279, 21966–21975.

Okochi, M., Walter, J., Koyama, A., Nakajo, S., Baba, M., Iwatsubo, T., Meijer, L., Kahle, P. J., and Haass, C. (2000). Constitutive phosphorylation of the Parkinson's disease associated alpha-synuclein. *J Biol Chem* 275, 390–397.

Outeiro, T. F., and Lindquist, S. (2003). Yeast cells provide insight into alpha-synuclein biology and pathobiology. *Science* 302, 1772–1775.

Payton, J. E., Perrin, R. J., Clayton, D. F., and George, J. M. (2001). Protein–protein interactions of alpha-synuclein in brain homogenates and transfected cells. *Brain Res Mol Brain Res* **95**, 138–145.

Payton, J. E., Perrin, R. J., Woods, W. S., and George, J. M. (2004). Structural determinants of PLD2 inhibition by alpha-synuclein. *J Mol Biol* **337**, 1001–1009.

Pepys, M. B., Hawkins, P. N., Booth, D. R., Vigushin, D. M., Tennent, G. A., Soutar, A. K., Totty, N., Nguyen, O., Blake, C. C., Terry, C. J. *et al.* (1993). Human lysozyme gene mutations cause hereditary systemic amyloidosis. *Nature* **362**, 553–557.

Perez, R. G., Waymire, J. C., Lin, E., Liu, J. J., Guo, F., and Zigmond, M. J. (2002). A role for alpha-synuclein in the regulation of dopamine biosynthesis. *J Neurosci* **22**, 3090–3099.

Periquet, M., Fulga, T., Myllykangas, L., Schlossmacher, M. G., and Feany, M. B. (2007). Aggregated alpha-synuclein mediates dopaminergic neurotoxicity *in vivo*. *J Neurosci* **27**, 3338–3346.

Perrin, R. J., Woods, W. S., Clayton, D. F., and George, J. M. (2000). Interaction of human alpha-Synuclein and Parkinson's disease variants with phospholipids, Structural analysis using site-directed mutagenesis. *J Biol Chem* **275**, 34393–34398.

Perrin, R. J., Woods, W. S., Clayton, D. F., and George, J. M. (2001). Exposure to long chain polyunsaturated fatty acids triggers rapid multimerization of synucleins. *J Biol Chem* **276**, 41958–41962.

Polymeropoulos, M. H., Higgins, J. J., Golbe, L. I., Johnson, W. G., Ide, S. E., Di Iorio, G., Sanges, G., Stenroos, E. S., Pho, L. T., Schaffer, A. A., Lazzarini, A. M., Nussbaum, R. L., and Duvoisin, R. C. (1996). Mapping of a gene for Parkinson's disease to chromosome 4q21-q23. *Science* **274**, 1197–1199.

Polymeropoulos, M. H., Lavedan, C., Leroy, E., Ide, S. E., Dehejia, A., Dutra, A., Pike, B., Root, H., Rubenstein, J., Boyer, R., Stenroos, E. S., Chandrasekharappa, S., Athanassiadou, A., Papapetropoulos, T., Johnson, W. G., Lazzarini, A. M., Duvoisin, R. C., Di Iorio, G., Golbe, L. I., and Nussbaum, R. L. (1997). Mutation in the alpha-synuclein gene identified in families with Parkinson's disease [see comments]. *Science* **276**, 2045–2047.

Poorkaj, P., Bird, T. D., Wijsman, E., Nemens, E., Garruto, R. M., Anderson, L., Andreadis, A., Wiederholt, W. C., Raskind, M., and Schellenberg, G. D. (1998). Tau is a candidate gene for chromosome 17 frontotemporal dementia. *Ann Neurol* **43**, 815–825.

Pronin, A. N., Morris, A. J., Surguchov, A., and Benovic, J. L. (2000). Synucleins are a novel class of substrates for G protein-coupled receptor kinases. *J Biol Chem* **275**, 26515–26522.

Ramakrishnan, M., Jensen, P. H., and Marsh, D. (2003). Alpha-synuclein association with phosphatidylglycerol probed by lipid spin labels. *Biochemistry* **42**, 12919–12926.

Rhoades, E., Ramlall, T. F., Webb, W. W., and Eliezer, D. (2006). Quantification of alpha-synuclein binding to lipid vesicles using fluorescence correlation spectroscopy. *Biophys J* **90**, 4692–4700.

Robertson, D. C., Schmidt, O., Ninkina, N., Jones, P. A., Sharkey, J., and Buchman, V. L. (2004). Developmental loss and resistance to MPTP toxicity of dopaminergic neurones in substantia nigra pars compacta of gamma-synuclein, alpha-synuclein and double alpha/gamma-synuclein null mutant mice. *J Neurochem* **89**, 1126–1136.

Saraiva, M. J., Birken, S., Costa, P. P., and Goodman, D. S. (1984). Amyloid fibril protein in familial amyloidotic polyneuropathy, Portuguese type. Definition of molecular abnormality in transthyretin (prealbumin). *J Clin Invest* **74**, 104–119.

Scherzer, C. R., Jensen, R. V., Gullans, S. R., and Feany, M. B. (2003). Gene expression changes presage neurodegeneration in a Drosophila model of Parkinson's disease. *Hum Mol Genet* **12**, 2457–2466.

Schluter, O. M., Fornai, F., Alessandri, M. G., Takamori, S., Geppert, M., Jahn, R., and Sudhof, T. C. (2003). Role of alpha-synuclein in 1-methyl-4-phenyl-1,2,3, 6-tetrahydropyridine-induced parkinsonism in mice. *Neuroscience* **118**, 985–1002.

Segrest, J. P., Jones, M. K., De Loof, H., Brouillette, C. G., Venkatachalapathi, Y. V., and Anantharamaiah, G. M. (1992). The amphipathic helix in the exchangeable apolipoproteins: A review of secondary structure and function. *J Lipid Res* **33**, 141–166.

Segrest, J. P., Jones, M. K., Klon, A. E., Sheldahl, C. J., Hellinger, M., De Loof, H., and Harvey, S. C. (1999). A detailed molecular belt model for apolipoprotein A-I in discoidal high density lipoprotein. *J Biol Chem* **274**, 31755–31758.

Selkoe, D. J. (1991). The molecular pathology of Alzheimer's disease. *Neuron* **6**, 487–498.

Serpell, L. C., Berriman, J., Jakes, R., Goedert, M., and Crowther, R. A. (2000). Fiber diffraction of synthetic alpha-synuclein filaments shows amyloid-like cross-beta conformation. *Proc Natl Acad Sci USA* **97**, 4897–4902.

Sharon, R., Bar-Joseph, I., Frosch, M. P., Walsh, D. M., Hamilton, J. A., and Selkoe, D. J. (2003). The formation of highly soluble oligomers of alpha-synuclein is regulated by fatty acids and enhanced in Parkinson's disease. *Neuron* **37**, 583–595.

Singleton, A. B., Farrer, M., Johnson, J., Singleton, A., Hague, S., Kachergus, J., Hulihan, M., Peuralinna, T., Dutra, A., Nussbaum, R., Lincoln, S., Crawley, A., Hanson, M., Maraganore, D., Adler, C., Cookson, M. R., Muenter, M., Baptista, M., Miller, D., Blancato, J., Hardy, J., and Gwinn-Hardy, K. (2003). Alpha-Synuclein locus triplication causes Parkinson's disease. *Science* 302, 841.

Spillantini, M. G., Schmidt, M. L., Lee, V. M., Trojanowski, J. Q., Jakes, R., and Goedert, M. (1997). Alpha-synuclein in Lewy bodies. *Nature* 388, 839–840.

Spillantini, M. G., Crowther, R. A., Jakes, R., Hasegawa, M., and Goedert, M. (1998a). alpha-Synuclein in filamentous inclusions of Lewy bodies from Parkinson's disease and dementia with lewy bodies. *Proc Natl Acad Sci USA* 95, 6469–6473.

Spillantini, M. G., Murrell, J. R., Goedert, M., Farlow, M. R., Klug, A., and Ghetti, B. (1998). Mutation in the tau gene in familial multiple system tauopathy with presenile dementia. *Proc Natl Acad Sci USA* 95, 7737–7741.

Sung, Y. H., and Eliezer, D. (2006). Secondary structure and dynamics of micelle bound beta- and gamma-synuclein. *Protein Sci* 15, 1162–1174.

Tawara, S., Nakazato, M., Kangawa, K., Matsuo, H., and Araki, S. (1983). Identification of amyloid prealbumin variant in familial amyloidotic polyneuropathy (Japanese type). *Biochem Biophys Res Commun* 116, 880–888.

Tofaris, G. K., Razzaq, A., Ghetti, B., Lilley, K. S., and Spillantini, M. G. (2003). Ubiquitination of alpha-synuclein in Lewy bodies is a pathological event not associated with impairment of proteasome function. *J Biol Chem* 278, 44405–44411.

Ueda, K., Fukushima, H., Masliah, E., Xia, Y., Iwai, A., Yoshimoto, M., Otero, D. A., Kondo, J., Ihara, Y., and Saitoh, T. (1993). Molecular cloning of cDNA encoding an unrecognized component of amyloid in Alzheimer disease. *Proc Natl Acad Sci USA* 90, 11282–11286.

Ulmer, T. S., and Bax, A. (2005). Comparison of structure and dynamics of micelle-bound human alpha-synuclein and Parkinson disease variants. *J Biol Chem* 280, 43179–43187.

Ulmer, T. S., Bax, A., Cole, N. B., and Nussbaum, R. L. (2005). Structure and dynamics of micelle-bound human alpha-synuclein. *J Biol Chem* 280, 9595–9603.

Uversky, V. N., Li, J., Souillac, P., Millett, I. S., Doniach, S., Jakes, R., Goedert, M., and Fink, A. L. (2002a). Biophysical properties of the synucleins and their propensities to fibrillate: Inhibition of alpha-synuclein assembly by beta- and gamma-synucleins. *J Biol Chem* 277, 11970–11978.

Uversky, V. N., Yamin, G., Souillac, P. O., Goers, J., Glaser, C. B., and Fink, A. L. (2002b). Methionine oxidation inhibits fibrillation of human alpha-synuclein *in vitro*. *FEBS Lett* 517, 239–244.

Volles, M. J., and Lansbury, P. T., Jr. (2007). Relationships between the sequence of alpha-synuclein and its membrane affinity, fibrillization propensity, and yeast toxicity. *J Mol Biol* 366, 1510–1522.

Walsh, D. M., Hartley, D. M., Kusumoto, Y., Fezoui, Y., Condron, M. M., Lomakin, A., Benedek, G. B., Selkoe, D. J., and Teplow, D. B. (1999). Amyloid beta-protein fibrillogenesis. Structure and biological activity of protofibrillar intermediates. *J Biol Chem* 274, 25945–25952.

Weinreb, P. H., Zhen, W., Poon, A. W., Conway, K. A., and Lansbury, P. T., Jr. (1996). NACP, a protein implicated in Alzheimer's disease and learning, is natively unfolded. *Biochemistry* 35, 13709–13715.

Wersinger, C., and Sidhu, A. (2003). Attenuation of dopamine transporter activity by alpha-synuclein. *Neurosci Lett* 340, 189–192.

Wersinger, C., Prou, D., Vernier, P., and Sidhu, A. (2003). Modulation of dopamine transporter function by alpha-synuclein is altered by impairment of cell adhesion and by induction of oxidative stress. *Faseb J* 17, 2151–2153.

Willingham, S., Outeiro, T. F., DeVit, M. J., Lindquist, S. L., and Muchowski, P. J. (2003). Yeast genes that enhance the toxicity of a mutant huntingtin fragment or alpha-synuclein. *Science* 302, 1769–1772.

Wright, P. E., and Dyson, H. J. (1999). Intrinsically unstructured proteins: Re-assessing the protein structure-function paradigm. *J Mol Biol* 293, 321–331.

Yavich, L., Tanila, H., Vepsalainen, S., and Jakala, P. (2004). Role of alpha-synuclein in presynaptic dopamine recruitment. *J Neurosci* 24, 11165–11170.

Yu, S., Li, X., Liu, G., Han, J., Zhang, C., Li, Y., Xu, S., Liu, C., Gao, Y., Yang, H., Ueda, K., and Chan, P. (2007). Extensive nuclear localization of alpha-synuclein in normal rat brain neurons revealed by a novel monoclonal antibody. *Neuroscience* 145, 539–555.

Zarranz, J. J., Alegre, J., Gomez-Esteban, J. C., Lezcano, E., Ros, R., Ampuero, I., Vidal, L., Hoenicka, J., Rodriguez, O., Atares, B., Llorens, V., Gomez Tortosa, E., del Ser, T., Munoz, D. G., and de Yebenes, J. G. (2004). The new mutation, E46K, of alpha-synuclein causes Parkinson and Lewy body dementia. *Ann Neurol* 55, 164–173.

Zhou, Y., Gu, G., Goodlett, D. R., Zhang, T., Pan, C., Montine, T. J., Montine, K. S., Aebersold, R. H., and Zhang, J. (2004). Analysis of alpha-synuclein-associated proteins by quantitative proteomics. *J Biol Chem* 279, 39155–39164.

Zhu, M., and Fink, A. L. (2003). Lipid binding inhibits alpha-synuclein fibril formation. *J Biol Chem* 278, 16873–16877.

Zhu, M., Li, J., and Fink, A. L. (2003). The association of alpha-synuclein with membranes affects bilayer structure, stability, and fibril formation. *J Biol Chem* 278, 40186–40197.

Zhu, M., Rajamani, S., Kaylor, J., Han, S., Zhou, F., and Fink, A. L. (2004). The flavonoid baicalein inhibits fibrillation of alpha-synuclein and disaggregates existing fibrils. *J Biol Chem*.

43

THE USE OF CELL-FREE SYSTEMS TO CHARACTERIZE PARKINSON'S DISEASE-RELATED GENE PRODUCTS

JEAN-CHRISTOPHE ROCHET AND JEREMY L. SCHIELER

Department of Medicinal Chemistry and Molecular Pharmacology, Purdue University, West Lafayette, IN, USA

Until the late 1990s, Parkinson's disease (PD) was viewed primarily as a disorder with little evidence of a role for genetic perturbations. However, in 1997 our understanding of PD was transformed by the discovery of a mutation in the gene encoding α-synuclein in a small number of patients with early-onset forms of the disease (Polymeropoulos *et al.*, 1997). Ten years after this initial discovery, seven additional genes involved in familial parkinsonism have been identified (Table 43.1) (Cookson, 2005). These genes are divided into two categories according to their pattern of inheritance. One category consists of genes involved in autosomal-dominant parkinsonism, including the genes encoding α-synuclein, ubiquitin carboxyl-terminal hydrolase L1 (UCH-L1), and leucine-rich repeat kinase 2 (LRRK2). Mutant forms of these genes are thought to undergo a gain of toxic function. A second category consists of genes involved in autosomal-recessive parkinsonism, including the genes encoding parkin, PTEN-induced kinase 1 (PINK1), DJ-1, and ATP13A2. Mutant forms of these genes generally exhibit a loss of protective function. Gene products from both categories have been the subject of intense research because it is thought that pathogenic mechanisms underlying familial forms

TABLE 43.1 Gene products involved in familial PD

Gene (locus)	Inheritance	Protein	Protein function
SNCA (PARK1/4)	AD	α-Synuclein	Regulation of synaptic vesicle release (?)
PRKN (PARK2)	AR	Parkin	E3 ubiquitin ligase
UCHL1 (PARK5)	AD	UCH-L1	Ubiquitin hydrolase
PINK1 (PARK6)	AR	PINK1	Serine/threonine kinase
DJ-1 (PARK7)	AR	DJ-1	Antioxidant Chaperone Anti-apoptotic function
LRRK2 (PARK8)	AD	LRRK2 (dardarin)	GTP-regulated kinase
ATP13A2 (PARK9)	AR	ATP13A2	Lysosomal ATPase

Parkinson's Disease: molecular and therapeutic insights from model systems

of PD also play a role in the more common, sporadic forms of the disease.

The objective of this chapter is to provide an overview of how cell-free systems are used to characterize the PD-related gene products listed above. For the purpose of this chapter, a "cell-free system" (or "test-tube model") is defined as any experimental setup involving the characterization of structural and functional properties of a *purified* protein. In turn, a "purified protein" can be a recombinant polypeptide generated in a heterologous expression system (e.g., *E. coli*, yeast, baculovirus) or a protein isolated from its native environment. The first two sections of the chapter focus on α-synuclein and DJ-1, two PD-related gene products that have been studied extensively in test-tube models. The next section describes the use of cell-free systems in studies of parkin, UCH-L1, PINK1, and LRRK2. The chapter concludes with a discussion of advantages and disadvantages of cell-free systems for PD research and how these models can be improved to become more predictive of pathogenic mechanisms *in vivo*.

α-SYNUCLEIN

α-Synuclein is a 14-kDa presynaptic protein involved in regulating neurotransmission (Maroteaux *et al.*, 1988; George *et al.*, 1995; Iwai *et al.*, 1995). The α-synuclein sequence spans three domains (Figure 43.1): (i) an N-terminal domain (residues 1–85), consisting of six repeats of the highly conserved hexamer sequence "KTK(E/Q)GV"; (ii) a central domain (residues 60–95; also referred to as the "non-Aβ component of AD amyloid" [NAC] domain), enriched with hydrophobic residues; and (iii) a C-terminal domain (residues 96–140), enriched with proline and acidic residues. In dilute aqueous buffer, α-synuclein lacks stable secondary structure and is referred to as "natively unfolded" (Weinreb *et al.*, 1996). However, in the presence of anionic phospholipid vesicles, the N-terminal repeat region of α-synuclein interacts with the lipid membrane and adopts an amphipathic α-helical structure (Davidson *et al.*, 1998; Jo *et al.*, 2000; Perrin *et al.*, 2000; Eliezer *et al.*, 2001; Bussell and Eliezer, 2003; Chandra *et al.*, 2003; Zhu and Fink, 2003; Jao *et al.*, 2004).

(a)

                                                                  *
1   MDVFMKGLSKAKEGVVAAAEKTKQGVAEAA

          *   *
31   GKTKEGVLYVGSKTKEGVVHGVATVAEKTK

61   EQVTNVGGAVVTGVTAVAQKTVEGAGSIAA

91   ATGFVKKDQLGKNEEGAPQEGILEDMPVDP

121 DNEAYEMPSEEGYQDYEPEA

(b)

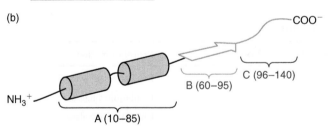

FIGURE 43.1 α-Synuclein is a modular protein with segments that exhibit different secondary structure propensities. (a) Amino-acid sequence of α-synuclein shown in one-letter code. The region that adopts an amphipathic α-helical structure in the presence of negatively charged lipid vesicles spans residues 1–85. The imperfect six-residue repeats (KTK[E/Q]GV) are shown in grey. The hydrophobic NAC segment spans residues 60–95 and the unfolded C-terminal region enriched with prolines and acidic residues spans residues 96–140. * indicate sites of amino-acid substitutions (A30P, E46K, and A53T) in mutant forms of the protein involved in familial PD. (b) Schematic showing the secondary structure preferences of the different segments along the α-helical polypeptide chain. Although the amphipathic α-helical region and NAC segment overlap, they are shown offset from each other to emphasize their different secondary structure preferences.

Research involving cell-free systems has yielded substantial insight into the role of α-synuclein in PD pathogenesis. Some of this research has focused on identifying discrete species on the α-synuclein aggregation pathway, including prefibrillar oligomers (collectively termed "protofibrils") and fibrils. Other studies have examined the influence of familial substitutions, post-translational modifications, and lipids on α-synuclein self-assembly. A number of biophysical analyses have also been carried out to characterize the structural properties of fibrillar α-synuclein. Not only have the data from these studies generated important hypotheses about how α-synuclein is involved in dopaminergic cell death, but they have also suggested new strategies for the treatment of PD.

Effects of Familial Mutations on α-Synuclein Self-assembly

Two types of mutations in the α-synuclein gene have been linked to early-onset, autosomal-dominant PD: (i) point mutations that encode the variants A30P (Kruger et al., 1998), E46K (Zarranz et al., 2004), and A53T (Polymeropoulos et al., 1997); and (ii) mutations that increase the copy number of the α-synuclein gene (Singleton et al., 2003; Chartier-Harlin et al., 2004). Moreover, fibrillar α-synuclein is a primary component of Lewy bodies and Lewy neurites, two neuropathological hallmarks of PD (Pollanen et al., 1993; Forno, 1996; Spillantini et al., 1997; Wakabayashi et al., 1997). These genetic and neuropathological findings led to the hypothesis that α-synuclein fibrillization is a key molecular event driving Lewy body formation in PD brain.

Shortly after a link was established between α-synuclein and PD, a number of groups examined whether the recombinant protein generated in an *E. coli* expression system forms fibrils in aqueous solution. If α-synuclein could be shown to undergo fibrillization in a simple test-tube model, then it might be possible to identify drug candidates that block α-synuclein fibril formation in this system. Early studies revealed that wild-type α-synuclein, A30P, and A53T produce fibrils in the test tube, and A53T undergoes self-assembly more rapidly than the wild-type protein or A30P (Conway et al., 1998; El-Agnaf et al., 1998; Hashimoto et al., 1998; Giasson et al., 1999; Narhi et al., 1999). In contrast, the fibrillization kinetics of A30P were not readily distinguished from those of the wild-type protein. Subsequently, the familial mutant E46K was found to undergo fibrillization more rapidly than wild-type α-synuclein (Choi et al., 2004; Greenbaum et al., 2005; Fredenburg et al., 2007). Fibrils formed by wild-type or mutant α-synuclein have various characteristic features of classic amyloid deposits (El-Agnaf et al., 1998; Narhi et al., 1999; Conway et al., 2000a; Rochet et al., 2000; Serpell et al., 2000). A 12-residue hydrophobic segment spanning residues 71–82 in the NAC region is strictly required for α-synuclein fibrillization (Giasson et al., 2001; Zibaee et al., 2007).

A closer examination of α-synuclein self-assembly revealed that this process involves the formation of protofibrils (Volles and Lansbury, 2003; Rochet et al., 2004). Far-UV circular dichroism (CD), atomic force microscopy (AFM), and electron microscopy (EM) analyses revealed that protofibrils consist of β-sheet-rich spheres, chains, and rings with heights of 3–5 nm (Conway et al., 2000a, b; Rochet et al., 2000, 2004; Volles et al., 2001; Ding et al., 2002; Lashuel et al., 2002b; Volles and Lansbury, 2003). In addition, kinetic data indicated that protofibrils appear transiently during α-synuclein fibrillization, reaching a maximum of ~15% of the total protein and abruptly disappearing upon the formation of mature fibrils (Conway et al., 2000b; Rochet et al., 2000; Lashuel et al., 2002b). Additional evidence from studies in low-pH solutions suggested that protofibril formation is preceded by the partial folding of the monomer to a β-sheet-rich conformation (Uversky et al., 2001).

As outlined above, α-synuclein binds phospholipid membranes, and ring-like structures are found in samples of the protofibrillar protein. These observations led to speculation that α-synuclein protofibrils form membrane-permeabilizing pores. In support of this idea, protofibrils, but not the monomer or fibrils, were found to bind phospholipid vesicles with high affinity and cause leakage of the vesicle contents (Volles et al., 2001; Ding et al., 2002; Volles and Lansbury, 2002). Other groups showed that oligomeric α-synuclein increases the conductance across a phospholipid bilayer (Kayed et al., 2004; Quist et al., 2005), although evidence from only one of these studies suggested that membrane disruption involves the formation of a discrete, selective pore (Quist et al., 2005). Recently, we reported that α-helical forms of the protein can form discrete ion channels in a planar lipid bilayer subjected to a trans-negative potential (Zakharov et al., 2007).

Together, these results imply that α-synuclein elicits toxicity by permeabilizing lipid membranes, perhaps via a pore-like mechanism. This permeabilizing activity could disrupt neuronal homeostasis by perturbing ionic gradients necessary for normal cellular function, including mitochondrial oxidative phosphorylation and the sequestration of DA in vesicles (Volles *et al.*, 2001; Volles and Lansbury, 2002; Volles and Lansbury, 2003).

A detailed study of the fibrillization kinetics of wild-type and mutant α-synuclein revealed that A53T forms fibrils more rapidly than wild-type α-synuclein, whereas A30P undergoes *less* rapid fibrillization compared to the wild-type protein (Conway *et al.*, 2000b). In contrast, A53T and A30P both form protofibrils more rapidly than the wild-type protein (Conway *et al.*, 2000b). Similar data from test-tube studies have been reported by other groups (Li *et al.*, 2001; Choi *et al.*, 2004; Hoyer *et al.*, 2004). These findings suggest that the early-onset disease associated with mutants is due not to the accelerated fibrillization of α-synuclein, but rather to accelerated protofibril formation. Additional studies in cell-free systems revealed that A30P and A53T have an enhanced propensity to form ring-like and/or tubular protofibrils (Lashuel *et al.*, 2002b), and both mutants exhibit a greater membrane-permeabilizing activity per mole of protein (Volles and Lansbury, 2002).

Test-tube models have also been used to study fibrillization in mixtures of different synuclein isoforms. These analyses provide insight into the self-assembly of one synuclein variant in the presence of another *in vivo* (e.g., in transgenic mice that express both human α-synuclein and the endogenous mouse protein). Data from one study indicated that α-synuclein fibrillization is inhibited, whereas protofibrils accumulate, in mixtures of the recombinant mouse and human proteins (Rochet *et al.*, 2000). These results imply that a non-productive interaction between mouse and human α-synuclein prevents the protofibril-to-fibril conversion, and they provide a rationale for why the α-synuclein-transgenic mice initially described by Masliah *et al.* (2000) only contain *non-fibrillar* neuronal inclusions. A subsequent *in vivo* study showed that A53T α-synuclein causes a more severe neurologic phenotype, characterized by motor dysfunction and degeneration of spinal cord motor neurons, in α-synuclein knockout mice than in animals with normal levels of the endogenous protein (Cabin *et al.*, 2005). These findings confirmed that the neurotoxicity of human α-synuclein

in transgenic mice is modulated by interactions between the human and the mouse proteins. However, the nature of this modulation appears to be context dependent – whereas toxicity is enhanced in dopaminergic nerve terminals (Masliah *et al.*, 2000), it is suppressed in spinal cord motor neurons (Cabin *et al.*, 2005).

In other studies, two groups showed that the fibrillization of α-synuclein is inhibited by β-synuclein, an α-synuclein homolog that lacks the critical 12-residue hydrophobic segment (see above) (Uversky *et al.*, 2002a; Park and Lansbury, 2003). The inhibitory effect of β-synuclein involves suppression of protofibril formation by α-synuclein (Park and Lansbury, 2003). These findings provide a rationale for the reduced neurodegeneration in doubly transgenic mice co-expressing α- and β-synuclein, compared to singly transgenic mice expressing only α-synuclein (Hashimoto *et al.*, 2001).

Effects of Post-translational Modifications on α-Synuclein Self-assembly

Various post-translational modifications are associated with fibrillar α-synuclein in patients with PD or other α-synuclein aggregation disorders (collectively referred to as "synucleinopathies"). Examples of these modifications include tyrosine nitration (Giasson *et al.*, 2000), phosphorylation at serine 129 (S129) (Fujiwara *et al.*, 2002; Anderson *et al.*, 2006), ubiquitylation (Hasegawa *et al.*, 2002; Anderson *et al.*, 2006), and C-terminal truncations (Baba *et al.*, 1998; Li *et al.*, 2005b; Liu *et al.*, 2005; Anderson *et al.*, 2006). Phosphorylated, nitrated, and oxidized forms of α-synuclein have also been detected in cellular and animal models of PD (Przedborski *et al.*, 2001; Neumann *et al.*, 2002; Smith *et al.*, 2005; Mirzaei *et al.*, 2006). Although these data reveal important neuropathological markers of PD and other synucleinopathies, they do not enable us to conclude whether post-translational modifications are a *cause* of enhanced α-synuclein aggregation and toxicity in these diseases. In contrast, this question can be addressed directly by examining the effects of the modifications on α-synuclein fibrillization in a cell-free system.

Data from test-tube studies indicate that α-synuclein oligomerization is stimulated by H_2O_2 and Fe^{++} or Cu^{++} (Hashimoto *et al.*, 1999; Paik *et al.*, 2000) and by the formation of dityrosine crosslinks under oxidizing conditions (Souza *et al.*,

2000; Krishnan *et al.*, 2003). α-synuclein modified by nitration (Norris *et al.*, 2003; Yamin *et al.*, 2003), metal-catalyzed oxidation (Cole *et al.*, 2005), reaction with a quinone derivative of DA (Conway *et al.*, 2001; Cappai *et al.*, 2005; Li *et al.*, 2005a; Bisaglia *et al.*, 2007), or oxidation by the lipid peroxidation product 4-hydroxy-2-nonenal (Qin *et al.*, 2007) does not form fibrils, but instead accumulates as soluble oligomers. Data from other studies indicate that DA oxidation products inhibit α-synuclein fibrillization (thereby promoting oligomer accumulation) and reduce the stability of preformed α-synuclein fibrils via non-covalent interactions (Li *et al.*, 2004; Norris *et al.*, 2005; Bisaglia *et al.*, 2007; Follmer *et al.*, 2007). Oxidized cholesterol metabolites have also been shown to accelerate the formation of α-synuclein protofibrils and fibrils in a cell-free system, apparently via a non-covalent mechanism (Bosco *et al.*, 2006).

Test-tube studies have revealed that the impact of oxidative modifications on α-synuclein aggregation depends on the ratio of oxidized to unoxidized protein. Although oligomerization and fibrillization are inhibited in pure solutions of α-synuclein oxidized at all four methionine residues, oligomers accumulate in mixtures of the methionine-oxidized and unoxidized protein (Conway *et al.*, 2001; Uversky *et al.*, 2002b; Hokenson *et al.*, 2004; Glaser *et al.*, 2005). Similarly, despite the inhibitory effect of nitration on α-synuclein fibril formation, nitrated monomeric or dimeric α-synuclein accelerates the fibrillization of the unmodified protein (Hodara *et al.*, 2004). Given the likelihood that α-synuclein is incompletely modified in the brain, data from studies of protein mixtures in cell-free systems may be highly relevant to α-synuclein aggregation *in vivo*.

Cell-free systems have also been used to examine the impact of phosphorylation and C-terminal truncation on α-synuclein self-assembly. The results of one study suggest that the phosphorylation of S129 promotes α-synuclein oligomerization and fibrillization in a test-tube model (Fujiwara *et al.*, 2002), whereas Feany and colleagues (Chen and Feany, 2005) reported that the S129D substitution (a mimic of S129 phosphorylation) has no effect on fibril formation by recombinant α-synuclein. The reason for this discrepancy is unclear, although differences in the experimental conditions may result in pronounced differences in the outcomes of fibrillization studies (Hoyer *et al.*, 2002; Chen and Feany, 2005). In another study, the phosphorylation

of α-synuclein on its three C-terminal tyrosine residues (Y125, Y133, and Y136) was found to inhibit fibrillization, although the effects of tyrosine phosphorylation on oligomer formation were not examined (Negro *et al.*, 2002). Finally, data from test-tube studies revealed that C-terminally deleted variants spanning residues 1 to ~120 of wild-type α-synuclein (similar to truncated species present in Lewy bodies) undergo more rapid fibrillization than the full-length protein (Crowther *et al.*, 1998; Murray *et al.*, 2003; Hoyer *et al.*, 2004; Li *et al.*, 2005b; Liu *et al.*, 2005; Anderson *et al.*, 2006).

Effects of Phospholipids on α-Synuclein Self-assembly

The analyses outlined above have focused on α-synuclein aggregation in aqueous solution. However, the behavior of the protein is likely to be more complex *in vivo* than in the test tube due to the interaction of α-synuclein with cellular membranes. To address this issue, several groups have monitored α-synuclein self-assembly in cell-free systems containing various types of phospholipids. Membrane-bound α-synuclein forms aggregates more rapidly than the cytosolic protein in incubated rat brain lysates (Lee *et al.*, 2002). This aggregation is inhibited by antioxidants, suggesting that the formation of membrane-associated oligomers depends on oxidative stress. α-synuclein also readily forms oligomers in the presence of long-chain polyunsaturated fatty acids (Perrin *et al.*, 2001; Sharon *et al.*, 2003) and undergoes accelerated fibrillization at the surface of anionic detergent micelles, phospholipid vesicles, or synaptosomal membranes (Necula *et al.*, 2003; Jo *et al.*, 2004). Other groups have reported that α-synuclein self-assembly is inhibited by excess synthetic vesicles (Narayanan and Scarlata, 2001; Zhu and Fink, 2003). However, α-synuclein forms fibrils more rapidly upon coincubation with vesicles at high protein/phospholipid ratios (Zhu and Fink, 2003).

α-Synuclein binds phospholipid membranes via its N-terminal α-helical domain, whereas the C-terminal domain remains unstructured and exposed to the aqueous environment (Davidson *et al.*, 1998; Jo *et al.*, 2000; Perrin *et al.*, 2000; Eliezer *et al.*, 2001; Bussell and Eliezer, 2003; Chandra *et al.*, 2003; Zhu and Fink, 2003; Jao *et al.*, 2004). However, data from studies of fluorescently labeled, recombinant α-synuclein suggest that

Ca^{2+}-induced structural rearrangements favor inter-actions of the C-terminal tail with phospholipids and an increase in β-sheet content (Tamamizu-Kato *et al.*, 2006). These data imply that the binding of monomeric α-synuclein to membranes may be an initiating event that can lead to the formation of high molecular weight complexes in response to some environmental "trigger" (Lee *et al.*, 2002; Necula *et al.*, 2003; Tamamizu-Kato *et al.*, 2006). Oxidative stress and metal ions may play a role in promoting the conversion of membrane-bound α-synuclein from benign, α-helical conformers to potentially toxic, β-sheet-rich aggregates.

Characterization of α-Synuclein Fibrillar Structure

The ability to generate α-synuclein oligomers and fibrils in test-tube models has enabled detailed structural analyses of these assemblies via elec-tron paramagnetic resonance (EPR) spectroscopy (Der-Sarkissian *et al.*, 2003), hydrogen exchange-mass spectrometry (HX-MS) (Del Mar *et al.*, 2005), and solid-state NMR (SS-NMR) (Heise *et al.*, 2005). The data from these studies reveal the presence of a folded core domain spanning residues ~30–100, flanked by an N-terminal, structurally heterogene-ous region and a C-terminal unfolded region. The core domain is organized into a series of strand-loop-strand motifs that project along the fibril axis to produce a multilayered structure (Figure 43.2) (Der-Sarkissian *et al.*, 2003; Del Mar *et al.*, 2005; Heise *et al.*, 2005). Each layer in this structure con-sists of an extended β-sheet with strands aligned parallel and in-register, similar to models described for fibrillar $A\beta_{1-40}$ and $A\beta_{1-42}$ (Luhrs *et al.*, 2005; Petkova *et al.*, 2006). The information from these structural analyses may have a significant impact on drug discovery by facilitating the rational design of molecules that inhibit α-synuclein fibrillization (Sato *et al.*, 2006; Sciarretta *et al.*, 2006).

Although the rational design of α-synuclein fibrillization inhibitors may yield new therapies for treating PD, this approach has an important limitation: namely, compounds that block fibril formation may cause a buildup of potentially toxic protofibrils. To rule out this possibility, one must verify that prefibrillar oligomers do not accumu-late upon incubating α-synuclein with an inhibitory compound (Conway *et al.*, 2001; Rochet *et al.*, 2004). At least some compounds designed on the

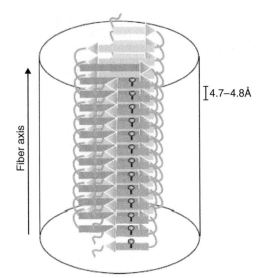

FIGURE 43.2 Structural model of fibrillar α-synuclein determined via EPR spectroscopy. The core domain of each α-synuclein molecule is organ-ized into a strand-loop-strand motif. Multiple cop-ies of this motif project along the fibril axis to form a multilayered structure. Each layer consists of a β-sheet with strands aligned parallel and in-register (the spacing between adjacent strands along the fibril axis is 4.7–4.8Å). The fibril has a characteristic cross-β structure in which individual strands run perpen-dicular to the fiber axis. The ovals represent spin labels introduced at identical sites in neighboring molecules of the fibrillar structure. This figure was generously provided by Dr. Ralf Langen (University of Southern California).

basis of the fibril structure should also inhibit protofibril formation, given that the two types of assemblies are likely to have common structural motifs (Kheterpal *et al.*, 2006). Ultimately, with an improved understanding of the conformational properties of protofibrillar α-synuclein, it may be possible to develop therapies that target protofibril formation directly.

Identification of Fibrillization Inhibitors via High-Throughput Screening

An important outcome of studying α-synuclein self-assembly in test-tube models has been the development of high-throughput screening (HTS) assays to identify fibrillization inhibitors. Screens carried out by various groups have led to the identification of antioxidant molecules that inhibit α-synuclein self-assembly and destabilize preformed fibrils. Examples include the flavonoid baicalein and polyphenols such as curcumin, myricetin, nordihydroguaiaretic acid, rosmarinic acid, and tannic acid (Zhu *et al.*, 2004; Masuda *et al.*, 2006; Ono and Yamada, 2006; Luk *et al.*, 2007). A number of therapeutic agents already used in the treatment of PD, including the DA agonist bromocriptine and the monoamine oxidase B inhibitor selegiline, have also been shown to block α-synuclein fibrillization and induce fibril dissociation (Ono *et al.*, 2007). A potential limitation of these findings is that the compounds may suppress fibrillization without altering the formation of toxic protofibrils. For example, the inhibitory effect of baicalein on α-synuclein fibrillization results in a buildup of soluble oligomers (Zhu *et al.*, 2004; Luk *et al.*, 2007). Accordingly, additional studies in cellular or animal models are required to determine whether baicalein and other fibrillization inhibitors truly have therapeutic potential *in vivo*.

Advantages and Limitations of Using Cell-Free Systems to Study α-Synuclein

Experiments in test-tube models have yielded critical insight into the role of α-synuclein in PD pathogenesis. Studies of the recombinant protein showed for the first time that prefibrillar oligomers or protofibrils are formed during α-synuclein fibrillization and may elicit toxicity by permeabilizing phospholipid vesicles. Comparisons of recombinant wild-type and mutant α-synuclein revealed that the familial mutants A30P, E46K, and A53T have a greater intrinsic propensity to form oligomers or protofibrils (but not necessarily fibrils) than the wild-type protein. Other data from cell-free systems demonstrated that post-translational modifications and phospholipid membranes induce α-synuclein aggregation (including protofibril and fibril formation) under specific conditions. Finally, structural studies of fibrillar α-synuclein provided clues about conformational motifs that may play a role in pathogenesis.

The use of cell-free systems to study α-synuclein pathobiology has several *advantages*. First, this approach *provides the opportunity to examine* *α-synuclein fibrillization in greater molecular detail than would be possible in cell culture or in vivo.* As a result, studies of recombinant α-synuclein have enabled the identification and characterization of protofibrillar forms of the protein. Evidence of protofibril formation in test-tube models has had a major impact on the interpretation of neuropathological data from the post-mortem brains of PD patients. In earlier studies, the presence of Lewy bodies enriched with fibrillar α-synuclein in PD brain was taken as evidence that fibrils contribute to dopaminergic cell death. However, this view was difficult to reconcile with the observation that the severity of neurodegeneration correlates poorly with Lewy body number (Braak *et al.*, 2004). The identification of protofibrils in test-tube models suggested an alternative hypothesis to account for this discrepancy – namely, that prefibrillar forms of α-synuclein may be the primary toxic species responsible for neurodegeneration, whereas fibrils might be relatively benign or even protective (Caughey and Lansbury, 2003; Volles and Lansbury, 2003). According to this "toxic protofibril" hypothesis, the presence of fibrillar α-synuclein aggregates in cell culture, animal models, or post-mortem tissue would be expected to correlate poorly with the severity of neurodegeneration because neurons containing these inclusions should be relatively spared, whereas neurons enriched with protofibrils would have a high probability to undergo cell death. This model is supported the observation that α-synuclein protofibrils (but not fibrils) permeabilize synthetic phospholipid vesicles, suggesting a rationale for the selective toxicity of protofibrillar α-synuclein (Volles *et al.*, 2001; Ding *et al.*, 2002; Volles and Lansbury, 2002). In addition, data from various PD animal models suggest that the expression of α-synuclein can trigger dopaminergic cell death and/or a loss of dopaminergic nerve terminals without causing a buildup of fibrillar, Lewy-like inclusions (Masliah *et al.*, 2000; Chen and Feany, 2005). An important advantage of studying protofibrils in test-tube models is that they can be generated from recombinant α-synuclein in sufficient purity and yield for biochemical and biophysical analyses. In contrast, similar species may be difficult to isolate from cells or brain tissue due to their low abundance and instability, and because they may destroy the neurons in which they accumulate (Caughey and Lansbury, 2003; Volles and Lansbury, 2003). In summary, the identification of α-synuclein protofibrils in test-tube models was a significant breakthrough because it suggested for the first time that these prefibrillar species may be involved in neurodegeneration and, therefore, should be considered potential therapeutic targets.

A second advantage is that *test-tube models are well suited to investigating the impact of perturbations relevant to PD pathogenesis on α-synuclein aggregation.* Examples of these perturbations include familial mutations and post-translational modifications such as methionine oxidation, tyrosine nitration, phosphorylation, and C-terminal truncation. The primary benefit to monitoring the impact of these perturbations in test-tube models is that one can analyze homogenous samples of recombinant α-synuclein with substitutions or modifications at specific sites in the protein sequence. As a result, these analyses provide insight into the effects of sequence-specific modifications on the *intrinsic* rate of α-synuclein aggregation. For example, in studies of the effects of methionine oxidation on α-synuclein aggregation, the recombinant protein was first incubated with H_2O_2 under conditions that converted all four methionine residues to the sulfoxide (as verified by mass spectrometry) (Uversky *et al.*, 2002b). The methionine-oxidized protein was then compared to unmodified α-synuclein with respect to kinetics of oligomerization and fibrillization. To determine the effects of oxidation at specific methionine residues on the rate of aggregation, the experiment was repeated using a series of engineered α-synuclein mutants with one or more methionines replaced by leucine (only the remaining methionine residues were available for oxidation by H_2O_2) (Hokenson *et al.*, 2004). Information about the effect of methionine oxidation on α-synuclein aggregation would have been more difficult to obtain using cellular or animal models because (i) post-translational modifications can influence the intracellular aggregation of α-synuclein indirectly (e.g., by modulating protein–protein or protein–membrane interactions) without necessarily affecting the intrinsic rate of self-assembly; and (ii) α-synuclein is subject to multiple, simultaneous modifications in cells, so that it is difficult to attribute effects on intracellular aggregation to a single type of modification. The use of test-tube models is also optimal to monitor the impact on α-synuclein aggregation of systematically varying "external" parameters such as phospholipid composition ("external" parameters are defined here as variables with potential modulatory effects

on α-synuclein self-assembly other than the primary structure of the protein itself). In general, it is more difficult to exert precise control over these types of parameters in cellular or animal models than in solutions of recombinant protein.

A third advantage is that *cell-free systems enable the identification of molecules that directly target α-synuclein aggregation.* Biophysical analyses have revealed important clues about the structures of protofibrils and fibrils formed from recombinant α-synuclein, and this information may facilitate the rational design of aggregation inhibitors. Moreover, test-tube assays have been used to identify inhibitors of α-synuclein self-assembly via HTS. Data from chemical screens indicate whether fibrillization inhibitors interfere with nucleation and/or elongation. This mechanistic insight (not easily deduced from screens in cellular or animal models) is critical to establish a prioritized list of "hits" in which compounds that block the formation of potentially toxic nuclei are ranked higher than molecules that slow elongation.

The use of cell-free systems to study α-synuclein pathobiology also has *limitations*. First, *oligomeric or fibrillar forms of α-synuclein generated in test-tube experiments may differ from aggregates formed in vivo.* One reason for this discrepancy may be that interactions between α-synuclein and other macromolecules inside the cell interfere with self-assembly mechanisms that occur in solutions of the recombinant protein. It is difficult to determine whether α-synuclein protofibrils generated in test-tube models are identical to oligomers formed *in vivo* because the latter cannot be easily isolated in sufficient purity or yield for detailed biophysical analyses. Instead, the α-synuclein oligomers isolated from cellular or animal models have been analyzed by low-resolution methods such as chromatography and electrophoresis (Mazzulli *et al.*, 2006). Moreover, although recombinant α-synuclein forms fibrils similar to those found in Lewy bodies upon prolonged incubation (Spillantini *et al.*, 1997; Conway *et al.*, 2000a; Serpell *et al.*, 2000), fibrillar α-synuclein generated in the test tube may be ultrastructurally different from fibrils actually formed *in vivo*. One explanation for this structural variation may be that the self-assembly conditions differ markedly between the two environments. Consistent with this idea, studies of fibrillization by recombinant α-synuclein (Hoyer *et al.*, 2002) and synthetic $A\beta_{1-40}$ (Petkova *et al.*, 2005) have shown that fibril morphology and structure are both highly sensitive to the incubation conditions.

The fact that α-synuclein oligomers or fibrils generated in cell-free systems may differ from aggregates formed *in vivo* has important implications for understanding PD pathogenesis and developing new therapies. By inferring neurotoxic mechanisms on the basis of test-tube data, one risks assigning a pathogenic role to species that differ from α-synuclein aggregates formed *in vivo*. Similarly, aggregation inhibitors that target oligomeric or fibrillar species assembled from recombinant α-synuclein may not suppress toxicity if these species are absent from the brain. Ideally, this problem could be addressed by demonstrating that α-synuclein protofibrils or fibrils formed in the test tube are nearly identical to aggregates generated in cellular or animal models. However, as discussed above, this approach is technically very challenging. Another strategy may be to characterize α-synuclein protofibrils and fibrils generated in cell-free systems under different incubation conditions. One would predict that structural motifs common to a range of aggregates formed from recombinant α-synuclein are more likely to contribute to neurotoxic mechanisms *in vivo* than motifs present in only a subset of test-tube assemblies. Accordingly, these common structural motifs may be excellent targets for rationally designed aggregation inhibitors. Similarly, small molecules identified via HTS that block the formation of various types of α-synuclein aggregates may be more likely to mitigate neurotoxicity *in vivo* than compounds that interfere with the formation of only a limited number of assemblies.

A second pitfall is that *test-tube models do not fully account for the biological complexity of α-synuclein in vivo*. As mentioned above, the relative simplicity of cell-free systems is an advantage for studying α-synuclein because it facilitates the characterization of intrinsic properties of the protein. On the other hand, this lack of complexity inevitably limits the scope of disease-related phenomena that can be observed in test-tube models, such that insights deduced from these systems may only reflect a subset of the elaborate network of molecular interactions underlying α-synuclein neurotoxicity. For example, studies of bacterially expressed α-synuclein have focused primarily on the unmodified protein or isoforms with only one type of post-translational modification (e.g., methionine oxidation, tyrosine nitration, phosphorylation) (Conway *et al.*, 2001; Fujiwara *et al.*, 2002; Negro *et al.*, 2002; Uversky *et al.*, 2002b; Norris *et al.*, 2003; Yamin *et al.*, 2003; Hokenson *et al.*, 2004;

Chen and Feany, 2005; Glaser *et al.*, 2005). In contrast, α-synuclein consists of a complex mixture of post-translationally modified isoforms in neuronal cells (Anderson *et al.*, 2006; Mirzaei *et al.*, 2006). Presumably, interactions involving the complete ensemble of modified species determine the overall pattern of α-synuclein aggregation and neurotoxicity in the brain. Test-tube studies typically do not account for all of these interactions, and, therefore, they provide only partial insight into the role of α-synuclein in PD pathogenesis. Additional examples of *in vivo* complexity that are difficult to model in cell-free systems include variations in post-translational modifications and protein–protein or protein–lipid interactions among different cellular compartments (e.g., cytosolic versus membrane-bound), cell types (e.g., neurons versus glia), and regions in the brain (e.g., *substantia nigra* versus cortex).

A third limitation is that *small-molecule inhibitors of α-synuclein self-assembly identified in cell-free systems may not have therapeutic value in vivo*. For example, compounds that block the aggregation of recombinant α-synuclein may be neurotoxic, or they may be sequestered by other molecular targets and, therefore, fail to suppress α-synuclein self-assembly inside the cell. To avoid these pitfalls, a better approach would be to identify therapeutic candidates by screening for compounds that suppress α-synuclein neurotoxicity in cellular models of PD. Not only does this approach effectively "filter out" toxic compounds, but it also enables the identification of chemical entities that prevent cell death by interacting with multiple targets (Lansbury, 2004). To ensure that hits from cellular screens are useful *in vivo*, these molecules must be tested in animal models of α-synuclein toxicity to eliminate compounds with unfavorable pharmacokinetic or pharmacodynamic properties or a weak ability to penetrate the blood–brain barrier.

DJ-I

A small number of cases of familial, early-onset, recessive PD have been linked to deletions or missense mutations in the gene encoding DJ-1 (Abou-Sleiman *et al.*, 2003; Bonifati *et al.*, 2003; Hering *et al.*, 2004; Annesi *et al.*, 2005). DJ-1 is a 19.9-kDa protein that has been shown to protect neurons from oxidative stress in a number of cell-culture and animal models (Yokota *et al.*, 2003; Canet-Aviles *et al.*, 2004; Martinat *et al.*, 2004; Taira *et al.*, 2004; Kim *et al.*, 2005b; Menzies *et al.*, 2005; Xu *et al.*, 2005; Yang *et al.*, 2005). Under oxidizing conditions, DJ-1 converts to a more acidic variant due to the oxidation of cysteine 106 to the sulfinic acid (Canet-Aviles *et al.*, 2004; Taira *et al.*, 2004). Studies using cell-free systems have revealed important clues about (i) how DJ-1 acts as a neuroprotective factor and (ii) how the stability and function of the protein are affected by familial mutations and post-translational modifications.

Characterization of DJ-1 Structure

The crystal structure of DJ-1, published by five independent groups (Honbou *et al.*, 2003; Huai *et al.*, 2003; Lee *et al.*, 2003; Tao and Tong, 2003; Wilson *et al.*, 2003), has provided valuable insight into the normal function of the protein and suggested reasons for the dysfunction of mutants linked to early-onset PD. In turn, these insights have stimulated new hypotheses about DJ-1 function that can be addressed in test-tube, cellular, and animal models. The structure reveals that DJ-1 exists as a homodimer with an α/β-fold closely resembling that of the ThiJ protein involved in thiamine biosynthesis (Figure 43.3a) (Bandyopadhyay and Cookson, 2004). Evidence from two of the published structures suggests that cysteine 106 is susceptible to oxidation to cysteine-sulfinic acid (Lee *et al.*, 2003; Wilson *et al.*, 2003; Canet-Aviles *et al.*, 2004).

Cysteine 106 is located near several acidic and basic residues that may promote its oxidation by stabilizing a reactive thiolate intermediate (Figure 43.3b) (Huai *et al.*, 2003; Tao and Tong, 2003; Wilson *et al.*, 2003; Canet-Aviles *et al.*, 2004). The presence of a reactive cysteine residue near acidic and basic residues is a characteristic feature of the active sites of peroxidases and peroxiredoxins, antioxidant enzymes that reduce H_2O_2 to water (Wilson *et al.*, 2003; Wood *et al.*, 2003). Accordingly, the fact that these molecular features are evident in the crystal structure led to speculation that DJ-1 might function as a peroxidase (Wilson *et al.*, 2003). Consistent with this hypothesis, DJ-1 exhibits a high degree of structural conservation with a domain essential for the activity of hydroperoxidase II and other catalases (Tao and Tong, 2003). The crystal structure of DJ-1 also reveals that the protein shares an evolutionarily conserved domain

with Hsp31, a related member of the DJ-1/ThiJ/PfpI superfamily with chaperone activity (Lee *et al.*, 2003). This observation, combined with evidence of the susceptibility to oxidation of cysteine 106, raised the possibility that DJ-1 might carry out its neuroprotective function by acting as a redox-sensitive chaperone (Bonifati *et al.*, 2004). Finally, although DJ-1 shares sequence and structural similarity with PfpI, a putative cysteine protease in the DJ-1/ThiJ/PfpI superfamily, the DJ-1 crystal structure reveals that the protein lacks a suitably arranged catalytic triad for proteolysis. This observation has led to the prediction that DJ-1 does not function as an intracellular protease (Wilson *et al.*, 2003). In contrast, others have argued that a catalytic diad consisting of cysteine 106 and histidine 126 could enable DJ-1 to carry out proteolysis via a caspase-like mechanism (Tao and Tong, 2003).

In addition to providing insight about the normal function of DJ-1, the three-dimensional structure suggests a rationale for the loss-of-function caused by familial substitutions. The residues at the subunit interface in the DJ-1 homodimer are highly conserved among various species, implying that the protein must adopt a dimeric structure to carry out its biological function (Honbou *et al.*, 2003). In support of this idea, the crystal structure of DJ-1 suggests that dimer formation is necessary for the assembly of two symmetry-related "active sites" at the subunit interface. Each active site consists of residues from both subunits (i.e., glutamate 15, glutamate 16, and glutamate 18 from one subunit; aspartate 24 and arginine 28 from the other) that form a pocket around cysteine 106 and promote the oxidation of this residue to the sulfinic acid (Tao and Tong, 2003; Wilson *et al.*, 2003). The L166P substitution is expected to interfere with dimerization due to the disruptive effect of proline 166 on α-helix G near the dimer interface (Honbou *et al.*, 2003; Huai *et al.*, 2003; Lee *et al.*, 2003; Tao and Tong, 2003; Wilson *et al.*, 2003). Another recently identified substitution, E163K (Annesi *et al.*, 2005), is also located in α-helix G and may have a similar effect on dimer stability as L166P. From the X-ray structure of DJ-1 (Honbou *et al.*, 2003; Huai *et al.*, 2003; Lee *et al.*, 2003; Tao and Tong, 2003; Wilson *et al.*, 2003) one can also rationalize how the M26I substitution might destabilize the dimer. Methionine 26 occurs in α-helix A, which promotes dimer formation via contacts with α-helix A from the opposite subunit. Perturbation of the structure

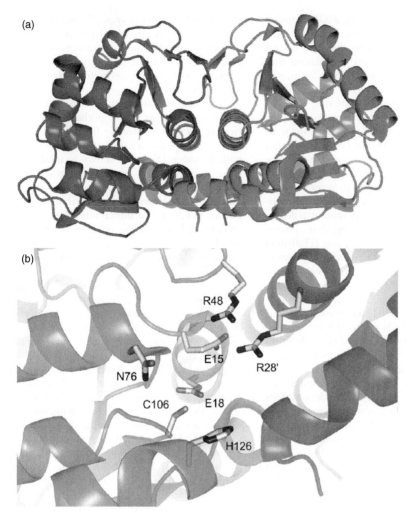

FIGURE 43.3 X-ray images of the subunit interface of wild-type DJ-1. (a) Image of homodimeric DJ-1. (b) Image showing that residues from both subunits contribute to the "active site" of the dimer (this active site is thought to promote the oxidation of C106 to the sulfinic acid). Residues E15, E18, R48, N76, C106, and H126 are from one monomer, and residue R28′ is from the opposite subunit. In (a) and (b), the two monomers are shown in different shades of grey. The images were generated using PDB code 1UCF (Honbou et al., 2003).

of this helix might destabilize the dimer, leading to unfolding, degradation, and/or aggregation of the protein (Hulleman et al., 2007).

Investigations of DJ-1 Function in Cell-Free Systems

Data from studies in cell culture (Yokota et al., 2003; Martinat et al., 2004; Taira et al., 2004; Kim et al., 2005b; Xu et al., 2005) and in animal models (Kim et al., 2005b; Menzies et al., 2005; Meulener et al., 2005; Park et al., 2005; Yang et al., 2005) suggest that DJ-1 protects dopaminergic neurons from oxidative stress. Moreover, in cells or animals exposed to oxidative insults, DJ-1 converts to acidic isoforms that reflect the oxidation of its own cysteine residues, primarily cysteine 106, but also to a lesser extent, C53 (Bandopadhyay et al., 2004;

Canet-Aviles *et al.*, 2004; Taira *et al.*, 2004; Betarbet *et al.*, 2006; Meulener *et al.*, 2006). These observations, together with evidence from the DJ-1 crystal structure (see above), led to early speculation that DJ-1 functions as a peroxidase or peroxiredoxin, preventing a buildup of reactive oxygen species (ROS) in dopaminergic neurons via a "buffering" mechanism involving the oxidation of cysteine 106. In one study, recombinant wild-type DJ-1 was shown to induce the elimination of H_2O_2 via a mechanism that depended on the stability of the homodimer (Taira *et al.*, 2004). The decrease in peroxide levels correlated with the conversion of DJ-1 to a more acidic isoform detectable by isoelectric focusing. In contrast to wild-type DJ-1, the C106A mutant failed to cause a depletion of H_2O_2, suggesting that cysteine 106 is necessary for peroxide reduction (Hulleman and Rochet, unpublished observations). If DJ-1 were acting as a peroxiredoxin, then reducing equivalents should regenerate the active-site cysteine residue for subsequent rounds of catalysis (Wood *et al.*, 2003). Peroxide elimination by DJ-1 occurred at similar rates and with similar yields in the absence or presence of thioredoxin, glutathione (GSH), or dithiothreitol (DTT), suggesting that the reduction of H_2O_2 by DJ-1 is coupled to a single round of oxidation at cysteine 106 (Hulleman and Rochet, unpublished observations) (Taira *et al.*, 2004, Andres-Mateos *et al.*, 2007). From these observations, it was inferred that DJ-1 does not function as a classic peroxiredoxin. Rather, the protein causes a noncatalytic, stoichiometric elimination of H_2O_2 upon being oxidized at cysteine 106, presumably as part of a redox signaling mechanism (discussed further below).

Examination of the DJ-1 crystal structure has stimulated divergent views about whether the protein could be proteolytically active. Whereas some investigators argue that DJ-1 is unlikely to have a proteolytic function due to the absence of a suitably arranged catalytic triad (Wilson *et al.*, 2003), others predict that DJ-1 may act as a protease because it has a Cys-His diad, a catalytic motif in the active sites of caspases (Tao and Tong, 2003). Several groups have reported that recombinant DJ-1 is inactive in proteolytic assays monitoring the cleavage of synthetic or natural polypeptides (Honbou *et al.*, 2003; Lee *et al.*, 2003; Wilson *et al.*, 2003; Shendelman *et al.*, 2004). In contrast, another group reported that DJ-1 exhibited weak proteolytic activity in a highly sensitive, fluorescence-based assay, and this activity was eliminated by the C106A

substitution (Olzmann *et al.*, 2004). Although studies addressing the potential proteolytic activity of recombinant DJ-1 have yielded inconsistent data, collectively the results imply that proteolysis is not a major part of the neuroprotective function of DJ-1 *in vivo*.

Evidence of structural similarity between DJ-1 and Hsp31 led to the hypothesis that DJ-1 carries out its neuroprotective function by acting as a chaperone (see above) (Lee *et al.*, 2003). Consistent with this idea, DJ-1 was shown to inhibit the heat-induced aggregation of citrate synthase and luciferase (Lee *et al.*, 2003). The ability of DJ-1 to function as a chaperone was attributed to a hydrophobic patch at the dimer interface, consisting of conserved, hydrophobic residues from α-helices G and H in each subunit (Lee *et al.*, 2003). In a second study, Abeliovich and colleagues (Shendelman *et al.*, 2004) demonstrated that recombinant, wild-type DJ-1 inhibited the heat-induced aggregation of citrate synthase, glutathione-*S*-transferase, and α-synuclein. Chaperone activity was abolished in this system by reducing disulfide linkages with DTT, and it was rescued upon exposure of the reduced protein to H_2O_2. Moreover, the mutant C53A exhibited no chaperone activity, whereas C106A suppressed protein aggregation with similar efficiency as the wild-type protein. These data suggested that DJ-1 suppresses α-synuclein aggregation via a mechanism involving disulfide formation by cysteine 53. In a third study, Fink and colleagues (Zhou *et al.*, 2006) reported that a modified form of wild-type DJ-1 in which cysteine 106 had been oxidized to the sulfinic acid (referred to as the "2O-DJ-1" isoform) suppressed α-synuclein fibrillization in a test-tube model. In contrast, unoxidized DJ-1 or variants of the protein oxidized more extensively than the 2O form failed to inhibit fibril formation. From these data, it was concluded that the chaperone function of DJ-1 is regulated by the oxidation of cysteine 106 to the sulfinic acid. The inhibition of α-synuclein aggregation by oxidized DJ-1 may account in part for the neuroprotective function of the protein.

Effects of Familial Substitutions and Oxidative Modifications on DJ-1 Stability and Function

Studies of DJ-1 in cell-free systems have provided insight into why mutant forms of the protein involved in early-onset, familial PD are dysfunctional. The X-ray structure of wild-type DJ-1 suggests that the protein must exist as a homodimer to carry out its biological activity (see above) (Honbou et al., 2003; Huai et al., 2003; Lee et al., 2003; Tao and Tong, 2003; Wilson et al., 2003). Evidence in support of a homodimeric structure has been obtained from gel filtration and sedimentation analyses of recombinant, wild-type DJ-1 (Wilson et al., 2003; Olzmann et al., 2004; Hulleman et al., 2007). In contrast, multiple lines of evidence from cell-free systems indicate that L166P has a dimerization defect (Moore et al., 2003; Gorner et al., 2004; Olzmann et al., 2004; Takahashi-Niki et al., 2004; Blackinton et al., 2005). In turn, this loss of dimer stability results in subunit dissociation followed by protein aggregation. Moreover, a thorough characterization of recombinant M26I by gel filtration, ultracentrifugation, and equilibrium-denaturation analyses revealed that this mutant forms a less stable dimer and has a greater propensity to aggregate than the wild-type protein (Hulleman et al., 2007). In contrast, data from test-tube studies suggest that the familial mutant E64D is at least as stable as (or perhaps even more stable than) wild-type DJ-1 (Gorner et al., 2004; Hulleman et al., 2007). This finding indicates that familial mutations linked to early-onset PD do not invariably disrupt DJ-1 function by causing a loss of protein stability.

Studies carried out in cell-free systems have also provided insight about the role of DJ-1 oxidation in PD pathogenesis. Emerging neuropathological data suggest that wild-type DJ-1 undergoes oxidative damage in sporadic PD and during aging (Choi et al., 2006; Meulener et al., 2006). In turn, these findings imply that oxidative modifications may disrupt the function of wild-type DJ-1 during aging, thereby increasing the risk for developing sporadic PD. Consistent with this idea, the 2O-DJ-1 isoform inhibits α-synuclein fibrillization in a test-tube model, whereas more extensively oxidized forms of recombinant DJ-1 have lost this chaperone activity (see above) (Zhou et al., 2006). Moreover, data from gel filtration, ultracentrifugation, and equilibrium-denaturation analyses indicate that over-oxidation destabilizes the recombinant, wild-type dimer, resulting in subunit dissociation and the formation of high molecular weight aggregates (Hulleman et al., 2007). The C106A mutant is more stable than wild-type DJ-1 after exposure to oxidizing conditions, suggesting that the conversion of cysteine 106 to the sulfonic acid (which cannot occur in C106A) contributes

significantly to destabilization of the wild-type protein upon over-oxidation (Hulleman et al., 2007).

Advantages and Limitations of Using Cell-Free Systems to Study DJ-1

Studies of DJ-1 in test-tube models have revealed important clues about the normal function of the protein and the role of DJ-1 dysfunction in PD pathogenesis. The data from these analyses suggest that (i) DJ-1 acts as a redox-sensitive molecular chaperone to suppress α-synuclein aggregation and (ii) familial substitutions and oxidative modifications cause DJ-1 dysfunction in PD by destabilizing the active dimer.

The use of cell-free systems to study DJ-1 has several *advantages*. First, this approach *is ideally suited for investigating intrinsic functional properties of DJ-1*. In contrast, this type of information would be more difficult to obtain in cellular or animal models because activities observed in these systems could be attributed to DJ-1-interacting proteins rather than DJ-1 itself. The three-dimensional structure initially provided valuable insight by suggesting testable hypotheses about DJ-1 function – namely, that the protein may act as a peroxidase, protease, or molecular chaperone. Subsequent analyses of recombinant DJ-1 failed to reveal a robust peroxidase or protease activity. Although the results of these studies did not corroborate hypotheses about DJ-1 function, they nevertheless added value by stimulating new investigations focused on alternative mechanisms. Notably, two groups reported that DJ-1 is partly localized to mitochondria, suggesting that it may modulate the oxidative stress response via effects on mitochondrial physiology (Canet-Aviles et al., 2004; Zhang et al., 2005). In one of these studies, it was suggested that DJ-1 carries out its antioxidant function by relocalizing from the cytosol to mitochondria under conditions of oxidative stress via a mechanism involving the oxidation of cysteine 106 to the sulfinic acid (Canet-Aviles et al., 2004). Another study revealed that DJ-1 stabilizes the transcription factor Nrf2 (nuclear factor erythroid 2-related factor 2), a master regulator of the antioxidant response (Clements et al., 2006), and this effect may play a key role in the neuroprotective activity of DJ-1 against oxidative insults (Liu et al., 2008).

The ability to characterize intrinsic protein activity in cell-free systems was critical for the discovery that DJ-1 suppresses α-synuclein aggregation via a redox-dependent chaperone mechanism (Shendelman et al., 2004; Zhou et al., 2006). Specifically, the use of test-tube models enabled controlled analyses of the chaperone function of DJ-1 by monitoring the effects of individual oxidized DJ-1 isoforms on the aggregation of a single substrate, α-synuclein. This exquisite control would not have been possible in cell-culture studies because intracellular DJ-1 consists of a mixture of modified isoforms (Bandopadhyay et al., 2004; Canet-Aviles et al., 2004; Taira et al., 2004; Betarbet et al., 2006; Choi et al., 2006; Meulener et al., 2006), and protein aggregation inside the cell typically involves the accumulation of multiple proteins in an inclusion body or "aggresome" (Johnston et al., 1998). Moreover, it would be difficult to prove that DJ-1 suppresses α-synuclein aggregation by acting directly as a chaperone in cell-culture experiments given that alternative intracellular mechanisms (e.g., inhibition of α-synuclein oxidation, transcriptional upregulation of other molecular chaperones) could account for this suppressive effect. In contrast, the inhibition of α-synuclein aggregation by 2O-DJ-1 in mixtures of the two recombinant proteins indicates that this DJ-1 isoform can function directly as a molecular chaperone.

A second advantage is that *studies of DJ-1 in cell-free systems provide detailed molecular insight into why mutant or oxidized forms of the protein are dysfunctional in PD.* Specifically, one can determine via test-tube experiments whether familial substitutions or oxidative modifications disrupt DJ-1 function by destabilizing the protein and/or interfering with one or more of its intrinsic activities. Structural and biochemical analyses of recombinant DJ-1 suggested that the M26I and L166P familial substitutions destabilize the native dimer by disrupting α-helical structure near the subunit interface (Honbou et al., 2003; Huai et al., 2003; Lee et al., 2003; Moore et al., 2003; Tao and Tong, 2003; Wilson et al., 2003; Gorner et al., 2004; Olzmann et al., 2004; Takahashi-Niki et al., 2004; Blackinton et al., 2005; Hulleman et al., 2007). Other test-tube data revealed that over-oxidized wild-type DJ-1 adopts a structure with a high propensity to undergo subunit dissociation and aggregation (Hulleman et al., 2007). It may be argued that a decrease in the stability of mutant DJ-1 or of wild-type DJ-1 under oxidized conditions can be easily inferred from a reduction in steady-state levels of the protein in cell-culture models. However,

these systems typically involve the co-expression of wild-type and mutant DJ-1, so that the reduced stability of the mutant may be masked by the formation of relatively stable heterodimers consisting of both types of subunits. Studies of recombinant DJ-1 have also shown that mutant and over-oxidized forms of the protein have a diminished intrinsic ability to suppress α-synuclein aggregation via a molecular chaperone function (Shendelman et al., 2004) (Hulleman and Rochet, unpublished observations). Evidence that familial DJ-1 variants have reduced stability and/or intrinsic activity enables one to conclude that the underlying genetic lesions are true loss-of-function mutations, rather than benign polymorphisms. This insight is critical given the small number of PD patients with mutations in the DJ-1 gene (Abou-Sleiman et al., 2003). Moreover, the observation that over-oxidized DJ-1 has decreased stability and function suggests that DJ-1 oxidation during aging is involved in sporadic PD and other age-related neurodegenerative disorders (Choi et al., 2006; Meulener et al., 2006). Not only does this information improve our understanding of PD pathogenesis, but also it justifies drug discovery efforts aimed at enhancing DJ-1 function in familial and sporadic forms of the disease.

A third advantage is that *insight about the structure and stability of recombinant DJ-1 may enable the development of new, rationally designed therapies for treating PD.* Evidence that two familial mutants (M26I and L166P) and the over-oxidized wild-type protein have a dimerization defect provides a strong rationale for rescuing DJ-1 function via dimer stabilization. An attractive approach would be to stabilize dimeric DJ-1 with chemical entities that bind at the dimer interface. The availability of the DJ-1 crystal structure makes it possible to identify these compounds via *in silico* screening. There is precedence for this type of structure-based approach in the discovery of therapeutic leads that stabilize the native, oligomeric forms of transthyretin and superoxide dismutatse-1 (Johnson et al., 2005; Ray et al., 2005; Rochet, 2007).

The use of cell-free systems to characterize DJ-1 also has *limitations.* First, *it is difficult to assess whether activities of DJ-1 identified in the test tube contribute to the neuroprotective effects of the protein in vivo.* For example, because there is little evidence of a robust physical interaction between DJ-1 and α-synuclein in cell-culture models (Jin et al., 2005; Moore et al., 2005), it is unclear whether DJ-1 suppresses α-synuclein aggregation via a classic molecular chaperone function *in vivo*. Conceivably, DJ-1 may be unavailable to interact with α-synuclein in the cytosol as it does in the test tube owing to its sequestration in multiprotein complexes (Baulac *et al.*, 2004; Moore *et al.*, 2005) or intracellular compartments (Canet-Aviles *et al.*, 2004; Junn *et al.*, 2005; Xu *et al.*, 2005; Zhang *et al.*, 2005; Zhong *et al.*, 2006). In addition, the α-synuclein conformer targeted by DJ-1 in cell-free systems may not be significantly populated *in vivo*. Moreover, the intrinsic chaperone activity of DJ-1 observed in test-tube models may play a minor role in preventing α-synuclein aggregation compared to the DJ-1-mediated upregulation of Hsp70 (Zhou and Freed, 2005, Liu *et al.*, 2008). Nevertheless, the results of test-tube studies clearly demonstrate that 2O-DJ-1 has the capacity to act directly as a molecular chaperone. Accordingly, if studies in cellular or animal models fail to provide evidence for such an activity, then it is imperative to identify reasons that account for differences in the behavior of DJ-1 in the test tube and *in vivo* (examples of such reasons are listed above). In turn, this information may prove vital to our understanding of normal DJ-1 function and PD pathogenesis.

A second limitation is that *studies of DJ-1 in cell-free systems may not account for parameters that modulate DJ-1 function or dysfunction in vivo*. As mentioned above, analyses of recombinant DJ-1 have failed to demonstrate that the protein acts as a peroxidase or protease. However, these catalytic functions may only be activated when DJ-1 interacts with a protein or cofactor that is absent from the test tube. For example, although DJ-1 consumes H_2O_2 by undergoing oxidation at cysteine 106, it lacks an intrinsic mechanism to convert the cysteine residue back to its reduced form (this conversion would be essential for additional rounds of catalysis). Instead, an as-of-yet undiscovered electron donor may be required to reactivate oxidized DJ-1 by reducing cysteine 106. Consistent with this idea, a number of peroxiredoxins are only activated upon the reduction of a catalytic cysteine residue by a specific electron donor such as glutathione-S-transferase π (Manevich *et al.*, 2004) or sulfiredoxin (Woo *et al.*, 2005). Data from test-tube studies may also yield an incomplete description of the functional properties of familial DJ-1 mutants. Even if a familial mutant is indistinguishable from the wild-type protein in a cell-free system, this does not prove that the underlying genetic

lesion is a benign polymorphism rather than a loss-of-function mutation. As one possibility, the mutant protein may be deficient in a functional property that is not modeled in the cell-free system. Alternatively, the mutant may only become inactive after undergoing destabilizing modifications (e.g., protein oxidation [Hulleman *et al.*, 2007]) that occur *in vivo* but not in the test tube. An example of a DJ-1 mutant with distinct behavior in the test tube versus in a cell-culture model is E64D. Although E64D exhibits similar structural properties as wild-type DJ-1 in a cell-free system (Gorner *et al.*, 2004; Hulleman *et al.*, 2007), the mutant has a decreased ability to protect against PD-related insults in primary mesencephalic cultures (Liu *et al.*, 2008).

A third limitation is that *many of the neuroprotective functions attributed to DJ-1 are difficult to model in cell-free systems*. As discussed above, the chaperone activity of DJ-1 has been successfully characterized in the test tube (Shendelman *et al.*, 2004; Zhou *et al.*, 2006). However, other proposed neuroprotective functions of DJ-1 involve modulatory effects on complex cellular pathways rather than a single protein target. For example, DJ-1 suppresses neuronal apoptosis by increasing signaling through the PI3K-Akt survival pathway (Kim *et al.*, 2005a; Yang *et al.*, 2005) and decreasing signaling through the DAXX-ASK1 (apoptosis signal-regulating kinase 1) or p53-glyceraldehyde-3-phosphate dehydrogenase (GAPDH)-BAX cell death pathway (Junn *et al.*, 2005; Bretaud *et al.*, 2007; Gorner *et al.*, 2007). DJ-1 also activates pro-survival genes by interacting with the transcriptional regulators p54nrb and pyrimidine tract-binding protein-associated splicing factor (PSF) (Xu *et al.*, 2005). Another study revealed that DJ-1 upregulates DA biosynthesis by antagonizing PSF, a transcriptional repressor of the TH gene (Zhong *et al.*, 2006). Limited (albeit meaningful) information about these and other DJ-1-regulated signaling pathways can be obtained from experiments in cell-free systems. For example, it may be possible to characterize binding interactions between recombinant DJ-1 and signaling molecules or transcription factors on these pathways (Junn *et al.*, 2005; Xu *et al.*, 2005; Gorner *et al.*, 2007). However, to achieve a comprehensive understanding of how DJ-1 modulates cell signaling, and how this modulation results in neuroprotection, one must conduct detailed analyses in cellular and animal models.

PARKIN, UCH-LI, PINKI, LRRK2

Cell-free systems have been used to explore the functional properties of parkin, UCH-L1, PINK1, and LRRK2, albeit less extensively than in the case of α-synuclein or DJ-1. The results of these studies have provided important insights into how each of these proteins is involved in PD pathogenesis.

Parkin

Mutations in the gene encoding parkin are the most frequent cause of early-onset, autosomal-recessive PD (Cookson, 2005). Parkin is a 51.5-kDa protein that functions as an E3 ubiquitin ligase. The N-terminal portion of the protein (the "ubiquitin-like" [Ubl] domain) shares homology with ubiquitin and is thought to play a role in the interaction of parkin with the proteasome (Cookson, 2005). The C-terminal domain is composed of two RING ("Really Interesting New Gene") finger domains and an IBR ("'In-Between RING finger") domain (Figure 43.4) (Cookson, 2005). The tripartite RING-IBR-RING motif is involved in recruiting the E2 ubiquitin conjugating enzyme (e.g., UBCH7 or UBCH8) and protein substrates targeted for ubiquitylation.

Experiments in test-tube models have advanced our understanding of the parkin E3 ubiquitin ligase activity and how this function is disrupted by familial mutations. The use of reconstituted cell-free systems to analyze parkin-dependent ubiquitylation led to the identification of various parkin substrates, including parkin itself (Zhang *et al.*, 2000), CDC-rel2 (SEPT5) (Choi *et al.*, 2003), cyclin E (Staropoli *et al.*, 2003), Pael-R (Imai *et al.*, 2001), synphilin-1 (Chung *et al.*, 2001), and a glycosylated form of α-synuclein (αSp22) found in human brain (Shimura *et al.*, 2001). In some of these studies, mutant forms of parkin associated with early-onset PD were shown to have reduced ubiquitin ligase activity in a cell-free system (Imai *et al.*, 2001;

Shimura *et al.*, 2001; Staropoli *et al.*, 2003). A potential limitation of these analyses is that the parkin variants were generated by *in vitro* translation or isolated from transfected cells via immunoprecipitation. Accordingly, the intrinsic activity of parkin in these assays may have been masked by contaminating E3 ligases or altered by other factors involved in regulating protein ubiquitylation (Hampe *et al.*, 2006; Matsuda *et al.*, 2006). To address this problem, two groups measured the auto-ubiquitylation activities of bacterially expressed, wild-type and mutant parkin in a reconstituted cell-free system (Hampe *et al.*, 2006; Matsuda *et al.*, 2006). Unexpectedly, the data indicated that (i) a substantial number of familial substitutions have no impact on intrinsic parkin activity and (ii) parkin has an intrinsic ability to catalyze monoubiquitylation rather than polyubiquitylation (Hampe *et al.*, 2006; Matsuda *et al.*, 2006).

Additional studies in cell-free systems revealed that parkin is covalently modified by oxidized DA (LaVoie *et al.*, 2005), nitrosylating agents (Chung *et al.*, 2004; Yao *et al.*, 2004), and the cyclin-dependent kinase, CDK5 (Avraham *et al.*, 2007). All three modifications were shown to cause a loss or dysregulation of parkin E3 ligase activity as measured by *in vitro* ubiquitylation assays. The NMR structure of the parkin IBR domain provides a rationale for the perturbation of ubiquitin ligase activity by S-nitrosylation and CDK5-mediated phosphorylation – namely, both modifications are predicted to destabilize the overall architecture of the RING–IBR–RING motif, thereby disrupting interactions with UBCH7 and UBCH8 (Beasley *et al.*, 2007).

UCH-LI

A heterozygous mutation in the gene encoding UCH-L1 (resulting in the I93M substitution) was identified in a kindred with familial, autosomal-dominant PD (Leroy *et al.*, 1998). In contrast, a polymorphic variant of UCH-L1 with the S18Y

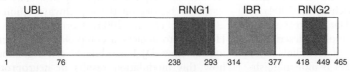

FIGURE 43.4 Domain structure of parkin showing the UBL ("ubiquitin-like"), RING ("really interesting new gene", and IBR ("in-between RING") domains. Adapted from von Coelln *et al.* (2004).

substitution is apparently associated with decreased susceptibility to PD (Maraganore *et al.*, 1999). UCH-L1 is a 24.8-kDa protein that belongs to the family of deubiquitinating enzymes, a group of proteases that catalyze the cleavage of ubiquitin from ubiquitylated polypeptides (Wilkinson *et al.*, 1989). Deubiquitinating enzymes play an essential role in recycling ubiquitin, making it available for conjugation to new protein substrates.

Biochemical analyses of recombinant UCH-L1 have shed light on possible mechanisms by which the I93M mutation and S18Y polymorphism might influence the risk of PD. Two groups reported that the I93M substitution causes an approximately twofold reduction in ubiquitin hydrolase activity compared to the wild-type enzyme (Leroy *et al.*, 1998; Nishikawa *et al.*, 2003). This loss of activity could elicit neurotoxicity by decreasing the pool of ubiquitin available for proteasomal degradation and/or by promoting the accumulation and subsequent aggregation of UCH-L1 substrates (Leroy *et al.*, 1998). In another study, evidence was reported that UCH-L1 has a ubiquityl ligase activity in addition to its more extensively characterized deubiquitinating function (Liu *et al.*, 2002). The ligase activity was found to correlate with the propensity of UCH-L1 to adopt a dimeric structure and resulted in the formation of multi-ubiquitylated, potentially neurotoxic forms of α-synuclein (Liu *et al.*, 2002). Intriguingly, S18Y formed less dimer and exhibited reduced ligase activity compared to the wild-type enzyme or I93M, suggesting a possible rationale for the protective effect of the S18Y polymorphism. The recently solved crystal structure of UCH-L1 indicates that the enzyme exists in an inactive state in the absence of ligand, but may undergo conformational changes coupled with activation upon the binding of an unknown substrate (Das *et al.*, 2006). The structure implies that the I93M substitution disrupts hydrolase activity by perturbing the arrangement of the catalytic cysteine residue in the active site. In contrast, the structure provides no rationale for how the S18Y substitution alters the enzymatic function of UCH-L1, raising the possibility that it instead affects the interaction of UCH-L1 with a binding partner (Das *et al.*, 2006).

PINK1

Homozygous mutations in the gene encoding PINK1 have been identified in patients with early-onset, autosomal-recessive PD (Valente *et al.*, 2004; Cookson, 2005). PINK1 is a 62.8-kDa protein with an N-terminal mitochondrial localization sequence, a serine/threonine protein kinase domain, and a C-terminal auto-regulatory domain. Most of the familial substitutions occur in the kinase domain and, to a lesser extent, in the mitochondrial localization sequence.

Test-tube analyses have provided insight into the enzymatic function of PINK1 and the deleterious effects of familial mutations. One study revealed that a GST fusion protein containing the PINK1 kinase domain catalyzes autophosphorylation, and this activity is dramatically reduced by mutating three residues predicted to play a role in phosphoryl transfer (Beilina *et al.*, 2005). Other groups observed that the PINK1 kinase domain catalyzes the phosphorylation of artificial protein substrates, including α-casein (Silvestri *et al.*, 2005) and histone H1 (Sim *et al.*, 2006). PINK1 was also shown to directly phosphorylate the mitochondrial molecular chaperone TNF receptor-associated protein 1 (TRAP1, also referred to as Hsp75) in a cell-free system (Pridgeon *et al.*, 2007). Collectively, the results of kinase assays carried out by various groups suggest that multiple PINK1 mutants associated with autosomal-recessive PD (e.g., A168P, G309D, L347P, G386A, G409V) have reduced kinase activity (Beilina *et al.*, 2005; Silvestri *et al.*, 2005; Sim *et al.*, 2006; Pridgeon *et al.*, 2007). In contrast, the truncation mutant W437X was reported to have enhanced or diminished kinase activity with α-casein or TRAP1 as the substrate, respectively (Silvestri *et al.*, 2005; Pridgeon *et al.*, 2007).

LRRK2

Mutations in the gene encoding leucine-rich repeat kinase 2 (LRRK2, also referred to as "dardarin") have been identified in patients with late-onset, autosomal-dominant PD or sporadic PD (Paisan-Ruiz *et al.*, 2004; Zimprich *et al.*, 2004; Mata *et al.*, 2006). LRRK2 has a molecular weight of 286 kDa and consists of multiple domains, including: (i) an N-terminal domain with ankyrin and armadillo repeats; (ii) a leucine-rich repeat (LRR) domain; (iii) a "Ras in complex proteins" (Roc) GTPase domain; (iv) a "C-terminal of Roc" (COR) domain; (v) a kinase domain with homology to tyrosine kinase-like (TKL) kinases; and (vi) a C-terminal WD40 domain (Figure 43.5) (Mata *et al.*, 2006). A striking feature of LRRK2 is the presence of a

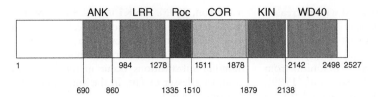

FIGURE 43.5   Domain structure of LRRK2 showing the N-terminal domain with ankyrin and armadillo repeats ("ANK"), leucine-rich repeat (LRR) domain, "Ras in complex proteins" ("Roc") GTPase domain, "C-terminal of Roc" ("COR") domain, kinase ("KIN") domain, and C-terminal WD40 domain. Adapted from Mata et al. (2006).

GTPase domain and a protein kinase domain within the same polypeptide chain. Moreover, the fact that LRKK2 contains various protein-interaction domains suggests that it acts as a scaffold for the assembly of multiprotein complexes involved in cell signaling (Mata *et al.*, 2006).

Studies in cell-free systems have provided important clues about the kinase function of LRKK2 and how this activity is affected by pathogenic mutations. Reports from several groups indicate that immunoprecipitated or affinity-purified LRKK2 catalyzes its autophosphorylation or the phosphorylation of artificial substrates (West *et al.*, 2005, 2007; Gloeckner *et al.*, 2006; Ito *et al.*, 2007). In addition, affinity-purified LRRK2 phosphorylates the actin-anchor protein moesin, a potential authentic substrate present in the brain (Jaleel *et al.*, 2007). Bacterially expressed LRKK2 fragments consisting of the kinase domain or a COR-kinase fusion also catalyze phosphorylation (Luzon-Toro *et al.*, 2007). Several groups have reported that a common pathogenic substitution in the kinase domain, G2019S, stimulates LRRK2-mediated phosphorylation (West *et al.*, 2005, 2007; Gloeckner *et al.*, 2006; Jaleel *et al.*, 2007; Luzon-Toro *et al.*, 2007), whereas another PD-linked mutation in this domain, I2012T, suppresses kinase activity (Jaleel *et al.*, 2007; West *et al.*, 2007). Studies of the effects of other familial substitutions (e.g., Y1699C, I2020T) have yielded inconsistent data (Gloeckner *et al.*, 2006; Jaleel *et al.*, 2007; West *et al.*, 2007).

A number of reports suggest that LRRK2 (or the isolated ROC domain) binds and hydrolyzes GTP in cell-free systems (Smith *et al.*, 2006; Ito *et al.*, 2007; Lewis *et al.*, 2007; Li *et al.*, 2007; West *et al.*, 2007). Collectively, the results of multiple groups indicate that GTP binding is stimulated or GTP hydrolysis is inhibited by familial substitutions in the ROC or COR domain (e.g., I1371V, R1441C, R1441G, Y1699C) (Lewis *et al.*, 2007; Li *et al.*, 2007; West *et al.*, 2007). Recombinant LRRK2 variants with mutations that disrupt or augment GTP binding have reduced or enhanced kinase activity, respectively (West *et al.*, 2005, 2007; Smith *et al.*, 2006; Ito *et al.*, 2007; Jaleel *et al.*, 2007). These findings suggest that LRRK2-mediated phosphorylation is regulated by the binding of GTP to the ROC-COR domains. Additional evidence from cell-culture studies indicates that GTP binding and protein kinase activity are both necessary for LRRK2 neurotoxicity (Smith *et al.*, 2006; West *et al.*, 2007).

Advantages and Limitations

The use of cell-free systems to study parkin, UCH-L1, PINK1, and LRRK2 has revealed a number of *advantages*. First, *this approach has made it possible to characterize intrinsic activities associated with each of these proteins*. In the absence of a cell-free system, it would have been challenging to uncover the monoubiquitylation activity of parkin or the ubiquityl ligase activity of UCH-L1 due to interference from other proteins that modulate ubiquitin conjugation *in vivo*. Similarly, it would not have been possible to characterize the GTP-binding and kinase functions of LRRK2 given that large numbers of other enzymes with similar activities are present inside the cell. Understanding the intrinsic activities of these proteins is critical to how we view their role in PD pathogenesis and how we design therapeutic strategies to overcome their dysfunction. For example, the fact that parkin catalyzes monoubiquitylation suggests that it modulates cellular processes such as transcription and endocytosis, rather than just ubiquitin-dependent proteasomal degradation (Hampe *et al.*, 2006; Matsuda *et al.*, 2006). In addition, evidence of a connection between the intrinsic GTP-binding and kinase functions of LRRK2 suggests that both activities are reasonable drug targets (West *et al.*, 2005, 2007; Smith *et al.*, 2006; Ito *et al.*, 2007; Jaleel *et al.*, 2007). A second advantage of working with cell-free systems is that this approach *has provided valuable insight into the effects of familial mutations and post-translational modifications on intrinsic activities of parkin, UCH-L1, PINK1, and LRRK2.* In the case of parkin, analyses of bacterially expressed mutants led to the discovery that not all familial substitutions disrupt the intrinsic ubiquitin ligase activity. In turn, this information has prompted new hypotheses about mechanisms underlying parkin dysfunction in familial PD (Hampe *et al.*, 2006; Matsuda *et al.*, 2006).

Conversely, the use of cell-free systems to characterize parkin, UCH-L1, PINK1, and LRRK2 has also revealed *limitations*. Notably, *it is unclear whether the intrinsic activities of these proteins identified in test-tube models play a significant role in vivo.* Each of these gene products is likely to interact with binding partners inside the cell, leading to the formation of complexes with overall activities markedly different from the intrinsic activities measured in the test tube. For example, the ubiquitin ligase activity of parkin may be affected by other proteins that modulate ubiquitin transfer, including E4-like factors (Matsuda *et al.*, 2006). Similarly, the function of the LRRK2 ROC–COR domains may be regulated by GTPase activating proteins (GAPs) and guanine nucleotide exchange factors (GEFs) (Mata *et al.*, 2006; Ito *et al.*, 2007; Li *et al.*, 2007). Another limitation is that *measurements of autophosphorylation or the phosphorylation of generic substrates may yield information about PINK1 or LRRK2 kinase activity with little relation to the catalytic properties of these enzymes in vivo.* To address this pitfall, future efforts must focus on developing biochemical assays to monitor PINK1- or LRRK2-mediated phosphorylation of authentic substrates. Finally, *test-tube experiments only provide partial insight into the biological properties of these PD-related gene products inside the cell.* This point is illustrated by the fact that LRRK2 has various protein-interaction domains which presumably serve as "adaptors" for the assembly of a multiprotein signaling complex (Mata *et al.*, 2006). Clearly, the integrated network

of signaling pathways triggered by such a complex would be difficult to model in a cell-free system.

CONCLUSIONS

Use of Cell-Free Systems to Explore Mechanisms of PD Pathogenesis

From the examples outlined in this chapter, it is evident that cell-free systems have played an essential role in PD research. Without the use of test-tube models, it would have been difficult to characterize protofibrils on the α-synuclein self-assembly pathway because these species are metastable and, presumably, lethal to neurons in which they accumulate. In contrast, protofibrils can be generated from recombinant α-synuclein in sufficient quantities for detailed biophysical and functional analyses. Studies in test-tube models have also enabled the characterization of PD-related gene products in terms of their *intrinsic* activities, including the redox-sensitive chaperone function of DJ-1 (Zhou *et al.*, 2006) and the GTP-regulated kinase activity of LRRK2 (Smith *et al.*, 2006). Intrinsic activities are difficult to characterize in cells or animals due to the presence of additional proteins with potentially dramatic effects on the phenotypic readout. Cell-free systems have also proven to be optimal for investigating effects of environmental insults (e.g., oxidative modifications) on the functional properties of PD-related gene products. Studying these effects in cellular or animal models is more challenging because environmental stresses inevitably impact many proteins in the cell rather than a single gene product of interest. Finally, high-resolution structural analyses of recombinant α-synuclein, DJ-1, and UCH-L1 have yielded new clues about the function and/or dysfunction of each of these proteins.

Although there are obvious advantages to studying PD in cell-free systems, this approach also has significant limitations. Because solutions of purified protein lack the complexity of the intracellular milieu, functional properties observed in a cell-free system may not play a major role in cellular or animal models. Conversely, the opposite pitfall is also true: because test-tube models are highly simplified, they may provide an incomplete description of how PD-related gene products function in neurons or the midbrain.

Use of Cell-Free Systems to Identify Therapeutic Targets and Drug Leads

Insights from test-tube models of PD have had a profound impact on the discovery of drug targets. An advantage of characterizing PD-linked gene products in test-tube models is that this approach enables the isolation of pathogenic, targetable species that would likely escape detection *in vivo* due to their metastability and/or neurotoxicity. This point is illustrated by the discovery that recombinant α-synuclein forms potentially toxic protofibrils, a finding which has led to the suggestion that protofibrils may be critical drug targets (Volles and Lansbury, 2003; Rochet, 2007). A second advantage of experiments involving cell-free systems is that these analyses reveal new insights about the intrinsic function or dysfunction of PD-related gene products (in contrast to studies in cells or animals – see above), and this information provides a rationale for targeting strategies. For example, data from studies of recombinant DJ-1 showed that the familial mutants M26I and L166P and the overoxidized wild-type protein have a dimerization defect (Hulleman *et al.*, 2007), suggesting that rescue of this defect with dimer "stabilizers" may be a reasonable approach to enhance DJ-1 function in familial and sporadic PD. In addition, evidence from cell-free systems has led to the hypothesis that LRRK2 can be targeted by inhibiting its GTP-binding and kinase activities (West *et al.*, 2005, 2007; Smith *et al.*, 2006; Ito *et al.*, 2007; Jaleel *et al.*, 2007). Without information about intrinsic activities, it is difficult to devise a strategy to rescue the function of PD-related gene products because the dysfunction of these proteins may occur via disparate mechanisms (and, therefore, one cannot be certain which functional deficit(s) should be targeted). For example, a gain of toxic function or loss of neuroprotective function could result from protein mislocalization and/or a disruption of interactions with binding partners, in addition to a perturbation of intrinsic activity. Importantly, the availability of high resolution structural data (obtained from X-ray or NMR analyses of purified proteins) facilitates the rational design of compounds that modulate the intrinsic activities of PD-related gene products.

A limitation associated with the use of cell-free systems to identify drug targets in PD is that protein conformers characterized in the test tube may differ markedly from the major species in neurons. For example, protofibrils generated in solutions of purified α-synuclein may be distinct from oligomers that accumulate in the brains of PD patients. Similarly, biological functions discovered in cell-free systems may differ from the predominant activities of PD-related gene products *in vivo*. Although α-synuclein has been shown to permeabilize synthetic lipid membranes in various biochemical assays (Volles *et al.*, 2001; Ding *et al.*, 2002; Lashuel *et al.*, 2002a, b; Volles and Lansbury, 2002; Kayed *et al.*, 2004; Quist *et al.*, 2005; Zakharov *et al.*, 2007), it has not been determined whether the protein disrupts complex biological membranes in the brain. Moreover, it is unclear whether DJ-1 inhibits α-synuclein aggregation *in vivo* via a chaperone mechanism similar to that observed in cell-free systems (Shendelman *et al.*, 2004; Zhou *et al.*, 2006). Another potential pitfall of working with recombinant proteins is that this strategy may lead one to overlook targets which play a critical role in pathogenesis but are difficult to reproduce in a test-tube model. This point is underscored by the observation that intracellular α-synuclein consists of a heterogeneous mixture of post-translationally modified forms (Mirzaei *et al.*, 2006), and the oligomers generated by these species are likely to be distinct from those formed in monodisperse solutions of the bacterially expressed protein. In addition, targetable activities such as the up- or downregulation of signaling pathways by DJ-1, PINK1, or LRRK2 cannot readily be modeled in cell-free systems. To address these limitations, it is essential to develop more complex test-tube models that replicate the intracellular milieu with higher fidelity (see below).

In principle, biochemical assays that report on the structure and/or function of drug targets relevant to PD can be used to identify new therapeutic leads. This point is illustrated by the discovery of small-molecule inhibitors of α-synuclein self-assembly using fluorescence-based methods to monitor fibrillization of the recombinant protein (Zhu *et al.*, 2004; Masuda *et al.*, 2006; Ono and Yamada, 2006; Luk *et al.*, 2007; Ono *et al.*, 2007). An advantage of using test-tube assays for drug discovery is that these methods can readily be adapted to high-throughput screens of chemical libraries. Moreover, "hit" compounds discovered using this approach are identified by screening against a known PD target (e.g., a gene product with a gain of toxic function such as α-synuclein or LRRK2). Assuming they have no off-target effects, these compounds are predicted to abrogate pathogenic mechanisms that occur strictly in diseased neurons (e.g., the formation of toxic α-synuclein aggregates) without disrupting normal signaling pathways in other types of cells. In contrast, small molecules identified by monitoring cell viability in tissue culture or animal models typically have less well-defined mechanisms of action. Although some compounds discovered using this approach may carry out their protective effect by interacting directly with a PD target, others may suppress cell death pathways activated in response to the target's dysfunction. A potential pitfall of molecules that inhibit downstream cell death mechanisms is that they may promote inappropriate cell survival or oncogenesis in peripheral tissues. Finally, an important advantage of working with recombinant, PD-related gene products is that if the structures of the proteins are known, then one can use this structural information to optimize initial drug leads.

A disadvantage of using biochemical assays for drug screening is that hits identified using this approach may not be neuroprotective in PD patients. For example, the compounds may be toxic *in vivo* due to off-target effects, or they may exhibit unfavorable pharmacokinetics or fail to penetrate the blood–brain barrier. Moreover, if a target identified in a cell-free system is ultimately found not to play a central role in PD pathogenesis, then small molecules that neutralize the target are unlikely to have an impact on dopaminergic cell death *in vivo*. A strategy to overcome these limitations is to screen for therapeutic compounds using cell-based assays. A limitation of this approach is that the mechanism of action of drug candidates identified via cellular screens may be poorly defined (see above). However, this potential drawback can be addressed by showing that the compounds suppress neurodegeneration and have minimal side-effects in animal models, even if their detailed mechanism of action is uncertain (Lansbury, 2004).

In summary, biochemical and cell-based assays have complementary advantages, and, therefore, both are reasonable to use in screens for PD drugs. Compounds identified using either method should be validated in animal models early in the drug discovery process to avoid retaining molecules with limited clinical potential.

Strategies to Validate Findings Obtained Using Cell-Free Systems

As indicated above, there are a number of limitations associated with using cell-free systems to investigate PD pathogenesis, characterize therapeutic targets, or identify new drugs. One way to address these pitfalls is to design test-tube models with greater complexity that more closely recapitulate the native intracellular environment. This strategy is illustrated by a recent study in which the authors tested the effects of a cytosolic fraction on the binding of recombinant α-synuclein to synthetic vesicles (Kim *et al.*, 2006). In the absence of cytosol, α-synuclein bound to the vesicles irreversibly via mainly electrostatic interactions. However, in the presence of cytosol the protein bound transiently via predominantly hydrophobic interactions, similar to its mode of interaction with intracellular membranes. In another study, the addition of brain cytosol was found to promote the release of α-synuclein from isolated synaptosomal membranes (Wislet-Gendebien *et al.*, 2006). These examples reveal a key benefit to "adding back" cellular fractions to a cell-free system – namely, one can use this approach to identify cellular components responsible for phenotypic differences between the test-tube and *in vivo* settings (Kim *et al.*, 2006; Wislet-Gendebien *et al.*, 2006).

The most convincing way to demonstrate that observations in a cell-free system are relevant to PD pathogenesis is to validate these findings in cellular or animal models. This approach is illustrated by a recent study in which mutations that enhance or suppress the fibrillization of recombinant α-synuclein were tested for their effects on α-synuclein aggregation and toxicity in transgenic *Drosophila* (Periquet *et al.*, 2007). Flies that expressed a deletion mutant lacking the hydrophobic stretch spanning residues 71–82 showed no evidence of aggregate formation or dopaminergic cell death, in contrast to flies that expressed wild-type α-synuclein (Periquet *et al.*, 2007). Conversely, flies expressing a C-terminally truncated variant exhibited increased oligomer and inclusion formation and more pronounced neurodegeneration. These data argue convincingly that test-tube models of α-synuclein can, in principle, predict key features of PD pathogenesis. More importantly, they demonstrate that α-synuclein aggregation is essential for dopaminergic cell death *in vivo*, although it remains unclear whether oligomers or fibrils are the primary neurotoxic species.

The validation of data from a test-tube model with parallel findings *in vivo* will justify the further use of this model to explore neurotoxic mechanisms in PD. If the data from a cell-free system are poorly predictive of *in vivo* phenomena, then it is imperative to determine the reasons for this discrepancy. In turn, this information may facilitate the identification of cellular factors that play essential roles in neurodegeneration or neuroprotection. Ultimately, a multipronged approach involving the use of test-tube, cellular, and animal models will be essential to advance our understanding of PD pathogenesis and identify new targets for therapeutic intervention.

ACKNOWLEDGEMENTS

We thank Dr. Ralf Langen (University of Southern California) for generously providing the image of a structural model of fibrillar α-synuclein in Figure 43.2 and Dr. Markus Lill (Purdue University) for assistance in preparing Figure 43.3.

REFERENCES

Abou-Sleiman, P. M., Healy, D. G., Quinn, N., Lees, A. J., and Wood, N. W. (2003). The role of pathogenic DJ-1 mutations in Parkinson's disease. *Ann Neurol* **54**, 283–286.

Anderson, J. P., Walker, D. E., Goldstein, J. M., de Laat, R., Banducci, K., Caccavello, R. J., Barbour, R., Huang, J., Kling, K., Lee, M., Diep, L., Keim, P. S., Shen, X., Chataway, T., Schlossmacher, M. G., Seubert, P., Schenk, D., Sinha, S., Gai, W. P., and Chilcote, T. J. (2006). Phosphorylation of Ser-129 is the dominant pathological modification of α-synuclein in familial and sporadic Lewy body disease. *J Biol Chem* **281**, 29739–29752.

Andres-Mateos, E., Perier, C., Zhang, L., Blanchard-Fillion, B., Greco, T. M., Thomas, B., Ko, H. S., Sasaki, M., Ischiropoulos, H., Przedborski, S., Dawson, T. M., and Dawson, V. L. (2007). DJ-1 gene deletion reveals that DJ-1 is an atypical peroxiredoxin-like peroxidase. *Proc Natl Acad Sci USA* **104**, 14807–14812.

Annesi, G., Savettieri, G., Pugliese, P., D'Amelio, M., Tarantino, P., Ragonese, P., La Bella, V., Piccoli, T., Civitelli, D., Annesi, F., Fierro, B., Piccoli, F., Arabia, G., Caracciolo, M., Ciro Candiano, I. C., and Quattrone, A. (2005). DJ-1 mutations and parkinsonism-dementia-amyotrophic lateral sclerosis complex. *Ann Neurol* **58**, 803–807.

Avraham, E., Rott, R., Liani, E., Szargel, R., and Engelender, S. (2007). Phosphorylation of parkin by the cyclin-dependent kinase 5 at the linker region modulates the E3 ubiquitin-ligase activity and parkin aggregation. *J Biol Chem* **282**, 12842–12850.

Baba, M., Nakajo, S., Tu, P.-H., Tomita, T., Nakaya, K., Lee, V. M.-Y., Trojanowski, J. Q., and Iwatsubo, T. (1998). Aggregation of α-synuclein in Lewy bodies of sporadic Parkinson's disease and dementia with Lewy bodies. *Am J Pathol* **152**, 879–884.

Bandopadhyay, R., Kingsbury, A. E., Cookson, M. R., Reid, A. R., Evans, I. M., Hope, A. D., Pittman, A. M., Lashley, T., Canet-Aviles, R., Miller, D. W., McLendon, C., Strand, C., Leonard, A. J., Abou-Sleiman, P. M., Healy, D. G., Ariga, H., Wood, N. W., De Silva, R., Revesz, T., Hardy, J. A., and Lees, A. J. (2004). The expression of DJ-1 (PARK7) in normal human CNS and idiopathic Parkinson's disease. *Brain* **127**, 420–430.

Bandyopadhyay, S., and Cookson, M. R. (2004). Evolutionary and functional relationships within the DJ1 superfamily. *BMC Evol Biol* **4**, 6.

Baulac, S., LaVoie, M. J., Strahle, J., Schlossmacher, M. G., and Xia, W. (2004). Dimerization of Parkinson's disease-causing DJ-1 and formation of high molecular weight complexes in human brain. *Mol Cell Neurosci* **27**, 236–246.

Beasley, S. A., Hristova, V. A., and Shaw, G. S. (2007). Structure of the Parkin in-between-ring domain provides insights for E3-ligase dysfunction in autosomal recessive Parkinson's disease. *Proc Natl Acad Sci USA* **104**, 3095–3100.

Beilina, A., Van Der Brug, M., Ahmad, R., Kesavapany, S., Miller, D. W., Petsko, G. A., and Cookson, M. R. (2005). Mutations in PTEN-induced putative kinase 1 associated with recessive parkinsonism have differential effects on protein stability. *Proc Natl Acad Sci USA* **102**, 5703–5708.

Betarbet, R., Canet-Aviles, R. M., Sherer, T. B., Mastroberardino, P. G., McLendon, C., Kim, J. H., Lund, S., Na, H. M., Taylor, G., Bence, N. F., Kopito, R., Seo, B. B., Yagi, T., Yagi, A., Klinefelter, G., Cookson, M. R., and Greenamyre, J. T. (2006). Intersecting pathways to neurodegeneration in Parkinson's disease: Effects of the pesticide rotenone on DJ-1, -synuclein, and the ubiquitin–proteasome system. *Neurobiol Dis* **22**, 404–420.

Bisaglia, M., Mammi, S., and Bubacco, L. (2007). Kinetic and structural analysis of the early oxidation products of dopamine. Analysis of the interactions with α-synuclein. *J Biol Chem* **282**, 15597–15605.

Blackinton, J., Ahmad, R., Miller, D. W., van der Brug, M. P., Canet-Aviles, R. M., Hague, S. M., Kaleem, M., and Cookson, M. R. (2005). Effects of DJ-1 mutations and polymorphisms on protein stability and subcellular localization. *Brain Res Mol Brain Res* **134**, 76–83.

Bonifati, V., Rizzu, P., van Baren, M. J., Schaap, O., Breedveld, G. J., Krieger, E., Dekker, M. C., Squitieri, F., Ibanez, P., Joosse, M., van Dongen, J. W., Vanacore, N., van Swieten, J. C., Brice, A., Meco, G., van Duijn, C. M., Oostra, B. A., and Heutink, P. (2003). Mutations in the DJ-1 gene associated with autosomal recessive early-onset parkinsonism. *Science* 299, 256–259.

Bonifati, V., Oostra, B. A., and Heutink, P. (2004). Linking DJ-1 to neurodegeneration offers novel insights for understanding the pathogenesis of Parkinson's disease. *J Mol Med* 82, 163–174.

Bosco, D. A., Fowler, D. M., Zhang, Q., Nieva, J., Powers, E. T., Wentworth, P., Jr., Lerner, R. A., and Kelly, J. W. (2006). Elevated levels of oxidized cholesterol metabolites in Lewy body disease brains accelerate α-synuclein fibrilization. *Nat Chem Biol* 2, 249–253.

Braak, H., Ghebremedhin, E., Rub, U., Bratzke, H., and Del Tredici, K. (2004). Stages in the development of Parkinson's disease-related pathology. *Cell Tissue Res* 318, 121–134.

Bretaud, S., Allen, C., Ingham, P. W., and Bandmann, O. (2007). p53-dependent neuronal cell death in a DJ-1-deficient zebrafish model of Parkinson's disease. *J Neurochem* 100, 1626–1635.

Bussell, R., Jr., and Eliezer, D. (2003). A structural and functional role for 11-mer repeats in α-synuclein and other exchangeable lipid binding proteins. *J Mol Biol* 329, 763–778.

Cabin, D. E., Gispert-Sanchez, S., Murphy, D., Auburger, G., Myers, R. R., and Nussbaum, R. L. (2005). Exacerbated synucleinopathy in mice expressing A53T SNCA on a Snca null background. *Neurobiol Aging* 26, 25–35.

Canet-Aviles, R. M., Wilson, M. A., Miller, D. W., Ahmad, R., McLendon, C., Bandyopadhyay, S., Baptista, M. J., Ringe, D., Petsko, G. A., and Cookson, M. R. (2004). The Parkinson's disease protein DJ-1 is neuroprotective due to cysteine-sulfinic acid-driven mitochondrial localization. *Proc Natl Acad Sci USA* 101, 9103–9108.

Cappai, R., Leck, S. L., Tew, D. J., Williamson, N. A., Smith, D. P., Galatis, D., Sharples, R. A., Curtain, C. C., Ali, F. E., Cherny, R. A., Culvenor, J. G., Bottomley, S. P., Masters, C. L., Barnham, K. J., and Hill, A. F. (2005). Dopamine promotes α-synuclein aggregation into SDS-resistant soluble oligomers via a distinct folding pathway. *FASEB J* 19, 1377–1379.

Caughey, B., and Lansbury, P. T. (2003). Protofibrils, pores, fibrils, and neurodegeneration: Separating the responsible protein aggregates from the innocent bystanders. *Annu Rev Neurosci* 26, 267–298.

Chandra, S., Chen, X., Rizo, J., Jahn, R., and Sudhof, T. C. (2003). A broken α-helix in folded α-synuclein. *J Biol Chem* 278, 15313–15318.

Chartier-Harlin, M. C., Kachergus, J., Roumier, C., Mouroux, V., Douay, X., Lincoln, S., Levecque, C., Larvor, L., Andrieux, J., Hulihan, M., Waucquier, N., Defebvre, L., Amouyel, P., Farrer, M., and Destee, A. (2004). α-Synuclein locus duplication as a cause of familial Parkinson's disease. *Lancet* 364, 1167–1169.

Chen, L., and Feany, M. B. (2005). α-Synuclein phosphorylation controls neurotoxicity and inclusion formation in a *Drosophila* model of Parkinson disease. *Nat Neurosci* 8, 657–663.

Choi, J., Sullards, M. C., Olzmann, J. A., Rees, H. D., Weintraub, S. T., Bostwick, D. E., Gearing, M., Levey, A. I., Chin, L. S., and Li, L. (2006). Oxidative damage of DJ-1 is linked to sporadic Parkinson and Alzheimer diseases. *J Biol Chem* 281, 10816–10824.

Choi, P., Snyder, H., Petrucelli, L., Theisler, C., Chong, M., Zhang, Y., Lim, K., Chung, K. K., Kehoe, K., D'Adamio, L., Lee, J. M., Cochran, E., Bowser, R., Dawson, T. M., and Wolozin, B. (2003). SEPT5_v2 is a parkin-binding protein. *Brain Res Mol Brain Res* 117, 179–189.

Choi, W., Zibaee, S., Jakes, R., Serpell, L. C., Davletov, B., Crowther, R. A., and Goedert, M. (2004). Mutation E46K increases phospholipid binding and assembly into filaments of human α-synuclein. *FEBS Lett* 576, 363–368.

Chung, K. K., Zhang, Y., Lim, K. L., Tanaka, Y., Huang, H., Gao, J., Ross, C. A., Dawson, V. L., and Dawson, T. M. (2001). Parkin ubiquitinates the α-synuclein-interacting protein, synphilin-1: Implications for Lewy-body formation in Parkinson disease. *Nat Med* 7, 1144–1150.

Chung, K. K., Thomas, B., Li, X., Pletnikova, O., Troncoso, J. C., Marsh, L., Dawson, V. L., and Dawson, T. M. (2004). S-nitrosylation of parkin regulates ubiquitination and compromises parkin's protective function. *Science* 304, 1328–1331.

Clements, C. M., McNally, R. S., Conti, B. J., Mak, T. W., and Ting, J. P. (2006). DJ-1, a cancer- and Parkinson's disease-associated protein, stabilizes the antioxidant transcriptional master regulator Nrf2. *Proc Natl Acad Sci USA* 103, 15091–15096.

Cole, N. B., Murphy, D. D., Lebowitz, J., Di Noto, L., Levine, R. L., and Nussbaum, R. L. (2005). Metal-catalyzed oxidation of α-synuclein: Helping to define the relationship between oligomers, protofilaments and filaments. *J Biol Chem* 280, 9678–9690.

Conway, K. A., Harper, J. D., and Lansbury, P. T., Jr. (1998). Accelerated *in vitro* fibril formation by a mutant α-synuclein linked to early-onset Parkinson disease. *Nat Med* 4, 1318–1320.

Conway, K. A., Harper, J. D., and Lansbury, P. T., Jr. (2000a). Fibrils formed *in vitro* from α-synuclein and two mutant forms linked to Parkinson's disease are typical amyloid. *Biochemistry* 39, 2552–2563.

Conway, K. A., Lee, S.-J., Rochet, J.-C., Ding, T. T., Williamson, R. E., and Lansbury, P. T., Jr. (2000b). Acceleration of oligomerization, not fibrillization, is a shared property of both α-synuclein mutations linked to early-onset Parkinson's disease: Implications for pathogenesis and therapy. *Proc Natl Acad Sci USA* 97, 571–576.

Conway, K. A., Rochet, J.-C., Bieganski, R. M., and Lansbury, P. T. (2001). Kinetic stabilization of the α-synuclein protofibril by a dopamine-α-synuclein adduct. *Science* 294, 1346–1349.

Cookson, M. R. (2005). The biochemistry of Parkinson's disease. *Annu Rev Biochem* 74, 29–52.

Crowther, R. A., Jakes, R., Spillantini, M. G., and Goedert, M. (1998). Synthetic filaments assembled from C-terminally truncated α-synuclein. *FEBS Lett* 436, 309–312.

Das, C., Hoang, Q. Q., Kreinbring, C. A., Luchansky, S. J., Meray, R. K., Ray, S. S., Lansbury, P. T., Ringe, D., and Petsko, G. A. (2006). Structural basis for conformational plasticity of the Parkinson's disease-associated ubiquitin hydrolase UCH-L1. *Proc Natl Acad Sci USA* 103, 4675–4680.

Davidson, W. S., Jonas, A., Clayton, D. F., and George, J. M. (1998). Stabilization of α-synuclein secondary structure upon binding to synthetic membranes. *J Biol Chem* 273, 9443–9449.

Del Mar, C., Greenbaum, E. A., Mayne, L., Englander, S. W., and Woods, V. L., Jr. (2005). Structure and properties of α-synuclein and other amyloids determined at the amino acid level. *Proc Natl Acad Sci USA* 102, 15477–15482.

Der-Sarkissian, A., Jao, C. C., Chen, J., and Langen, R. (2003). Structural organization of α-synuclein fibrils studied by site-directed spin labeling. *J Biol Chem* 278, 37530–37535.

Ding, T. T., Lee, S.-J., Rochet, J.-C., and Lansbury, P. T., Jr. (2002). Annular α-synuclein protofibrils are produced when spherical protofibrils are incubated in solution or bound to brain-derived membranes. *Biochemistry* 41, 10209–10217.

El-Agnaf, O. M., Jakes, R., Curran, M. D., and Wallace, A. (1998). Effects of the mutations Ala30 to Pro and Ala53 to Thr on the physical and morphological properties of α-synuclein protein implicated in Parkinson's disease. *FEBS Lett* 440, 67–70.

Eliezer, D., Kutluay, E., Bussell, R., Jr., and Browne, G. (2001). Conformational properties of α-synuclein in its free and lipid-associated states. *J Mol Biol* 307, 1061–1073.

Follmer, C., Romao, L., Einsiedler, C. M., Porto, T. C., Lara, F. A., Moncores, M., Weissmuller, G., Lashuel, H. A., Lansbury, P., Neto, V. M., Silva, J. L., and Foguel, D. (2007). Dopamine affects the stability, hydration, and packing of protofibrils and fibrils of the wild type and variants of α-synuclein. *Biochemistry* 46, 472–482.

Forno, L. S. (1996). Neuropathology of Parkinson's disease. *J Neuropathol Exp Neurol* 55, 259–272.

Fredenburg, R. A., Rospigliosi, C., Meray, R. K., Kessler, J. C., Lashuel, H. A., Eliezer, D., and Lansbury, P. T., Jr. (2007). The impact of the E46K mutation on the properties of α-synuclein in its monomeric and oligomeric states. *Biochemistry* 46, 7107–7118.

Fujiwara, H., Hasegawa, M., Dohmae, N., Kawashima, A., Masliah, E., Goldberg, M. S., Shen, J., Takio, K., and Iwatsubo, T. (2002). α-Synuclein is phosphorylated in synucleinopathy lesions. *Nat Cell Biol* 4, 160–164.

George, J. M., Jin, H., Woods, W. S., and Clayton, D. F. (1995). Characterization of a novel protein regulated during the critical period for song learning in the zebra finch. *Neuron* 15, 361–372.

Giasson, B. I., Uryu, K., Trojanowski, J. Q., and Lee, V. M.-Y. (1999). Mutant and wild type human α-synucleins assemble into elongated filaments with distinct morphologies *in vitro*. *J Biol Chem* 274, 7619–7622.

Giasson, B. I., Duda, J. E., Murray, I. V., Chen, Q., Souza, J. M., Hurtig, H. I., Ischiropoulos, H., Trojanowski, J. Q., and Lee, V. M. (2000). Oxidative damage linked to neurodegeneration by selective α-synuclein nitration in synucleinopathy lesions. *Science* 290, 985–989.

Giasson, B. I., Murray, I. V., Trojanowski, J. Q., and Lee, V. M. (2001). A hydrophobic stretch of 12 amino acid residues in the middle of α-synuclein is essential for filament assembly. *J Biol Chem* 276, 2380–2386.

Glaser, C. B., Yamin, G., Uversky, V. N., and Fink, A. L. (2005). Methionine oxidation, α-synuclein and Parkinson's disease. *Biochim Biophys Acta* 1703, 157–169.

Gloeckner, C. J., Kinkl, N., Schumacher, A., Braun, R. J., O'Neill, E., Meitinger, T., Kolch, W., Prokisch, H., and Ueffing, M. (2006). The Parkinson disease causing LRRK2 mutation I2020T is associated with increased kinase activity. *Hum Mol Genet* 15, 223–232.

Gorner, K., Holtorf, E., Odoy, S., Nuscher, B., Yamamoto, A., Regula, J. T., Beyer, K., Haass, C., and Kahle, P. J. (2004). Differential effects of Parkinson's disease-associated mutations on stability and folding of DJ-1. *J Biol Chem* 279, 6943–6951.

Gorner, K., Holtorf, E., Waak, J., Pham, T. T., Vogt-Weisenhorn, D. M., Wurst, W., Haass, C., and Kahle, P. J. (2007). Structural determinants of the C-terminal helix-kink-helix motif essential for protein stability and survival promoting activity of DJ-1. *J Biol Chem* 282, 13680–13691.

Greenbaum, E. A., Graves, C. L., Mishizen-Eberz, A. J., Lupoli, M. A., Lynch, D. R., Englander, S. W., Axelsen, P. H., and Giasson, B. I. (2005). The E46K mutation in α-synuclein increases amyloid fibril formation. *J Biol Chem* **280**, 7800–7807.

Hampe, C., Ardila-Osorio, H., Fournier, M., Brice, A., and Corti, O. (2006). Biochemical analysis of Parkinson's disease-causing variants of Parkin, an E3 ubiquitin–protein ligase with monoubiquitylation capacity. *Hum Mol Genet* **15**, 2059–2075.

Hasegawa, M., Fujiwara, H., Nonaka, T., Wakabayashi, K., Takahashi, H., Lee, V. M., Trojanowski, J. Q., Mann, D., and Iwatsubo, T. (2002). Phosphorylated α-synuclein is ubiquitinated in α-synucleinopathy lesions. *J Biol Chem* **277**, 49071–49076.

Hashimoto, M., Hsu, L. J., Sisk, A., Xia, Y., Takeda, A., Sundsmo, M., and Masliah, E. (1998). Human recombinant NACP/α-synuclein is aggregated and fibrillated *in vitro*: Relevance for Lewy body disease. *Brain Res* **799**, 301–306.

Hashimoto, M., Hsu, L. J., Xia, Y., Takeda, A., Sisk, A., Sundsmo, M., and Masliah, E. (1999). Oxidative stress induces amyloid-like aggregate formation of NACP/α-synuclein *in vitro*. *Neuroreport* **10**, 717–721.

Hashimoto, M., Rockenstein, E., Mante, M., Mallory, M., and Masliah, E. (2001). β-Synuclein inhibits α-synuclein aggregation: a possible role as an antiparkinsonian factor. *Neuron* **32**, 213–223.

Heise, H., Hoyer, W., Becker, S., Andronesi, O. C., Riedel, D., and Baldus, M. (2005). Molecular-level secondary structure, polymorphism, and dynamics of full-length α-synuclein fibrils studied by solid-state NMR. *Proc Natl Acad Sci USA* **102**, 15871–15876.

Hering, R., Strauss, K. M., Tao, X., Bauer, A., Woitalla, D., Mietz, E. M., Petrovic, S., Bauer, P., Schaible, W., Muller, T., Schols, L., Klein, C., Berg, D., Meyer, P. T., Schulz, J. B., Wollnik, B., Tong, L., Kruger, R., and Riess, O. (2004). Novel homozygous p, E64D mutation in DJ1 in early onset Parkinson disease (PARK7). *Hum Mutat* **24**, 321–329.

Hodara, R., Norris, E. H., Giasson, B. I., Mishizen-Eberz, A. J., Lynch, D. R., Lee, V. M., and Ischiropoulos, H. (2004). Functional consequences of α-synuclein tyrosine nitration: diminished binding to lipid vesicles and increased fibril formation. *J Biol Chem* **279**, 47746–47753.

Hokenson, M. J., Uversky, V. N., Goers, J., Yamin, G., Munishkina, L. A., and Fink, A. L. (2004). Role of individual methionines in the fibrillation of methionine-oxidized α-synuclein. *Biochemistry* **43**, 4621–4633.

Honbou, K., Suzuki, N. N., Horiuchi, M., Niki, T., Taira, T., Ariga, H., and Inagaki, F. (2003). The crystal structure of DJ-1, a protein related to male fertility and Parkinson's disease. *J Biol Chem* **278**, 31380–31384.

Hoyer, W., Antony, T., Cherny, D., Heim, G., Jovin, T. M., and Subramaniam, V. (2002). Dependence of α-synuclein aggregate morphology on solution conditions. *J Mol Biol* **322**, 383–393.

Hoyer, W., Cherny, D., Subramaniam, V., and Jovin, T. M. (2004). Impact of the acidic C-terminal region comprising amino acids 109-140 on α-synuclein aggregation *in vitro*. *Biochemistry* **43**, 16233–16242.

Huai, Q., Sun, Y., Wang, H., Chin, L. S., Li, L., Robinson, H., and Ke, H. (2003). Crystal structure of DJ-1/RS and implication on familial Parkinson's disease. *FEBS Lett* **549**, 171–175.

Hulleman, J. D., Mirzaei, H., Guigard, E., Taylor, K. L., Ray, S. S., Kay, C. M., Regnier, F. E., and Rochet, J. C. (2007). Destabilization of DJ-1 by familial substitution and oxidative modifications: Implications for Parkinson's disease. *Biochemistry* **46**, 5776–5789.

Imai, Y., Soda, M., Inoue, H., Hattori, N., Mizuno, Y., and Takahashi, R. (2001). An unfolded putative transmembrane polypeptide, which can lead to endoplasmic reticulum stress, is a substrate of Parkin. *Cell* **105**, 891–902.

Ito, G., Okai, T., Fujino, G., Takeda, K., Ichijo, H., Katada, T., and Iwatsubo, T. (2007). GTP binding is essential to the protein kinase activity of LRRK2, a causative gene product for familial Parkinson's disease. *Biochemistry* **46**, 1380–1388.

Iwai, A., Masliah, E., Yoshimoto, M., Ge, N., Flanagan, L., de Silva, H. A., Kittel, A., and Saitoh, T. (1995). The precursor protein of non-A beta component of Alzheimer's disease amyloid is a presynaptic protein of the central nervous system. *Neuron* **14**, 467–475.

Jaleel, M., Nichols, R. J., Deak, M., Campbell, D. G., Gillardon, F., Knebel, A., and Alessi, D. R. (2007). LRRK2 phosphorylates moesin at threonine-558: Characterization of how Parkinson's disease mutants affect kinase activity. *Biochem J* **405**, 307–317.

Jao, C. C., Der-Sarkissian, A., Chen, J., and Langen, R. (2004). Structure of membrane-bound α-synuclein studied by site-directed spin labeling. *Proc Natl Acad Sci USA* **101**, 8331–8336.

Jin, J., Meredith, G. E., Chen, L., Zhou, Y., Xu, J., Shie, F. S., Lockhart, P., and Zhang, J. (2005). Quantitative proteomic analysis of mitochondrial proteins: Relevance to Lewy body formation and Parkinson's disease. *Brain Res Mol Brain Res* **134**, 119–138.

Jo, E., McLaurin, J., Yip, C. M., St. George-Hyslop, P., and Fraser, P. E. (2000). α-Synuclein membrane interactions and lipid specificity. *J Biol Chem* **275**, 34328–34334.

Jo, E., Darabie, A. A., Han, K., Tandon, A., Fraser, P. E., and McLaurin, J. (2004). α-Synuclein-synaptosomal membrane interactions: Implications for fibrillogenesis. *Eur J Biochem* **271**, 3180–3189.

Johnson, S. M., Wiseman, R. L., Sekijima, Y., Green, N. S., Adamski-Werner, S. L., and Kelly, J. W. (2005). Native state kinetic stabilization as a strategy to ameliorate protein misfolding diseases: A focus on the transthyretin amyloidoses. *Acc Chem Res* **38**, 911–921.

Johnston, J. A., Ward, C. L., and Kopito, R. R. (1998). Aggresomes: A cellular response to misfolded proteins. *J Cell Biol* **143**, 1883–1898.

Junn, E., Taniguchi, H., Jeong, B. S., Zhao, X., Ichijo, H., and Mouradian, M. M. (2005). Interaction of DJ-1 with Daxx inhibits apoptosis signal-regulating kinase 1 activity and cell death. *Proc Natl Acad Sci USA* **102**, 9691–9696.

Kayed, R., Sokolov, Y., Edmonds, B., McIntire, T. M., Milton, S. C., Hall, J. E., and Glabe, C. G. (2004). Permeabilization of lipid bilayers is a common conformation-dependent activity of soluble amyloid oligomers in protein misfolding diseases. *J Biol Chem* **279**, 46363–46366.

Kheterpal, I., Chen, M., Cook, K. D., and Wetzel, R. (2006). Structural differences in Abeta amyloid protofibrils and fibrils mapped by hydrogen exchange – mass spectrometry with on-line proteolytic fragmentation. *J Mol Biol* **361**, 785–795.

Kim, R. H., Peters, M., Jang, Y., Shi, W., Pintilie, M., Fletcher, G. C., Deluca, C., Liepa, J., Zhou, L., Snow, B., Binari, R. C., Manoukian, A. S., Bray, M. R., Liu, F. F., Tsao, M. S., and Mak, T. W. (2005a). DJ-1, a novel regulator of the tumor suppressor PTEN. *Cancer Cell* **7**, 263–273.

Kim, R. H., Smith, P. D., Aleyasin, H., Hayley, S., Mount, M. P., Pownall, S., Wakeham, A., You-Ten, A. J., Kalia, S. K., Horne, P., Westaway, D., Lozano, A. M., Anisman, H., Park, D. S., and Mak, T. W. (2005b). Hypersensitivity of DJ-1-deficient mice to 1-methyl-4-phenyl-1,2,3,6-tetrahydropyridine (MPTP) and oxidative stress. *Proc Natl Acad Sci USA* **102**, 5215–5220.

Kim, Y. S., Laurine, E., Woods, W., and Lee, S. J. (2006). A novel mechanism of interaction between α-synuclein and biological membranes. *J Mol Biol* **360**, 386–397.

Krishnan, S., Chi, E. Y., Wood, S. J., Kendrick, B. S., Li, C., Garzon-Rodriguez, W., Wypych, J., Randolph, T. W., Narhi, L. O., Biere, A. L., Citron, M., and Carpenter, J. F. (2003). Oxidative dimer formation is the critical rate-limiting step for Parkinson's disease α-synuclein fibrillogenesis. *Biochemistry* **42**, 829–837.

Kruger, R., Kuhn, W., Muller, T., Woitalla, D., Graeber, M., Kosel, S., Przuntek, H., Epplen, J. T., Schols, L., and Riess, O. (1998). Ala30Pro mutation in the gene encoding α-synuclein in Parkinson's disease. *Nat Genet* **18**, 106–108.

Lansbury, P. T., Jr. (2004). Back to the future: the 'old-fashioned' way to new medications for neurodegeneration. *Nat Med* **10**, Suppl, S51–S57.

Lashuel, H. A., Hartley, D., Petre, B. M., Walz, T., and Lansbury, P. T., Jr. (2002a). Neurodegenerative disease: Amyloid pores from pathogenic mutations. *Nature* **418**, 291.

Lashuel, H. A., Petre, B. M., Wall, J., Simon, M., Nowak, R. J., Walz, T., and Lansbury, P. T., Jr. (2002b). α-Synuclein, especially the Parkinson's disease-associated mutants, forms pore-like annular and tubular protofibrils. *J Mol Biol* **322**, 1089–1102.

LaVoie, M. J., Ostaszewski, B. L., Weihofen, A., Schlossmacher, M. G., and Selkoe, D. J. (2005). Dopamine covalently modifies and functionally inactivates parkin. *Nat Med* **11**, 1214–1221.

Lee, H.-J., Choi, C., and Lee, S.-J. (2002). Membrane-bound α-synuclein has a high aggregation propensity and the ability to seed the aggregation of the cytosolic form. *J Biol Chem* **277**, 671–678.

Lee, S. J., Kim, S. J., Kim, I. K., Ko, J., Jeong, C. S., Kim, G. H., Park, C., Kang, S. O., Suh, P. G., Lee, H. S., and Cha, S. S. (2003). Crystal structures of human DJ-1 and *Escherichia coli* Hsp31, which share an evolutionarily conserved domain. *J Biol Chem* **278**, 44552–44559.

Leroy, E., Boyer, R., Auburger, G., Leube, B., Ulm, G., Mezey, E., Harta, G., Brownstein, M. J., Jonnalagada, S., Chernova, T., Dehejia, A., Lavedan, C., Gasser, T., Steinbach, P. J., Wilkinson, K. D., and Polymeropoulos, M. H. (1998). The ubiquitin pathway in Parkinson's disease. *Nature* **395**, 451–452.

Lewis, P. A., Greggio, E., Beilina, A., Jain, S., Baker, A., and Cookson, M. R. (2007). The R1441C mutation of LRRK2 disrupts GTP hydrolysis. *Biochem Biophys Res Commun* **357**, 668–671.

Li, H. T., Lin, D. H., Luo, X. Y., Zhang, F., Ji, L. N., Du, H. N., Song, G. Q., Hu, J., Zhou, J. W., and Hu, H. Y. (2005a). Inhibition of α-synuclein fibrillization by dopamine analogs via reaction with the amino groups of α-synuclein. Implication for dopaminergic neurodegeneration. *FEBS J* **272**, 3661–3672.

Li, J., Uversky, V. N., and Fink, A. L. (2001). Effect of familial Parkinson's disease point mutations A30P and A53T on the structural properties, aggregation, and fibrillation of human α-synuclein. *Biochemistry* **40**, 11604–11613.

Li, J., Zhu, M., Manning-Bog, A. B., Di Monte, D. A., and Fink, A. L. (2004). Dopamine and L-dopa disaggregate amyloid fibrils: implications for Parkinson's and Alzheimer's disease. *FASEB J* **18**, 962–964.

Li, W., West, N., Colla, E., Pletnikova, O., Troncoso, J. C., Marsh, L., Dawson, T. M., Jakala, P., Hartmann, T., Price, D. L., and Lee, M. K. (2005b). Aggregation promoting C-terminal truncation of α-synuclein is a normal cellular process and is enhanced by the familial Parkinson's disease-linked mutations. *Proc Natl Acad Sci USA* **102**, 2162–2167.

Li, X., Tan, Y. C., Poulose, S., Olanow, C. W., Huang, X. Y., and Yue, Z. (2007). Leucine-rich repeat kinase 2 (LRRK2)/PARK8 possesses GTPase activity that is altered in familial Parkinson's disease R1441C/G mutants. *J Neurochem* 103, 238–247.

Liu, C. W., Giasson, B. I., Lewis, K. A., Lee, V. M., Demartino, G. N., and Thomas, P. J. (2005). A precipitating role for truncated α-synuclein and the proteasome in α-synuclein aggregation: Implications for pathogenesis of Parkinson disease. *J Biol Chem* 280, 22670–22678.

Liu, F., Nguyen, J. L., Hulleman, J. D., Li, L., and Rochet, J. C., (2008). Mechanisms of DJ-1 neuroprotection in a cellular model of Parkinsons's disease. *J Neurochem* 105, 2435–2453.

Liu, Y., Fallon, L., Lashuel, H. A., Liu, Z., and Lansbury, P. T., Jr. (2002). The UCH-L1 gene encodes two opposing enzymatic activities that affect α-synuclein degradation and Parkinson's disease susceptibility. *Cell* 111, 209–218.

Luhrs, T., Ritter, C., Adrian, M., Riek-Loher, D., Bohrmann, B., Dobeli, H., Schubert, D., and Riek, R. (2005). 3D structure of Alzheimer's amyloid-beta(1-42) fibrils. *Proc Natl Acad Sci USA* 102, 17342–17347.

Luk, K. C., Hyde, E. G., Trojanowski, J. Q., and Lee, V. M. (2007). Sensitive fluorescence polarization technique for rapid screening of α-synuclein oligomerization/fibrillization inhibitors. *Biochemistry* 46, 12522–12529.

Luzon-Toro, B., de la Torre, E. R., Delgado, A., Perez-Tur, J., and Hilfiker, S. (2007). Mechanistic insight into the dominant mode of the Parkinson's disease-associated G2019S LRRK2 mutation. *Hum Mol Genet* 16, 2031–2039.

Manevich, Y., Feinstein, S. I., and Fisher, A. B. (2004). Activation of the antioxidant enzyme 1-CYS peroxiredoxin requires glutathionylation mediated by heterodimerization with pi GST. *Proc Natl Acad Sci USA* 101, 3780–3785.

Maraganore, D. M., Farrer, M. J., Hardy, J. A., Lincoln, S. J., McDonnell, S. K., and Rocca, W. A. (1999). Case–control study of the ubiquitin carboxyterminal hydrolase L1 gene in Parkinson's disease. *Neurology* 53, 1858–1860.

Maroteaux, L., Campanelli, J. T., and Scheller, R. H. (1988). Synuclein: A neuron-specific protein localized to the nucleus and presynaptic nerve terminal. *J Neurosci* 8, 2804–2815.

Martinat, C., Shendelman, S., Jonason, A., Leete, T., Beal, M. F., Yang, L., Floss, T., and Abeliovich, A. (2004). Sensitivity to oxidative stress in DJ-1-deficient dopamine neurons: An ES-derived cell model of primary parkinsonism. *PLoS Biol* 2, e327.

Masliah, E., Rockenstein, E., Veinbergs, I., Mallory, M., Hashimoto, M., Takeda, A., Sagara, Y., Sisk, A., and Mucke, L. (2000). Dopaminergic loss and inclusion body formation in α-synuclein mice: Implications for neurodegenerative disorders. *Science* 287, 1265–1269.

Masuda, M., Suzuki, N., Taniguchi, S., Oikawa, T., Nonaka, T., Iwatsubo, T., Hisanaga, S., Goedert, M., and Hasegawa, M. (2006). Small molecule inhibitors of α-synuclein filament assembly. *Biochemistry* 45, 6085–6094.

Mata, I. F., Wedemeyer, W. J., Farrer, M. J., Taylor, J. P., and Gallo, K. A. (2006). LRRK2 in Parkinson's disease: Protein domains and functional insights. *Trends Neurosci* 29, 286–293.

Matsuda, N., Kitami, T., Suzuki, T., Mizuno, Y., Hattori, N., and Tanaka, K. (2006). Diverse effects of pathogenic mutations of Parkin that catalyze multiple monoubiquitylation *in vitro*. *J Biol Chem* 281, 3204–3209.

Mazzulli, J. R., Mishizen, A. J., Giasson, B. I., Lynch, D. R., Thomas, S. A., Nakashima, A., Nagatsu, T., Ota, A., and Ischiropoulos, H. (2006). Cytosolic catechols inhibit α-synuclein aggregation and facilitate the formation of intracellular soluble oligomeric intermediates. *J Neurosci* 26, 10068–10078.

Menzies, F. M., Yenisetti, S. C., and Min, K. T. (2005). Roles of *Drosophila* DJ-1 in survival of dopaminergic neurons and oxidative stress. *Curr Biol* 15, 1578–1582.

Meulener, M., Whitworth, A. J., Armstrong-Gold, C. E., Rizzu, P., Heutink, P., Wes, P. D., Pallanck, L. J., and Bonini, N. M. (2005). Drosophila DJ-1 mutants are selectively sensitive to environmental toxins associated with Parkinson's disease. *Curr Biol* 15, 1572–1577.

Meulener, M. C., Xu, K., Thompson, L., Ischiropoulos, H., and Bonini, N. M. (2006). Mutational analysis of DJ-1 in *Drosophila* implicates functional inactivation by oxidative damage and aging. *Proc Natl Acad Sci USA* 103, 12517–12522.

Mirzaei, H., Schieler, J. L., Rochet, J.-C., and Regnier, F. (2006). Identification of rotenone-induced modifications in α-synuclein using affinity pull-down and tandem mass spectrometry. *Anal Chem* 78, 2422–2431.

Moore, D. J., Zhang, L., Dawson, T. M., and Dawson, V. L. (2003). A missense mutation (L166P) in DJ-1, linked to familial Parkinson's disease, confers reduced protein stability and impairs homooligomerization. *J Neurochem* 87, 1558–1567.

Moore, D. J., Zhang, L., Troncoso, J., Lee, M. K., Hattori, N., Mizuno, Y., Dawson, T. M., and Dawson, V. L. (2005). Association of DJ-1 and parkin mediated by pathogenic DJ-1 mutations and oxidative stress. *Hum Mol Genet* 14, 71–84.

Murray, I. V., Giasson, B. I., Quinn, S. M., Koppaka, V., Axelsen, P. H., Ischiropoulos, H., Trojanowski, J. Q., and Lee, V. M. (2003). Role of α-synuclein carboxyterminus on fibril formation *in vitro*. *Biochemistry* 42, 8530–8540.

Narayanan, V., and Scarlata, S. (2001). Membrane binding and self-association of α-synucleins. *Biochemistry* 40, 9927–9934.

Narhi, L., Wood, S. J., Steavenson, S., Jiang, Y., Wu, G. M., Anafi, D., Kaufman, S. A., Martin, F., Sitney, K., Denis, P., Louis, J. C., Wypych, J., Biere, A. L., and Citron, M. (1999). Both familial Parkinson's disease mutations accelerate α-synuclein aggregation. *J Biol Chem* **274**, 9843–9846.

Necula, M., Chirita, C. N., and Kuret, J. (2003). Rapid anionic micelle-mediated α-synuclein fibrillization *in vitro. J Biol Chem* **278**, 46674–46680.

Negro, A., Brunati, A. M., Donella-Deana, A., Massimino, M. L., and Pinna, L. A. (2002). Multiple phosphorylation of α-synuclein by protein tyrosine kinase Syk prevents eosin-induced aggregation. *FASEB J* **16**, 210–212.

Neumann, M., Kahle, P. J., Giasson, B. I., Ozmen, L., Borroni, E., Spooren, W., Muller, V., Odoy, S., Fujiwara, H., Hasegawa, M., Iwatsubo, T., Trojanowski, J. Q., Kretzschmar, H. A., and Haass, C. (2002). Misfolded proteinase K-resistant hyper-phosphorylated α-synuclein in aged transgenic mice with locomotor deterioration and in human α-synucleinopathies. *J Clin Invest* **110**, 1429–1439.

Nishikawa, K., Li, H., Kawamura, R., Osaka, H., Wang, Y. L., Hara, Y., Hirokawa, T., Manago, Y., Amano, T., Noda, M., Aoki, S., and Wada, K. (2003). Alterations of structure and hydrolase activity of parkinsonism-associated human ubiquitin carboxyl-terminal hydrolase L1 variants. *Biochem Biophys Res Commun* **304**, 176–183.

Norris, E. H., Giasson, B. I., Ischiropoulos, H., and Lee, V. M. (2003). Effects of oxidative and nitrative challenges on α-synuclein fibrillogenesis involve distinct mechanisms of protein modifications. *J Biol Chem* **278**, 27230–27240.

Norris, E. H., Giasson, B. I., Hodara, R., Xu, S., Trojanowski, J. Q., Ischiropoulos, H., and Lee, V. M. (2005). Reversible inhibition of α-synuclein fibrilliza-tion by dopaminochrome-mediated conformational alterations. *J Biol Chem* **280**, 21212–21219.

Olzmann, J. A., Brown, K., Wilkinson, K. D., Rees, H. D., Huai, Q., Ke, H., Levey, A. I., Li, L., and Chin, L. S. (2004). Familial Parkinson's disease-associated L166P mutation disrupts DJ-1 protein folding and function. *J Biol Chem* **279**, 8506–8515.

Ono, K., and Yamada, M. (2006). Antioxidant compounds have potent anti-fibrillogenic and fibril-destabilizing effects for α-synuclein fibrils *in vitro. J Neurochem* **97**, 105–115.

Ono, K., Hirohata, M., and Yamada, M. (2007). Anti-fibrillogenic and fibril-destabilizing activities of anti-Parkinsonian agents for α-synuclein fibrils *in vitro. J Neurosci Res* **85**, 1547–1557.

Paik, S. R., Shin, H. J., and Lee, J. H. (2000). Metal-catalyzed oxidation of α-synuclein in the presence of copper(II) and hydrogen peroxide. *Arch Biochem Biophys* **378**, 269–277.

Paisan-Ruiz, C., Jain, S., Evans, E. W., Gilks, W. P., Simon, J., van der Brug, M., Lopez de Munain, A., Aparicio, S., Gil, A. M., Khan, N., Johnson, J., Martinez, J. R., Nicholl, D., Carrera, I. M., Pena, A. S., de Silva, R., Lees, A., Marti-Masso, J. F., Perez-Tur, J., Wood, N. W., and Singleton, A. B. (2004). Cloning of the gene containing mutations that cause PARK8-linked Parkinson's disease. *Neuron* **44**, 595–600.

Park, J., Kim, S. Y., Cha, G. H., Lee, S. B., Kim, S., and Chung, J. (2005). *Drosophila* DJ-1 mutants show oxi-dative stress-sensitive locomotive dysfunction. *Gene* **361**, 133–139.

Park, J. Y., and Lansbury, P. T., Jr. (2003). β-Synuclein inhibits formation of α-synuclein protofibrils: A pos-sible therapeutic strategy against Parkinson's disease. *Biochemistry* **42**, 3696–3700.

Periquet, M., Fulga, T., Myllykangas, L., Schlossmacher, M. G., and Feany, M. B. (2007). Aggregated α-synuclein mediates dopaminergic neuro-toxicity *in vivo. J Neurosci* **27**, 3338–3346.

Perrin, R. J., Woods, W. S., Clayton, D. F., and George, J. M. (2000). Interaction of human α-synuclein and Parkinson's disease variants with phospholipids. Structural analysis using site-directed mutagenesis. *J Biol Chem* **275**, 34393–34398.

Perrin, R. J., Woods, W. S., Clayton, D. F., and George, J. M. (2001). Exposure to long chain polyun-saturated fatty acids triggers rapid multimerization of synucleins. *J Biol Chem* **276**, 41958–41962.

Petkova, A. T., Leapman, R. D., Guo, Z., Yau, W. M., Mattson, M. P., and Tycko, R. (2005). Self-propagating, molecular-level polymorphism in Alzheimer's beta-amyloid fibrils. *Science* **307**, 262–265.

Petkova, A. T., Yau, W. M., and Tycko, R. (2006). Experimental constraints on quaternary structure in Alzheimer's beta-amyloid fibrils. *Biochemistry* **45**, 498–512.

Pollanen, M. S., Dickson, D. W., and Bergeron, C. (1993). Pathology and biology of the Lewy body. *J Neuropath Exp Neurol* **52**, 183–191.

Polymeropoulos, M. H., Lavedan, C., Leroy, E., Ide, S. E., Dehejia, A., Dutra, A., Pike, B., Root, H., Rubenstein, J., Boyer, R., Stenroos, E. S., Chandradekharappa, S., Athanassiadou, A., Papapetropoulos, T., Johnson, W. G., Lazzarini, A. M., Duvoisin, R. C., Di Iorio, G., Golbe, L. I., and Nussbaum, R. L. (1997). Mutation in the α-synuclein gene identified in families with Parkinson's disease. *Science* **276**, 2045–2047.

Pridgeon, J. W., Olzmann, J. A., Chin, L. S., and Li, L. (2007). PINK1 protects against oxidative stress by phosphorylating mitochondrial chaperone TRAP1. *PLoS Biol* **5**, e172.

Przedborski, S., Chen, Q., Vila, M., Giasson, B. I., Djaldatti, R., Vukosavic, S., Souza, J. M., Jackson-Lewis, V., Lee, V. M., and Ischiropoulos, H. (2001). Oxidative post-translational modifications of α-synuclein in the 1-methyl-4-phenyl-1,2,3,6-tetrahydropyridine (MPTP) mouse model of Parkinson's disease. *J Neurochem* **76**, 637–640.

Qin, Z., Hu, D., Han, S., Reaney, S. H., Di Monte, D. A., and Fink, A. L. (2007). Effect of 4-hydroxy-2-nonenal modification on α-synuclein aggregation. *J Biol Chem* **282**, 5862–5870.

Quist, A., Doudevski, I., Lin, H., Azimova, R., Ng, D., Frangione, B., Kagan, B., Ghiso, J., and Lal, R. (2005). Amyloid ion channels: A common structural link for protein-misfolding disease. *Proc Natl Acad Sci USA* **102**, 10427–10432.

Ray, S. S., Nowak, R. J., Brown, R. H., Jr., and Lansbury, P. T., Jr. (2005). Small-molecule-mediated stabilization of familial amyotrophic lateral sclerosis-linked superoxide dismutase mutants against unfolding and aggregation. *Proc Natl Acad Sci USA* **102**, 3639–3644.

Rochet, J. C. (2007). Novel therapeutic strategies for the treatment of protein-misfolding diseases. *Expert Rev Mol Med* **9**, 1–34.

Rochet, J. C., Conway, K. A., and Lansbury, P. T., Jr. (2000). Inhibition of fibrillization and accumulation of prefibrillar oligomers in mixtures of human and mouse α-synuclein. *Biochemistry* **39**, 10619–10626.

Rochet, J. C., Outeiro, T. F., Conway, K. A., Ding, T. T., Volles, M. J., Lashuel, H. A., Bieganski, R. M., Lindquist, S. L., and Lansbury, P. T. (2004). Interactions among α-synuclein, dopamine, and biomembranes: Some clues for understanding neurodegeneration in Parkinson's disease. *J Mol Neurosci* **23**, 23–34.

Sato, T., Kienlen-Campard, P., Ahmed, M., Liu, W., Li, H., Elliott, J. I., Aimoto, S., Constantinescu, S. N., Octave, J. N., and Smith, S. O. (2006). Inhibitors of amyloid toxicity based on beta-sheet packing of Abeta40 and Abeta42. *Biochemistry* **45**, 5503–5516.

Sciarretta, K. L., Boire, A., Gordon, D. J., and Meredith, S. C. (2006). Spatial separation of beta-sheet domains of beta-amyloid: Disruption of each beta-sheet by N-methyl amino acids. *Biochemistry* **45**, 9485–9495.

Serpell, L. C., Berriman, J., Jakes, R., Goedert, M., and Crowther, R. A. (2000). Fiber diffraction of synthetic α-synuclein filaments shows amyloid-like cross-b conformation. *Proc Natl Acad Sci USA* **97**, 4897–4902.

Sharon, R., Bar-Joseph, I., Frosch, M. P., Walsh, D. M., Hamilton, J. A., and Selkoe, D. J. (2003). The formation of highly soluble oligomers of α-synuclein is regulated by fatty acids and enhanced in Parkinson's disease. *Neuron* **37**, 583–595.

Shendelman, S., Jonason, A., Martinat, C., Leete, T., and Abeliovich, A. (2004). DJ-1 is a redox-dependent molecular chaperone that inhibits α-synuclein aggregate formation. *PLoS Biol* **2**, e362.

Shimura, H., Schlossmacher, M. G., Hattori, N., Frosch, M. P., Trockenbacher, A., Schneider, R., Mizuno, Y., Kosik, K. S., and J., S. D. (2001). Ubiquitination of a new form of α-synuclein by parkin from human brain: Implications for Parkinson's disease. *Science* **293**, 263–269.

Silvestri, L., Caputo, V., Bellacchio, E., Atorino, L., Dallapiccola, B., Valente, E. M., and Casari, G. (2005). Mitochondrial import and enzymatic activity of PINK1 mutants associated to recessive parkinsonism. *Hum Mol Genet* **14**, 3477–3492.

Sim, C. H., Lio, D. S., Mok, S. S., Masters, C. L., Hill, A. F., Culvenor, J. G., and Cheng, H. C. (2006). C-terminal truncation and Parkinson's disease-associated mutations down-regulate the protein serine/threonine kinase activity of PTEN-induced kinase-1. *Hum Mol Genet* **15**, 3251–3262.

Singleton, A. B., Farrer, M., Johnson, J., Singleton, A., Hague, S., Kachergus, J., Hulihan, M., Peuralinna, T., Dutra, A., Nussbaum, R., Lincoln, S., Crawley, A., Hanson, M., Maraganore, D., Adler, C., Cookson, M. R., Muenter, M., Baptista, M., Miller, D., Blancato, J., Hardy, J., and Gwinn-Hardy, K. (2003). α-Synuclein locus triplication causes Parkinson's disease. *Science* **302**, 841.

Smith, W. W., Margolis, R. L., Li, X., Troncoso, J. C., Lee, M. K., Dawson, V. L., Dawson, T. M., Iwatsubo, T., and Ross, C. A. (2005). α-Synuclein phosphorylation enhances eosinophilic cytoplasmic inclusion formation in SH-SY5Y cells. *J Neurosci* **25**, 5544–5552.

Smith, W. W., Pei, Z., Jiang, H., Dawson, V. L., Dawson, T. M., and Ross, C. A. (2006). Kinase activity of mutant LRRK2 mediates neuronal toxicity. *Nat Neurosci* **9**, 1231–1233.

Souza, J. M., Giasson, B. I., Chen, Q., Lee, V. M., and Ischiropoulos, H. (2000). Dityrosine cross-linking promotes formation of stable α-synuclein polymers. Implication of nitrative and oxidative stress in the pathogenesis of neurodegenerative synucleinopathies. *J Biol Chem* **275**, 18344–18349.

Spillantini, M. G., Schmidt, M. L., Lee, V. M.-Y., Trojanowski, J. Q., Jakes, R., and Goedert, M. (1997). α-Synuclein in Lewy bodies. *Nature* **388**, 839–840.

Staropoli, J. F., McDermott, C., Martinat, C., Schulman, B., Demireva, E., and Abeliovich, A. (2003). Parkin is a component of an SCF-like ubiquitin ligase complex and protects postmitotic neurons from kainate excitotoxicity. *Neuron* **37**, 735–749.

Taira, T., Saito, Y., Niki, T., Iguchi-Ariga, S. M., Takahashi, K., and Ariga, H. (2004). DJ-1 has a role in antioxidative stress to prevent cell death. *EMBO Rep* 5, 213–218.

Takahashi-Niki, K., Niki, T., Taira, T., Iguchi-Ariga, S. M., and Ariga, H. (2004). Reduced anti-oxidative stress activities of DJ-1 mutants found in Parkinson's disease patients. *Biochem Biophys Res Commun* 320, 389–397.

Tamamizu-Kato, S., Kosaraju, M. G., Kato, H., Raussens, V., Ruysschaert, J. M., and Narayanaswami, V. (2006). Calcium-triggered membrane interaction of the α-synuclein acidic tail. *Biochemistry* 45, 10947–10956.

Tao, X., and Tong, L. (2003). Crystal structure of human DJ-1, a protein associated with early onset Parkinson's disease. *J Biol Chem* 278, 31372–31379.

Uversky, V. N., Li, J., and Fink, A. L. (2001). Evidence for a partially folded intermediate in α-synuclein fibril formation. *J Biol Chem* 276, 10737–10744.

Uversky, V. N., Li, J., Souillac, P., Millett, I. S., Doniach, S., Jakes, R., Goedert, M., and Fink, A. L. (2002a). Biophysical properties of the synucleins and their propensities to fibrillate – Inhibition of α-synuclein assembly by β- and γ-synucleins. *J Biol Chem* 277, 11970–11978.

Uversky, V. N., Yamin, G., Souillac, P. O., Goers, J., Glaser, C. B., and Fink, A. L. (2002b). Methionine oxidation inhibits fibrillation of human α-synuclein *in vitro*. *FEBS Lett* 517, 239–244.

Valente, E. M., Abou-Sleiman, P. M., Caputo, V., Muqit, M. M., Harvey, K., Gispert, S., Ali, Z., Del Turco, D., Bentivoglio, A. R., Healy, D. G., Albanese, A., Nussbaum, R., Gonzalez-Maldonado, R., Deller, T., Salvi, S., Cortelli, P., Gilks, W. P., Latchman, D. S., Harvey, R. J., Dallapiccola, B., Auburger, G., and Wood, N. W. (2004). Hereditary early-onset Parkinson's disease caused by mutations in PINK1. *Science* 304, 1158–1160.

Volles, M. J., and Lansbury, P. T., Jr. (2002). Vesicle permeabilization by protofibrillar α-synuclein is sensitive to Parkinson's disease-linked mutations and occurs by a pore-like mechanism. *Biochemistry* 41, 4595–4602.

Volles, M. J., and Lansbury, P. T., Jr. (2003). Zeroing in on the pathogenic form of α-synuclein and its mechanism of neurotoxicity in Parkinson's disease. *Biochemistry* 42, 7871–7878.

Volles, M. J., Lee, S.-J., Rochet, J.-C., Shtilerman, M. D., Ding, T. T., Kessler, J. C., and Lansbury, P. T., Jr. (2001). Vesicle permeabilization by protofibrillar α-synuclein: Implications for the pathogenesis and treatment of Parkinson's disease. *Biochemistry* 40, 7812–7819.

von Coelln, R., Dawson, V. L., and Dawson, T. M. (2004). Parkin-associated Parkinson's disease. *Cell Tissue Res* 318, 175–184.

Wakabayashi, K., Matsumoto, K., Takayama, K., Yoshimoto, M., and Takahashi, H. (1997). NACP, a presynaptic protein, immunoreactivity in Lewy bodies in Parkinson's disease. *Neurosci Lett* 239, 45–48.

Weinreb, P. H., Zhen, W., Poon, A. W., Conway, K. A., and Lansbury, P. T., Jr. (1996). NACP, a protein implicated in Alzheimer's disease and learning, is natively unfolded. *Biochemistry* 35, 13709–13715.

West, A. B., Moore, D. J., Biskup, S., Bugayenko, A., Smith, W. W., Ross, C. A., Dawson, V. L., and Dawson, T. M. (2005). Parkinson's disease-associated mutations in leucine-rich repeat kinase 2 augment kinase activity. *Proc Natl Acad Sci USA* 102, 16842–16847.

West, A. B., Moore, D. J., Choi, C., Andrabi, S. A., Li, X., Dikeman, D., Biskup, S., Zhang, Z., Lim, K. L., Dawson, V. L., and Dawson, T. M. (2007). Parkinson's disease-associated mutations in LRRK2 link enhanced GTP-binding and kinase activities to neuronal toxicity. *Hum Mol Genet* 16, 223–232.

Wilkinson, K. D., Lee, K. M., Deshpande, S., Duerksen-Hughes, P., Boss, J. M., and Pohl, J. (1989). The neuron-specific protein PGP 9.5 is a ubiquitin carboxyl-terminal hydrolase. *Science* 246, 670–673.

Wilson, M. A., Collins, J. L., Hod, Y., Ringe, D., and Petsko, G. A. (2003). The 1.1-A resolution crystal structure of DJ-1, the protein mutated in autosomal recessive early onset Parkinson's disease. *Proc Natl Acad Sci USA* 100, 9256–9261.

Wislet-Gendebien, S., D'Souza, C., Kawarai, T., St George-Hyslop, P., Westaway, D., Fraser, P., and Tandon, A. (2006). Cytosolic proteins regulate α-synuclein dissociation from presynaptic membranes. *J Biol Chem* 281, 32148–32155.

Woo, H. A., Jeong, W., Chang, T. S., Park, K. J., Park, S. J., Yang, J. S., and Rhee, S. G. (2005). Reduction of cysteine sulfinic acid by sulfiredoxin is specific to 2-cys peroxiredoxins. *J Biol Chem* 280, 3125–3128.

Wood, Z. A., Schroder, E., Robin Harris, J., and Poole, L. B. (2003). Structure, mechanism and regulation of peroxiredoxins. *Trends Biochem Sci* 28, 32–40.

Xu, J., Zhong, N., Wang, H., Elias, J. E., Kim, C. Y., Woldman, I., Pifl, C., Gygi, S. P., Geula, C., and Yankner, B. A. (2005). The Parkinson's disease-associated DJ-1 protein is a transcriptional co-activator that protects against neuronal apoptosis. *Hum Mol Genet* 14, 1231–1241.

Yamin, G., Uversky, V. N., and Fink, A. L. (2003). Nitration inhibits fibrillation of human α-synuclein *in vitro* by formation of soluble oligomers. *FEBS Lett* 542, 147–152.

Yang, Y., Gehrke, S., Haque, M. E., Imai, Y., Kosek, J., Yang, L., Beal, M. F., Nishimura, I., Wakamatsu, K., Ito, S., Takahashi, R., and Lu, B. (2005). Inactivation of Drosophila DJ-1 leads to impairments of oxidative stress response and phosphatidylinositol 3-kinase/Akt signaling. *Proc Natl Acad Sci USA* 102, 13670–13675.

Yao, D., Gu, Z., Nakamura, T., Shi, Z. Q., Ma, Y., Gaston, B., Palmer, L. A., Rockenstein, E. M., Zhang, Z., Masliah, E., Uehara, T., and Lipton, S. A. (2004). Nitrosative stress linked to sporadic Parkinson's disease: S-nitrosylation of parkin regulates its E3 ubiquitin ligase activity. *Proc Natl Acad Sci USA* 101, 10810–10814.

Yokota, T., Sugawara, K., Ito, K., Takahashi, R., Ariga, H., and Mizusawa, H. (2003). Down regulation of DJ-1 enhances cell death by oxidative stress, ER stress, and proteasome inhibition. *Biochem Biophys Res Commun* 312, 1342–1348.

Zakharov, S. D., Hulleman, J. D., Dutseva, E. A., Antonenko, Y. N., Rochet, J. C., and Cramer, W. A. (2007). Helical α-synuclein forms highly conductive ion channels. *Biochemistry* 46, 14369–14379.

Zarranz, J. J., Alegre, J., Gomez-Esteban, J. C., Lezcano, E., Ros, R., Ampuero, I., Vidal, L., Hoenicka, J., Rodriguez, O., Atares, B., Llorens, V., Gomez Tortosa, E., del Ser, T., Munoz, D. G., and de Yebenes, J. G. (2004). The new mutation, E46K, of α-synuclein causes Parkinson and Lewy body dementia. *Ann Neurol* 55, 164–173.

Zhang, L., Shimoji, M., Thomas, B., Moore, D. J., Yu, S. W., Marupudi, N. I., Torp, R., Torgner, I. A., Ottersen, O. P., Dawson, T. M., and Dawson, V. L. (2005). Mitochondrial localization of the Parkinson's disease related protein DJ-1: Implications for pathogenesis. *Hum Mol Genet* 14, 2063–2073.

Zhang, Y., Gao, J., Chung, K. K., Huang, H., Dawson, V. L., and Dawson, T. M. (2000). Parkin functions as an E2-dependent ubiquitin–protein ligase and promotes the degradation of the synaptic vesicle-associated protein, CDCrel-1. *Proc Natl Acad Sci USA* 97, 13354–13359.

Zhong, N., Kim, C. Y., Rizzu, P., Geula, C., Porter, D. R., Pothos, E. N., Squitieri, F., Heutink, P., and Xu, J. (2006). DJ-1 transcriptionally up-regulates the human tyrosine hydroxylase by inhibiting the sumoylation of pyrimidine tract-binding protein-associated splicing factor. *J Biol Chem* 281, 20940–20948.

Zhou, W., and Freed, C. R. (2005). DJ-1 up-regulates glutathione synthesis during oxidative stress and inhibits A53T α-synuclein toxicity. *J Biol Chem* 280, 43150–43158.

Zhou, W., Zhu, M., Wilson, M. A., Petsko, G. A., and Fink, A. L. (2006). The oxidation state of DJ-1 regulates its chaperone activity toward α-synuclein. *J Mol Biol* 356, 1036–1048.

Zhu, M., and Fink, A. L. (2003). Lipid binding inhibits α-synuclein fibril formation. *J Biol Chem* 278, 16873–16877.

Zhu, M., Rajamani, S., Kaylor, J., Han, S., Zhou, F., and Fink, A. L. (2004). The flavonoid baicalein inhibits fibrillation of α-synuclein and disaggregates existing fibrils. *J Biol Chem* 279, 26846–26857.

Zibaee, S., Jakes, R., Fraser, G., Serpell, L. C., Crowther, R. A., and Goedert, M. (2007). Sequence determinants for amyloid fibrillogenesis of human α-synuclein. *J Mol Biol* 374, 454–464.

Zimprich, A., Biskup, S., Leitner, P., Lichtner, P., Farrer, M., Lincoln, S., Kachergus, J., Hulihan, M., Uitti, R. J., Calne, D. B., Stoessl, A. J., Pfeiffer, R. F., Patenge, N., Carbajal, I. C., Vieregge, P., Asmus, F., Muller-Myhsok, B., Dickson, D. W., Meitinger, T., Strom, T. M., Wszolek, Z. K., and Gasser, T. (2004). Mutations in LRRK2 cause autosomal-dominant parkinsonism with pleomorphic pathology. *Neuron* 44, 601–607.

INDEX

Printed and bound by CPI Group (UK) Ltd, Croydon, CR0 4YY

08/05/2025

01865016-0001